STANDARD 9: Geometry and Spatial Sense

In grades K-4, the mathematics curriculum should include two- and three-dimensional geometry so that students can—

- describe, model, draw, and classify shapes;
- Investigate and predict the results of combining, subdividing, and changing shapes;
- develop spatial sense;
- relate geometric ideas to number and measurement ideas;
- recognize and appreciate geometry in their world.

STANDARD 10: Measurement

In grades K-4, the mathematical curriculum should include measurement so that students can—

- understand the attributes of length, capacity, weight, area, volume, time, temperature, and angle;
- develop the process of measuring and concepts related to units of measurement;
- make and use estimates of measurement;
- make and use measurements in problem and everyday situations.

STANDARD 11: Statistics and Probability

In grades K-4, the mathematics curriculum should include experiences with data analysis and probability so that students can—

-
-
-
- explore concepts of chance.

STANDARD 12: Fractions and Decimals

In grades K-4, the mathematics curriculum should include fractions and decimals so that students can—

- develop concepts of fractions, mixed numbers, and decimals;
- develop number sense for fractions and decimals;
- use models to relate fractions to decimals and to find equivalent fractions;
- use models to explore operations on fractions and decimals;
- apply fractions and decimals to problem situations.

STANDARD 13: Patterns and Relationships

In grades K-4, the mathematics curriculum should include the study of patterns and relationships so that students can—

- recognize, describe, extend, and create a wide variety of patterns;
- represent and describe mathematical relationships;
- explore the use of variables and open sentences to express relationships.

CURRICULUM STANDARDS FOR GRADES 5-8

STANDARD 1: Mathematics as Problem Solving

In grades 5-8, the mathematics curriculum should include numerous and varied experiences with problem solving as a method of inquiry and application so that students can—

- use problem-solving approaches to investigate and understand mathematical content;
- formulate problems from situations within and outside mathematics;
- develop and apply a variety of strategies to solve problems, with emphasis on multistep and nonroutine problems;
- verify and interpret results with respect to the original problem situation;
- generalize solutions and strategies to new problem situations;
- acquire confidence in using mathematics meaningfully.

STANDARD 2: Mathematics as Communication

In grades 5-8, the study of mathematics should include opportunities to communicate so that students can—

- model situations using oral, written, concrete, pictorial, graphical, and algebraic methods;
- reflect on and clarify their own thinking about mathematical ideas and situations;
- develop common understandings of mathematical ideas, including the role of definitions;
- use the skills of reading, listening, and viewing to interpret and evaluate mathematical ideas;
- discuss mathematical ideas and make conjectures and convincing arguments;
- appreciate the value of mathematical notation and its role in the development of mathematical ideas.

STANDARD 3: Mathematics as Reasoning

In grades 5-8, reasoning shall permeate the mathematics curriculum so that students can—

- recognize and apply deductive and inductive reasoning;
- understand and apply reasoning processes, with special attention to spatial reasoning and reasoning with proportions and graphs;
- make and evaluate mathematical conjectures and arguments;
- validate their own thinking;
- appreciate the pervasive use and power of reasoning as a part of mathematics.

Curriculum Standards for Grades 5-8 are continued on inside back cover.

A Problem Solving Approach to
MATHEMATICS FOR ELEMENTARY SCHOOL TEACHERS

FIFTH EDITION

A Problem Solving Approach to
MATHEMATICS FOR ELEMENTARY SCHOOL TEACHERS

FIFTH EDITION

Rick Billstein
University of Montana
Missoula, Montana

Shlomo Libeskind
University of Oregon
Eugene, Oregon

Johnny W. Lott
University of Montana
Missoula, Montana

▼▲
▲▼ **ADDISON-WESLEY PUBLISHING COMPANY**

Reading, Massachusetts ■ Menlo Park, California ■ New York ■ Don Mills, Ontario
Wokingham, England ■ Amsterdam ■ Bonn ■ Sydney ■ Singapore ■ Tokyo
Madrid ■ San Juan ■ Milan ■ Paris

Sponsoring Editor ■ *Bill Poole*
Associate Editor ■ *Kathleen Manley*
Developmental Editor ■ *Shelley Parlante*
Managing Editor ■ *Karen M. Guardino*
Production Supervisor ■ *Jennifer Brownlow Bagdigian*
Production Coordinator ■ *Patricia A. Oduor*
Text Designer ■ *Vanessa Piñeiro, Piñeiro Design Associates*
Chapter Opener Designer ■ *Martha Podren, Podren Design*
Copyeditor ■ *Carmen Wheatcroft*
Proofreader ■ *Laura K. Michaels, Michaels Communications*
Math Marketing Manager ■ *Sylvia Herrera-Alaniz*
Senior Manufacturing Manager ■ *Roy Logan*
Manufacturing Coordinator ■ *Evelyn Beaton*
Cover Designer ■ *Peter Blaiwas*
Composition/Prepress Services ■ *TSI Graphics*
Illustrations ■ *Monotype Composition Company, Inc.*
Printer ■ *Von Hoffmann Press*

Library of Congress Cataloging-in-Publication Data

Billstein, Rick.
    A problem solving approach to mathematics for elementary school teachers / Rick Billstein, Shlomo Libeskind, Johnny W. Lott.—5th ed.
      p.   cm.
    ISBN 0-201-52565-8
    1. Mathematics—Study and teaching (Elementary)   2. Problem solving. I. Libeskind, Shlomo. II. Lott, Johnny W., 1944-   .
III. Title.
QA135.5.B49   1993b
372.7—dc20
                                                                         92–32219
                                                                          CIP

Reprinted with corrections, November 1993.

Copyright © 1993 by Addison-Wesley Publishing Company, Inc.

All rights reserved. No part of this publication may be reproduced, stored in a retrieval system, or transmitted, by any form or by any means, electronic, mechanical, photocopying, recording, or otherwise, without the prior written permission of the publisher. Printed in the United States of America.

7 8 9 10 – VH – 99 98 97 96

To Jane, Molly, and Karly—R.B.

To my mother, Genia, and to the memory of my father, Mendel—S.L.

To John for courage, Carolyn for understanding, and Ouidamai for support—J.W.L.

Cover art: Patchwork Sampler quilt by Margit Echols © 1983; Photo: Schecter Lee. All quotations from *Curriculum and Evaluation Standards for School Mathematics* ('Standards') and *Professional Standards for Teaching Mathematics* ('Teaching Standards') are reprinted with permission of the National Council of Teachers of Mathematics, Inc.; All student pages from Eicholz, et al, *Addison-Wesley Mathematics Series,* are reprinted with permission of Addison-Wesley Publishing Company; Pages 2, 50, 174, 301, 411, 423, 429, 431, and 438, PEANUTS reprinted by permission of UFS, Inc.; Page 213, CALVIN & HOBBES © 1987 Watterson. Reprinted with permission of Universal Press Syndicate. All rights reserved; Page 242, THE FAR SIDE cartoon by Gary Larson is reprinted by permission of Chronicle Features, San Francisco, CA; Page 282, student page from HEATH MATHEMATICS CONNECTIONS, Grade 5, © 1992. Reprinted with permission of D.C. Heath and Company; Page 296, student page from HEATH MATHEMATICS CONNECTIONS, Grade 6, © 1992. Reprinted with permission of D.C. Heath and Company; Page 352, BORN LOSER reprinted by permission of NEA, Inc.; Page 420, cartoon courtesy of Carole Cable; Page 443, student page from EXPLORING MATHEMATICS, Grade 6 (1991). Reprinted with permission of Scott, Foresman and Company; Page 485, THE FAR SIDE © 1985 Far Works, Inc. Reprinted with permission of Universal Press Syndicate. All rights reserved; Pages 500, 636, and 696, B.C. cartoons are reprinted by permission of Johnny Hart and Creators Syndicate, Inc.; Page 505, student page reprinted with permission of Macmillan/McGraw-Hill School Publishing Company from MATHEMATICS IN ACTION, Grade 5 by Hoffer, et al. Copyright 1991; Page 543, student page from HEATH MATHEMATICS CONNECTIONS, Grade 6, © 1992. Reprinted with permission of D.C. Heath and Company; Page 604, student page from MATHEMATICS: EXPLORING YOUR WORLD, Grade 7, © 1991 Silver, Burdett & Ginn. Reprinted with permission; Page 613, HAGAR THE HORRIBLE cartoon is reprinted with special permission of King Features Syndicate, Inc.; Page 656, pottery fragment, source unknown; Page 661, Chevrolet logo courtesy of Chevrolet Motor Division. Chrysler logo courtesy of Chrysler Corporation. Bell Symbol courtesy of Pacific Telesis Group. Volkswagen of America logo courtesy of Volkswagen of America; Page 664, lizard tessellation © 1939 M.C. Escher/Cordon Art - Baarn - Holland; Page 668, bird tessellation © 1955 M.C. Escher/Cordon Art - Baarn - Holland; Page 677, fractal from Peitgen & Richter, BEAUTY OF FRACTALS, © 1986. Reprinted with permission of Springer-Verlag; Page 719, student page from HOUGHTON MIFFLIN MATHEMATICS by Duncan, et al. Copyright © 1991 by Houghton Mifflin Company. Reprinted by permission of Houghton Mifflin Company; Page 775, THE FAR SIDE © 1985 Far Works, Inc. Reprinted with permission of Universal Press Syndicate. All rights reserved.

PREFACE

he fifth edition of *A Problem Solving Approach to Mathematics for Elementary School Teachers* retains the goals of the previous editions, but extends them to meet the requests of users of the fourth edition as well as to meet new challenges of the 1990s.

Standards of the NCTM

In particular, we focus on two National Council of Teachers of Mathematics (NCTM) publications, the 1989 *Curriculum and Evaluation Standards of School Mathematics* (hereafter referred to as the *Standards*), and the 1991 *Professional Standards for Teaching Mathematics* (hereafter referred to as the *Teaching Standards*). Of primary importance to us is the standard (*Standards*, p. 253): "Prospective teachers must be taught in a manner similar to how they are to teach—by exploring, conjecturing, communicating, reasoning, and so forth." In addition, "all teachers need an understanding of both the historical development and current applications of mathematics. Furthermore, they should be familiar with the power of technology." In particular, the *Teaching Standards* (p. 3) emphasize the need for the following shifts in teaching of mathematics.

- Toward logic and mathematical evidence as verification—away from the teacher as the sole authority for right answers;
- toward mathematical reasoning—away from merely memorizing procedures;
- toward conjecturing, inventing, and problem solving—away from an emphasis on mechanistic answer-finding;
- toward connecting mathematics, its ideas, and its applications—away from treating mathematics as a body of isolated concepts and procedures.

To achieve the aforementioned shifts, the fifth edition allows instructors a variety of approaches to teaching and encourages "doing" as well as lecturing. We kept the future elementary mathematics student in mind and produced a text which allows prospective teachers to examine, represent, transform, conjecture, justify, apply, solve problems, and communicate in mathematics.

Continuing Goals

In the fifth edition our goals remain:

- To present appropriate mathematics in an intellectually honest and mathematically correct manner.
- To use the heuristics of problem solving as an integral part of mathematics.
- To approach mathematics in a sequence which instills confidence, then challenges students.

New Goals

For this edition, we have additional goals:

- To present the topics in the context of the *Standards*.
- To revise the coverage of most topics, in order to make them more accessible.
- To identify and use the various problem solving strategies.
- To provide communication problems so that future teachers can develop writing skills and practice the explanation of their thinking.

Problem Solving in the Fifth Edition

We showcase problem solving skills by:

- Devoting an expanded Chapter 1 to problem solving.
- Using a four-step problem solving process to solve problems in each chapter. The four steps are: Understanding the Problem, Devising a Plan, Carrying out the Plan, and Looking Back.
- Beginning Chapters 1 through 14 with a preliminary problem which poses a question that students can answer with the skills mastered from the chapter.

We encourage teachers to introduce the preliminary problem at the beginning of the chapter, then spiral back after covering the chapter to show how the techniques therein are necessary to solve the problem.

Features Retained in this Edition

Wherever possible, we present topics in ways that could be used in actual classrooms. Further, we have incorporated various study aids and features to facilitate learning.

- Historical notes add context and humanize the mathematics.
- Brain Teasers provide a different avenue for problem solving. They are solved in the Instructor's Guide and may be assigned or used by the teacher to challenge students.
- Computer Corners, which provide additional information on using the computer in the classroom, are set off from the body of the text. This feature includes greater coverage of LOGO than in the past edition. In addition, problems emphasizing computer usage are denoted by a computer symbol 💻.

- Laboratory Activities are integrated throughout the book to provide hands-on learning exercises. In each non-optional geometry section, van Hiele-type laboratory activities are included.
- Cartoons teach or emphasize important material and add levity.
- Key terms are presented in the margins for quick review.
- Definitions are either set off in text or presented as key terms in the margin for quick review.
- Review problems are included at the end of each non-optional section.
- Optional sections as well as problems based on these sections are marked with an asterisk (*); more difficult problems are marked with a star (★). Problems numbered in color have answers at the back of the book.
- Questions from the Classroom have been expanded. We strongly recommend that instructors use these questions posed by actual students in actual classrooms when building a course syllabus. Instructors may require students to write two answers to the questions—one mathematical and one pedagogical—using student texts and professional journals for research.
- Chapter outlines at the end of each chapter help students review the chapter.
- Chapter tests at the end of each chapter allow students to test themselves.
- Selected Bibliographies have been updated and revised. They are at the end of each chapter.

Features New to this Edition

- Problem sets are reorganized first to follow the order of the chapter and second to proceed from easy to difficult.
- Problem sets have been extended to include many more relevant and realistic problems.
- Problem solving strategies are often highlighted in italics and indicated by the strategy symbol ■ in the margin.
- Relevant quotes from the *Standards* and *Teaching Standards* are incorporated throughout the text and marked by the standard icon ▲.
- In view of the *Standards'* emphasis on communicating mathematical ideas, problem sets contain numerous problems in which students are asked to explain or justify their answers. Communication problems are noted by the ▶ symbol.
- Full color has been added for pedagogical reasons and consequently figures are more attractive and easier to follow. Additionally, all of the pages taken from elementary math texts are now presented in full color.

Content

The material has been rewritten and the problem sets revised so that students with diverse backgrounds will find the material accessible. Flexibility is built in for instructors who wish to adapt the text to a variety of course lengths and organizations. Sections marked with asterisks (*) are optional and may be omitted without loss of continuity.

Geometry

Special attention has been paid to revising the geometry chapters emphasizing more intuitive approaches and informal justifications.

Chapter 10: Introduces material necessary to study geometry including topological notions, and sets the stage for van Hiele-type activities.

Chapter 11: Includes various types of construction activities involving congruent and similar figures. Activities use tracing paper, paper-folding, Mira, and compass and straightedge.

Chapter 12: Has been reorganized and rewritten to develop a transformational approach to congruence and similarity.

Chapter 13: Presents the basics of the English and metric measuring systems and their uses.

Chapter 14: Has been reorganized to cover coordinate geometry and the relationship between algebra and geometry.

Appendix II: Covers basics of the Logo computer language, a tool for learning geometry. Optional sections and problems in each chapter also use Logo.

Calculator usage

As prescribed in the *Standards,* coverage of calculators is necessary and timely. Calculators are introduced in Chapter 1. A discussion of use of scientific/fraction calculators has been added. Problems involving calculators are marked with an icon (▦) and appear in most sections throughout the book.

Numerations systems, whole numbers, and integers

Historical significance of numerations systems and conceptual development of whole numbers and integers are highlighted.

Chapters 3 and 4: Emphasize use of different models for presenting the mathematics and include sections on mental mathematics and estimation. Additional material has been added on translating word problems into algebraic expressions.

Sets, relations, functions, and logic

These topics, developed in Chapter 2, provide a background for later formal development of mathematics.

Number theory

Number theory has long been one of the most exciting areas in mathematics.

Chapter 5: Develops divisibility rules and presents unsolved problems. More has been added on the application of calculators to number theory.

Real numbers

Decimals, fractions, and computation with these numbers are typically among the most difficult concepts elementary teachers are requird to teach.

> **Chapters 6 and 7:** Provide necessary mathematical tools for understanding relationships among these sets of numbers. New material has been added to Chapter 7 on the use and misuse of percent.

Probability

Topics in probability are now common in elementary school texts.

> **Chapter 8:** Includes geometric probability and more on simulations, as well as tree diagrams and Monte Carlo methods, to aid in an understanding of probability topics. The coverage of conditional probability using formulas has been deleted in the 5th edition.

Statistics

Understanding statistics is key to becoming an educated person.

> **Chapter 9:** Presents new and expanded material using such techniques as stem and leaf graphs, box plots, and line plots for presenting and understanding data. Additional material has been added on the misuse of statistics.

The content of the book is compatible with the recommendations from the *Standards* and the *Teaching Standards*. We strongly believe that most people learn mathematics first intuitively and then formally. We also recognize that prospective teachers must learn both the content and the methods of teaching at the same time. As a result, we present the material with motivation for what a concept is, why it is important and how it may be used. To this same end, we encourage users to incorporate laboratory activities and to use the accompanying activities book by Dolan and Williamson.

Supplements for the Student

- *Mathematics Activities for Elementary School Teachers: A Problem Solving Approach*, 2nd Edition, by Daniel Dolan, James Williamson, and Mari Muri. This new edition features both short activities with possible extensions as well as longer activities. Activities are organized from lowest to highest grade level within each chapter.
- Student's Solutions Manual, by Louis Levy and Edward Fritz, new to this edition, contains solutions to all odd-numbered exercises.

Supplements for the Instructor

- Instructor's Guide includes: Answers to all problems; Two sample chapter tests with answers for each chapter; Suggested answers to Questions from the Classroom; Solutions to Brain Teasers.
- Instructor's Solutions Manual, by Louis Levy and Edward Fritz, new to this edition, contains solutions to all exercises including communication exercises.

- Instructor's Guide to *Mathematics Activities for Elementary School Teachers: A Problem Solving Approach, 2nd Edition* by Daniel Dolan, Jim Williamson, and Mari Muri, new to this edition, contains answers to all activities and key ideas and extensions of activities.
- OmniTest II (IBM 5.25 or 3.5 inch disks) is a powerful new testing system developed exclusively for Addison-Wesley. It is an algorithm-driven system that allows the user to create up to 99 perfectly parallel forms of any test effortlessly. It also allows the user to add his or her own test items and edit existing items with an easy to use, on screen "What-You-See-Is-What-You-Get" interface.

Acknowledgements

Our students and users of this text from across the nation have provided us with valuable feedback in revising this text. We have welcomed and continue to welcome constructive comments concerning topics in the text. Comments have led to many positive changes, particularly those of the many reviewers who have contributed time to helping us with the book.

Rick Billstein
Shlomo Libeskind
Johnny W. Lott

Reviewers of this edition and previous editions

Leon J. Ablon
G. L. Alexanderson
Jane Barnard
Joann Becker
Cindy Bernlohr
Jim Boone
Maurice Burke
David Bush
Laura Cameron
Louis J. Chatterley
Donald J. Dessart
Ronald Dettmers
Jackie Dewar
Amy Edwards
Margaret Ehringer
Albert Filano
Marjorie Fitting
Michael Flom
Sandy Geiger
Glenadine Gibb
Elizabeth Gray
Alice Guckin
Boyd Henry
Alan Hoffer
E. John Hornsby Jr.
Jerry Johnson
Wilburn C. Jones
Robert Kalin
Herbert E. Kasube
Sarah Kennedy
Steven D. Kerr
Leland Knauf
Don Loftsgaarden
Stanley Lukawecki
Barbara Moses
Charles Nelson
Glenn Nelson
Keith Peck
Barbara Pence
Glenn L. Pfeifer
Jack Porter
Edward Rathnell
Helen R. Santiz
M. Geralda Schaefer
Jane Schielack
Barbara Shabell
Gwen Shufelt
Ron Smit
Joe K. Smith
William Sparks
Virginia Strawderman
Viji Sundar
C. Ralph Verno
Hubert Voltz
John Wagner
Virginia Warfield
Mark F. Weiner
Grayson Wheatley
Jerry L. Young

BRIEF CONTENTS

CHAPTER 1   Tools for Problem Solving   1
CHAPTER 2   Sets, Functions, and Logic   49
CHAPTER 3   Numeration Systems and Whole Numbers   103
CHAPTER 4   The Integers   173
CHAPTER 5   Number Theory   219
CHAPTER 6   Rational Numbers as Fractions   265
CHAPTER 7   Decimals   327
CHAPTER 8   Probability   383
CHAPTER 9   Statistics: An Introduction   437
CHAPTER 10 Introductory Geometry   497
CHAPTER 11 Constructions, Congruence, and Similarity   569
CHAPTER 12 Motion Geometry and Tessellations   625
CHAPTER 13 Concepts of Measurement   685
CHAPTER 14 Coordinate Geometry   763
             Appendix I: An Introduction to BASIC   815
             Appendix II: An Introduction to Logo Turtle Graphics   833
             Answers to Selected Problems   857
             Index   897

Solutions to the cover puzzles follow the index.

CONTENTS

1 Tools For Problem Solving   1

Preliminary Problem   1
1-1   Exploration with Patterns   4
1-2   Using the Problem-solving Process   17
1-3   Using a Calculator as a Problem-solving Tool   37
Solution to the Preliminary Problem   44
Questions from the Classroom   45
Chapter Outline   46
Chapter Test   46
Selected Bibliography   47

2 Sets, Functions, and Logic   49

Preliminary Problem   49
2-1   Describing Sets   51
2-2   Other Set Operations and Their Properties   60
2-3   Relations and Functions   69
*__2-4__   Logic: An Introduction   83
*__2-5__   Conditionals and Biconditionals   88
Solution to the Preliminary Problem   96
Questions from the Classroom   98
Chapter Outline   98
Chapter Test   99
Selected Bibliography   101

3 Numeration Systems and Whole Numbers 103

Preliminary Problem 103
- **3-1** Numeration Systems 104
- **3-2** Addition and Subtraction of Whole Numbers 111
- **3-3** Multiplication and Division of Whole Numbers 122
- **3-4** Algorithms, Mental Math, and Estimation for Whole-number Addition and Subtraction 133
- **3-5** Algorithms for Whole-number Multiplication and Division 148
- ***3-6** Other Number Bases 158

Solution to the Preliminary Problem 167
Questions from the Classroom 169
Chapter Outline 169
Chapter Test 170
Selected Bibliography 171

4 The Integers 173

Preliminary Problem 173
- **4-1** Integers and the Operations of Addition and Subtraction 175
- **4-2** Multiplication and Division of Integers 188
- **4-3** Solving Equations and Inequalities 197

Solution to the Preliminary Problem 214
Questions from the Classroom 216
Chapter Outline 216
Chapter Test 217
Selected Bibliography 218

5 Number Theory 219

Preliminary Problem 219
- **5-1** Divisibility 220
- **5-2** Prime and Composite Numbers 231
- **5-3** Greatest Common Divisor and Least Common Multiple 244
- ***5-4** Clock and Modular Arithmetic 254

Solution to the Preliminary Problem 259
Questions from the Classroom 261
Chapter Outline 261
Chapter Test 262
Selected Bibliography 263

6 Rational Numbers as Fractions 265

Preliminary Problem 265
6-1 Fractions and Rational Numbers 267
6-2 Addition and Subtraction of Rational Numbers 276
6-3 Multiplication and Division of Rational Numbers 289
6-4 Ordering Rational Numbers 302
6-5 Ratio and Proportion 312
6-6 Exponents Revisited 318
Solution to the Preliminary Problem 323
Questions from the Classroom 323
Chapter Outline 324
Chapter Test 325
Selected Bibliography 326

7 Decimals 327

Preliminary Problem 327
7-1 Decimals and Decimal Operations 328
7-2 Decimals and Their Properties 341
7-3 Percents 351
*7-4 Computing Interest 363
7-5 Real Numbers 368
*7-6 Radicals and Rational Exponents 375
Solution to the Preliminary Problem 377
Questions from the Classroom 378
Chapter Outline 379
Chapter Test 380
Selected Bibliography 381

8 Probability 383

Preliminary Problem 383
8-1 How Probabilities Are Determined 384
8-2 Multistage Experiments with Tree Diagrams and Geometric Probabilities 396
8-3 Using Simulations in Probability 408
8-4 Odds and Expected Value 417
*8-5 Methods of Counting 424
Solution to the Preliminary Problem 432
Questions from the Classroom 433
Chapter Outline 434
Chapter Test 435
Selected Bibliography 436

9 Statistics: An Introduction 437

Preliminary Problem 437
9-1 Statistical Graphs 439
9-2 Measures of Central Tendency and Variation 457
*****9-3** Normal Distributions 474
*****9-4** Abuses of Statistics 484
Solution to the Preliminary Problem 491
Questions from the Classroom 492
Chapter Outline 493
Chapter Test 493
Selected Bibliography 495

10 Introductory Geometry 497

Preliminary Problem 497
10-1 Basic Notions 499
10-2 Polygonal Curves 517
10-3 More about Angles 530
10-4 Geometry in Three Dimensions 541
*****10-5** Networks 553
*****10-6** Introducing Logo as a Tool in Geometry 559
Solution to the Preliminary Problem 562
Questions from the Classroom 564
Chapter Outline 564
Chapter Test 565
Selected Bibliography 567

11 Constructions, Congruence, and Similarity 569

Preliminary Problem 569
11-1 Congruence through Constructions 570
11-2 Other Congruence Properties 581
11-3 Other Constructions 587
11-4 Circles and Spheres 597
11-5 Similar Triangles and Similar Figures 603
*****11-6** More on Geometry and Logo 615
Solution to the Preliminary Problem 619
Questions from the Classroom 620
Chapter Outline 621
Chapter Test 622
Selected Bibliography 624

12 Motion Geometry and Tessellations 625

Preliminary Problem 625
12-1 Translations and Rotations 626
12-2 Reflections and Glide Reflections 636
12-3 Size Transformations 648
12-4 Symmetries 654
12-5 Tessellations of the Plane 663
*12-6 Escher-like Logo-type Tessellations 670
Solution to the Preliminary Problem 678
Questions from the Classroom 679
Chapter Outline 679
Chapter Test 680
Selected Bibliography 682

13 Concepts of Measurement 685

Preliminary Problem 685
13-1 Units of Length 686
13-2 Areas of Polygons and Circles 698
13-3 The Pythagorean Relationship 714
13-4 Applications Involving Areas 724
13-5 Volume Measure and Volumes 735
*13-6 Mass and Temperature 747
*13-7 Using Logo to Draw Circles 752
Solution to the Preliminary Problem 755
Questions from the Classroom 758
Chapter Outline 758
Chapter Test 760
Selected Bibliography 762

14 Coordinate Geometry 763

Preliminary Problem 763
14-1 Coordinate System in a Plane 764
14-2 Equations of Lines 776
14-3 Systems of Linear Equations 794
*14-4 Coordinate Geometry and Logo 803
Solution to the Preliminary Problem 809
Questions from the Classroom 810
Chapter Outline 811
Chapter Test 812
Selected Bibliography 813

Appendix I

An Introduction to BASIC 815
AI-1 BASIC: Variables and Operations 815
AI-2 Branching 819

Appendix II

An Introduction to Logo Turtle Graphics 833
AII-1 Introducing the Turtle 833
AII-2 Using Recursion as a Problem-solving Tool 850

Answers to Selected Problems 857

Index 897

Solutions to the cover puzzles follow the index.

1 Tools for Problem Solving

Preliminary Problem

A landscape architect needs to design a path for the Museum of Modern Art. The path is to be made of black and red tiles in the following pattern: 1 black, 1 red, 2 black, 2 red, 3 black, 3 red, 4 black, 4 red, ... The pattern of alternating colors is to continue so that each block of a single color has one more tile than the preceding block of the same color. The museum curator estimated that 5000 tiles will be needed for the project. She ordered 2500 black tiles and 2500 red ones. If the landscape architect wants to use all of these tiles and have the path end with a block of black tiles followed by the same number of red tiles, what is the smallest number of additional black and red tiles that should be ordered?

The National Council of Teachers of Mathematics (NCTM) publication *Curriculum and Evaluation Standards for School Mathematics* (1989), called simply the *Standards* hereafter, addresses *problem solving* at each of the three levels (K–4, 5–8, and 9–12). For example, in Standard 1 for grades K–4 (p. 23) we find the following quote:

> Problem solving should be the central focus of the mathematics curriculum. As such, it is a primary goal of all mathematics instruction and an integral part of all mathematical activity. Problem solving is not a distinct topic but a process that should permeate the entire program and provide the context in which concepts and skills can be learned.

Most textbook problems are designed to give students practice with content presented in the lesson. These types of problems are exercises involving routine procedures for finding solutions. For example, $12 \times 4 = \square$ is an exercise for most eighth graders but may be a problem for most second graders. Exercises serve their purpose in mathematics, but students are not exposed to problem solving through exercises alone. In the accompanying cartoon, Peppermint Patty tries to substitute one kind of exercise for a problem.

PEANUTS reprinted by permission of UFS, Inc.

In "The Heart of Mathematics," Paul Halmos wrote, "It is the duty of all teachers, and of teachers of mathematics in particular, to expose their students to problems much more than to facts." A good problem solver needs to experience a variety of problem-solving situations in which the focus is on the problem-solving process rather than just on obtaining the correct answer. Problem solving may require the use of many skills or strategies.

Four-step Problem-solving Process

George Polya described the experience of problem solving in his book, *How to Solve It* (p. v):

A great discovery solves a great problem but there is a grain of discovery in the solution of any problem. Your problem may be modest; but if it challenges your curiosity and brings into play your inventive faculties, and if you solve it by your own means, you may experience the tension and enjoy the triumph of discovery.

As part of his work on problem solving, Polya developed a four-step problem-solving process similar to the following:

1. Understanding the Problem
   (a) Can you state the problem in your own words?
   (b) What are you trying to find or do?
   (c) What are the unknowns?
   (d) What information do you obtain from the problem?
   (e) What information, if any, is missing or not needed?
2. Devising a Plan
   The following list of strategies, although not exhaustive, is very useful:
   (a) Look for a pattern.
   (b) Examine related problems and determine if the same technique can be applied.
   (c) Examine a simpler or special case of the problem to gain insight into the solution of the original problem.
   (d) Make a table.
   (e) Make a diagram.
   (f) Write an equation.
   (g) Use guess and check.
   (h) Work backward.
   (i) Identify a subgoal.
3. Carrying Out the Plan
   (a) Implement the strategy or strategies in step 2 and perform any necessary actions or computations.
   (b) Check each step of the plan as you proceed. This may be intuitive checking or a formal proof of each step.
   (c) Keep an accurate record of your work.
4. Looking Back
   (a) Check the results in the original problem. (In some cases, this will require a proof.)
   (b) Interpret the solution in terms of the original problem. Does your answer make sense? Is it reasonable?
   (c) Determine whether there is another method of finding the solution.
   (d) If possible, determine other related or more general problems for which the techniques will work.

These and other general mathematics problem-solving strategies, or rules of thumb for successful problem solving, are called *heuristics*.

> **HISTORICAL NOTE**
>
> George Polya (1887–1985) was born in Hungary and received his Ph.D. in 1912 from the University of Budapest for research on geometric probability. Polya studied law and literature before turning to mathematics. In 1940 he came to the United States and taught at Brown University before moving to Stanford University in 1942. Polya authored over 250 articles, 10 books, and numerous monographs. His book *How to Solve It,* published in 1945, has been translated into at least 17 other languages, has sold over 1 million copies, and has never been out of print. *How to Solve It* was followed in 1954 by two volumes of *Mathematics and Plausible Reasoning* and in 1962 and 1965 by the two volumes of *Mathematical Discovery.* Polya was renowned for his ability to describe abstract concepts in concrete terms.

We urge you to attempt the preliminary problem. If you try to solve the problem but are unable to do so, you can look at the solution presented at the end of the chapter. If you have not solved the problem, read only enough of the solution to get a hint and then try to complete the solution on your own.

In the following sections, we present a variety of problems and use the four-step problem-solving process to solve many of them. The four-step process does not ensure a solution to a problem, but it gives valuable guidelines when there is no obvious way to proceed.

One of the strategies of problem solving, looking for a pattern, is used so often it is discussed in a separate section. In addition, a separate section is devoted to choosing and using a calculator, an indispensable tool in performing routine computations and an invaluable aid in many problem-solving situations.

Section 1-1 Exploration with Patterns

G. H. Hardy (1877–1947), the eminent British mathematician, described mathematicians as makers of patterns of ideas. René Descartes (1596–1650), whom Polya credits with laying the foundation of his work on problem solving, wrote, "Each problem that I solved became a rule which served afterwards to solve other problems." (*Discours de la Méthode,* pp. 20–21). To a great extent, the purpose of mathematics is to create order out of chaos.

This section focuses on *searching for a pattern* within a given problem as an important strategy in problem solving. This strategy is emphasized in the *Standards* (p. 29, 98).

In grades K–4, the study of mathematics should emphasize reasoning so that students can use patterns and relationships to analyze mathematical situations. In Grades 5–8, the mathematics curriculum should include explorations of patterns and functions so that students can describe, extend, analyze, and create a wide variety of patterns, describe and represent relationships with tables, graphs, and rules.

Police investigators study case files to find the modus operandi, or pattern of operation, when a series of crimes is committed. Their discovery of a pattern, sometimes by using a computer, does not necessarily find the criminal, but it may provide the necessary clues to do so. Similarly, in science and mathematics we try to find solutions to problems by studying patterns and searching for clues. The patterns may or may not provide solutions.

Patterns can be surprising and aesthetically pleasing. For example, consider the following:

$$1 \times 9 + 2 = 11$$
$$12 \times 9 + 3 = 111$$
$$123 \times 9 + 4 = 1111$$
$$1234 \times 9 + 5 = 11111$$
$$12345 \times 9 + 6 = 111111$$
$$123456 \times 9 + 7 = 1111111$$

Here, the challenge would be to find why the pattern works. Sometimes, discovering a pattern among numbers can be challenging. Upon completion of this section, you should be able to find a pattern in the following sequence:

$$1, 5, 14, 30, 55, 91, \ldots .$$

Example 1-1 shows that based on a given set of data, more than one pattern is possible.

E X A M P L E 1 - 1   Find the next three terms to complete a pattern based on the following sequence:

$$1, 2, 4, \underline{}, \underline{}, \underline{}.$$

Solution   The difference between the first two terms is 1, and the difference between the second two terms is 2; consequently, the difference between the next two terms might be 3, then 4, and so on. Thus we conjecture that the completed sequence might appear as follows:

$$1, 2, 4, \underline{7}, \underline{11}, \underline{16}.$$

Another property that 1, 2, and 4 share is that 2 is twice 1 and 4 is twice 2. Thus the next terms could be 8, 16, and 32. Hence, the completed sequence might appear as follows:

$$1, 2, 4, \underline{8}, \underline{16}, \underline{32}.$$

It is evident that more than one pattern is possible, based on the given information. ∎

E X A M P L E 1 - 2   Find the next three terms to complete a pattern.

□, △, △, □, △, △, □, __, __, __.

Solution   Notice that between any two squares there are two consecutive triangles. Based on this observation, the next three terms are two triangles followed by a square. Thus the completed sequence might appear as follows:

□, △, △, □, △, △, □, △̲, △̲, □̲, ∎

CHAPTER 1 TOOLS FOR PROBLEM SOLVING

Inductive Reasoning

inductive reasoning

conjecture

counterexample

Reasoning based on examining a variety of cases or sets of data, discovering patterns, and forming conclusions is called **inductive reasoning.** Scientists use inductive reasoning when they perform experiments to discover various laws of nature. Statisticians use inductive reasoning when they form conclusions based on collected data. Inductive reasoning may lead to a **conjecture,** a statement thought to be true but not yet proved to be either true or false. Inductive reasoning should be used cautiously because conjectures developed using inductive reasoning may be false, as a **counterexample** can show. Sometimes it is necessary to test a large number of cases to find that a conjectured pattern does not continue.

The following discussion illustrates the danger of making a conjecture based on a few cases. In Figure 1-1, we choose points on a circle and connect them to form distinct, nonoverlapping regions. In this figure, we see that 2 points determine 2 regions, 3 points determine 4 regions, and 4 points determine 8 regions. What is the maximum number of regions that would be determined by 12 points and, in general, n points?

FIGURE 1-1

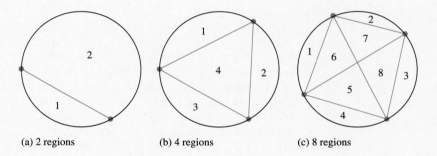

(a) 2 regions   (b) 4 regions   (c) 8 regions

The data from Figure 1-1 are recorded in Table 1-1. It appears that each time we increase the number of points by 1, we double the number of regions. If this were true, then for 5 points we would have 16, or 2^4, regions, for 6 points we would have 32, or 2^5, regions, and so on. If we base our solution on this pattern, we would have 2^{11}, or 2048 regions for 12 points and 2^{n-1} regions for n points.

TABLE 1-1

Number of points	2	3	4	5	6	...	12	...	n
Number of regions	$2 = 2^1$	$4 = 2^2$	$8 = 2^3$				?		?

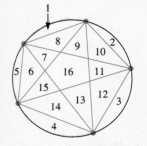

FIGURE 1-2

Before reading on, check to see whether we obtain 16 regions for 5 points. We obtain a figure similar to Figure 1-2, and our guess of 16 regions is verified. For 6 points, the pattern seems to predict that the number of regions will be 32. Choose the points so that they are not equally spaced and count the regions carefully. You should obtain 31 regions, and not 32 regions as predicted. Do you see that no matter how the points are located on the circle, our guess of 2^{n-1} for n points is not true for all n? (Checking further will verify that it is not true for the numbers 7 through 12.) This may be shocking to most of us who put our faith in the continuation of a simple pattern once we have been fortunate enough to discover it. But it should be noted that a pattern to the sequence generated by 1, 2, 4, 8, 16, 31, . . . does exist; the nth term is given by

$$\frac{n(n-1)(n-2)(n-3)}{24} + \frac{n(n-1)}{2} + 1.$$

Arithmetic Sequence

sequence

In the previous examples, the terms were given in an ordered arrangement. The word **sequence** is used to describe terms given in such a way that they can be thought of as being numbered, that is, there is a first, a second, and so on. The first three sequences below have a common property that the fourth sequence does not share:

(a) 1, 2, 3, 4, 5, 6, . . .

(b) 0, 5, 10, 15, 20, 25, . . .

(c) 2, 6, 10, 14, 18, 22, . . .

(d) 1, 11, 111, 1111, 11111, 111111, . . .

difference
arithmetic sequence

In the first sequence, each term—starting from the second—is obtained from the preceding one by adding 1. In other words, the difference between each number and the preceding one is always 1. In the second sequence, the difference between each term and the preceding one is always 5, and in the third sequence, it is always 4. However, in the fourth sequence the differences are $11 - 1 = 10$, $111 - 11 = 100$, $1111 - 111 = 1000$, and so forth. In the first three sequences, the difference between each term and the preceding one does not change, that is, it is fixed within the sequence. This is not the case in the fourth sequence. Sequences such as the first three, in which each successive term is obtained from the previous term by the addition of a fixed number, the **difference,** is an **arithmetic sequence.**

Neither pattern in Example 1-1 is an arithmetic sequence because no fixed number has been added.

E X A M P L E 1 - 3 Find a pattern in the number of matchsticks required to continue the pattern shown in Figure 1-3. Assume that the matchsticks are arranged so that each figure has one more square than the preceding figure.

FIGURE 1-3

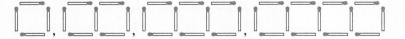

Solution The number of matchsticks required to make the successive figures are 4, 7, 10, and 13. As seen below, each term after the first is 3 units greater than the previous term:

$$
\begin{array}{lccccc}
\text{Sequence} & 4 & 7 & 10 & 13 & \_\_ \\
 & \diagdown\diagup & \diagdown\diagup & \diagdown\diagup & \diagdown\diagup & \\
\text{Difference} & 3 & 3 & 3 & 3 &
\end{array}
$$

If this pattern continues, the next three terms will be 16, 19, and 22, which indicates that the number of matchsticks in each of the next three terms will be 16, 19, and 22, respectively.

We can show that our answers are correct by demonstrating that this pattern of adding 3 each time continues. Observe that the addition of another square requires 3 matchsticks to be added to form the 3 sides of the right-most square. The fourth side is determined by an existing matchstick. Thus the sequence is an arithmetic sequence with a difference of 3. ■

It is often useful to predict the terms in a sequence. For example, suppose we want to find the number of matchsticks in the 100th figure. The strategy of *making a table* is a helpful problem-solving aid for finding such values. Table 1-2 shows the sequence in Example 1-3. The column headed "Number of Term" refers to the order of the term in the sequence. The column headed "Term" lists the accompanying terms of the sequence. We use an **ellipsis,** denoted by three dots, to indicate that the sequence continues in the same manner.

TABLE 1-2

Number of Term	Term
1	4
2	$7 = 4 + 3 = 4 + 1 \cdot 3$
3	$10 = 4 + 3 + 3 = 4 + 2 \cdot 3$
4	$13 = 4 + 3 + 3 + 3 = 4 + 3 \cdot 3$
.	.
.	.
.	.

Notice that the number of 3's in each term is one less than the number of the term. If we assume that this pattern continues, the 10th term is $4 + 9 \cdot 3$, or 31, and the 100th term is $4 + 99 \cdot 3$, or 301. The general, or nth, term of a sequence enables us to find any term, given the number of the term. In the preceding sequence, the 10th term was $4 + (10 - 1) \cdot 3$ and the 100th term was $4 + (100 - 1) \cdot 3$. Hence, the nth term should be $4 + (n - 1) \cdot 3$. For example, the 200th term for this sequence can be obtained by substituting 200 for n to obtain $4 + (200 - 1) \cdot 3$, or $4 + 199 \cdot 3$, or 601.

Some students may approach finding the 100th and nth term in Example 1-3 differently. The *Standards* recommends such an activity, which suggests increased attention to "The active involvement of students in constructing and applying mathematical ideas" (p. 129). A different approach might be as follows: If the matchstick figure has 100 squares, we could find the total number of matchsticks by adding the number of horizontal and the number of vertical sticks. There are $2 \cdot 100$ placed horizontally. Why? Notice that in the 1st figure, there are 2 matchsticks placed vertically; in the 2nd, 3; and in the 3rd, 4. In the 100th figure, there should $100 + 1$ vertical matchsticks. Altogether there will be $2 \cdot 100 + (100 + 1)$, or 301 matchsticks in the 100th figure. Similarly, in the nth figure, there would be $2n$ horizontal and $(n + 1)$ vertical matchsticks for a total of $3n + 1$. Determine if you can use algebra to show that this answer is equivalent to the one previously obtained, that is, $4 + (n - 1) \cdot 3 = 3n + 1$.

In the sequence in Table 1-2, the nth term is $4 + (n - 1)3$. We could use this expression to find the number of a term if we were given the value of the term. For example, given the term 1798 we know that $4 + (n - 1)3 = 1798$. Hence, $(n - 1)3 = 1794$ and

$n - 1 = 598$. Thus $n = 599$. Consequently, the 599th term is 1798. We would obtain the same answer by solving the equation $3n + 1 = 1798$.

A related problem is to generate the sequence if given the nth term.

EXAMPLE 1-4   Find the 1st four terms of a sequence whose nth term is given by the following:

(a) $4n + 3$ (b) $n^2 - 1$

Solution   (a) To find the 1st term, we substitute $n = 1$ in the formula $4n + 3$ to obtain $4 \cdot 1 + 3$, or 7. Similarly, substituting $n = 2, 3, 4$, we obtain $4 \cdot 2 + 3$, or 11, $4 \cdot 3 + 3$, or 15, and $4 \cdot 4 + 3$, or 19, respectively. Hence, the 1st four terms of the arithmetic sequence are 7, 11, 15, 19.

(b) Substituting $n = 1, 2, 3, 4$ in the formula $n^2 - 1$, we obtain $1^2 - 1$, or 0, $2^2 - 1$, or 3, $3^2 - 1$, or 8, $4^2 - 1$, or 15, respectively. Thus the 1st four terms of the sequence are 0, 3, 8, 15. This sequence is not arithmetic. ■

It is possible to generalize our work with arithmetic sequences. Suppose the 1st term in an arithmetic sequence is a and the difference is d. The strategy of *making a table* can be used to investigate the general term for the sequence $a, a + d, a + 2d, a + 3d, \ldots$, as shown in Table 1-3. We see that *the nth term of any sequence with first term a and difference d is given by $a + (n - 1)d$*. For example, in the sequence 5, 9, 13, 17, 21, 25, $\ldots$, the 1st term is 5 and the difference is 4. Thus the nth term is given by $a + (n - 1)d = 5 + (n - 1)4$. Simplifying algebraically, we obtain $5 + 4n - 4 = 4n + 1$.

TABLE 1-3

Number of Term	Term
1	a
2	$a + d$
3	$a + 2d$
4	$a + 3d$
5	$a + 4d$
.	.
.	.
.	.
n	$a + (n - 1)d$

Often, more than one strategy is used in solving a given problem, as the following example shows.

EXAMPLE 1-5   Find the 1st two terms of an arithmetic sequence in which the 3rd term is 13 and the 30th term is 121.

Solution   For better understanding of the problem, we use the strategy of *making a table*. We construct a table similar to Table 1-3. We do not know what the 1st two terms are, but the 3rd term is 13. Because it is given that the sequence is arithmetic, the common difference is fixed. If this fixed difference is denoted by d, then the 4th term is $13 + d$ and so on, as in Table 1-4.

TABLE 1-4

Number of Term	Term
1	?
2	?
3	13
4	$13 + d$
5	$13 + 2d$
6	$13 + 3d$
.	.
.	.
.	.
30	$13 + \square \cdot d$

 To find the 1st two terms, we need to find only d because we could subtract d from 13, the 3rd term to obtain the 2nd term, and again subtract d to obtain the 1st term. Thus we have *identified a subgoal,* which is to find d.

According to Table 1-4, it appears that the 30th term is in the form $13 + \square \cdot d$. If we could find the number of d's in this expression, we could write the equation

$$13 + \square \cdot d = 121$$

and solve it for d. Thus our new *subgoal* is to find the number of d's in the above expression.

Because the 4th term is $13 + 1d$, the 5th term is $13 + 2d$, and the 6th term is $13 + 3d$, and because we have an arithmetic sequence, we see that the number of d's in each case is always 3 less than the number of the term. Therefore the 30th term is given by $13 + 27d$. Because the 30th term is 121, we have the following:

$$13 + 27d = 121$$
$$27d = 108$$
$$d = 4$$

Now we see that the 2nd term is $13 - 4$, or 9, and the 1st term is $9 - 4$, or 5. ∎

Geometric Sequence

geometric sequence

ratio

Suppose a child in a family has 2 parents, 4 grandparents, 8 great-grandparents, 16 great-great-grandparents, and so on. We see that the numbers of ancestors form the sequence 2, 4, 8, 16, 32, ... This type of sequence is called a **geometric sequence.** Each successive term of a geometric sequence is obtained from its predecessor by multiplying by a fixed number called the **ratio.** In this example, both the 1st term and the ratio are 2. To find the nth term, examine Table 1-5.

Table 1-5 reveals a pattern: When the given term is written as a power of 2, the number of the term is the exponent of 2. Following this pattern, the 10th term is 2^{10}, or 1024, the 100th term is 2^{100}, and the nth term is 2^n. Thus the number of ancestors in the nth previous generation is 2^n.

The sequence 2, 4, 8, 16, 32, ... is a *doubling* sequence, and the nth term is 2^n. In the sequence 2, 6, 18, 54, . . . , each term is obtained from the previous one by multi-

TABLE 1-5

Number of Term	Term
1	$2 = 2^1$
2	$4 = 2 \cdot 2 = 2^2$
3	$8 = (2 \cdot 2) \cdot 2 = 2^3$
4	$16 = (2 \cdot 2 \cdot 2) \cdot 2 = 2^4$
5	$32 = (2 \cdot 2 \cdot 2 \cdot 2) \cdot 2 = 2^5$
.	.
.	.
.	.

plication by 3, as shown in Table 1-6. Following this pattern, the 10th term is $2 \cdot 3^9$, the 100th term is $2 \cdot 3^{99}$, and the nth term is $2 \cdot 3^{n-1}$.

TABLE 1-6

Number of Term	Term
1	2
2	$6 = 2 \cdot 3$
3	$18 = (2 \cdot 3) \cdot 3 = 2 \cdot 3^2$
4	$54 = (2 \cdot 3^2) \cdot 3 = 2 \cdot 3^3$
.	.
.	.
.	.

A similar approach will work for finding the nth term of any geometric sequence when given the 1st term and the common ratio. If the 1st term is a and the ratio is r, then the terms are as given in Table 1-7. We see that the nth term of any geometric sequence where the 1st term is a and the ratio is r is given by the formula ar^{n-1}. Notice that for $n = 1$, we have $ar^{1-1} = ar^0$. If $r \neq 0$, then $r^0 = 1$. (This is discussed in Chapter 6.) Thus when $n = 1$ and $r \neq 0$, we have $ar^0 = a(1) = a$. If we are given the geometric sequence 3, 12, 48, 192, . . . , the 1st term is 3 and the ratio is 4. Thus the nth term is given by $ar^{n-1} = 3 \cdot 4^{n-1}$.

TABLE 1-7

Number of Term	Term
1	a
2	ar
3	ar^2
4	ar^3
5	ar^4
.	.
.	.
.	.
n	ar^{n-1}

Other Sequences

figurate numbers

Figurate numbers provide examples of sequences that are neither arithmetic nor geometric. They can be represented by dots arranged in the shape of certain geometric figures. The number 1 is the beginning of most patterns of figurate numbers. Consider the arrays in Figure 1-4, which represent the 1st four **square numbers,** or *perfect squares*. The square numbers pictured may be written as $1^2, 2^2, 3^2$, and 4^2. The number of dots in the 10th array is 10^2, the number of dots in the 100th array is 100^2, and the number of dots in the nth array is n^2.

square numbers

FIGURE 1-4

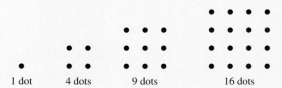

1 dot   4 dots   9 dots   16 dots

As the following diagram shows, the square numbers do not generate an arithmetic sequence because there is no common difference.

$$1 \quad 4 \quad 9 \quad 16 \quad 25 \quad 36 \quad \underline{}$$
(1st difference) $\quad 3 \quad 5 \quad 7 \quad 9 \quad 11 \quad \underline{}$

However, the sequence of first differences, 3, 5, 7, 9, 11, . . . , does form an arithmetic sequence with common difference 2.

$$3 \quad 5 \quad 7 \quad 9 \quad 11 \quad \underline{}$$
(2nd difference) $\quad 2 \quad 2 \quad 2 \quad 2 \quad 2$

Because the 2nd differences are all 2, the next 1st difference is 13. Using this information, we can determine that the next term in the original sequence is 36 + 13, or 49. Additional terms in the sequence can be generated in a similar matter.

EXAMPLE 1-6   The arrays shown in Figure 1-5 represent the first four terms of a sequence of numbers called *triangular numbers*. What is the 10th term? What is the 100th term? What is the nth term?

FIGURE 1-5

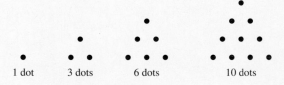

1 dot   3 dots   6 dots   10 dots

We see that the 1st four terms of the sequence are 1, 3, 6, 10. This sequence is not an arithmetic sequence because there is no common difference.

Solution   Table 1-8 suggests the sequence of numbers and a pattern for finding the desired terms. From Table 1-8, we see that the 2nd term is obtained from the 1st term by adding 2, the 3rd term is obtained from the 2nd by adding 3, and so on. In general, because the nth triangular number has n dots in the nth row, it is equal to the sum of the dots in the previous triangular number (the $(n-1)$th one) plus the n dots in the nth row. Following this pattern, the 10th term is $1 + 2 + 3 + 4 + 5 + 6 + 7 + 8 + 9 + 10$, or 55, the 100th term is $1 + 2 + 3 + 4 + 5 + \cdots + 99 + 100$, and the nth term is $1 + 2 + 3 + 4 + 5 + \cdots + (n-1) + n$.

TABLE 1-8

Number of Term	Term
1	1
2	$3 = 1 + 2$
3	$6 = 1 + 2 + 3$
4	$10 = 1 + 2 + 3 + 4$
5	$15 = 1 + 2 + 3 + 4 + 5$
.	.
.	.
.	.
10	$55 = 1 + 2 + 3 + 4 + 5 + 6 + 7 + 8 + 9 + 10$

Another way of looking at the sequence of triangular numbers is to consider differences, as we did for the square numbers.

$$
\begin{array}{ccccccccccc}
& 1 & & 3 & & 6 & & 10 & & 15 & \\
& & \searrow\!\!\!\swarrow & & \searrow\!\!\!\swarrow & & \searrow\!\!\!\swarrow & & \searrow\!\!\!\swarrow & & \searrow\!\!\!\swarrow \\
\text{(1st difference)} & & 2 & & 3 & & 4 & & 5 & & 6 \\
& & & \searrow\!\!\!\swarrow & & \searrow\!\!\!\swarrow & & \searrow\!\!\!\swarrow & & \searrow\!\!\!\swarrow & \\
\text{(2nd difference)} & & & 1 & & 1 & & 1 & & 1 &
\end{array}
$$

Using the idea of differences, we see that the next triangular number after 15 is $15 + 6$, or 21.

The next example involves sequences for which it is helpful to take more than one successive difference to find a pattern.

EXAMPLE 1-7 Assuming that the pattern you discover continues, find the 7th term in each of the following sequences:

(a) 5, 6, 14, 29, 51, 80, . . .

(b) 2, 3, 9, 23, 48, 87, . . .

Solution (a) The pattern for the differences between successive terms is not easily recognizable.

$$
\begin{array}{ccccccccccc}
& 5 & & 6 & & 14 & & 29 & & 51 & & 80 \\
& & \searrow\!\!\!\swarrow & & \searrow\!\!\!\swarrow & & \searrow\!\!\!\swarrow & & \searrow\!\!\!\swarrow & & \searrow\!\!\!\swarrow & \\
\text{(1st difference)} & & 1 & & 8 & & 15 & & 22 & & 29
\end{array}
$$

To discover a pattern for the original sequence, we try to find a pattern for the sequence of differences 1, 8, 15, 22, 29, ... This sequence is an arithmetic sequence with fixed difference 7.

$$
\begin{array}{ccccccccccc}
& 5 & & 6 & & 14 & & 29 & & 51 & & 80 \\
& & \searrow\!\!\!\swarrow & & \searrow\!\!\!\swarrow & & \searrow\!\!\!\swarrow & & \searrow\!\!\!\swarrow & & \searrow\!\!\!\swarrow & \\
\text{(1st difference)} & & 1 & & 8 & & 15 & & 22 & & 29 \\
& & & \searrow\!\!\!\swarrow & & \searrow\!\!\!\swarrow & & \searrow\!\!\!\swarrow & & \searrow\!\!\!\swarrow & \\
\text{(2nd difference)} & & & 7 & & 7 & & 7 & & 7 &
\end{array}
$$

Thus the 6th term in the first difference row is 29 + 7, or 36 and the 7th term in the original sequence is 80 + 36, or 116. What number follows 116?

(b) Because the 2nd difference is not a fixed number, we go on to the 3rd difference, as shown.

$$\begin{array}{cccccc} 2 & 3 & 9 & 23 & 48 & 87 \\ \end{array}$$

(1st difference) 1   6   14   25   39

(2nd difference)   5   8   11   14

(3rd difference)   3   3   3

The 3rd difference is a fixed number, therefore the 2nd difference is an arithmetic sequence. The 5th term in the 2nd-difference sequence is 14 + 3, or 17, the 6th term in the first-difference sequence is 39 + 17, or 56, and the 7th term in the original sequence is 87 + 56, or 143. ∎

When asked to find a pattern for a given sequence, first look for some easily recognizable pattern. If none exists, determine whether the sequence is either arithmetic or geometric. If a pattern is still unclear, taking successive differences may help. *It is possible that none of the methods described will reveal a pattern.*

PROBLEM SET 1-1

1. For each of the following sequences of figures, determine a possible pattern and show what figure would be next according to that pattern.

(a)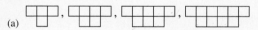

(b) △, △△, △△△, △△△△

(c)

(d)

(e) ⊠, ⊠, ⊠, ⊠

2. ▶For each of the following, list three more terms that continue a possible pattern. Describe the pattern in each case.
   (a) $1 \times 2, 2 \times 3, 3 \times 4, 4 \times 5$
   (b) 61, 57, 53, 49
   (c) 5, 6, 8, 11
   (d) 2, 5, 10, 17
   (e) X, Y, X, X, Y, X, X
   (f) 1, 3, 1, 8, 1, 13
   (g) 1, 1, 2, 3, 5, 8, 13, 21
   (h) 1, 11, 111, 1111, 11111
   (i) 1, 12, 123, 1234, 12345
   (j) $1 \times 2, 2 \times 2^2, 3 \times 2^3, 4 \times 2^4, 5 \times 2^5$
   (k) $2, 2^2, 2^4, 2^8, 2^{16}$
   (l) 5, 10, _____, 20, _____, 30, 35, _____
   (m) 0, 7, 14, _____, 28, _____, 42, _____
   (n) 0, $\frac{1}{2}$, 1, _____, 2, _____, 3, _____
   (o) 0, 22, _____, 66, _____, _____

3. In each of the following, list terms that continue a possible pattern. Which of the following sequences are arithmetic, which are geometric, and which are neither?
   (a) 1, 3, 5, 7, 9
   (b) 0, 50, 100, 150, 200
   (c) 3, 6, 12, 24, 48
   (d) 10, 100, 1000, 10,000, 100,000
   (e) $5^2, 5^3, 5^4, 5^5, 5^6$
   (f) 11, 22, 33, 44, 55
   (g) $2^1, 2^3, 2^5, 2^7, 2^9$
   (h) 9, 13, 17, 21, 25, 29
   (i) 1, 8, 27, 64, 125

4. Fill in the circles by making use of the patterns in a calendar.

(a)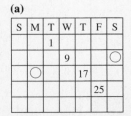

(b)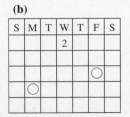

5. The following geometric arrays suggest a sequence of numbers:

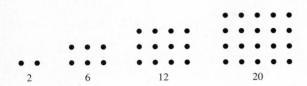

2   6   12   20

(a) Find the next three terms.
(b) Find the 100th term.
(c) Find the nth term.

6. In the following pattern, one hexagon takes 6 toothpicks to build, two hexagons take 11 toothpicks to build, and so on. How many toothpicks would it take to build (a) 10 hexagons? (b) n hexagons?

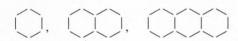

7. The first windmill takes 5 squares to build, the second takes 9 squares to build, and the third takes 13 squares to build, as shown. How many squares will it take to build (a) the 10th windmill? (b) the nth windmill?

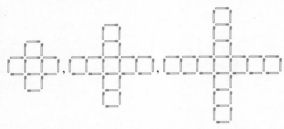

8. The school population for a certain school was predicted to increase by 50 students a year for the next 10 yrs. If the current enrollment is 700 students, what will the enrollment be after 10 yrs?

9. A tank contains 15,360 L of water. At the end of each day, half of the water is removed and not replaced. How much water is left in the tank after 10 days?

10. A well driller charges $10 a foot for the first 10 ft, $10.50 a foot for the next 10 ft, $11 a foot for the next 10 ft, and so on, increasing the price by 50¢ for each 10 ft. What is the cost of drilling a 100-ft well?

11. An employee is paid $1200 at the end of the first month on the job. Each month after that, the worker is paid $20 more than in the preceding month.
    (a) What is the employee's monthly salary at the end of the second year on the job?
    (b) How much will the employee have earned after 6 mo?
    (c) After how many months will the employee's monthly salary be $3240?

12. A commuter train picks up passengers at 7:30 A.M. If 1 person gets on at the 1st stop, 3 at the 2nd stop, 5 at the 3rd stop, and so on in this manner, how many people get on at the 10th stop?

13. Joe's annual income has been increasing each year by the same amount. The first year his income was $24,000, and the ninth year his income was $31,680. In which year was his income $45,120?

14. The following sequence of figures is made from matchsticks.

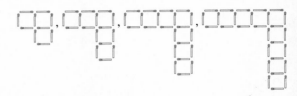

(a) Find the number of matchsticks required to construct the 100th figure.
(b) Given 500 matchsticks, what is the largest number of matchsticks that can be used to construct a single figure in the sequence?
(c) How many squares will the figure in part (b) contain?

15. Each of the following figures is made of smaller triangles like the first one in the sequence. (The second figure is made of 4 triangles.) Make a conjecture concerning the number of smaller triangles needed to make (a) the 100th and (b) the nth figure.

16. In the following sequence, the figures are made of cubes that are glued together. If the exposed surface needs to be painted, how many squares will be painted in (a) the 10th figure? (b) the nth figure?

17. The 1st difference of a sequence is 2, 4, 6, 8, ... Find the 1st six terms of the original sequence in each of the following cases:
    (a) The 1st term of the original sequence is 3.
    (b) The sum of the 1st two terms of the original sequence is 10.
    (c) The 5th term of the original sequence is 35.

18. List the next three terms to continue a pattern in each of the following. (Finding differences may be helpful.)
    (a) 5, 6, 14, 32, 64, 115, 191
    (b) 0, 2, 6, 12, 20, 30, 42
    ★(c) 10, 8, 3, 0, 4, 20, 53

19. How many terms are there in the following sequence?
    (a) 51, 52, 53, 54, . . . , 151
    (b) 1, 2, 2^2, 2^3, . . . , 2^{60}
    (c) 10, 20, 30, 40, . . . , 2000
    (d) 9, 13, 17, 21, 25, . . . , 353
    (e) 1, 2, 4, 8, 16, 32, . . . , 1024

(f) $3, 3\cdot 5, 3\cdot 5^2, 3\cdot 5^3, 3\cdot 5^4, 3\cdot 5^5, \ldots, 3\cdot 5^{20}$
(g) $3\cdot 4, 4\cdot 5, 5\cdot 6, 6\cdot 7, 7\cdot 8, \ldots, 100\cdot 101$

20. ▶(a) If a fixed number is added to each term of an arithmetic sequence, is the resulting sequence an arithmetic sequence? Justify your answer.
    (b) If each term of an arithmetic sequence is multiplied by a fixed number, will the resulting sequence always be an arithmetic sequence? Justify your answer.

21. ▶Answer the questions in Problem 20 for a geometric sequence.

22. ▶The following is an example of term-by-term addition of two arithmetic sequences:

    $$\begin{array}{r}1, 3, 5, 7, 9, 11, \ldots\\ +2, 4, 6, 8, 10, 12, \ldots\\ \hline 3, 7, 11, 15, 19, 23, \ldots\end{array}$$

    Notice that the resulting sequence is also arithmetic. Investigate whether this happens again, by trying two other examples. Will the resulting sequence always be an arithmetic sequence? Explain your answer.

23. ▶Will a term-by-term addition of two different geometric sequences ever be a geometric sequence? Justify your answer.

24. ▶Will a term-by-term multiplication of two geometric sequences always be a geometric sequence? Justify your answer.

25. The sequence $32, a, b, c, 162$ is a geometric sequence. Find a, b, c.

26. (a) Find the 1st two terms of an arithmetic sequence in which the 4th term is 24 and the 50th term is 300.
    (b) The 3rd term of an arithmetic sequence is 9 and the 6th term is 27. What is the 1st term?

27. Find the 1st five terms of the sequence whose nth term is as follows:
    (a) $n^2 + 2$ (b) $5n - 1$ (c) $10^n - 1$ (d) $3n + 2$

28. The sequence $1, 1, 2, 3, 5, 8, 13, 21, \ldots$, in which each term starting with the 3rd one is the sum of the two preceding terms, is called a *Fibonacci sequence*. This sequence is named after the great Italian mathematician Leonardo Fibonacci, who lived in the twelfth and thirteenth centuries.
    (a) Write the first 12 terms of the sequence.
    (b) Notice that the sum of the first three terms in the sequence is one less than the fifth term of the sequence. Does a similar relationship hold for the sum of the first four terms, five terms, and six terms?
    (c) Guess the sum of the first 10 terms of the sequence.
    ★(d) Make a conjecture concerning the sum of the first n terms of the sequence.

29. The Fibonacci sequence defined in Problem 28 can start with arbitrary 1st and 2nd terms. If the 1st term is 2 and the 2nd is 4, then the sequence is $2, 4, 6, 10, 16, 26, 42, \ldots$. Answer the questions in Problem 28 for this sequence.

30. Find the 100th term and the nth term in each of the sequences of Problem 3.

31. (a) Consider the following geometric arrays of pentagonal numbers. The numbers are formed by counting the dots. Find the first five numbers suggested by this sequence.
    ★(b) What is 100th pentagonal number?

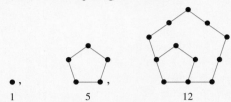

32. Consider the following sequences:
    (a) $300, 500, 700, 900, 1100, 1300, \ldots$
    (b) $2, 4, 8, 16, 32, 64, \ldots$
    Find the number of the term in which the geometric sequence becomes greater than the arithmetic sequence.

33. Start with a piece of paper. Cut it into 5 pieces. Take any one of the pieces and again cut it into 5 pieces, and so on.
    (a) What number of pieces can be obtained in this way?
    (b) What is the number of pieces after the nth experiment?

34. The "triangle" pictured below is called **Pascal's triangle,** after the French mathematician Blaise Pascal. Each number in the triangle, except the ones on the "boundary," equals the sum of two immediate neighboring numbers in the preceding row.

													Row
						1							(0)
					1		1						(1)
				1		2		1					(2)
			1		3		3		1				(3)
		1		4		6		4		1			(4)
	1		5		10		10		5		1		(5)
1		6		15		20		15		6		1	(6)
1	7		21		35		35		21		7	1	(7)

    (a) Continue the triangle by finding two more rows.
    (b) Find the sums of the numbers in the 1st row, the 2nd row, the 3rd row, and the 4th row. Do you notice a pattern? Can you predict the sum of the numbers in the 10th row? Make a general conjecture for the sum of the numbers in the nth row.
    (c) Find the "alternate" sum of numbers in each row after row 1; that is,

    $1 - 1$
    $1 - 2 + 1$
    $1 - 3 + 3 - 1$
    $1 - 4 + 6 - 4 + 1$

    (d) Find other patterns among the numbers in Pascal's triangle.

BRAIN TEASER

Find the patterns in each of the following:
(a) Find the next three terms in the following sequence:

O, T, T, F, F, S, S, E, _____, _____, _____

(b) Determine a pattern for placing letters above or below the horizontal line in the following diagram:

```
A     EF  HI KLMN        T  VWXYZ
────────────────────────────────
 BCD     G   J       OPQRS  U
```

Section 1-2 Using the Problem-solving Process

If you follow only certain patterns in attacking problems, you may risk forming a mind-set. A mind-set occurs when you approach a problem in only one way. For example, consider the following children's nursery rhyme:

As I was going to St. Ives
I met a man with seven wives.
Every wife had seven sacks,
Every sack had seven cats,
Every cat had seven kits,
Kits, cats, sacks, and wives,
How many were going to St. Ives?

Without carefully reading the rhyme, you may start counting the number of wives, sacks, cats, and kits. If you do, you have a mind-set. Reread the rhyme. There is only one person going to St. Ives. Could you solve the problem if the question were, "How many were coming from St. Ives?"

Other common mind-sets follow: Spell the word "spot" three times aloud. "S-P-O-T! S-P-O-T! S-P-O-T!" Now answer the question, "What do you do when you come to a green light?" Write your answer. If you answered "Stop," you may be guilty of forming a mind-set. You do not stop at a *green* light.

Consider the following problem: "A man had 36 sheep. All but 10 died. How many lived?"

Did you answer "10"? If you did, you are catching on and are ready to try some problems. If you did not answer "10," then you did not understand the question. *Understanding the problem* is the first step in the four-step process mentioned in the introduction. Using the four-step process does not guarantee a solution to a problem, but it does provide a systematic means of attacking problems. We now discuss the process in more detail. In the various steps of the process, we follow the recommendations of the NCTM publication *Professional Standards for Teaching Mathematics* (1991), hereafter called *The Teaching Standards*. In *Standard 5* (p. 23) it is pointed out that:

Teachers should engage students in mathematical discourse about problem solving. This includes discussing different solutions and solution strategies for a given problem, how solutions can be extended and generalized, and different kinds of problems that can be created from a given situation.

Step 1: Understanding the Problem

Understanding the Problem involves not only applying the skills necessary for literary reading but determining what is being asked, what information is known, what information is extraneous, and what information is missing or not known. Consider the following example:

Bo and Jojo went to a football game. The tickets cost $4.00 each. Bo gave the cashier $10.00 and received $2.00 in change. At the concession stand, Jojo bought two cans of juice at $1.25 each and two containers of popcorn at $0.80 each. Three minutes before halftime, Bo bought a hot dog for $0.75 and a can of juice. They left the game with 4 minutes and 13 seconds left to play. How much did they spend on juice?

In this problem, the question is clear. However, there is a lot of extraneous information. The only important information is that three cans of juice were bought and each one cost $1.25, for a total of $3.75. Another example follows:

Thanksgiving was on November 24, 1983. Memorial Day was on May 28, 1984. How many days were there between the two holidays?

In this problem, there is no extraneous information but additional information is needed. We have to know how many days there are in the months of November, December, January, February, March, and April. In addition, we have to recognize that 1984 was a leap year and that February has an additional day in a leap year. Finally, we have to know what *between* means. In mathematics, *between* is not inclusive, which means that we do not count November 24 and May 28.

Step 2: Devising a Plan

Devising a Plan involves finding a strategy to aid in solving a problem. In his book, *How to Solve It*, Polya emphasized the importance of this step when he wrote "the main achievement in the solution of a problem is to conceive the idea of a plan."

Students often ask which strategy to use for a specific problem. There is no definite answer to this question. However, being aware of and practicing general strategies for problem solving should be helpful in determining an appropriate strategy for a particular problem. A specific strategy is learned by practicing it. Once learned, strategies are simply tools to aid in the problem-solving process. The notion here is similar to that contained in the following ancient proverb:

If you give a person a loaf of bread, you feed the person for a day;
If you teach the person to bake, you feed the person for a lifetime.

We consider several strategies in detail later in this section.

Step 3: Carrying Out the Plan

Carrying out the Plan involves attempting to solve the problem with some chosen strategy. If the chosen strategy does not work, we try to devise a new strategy. Here we also perform any necessary arithmetic or algebraic computations. An important tool for performing arithmetic operations is the calculator.

Step 4: Looking Back

The looking back step is where we check the solution in terms of the original problem. We begin by asking if the answer is reasonable and if it answers the required question or questions. In this step, we also consider related problems and other ways to solve the problem.

Strategies for Problem Solving

In a problem-solving situation, a student has a goal to achieve but may not have the means to achieve it immediately. *Solving the problem* consists of constructing the means, or discovering them. This process sometimes involves unanswered questions, false starts, and dead ends. However, any of these could provide a tool to solve the given problem or some other problems. *Strategies* are tools that might be useful in discovering or constructing means to achieve a goal. For each strategy described, a problem is given that can be solved by using that strategy. Read each problem and try to solve it before reading the solution. If you need a hint, read only enough of the solution to help you get started. After you have solved the problem, compare your solution with the one in the text. Often, problems can be solved in more than one way. Your strategy may be different from that in the text, but this is all right.

Strategy—Look for a Pattern

The strategy of looking for a pattern was examined in the previous section, where we concentrated on sequences of numbers. We continue that investigation here.

PROBLEM 1

When the famous German mathematician Carl Gauss was a child, his teacher required the students to find the sum of the first 100 natural numbers. The teacher expected this problem to keep the class occupied for some time. Gauss gave the answer almost immediately. Can you?

natural numbers   **Understanding the Problem.** The **natural numbers** are 1, 2, 3, 4, ... Thus the problem is to find the sum $1 + 2 + 3 + 4 + \ldots + 100$.

Devising a Plan. One possible strategy is that of *looking for a pattern*. The obvious way of adding the numbers in the order they appear does not reveal a recognizable pattern. However, by considering $1 + 100, 2 + 99, 3 + 98, \ldots, 50 + 51$, it is evident that there are 50 pairs of numbers, each with a sum of 101, as shown in Figure 1-6.

FIGURE 1-6

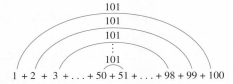

Carrying Out the Plan. There are 50 pairs, each with the sum 101. Thus the total sum is 50(101), or 5050.

Looking Back. The method is mathematically correct because addition can be performed in any order, and multiplication is repeated addition. A more general problem is to find the sum of the first n numbers, $1 + 2 + 3 + 4 + 5 + \cdots + n$, where n is any natural number. We use the same plan as before and notice the relationship in Figure 1-7. If n is an even natural number, there are $n/2$ pairs of numbers. The sum of each pair is $n + 1$. Therefore the sum $1 + 2 + 3 + \cdots + n$ is given by $(n/2)(n + 1)$. How would you find the sum if $n = 101$ and, in general, if n is odd? Does the same formula work if n is odd?

FIGURE 1-7

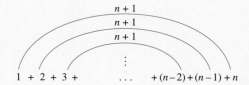

A different strategy for finding the sum $1 + 2 + 3 + \cdots + n$ involves thinking of the sum geometrically as a stack of blocks. To find the sum, we might consider the stack in Figure 1-8(a) and a stack of the same size placed differently, as in Figure 1-9(b). The total number of blocks in the stack in Figure 1-8(b) is $n(n + 1)$, which is twice the desired sum. Thus the desired sum is $n(n + 1)/2$.

FIGURE 1-8

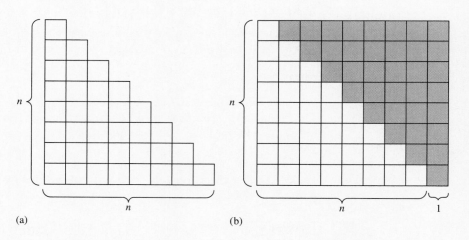

HISTORICAL NOTE

Carl Gauss (1777–1855) is regarded as the greatest mathematician of the nineteenth century and one of the greatest mathematicians of all time. Born to humble parents in Brunswick, Germany, he was an infant prodigy who, it is said, at age 3 corrected an arithmetic error in his father's bookkeeping. Gauss used to claim that he could figure before he could talk.

Gauss made contributions in the areas of astronomy, geodesy, and electricity. After Gauss's death, the King of Hanover ordered a commemorative medal prepared in his honor. On the medal was an inscription referring to Gauss as the "Prince of Mathematics," a title that has stayed with his name.

Strategy—Make a Table

A table can be used to summarize data or to help us see a pattern. It also helps us consider all possible cases in a given problem.

PROBLEM 2

How many ways are there to make change for a quarter, using only dimes, nickels, and pennies?

Understanding the Problem. There are no special limits on the number of coins that may be used to make change for a quarter. Nickels, dimes, and pennies need not all be used; that is, 25 pennies is an acceptable answer, as is 2 dimes and 1 nickel.

Devising a Plan. In this problem, the strategy of *making a table* is used to keep a record of all possibilities as they are examined.

Carrying Out the Plan. First, consider the possibilities when the number of nickels and dimes is zero and the number of pennies is 25. Continue the chart by trading nickels for pennies, as shown in Table 1-9. Are there other combinations? What about dimes? To finish the problem, consider all possibilities using dimes. Start with combinations using one dime. With one dime, the greatest number of pennies possible is 15. Next, trade nickels for pennies, as shown in Table 1-10. The last case to consider is possibilities with 2 dimes. Proceeding as before, we obtain Table 1-11. Thus there are 6 + 4 + 2 = 12 ways to make change for a quarter using only dimes, nickels, and pennies.

TABLE 1-9

D	N	P
0	0	25
0	1	20
0	2	15
0	3	10
0	4	5
0	5	0

(6 ways using 0 dimes)

TABLE 1-11

D	N	P
2	0	5
2	1	0

(2 ways using 2 dimes)

TABLE 1-10

D	N	P
1	0	15
1	1	10
1	2	5
1	3	0

(4 ways using 1 dime)

Looking Back. Check each row of each table to confirm that it shows change for a quarter. The systematic listing used in the tables shows that all cases have been considered. The problem can be extended easily by starting with an initial amount other than one quarter.

Another interesting, related problem is as follows. Given the number of coins it takes to make change for a quarter, is it possible to determine exactly which coins they are? (*Hint:* Look at the tables listing the 12 different combinations. Is the number of coins in each combination different?) If you think you know the answer, try it with a friend to see if it works. ∎

Strategy—Examine a Simpler Case

Sometimes it is possible to devise a strategy for solving a complex problem by first *solving a simpler case* of the problem.

PROBLEM 3

In a portion of a large city, the streets divide the city into square blocks of equal size, as shown in Figure 1-9. Eugene drives his taxi daily from the train depot (T) to the bus depot (B). One day, he drove due east from the train depot to the courthouse (C) along First Street and then due north to the bus depot along Seventh Avenue, covering a distance of 11 blocks in all. To avoid boredom, Eugene varies his route, but to save gas he does not want to travel any unnecessary distance. How many possible routes are there from the train depot to the bus depot?

FIGURE 1-9

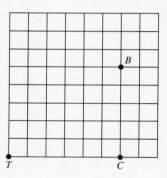

Understanding the Problem. To travel the minimum distance, Eugene should go only north (upward) and east (to the right). Two routes are shown in Figure 1-10. For each route, the total length of the horizontal segments is 6 blocks. Similarly, the total length of the vertical segments is 5 blocks. Thus the length of each of the taxi driver's routes from T to B equals 11 blocks.

FIGURE 1-10

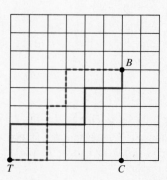

Devising a Plan. One way to solve the problem is by drawing all possible routes from T to B and counting them. Because this is a difficult task, we examine some *simpler cases*. In Figure 1-11, there is only one possible route from T to D and only one route from T to F. In fact, all points due east or all points due north of T can be reached by only one (shortest) route.

FIGURE 1-11

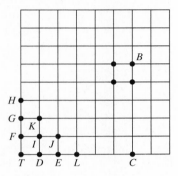

Next, we examine the routes from T to I. Only two are possible, T-D-I and T-F-I. From T to J, there are three routes, namely, T-D-I-J, T-D-E-J, and T-F-I-J. Similarly, the number of routes to various points from T can be counted. Figure 1-12 shows the number of possible routes to various points, starting from T.

FIGURE 1-12

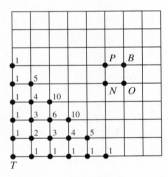

It appears that the number of routes to any given point from T is the sum of the number of routes to each of its two neighboring points, one immediately to the left and the other immediately below. If this pattern continues, it would be easy to work from point to point until we reach point B.

Carrying Out the Plan. Following the discovered pattern, we can find the number of routes to the various points in Figure 1-13. We find that there are 462 routes from T to B, as shown in Figure 1-13.

The pattern can be justified as follows. Any route from T to B in Figure 1-12 must pass through either O or P. The number of routes from T to B that passes through P is the same as that from T to P because for each route from T to P, there is one single route to B that passes through P. Similarly, the number of routes from T to B that passes through O is the same as that from T to O. Thus the number of routes from T to B is the sum of the routes from T to P and from T to O. Consequently, the pattern in Figure 1-12 continues, and there are 462 routes from T to B, as shown in Figure 1-13.

FIGURE 1-13

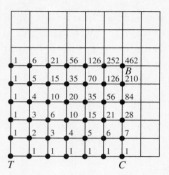

A related problem is to find the number of shortest routes from any intersection point to any other intersection point. Still another problem is to find the number of shortest routes from T to B that pass through a given intersection point.

The problem-solving strategies discussed in this text are used in elementary schools. On page 25 is an example of how the strategy of examining a simpler problem is used in *Addison-Wesley Mathematics*, 1991, Grade 5.

STRATEGY—IDENTIFY A SUBGOAL

In an attempt to devise a plan for solving some problems, it may become apparent that the problem could be solved if the solution to a somewhat easier or more familiar problem could be found. In such a case, finding the solution to the easier problem may become a *subgoal* of the primary goal of solving the original problem. Problem 4 shows an example of this.

PROBLEM 4

Arrange the numbers 1 through 9 into a square subdivided into nine smaller squares like the one shown in Figure 1-14, so that the sum of every row, column, and main diagonal is the same. (The result is called a *magic square*.)

FIGURE 1-14

Understanding the Problem. We need to put each of the nine numbers 1, 2, 3, . . . , 9 in the small squares, a different number in each square, so that the sum of the numbers in each row, in each column, and in each of the two diagonals is the same.

Devising a Plan. If we know the fixed sum of the numbers in each row, column, and diagonal, we would have a better idea of which numbers can appear together in a single row, column, or diagonal. Thus our *subgoal* is to find that fixed sum. The sum of the nine numbers, $1 + 2 + 3 + \ldots + 9$ equals 3 times the sum in one row. Consequently, the fixed sum is obtained by dividing $1 + 2 + 3 + \ldots + 9$ by 3. We obtain $45 \div 3 = 15$; hence, the sum in each row, column, and diagonal must be 15. We need to decide what numbers could occupy the various squares. The number in the center space will appear in 4 sums,

Problem Solving
Solve a Simpler Problem

LEARN ABOUT IT

Sometimes you can find the answer to a problem by solving a problem that is like it but has smaller numbers. This problem solving strategy is called **Solve a Simpler Problem.**

Roger is helping to set up 20 small tables for an exhibit of Eskimo sculpture at his school. Each table can have one sculpture on each side. If Roger pushes all the tables together to make one long table, how many sculptures can he display?

> Instead of drawing 20 tables, I'll solve a simpler problem using 2, 3, and 4 tables. That may help me solve the more difficult problem.

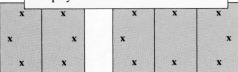

> I see! Two sculptures can be displayed at each of the 20 tables. Then 2 more sculptures can be placed at each end.

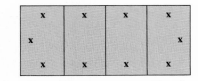

$2 \times 20 = 40$ $40 + 2 = 42$

42 sculptures can be displayed.

TRY IT OUT

Thelma was in charge of planning a dinner for visiting students from Anchorage, Alaska. She made a large square table from 25 small square tables. Each small table can seat one person on a side. How many students can sit at the large table? (Hint: Try 4 small tables. Then try 9.)

- How many students can sit on each side of a small square table?
- What shape does Thelma want the large table to be?
- Copy and complete the drawings to solve the problem.

each adding to 15 (2 diagonals, the second row and the second column). Each number in the corners will appear in three sums of 15. (Do you see why?) If we write 15 as a sum of three different numbers 1 through 9 in all possible ways, we could then count how many sums contain each of the numbers 1 through 9. The numbers that appear in at least four sums are candidates for placement in the center square, whereas the numbers that appear in at least three sums are candidates for the corner squares. Thus our new *subgoal* is to write 15 in as many ways as possible as a sum of three different numbers from the set: {1, 2, 3, . . . , 9}.

Carrying Out the Plan. We see that 15 can be written systematically as follows:

$$9 + 5 + 1$$
$$9 + 4 + 2$$
$$8 + 6 + 1$$
$$8 + 5 + 2$$
$$8 + 4 + 3$$
$$7 + 6 + 2$$
$$7 + 5 + 3$$
$$6 + 5 + 4$$

Notice that the order in each sum is not important. (Do you see why?) Hence, $1 + 5 + 9$ and $5 + 1 + 9$, for example, are counted as the same. Notice that 1 appears in only two sums, 2 in three sums, 3 in two sums, and so on. Table 1-12 summarizes this pattern:

TABLE 1-12

Number	1	2	3	4	5	6	7	8	9
Number of sums containing the number	2	3	2	3	4	3	2	3	2

We see that the only number that appears in four sums is 5; hence, 5 must be in the center of the square. (Do you see why?) Because 2, 4, 6, and 8 appear three times each, they must go in the corners. Suppose we choose 2 for the upper left corner. Then 8 must be in the lower right corner. (Why?) This is shown in Figure 1-15(a). Now we could place 6 in the lower left corner or upper right corner. If we choose the upper right corner, we obtain the result in Figure 1-15(b). The magic square can now be completed, as shown in Figure 1-15(c).

FIGURE 1-15

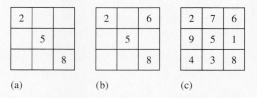

(a) (b) (c)

Looking Back. We have seen that 5 was the only number among the given numbers that could appear in the center. However, we had various choices for a corner, and hence it seems that the magic square we found is not the only one possible. Can you find all the others?

Another way to see that 5 must be in the center square is to consider the sums $1 + 9$, $2 + 8$, $3 + 7$, $4 + 6$, as shown in Figure 1-16. We could add 5 to each to obtain 15. ∎

FIGURE 1-16

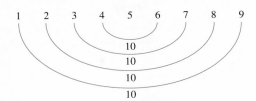

Strategy—Examine a Related Problem

Sometimes we find that the problem we are working on is similar to a previously encountered problem. If so, it is often possible to apply a similar approach to solve the new problem. Such is the case in Problem 5.

PROBLEM 5

Ryan is building matchstick square sequences, as shown in Figure 1-17. He used 67 matchsticks to form the last figure in his sequence. How many matchsticks will he use for the entire project?

FIGURE 1-17

Understanding the Problem. From our experience with patterns in Example 1-3, we recognize the sequence generated by the matchsticks as 4, 7, 10, 13, . . . , 67. The last number is 67, because Ryan used 67 matchsticks to form the last figure in the sequence. This is an arithmetic sequence with difference 3. We are to find the sum of the numbers in this sequence.

Devising a Plan. A *related problem* is Gauss's problem of finding the sum $1 + 2 + 3 + 4 + \ldots + 100$. In that problem, we paired 1 with 100, 2 with 99, 3 with 98, and so on, and observed that there were 50 pairs of numbers, each with a sum of 101. A similar approach in the present problem yields a sum of 71. To find the total, we need to know the number of pairs in Figure 1-18.

FIGURE 1-18

   To find the number of pairs, we need the number of terms in the sequence. Thus we have identified a *subgoal,* which is to find the number of terms in the sequence. In Example 1-3, we found the nth term of this sequence to be $4 + (n - 1)3$. To find the number of the term corresponding to 67, we solve the equation $4 + (n - 1)3 = 67$ and obtain $n = 22$. Thus there are 22 terms in the given sequence.

Carrying Out the Plan. Because the number of terms is 22, we have 11 pairs of matchstick figures whose sum is 71 matchsticks each. Therefore the total is 11×71, or 781 matchsticks.

Looking Back. Using the outlined procedure, we should be able to find the sum of any arithmetic sequence in which we know the first two terms and the last term.

Strategy—Work Backward

In some problems, it is easier to start with what might be considered the final result and to work backward. This strategy is used in Problem 6.

PROBLEM 6

Charles and Cynthia are playing a game called NIM. Each has a box of matchsticks. They take turns putting 1, 2, or 3 matchsticks in a common pile. The person who is able to add a number of matchsticks to the pile to make a total of 24 wins the game. What should be Charles' strategy to ensure he wins the game?

Understanding the Problem. Each player chooses 1, 2, or 3 matchsticks to place in the pile. If Charles puts 3 matchsticks in the pile, Cynthia may put 1, 2, or 3 matchsticks in the pile, which makes a total of 4, 5, or 6. It is now Charles' turn. Whoever makes a total of 24 wins the game.

Devising a Plan. Here the strategy of *working backward* can be used. If there are 21, 22, or 23 matchsticks in the pile, Charles would like it to be his turn because he can win by adding 3, 2, or 1 matchsticks, respectively. However, if there are 20 matchsticks in the pile, Charles would like for it to be Cynthia's turn because she must add 1, 2, or 3, which would give a total of 21, 22, or 23. A *subgoal* for Charles is to reach 20 matchsticks, which forces Cynthia's total to be 21, 22, or 23. The subgoal of 20 matchsticks can be reached if there are 17, 18, or 19 matchsticks in the pile when Cynthia has completed her turn. For this to happen, there should be 16 matchsticks in the pile when Charles has completed his turn. Hence, a new subgoal for Charles is to reach 16 matchsticks. By similar reasoning, we see that Charles's additional subgoals are to reach 12, 8, and 4 matchsticks.

Carrying Out the Plan. Using the reasoning developed in Devising a Plan, we see that the winning strategy for Charles is to have a total of 4 matchsticks and then reach the totals of 8, 12, 16, 20, and 24 on successive turns. To do this, Charles should play second; if Cynthia puts 1, 2, or 3 matchsticks in the pile, Charles should add 3, 2, or 1, respectively, to make a total of 4. The totals 8, 12, 16, 20, and 24 can be achieved in a similar fashion.

Looking Back. A related problem is to solve the game in which the person who reaches 24 or more matchsticks loses. Now what is the winning strategy? Other related games can be examined in which different numbers are used as goals or different numbers of matchsticks are allowed to be added. For example, suppose that the goal is 21 and that 1, 3, or 5 matchsticks can be added each time.

Strategy—Write an Equation

A problem-solving strategy commonly used in algebra consists of writing an equation. We discuss how to write equations and solve them in Chapter 4. Here, we discuss the strategy for problems in which the solutions to the corresponding equations require little or no algebra.

PROBLEM 7

A farmer wants to enclose a rectangular grazing area using an existing fence on one side and 244 ft of new fencing on three sides. If the length of the rectangle is supposed to be twice as long as it is wide, what are the dimensions of the enclosure?

Understanding the Problem. We need to find the width and length of the rectangular enclosure shown in Figure 1-19(a). The length is to be twice the width, and the total new fencing is 244 ft.

FIGURE 1-19

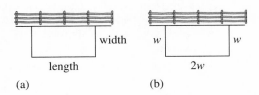

Devising a Plan. If we designate the width by w, then the length must be $2 \cdot w$, as shown in Figure 1-19(b). To find w, we use the strategy of *writing an equation*. The fencing is 244 ft. The total new fencing of 244 ft is the sum of the lengths of the three sides of the rectangle shown in Figure 1-19(b), that is, $w + w + 2w$ or $4w$. Thus $4w = 244$. It remains to solve this equation for w.

Carrying Out the Plan. $4 \cdot w = 244$ implies $w = \dfrac{244}{4} = 61$. Because the width is 61 ft, the length is $2 \cdot w = 2 \cdot 61$ ft $= 122$ ft.

Looking Back. It is important to check the solution in relation to the original problem. Because the width is 61 ft and the length is 122 ft, the total fencing should be $61 + 61 + 122$, or 244, ft, which agrees with what is given in the problem. Does the problem have a unique solution without the restriction that the length is twice the width? ∎

Strategy—Draw a Diagram

It has often been said that a picture is worth a thousand words. This is particularly true in problem solving. In geometry, drawing a picture often provides the insight necessary to solve a problem. A nongeometric problem that can be solved by drawing a diagram is offered in Problem 8.

PROBLEM 8

On the first day of math class, 20 people are present in the room. To become acquainted with one another, each person shakes hands just once with everyone else. How many handshakes take place.

Understanding the Problem. There are 20 people in the room, and each person shakes hands with each other person only once. It takes 2 people for 1 handshake; that is, if Maria shakes hands with John and John shakes hands with Maria, this counts as 1 handshake, not 2. The problem is to find the number of handshakes that take place.

Devising a Plan. One plan that would certainly work is to take 20 people and actually count the handshakes. Although this plan provides a solution, it would be nice to find a less elaborate one. One way of investigating this problem is to use the strategy of *drawing a diagram*. A diagram showing a handshake between persons A and B can be indicated by a line segment connecting A and B.

Diagrams showing handshakes for 3, 4, and 5 people are given in Figure 1-20. From the diagrams, we see that the problem becomes one of counting the different line segments needed to connect various numbers of points. In looking at the problem for 5 people (Figure 1-20c), we see that A shakes hands with persons B, C, D, and E (4 handshakes). Also, B shakes hands with A, C, D, and E (4 handshakes). In fact, each person shakes hands with 4 other people. Therefore it appears that there are 5 · 4, or 20, handshakes. However, notice that the handshake between A and B has been counted twice. This dual counting occurs for all 5 people. Consequently, each handshake was counted twice; thus to obtain the answer, we must divide by 2. The answer is (5 · 4)/2, or 10. This approach can be generalized for any number of people.

FIGURE 1-20

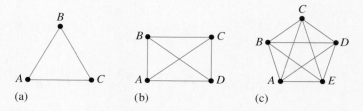

(a)   (b)   (c)

Carrying Out the Plan. Using the outlined strategy, we see that with 20 people there are (20 · 19)/2, or 190, handshakes.

Looking Back. Our answer can be checked by solving the problem by means of a different strategy. We try the strategy of *looking at a simpler problem*. With one person in the room, there are no handshakes. If a second person enters the room, there is 1 handshake (remember, 2 people shaking hands counts as 1 handshake). If a third person enters the room, he or she shakes hands with each of the other persons present, so there are 2 additional handshakes, for a total of 1 + 2. If a fourth person enters the room, he or she shakes hands with each of the other three members present, so there is an addition of 3 shakes for a total of 1 + 2 + 3. If a fifth person enters the room, an additional 4 shakes take place.

TABLE 1-13

Number of People	Number of Handshakes
1	0
2	1
3	3 = 1 + 2
4	6 = 1 + 2 + 3
5	10 = 1 + 2 + 3 + 4

In Table 1-13, we record the number of handshakes. Notice that the last number in the expression 1 + 2 + 3 + 4 is one less than the number of people shaking hands. Following this pattern, the answer for 20 people is given by 1 + 2 + 3 + 4 + · · · + 19. A *related problem* used by Gauss (Problem 1) to find sums of consecutive natural numbers is very useful in completing the problem. Applying this technique, we have the following:

$$1 + 2 + 3 + \cdots + 19 = \frac{19(20)}{2} = 190.$$

Strategy—Guess and Check

In the strategy of guess and check, we first guess at a solution using as reasonable a guess as possible. We then check to see if the guess is correct. If not, the next step is to learn as much as possible about the solution based on the guess before we make the next guess. This strategy can be regarded as a form of trial and error, where the information about the error helps us choose what trial to make next. The guess and check strategy is seen in Problem 9.

PROBLEM 9

Marques, a fourth grader, says to Mr. Treacher, "I'm thinking of a number less than or equal to 1000. Can you guess my number?" Mr. Treacher replies, "Not only can I guess your number, but I can guess it within no more than 10 questions, provided your answers to my questions are yes or no and are truthful." How can Mr. Treacher be so positive about the maximum number of questions he will have to ask?

Understanding the Problem. To guarantee that Mr. Treacher can make good his statement, we need to know that Marques is thinking of a counting number. There are 1000 possibilities for the number. What types of questions could Mr. Treacher ask? Suppose he asked, "Is the number 47?" With this type of question, it seems impossible to determine Marques's number within ten or fewer questions. He needs to ask questions of a form such that he learns information about more than one number each time. For example, if he asked if the number is even, then no matter what the answer was, he would have only 500 numbers left to worry about with his next question.

Devising a Plan. What types of questions should Mr. Treacher ask? A strategy that he could use here is *guess and check*, where successive guess is based on the information learned from Marques's answer to his previous question. In his questions, Mr. Treacher should try not only to narrow the number of numbers left to choose from but also to determine how far apart they are, that is, to find the range of the numbers. For example, if his first question is "Is the number less than 500?" and Marques answers yes, his number is in the range from 1 to 499. If he answers no, then his number is in the range from 500 to 1000. An equally good question is, "Is the number greater than 500?" Questions like this are better than "Is it an even number?" because if the answer to this question is yes, then there are 500 numbers left, but the range is from 2 to 1000. By successively asking questions such as, "Is the number less than 500?", where each question determines the range for the next question, Mr. Treacher successively halves both the field of numbers and the range in which the number lies.

Carrying Out the Plan. Suppose Marques chose 38 for his number. The questions and answers might be as in Table 1-14.

TABLE 1-14

Mr. Treacher's Questions	Marques's Answers
Is the number less than 500?	Yes
Is the number less than 250?	Yes
Is the number less than 125?	Yes
Is the number less than 62?	Yes
Is the number less than 31?	No
Is the number less than 46?	Yes
Is the number less than 38?	No
Is the number less than 42?	Yes
Is the number less than 40?	Yes
Is the number less than 39?	Yes

Consequently, Marques's number is 38.

Each time Mr. Treacher asks a question, he halves the range of numbers he needs to consider. When the range contains only one number, Mr. Treacher knows what Marques's number is. Because 1000 is between $2^9 = 512$ and $2^{10} = 1024$, it takes 10 times to repeatedly halve 1000 and the successive answers to obtain a range of numbers containing 1 number.

Looking Back. The guess and check strategy, in which we utilize the information gleaned from a question to determine the next guess, appears to be the most efficient way to attack the problem. We could change the range for Marques's number in order to change the problem. We also could use a calculator to find the values to ask about in the questions. If Marques's number were between 1 and 1,000,000, what would be the maximum number of questions that Mr. Treacher would have to ask to determine the number? If Mr. Treacher could ask 30 questions, what would be the largest allowable range for Marques's number within which Mr. Treacher could determine the chosen number? ∎

Strategy—Use Indirect Reasoning

To show that a statement is true, it is sometimes easier to show that it is impossible for the statement to be false. This can be done by showing that if the statement were false, something contradictory or impossible would follow. This approach is useful when it is difficult to start a direct argument and, when negating the given statement, gives us something tangible with which to work. An example follows.

PROBLEM 10

In Figure 1-21, you are given a checkerboard with the two squares on opposite corners removed and a set of dominoes such that each domino can cover 2 squares on the board. Can the dominoes be arranged in such a way that all the remaining squares on the board can be covered? If not, why not?

Understanding the Problem. Two red spaces on opposite corners were removed from the checkerboard in Figure 1-21. We are asked whether it is possible to cover the remaining 62 squares with dominoes the size of 2 squares.

FIGURE 1-21

Devising a Plan. If we try to cover the board in Figure 1-21 with dominoes, we will find the dominoes do not fit and some squares will remain uncovered. To show that there is no way to cover the board with dominoes, we use *indirect reasoning*. If the remaining 62 squares could be covered with dominoes, it would take 31 dominoes to accomplish the task. We want to show now that this implies something impossible.

Carrying Out the Plan. Each domino must cover 1 black and 1 red square. Hence, 31 dominoes would cover 31 red and 31 black squares. This is impossible, however, because the board in Figure 1-21 has 30 red and 32 black squares. Consequently, our assumption that the board in Figure 1-21 can be covered with dominoes is wrong.

Looking Back. The counting of black and red squares implies that if we remove any number of squares from a checkerboard so that the number of remaining red squares differs from the number of remaining black squares, the board cannot be covered with dominoes. (Do you see why?) We could also investigate what happens when two squares of the same color are removed from an 8-by-7 board and other-sized boards. Also, is it always possible to cover the remaining board if two squares of opposite colors are removed? ∎

Before you attempt the problems in the problem set, try the following puzzles. These puzzles have been around in one form or another for many years. They should help you begin to think and to understand what is really being asked in a problem.

1. How much dirt is in a hole 2 ft long, 3 ft wide, and 2 ft deep?
2. Two U.S. coins have a total value of 55¢. One coin is not a nickel. What are the two coins?
3. Walter had a dozen apples in his office. He ate all but 4. How many are left?
4. Sal owns 20 blue and 20 brown socks, which he keeps in a drawer in complete disorder. What is the minimum number of socks that he must pull out of the drawer on a dark morning to be sure he has a matching pair?
5. You have 8 sticks. Four of them are exactly half the length of the other 4. Enclose exactly 3 squares of equal size with them.
6. Suppose you have only one 5-L container and one 3-L container. How can you measure exactly 4 L of water if neither container is marked for measuring?
7. What is the minimum number of pitches possible for a pitcher to make in a major league baseball game, assuming that he plays the entire 9-inning game and it is not called prior to completion?

8. Consider the following banking transaction. Deposit $50 and withdraw it as follows:

Withdraw	$20	Leaving	$30
Withdraw	15	Leaving	15
Withdraw	9	Leaving	6
Withdraw	6	Leaving	0
	$50		$51

   Where did the extra dollar come from? To whom does it belong?

9. A businessperson bought 4 pieces of solid-gold chain, each consisting of 3 links.

   He wanted to keep them as an investment, but his wife felt that joined together, the pieces would make a lovely necklace. A jeweler charges $10.50 to break a link and $10.50 to rejoin it. What is the minimum charge possible to form a necklace using all the pieces?

10. Two people played 5 games of checkers. Each won 3 games. How is that possible?

11. How many animals of each species did Adam take with him on the ark?

12. There are four volumes of Shakespeare's collected works on a shelf. The volumes are in order from left to right. The pages of each volume are exactly 2 in. thick. The covers are each $\frac{1}{6}$ in. thick. A bookworm started eating at page 1 of Volume I and ate through to the last page of Volume IV. What is the distance the bookworm traveled?

PROBLEM SET 1-2

1. Use Gauss's approach in Problem 1, p. 19, to find the following sums (do not use formulas):
   (a) $1 + 2 + 3 + 4 + \ldots + 99$
   (b) $1 + 2 + 3 + 4 + \ldots + n$ where n is odd
   (c) $1 + 3 + 5 + 7 + \ldots + 1001$

2. ▶An alternate version of the story of Gauss computing $1 + 2 + 3 + \cdots + 100$ reports that he simply listed the numbers in the following way to discover the sum:
   $$1 + 2 + 3 + 4 + 5 + \cdots + 98 + 99 + 100$$
   $$100 + 99 + 98 + 97 + 96 + \cdots + 3 + 2 + 1$$
   $$\overline{101 + 101 + 101 + 101 + 101 + \cdots + 101 + 101 + 101}$$

   Does this method give the same answer? Discuss the advantages of this method over the one described in the text.

3. In the array below, there are 6 rows of 6 squares each. Count the number of squares in two ways. In one way, the number of squares is $6 \cdot 6$, or 6^2. In the other way, count the number of squares starting with the single square in the upper left corner and then along the L-shaped routes. Consequently, notice that $1 + 3 + 5 + 7 + 9 + 11 =$ $36 = 6^2$. Generalize this result to an n-by-n array of squares.

   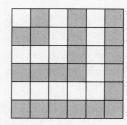

4. How many different squares are in the following figure?

   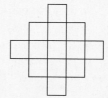

SECTION 1-2 EXERCISES 35

5. What is the largest sum of money—all in coins and no silver dollars—that I could have in my pocket without being able to give change for a dollar, a half-dollar, a quarter, a dime, or a nickel?

6. ▶(a) Using the existing lines on the checkerboard shown in the figure, how many different squares are there? Justify your answer.

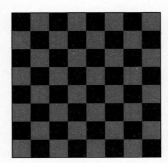

(b) If the number of rows and columns of the checkerboard is doubled, is the number of different squares doubled? Justify your answer.

7. How can you cook an egg for exactly 15 min if all you have are a 7-min and an 11-min timer?

8. How many different ways can you make change for a $50 bill using $5, $10, and $20 bills?

9. How many four-digit numbers have the same digits as 1993?

★10. Looking out in the backyard one day, I saw an assortment of boys and dogs. Counting heads, I got 22. Counting feet, I got 68. How many boys and how many dogs were in the yard?

11. A compass and a ruler together cost $4. The compass costs 90¢ more than the ruler. How much does the compass cost?

12. A cat is at the bottom of an 18-ft well. Each day it climbs up 3 ft, and each night it slides back 2 ft. How long will it take the cat to get out of the well?

13. Two houses on the same street are separated by a large, empty field. The first house is numbered 29, and the other house is numbered 211. An architect is designing 13 new houses to be built between the two existing houses.
    (a) What should the numbers of the new houses be if along with the existing houses, the numbers need to form an arithmetic sequence?
    (b) What is the difference of this sequence?

14. Same-sized cubes are glued together to form a staircase-like sequence of solids as shown.

(a) How many cubes will it take to build a staircase that is 25 cubes high?

(b) All of the faces of cubes not glued together need to be painted. How many squares will need to be painted in (i) the 100th solid? (ii) the nth solid?

15. You are standing on the middle rung of a ladder. If you first move up 3 rungs, then move down 5 rungs, and then climb up 10 rungs to get onto the roof, how many rungs are on the ladder?

16. David, Judy, and Jacobo are playing with marbles. Each has a different number of marbles, and they decide to lend each other some. David lends Judy and Jacobo as many marbles as they already have. Then Judy lends David and Jacobo as much as they already have. Finally, Jacobo does the same; that is, he lends David and Judy as many marbles as they already have. Each child now has 48 marbles. How many did each have originally?

17. Jacobo decides to give half of his marbles to Judy and 2 to David. From what remains, he gives half to another friend and 2 more to David. From what remains now, he gives one third to another friend and 20 more to David. If he is finally left with 100 marbles, how many marbles did Jacobo have at the start?

18. ▶Eight marbles look alike, but one is slightly heavier than the others. Using a balance scale, explain how you can determine the heavier one in exactly
    (a) 3 weighings.   (b) 2 weighings.

19. Marc goes to the store with exactly $1.00 in change. He has at least one of each coin less than a half-dollar coin, but he does not have a half-dollar coin.
    (a) What is the least number of coins he could have?
    (b) What is the greatest number of coins he could have?

20. A farmer needs to fence a rectangular piece of land. She wants the length of the field to be 80 ft longer than the width. If she has 1080 ft of fencing material, what should the length and the width of the field be?

21. Find a 3-by-3 magic square using the numbers 3, 5, 7, 9, 11, 13, 15, 17, 19.

22. ▶Explain why it is impossible to have a 3×3 magic square with numbers 1, 3, 4, 5, 6, 7, 8, 9, 10.

23. ▶Use the strategy of *indirect reasoning* to justify the following:
    (a) If the product of two positive numbers is greater than 82, then at least one of the numbers is greater than 9.
    (b) If the product of two positive numbers is greater than 81, then at least one of the numbers is greater than 9.

24. Find the following sums:
    (a) $2 + 4 + 6 + 8 + 10 + \cdots + 1020$
    (b) $1 + 6 + 11 + 16 + 21 + \cdots + 1001$
    (c) $3 + 7 + 11 + 15 + 19 + \cdots + 403$

25. There were 20 people at a round table for dinner each of whom shook hands with the person on his or her immediate right and left. At the end of the dinner, each person got up and shook hands with everybody except with the people who sat to his or her immediate right or left at dinner. Find

the number of handshakes that took place after dinner.

26. A student has a sheet of $8\frac{1}{2}$- by 11-in. paper. She needs to measure exactly 6 in. Can she do it using the sheet of paper?

27. In the game of Life, José had to pay $1500 when he was married; then he lost half the money he had left. Next, he paid half the money he had for a house. Then the game was stopped, and he had $3000 left. With how much money did he start?

28. ▶Three friends play a game in which the loser is to double the money of each of the other two. After three games, each has lost exactly once and each ended up with $48. How much did each have initially? Explain your reasoning.

29. Two distinct lines may intersect in at most one point. The following figure shows that three distinct lines may intersect in at most three points. To determine the maximum number of intersection points given any number of lines, answer each of the following:

(a) Use the strategy and an approach similar to the one in the solution to Problem 8, p. 29, to find the maximum number of intersection points for 20 lines. Explain your reasoning.

▶(b) What is the maximum number of intersection points for n lines? Explain your reasoning.

★30. Ten women are fishing all in a row in a boat. One seat in the center of the boat is empty. The 5 women in the front of the boat want to change seats with the 5 women in the back of the boat. A person can move from her seat to the next empty seat or she can step over one person without capsizing the boat. What is the minimum number of moves needed for the 5 women in front to change places with the 5 in back?

Review Problems

31. List three more terms to complete a possible pattern.
    (a) 3, 6, 9, 12, 15, 18,
    (b) 1, 2, 3, 2, 9, 2, 27, 2, 81, 2,
32. Find the nth term for the sequence 22, 32, 42, 52, ...
33. How many terms are in the sequence 3, 7, 11, 15, 19, . . . , 83.
34. Find the sums of the terms in the sequence in Problem 33.

▼ BRAIN TEASER

What day follows the day before yesterday if 2 days from now it will be Sunday?

LABORATORY ACTIVITY

Place a half-dollar, a quarter, and a nickel in position A, as shown in the figure below. Try to move these coins, one at a time, to position C. At no time may a larger coin be placed on a smaller coin. Coins may be placed in position B. How many moves does it take? Now, add a penny to the pile and see how many moves it takes. This is a simple case of the famous Tower of Hanoi problem, in which ancient Brahman priests were required to move a pile of 64 disks of decreasing size, after which the world would end. How long would this take at a rate of one move per second?

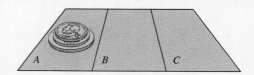

Section 1-3 Using a Calculator as a Problem-solving Tool

At the K–4 level, the *Standards* (p. 19) makes the following points:

The K–4 curriculum should make appropriate and ongoing use of calculators and computers. Calculators must be accepted at the K–4 level as valuable tools for learning mathematics. Calculators enable children to explore number ideas and patterns, to have valuable concept-development experiences, to focus on problem-solving processes, and to investigate realistic applications. The thoughtful use of calculators can increase the quality of the curriculum as well as the quality of children's learning.

Calculators do not replace the need to learn basic facts, to compute mentally, or to do reasonable paper-and-pencil computation. Classroom experience indicates that young children take a commonsense view about calculators and recognize the importance of not relying on them when it is more appropriate to compute in other ways. The availability of calculators means, however, that educators must develop a broader view of the various ways computation can be carried out and must place less emphasis on complex paper-and-pencil computation. Calculators also highlight the importance of teaching children to recognize whether computed results are reasonable.

The *Standards* assumes that a calculator with features similar to the scientific calculator described next will be available at all times for all students in grades 4–8. In this section, we discuss features of calculators and give some examples and problems appropriate for calculator use.

Recommended Features for Elementary School Calculators

At the upper elementary level, calculators should be able to do more than the basic four operations. For these grades and for this text, more advanced calculators, called *scientific calculators,* should be used. Scientific calculators differ greatly in the keys available, but we recommend that they have the following: $\boxed{y^x}$ (power), $\boxed{\sqrt{}}$ (square root), $\boxed{x!}$ (factorial), $\boxed{+/-}$ (change of sign), $\boxed{\pi}$ (pi), and $\boxed{1/x}$ (reciprocal). Further, they should be able to handle scientific notation and the correct order of operations, and they should have memory keys. We also recommend using solar-powered calculators so that batteries are not a problem. Figure 1-22 shows a scientific calculator.

The calculator should have an easily accessible on-off switch. The position of the keys on the keyboard may vary, but the keys should be adequately spaced and large enough so that the user's finger will press no more than one key at a time. Separate $\boxed{\text{CLEAR}}$ and $\boxed{\text{CLEAR ENTRY}}$ keys are desirable. These keys are usually abbreviated as $\boxed{\text{C}}$ and $\boxed{\text{CE}}$.

FIGURE 1-22

Types of Logic

The types of logic built into a calculator determines how a computation is performed. With different types of logic, entering a computation as it is written may or may not produce the correct result. The order in which operations are done in mathematics is very important, and you must be familiar with how your calculator works if you are to compute correctly. Be sure to retain and study the owner's manual that comes with your calculator. Many inexpensive four-function calculators process operations in the order in which they are entered. For example,

$$\boxed{2}\,\boxed{+}\,\boxed{4}\,\boxed{\times}\,\boxed{5}\,\boxed{-}\,\boxed{4}\,\boxed{\div}\,\boxed{2}\,\boxed{=}$$

would be evaluated as

$$6 \cdot 5 - 4 \div 2$$

then as

$$30 - 4 \div 2$$

and, finally, as

$$26 \div 2 = 13.$$

However, multiplications and divisions should be done in order from left to right before additions and subtractions. Thus the correct solution to the problem $2 + 4 \cdot 5 - 4 \div 2$ is $2 + 20 - 2 = 22 - 2 = 20$. This logic is especially disturbing when we try to perform a computation like $\frac{1}{2} + \frac{1}{4}$ by pressing the keys in the order given, that is, $\boxed{1}\,\boxed{\div}\,\boxed{2}\,\boxed{+}\,\boxed{1}\,\boxed{\div}\,\boxed{4}\,\boxed{=}$. A calculator with this type of logic will evaluate $\frac{1}{2} + \frac{1}{4}$ as 0.375 rather than as 0.75. (Do you see why?)

algebraic operating system

Many calculators, such as the one in Figure 1-22, include a desirable feature called an **algebraic operating system.** It evaluates expressions inside parentheses first, then multiplications and divisions, and then additions and subtractions. For example, in the computation $2 + 4 \cdot 5 - 4 \div 2$, if $\boxed{2}\,\boxed{+}\,\boxed{4}$ is entered the calculator will not perform the addition. If $\boxed{2}\,\boxed{+}\,\boxed{4}\,\boxed{\times}$ is entered, no calculations will be completed. After $\boxed{2}\,\boxed{+}\,\boxed{4}\,\boxed{\times}\,\boxed{5}\,\boxed{-}$ is entered, the display will show 22. In other words, the calculator performs 4×5 before adding 2. A calculator with the algebraic operating system feature will complete the original problem $2 + 4 \cdot 5 - 4 \div 2$ and give the desired answer of 20 if the computation is entered in the order in which it is written and the $\boxed{=}$ key is pressed. If the calculator has parentheses, these keys can be used to ensure that the operations are performed in the desired order.

Decimal Notation

The calculator should have a floating decimal point or a method of allowing the user to input the number of decimal places in a computation. For example, when $\boxed{1}\,\boxed{\div}\,\boxed{3}\,\boxed{=}$ is entered, the display should show 0.3333333 rather than 0.33, as some displays do on fixed-point machines. Be aware of how a calculator rounds decimals, if it does. For example, in $\boxed{2}\,\boxed{\div}\,\boxed{3}\,\boxed{=}$, the display with a floating decimal may show 0.6666666 or 0.6666667. If the display shows 0.6666667, the round-off is apparent. If it shows 0.6666666, then we can multiply by 3 and observe the result, which may be either 1.9999998, 1.9999999, or 2. If 1.9999998 appears, there is a truncation, rather than round-off, by the calculator. If 1.9999999 or 2 appears, there is an internal round-off.

Also, when considering decimal notation, we should determine whether or not the calculator uses scientific notation and, if so, how it works. In scientific notation, a number like 238,000 is written as a product of a number greater than or equal to 1 but less than 10 times a power of 10. Thus $238{,}000 = 2.38 \times 10^5$. Consult your owner's manual to see how scientific notation is displayed.

Error Indicator

There should be some signal on the calculator to indicate when an "illegal" operation is entered. For example, $\boxed{1} \boxed{\div} \boxed{0} \boxed{=}$ should cause the display to show an error. This indicator also should show when the computing limit of the calculator has been exceeded.

Special Keys

Several special keys are convenient. The first of these is the constant key, $\boxed{K}$, which allows an operation to be repeated without your having to press all the keys each time. For example, the calculator might be designed so that if $\boxed{7} \boxed{+} \boxed{K}$ is entered, then 7 is added to whatever appears on the display each time $\boxed{=}$ is pressed. Some machines have automatic constants built into them, rather than a separate constant key.

For example, if the machine has an automatic constant, we simply enter $\boxed{2} \boxed{+} \boxed{=} \boxed{=} \boxed{=} \boxed{=}$ . . . and the multiples of 2 are displayed. Some models require that you press the operation key twice to activate the constant feature. You also should check to see if the other basic operations can be used with the constant feature. If you experience difficulties with the constant feature, consult your owner's manual.

Another special key is the change-of-sign key, $\boxed{+/-}$. This allows for the entry of negative numbers. Normally, each number entered into a calculator is positive. Pressing $\boxed{3} \boxed{+/-}$ changes 3 to $^-3$. It is desirable that the negative sign immediately precede a number to denote a negative number, rather than leaving a space between the sign and the number, as is done on some calculators.

Still another special key is the percent key, $\boxed{\%}$. This key may operate in a variety of ways, depending on the calculator. It may change a percent to a decimal. For example, pressing $\boxed{6} \boxed{\%}$ may give 0.06 on the display. On other machines, pressing $\boxed{2} \boxed{\times} \boxed{3} \boxed{\%}$ may yield 0.06 without your using the $\boxed{=}$ key. If the $\boxed{=}$ key is used, the display might show 2.06, which is 2 + 2(3%). You should carefully check how the $\boxed{\%}$ key operates on your calculator.

For work in this text, the y-to-the-x-power key, $\boxed{y^x}$, is very important. It raises y to the power of x. For example, pressing $\boxed{2} \boxed{y^x} \boxed{1} \boxed{0} \boxed{=}$ yields 1024, which is 2^{10}. To perform 2^{-3}, we key $\boxed{2} \boxed{y^x} \boxed{3} \boxed{+/-} \boxed{=}$.

Other keys that may be convenient are $\boxed{\sqrt{}}$, the square-root key, and $\boxed{x^2}$, the squaring key. Numerous other keys are available for very little cost and may be useful, depending on individual needs.

Texas Instruments introduced the *Math Explorer*™ calculator shown in Figure 1-23. It can perform the four basic operations on fractions, change decimals to fractions, and convert improper fractions to mixed numbers and back. The calculator also can perform integer division by displaying the integer part and the remainder. The key $\boxed{/}$ is used for the fraction bar. For example, to compute $\frac{1}{4} + \frac{3}{8}$, press $\boxed{1} \boxed{/} \boxed{4} \boxed{+} \boxed{3} \boxed{/} \boxed{8}$ and the calculator will display 5/8. To convert the displayed fraction to a decimal, press $\boxed{F \leftrightarrow D}$ and 0.625 will be displayed. Press the $\boxed{F \leftrightarrow D}$ key again and the display will show 625/1000. The $\boxed{Simp}$ is used to simplify or reduce fractions. Pressing $\boxed{Simp}$ followed by a number will divide the numerator and denominator of a displayed fraction by that number. For example, if the fraction $\boxed{18/24}$ is displayed and you press $\boxed{Simp}$ 3 $\boxed{=}$, the calculator will display 6/8 along with N/D $\rightarrow$ n/d, which indicates that the fraction can be further simplified. If you now press $\boxed{Simp}$ 2 $\boxed{=}$, the calculator will display 3/4. Now N/D $\rightarrow$ n/d will no longer be displayed, indicating that the fraction is in the

FIGURE 1-23

simplest form. You may also let the calculator choose the factor by which a fraction is reduced. For example, 9/12 [Simp] [=] yields 3/4. The key [x⌒y] can be used to reveal the factor by which the fraction was reduced. Thus pressing [x⌒y] now yields 3. The [INT÷] key performs division with a remainder. For example, 154 [INT÷] 6 [=] displays ⌊—Q—25—⌋ ⌊—R—4—⌋, which shows that the quotient is 25 and the remainder is 4.

PROBLEM 11

Would you rather work for a month (31 days) and get $1,000,000 or be paid 1¢ the first day, 2¢ the second day, 4¢ the third day, and so on, but be allowed to keep only the amount that would be paid on the 31st day?

Understanding the Problem. Because we know that the wages are $1,000,000 for 31 days' work under the first option, we must compute the amount of pay under the second option. If 1¢ is paid for the first day, 2¢ for the second day, 4¢ for the third day, and so on, we need to find the amount paid on the 31st day. Then we can determine the better option.

Devising a Plan. One strategy is to *make a table* and look for a pattern for the amount of pay for each day. Table 1-15 shows a pattern for the second pay plan. From Table 1-15, we see that the pay for consecutive days generates a geometric sequence with ratio 2. The exponent in each case is one less than the number of the day. Thus the amount of money for the 31st day is 2^{30} cents. To see how great a number 2^{30} is, we could use a calculator. Then we could convert this number into dollars and compare it with $1,000,000 to determine which is greater.

TABLE 1-15

Day	Amount of Pay in Cents
1	1
2	$2 = 2^1$
3	$4 = 2^2$
4	$8 = 2^3$
5	$16 = 2^4$
6	$32 = 2^5$
.	.
.	.
.	.
31	?

Carrying Out the Plan. We can determine the value of 2^{30} in various ways on a calculator. If the calculator has a [y^x] key, which allows the user to raise numbers to powers, then 2^{30} could be determined by pressing [2] [y^x] [3] [0] [=]. If the [y^x] key is not present and the calculator has a constant feature, then the latter feature could be used. Another approach is to use the calculator to compute $2^{10} = 1024$ and then compute $2^{30} = 2^{10} \cdot 2^{10} \cdot 2^{10} = 1024 \cdot 1024 \cdot 1024 = 1,073,741,824$. Depending on the calculator, this result may be displayed in scientific notation; for example, the calculator might read 1.0737 09, which means $1.0737 \cdot 10^9$, or 1,073,700,000. Notice that numbers in scientific notation are rounded. To convert this number of cents into dollars, we divide by 100 and see that the rounded amount received on the 31st day is much greater than $1,000,000; hence, the second option is better.

On some calculators, you may see the word Error or E displayed when the number is too large for the display. Such calculators do not do exponential notation.

Looking Back. An alternate problem might be to consider which option is better if we keep only the money on the 25th day. How many days are needed before the second option is more attractive than $1,000,000? What if we are allowed to keep all the money from each day? How do the preceding answers change? Try to estimate which option is better without using a calculator. ∎

PROBLEM 12

Sara and David were reading the same novel. When Sara asked David what page he was reading, he replied that the product of the number of the page he was reading and the next page number was 98,282. What page was David reading?

Understanding the Problem. We know that the product of the page number of the page David was reading and the next page number is 98,283. We are asked to find the number of the page David was reading.

Devising a Plan. Adjacent pages must have consecutive numbers. If we denote the page number David was on by x, then the next page number is $x + 1$. The product of these page numbers is 98,282, so we *write the equation* as $x(x + 1) = 98,282$. To solve the equation, we use the *guess and check* strategy. We use a calculator to multiply various consecutive numbers, trying to obtain the product 98,282. Each new guess should be based on the information obtained from previous trials.

Carrying Out the Plan. Table 1-16 shows a series of guesses. From Table 1-16, we see that the desired page number must be closer to 300 than to 400. Checking $x = 310$ yields $310 \cdot 311 = 96,410$, which shows that 310 is too small for the solution. Successive trials reveal that $313 \cdot 314 = 98,282$, so David was reading page 313.

TABLE 1-16

x	$x + 1$	$x(x + 1)$
100	101	$100 \cdot 101$, or 10,100
200	201	$200 \cdot 201$, or 40,200
300	301	$300 \cdot 301$, or 90,300
400	401	$400 \cdot 401$, or 160,400

Looking Back. An alternate solution involves using the concept of square root. The desired page number is close to the number that when multiplied by itself, yields the product 98,282. This number is called the *square root* of 98,282. Using a calculator, we press the keys $\boxed{9}\,\boxed{8}\,\boxed{2}\,\boxed{8}\,\boxed{2}\,\boxed{\sqrt{}}$. This yields 313.4996. Thus a good guess for the desired page number is 313. ∎

CALCULATOR TIME OUT

Many words can be formed when the calculator display is turned upside down. For example, to become better acquainted with your calculator, press 0.7734 and turn the calculator upside down. The digits and the letters they represent follow:

$$0 \rightarrow O \quad 3 \rightarrow E \quad 7 \rightarrow L$$
$$1 \rightarrow I \quad 4 \rightarrow H \quad 8 \rightarrow B$$
$$2 \rightarrow Z \quad 5 \rightarrow S \quad 9 \rightarrow G$$

A small vocabulary list follows:

$$37818 \rightarrow \text{BIBLE} \qquad 3781937 \rightarrow \text{LEGIBLE}$$
$$379908 \rightarrow \text{BOGGLE} \qquad 35007 \rightarrow \text{LOOSE}$$
$$37819173 \rightarrow \text{ELIGIBLE} \qquad 35380 \rightarrow \text{OBESE}$$
$$35339 \rightarrow \text{GEESE} \qquad 372215 \rightarrow \text{SIZZLE}$$
$$5379919 \rightarrow \text{GIGGLES} \qquad 491375 \rightarrow \text{SLEIGH}$$
$$378809 \rightarrow \text{GOBBLE} \qquad 45075 \rightarrow \text{SLOSH}$$

It is possible to make up word problems to make your calculator talk; see the following three examples. After studying these examples, make up three new words and three new word problems.

1. Is 13,632 greater than or less than 19,169? Work the following problem, turn the calculator upside down, and read the display.

$$19{,}169 - 13{,}632 =$$

2. In the Winter Olympics, one of the most dangerous events is the ____ race. Complete the following problem, turn the calculator upside down, and read the display.

$$144 \times 349{,}832 =$$

3. The fire in the burning building reached 418°; 128 firefighters showed up to battle the blaze but were not successful because they forgot their ____. Multiply these numbers and turn your calculator upside down to find out what they forgot.

$$418 \times 128 =$$

BRAIN TEASER

What holiday does the following array suggest?

A	B	C	D	E
F	G	H	I	J
K	M	N	O	P
Q	R	S	T	U
V	W	X	Y	Z

PROBLEM SET 1-3

1. (a) Place the digits 1, 2, 4, 5, and 7 in the boxes so that in (i), the greatest product is obtained and in (ii) the greatest quotient is obtained.

    (i) ☐☐☐
        ×☐☐   (ii) ☐☐)☐☐☐

    (b) Use the same digits as in (a) to obtain (i) the least product and (ii) the least quotient.

2. Which of the following savings plans yields the greatest amount of money?
    (a) $10 a day for a year
    (b) $120 a week for a year
    (c) 25¢ an hour for a year
    (d) 1¢ a minute for a year

3. Vera spent $16.33 for three of the following items. Which three did she buy?
   $5.77, $3.99, $4.33, $5.87, $6.47

4. ▶Pick your favorite single-digit number greater than zero. Multiply it by 259. Now multiply your result by 429. What is your answer? Try it with other numbers. Why does it work?

5. Use your calculator's constant feature, if it has one, to count the number of terms in the following sequence:
   1, 8, 15, 22, . . . , 113

6. If 0.2 oz of catsup is used on each of 22 billion hamburgers, how many 16-oz bottles are needed?

7. How many natural numbers that are evenly divisible by 5,230,010 can be displayed on your calculator without using scientific notation?

8. Suppose the $\boxed{7}$, $\boxed{8}$, $\boxed{9}$, and $\boxed{\div}$ keys on your calculator do not work. Devise ways to perform the following computations on your calculator:
    (a) 756 + 183   (b) 155 ÷ 31

9. Suppose that your $\boxed{7}$, $\boxed{8}$, and $\boxed{+}$ keys are broken. How could you make your calculator display 73?

10. An integer is selected to be the target number. Suppose that only one numeral key works on your calculator. You want to display your target number in the fewest number of key presses. Determine what is the fewest number of key presses and what keys are used for each of the following:
    (a) The working key is 7 and the target number is 301.
    (b) The working key is 2 and the target number is 3246.

11. Suppose you could spend $10 every minute, night and day. How much could you spend in a year? (Assume there are 365 days in a year.)

12. How many times does your heart beat in each of the following?
    (a) One minute   (b) One hour
    (c) One day   (d) One week
    (e) One year ($365\frac{1}{4}$ days)

13. Suppose that a number is entered on the calculator. Then it is divided by 25; 18 is subtracted from it; and it is multiplied by 37. If the answer is 259, what is the original number?

14. The number 5! (read "five factorial") is defined to be $5 \cdot 4 \cdot 3 \cdot 2 \cdot 1$, and $4! = 4 \cdot 3 \cdot 2 \cdot 1$. Evaluate 10!. If your calculator has a factorial key, $\boxed{x!}$, work the exercise with and without using the key.

15. ▶(a) Multiply several two-digit numbers by 99 and study the products. What do you notice?
    ▶(b) Multiply several two-digit numbers by 999. What do you notice?

16. If your calculator displays 0.3333333 when 1 is divided by 3, what other division could be performed to yield a display of 0.0333333?

17. The distance around the world is approximately 40,000 km. Approximately how many people holding hands would it take to stretch around the world?

18. ▶Find two positive integers whose product is 5459. Find the most efficient way to answer the question. Explain your method.

19. The following is one version of a game called NIM. Two players and one calculator are needed. Player 1 presses $\boxed{1}$ and $\boxed{+}$ or $\boxed{2}$ and $\boxed{+}$. Player 2 does the same. The players take turns until the target number of 21 is reached. The first player to make the display read 21 is the winner. Determine a strategy for deciding who always wins.

20. ▶Try a game of NIM (see Problem 19) using the digits 1, 2, 3, and 4, with a target number of 104. The first player to reach 104 wins. What is the winning strategy?

21. ▶Try a game of NIM using the digits 3, 5, and 7, with a target number of 73. The first player to exceed 73 loses. What is the winning strategy?

22. ▶The game of NIM in Problem 19 requires two players and one calculator. Player 1 presses $\boxed{1}$ and $\boxed{+}$ or $\boxed{2}$ and $\boxed{+}$. Player 2 does the same. Both try to reach the target number of 21. Now play Reverse NIM. Instead of $\boxed{+}$, use $\boxed{-}$. Put 21 on the display. Let the new target number be 0. Determine a strategy for winning Reverse NIM.

23. ▶Try Reverse NIM, using the digits 1, 2, and 3 and starting with 24 on the display. (See Problem 22.) The target number is 0. What is the winning strategy?

24. ▶Try Reverse NIM, using the digits 3, 5, and 7 and starting with 73 on the display. The first player to display a negative number loses. What is the winning strategy?

25. Based on the following pattern, answer questions (a) through (c):

$$1 = 2^1 - 1$$
$$1 + 2 = 2^2 - 1$$
$$1 + 2 + 2^2 = 2^3 - 1$$
$$1 + 2 + 2^2 + 2^3 = 2^4 - 1$$
$$1 + 2 + 2^2 + 2^3 + 2^4 = 2^5 - 1$$

▶ (a) Write a simpler expression for $1 + 2 + 2^2 + 2^3 + 2^4 + 2^5$. Justify your answer.
(b) Write a simpler expression for the sum in the nth row in the above pattern.
(c) Use a calculator to check your answer in (b) for $n = 15$.

Review Problems

26. List three additional terms to continue a possible pattern in the following sequences:
    (a) 7, 14, 21, 28
    (b) 4, 1, 8, 1, 12
27. Find the nth term for the following sequence:
    12, 32, 52, 72.
28. How many terms are in the following sequence?
    6, 10, 14, 18, . . . , 86
29. In how many ways can you make change for $.21?

Solution to the Preliminary Problem

Understanding the Problem. An architect needs to design a path made of black (B) and red (R) tiles in the following pattern:

$$\text{BRBBRRBBBRRRBBBBRRRR} \ldots$$

Assuming that the existing supply of 2500 black and 2500 red tiles must be used, we need to determine the least number of tiles of each color to be purchased so that the path ends with a block of black tiles followed by the same number of red tiles.

Devising a Plan. We see that the number of tiles follows the pattern.

$$1, 1, 2, 2, 3, 3, 4, 4, 5, 5, \ldots,$$

where each number repeats itself and the first number in each pair indicates the number of black tiles and the second, the number of red tiles. We *examine a simpler case* first. Suppose we had only 12 tiles: 6 black and 6 red. Would we need any extra tiles to fulfill the requirements in the problem? We would find that $1 + 1 + 2 + 2 + 3 + 3 = 12$, and no extra tiles would be needed. However, if we had 14 tiles (7 black and 7 red), we would continue the pattern to include the next pair of equal numbers and obtain $1 + 1 + 2 + 2 + 3 + 3 + 4 + 4 = 20$. Thus the museum would have to order $20 - 14$, or 6 tiles (3 black and 3 red). To solve the original problem, we could continue adding the pairs until we reach a sum equal to or greater than 5000. The excess over 5000 will be the number of tiles to be ordered. Because this method is too cumbersome, we look for a more efficient approach.

If we could find how far we must add to obtain a sum greater or equal to 5000, we could use a *related problem,* that is, Gauss's sum in Problem 1, and sum the numbers, then find the excess.

To figure how far to go, we need only find the last pair that must be considered. Doing this becomes our *subgoal.*

Notice that the nth pair is n, n. (Do you see why?) Thus we need to find the least n for which the sum

$$1 + 1 + 2 + 2 + 3 + 3 + 4 + 4 + \ldots + n + n$$

is greater than or equal to 5000. Recall Gauss's sum in Problem 1:

$$1 + 2 + 3 + 4 + \ldots + n = \frac{n(n + 1)}{2}.$$

Because our sum is twice as large (why?), we have

$$1 + 1 + 2 + 2 + 3 + 3 + 4 + 4 + \ldots + n + n = n(n + 1).$$

Consequently, we need only find the least n for which $n(n + 1)$ is greater than or equal to 5000. This task is our new *subgoal*.

Carrying Out the Plan. To find the least n for which $n(n + 1)$ is greater than or equal to 5000, we could use the *guess and check* strategy. Table 1-17 shows a series of guesses. From this table, we see that the first n that makes $n(n + 1)$ greater than 5000 is 71. The corresponding sum is 5112; hence, the excess needed is 112. Consequently, the museum needs to order 56 black and 56 red tiles.

TABLE 1-17

n	n + 1	n(n + 1)
60	61	3660
70	71	4970
71	72	5112

Looking Back. An alternate approach involves the concept of square root. The desired value of n is close to the number that when multiplied by itself, yields the product 5000. This number is called the *square root* of 5000. Using a calculator, we press the keys $\boxed{5}\,\boxed{0}\,\boxed{0}\,\boxed{0}\,\boxed{\sqrt{}}$. This yields 70.7106. Thus a good guess for the desired value of n is 71.

QUESTIONS FROM THE CLASSROOM

1. A student claims that she checked that $n^{50} > 2^n$ (the symbol $>$ designates "greater than") for $n = 1, 2, 3, \ldots, 50$. Hence, she claims that $n^{50} > 2^n$ should be true for all values of n. How do you respond?

2. A student says that many arithmetic sequences can be expressed using geometric figures and hence, is wondering why arithmetic sequences are not called "geometric." How do you respond?

3. A student says that she read that Thomas Robert Malthus (1766–1834), a renowned British economist and demographer, claimed that the increase of population will take place, if unchecked, in a geometric sequence, while the supply of food will increase only in an arithmetic sequence. This theory implies that population increases faster than production of food. The student is wondering why. How do you respond?

4. A student found that in the sequence

$$1, 2, 2^2, 2^3, 2^4, \ldots,$$

the sum of the first n terms is approximately equal to the $(n + 1)$st term. The student wonders whether this is always true and if so, why. How do you respond?

5. A student notices that when she enters 0.3333333×3 on her calculator and then presses the $\boxed{=}$ key, the calculator displays 0.9999999. When she enters $\boxed{1}\,\boxed{\div}\,\boxed{3}\,\boxed{=}$, the calculator displays 0.3333333. She then enters $\boxed{\times}\,\boxed{3}$ and the calculator displays 1 rather than 0.9999999 as before. She wonders where she made a mistake. How do you respond?

CHAPTER OUTLINE

I. Mathematical patterns
   A. Patterns are an important part of problem solving.
   B. Patterns are used in **inductive reasoning** to form conjectures. A **conjecture** is a statement that is thought to be true but has not yet been proved to be true or false.
   C. A **sequence** is a group of terms in a definite order.
      1. **Arithmetic sequence:** Each successive term is obtained from the previous one by the addition of a fixed number called the **difference.** The nth term is given by $a + (n - 1)d$, where a is the first term and d is the difference.
      2. **Geometric sequence:** Each successive term is obtained from its predecessor by multiplying it by a fixed number called the **ratio.** The nth term is given by ar^{n-1}, where a is the first term and r is the ratio.
      3. $a^n = \underbrace{a \cdot a \cdot a \cdot a \cdot a \cdot \ldots \cdot a}_{n \text{ terms}}$
      4. $a^0 = 1$, where a is a natural number.
      5. Finding differences for a sequence is one technique for finding the next terms.

II. Problem Solving
   A. Problem solving can be guided by the following four-step process:
      1. Understanding the Problem
      2. Devising a Plan
      3. Carrying Out the Plan
      4. Looking Back
   B. Important problem-solving strategies include:
      1. Look for a pattern.
      2. Make a table.
      3. Examine a simpler or special case of the problem to gain insight into the solution of the original problem.
      4. Identify a subgoal.
      5. Examine related problems and determine if the same technique can be applied.
      6. Work backward.
      7. Write an equation.
      8. Make a diagram.
      9. Use guess and check.
      10. Use indirect reasoning.
   C. Beware of mind-sets!

III. Features of calculators
   A. Types of logic
      1. Without algebraic operating system
      2. With algebraic operating system
   B. Special keys
      1. Constant key
      2. Change-of-sign key
      3. Parentheses keys
      4. Percent key
      5. Power key
      6. Square-root key
      7. Memory keys
      8. Fraction bar key
      9. Conversion from decimal to fraction and back key
      10. Reducing fractions key
      11. Division with quotient and remainder key

CHAPTER TEST

1. List three more terms that complete a possible pattern in each of the following:
   (a) 0, 1, 3, 6, 10,
   (b) 52, 47, 42, 37,
   (c) 6400, 3200, 1600, 800,
   (d) 1, 2, 3, 5, 8, 13,
   (e) 2, 5, 8, 11, 14,
   (f) 1, 4, 16, 64,
   (g) 0, 4, 8, 12,
   (h) 1, 8, 27, 64,

2. Classify each sequence in Problem 1 as arithmetic, geometric, or neither.

3. Find the nth term in each of the following:
   (a) 5, 8, 11, 14, . . .   (b) 1, 8, 27, 64, . . .
   (c) 3, 9, 27, 81, 243, . . .

4. Find the first five terms of the sequences whose nth term is given as follows:
   (a) $3n + 2$ (b) $n^2 + n$ (c) $4n - 1$

5. Find the following sums:
   (a) $2 + 4 + 6 + 8 + 10 + \cdots + 200$
   (b) $51 + 52 + 53 + 54 + \cdots + 151$

6. (a) Determine a possible pattern in the sequence

   1, 12, 123, 1234, 12345, ...

   ▶(b) If the tenth term of the sequence in (a) is supposed to have 10 digits, what is the tenth term? Explain your reasoning.

7. Complete the following magic square; that is, complete the square so that the sum in each row, column, and diagonal is the same.

16	3	2	13
	10		
9		7	12
4		14	

8. How many years are there between the fifth day of the year 45 B.C. and the fifth day of the year A.D. 45?
9. A worm is at the bottom of a glass that is 20 cm deep. Each day the worm crawls up 3 cm and each night it slides back 1 cm. How long will it take the worm to climb out of the glass?
10. How many people can be seated at 12 square tables lined up end to end if each table used individually holds four persons?
11. A shirt and a tie sell for $9.50. The shirt costs $5.50 more than the tie. What is the cost of the tie?
12. If fence posts are to be placed in a row 5 m apart, how many posts are needed for 100 m of fence?
13. A total of 129 players entered a single-elimination handball tournament. In the first round of play, the top-seeded player received a bye, and the remaining 128 players played in 64 matches; thus 65 players entered the second round of play. How many matches must be played to determine the tournament champion?
14. Given the six numbers 3, 5, 7, 9, 11, and 13, pick five of them that when multiplied, give 19,305.
15. If a complete turn of a car tire moves a car forward 6 ft, how many turns of a tire occur before a tire goes off its 50,000-mi warranty?
16. The members of Mrs. Grant's class are standing in a circle; they are evenly spaced and are numbered in order. The student with number 7 is standing directly across from the student with number 17. How many students are in the class?
17. A carpenter has three separate large boxes. Inside each large box are two medium-sized boxes. Inside each medium-sized box are five separate small boxes. How many boxes are there altogether?

18. ▶How many different triangles are there in the following figure? Explain your reasoning.

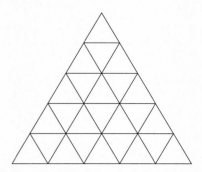

19. Study the following pattern and find an expression for the given sums:

$$1 = \frac{9}{9} = \frac{10^1 - 1}{9}$$

$$1 + 10 = 11 = \frac{99}{9} = \frac{10^2 - 1}{9}$$

$$1 + 10 + 10^2 = 111 = \frac{999}{9} = \frac{10^3 - 1}{9}$$

$$1 + 10 + 10^2 + 10^3 = 1111 = \frac{9999}{9} = \frac{10^4 - 1}{9}$$

(a) $1 + 10 + 10^2 + 10^3 + \ldots + 10^{12}$
(b) $1 + 10 + 10^2 + 10^3 + \ldots + 10^n$
(c) $1 + 10 + 10^2 + 10^3 + \ldots + 10^{n-1}$

20. Mary left her home and averaged 16 km/hr riding her bicycle on an uphill trip to Larry's house. On the return trip over the same route, she averaged 20 km/hr. If it took 4 hr to make the return trip, how much cycling time did the entire trip take?

21. ▶The perimeter of a rectangle is 68 ft, and the length of the rectangle is 4 ft more than twice the width. Find the length and width of the rectangle. Explain your reasoning.

SELECTED BIBLIOGRAPHY

An excellent bibliography is available in *Problem Solving in School Mathematics*, 1980, yearbook published by the National Council of Teachers of Mathematics, S. Krulick and R. Reys, eds.

Billstein, R. "Checkerboard Mathematics." *Mathematics Teacher* 86 (December 1975): 640–646.

Bledsoe, G. "Hook Your Students on Problem Solving." *Arithmetic Teacher* 37 (December 1989): 16–20.

Brown, S., and M. Walter. *The Art of Problem Posing.* Philadelphia: Franklin Institute Press, 1991.

Campbell, P. "Implementing the Standards. The Vision of Problem Solving in the Standards." *Arithmetic Teacher* 37 (May 1990): 14–17.

Dick, T. "The Continuing Calculator Controversy." *Arithmetic Teacher* 35 (April 1988): 37–41.

Demana, F., and A. Osborne. "Chasing a Calculator: Four Function Foul-ups." *Arithmetic Teacher* 35 (March 1988): 2–3.

Halmos, P. R. "The Heart of Mathematics." *American Mathematical Monthly* 87 (August–September 1980): 519–524.

Johnson, M., and T. Offerman. "How Do You Evaluate Problem Solving." *Arithmetic Teacher* 35 (April 1988): 49–51.

Lester, F. "Research Into Practice." *Arithmetic Teacher* 37 (November 1989): 33–35.

Moody, W. "A Program in Middle School Problem Solving." *Arithmetic Teacher* 38 (December 1990): 6–11.

Norman, F. "Figurate Numbers in the Classroom." *Arithmetic Teacher* 38 (March 1991): 42–45.

Polya, G. *How to Solve It*. Princeton, N.J.: Princeton University Press, 1957.

Seymour, D., and E. Beardslee. *Critical Thinking Activities*. Palo Alto, Calif.: Dale Seymour, 1988.

Talton, C. "Let's Solve the Problem Before We Find the Answer." *Arithmetic Teacher* 36 (September 1988): 40–45.

Thompson, A. "On Patterns, Conjectures, and Proof: Developing Students." *Arithmetic Teacher* 33 (September 1985): 20–23.

2 Sets, Functions, and Logic

Preliminary Problem

The Red Cross looks for three types of antigens in blood tests: A, B, and Rh. When the antigen A or B is present, it is listed, but if both these antigens are absent, the blood is type O. If the Rh antigen is present, the blood is positive; otherwise, it is negative. If a laboratory technician reports the following results after testing the blood samples of 100 people, how many were classified as O negative?

Number of Samples	Antigens in Blood
40	A
18	B
82	Rh
5	A and B
31	A and Rh
11	B and Rh
4	A, B, and Rh

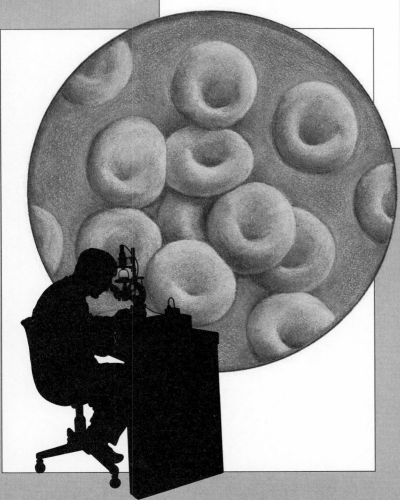

eorge Cantor, in the years 1871–1884, created a new area of mathematics called *set theory*. His theories had a profound effect on research and on mathematics teaching.

HISTORICAL NOTE

Georg Cantor, 1845–1918, a German mathematician, was born of Danish parents at St. Petersburg, Russia. His family moved to Frankfurt when he was 11. Against his father's advice, Cantor pursued a career in mathematics and obtained his doctorate in Berlin at age 22. Most of his academic work was spent at the University of Halle. His hope of becoming a professor at the University of Berlin did not materialize, as his work gained little recognition during his lifetime.

Cantor suffered from nervous breakdowns and died in a mental hospital. His work was praised as an "astonishing product of mathematical thought, one of the most beautiful realizations of human activity . . . ."

The language of set theory was introduced into schools in the 1960s in the post-Sputnik era. In the 1970s, numerous people felt that the new language and symbolism caused confusion for children and adults. The following cartoon illustrates the feelings of many of these people.

PEANUTS reprinted by permission of UFS, Inc.

However, the basic set concepts clarify many mathematical ideas and are used in elementary school texts. The *Teaching Standards* (p. 136) assert, *Teachers need to experience the development of mathematical language and symbolism and how these have influenced the way we communicate mathematical ideas*. In this chapter, we discuss set notation, relations between sets, and set operations and their properties. We also use the concept of a set to define relations and functions. In the last two (optional) sections, we introduce the fundamentals of logic.

Section 2-1 Describing Sets

set
elements / members

A **set** is any collection of objects. Individual objects in a set are **elements,** or **members,** of the set. For example, each letter is an element of the set of letters in the English language. We use braces to enclose the elements of a set and label the set with a capital letter. The set of letters of the English alphabet can be written as

$$A = \{a, b, c, d, e, f, g, h, i, j, k, l, m, n, o, p, q, r, s, t, u, v, w, x, y, z\}.$$

The order in which the elements are written makes no difference, and *each element is listed only once*. For example, the set of letters in the word *book* could be written as $\{b, o, k\}$, $\{o, b, k\}$, or $\{k, o, b\}$.

We symbolize an element belonging to a set by using the symbol $\in$. For example, $b \in A$. The fact that A does not contain α is written as $\alpha \notin A$.

well defined

For a given set to be useful in mathematics, it must be **well defined.** If we are given a set and some particular object, the object does or does not belong to the set. For example, the set of all citizens of Portland, Oregon, who ate rice on January 1, 1993, is well defined. We may not know if a particular resident of Portland ate rice or not, but we do know that that person either did or did not. On the other hand, the set of all large numbers is not well defined because we do not know which particular numbers qualify as large numbers.

natural numbers
counting numbers

We may use sets to define mathematical terms. For example, the set of **natural,** or **counting, numbers** is defined by the following:

$$N = \{1, 2, 3, 4, \ldots\}.$$

set-builder notation

Sometimes the individual elements of a set are not known or they are too numerous to list. In these cases, the elements are indicated by using **set-builder notation.** The set of animals in the San Diego Zoo can be written as

$$Z = \{x | x \text{ is an animal in the San Diego Zoo}\}.$$

This is read "Z is the set of all elements x such that x is an animal in the San Diego Zoo." The vertical line is read "such that."

EXAMPLE 2-1   Write the following sets using set-builder notation:

(a) $\{51, 52, 53, 54, \ldots, 498, 499\}$ (b) $\{2, 4, 6, 8, 10, \ldots\}$
(c) $\{1, 3, 5, 7, \ldots\}$ (d) $\{1^2, 2^2, 3^2, 4^2, \ldots\}$

Solution (a) $\{x | x \text{ is a natural number greater than 50 and less than 500}\}$, or $\{x | 50 < x < 500, x \in N\}$

(b) $\{x | x \text{ is an even natural number}\}$, or $\{x | x = 2n, n \in N\}$

(c) $\{x | x \text{ is an odd natural number}\}$, or $\{x | x = 2n - 1, n \in N\}$

(d) $\{x | x \text{ is a square of a natural number}\}$ or $\{x | x = n^2, n \in N\}$ ∎

equal sets

Two sets are **equal** if, and only if, they contain exactly the same elements. The order in which the elements are listed does not matter. If A and B are equal, written $A = B$, then every element of A is an element of B, and every element of B is an element of A. If A does not equal B, we write $A \neq B$. Consider sets $D = \{1, 2, 3\}$, $E = \{2, 5, 1\}$, and $F = \{1, 2, 5\}$. Sets D and E are not equal; sets E and F are equal.

One-to-one Correspondence

Consider the set of people $P = \{$Tomas, Dick, Mari$\}$ and the set of swimming lanes $S = \{1, 2, 3\}$. Suppose each person in P is to swim in a lane numbered 1, 2, or 3 so that no two people swim in the same lane. Such a person-lane pairing is a **one-to-one correspondence.** One way to exhibit this one-to-one correspondence is: Tomas $\leftrightarrow$ 1, Dick $\leftrightarrow$ 2, Mari $\leftrightarrow$ 3.

one-to-one correspondence

FIGURE 2-1

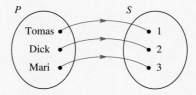

Other possible one-to-one correspondences exist between the sets P and S. There are several schemes for exhibiting them. For example, all six possible one-to-one correspondences for sets P and S can be listed as follows:

(1) Tomas $\leftrightarrow$ 1	(2) Tomas $\leftrightarrow$ 1	(3) Tomas $\leftrightarrow$ 2
Dick $\leftrightarrow$ 2	Dick $\leftrightarrow$ 3	Dick $\leftrightarrow$ 1
Mari $\leftrightarrow$ 3	Mari $\leftrightarrow$ 2	Mari $\leftrightarrow$ 3
(4) Tomas $\leftrightarrow$ 2	(5) Tomas $\leftrightarrow$ 3	(6) Tomas $\leftrightarrow$ 3
Dick $\leftrightarrow$ 3	Dick $\leftrightarrow$ 1	Dick $\leftrightarrow$ 2
Mari $\leftrightarrow$ 1	Mari $\leftrightarrow$ 2	Mari $\leftrightarrow$ 1

DEFINITION OF ONE-TO-ONE CORRESPONDENCE

If the elements of sets P and S can be paired so that for each element of P there is exactly one element of S and for each element of S there is exactly one element of P, then the two sets P and S are said to be in **one-to-one correspondence** (or matched).

Another method of demonstrating one-to-one correspondence is to use a chart, as seen in Table 2-1, where the lane numbers are listed across the top of the chart and the possible pairings of swimmers to lanes are listed below.

TABLE 2-1

1	2	3
Tomas	Dick	Mari
Tomas	Mari	Dick
Dick	Tomas	Mari
Dick	Mari	Tomas
Mari	Tomas	Dick
Mari	Dick	Tomas

Finally, a tree diagram can be used to list the possible one-to-one correspondences, as Figure 2-2 shows. To read the tree diagram and see the one-to-one correspondences, follow each branch. The person occupying a specific lane in a correspondence is listed

FIGURE 2-2

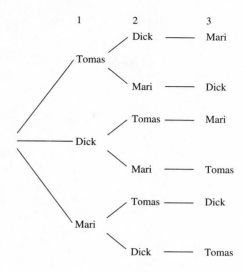

below the lane number. For example, the top branch gives the pairing (Tomas, 1), (Dick, 2), and (Mari, 3). Observe that for lane 1, there are actually three choices at the start, then once that lane is filled, there are only two choices for lane 2, and finally only one choice for lane 3. Hence, there are $3 \cdot 2 \cdot 1$, or 6, possible one-to-one correspondences. The counting argument used to find the number of possible one-to-one correspondences is an example of the *Fundamental Counting Principle*.

> **PROPERTY**
>
> **Fundamental Counting Principle** If event M can occur in m ways and, after it has occurred, event N can occur in n ways, then event M followed by event N can occur in $m \cdot n$ ways.

> **REMARK** The Fundamental Counting Principle can be extended to any number of events.

Equivalent Sets

equivalent sets

Suppose a room contains 20 chairs and one student is sitting in each chair with no one standing. There is a one-to-one correspondence between the set of chairs and the set of students in the room. In this case, the set of chairs and the set of students are **equivalent sets.**

> **DEFINITION OF EQUIVALENT SETS**
>
> Two sets A and B are **equivalent,** written $A \sim B$, if and only if there exists a one-to-one correspondence between the sets.

The term *equivalent* should not be confused with *equal*. The difference should be made clear by the following example.

EXAMPLE 2-2   Let

$$A = \{p, q, r, s\}, B = \{a, b, c\}, C = \{x, y, z\}, D = \{b, a, c\}.$$

Compare the sets, using the terms *equal* and *equivalent*.

Solution   Sets A and B are not equivalent ($A \not\sim B$) and not equal ($A \neq B$).
Sets A and C are not equivalent ($A \not\sim C$) and not equal ($A \neq C$).
Sets A and D are not equivalent ($A \not\sim D$) and not equal ($A \neq D$).
Sets B and C are equivalent ($B \sim C$), but not equal ($B \neq C$).
Sets B and D are equivalent ($B \sim D$) and equal ($B = D$).
Sets C and D are equivalent ($C \sim D$), but not equal ($C \neq D$). ■

> **REMARK**   Note that if two sets are equal, they are equivalent; however, if two sets are equivalent, they are not necessarily equal.

Cardinal Numbers

The five sets $\{a, b\}$, $\{1, 2\}$, $\{x, y\}$, $\{b, a\}$, and $\{*, \#\}$ are equivalent to one another and share the property of "twoness." These sets have the same cardinal number, namely, 2. The

cardinal number   **cardinal number** of a set X, denoted by $n(X)$, indicates the number of elements in the set X. If $D = \{a, b\}$, the cardinal number of D is 2, and we write $n(D) = 2$.

If A is equivalent to, or equal to, B, then A and B have the same cardinal number; that is, $n(A) = n(B)$. If $n(A) = n(B)$, sets A and B are equivalent, but not necessarily equal.

finite set   A set is a **finite set** if the number of elements in the set is zero or a natural number. For example, the set of letters in the English alphabet is a finite set because it contains exactly 26 elements. Another way to think of this is that the set of letters in the English alphabet can be put into a one-to-one correspondence with the set $\{1, 2, 3, \ldots, 26\}$. The set of

infinite set   natural numbers N is an example of an **infinite set,** a set that is not finite.

More about Sets

empty set   A set that contains no elements has cardinal number 0 and is called an **empty,** or **null, set.**
null set   The empty set is designated by the symbol $\varnothing$ or $\{\ \}$. Two examples of sets with no elements are the following:

$$C = \{x \mid x \text{ was a state of the United States before 1200}\}$$

$$D = \{x \mid x \text{ is a natural number less than 1}\}$$

> **REMARK**   The empty set is often incorrectly recorded as $\{\varnothing\}$. This set is not empty but contains one element. Likewise, $\{0\}$ does not represent the empty set.

SECTION 2-1 DESCRIBING SETS 55

universal set / universe

The **universal set,** or the **universe,** denoted by U, is the set that contains all elements being considered in a given discussion. For this reason, you should be aware of what the universal set is in any given problem. Suppose $U = \{x \mid x \text{ is a person living in California}\}$ and $F = \{x \mid x \text{ is a female living in California}\}$. The universal set and set F can be represented by a diagram, as in Figure 2-3(a). The universal set is usually indicated by a large rectangle, and particular sets are indicated by geometric figures inside the rectangle, as shown in Figure 2-3(a). This figure is an example of a **Venn diagram,** named after the Englishman John Venn, who used such diagrams to illustrate ideas in logic. The set of elements in the universe that are not in F is the set of males living in California and is the **complement** of F. It is represented by the shaded region in Figure 2-3(b).

Venn diagram

complement

FIGURE 2-3

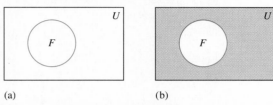

(a)   (b)

DEFINITION OF SET COMPLEMENT

The **complement** of a set F, written $\overline{F}$, is the set of all elements in the universal set U that are not in F. $\overline{F} = \{x \mid x \in U \text{ and } x \notin F\}$.

The key word in the definition of complement is *not*.

EXAMPLE 2-3

(a) If $U = \{a, b, c, d\}$ and $B = \{c, d\}$, find: (i) $\overline{B}$; (ii) $\overline{U}$; (iii) $\overline{\varnothing}$.

(b) If $U = \{x \mid x \text{ is an animal in the zoo}\}$ and $S = \{x \mid x \text{ is a snake in the zoo}\}$, find $\overline{S}$.

Solution (a) (i) $\overline{B} = \{a, b\}$ (ii) $\overline{U} = \varnothing$ (iii) $\overline{\varnothing} = U$.

(b) Because the individual animals in the zoo are not known, $\overline{S}$ must be described using set-builder notation.

$\overline{S} = \{x \mid x \text{ is an animal in the zoo that is not a snake}\}$ ∎

Subsets

Consider the sets $A = \{1, 2, 3, 4, 5, 6\}$ and $B = \{2, 4, 6\}$. All the elements of B are contained in A and B is a **subset** of A. We write $B \subseteq A$. In general, we have the following definition.

subset

DEFINITION OF SUBSET

B is a **subset** of A, written $B \subseteq A$, if and only if every element of B is an element of A.

This definition allows B to be equal to A. The definition is written with the phrase "if and only if," which means "if B is a subset of A, then every element of B is an element of

proper subset

A, and if every element of B is an element of A, then B is a subset of A." If both $A \subseteq B$ and $B \subseteq A$, then $A = B$.

If B is a subset of A and B is not equal to A, then B is a **proper subset** of A, written $B \subset A$. This means that every element of B is contained in A and there is at least one element of A that is not in B.

EXAMPLE 2-4 Given that $U = \{1, 2, 3, 4, 5\}$, $D = \{1, 3, 5\}$, and $E = \{1, 3\}$:

(a) Which sets are subsets of each other?

(b) Which sets are proper subsets of each other?

Solution (a) $D \subseteq U$, $E \subseteq U$, $E \subseteq D$, $D \subseteq D$, $E \subseteq E$, and $U \subseteq U$.

(b) $D \subset U$, $E \subset U$, and $E \subset D$. ∎

Suppose $A \subset B$. Can we always conclude that $A \subseteq B$? If $A \subseteq B$, does it follow that $A \subset B$? These and related questions are explored in the problem set.

When a set A is not a subset of another set B, we write $A \not\subseteq B$. To show that $A \not\subseteq B$, we must find at least one element of A that is not in B. If $A = \{1, 3, 5\}$ and $B = \{1, 2, 3\}$, then A is not a subset of B because 5 is an element of A but not of B. Likewise, $B \not\subseteq A$ because 2 belongs to B but not to A.

It is not obvious how the empty set fits the definition of a subset because no elements in the empty set are elements of another set. To investigate this problem, we use the strategies of *indirect reasoning* and *looking at a special case*.

For the set $\{1, 2\}$, either $\varnothing \subseteq \{1, 2\}$ or $\varnothing \not\subseteq \{1, 2\}$. Suppose $\varnothing \not\subseteq \{1, 2\}$. Then there must be some element in $\varnothing$ that is not in $\{1, 2\}$. Because the empty set has no elements, there cannot be an element in the empty set that is not in $\{1, 2\}$. Consequently, $\varnothing \not\subseteq \{1, 2\}$ is false and $\varnothing \subseteq \{1, 2\}$ is true. The same reasoning can be applied in the case of the empty set and any other set. In particular, we note that the empty set is a subset of itself and a proper subset of any set other than itself.

Subsets and elements of sets are often confused. We say that $2 \in \{1, 2, 3\}$. But because 2 is not a set, we cannot substitute the symbol $\subset$ for $\in$. However, $\{2\} \subseteq \{1, 2, 3\}$ and $\{2\} \subset \{1, 2, 3\}$. Conversely, the symbol $\in$ should not be used between $\{2\}$ and $\{1, 2, 3\}$.

Inequalities

The notion of a proper subset can be used to define the concept of "less than" among natural numbers. The set $\{a, b, c\}$ has *fewer* elements than the set $\{x, y, z, w\}$ because when we try to pair the elements of the two sets, as in

$$\{a, b, c\}$$
$$| \quad | \quad |$$
$$\{x, y, z, w\},$$

less than

we see that there is an element of the second set that is not paired with any element of the first set. The set $\{a, b, c\}$ is equivalent to a proper subset of the set $\{x, y, z, w\}$. In general, if A and B are finite sets, A has fewer elements than B, or $n(A) < n(B)$, if A is equivalent to a proper subset of B. This leads us to the following definition of **less than** for natural numbers.

DEFINITION OF LESS THAN

For natural numbers a and b, a is **less than** b, written $a < b$, if and only if for sets A and B with $n(A) = a$ and $n(B) = b$, there exists a proper subset of B equivalent to A.

greater than

We say that a is **greater than** b, written $a > b$, if and only if $b < a$. Defining the concept of "less than or equal" in a similar way is explored in the problem set.

PROBLEM 1

A committee of senators consists of Abel, Baro, Carni, and Davis. Suppose each member of the committee has one vote and a simple majority is needed either to pass or reject any measure. A measure that is neither passed nor rejected is considered to be blocked and will be voted on again. Determine the number of ways a measure could be passed or rejected and the number of ways a measure could be blocked.

Understanding the Problem. A Senate committee consisting of four members—Abel, Baro, Carni, and Davis—requires a simple majority of votes either to pass or reject a measure. We are asked to determine how many ways the committee could pass or reject a proposal and how many ways the committee could block a proposal. To pass or reject a proposal requires a winning coalition, that is, a group of senators that can pass or reject the proposal, regardless of what others do. To block a proposal, there must be a blocking coalition, that is, a group that can prevent any proposal from passing but that cannot reject the measure.

Devising a Plan. To solve the problem, we need to *make a list* of subsets of the set of senators. Any subset of the set of senators with three or four members will form a winning coalition. Any subset of the set of senators that contains exactly two members will form a blocking coalition.

Carrying Out the Plan. To solve the problem, we must find all subsets of the set $S = \{\text{Abel, Baro, Carni, Davis}\}$ that have at least three elements and all subsets that have exactly two elements. For ease, we identify the members as follows: A—Abel, B—Baro, C—Carni, D—Davis. All the subsets are given below:

$\varnothing$	$\{A, B\}$	$\{A, B, C\}$	$\{A, B, C, D\}$
$\{A\}$	$\{A, C\}$	$\{A, B, D\}$	
$\{B\}$	$\{A, D\}$	$\{A, C, D\}$	
$\{C\}$	$\{B, C\}$	$\{B, C, D\}$	
$\{D\}$	$\{B, D\}$		
	$\{C, D\}$		

Thus we see that there are five subsets with at least three members that can form a winning coalition and pass or reject a measure and six subsets with exactly two members that can block a measure.

Looking Back. From the preceding, we know that there are five winning coalitions and six blocking coalitions. Other questions that might be considered include the following:

1. How many minimal winning coalitions are there? In other words, how many subsets are there of which no proper subset could pass a measure?

2. Devise a method to solve this problem without listing all subsets.
3. Solve the problem if the committee has six members.

Problem 1 suggests the general problem of finding the number of subsets that a set containing n elements has. To obtain a general formula, we use the strategy of *trying simpler cases* first.

1. If $B = \{a\}$, then B has 2 subsets, $\emptyset$ and $\{a\}$.
2. If $C = \{a, b\}$, then C has 4 subsets, $\emptyset$, $\{a\}$, $\{b\}$, and $\{a, b\}$.
3. If $D = \{a, b, c\}$, then D has 8 subsets, $\emptyset$, $\{a\}$, $\{b\}$, $\{c\}$, $\{a, b\}$, $\{a, c\}$, $\{b, c\}$, and $\{a, b, c\}$.

Using the information from these cases, *we make a table and search for a pattern*, as seen in Table 2-2.

TABLE 2-2

Number of Elements	Number of Subsets
1	2, or 2^1
2	4, or 2^2
3	8, or 2^3
.	.
.	.
.	.

Table 2-2 suggests that for 4 elements, there are 2^4, or 16, subsets. Is this guess correct? If $E = \{a, b, c, d\}$, then all the subsets of $D = \{a, b, c\}$ are also subsets of E. Eight new subsets are also formed by adjoining the element d to each of the 8 subsets of D. The 8 new subsets are $\{d\}$, $\{a, d\}$, $\{b, d\}$, $\{c, d\}$, $\{a, b, d\}$, $\{a, c, d\}$, $\{b, c, d\}$, and $\{a, b, c, d\}$. Thus there are twice as many subsets of set E (with 4 elements) as there are of set D (with 3 elements). Consequently, there are indeed 16, or 2^4, subsets of a set with 4 elements. In a similar way, we can argue that there are $2^4 + 2^4$—that is, $2 \cdot 2^4$ or 2^5—subsets of a set with 5 elements. Notice that in each case the number of elements and the power of 2 used to obtain the number of subsets match exactly. In general it can be shown that, *if there are n elements in a set, there are 2^n subsets that can be formed*. This result can also be justified by using the Fundamental Counting Principle.

The formula 2^n for the number of subsets of a set with n elements is based on the observation that adding one more element to a set doubles the number of possible subsets of the new set. If we apply this formula to the empty set—that is, when $n = 0$—then we have $2^0 = 1$ because the empty set has only one subset—itself. The fact that $a^0 = 1$, where a is a natural number, is investigated in Chapter 6.

▼ **BRAIN TEASER**

Bertrand Russell's antinomy, "Is the set of all sets that are not members of themselves a member of itself?" has become popularized in the following paradox: The town barber shaves all those males, and only those males, who do not shave themselves. Assuming the barber is a male who shaves, who shaves the barber?

PROBLEM SET 2-1

1. ▶Which of the following sets are well defined? Explain your answers.
   (a) The set of wealthy school teachers.
   (b) The set of great books.
   (c) The set of natural numbers greater than 100.
   (d) The set of subsets of {1, 2, 3, 4, 5, 6}.

2. Write the following sets by listing the members or using set-builder notation:
   (a) The set of letters in the word *mathematics*.
   (b) The set of states in the continental United States.
   (c) The set of months whose names begin with J.
   (d) The set of natural numbers greater than 20.
   (e) The set of states in the United States.
   (f) The set of days in a week starting with the letter P.
   (g) The set of states in the United States that border the Pacific Ocean.

3. Rewrite the following statements using mathematical symbols:
   (a) B is equal to the set whose elements are x, y, z, and w.
   (b) 3 is not an element of set B.
   (c) The set consisting of the elements 1 and 2 is a proper subset of the set consisting of the elements 1, 2, 3, and 4.
   (d) Set D is not a subset of set E.
   (e) Set A is not a proper subset of set B.
   (f) 0 is not an element of the empty set.
   (g) The set whose only element is 0 is not equal to the empty set.

4. ▶(a) Describe three sets of which you are a member.
   (b) Describe three sets that have no members.

5. Which of the following pairs of sets can be placed in one-to-one correspondence?
   (a) {1, 2, 3, 4, 5} and {m, n, o, p, q}
   (b) {m, a, t, h} and {f, u, n}
   (c) {a, b, c, d, e, f, . . . , m} and {1, 2, 3, 4, 5, 6, . . . , 13}
   (d) {$x|x$ is a letter in the word *mathematics*} and {1, 2, 3, 4, . . . , 11}
   (e) {○, △} and {2}

6. Show all possible one-to-one correspondences between the sets A and B if $A = \{1, 2\}$ and $B = \{a, b\}$.

7. How many different one-to-one correspondences are there between two sets with
   (a) four elements each?
   (b) five elements each?
   (c) n elements each?

8. ▶Describe five subsets of a set of college professors.

9. Which of the following represent equal sets?
   $A = \{a, b, c, d\}$
   $B = \{x, y, z, w\}$
   $C = \{c, d, a, b\}$
   $D = \{x|x$ is one of the first four letters of the English alphabet$\}$
   $E = \emptyset$
   $F = \{\emptyset\}$
   $G = \{0\}$
   $H = \{\ \}$

10. If U is the set of all college students and A is the set of all college students with a straight-A average, describe $\overline{A}$.

11. Suppose B is a proper subset of C.
    (a) If $n(C) = 8$, what is the maximum number of elements in B?
    (b) What is the least possible number of elements in C?

12. Suppose C is a subset of D, and D is a subset of C.
    (a) If $n(C) = 5$, find $n(D)$.
    (b) What other relationship exists between sets C and D?

13. ▶Is $\emptyset$ a proper subset of every set? Why?

14. Indicate which symbol, $\in$, or $\notin$, makes each of the following statements true:
    (a) 3 _____ {1, 2, 3}
    (b) 0 _____ $\emptyset$
    (c) {1} _____ {1, 2}
    (d) $\emptyset$ _____ $\emptyset$
    (e) {1, 2} _____ {1, 2}

15. Indicate which symbol, $\subseteq$ or $\not\subseteq$, makes each part of Problem 14 true.

16. Is it always true that $A \not\subseteq B$ implies $B \subseteq A$? Why?

17. Classify each of the following as true or false. If you answer "false," tell why.
    (a) If $A = B$, then $A \subseteq B$.
    (b) If $A \subseteq B$, then $A \subset B$.
    (c) If $A \subset B$, then $A \subseteq B$.
    (d) If $A \subseteq B$, then $A = B$.

18. Use the definition of *less than* to show each of the following:
    (a) $2 < 4$ (b) $3 < 100$ (c) $0 < 3$

19. ▶Define *less than or equal to* in a way similar to the definition of *less than*.

20. (a) If $A = \{a, b, c, d, e, f\}$, how many proper subsets does A have?
    (b) If a set B has n elements where n is some natural number, how many proper subsets does B have?

21. On a certain Senate committee there are seven senators: Abel, Brooke, Cox, Dean, Eggers, Funk, and Gage. Three of these members are to be appointed to a subcommittee. How many possible subcommittees are there?

Section 2-2 Other Set Operations and Their Properties

Finding the complement of a set is an operation that acts on only one set at a time. In this section, we consider operations on two sets.

Set Intersection

Suppose that during the fall quarter, a college wants to mail a survey to all its students who are enrolled in both art and biology classes. To do this, the school officials must identify those students who are taking both classes. If A and B are the sets of students taking art courses and the set of students taking biology courses during the fall quarter, respectively, then the desired set of students includes those common to A and B, or the **intersection** of A and B. The intersection of sets A and B is pictured in Figure 2-4.

intersection

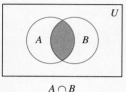

$A \cap B$

FIGURE 2-4

DEFINITION OF SET INTERSECTION

The **intersection** of two sets A and B, written $A \cap B$, is the set of all elements common to both A and B. $A \cap B = \{x \mid x \in A \text{ and } x \in B\}$.

The key word in the definition of intersection is the word *and*. In everyday language, as in mathematics, *and* implies that both conditions must be met. In the above example, the desired set is the set of those students enrolled in both art and biology.

disjoint sets

If sets such as A and B have no elements in common, we call them **disjoint sets.** In other words, two sets A and B are disjoint if, and only if, $A \cap B = \emptyset$. For example, the set of males taking biology and the set of females taking biology are disjoint.

Set Union

If A is the set of students taking art courses during the fall quarter and B is the set of students taking biology courses during the fall quarter, then the set of students taking art or biology or both is the **union** of sets A and B. The union of sets A and B is pictured in Figure 2-5.

union

FIGURE 2-5

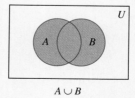

$A \cup B$

DEFINITION OF SET UNION

The **union** of two sets A and B, written $A \cup B$, is the set of all elements in A or in B. $A \cup B = \{x \mid x \in A \text{ or } x \in B\}$.

The key word in the definition of union is *or*. In mathematics, *or* usually means "one or the other or both." This usage is known as the *inclusive or*.

Set Difference

complement of A relative to B / set difference

If A is the set of students taking art classes during the fall quarter and B is the set of students taking biology classes, then the set of all students taking biology classes but not art classes is called the **complement of A relative to B** or the **set difference** of B and A.

DEFINITION OF COMPLEMENT

The **complement of A relative to B,** written $B - A$, is the set of all elements in B that are not in A. $B - A = \{x | x \in B \text{ and } x \notin A\}$.

A Venn diagram representing $B - A$ is shown in Figure 2-6(a). A Venn diagram for $B \cap \overline{A}$ is given in Figure 2-6(b). These diagrams imply $B - A = B \cap \overline{A}$.

FIGURE 2-6

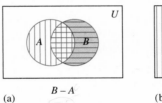
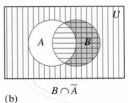

(a) $B - A$ (b) $B \cap \overline{A}$

EXAMPLE 2-5 If $U = \{a, b, c, d, e, f, g\}$, $A = \{d, e, f\}$, and $B = \{a, b, c, d, e\}$, find each of the following:
(a) $A \cup B$ (b) $A \cap B$ (c) $A - B$
(d) $B - A$ (e) $B \cup A$ (f) $B \cap A$

Solution
(a) $A \cup B = \{a, b, c, d, e, f\}$
(b) $A \cap B = \{d, e\}$
(c) $A - B = \{f\}$
(d) $B - A = \{a, b, c\}$
(e) $B \cup A = \{a, b, c, d, e, f\}$
(f) $B \cap A = \{d, e\}$

Properties of Set Operations

From Example 2-5(a) and (e), we see that $A \cup B$ is equal to $B \cup A$. This is an example of the commutative property of set union. It does not matter in which order we write the sets when the union of two sets is involved. Also, from Example 2-5(b) and (f), we see that $A \cap B = B \cap A$. This is an example of the commutative property of set intersection. The following properties are true for any sets.

PROPERTIES

1. **Commutative Properties** For all sets A and B:
   (a) $A \cup B = B \cup A$ Commutative property of set union.
   (b) $A \cap B = B \cap A$ Commutative property of set intersection.
2. **Associative Properties** For all sets A, B, and C:
   (a) $(A \cap B) \cap C = A \cap (B \cap C)$ Associative property of set intersection.
   (b) $(A \cup B) \cup C = A \cup (B \cup C)$ Associative property of set union.
3. **Identity Properties** For every set A and universe U:
   (a) $A \cap U = U \cap A = A$ U is the identity for set intersection.
   (b) $A \cup \emptyset = \emptyset \cup A = A$ $\emptyset$ is the identity for set union.
4. **Complement Properties** For every set A and universe U:
   (a) $\overline{U} = \emptyset$ (b) $\overline{\emptyset} = U$ (c) $A \cap \overline{A} = \emptyset$
   (d) $A \cup \overline{A} = U$ (e) $\overline{\overline{A}} = A$

Is grouping important when two different set operations are involved? For example, is it true that $A \cap (B \cup C) = (A \cap B) \cup C$? To investigate this, let $A = \{a, b, c, d\}$, $B = \{c, d, e\}$, and $C = \{d, e, f, g\}$.

$$A \cap (B \cup C) = \{a, b, c, d\} \cap (\{c, d, e\} \cup \{d, e, f, g\})$$
$$= \{a, b, c, d\} \cap \{c, d, e, f, g\}$$
$$= \{c, d\}$$

$$(A \cap B) \cup C = (\{a, b, c, d\} \cap \{c, d, e\}) \cup \{d, e, f, g\}$$
$$= \{c, d\} \cup \{d, e, f, g\}$$
$$= \{c, d, e, f, g\}$$

In this case, $A \cap (B \cup C) \neq (A \cap B) \cup C$. We have found a counterexample, an example illustrating that the general statement is not always true.

To discover an expression that is equal to $A \cap (B \cup C)$, consider the Venn diagram for $A \cap (B \cup C)$ shown by the shaded region in Figure 2-7. In the figure, $A \cap C$ and $A \cap B$ are subsets of the shaded region. The union of $A \cap C$ and $A \cap B$ is the entire shaded region. Thus $A \cap (B \cup C) = (A \cap B) \cup (A \cap C)$. A similar approach illustrates that $A \cup (B \cap C) = (A \cup B) \cap (A \cup C)$. These properties relate intersection and union and are called **distributive properties**.

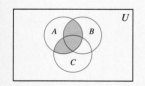

FIGURE 2-7

distributive properties

PROPERTIES

Distributive Properties For all sets A, B, and C:
1. **Distributive property of set intersection over union.**
   $A \cap (B \cup C) = (A \cap B) \cup (A \cap C)$
2. **Distributive property of set union over intersection.**
   $A \cup (B \cap C) = (A \cup B) \cap (A \cup C)$

EXAMPLE 2-6 If $A = \{a, b, c\}$, $B = \{b, c, d\}$, and $C = \{d, e, f, g\}$, check the distributive property of intersection over union for these sets.

Solution
$$A \cap (B \cup C) = \{a, b, c\} \cap (\{b, c, d\} \cup \{d, e, f, g\})$$
$$= \{a, b, c\} \cap \{b, c, d, e, f, g\}$$
$$= \{b, c\}$$

$$(A \cap B) \cup (A \cap C) = (\{a, b, c\} \cap \{b, c, d\}) \cup (\{a, b, c\} \cap \{d, e, f, g\})$$
$$= \{b, c\} \cup \varnothing$$
$$= \{b, c\}$$

Thus $A \cap (B \cup C) = (A \cap B) \cup (A \cap C)$.

Using Venn Diagrams as a Problem-solving Tool

Venn diagrams can be used as a problem-solving tool for modeling information, as this section shows.

EXAMPLE 2-7 Suppose M is the set of all students taking mathematics and E is the set of all students taking English. Identify the students described by each region in Figure 2-8.

FIGURE 2-8

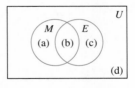

Solution Region (a) contains all students taking mathematics but not English.
Region (b) contains all students taking both mathematics and English.
Region (c) contains all students taking English but not mathematics.
Region (d) contains all students taking neither mathematics nor English.

EXAMPLE 2-8 Use set notation to describe the shaded portions of the Venn diagrams in Figure 2-9(a) and (b).

FIGURE 2-9

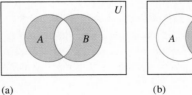 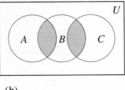

(a)   (b)

Solution The solutions can be described in many different, but equivalent, forms. The following are possible answers:

(a) $(A \cup B) \cap \overline{(A \cap B)}$, or $(A \cup B) - (A \cap B)$
(b) $(A \cap B) \cup (B \cap C)$, or $B \cap (A \cup C)$

EXAMPLE 2-9   In a survey of 110 college freshmen investigating their high school backgrounds, the following information was gathered:

>25 took physics
>45 took biology
>48 took mathematics
>10 took physics and mathematics
>8 took biology and mathematics
>6 took physics and biology
>5 took all three subjects

How many students took biology, but neither physics nor mathematics?
How many did not take any of the three subjects?

Solution   To solve this problem, we *build a model using sets*. Because there are three distinct subjects, we should use three circles. The maximum number of regions of a Venn diagram determined by three circles is 8. In Figure 2-10, P is the set of students taking physics, B is the set taking biology, and M is the set taking mathematics. The shaded region represents the 5 students who took all three subjects. The lined region represents the students who took physics and mathematics, but who did not take biology.

One mind-set to beware of in this problem is thinking that the 25 who took physics, for example, took only physics. That is not necessarily the case. If those students had been taking only physics, then we should have been told so.

Because a total of 10 students took physics and mathematics and because 5 of those also took biology, $10 - 5$, or 5, students took physics and math, but not biology. The other numbers in the diagram were derived by means of similar reasoning. After completing the diagram, we interpret the results. Of all the students, 36 took only biology, but neither physics nor mathematics; 11 did not take any of the three subjects. ∎

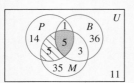

FIGURE 2-10

Cartesian Products

Cartesian product   Another way to produce a set from two given sets is by forming the **Cartesian product.** This formation pairs the elements of one set with the elements of another set in a specific way. Suppose a person has three pairs of pants, $P = \{$blue, white, green$\}$, and two shirts, $S = \{$blue, red$\}$. According to the Fundamental Counting Principle, there are $3 \cdot 2$, or 6, possible different pant-and-shirt pairs.

The pairs of pants and shirts form a set of all possible pairs in which the first member of the pair is an element of set P and the second member is an element of set S. The set of all possible pairs is given in Figure 2-11. Because the first component in each pair represents pants and the second component in each pair represents shirts, the order in which the components are written is important. Thus (green, blue) represents green pants and a blue shirt, whereas (blue, green) represents blue pants and a green shirt. Therefore the two pairs represent different outfits. Because the order in each pair is important, the pairs are called *ordered pairs* **ordered pairs.** The positions that the ordered pairs occupy within the set of outfits is *components* immaterial. Only the order of the **components** within each pair is significant. By the definition of equality for ordered pairs, $(x,y) = (m,n)$ if, and only if, the first components are equal and the second components are equal. A set consisting of ordered pairs such as the ones in the pants-and-shirt example is the Cartesian product of the set of pants and the set of shirts.

FIGURE 2-11

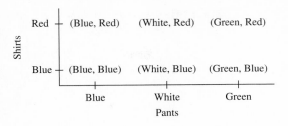

DEFINITION OF CARTESIAN PRODUCT

For any sets A and B, the **Cartesian product** of A and B, written $A \times B$, is the set of all ordered pairs such that the first element of each pair is an element of A and the second element of each pair is an element of B.

$$A \times B = \{(x, y) \mid x \in A \text{ and } y \in B\}$$

REMARK $A \times B$ is commonly read as "A cross B."

A third way of obtaining a Cartesian product is by using a tree diagram, as in Figure 2-12. The elements of the Cartesian product are formed by following each of the branches of the tree.

FIGURE 2-12

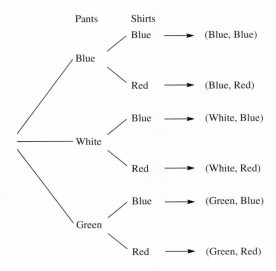

EXAMPLE 2-10 If $A = \{a, b, c\}$ and $B = \{1, 2, 3\}$, find each of the following:
(a) $A \times B$ (b) $B \times A$ (c) $A \times A$

Solution (a) $A \times B = \{(a, 1), (a, 2), (a, 3), (b, 1), (b, 2), (b, 3), (c, 1), (c, 2), (c, 3)\}$

(b) $B \times A = \{(1, a), (1, b), (1, c), (2, a), (2, b), (2, c), (3, a), (3, b), (3, c)\}$

(c) $A \times A = \{(a, a), (a, b), (a, c), (b, a), (b, b), (b, c), (c, a), (c, b), (c, c)\}$

It is possible to form a Cartesian product involving the null set. Suppose $A = \{1, 2\}$. Because there are no elements in $\varnothing$, no ordered pairs (x, y) with $x \in A$ and $y \in \varnothing$ are possible, so $A \times \varnothing = \varnothing$. This is true for all sets A. Similarly, $\varnothing \times A = \varnothing$ for all sets A.

PROBLEM SET 2-2

1. Suppose $U = \{e, q, u, a, l, i, t, y\}$, $A = \{l, i, t, e\}$, $B = \{t, i, e\}$, and $C = \{q, u, e\}$. Decide whether the following pairs of sets are equal:
   (a) $A \cap B$ and $B \cap A$
   (b) $A \cup B$ and $B \cup A$
   (c) $A \cup (B \cup C)$ and $(A \cup B) \cup C$
   (d) $A \cup \varnothing$ and A
   (e) $(A \cap A)$ and $(A \cap \varnothing)$
   (f) $\overline{\overline{C}}$ and C

2. Tell whether each of the following is true or false for all sets A, B, or C. If false, give a counterexample.
   (a) $A \cup \varnothing = A$
   (b) $A - B = B - A$
   (c) $A \cup A = A$
   (d) $\overline{A \cap B} = \overline{A} \cap \overline{B}$
   (e) $A \cap B = B \cap A$
   (f) $(A \cup B) \cup C = A \cup (B \cup C)$
   (g) $A - \varnothing = A$

3. If $B \subseteq A$, find a simpler expression for each.
   (a) $A \cap B$ (b) $A \cup B$

4. For each of the following, shade the portion of the Venn diagram that illustrates the set:
   (a) $A \cup B$
   (b) $A \cap \overline{B}$
   (c) $\overline{A \cap B}$
   (d) $(A \cap B) \cup (A \cap C)$
   (e) $\overline{A \cap B}$
   (f) $(A \cup B) \cap \overline{C}$
   (g) $(A \cap B) \cup C$
   (h) $(\overline{A} \cap B) \cup C$

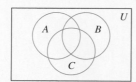

5. If S is a subset of universe U, find each of the following:
   (a) $S \cup \overline{S}$ (b) $S \cup U$ (c) $\varnothing \cup S$
   (d) $\overline{U}$ (e) $S \cap U$ (f) $\overline{\varnothing}$
   (g) $S \cap \overline{S}$ (h) $S - \overline{S}$ (i) $U \cap \overline{S}$
   (j) $\overline{\overline{S}}$ (k) $\varnothing \cap S$ (l) $U - S$

6. ▶ Answer each of the following and justify your answer:
   (a) If $a \in A \cap B$, is it true that $a \in A \cup B$?
   (b) If $a \in A \cup B$, is it true that $a \in A \cap B$?

7. For each of the following conditions, find $A - B$.
   (a) $A \cap B = \varnothing$
   (b) $B = U$
   (c) $A = B$
   (d) $A \subseteq B$

8. Use set notation to identify each of the following shaded regions.

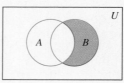
(a)

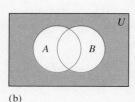
(b)

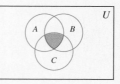
(c)

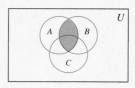
(d)

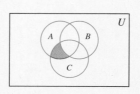
(e)   (f)

9. Shade the portion of the diagram that represents the given sets.

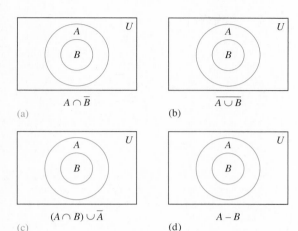

10. Use Venn diagrams to determine if each of the following is true:
    (a) $A \cup (B \cap C) = (A \cup B) \cap C$
    (b) $A \cap (B \cup C) = (A \cap B) \cup C$
    (c) $A - B = B - A$
    (d) $A - (B - C) = (A - B) - C$
    (e) $A - (B \cup C) = (A - B) \cup (A - C)$

11. (a) If A has 3 elements and B has 2 elements, what is the greatest number of elements possible in (i) $A \cup B$? (ii) $A \cap B$?
    (b) If A has n elements and B has m elements, what is the greatest number of elements possible in (i) $A \cup B$? (ii) $A \cap B$?

12. ▶The primary colors are red, blue, and yellow. If each is considered a set, write an explanation of what we would expect to get with the intersection of each of these sets from the regions pictured.

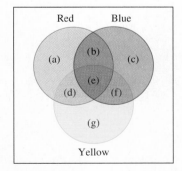

13. If E is the set of even natural numbers and O is the set of odd natural numbers, describe each of the following:
    (a) The union of sets E and O
    (b) The intersection of sets E and O
    (c) The set difference of E and O
    (d) The set difference of O and E

14. The equations $\overline{A \cup B} = \overline{A} \cap \overline{B}$ and $\overline{A \cap B} = \overline{A} \cup \overline{B}$ are referred to as *DeMorgan's Laws* in honor of the famous British mathematician who first discovered them. Use Venn diagrams to show each of the following:
    (a) $\overline{A \cup B} = \overline{A} \cap \overline{B}$ (b) $\overline{A \cap B} = \overline{A} \cup \overline{B}$
    (c) Verify (a) and (b) for specific sets A and B.

15. If $A \cap B = A \cup B$, how are A and B related?

16. Given that the universe is the set of all humans, $B = \{x \mid x$ is a college basketball player$\}$, and $S = \{x \mid x$ is a college student more than 200 cm tall$\}$, describe each of the following in words:
    (a) $B \cap S$ (b) $\overline{S}$ (c) $B \cup S$
    (d) $\overline{B \cup S}$ (e) $\overline{B} \cap S$ (f) $B \cap \overline{S}$

17. Suppose P is the set of all eighth-grade students at the Paxson school, with B the set of all students in the band and C the set of all students in the choir. Identify in words the students described by each region of the diagram.

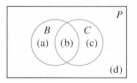

18. Of the eighth graders at the Paxson school, there were 7 who played basketball, 9 who played volleyball, 10 who played soccer, 1 who played basketball and volleyball only, 1 who played basektball and soccer only, 2 who played volleyball and soccer only, and 2 who played volleyball, basketball, and soccer. How many played one or more of the three sports?

19. In a fraternity with 30 members, 18 take mathematics, 5 take both mathematics and biology, and 8 take neither mathematics nor biology. How many take biology but not mathematics?

20. Write the letters in the appropriate sections of the Venn diagram using the directions below:

    Set A contains the letters in the word *Iowa*.
    Set B contains the letters in the word *Hawaii*.
    Set C contains the letters in the word *Ohio*.
    The universal set contains the letters in the word *Washington*.

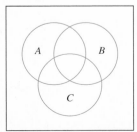

21. In Paul's bicycle shop, 40 bicycles are inspected. If 20 needed new tires and 30 needed gear repairs, answer the following:
    (a) What is the greatest number of bikes that could have needed both?
    (b) What is the least number of bikes that could have needed both?
    (c) What is the greatest number of bikes that could have needed neither?

22. Classify the following as true or false. If false, give a counterexample. The first statement is true.
    (a) $n(A) = n(B)$. $A = B$.
    (b) $A \sim B$. $A \cup B$ is not equivalent to B.
    (c) $A - B = \varnothing$. $A = B$.
    (d) $B - A = \varnothing$. $B \subseteq A$.
    (e) $A \subset B$. $n(A) < n(B)$.
    (f) $n(A) < n(B)$. $A \subset B$.

23. Three announcers each tried to predict the winners of Sunday's professional football games. The only team not picked that is playing Sunday was the Giants. The choices for each person were as follows:

    Phyllis: Cowboys, Steelers, Vikings, Bills
    Paula: Steelers, Packers, Cowboys, Redskins
    Rashid: Redskins, Vikings, Jets, Cowboys

    If the only teams playing Sunday are those just mentioned, which teams will play which other teams?

24. Let $A = \{x, y\}$, $B = \{a, b, c\}$, and $C = \{0\}$. Find each of the following:
    (a) $A \times B$
    (b) $B \times A$
    (c) $B \times \varnothing$
    (d) $C \times C$
    (e) $\varnothing \times C$
    (f) $(A \times B) \cup (A \times C)$
    (g) $(A \cup B) \times C$
    (h) $A \cup (B \times C)$

25. ▶(a) Is the operation of forming Cartesian products commutative? Explain why or why not.
    ▶(b) Is the operation of forming Cartesian products associative? Explain why or why not.

26. For each of the following, the Cartesian product $C \times D$ is given by the following sets. Find C and D.
    (a) $\{(a, b), (a, c), (a, d), (a, e)\}$
    (b) $\{(1, 1), (1, 2), (1, 3), (2, 1), (2, 2), (2, 3)\}$
    (c) $\{(0, 1), (0, 0), (1, 1), (1, 0)\}$

27. Answer each of the following:
    (a) If A has 5 elements and B has 4 elements, how many elements are in $A \times B$?
    (b) If A has m elements and B has n elements, how many elements are in $A \times B$?
    (c) If A has m elements, B has n elements, and C has p elements, how many elements are in $(A \times B) \times C$?

28. If $A = \{1, 2, 3\}$, $B = \{0\}$, and $C = \varnothing$, find the number of elements in each of the following:
    (a) $A \times C$
    (b) $B \times C$
    (c) $C \times C$

29. If the number of elements in set B is 3 and the number of elements in $(A \cup B) \times B$ is 24, what is the number of elements in A if $A \cap B = \varnothing$?

30. If A and B are nonempty sets such that $A \times B = B \times A$, does $A = B$?

31. If there are 6 teams in the Alpha league and 5 teams in the Beta league and if each team from one league plays each team from the other league exactly once, how many games are played?

32. José has 4 pairs of slacks, 5 shirts, and 3 sweaters. From how many different combinations can he choose if he chooses a pair of slacks, a shirt, and a sweater each time?

★33. At the end of a tour of the Grand Canyon, several guides were talking about the people on the latest British-American tour. The guides could not remember the total number in the group; however, together they compiled the following statistics about the group. It contained 26 British females, 17 American women, 17 American males, 29 girls, 44 British citizens, 29 women, and 24 British adults. Find the total number of people in the group.

34. ▶The drawing below, taken from *Professional Standards for Teaching Mathematics*, was used to discuss the relationship among the knowledge bases for teachers of grades K-4, 5-8, and 9-12. Explain how this might have worked.

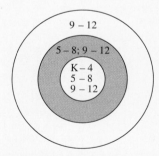

35. ▶A pollster interviewed 500 university seniors who owned credit cards. She reported that 240 owned Goldcard, 290 had Supercard, and 270 had Thriftcard. Of these seniors, the report said that 80 owned only a Goldcard and a Supercard, 70 owned only a Goldcard and a Thriftcard, 60 owned only a Supercard and a Thriftcard, and 50 owned all three cards. When the report was submitted for publication in the local campus newspaper, the editor refused to publish it, claiming that the poll was not accurate. Was the editor right? Why or why not?

Review Problems

36. List all the subsets of $A = \{a, b, c\}$.

37. Are the following two sets equal?

    $\{2, 4, 6, 8, 10, \ldots\}$
    $\{x \mid x = 2n, n \in N\}$

38. Given $B = \{p, q, r, s\}$, list the nonempty, proper subsets of B.

39. Write the set of states of the United States that begin with the letter M by
    (a) listing them,
    (b) using set-builder notation.

▼ BRAIN TEASER

Simon and Gaussfunbel were arguing over how big Missoula, Montana, actually is. Simon insisted it must have at least 60,000 people since it was home to 10,000 University of Montana students. Gaussfunbel, however, was adamant that Missoula was tiny. He argued that it was so small that no two people in the town could not have different initials. Who was correct?

LABORATORY ACTIVITY

A set of attribute blocks consists of 32 blocks. Each block is identified by its own shape, size, and color. The four shapes in a set are square, triangle, rhombus, and circle; the four colors are red, yellow, blue, and green; the two sizes are large and small. In addition to the blocks, each set contains a group of 20 cards. Ten of the cards specify one of the attributes of the blocks (for example, red, large, square). The other 10 cards are negation cards and specify the lack of an attribute (for example, not green, not circle). Many set-type problems can be studied with these blocks. For example, let A be the set of all green blocks and B be the set of all large blocks. Using the set of all blocks as the universal set, describe elements in each set listed below to determine which are equal:

1. $A \cup B$; $B \cup A$
2. $\overline{A \cup B}$; $\overline{A} \cap \overline{B}$
3. $\overline{A \cap B}$; $\overline{A} \cup \overline{B}$
4. $A - B$; $A \cap \overline{B}$

Section 2-3 Relations and Functions

Relations

relation   A subset of a Cartesian product is called a **relation.** Before formally examining this mathematical concept, let us examine some relations. The word *relations* brings to mind parents, brothers, sisters, grandfathers, aunts, and so on. For example, "is the brother of" may express a relation between Billy and Jimmy.

To illustrate relations, a diagram like Figure 2-13 is useful. Suppose each point in the figure represents a child on a playground, the letters represent their names, and an arrow going from I to J means that I "is the sister of" J.

FIGURE 2-13

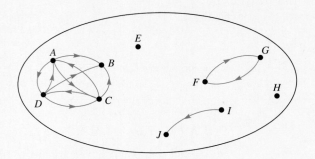

The information in Figure 2-13 indicates that *A*, *C*, *D*, *F*, *G*, and *I* are definitely girls and that *B* and *J* are definitely boys. Why? It also indicates that *H* and *E* have no sisters on the playground, but it does not indicate the gender of *H* and *E*.

Another way to exhibit the relation "is a sister of" is by using the same set twice, with arrows, as in Figure 2-14.

FIGURE 2-14

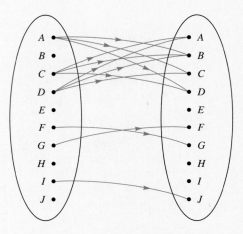

Still another way to show the relation "is a sister of" is to write the relation "*A* is a sister of *B*" as an ordered pair (*A*, *B*). For the set of children on the playground, we may describe the relation in ordered pairs as follows:
{(*A*, *B*), (*A*, *C*), (*A*, *D*), (*C*, *A*), (*C*, *B*), (*C*, *D*), (*D*, *A*), (*D*, *B*), (*D*, *C*), (*F*, *G*), (*G*, *F*), (*I*, *J*)}.

EXAMPLE 2-11 The pairs (Helena, Montana), (Denver, Colorado), Springfield, Illinois), and (Juneau, Alaska) are included in some relation. Give a rule that describes the relation.

Solution One possible rule is that the ordered pair (*x*, *y*) indicates that *x* is the capital of *y*. ∎

In Example 2-11, the first components of the ordered pairs are state capitals; the second components are states of the United States. Each ordered pair in the example is an element of the Cartesian product $A \times B$, where *A* is the set of state capitals and *B* is the set of states in the United States.

SECTION 2-3 RELATIONS AND FUNCTIONS

> **DEFINITION OF RELATION**
>
> Given any two sets A and B, a **relation** from A to B is a subset of $A \times B$; that is, if R is a relation, then $R \subseteq A \times B$. If $A = B$, we say that the **relation is on A**.

Properties of Relations

Figure 2-15 represents a set of children in a small group. They have drawn all possible arrows representing the relation "has the same first letter in his or her name as." Notice that the children were very careful to observe that each child in the group has the same first initial as himself or herself. Figure 2-15 shows three properties of relations.

FIGURE 2-15

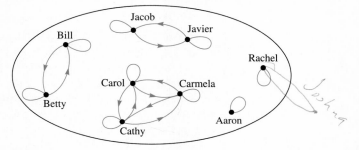

In Figure 2-15, there is a loop at every point. For example, Rachel has the same first initial as herself, namely, R. The relation has the reflexive property.

> **PROPERTY**
>
> **The Reflexive Property** A relation R on a set X is reflexive if, and only if, for every element, $a \in X$, a is related to a. That is, for every $a \in X$, $(a, a) \in R$.

A relation such as "is taller than" is not reflexive because people cannot be taller than themselves.

In Figure 2-15, every pair of points that has an arrow headed in one direction also has a return arrow. For example, if Bill has the same first initial as Betty, then Betty has the same first initial as Bill. This relation has the symmetric property.

> **PROPERTY**
>
> **The Symmetric Property** A relation R on a set X is symmetric if, and only if, for all elements a and b in X, whenever a is related to b, then b also is related to a; that is, if $(a, b) \in R$, then $(b, a) \in R$.

A relation such as "is a brother of" is not symmetric, since Dick can be a brother of Jane, but Jane cannot be a brother of Dick.

In Figure 2-15, every connected portion satisfies the transitive property. For example, if Carol has the same first intial as Carmela, and Carmela has the same first initial as Cathy, then Carol has the same first initial as Cathy. The relation has the transitive property.

> ## PROPERTY
>
> **The Transitive Property** A relation R on a set X is transitive if, and only if, for all elements a, b, and c of X, whenever a is related to b and b is related to c, then a is related to c; that is, if $(a, b) \in R$ and $(b, c) \in R$, then $(a, c) \in R$.

> **REMARK** a, b, and c do not have to be different. Three symbols are used to allow for difference.

A relation such as "is the father of" is not transitive since if Tom Jones, Sr., is the father of Tom Jones, Jr., and Tom Jones, Jr., is the father of Joe Jones, then Tom Jones, Sr., is not the father of Joe Jones. He is instead the grandfather.

The relation "is the same color as" is reflexive, symmetric, and transitive. The common relation "is equal to" also satisfies all three properties. In general, relations that satisfy all three properties are called **equivalence relations.**

equivalence relations

> ## DEFINITION OF EQUIVALENCE RELATION
>
> An **equivalence relation** on a set A is any relation R on A that satisfies the reflexive, symmetric, and transitive properties.

The most natural mathematical equivalence relation encountered in elementary school is "is equal to" on the set of all numbers.

Suppose P is the set consisting of all persons attending the State University and consider the relation "is the same sex as." This relation is an equivalence relation. The relation partitions the persons at State into two classes: females and males. Any equivalence relation defined on a set has the effect of partitioning the set into disjoint subsets, called **equivalence classes.** In this example, the class of females can be described as the set of all students who are the same sex as Jane, a student at State, and the set can be called Jane's equivalence class. This class also can be called Mary's equivalence class as long as Mary is a student at State. An equivalence class can be named after any of its members.

equivalence classes

EXAMPLE 2-12 Tell whether the following relations are reflexive, symmetric, or transitive on the set of all people:

(a) "Is older than"

(b) "Sits in the same row as"

(c) "Is heavier than"

Solution The following table gives the answers.

Relation	Reflexive	Symmetric	Transitive
(a) "Is older than"	No	No	Yes
(b) "Sits in the same row as"	Yes	Yes	Yes
(c) "Is heavier than"	No	No	Yes

Note that "sits in the same row as" is an equivalence relation.

Functions

In the *Teaching Standards (p. 136)*, we find:

To build bridges for their students to the mathematics that comes later in the school curriculum, teachers must have a basic understanding of the concepts of functions and their use in the growth of mathematical ideas. Understanding different representations of functions (tabular, graphical, symbolic, verbal), how to move among these representations, and the strengths and limitations of each is fundamental.

The following is an example of a game called "guess my rule," often used to introduce the concept of a function.

When Tom said 2, Noah said 5. When Dick said 4, Noah said 7. When Mary said 10, Noah said 13. When Liz said 6, what did Noah say? What is Noah's rule?

The answer to the first question may be 9, and the rule could be, "Take the original number and add 3"; that is, for any number n, Noah's answer is $n + 3$.

EXAMPLE 2-13 Guess the teacher's rule for the following responses:

(a) You	Teacher	(b) You	Teacher	(c) You	Teacher
1	3	2	5	2	0
0	0	3	7	4	0
4	12	5	11	7	1
10	30	10	21	21	1

Solution (a) The teacher's rule could be, "Multiply the given number n by 3," that is, $3n$.

(b) The teacher's rule could be, "Double the original number n and add 1," that is, $2n + 1$.

(c) The teacher's rule could be, "If the number n is even, answer 0; if the number is odd, answer 1."

HISTORICAL NOTE

The Babylonians (ca. 2000 B.C.) probably had a working idea of what a function was. To them, it was a table or a correspondence. René Descartes (1637), Gottfried Wilhelm von Leibnitz (1692), Johann Bernoulli (1718), Leonhard Euler (1750), Joseph Louis Lagrange (1800), and Jean Joseph Fourier (1822) were among the mathematicians contributing to the notion of function. Leonhard Euler in 1734 first used the notation $f(x)$. In the late 1800s, Georg Cantor and others began to use the modern definition.

function
domain
range

Another way to prepare students for the idea of a **function** is by using a "function machine." The inputs are from a set called the **domain,** and the outputs are from a set called the **range.** The function machine is sometimes pictured as in Figure 2-16.

FIGURE 2-16

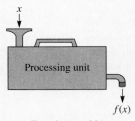

Function machine

The processing unit normally consists of some rule to assign a value from the domain to a value of the range.

In a function machine, for any input element x there is an output element denoted by $f(x)$, read "f of x." A function machine associates *exactly one output with each input*. If you enter some number x as input and obtain some number $f(x)$ as output, then *every* time you enter that same x as input, you will obtain that same $f(x)$ as output. The idea of a function machine associating exactly one output with each input according to some rule leads us to the following definition.

DEFINITION OF FUNCTION

A **function** from A to B is a relation from A to B in which each element of A is paired with one, *and only one,* element of B.

REMARK If a function is represented as a set of ordered pairs, then the set of all first components is the domain and the set of the second components is the range.

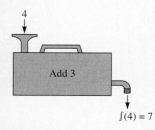

FIGURE 2-17

EXAMPLE 2-14 Consider the function machine in Figure 2-17. What will happen if the numbers 0, 1, 3, and 6 are entered?

Solution If the numbers output are denoted by $f(x)$, the corresponding values can be described using Table 2-3.

TABLE 2-3

x	f(x)
0	3
1	4
3	6
6	9

We can write an equation to depict the rule in Example 2-14 as follows. If the input is x, the output is $x + 3$; that is, $f(x) = x + 3$. The output values can be obtained by substituting the values 0, 1, 3, 4, and 6 for x in $f(x) = x + 3$, as shown.

$$f(0) = 0 + 3 = 3$$
$$f(1) = 1 + 3 = 4$$
$$f(3) = 3 + 3 = 6$$
$$f(4) = 4 + 3 = 7$$
$$f(6) = 6 + 3 = 9$$

Normally, if no domain is given to describe a function, then the domain is assumed to be the largest set for which the rule is meaningful.

A calculator is a function machine. Suppose a student enters $\boxed{9}$ $\boxed{\times}$ $\boxed{K}$ on the calculator, using the constant key, $\boxed{K}$. The student then presses $\boxed{0}$ and hands the calculator to another student. The other student is to determine the rule by entering various numbers followed by the $\boxed{=}$ key. Machines with an automatic constant feature can also be used.

Other buttons on a calculator are function buttons. For example, the $\boxed{\pi}$ button always displays an approximation for π, such as 3.1415927; the $\boxed{+/-}$ button displays a negative sign in front of a number or removes it; and the $\boxed{x^2}$ and $\boxed{\sqrt{}}$ buttons square and take the square root of numbers, respectively.

Are all input-output machines function machines? Consider the machine in Figure 2-18. For any natural-number input x, the machine outputs a number that is less than x. If, for example, you input the number 10, the machine may output 9, since 9 is less than 10. If you input 10 again, the machine may output 3, since 3 is less than 10. This clearly violates the definition of a function, since 10 can be paired with more than one element. The machine is not a function machine.

Consider the relations described in Figure 2-19. Do they illustrate functions?

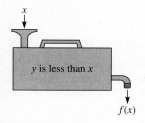

FIGURE 2-18

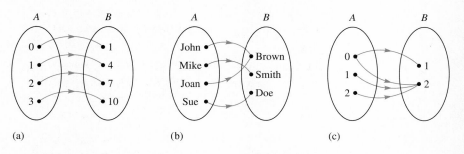

FIGURE 2-19

In Figure 2-19(a), for every element belonging to the domain A, there is one, and only one, element belonging to *B*. Thus this relation is a function from *A* to *B*. Figure 2-19(b) illustrates a function, since there is only one arrow leaving each element in *A*. It does not matter that an element of set *B*, Brown, has two arrows pointing to it. Figure 2-19(c) does not define a function because 0 is paired with more than one element.

EXAMPLE 2-15   Which, if any, of the parts of Figure 2-20 exhibits a function from *A* to *B*?

FIGURE 2-20

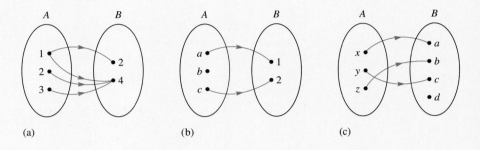

Solution   (a) Figure 2-20(a) does not define a function from *A* to *B*, since the element 1 is paired with both 2 and 4.

(b) Figure 2-20(b) does not define a function from *A* to *B*, since the element *b* is not paired with any element of *B*. (It is a function from a subset of *A* to *B*.)

(c) Figure 2-20(c) does define a function from *A* to *B*, since there is one and only one arrow leaving each element of *A*. The fact that *d*, an element of *B*, is not paired with any element in the domain does not violate the definition. ■

EXAMPLE 2-16   Determine whether the following relations are functions from the set of first components to the set of second components:

(a) $\{(1, 2), (2, 5), (3, 7), (1, 4), (4, 8)\}$

(b) $\{(1, 2), (2, 2), (3, 2), (4, 2)\}$

Solution   (a) This is not a function, since the first component, 1, is associated with two different second components, namely, 2 and 4.

(b) This is a function from $\{1, 2, 3, 4\}$ to $\{2\}$, since each first components is associated with exactly one second component. The fact that the second component, 2, is associated with more than one first component does not violate the definition of a function. ■

EXAMPLE 2-17   Explain why a telephone company would not set rates for telephone calls as depicted on the graph in Figure 2-21.

FIGURE 2-21

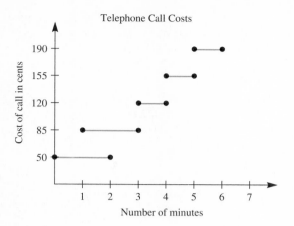

Solution The graph does not depict a function. For example, a customer could be charged either 50¢ or 85¢ for the second minute of conversation. ∎

Operations on Functions

Consider the function machines in Figure 2-22. If 2 is entered in the top machine in Figure 2-22, then $f(2) = 2 + 4 = 6$. Six is then entered in the second machine and $g(6) = 2 \cdot 6 = 12$. The functions in Figure 2-22 illustrate the **composition of two functions.** In the composition of two functions, the range of the first function becomes the domain of the second function.

composition of two functions

FIGURE 2-22

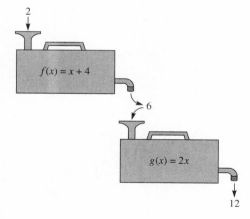

If the first function f is followed by a second function g, as in Figure 2-22, then we symbolize the composition of the functions as $g \circ f$. If we input 3 in the function machines of Figure 2-22, then the output is symbolized by $(g \circ f)(3)$. Because f acts first on 3, to compute $(g \circ f)(3)$, we find $f(3) = 3 + 4 = 7$ and then $g(7) = 2 \cdot 7 = 14$. Hence, $(g \circ f)(3) = 14$. Notice that $(g \circ f)(3) = g(f(3))$.

Elementary school math texts introduce composition of two functions, one followed by the other, as seen on the student page from *Addison-Wesley Mathematics,* 1991, Grade 4.

The output of function machine A is the input for function machine B. Study the example. Then figure out the two function rules and complete the table.

Input A	Output B
24	5
72	11
8	3
32					
40					
48					

EXAMPLE 2-18 If $f(x) = 2x + 3$ and $g(x) = x - 3$, find the following:

(a) $(f \circ g)(3)$ (b) $(g \circ f)(3)$

Solution (a) $(f \circ g)(3) = f(0) = 2 \cdot 0 + 3 = 3$
(b) $(g \circ f)(3) = g(9) = 9 - 3 = 6$ ∎

REMARK Example 2-18 shows that composition of functions is not commutative since $(f \circ g)(3) \neq (g \circ f)(3)$.

EXAMPLE 2-19 Find the range of $g \circ f$ for each of the following where the domain of $g \circ f$ is the set $\{1, 2, 3\}$:

(a) $f(x) = 2x$; $g(x) = 3x$
(b) $f(x) = x + 2$; $g(x) = x - 2$

Solution (a) The composition follows in Figure 2-23.

FIGURE 2-23

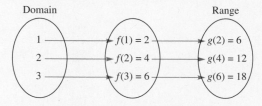

Thus the range of $g \circ f$ is the set $\{6, 12, 18\}$.

(b) The composition follows in Figure 2-24.

FIGURE 2-24

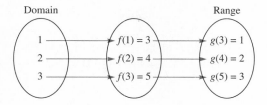

Thus the range of $g \circ f$ is the set $\{1, 2, 3\}$. ∎

identity function

> **REMARK** Under the function $g \circ f$ in Example 2-19(b), the image of every element in the domain is the element itself, that is $(g \circ f)(1) = 1$, $(g \circ f)(2) = 2$, and $(g \circ f)(3) = 3$. Such a function is called an **identity function.**

Applications of Functions

Functions have many real-life applications. For example, on direct-dial, long-distance calls, you pay only for the minutes you talk. The initial rate period is 1 min. Suppose the weekday rate for a long-distance phone call from Missoula, Montana, to Butte, Montana, is 50¢ for the first minute and 30¢ for each additional minute or part of a minute. We have seen that one way to describe a function is by writing an equation. Based on the information in Table 2-4, the equation relating time to cost is $C = 50 + (t - 1)30$. This also could be written as $f(t) = 50 + 30(t - 1)$, where $f(t)$ is the cost of the call. If we restrict the time in minutes to the first five natural numbers, the function can be described as the set of ordered pairs $\{(1, 50), (2, 80), (3, 110), (4, 140), (5, 170)\}$.

The information in the first five rows in Table 2-4 might also be shown on a *lattice graph,* as in Figure 2-25. The lattice graph is formed by taking the Cartesian product of the sets $\{1, 2, 3, 4, 5\}$ and $\{50, 80, 110, 140, 170\}$ and plotting points corresponding to the ordered pairs. If (a, b) is in the Cartesian product, then the corresponding point in the lattice is found by starting at the lower left corner and moving first horizontally a units, and

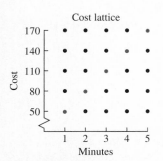

FIGURE 2-25

TABLE 2-4

Number of Minutes Talked	Total Cost in Cents
1	50
2	$50 + 1 \cdot 30 = 80$
3	$50 + 2 \cdot 30 = 110$
4	$50 + 3 \cdot 30 = 140$
5	$50 + 4 \cdot 30 = 170$
.	.
.	.
.	.
t	$50 + (t - 1) \cdot 30$

then vertically b units. Functions and their graphs are frequently used in the business world, as the following example illustrates.

EXAMPLE 2-20 In Figure 2-26, the blue graph shows the cost C in dollars of producing a given number of tee shirts. The red graph shows the revenue R in dollars from selling any number of tee shirts.

FIGURE 2-26

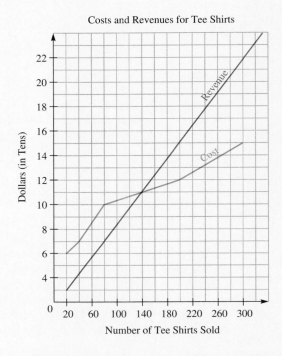

The graphs technically should be lattice graphs because a factory produces only a whole number of tee shirts. However, for visual effect, it is customary to connect the points to produce continuous segments. Based on the information in the graphs, find the following:

(a) The cost of producing the first 300 tee shirts

(b) The revenue from the sale of the first 80 tee shirts

(c) The profit or loss if the first 80 tee shirts are produced and sold

(d) The break-even point, that is, the number of items that must be produced and sold in order for the net profit to be $0.

Solution (a) From the blue graph in Figure 2-26, we see that the cost corresponding to 300 tee shirts is $150.00.

(b) From the red graph, we see that the revenue from the sale of the first 80 tee shirts is $70.00.

(c) The cost of producing 80 tee shirts is $100. Because the profit is the difference between the cost and the revenue, we find that the loss in this case is $100.00 to $70.00, or $30.00.

(d) The break-even point is at point C, where the graphs intersect. At that point, the cost and the revenue are the same. The number of tee shirts corresponding to point C is 140.

PROBLEM SET 2-3

1. Each of the following gives pairs that are included in some relation. Give a rule or phrase that could describe each relation and list two more pairs that could be included in the relation.
   (a) (1, 1), (2, 4), (3, 9), (4, 16)
   (b) (Blondie, Dagwood), (Martha, George), (Rosalyn, Jimmy), (Flo, Andy), (Scarlett, Rhett), (Barbara, George), (Dan, Marilyn)
   (c) (a, A), (b, B), (c, C), (d, D)
   (d) (3 candies, 10¢), (6 candies, 20¢)

2. ▶Let $X = \{a, b, c\}$ and $Y = \{m, n\}$ and suppose X and Y represent two sets of students. The students in X point to the shorter students in Y. The following diagrams list two possibilities. Tell as much as you can about the students depicted in each diagram.

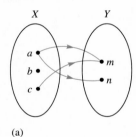

(a)

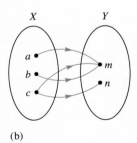

(b)

3. Write three ordered pairs that satisfy the relation "is owned by," from a set of cats to a set of people.

4. The following are the ages of the children in a family: Bill, 17; Becky, 14; John, 9; Abby, 3; Karly, 1. Draw an arrow diagram showing the names of the children and the relation "is younger than."

5. Tell whether each of the following is reflexive, symmetric, or transitive on the set of all people. Which are equivalence relations?
   (a) "Is a parent of"
   (b) "Is the same age as"
   (c) "Has the same last name as"
   (d) "Is the same height as"
   (e) "Is married to"
   (f) "Lives within 10 mi of"
   (g) "Is older than"

6. Tell whether each of the following is reflexive, symmetric, or transitive on the set of subsets of a nonempty set. Which are equivalence relations?
   (a) "Is equal to"   (b) "Is a proper subset of"
   (c) "Is not equal to"

7. ▶For each of the following, guess the teacher's rule. Explain your answers.

(a)
You	Teacher
3	8
4	11
5	14
10	29

(b)
You	Teacher
0	1
3	10
5	26
8	65

(c)
You	Teacher
6	42
0	0
8	72
2	6

8. The following sets of ordered paris are functions. Give a rule that describes each function.
   (a) {(2, 4), (3, 6), (9, 18), (12, 24)}
   (b) {(5, 3), (7, 5), (11, 9), (14, 12)}
   (c) {(2, 8), (5, 11), (7, 13), (4, 10)}
   (d) {(2, 5), (3, 10), (4, 17), (5, 26)}

9. Following are five relations from the set {1, 2, 3} to the set {a, b, c, d}. Which are functions? If the relation is not a function, tell why it is not.
   (a) {(1, a), (2, b), (3, c), (1, d)}
   (b) {(1, c), (3, d)}
   (c) {(1, a), (2, b), (3, a)}
   (d) {(1, a), (1, b), (1, c)}

10. Write a function rule to describe the number of wheels on a normal car.

11. ▶Is a one-to-one correspondence a function? Give an example and write an explanation of your answer.

12. ▶Does the diagram define a function from A to B? Why or why not?

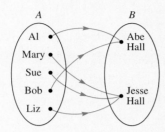

13. If $g(x) = 3x + 5$, find each value.
    (a) $g(0)$ (b) $g(2)$ (c) $g(10)$ (d) $g(a)$
14. (a) Draw a diagram of a function with domain {1, 2, 3, 4, 5} and range {a, b}.
    (b) How many possible functions are there in part (a)?
15. Suppose that $f(x) = 2x + 1$ and the domain is {0, 1, 2, 3, 4}. Describe the function in the following ways:
    (a) Draw an arrow diagram involving two sets.
    (b) Use ordered pairs.
    (c) Make a table.
    (d) Draw a lattice graph to depict the function.
16. Consider two function machines that are placed as shown.

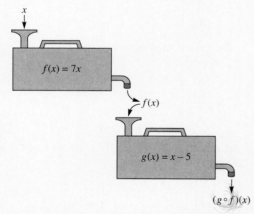

Find the final output for each of the following inputs:
    (a) 0 (b) 3 (c) 10
17. (a) Write a rule for computing the cost of mailing a first-class letter based upon its weight.
    (b) Find the cost of mailing a 3-oz letter.
18. According to wildlife experts, the rate at which crickets chirp is a function of the temperature; that is, $C = T - 40$, where C is the number of chirps every 15 sec and T is the temperature in degrees Fahrenheit.
    (a) How many chirps does the cricket make per second if the temperature is 70°F?
    (b) What is the temperature if the cricket chirps 40 times in 1 min?
19. If taxi fares are $2.50 for the first $\frac{1}{2}$ mi and 50¢ for each additional $\frac{1}{4}$ mi, answer the following:
    (a) What is the fare for a 2-mi trip?
    (b) Write a rule for computing the fare for an *n*-mile trip by taxi.
20. Work the problem on the student page of this section.
21. The following graph shows arithmetic achievement-test scores for students of a sixth-grade class. From the graph, estimate the following:
    (a) The frequency of the score made most often
    (b) The highest score obtained
    (c) The number of boys who would have to score 54 on the test in order to match the number of girls scoring 54

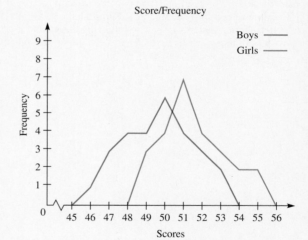

22. Find a rule for each of the following sequences, whose domains are the natural numbers:
    (a) {3, 8, 13, 18, 23, ... }
    (b) {3, 9, 27, 81, 243, ... }
    (c) {2, 4, 6, 8, 10, ... }
23. (a) Is the rule "has as mother" a function whose domain is the set of all people?
    (b) Is the relation "has as brother" a function on the set of all boys?

Review Problems

24. Draw Venn diagrams to verify the following:
    (a) $A - (B \cup C) = (A - B) \cap (A - C)$
    (b) If $A \subseteq B$, then $A \cup B = B$.
25. ▶Is the set of all rich men well defined? Why or why not?

26. Suppose U is the set of natural numbers, $\{1, 2, 3, 4, \ldots\}$. Write each of the following, using set-builder notation:
    (a) The set of even numbers greater than 12.
    (b) The set of number less than 14.
27. If $U = \{a, b, c, d\}$, $A = \{a, b, c\}$, $B = \{b, c\}$, and $C = \{d\}$, find each of the following:
    (a) $A \cup \overline{B}$
    (b) $\overline{A \cap B}$
    (c) $A \cap \emptyset$
    (d) $B \cap C$
    (e) $B - A$
28. Write two sets with three elements each and establish a one-to-one correspondence between them.
29. Write a set that is equivalent, but not equal, to the set $\{5, 6, 7, 8\}$.
30. (a) How many different one-to-one correspondences are possible between $A = \{a, b, c\}$ and $B = \{1, 2, 3\}$?
    (b) How many elements are there in $A \times B$?
31. Illustrate the associative property of set union with the sets $U = \{h, e, l, p, m, n, o, w\}$, $A = \{h, e, l, p\}$, $B = \{m, e\}$, and $C = \{n, o, w\}$.

▼ BRAIN TEASER

Only 10 rooms were vacant in the Village Hotel. Eleven men went into the hotel at the same time, each wanting a separate room. The clerk, settling the argument, said, "I'll tell you what I'll do. I'll put two men in Room 1 with the understanding that I will come back and get one of them a few minutes later." The men agreed to this. The clerk continued, "I will put the rest of you men in rooms as follows: the 3rd man in Room 2, the 4th man in Room 3, the 5th man in Room 4, the 6th man in Room 5, the 7th man in Room 6, the 8th man in Room 7, the 9th man in Room 8, and the 10th man in Room 9." Then the clerk went back and got the extra man he had left in Room 1 and put him in Room 10. Everybody was happy. What is wrong with this plan?

*Section 2-4 Logic: An Introduction

statement   Logic is a tool used in mathematical thinking and problem solving. In logic, a **statement** *is a sentence that is either true or false, but not both.*

The following expressions are not statements because their truth value cannot be determined without more information.

1. She has blue eyes.
2. $x + 7 = 18$
3. $2y + 7 > 1$
4. $2 + 3$

Expressions (1), (2), and (3) become statements if, for (1), "she" is identified, and for (2) and (3), values are assigned to x and y, respectively. However, an expression involving *he* or *she* or x or y may already be a statement. For example, "If he is over 210 cm tall, then he is over 2 m tall," and "$2(x + y) = 2x + 2y$" are both statements because they are true no matter who *he* is or what the numerical values of x and y are.

From a given statement, it is possible to create a new statement by forming a **negation.** The negation of a statement is a statement with the opposite truth value of the given

negation

statement. If a statement is true, its negation is false, and if a statement is false, its negation is true. Consider the statement "It is snowing." The negation of this statement may be stated simply as "It is not snowing."

EXAMPLE 2-21   Negate each of the following statements:

(a) $2 + 3 = 5$

(b) A hexagon has six sides.

(c) Today is not Monday.

Solution   (a) $2 + 3 \neq 5$

(b) A hexagon does not have six sides.

(c) Today is Monday.   ∎

The statements "The shirt is blue" and "The shirt is green" are not negations of each other. A statement and its negation must have opposite truth values. If the shirt is actually red, then both of the above statements are false and, hence, cannot be negations of each other. However, the statements "The shirt is blue" and "The shirt is not blue" are negations of each other because they have opposite truth values no matter what color the shirt really is.

quantifiers   Some statements involve **quantifiers** and are more complicated to negate. Quantifiers include words such as *all, some, every,* and *there exists*.

The quantifiers *all, every,* and *no* refer to each and every element in a set and are *universal quantifiers*   **universal quantifiers.** The quantifiers *some* and *there exists at least one* refer to one or more, or possibly all, of the elements in a set. *Some* and *there exists* are called **existential** *existential quantifiers*   **quantifiers.** Examples with universal and existential quantifiers follow:

1. All roses are red. [universal]
2. Every student is important. [universal]
3. For each counting number x, $x + 0 = x$. [universal]
4. Some roses are red. [existential]
5. There exists at least one even counting number less than 3. [existential]
6. There are women who are taller than 200 cm. [existential]

Venn diagrams can be used to picture statements involving quantifiers. For example, Figure 2-27(a) and (b) picture statements (1) and (4). The x in Figure 2-27(b) can be used to show that there must be at least one element of the set of roses that is red.

FIGURE 2-27

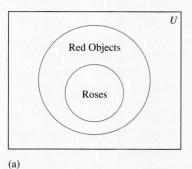
(a)

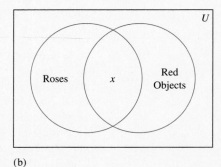
(b)

SECTION 2-4 LOGIC: AN INTRODUCTION

Consider the following statement involving the existential quantifier *some*. "Some professors at Paxson University have blue eyes." This means that at least one professor at Paxson University has blue eyes. It does not rule out the possibilities that all the Paxson professors have blue eyes or that some of the Paxson professors do not have blue eyes. Because the negation of a true statement is false, neither "Some professors at Paxson University do not have blue eyes" nor "All professors at Paxson have blue eyes" are negations of the original statement. One possible negation of the original statement is "No professors at Paxson University have blue eyes."

To discover if one statement is a negation of another, we use arguments similar to the preceding one to determine if they have opposite truth values in all possible cases. Some general forms of quantified statements with their negations follow:

Statement	Negation
Some *a* are *b*.	No *a* is *b*.
Some *a* are not *b*.	All *a* are *b*.
All *a* are *b*.	Some *a* are not *b*.
No *a* is *b*.	Some *a* are *b*.

EXAMPLE 2-22 Negate each of the following statements:

(a) All students like hamburgers.

(b) Some people like mathematics.

(c) There exists a counting number x such that $3x = 6$.

(d) For all counting numbers, $3x = 3x$.

Solution
(a) Some students do not like hamburgers.

(b) No people like mathematics.

(c) For all counting numbers x, $3x \neq 6$.

(d) There exists a counting number x such that $3x \neq 3x$. ■

TABLE 2-5

Statement p	Negation $\sim p$
T	F
F	T

truth table

There is a symbolic system defined to help in the study of logic. If p represents a statement, the negation of the statement p is denoted by $\sim p$. **Truth tables** are often used to show all possible true-false patterns for statements. Table 2-5 summarizes the truth tables for p and $\sim p$.

Observe that p and $\sim p$ are analogous to sets P and $\overline{P}$. If x is an element of P, then x is not an element of $\overline{P}$.

Compound Statements

compound statement

From two given statements, it is possible to create a new, **compound statement** by using a connective such as *and*. For example, "It is snowing" and "The ski run is open" together with *and* give "It is snowing and the ski run is open." Other compound statements can be obtained by using the connective *or*. For example, "It is snowing or the ski run is open."

The symbols $\wedge$ and $\vee$ are used to represent the connectives *and* and *or*, respectively. For example, if p represents "It is snowing," and if q represents "The ski run is open," then "It is snowing and the ski run is open" is denoted by $p \wedge q$. Similarly, "It is snowing or the ski run is open" is denoted by $p \vee q$.

The truth value of any compound statement, such as $p \wedge q$, is defined using the truth table of each of the simple statements. Because each of the statements p and q may be either true or false, there are four distinct possibilities for the truth of $p \wedge q$, as shown in Table 2-6. The compound statement $p \wedge q$ is the **conjunction** of p and q and is defined to be true if, and only if, both p and q are true. Otherwise, it is false.

We can find similarities between conjunction and set intersection. Consider Table 2-7, which shows all possibilities of whether an element is a member of sets, P, Q, and $P \cap Q$. If we consider $\in$ analogous to T and $\notin$ analogous to F, we see that Tables 2-6 and 2-7 are equivalent. The language involving set intersection and the language involving *and* in logic are equivalent. Thus for every property involving set intersection, there should be an equivalent property involving *and*.

conjunction

TABLE 2-6

p	q	Conjunction $p \wedge q$
T	T	T
T	F	F
F	T	F
F	F	F

TABLE 2-7

P	Q	$P \cap Q$
$\in$	$\in$	$\in$
$\in$	$\notin$	$\notin$
$\notin$	$\in$	$\notin$
$\notin$	$\notin$	$\notin$

disjunction

The compound statement $p \vee q$—that is, p *or* q—is a **disjunction.** In everyday language, *or* is not always interpreted in the same way. In logic, we use an *inclusive or*. The statement "I will go to a movie or I will read a book" means that I will either go to a movie, or read a book, or do both. Hence, in logic, p or q, symbolized as $p \vee q$, is defined to be false if both p and q are false and true in all other cases. This is summarized in Table 2-8.

TABLE 2-8

p	q	Disjunction $p \vee q$
T	T	T
T	F	T
F	T	T
F	F	F

EXAMPLE 2-23 Given the following statements, classify each of the conjunctions and disjunctions as true or false:

p: $2 + 3 = 5$ r: $5 + 3 = 9$
q: $2 \cdot 3 = 6$ s: $2 \cdot 4 = 9$

(a) $p \wedge q$ (b) $p \wedge r$ (c) $s \wedge q$ (d) $r \wedge s$
(e) $\sim p \wedge q$ (f) $\sim (p \wedge q)$ (g) $p \vee q$ (h) $p \vee r$
(i) $s \vee q$ (j) $r \vee s$ (k) $\sim p \vee q$ (l) $\sim (p \vee q)$

Solution   (a) p is true and q is true, so $p \wedge q$ is true.
(b) p is true and r is false, so $p \wedge r$ is false.
(c) s is false and q is true, so $s \wedge q$ is false.
(d) r is false and s is false, so $r \wedge s$ is false.
(e) $\sim p$ is false and q is true, so $\sim p \wedge q$ is false.
(f) $p \wedge q$ is true [part (a)], so $\sim (p \wedge q)$ is false.
(g) p is true and q is true, so $p \vee q$ is true.
(h) p is true and r is false, so $p \vee r$ is true.

(i) s is false and q is true, so $s \lor q$ is true.
(j) r is false and s is false, so $r \lor s$ is false.
(k) $\sim p$ is false and q is true, so $\sim p \lor q$ is true.
(l) $p \lor q$ is true [part (g)], so $\sim(p \lor q)$ is false.

Not only are truth tables used to summarize the truth values of compound statements, they also are used to determine if two statements are logically equivalent. Two statements are **logically equivalent** if, and only if, they have the same truth values. For example, we could show that $p \land q$ is logically equivalent to $q \land p$ by using truth tables as in Table 2-9.

logically equivalent

TABLE 2-9

p	q	$p \land q$	$q \land p$
T	T	T	T
T	F	F	F
F	T	F	F
F	F	F	F

Table 2-9 shows that *and* is commutative. Just as *set intersection* and *and* are analogous, so are *set union* and *or*. Similarly for all properties involving set union, *there are corresponding properties involving* or, *such as the commutative and associative properties*. Expressions involving statements inside parentheses are treated similarly to expressions involving set unions and intersections.

EXAMPLE 2-24 Use a truth table to determine if $\sim p \lor \sim q$ and $\sim(p \land q)$ are logically equivalent.

Solution Table 2-10 shows headings and the four distinct possibilities for p and q. In the column headed $\sim p$, we write the negations of the p column. In the $\sim q$ column, we write the negation of the q column. Next, we use the values in the $\sim p$ and the $\sim q$ columns to construct the $\sim p \lor \sim q$ column. To find the truth values for $\sim(p \land q)$, we use the p and q columns to find the truth value for $p \land q$ and then negate $p \land q$.

TABLE 2-10

p	q	$\sim p$	$\sim q$	$\sim p \lor \sim q$	$p \land q$	$\sim(p \land q)$
T	T	F	F	F	T	F
T	F	F	T	T	F	T
F	T	T	F	T	F	T
F	F	T	T	T	F	T

PROBLEM SET 2-4

1. Determine which of the following are statements, and then classify each statement as true or false:
   (a) $2 + 4 = 8$
   (b) Shut the window.
   (c) Los Angeles is a state.
   (d) He is in town.
   (e) What time is it?
   (f) $5x = 15$
   (g) $3 \cdot 2 = 6$
   (h) $2x^2 > x$
   (i) This statement is false.
   (j) Stay put!

2. Use quantifiers to make each of the following true where x is a natural number:
   (a) $x + 8 = 11$
   (b) $x + 0 = x$
   (c) $x^2 = 4$
   (d) $x + 1 = x + 2$

3. Use quantifiers to make each equation in Problem 2 false.

4. Write the negation for each of the following statements:
   (a) The book has 500 pages.
   (b) Six is less than eight.
   (c) $3 \cdot 5 = 15$
   (d) Some people have blond hair.
   (e) All dogs have four legs.
   (f) Some cats do not have nine lives.
   (g) All squares are rectangles.
   (h) Not all rectangles are squares.

(i) For all natural numbers x, $x + 3 = 3 + x$.
(j) There exists a natural number x such that $3 \cdot (x + 2) = 12$.
(k) Every counting number is divisible by itself and 1.
(l) Not all natural numbers are divisible by 2.
(m) For all natural numbers x, $5x + 4x = 9x$.

5. Complete each of the following truth tables:

(a)

p	$\sim p$	$\sim(\sim p)$
T		
F		

(b)

p	$\sim p$	$p \vee \sim p$	$p \wedge \sim p$
T			
F			

(c) Based on part (a), is p logically equivalent to $\sim(\sim p)$?
(d) Based on part (b), is $p \vee \sim p$ logically equivalent to $p \wedge \sim p$?

6. If q stands for "This course is easy" and r stands for "Lazy students do not study," write each of the following in symbolic form:
(a) This course is easy and lazy students do not study.
(b) Lazy students do not study or this course is not easy.
(c) It is false that both this course is easy and lazy students do not study.
(d) This course is not easy.

7. If p is false and q is true, find the truth values for each of the following:
(a) $p \wedge q$ (b) $p \vee q$
(c) $\sim p$ (d) $\sim q$
(e) $\sim(\sim p)$ (f) $\sim p \vee q$
(g) $p \wedge \sim q$ (h) $\sim(p \vee q)$
(i) $\sim(\sim p \wedge q)$ (j) $\sim q \wedge \sim p$

8. Find the truth value for each statement in Problem 7 if p is false and q is false.

9. For each of the following, is the pair of statements logically equivalent?
(a) $\sim(p \vee q)$ and $\sim p \vee \sim q$
(b) $\sim(p \vee q)$ and $\sim p \wedge \sim q$
(c) $\sim(p \wedge q)$ and $\sim p \wedge \sim q$
(d) $\sim(p \wedge q)$ and $\sim p \vee \sim q$

10. ▶(a) Write two logical equivalences discovered in parts 9(a)-(d). These equivalences are called DeMorgan's Laws for "*and*" and "*or*."
(b) Write an explanation of the analogy between DeMorgan's Laws for sets and those found in part (a).

11. Complete the following truth table:

p	q	$\sim p$	$\sim q$	$\sim p \vee q$
T	T			
T	F			
F	T			
F	F			

12. Restate the following in a logically equivalent form:
(a) It is not true that both today is Wednesday and the month is June.
(b) It is not true that yesterday I both ate breakfast and watched television.
(c) It is not raining or it is not July.

▼ BRAIN TEASER

On an island inhabited by two tribes, the Abes and the Babes, Abes always tell the truth and Babes always lie. A traveler met three natives on the shore. He asked the first native to name his tribe and the native responded in his native tongue, which the traveler did not understand. The second native stated that the first native said that he was an Abe. The third native then stated that the first native had said he was a Babe. To what tribes do the second and third natives belong?

---

*Section 2-5 Conditionals and Biconditionals

conditionals
implications

Statements expressed in the form "if p, then q" are called **conditionals,** or **implications,** and are denoted by $p \rightarrow q$. Such statements also can be read "p implies q." The "if" part

hypothesis of a conditional is called the **hypothesis** of the implication and the "then" part is called the
conclusion **conclusion.**

Many types of statements can be put in "if-then" form; an example follows:

Statement: All first-graders are 6 years old.
If-then form: If a child is a first-grader, then the child is 6 years old.

An implication may also be thought of as a promise. Suppose Betty makes the promise, "If I get a raise, then I will take you to dinner." If Betty keeps her promise, the implication is true; if Betty breaks her promise, the implication is false. Consider the following four possibilities:

	p	q	
(1)	T	T	Betty gets the raise; she takes you to dinner.
(2)	T	F	Betty gets the raise; she does not take you to dinner.
(3)	F	T	Betty does not get the raise; she takes you to dinner.
(4)	F	F	Betty does not get the raise; she does not take you to dinner.

TABLE 2-11

p	q	Implication $p \rightarrow q$
T	T	T
T	F	F
F	T	T
F	F	T

The only case in which Betty breaks her promise is when she gets her raise and fails to take you to dinner, case (2). If she does not get the raise, she can either take you to dinner or not without breaking her promise. The definition of implication is summarized in Table 2-11. Observe that the only case for which the implication is false is when p is true and q is false.

An implication may be worded in several equivalent ways, as follows:

1. If the sun shines, then the swimming pool is open. (If p, then q.)
2. If the sun shines, the swimming pool is open. (If p, q.)
3. The swimming pool is open if the sun shines. (q if p.)
4. The sun shines implies the swimming pool is open. (p implies q.)
5. The sun is shining only if the pool is open. (p only if q.)
6. The sun's shining is a sufficient condition for the swimming pool to be open. (p is a sufficient condition for q.)
7. The swimming pool's being open is a necessary condition for the sun to be shining. (q is a necessary condition for p.)

Any implication $p \rightarrow q$ has three related implication statements, as follows:

Statement:	If p, then q.	$p \rightarrow q$
Converse:	If q, then p.	$q \rightarrow p$
Inverse:	If not p, then not q.	$\sim p \rightarrow \sim q$
Contrapositive:	If not q, then not p.	$\sim q \rightarrow \sim p$

EXAMPLE 2-25 Write the converse, the inverse, and the contrapositive for each of the following statements:

(a) If $2x = 6$, then $x = 3$.

(b) If I am in San Francisco, then I am in California.

Solution (a) *Converse:* If $x = 3$, then $2x = 6$.
Inverse: If $2x \neq 6$, then $x \neq 3$.
Contrapositive: If $x \neq 3$, then $2x \neq 6$.

(b) *Converse:* If I am in California, then I am in San Francisco.
Inverse: If I am not in San Francisco, then I am not in California.
Contrapositive: If I am not in California, then I am not in San Francisco. ■

Table 2-12 shows that an implication and its converse do not always have the same truth value. However, an implication and its contrapositive do always have the same truth value. Also, the converse and inverse of a conditional statement are logically equivalent.

TABLE 2-12

p	q	$\sim p$	$\sim q$	Implication $p \rightarrow q$	Converse $q \rightarrow p$	Inverse $\sim p \rightarrow \sim q$	Contrapositive $\sim q \rightarrow \sim p$
T	T	F	F	T	T	T	T
T	F	F	T	F	T	T	F
F	T	T	F	T	F	F	T
F	F	T	T	T	T	T	T

biconditional

Connecting a statement and its converse with the connective *and* gives $(p \rightarrow q) \land (q \rightarrow p)$. This compound statement can be written as $p \leftrightarrow q$ and usually is read "p if and only if q." The statement "p if and only if q" is a **biconditional**. A truth table for $p \leftrightarrow q$ is given in Table 2-13. Observe that $p \leftrightarrow q$ is true if and only if both statements are true or both are false.

TABLE 2-13

p	q	$p \rightarrow q$	$q \rightarrow p$	Biconditional $(p \rightarrow q) \land (q \rightarrow p)$ or $p \leftrightarrow q$
T	T	T	T	T
T	F	F	T	F
F	T	T	F	F
F	F	T	T	T

EXAMPLE 2-26 Given the following statements, classify each of the biconditionals as true or false:

p: $2 = 2$ r: $2 = 1$
q: $2 \neq 1$ s: $2 + 3 = 1 + 3$

(a) $p \leftrightarrow q$ (b) $p \leftrightarrow r$ (c) $s \leftrightarrow q$ (d) $r \leftrightarrow s$

Solution (a) $p \to q$ is true and $q \to p$ is true, so $p \leftrightarrow q$ is true.
(b) $p \to r$ is false and $r \to p$ is true, so $p \leftrightarrow r$ is false.
(c) $s \to q$ is true and $q \to s$ is false, so $s \leftrightarrow q$ is false.
(d) $r \to s$ is true and $s \to r$ is true, so $r \leftrightarrow s$ is true. ∎

In the previous section, we discussed analogies between the conjunction $p \wedge q$ and set intersection and between the disjunction $p \vee q$ and set union. Similar analogies exist for implication. Consider the implication "If a flower is a violet, then it is blue." The set of violets is a subset of the set of blue objects. In general, the implication $p \to q$ is analogous to $P \subseteq Q$. In fact, the definition of $P \subseteq Q$ tells us that $x \in P$ implies $x \in Q$. Thus for every property involving set inclusion, we should have a corresponding property involving implications.

Now consider the following statement:

It is raining or it is not raining.

This statement, which can be modeled as $p \vee (\sim p)$, is always true, as shown in Table 2-14. A statement that is always true is called a **tautology**. One way to make a tautology is to take two logically equivalent statements such as $p \to q$ and $\sim q \to \sim p$ (from Table 2-12) and form them into a biconditional as follows:

$$(p \to q) \leftrightarrow (\sim q \to \sim p)$$

Because $p \to q$ and $\sim q \to \sim p$ have the same truth values, then $(p \to q) \leftrightarrow (\sim q \to \sim p)$ is a tautology.

TABLE 2-14

p	$\sim p$	$p \vee (\sim p)$
T	F	T
F	T	T

Valid Reasoning

In problem solving, the reasoning used is said to be **valid** if the conclusion follows unavoidably from the hypotheses. Consider the following example:

Hypotheses: All roses are red.
             This flower is a rose.
Conclusion: Therefore this flower is red.

The statement "All roses are red" can be written as the implication "If a flower is a rose, then it is red" and pictured with the Venn diagram in Figure 2-28(a).

FIGURE 2-28

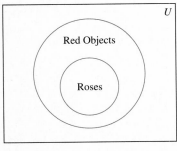
(a)

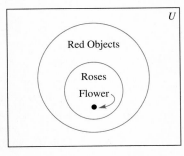
(b)

The information "This flower is a rose" implies that this flower must belong to the circle containing roses, as pictured in Figure 2-28(b). This flower also must belong to the circle containing red objects. Thus the reasoning is valid because it is impossible to draw a picture satisfying the hypotheses and contradicting the conclusion.

Consider the following argument:

Hypotheses: All elementary school teachers are mathematically literate.
              Some mathematically literate people are not childern.
Conclusion: Therefore no elementary school teacher is a child.

Let E be the set of elementary school teachers, M be the set of mathematically literate people, and C be the set of children. Then the statement "All elementary school teachers are mathematically literate" can be pictured as in Figure 2-29(a). The statement "Some mathematically literate people are not children" can be pictured in several ways. Three of these are illustrated in Figure 2-29(b)-(d).

According to Figure 2-29(d), it is possible that some elementary school teachers are children, and yet the given statements are satisfied. Therefore the conclusion that "No elementary school teacher is a child" does not follow from the given hypotheses. Hence, the reasoning is not valid.

FIGURE 2-29

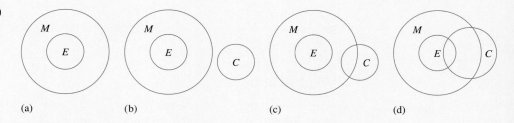

(a)   (b)   (c)   (d)

If a single picture can be drawn to satisfy the hypotheses of an argument and contradict the conclusion, the argument is not valid. However, to show that an argument is valid, *all* possible pictures must be considered to show that there are no contradictions. There must be no way to satisfy the hypotheses and contradict the conclusion if the argument is valid.

EXAMPLE 2-27   Determine if the following argument is valid:

Hypotheses: In Washington, D. C., all senators wear power ties.
              No one in Washington, D. C., over 6 ft tall wears a power tie.
Conclusion: Persons over 6 ft tall are not senators in Washington, D. C.

Solution   If S represents the set of senators and P represents the set of people who wear power ties, the first hypothesis is pictured as shown in Figure 2-30(a). If T represents the set of people in Washington, D.C., over 6 ft tall, the second hypothesis is pictured in Figure 2-30(b). Because people over 6 ft tall are outside the circle representing power tie wearers and senators are inside the circle P, the conclusion is valid and no person over 6 ft tall can be a senator in Washington, D. C.

FIGURE 2-30

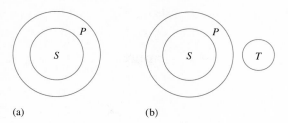

(a)   (b)

direct reasoning
Modus Ponens

A different method for determining if an argument is valid uses **direct reasoning** and a form of argument called the Law of Detachment (or **Modus Ponens**). For example, consider the following true statements:

> If the sun is shining, then we shall take a trip.
> The sun is shining.

Law of Detachment

Using these two statements, we can conclude that we shall take a trip. In general, the **Law of Detachment** is stated as follows:

If a statement is in the form "if p, then q" is true, and p is true, then q must also be true.

The Law of Detachment is sometimes described schematically as follows, where all statements above the horizontal line are true and the statement below the horizontal line is the conclusion.

$$\begin{array}{c} p \rightarrow q \\ \underline{p} \\ q \end{array}$$

The Law of Detachment follows from the truth table for $p \rightarrow q$ given in Table 2-11. The only case in which both p and $p \rightarrow q$ are true is when q is true (line 1 in the table).

EXAMPLE 2-28 Determine if each of the following arguments is valid:

(a) Hypotheses: If you eat spinach, then you will be strong.
                 You eat spinach.
    Conclusion: Therefore you will be strong.
(b) Hypotheses: If Claude goes skiing, he will break his leg.
                 If Claude breaks his leg, he cannot enter the dance contest.
                 Claude goes skiing.
    Conclusion: Therefore Claude cannot enter the dance contest.

Solution (a) Using the Law of Detachment, we see that the conclusion is valid.
              (b) By using the Law of Detachment twice, we see that the conclusion is valid.

indirect reasoning

A different type of reasoning, **indirect reasoning,** uses a form of argument called

Modus Tollens **Modus Tollens.** For example, consider the following true statements:

If Chicken Little had been hit by a jumping frog, he would have thought the earth was rising.
Chicken Little did not think the earth was rising.

What is the conclusion? The conclusion is that Chicken Little did not get hit by a jumping frog. This leads us to the general form of Modus Tollens:

If we have a conditional accepted as true, and we know the conclusion is false, then the hypothesis must be false.

Modus Tollens is sometimes schematically described as follows:

$$\begin{array}{c} p \to q \\ \sim q \\ \hline \sim p \end{array}$$

The validity of Modus Tollens also follows from the truth table for $p \to q$ given in Table 2-11. The only case in which both $p \to q$ is true and q is false is when p is false (line 4 in the table). The validity of Modus Tollens also can be established from the fact that an implication and its contrapositive are equivalent.

EXAMPLE 2-29 Determine conclusions for each of the following sets of true statements:

(a) If an old woman lives in a shoe, then she does not know what to do. Mrs. Pumpkin Eater, an old woman, knows what to do.

(b) If Jack is nimble, he will not get burned. Jack was burned.

Solution (a) Mrs. Pumpkin Eater does not live in a shoe.

(b) Jack was not nimble. ∎

Chain Rule The final reasoning argument to be considered here involves the **Chain Rule.** Consider the following statements:

If my wife works, I will retire early.
If I retire early, I will become lazy.

What is the conclusion? The conclusion is that if my wife works, I will become lazy. In general, the Chain Rule can be stated as follows:

If "if p, then q" and "if q, then r" are true, then "if p, then r" is true.

The Chain Rule is sometimes schematically described as follows:

$$\begin{array}{c} p \to q \\ q \to r \\ \hline p \to r \end{array}$$

Notice that the chain rule shows that implication is a transitive relation.

People often make invalid conclusions based on advertising or other information. Consider, for example, the statement "Healthy people eat Super-Bran cereal." Are the following conclusions valid?

If a person eats Super-Bran cereal, then the person is healthy.
If a person is not healthy, the person does not eat Super-Bran cereal.

If the original statement is denoted by $p \to q$, where p is "a person is healthy" and q is "a person eats Super-Bran cereal," then the first conclusion is the converse of $p \to q$—that is, $q \to p$—and the second conclusion is the inverse of $p \to q$—that is, $\sim p \to \sim q$. Table 2-12 points out that neither the converse nor the inverse are logically equivalent to the original statement, and consequently the conclusions are not necessarily true.

EXAMPLE 2-30 Determine conclusions for each of the following sets of true statements.
(a) If Alice follows the White Rabbit, she falls into a hole. If she falls into a hole, she goes to a tea party.
(b) If Chicken Little is hit by an acorn, we think the sky is falling. If we think the sky is falling, we will go to a fallout shelter. If we go to a fallout shelter, we will stay there a month.

Solution (a) If Alice follows the White Rabbit, she goes to a tea party.
(b) If Chicken Little is hit by an acorn, we will stay in a fallout shelter for a month.

REMARK Note that in Example 2-30(b), the Chain Rule can be extended to contain several implications.

PROBLEM SET 2-5

1. Write each of the following in symbolic form if p is the statement "It is raining" and q is the statement "The grass is wet."
   (a) If it is raining, then the grass is wet.
   (b) If it is not raining, then the grass is wet.
   (c) If it is raining, then the grass is not wet.
   (d) The grass is wet if it is raining.
   (e) The grass is not wet implies that it is not raining.
   (f) The grass is wet if, and only if, it is raining.

2. For each of the following implications, state the converse, inverse, and contrapositive:
   (a) If you eat Meaties, then you are good in sports.
   (b) If you do not like this book, then you do not like mathematics.
   (c) If you do not use Ultra Brush toothpaste, then you have cavities.
   (d) If you are good at logic, then your grades are high.

3. Construct a truth table for each of the following:
   (a) $p \to (p \lor q)$ (b) $(p \land q) \to q$
   (c) $p \leftrightarrow \sim(\sim p)$ (d) $\sim(p \to q)$

4. If p is true and q is false, find the truth values for each of the following:
   (a) $\sim p \to \sim q$ (b) $\sim(p \to q)$
   (c) $(p \lor q) \to (p \land q)$ (d) $p \to \sim p$
   (e) $(p \lor \sim p) \to p$ (f) $(p \lor q) \leftrightarrow (p \land q)$

5. If p is false and q is false, find the truth values for each of the statements in Problem 4.

6. ▶Can an implication and its converse both be false? Explain your answer.

7. Iris makes the true statement, "If it rains, then I am going to the movies." Does it follow logically that if it does not rain, then Iris does not go to the movies?

8. Consider the statement "If every digit of a number is 6, then the number is divisible by 3." Which of the following is logically equivalent to the statement?
   (a) If every digit of a number is not 6, then the number is not divisible by 3.
   (b) If a number is not divisible by 3, then some digit of the number is not 6.
   (c) If a number is divisible by 3, then every digit of the number is 6.

9. Write a statement logically equivalent to the statement "If a number is a multiple of 8, then it is a multiple of 4."

10. Use truth tables to prove that the following are tautologies:
    (a) $(p \rightarrow q) \rightarrow [(p \wedge r) \rightarrow q]$ Law of Added Hypothesis
    (b) $[(p \rightarrow q) \wedge p] \rightarrow q$ Law of Detachment
    (c) $[(p \rightarrow q) \wedge (\sim q)] \rightarrow \sim p$ Modus Tollens
    (d) $[(p \rightarrow q) \wedge (q \rightarrow r)] \rightarrow (p \rightarrow r)$ Chain Rule

11. ▶(a) If $p \rightarrow q$ is true but $q \rightarrow p$ is false, what is the analogous relation between sets?
    ▶(b) Suppose that $p \rightarrow q$ and $q \rightarrow p$ are true. What is the analogous relation between sets?
    ▶(c) Suppose that $A \subseteq B$ and $\overline{A} \subseteq \overline{B}$. What are the analogous statements in logic? What can you conclude about A and B?

12. (a) Suppose that $p \rightarrow q$, $q \rightarrow r$, and $r \rightarrow s$ are all true, but s is false. What can you conclude about the truth value of p?
    (b) Suppose that $(p \wedge q) \rightarrow r$ is true, r is false, and q is true. What can you conclude about the truth value of p?
    (c) Suppose that $p \rightarrow q$ is true and $q \rightarrow p$ is false. Can q be true? Why or why not?

13. ▶Translate the following statements into symbolic form. Give the meanings of the symbols that you use.
    (a) If Mary's little lamb follows her to school, then its appearance there will break the rules and Mary will be sent home.
    (b) If it is not the case that Jack is nimble and quick, then Jack will not make it over the candlestick.
    (c) If the apple had not hit Isaac Newton on the head, then the laws of gravity would not have been discovered.

14. Investigate the validity of each of the following arguments:
    (a) All women are mortal.
        Hypatia was a woman.
        Therefore Hypatia was mortal.
    (b) All squares are quadrilaterals.
        All quadrilaterals are polygons.
        Therefore all squares are polygons.
    (c) All teachers are intelligent.
        Some teachers are rich.
        Therefore some intelligent people are rich.
    (d) If a student is a freshman, then she takes mathematics.
        Jane is a sophomore.
        Therefore Jane does not take mathematics.

15. For each of the following, form a conclusion that follows logically from the given statements:
    (a) All college students are poor.
        Helen is a college student.
    (b) Some freshmen like mathematics.
        All people who like mathematics are intelligent.
    (c) If I study for the final, then I will pass the final.
        If I pass the final, then I will pass the course.
        If I pass the course, then I will look for a teaching job.
    (d) Every equilateral triangle is isosceles.
        There exist triangles that are isosceles.

16. Write the following in if-then form:
    (a) Every figure that is a square is a rectangle.
    (b) All integers are rational numbers.
    (c) Figures with exactly three sides may be triangles.
    (d) It rains only if it is cloudy.

Solution to the Preliminary Problem

Understanding the Problem.   The Red Cross looks for A, B, and Rh antigens in blood tests. If A or B is present it is listed; if both are absent, the blood is typed as O. If the Rh antigen is present, the blood type is positive; otherwise, the type is negative. The following are results of tests on 100 samples:

Number of Samples	Antigens in Blood
40	A
18	B
82	Rh
5	A and B
31	A and Rh
11	B and Rh
4	A, B, and Rh

We are to find how many of the 100 samples are classified as O negative. One of the first

SOLUTION TO THE PRELIMINARY PROBLEM 97

tasks is to determine how many blood types there are and then to sort them to answer the question.

Devising a Plan. An antigen is either present or not in any given sample, and we know that there are three types of antigens, so from the Fundamental Counting Principle, there should be $2 \cdot 2 \cdot 2$, or 8 choices of blood types. Using the strategy of *drawing a picture* with each antigen as a circle as in Figure 2-31, the Venn diagram depicts the 8 different blood types. Circle A includes samples with blood type A, circle B includes samples with blood type B, and each sample in circle Rh includes a "$+$". Each type outside circle Rh includes a "$-$". We are to determine the number of samples of O^-. This may be done by finding the cardinal number of each section of the Venn diagram.

FIGURE 2-31

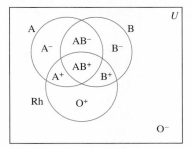

Carrying Out the Plan. Because there are 4 samples containing all three antigens and 11 containing B and Rh, then there must be $11 - 4$, or 7, others containing B but neither A nor Rh. Similarly, there must be $31 - 4$, or 27, containing A but neither B nor Rh and $5 - 4$, or 1, containing A but neither B nor Rh. Because there are 40 samples containing the A antigen, 1 containing A and B but not Rh, 4 containing all three, and 27 containing A and Rh but not B, there must be $40 - 1 - 4 - 27$, or 8, containing A alone. Similarly, there are 6 samples containing B alone and 44 containing Rh alone. This information is summarized in Figure 2-32.

FIGURE 2-32

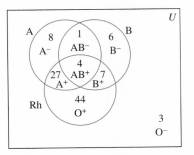

From Figure 2-32, we have a total of $8 + 1 + 6 + 27 + 4 + 7 + 44$, or 97, samples accounted for in circles A, B, and Rh, so we must have $100 - 97$, or 3, samples outside the circles. Therefore there are 3 samples of O^- in the set of samples.

Looking Back. The problem can be altered by changing the numbers of each type listed or by changing the sample size. However, the information in the problem is comparable to data gathered by the Red Cross in the 1970s. Would you expect to obtain comparable numbers today?

In the 1970s, what was the rarest blood type? Which type do you think is considered to be a universal donor? Why?

QUESTIONS FROM THE CLASSROOM

1. A student argues that $\{\emptyset\}$ is the proper notation for the empty set. What is your response?
2. A student asks, "If $A = \{a, b, c\}$ and $B = \{b, c, d\}$, why isn't it true that $A \cup B = \{a, b, c, b, c, d\}$?" What is your response?
3. A student says that she can show that if $A \cap B = A \cap C$, then it is not necessarily true that $B = C$; but she thinks that whenever $A \cap B = A \cap C$ and $A \cup B = A \cup C$, then $B = C$. What is your response?
4. A student claims that a finite set of numbers is any set that has a largest element. Do you agree?
5. A student claims that the complement bar can be broken over the operation of intersections; that is, $\overline{A \cap B} = \overline{A} \cap \overline{B}$. What is your response?
6. A student claims that $\overline{A} \cap \overline{B}$ includes all elements that are not in A. What is your response?
7. A student asks whether a formula and a function are the same. What is your response?
8. A student states that either $A \subseteq B$ or $B \subseteq A$. Is the student correct?
9. A student is asked to find all one-to-one correspondences between two given sets. He finds the Cartesian product of the sets and claims that his answer is correct because it includes all possible pairings between the elements of the sets. How do you respond?
10. A student argues that adding two sets A and B, or $A + B$, and taking the union of two sets, $A \cup B$, is the same thing. How do you respond?

CHAPTER OUTLINE

I. Set definitions and notation
   A. A **set** can be described as any collection of objects.
   B. Sets should be **well defined** so that an object either does or does not belong to the set.
   C. An **element** is any **member** of a set.
   D. Sets can be specified by either listing all the elements or using **set-builder notation**.
   E. The **empty set,** written $\emptyset$, contains no elements.
   F. The **universal set** contains all the elements being discussed.

II. Relationships and operations on sets
   A. Two sets are **equal** if, and only if, they have exactly the same elements.
   B. Two sets A and B are in **one-to-one correspondence** if, and only if, each element of A can be paired with exactly one element of B and each element of B can be paired with exactly one element of A.
   C. The **Fundamental Counting Principle:** If event M can occur in m ways and after it has occurred, event N can occur in n ways, then event M followed by event N can occur in $m \cdot n$ ways.
   D. Two sets are **equivalent** if, and only if, their elements can be placed into one-to-one correspondence (written $A \sim B$).
   E. Set A is a **subset** of set B if, and only if, every element of A is an element of B (written $A \subseteq B$).
   F. Set A is a **proper subset** of set B if, and only if, every element of A is an element of B and there is at least one element of B that is not in A (written $A \subset B$).
   G. The **union** of two sets A and B is the set of all elements in A, in B, or in both A and B (written $A \cup B$).
   H. The **intersection** of two sets A and B is the set of all elements belonging to both A and B (written $A \cap B$).
   I. The **cardinal number** of a set indicates the number of elements in the set.
   J. A set is **finite** if the number of elements in the set is zero or a natural number. Otherwise, the set is **infinite**.
   K. Two sets A and B are **disjoint** if they have no elements in common.
   L. The **complement** of a set A is the set consisting of the elements of the universal set not in A (written $\overline{A}$).
   M. The **complement of set A relative to set B** (set difference) is the set of all elements in B that are not in A (written $B - A$).
   N. The **Cartesian product** of sets A and B is the set of all ordered pairs such that the first element of each pair is an element of A and the second element of each pair is an element of B (written $A \times B$).

III. Properties of set operations
   For all sets A, B, C, and universal set U, the following properties hold:
   A. $A \cap B = B \cap A$; **commutative property of set intersection.**
   B. $A \cup B = B \cup A$; **commutative property of set union.**
   C. $(A \cap B) \cap C = A \cap (B \cap C)$; **associative property of set intersection.**

D. $(A \cup B) \cup C = A \cup (B \cup C)$; **associative property of set union.**
E. $A \cap (B \cup C) = (A \cap B) \cup (A \cap C)$; **distributive property of set intersection over union.**
F. $A \cup (B \cap C) = (A \cup B) \cap (A \cup C)$; **distributive property of set union over intersection.**
G. $A \cap U = U \cap A = A$; U is the **identity for set intersection.**
H. $A \cup \emptyset = \emptyset \cup A = A$; $\emptyset$ is the **identity for set union.**
I. $\overline{U} = \emptyset$; $\overline{\emptyset} = U$; $A \cap \overline{A} = \emptyset$; $A \cup \overline{A} = U$; $A = \overline{\overline{A}}$; **complement properties.**

IV. Relations and functions
  A. A **relation** R from set A to set B is a subset of $A \times B$; that is, if R is a relation, then $R \subseteq A \times B$.
  B. Properties of relations
    1. A relation R on a set X is **reflexive** if, and only if, for every element a of X, a is related to a.
    2. A relation R on a set X is **symmetric** if, and only if, for all elements a and b of X, whenever a is related to b, then b is related to a.
    3. A relation R on a set X is **transitive** if, and only if, for all elements a, b, and c of X, whenever a is related to b and b is related to c, then a is related to c.
  C. An **equivalence relation** is any relation R that satisfies the reflexive, symmetric, and transitive properties.
  D. A **function** from set A to set B is a relation from A to B in which each element of A is paired with one, and only one, element of B. A is called the **domain** of the function; B is called the **range** of the function.

*V. Logic
  A. A **statement** is a sentence that is either true or false, but not both.
  B. The **negation** of a statement is a statement with the opposite truth value of the given statement.
  C. **Universal quantifiers** refer to each and every element in a set.
  D. **Existential quantifiers** refer to one or more, or possibly all, of the elements in a set.
  E. The **compound statement** $p \wedge q$ is called the **conjunction** of p and q and is defined to be true if, and only if, both p and q are true.
  F. The compound statement $p \vee q$ is called the **disjunction** of p and q and is true if either p or q or both are true.
  G. Statements of the form "if p, then q" are called **conditionals** or **implications** and are false only if p is true and q is false.
  H. Given the conditional $p \to q$, the following can be found:
    1. **Converse:** $q \to p$
    2. **Inverse:** $\sim p \to \sim q$
    3. **Contrapositive:** $\sim q \to \sim p$
  I. Two statements are **logically equivalent** if, and only if, they have the same truth value.
  J. The statement "p if and only if q" is called a **biconditional.** It is true only if p and q have the same truth values.
  K. A **tautology** is a statement that is always true.
  L. Laws to determine the validity of arguments include the **Law of Detachment, Modus Tollens,** and the **Chain Rule.**

CHAPTER TEST

1. Write the set of letters of the Greek alphabet using set-builder notation.
2. List all the subsets of $\{m, a, t, h\}$.
3. ▶Let
   $U = \{x | x \text{ is a person living in Montana}\}$
   $A = \{x | x \text{ is a person 30 yr or older}\}$
   $B = \{x | x \text{ is a person less than 30 yr old}\}$
   $C = \{x | x \text{ is a person who owns a pickup truck}\}$
   Describe in words a member of each of the following sets:
   (a) $\overline{A}$ (b) $A \cap C$ (c) $A \cup B$
   (d) $\overline{C}$ (e) $\overline{A \cap C}$ (f) $A - C$
4. Let $U = \{u, n, i, v, e, r, s, a, l\}$
   $A = \{r, a, v, e\}$ $C = \{l, i, n, e\}$
   $B = \{a, r, e\}$ $D = \{s, a, l, e\}$
   Find each of the following:
   (a) $A \cup B$ (b) $C \cap D$
   (c) $\overline{D}$ (d) $A \cap \overline{D}$
   (e) $\overline{B \cup C}$ (f) $(B \cup C) \cap D$
   (g) $(\overline{A} \cup B) \cap (C \cap \overline{D})$ (h) $(C \cap D) \cap A$
   (i) $n(\overline{C})$ (j) $n(C \times D)$
5. Indicate the following sets by shading:

   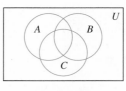   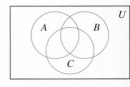

   (a) $A \cap (B \cup C)$ (b) $\overline{(A \cup B)} \cap C$

6. Let $A = \{s, e, t\}$ and $B = \{i, d, e, a\}$. Find each of the following:
   (a) $B \times A$ (b) $A \times A$
   (c) $n(A \times \emptyset)$ (d) $n(B - A)$
7. Suppose you are playing a word game with seven letters. How many possible seven-letter words could there be?
8. (a) Show one possible one-to-one correspondence

between sets D and E if $D = \{t, h, e\}$ and $E = \{e, n, d\}$.
   (b) How many different one-to-one correspondences between sets D and E are possible?
9. Use a Venn diagram to determine whether $A \cap (B \cup C) = (A \cap B) \cup C$ for all sets A, B, and C.
10. Describe, using symbols, the shaded portion in each of the following:

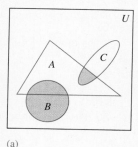
(a)

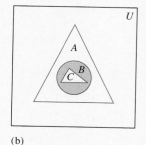
(b)

11. (a) If $A = \{1, 2, 3\}$, $B = \{2, 3, 4, 5\}$, and $C = \{3, 4, 5, 6, 7\}$, illustrate the associative property of intersection of sets.
    (b) Using sets A and B, illustrate the commutative property of union of sets.
12. Classify each of the following as true or false. If false, tell why.
    (a) For all sets A and B, either $A \subseteq B$ or $B \subseteq A$.
    (b) The empty set is a proper subset of every set.
    (c) For all sets A and B, if $A \sim B$, then $A = B$.
    (d) The set $\{5, 10, 15, 20, \ldots\}$ is a finite set.
    (e) No set is equivalent to a proper subset of itself.
    (f) If A is an infinite set and $B \subseteq A$, then B also is an infinite set.
    (g) For all finite sets A and B, if $A \cap B \neq \emptyset$, then $n(A \cup B) \neq n(A) + n(B)$.
    (h) If A and B are sets such that $A \cap B = \emptyset$, then $A = \emptyset$ or $B = \emptyset$.
13. Suppose P and Q are equivalent sets and $n(P) = 17$.
    (a) What is the minimum number of elements in $P \cup Q$?
    (b) What is the maximum number of elements in $P \cup Q$?
    (c) What is the minimum number of elements in $P \cap Q$?
    (d) What is the maximum number of elements in $P \cap Q$?
14. Case Eastern Junior College awarded 26 varsity letters in crew, 15 in swimming, and 16 in soccer. If awards went to 46 students and only 2 lettered in all sports, how many students lettered in exactly two of the three sports?
15. ▶Consider the set of northwestern states or provinces {Montana, Washington, Idaho, Oregon, Alaska, British Columbia, Alberta}. If a person chooses one element, show that in three "yes" or "no" questions we can determine the element.
16. According to a student survey, 16 students liked history, 19 liked English, 18 liked mathematics, 8 liked mathematics and English, 5 liked history and English, 7 liked history and mathematics, 3 liked all three subjects, and every student liked at least one of the subjects. Draw a Venn diagram describing this information and answer the following questions:
    (a) How many students were in the survey?
    (b) How many students liked only mathematics?
    (c) How many students liked English and mathematics but not history?
17. Which of the following relations are functions from the set of first components to the set of second components?
    (a) $\{(a, b), (c, d), (e, a), (f, g)\}$
    (b) $\{(a, b), (a, c), (b, b), (b, c)\}$
    (c) $\{(a, b), (b, a)\}$
18. If $f(x) = 3x + 7$, find each of the following:
    (a) $f(0)$ (b) $f(8)$ (c) $f(10)$
19. Consider the set of post offices, P, in the United States and the set of ZIP codes, Z, in the United States.
    (a) Describe a function from P to Z.
    (b) The first digit of a ZIP code divides the country into 10 sets of states; each state is divided into an average of 10 smaller geographic areas, identified by the second and third digits of the code; the fourth and fifth digits identify a local delivery area. If I have the ZIP code 59801, describe a series of set operations the U.S. Postal Service might use to determine exactly where I live.
20. Given the following function rules and the domains, find the associated ranges:
    (a) $f(x) = x + 3$ domain = $\{0, 1, 2, 3\}$
    (b) $f(x) = 3x - 1$ domain = $\{5, 10, 15, 20\}$
    (c) $f(x) = x^2$ domain = $\{0, 1, 2, 3, 4\}$
    (d) $f(x) = x^2 + 3x + 5$ domain = $\{0, 1, 2\}$
21. What properties do each of the following relations, defined on the set of all people, have?
    (a) Belongs to the same book club as
    (b) Is thinner than
    (c) Is married to
    (d) Is the father of
*22. Which of the following are statements?
    (a) The moon is inhabited.
    (b) $3 + 5 = 8$
    (c) $x + 7 = 15$
    (d) Some women have Ph. Ds in mathematics.
*23. Negate each of the following:
    (a) Some women smoke.
    (b) $3 + 5 = 8$
    (c) All heavy metal rock is loud.
    (d) Beethoven wrote only classical music.

*24. Write truth tables for each of the following:
   (a) $[p \vee (\sim q)] \wedge p$
   (b) $[p \to (\sim q)] \vee q$
   (c) $[p \to (\sim q)] \wedge [(\sim q) \to p]$
   (d) $[(\sim p) \vee (\sim q)] \to (q \wedge p)$

*25. Decide whether the following are equivalent:
   (a) $p \wedge (q \vee r)$; $(p \wedge q) \vee (p \wedge r)$
   (b) $p \to q$; $q \to p$

*26. Write the converse, inverse, and contrapositive of the following: If we have a rock concert, someone will faint.

*27. Find valid conclusions for the following arguments:
   (a) All Americans love Mom and apple pie.
       Joe Czernyu is an American.
   (b) Steel eventually rusts.
       The Statue of Liberty has a steel structure.
   (c) Albertina will pass Math 100 or be a dropout.
       Albertina is not a dropout.

*28. Write the following argument symbolically and then determine its validity:
   If you are fair-skinned, you will sunburn.
   If you sunburn, you will not go to the dance.
   If you do not go to the dance, your parents will want to know why you didn't go to the dance.
   Your parents do not want to know why you didn't go to the dance.
   Therefore you are not fair-skinned.

*29. Determine whether each of the following arguments is valid:
   (a) All diets are ridiculous.
       All diets are degrading.
       Therefore all ridiculous diets are degrading.
   (b) Blondes have more fun.
       Some blondes are really brunettes.
       Therefore some brunettes have more fun.
   (c) If Gloria goes fishing, she does not use flies.
       If Gloria does not use flies, then she is not a real fisherwoman.
       Gloria goes fishing.
       Therefore Gloria is not a real fisherwoman.

SELECTED BIBLIOGRAPHY

Artzt, A. and C. Newman. "Equivalence: A Unifying Concept." *Mathematics Teacher* 84 (February 1991): 128–132.

Coltharp, F. "Mathematical Aspects of the Attribute Games." *Arithmetic Teacher* 21 (March 1974): 246–251.

Johnston, A. "Introducing Function and Its Notation." *Mathematics Teacher* 80 (October 1987): 558–560.

McGinty, R. and J. Van Beynen. "Deductive and Analytical Thinking." *Mathematics Teacher* 78 (March 1985): 188–194.

O'Daffer, P. "Inductive and Deductive Reasoning." *Mathematics Teacher* 83 (May 1990): 378–384.

O'Regan, P. "Intuition and Logic." *Mathematics Teacher* 81 (November 1988): 664–668.

Sanders, W. and R. Antes. "Teaching Logic with Logic Boxes." *Mathematics Teacher* 81 (November 1988): 643–647.

Spence, L. "How Many Elements Are in a Union of Sets?" *Mathematics Teacher* 80 (November 1987): 666–670, 681.

3 Numeration Systems and Whole Numbers

Preliminary Problem

Pennies are placed in stacks on a checkerboard as shown, with 1 penny on the first square, 2 pennies on the second square, 4 on the third square, and 8 on the fourth square. The number of pennies in a stack doubles with each consecutive square. How high will the stack be on square number 64?

n the K–4 *Standards* (p. 41), we find the following:

> *Understanding the fundamental operations of addition, subtraction, multiplication, and division is central to knowing mathematics. One essential component of what it means to understand an operation is recognizing conditions in real-world situations that indicate that the operation would be useful in those situations. Other components include building an awareness of models and the properties of an operation, seeing relationships among operations, and acquiring insight into the effects of an operation on a pair of numbers. These four components are aspects of operation sense. Children with good operation sense are able to apply operations meaningfully and with flexibility.*

Also in the *Teaching Standards* (p. 136), we find, *Teachers of mathematics should have a well-defined number sense (including mental mathematics, estimation, and reasonableness of results) and an understanding of the use of number concepts, operations, and properties (including basic number theory), of the role of algorithms, and of place value.*

Our goal in this chapter is to help you develop number sense and operation sense. You may be so familiar with our numeration system and with operations such as addition and multiplication that you take them for granted. As a teacher, however, you will work with students who are encountering them for the first time. When you have finished this chapter, you will have many new tools for helping students understand what they are doing when they count to 10, add 2 and 3, or divide 16 by 4.

First we will offer fresh insight into how our system works by comparing our numeration system with other systems developed throughout history. Then we will present ways in which to sharpen your operation sense; for example, how to do mathematical operations, which ones to use when, and why they work. In this chapter we will look specifically at algorithms for addition, subtraction, multiplication, and division of whole numbers and at models for these operations. Then, building on operation sense, we will present techniques for using mental math and estimation in connection with computation. Finally, building on whole number operations knowledge, we will look at number bases other than ten to develop a better understanding of our present system.

Section 3-1 Numeration Systems

numerals

Early methods of "writing down" numbers included making notches or strokes on stone or wood and tying knots in a cord. The recorded symbols usually represented numbers of animals. Since that time, numerals have changed extensively. **Numerals** are written symbols used to represent quantities or numbers.

Table 3-1 shows some ways that numbers have been recorded. The Babylonians used wedge-shaped marks pressed in wet clay. The Egyptians used papyrus and ink-filled brushes, basing their system on tally marks. The Mayans introduced a symbol for zero. The numerals we use today are called *Hindu-Arabic* numerals. Different symbols can be used to represent the same quantity. For example, in Table 3-1 we see that 3 and III represent the quantity we call *three*.

TABLE 3-1

Babylonian	▼	▼▼	▼▼▼	▼▼▼▼	▼▼▼/▼▼	▼▼▼/▼▼▼	▼▼▼▼/▼▼▼	▼▼▼▼/▼▼▼▼	▼▼▼▼▼/▼▼▼▼	<
Egyptian	I	II	III	IIII	III/II	III/III	IIII/III	IIII/IIII	III/III/III	∩
Mayan	👁	•	••	•••	••••	—	•̇	•• / —	••• / —	•••• / —	=
Greek	α	β	γ	δ	ε	φ	ζ	η	θ	ι
Chinese	一	二	三	四	五	六	七	八	九	十
Roman	I	II	III	IV	V	VI	VII	VIII	IX	X
Hindu	0	1	2	3	8	9	6	ʌ	8	9
Arabic	•	١	٢	٣	٤	٥	٦	٧	٨	٩
Hindu–Arabic	0	1	2	3	4	5	6	7	8	9	10

Egyptian Numeration System

The Egyptian system, dating back to about 3400 B.C., used *tally marks*. Tally marks are scratches or marks that represent the items being counted. In a *tally numeration system*, there is a one-to-one correspondence between the marks and the items being counted. The first nine numerals in the Egyptian system in Table 3-1 show the use of tally marks. The Egyptians improved on the system based only on tally marks by developing a *grouping system*. In a grouping system, new numerals are used to represent certain sets of numbers, which make the numbers easier to record. The Egyptians drew on their environment for their symbols. For example, the Egyptians used a heel bone symbol, ∩, to stand for a grouping of ten tally marks.

$$|||||||||| \rightarrow \cap$$

Table 3-2 shows other numerals that the Egyptians used in their system.

TABLE 3-2

Egyptian Numeral	Description	Hindu-Arabic Equivalent
I	Vertical staff	1
∩	Heel bone	10
𓏲	Scroll	100
𓆼	Lotus flower	1000
𓂀	Pointing finger	10,000
𓆐	Polliwog or burbot	100,000
𓀀	Astonished man	1,000,000

106 CHAPTER 3 NUMERATION SYSTEMS AND WHOLE NUMBERS

additive property

The Egyptian system involved an **additive property;** that is, the value of a number was the sum of the values of the numerals. The Egyptians customarily wrote the numerals in decreasing order from left to right. An example follows:

ꜣ	represents	100,000	
999	represents	300	(100 + 100 + 100)
∩∩	represents	20	(10 + 10)
∥	represents	2	(1 + 1)
ꜣ999∩∩∥	represents	100,322	

Babylonian Numeration System

The Babylonian system was developed at about the same time as the Egyptian system. The symbols shown in Table 3-3 were made using a stylus either vertically or horizontally.

TABLE 3-3

Babylonian Numeral	Hindu-Arabic Equivalent
▼	1
＜	10

The Babylonian numerals 1 through 59 were similar to the Egyptian numerals, but the staff and the heel bone were replaced by the symbols shown in Table 3-3.

For example, ≪ ▼▼ represented 22. For numbers greater than 59, the Babylonians used **place value.** The value of a symbol in a given numeral depended on the placement of the symbol with respect to other symbols in the numeral. The ＜ symbols appear to the left of the ▼ symbols for numbers between 10 and 60. Numbers greater than 59 were represented by repeated groupings of 60, much as we use groupings of 10 today. For example ▼▼ ≪ represented $2 \cdot 60 + 20$, or 140. The space indicates that ▼▼ represents $2 \cdot 60$ rather than 2. In the number ▼▼ ≪, the place value of ▼▼ is 60 and the **face value** of ▼▼ is 2, which tells how many groupings of 60 are indicated. The Babylonians chose to work with 60 because it can be evenly divided by many numbers. This simplifies division and operations with fractions. The size of the basic groups in a numeration system is called the **number base** of the numeration system. Thus the system that we use today is referred to as a *base-ten system*.

place value

face value

number base

The initial Babylonian system contained inadequacies. For example, the symbol ▼▼ could have represented 2 or $2 \cdot 60$ because the Babylonian system lacked a symbol for zero until after 300 B.C.

Numerals to the left of a second space have a value $60 \cdot 60$ times their face value, and so on.

≪ ▼	represents	$20 \cdot 60 + 1$, or 1201
＜▼ ＜▼ ▼	represents	$11 \cdot 60 \cdot 60 + 11 \cdot 60 + 1$, or 40,261
▼ ＜▼ ＜▼ ▼	represents	$1 \cdot 60 \cdot 60 \cdot 60 + 11 \cdot 60 \cdot 60 + 11 \cdot 60 + 1$, or 256,261

expanded form

The representation of ＜▼ ＜▼ ▼ as $11 \cdot 60 \cdot 60 + 11 \cdot 60 + 1$ is called the **expanded form** of the number. The expanded form consists of the sum of the products resulting from multiplying face values by place values. Using exponents, a product such as $60 \cdot 60$ can be

SECTION 3-1 NUMERATION SYSTEMS 107

factor written as 60^2, in which case 60 is called a **factor** of the product. Thus <▼ <▼ ▼ can be written in expanded form using exponents as $11 \cdot 60^2 + 11 \cdot 60 + 1$. The notion of exponents can be generalized as follows.

DEFINITION OF a^n

If a is any number and n is any natural number, then a^n is defined by the following equation:
$$a^n = \underbrace{a \cdot a \cdot a \cdot \ldots \cdot a}_{n \text{ factors}}.$$

---

nth power of a In this definition, a^n is the **nth power of a,** n is the **exponent,** and a is the **base.**
exponent/base   Chapter 6 discusses the definitions and properties of exponents in detail. The chapter also discusses why it is useful to define a^0 as 1 if $a \neq 0$.

Computations involving exponents can be made using a calculator. If the calculator has a constant key, the following steps will compute $5^4 = 625$:

$$\boxed{5} \; \boxed{\times} \; \boxed{K} \; \boxed{=} \; \boxed{=} \; \boxed{=} \; .$$

On a calculator with an automatic constant, the same type of computation is possible by pressing $\boxed{5} \; \boxed{\times} \; \boxed{5} \; \boxed{=} \; \boxed{=} \; \boxed{=}$. Some calculators have an exponential key such as $\boxed{y^x}$. This function computes the xth power of y; for example, to compute 5^4 we press $\boxed{5}$ $\boxed{y^x} \; \boxed{4} \; \boxed{=}$.

Mayan Numeration System

At some point in numeration history, people began using parts of their bodies to count. Fingers could be matched to objects to stand for one, two, three, four, or five objects. Two hands could then stand for a set of ten objects. In warmer climates where people went barefoot, people may have used their toes as well as their fingers for counting. The Mayans introduced a new attribute that was present neither in the Egyptian nor in the Babylonian systems, namely, a symbol for zero. The Mayan system used only three symbols, which Table 3-4 shows.

TABLE 3-4

Mayan Numeral	Hindu-Arabic Equivalent
•	1
—	5
👁	0

The symbols for the first ten numerals are shown in Table 3-1. Notice the groupings of five in Table 3-1, where each horizontal bar represents a group of five. Thus the symbol for 19 was ≣, or 3 fives and 4 ones. The symbol for 20 was 👁̇, which represents 1 group of twenty plus 0 ones. The Mayans wrote numbers vertically with the greatest value on top. Several examples of the Mayan system, using units and twenties, are given next. In Figure 3-1(a), we have $2 \cdot 5 + 3 \cdot 1$, or 13 groups of twenty plus $2 \cdot 5 + 1 \cdot 1$, or 11 ones, for

FIGURE 3-1

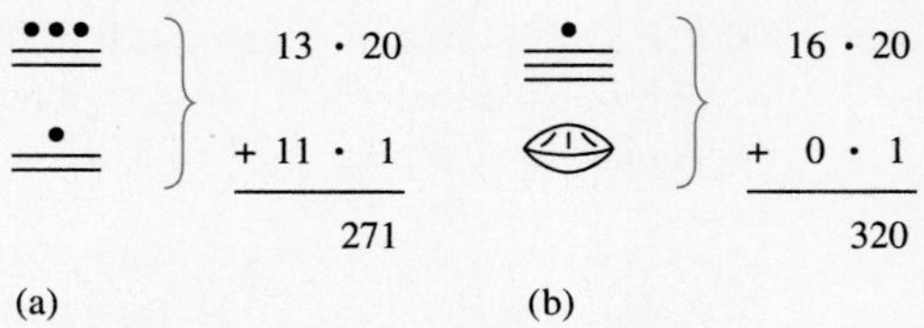

(a) 13 · 20 + 11 · 1 = 271

(b) 16 · 20 + 0 · 1 = 320

a total of 271. In Figure 3-1(b), we have $3 \cdot 5 + 1 \cdot 1$, or 16 groups of twenty and 0 ones, for a total of 320.

In a true base-twenty system, the value of the symbols in the third position vertically from the bottom should be 20^2, or 400. However, it is conjectured that the Mayans used $20 \cdot 18$, or 360, instead of 400. (The number 360 is an approximation of the length of a calendar year, which consisted of 18 months of 20 days each, plus 5 "unlucky" days.) Thus, instead of place values of 1, 20, 20^2, 20^3, 20^4, and so on, the Mayans used 1, 20, $20 \cdot 18$, $20^2 \cdot 18$, $20^3 \cdot 18$, and so on. For example, in Figure 3-2(a), we have $5 + 1$ (or 6) groups of 360, plus $5 + 5 + 2$ (or 12) groups of 20, plus $5 + 4$ (or 9) groups of 1, for a total of 2409. In Figure 3-2(b), we have $2 \cdot 5$ (or 10) groups of 360, plus 0 groups of 20, plus 2 ones, for a total of 3602. Spacing is important in the Mayan system. For example, if two horizontal bars are placed close together, as in ≡, the symbols represent $5 + 5 = 10$. If the bars are spaced apart, as in ═, then the value is $5 \cdot 20 + 5 \cdot 1 = 105$.

FIGURE 3-2

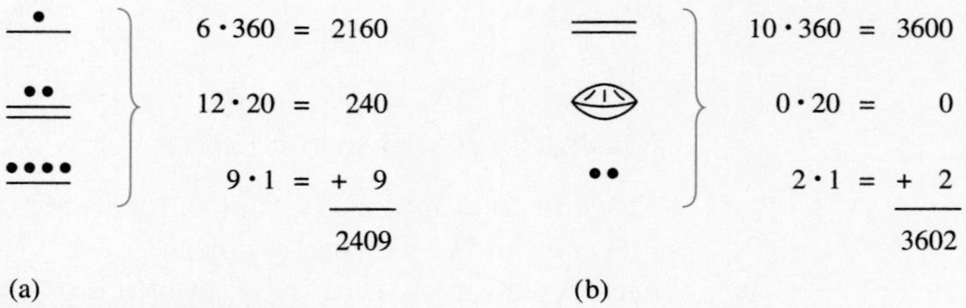

(a) 6 · 360 = 2160; 12 · 20 = 240; 9 · 1 = + 9; 2409

(b) 10 · 360 = 3600; 0 · 20 = 0; 2 · 1 = + 2; 3602

Roman Numeration System

The Roman numeration system remains in use today, as seen on cornerstones, on the opening pages of books, and on the faces of clocks. The basic Roman numerals are pictured in Table 3-5.

TABLE 3-5

Roman Numeral	Hindu-Arabic Equivalent
I	1
V	5
X	10
L	50
C	100
D	500
M	1000

Roman numerals can be combined by using an additive property. For example, MDCLXVI represents $1000 + 500 + 100 + 50 + 10 + 5 + 1 = 1666$, CCCXXVIII represents 328, and VI represents 6.

subtractive property

To avoid repeating a symbol more than three times, as in IIII, a **subtractive property** was introduced in the Middle Ages. For example, I is less than V, so if it is to the left of V, it is subtracted. Thus, IV has a value of $5 - 1$, or 4, and XC represents $100 - 10$, or 90.

Some extensions of the subtractive property could lead to ambiguous results. For example, IXC could be 91 or 89. By custom, 91 is written XCI, and 89 is written LXXXIX. In general, only one smaller number symbol can be to the left of a larger number symbol, and the pair must be one of those listed in Table 3-6.

TABLE 3-6

Roman Numeral	Hindu-Arabic Equivalent
IV	$5 - 1$, or 4
IX	$10 - 1$, or 9
XL	$50 - 10$, or 40
XC	$100 - 10$, or 90
CD	$500 - 100$, or 400
CM	$1000 - 100$, or 900

multiplicative property

The Romans adopted the use of bars to write large numbers. The use of bars is based on a **multiplicative property.** A bar over a symbol or symbols indicates that the value is multiplied by 1000. For example, $\overline{V}$ represents $5 \cdot 1000$, or 5000, and $\overline{CDX}$ represents $410 \cdot 1000$, or 410,000. To indicate even greater numbers, more bars appear. For example, $\overline{\overline{V}}$ represents $(5 \cdot 1000) \cdot 1000$, or 5,000,000; $\overline{\overline{\overline{CXI}}}$ represents $111 \cdot 1000^3$, or 111,000,000,000; and $\overline{CXI}$ represents $110 \cdot 1000 + 1$, or 110,001.

Several properties might be used to represent some numbers, for example:

$$\overline{DCLIX} = \underbrace{(500 \times 1000)}_{\text{Multiplicative}} + \underbrace{(100 + 50)}_{\text{Additive}} + \underbrace{(10 - 1)}_{\text{Subtractive}} = 500{,}159$$

Hindu-Arabic Numeration System

digits

decimal system

The Hindu-Arabic numeration system we use today is a base-ten system. The ten basic symbols are called **digits:** 0, 1, 2, 3, 4, 5, 6, 7, 8, 9. The word digit is derived from the Latin word *digitus* for finger. Each place in a Hindu-Arabic numeral represents a power of 10. Thus, the system is a **decimal system,** after the Latin word *decem* for ten.

The Hindu-Arabic system has the following important characteristics:

1. All numerals are constructed from the ten basic digits.
2. The system uses place value based on repeated groupings of ten.
3. There is a symbol for zero.

Each digit in a numeral has two functions:

1. Its position in the numeral names its *place value*.
2. The digit itself names its *face value;* that is, it tells how many groupings of ten are indicated.

For example, in the numeral 5984, the 5 has place value "thousands," the 9 has place value "hundreds," the 8 has place value "tens," and the 4 has place value "units," as

FIGURE 3-3

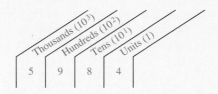

shown in Figure 3-3. Hence, we can write 5984 in expanded form as $5 \cdot 1000 + 9 \cdot 100 + 8 \cdot 10 + 4 \cdot 1$, or $5 \cdot 10^3 + 9 \cdot 10^2 + 8 \cdot 10 + 4 \cdot 1$.

PROBLEM SET 3-1

1. For each of the following, tell which numeral represents the greater number and why:
   (a) $\overline{\text{MCDXXIV}}$ and $\overline{\overline{\text{MCDXXIV}}}$
   (b) 4632 and 46,032
   (c) ◁▼▼ and ◁ ▼▼
   (d) 999∩∩|| and 𓐍∩|
   (e) ⋮ and 👁
2. For each of the following, name both the succeeding and preceding numerals (one more and one less):
   (a) MCMXLIX   (b) $\overline{\text{MI}}$ (c) CMXCIX
   (d) ◁◁ ◁▼   (e) 𓐍99   (f) ⋮
3. The cornerstone on the building reads MCMXXII. When was this building built?
4. ▶For each of the following systems, discuss how you might add 245 and 989:
   (a) Babylonian   (b) Egyptian
5. ▶How might you perform the following subtraction problem using Egyptian numerals?

   9∩∩|||
   −∩∩∩|||||

6. Write each of the following in Roman symbols:
   (a) 121   (b) 42   (c) 89   (d) 5282
7. Write each of the following in Egyptian symbols:
   (a) 52   (b) 103   (c) 100,003   (d) 38
8. Complete the following table, which compares symbols for numbers in different numeration systems:

Hindu-Arabic	Babylonian	Egyptian	Roman	Mayan
(a) 72				
(b)	◁ ▼▼			
(c)		𓐍99∩∩|||		

9. ▶List three places where Roman numerals are used today.
10. ▶Ben claims that zero is the same as nothing. How do you respond?
11. ▶In your opinion, what are the major drawbacks of the following systems?
    (a) Egyptian   (b) Babylonian   (c) Roman
12. ▶(a) Create a numeration system of your own with unique symbols, and write a paragraph explaining the properties of your system.
    (b) Complete the table using your system.

Hindu-Arabic Numeral	Your System Numeral	Hindu-Arabic Numeral	Your System Numeral
1		100	
5		5,000	
10		10,000	
50		15,280	

13. For each of the following decimal numerals, give the place value of the underlined numeral:
    (a) 827,367   (b) 8,421,000
    (c) 97,998   (d) 810,485
14. Rewrite each of the following as a base-ten numeral:
    (a) $3 \cdot 10^6 + 4 \cdot 10^3 + 5$
    (b) $2 \cdot 10^4 + 1$
    (c) $3 \cdot 10^3 + 5 \cdot 10^2 + 6 \cdot 10$
    (d) $9 \cdot 10^6 + 9 \cdot 10 + 9$
15. ▶Why are large numbers written with commas separating groups of three digits?
16. Study the counting frame shown next. In the frame, the value of each dot is represented by the number in the box below the dot. For example, the following figure represents the number 154:

••	•••	••
64	8	1

    What numbers are represented in the frames in (a) and (b)?

    (a)    (b)

17. A certain three-digit whole number has the following properties: The hundreds digit is greater than 7; the tens digit is an odd number; and the sum of the digits is 10. What is the number?

18. ▶For the last 600 yr, no significant changes have taken place in the Hindu-Arabic system. What suggestions for change might make the system better or easier to use?

19. 🖩 Use the constant feature on a calculator to determine the value of $9 \cdot 9 \cdot 9 \cdot 9 \cdot 9 \cdot 9 \cdot 9$, or 9^7.

20. 🖩 Use only the keys $\boxed{1}, \boxed{2}, \boxed{3}, \boxed{4}, \boxed{5}, \boxed{6}, \boxed{7}, \boxed{8}$, and $\boxed{9}$, for each of the following:
    (a) Fill the display to show the greatest number possible; each key may be used only once.
    (b) Fill the display to show the least number possible; each key may be used only once.
    (c) Fill the display to show the greatest number possible if a key may be used more than once.
    (d) Fill the display to show the least number possible if a key may be used more than once.

21. 🖩 In a game called WIPEOUT, we are to "wipe out" digits from the calculator display without changing any of the other digits. "Wipeout" in this case means replace the chosen digit(s) with a 0. For example, if the initial number is 54,321 and we are to wipe out the 4, we could subtract 4000 to obtain 50,321. Complete the following two problems and then try other numbers or challenge another person to wipe out a number from the number you have placed on the screen.
    (a) Wipe out the 2s from 32,420.
    (b) Wipe out the 5 from 67,357.

▼ BRAIN TEASER

There are 3 nickels and 3 dimes concealed inside three boxes. Two coins are placed in each of the boxes, which are labeled 10¢, 15¢, and 20¢. The coins are placed in such a way that no box contains the amount of money showing on its label; for example, the box labeled 10¢ does not really have a total of 10¢ in it. What is the minimum number of coins that you would have to remove from a box, and from which box or boxes, to determine which coins are in which boxes?

Section 3-2   Addition and Subtraction of Whole Numbers

whole numbers

Zero is important in the evolution of numeration systems. In terms of set theory, zero can be defined as the cardinal number of the empty set: $n(\emptyset) = 0$. When zero is joined with the set of natural numbers, $N = \{1, 2, 3, 4, 5, \ldots\}$, we have the set of numbers called **whole numbers,** denoted by $W = \{0, 1, 2, 3, 4, 5, \ldots\}$. Another way of defining a whole number is to define it as the cardinal number of a finite set.

The K-4 *Standards* (p. 44) addresses the importance of teaching children a variety of ways to compute with whole numbers as well as the usefulness of calculators in solving problems. In this section, we provide a variety of models for teaching computational skills.

Addition of Whole Numbers

Addition is a *binary operation,* that is, a function that involves using two numbers at a time to produce a single number. The addition of whole numbers can be modeled in several ways. We present the set model and the number-line model.

SET MODEL

Suppose Jane has 4 pencils in one pile and 3 pencils in another. If she combines the two groups of pencils, how many pencils are there in the combined group? Figure 3-4 shows

FIGURE 3-4

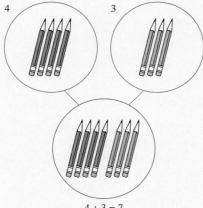

$4 + 3 = 7$

the solution as it might appear in an elementary school text. The combined set of pencils is the union of the set of 4 pencils and the set of 3 pencils. After the sets have been combined, children count the pencils to determine that there are 7 pencils in all. Note the importance of the sets having no elements in common. If the sets have common elements, then an incorrect conclusion can be drawn.

DEFINITION OF ADDITION OF WHOLE NUMBERS

Let A and B be two disjoint finite sets. If $n(A) = a$ and $n(B) = b$, then $a + b = n(A \cup B)$.

addends The numbers a and b in this definition are the **addends;** and $(a + b)$ is the **sum.**
sum

HISTORICAL NOTE

The symbol "+" first appeared in a 1417 manuscript and was a short way of writing the Latin word *et,* which means "and." However, Johann Widmann wrote a book in 1498 that made use of the + and − symbols for addition and subtraction. The word *minus* means "less" in Latin; at first it was written as an *m,* which was later shortened to a horizontal bar.

NUMBER-LINE MODEL

A number line may be used to model whole-number addition. Any line marked with two fundamental points, one representing 0 and the other representing 1, can be turned into a number line. The points representing 0 and 1 mark the ends of a *unit segment*. Other points may be marked and labeled as shown in Figure 3-5. Any two consecutive points in Figure 3-5 mark the ends of a segment that has the same length as the unit segment.

Addition problems can be modeled using directed arrows on the number line. For example, the sum of $4 + 3$ is shown in Figure 3-5. Arrows representing the addends, 4 and 3, are combined into one arrow representing the sum.

FIGURE 3-5

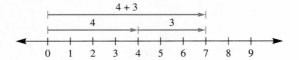

Greater-than and Less-than Relations

greater-than / less-than

A number line can also be used to describe **greater-than** and **less-than** relations on the set of whole numbers. For example, in Figure 3-5, notice that 4 is to the left of 7 on the number line. We say, "four is less than seven," and we write $4 < 7$. We can also say "seven is greater than four" and write $7 > 4$. Since 4 is to the left of 7, there is a natural number that can be added to 4 to get 7, namely, 3. Thus $4 < 7$ because $4 + 3 = 7$. (We can generalize this discussion to form a definition for *less than*.)

> **DEFINITION OF LESS THAN**
>
> For any whole numbers a and b, a is **less than** b, written $a < b$, if and only if, there exists a natural number k such that $a + k = b$.

greater than or equal to
less than or equal to

Sometimes equality is combined with the inequalities greater than and less than to give the relations **greater than or equal to** and **less than or equal to,** denoted by $\geq$ and $\leq$. The emphasis with respect to these symbols has to be on the *or*. Observe that "$3 < 5$ or $3 = 5$" is a true statement, so $3 \leq 5$ is true. Both $5 \geq 3$ and $3 \geq 3$ are true statements.

Whole-number Addition Properties

We now examine properties of whole-number addition. These properties will be used to develop algorithms for more complicated addition. When two whole numbers are added, the sum is a unique whole number. This property is the *closure property of addition of whole numbers,* and we say, "The set of whole numbers is closed under addition."

> **PROPERTY**
>
> **Closure Property of Addition of Whole Numbers** If a and b are any whole numbers, then $a + b$ is a unique whole number.

Note that this property guarantees both the existence and the uniqueness of the sum. With some sets, the addition of two numbers from the set results in a number that does not belong to the set. In this case, we say the set is *not closed* under addition. For example, the set $\{0, 1, 2, 3\}$ is not closed under addition because a sum such as $2 + 3$ is not an element of the set.

EXAMPLE 3-1   Which of these sets are closed under addition?

(a) $\{0, 1\}$ (b) $\{0\}$ (c) $\{1, 3, 5, 7, 9, \ldots\}$

Solution (a) This set is not closed under addition. Although the sum of two different numbers such as 0 and 1 belongs to the set {0, 1}, it is not true for all sums involving numbers from the set. For example, $1 + 1 \notin \{0, 1\}$.

(b) This set is closed under addition because $0 + 0$ belongs to the set.

(c) This set is not closed under addition because, for example, $5 + 3$ is not an element of the original set. ∎

If $a \in W$ and $b \in W$, then, by the closure property of addition, $a + b \in W$. If $c \in W$, it follows that $(a + b) + c \in W$. This reasoning can be extended to more than three whole numbers.

Figure 3-6 shows two additions. Pictured above the number line is $3 + 5$, and below the number line is $5 + 3$. The sums are exactly the same. This demonstrates that $3 + 5 = 5 + 3$. This idea that two whole numbers can be added in either order is true for any two whole numbers a and b. (We can see this by using the definition of addition of whole numbers and the fact that $A \cup B = B \cup A$.) This property is called the *commutative property of addition of whole numbers*, and we say, "Addition of whole numbers is commutative." The root word of "commutative" is "commute," which implies an interchange.

FIGURE 3-6

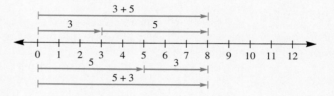

PROPERTY

Commutative Property of Addition of Whole Numbers If a and b are any whole numbers, then $a + b = b + a$.

REMARK The set model can also be used to show commutative property simply by interchanging the sets.

The commutative property of addition of whole numbers is not obvious to many young children. They may be able to find the sum $9 + 2$ and not be able to find the sum $2 + 9$. This is because one of the techniques used to teach addition is *counting on*. Using this technique, $9 + 2$ can be computed by starting at 9 and then counting on two more, as "ten" and "eleven." To compute $2 + 9$, the *counting on* is more involved. Students need to understand that $2 + 9$ is another name for $9 + 2$.

Another property of addition is demonstrated when we select the order in which to add three or more numbers. For example, we could compute $24 + 8 + 2$ by grouping the 24 and the 8 together: $(24 + 8) + 2 = 32 + 2 = 34$. (The parentheses indicate that the first two numbers are grouped together.) We might also recognize that it is easy to add any number to 10 and compute it as $24 + (8 + 2) = 24 + 10 = 34$. This example illustrates the *associative property of addition of whole numbers*. In many elementary school texts in the lower grades, this property is referred to as the *grouping property for addition*.

> **PROPERTY**
>
> **Associative Property of Addition of Whole Numbers** If a, b, and c are any whole numbers, then $(a + b) + c = a + (b + c)$.

When several numbers are being added, the parentheses are usually omitted, since the grouping does not alter the result. The commutative and associative properties of addition are often used together. For example, to find the sum $20 + 5 + 60 + 4$, we group the addends as $(20 + 60) + (5 + 4)$ to obtain $80 + 9$, or 89.

Another property of addition of whole numbers is seen when one addend is 0. In Figure 3-7, set A has 5 blocks and set B has 0 blocks. The union of sets A and B has only 5 blocks.

FIGURE 3-7

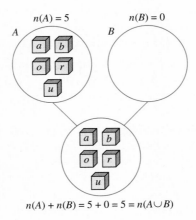

$n(A) + n(B) = 5 + 0 = 5 = n(A \cup B)$

This example illustrates the following property of whole numbers.

> **PROPERTY**
>
> **Identity Property of Addition of Whole Numbers** There is a unique whole number 0, called the **additive identity**, such that for any whole number a, $a + 0 = a = 0 + a$.

REMARK The identity property can be justified by using set theory, since $0 = n(\emptyset)$ and, if $a = n(A)$, then $a + 0 = n(A) + n(\emptyset) = n(A \cup \emptyset) = n(A) = a$. A similar argument shows that $0 + a = a$.

EXAMPLE 3-2   Which properties justify each of the following?

(a) $5 + 7 = 7 + 5$
(b) $1001 + 733$ is a whole number.
(c) $(3 + 5) + 7 = (5 + 3) + 7$
(d) $(8 + 5) + 2 = 8 + (5 + 2)$
(e) $(10 + 5) + (10 + 3) = (10 + 10) + (5 + 3)$

Solution
(a) Commutative property of addition
(b) Closure property of addition
(c) Commutative property of addition
(d) Associative property of addition
(e) Commutative and associative properties of addition

Mastering Basic Addition Facts

Certain mathematical facts are called *basic addition facts*. Basic addition facts are those involving a single digit plus a single digit. As the K-4 *Standards* (p. 19) point out, *Calculators do not replace the need to learn basic facts, to compute mentally, or to do reasonable paper-and-pencil computation*. One method of learning the basic facts is to organize them according to different strategies.

1. *Counting On*. The strategy of *counting on* from the *greater* of the addends is usually used when the other addend is 1, 2, or 3. For example, $5 + 3$ can be computed by starting at 5 and then counting on 6, 7, and 8. Likewise, $2 + 8$ would be computed by starting at 8 and then counting 9 and 10.

2. *Doubles*. The next strategy considered involves the use of *doubles*. Doubles such as $4 + 4$ and $6 + 6$ receive special attention with students. After doubles are mastered, *doubles* + *1* and *doubles* + *2* can be easily learned. For example, if a student knows $6 + 6 = 12$, then $6 + 7$ is $(6 + 6) + 1$, or one more than the double of 6, or 13. Likewise, $7 + 9$ is $(7 + 7) + 2$, or two more than the double of 7, or 16.

3. *Making 10*. Another strategy is that of *making 10* and then adding any leftover. For example, we could think of $8 + 5$ as shown in Figure 3-8. Notice that we are really using the associative property of addition.

FIGURE 3-8

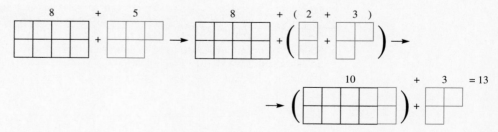

Many basic facts might be classified under more than one strategy. For example, we could find $9 + 8$ using *making 10* as $9 + (1 + 7) = (9 + 1) + 7 = 10 + 7 = 17$, or using a *double plus 1* as $(8 + 8) + 1$.

Subtraction of Whole Numbers

Subtraction of whole numbers can be modeled in several ways: the set (take-away) model, the missing-addend model, the comparison model, and the number-line model.

TAKE-AWAY MODEL

One way to think about subtraction is this: Instead of imagining a second set of objects as being joined to a first set (as in addition), consider the second set as being *taken away* from a first set. For example, suppose we have 8 blocks and we take away 3 of them, as shown in Figure 3-9. We record this process as $8 - 3 = 5$.

FIGURE 3-9

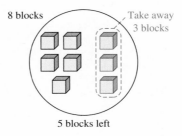

> **REMARK** Figure 3-9 shows that $a - b$ could be defined in terms of sets as follows. If A and B are sets such that $n(A) = a$, $n(B) = b$, and $B \subseteq A$, then $a - b = n(A - B)$.

MISSING-ADDEND MODEL

A second model for subtraction, the *missing-addend* model, relates subtraction and addition. Recall that in Figure 3-9, $8 - 3$ is pictured with blocks as 8 blocks "take away" 3 blocks. The number of blocks left is the number $8 - 3$, or 5. This can also be thought of as the number of blocks that could be added to 3 blocks in order to get 8 blocks, that is,

$$\boxed{8 - 3} + 3 = 8.$$

missing addend   The number $8 - 3$, or 5, is called the **missing addend** in the equation

$$\square + 3 = 8.$$

The relationships between addition and subtraction are further illustrated for elementary students in Figure 3-10. Cashiers often use this model. For example, if the bill for a movie is $8 and you pay $10, the cashier might say "eight and two is ten." This idea can be generalized for whole numbers a and b, as shown next.

FIGURE 3-10

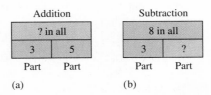

DEFINITION OF SUBTRACTION OF WHOLE NUMBERS

For any whole numbers a and b, such that $a > b$, $a - b$ is the unique whole number c such that $a = b + c$.

minuend
subtrahend / difference

REMARK The notation $a - b$ is read "a minus b." The number a is called the **minuend**; b is called the **subtrahend**; c is called the **difference**.

COMPARISON MODEL

A third way to consider subtraction is by using a *comparison* model. Suppose we have 8 blocks and 3 balls and we would like to know how many more blocks we have than balls. We can pair the blocks and balls, as shown in Figure 3-11, and determine that there are 5 more blocks than balls. We also write this as $8 - 3 = 5$.

FIGURE 3-11

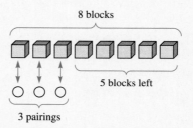

Subtraction can also be defined in terms of set theory. This will be investigated in Problem Set 3-2.

NUMBER-LINE MODEL

We can also model subtraction by using a number line. For example, $5 - 3$ is shown on a number line in Figure 3-12. Observe that an arrow extends 5 units to the right from 0. Because the operation is subtraction, the second arrow extends 3 units to the left from the end of the first arrow. Thus, we see that $5 - 3 = 2$.

FIGURE 3-12

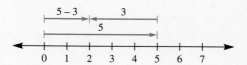

An alternate way to find $5 - 3$ by using a number line is to determine the missing addend that must be added to 3 to obtain the sum of 5. In Figure 3-13, the missing addend is 2, and we have $5 - 3 = 2$.

FIGURE 3-13

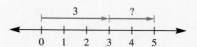

SECTION 3-2 ADDITION AND SUBTRACTION OF WHOLE NUMBERS

Many elementary school texts refer to the relationship between addition and subtraction in terms of "fact families." Notice the fact family on the student page from *Addison-Wesley Mathematics*, 1991, Grade 4. Snap cubes are used as manipulatives to develop the concept and the *TALK ABOUT IT* section asks two questions for students to discuss. The *Standards* call for more communications in mathematics in the elementary school.

Using Addition to Subtract

LEARN ABOUT IT

EXPLORE Use Snap Cubes

Since addition and subtraction are related, you can use addition to find differences. Work in groups. Use red and blue snap cubes to show the parts of this fact family.

Now use 10 cubes with another combination of red and blue to show a different fact family.

Fact Family
7 + 3 = 10
10 − 3 = 7
3 + 7 = 10
10 − 7 = 3

TALK ABOUT IT

1. Look at one of your fact families. What action did you use to show each fact?

2. Use the 10 cubes and a fact family to explain how addition and subtraction are related.

Math Point
When 0 is subtracted from a number, the difference is that number.
6 − 0 = 6
When a number is subtracted from itself, the difference is 0.
6 − 6 = 0

An addition fact shows two addends and their sum. The related subtraction fact shows the sum and one addend. You use the sum and this addend to find the other addend.

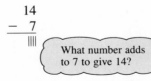

12 − 9 = ||||

The missing addend is 3.   The missing addend is 7.

Properties of Subtraction

Consider the difference $3 - 5$. Using the definition of subtraction, $3 - 5 = c$ means $c + 5 = 3$. Since there is no whole number c that satisfies the equation, the solution for $3 - 5$ cannot be found in the set of whole numbers. This means that the set of whole numbers is *not* closed under subtraction. We also see that, for whole numbers a and b, $a - b$ is meaningful if, and only if, $a - b$ is a whole number, that is, if a is greater than or equal to b. Later, we consider the set of integers, which is closed under subtraction. Showing that subtraction of whole numbers is not commutative, is not associative, and has no identity property is left as an exercise in Problem Set 3-2.

PROBLEM SET 3-2

1. ▶Explain why $5 < 7$ and why $6 > 3$ by finding natural numbers k such that each is true.
   (a) $5 + k = 7$ (b) $6 = 3 + k$
2. ▶In the definition of *less than*, can the natural number k be replaced by the whole number k? Why or why not?
3. Give an example to show why, in the definition of addition, sets A and B must be disjoint.
4. Use the number-line model to illustrate the following:
   (a) $6 + 3 = 9$ (b) $11 - 3 = 8$
5. For each of the following, find whole numbers to make the statements true, if possible.
   (a) $2 + \square = 7$ (b) $\square + 4 = 6$
   (c) $3 + \square \le 5$ (d) $\square + 6 \ge 9$
6. For each of the following, find whole numbers to make the statements true, if possible, where $a \in W$.
   (a) $8 - 5 = \square$ (b) $\square - 4 = 9$
   (c) $a - 0 = \square$ (d) $a - \square = a$
   (e) $\square - 3 \le 6$ (f) $\square - 3 > 6$
7. Tell whether the following sets are closed under addition. If not, give a counterexample.
   (a) $B = \{0\}$
   (b) $T = \{0, 3, 6, 9, 12, \ldots\}$
   (c) $N = \{1, 2, 3, 4, 5, \ldots\}$
   (d) $V = \{3, 5, 7\}$
   (e) $\{x | x \in W, x > 10\}$
8. Rewrite each of the following subtraction problems as an equivalent addition problem.
   (a) $x - 119 = 213$
   (b) $213 - x = 119$
   (c) $213 - 119 = x$
9. Each of the following is an example of one of the properties for addition of whole numbers. Identify the property illustrated.
   (a) $6 + 3 = 3 + 6$
   (b) $(6 + 3) + 5 = 6 + (3 + 5)$
   (c) $(6 + 3) + 5 = (3 + 6) + 5$
10. Use the digits 2, 3, 8, and 0 to fill in the blanks to make each statement true. Use each digit only once. List all possibilities for each case.
    (a) $3280 < \underline{} < 8032$
    (b) $2803 < \underline{} < 3820$
11. Find the next three terms in each of the following sequences:
    (a) 8, 13, 18, 23, 28
    (b) 98, 91, 84, 77, 70, 63
12. Illustrate $9 - 2$, using each of the following models:
    (a) Comparison model
    (b) Number-line model
13. Suppose $A \subseteq B$. If $n(A) = a$ and $n(B) = b$, then $b - a$ could be defined as $n(B - A)$. Choose two sets A and B, and illustrate this definition.
14. Give a counterexample to show that each of the following is false in the set of whole numbers:
    (a) $a - b = b - a$
    (b) $(a - b) - c = a - (b - c)$
    (c) $a - 0 = 0 - a = a$
15. ▶If A, B, and C each stand for a different single digit from 1 to 9, answer the following if
    $$A + B = C.$$
    (a) What is the greatest digit that C could be? Why?
    (b) What is the greatest digit that A could be? Why?
    (c) What is the smallest digit that C could be? Why?
    (d) If A, B, and C are even, what number(s) could C be? Why?
    (e) If C is 5 more than A, what number(s) could B be? Why?
    (f) If A is 3 times as great as B, then what number(s) could C be? Why?
    (g) If A is odd and A is 5 more than B, what number(s) could C be? Why?

16. If *A*, *B*, *C*, and *D* each stand for a different single digit from 1 to 9, answer each of the following if

$$\begin{array}{r} A \\ + B \\ \hline CD \end{array}.$$

▶ (a) What is the value of *C*? Why?
▶ (b) Can *D* be 1? Why?
 (c) If *D* is 7, what values can *A* be?
 (d) If *A* is 6 greater than *B*, then what is the value of *D*?

17. Find the total of the terms in the 50th row in the following figure:

1	1st row
1 − 1	2nd row
1 − 1 + 1	3rd row
1 − 1 + 1 − 1	4th row
1 − 1 + 1 − 1 + 1	5th row

18. Make each of the following a magic square:

(a)
1	6	
	5	7
4		2

(b)
17	10	
		14
13	18	

19. Place whole numbers in the four squares so that each pair has the sum shown. Note that one diagonal sum must be 20 and the other diagonal sum must be 7.

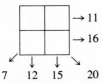

20. Switch two numbers in the squares below so that each row and column will sum to 15.

1	5	3
6	7	2
8	9	4

21. Place the numbers 1, 2, 3, 4, 5, 6, and 7 in the seven boxes so that the sum along each line segment is 13; that is, the three sets of connected boxes should all sum to 13.

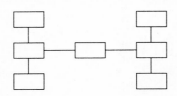

22. Fill in the boxes to obtain a magic square.

32 + 35	240 − 180	29 + 36
96 − 34	☐ + 20	78 − ☐
☐ + 2	80 − 12	14 + 47

23. (a) Place the numbers 1, 2, 3, 4, 5, and 6 in the boxes so that no square has a number greater than the one directly below it or directly to the right of it.

 (b) Can you find more than one way to place the numbers?

24. A domino set contains all number pairs from double-zero to double-six, with each number pair occurring only once; that is, the following domino counts as two-four and four-two. How many dominoes are in the set?

25. Millie and Samantha began saving money at the same time. Millie plans to save $3 a month, and Samantha plans to save $5 a month. After how many months will Samantha have exactly $10 more than Millie?

26. String art is formed by connecting evenly spaced nails on the vertical and horizontal axes by line segments. Connect the nail farthest from the origin on the vertical axis with the nail closest to the origin on the horizontal axis. Continue until all nails are connected, as shown in the figure that follows. How many intersection points are created with 10 nails on each axis?

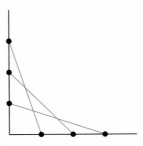

3 nails per axis
3 intersections

27. At a certain party, when the doorbell rang the first time, one guest arrived. On each successive ring, the number of arriving guests was 2 more than the number that had arrived on the previous ring. After 20 rings, how many guests had arrived?
28. ▶Why is it important that elementary students learn the various properties of addition of whole numbers?
29. ▶Why is it important that elementary students learn more than one model for performing the operations of addition and subtraction?
30. ▶Do elementary students still have to learn their basic facts when the calculator is a part of the curriculum? Why or why not?
31. ▶How can the commutative property be used to help a child find the answer to 3 + 9?
32. ▦ Make a calculator display numbers that have the following values:
    (a) Seven tens
    (b) Nine thousands
    (c) Eleven hundreds
    (d) Fifty-six tens
    (e) Three hundred forty-seven tens
33. ▦ Make a calculator count to 100. (Use a constant operation if available.)
    (a) By ones   (b) By twos   (c) By fives
34. ▦ Make a calculator count backward from 27 to 0. (Use a constant operation if possible.)
    (a) By ones   (b) By threes   (c) By nines
35. ▦ If a calculator is made to count by twos starting at 2, what is the thirteenth number in the sequence?

Review Problems

36. Write the number that precedes each of the following:
    (a) CMLX   (b) XXXIX
37. ▶What are the advantages of the Babylonian system over the Egyptian system?
38. Write 5286 in expanded form.

▼ B R A I N T E A S E R

Design an *unmagic square;* that is, use each of the digits 1, 2, 3, 4, 5, 6, 7, 8, and 9 exactly once so that every column, row, and diagonal adds to a different sum.

Section 3-3 Multiplication and Division of Whole Numbers

Multiplication of Whole Numbers

In this section, we use three models to discuss multiplication: the repeated-addition model, the array model, and the Cartesian-product model.

REPEATED-ADDITION MODEL

Suppose we have a classroom with 5 rows of 4 chairs each, as shown in Figure 3-14. How many chairs are there altogether? We can think of this as combining 5 sets of 4 objects into a single set.

FIGURE 3-14

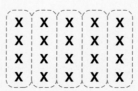

The 5 rows of 4 suggest the following addition:

$$\underbrace{4 + 4 + 4 + 4 + 4}_{\text{five 4s}} = 20$$

Instead of writing the sum of 4s just shown, we use the notation 5×4, or $5 \cdot 4$, to mean that five 4s are added. Thus, multiplication can be defined in terms of repeated addition.

DEFINITION OF MULTIPLICATION OF WHOLE NUMBERS

For any whole numbers a and $n \neq 0$,

$$n \cdot a = \underbrace{a + a + a + \cdots + a}_{n \text{ terms}}.$$

If $n = 0$, then $0 \cdot a = 0$.

REMARK In terms of set theory, $n \cdot a$ can be thought of as the union of n equivalent, disjoint sets, each with a elements.

HISTORICAL NOTE

William Oughtred (1575–1660), an English mathematician, placed emphasis on mathematical symbols. He first introduced the use of "St. Andrew's cross" ($\times$) as the symbol for multiplication. This symbol was not readily adopted because, as Gottfried Wilhelm von Leibnitz (1646–1716) objected, it was too easily confused with the letter x. Leibnitz adopted the use of the dot ($\cdot$) for multiplication, which then became commonly used.

NUMBER-LINE MODEL

Multiplication can be modeled on a number line. For example, the number line for $5 \cdot 4$ is shown in Figure 3-15.

FIGURE 3-15

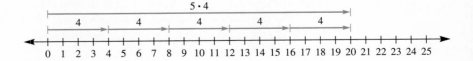

ARRAY MODEL

Another model that is useful in exploring multiplication of whole numbers is the *array model*. We introduce this model by crossing sticks to create intersection points, forming an array. For example, to show 2 · 3, we place three sticks side by side and then cross them with two sticks, as shown in Figure 3-16(a). The product of 2 · 3 is the number of intersection points. The product of 4 · 3 is modeled in Figure 3-16(b). In Figure 3-16(c), we see that the product of 2 · 0 is 0 because there are no intersection points.

FIGURE 3-16

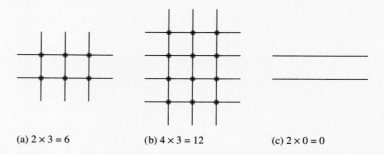

(a) 2 × 3 = 6   (b) 4 × 3 = 12   (c) 2 × 0 = 0

CARTESIAN-PRODUCT MODEL

Suppose you can order a soyburger on light or dark bread with one condiment: mustard, mayonnaise, or horseradish. To show the number of different soyburger orders that a waiter could write for the cook, we use a *tree diagram*. The ways of writing the order are listed in Figure 3-17, where the bread is chosen from the set $B = \{$light, dark$\}$ and the condiment is chosen from the set $C = \{$mustard, mayonnaise, horseradish$\}$.

FIGURE 3-17

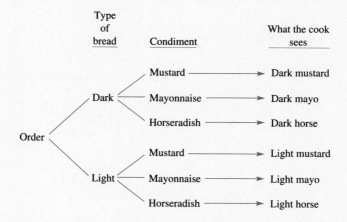

Each order can be written as an ordered pair, for example, (dark, mustard). The set of ordered pairs forms the Cartesian product $B \times C$. The Fundamental Counting Principle tells us that the number of ordered pairs in $B \times C$ is 2 · 3.

The preceding discussion demonstrates how multiplication can be defined in terms of Cartesian products. This alternate definition follows.

ALTERNATE DEFINITION OF MULTIPLICATION OF WHOLE NUMBERS

For finite sets A and B, if $n(A) = a$ and $n(B) = b$, then $a \cdot b = n(A \times B)$.

product / factors

REMARK In this definition, sets A and B do not have to be disjoint. The expression $a \cdot b$ is the **product** of a and b, and a and b are **factors**. Also, note that $A \times B$ indicates the Cartesian product, not multiplication. We multiply numbers, not sets.

Properties of Whole-number Multiplication

As with addition, multiplication on the set of whole numbers has the closure, commutative, associative, and identity properties.

> **PROPERTIES OF MULTIPLICATION OF WHOLE NUMBERS**
>
> **Closure Property of Multiplication of Whole Numbers** For any whole numbers a and b, $a \cdot b$ is a unique whole number.
> **Commutative Property of Multiplication of Whole Numbers** For any whole numbers a and b, $a \cdot b = b \cdot a$.
> **Associative Property of Multiplication of Whole Numbers** For any whole numbers a, b, and c, $(a \cdot b) \cdot c = a \cdot (b \cdot c)$.
> **Identity Property of Multiplication of Whole Numbers** There is a unique whole number 1 such that for any whole number a, $a \cdot 1 = a = 1 \cdot a$.
> **Zero Multiplication Property of Whole Numbers** For any whole number a, $a \cdot 0 = 0 = 0 \cdot a$.

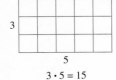

$3 \cdot 5 = 15$

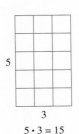

$5 \cdot 3 = 15$

FIGURE 3-18

The commutative property of multiplication of whole numbers is easily illustrated by building a 3×5 grid and then turning it sideways, as shown in Figure 3-18. We see that the number of 1×1 squares present in either case is 15, that is, $3 \cdot 5 = 15 = 5 \cdot 3$. The commutative property can be verified by recalling that $n(A \times B) = n(B \times A)$.

The associative property of multiplication of whole numbers can be illustrated as follows. Suppose $a = 3$, $b = 5$, and $c = 4$. In Figure 3-19(a), we see a picture of $3 \cdot (5 \cdot 4)$ blocks. In Figure 3-19(b), we see the same blocks, this time arranged as $(3 \cdot 5) \cdot 4$. Because both sets of blocks in Figure 3-19(a) and (b) compress to the set shown in Figure 3-19(c), we see that $3 \cdot (5 \cdot 4) = (3 \cdot 5) \cdot 4$. The associative property is useful in computations such as the following:

$$3 \cdot 40 = 3 \cdot (4 \cdot 10) = (3 \cdot 4) \cdot 10 = 12 \cdot 10 = 120.$$

FIGURE 3-19

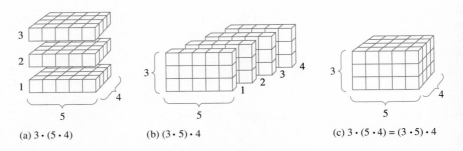

(a) $3 \cdot (5 \cdot 4)$ (b) $(3 \cdot 5) \cdot 4$ (c) $3 \cdot (5 \cdot 4) = (3 \cdot 5) \cdot 4$

The multiplicative identity for whole numbers is 1. For example, $3 \cdot 1 = 1 + 1 + 1 = 3$. In general, for any whole number a,

$$a \cdot 1 = \underbrace{1 + 1 + 1 + \cdots + 1}_{a \text{ terms}} = a.$$

Thus $a \cdot 1 = a$, which, along with the commutative property for multiplication, implies that $a \cdot 1 = a = 1 \cdot a$. Cartesian products can also be used to show that $a \cdot 1 = a = 1 \cdot a$.

Next, consider multiplication involving 0. For example, $6 \cdot 0 = 0 + 0 + 0 + 0 + 0 + 0 = 0$. Thus, we see that multiplying 0 by 6 yields a product of 0 and, by commutativity, $0 \cdot 6 = 0$. This is an example of the Zero Multiplication Property, which can also be verified by using the definition of multiplication in terms of Cartesian products. Let A be any set such that $n(A) = a$. Then, $a \cdot 0 = n(A \times \emptyset) = n(\emptyset) = 0$. Similarly, we can show that $0 \cdot a = 0$.

The Distributive Property of Multiplication over Addition

The next property we investigate is the basis for understanding multiplication algorithms. In Figure 3-20, $5 \cdot (3 + 4) = (5 \cdot 3) + (5 \cdot 4)$.

FIGURE 3-20

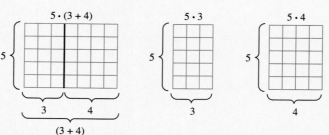

The properties of addition and multiplication also can be used to justify this result.

$5 \cdot (3 + 4) = \underbrace{(3 + 4) + (3 + 4) + (3 + 4) + (3 + 4) + (3 + 4)}_{5 \text{ terms}}$

Definition of multiplication

$= (3 + 3 + 3 + 3 + 3) + (4 + 4 + 4 + 4 + 4)$

Commutative and associative properties of addition

$= 5 \cdot 3 + 5 \cdot 4$

Definition of multiplication

This example illustrates the *distributive property of multiplication over addition* for whole numbers, which is stated in general as follows.

PROPERTY

Distributive Property of Multiplication over Addition for Whole Numbers For any whole numbers a, b, and c,

$$a \cdot (b + c) = a \cdot b + a \cdot c.$$

SECTION 3-3 MULTIPLICATION AND DIVISION OF WHOLE NUMBERS

> **REMARK** Because the commutative property of multiplication of whole numbers holds, the distributive property of multiplication over addition can be rewritten as $(b + c) \cdot a = b \cdot a + c \cdot a$. The distributive property can be generalized to any finite number of terms. For example, $a \cdot (b + c + d) = a \cdot b + a \cdot c + a \cdot d$.

Students find the distributive property of multiplication over addition useful when doing *mental mathematics*. For example,

$$11 \cdot 17 = (10 + 1) \cdot 17 = 10 \cdot 17 + 1 \cdot 17 = 170 + 17 = 187.$$

The distributive property is used to combine like terms when we work with variables; for example, $3ab + 2ab = (3 + 2) \cdot ab = 5ab$.

EXAMPLE 3-3 Rename each of the following using the distributive property:

(a) $3 \cdot (x + y)$
(b) $(x + 1) \cdot x$
(c) $3 \cdot (2x + y + 3)$
(d) $a \cdot x + a \cdot y$
(e) $a \cdot x + a$
(f) $(x + 2) \cdot 5 + (x + 2) \cdot a$

Solution
(a) $3 \cdot (x + y) = 3 \cdot x + 3 \cdot y = 3x + 3y$
(b) $(x + 1) \cdot x = x \cdot x + 1 \cdot x = x^2 + x$
(c) $3 \cdot (2x + y + 3) = 3 \cdot (2x) + 3 \cdot y + 3 \cdot 3 = (3 \cdot 2) \cdot x + 3 \cdot y + 9$
$= 6x + 3y + 9$
(d) $a \cdot x + a \cdot y = a \cdot (x + y) = a(x + y)$
(e) $a \cdot x + a = a \cdot x + a \cdot 1 = a \cdot (x + 1) = a(x + 1)$
(f) $(x + 2) \cdot 5 + (x + 2) \cdot a = (x + 2) \cdot (5 + a) = (x + 2)(5 + a)$ ∎

EXAMPLE 3-4 If $a \in W$ and $b \in W$, use the distributive property to write $(a + b)^2$ as a sum without parentheses.

Solution By the definition of exponents, $(a + b)^2 = (a + b)(a + b)$. We consider the first term, $(a + b)$, as a single whole number and apply the distributive property. Then, applying the commutative, associative, and distributive properties, we obtain the following:

$$\begin{aligned}(a + b)(a + b) &= (a + b)a + (a + b)b \\ &= (aa + ba) + (ab + bb) \\ &= (a^2 + ba) + (ab + b^2) \\ &= a^2 + (ba + ab) + b^2 \\ &= a^2 + (ab + ab) + b^2 \\ &= a^2 + (1 \cdot ab + 1 \cdot ab) + b^2 \\ &= a^2 + (1 + 1)ab + b^2 \\ &= a^2 + 2ab + b^2\end{aligned}$$
∎

The result in Example 3-4 can be demonstrated geometrically by using the fact that the area A of a rectangle is given by $A = l \cdot w$, where l is the length of the rectangle and w is the

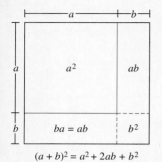

FIGURE 3-21

width. If we build a rectangle of length and width $(a + b)$ units, as shown in Figure 3-21, then it has area $(a + b)^2$. Notice that this figure is divided into four disjoint regions that make up the large square. From the figure we see that $(a + b)^2 = a^2 + ab + ab + b^2 = a^2 + 2ab + b^2$. Students often mistakenly think that $(a + b)^2 = a^2 + b^2$. From Figure 3-21, we see that this is not the case.

Order of Operations

Difficulties involving the order of arithmetic operations sometimes arise. For example, many students will treat $2 + 3 \cdot 6$ as $(2 + 3) \cdot 6$, while others will treat it as $2 + (3 \cdot 6)$. In the first case, the value is 30; in the second case, the value is 20. In order to avoid confusion, mathematicians agree that when no parentheses are present, multiplications are performed *before* additions. Thus, $2 + 3 \cdot 6 = 2 + 18 = 20$. This order of operations is not built into calculators that display an incorrect answer of 30.

Division of Whole Numbers

We discuss division using three models: the *set (partition)* model, the *missing-factor* model, and the *repeated-subtraction* model.

SET (PARTITION) MODEL

Suppose we have 18 cookies and want to give an equal number of cookies to each of three friends, Bob, Dean, and Charlie. How many should each person receive? If we draw a picture, we see that we can divide (or partition) the 18 cookies into three *sets*, with an equal number of cookies in each set. Figure 3-22 shows that each friend received 6 cookies.

FIGURE 3-22

The answer may be symbolized as $18 \div 3 = 6$. Thus, $18 \div 3$ is the number of cookies in each of three disjoint sets whose union has 18 cookies. In this approach to division, we partition a set into a number of equivalent subsets.

MISSING-FACTOR MODEL

Another strategy for dividing 18 cookies among 3 friends is to use the *missing-factor* model. If each friend receives c cookies, then the three friends receive $3 \cdot c$ cookies, or 18 cookies. Hence, $3 \cdot c = 18$. Since $3 \cdot 6 = 18$, then $c = 6$. We have answered the division computation by using multiplication. This leads us to the following definition of division of whole numbers.

DEFINITION OF DIVISION OF WHOLE NUMBERS

For any whole numbers a and b, with $b \neq 0$, $a \div b = c$ if, and only if, c is the unique whole number such that $b \cdot c = a$.

SECTION 3-3 MULTIPLICATION AND DIVISION OF WHOLE NUMBERS

dividend / divisor
quotient

> **REMARK** The number a is the **dividend,** b is the **divisor,** and c is the **quotient.** Note that $a \div b$ can also be written as $\frac{a}{b}$ or $b\overline{)a}$.

REPEATED-SUBTRACTION MODEL

Suppose we have 18 cookies and want to package them in cookie boxes that hold 6 cookies each. How many boxes are needed? We could reason that if one box is filled, then we would have $18 - 6$ (or 12) cookies left. If one more box is filled, then there are $12 - 6$ (or 6) cookies left. Finally, we could place the last 6 cookies in a third box. This discussion can be summarized by writing $18 - 6 - 6 - 6 = 0$. We have found by repeated subtraction that $18 \div 6 = 3$.

Calculators can be used to show that division of whole numbers can be thought of as repeated subtraction. For example, consider $135 \div 15$. If the calculator has a constant key, press $\boxed{1}\,\boxed{5}\,\boxed{-}\,\boxed{K}\,\boxed{1}\,\boxed{3}\,\boxed{5}\,\boxed{=}\ldots$, and then count how many times you must press the $\boxed{=}$ key in order to make the display read 0. Calculators with a different constant feature may require a different sequence of entries. For example, if the calculator has an automatic constant, we can press $\boxed{1}\,\boxed{3}\,\boxed{5}\,\boxed{-}\,\boxed{1}\,\boxed{5}\,\boxed{=}$ and then count the number of times we press the $\boxed{=}$ key to make the display read 0. Compare your answer with the one achieved by pressing this sequence of keys:

$$\boxed{1}\,\boxed{3}\,\boxed{5}\,\boxed{\div}\,\boxed{1}\,\boxed{5}\,\boxed{=}.$$

Division Algorithm

Figure 3-23 provides a visual model to show relationships among the four basic whole-number operations. Just as subtraction of whole numbers is not always meaningful, division of whole numbers is not always meaningful. For example, to find $383 \div 57$ we look for a whole number c such that $57 \cdot c = 383$.

FIGURE 3-23

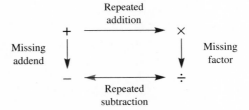

Table 3-7 shows several products of whole numbers times 57. Since 383 is between 342 and 399, there is no whole number c such that $57 \cdot c = 383$. Because no whole number c satisfies this equation, we see that $383 \div 57$ has no meaning in the set of whole numbers, and the set of whole numbers is not closed under division.

TABLE 3-7

$57 \cdot 1$	$57 \cdot 2$	$57 \cdot 3$	$57 \cdot 4$	$57 \cdot 5$	$57 \cdot 6$	$57 \cdot 7$
57	114	171	228	285	342	399

remainder

division algorithm

However, if 383 apples were to be divided among 57 students, each student would receive 6 apples, and 41 apples would remain. The number 41 is the **remainder.** Thus, 383 contains six 57s with a remainder of 41. Observe that the remainder is a whole number less than 57. The concept illustrated is the **division algorithm.**

DIVISION ALGORITHM

Given any whole numbers a and b with $b \neq 0$, there exist unique whole numbers q (quotient) and r (remainder) such that

$$a = b \cdot q + r \quad \text{with} \quad 0 \leq r < b.$$

REMARK The quotient q is the greatest whole number of b's in a.

EXAMPLE 3-5 If 123 is divided by a number and the remainder is 13, what are the possible divisors?

Solution From the division algorithm, we have

$$123 = b \cdot q + 13 \quad \text{and} \quad b > 13.$$

Using the definition of subtraction, we have $bq = 123 - 13$, and hence $110 = b \cdot q$. Now we are looking for two numbers whose product is 110, where one number is greater than 13. Table 3-8 shows the pairs of factors of 110.

We see that 110, 55, and 22 are possible divisors because each is greater than 13. The numbers 1, 2, 5, 10, and 11 cannot be divisors. ∎

TABLE 3-8

1	110
2	55
5	22
10	11

Division by 0 and 1

The whole numbers 0 and 1 deserve special attention with respect to division of whole numbers. Before reading on, try to find the values of the following three expressions:

1. $3 \div 0$ 2. $0 \div 3$ 3. $0 \div 0$

Consider the following explanations:

1. By definition, $3 \div 0 = c$ if there is a unique number c such that $0 \cdot c = 3$. Since the zero property of multiplication states that $0 \cdot c = 0$ for any whole number c, there is no whole number c such that $0 \cdot c = 3$. Thus, $3 \div 0$ is undefined because there is no answer to the equivalent multiplication problem.

2. By definition, $0 \div 3 = c$ if there exists a unique number c such that $3 \cdot c = 0$. The zero property of multiplication states that any number times 0 is 0. Since $3 \cdot 0 = 0$, then $c = 0$ and $0 \div 3 = 0$. Note that $c = 0$ is the only number that satisfies $3 \cdot c = 0$.

3. By definition, $0 \div 0 = c$ if there is a unique whole number c such that $0 \cdot c = 0$. Notice that, for *any* c, $0 \cdot c = 0$. According to the definition of division, c must be unique. Since there is no *unique* number c such that $0 \cdot c = 0$, it follows that $0 \div 0$ is indeterminate, or undefined.

Division involving 0 may be summarized as follows.
Let n be any natural number. Then

1. $n \div 0$ is undefined;
2. $0 \div n = 0$; and
3. $0 \div 0$ is indeterminate, or undefined.

Recall that $n \cdot 1 = n$ for any whole number n. Thus, by the definition of division, $n \div 1 = n$. For example, $3 \div 1 = 3$, $1 \div 1 = 1$, and $0 \div 1 = 0$.

PROBLEM SET 3-3

1. Use the number-line model to illustrate why $3 \cdot 5 = 15$.
2. Each ticket to the band concert costs $2.00. Each ticket to the football game costs $5.00. Jim bought 5 tickets to each event. What was his total bill?
3. Tell whether the following sets are closed under multiplication. If not, give a counterexample.
   (a) $\{0,1\}$
   (b) $\{0\}$
   (c) $\{2, 4, 6, 8, 10, \ldots\}$
   (d) $\{1, 3, 5, 7, 9, \ldots\}$
   (e) $\{1, 4, 7, 10, 13, \ldots\}$
   (f) $\{0, 1, 2\}$
4. ▶(a) If 5 is removed from the set of whole numbers, is the set closed with respect to addition?
   (b) If 5 is removed from the set of whole numbers, is the set closed with respect to multiplication?
   (c) Answer the same questions as (a) and (b) if 6 is removed from the set of whole numbers.
5. ▶Discuss the claim "If a set of whole numbers is closed under multiplication, then it is closed under addition."
6. Use the distributive property to describe how you might find the product $8 \cdot 3$ if you know only the addition table and the 2 and 6 multiplication facts.
7. Identify the property being illustrated in each of the following:
   (a) $3 \cdot 2 = 2 \cdot 3$
   (b) $3(2 \cdot 4) = (3 \cdot 2)4$
   (c) $3(2 + 3) = 3(3 + 2)$
   (d) $8 \cdot 0 = 0 = 0 \cdot 8$
   (e) $1 \cdot 8 = 8 = 8 \cdot 1$
   (f) $6(3 + 5) = (3 + 5)6$
   (g) $6(3 + 5) = 6 \cdot 3 + 6 \cdot 5$
   (h) $(3 + 5)6 = 3 \cdot 6 + 5 \cdot 6$
8. For each of the following, find, if possible, the whole numbers that make the equations true.
   (a) $3 \cdot \square = 15$
   (b) $18 = 6 + 3 \cdot \square$
   (c) $\square \cdot (5 + 6) = \square \cdot 5 + \square \cdot 6$
9. Rename each of the following, using the distributive property for multiplication over addition so that there are no parentheses in the final answer:
   (a) $(a + b)(c + d)$
   (b) $3(x + y + 5)$
   (c) $\square(\triangle + \bigcirc)$
   (d) $(x + y)(x + y + z)$
10. Perform each of the following computations.
    (a) $2 \cdot 3 + 5$
    (b) $2(3 + 5)$
    (c) $2 \cdot 3 + 2 \cdot 5$
    (d) $3 + 2 \cdot 5$
11. Place parentheses, if needed, to make each of the following equations true:
    (a) $4 + 3 \times 2 = 14$
    (b) $9 \div 3 + 1 = 4$
    (c) $5 + 4 + 9 \div 3 = 6$
    (d) $3 + 6 - 2 \div 1 = 7$
12. The generalized distributive property for three terms states that, for any whole numbers a, b, c, and d, $a(b + c + d) = ab + ac + ad$. Justify this property, using the distributive property for two terms.
13. Which of the following, if any, are true?
    (a) $(3 \cdot 5) \cdot 7 = 3 \cdot (5 \cdot 7)$
    (b) $3 \cdot (7 + 8) = (3 \cdot 7) + 8$
    (c) $3 \cdot (7 + 5) = 3 \cdot 7 + 5$
    (d) $5 \cdot (3 \cdot 7) = (5 \cdot 3) \cdot (5 \cdot 7)$
14. The FOIL method is often used as a shortcut to multiply expressions like $(m + n)(x + y)$.

    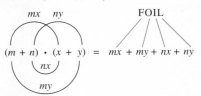

    where F stands for product of the *first* terms, mx
    O stands for product of the *outer* terms, my
    I stands for product of the *inner* terms, nx
    L stands for product of the *last* terms, ny.

    Use the distributive property to show why the FOIL method works.
15. For each of the following, find whole numbers to make the statement true, if possible.
    (a) $18 \div 3 = \square$
    (b) $\square \div 76 = 0$
    (c) $28 \div \square = 7$
16. Illustrate geometrically each of the following, using the concept of area.
    (a) $a \cdot (b + c) = ab + ac$
    (b) $(a + b) \cdot (c + d) = ac + ad + bc + bd$

17. Rewrite each of the following division problems as a multiplication problem.
    (a) $40 \div 8 = 5$
    (b) $326 \div 2 = x$
    (c) $48 \div x = 16$
    (d) $x \div 5 = 17$
18. ▶Think of a number. Multiply it by 2. Add 2. Divide by 2. Subtract 1. How does the result compare with your original number? Will this work all the time? Explain your answer.
19. Write $+, -, \times$, or $\div$ in the circles to make each equation true.
    (a) $(5 \bigcirc 2) \bigcirc 6 = 16$
    (b) $(5 \bigcirc 3) \bigcirc 5 = 40$
    (c) $(15 \bigcirc 3) \bigcirc 4 = 1$
    (d) $(6 \bigcirc 3) \bigcirc (5 \bigcirc 3) = 4$
20. Show that, in general, each of the following is false if a, b, and c are whole numbers:
    (a) $a \div b = b \div a$
    (b) $(a \div b) \div c = a \div (b \div c)$
    (c) $a \div (b + c) = (a \div b) + (a \div c)$
    (d) $a \div b$ is a whole number.
21. Because the Jones' water meter was stuck, they were billed the same amount for water each month for 5 mo. If they paid $160, what was the monthly bill?
22. There were 17 sandwiches for 7 people on a picnic. How many whole sandwiches were there for each person if they were divided equally? How many were left over?
23. If it takes 1 min per cut, how long will it take to cut a 10-ft log into 10 equal pieces?
24. For each of the following, name all the possible whole-number replacements for □ and △:
    (a) $34 = \square \cdot 8 + \triangle$ (b) $\triangle = 4 \cdot 16 + 2$
25. Find all the pairs of whole numbers whose product is 36.
26. A new model of a car is available in 4 different exterior colors and 3 different interior colors. How many different color schemes are possible for the car?
27. Tony has 5 ways to get from his home to the park. He has 6 ways to get from the park to the school. How many ways can Tony get from his home to school by way of the park?
28. Students were divided into 8 teams with 9 on each team. Later the same students were divided into teams with 6 on each team. How many teams were there then?
29. ▶How is division related to multiplication?
30. ▶Sue claims the following is true by the distributive law, where a and b are whole numbers:
    $$3(ab) = (3a)(3b).$$
    How might you help her?

★31. The binary operation $\odot$ is defined on the set $S = \{a, b, c\}$, as shown in the following table. For example, $a \odot b = b$ and $b \odot a = b$.

$\odot$	a	b	c
a	a	b	c
b	b	a	c
c	c	c	c

   (a) Is S closed with respect to $\odot$?
   (b) Is $\odot$ commutative on S?
   (c) Is there an identity for $\odot$ on S? If yes, what is it?
   (d) Is $\odot$ associative on S?

32. 🖩 To find $7 \div 5$ on the calculator, press $\boxed{7}\boxed{\div}\boxed{5}\boxed{=}$, which yields 1.4. To find the whole-number remainder, ignore the decimal portion of 1.4, multiply $5 \cdot 1$, and subtract this product from 7. The result is the remainder. Use a calculator to find the whole-number remainder for each of the following divisions:
    (a) $28 \div 5$
    (b) $32 \div 10$
    (c) $29 \div 3$
    (d) $41 \div 7$
    (e) $49{,}382 \div 14$
33. 🖩 In the problems that follow, use only the designated number keys. Use any function keys on the calculator.
    (a) Use the keys $\boxed{1}, \boxed{9}$, and $\boxed{7}$ exactly once each in any order, and use any operations available to write as many of the whole numbers as possible from 1 to 20. For example, $9 - 7 - 1 = 1$ and $1 \cdot 9 - 7 = 2$.
    (b) Use the $\boxed{4}$ key as many times as desired with any operations to display 13.
    (c) Use the $\boxed{2}$ key three times with any operations to display 24.
    (d) Use the $\boxed{1}$ key five times with any operations to display 100.

Review Problems

34. Write 75, using Egyptian, Roman, and Babylonian numerals.
35. Write 35,206 in expanded form.
36. Give a set that is not closed under addition.
37. Are the whole numbers commutative under subtraction? If not, give a counterexample.
38. Illustrate $11 - 3$, using a number-line model.

BRAIN TEASER

Rosalie bought a bike for $50 and sold it for $60. Then she bought it back for $70 and sold it for $80. What is the financial outcome of these transactions?

LABORATORY ACTIVITY

Enter a number less than 20 on the calculator. If the number is even, divide it by 2; if it is odd, multiply it by 3 and add 1. Next, use the number on the display. Follow the given directions. Repeat the process again.

1. Will the display eventually reach 1?
2. Which number less than 20 takes the most steps before reaching 1?
3. Do even or odd numbers reach 1 more quickly?
4. Investigate what happens with numbers greater than 20.

Section 3-4 Algorithms, Mental Math, and Estimation for Whole-number Addition and Subtraction

It is no longer necessary or advisable to devote large portions of instructional time to paper-and-pencil computation. Computational ability is still important, but the role of technology, mental mathematics, and estimation must be considered. Knowledge of basic facts is necessary; for example, students still need to know that $9 + 8 = 17$ in order to compute $900 + 800$ mentally or to estimate $889 + 797$. When computing taxes, however, a person should probably use a calculator or computer. In determining a tip in a restaurant, mental mathematics is appropriate. If you need to find the total number of students in four sixth-grade classes, pencil and paper might be the most appropriate method.

In previous sections, the definitions of addition and subtraction were introduced. These definitions, along with a knowledge of basic facts and properties, are necessary to perform more complex additions and subtractions. More complex operations are commonly done by applying various algorithms. An **algorithm** (named for the ninth-century Arabian mathematician Mohammed al-Khowârizmî) is a step-by-step systematic procedure used to accomplish an operation. Every prospective elementary school teacher should know more than one algorithm for doing operations. Not all students learn in the same manner, and the shortest, most efficient algorithms may not be the best for every individual. In the *Standards* (p. 8), we find the following with regard to algorithms: *Similarly, the availability of calculators does not eliminate the need for students to learn algorithms. Some proficiency with paper-and-pencil computational algorithms is important, but such knowledge should grow out of problem situations that have given rise to the need for such algorithms.*

algorithm

Addition Algorithms

Paper-and-pencil algorithms need to be taught developmentally; that is, they must proceed from the concrete stage to the abstract stage at appropriate times. The use of concrete teaching aids—such as chips, bean sticks, an abacus, or base-ten blocks—helps provide insight into the creation of algorithms for addition. A set of base-ten blocks, shown in Figure 3-24, consists of *units, longs, flats,* and *blocks*, representing 1, 10, 100, and 1000, respectively.

FIGURE 3-24

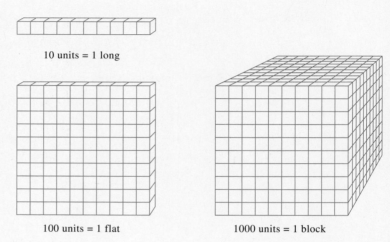

Students trade blocks by regrouping; that is, they take a set of base-ten blocks representing a number and trade them until they have the fewest possible pieces representing the same number. For example, suppose you have 58 units and want to trade them. What pieces do you have if you have the smallest number of pieces you can receive in a fair exchange? The units can be grouped into tens to form longs. Five sets of 10 units each can be traded for 5 longs. Thus, 58 units can be traded for 5 longs and 8 units. In terms of numbers, this is analogous to rewriting 58 as $5 \cdot 10 + 8$. In this case, you cannot receive flats or blocks.

EXAMPLE 3-6 What is the fewest number of pieces you can receive in a fair exchange for 11 flats, 17 longs, and 16 units?

Solution The 16 units can be traded for 1 long and 6 units.

11 flats	17 longs	16 units	(16 units = 1 long and 6 units)
	1 long	6 units	(Trade)
11 flats	18 longs	6 units	(After the first trade)
11 flats	18 longs	6 units	(18 longs = 1 flat and 8 longs)
1 flat	8 longs		(Trade)
12 flats	8 longs	6 units	(After the second trade)
	12 flats	8 longs	6 units	(12 flats = 1 block and 2 flats)
1 block	2 flats			(Trade)
1 block	2 flats	8 longs	6 units	(After the third trade)

SECTION 3-4 ALGORITHMS FOR WHOLE-NUMBER ADDITION AND SUBTRACTION

In terms of numbers, this is analogous to rewriting $11 \cdot 10^2 + 17 \cdot 10 + 16$ as $1 \cdot 10^3 + 2 \cdot 10^2 + 8 \cdot 10 + 6$, which implies that there are 1286 units. ∎

We now use base-ten blocks to help develop an algorithm for whole-number addition. Suppose we wish to add $14 + 23$. We show this computation with a concrete model in Figure 3-25(a), with an introductory algorithm in Figure 3-25(b), and with the familiar algorithm in Figure 3-25(c).

FIGURE 3-25

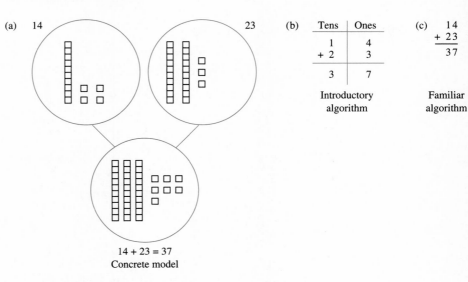

A formal justification for this addition is the following:

$14 + 23 = (1 \cdot 10 + 4) + (2 \cdot 10 + 3)$	Expanded form
$= (1 \cdot 10 + 2 \cdot 10) + (4 + 3)$	Commutative and associative properties of addition
$= (1 + 2) \cdot 10 + (4 + 3)$	Distributive property of multiplication over addition
$= 3 \cdot 10 + 7$	Single-digit addition facts
$= 37$	Place value

> **REMARK** Some texts show the partial sums in the computation as follows:
>
> $$\begin{array}{r} 14 \\ +\ 23 \\ \hline 7 \\ 30 \\ \hline 37 \end{array}$$

Although this mathematical justification is not usually presented at the elementary school level, the ideas and properties shown are necessary to understand why the algorithm works. Some problems are more involved than this one because they involve "regrouping'' or ''carrying,'' in which students trade by regrouping. This is described in terms of the base-ten blocks on a student page from *Addison-Wesley Mathematics,* 1991, Grade 2.

Adding
Trading 10 Tens for 1 Hundred

Find the sum of 143 and 82. Use blocks to help.

1. Add the ones. Can you trade? yes no

   What you do.   What you write.

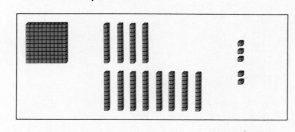 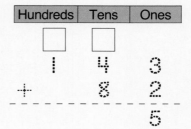

2. Add the tens. Can you trade? yes no

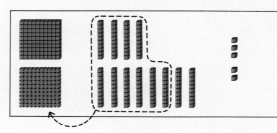

3. Add the hundreds.

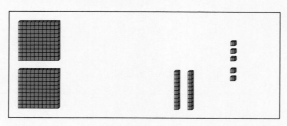

SECTION 3-4 ALGORITHMS FOR WHOLE-NUMBER ADDITION AND SUBTRACTION

After using concrete aids, children are ready to complete a computation such as 28 + 34. Figure 3-26(a) and (b) show introductory algorithms, whereas Figure 3-26(c) shows the traditional algorithm.

FIGURE 3-26

(a)
Tens	Ones
2	8
+3	4
5	$\cancel{1}2$	(Add)
+1	2	(Regroup)
6	2

(b)
```
   2 8
 + 3 4
 ─────
   1 2    (Sum of ones)
 + 5      (Sum of tens)
 ─────
   6 2
```

(c)
```
    1
   2 8
 + 3 4
 ─────
   6 2
```

Scratch Addition

scratch addition   The **scratch addition** algorithm allows students to do complicated additions by doing a series of additions involving only two single digits. An example follows.

1. $\begin{array}{r} 87 \\ 65_2 \\ + 49 \\ \hline \end{array}$ Add the numbers in the units place starting at the top. When the sum is 10 or more, record this sum by scratching a line through the last digit added and writing the number of units next to the scratched digit. For example, since $7 + 5 = 12$, the "scratch" represents 10 and the 2 represents the units.

2. $\begin{array}{r} 87 \\ 65_2 \\ + 49_1 \\ \hline \end{array}$ Continue adding the units, including any new digits written down. When the addition again results in a sum of 10 or more, as with $2 + 9 = 11$, repeat the process described in (1).

3. $\begin{array}{r} ^2 87 \\ 65_2 \\ + 49_1 \\ \hline 1 \end{array}$ When the first column of additions is completed, write the number of units, 1, below the addition line. Count the number of scratches, 2, and add this number to the second column.

4. $\begin{array}{r} ^2 8_0 7 \\ 6\ 5_2 \\ 4_0 9_1 \\ \hline 2\ 0\ 1 \end{array}$ Repeat the procedure for each successive column.

E X A M P L E   3 - 7   Compute these additions using the scratch algorithm:

(a)
```
   296
   840
 +  27
```

(b)
```
   1369
   4813
   5879
 + 6183
```

138   CHAPTER 3   NUMERATION SYSTEMS AND WHOLE NUMBERS

Solution   (a) ${}^{1}2\ {}^{1}\cancel{9}0\ 6$ (b) ${}^{2}1\ {}^{2}3\ {}^{2}6\ 9$
$\cancel{8}_1\ 4\ 04\ \cancel{8}\ 1\ \cancel{3}_2$
$+2\ \cancel{7}_3\cancel{5}\ \cancel{8}^3\ \cancel{7}\ \cancel{9}_1$
$\overline{1\ 1\ 6\ 3}+6^2\ 1^1\ \cancel{8}_4^6\ 3$
$\overline{1\ 8\ 2\ 4\ 4}$ ■

Mental Mathematics: Addition

Mental mathematics is an important tool in elementary schools. *Mental mathematics is the process of producing an exact answer to a computation without using external computational aids.* As stated in the *Standards* (p. 45). *Children need more time to explore and to invent alternative strategies for computing mentally. Both mental computation and estimation should be ongoing emphases that are integrated throughout all computational work.*

Several ways of performing mental addition are given next.

1. *Adding from the left*

   (a) $\begin{array}{r}67\\+36\\\hline\end{array}$ $\begin{array}{l}60+30=90\\7+6=13\\90+13=103\end{array}$ Add the tens.
   Add the units.
   Add the two sums.

   (b) $\begin{array}{r}36\\+36\\\hline\end{array}$ $\begin{array}{l}30+30=60\\6+6=12\\60+12=72\end{array}$ Double 30.
   Double 6.
   Add the doubles.

2. *Breaking up and bridging*

   $\begin{array}{r}67\\+36\\\hline\end{array}$ $\begin{array}{l}67+30=97\\97+6=103\end{array}$ Add the first number to the tens in the second number.
   Add this sum to the units in the second number.

3. *Trading off*

   (a) $\begin{array}{r}67\\+36\\\hline\end{array}$ $\begin{array}{l}67+3=70\\36-3=33\\70+33=103\end{array}$ Add 3 to make a multiple of 10.
   Subtract 3 to compensate for the 3 that was added.
   Add the two numbers.

   (b) $\begin{array}{r}67\\+29\\\hline\end{array}$ $\begin{array}{l}67+30=97\\97-1=96\end{array}$ Add 30 (next multiple of 10 greater than 29).
   Subtract 1 to compensate for the extra 1 that was added.

4. *Using compatible numbers*

   Compatible numbers are numbers whose sums are easy to calculate mentally.

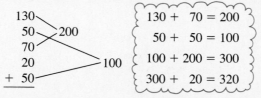

5. *Making compatible numbers*

$$\begin{array}{r} 25 \\ +79 \\ \hline \end{array} \quad \begin{array}{l} 25 + 75 = 100 \\ 100 + 4 = 104 \end{array} \quad \begin{array}{l} 25 + 75 \text{ adds to } 100. \\ \text{Add 4 more units.} \end{array}$$

PROBLEM 1

A wise man received a reward from a wealthy king. The king announced he would grant the man any wish. The wise man requested the amount of wheat that would be required to cover a checkerboard beginning by placing 1 grain on the first square, 2 grains on the second square, 4 grains on the third square, 8 grains on the fourth square, and so on, always doubling the number of grains on successive squares. The king quickly agreed to this modest-sounding request. How many grains of wheat did the wise man request?

Understanding the Problem. We must determine the total number of grains of wheat obtained by placing 1 grain on the first square of a checkerboard, 2 grains on the second square, 4 grains on the third square, and so on, doubling the number of grains each time, until all 64 squares on the checkerboard have been accounted for.

Devising a Plan. We *reduce the problem to some simpler cases* of boards with 4 and 9 squares, as shown in Figure 3-27. We hope to gain insight into how to solve the more difficult problem.

FIGURE 3-27

The 2×2 board holds $1 + 2 + 2^2 + 2^3 = 15$ grains, and the 3×3 board holds $1 + 2 + 2^2 + 2^3 + 2^4 + 2^5 + 2^6 + 2^7 + 2^8 = 511$ grains. Because 15 is one less than 16 (or 2^4) and 511 is one less than 512 (or 2^9), it appears that the number of grains on the 8×8 checkerboard is $1 + 2 + 2^2 + 2^3 + 2^4 + \cdots + 2^{63} = 2^{64} - 1$.

Carrying Out the Plan. The number of grains of wheat to be calculated is $1 + 2 + 2^2 + 2^3 + 2^4 + \cdots + 2^{63}$. A justification that this sum is $2^{64} - 1$ follows:

$$S = 1 + 2 + 2^2 + 2^3 + \cdots + 2^{62} + 2^{63}.$$

Multiply both sides of this equation by 2 to obtain

$$2S = 2 + 2^2 + 2^3 + 2^4 + \cdots + 2^{63} + 2^{64}.$$

Subtracting S from $2S$, we obtain the following:

$$\begin{aligned} 2S - S &= (2 + 2^2 + 2^3 + \cdots + 2^{64}) - (1 + 2 + 2^2 + \cdots + 2^{63}) \\ &= (2 - 2) + (2^2 - 2^2) + (2^3 - 2^3) + \cdots + (2^{63} - 2^{63}) + 2^{64} - 1 \\ &= 2^{64} - 1 \end{aligned}$$

How large a number is this? For estimating, a useful fact is that 2^{10} is approximately 1000. Therefore $2^{64} = 2^4 \cdot 2^{60} = 2^4 \cdot (2^{10})^6$, which is approximately $2^4 \cdot (10^3)^6$, or $16 \cdot 10^{18}$, which is 16 billion billion, or 16 quintillion. On a calculator, if we find 2^{64} using the $\boxed{y^x}$ key, the value $\boxed{1.8447 \quad 19}$ is displayed. Remember, this number is in scientific notation and represents $1.8447 \cdot 10^{19}$ or $18,447,000,000,000,000,000$. The exact value of 2^{64} is 18,446,744,073,709,551,616. *We obtain only an approximation on the calculator.* Assuming that the current production remains the same, it would take the United States approximately 14,000 yr to produce the amount of wheat necessary to place on the checkerboard.

Looking Back. The size of the checkerboard could be varied to introduce more problems. Questions concerning the length of time required to count the grains or the amount of storage space required to store the grain could also be asked. We might also compute the number of grains if each square contains three times as many grains as the previous square, or, in general, n times as many as the previous square, where n is a positive integer. ∎

Subtraction Algorithms

As with addition, base-ten blocks can provide a concrete model for subtraction. Consider $36 - 24$. We do this computation in Figure 3-28(a), using base-ten blocks; in Figure 3-28(b), using an introductory algorithm based on the blocks; and finally in Figure 3-28(c), using the familiar algorithm. The slashes through the blocks show the ones taken away.

FIGURE 3-28

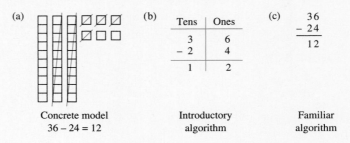

Notice that this subtraction problem can be checked by using the definition of subtraction: $36 - 24 = 12$, because $12 + 24 = 36$.

Subtractions become more involved when regrouping is necessary, as in $56 - 29$. In the concrete model, 9 units cannot be taken from 6 units so 1 long must be traded for 10 units, giving a total of 16. The three stages for working this problem are shown in Figure 3-29.

FIGURE 3-29

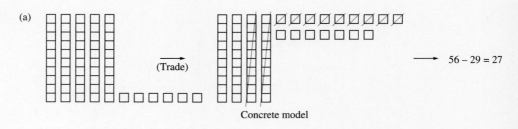

Mental Mathematics: Subtraction

1. *Subtracting in parts*

   $\begin{array}{r} 67 \\ -36 \\ \hline \end{array}$ $67 - 30 = 37$ Subtract the tens in the second number from the first number.

   $37 - 6 = 31$ Subtract the units in the second number from the difference.

2. *Making an easier problem*

   $\begin{array}{r} 71 \\ -39 \\ \hline \end{array}$ $(71 + 1) = 72; (39 + 1) = 40$ Add 1 to both numbers. Perform the subtraction, which is easier than the original problem.

   $(72 - 40) = 32$

   (Notice that adding 1 to both numbers does not change the answer. Why?)

3. *Drop the zeros*

   $\begin{array}{r} 8700 \\ -500 \\ \hline \end{array}$ $87 - 5 = 82$ Notice that there are two zeros in each number. Drop these zeros and perform the computation. Then replace the two zeros to obtain proper place value.

   $82 \rightarrow 8200$

Another mental mathematics technique for subtraction is called "adding up." This method is based on the *missing addend* approach and is sometimes referred to as the "cashier's algorithm." An example of the cashier's algorithm follows.

EXAMPLE 3-8   Noah owed $11 for his groceries. He used a $50 check to pay the bill. While handing Noah the change, the cashier said, "$11, $12, $13, $14, $15, $20, $30, $50." How much change did Noah receive?

Solution   Table 3-9 shows what the cashier said and how much money Noah received each time. Since $11 plus $1 is $12, Noah must have received $1 when the cashier said $12. The same reasoning follows for $13, $14, and so on. Thus, the total amount of change that Noah received is given by $1 + $1 + $1 + $1 + $5 + $10 + $20 = $39. In other words, $50 - $11 = $39 because $39 + $11 = $50.

TABLE 3-9

What the cashier said	$11	$12	$13	$14	$15	$20	$30	$50
Amount of money Noah received each time	0	$1	$1	$1	$1	$5	$10	$20

Computational Estimation

Computational estimation is the process of forming an approximate answer to a numerical problem. Computational estimation is useful in determining whether an answer is reasonable when the computation is done on a calculator. As reflected in the *Standards* (p. 37), *Continual emphasis on computational estimation helps children develop creative and flexible thought processes and fosters in them a sense of mathematical power.*

142 CHAPTER 3 NUMERATION SYSTEMS AND WHOLE NUMBERS

The *front-end* computational estimation strategy is demonstrated on the student page from *Addison-Wesley Mathematics*, 1991, Grade 5.

Front-End Estimation

LEARN ABOUT IT

You can use mental math to estimate sums.

EXPLORE Solve to Understand
- How many pairs of numbers can you find on the bulletin board that have a sum within 10 of the target number 100?
- List the target pairs and tell whether each sum is more or less than 100.

TALK ABOUT IT

1. How would you find a number that forms a target pair with 32?
2. How can you see that 55 and 59 do not form a target pair?
3. Can you look only at the tens digits to find target pairs? Explain.

Front-end estimation is a method of estimation where we add the digits with the greatest place value (the front-end digits) to get a rough estimate. Then we use the rest of the numbers to adjust the estimate.

Add the front-end digits. → Adjust using the other digits.

$$\begin{array}{r}3\!\!\mid\!96\\2\!\!\mid\!63\\+5\!\!\mid\!37\end{array}\quad 1{,}000+\quad\begin{array}{r}3\,9\!\!\mid\!6\\2\,6\!\!\mid\!3\\+5\,3\!\!\mid\!7\end{array}\quad\text{about 200 more}$$

Estimate: 1,200

TRY IT OUT

Use the front-end digits to estimate. Tell whether the exact sum or difference is more than or less than your estimate.

1. 475
   +420

2. 934
   −306

3. 508
   −259

4. 820
   470
   +381

A summary of some of the common estimation strategies is given next.

1. *Front-end strategy*

   Suppose we are to add the following column of figures. To obtain an estimate, we focus on the "front-end," or left-most, digits, which are the most significant. Front-end estimation is a two-step process.

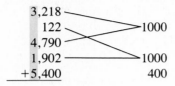

   1. Find the total of the most important lead digits, and use place value.

      $3 + 4 + 1 + 5 = 13$

      Place value gives 13,000.

   2. Adjust the estimate: $218 + 790$ is about 1000; $122 + 902$ is about 1000; plus 400, so about 2400.

   Hence, an estimate would be $13,000 + 2400$, or 15,400.

2. *Grouping to nice numbers strategy*

   The strategy used to obtain the adjustment in the preceding example is the *grouping to nice numbers* strategy, which means that numbers that "nicely" fit together are grouped. Another example is given here.

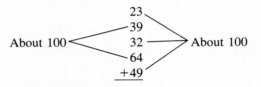

   Therefore, the sum is about $100 + 100$, or 200.

3. *Clustering strategy*

   Clustering is used when a group of numbers cluster around a common value. This strategy is limited to certain kinds of computations. In the next example, the numbers seem to cluster around 6000.

   ```
   6200
   5842
   6512
   5521
   +6319
   ```

   1. Estimate the "average"—about 6000.
   2. Multiply the "average" by the number of values to obtain

      $5 \cdot 6000 = 30,000$.

4. *Rounding strategy*

   Rounding is a way of cleaning up numbers so that they are easier to handle. Rounding enables us to find approximate answers to calculations, as follows:

   (a) 4724   5000   Round 4724 to 5000.
       +3192   +3000   Round 3192 to 3000.
                    8000   Add the rounded numbers.

   (b) 1267   1300   Round 1267 to 1300.
       - 510   - 500   Round 510 to 500.
                    800   Subtract the rounded numbers.

   Performing estimations requires a knowledge of place value and rounding techniques. We illustrate a rounding procedure that can be generalized to all rounding situations. For example, suppose we wish to round 4724 to the nearest thousand. We may proceed in four steps (see also Figure 3-30).

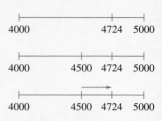

1. Determine between which two consecutive thousands the number lies.
2. Determine the midpoint between the thousands.
3. Determine which thousand the number is closer to, by observing whether it is greater than or less than the midpoint. (Not all texts use the same rule for rounding when a number falls at a midpoint.)

FIGURE 3-30

4. If the number to be rounded is greater than or equal to the midpoint, round the given number to the greater thousand; otherwise, round to the lesser thousand. In this case, we round 4724 to 5000.
5. *Range Strategy*
   It is often useful to know into what *range* an answer falls. The range is determined by finding a low estimate and a high estimate and reporting the answer falls in this interval. An example follows:

Problem	Low Estimate	High Estimate
378	300	400
+524	+500	+600
	800	1000

Thus a range for this problem is from 800 to 1000.

We have discussed calculators, pencil-and-paper computations, mental mathematics, and estimation. How are these related? The *Standards* gives a chart similar to Figure 3-31 to show how decisions about calculation procedures in numerical problems may be made.

FIGURE 3-31

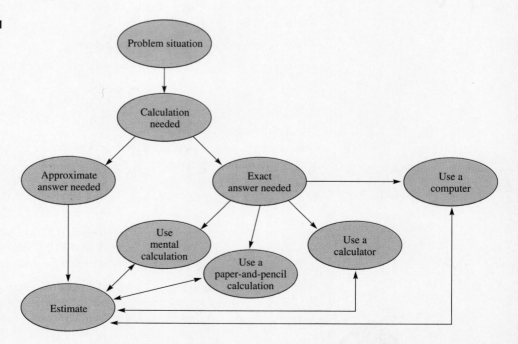

BRAIN TEASER

The number on a license plate consists of five digits. When the license plate is looked at upside down, you can still read it, but the value of the upside-down number is 78,633 greater than the real license number. What is the license number?

PROBLEM SET 3-4

1. Perform the following additions, using both the scratch and conventional algorithms. Use estimations to determine if your answers are reasonable.
   (a) 3789
        9296
       +6843

   (b) 524
        328
        567
       +135

2. ▶Explain why the scratch addition algorithm works.

3. ▶An addition algorithm from an elementary school text follows. Explain why it works.

2	7
+6	8
1	5
8	
9	5

4. Find the missing numbers in each of the following:
   (a) _ _ 1
       + 4 2 _
        _ 4 0 2

   (b) _ 0 2 5
        1 1 _ 6
       +3 1 4 8
        6 _ 6 _

   (c) 1 _ 6 9
        2 _ 9 4
        9 5 4 6
        9 _ _ 3
       + 7 _ 6 4
        2 8 7 7 6

   (d) 2 _ 1
        4 5 _
       + _ 8 4
        1 3 2 6

5. Find the missing numbers in each of the following:
   (a) 8 7 6 9 3
       - _ _ _ _ _
        4 1 2 7 9

   (b) 8 1 3 5
       -4 6 8 2
        _ _ _ _

   (c) 3 _ _
       -1 5 9
        _ 2 4

   (d) 1 _ _ _ 6
       -   8 3 0 9
            4 9 8 7

6. Place the digits 7, 6, 8, 3, 5, and 2 in the boxes to obtain:
   (a) the greatest sum;   (b) the least sum.

   ☐ ☐ ☐
   +☐ ☐ ☐

7. Place the digits 7, 6, 8, 3, 5, and 2 in the boxes to obtain:
   (a) the greatest difference;   (b) the least difference.

   ☐ ☐ ☐
   -☐ ☐ ☐

8. At the beginning of the year, the library had 15,282 books. During fall quarter, 125 books were added; during winter quarter, 137 were added; and during spring quarter, 238 were added. How many books did the library have at the end of the school year?

9. Find the next three numbers in each of the sequences given below:
   (a) 9, 14, 19, 24, 29
   (b) 97, 94, 91, 88, 85

10. Maria goes into a store with 87¢. If she buys a granola bar for 25¢, a balloon for 15¢, and a comb for 17¢, how much money does she have left?

11. Tom's diet allows only 1500 calories per day. For breakfast, Tom had skim milk (90 calories), a waffle with no syrup (120 calories), and a banana (119 calories). For lunch, he had $\frac{1}{2}$ cup of salad (185 calories) with mayonnaise (110 calories), and tea (0 calories), and then he "blew it" with pecan pie (570 calories). Can he have dinner consisting of steak (250 calories), a salad with no mayonnaise, and tea?

12. With three girls on a large scale, the scale read 170 pounds. When Molly stepped off, the scale read 115 pounds. When Karly stepped off, leaving only Samantha, the scale read 65 pounds. What is the weight of each of the three girls?

13. Wally the waiter kept track of last week's money transactions. His salary was $150 plus $54 in overtime and $260 in tips. His transportation expenses were $22, his food expenses were $60, his laundry costs were $15, his entertainment expenditures amounted to $58, and his rent was $185. Did he save any money last week? If so, how much?

14. In the problem below, the sum is correct but the order of the numbers in each addend has been scrambled. Correct the addends to obtain the correct sum.

    2 8 3 4
   +6 3 1 5
    9 0 5 9

    ☐ ☐ ☐ ☐
   +☐ ☐ ☐ ☐
    9 0 5 9

15. (a) Would the clustering strategy of estimation be a good one to use in each of the following cases? Why or why not?

    (i)   474
          1467
            64
         +2445

    (ii)   483
           475
           530
           503
          +528

(b) Estimate each part of 15(a), using:
  (i) The front-end method.
  (ii) Grouping to nice numbers.
  (iii) Rounding.

16. Dana obtained the following results for boxes of Girl Scout cookies sold for the week. She estimated total sales at 400 boxes. Do you think her estimate is too high or too low? Why?

Monday	38
Tuesday	92
Wednesday	74
Thursday	17
Friday	130

17. If 1 mo is approximately 4 wk and 1 yr is approximately 365 days or 52 wk, answer the following:
  (a) Lewis and Clark spent approximately 2 yr, 4 mo, and 9 days exploring the territory in the Northwest. What is this time in weeks?
  (b) It took Magellan 1126 days to circle the world. How many years is this?
  (c) How many seconds old are you?
  (d) Approximately how many times does your heart beat in one year?

18. The Hawks played the Warriors in a basketball game. Based on the information below, complete the scoreboard showing the number of points scored by each team during each quarter and the final score of the game.

TEAMS	QUARTERS				FINAL SCORE
	1	2	3	4	
Hawks					
Warriors					

  (a) The Hawks scored 15 points in the first quarter.
  (b) The Hawks were behind by 5 points at the end of the first quarter.
  (c) The Warriors scored 5 more points in the second quarter than they did in the first quarter.
  (d) The Hawks scored 7 more points than the Warriors in the second quarter.
  (e) The Warriors outscored the Hawks by 6 points in the fourth quarter.
  (f) The Hawks scored a total of 120 points in the game.
  (g) The Hawks scored twice as many points in the third quarter as the Warriors did in the first quarter.
  (h) The Warriors scored as many points in the third quarter as the Hawks did in the first two quarters combined.

19. ▶Explain what algorithms are and why they are taught.
20. ▶Why is a front-end estimate before the adjustment always less than the exact sum?
21. ▶Give several examples where an estimate, rather than an exact answer, is close enough.

22. Janet worked her addition problems by placing the partial sums as shown below:

$$\begin{array}{r} 569 \\ +645 \\ \hline 14 \\ 10 \\ 11 \\ \hline 1214 \end{array}$$

  (a) Use this method to work the following:
   (i) 687
       +549
   (ii) 359
       +673
  ▶(b) Explain why this algorithm works.

23. ▶Discuss the advantages and disadvantages of lattice multiplication.

24. ▶Analyze the following errors. Explain what is wrong in each case.
  (a) 135   (b) 87   (c) 57   (d) 56
     + 47   +25   -38   -18
     ───   ───   ──   ──
      172   1012   21   48

25. 🔲 A palindrome is any number that reads the same backward and forward, for example, 121 and 2332. Try the following. Begin with any number. Is it a palindrome? If not, reverse the digits and add this new number to the original number. Is this a palindrome? If not, repeat the above procedure until a palindrome is obtained. For example, start with 78. Because 78 is not a palindrome, we add: 78 + 87 = 165. Because 165 is not a palindrome, we add: 165 + 561 = 726. Again, 726 is not a palindrome, so we add 726 + 627 to obtain 1353. Finally, 1353 + 3531 yields 4884, which is a palindrome.
  (a) Try this method with the following numbers.
   (i) 93   (ii) 588   (iii) 2003
  (b) Find a number for which the procedure described takes more than five steps to form a palindrome.

26. 🔲 Given the following addition problem, replace nine digits with 0s so that the sum of the numbers is 1111.

$$\begin{array}{r} 999 \\ 777 \\ 555 \\ 333 \\ 111 \\ \hline \end{array}$$

27. 🔲 Arrange eight 8's so that the sum is 1000.

28. 🔲 (a) Place the numbers 24 through 32 in the following circles so that the sums are the same in each direction.

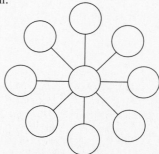

(b) How many different numbers can be placed in the middle to obtain a solution?

29. ▦ Andrew's calculator was not functioning properly. When he pressed 8 + 6 =, the numeral 20 appeared on the display. When he pressed 5 + 4 =, 13 was displayed. When he pressed 1 5 − 3 =, 9 was displayed. What do you think Andrew's calculator was doing?

30. ▦ The following is a supermagic square taken from an engraving called *Melancholia* by Dürer (1514).

16	3	2	13
5	10	11	8
9	6	7	12
4	15	14	1

(a) Find the sum of each row, the sum of each column, and the sum of each diagonal.
(b) Find the sum of the four numbers in the center.
(c) Find the sum of the four numbers in each corner.
(d) Add 11 to each number in the square. Is the square still a magic square? Explain your answer.
(e) Subtract 11 from each number in the square. Is the square still a magic square?

Review Problems

31. Write 5280 in expanded form.
32. Give an example of the associative property of addition for whole numbers.
33. Illustrate $11 + 8$ using a number-line model.
34. What is the value of $\overline{MCDX}$ in Hindu-Arabic numerals?
35. Rename the following using the distributive property of multiplication over addition:
    (a) $ax + a$ (b) $3(x + y) + a(x + y)$
36. Jim has 5 new shirts and 3 new pairs of pants. How many combinations of new shirts and pants does he have?

LABORATORY ACTIVITY

1. For each of the following, subtract the numbers in each row and column, as shown in Figure 3-32. Investigate why this works.

FIGURE 3-32

(a)
16	3	
8	1	

(b)
28	7	
15	12	

12	7	5
5	4	1
7	3	4

2. The Chinese abacus, *suan pan* (see Figure 3-33), is still in use today. A bar separates two sets of bead counters. Each counter above the bar represents five times the counter below the bar. Numbers are illustrated by moving the counter toward the bar. The number 7362 is pictured. Practice demonstrating numbers and adding on the *suan pan*.

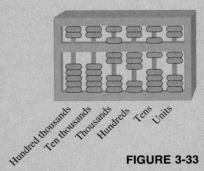

FIGURE 3-33

Section 3-5 Algorithms for Whole-number Multiplication and Division

Multiplication Algorithms

To develop algorithms for multiplying multidigit whole numbers, we use the strategy of *examining simpler computations first*. Consider $4 \cdot 12$. This computation could be pictured, as in Figure 3-34 with 4 rows of 12 dots, or 48 dots. The dots in Figure 3-34 can also be partitioned to show that $4 \cdot 12 = 4 \cdot (10 + 2) = 4 \cdot 10 + 4 \cdot 2$. The numbers $4 \cdot 10$ and $4 \cdot 2$ are *partial products*.

FIGURE 3-34

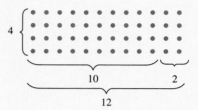

Figure 3-34 illustrates the distributive property of multiplication over addition on the set of whole numbers. The process leading to an algorithm for multiplying $4 \cdot 12$ is as follows:

Tens	Ones
1	2
×	4

$\longrightarrow$
$$\begin{array}{r} 10 + 2 \\ \times 4 \\ \hline 40 + 8 \end{array}$$
$\longrightarrow$
$$\begin{array}{r} 12 \\ \times 4 \\ \hline 8 \\ 40 \\ \hline 48 \end{array}$$
$\longrightarrow$
$$\begin{array}{r} 12 \\ \times 4 \\ \hline 48 \end{array}$$

To compute products involving powers of 10, such as $3 \cdot 200$, we proceed as follows:

$$\begin{aligned} 3 \cdot 200 &= 3 \cdot (2 \cdot 10^2) \\ &= (3 \cdot 2) \cdot 10^2 \\ &= 6 \cdot 10^2 \\ &= 6 \cdot 10^2 + 0 \cdot 10^1 + 0 \cdot 1 \\ &= 600 \end{aligned}$$

We see that multiplying 6 by 10^2 results in annexing two zeros to 6. This idea can be generalized to the statement that *multiplication of any natural number by 10^n, where n is a natural number, results in annexing n zeros to the number*.

When multiplying powers of 10, an extension of the definition of exponents is used. For example, $10^2 \cdot 10^1 = (10 \cdot 10) \cdot 10 = 10^3$. Observe that $10^3 = 10^{2+1}$. In general, where a is a natural number and m and n are whole numbers, $a^m \cdot a^n$ is given by the following:

$$a^m \cdot a^n = \underbrace{(a \cdot a \cdot a \cdot \ldots \cdot a)}_{m \text{ factors}} \cdot \underbrace{(a \cdot a \cdot a \cdot \ldots \cdot a)}_{n \text{ factors}}$$

$$= \underbrace{a \cdot a \cdot a \cdot \ldots \cdot a}_{m + n \text{ factors}} = a^{m+n}$$

Consequently, $a^m \cdot a^n = a^{m+n}$.

SECTION 3-5 ALGORITHMS FOR WHOLE-NUMBER MULTIPLICATION AND DIVISION

EXAMPLE 3-9 Find each of the following products:

(a) $10^5 \cdot 36$
(b) $10^3 \cdot 279$
(c) $10^{13} \cdot 10^8$
(d) $7 \cdot 200$

Solution

(a) $10^5 \cdot 36 = 3,600,000$
(b) $10^3 \cdot 279 = 279,000$
(c) $10^{13} \cdot 10^8 = 10^{13+8} = 10^{21}$
(d) $7 \cdot 200 = 7 \cdot (2 \cdot 10^2) = (7 \cdot 2) \cdot 10^2 = 14 \cdot 10^2 = 1400$ ∎

Next we consider computations with two-digit factors, such as $14 \cdot 23$. One possibility is to use the distributive property of multiplication over addition to write out all the partial products and add, as shown.

$$
\begin{array}{r}
14 \\
\times\, 23 \\
\hline
12 \quad (3 \times 4) \\
30 \quad (3 \times 10) \\
80 \quad (20 \times 4) \\
+\,200 \quad (20 \times 10) \\
\hline
322
\end{array}
$$

Another approach is to write 14 as $10 + 4$ and use the distributive property of multiplication over addition, as follows:

$$
\begin{aligned}
14 \cdot 23 &= (10 + 4) \cdot 23 \\
&= 10 \cdot 23 + 4 \cdot 23 \\
&= 230 + 92 \\
&= 322
\end{aligned}
$$

This last approach leads to an algorithm for multiplication.

$$
\begin{array}{r}
23 \\
\times\, 14 \quad (10 + 4) \\
\hline
92 \quad (4 \cdot 23) \\
230 \quad (10 \cdot 23) \\
\hline
322
\end{array}
\quad \text{or} \quad
\begin{array}{r}
23 \\
\times\, 14 \\
\hline
92 \\
23 \\
\hline
322
\end{array}
$$

We are accustomed to seeing the partial product 230 written without the zero, as 23. The placement of 23 with 3 in the tens column obviates having to write the 0 in the units column. When children first learn multiplication algorithms, they should be encouraged to include the zero, in order to avoid errors and promote better understanding. Children should also be encouraged to *estimate* whether their answers are reasonable. In this exercise, we know that the answer must be between $10 \cdot 20 = 200$ and $20 \cdot 30 = 600$ because $10 < 14 < 20$ and $20 < 23 < 30$. Because 322 is between 200 and 600, the answer is reasonable.

Lattice Multiplication

lattice multiplication — The **lattice multiplication** algorithm for multiplying 14 and 23 follows. (Determining the reasons why lattice multiplication works is left as an exercise.)

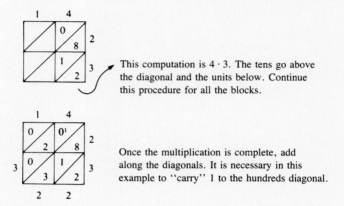

This computation is 4 · 3. The tens go above the diagonal and the units below. Continue this procedure for all the blocks.

Once the multiplication is complete, add along the diagonals. It is necessary in this example to "carry" 1 to the hundreds diagonal.

Mental Mathematics: Multiplication

As with addition and subtraction, mental mathematics is useful for multiplication. Several examples are given next.

1. *Front-end multiplying*

   $\quad$ 64
   $\times$ 5

   $60 \times 5 = 300$ — Multiply the number of tens in the first number by 5.
   $4 \times 5 = 20$ — Multiply the number of units in the first number by 5.
   $300 + 20 = 320$ — Add the two products.

2. *Compatible number multiplication*

   $2 \times 9 \times 5 \times 20 \times 5$

   Rearrange as $9 \times (2 \times 5) \times (20 \times 5) = 9 \times 10 \times 100 = 9000$

3. *Thinking money*

   (a) $\quad$ 64
   $\quad\quad \times 5$

   Think of the product as 64 nickels, which can be thought of as 32 dimes, which is $32 \times 10 = 320$ cents.

   (b) $\quad$ 64
   $\quad\quad \times 50$

   Think of the product as 64 half-dollars, which is 32 dollars, or 3200 cents.

   (c) $\quad$ 64
   $\quad\quad \times 25$

   Think of the product as 64 quarters, which is 32 half-dollars, or 16 dollars. Thus, we have 1600 cents.

Division Algorithms

Algorithms for division can be developed by using repeated subtraction. Consider the following:

A shopkeeper is packaging juice in cartons that hold 6 bottles each. She has 726 bottles. How many cartons does she need?

We might reason that if 1 carton holds 6 bottles, then 10 cartons hold 60 bottles and 100 cartons hold 600 bottles. If 100 cartons are filled, there are $726 - 100 \cdot 6$, or 126, bottles remaining. If 10 more cartons are filled, then $126 - 10 \cdot 6$, or 66, bottles remain. Similarly, if 10 more cartons are filled, $66 - 10 \cdot 6$, or 6, bottles remain. Finally, 1 carton will hold the remaining 6 bottles. The total number of cartons necessary is $100 + 10 + 10 + 1$, or 121. This procedure is summarized in Figure 3-35(a). A more efficient method is shown in Figure 3-35(b).

FIGURE 3-35

(a)
```
    _____
  6)726
  -600      100 sixes
   ___
   126
   - 60     10 sixes
   ___
    66
   - 60     10 sixes
   ___
     6
   -  6     1 six
   ___
     0      121 sixes
```

(b)
```
    _____
  6)726
  -600      100 sixes
   ___
   126
  - 120     20 sixes
   ___
     6
   -  6     1 six
   ___
     0      121 sixes
```

Divisions such as the one in Figure 3-35 are usually shown in elementary school texts in the most efficient form, as in Figure 3-36(b), in which the numbers in color in Figure 3-36(a) are omitted. The technique used in Figure 3-36(a) is often called "scaffolding" and may be used as a preliminary step to Figure 3-36(b).

FIGURE 3-36

(a)
```
        121
          1
         20
        100
      _____
    6)726
    -600
     ___
     126
    -120
     ___
       6
    -  6
     ___
       0
```

(b)
```
        121
      _____
    6)726
    - 6
     ___
      12
     -12
     ___
       6
     - 6
     ___
       0
```

Division in most elementary texts is taught using a four-step algorithm: *estimate, multiply, subtract,* and *compare*. This is demonstrated on the student page from *Addison-Wesley Mathematics*, 1991, Grade 3, shown on p. 152.

Dividing
Finding Quotients and Remainders

LEARN ABOUT IT

EXPLORE Think About the Process

Amber took a roll of 32 pictures on her vacation. If she puts 6 pictures on each page of her photo book, how many pages can Amber fill? How many pictures will be left?

You divide because you are sharing equally.

TALK ABOUT IT

1. Why is $30 \div 6$ a good way to estimate?
2. Is the remainder less than the divisor?
3. Use complete sentences to answer the problem.

An example of division by a divisor of more than one digit is given next. Consider $32 \overline{)2618}$.

1. Estimate the quotient in $32 \overline{)2618}$. Because $1 \cdot 32 = 32$, $10 \cdot 32 = 320$, $100 \cdot 32 = 3200$, we see that the quotient is between 10 and 100.
2. Find the number of tens in the quotient. Because $26 \div 3$ is approximately 8, then 26 hundreds divided by 3 tens is approximately 8 tens. We then write the 8 in the tens

place, as shown.

$$\begin{array}{r} 80 \\ 32\overline{)2618} \\ -2560 \quad (32\cdot 80) \\ \hline 58 \end{array}$$

3. Find the number of units in the quotient. Because $5 \div 3$ is approximately 1, then 5 tens divided by 3 tens is approximately 1. We have the following:

$$\begin{array}{r} 81 \\ \overline{1} \\ 80 \\ 32\overline{)2618} \\ -2560 \\ \hline 58 \\ -32 \quad (32\cdot 1) \\ \hline 26 \end{array}$$

4. Check: $32 \cdot 81 + 26 = 2618$.

Normally in grade-school books, we see the following format, with the remainder written by the quotient:

$$\begin{array}{r} 81 \text{ R}26 \\ 32\overline{)2618} \\ -256 \\ \hline 58 \\ -32 \\ \hline 26 \end{array}$$

Because of the advent of calculators, many mathematics educators are suggesting that division by divisors of more than two digits should not be taught. What do you think?

The process just described is usually referred to as "long" division. Another technique, called "short" division, can be used when the divisor is a one-digit number and most of the work is done mentally. An example of short division is given next.

Decide where to start.	Divide the hundreds. Write the remainder by the tens.	Divide the tens. Write the remainder by the ones.	Divide the ones.
$\begin{array}{r} 5 \\ 5\overline{)2\,8\,8\,0} \end{array}$	$\begin{array}{r} 5 \\ 5\overline{)2\,8^3 8\,0} \end{array}$	$\begin{array}{r} 5\,7 \\ 5\overline{)2\,8^3 8^3 0} \end{array}$	$\begin{array}{r} 5\,7\,6 \\ 5\overline{)2\,8^3 8^3 0} \end{array}$
Not enough thousands, $5 > 2$. $5 < 28$, so divide the hundreds.	$28 \div 5 = 5 \text{ R}3$	$38 \div 5 = 7 \text{ R}3$	$30 \div 5 = 6 \text{ R}0$

154 CHAPTER 3 NUMERATION SYSTEMS AND WHOLE NUMBERS

Mental Mathematics: Division

1. *Breaking up the dividend*

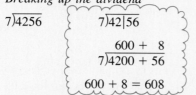

Break up the dividend into parts.

Divide both parts by 7.

Add the answers together.

2. *Compatible numbers in division*

(a) $3\overline{)105}$

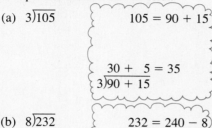

Look for numbers that you recognize as divisible by 3 and having a sum of 105.

Divide both parts and add the answers.

(b) $8\overline{)232}$

Look for numbers that are easily divisible by 8 and whose difference is 232.

Divide both parts and take the difference.

Estimation: Multiplication and Division

Examples of estimation strategies for multiplication and division are given next.

1. *Front-end multiplication*

$$\begin{array}{r} 524 \\ \times\ 8 \end{array}$$

$500 \times 8 = 4000$ Start multiplying at the front to obtain a first estimate.

$20 \times 8 = 160$ Multiply 8 times the next important digit.

$4000 + 160 = 4160$ Adjust the first estimate by adding the two numbers.

2. *Compatible numbers*

$5\overline{)4163}$ 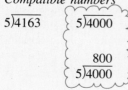   Change 4163 to a number close to it that you know is divisible by 5.

Carry out the division and obtain the first estimate of 800. Various techniques can be used to adjust the first estimate.

PROBLEM SET 3-5

1. Perform the following multiplications, using both the conventional and the lattice multiplication algorithms.

   (a) 728
        × 94

   (b) 306
        × 24

2. ▶Explain why the lattice multiplication algorithm works.

3. Fill in the missing numbers in each of the following.

   (a)   4_6
        ×783
        1_78
        3408
        _982
        3335_8

   (b)   327
        ×9_1
         327
        1_08
        _9_3
        30__07

4. The following chart gives average water usage for one person for one day.

Use	Average Amount
Taking bath	110 l (liters)
Taking shower	75 l
Flushing toilet	22 l
Washing hands, face	7 l
Getting a drink	1 l
Brushing teeth	1 l
Doing dishes (one meal)	30 l
Cooking (one meal)	18 l

(a) Use the chart to calculate how much water you use each day.
(b) The average American uses approximately 200 l of water per day. Are you average?
(c) If there are 215,000,000 people in the United States, approximately how much water is used in the United States per day?

5. Simplify each of the following, using properties of exponents. Leave answers as powers.
(a) $5^7 \cdot 5^{12}$
(b) $6^{10} \cdot 6^2 \cdot 6^3$
(c) $10^{296} \cdot 10^{17}$
(d) $2^7 \cdot 10^5 \cdot 5^7$

6. ▶(a) Which is greater, $2^{80} + 2^{80}$ or 2^{100}? Why?
▶(b) Which is the greatest, 2^{101}, $3 \cdot 2^{100}$, or 2^{102}? Why?

7. The given model illustrates that
$23 \cdot 14 = (20 + 3) \cdot (10 + 4)$
or
$20 \cdot 10 + 20 \cdot 4 + 3 \cdot 10 + 3 \cdot 4$.
Draw similar models illustrating each of the following.
(a) $6 \cdot 23$
(b) $18 \cdot 25$

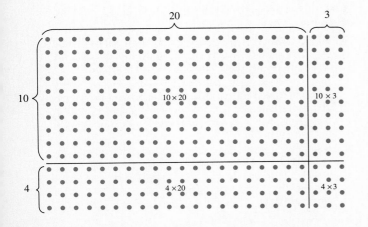

8. Consider the following:
```
    476
   ×293
    952   (2 · 476)
   4284   (9 · 476)
   1428   (3 · 476)
 139468
```
(a) Show, by using the conventional algorithm, that the answer is correct.
(b) Explain why the algorithm works.
(c) Try the method to multiply 84×363.

9. ▶The Russian peasant algorithm for multiplying 27×68 follows. (Disregard remainders when halving.)

	Halves		Doubles	
→	27 ×		68	
Halve 27 →	13		136	Double 68.
Halve 13	6		272	Double 136.
Halve 6 →	3		544	Double 272.
Halve 3 →	1		1088	Double 544.

In the "Halves" column, choose the odd numbers. In the "Doubles" column, circle the numbers paired with the odds from the "Halves" column. Add the circled numbers.

```
   68
  136
  544
 1088
 1836
```
This is the product $27 \cdot 68$.
Try this algorithm for $17 \cdot 63$ and other numbers.

10. Find the greatest possible whole-number value of n such that
(a) $14n < 300$ (b) $21n \leq 7459$
(c) $7n \leq 2134$ (d) $483n < 79485$

11. Find the least possible whole-number such that
(a) $14n > 300$ (b) $23n \geq 4369$
(c) $123n > 782$ (d) $222n > 8654$

12. Use the distributive property of multiplication over addition or subtraction to compute mentally each of the following.
(a) $15 \cdot 12$ (b) $14 \cdot 102$ (c) $30 \cdot 99$

13. Complete the following table:

a	b	$a \cdot b$	$a + b$
	56	3752	
32			110
		270	33

14. Answer the following questions based on the activity chart given next.

Activity	Calories Burned per Hour
Playing tennis	462
Snowshoeing	708
Cross-country skiing	444
Playing volleyball	198

    (a) How many calories are burned during 3 hr of cross-country skiing?

    (b) Jane played tennis for 2 hr while Carolyn played volleyball for 3 hr. Who burned more calories, and how many more?

    (c) Lyle went snowshoeing for 3 hr and Maurice went cross-country skiing for 5 hr. Who burned more calories, and how many more?

15. On a 14-day vacation, Glenn increased his caloric intake by 1500 calories per day. He also worked out more than usual by swimming 2 hr a day. Swimming burns 666 calories per hour, and a net gain of 3500 calories adds 1 lb of weight. Did Glenn gain at least 1 lb during his vacation?

16. Sue purchased a $30,000 life-insurance policy at the price of $24 for each $1000 of coverage. If she pays the premium in 12 monthly installments, how much is each installment?

17. Perform each of the following divisions, using both the repeated-subtraction and familiar algorithms.
    (a) $8\overline{)623}$ (b) $36\overline{)298}$ (c) $391\overline{)4001}$

18. Place the digits 7, 6, 8, and 3 in the boxes to obtain:
    (a) the greatest quotient;   (b) the least quotient.

    $1\overline{)\square\square\square}$

19. Rudy is buying a new car that costs $8600. The car salesman said Rudy could pay cash or pay $1500 down and $450 a month for 2 yr.
    (a) Which option is more expensive?
    (b) How much more expensive?

20. A 1K computer memory chip can store 1024 bits of information. How many bits of information can be stored in a 64K chip?

21. Using a calculator, Ralph multiplied by 10 when he should have divided by 10. The display read 300. What should the correct answer be?

22. Twenty members of the band plan to attend a festival. The band members washed 245 cars at $2 per car to help cover expenses. The school will match every dollar that the band raises with a dollar from the school budget. The cost of renting the bus to take the band is 72¢ per mile and the round trip is 350 mi. The band members can stay in the dorm for 2 nights at $5 per person per night. Meals for the trip will cost $28 per person. Has the band raised enough money yet? If not, how many more cars do they have to wash?

23. The following figure shows four function machines. The output from one machine becomes the input for the one below it. Complete the accompanying chart.

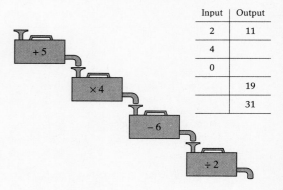

Input	Output
2	11
4	
0	
	19
	31

24. Choose three different digits.
    (a) Form six different two-digit numbers from the numbers you chose. Each number can be used only once.
    (b) Add the six numbers.
    (c) Add the three digits you chose.
    (d) Divide the answer in (b) by the answer in (c).
    (e) Repeat (a)–(d) with three different numbers.
    (f) Is the final result always the same? Why?

25. Consider the following multiplications. Notice that when the digits in the factors are reversed, the products are the same.

    $$\begin{array}{r}36\\\times 42\\\hline 1512\end{array} \qquad \begin{array}{r}63\\\times 24\\\hline 1512\end{array}$$

    (a) Find other multiplications where this procedure works.
    (b) Find a pattern for the numbers that work in this way.

26. In a certain book, 2981 digits were used to print the page numbers. How many pages are in the book?

27. ▶Pick a number. Double it. Multiply the result by 3. Add 24. Divide by 6. Subtract your original number. Is the result always the same? Discuss why or why not.

28. Molly read 160 pages in her book in 4 hr. Her sister Karly took 4 hr to read 100 pages in the same book. If the book is 200 pages long and if the two girls continued to read at these rates, how much longer would it take Karly to read the book than Molly?

29. ▶Sami has a paper route with 38 customers. Each customer is charged $12 a month. She estimates that if she makes all her collections, she will collect about $600. Is her estimate low or high? Why? Discuss how she could get a closer estimate.

30. ▶Discuss possible error patterns in each of the following:

(a) $\begin{array}{r} 34 \\ \times 8 \\ \hline 2432 \end{array}$ (b) $\begin{array}{r} 35 \\ \times 26 \\ \hline 90 \end{array}$ (c) $\begin{array}{r} 34 \\ \times 6 \\ \hline 114 \end{array}$ (d) $\begin{array}{r} 5\,3 \\ 5\overline{)2515} \\ -25 \\ \hline 15 \\ -15 \\ \hline 0 \end{array}$

31. A dog and a cat are 100 ft apart. The cat can run 30 ft/sec and the dog can run 20 ft/sec. If they start running at the same time with the cat chasing the dog, how long will it take for the cat to catch the dog?

32. Mira is saving to buy a new computer. She has saved $356, and the total cost of the machine is $980. Each week, she can save $30. How long will it take until she can purchase her computer?

33. To transport the complete student body of 1672 students to a talk given by the governor, the school plans to rent buses that can hold 29 students each. How many buses are needed? Will all the buses be full?

34. Larry's new car holds 40 l of gas. He drove 396 km and had 4 l left. How many kilometers does Larry's car get per liter?

35. ▶What happens when you multiply any two-digit number by 101? Discuss why this happens.

36. Cynthia can buy skis, bindings, poles, and boots, or she can rent them each time she goes skiing. She can buy or rent individual items or she can buy or rent complete packages, as shown below.

	Buy	Rent
Skis	$200	$25
Bindings	80	0
Poles	20	5
Boots	100	10
Complete ski package	330	30

(a) How much does she save on the complete ski package if she
   (i) buys?   (ii) rents?
(b) If she buys her own equipment, how many trips must she make so that buying the whole package costs less than renting?

37. ▦ When multiplying $12 \times 483 = 5796$, every digit 1 through 9 is used either in one of the factors or in the product.
(a) Show that this also happens in the following:
   (i) 27×198 (ii) 48×159 (iii) 39×186
(b) Find other examples where all digits are used that have the following as factors:
   (i) 1963   (ii) 483   (iii) 297
▶(c) Which whole numbers 1 through 9, if any, cannot be the units digit of a factor when using every digit in multiplication as described above. Why?

38. ▦ Place the digits 7, 6, 8, and 3 in the boxes to obtain:
(a) the greatest product;   (b) the least product.

□□□
× □

39. ▦ Place the digits 7, 6, 8, 3, and 2 in the boxes to obtain:
(a) the greatest product;   (b) the least product.

□□□
× □□

40. ▦ If a cow produces 700 lb of hamburger and there are 4 Quarter Pounders to a pound, how many cows would it take to produce 21 billion Quarter Pounders?

41. ▦ Use a calculator to find the missing numbers.

(a) $\begin{array}{r} 37 \\ \times 43 \\ \hline \_\_\_\_ \\ \_591 \end{array}$ (b) $\begin{array}{r} \_\_ \\ \times 36 \\ \hline 558 \\ 2790 \\ \hline \_\_\_\_ \end{array}$ (c) $\begin{array}{r} \_)\overline{123} \\ -9 \\ \hline 33 \\ -27 \\ \hline 6 \end{array}$

42. ▦ Find the products of the following, and describe the pattern that emerges:

(a) 1×1 (b) 99×99
11×11 999×999
111×111 9999×9999
1111×1111

43. ▦ Suppose a person can spend $1 per second. How much can that person spend in a minute? An hour? A day? A week? A month? A year? 20 years?

44. ▦ Suppose a friend chooses a number between 250,000 and 1,000,000. What is the least number of questions you must ask in order to guess the number if the friend answers only "yes" or "no" to the questions?

Review Problems

45. Write the number succeeding 673 in Egyptian numerals.
46. Write $3 \cdot 10^5 + 2 \cdot 10^2 + 6 \cdot 10$ as a Hindu-Arabic numeral.
47. Illustrate the identity property of addition for whole numbers.
48. Rename each of the following, using the distributive property of multiplication over addition:
(a) $ax + bx + 2x$
(b) $3(a + b) + x(a + b)$
49. At the beginning of a trip, the odometer registered 52,281. At the end of the trip, the odometer registered 59,260. How many miles were traveled on this trip?
50. The registration for the computer conference was 192 people on Thursday, 215 on Friday, and 317 on Saturday. What was the total registration?

BRAIN TEASER

For each of the following, replace the letters with digits in such a way that the computation is correct. Each letter may represent only one digit.

(a) $\quad$ LYNDON
$\quad\quad\quad \underline{\times B}$
$\quad\quad$ JOHNSON

(b) $\quad$ MA
$\quad\quad$ MA
$\quad\quad \underline{+MA}$
$\quad\quad$ EEL

LABORATORY ACTIVITY

Finger multiplication has long been popular in many parts of the world. Multiplication of single digits by 9 is very simple using the following steps.

1. Place your hands next to each other as shown.

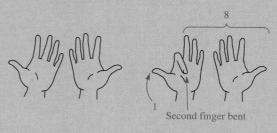

2. To multiply 2 by 9, bend down the second finger from the left. The remaining fingers show the product.

3. Similarly, to multiply 3 by 9, bend down the third finger from the left. The remaining fingers will show the product $3 \times 9 = 27$. Try this procedure with other multiplications by 9.

*Section 3-6 Other Number Bases

We study other number bases for several reasons, as follows:

1. To deepen our understanding of base ten by comparing it with other systems.
2. To allow us to experience learning in an unfamiliar system so we can better understand elementary students' learning problems.
3. To see the practical value of different number bases. For example, base two and base sixteen are used with computers, and base twelve is used in packaging items by units—twelves—and by the gross. Base sixty is used with units of time and angle measure.

Mathematical historians believe that one reason the majority of the world uses the base-ten system, with the ten digits 0 through 9, is that most people have ten fingers. Suppose you can use only one hand and the digits available for counting are 0, 1, 2, 3, and 4. In the "one-hand system," you count 1, 2, 3, 4, 10, where 10 represents one hand and no fingers. The one-hand system is a base-five system. Counting in base five proceeds as shown in Figure 3-37. We write the small "five" below the numeral as a reminder that the number is written in base five.

SECTION 3-6 OTHER NUMBER BASES

no fingers. The one-hand system is a base-five system. Counting in base five proceeds as shown in Figure 3-37. We write the small "five" below the numeral as a reminder that the number is written in base five.

FIGURE 3-37

Base-Five Symbol	Base-Five Grouping	One-Hand System
0_{five}		0 fingers
1_{five}	x	1 finger
2_{five}	xx	2 fingers
3_{five}	xxx	3 fingers
4_{five}	xxxx	4 fingers
10_{five}	(xxxxx)	1 hand and 0 fingers
11_{five}	(xxxxx) x	1 hand and 1 finger
12_{five}	(xxxxx) xx	1 hand and 2 fingers
13_{five}	(xxxxx) xxx	1 hand and 3 fingers
14_{five}	(xxxxx) xxxx	1 hand and 4 fingers
20_{five}	(xxxxx)(xxxxx)	2 hands and 0 fingers
21_{five}	(xxxxx)(xxxxx) x	2 hands and 1 finger

What number follows 44_{five}? There are no more two-digit numbers in the system after 44_{five}. In base ten, the same situation occurs at 99. We use 100 to represent ten tens or one hundred. In the base-five system, we need a symbol to represent five fives. To continue the analogy with base ten, we use 100_{five} to represent 1 group of five fives, 0 groups of five, and 0 units. To distinguish from "one hundred" in base ten, the name for 100_{five} is "one-zero-zero base five." The number 100 means $1 \cdot 10^2 + 0 \cdot 10^1 + 0$, whereas the number 100_{five} means $(1 \cdot 10^2 + 0 \cdot 10^1 + 0)_{five}$, or $(1 \cdot 5^2 + 0 \cdot 5^1 + 0)_{ten}$ or 25.

EXAMPLE 3-10 Convert 11244_{five} to base ten.

Solution
$$11244_{five} = 1 \cdot 5^4 + 1 \cdot 5^3 + 2 \cdot 5^2 + 4 \cdot 5 + 4 \cdot 1$$
$$= 1 \cdot 625 + 1 \cdot 125 + 2 \cdot 25 + 4 \cdot 5 + 4 \cdot 1$$
$$= 625 + 125 + 50 + 20 + 4$$
$$= 824$$

Example 3-10 suggests a method for changing a base-ten number to a base-five number using powers of five. To convert 824 to base five, we divide by successive powers of five. A shorthand method for illustrating this conversion is shown on the following page.

```
625 ) 824  1     How many groups of 625 in 824?
     -625
125 ) 199  1     How many groups of 125 in 199?
     -125
 25 )  74  2     How many groups of 25 in 74?
      -50
  5 )  24  4     How many groups of 5 in 24?
      -20
  1 )   4  4     How many 1's in 4?
       -4
        0
```

Thus, $824 = 11244_{\text{five}}$.

A different method of converting 824 to base five is shown next using successive divisions by 5. The quotient in each case is placed below the dividend and the remainder is placed on the right, on the same line with the quotient. The answer is read from bottom to top, that is, as 11244_{five}. Why does it work?

```
5 ) 824
  5 ) 164   4
    5 ) 32  4
      5 ) 6  2
          1  1
```

binary system

Historians tell of early tribes that used base two. Some Australian tribes still count "one, two, two and one, two twos, two twos and one, ..." Because base two has only two digits, it is called the **binary system.** Base two is especially important because of its use in computers. One of the two digits is represented by the presence of an electrical signal and the other by the absence of an electrical signal. Although base two works well for computers, it is inefficient for everyday use because multidigit numbers are reached very rapidly in counting in this system.

Conversions from base two to base ten, and vice versa, may be accomplished in a manner similar to that used for base-five conversions.

EXAMPLE 3-11   (a) Convert 10111_{two} to base ten.
(b) Convert 27 to base two.

Solution   (a) $10111_{\text{two}} = 1 \cdot 2^4 + 0 \cdot 2^3 + 1 \cdot 2^2 + 1 \cdot 2^1 + 1$
$= 16 + 0 + 4 + 2 + 1$
$= 23$

(b)
```
  16 | 27  | 1     How many groups of 16 in 27?
      -16
   8  | 11  | 1    How many groups of 8 in 11?
       -8
   4  |  3  | 0    How many groups of 4 in 3?
       -0
   2  |  3  | 1    How many groups of 2 in 3?
       -2
   1  |  1  | 1    How many 1's in 1?
       -1
        0
```

Alternate Solution
```
2 | 27
2 | 13 | 1
2 |  6 | 1
2 |  3 | 0
    1    1
```

Thus, 27 is equivalent to 11011_{two}.

Another commonly used number base system is the base-twelve, or duodecimal, system, known popularly as the "dozens" system. Eggs are bought by the dozen, and pencils are bought by the *gross* (a dozen dozens). In base twelve, there are twelve digits, just as there are ten digits in base ten, five digits in base five, and two digits in base two. In base twelve, new symbols are needed to represent the following groups of x's.

$$\underbrace{x\,x\,x\,x\,x\,x\,x\,x\,x\,x}_{10\ x\text{'s}} \quad \text{and} \quad \underbrace{x\,x\,x\,x\,x\,x\,x\,x\,x\,x\,x}_{11\ x\text{'s}}$$

The new symbols chosen are T and E, respectively, so that the base-twelve digits are 0, 1, 2, 3, 4, 5, 6, 7, 8, 9, T, E. Thus, in base twelve we count "1, 2, 3, 4, 5, 6, 7, 8, 9, T, E, 10, 11, 12, . . . , 17, 18, 19, $1T$, $1E$, 20, 21, 22, . . . , 28, 29, $2T$, $2E$, 30, ..."

EXAMPLE 3-12
(a) Convert $E2T_{twelve}$ to base ten. (b) Convert 1277 to base twelve.

Solution

(a) $E2T_{twelve} = 11 \cdot 12^2 + 2 \cdot 12^1 + 10$
$= 11 \cdot 144 + 24 + 10$
$= 1584 + 24 + 10$
$= 1618$

(b)
```
 144 | 1277 | 8     How many groups of 144 in 1277?
      -1152
  12 |  125 | T     How many groups of 12 in 125?
       -120
   1 |    5 | 5     How many 1s in 5?
        -5
         0
```

Thus, $1277 = 8T5_{twelve}$.

Addition and Subtraction in Different Bases

Before studying algorithms in base ten, we assumed a knowledge of the basic addition and multiplication facts. The same is true for other bases. Using a number line such as in Figure 3-38, which shows $4_{five} + 3_{five} = 12_{five}$, we construct the base-five addition table shown in Table 3-10.

FIGURE 3-38

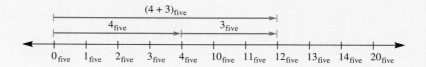

TABLE 3-10 Addition Table (Base Five)

+	0	1	2	3	4
0	0	1	2	3	4
1	1	2	3	4	10
2	2	3	4	10	11
3	3	4	10	11	12
4	4	10	11	12	13

Using the addition facts in Table 3-10, we develop algorithms for base-five addition similar to those for base-ten addition. Concrete teaching aids, such as multibase blocks, chip trading, and bean sticks, can be used to develop these algorithms.

Suppose we wish to add $12_{five} + 31_{five}$. We show the computation using a concrete model in Figure 3-39(a), using an introductory algorithm in Figure 3-39(b), and using the familiar algorithm in Figure 3-39(c). Additions in other number bases can be handled similarly.

FIGURE 3-39

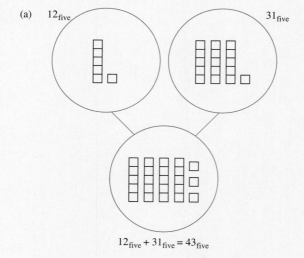

$12_{five} + 31_{five} = 43_{five}$

Subtraction such as $12_{five} - 4_{five}$ can be modeled using a number line, as shown in Figure 3-40. Thus, we see that $12_{five} - 4_{five} = 3_{five}$. The subtraction facts for base five can also be derived from the addition-facts table, by using the definition of subtraction. For

FIGURE 3-40

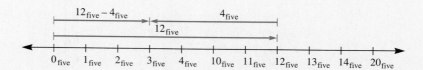

example, to find $(12 - 4)_{\text{five}}$, recall that $(12 - 4)_{\text{five}} = c_{\text{five}}$ if, and only if, $(c + 4)_{\text{five}} = 12_{\text{five}}$. From Table 3-10, we see that $c = 3_{\text{five}}$. An example of subtraction involving regrouping, $32_{\text{five}} - 14_{\text{five}}$, is developed in Figure 3-41.

FIGURE 3-41

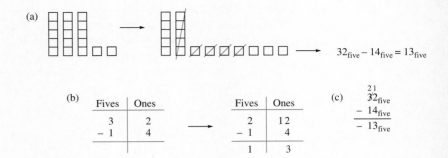

Multiplication and Division in Different Bases

As with addition and subtraction, we need to identify the basic facts of multiplication before we can use algorithms. The multiplication facts for base five are given in Table 3-11. These facts can be derived by using repeated addition.

TABLE 3-11 Multiplication Table (Base Five)

×	0	1	2	3	4
0	0	0	0	0	0
1	0	1	2	3	4
2	0	2	4	11	13
3	0	3	11	13	22
4	0	4	13	22	31

There are various ways to do the multiplication $21_{\text{five}} \cdot 3_{\text{five}}$.

Five	Ones
2	1
×	3

$$\begin{array}{r} (20 + 1)_{\text{five}} \\ \times \quad 3_{\text{five}} \\ \hline (110 + 3)_{\text{five}} \end{array} \longrightarrow \begin{array}{r} 21_{\text{five}} \\ \times \quad 3_{\text{five}} \\ \hline 3 \\ 110 \\ \hline 113_{\text{five}} \end{array} \longrightarrow \begin{array}{r} 21_{\text{five}} \\ \times \quad 3_{\text{five}} \\ \hline 113_{\text{five}} \end{array}$$

The multiplication of a two-digit number by a two-digit number is developed next.

$$\begin{array}{r} 23_{\text{five}} \\ \times \; 14_{\text{five}} \\ \hline 22 \\ 130 \\ 30 \\ 200 \\ \hline 432_{\text{five}} \end{array} \quad \begin{array}{l} (10 + 4)_{\text{five}} \\ (4 \cdot 3)_{\text{five}} \\ (4 \cdot 20)_{\text{five}} \\ (10 \cdot 3)_{\text{five}} \\ (10 \cdot 20)_{\text{five}} \end{array} \quad \begin{array}{r} 23_{\text{five}} \\ \times \; 14_{\text{five}} \\ \hline 202 \\ 230 \\ \hline 432_{\text{five}} \end{array}$$

Lattice multiplication can also be used to multiply numbers in various number bases. This will be explored in Problem Set 3-6.

Division in different bases can be performed using the multiplication facts and the definition of division. For example, $22_{five} \div 3_{five} = c$ if, and only if, $c \cdot 3_{five} = 22_{five}$. From Table 3-11, we see that $c = 4_{five}$. As in base ten, computing multidigit divisions efficiently in different bases requires practice. The ideas behind the algorithms for division can be developed by using repeated subtraction, just as they were for base ten. For example, $3241_{five} \div 43_{five}$ is computed by means of the repeated-subtraction technique in Figure 3-42(a) and by means of the conventional algorithm in Figure 3-42(b). Thus $3241_{five} \div 43_{five} = 34_{five}$ with remainder 14_{five}.

FIGURE 3-42

(a)

$$
\begin{array}{rrl}
43_{five}\,\overline{)\,3241_{five}} & & \\
-430 & & (10 \cdot 43)_{five} \\
\hline
2311 & & \\
-430 & & (10 \cdot 43)_{five} \\
\hline
1331 & & \\
-430 & & (10 \cdot 43)_{five} \\
\hline
401 & & \\
-141 & & (2 \cdot 43)_{five} \\
\hline
210 & & \\
-141 & & (2 \cdot 43)_{five} \\
\hline
14 & & (34 \cdot 43)_{five}
\end{array}
$$

(b)

$$
\begin{array}{r}
34_{five} \\
43_{five}\,\overline{)\,3241_{five}} \\
-234 \\
\hline
401 \\
-332 \\
\hline
14
\end{array}
$$

Computations involving base two are demonstrated in Example 3-13.

EXAMPLE 3-13

(a) Add:

$$
\begin{array}{r}
101_{two} \\
111_{two} \\
+110_{two} \\
\hline
\end{array}
$$

(b) Subtract:

$$
\begin{array}{r}
1010_{two} \\
-\ 111_{two} \\
\hline
\end{array}
$$

(c) Multiply:

$$
\begin{array}{r}
101_{two} \\
\times\ 11_{two} \\
\hline
\end{array}
$$

(d) Divide:

$$
101_{two}\,\overline{)\,110110_{two}}
$$

Solution

(a)

$$
\begin{array}{r}
{\scriptstyle 11}\phantom{1_{two}} \\
101_{two} \\
111_{two} \\
+\ \ 110_{two} \\
\hline
10010_{two}
\end{array}
$$

(b)

$$
\begin{array}{r}
1010_{two} \\
-\ 111_{two} \\
\hline
11_{two}
\end{array}
$$

(c)

$$
\begin{array}{r}
101_{two} \\
\times\ \ 11_{two} \\
\hline
101 \\
101 \\
\hline
1111_{two}
\end{array}
$$

(d)

$$
\begin{array}{r}
1010_{two} \\
101_{two}\,\overline{)\,110110_{two}} \\
-101 \\
\hline
111 \\
-101 \\
\hline
100
\end{array}
$$

■

PROBLEM SET 3-6

1. Write the first 15 counting numbers for each of the following bases:
   (a) Base two
   (b) Base three
   (c) Base four
   (d) Base eight
2. How many different digits are needed for base twenty?
3. Write 2032_{four} in expanded base-four notation.
4. What is the greatest three-digit number in each base?
   (a) Base two
   (b) Base six
   (c) Base ten
   (d) Base twelve
5. Find the numbers preceding and succeeding each of the following:
   (a) $EE0_{twelve}$
   (b) 100000_{two}
   (c) 555_{six}
   (d) 100_{seven}
   (e) 1000_{five}
   (f) 110_{two}
6. What, if anything, is wrong with the numerals below?
   (a) 204_{four}
   (b) 607_{five}
   (c) $T12_{three}$
7. Convert each of the following base-ten numbers to numbers in the indicated bases:
   (a) 432 to base five
   (b) 1963 to base twelve
   (c) 404 to base four
   (d) 37 to base two
   (e) $4 \cdot 10^4 + 3 \cdot 10^2$ to base twelve
8. Change 42_{eight} to base two.
9. Write each of the following numbers in base ten:
   (a) 432_{five}
   (b) 101101_{two}
   (c) $92E_{twelve}$
   (d) $T0E_{twelve}$
   (e) 111_{twelve}
   (f) 346_{seven}
10. Suppose you have two quarters, four nickels, and two pennies. What is the value of your money in cents? Write a base-five representation to indicate the value of your fortune.
11. You are asked to distribute $900 in prize money. The dollar amounts for the prizes are $625, $125, $25, $5, and $1. How should this $900 be distributed in order to give the fewest number of prizes?
12. What is the minimum number of quarters, nickels, and pennies necessary to make 97¢?
13. Convert each of the following.
    (a) 58 days to weeks and days
    (b) 54 mo to years and months
    (c) 29 hr to days and hours
    (d) 68 in. to feet and inches
14. A bookstore ordered 11 gro, 6 dz, and 6 pencils. Express the number of pencils in base twelve and in base ten.
15. For each of the following, find b:
    (a) $b2_{seven} = 44_{ten}$
    (b) $5b2_{twelve} = 734_{ten}$
    (c) $23_{ten} = 25_b$
16. George is cooking an elaborate meal for Thanksgiving. He can only cook one thing at a time in his microwave oven. His turkey takes 75 min; the pumpkin pie takes 18 min; rolls take 45 sec; and a cup of coffee takes 30 sec to heat. How much time does he need to cook the meal?
17. An inspector of weights and measures uses a special set of weights to check the accuracy of scales. Various weights are placed on a scale to check accuracy of any amount from 1 oz through 15 oz. What is the least number of weights the inspector needs? What weights are needed to check the accuracy of scales from 1 oz through 15 oz? From 1 through 31 oz?
18. (a) Anna's bank contains only pennies, nickels, and quarters. What is the minimum number of coins she could trade for 117 pennies?
    (b) If she trades 2 quarters, 4 nickels, and 3 pennies for pennies, how many pennies will she have?
19. Perform each of the following operations, using the bases shown:
    (a) $43_{five} + 23_{five}$
    (b) $43_{five} - 23_{five}$
    (c) $432_{five} + 23_{five}$
    (d) $42_{five} - 23_{five}$
    (e) $110_{two} + 11_{two}$
    (f) $10001_{two} - 111_{two}$
20. Construct addition and multiplication tables for base eight.
21. Perform each of the following operations:
    (a) 3 hr 36 min 58 sec
        +5 hr 56 min 27 sec
    (b) 5 hr 36 min 38 sec
        −3 hr 56 min 58 sec
22. Perform each of the following operations (2 c = 1 pt, 2 pt = 1 qt, 4 qt = 1 gal):
    (a) 1 qt 1 pt 1 c
        +   1 pt 1 c
    (b) 1 qt 1 c
        −1 pt 1 c
    (c) 1 gal 3 qt 1 c
        −   4 qt 2 c
23. Mari is going to invite 20 friends to a party. She would like to have at least 2 c of cider for each guest. If cider is sold only by the gallon, how many gallons should she buy?
24. Use scratch addition to perform the following:
    32_{five}
    13_{five}
    22_{five}
    43_{five}
    23_{five}
    $+12_{five}$
25. Perform each of the following operations:
    (a)   4 gro 4 dz 6 ones
        −   5 dz 9 ones
    (b)   2 gro 9 dz 7 ones
        +3 gro 5 dz 9 ones
26. In a small rural community, the elementary school has no refrigerators. Through a federally financed program, the school provides 1 c of milk per day for each student. Milk for the day is purchased at the local store each morning

and the school buys the exact amount necessary. The milk is available in gallons, half-gallons, quarts, pints, or cups, and the larger containers are better buys.
  (a) If 1 gal, 1 qt, and 1 pt of milk were purchased on Tuesday, how many students were at school that day?
  (b) On Wednesday, 31 students were at school. How much milk was purchased that day to make the best buy?

27. A *score* is equal to 20. Indicate each of the following as a base-ten number:
  (a) Three score and ten
  (b) Four score and seven

28. What is wrong with the following?

   22_{five}
   $+33_{five}$
   $\overline{55_{five}}$

29. Fill in the missing numbers in each of the following:
  (a) $\quad 2\_\_{}_{five}$
      $\quad -\ 2\ 2_{five}$
      $\quad \overline{\_ 0\ 3_{five}}$
  (b) $\quad 2\ 0\ 0\ 1\ 0_{three}$
      $\quad -\ 2\_ 2\_{}_{three}$
      $\quad \overline{1\_ 2\_ 1_{three}}$

30. Perform each of the following operations using the bases shown:
  (a) $32_{five} \cdot 4_{five}$
  (b) $32_{five} \div 4_{five}$
  (c) $43_{five} \cdot 23_{five}$
  (d) $143_{five} \div 3_{five}$
  (e) $13_{eight} \cdot 5_{eight}$
  (f) $67_{eight} \div 4_{eight}$
  (g) $10010_{two} \div 11_{two}$
  (h) $10110_{two} \cdot 101_{two}$

31. For what possible bases are each of the following computations correct?
  (a) $\quad 213$
      $\quad +308$
      $\quad \overline{522}$
  (b) $\quad 322$
      $\quad -233$
      $\quad \overline{23}$
  (c) $\quad\ \ 213$
      $\quad \times\ 32$
      $\quad \overline{\ \ 430}$
      $\quad\ 1043$
      $\quad \overline{11300}$
  (d) $\quad\quad 101$
      $\quad 11\overline{)1111}$
      $\quad\quad \underline{-11}$
      $\quad\quad\ \ 11$
      $\quad\quad \underline{-11}$
      $\quad\quad\quad 0$

32. Use lattice multiplication to compute $(323_{five}) \cdot (42_{five})$.

33. Find the smallest values of a and b so that $32_a = 23_b$.

34. ▶Discuss some applications of number bases other than base ten that people encounter in their lives.

LABORATORY ACTIVITY

1. Messages can be coded on paper tape in base two. A hole in the tape represents 1, whereas a space represents 0. The value of each hole depends on its position; from left to right, 16, 8, 4, 2, 1 (all powers of 2). Letters of the alphabet may be coded in base two according to their position in the alphabet. For example, G is the seventh letter. Since $7 = 1 \cdot 4 + 1 \cdot 2 + 1$, the holes appear as they do in the following figure:

   ○ ○ ○
   16 8 4 2 1

  (a) Decode the following message.

  (b) Write your name on a tape, using base two.

2. The following number game uses base-two arithmetic:

Card E		Card D		Card C		Card B		Card A	
16	24	8	24	4	20	2	18	1	17
17	25	9	25	5	21	3	19	3	19
18	26	10	26	6	22	6	22	5	21
19	27	11	27	7	23	7	23	7	23
20	28	12	28	12	28	10	26	9	25
21	29	13	29	13	29	11	27	11	27
22	30	14	30	14	30	14	30	13	29
23	31	15	31	15	31	15	31	15	31

(a) Suppose a person's age appears on cards E, C, and B. Then, the person is 22. Can you discover how this works and why?

(b) Design card F so that the numbers 1–63 can be used in the game. Note that cards A–E must also be changed.

Solution to the Preliminary Problem

Understanding the Problem. There are to be 64 stacks of pennies placed on a checkerboard with the number of pennies in each stack determined by placing 1 penny on the first square of a checkerboard, 2 pennies on the second square, 4 pennies on the third square and so on, doubling the number of pennies on each consecutive square. We are to find the height of the stack of pennies on square number 64.

Devising a Plan. Since the problem deals with the height of stacks of pennies, we must determine the number of pennies in a particular unit of length. Using a ruler, we see that it takes approximately 17 pennies to reach a 1-in. height. To determine the number of pennies in each stack, we use the strategy of *building a table,* as in Table 3-12.

TABLE 3-12

Number of Square	Number of Pennies
1	1
2	2
3	4
4	8
5	16
6	32
7	64
8	128

It would be cumbersome to carry the table out to 64 squares, so we must find a way to determine the number of pennies on a square without extending the table. The number of pennies form a geometric sequence. If we can find a formula for computing the nth term of this sequence, then we can find the number of pennies on the 64th square and, in turn, compute the height.

Carrying Out the Plan. We see from Table 3-12 that the geometric sequence has first term 1 and fixed ratio 2, so the nth term is given by $1 \times 2^{n-1}$ or 2^{n-1}. Therefore, the number of pennies on square 64 is 2^{63}. To find the height of the stacks of pennies in inches, we divide by 17 (17 pennies in 1 in.), then convert to feet by dividing by 12 (1 ft = 12 in.), and finally, convert to miles by dividing by 5280 (1 mi = 5280 ft).

Using a calculator with an $\boxed{x^y}$ key, we see that 2^{63} is approximately 9×10^{18}, which is a stack approximately 8×10^{12}, or 8 trillion mi, tall. Who said we didn't know how to stretch a penny?

Looking Back. The preceding answers for the 64th square are approximate and depend on the calculator being used and the rounding involved in the calculations. Related problems might include the following:

1. If a person could travel up the stack of pennies at 60 mph, how long would it take to reach the top of the stack on the 32nd and 64th squares?
2. What is the approximate value of the pennies on the 32nd square?
3. Could the stack of pennies on the 64th square reach the moon?

QUESTIONS FROM THE CLASSROOM

1. A student asks "Does $2 \cdot (3 \cdot 4)$ equal $(2 \cdot 3) \cdot (2 \cdot 4)$?" Is there a distributive property of multiplication over multiplication?
2. Since $39 + 41 = 40 + 40$, is it true that $39 \cdot 41 = 40 \cdot 40$?
3. The division algorithm, $a = bq + r$, holds for $a > b$; $a, b, q, r \in W$ and $b \neq 0$. Does the algorithm hold when $a < b$?
4. A student asks if 5 times 4 is the same as 5 multiplied by 4. How do you respond?
5. Can we define $0 \div 0$ as 1? Why or why not?
6. A student divides as follows. How do you help?

$$\begin{array}{r} 15 \\ 6\overline{)36} \\ \underline{6} \\ 30 \\ \underline{30} \end{array}$$

7. When using Roman numerals, a student asks whether it is correct to write $\overline{\text{II}}$, as well as MI, for 1001. How do you respond?
8. A student says that $(x + 7) \div 7 = x + 1$. What is that student doing wrong?
9. A student says $x \div x$ is always 1. Is the student correct?
10. A student claims that the expressions $(2^3)^2$ and $2^{(3^2)}$ are equal. How do you respond?
11. A student asks if division on the set of whole numbers is distributive over subtraction. How do you respond?
12. A student says that 0 is the identity for subtraction. How do you respond?
13. A student claims that, on the following number line, the arrow doesn't really represent 3 because the end of the arrow does not start at 0. How do you respond?

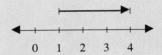

CHAPTER OUTLINE

I. Numeration systems
   A. Properties of numeration systems give basic structure to the systems.
      1. Additive property
      2. Place-value property
      3. Subtractive property
      4. Multiplicative property

II. Exponents
   A. For any whole number a and any natural number n,
   $$a^n = \underbrace{(a \cdot a \cdot a \cdots a)}_{n \text{ factors}}$$
   where a is the **base** and n is the **exponent**.
   B. $a^0 = 1$, $a \in N$
   C. For any natural number a, with whole numbers m and n, $a^m \cdot a^n = a^{m+n}$.

III. Whole numbers
   A. The set of **whole numbers** W is $\{0, 1, 2, 3, \ldots\}$.
   B. The basic operations for whole numbers are addition, subtraction, multiplication, and division.
      1. Addition: If $n(A) = a$ and $n(B) = b$, where $A \cap B = \varnothing$, then $a + b = n(A \cup B)$. The numbers a and b are **addends** and $a + b$ is the **sum**.
      2. Subtraction: If a and b are any whole numbers, then $a - b$ is the unique whole number c such that $a = b + c$. The number a is the **minuend**, b is the **subtrahend**, and c is the **difference**.
      3. Multiplication: If a and b are any whole numbers,
      $$a \cdot b = \underbrace{b + b + b + \cdots + b}_{a \text{ terms}}$$
      a and b are called **factors** and $a \cdot b$ is the **product**.
      4. Multiplication: If A and B are sets such that $n(A) = a$ and $n(B) = b$, then $a \cdot b = n(A \times B)$.
      5. Division: If a and b are any whole numbers with $b \neq 0$, $a \div b$ is the unique whole number c such that $b \cdot c = a$. The number a is the **dividend**, b is the **divisor**, and c is the **quotient**.
      6. **Division algorithm:** Given any whole numbers a and b, with $b \neq 0$, there exist unique whole numbers q and r such that $a = b \cdot q + r$, with $0 \leq r < b$.

170   CHAPTER 3   NUMERATION SYSTEMS AND WHOLE NUMBERS

C. Properties of addition and multiplication of whole numbers.
1. Closure: If $a, b \in W$, then $a + b \in W$ and $a \cdot b \in W$.
2. Commutative: If $a, b \in W$, then $a + b = b + a$ and $a \cdot b = b \cdot a$.
3. Associative: If $a, b, c \in W$, then $(a + b) + c = a + (b + c)$ and $a \cdot (b \cdot c) = (a \cdot b) \cdot c$.
4. Identity: 0 is the unique identity element for addition of whole numbers; 1 is the unique identity element for multiplication.
5. Distributive property of multiplication over addition: If $a, b, c \in W$, then $a \cdot (b + c) = a \cdot b + a \cdot c$.
6. Zero multiplication property
   For any whole number a, $a \cdot 0 = 0 = 0 \cdot a$.

D. Relations on whole numbers
1. $a < b$ if and only if there is a natural number c such that $a + c = b$.
2. $a > b$ if and only if $b < a$.

CHAPTER TEST

1. Convert each of the following to base ten:
   (a) $\overline{\text{CDXLIV}}$ *(b) 432_{five} *(c) $ET0_{\text{twelve}}$
   *(d) 1011_{two} *(e) 4136_{seven}

2. Convert each of the following numbers to numbers in the indicated system:
   (a) 999 to Roman
   (b) 86 to Egyptian
   (c) 123 to Mayan
   *(d) 346_{ten} to base five
   *(e) 1728_{ten} to base twelve
   *(f) 27_{ten} to base two
   *(g) 928_{ten} to base nine
   *(h) 13_{eight} to base two

3. Simplify each of the following, if possible. Write your answers in exponential form, a^b.
   (a) $3^4 \cdot 3^7 \cdot 3^6$
   (b) $2^{10} \cdot 2^{11}$
   (c) $3^4 + 2 \cdot 3^4$
   (d) $2^{80} + 3 \cdot 2^{80}$

4. For each of the following, identify the properties of the operation(s) for whole numbers illustrated:
   (a) $3 \cdot (a + b) = 3 \cdot a + 3 \cdot b$
   (b) $2 + a = a + 2$
   (c) $16 \cdot 1 = 1 \cdot 16 = 16$
   (d) $6 \cdot (12 + 3) = 6 \cdot 12 + 6 \cdot 3$
   (e) $3 \cdot (a \cdot 2) = 3 \cdot (2 \cdot a)$
   (f) $3 \cdot (2 \cdot a) = (3 \cdot 2) \cdot a$

5. Using the definitions of less than or greater than, prove that each of the following inequalities is true:
   (a) $3 < 13$ (b) $12 > 9$

6. ▶Explain why the product of $1000 \cdot 483$, namely, 483,000, has 0 for the hundreds, tens, and units digits.

7. Use both the scratch and the traditional algorithms to perform each of the following:
   (a)   316
         712
       + 91
   *(b) 316_{twelve}
         712_{twelve}
       + 913_{twelve}

8. Use both the traditional and the lattice multiplication algorithms to perform each of the following:
   (a)   613
       × 98
   *(b) 216_{eight}
       × 54_{eight}

9. Use both the repeated-subtraction and the conventional algorithms to perform each of the following:
   (a) $912 \overline{)4803}$
   (b) $11 \overline{)1011}$
   *(c) $23_{\text{five}} \overline{)3312_{\text{five}}}$
   *(d) $11_{\text{two}} \overline{)1011_{\text{two}}}$

10. Use the division algorithm to check your answers in Problem 9.

11. For each of the following base-ten numbers, tell the place value for each of the circled digits:
    (a) 4③2 (b) ③432 (c) 19③24

12. For each of the following, find all possible replacements to make the following statements true for whole numbers:
    (a) $4 \cdot \square - 37 < 27$
    (b) $398 = \square \cdot 37 + 28$
    (c) $\square \cdot (3 + 4) = \square \cdot 3 + \square \cdot 4$
    (d) $42 - \square \geq 16$

13. Use a number line to perform each of the following operations:
    (a) $27 - 15$
    (b) $17 + 2$
    *(c) $3_{\text{five}} + 11_{\text{five}}$
    *(d) $12_{\text{three}} + 2_{\text{three}}$

14. Use the distributive property of multiplication and addition facts, if possible, to rename each of the following.
    (a) $3a + 7a + 5a$
    (b) $3x^2 + 7x^2 - 5x^2$
    (c) $x(a + b + y)$
    (d) $(x + 5)3 + (x + 5)y$

15. If A, B, C, and D each stand for different single digits from 1 to 9, answer the following if
    $$A + B + C = D.$$
    (a) If $A = 2$, $B = 4$, and $D = 7$, then what is the value of C?

(b) What is the smallest number D could be?
(c) If B is 2 greater than A and 2 less than C, then what is the value of D?

16. You had a balance in your checking account of $720 before writing checks for $162, $158, and $33 and making a deposit of $28. What is your new balance?

17. Jim was paid $320 a month for 6 mo and $410 a month for 6 mo. What were his total earnings for the year?

18. A soft drink manufacturer produces 15,600 cans of his product each hour. Cans are packed 24 to a case. How many cases are produced in 4 hr?

19. A limited partnership of 120 investors sold a piece of land for $461,040. How much did each investor receive?

20. (a) Use each of the digits 1 through 9 to obtain a correct sum in the boxes below.
    ▶(b) Is only one correct answer possible? Why?

21. Merle took a 2040-mi trip, which took 10 days. Each day he drove 30 mi more than he had the day before. How many miles did he cover on the first day?

22. How many 12-oz cans of juice would it take to give 60 people one 8-oz serving each?

23. Heidi has a brown and a gray pair of slacks; a brown, a yellow, and a white blouse; and a blue and a white sweater. How many different outfits does she have if she wears slacks, a blouse, and a sweater?

24. Complete the following base-ten addition table:

+			9
8		15	
	16		
		26	30

25. I am thinking of a whole number. If I divide it by 13, then multiply the answer by 12, then subtract 20, and then add 89, I end up with 93. What was my original number?

26. Apples normally sell for 32¢ each. They go on sale for 3 for 69¢. How much money is saved if we purchase 2 dz apples?

27. A ski resort offers a weekend ski package for $80 per person or $6000 for a group of 80 people. Which would be the cheaper option for a group of 80?

28. The owner of a bicycle shop reported his inventory of bicycles and tricycles in an unusual way. He said he counted 126 wheels and 108 pedals. How many bikes and how many trikes did he have?

29. Josi has a job in which she works 30 hr/wk and gets paid $5/hr. If she works over 30 hr in a week, she receives $8/hr for each hour over 30 hr. If she worked 38 hr this week, how much did she earn?

30. In a television game show, there are five questions to answer. Each question is worth twice as much as the previous question. If the last question was worth $6400, what was the first question worth?

SELECTED BIBLIOGRAPHY

Bates, T., and L. Rousseau. "Will the Real Division Algorithm Please Stand Up?" *Arithmetic Teacher* 33 (March 1987): 42–46.

Bobis, J. "Using a Calculator to Develop Number Sense." *Arithmetic Teacher* 38 (January 1991): 42–45.

Bradbent, F. "Lattice Multiplication and Division." *Arithmetic Teacher* 34 (January 1987): 28–31.

Feinberg, M. "Using Patterns to Practice Basic Facts." *Arithmetic Teacher* 37 (April 1990): 38–41.

Gluck, D. "Helping Students Understand Place Value." *Arithmetic Teacher* 38 (March 1991): 10–13.

Hope, J. "Promoting Number Sense in School." *Arithmetic Teacher* 36 (February 1989): 12–16.

Hope, J., B. Reys, and R. Reys. *Mental Math in the Middle Grades*. Palo Alto: Dale Seymour Publishing, 1987.

Huinker, D. "Multiplication and Division Word Problems: Improving Students Understanding." *Arithmetic Teacher* 37 (October 1989): 8–12.

Kami, C., and L. Joseph. "Teaching Place Value and Double-Column Addition." *Arithmetic Teacher* 35 (February 1988): 48–52.

Moore, T. "More on Mental Computation." *Mathematics Teacher* 79 (March 1987): 168–169.

Reys, R. *Computational Estimation (Grades 6, 7, and 8)*. Palo Alto: Dale Seymour Publishing, 1987.

Sowder, J. "Mental Computation and Number Sense." *Arithmetic Teacher* 37 (March 1990): 18–20.

Stanic, G., and W. McKillip. "Developmental Algorithms Have a Place in Elementary School Mathematics." *Arithmetic Teacher* 36 (January 1989): 14–16.

Sundar, V. "Thou Shalt Not Divide by Zero." *Arithmetic Teacher* 37 (March 1991): 50–51.

Thornton, C., and P. Smith. "Action Research: Strategies for Learning Subtraction Facts." *Arithmetic Teacher* 35 (April 1988): 8–12.

Trafton, P., and J. Zawojewski. "Meanings of Operations." *Arithmetic Teacher* 38 (November 1990): 18–22.

Vande Walle, J. "Redefining Computation." *Arithmetic Teacher* 38 (January 1991): 46–51.

Wood, E. "More Magic with Magic Squares." *Arithmetic Teacher* 37 (December 1989): 42–46.

4 The Integers

Preliminary Problem

Maria noticed that every 30 seconds the temperature of a chemical reaction in her lab was decreasing by the same number of degrees. Initially, she measured the temperature as 28°C and 5 minutes later as ⁻12°C. In a second experiment, Maria noticed that the temperature of the chemical reaction was initially ⁻57°C and was decreasing by 3°C every minute. If she started the two experiments at the same time, when were the temperatures of the experiments the same? What was that temperature?

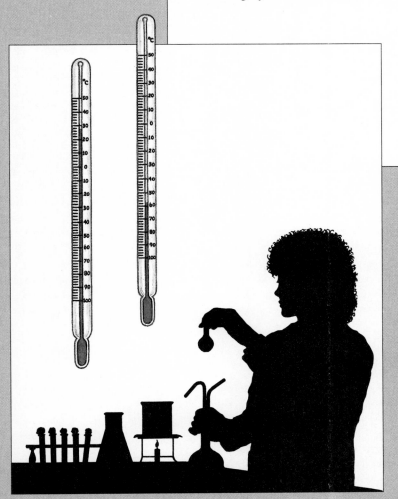

egative numbers are useful in everyday life. For example, Mount Everest (the highest point on earth) is 29,028 ft above sea level, but the Dead Sea (the lowest point on earth) is 1293 ft below sea level. We may say that the elevation of Mount Everest is 29,028 ft, and the elevation of the Dead Sea is ⁻1293 ft.

In mathematics, the need for negative numbers arises because subtractions cannot always be performed using only the set of whole numbers. The cartoon depicts Linus attempting a subtraction using only whole numbers.

PEANUTS reprinted by permission of UFS, Inc.

To compute $4 - 6$ using the definition of subtraction for whole numbers, a whole number a must be found such that $6 + a = 4$. Because there is no such whole number a, Linus's subtraction is not possible using only whole numbers. To perform the computation, we must invent a new number. This new number is called a *negative integer*. This chapter deals with the creation of negative integers, and with properties of integers.

The last section of the chapter deals with prealgebra and algebra skills. This section should help prepare teachers to meet the recommendations of the *Standards* (p. 102), which states that students in grades 5–8 should be able to *apply algebraic methods to examine and solve a variety of real-world and mathematical problems* and that students should be able to *develop confidence in solving linear equations using concrete, informal, and formal methods* and to *investigate inequalities and nonlinear equations informally*.

■ HISTORICAL NOTE

The Chinese used red rods for positive numbers and black rods for negative numbers in calculations possibly as early as 500 B.C. Brahmagupta, a seventh-century Hindu mathematician, wrote, "Positive divided by positive, or negative by negative, is affirmative." The Italian mathematician Girolamo Cardano (1501–1576) provided the first significant treatment of negative numbers (which he called "false numbers"). However, as late as the eighteenth century, some mathematicians worried whether two negative numbers could be multiplied, and many textbooks categorically denied the possibility of multiplying two negative numbers.

Section 4-1 Integers and the Operations of Addition and Subtraction

If we attempt to calculate $4 - 6$ on a number line, as we did with whole numbers, we must draw intervals to the left of 0. In Figure 4-1, $4 - 6$ is pictured as an arrow that starts at 0 and ends 2 units to the left of 0. The new number that corresponds to a point 2 units to the left of 0 is *negative two,* symbolized by $^-2$. Other numbers to the left of 0 are created similarly. The new set of numbers, $\{^-1, ^-2, ^-3, ^-4, ^-5, \ldots\}$ is called the set of **negative integers**.

negative integers

FIGURE 4-1

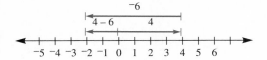

Unfortunately, the symbol "$-$" is used to indicate both a subtraction and a negative sign. To reduce confusion between the uses of this symbol, a raised "$^-$" sign is used for negative numbers, as in $^-2$, in contrast to the lower sign for subtraction. To emphasize that an integer is positive, some people use a raised plus sign, as in $^+3$. In this text, we use the plus sign for addition only and write $^+3$ simply as 3.

The union of the set $\{^-1, ^-2, ^-3, ^-4, ^-5, \ldots\}$ and the set of whole numbers, $\{0, 1, 2, 3, \ldots\}$, is called the set of **integers.** The set of integers is denoted by I:

integers

$$I = \{\ldots, ^-5, ^-4, ^-3, ^-2, ^-1, 0, 1, 2, 3, 4, 5, \ldots\}.$$

The set of integers can be partitioned into three subsets $\{1, 2, 3, 4, \ldots\}$, $\{0\}$, and $\{^-1, ^-2, ^-3, ^-4, \ldots\}$, which are the *positive integers, zero,* and the *negative integers,* respectively. *Zero is neither a positive nor a negative number*.

The negative integers are mirror images of the positive integers (Figure 4-1). For example, the mirror image of 5 is $^-5$, and the mirror image of 0 is 0. Similarly, the positive integers are mirror images of the negative integers. For example, 4 is the mirror image of $^-4$. Another term for "mirror image of" is "**opposite** of." Thus, the opposite of 4 is denoted by $^-4$, and the opposite of $^-4$ can be denoted by $^-(^-4)$, or 4. In general, we have the following definition.

opposite

DEFINITION OF OPPOSITE

If n is an integer, then the unique integer ^-n is called the **opposite** of n if $n + (^-n) = 0 = (^-n) + n$.

In the set of integers I, every element has an opposite that is also in I. This is not the case for the set W of whole numbers. If a is a nonzero element of W, its opposite ^-a is not in the set W.

HISTORICAL NOTE

The dash has not always been used for both the subtraction operation and the negative sign. Other notations were developed but never adopted. One such notation was used by Mohammed al-Khowârizmî (ca. 825), who indicated a negative number by placing a small circle over it. For example, $^-4$ was recorded as $\overset{\circ}{4}$. The Hindus denoted a negative number by enclosing it in a circle; for example, $^-4$ was recorded as ④. The symbols $+$ and $-$ first appeared in print in European mathematics in the late fifteenth century. The symbols referred not to addition or subtraction or to positive or negative numbers, but to surpluses and deficits in business problems.

EXAMPLE 4-1   For each of the following, find the opposite of x:

(a) $x = 3$ (b) $x = {}^-5$ (c) $x = 0$

Solution   (a) $^-x = {}^-3$ (b) $^-x = {}^-({}^-5) = 5$ (c) $^-x = {}^-0 = 0$

The value of ^-x in Example 4-1(b) is 5. *The term ^-x does not necessarily represent a negative integer.* In other words, x is a variable that can be replaced by some number, either positive, zero, or negative. For this reason, many teachers encourage students to refer to ^-x as "the opposite of x" and not "minus x" or "negative x."

Absolute Value

Because 4 and $^-4$ are opposites of each other, they are on opposite sides of 0 on the number line and are the same distance (4 units) from 0, as shown in Figure 4-2.

FIGURE 4-2

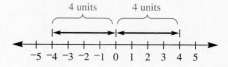

absolute value

Distance is always a positive number or zero. The distance between the points corresponding to an integer and 0 is called the **absolute value** of the integer. Thus, the absolute value of both 4 and $^-4$ is 4, written as $|4| = 4$ and $|{}^-4| = 4$, respectively. (A more formal definition of absolute value as a function is given in Problem Set 4-1.)

EXAMPLE 4-2   Evaluate each of the following:

(a) $|20|$ (b) $|{}^-5|$ (c) $|0|$ (d) $^-|{}^-3|$ (e) $|2 - 5|$

Solution   (a) $|20| = 20$ (b) $|{}^-5| = 5$ (c) $|0| = 0$ (d) $^-|{}^-3| = {}^-3$
(e) $|2 - 5| = |{}^-3| = 3$

EXAMPLE 4-3 Solve each of the following for x, where x is an integer:
(a) $|x| = 4$ (b) $|^-x| = 4$ (c) $|x - 3| = 4$ (d) $^-|^-x| = 4$
(e) $|x| = |^-x|$ (f) $|x| = x$ (g) $|x| = ^-x$

Solution (a) The only two points on the number line at a distance of 4 away from 0 are 4 and $^-4$. Thus, $x = 4$ or $x = ^-4$.

(b) If $|^-x| = 4$, then the only two points on the number line at a distance 4 from 0 are 4 and $^-4$. Therefore, $^-x = 4$ or $^-x = ^-4$. Thus, $x = ^-4$ or $x = 4$.

(c) Since both 4 and $^-4$ are at a distance 4 from 0, we have $x - 3 = 4$ or $x - 3 = ^-4$. Thus, $x = 7$ or $x = ^-1$.

(d) Since $^-|^-x| = 4$, then $|^-x| = ^-4$. Because there are no points on the number line at a distance $^-4$ from 0, this equation has no solution, or the solution set is the empty set.

(e) For all integers x, ^-x and x are the same distance from 0. Therefore, the statement is true for all integers.

(f) If x is positive or 0, then the distance from x to 0 is x. Hence, the solution is any element in the set $W = \{0, 1, 2, 3, \ldots\}$.

(g) $|x|$ is positive or 0 and $|x| = ^-x$. This means ^-x must be positive or 0, and x is negative or 0. Every negative integer or 0 is a solution, so the solution is any element in the set $\{0, ^-1, ^-2, ^-3, \ldots\}$. ∎

Integer Addition

Absolute value can be used to define addition of integers. We consider more informal approaches.

CHIP MODEL

In this model, positive integers are represented by black chips and negative integers by red chips. One red chip "cancels" one black chip. Hence, the integer $^-1$ can be represented by one red chip, or 2 red and 1 black, or 3 red and 2 black, and so on. Similarly, every integer can be represented in many ways using chips. Figure 4-3 shows a chip model for the addition $^-4 + 3$. We put four red chips together with 3 black chips. Because 3 red chips "cancel" 3 black ones, Figure 4-3 represents the equivalent of 1 red chip, or $^-1$.

FIGURE 4-3

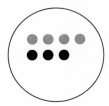

The chip model is frequently used in elementary school, as seen in the student page from *Addison-Wesley Mathematics*, 1991, Grade 7.

Adding Integers

LEARN ABOUT IT

EXPLORE **Use Counting Chips**

Each white chip represents $^+1$. Each red chip represents $^-1$. Since they are opposites, they have a sum of 0 and are said to "cancel each other." It's easy to see that the chips in box A represent $^+2$ if you first "cancel" the pairs of opposites.

Work with a partner. One of you selects some white chips and records the integer the chips represent. The other selects some red chips and records the integer those chips represent. Combine both sets of chips and record the integer they represent. Repeat the activity several times using different numbers of red and white chips.

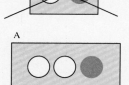

A

TALK ABOUT IT

1. When you and your partner combined chips, how did you decide what integer they represented?

2. What happens if you choose the same number of red and white counters?

You can use counting chips to model addition with integers.

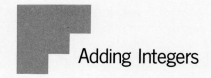

$^+3 + {}^+4$ means ○○ ○○ ○○ → ○○○ ○○○○ so $^+3 + {}^+4 = {}^+7$.

$^-2 + {}^-3$ means ●● ●●● → ●● ●●● so $^-2 + {}^-3 = {}^-5$.

$^+4 + {}^-1$ means ○○ ○○ ● → ○⊘ ○○ so $^+4 + {}^-1 = {}^+3$.

$^+7 + ({}^+3 + {}^-3)$ means ○○○ ○○○○ ○○○ ●●● → ○○○ ○○○○ so $^+7 + ({}^+3 + {}^-3) = {}^+7$.

TRY IT OUT

Use chips to model each sum. Write a complete equation.

1. $^-2 + {}^-2$ 2. $^+6 + {}^+5$ 3. $^+5 + {}^-3$ 4. $^-2 + {}^+6$
5. $^+3 + {}^-3$ 6. $^-8 + 0$ 7. $^+3 + ({}^-7 + {}^-7)$ 8. $^-6 + ({}^+5 + {}^+6)$

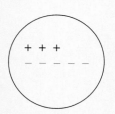

FIGURE 4-4

CHARGED-FIELD MODEL

A model similar to the chip model uses positive and negative charges. A field has 0 charge if it has the same number of positive (+) and negative (−) charges. As in the chip model, a given integer can be represented in many ways using the charged-field model. Figure 4-4 uses the model for $3 + {}^-5$. Because 3 positive charges "cancel" 3 negative charges, the net result is 2 negative ones. Hence, $3 + {}^-5 = {}^-2$.

NUMBER-LINE MODEL

Another model for addition of integers involves a number line used with a toy car. The car starts at 0, facing in a positive direction (to the right). To represent a positive integer, the car moves forward, and to represent a negative integer, it moves in reverse. For example, Figure 4-5(a)-(d) illustrates four different additions.

FIGURE 4-5

(a)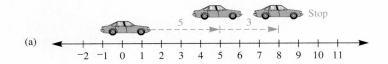

5 + 3 is seen as moving the car forward 5 units and then 3 units for a net move of 8 units to the right from 0. Thus, 5 + 3 = 8.

(b)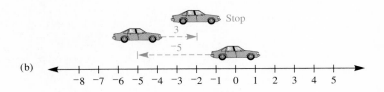

$^-5 + 3$ is seen as moving the car 5 units in reverse and then moving it forward 3 units, for a net move of 2 units to the left from 0. Thus, $^-5 + 3 = {}^-2$.

(c)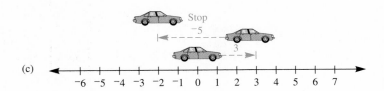

$3 + {}^-5$ is seen as moving the car 3 units forward and then moving it 5 units in reverse, for a net move of 2 units to the left from 0. Thus, $3 + {}^-5 = {}^-2$.

(d)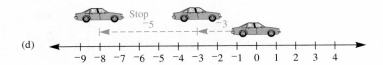

$^-3 + {}^-5$ is seen as moving the car 3 units in reverse and 5 units in reverse, for a net move of 8 units to the left from 0. Thus, $^-3 + {}^-5 = {}^-8$.

Without the car, $^-3 + {}^-5$ can be pictured as in Figure 4-6.

FIGURE 4-6

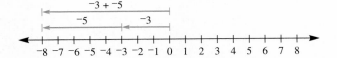

Figure 4-7 similarly depicts integer addition of 3 + ⁻5.

FIGURE 4-7

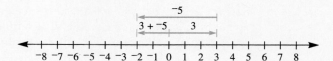

We can describe addition of integers as finding the difference or the sum of the absolute values of these integers and attaching an appropriate sign. Problem 40 in Problem Set 4-1 explores this.

E X A M P L E   4 - 4   Find each of the following sums:

(a) 9 + ⁻4   (b) ⁻5 + ⁻8   (c) ⁻5 + 5   (d) 3 + ⁻10

Solution   (a) 9 + ⁻4 = 5   (b) ⁻5 + ⁻8 = ⁻13
(c) ⁻5 + 5 = 0   (d) 3 + ⁻10 = ⁻7

Example 4-5 involves a thermometer with a scale in the form of a vertical number line.

E X A M P L E   4 - 5   The temperature was ⁻4°C. In an hour, it rose 10°C. What is the new temperature?

Solution   Figure 4-8 shows that the new temperature is 6°C and that ⁻4 + 10 = 6.

Properties of Integer Addition

Integer addition has all the properties of whole-number addition. These properties are summarized next.

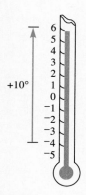

FIGURE 4-8

> **PROPERTIES**
>
> Given integers a, b, and c:
>
> **Closure Property of Addition of Integers** $a + b$ is a unique integer.
> **Commutative Property of Addition of Integers** $a + b = b + a$
> **Associative Property of Addition of Integers** $(a + b) + c = a + (b + c)$
> **Identity Element of Addition of Integers**   0 is the unique integer such that, for all integers a, $0 + a = a = a + 0$.

additive inverse   We have seen that every integer has an opposite. This opposite is also called the **additive inverse**. The fact that each integer has a unique (one and only one) additive inverse is recorded below.

UNIQUENESS PROPERTY OF ADDITIVE INVERSE

For every integer a, there exists a unique integer ^-a, called the additive inverse of a, such that $a + {}^-a = 0 = {}^-a + a$.

Observe that the additive inverse of ^-a can be written as $^-({}^-a)$, or a. Because the additive inverse of ^-a must be unique, we have $^-({}^-a) = a$.

The additive inverse property and other properties of integer addition make it possible to prove that $^-a + {}^-b = {}^-(a + b)$. (The proof is left as an exercise.) We summarize the properties of the additive inverse below.

PROPERTIES OF ADDITIVE INVERSE

For any integers, a and b:
(1) $^-({}^-a) = a$
(2) $^-a + {}^-b = {}^-(a + b)$

EXAMPLE 4-6 Find the additive inverse of each of the following:
(a) $^-(3 + x)$ (b) $(a + {}^-4)$ (c) $^-3 + ({}^-x)$

Solution   (a) $3 + x$
(b) $^-(a + {}^-4)$, which can be written as $^-(a) + {}^-({}^-4)$, or $^-a + 4$.
(c) $^-[{}^-3 + ({}^-x)]$, which can be written as $^-({}^-3) + {}^-({}^-x)$, or $3 + x$. ∎

Integer Subtraction

As with integer addition, we explore several models for integer subtraction.

CHIP MODEL

To find $3 - {}^-2$ we want to subtract $^-2$ (or remove 2 red chips) from 3 black chips. We need to represent 3 so that at least 2 red chips are present. In Figure 4-9, 3 is represented using 2 red and 5 black chips. When the 2 red chips are removed, 5 black ones are left and, hence, $3 - {}^-2 = 5$.

FIGURE 4-9

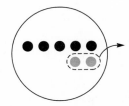

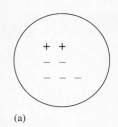
(a)

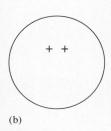
(b)

FIGURE 4-10

CHARGED-FIELD MODEL

Integer subtraction can be modeled with a charged field. For example, consider $^-3 - {}^-5$. To subtract $^-5$ from $^-3$, we must represent $^-3$ so that at least 5 negative charges are present. An example is shown in Figure 4-10(a). To subtract $^-5$, remove the 5 negative charges, leaving 2 positive charges, as in Figure 4-10(b). Hence, $^-3 - {}^-5 = 2$.

PATTERNS MODEL

We may find the difference of two integers by considering the following patterns, where we start with subtractions that we already know how to do. Both the pattern on the left and the pattern on the right start with $3 - 2 = 1$.

$$
\begin{array}{ll}
3 - 2 = 1 & 3 - 2 = 1 \\
3 - 3 = 0 & 3 - 1 = 2 \\
3 - 4 = ? & 3 - 0 = 3 \\
3 - 5 = ? & 3 - {}^-1 = ?
\end{array}
$$

In the pattern on the left, the difference decreases by 1. If we continue the pattern, we have $3 - 4 = {}^-1$ and $3 - 5 = {}^-2$. In the pattern on the right, the difference increases by 1. If we continue the pattern, we have $3 - {}^-1 = 4$ and $3 - {}^-2 = 5$.

NUMBER-LINE MODEL

The number-line model used for integer addition may also be used to model integer subtraction. In Figure 4-11, the car starts at 0 and is pointed in a positive direction (to the right). In this model, the operation of subtraction corresponds to facing the car in a negative direction. We subtract a positive integer by moving the car forward and a negative integer by moving the car in reverse. Figure 4-11 shows some examples.

Subtraction as the Inverse of Addition

Subtraction of integers, like subtraction of whole numbers, can be defined in terms of addition. Recall that $5 - 3$ can be computed by finding a whole number n as follows:

$$5 - 3 = n \quad \text{if and only if} \quad 5 = 3 + n.$$

Because $3 + 2 = 5$, then $n = 2$.

Similarly, we compute $3 - 5$ as follows:

$$3 - 5 = n \quad \text{if and only if} \quad 3 = 5 + n.$$

Because $5 + {}^-2 = 3$, then $n = {}^-2$. In general, for integers a and b, we have the following definition of *subtraction*.

DEFINITION OF SUBTRACTION

For integers a and b, $a - b$ is the unique integer n such that $a = b + n$.

From our previous work with addition of integers, we know that $3 + {}^-5 = {}^-2$ and $3 - 5 = {}^-2$. Hence, $3 - 5 = 3 + {}^-5$. In general, the following is true.

PROPERTY

For all integers a and b, $a - b = a + (^-b)$.

FIGURE 4-11

(a)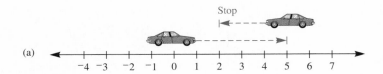

$5 - 3$ first tells you to move the car forward 5 units; then the subtraction sign tells you to face the car in the negative direction, finally move forward 3 units, for a net move of 2 units to the right from 0. Thus, $5 - 3 = 2$.

(b)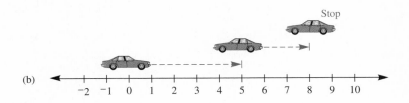

$5 - {^-3}$ first tells you to move the car forward 5 units and then face it in a negative direction and move it in reverse 3 units. The net move is 8 units to the right of 0. Thus, $5 - {^-3} = 8$.

(c)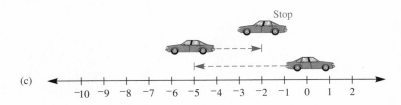

In $^-5 - {^-3}$ move the car 5 units in reverse then face it in a negative direction and move it in reverse 3 units. The net move is 2 units to the left of 0. Thus, $^-5 - {^-3} = {^-2}$.

EXAMPLE 4-7 Use the definition of subtraction to compute the following:

(a) $3 - 10$ (b) $^-2 - 10$

Solution (a) Let $3 - 10 = n$. Then $10 + n = 3$, so $n = {^-7}$. Therefore $3 - 10 = {^-7}$.
(b) Let $^-2 - 10 = n$. Then $10 + n = {^-2}$, so $n = {^-12}$. Therefore $^-2 - 10 = {^-12}$. ∎

Many calculators have a change-of-sign key, either [CHS] or [+/−], that allows computation with integers. For example, to compute $8 - (^-3)$, we would press [8] [−] [3] [+/−] [=].

EXAMPLE 4-8   Compute each of the following, using the fact that $a - b = a + (^-b)$:

(a) $2 - 8$ (b) $2 - (^-8)$ (c) $^-12 - (^-5)$ (d) $^-12 - 5$

Solution   (a) $2 - 8 = 2 + {^-8} = {^-6}$
(b) $2 - (^-8) = 2 + {^-(^-8)} = 2 + 8 = 10$
(c) $^-12 - (^-5) = {^-12} + {^-(^-5)} = {^-12} + 5 = {^-7}$
(d) $^-12 - 5 = {^-12} + {^-5} = {^-17}$

EXAMPLE 4-9   Write an expression equal to $^-(b + {^-c})$ that contains no parentheses.

Solution $^-(b + {^-c}) = {^-b} + {^-(^-c)} = {^-b} + c$

EXAMPLE 4-10   Simplify each of the following:

(a) $2 - (5 - x)$ (b) $5 - (x - 3)$ (c) $^-(x - y) - y$

Solution   (a) $2 - (5 - x) = 2 + {^-(5 + {^-x})}$
$= 2 + {^-5} + {^-(^-x)}$
$= 2 + {^-5} + x$
$= {^-3} + x$
(b) $5 - (x - 3) = 5 + {^-(x + {^-3})}$
$= 5 + {^-x} + {^-(^-3)}$
$= 5 + {^-x} + 3$
$= 8 + {^-x}$
$= 8 - x$
(c) $^-(x - y) - y = ({^-x} + y) - y$
$= ({^-x} + y) + {^-y}$
$= {^-x} + (y + {^-y})$
$= {^-x} + 0$
$= {^-x}$

Order of Operations

Subtraction on the set of integers is neither commutative nor associative, as illustrated in these counterexamples.

$$5 - 3 \neq 3 - 5 \quad \text{because} \quad 2 \neq {^-2}$$
$$(3 - 15) - 8 \neq 3 - (15 - 8) \quad \text{because} \quad {^-20} \neq {^-4}$$

Remember, any computations within parentheses must be completed before other computations.

An expression such as $3 - 15 - 8$ is ambiguous unless we know in which order to perform the subtractions. Mathematicians agree that $3 - 15 - 8$ means $(3 - 15) - 8$; that is, the subtractions in $3 - 15 - 8$ are performed in order from left to right. Similarly, $3 - 4 + 5$ means $(3 - 4) + 5$ and not $3 - (4 + 5)$. Thus, $(a - b) - c$ may be written without parentheses as $a - b - c$.

EXAMPLE 4-11 Compute each of the following:

(a) $2 - 5 - 5$ (b) $3 - 7 + 3$ (c) $3 - (7 - 3)$

Solution (a) $2 - 5 - 5 = {}^-3 - 5 = {}^-8$
(b) $3 - 7 + 3 = {}^-4 + 3 = {}^-1$
(c) $3 - (7 - 3) = 3 - 4 = {}^-1$

PROBLEM SET 4-1

1. Find the opposite of each of the following integers. Write your answer in the simplest possible form.
   (a) 2   (b) $^-5$ (c) m
   (d) 0   (e) ^-m (f) $a + b$
2. Simplify each of the following:
   (a) $^-(^-2)$ (b) $^-(^-m)$ (c) $^-0$
3. Evaluate each of the following:
   (a) $|^-5|$ (b) $|10|$ (c) $^-|^-5|$ (d) $^-|5|$
4. Demonstrate each of the following additions, using the charged-field model:
   (a) $5 + {}^-3$ (b) $^-2 + 3$ (c) $^-3 + 2$
   (d) $^-3 + {}^-2$
5. Demonstrate each of the additions in Problem 4, using the colored-chips model.
6. Demonstrate each of the additions in Problem 4, using a number-line model.
7. Add each of the following:
   (a) $10 + {}^-3$ (b) $10 + {}^-12$
   (c) $10 + {}^-10$ (d) $^-10 + 10$
   (e) $^-2 + {}^-8$ (f) $(^-2 + {}^-3) + 7$
   (g) $^-2 + (^-3 + 7)$
8. ▶Write an addition fact corresponding to each of the following sentences, and then answer the question:
   (a) A certain stock dropped 17 points and the following day gained 10 points. What was the net change in the stock's worth?
   (b) The temperature was $^-10°C$ and then it rose by 8°C. What is the new temperature?
   (c) The plane was at 5000 ft and dropped 100 ft. What is the new altitude of the plane?
   (d) A visitor in a Las Vegas casino lost $200, won $100, and then lost $50. What is the change in the gambler's net worth?
   (e) In four downs, the football team lost 2 yd, gained 7 yd, gained 0 yd, and lost 8 yd. What is the total gain or loss?
9. On January 1, Jane's bank balance was $300. During the month, she wrote checks for $45, $55, $165, $35, and $100 and made deposits of $75, $25, and $400.
   (a) If a check is represented by a negative integer and a deposit by a positive integer, express Jane's transactions as a sum of positive and negative integers.
   (b) What was the balance in Jane's account at the end of the month?
10. Use the charged-field model to show each of the following:
    (a) $3 - {}^-2 = 5$ (b) $^-3 - 2 = {}^-5$
    (c) $^-3 - {}^-2 = {}^-1$
11. Use the car model to find the following:
    (a) $^-4 - {}^-1$ (b) $^-4 - {}^-3$
12. Use patterns to show the following:
    (a) $^-4 - {}^-1 = {}^-3$ (b) $^-2 - 1 = {}^-3$
13. Write the integer suggested by each of the following chip sequences:
    (a) ○, ●●○, ●●●●○, ●●●●●●○, . . .
    (b) ●, ●●○, ●●●●○, ●●●●●●○, . . .
    (c) ●●○, ●●●●○, ●●●●●●○, ●●●●●●●●○, . . .
14. Evaluate each of the following, using the definition of subtraction:
    (a) $2 - 11$
    (b) $^-3 - 7$
    (c) $5 - (^-8)$
    (d) $0 - 4$
15. Perform each of the following:
    (a) $^-2 + (3 - 10)$
    (b) $[8 - (^-5)] - 10$
    (c) $(^-2 - 7) + 10$
    (d) $^-2 - (7 + 10)$
    (e) $8 - 11 - 10$
    (f) $^-2 - 7 + 3$
16. ▶In each of the following, write a subtraction problem that corresponds to the question and an addition problem that corresponds to the question and then answer the question.
    (a) The temperature is 55°F, and it is supposed to drop 60°F by midnight. What is the expected midnight temperature?
    (b) Moses has overdraft privileges at his bank. If he had $200 in his checking account and he wrote a $220 check, what is his balance?

186 CHAPTER 4 THE INTEGERS

17. Simplify each of the following as much as possible:
    (a) $3 - (2 - 4x)$
    (b) $x - (^-x - y)$
    (c) $4x - 2 - 3x$

18. Find all integers x such that the following are true:
    (a) ^-x is positive.
    (b) ^-x is negative.
    (c) $^-x - 1$ is positive.
    (d) $|x| = 2$
    (e) $^-|x| = 2$
    (f) $^-|x|$ is negative.
    (g) $^-|^-x|$ is positive.

19. Columbus discovered America in 1492. Rome was founded 2275 yr before that. When was Rome founded?

20. Let W stand for the set of whole numbers, I the set of integers, I^+ the set of positive integers, and I^- the set of negative integers. Find each of the following:
    (a) $W \cup I$
    (b) $W \cap I$
    (c) $I^+ \cup I^-$
    (d) $I^+ \cap I^-$
    (e) $W - I$
    (f) $I - W$
    (g) $W - I^+$
    (h) $W - I^-$
    (i) $I \cap I$

21. Complete the magic square, using the following integers: $^-13, ^-10, ^-7, ^-4, 2, 5, 8, 11$.

22. Place the integers 1 through 8 in the following boxes so that no two consecutive integers are in boxes that share a common side or vertex (corner).

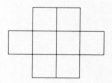

23. Donna picked the Knicks basketball team to win by 12 points. Instead, they lost by 21. By how many points did Donna misjudge the score?

24. Answer each of the following:
    (a) In a game of Triominoes, Jack's scores in five successive turns are 17, $^-8$, $^-9$, 14, and 45. What is his total at the end of five turns?
    (b) The largest bubble chamber in the world is 15 ft in diameter and contains 7259 gal of liquid hydrogen at a temperature of $^-247°C$. If the temperature is dropped by 11°C per hour for 2 consecutive hours, what is the new temperature?
    (c) The greatest recorded temperature ranges in the world are around the "cold pole" in Siberia. Temperatures in Verkhoyansk have varied from $^-94°F$ to 98° F. What is the difference between the high and low temperatures in Verkhoyansk?
    (d) ▶A turnpike driver had car trouble. He knew that he had driven 12 mi from milepost 68 before the trouble started. Assuming he is confused and disoriented when he calls on his CB for help, how can he determine his possible location? Explain.

25. The daily changes in Dolores's favorite stock were recorded as follows: 5, $^-10$, 8, $^-2$, 3, $^-1$, $^-1$. What was the change for the week?

26. Jim recorded his weight gains and losses during 8 wk as follows: $^-2$, $^-4$, 3, 0, $^-2$, $^-3$, 1, 3. How much did Jim gain or lose?

27. Motor oils protect car engines over a range of temperatures. These oils have names like 10W-40 or 5W-30. The graph shows the temperatures, in degrees Fahrenheit, at which the engine is protected by a particular oil. Using the graph below, find which oils can be used for the following temperatures.
    (a) Between $^-5°$ and 90°
    (b) Below $^-20°$
    (c) Between $^-10°$ and 50°
    (d) From $^-20°$ to over 100°
    (e) From $^-8°$ to 90°

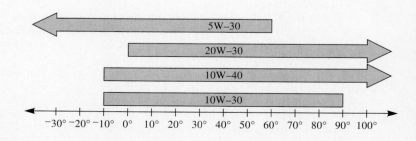

28. Let $f(x) = {}^-x - 1$. Find the following:
    (a) $f({}^-1)$ (b) $f(100)$ (c) $f({}^-2)$
    (d) For which values of x will the output be 3?
29. Let $f(x) = |1 - x|$. Find the following:
    (a) $f(10)$ (b) $f({}^-1)$
    (c) All the inputs for which the output is 1
    (d) The range
30. The following is the definition for the absolute value function, where the domain is the set of integers:

    If x is a positive integer or 0, then $|x| = x$.
    If x is a negative integer, then $|x| = {}^-x$.

    (a) What is the range of this function?
    (b) Use the definition to evaluate each of the following:
        (i) $|5|$ (ii) $|{}^-5|$
        (iii) $|0|$ (iv) $-|{}^-7|$
31. How many integers are there between the given integers (not including the given integers)?
    (a) 10 and 100 (b) $^-30$ and $^-10$
    (c) $^-10$ and 10
    (d) x and y (if $x < y$)
32. Suppose $a = 6$, $b = 5$, $c = 4$, and $d = {}^-3$. Insert parentheses in the expression $a - b - c - d$ to obtain the greatest possible and the least possible values. What are these values?
33. An arithmetic sequence may have a positive or negative difference. In each of the following arithmetic sequences, find the difference and write the next two terms.
    (a) 0, $^-3$, $^-6$, $^-9$
    (b) 7, 3, $^-1$, $^-5$
    (c) $x + y$, x, $x - y$
    (d) $1 - 3x$, $1 - x$, $1 + x$
34. Find the sums of the following arithmetic sequences:
    (a) $^-20 + {}^-19 + {}^-18 + \ldots + 18 + 19 + 20$
    (b) $100 + 99 + 98 + \ldots + {}^-50$
    (c) $100 + 98 + 96 + \ldots + {}^-6$

35. Find the opposites for each of the following, using the $+/-$ key on a calculator.
    (a) 14 (b) 24 (c) $^-2$ (d) $^-5$
36. Complete each of the following integer arithmetic problems on the calculator, making use of the $+/-$ key. For example, to find $^-5 + {}^-4$, press $5\ +/-\ +\ 4\ +/-\ =$.
    (a) $^-12 + {}^-6$ (b) $^-7 + {}^-99$
    (c) $^-12 + 6$ (d) $27 + {}^-5$
    (e) $3 + {}^-14$ (f) $^-7 - {}^-9$
    (g) $^-12 - 6$ (h) $16 - {}^-7$
37. Estimate each of the following and then use a calculator to find the actual answer.
    (a) $343 + {}^-42 - 402$
    (b) $^-1992 + 3005 - 497$
    (c) $992 - {}^-10003 - 101$
    (d) $^-301 - {}^-1303 + 4993$
38. ▶ Explain why $b - a$ and $a - b$ are opposites of each other.
39. ▶ (a) Show that when $a + b$ is added to $^-a + {}^-b$, the result is 0.
    ▶ (b) Use part (a) and the definition of additive inverse to explain why $^-a + {}^-b = {}^-(a + b)$.
40. ▶ Addition of integers with like signs can be described using absolute values as follows:

    To add integers with like signs, add the absolute values of the integers. The sum has the same sign as the integers.

    Describe in a similar way how to add integers with unlike signs.
41. Classify each of the following as true or false. If false, give a counterexample.
    (a) $|{}^-x| = |x|$
    (b) $|x - y| = |y - x|$
    (c) $|{}^-x + {}^-y| = |x + y|$
    (d) $|x^2| = x^2$
    (e) $|x^3| = x^3$
    (f) $|x^3| = x^2|x|$

▼ BRAIN TEASER

If the digits 1 through 9 are written in order, it is possible to place plus and minus signs between the numbers or to use no operation symbol at all to obtain a total of 100. For example,

$1 + 2 + 3 + {}^-4 + 5 + 6 + 78 + 9 = 100.$

Can you obtain a total of 100 using fewer plus or minus signs than in the given example? Notice that digits, such as 7 and 8, may be combined.

COMPUTER CORNER

Type the given Logo programs on your computer.

```
TO ABS :X
    IF :X < 0 THEN OUTPUT ( -:X )
    OUTPUT :X
END
```

(*In LCSI replace IF :X < O THEN OUTPUT (−:X) with IF :X < 0 [OUTPUT −:X].*)

Run this program and input the following values:

(a) ⁻7   (b) 0   (c) 140   (d) ⁻21

Section 4-2 Multiplication and Division of Integers

We may approach multiplication of integers through a variety of models.

PATTERNS MODEL

First, we may approach multiplication of integers by using repeated addition. For example, if E. T. Simpson lost 2 yd on each of three carries in a football game, then he had a net loss of ⁻2 + ⁻2 + ⁻2, or ⁻6, yards. Since ⁻2 + ⁻2 + ⁻2 can be written as 3 · (⁻2), using repeated addition, we have 3 · (⁻2) = ⁻6.

Consider (⁻2) · 3. It is meaningless to say that there are ⁻2 threes in a sum. The following pattern can help develop a feeling for what (⁻2) · 3 should be:

$$4 \cdot 3 = 12$$
$$3 \cdot 3 = 9$$
$$2 \cdot 3 = 6$$
$$1 \cdot 3 = 3$$
$$0 \cdot 3 = 0$$
$$^{-}1 \cdot 3 = ?$$
$$^{-}2 \cdot 3 = ?$$

The first five products, 12, 9, 6, 3, and 0, are terms of an arithmetic sequence with fixed difference ⁻3. If the pattern continues, the next two terms in the sequence are ⁻3 and ⁻6. Thus it appears that (⁻2) · 3 = ⁻6. Recall that 3 · (⁻2) also equals ⁻6. Hence, if (⁻2) · 3 = ⁻6, we have (⁻2) · 3 = 3 · (⁻2). This result is consistent with the commutative property of multiplication for whole numbers.

Next, consider (⁻2) · (⁻3). Using the previous results, we can develop the pattern that follows:

SECTION 4-2 MULTIPLICATION AND DIVISION OF INTEGERS

$$(^-2) \cdot 3 = {^-6}$$
$$(^-2) \cdot 2 = {^-4}$$
$$(^-2) \cdot 1 = {^-2}$$
$$(^-2) \cdot 0 = 0$$
$$(^-2) \cdot (^-1) = ?$$
$$(^-2) \cdot (^-2) = ?$$
$$(^-2) \cdot (^-3) = ?$$

The first four products, $^-6$, $^-4$, $^-2$, and 0, are terms in an arithmetic sequence with fixed difference 2. If the pattern continues, the next three terms in the sequence are 2, 4, and 6. Thus, it appears that $(^-2) \cdot (^-3) = 6$.

CHARGED-FIELD MODEL AND CHIP MODELS

The charged-field and chip models can be used to illustrate multiplication of integers, although an interpretation must be given to the signs. Consider Figure 4-12, where $3 \cdot (^-2)$ is pictured using a chip model.

FIGURE 4-12

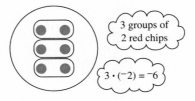

To find $^-3 \cdot (^-2)$ using the charged-field model, remove 3 groups of 2 negative charges from 0 charge field. For that purpose, start with a 0 charge field including at least 6 negative charges, as shown in Figure 4-13.

FIGURE 4-13

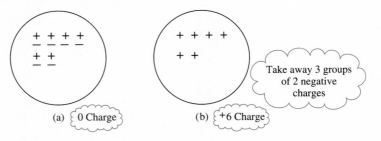

The result is a charge of positive 6, so $^-3 \circ (^-2) = 6$.

NUMBER-LINE MODEL

As for addition and subtraction, we use a car moving along a number line, according to the following rules:

1. Traveling to the left (west) means moving in the negative direction, and traveling to the right (east) means moving in the positive direction.

2. Time in the future is denoted by a positive value, and time in the past is denoted by a negative value.

Consider the number line shown in Figure 4-14. Various cases using this number line are given next.

FIGURE 4-14

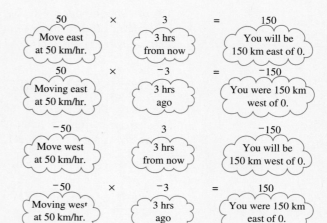

1. If you are now at 0 and move east at 50 km/hr, where will you be 3 hrs from now?

   50 (Move east at 50 km/hr.) × 3 (3 hrs from now) = 150 (You will be 150 km east of 0.)

2. If you are now at 0, moving east at 50 km/hr, where were you 3 hrs ago?

   50 (Moving east at 50 km/hr.) × ⁻3 (3 hrs ago) = ⁻150 (You were 150 km west of 0.)

3. If you are now at 0 and move west at 50 km/hr, where will you be 3 hrs from now?

   ⁻50 (Move west at 50 km/hr.) × 3 (3 hrs from now) = ⁻150 (You will be 150 km west of 0.)

4. If you are now at 0, moving west at 50 km/hr, where were you 3 hrs ago?

   ⁻50 (Moving west at 50 km/hr.) × ⁻3 (3 hrs ago) = 150 (You were 150 km east of 0.)

These models illustrate the following definition of *multiplication of integers*.

DEFINITION OF MULTIPLICATION OF INTEGERS

For any whole numbers a and b:
1. $(^-a)(^-b) = ab$
2. $(^-a)b = b(^-a) = {}^-(ab)$

HISTORICAL NOTE

Leonard Euler, in his book *Anleitung zur Algebra* (1770), was one of the first mathematicians to attempt to prove that $(^-1) \cdot (^-1) = 1$. He reasoned that the product must be either 1 or ⁻1. It was already known that $(1) \cdot (^-1) = {}^-1$, so he reasoned that $(^-1) \cdot (^-1) = 1$.

Properties of Integer Multiplication

The set of integers has properties under multiplication analogous to those of the set of whole numbers under multiplication. These properties are summarized next.

PROPERTIES OF INTEGER MULTIPLICATION

The set of integers I satisfies the following properties of multiplication for all integers a, b, $c \in I$:

Closure Property of Multiplication of Integers ab is a unique integer.
Commutative Property of Multiplication of Integers $ab = ba$.
Associative Property of Multiplication of Integers $(ab)c = a(bc)$
Multiplicative Identity Property 1 is the unique integer such that for all integers a, $1 \cdot a = a = a \cdot 1$.
Distributive Properties of Multiplication over Addition for Integers $a(b + c) = ab + ac$ and $(b + c)a = ba + ca$
Zero Multiplication Property of Integers 0 is the unique integer such that, for all integers a, $a \cdot 0 = 0 = 0 \cdot a$

HISTORICAL NOTE

Emmy Noether (1882–1935) is one of the outstanding mathematicians of the 20th century. She made lasting contributions to the study of *rings,* algebraic systems among which is the set of integers. When she entered the University of Erlanger (Germany) in 1900, Emmy Noether was one of only two women enrolled. After completing her doctorate in 1907, she could not find a suitable job, despite her outstanding achievements, because she was a woman. In 1919, she got a university appointment without pay and, only later, a very modest salary. In 1933, along with many other scholars, she was dismissed from the University at Göttingen because she was Jewish. She immigrated to the United States and taught at Bryn Mawr College until her untimely death only 18 months after arriving in the United States.

A mathematical approach to showing that $(^-2) \cdot 3 = {}^-(2 \cdot 3)$ uses the uniqueness property of additive inverses. If we can show that $(^-2) \cdot 3$ and $^-(2 \cdot 3)$ are additive inverses of the same number, then they must be equal. By definition, the additive inverse of $(2 \cdot 3)$ is $^-(2 \cdot 3)$. That $(^-2) \cdot 3$ is also the additive inverse of $2 \cdot 3$ can be proved by showing that $(^-2) \cdot 3 + 2 \cdot 3 = 0$. The proof follows.

$$(^-2) \cdot 3 + 2 \cdot 3 = (^-2 + 2) \cdot 3 \quad \text{Distributive property of multiplication over addition}$$
$$= 0 \cdot 3 \quad \text{Additive inverse}$$
$$= 0 \quad \text{Zero multiplication}$$

Because $(^-2) \cdot 3$ and $^-(2 \cdot 3)$ are both additive inverses of $(2 \cdot 3)$ and because the additive inverse must be unique, $(^-2) \cdot 3 = {}^-(2 \cdot 3)$.

The preceding proof holds for any integers a and b.

PROPERTY

For any integers a and b, $(^-a)b = {}^-(ab)$.

Similarly, we can prove the following.

> **PROPERTY**
>
> For any integers a and b, $(^-a)(^-b) = ab$.

> **REMARK** Note that there is no restriction that a must be positive or that ^-a must be negative. If we substitute $a = 1$ in $(^-a) \cdot b = {^-(ab)}$, we obtain $(^-1) \cdot b = {^-b}$. This says that multiplication of an integer by $^-1$ results in the opposite of the integer.

EXAMPLE 4-12 Find each of the following:

(a) $(^-3) \cdot (^-15)$ (b) $(^-5) \cdot 7$ (c) $0 \cdot (^-3)$
(d) $0 \cdot (^-n)$, $n \in W$ (e) $(^-5)^4$

Solution (a) $(^-3) \cdot (^-15) = 45$ (b) $(^-5) \cdot 7 = {^-35}$
(c) $0 \cdot (^-3) = 0$ (d) $0 \cdot (^-n) = 0$
(e) $(^-5)^4 = (^-5) \cdot (^-5) \cdot (^-5) \cdot (^-5) = 625$ ∎

The distributive property of multiplication over subtraction follows from the distributive property of multiplication over addition:

$$\begin{aligned} a(b - c) &= a(b + {^-c}) \\ &= ab + a(^-c) \\ &= ab + {^-(ac)} \\ &= ab - ac \end{aligned}$$

Consequently, $a(b - c) = ab - ac$. Similarly, we can show that $(b - c)a = ba - ca$.

> **PROPERTY**
>
> **Distributive Property of Multiplication over Subtraction for Integers** For any integers a, b, and c,
>
> $$a(b - c) = ab - ac$$
> $$(b - c)a = ba - ca$$

EXAMPLE 4-13 Simplify each of the following so that there are no parentheses in the final answer:

(a) $(^-3)(x - 2)$ (b) $(a + b)(a - b)$

Solution (a) $(^-3)(x - 2) = (^-3)x - (^-3)(2) = ^-3x - (^-6) = ^-3x + 6$
(b) $(a + b)(a - b) = (a + b)a - (a + b)b$
$= (a^2 + ba) - (ab + b^2)$
$= a^2 + ab - ab - b^2$
$= a^2 - b^2$
Thus, $(a + b)(a - b) = a^2 - b^2$.

difference of squares   The result $(a + b)(a - b) = a^2 - b^2$ in Example 4-13(b) is commonly called the **difference-of-squares** formula.

EXAMPLE 4-14   Use the difference-of-squares formula to aid in simplifying the following:
(a) $22 \cdot 18$ (b) $(4 + b)(4 - b)$ (c) $(^-4 + b)(^-4 - b)$ (d) $24 \cdot 26$

Solution (a) $22 \cdot 18 = (20 + 2)(20 - 2) = 20^2 - 2^2 = 400 - 4 = 396$
(b) $(4 + b)(4 - b) = 4^2 - b^2 = 16 - b^2$
(c) $(^-4 + b)(^-4 - b) = (^-4)^2 - b^2 = 16 - b^2$
(d) $24 \cdot 26 = (25 - 1)(25 + 1) = (25)^2 - 1^2 = 625 - 1 = 624$

Both the difference-of-squares formula and the distributive properties of multiplication over addition and subtraction can be used for factoring.

EXAMPLE 4-15   Factor each of the following completely:
(a) $x^2 - 9$ (b) $(x + y)^2 - z^2$ (c) $^-3x + 5xy$ (d) $3x - 6$

Solution (a) $x^2 - 9 = x^2 - 3^2 = (x + 3)(x - 3)$
(b) $(x + y)^2 - z^2 = (x + y + z)(x + y - z)$
(c) $^-3x + 5xy = x(^-3 + 5y)$
(d) $3x - 6 = 3(x - 2)$

Integer Division

In the set of whole numbers, $a \div b$, where $b \neq 0$, is the unique whole number c such that $a = bc$. If such a whole number c does not exist, then $a \div b$ is undefined. Division on the set of integers is defined analogously.

DEFINITION OF INTEGER DIVISION

If a and b are any integers, with $b \neq 0$, then $a \div b$ is the unique integer c, if it exists, such that $a = bc$.

EXAMPLE 4-16   Use the definition of division to evaluate each of the following if possible:
(a) $12 \div (^-4)$ (b) $^-12 \div 4$ (c) $^-12 \div (^-4)$ (d) $^-12 \div 5$

Solution (a) Let $12 \div (^-4) = c$. Then, $12 = ^-4c$, and consequently, $c = ^-3$. Thus $12 \div (^-4) = ^-3$.

(b) Let $^-12 \div 4 = c$. Then, $^-12 = 4c$, and therefore, $c = ^-3$. Thus $^-12 \div 4 = ^-3$.

(c) Let $^-12 \div (^-4) = c$. Then, $^-12 = ^-4c$, and consequently, $c = 3$. Thus $^-12 \div (^-4) = 3$.

(d) Let $^-12 \div 5 = c$. Then, $^-12 = 5c$. Because no integer c exists to satisfy this equation, $^-12 \div 5$ is undefined. ∎

Example 4-16 suggests that, *if it exists, the quotient of two negative integers is a positive integer and, if it exists, the quotient of a positive and a negative integer or of a negative and a positive integer is negative.*

Order of Operations on Integers

The following rules apply to the order in which arithmetic operations are performed. Recall that, when addition and multiplication appear in a problem without parentheses, multiplication is done first.

When addition, subtraction, multiplication, and division appear without parentheses, multiplications and divisions are done first, in the order of their appearance from left to right, and then additions and subtractions are done, in the order of their appearance from left to right. Any arithmetic operation appearing inside parentheses must be done first.

EXAMPLE 4-17 Evaluate each of the following:

(a) $2 - 5 \cdot 4 + 1$
(b) $(2 - 5) \cdot 4 + 1$
(c) $2 - 3 \cdot 4 + 5 \cdot 2 - 1 + 5$
(d) $2 + 16 \div 4 \cdot 2 + 8$
(e) $(^-3)^4$
(f) $^-3^4$

Solution (a) $2 - 5 \cdot 4 + 1 = 2 - 20 + 1 = ^-18 + 1 = ^-17$

(b) $(2 - 5) \cdot 4 + 1 = ^-3 \cdot 4 + 1 = ^-12 + 1 = ^-11$

(c) $2 - 3 \cdot 4 + 5 \cdot 2 - 1 + 5 = 2 - 12 + 10 - 1 + 5 = 4$

(d) $2 + 16 \div 4 \cdot 2 + 8 = 2 + 4 \cdot 2 + 8 = 2 + 8 + 8 = 10 + 8 = 18$

(e) $(^-3)^4 = (^-3)(^-3)(^-3)(^-3) = 81$

(f) $^-3^4 = ^-(3^4) = ^-(81) = ^-81$ ∎

> **REMARK** Notice that from Example 4-17(e) and (f), we have $(^-3)^4 \neq ^-3^4$. By convention, $(^-3)^4$ means $(^-3)(^-3)(^-3)(^-3)$ and $^-3^4$ means $^-(3^4)$, or $^-(3 \cdot 3 \cdot 3 \cdot 3)$.

BRAIN TEASER

Express each of the numbers from 1 through 10 using four 4s and any operations. For example,

$1 = 44 \div 44$ or
$1 = (4 \div 4)^{44}$ or
$1 = {}^-4 + 4 + (4 \div 4)$.

PROBLEM SET 4-2

1. Use patterns to show that $(^-1)(^-1) = 1$.
2. Use the charged-field model to show that $(^-4)(^-2) = 8$.
3. Use the number-line model to show that $(^-4) \cdot 2 = {}^-8$.
4. The number of students eating in the school cafeteria has been decreasing at the rate of 20 per year. Assuming that this trend continues, write a multiplication problem that describes the change in the number of students eating in the school cafeteria for each of the following:
   (a) The change over the next 4 years
   (b) The situation 4 years ago
   (c) The change over the next n years
   (d) The situation n years ago
5. Evaluate each of the following:
   (a) $^-3(^-4)$
   (b) $3(^-5)$
   (c) $(^-5) \cdot 3$
   (d) $^-5 \cdot 0$
   (e) $^-2(^-3 \cdot 5)$
   (f) $[^-2(^-5)](^-3)$
   (g) $(^-4 + 4)(^-3)$
   (h) $(^-5 - {}^-3)(^-5 - 3)$
6. ▶Use the definition of division to find each quotient (if possible). If a quotient is not defined, explain why.
   (a) $^-40 \div {}^-8$
   (b) $143 \div (^-11)$
   (c) $^-143 \div 13$
   (d) $0 \div (^-5)$
   (e) $^-5 \div 0$
   (f) $0 \div 0$
7. Evaluate each of the following (if possible).
   (a) $(^-10 \div {}^-2)(^-2)$
   (b) $(^-40 \div 8)8$
   (c) $(a \div b)b$
   (d) $(^-10 \cdot 5) \div 5$
   (e) $(ab) \div b$
   (f) $(^-8 \div {}^-2)(^-8)$
   (g) $(^-6 + {}^-14) \div 4$
   (h) $(^-8 + 8) \div 8$
   (i) $^-8 \div (^-8 + 8)$
   (j) $(^-23 - {}^-7) \div 4$
   (k) $(^-6 + 6) \div (^-2 + 2)$
   (l) $^-13 \div (^-1)$
   (m) $(^-36 \div 12) \div 3$
   (n) $|^-24| \div (3 - 15)$
8. In a lab, the temperature of various chemical reactions was changing by a fixed number of degrees per minute. Write a multiplication problem that describes each of the following:
   (a) The temperature at 8:00 P.M. was 32°C. If it dropped 3°C per minute, what is the temperature at 8:30 P.M.?
   (b) The temperature at 8:20 P.M. was 0°C. If it dropped 4°C per minute, what is the temperature at 7:55 P.M.?
   (c) The temperature at 8:00 P.M. was $^-20$°C. If it dropped 4°C per minute, what is the temperature at 7:30 P.M.?
   (d) The temperature at 8:00 P.M. was 25°C. If it increased every minute by 3°C, what is the temperature at 7:40 P.M.?
   (e) The temperature at 8:00 P.M. was 0°C. If it dropped d degrees per minute, what was the temperature m minutes before?
   (f) The temperature at 8:00 P.M. was 20°C. If it increased every minute by d degrees, what was the temperature m minutes before?
9. (a) On each of four consecutive plays in a football game, Foo University lost 11 yd. If lost yardage is interpreted as a negative integer, write the information as a product of integers and determine the total number of yards lost.
   (b) If Jack Jones lost a total of 66 yd in 11 plays, how many yards, on the average, did he lose on each play?
10. In 1979, it was predicted that the farmland acreage lost to family dwellings over the next 9 yr would be 12,000 acres per year. If this prediction were true, and if this pattern were to continue, how much acreage would be lost to homes by the end of 1992?
11. Show that the distributive property of multiplication over addition, $a(b + c) = ab + ac$ is true for each of the following values of a, b, and c:
    (a) $a = {}^-1$, $b = {}^-5$, $c = {}^-2$
    (b) $a = {}^-3$, $b = {}^-3$, $c = 2$
    (c) $a = {}^-5$, $b = 2$, $c = {}^-6$
12. Compute each of the following:
    (a) $(^-2)^3$
    (b) $(^-2)^4$
    (c) $(^-10)^5 \div (^-10)^2$
    (d) $(^-3)^5 \div (^-3)$
    (e) $(^-1)^{10}$
    (f) $(^-1)^{15}$
    (g) $(^-1)^{50}$
    (h) $(^-1)^{151}$
13. Compute each of the following:
    (a) $^-2 + 3 \cdot 5 - 1$
    (b) $10 - 3 \cdot 7 - 4(^-2) + 3$

196   CHAPTER 4   THE INTEGERS

    (c) $10 - 3 - 12$ (d) $10 - (3 - 12)$
    (e) $(^-3)^2$ (f) $^-3^2$
    (g) $^-5^2 + 3(^-2)^2$ (h) $^-2^3$
    (i) $(^-2)^5$ (j) $^-2^4$

14. If x is an integer and $x \neq 0$, which of the following are always positive and which are always negative?
    (a) $^-x^2$ (b) x^2 (c) $(^-x)^2$ (d) $^-x^3$
    (e) $(^-x)^3$ (f) $^-x^4$ (g) $(^-x)^4$ (h) x^4
    (i) x (j) ^-x

15. Which of the expressions in Problem 14 are equal to each other for all values of x except 0?

16. Identify the property of integers being illustrated in each of the following:
    (a) $(^-3) \cdot (4 + 5) = (4 + 5) \cdot (^-3)$
    (b) $^-4 + ^-7 \in I$
    (c) $5 \cdot [4 \cdot (^-3)] = (5 \cdot 4) \cdot (^-3)$
    (d) $(^-9) \cdot [5 + (^-8)] = (^-9) \cdot 5 + (^-9) \cdot (^-8)$

17. Identify the property of integers used in each of the following:
    (a) $^-x \cdot (y + ^-y) = (^-x) \cdot y + (^-x) \cdot (^-y)$
    (b) $a \cdot (b - 1) = a \cdot (b + ^-1)$
    (c) $a \cdot (b + c) = (b + c) \cdot a$
    (d) $a \cdot (b + c) = a \cdot (c + b)$
    (e) $xy + ^-x = xy - x$

18. Simplify each of the following:
    (a) $(^-x)(^-y)$ (b) $^-2x(^-y)$
    (c) $^-(x + y) + x + y$ (d) $^-1 \cdot x$
    (e) $x - 2(^-y)$ (f) $a - (a - b)$
    (g) $y - (y - x)$ (h) $^-(x - y) + x$

19. Find all integers x (if possible) that make each of the following true:
    (a) $^-3x = 6$ (b) $^-3x = ^-6$
    (c) $^-2x = 0$ (d) $5x = ^-30$
    (e) $x \div 3 = ^-12$ (f) $x \div (^-3) = ^-2$
    (g) $x \div (^-x) = ^-1$ (h) $0 \div x = 0$
    (i) $x \div 0 = 1$ (j) $x^2 = 9$
    (k) $x^2 = ^-9$ (l) $^-x \div ^-x = 1$
    (m) $^-x^2$ is negative.   (n) $^-(1 - x) = x - 1$
    (o) $x - 3x = ^-2x$

20. Multiply each of the following and simplify.
    (a) $^-2(x - 1)$ (b) $^-2(x - y)$
    (c) $x(x - y)$ (d) $^-x(x - y)$
    (e) $^-2(x + y - z)$ (f) $^-x(x - y - 3)$
    (g) $(^-5 - x)(5 + x)$
    (h) $(x - y - 1)(x + y + 1)$
    (i) $(^-x^2 + 2)(x^2 - 1)$

21. Use the difference-of-squares formula to simplify each of the following, if possible.
    (a) $52 \cdot 48$ (b) $(5 - 100)(5 + 100)$
    (c) $(^-x - y)(^-x + y)$ (d) $(2 + 3x)(2 - 3x)$

    (e) $(x - 1)(1 + x)$ (f) $213^2 - 13^2$

22. ▶Can $(^-x - y)(x + y)$ be multiplied by using the difference-of-squares formula? Explain why or why not.

23. Factor each of the following expressions completely and then simplify, if possible.
    (a) $3x + 5x$ (b) $ax + 2x$
    (c) $xy + x$ (d) $ax - 2x$
    (e) $x^2 + xy$ (f) $3x - 4x + 7x$
    (g) $3xy + 2x - xz$ (h) $3x^2 + xy - x$
    (i) $abc + ab - a$
    (j) $(a + b)(c + 1) - (a + b)$
    (k) $16 - a^2$ (l) $x^2 - 9y^2$
    (m) $4x^2 - 25y^2$ (n) $(x^2 - y^2) + x + y$

24. (a) Develop a formula for $(a - b)^2$.
    (b) Use your results from part (a) to compute each of the following in your head:
        (i) 98^2 (*Hint:* Write $98 = 100 - 2$.)
        (ii) 99^2
        (iii) 997^2

25. ▶Kahlil said that using the formula $(a + b)^2 = a^2 + 2ab + b^2$, he can find a similar formula for $(a - b)^2$. Examine his argument. If it is correct, supply any missing steps or justifications; if it is incorrect, point out why.
$(a - b)^2 = [a + (^-b)]^2$
         $= a^2 + 2a(^-b) + (^-b)^2$
         $= a^2 - 2ab + b^2$

26. If x is positive and $y = ^-x$, then which of the following statements is false?
    (a) $x^2y > 0$ (b) $x + y = 0$
    (c) xy is negative   (d) xy^2 is positive

27. (a) Given a calendar for any month of the year, such as the one shown, pick several 3×3 groups of numbers and find the sum of these numbers. How are the obtained sums related to the middle number?

		JULY				
S	M	T	W	T	F	S
		1	2	3	4	5
6	7	8	9	10	11	12
13	14	15	16	17	18	19
20	21	22	23	24	25	26
27	28	29	30	31		

★ (b) Prove that the sum of any 9 digits in any 3×3 set of numbers selected from a monthly calendar will always be equal to 9 times the middle number.

28. In each case, find the next two terms. If a sequence is arithmetic or geometric, find its difference or ratio and the nth term.
    (a) $^-10, ^-7, ^-4, ^-1, 2, 5, \ldots$
    (b) $10, 7, 4, 1, ^-2, ^-5, \ldots$

(c) $^-2, ^-4, ^-8, ^-16, ^-32, ^-64, \ldots$
(d) $^-2, 4, ^-8, 16, ^-32, 64, \ldots$
(e) $2, ^-2^2, 2^3, ^-2^4, 2^5, ^-2^6, \ldots$
(f) $3 \cdot 2^2, ^-4 \cdot 2^3, 5 \cdot 2^4, ^-6 \cdot 2^5, 7 \cdot 2^6, \ldots$

29. Find the sum of the first 100 terms in parts (a) and (b) of Problem 28.

30. Find the first five terms of the sequences whose nth term is as follows:
(a) $n^2 - 10$ (b) $^-5n + 3$ (c) $(^-2)^n - 1$
(d) $(^-2)^n + 2^n$ (e) $n^2(^-1)^n$ (f) $^-n(^-2)^n$
(g) $|10 - n|$ (h) $[1 + (^-1)^n] \cdot 2^n$

31. Find the first two terms of an arithmetic sequence in which the fourth term is $^-8$ and the 101st term is $^-493$.

★32. Use the distributive property, the definition of additive inverse and other properties of integers to justify each of the following:
(a) $(^-a)b = ^-(ab)$ (Hint: Show that $(^-a)b + ab = 0$.)
(b) $(^-a)(^-b) = ab$ (Hint: Use part (a) to show that $(^-a)(^-b) + ^-(ab) = 0$.)

33. ▶Seventh grader Nancy gave the following argument to show that $(^-a)b = ^-(ab)$ for all integers a and b. If the argument is valid, complete its details; if it is not valid, explain why not. *I first show that $(^-1)a = ^-a$ by showing that $(^-1)a + a = 0$.*
Now $(^-a)b = [(^-1)a] \cdot b$
$= (^-1) \cdot (ab)$
$= ^-(ab)$

34. ▶Hosni gave the following argument that $^-(a + b) = ^-a + ^-b$ for all integers a and b. If the argument is correct, supply the missing reasons. If it is incorrect, explain why not.
$^-(a + b) = (^-1) \cdot (a + b)$
$= (^-1)a + (^-1)b$
$= ^-a + ^-b$

35. 🖩 Use the ⁺⁄₋ key on the calculator to compute each of the following:
(a) $^-27 \times 3$ (b) $^-46 \times ^-4$
(c) $^-26 \div 13$ (d) $^-26 \div ^-13$

Review Problems

36. Compute each of the following:
(a) $3 - 6$ (b) $8 + ^-7$
(c) $5 - ^-8$ (d) $^-5 - ^-8$
(e) $^-8 + 5$ (f) $^-8 + ^-5$

37. Illustrate $^-8 + ^-5$ on a number line.

38. Find the opposite of each of the following:
(a) $^-5$ (b) 7 (c) 0

39. Compute each of the following:
(a) $|^-14|$ (b) $|^-14| + 7$
(c) $8 - |^-12|$ (d) $|11| + |^-11|$

40. In the 1400s, European merchants used positive and negative numbers to label barrels of flour. For example, a barrel labeled $^+3$ meant that the barrel was 3 lb overweight, whereas a barrel labeled $^-5$ meant that the barrel was 5 lb underweight. If the following numbers were found on 100-lb barrels, what was the total weight of the barrels?

▼ BRAIN TEASER

If $a, \ldots, z$ are integers, find the product
$$(x - a)(x - b)(x - c) \ldots (x - z).$$

Section 4-3 Solving Equations and Inequalities

The topic of this section is writing and solving equations and inequalities. The *Standards* (p 102), point out that:

In Grades 5–8, the mathematics curriculum should include exploration of algebraic concepts and processes so that students can

- understand the concepts of variable, expression, and equation;
- develop confidence in solving linear equations using concrete, informal, and formal methods;
- apply algebraic methods to solve a variety of real-world and mathematical problems.

Properties of Equations

To solve equations, we need several properties of equality. Children can discover many of these by using a balance scale. For example, consider two weights of amounts a and b on the balances, as in Figure 4-15(a). If the balance is level, then $a = b$. When we add an equal amount of weight c to both sides, the balance is still level, as in Figure 4-15(b).

FIGURE 4-15

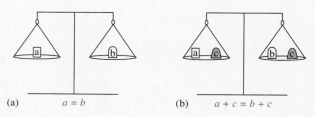

(a) $a = b$ (b) $a + c = b + c$

This demonstrates that, if $a = b$, then $a + c = b + c$.

Similarly, if the scale is balanced with amounts a and b, as in Figure 4-16(a), and we put additional a's on one side and an equal number of b's on the other side, the scale remains level, as in Figure 4-16(b).

FIGURE 4-16

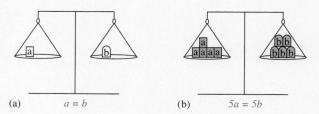

(a) $a = b$ (b) $5a = 5b$

This demonstrates that if $a = b$, then $ac = bc$. The above properties are summarized below.

PROPERTIES

The Addition Property of Equality For any integers a, b, and c, if $a = b$, then $a + c = b + c$.

The Multiplication Property of Equality For any integers a, b, and c, if $a = b$, then $ac = bc$.

The properties imply that we may add the same integer to both sides of an equation or multiply both sides of an equation by the same integer without affecting the equality. (Multiplication of both sides by zero is rarely used.)

Equality is not affected if we substitute a number for its equal. This property is referred to as the **substitution property.** Examples of substitution follow:

substitution property

1. If $a + b = c + d$ and $d = 5$, then $a + b = c + 5$.
2. If $a + b = c + d$, if $b = e$, and if $d = f$, then $a + e = c + f$.

The addition property of equality is formulated as follows. For any integers, a, b, and c, if $a = b$, then $a + c = b + c$. A new statement results from reversing the order of the *if* and *then* parts of this property. The new statement is called the *converse* of the original statement. In the case of the addition property, the converse is a true statement. The converse of the multiplication property of equality is also true when $c \neq 0$. These properties are summarized below.

CANCELLATION PROPERTIES OF EQUALITY

1. For any integers a, b, and c, if $a + c = b + c$, then $a = b$.
2. For any integers a, b, and c, with $c \neq 0$, if $ac = bc$, then $a = b$.

Properties of Inequalities

Before we consider solving equations and inequalities, we develop additional properties of inequalities for integers. Recall that for whole numbers, $5 > 3$ because 5 is to the right of 3 on the number line. Similarly, $^-3 > {^-5}$ because $^-3$ is to the right of $^-5$, as shown in Figure 4-17(a). Because 5 is to the right of 3, there is a positive integer that can be added to 3 to obtain 5, namely 2. Similarly, since $^-3$ is to the right of $^-5$, there is a positive integer that can be added to $^-5$ to obtain $^-3$, namely 2, as shown in Figure 4-17(b). We have $^-5 < {^-3}$, since $^-5 + 2 = {^-3}$ and, in general, we have the following definition.

FIGURE 4-17

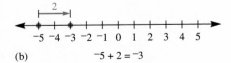

(a) $3 < 5$ and $^-5 < {^-3}$

(b) $^-5 + 2 = {^-3}$

DEFINITION OF GREATER THAN AND LESS THAN FOR INTEGERS

For any integers a and b, a is **greater than** b, written $a > b$, if and only if there exists a positive integer k such that $a = b + k$. Also, b is **less than** a, written $b < a$, if and only if $a > b$.

By the definition of *greater than*, $a > b$ if, and only if, there exists a positive integer k such that $a = b + k$. By the definition of subtraction, $a = b + k$ if, and only if, $a - b = k$. Because k is positive, $a - b > 0$. We summarize this discussion as follows.

For any two integers a and b, $a > b$ if, and only if, $a - b > 0$; that is, $a > b$ if, and only if, $a - b$ is positive.

Now we explore the properties of inequality relations by comparing them to properties of the equality relations. Recall that $a = b$ implies $a + c = b + c$ for all integers c. Analogously, $a > b$ implies that $a + c > b + c$ for all integers c.

Next, consider the multiplication property of equality: $a = b$ implies $ac = bc$ for any integer c. Is an analogous property true for inequalities? If $5 > 3$, then $5 \cdot 2 > 3 \cdot 2$, and $5 \cdot 100 > 3 \cdot 100$, but $5 \cdot (^-1) < 3 \cdot (^-1)$ and $5 \cdot (^-2) < 3 \cdot (^-2)$. In general, $a > b$ and $c > 0$ implies $ac > bc$, but $a > c$ and $c < 0$ implies $ac < bc$. Table 4-1 summarizes these properties and compares them to properties of the equality relation. (The variables a, b, and c represent integers.)

TABLE 4-1

Property	Equality	Inequality
Addition	$a = b$ implies $a + c = b + c$	$a > b$ implies $a + c > b + c$ $a < b$ implies $a + c < b + c$
Multiplication	$a = b$ implies $ac = bc$	$a > b$ and $c > 0$ implies $ac > bc$ $a > b$ and $c < 0$ implies $ac < bc$ $a < b$ and $c > 0$ implies $ac < bc$ $a < b$ and $c < 0$ implies $ac > bc$

All of these properties can be proved using the definition of ">" and "<" or using the property that $a > b$ if, and only if, $a - b$ is a positive integer. It is possible to combine properties of equality and inequality by using $\geq$ (greater than or equal) or $\leq$ (less than or equal).

Examples of the addition property of greater than follow:

$$5 > 2 \quad \text{implies} \quad 5 + 10 > 2 + 10$$
$$^-2 > ^-5 \quad \text{implies} \quad ^-2 + 2 > ^-5 + 2$$
$$x > 3 \quad \text{implies} \quad x + 2 > 3 + 2$$
$$x - 3 > 5 \quad \text{implies} \quad x - 3 + 3 > 5 + 3$$

The following are examples of the multiplication property:

$$5 > 3 \quad \text{implies} \quad 5 \cdot 2 > 3 \cdot 2, \text{ but } 5(^-2) < 3(^-2)$$
$$^-3 > ^-5 \quad \text{implies} \quad (^-3)2 > (^-5)2, \text{ but } (^-3)(^-2) < (^-5)(^-2)$$
$$x > 3 \quad \text{implies} \quad 2x > 2 \cdot 3, \text{ but } ^-2x < ^-2 \cdot 3$$

Properties for subtraction and division of inequalities follow from the addition and multiplication properties of inequality.

PROPERTIES

Subtraction and Division Properties of Inequalities:
If a, b, and c are any integers, then:

1. $a > b$ implies $a - c > b - c$;
2. $a > b$ and $c > 0$ implies $a \div c > b \div c$, provided that the divisions are defined;
3. $a > b$ and $c < 0$ implies $a \div c < b \div c$, provided that the divisions are defined.

EXAMPLE 4-18 Justify each of the following:

(a) $^-2 > {}^-5$ implies $^-7 > {}^-10$. (b) $10 > 6$ implies $5 > 3$.
(c) $10 > 6$ implies $^-5 < {}^-3$. (d) $2x > 6$ implies $x > 3$.
(e) $^-2x > {}^-6$ implies $x < 3$.

Solution (a) By the subtraction property of inequality, $^-2 > {}^-5$ implies $^-2 - 5 > {}^-5 - 5$; that is, $^-7 > {}^-10$.

(b) By the division property of inequality, $10 > 6$ implies $10 \div 2 > 6 \div 2$; that is, $5 > 3$.

(c) By the division property of inequality, $10 > 6$ implies $10 \div {}^-2 < 6 \div {}^-2$; that is, $^-5 < {}^-3$.

(d) By the division property of inequality, $2x > 6$ implies $2x/2 > 6/2$; that is, $x > 3$.

(e) By the division property of inequality, $^-2x > {}^-6$ implies $^-2x/({}^-2) < {}^-6/({}^-2)$; that is $x < 3$. ∎

Solving Equations and Inequalities

Part of the study of algebra involves operations on numbers and other elements represented by symbols. Finding solutions to equations and inequalities is one part of algebra.

HISTORICAL NOTE

The word *algebra* comes from the Arabic book *Al-jabr wa'l muqabalah* written by Mohammed al-Khowârizmî (ca. 825). *Al-jabr* means restoring the balance in an equation by putting on one side of an equation a term that has been removed from the other side. Algebra was introduced in Europe in the thirteenth and fourteenth centuries by Leonardo of Pisa (also called Fibonacci). Algebra was occasionally referred to as *Ars Magna*, or "the great art." Both Diophantus (ca. A.D. 250) and Francois Viète (1540–1603) have been called "fathers of algebra." Little is known about Diophantus, a Greek, except that he is supposed to have lived to be 84 years old and that he wrote *Arithmetica*, a treatise originally in 13 books. Viète was a French lawyer who devoted his leisure time to mathematics. Not liking the word *algebra*, he referred to the subject as "the analytic art."

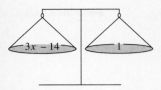

FIGURE 4-18

In order to solve equations, we may use the properties of equality developed earlier. A balance scale can be used to demonstrate solving equations. Consider $3x - 14 = 1$. Put the equal expressions on the opposite pans of the balance scale. Since the expressions are equal, the pans should be level, as in Figure 4-18.

To solve for x, we use the properties of equality to manipulate the expressions on the scale so that, after each step, the scale remains level and, at the final step, only an x remains on one side of the scale. The number on the other side of the scale represents the solution to the original equation. To find x in the equation of Figure 4-18, consider the scales pictured in successive steps in Figure 4-19. In Figure 4-19, each successive scale represents an equation that is equivalent to the original equation; that is, each has the same solution as the original. The last scale shows $x = 5$. To check that 5 is the correct solution, we substitute 5 into the original equation for x. Because $3 \cdot 5 - 14 = 1$ is a true statement, 5 is the solution to the original equation.

FIGURE 4-19

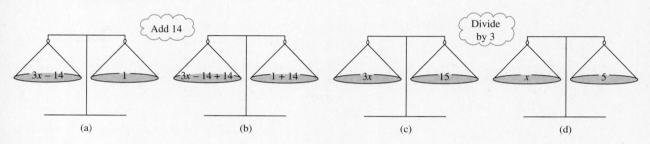

EXAMPLE 4-19 Solve each of the following for x, where x is an integer:

(a) $x + 4 = {}^-6$ (b) $x + 4 > {}^-6$
(c) ${}^-x - 5 = 8$ (d) ${}^-x - 5 \geq 8$
(e) ${}^-2x + 3 = {}^-11$ (f) ${}^-2x + 3 > {}^-11$

Solution

(a)
$$x + 4 = {}^-6$$
$$(x + 4) + {}^-4 = {}^-6 + {}^-4$$
$$x + (4 + {}^-4) = {}^-6 + {}^-4$$
$$x + 0 = {}^-10$$
$$x = {}^-10$$

(b)
$$x + 4 > {}^-6$$
$$(x + 4) + {}^-4 > {}^-6 + {}^-4$$
$$x + (4 + {}^-4) > {}^-6 + {}^-4$$
$$x + 0 > {}^-10$$
$$x > {}^-10, x \in I$$

(c)
$${}^-x - 5 = 8$$
$$({}^-x - 5) + 5 = 8 + 5$$
$${}^-x = 13$$
$$({}^-x)({}^-1) = 13({}^-1)$$
$$x = {}^-13$$

(d)
$${}^-x - 5 \geq 8$$
$$({}^-x - 5) + 5 \geq 8 + 5$$
$${}^-x \geq 13$$
$$({}^-x)({}^-1) \leq 13({}^-1)$$
$$x \leq {}^-13, x \in I$$

(e)
$${}^-2x + 3 = {}^-11$$
$$({}^-2x + 3) + {}^-3 = {}^-11 + {}^-3$$
$${}^-2x = {}^-14$$
$$({}^-2x) \div {}^-2 = {}^-14 \div {}^-2$$
$$x = 7$$

(f)
$${}^-2x + 3 > {}^-11$$
$$({}^-2x + 3) + {}^-3 > {}^-11 + {}^-3$$
$${}^-2x > {}^-14$$
$$({}^-2x) \div {}^-2 < ({}^-14) \div {}^-2$$
$$x < 7, x \in I$$ ∎

SECTION 4-3 SOLVING EQUATIONS AND INEQUALITIES

Developing Algebra Skills

To apply algebra in solving problems, we frequently need to translate given information into a symbolic expression involving quantities designated by letters. Consider the following examples:

EXAMPLE 4-20 In each of the following, translate the given information into a symbolic expression involving quantities designated by letters.

(a) One weekend, a music store sold twice as many CDs as cassettes and 25 fewer records than CDs. If the music store sold x number of cassettes, how many records and CDs did it sell?

(b) French fries have about 12 calories apiece. A hamburger with a bun and relish has about 600 calories. Akiva is on a diet of 2000 calories per day. If he ate x french fries and one hamburger, how many more calories can he consume that day?

(c) First class postage in 1991 costs 29¢ for the first ounce and 23¢ for each additional ounce. At this rate, what is the cost of mailing a letter that weighs x ounces?

Solution

(a) Because x cassettes were sold, twice as many CDs as cassettes implies $2x$ CDs. Thus, 25 fewer records than CDs implies $2x - 25$ records.

(b) First, we find in terms of x how many calories Akiva consumed eating x french fries and one hamburger. Then, to find how many more calories he can consume, we subtract this expression from 2000.

| 1 french fry | 12 calories |
| x french fries | $12x$ calories |

Number of calories in x french fries and one hamburger:

$$600 + 12x$$

The number of calories left for the day is $2000 - (12x + 600)$, or $2000 - 12x - 600$, or $1400 - 12x$.

(c) If a letter weighs x ounces, the first ounce costs 29¢ and the remaining $x - 1$ ounces cost 23¢ each. The cost of the $x - 1$ ounces therefore is $23(x - 1)$ cents. We have:

Cost of mailing an x-ounce letter = cost of first ounce + cost of next $x - 1$ ounces
$= 29 + 23(x - 1)$ cents ∎

REMARK Note that in each part of Example 4-20, the answer is a function of x.

EXAMPLE 4-21 A teacher instructed her class as follows:

Take any number and add 15 to it. Now multiply that sum by 4. Next subtract 8, and divide the difference by 4. If you now subtract 12 from the quotient and tell me the number, I will tell you the number you started with.

Analyze the instructions to see how the teacher was able to determine the original number.

Solution We translate the information into an algebraic form.

Instructions	Discussion	Symbols
Take any number.	Since any number is used, we need a variable to represent the number. Let n be that variable.	n
Add 15 to it.	We are told to add 15 to "it." "It" refers to the variable n.	$n + 15$
Multiply that sum by 4.	We are told to multiply "that sum" by 4. "That sum" is $n + 15$.	$4(n + 15)$
Subtract 8.	Now we subtract 8 from the product.	$4(n + 15) - 8$
Divide the difference by 4.	The difference is $4(n + 15) - 8$. We divide it by 4.	$\dfrac{4(n + 15) - 8}{4}$
Subtract 12 from the quotient, and tell me the answer.	We subtract 12 from the quotient.	$\dfrac{4(n + 15) - 8}{4} - 12$

We simplify the last expression to see how the teacher could give the answer quickly.

$$\frac{4(n + 15) - 8}{4} = \frac{4(n + 15)}{4} - \frac{8}{4}$$
$$= n + 15 - 2$$
$$= n + 13$$

Consequently,

$$\frac{4(n + 15) - 8}{4} - 12 = n + 13 - 12$$
$$= n + 1.$$

The answer the student gives the teacher is 1 greater than the student's original number. To tell the student's number, the teacher merely subtracts 1 from the student's answer. ∎

Some common English phrases used in word problems, together with their symbolic translations, are given in Table 4-2, where n and a represent variables.

TABLE 4-2

Greater than n by a	$n + a$
Less than n by a	$n - a$
a times n	an
The difference of n and a	$n - a$
The sum of n and a	$n + a$
The square of a subtracted from the square of n	$n^2 - a^2$
The square of the difference of a and n	$(a - n)^2$

REMARK It should be noted that the difference of n and a could also be written as $a - n$. The context is frequently the only guide as to which is intended.

HISTORICAL NOTE

Mary Fairfax Somerville (1780–1872) was born in Scotland of upper-class parents. Her introduction to algebra came at about age 13, while reading a ladies' fashion magazine which contained some puzzles. Though not allowed to study mathematics formally, at age 27, widowed and with two children, she bought and studied a set of mathematics books. In her autobiography, she wrote, "I was sometimes annoyed when in the midst of a difficult problem someone would enter and say, I have come to spend a few hours with you." Shortly before her death, she wrote, "I am now in my ninety-second year, . . . , I am extremely deaf, and my memory of ordinary events, and especially of the names of people, is failing, but not for mathematical and scientific subjects. I am still able to read books on the higher algebra for four or five hours in the morning and even to solve the problems. Sometimes I find them difficult, but my old obstinacy remains, for if I do not succeed today, I attack them again tomorrow."

We can use algebra to solve many types of problems. Naturally, the problems we attempt in elementary school are not the complex problems involved with topics such as world economics or space travel, but they can help develop competence in problem solving. The following simple model demonstrates a method for solving word problems: Formulate the word problem as a mathematical problem, solve the mathematical problem, and then interpret the solution in terms of the original problem.

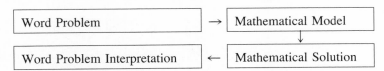

At the third-grade level, an example of this model appears as follows:

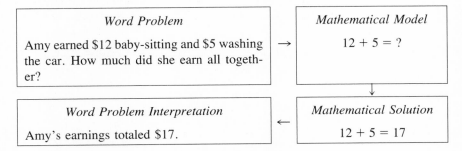

We can apply Polya's four-step problem-solving process to solving word problems in which the use of algebra is appropriate.

In *Understanding the Problem*, we identify what is given and what is to be found. In *Devising a Plan,* we assign letters to the unknown quantities and translate the information in the problem into a model involving equations or inequalities. In *Carrying Out the Plan*, we solve the equations or inequalities. In *Looking Back,* we interpret and check the solution in terms of the original problem. This process is demonstrated on the following, a student page from *Addison-Wesley Mathematics*, 1991, Grade 8.

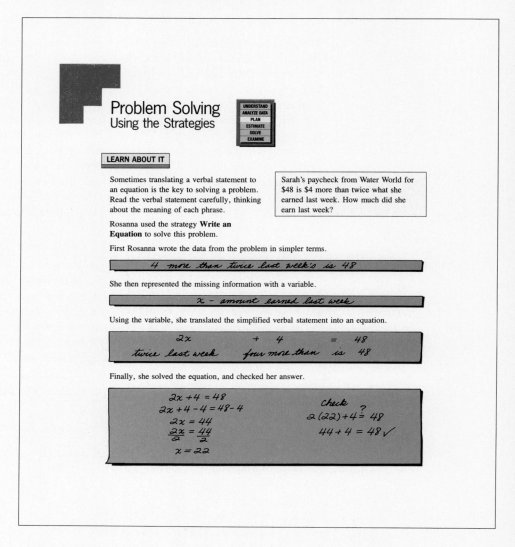

PROBLEM 1

David and Mona Liu bought some new household appliances. The total cost, with tax and interest, was $2450. They made a down payment of $350. The balance was to be paid in 12 equal installments. How much was each payment?

Understanding the Problem. The total cost of the new appliances was $2450. The down payment was $350. The balance was to be paid in 12 equal payments. We need to find the amount of each payment.

Devising a Plan. Let x be the amount of each payment. The given information suggests the strategy of *writing an equation:*

Total cost = down payment + 12 equal payments.

Hence,

$$2450 = 350 + 12x.$$

Carrying Out the Plan. Solve the following equation:

$$2450 = 350 + 12x.$$
$$2450 - 350 = 350 - 350 + 12x$$
$$2100 = 12x$$
$$\frac{2100}{12} = x$$
$$175 = x$$

Thus each payment was $175.

Looking Back. To check that $175 is the correct solution, follow the written information. Twelve payments of $175 each results in 12 · 175, or $2100. Adding the $350 down payment results in $2100 + $350, or $2450, which was the total cost as given. So the answer is correct.

We can also solve this problem without algebra. After the $350 down payment, David and Mona still owe $2450 − $350, or $2100. This amount must be paid in 12 equal payments. Hence, each payment is $2100 ÷ 12, or $175. ■

PROBLEM 2

Bruno has five books overdue at the library. The fine is 10¢ a day per book. He remembers that he checked out an astronomy book a week earlier than the four novels. If his total fine was $8.70, how long was each book overdue?

Understanding the Problem. Bruno has five books overdue. He checked out an astronomy book 7 days earlier than the four novels so the astronomy book is overdue 7 days more than the novels. The fine per day for each book is 10¢, and the total fine was $8.70. We need to find out how many days each book is overdue.

Devising a Plan. Let x be the number of days that each of the four novels is overdue. The astronomy book is overdue 7 days longer, that is, $x + 7$ days. To *write an equation* for x, we express the total fine in two different ways. We know that the total fine is $8.70. This fine equals the fine for the astronomy book plus the fine for the 4 novels. We have (in cents):

Fine for each of the novels = fine per day times the number of overdue days

10 · x

= $10x$

Fine for the four novels = $\underbrace{\text{one day fine for 4 novels}}_{4 \cdot 10}$ times $\underbrace{\text{number of overdue days}}_{x}$

$$= (4 \cdot 10)x$$
$$= 40x$$

Fine for the astronomy book = $\underbrace{\text{fine per day}}_{10}$ times $\underbrace{\text{the number of overdue days}}_{(x+7)}$

$$= 10 \cdot (x + 7) \quad \text{(in cents)}$$

Because each of the above expressions is in cents, we need to write the total fine of $8.70 as 870¢ to produce the following:

Fine for the four novels + fine for the astronomy book = total fine
$\quad\quad 40x \quad\quad\quad + \quad\quad 10(x+7) \quad\quad\quad = 870$

Carrying Out the Plan. We must solve the equation for x.

$$40x + 10(x + 7) = 870$$
$$40x + 10x + 70 = 870$$
$$50x = 870 - 70$$
$$50x = 800$$
$$x = 16$$

Thus, each of the four novels was 16 days overdue, and the astronomy book was overdue $x + 7$, or 23, days.

Looking Back. To check the answer, we follow the original information. We found that each of the four novels was 16 days overdue, and the astronomy book was 23 days overdue. Because the fine was 10¢ per day per book, the fine for each of the novels was $16 \cdot 10¢$, or 160¢. Hence, the fine for all four novels was $4 \cdot 160¢$, or 640¢. The fine for the astronomy book was $23 \cdot 10¢$, or 230¢. Consequently, the total fine was 640¢ + 230¢, or 870¢, which agrees with the given information of $8.70 as the total fine.

Rather than letting x be the number of days that each of the four novels was overdue, we could designate the number of days that the astronomy book was overdue by x. Then the number of days that each of the novels was overdue would be 7 days less, that is, $x - 7$. Then, the total fine would be $10x + 40(x - 7) = 870$. Solving this equation, we would find $x = 23$ and $x - 7 = 16$, that is, 23 overdue days for the astronomy book and 16 days for each of the four novels. ∎

PROBLEM 3

Beans that cost 75¢ per pound are mixed with beans that cost 95¢ per pound to produce a 20-lb mixture that costs 80¢ per pound. How many pounds of the beans costing 75¢ per pound are used?

Understanding the Problem. We need to find how many pounds of beans costing 75¢ per pound are necessary to make 20 lb of a mixture costing 80¢ per pound. To make the 20-lb mixture, beans costing 95¢ per pound are mixed with beans costing 75¢ per pound. Thus, the number of pounds of beans costing 95¢ per pound is 20 minus the number of pounds of beans costing 75¢ per pound. Also, the total cost of the beans costing 75¢ per pound and the beans costing 95¢ per pound must be the cost of 20 lbs of beans costing 80¢ per pound.

Devising a Plan. The information suggests *writing an equation*. Let x be the number of pounds of beans costing 75¢ per pound. Using this symbolism, we know that the number of pounds of beans costing 95¢ per pound is $20 - x$ pounds. Since the rest of the given information involves cost, we need the cost of each type of beans. The cost of x pounds of beans costing 75¢ per pound is $75x$ (in cents). Similarly, the cost of $20 - x$ pounds of beans costing 95¢ per pound is $95(20 - x)$ (in cents). The total mixture, 20 lbs, costs 80¢ per pound or $80 \cdot 20$ cents. We use this information to write the following equation:

$$75x + 95(20 - x) = 80 \cdot 20.$$

Carrying Out the Plan. We must solve the equation for x. This will give the number of pounds of beans costing 75¢ per pound.

$$75x + 95(20 - x) = 80 \cdot 20$$
$$75x + 1900 - 95x = 1600$$
$$^{-}20x + 1900 = 1600$$
$$^{-}20x = {}^{-}300$$
$$x = 15$$

Thus, 15 lbs of beans costing 75¢ per pound are required. Because there are 20 lbs of the final mixture, of which 15 lbs consist of beans costing 75¢ per pound, $20 - 15$, or 5, pounds of beans costing 95¢ per pound are used.

Looking Back. The solution should be checked in the original problem. The cost of 15 lbs of beans costing 75¢ per pound is $15 \cdot 75¢$, or \$11.25. The cost of 5 lbs of beans costing 95¢ per pound is $5 \cdot 95¢$, or \$4.75. The cost of 20 lbs of the final mixture at 80¢ per pound is $20 \cdot 80¢$, or \$16.00. The conditions of the problem are satisfied because \$11.25 + \$4.75 = \$16.00.

A different method of solving the problem involves using two unknowns. We let x be the number of pounds of 75¢ beans and y the number of pounds of 95¢ beans. Next, we translate the information from the problem into mathematical statements. There are 20 lbs in the blend, so we have $x + y = 20$. The remaining information tells us about the cost per pound of each type of bean. To produce an equation using this information, notice that the value of beans costing 75¢ per pound plus the value of beans costing 95¢ per pound equals the value of the 20-lb mixture of beans.

Cost 75¢ per pound beans plus cost of 95¢ per pound beans = cost of mixture
          $75x$ + $95y$ = $80 \cdot 20$

The two equations obtained are as follows:

$$x + y = 20$$
$$75x + 95y = 80 \cdot 20$$

We know how to *solve a related problem*, that is equations with one unknown, so we try to combine these two equations into one equation with one unknown. This can be achieved by solving one of the equations for y and then substituting the expression for y in the other equation. Because $x + y = 20$ implies $y = 20 - x$, we substitute $20 - x$ for y in the second equation and solve for x. We then use the value of x obtained to find the value of y. ∎

PROBLEM 4

In a small town, three youngsters deliver all the newspapers. Abby delivers 3 times as many papers as Bob, and Connie delivers 13 more than Abby. If the three youngsters delivered a total of 496 papers, how many papers does each deliver?

Understanding the Problem. The problem asks for the number of papers that each youngster delivers. It gives information that compares the number of papers that each youngster delivers as well as the total number of papers delivered in the town.

Devising a Plan. Let a, b, and c be the number of papers delivered by Abby, Bob, and Connie, respectively. We translate the given information into *equations* as follows:

Abby delivers 3 times as many papers as Bob: $a = 3b$
Connie delivers 13 more papers than Abby: $c = a + 13$
Total delivery is 496: $a + b + c = 496$

In order to reduce the number of variables, substitute $3b$ for a in the second and third equations:

$$c = a + 13 \text{ becomes} \qquad c = 3b + 13$$
$$a + b + c = 496 \text{ becomes } 3b + b + c = 496$$

Next, make an equation in one variable, b, by substituting $3b + 13$ for c in the equation $3b + b + c = 496$, solve for b, and then find a and c.

Carrying Out the Plan.

$$3b + b + 3b + 13 = 496$$
$$7b + 13 = 496$$
$$7b = 483$$
$$b = 69$$

Thus, $a = 3b = 3 \cdot 69 = 207$. Also, $c = a + 13 = 207 + 13 = 220$. So, Abby delivers 207 papers, Bob delivers 69 papers, and Connie delivers 220 papers.

Looking Back. To check the answers, follow the original information, using $a = 207$, $b = 69$, and $c = 220$. The information in the first sentence, "Abby delivers 3 times as many papers as Bob" checks, since $207 = 3 \cdot 69$. The second sentence, "Connie delivers 13 more papers than Abby" is true, because $220 = 207 + 13$. The information on the total delivery checks, since $207 + 69 + 220 = 496$.

An alternate solution involves expressing all quantities in terms of a single variable. Let x be the number of papers that Bob delivers and then express the number of papers that Abby and Connie deliver in terms of x.

Information	Mathematical Translation
The number of papers that Bob delivers.	x
Abby delivers 3 times as many as Bob.	$3x$
Connie delivers 13 more papers than Abby.	$3x + 13$
The total delivery is 496.	$x + 3x + (3x + 13) = 496$

Solve the equation.

$$x + 3x + (3x + 13) = 496$$
$$7x + 13 = 496$$
$$7x = 483$$
$$x = 69$$

Hence, Bob delivers 69 papers. Because $3x = 3 \cdot 69 = 207$, Abby delivers 207 papers; $3x + 13 = 207 + 13 = 220$, so Connie delivers 220 papers.

If we had let x be the number of papers that Abby delivers, the problem would have been more complicated to solve, because, according to that, Bob delivers $x \div 3$ papers. If we had let x be the number of papers that Connie delivers, how would Abby's and Bob's delivered papers have been expressed? ∎

PROBLEM 5

A freight train leaves Rio de Janeiro for Brasilia. At 8:10 A.M., when it passed a suburb of Rio, the train reached a speed of 210 kilometers per hour. Later, a passenger train left Rio for Brasilia. At 9:10 A.M., when it passed the same suburb, this train reached a speed of 240 kilometers per hour. If the trains continue at their respective speeds, at what time will the passenger train catch the freight train?

Understanding the Problem. Two trains leave from the same place (Rio) and pass the same suburb of Rio at known times. The freight train travels at 210 kilometers per hour, and the passenger train passes the suburb 1 hr later traveling at 240 kilometers per hour. We need to determine when the passenger train will catch up with the freight train.

Devising a Plan. Let t be the time in hours that it takes the passenger train to meet the freight train. Since the passenger train leaves 1 hr later, it will travel 1 hr less than the freight train, that is, $t - 1$ hours. We know that each train traveled the same distance from the suburb to the meeting place. Each distance can be expressed in terms of t. Equating the distances will give us an *equation* in t.

In 1 hr, the freight train travels 210 km; therefore, in t hours it travels $210 \cdot t$ kilometers. Similarly, the passenger train travels in $t - 1$ hours $240 \cdot (t - 1)$ kilometers. Thus, we have:

$$\text{Freight train's distance} = \text{Passenger train's distance}$$
$$210t = 240(t - 1)$$

Carrying Out the Plan. We solve the equation.

$$210t = 240(t - 1)$$
$$210t = 240t - 240$$
$$210t - 240t = 240t - 240t - 240$$
$$^{-}30t = {}^{-}240$$
$$t = \frac{{}^{-}240}{{}^{-}30}$$
$$t = 8$$

Thus, the trains met 8 hr after 8:10 A.M. (the time when the freight train passed the suburb), that is, at 4:10 P.M.

Looking Back. At 4:10 P.M., the freight train has traveled for 8 hr, or a distance of 8 · 210, or 1680, kilometers. The passenger train passes the suburb 1 hr later, so at 4:10 P.M. it has traveled for 7 hr, or a distance of 7 · 240, or 1680, kilometers. Because the distances they have traveled by 4:10 P.M. are the same, the trains indeed meet at that time.

We could have designated the time that the passenger train travels to the meeting point by t. Then the freight train's time would have been $t + 1$. This approach is left for you the reader to complete. ∎

▼ BRAIN TEASER

The following is an argument showing that an ant weighs as much as an elephant. What is wrong?

Let e be the weight of the elephant and a the weight of the ant. Let $e - a = d$. Consequently, $e = a + d$. Multiply each side of $e = a + d$ by $e - a$. Then simplify.

$$e(e - a) = (a + d)(e - a)$$
$$e^2 - ea = ae + de - a^2 - da$$
$$e^2 - ea - de = ae - a^2 - da$$
$$e(e - a - d) = a(e - a - d)$$
$$e = a$$

Thus, the weight of the elephant equals the weight of the ant.

PROBLEM SET 4-3

1. Write each of the following lists of numbers in increasing order:
   (a) $^-13, ^-20, ^-5, 0, 4, ^-3$
   (b) $^-5, ^-6, 5, 6, 0$
   (c) $^-20, ^-15, ^-100, 0, ^-13$
   (d) $13, ^-2, ^-3, 5$

2. Show that each of the following is true:
   (a) $^-3 > ^-5$ (b) $^-6 < 0$
   (c) $^-8 > ^-10$ (d) $^-5 < 4$

3. Solve each of the following if x is an integer:
   (a) $x + 3 = ^-15$ (b) $x + 3 > ^-15$
   (c) $3 - x = ^-15$ (d) $^-x + 3 > ^-15$
   (e) $^-x - 3 = 15$ (f) $^-x - 3 \geq 15$
   (g) $3x + 5 = ^-16$ (h) $3x + 5 < ^-16$
   (i) $^-3x + 5 = 11$ (j) $^-3x + 5 \leq 11$
   (k) $5x - 3 = 7x - 1$ (l) $5x - 3 > 7x - 1$
   (m) $3(x + 5) = ^-4(x + 5) + 21$
   (n) $^-5(x + 3) > 0$

4. Which of the following are true for all possible integer values of x?
   (a) $3(x + 1) = 3x + 3$
   (b) $x - 3 = 3 - x$
   (c) $x + 3 = 3 + x$
   (d) $2(x - 1) + 2 = 3x - x$
   (e) $x^2 + 1 > 0$
   (f) $3x > 4x - x$

5. For each of the following, which elements of the given set, if any, satisfy the equation or inequality?
   (a) $x^3 + x^2 = 2x$, $\{1, ^-1, ^-2, 0\}$
   (b) $3x - 3 = 24$, $\{^-9, 9\}$
   (c) $^-x \geq 5$, $\{6, ^-6, 7, ^-7, 2\}$
   (d) $x^2 < 16$, $\{^-4, ^-3, ^-2, ^-1, 0, 1, 2, 3, 4\}$

6. Solve each of the following equations. Check your answers by substituting in the given equation. Assume that x, y, and z represent integers.
   (a) $^-2x + ^-11 = 3x + 4$ (b) $5(^-x + 1) = 5$
   (c) $^-3y + 4 = y$ (d) $^-3(z - 1) = 8z + 3$

7. Translate each of the following expressions into symbolic expressions, where n represents the unknown number:
   (a) The difference of 6 and another number.
   (b) The sum of a number and 14.
   (c) Seven less than four times n.
   (d) Eight greater than three times n.
   (e) The number increased by 10.
   (f) The number multiplied by 4.
   (g) Thirteen decreased by the number.
   (h) The number less four.

Calvin and Hobbes

by Bill Watterson

CALVIN & HOBBES © 1987 Watterson. Reprinted with permission of Universal Press Syndicate. All rights reserved.

8. Write an expression in terms of the given variable that represents the indicated quantity (that is, write the quantity as a function of the given variable).
   (a) The distance traveled at a constant speed of 60 miles/hour during t hours.
   (b) The total amount of money spent on a trip by Dan, Giora, and Yael if Dan spent as much as Giora and Yael combined, and Yael spent $15 more than Giora, who spent g dollars.
   (c) The cost of having a plumber spend x hours at your house if the plumber charges $20 for coming to the house and $25 per hour for labor.
   (d) The amount of money in cents in a jar containing d dimes and some nickels and quarters, if there are three times as many nickels as dimes and twice as many quarters as nickels.
   (e) The sum of three consecutive integers if the smallest integer is x.
   (f) The sum of three consecutive odd integers if the smallest integer is x.
   (g) The sum of three consecutive integers if the middle integer is m.
   (h) The product of three consecutive integers if the middle integer is m.
   (i) The amount of bacteria after n minutes if the initial amount of bacteria is q and the amount of bacteria doubles every minute. (*Hint:* The answer should contain q as well as n.)
   (j) The temperature after t hours if the initial temperature is 40°F and each hour it drops by 3°F.
   (k) Pawel's salary after 3 yr if the first year his salary was s dollars, the second year $5000 higher, and the third year twice as much as the second year.

9. The area of Asia is 982,000 sq mi more than twice that of North America. The area of North America exceeds that of South America by 1,186,000 sq mi and exceeds that of Europe by 4,383,000 sq mi. The total area of the four continents is 35,692,000 sq mi. Find the area of each.

10. If you multiply Tom's age by 3 and add 4, the result is more than 37. What can you tell about Tom's age?

11. If you multiply a number by ⁻6 and then add 20 to the product, the result is 50. What is the number?

12. David has three times as much money as Rick. Together, they have $400. How much does each have?

13. Help Calvin in the cartoon above by translating the information in the first panel of the cartoon into symbols. Assume that A, B, and C lie in a straight line.

14. Factory A produces twice as many cars per day as factory B. Factory C produces 300 cars more per day than factory A. If the total production in the three factories is 7300 cars per day, how many cars per day are produced in each factory?

15. Tea that costs 60¢ per pound is mixed with tea that costs 45¢ per pound to produce a 100-lb blend that costs 51¢ per pound. How much of each kind of tea is used?

16. For a certain event, 812 tickets were sold, totaling $1912. If students paid $2 per ticket and non-students paid $3 per ticket, how many student tickets were sold?

17. The sum of three consecutive integers is 237. Find the three integers.

18. The sum of three consecutive even integers is 240. Find the three integers.

19. The sum of two integers is 21. The first number is twice the second number. Find the integers.

20. A man left an estate of $64,000 to three children. The eldest child received three times as much as the youngest. The middle child received $14,000 more than the youngest. How much did each child receive?

21. There are three consecutive even integers. Seven times the smallest equals five times the largest. What are the three integers?

22. Tickets to the Bach Festival sold for $16 for adults and $6 for students. If there were four times as many adult tickets sold as student tickets, and the total from all ticket sales was $14,000, how many of each kind of tickets were sold?

23. ▶The Race Across America (RAAM) is an annual bicycle race from California to the East Coast. Winning riders average over 500 km per day and often ride for more than 18 hr per day. Pete leaves Albuquerque early one morning riding 40 kph, and Ron passes the same point 2 hr later, riding at 45 kph. How far from Albuquerque will Ron overtake Pete? Explain your reasoning.

24. ▶Two Explorer troops plan to meet at 8:00 A.M. to hike around Clear Lake, a distance of 22 mi. The troop with the picnic lunch arrives on time and decides to leave and hike clockwise around the lake, even though the other troop has not yet arrived. Since they are packing all the food, this troop will be able to hike at only 3 mph. They know that the other troop will be able to hike at 5 mph. They leave a note for the other troop telling of their plans. The second troop shows up, reads the note, and is ready to leave at 10:00 A.M. They are hungry and want to eat as soon as possible. Should they start clockwise or counterclockwise around the lake? Explain your reasoning.

25. ▶Berna left on her bicycle at 7:00 A.M. from her home in Aurora to visit her grandparents in Zenith. She had a tail wind and was able to average 16 mph. She stayed at her grandparents' house for 5 hr and then rode home. Since she then had to ride into the wind, she was able to average only 12 mph. She arrived home at 7:00 P.M. How far is it from Aurora to Zenith? Explain your reasoning.

26. ▶(a) Find all integers x such that $|x| < 4$.
    ▶(b) Find all integers y such that $|y| \leq 3$.
    ▶(c) If x and y satisfy the conditions in (a) and (b), find the smallest and largest values of $|x + y|$ and $|x - y|$. Explain your reasoning.

★27. (a) Is it always true that, for any integers x and y, $x^2 + y^2 \geq 2xy$? Prove your answer.
    (b) For which integers x and y is $x^2 + y^2 = 2xy$?

★28. If $0 < a < b$, where a and b are integers, prove that $a^2 < b^2$.

★29. If $a < b$, where a and b are integers, is it always true that $a^2 < b^2$?

★30. If $a < b$, where a and b are integers, prove that $c - b < c - a$, if c is an integer.

★31. For each of the following, find all integers x such that the statement is true:
    (a) $x + 1 < 3$ and $^-x + 1 < 5$
    (b) $2x < {}^-6$ or $1 + x < 0$

Review Problems

32. Find the additive inverse of each of the following:
    (a) $^-7$ (b) 5
    (c) $^-(^-3)$ (d) $3 - (^-7)$

33. Compute each of the following:
    (a) $^-3 + {}^-7$ (b) $^-3 - {}^-7$
    (c) $3 + {}^-7$ (d) $3 - {}^-7$
    (e) $^-3 \cdot 7$ (f) $^-3 \cdot {}^-7$
    (g) $3 - 7$ (h) $^-21 \div 7$
    (i) $^-21 \div {}^-7$ (j) $7 - 3 - 8$
    (k) $8 + 2 \cdot 3 - 7$ (l) $^-8 - 7 - 2 \cdot 3$
    (m) $|^-7| \cdot |^-3|$ (n) $|^-7| \cdot |^-8|$
    (o) $|^-7| + 8$ (p) $|^-7| - |^-8|$

34. Compute $^-7 + (^-3)$, using a number line.

▼ BRAIN TEASER

Mary, a 10-year-old calculator genius, announced a discovery to her classmates one day. She said, "I have found a special five-digit number I call *abcde*. If I enter 1 and then the number on my calculator and then multiply by 3, the result is the number with 1 on the end!" Can you find her number?

Solution to the Preliminary Problem

Understanding the Problem. The problem asks when the temperatures of two experiments will be the same and what that temperature will be. The initial temperature of the first experiment was 28°F and 5 min later it was $^-12$°F. The temperature decreased by the same number of degrees every minute. In a second experiment, the initial temperature was $^-57°F$ and was decreasing by 3°F every minute.

Devising a Plan. We need to find out how many minutes will have elapsed when the temperatures of the experiments are the same. Let n be that number of minutes and express the temperature of each experiment after n minutes in terms of n. By equating these temperatures, we would then have an *equation* in terms of n. It seems easier to find the temperature in the second experiment. (Why?) Because the temperature decreased by 3°F every minute, in n minutes the temperature decreased $3n$ degrees. Since the initial temperature was $^-57°F$, after n minutes the temperature was $^-57 - 3n$.

We know that in the first experiment, the temperature decreased by the same number every minute. If we know that number, we could find the temperature after n minutes in terms of n. Thus our *subgoal* is to find the decrease in temperature per minute in the first experiment.

Let d be the decrease in temperature per minute in the first experiment. Because the initial temperature was 28°F and after 5 min it was $^-12°F$, we can *write an equation* in d. Because the temperature decreased each minute by d degrees, after 5 min the temperature was $28 - 5d$. Because we know that the temperature was actually $^-12°F$, we have $28 - 5d = ^-12$.

Carrying Out the Plan. We solve the equation.

$$28 - 5d = {}^-12$$
$$28 - 5d + 5d = {}^-12 + 5d$$
$$28 = {}^-12 + 5d$$
$$12 + 28 = 12 + ({}^-12) + 5d$$
$$40 = 5d$$
$$8 = d$$

Thus, the temperature decreased by 8°F per minute. In n minutes, the temperature will have decreased by $8n$ degrees and because the initial temperature was 28°F, after n minutes the temperature can be expressed as $28 - 8n$. We have already found that the temperature of the second experiment after n minutes was $^-57 - 3n$. Equating the temperatures, we have:

$$28 - 8n = {}^-57 - 3n$$
$$28 - 8n + 8n = {}^-57 - 3n + 8n$$
$$28 = {}^-57 + 5n$$
$$57 + 28 = 57 + ({}^-57) + 5n$$
$$85 = 5n$$
$$17 = n$$

Consequently, the temperatures were the same after 17 min. To find the temperature then, we substitute 17 in $28 - 8n$ (or $^-57 - 3n$) and obtain $28 - 8 \cdot 17 = {}^-108$. Hence, the common temperature is $^-108°F$.

Looking Back. We substituted 17 in $28 - 8n$ and obtained $^-108$. We should get the same answer if we substitute 17 in $^-57 - 3n$. Indeed, $^-57 - 3 \cdot 17 = {}^-108$. Therefore, if our original equation is correct, the final answer is correct.

QUESTIONS FROM THE CLASSROOM

1. A fourth-grade student devised the following subtraction algorithm for subtracting $84 - 27$.
   Four minus seven equals negative three.

   $$\begin{array}{r} 84 \\ -27 \\ \hline {}^-3 \end{array}$$

   Eighty minus twenty equals sixty.

   $$\begin{array}{r} 84 \\ -27 \\ \hline {}^-3 \\ 60 \end{array}$$

   Sixty plus negative three equals fifty-seven.

   $$\begin{array}{r} 84 \\ -27 \\ \hline {}^-3 \\ +60 \\ \hline 57 \end{array}$$

   Thus, the answer is 57. What is your response as a teacher?

2. A seventh-grade student does not believe that $^-5 < {}^-2$. The student argues that a debt of \$5 is greater than a debt of \$2. How do you respond?

3. An eighth-grade student claims she can prove that subtraction of integers is commutative. She points out that if a and b are integers, then $a - b = a + {}^-b$. Since addition is commutative, so is subtraction. What is your response?

4. A student claims that, since $(a \cdot b)^2 = a^2 \cdot b^2$, it must also be true that $(a + b)^2 = a^2 + b^2$. How do you respond?

5. A student solves $1 - 2x > x - 5$, where x is an integer, and reports the solution as $x < 2$. The student asks if it is possible to check the answer in a way similar to the method of substitution for equations. What is your response?

6. A student computes $^-8 - 2(^-3)$ by writing $^-10(^-3) = 30$. How would you help this student?

7. A student says that his father showed him a very simple method for dealing with expressions like $^-(a - b + 1)$ and $x - (2x - 3)$. The rule is, if there is a negative sign before the parentheses, change the signs of the expressions inside the parentheses. Thus, $^-(a - b + 1) = {}^-a + b - 1$ and $x - (2x - 3) = x - 2x + 3$. What is your response?

8. A student solving word problems always checks her solutions by substituting in equations rather than by following the written information. Is this an accurate check for the word problem?

9. A student shows you the following proof that $(^-1)(^-1) = 1$: There are two possibilities, either $(^-1)(^-1) = 1$ or $(^-1)(^-1) = {}^-1$. Suppose $(^-1)(^-1) = {}^-1$. Since $^-1 = (^-1) \cdot 1$, then $(^-1)(^-1) = {}^-1$ can be written as $(^-1)(^-1) = (^-1) \cdot 1$. By the cancellation property of multiplication, it follows that $^-1 = 1$, which is impossible. Hence, $(^-1)(^-1)$ cannot equal $^-1$ and must, therefore, equal 1. What is your reaction?

10. A student had the following picture of an integer and its opposite. Other students in the class objected, saying that ^-a should be to the left of 0. How do you respond?

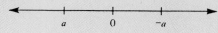

CHAPTER OUTLINE

I. Basic concepts of integers
   A. The set of **integers**, I, is $\{\ldots, {}^-3, {}^-2, {}^-1, 0, 1, 2, 3, \ldots\}$.
   B. The distance from any integer to 0 is called the **absolute value** of the integer. The absolute value of an integer x is denoted $|x|$.
   C. Operations with integers
      1. **Addition:** For any integers a and b:
         $^-a + {}^-b = {}^-(a + b)$
      2. **Subtraction:**
         (a) If a and b are any integers, then $a - b = n$ if and only if $a = b + n$.
         (b) For all integers a and b, $a - b = a + {}^-b$.
      3. **Multiplication:** For any integers a and b:
         (a) $(^-a) \cdot (^-b) = ab$
         (b) $(^-a) \cdot b = b \cdot (^-a) = {}^-(ab)$
      4. **Division:** If a and b are any integers with $b \neq 0$, then $a \div b$ is the unique integer c, if it exists, such that $a = bc$.

5. **Order of operations:** When addition, subtraction, multiplication, and division appear without parentheses, multiplications and divisions are done first, in the order of their appearance from left to right, and then additions and subtractions are done, in the order of their appearance from left to right. Any arithmetic in parentheses is done first.

II. **The system of integers**
   A. The set of integers, $I = \{\ldots, ^{-}3, ^{-}2, ^{-}1, 0, 1, 2, 3, \ldots\}$, along with the operations of addition and multiplication, satisfy the following properties:

Property	+	×
Closure	Yes	Yes
Commutative	Yes	Yes
Associative	Yes	Yes
Identity	Yes, 0	Yes, 1
Inverse	Yes	No
Distributive property of multiplication over addition		

   B. **Zero multiplication property of integers:** $a \cdot 0 = 0 = 0 \cdot a$.

   C. **Addition property of equality:** For any integers a, b, and c, if $a = b$, then $a + c = b + c$.
   D. **Multiplication property of equality:** For any integers a, b and c, if $a = b$, then $ac = bc$.
   E. **Substitution property:** Any number may be substituted for its equal.
   F. **Cancellation properties of equality:**
      (a) For any integers a, b, and c, if $a + c = b + c$, then $a = b$.
      (b) For any integers a, b, and c, if $c \neq 0$ and $ac = bc$, then $a = b$.
   G. For all integers a, b, and c:
      1. $^{-}(^{-}a) = a$
      2. $a - (b - c) = a - b + c$
      3. $(a + b)(a - b) = a^2 - b^2$ (**Difference-of-squares formula**)

III. **Inequalities**
   A. $a > b$ if, and only if, there exists a positive integer k such that $a = b + k$. $b < a$ if, and only if, $a > b$.
   B. Let a and b be any two integers. Then, $a > b$ if, and only if, $a - b > 0$.
   C. Properties of inequalities:
      1. **Addition property:** If $a > b$ and c is any integer, then $a + c > b + c$.
      2. **Multiplication properties:**
         (a) If $a > b$ and $c > 0$, then $ac > bc$.
         (b) If $a > b$ and $c < 0$, then $ac < bc$.

CHAPTER TEST

1. Find the additive inverse of each of the following:
   (a) 3 (b) ^{-}a (c) 0
   (d) $x + y$ (e) $^{-}x + y$ (f) $(^{-}2)^5$ (g) $^{-}2^5$

2. Perform each of the following operations:
   (a) $(^{-}2 + ^{-}8) + 3$ (b) $^{-}2 - (^{-}5) + 5$
   (c) $^{-}3(^{-}2) + 2$ (d) $^{-}3(^{-}5 + 5)$
   (e) $^{-}40 \div (^{-}5)$ (f) $(^{-}25 \div 5)(^{-}3)$

3. For each of the following, find all integer values of x (if there are any) that make the given equation true:
   (a) $^{-}x + 3 = 0$ (b) $^{-}2x = 10$
   (c) $0 \div (^{-}x) = 0$ (d) $^{-}x \div 0 = ^{-}1$
   (e) $3x - 1 = ^{-}124$ (f) $^{-}2x + 3x = x$

4. ▶Use a pattern approach to explain why $(^{-}2)(^{-}3) = 6$.

5. In each of the following chip models, the encircled chips are removed. Write the corresponding integer problem with its solution.

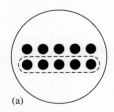

(a)

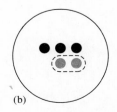

(b)

6. ▶(a) Show that $(x - y)(x + y) = x^2 - y^2$.
   (b) Use the result in (a) to compute $(^{-}2 - x)(^{-}2 + x)$.

7. Simplify each of the following expressions:
   (a) $^{-}1x$ (b) $(^{-}1)(x - y)$
   (c) $2x - (1 - x)$ (d) $(^{-}x)^2 + x^2$
   (e) $(^{-}x)^3 + x^3$ (f) $(^{-}3 - x)(3 + x)$

8. Factor each of the following expressions and then simplify, if possible:
   (a) $x - 3x$
   (b) $x^2 + x$
   (c) $x^2 - 36$
   (d) $81y^6 - 16x^4$
   (e) $5 + 5x$
   (f) $(x - y)(x + 1) - (x - y)$

9. Solve each of the following for x, if x is an integer:
   (a) $^{-}3x + 7 = ^{-}x + 11$
   (b) $|x| = 5$
   (c) $^{-}2x + 1 < 0$
   (d) $^{-}2(^{-}3x + 7) < ^{-}2(^{-}x + 11)$

10. ▶Classify each of the following as true or false (all letters represent integers). Justify your answers.
    (a) $|x|$ always is positive.
    (b) For all x and y, $|x + y| = |x| + |y|$.

(c) If $a < {}^-b$, then $a < 0$.
(d) For all x and y, $(x - y)^2 = (y - x)^2$.
(e) $({}^-a)({}^-b)$ is the additive inverse of ab.

11. Find a counterexample to disprove each of the properties on the set of integers.
    (a) Commutative property of division
    (b) Associative property of subtraction
    (c) Closure property for division
    (d) Distributive property of division over subtraction

12. Use the car model to illustrate each of the following computations:
    (a) ${}^-3 - {}^-4$ (b) ${}^-3 + {}^-4$ (c) ${}^-3({}^-4)$

13. If the temperature was ${}^-16°C$ and it rose by $9°C$, what is the new temperature?

14. A truck contains 150 small packages, some weighing 1 kg each and some weighing 2 kg each. How many packages of each weight are in the truck if the total weight of the packages is 265 kg?

15. John has a collection of nickels and dimes. He has three more dimes than twice the number of nickels. If he has $2.05, how many of each type of coin does he have?

16. A certain college has 5715 undergraduates. There are 115 more seniors than juniors. The number of sophomores is twice the number of seniors, and the number of freshmen is twice the number of juniors. How many freshmen, sophomores, juniors, and seniors attend the college?

17. Two kegs contain equal quantities of beer. From one keg 37 gal are drawn, and from the other, 7 gal are drawn. The quantity now remaining in one keg is seven times that remaining in the other. How much did each keg contain at first?

18. A mail order firm is preparing 1-lb gift packages of mixed nuts. The company is gearing up to make an initial batch of 10,000 packages. They are using Spanish peanuts that cost $2.20 per pound, cashews that cost $3.00 per pound, and pecans that cost $3.20 per pound. Market research has shown that the firm should use twice as many pounds of cashews as pecans. The accounting department wants the mixture to cost $2.72 per pound. To cover taxes and shipping costs and to make a profit, they will then retail the bags at $4.00 apiece. How many pounds of each kind of nut should they use to please both the marketing and accounting departments?

SELECTED BIBLIOGRAPHY

Ballowe, J. "Teaching Difficult Problems Involving Absolute-value Signs." *Mathematics Teacher* 81 (May 1988): 373–374.

Battista, M. "A Complete Model for Operations on Integers." *Arithmetic Teacher* 30 (May 1983): 26–31.

Billstein, R. "Teach a Turtle to Add and Subtract." *The Computing Teacher* 14 (May 1987): 47–50.

Brumfiel, C. "Teaching the Absolute Value Function." *Mathematics Teacher* 73 (January 1980): 24–30.

Charles, R. "Get the Most Out of Word Problems." *Arithmetic Teacher* 29 (November 1981): 39–40.

Crowley, M. and K. Dunn. "On Multiplying Negative Numbers." *Mathematics Teacher* 78 (April 1985): 252–256.

Kohn, J. "A Physical Model for Operations with Integers." *Mathematics Teacher* 71 (December 1978): 734–736.

Morrow, L. "Flow Charts for Equation Solving and Maintenance of Skills." *Mathematics Teacher* 66 (October 1973): 499–506.

Peterson, J. "Fourteen Different Strategies for Multiplication of Integers, or Why $({}^-1)({}^-1) = ({}^+1)$." *Arithmetic Teacher* 19 (May 1972): 396–403.

Richardson, L. "The Role of Strategies for Teaching Pupils to Solve Verbal Problems." *Arithmetic Teacher* 22 (May 1975): 414–421.

Schoenfeld, A. and A. Arcavi. "On the Meaning of Variable." *Mathematics Teacher* 81 (September 1988): 420–427.

Usiskin, Z. "Why Elementary Algebra Can, Should, and Must Be an Eighth-Grade Course for Average Students." *Mathematics Teacher* 80 (September 1987): 428–438.

5 Number Theory

Preliminary Problem

In a national high school essay contest, winners are ranked without ties to receive 1000 $25 U.S. Savings Bonds in such a way that the person with the top ranking receives one more bond than the second-ranked person. The second-ranked person receives one more bond than the third-ranked person, and so on. If all bonds are distributed, what is the minimum number of winners (more than one) for which this scheme will work?

Number theory, concerned primarily with relationships among integers, is associated with names like Pythagoras (500 B.C.), Euclid (300 B.C.), and Diophantus (A.D. 250). As a field of study, number theory flourished in the seventeenth century with the work of Pierre de Fermat. Such number theory topics as multiples, factors, divisibility tests, prime numbers, prime factorizations, greatest common divisors, and least common multiples are incorporated in the elementary mathematics curriculum. A major use of topics in number theory arises in operations on rational numbers written as fractions. The *Standards* (p. 91) offers the following remarks:

Number theory provides rich opportunities for explorations that are interesting, enjoyable, and useful. The fruits of these explorations have payoffs in problem solving, in understanding and developing other mathematical concepts, and in illustrating the beauty of mathematics.

The number theory topic of congruences, introduced by Karl Gauss (1777–1855), is also incorporated into the elementary school curriculum through clock and modular arithmetic. In the *Teaching Standards* (p. 137), we find the following:

Investigations of selected algebraic structures should include concrete examples such as clock arithmetic, modular systems, and matrices.

Section 5-1 Divisibility

In a division such as $12 \div 3 = 4$, all of the statements in the left-hand column below are true. Each statement in the left-hand column can be written as $3|12$.

Example	*General Statement*
12 is divisible by 3.	a is divisible by b.
3 is a divisor of 12.	b is a divisor of a.
12 is a multiple of 3.	a is a multiple of b.
3 is a factor of 12.	b is a factor of a.
3 divides 12.	b divides a.

In general, if $a \div b = c$, where a, b, and c are integers, then the statements in the right-hand column are true. Each statement in the right-hand column can be written as $b|a$. The expression $b|a$ is usually read "b **divides** a."

divides

> **DEFINITION**
>
> If a and b are any integers, then b divides a, written $b|a$, if and only if there is an integer c such that $a = bc$.

factor
divisor
multiple

If $b|a$, then b is a **factor**, or a **divisor**, of a, and a is a **multiple** of b.

Do not confuse $b|a$ with b/a, which is interpreted as $b \div a$. The former, a relation, is either true or false. The latter, an operation, has a numerical value. To compare $0 \div 0$ and

$0 | 0$, recall that $0 \div 0$ is undefined. However, $0 | 0$ is a true statement because $0 = 0 \cdot a$ for all integers a.

HISTORICAL NOTE

Pierre de Fermat (1601–1665) was a lawyer and a magistrate who served in the provincial parliament in Toulouse, France. He devoted his leisure time to mathematics—a subject in which he had no formal training. After his death, his son decided to publish a new edition of Diophantus' *Arithmetica* with Fermat's notes. One of the notes in the margin of Fermat's copy asserted that the equation $x^n + y^n = z^n$ has no positive integer solutions if n is an integer greater than 2 and commented, "I have found an admirable proof of this, but the margin is too narrow to contain it." Many great mathematicians spent years trying to prove Fermat's assertion, now called "Fermat's Last Theorem." With the help of a computer, Fermat's Last Theorem has been proved for all exponents up to 125,000. In 1983, a 29-year-old West German mathematician, Gerd Falting, made major progress toward the solution of the problem, for which he received the Field's Medal in Mathematics. However, the original Fermat's Last Theorem remains unproved after more than 300 years.

We write $5 \nmid 12$ to symbolize that 12 is not divisible by 5, or that 5 does not divide 12. The notation $5 \nmid 12$ is also used to indicate that 12 is not a multiple of 5 and 5 is not a factor of 12.

EXAMPLE 5-1 Classify each of the following as true or false. Explain your answers.

(a) $^-3 | 12$ (b) $0 | 3$ (c) $3 | 0$ (d) If $3 | a$, then $3 | na$, where n is any integer
(e) $8 \nmid 2$ (f) For all integers, a, $1 | a$ (g) For all integers a, $^-1 | a$

Solution

(a) $^-3 | 12$ is true because $12 = ^-4(^-3)$.

(b) $0 | 3$ is false because there is no integer c such that $3 = c \cdot 0$.

(c) $3 | 0$ is true because $0 = 0 \cdot 3$.

(d) $3 | na$ is true. If $3 | a$, then there is an integer k so that $a = 3k$. Multiplying both sides of the equation by n, we have $an = (3k)n$. By the commutative, associative, and closure properties of multiplication of integers, we have $na = 3(nk)$, where nk is an integer, so $3 | na$.

(e) $8 \nmid 2$ is true because there is no integer c such that $2 = c \cdot 8$.

(f) $1 | a$ is true for all integers a because $a = a \cdot 1$.

(g) $^-1 | a$ is true for all integers a because $a = (^-a)(^-1)$. ∎

In Example 5-1(d), we see that if 3 divides a, 3 divides any integer multiple of a. This may be further generalized as Theorem 5-1 shows.

THEOREM 5-1

For any integers a and d, if $d | a$ and n is any integer, then $d | na$.

FIGURE 5-1

Theorem 5-1 can be proved using the definition of divides and the associative property of multiplication for integers.

We can construct other notions of divisibility from everyday models. Consider two packages of chewing gum with five pieces per package, as in Figure 5-1. We can evenly divide each package of gum among five students. In addition, if we opened both packages and put all of the pieces in a bag, we could still evenly divide the pieces of gum among the five students. To generalize this notion, if we buy gum in larger packages with a pieces in one package and b pieces in a second package with both a and b divisible by 5, we could record the preceding discussion as follows:

$$\text{If } 5|a \text{ and } 5|b, \text{ then } 5|(a+b).$$

If the number, a, of pieces of gum in one package is divisible by five, but the number, b, of pieces in the other package is not, then the total, $a + b$, cannot be divided evenly among the five students. This could be recorded as follows:

$$\text{If } 5|a \text{ and } 5 \nmid b \text{ then } 5 \nmid (a+b).$$

What, if anything, can you conclude if $5 \nmid a$ and $5 \nmid b$?

Since subtraction is defined in terms of addition, similar results hold for subtraction. These ideas may be generalized in the following theorem.

THEOREM 5-2

For any integers a, b, and d,

(a) If $d|a$ and $d|b$, then $d|(a+b)$.
(b) If $d|a$ and $d \nmid b$, then $d \nmid (a+b)$.
(c) If $d|a$ and $d|b$, then $d|(a-b)$.
(d) If $d|a$ and $d \nmid b$, then $d \nmid (a-b)$.

The proofs of most theorems in this section are left as exercises, but the proof of Theorem 5-2(a) is given as an illustration.

Proof. To show that $d|(a+b)$, we must show that $(a+b) = d \cdot k$, for some $k \in I$. To do this, we proceed as follows:

$$d|a \quad \text{implies} \quad a = m \cdot d, \quad m \in I$$
$$d|b \quad \text{implies} \quad b = n \cdot d, \quad n \in I$$

Substituting, we obtain

$$a + b = md + nd$$
$$= (m+n)d.$$

Thus, $a + b = (m+n)d$.

Because $m + n$ is an integer, d divides $a + b$, and the proof is complete. ∎

EXAMPLE 5-2 Classify each of the following as true or false, where x, y, and z are integers. If a statement is true, prove it. If a statement is false, provide a counterexample.

(a) If $3|x$ and $3|y$, then $3|xy$.

(b) If $3|(x + y)$, then $3|x$ and $3|y$.

(c) If $9 \nmid a$, then $3 \nmid a$.

Solution (a) True. By Theorem 5-1, if $3|x$, then, for any integer k, $3|kx$. If $k = y$, then $3|yx$ or $3|xy$. (Notice that $3|xy$, regardless of whether $3|y$ or $3 \nmid y$.)

(b) False. For example, $3|(7 + 2)$, but $3 \nmid 7$ and $3 \nmid 2$.

(c) False. For example, $9 \nmid 21$, but $3|21$.

EXAMPLE 5-3 Five students found a padlocked money box, which had a deposit slip attached to it. The deposit slip was water-spotted, so the currency total appeared as shown in Figure 5-2. One student remarked that if the money listed on the deposit slip was in the box, it could easily be divided equally among the five students without using coins. How did the student know this?

FIGURE 5-2

Solution Because the units digit of the amount of the currency is zero, the solution to the problem becomes one of determining whether any natural number whose units digit is 0 is divisible by 5. One method for attacking this problem is to *look for a pattern*. Natural numbers whose units digit is zero form a pattern, that is, 10, 20, 30, 40, 50, ... These numbers are multiples of 10. We are to determine whether 5 divides all multiples of 10. Since $5|10$, by Theorem 5-1, 5 divides any multiple of 10. Hence, 5 divides the amount of money in the box, and the student is correct.

Divisibility Rules

Elementary texts frequently state divisibility rules for determining whether one number divides another. Today, however, divisibility rules have limited use except for mental arithmetic. It is possible to determine whether or not 1734 is divisible by 17, either by using pencil and paper or a calculator. To check divisibility and avoid decimals, we can use a calculator with an integer division button, $\boxed{\text{INT} \div}$. On such a calculator, integer division may be performed using the sequence of buttons listed below:

$\boxed{1}\boxed{7}\boxed{3}\boxed{4}\boxed{\text{INT} \div}\boxed{1}\boxed{7}\boxed{=}$ to obtain the display $\underbrace{\lfloor 1\ 0\ 2 \rfloor}_{Q}$ $\underbrace{\lfloor 0 \rfloor}_{R}$.

This implies that $1734/17 = 102$ with a remainder of 0, which, in turn, implies that $17|1734$.

We could have determined this same result mentally by considering the following:

$$1734 = 1700 + 34.$$

Because $17|1700$ and $17|34$, by Theorem 5-2(a), we have $17|(1700 + 34)$, or $17|1734$. Similarly, we could determine mentally that $17 \nmid 1735$.

To determine mentally whether or not a given integer n is divisible by another integer d, we think of n as the sum or difference of two integers where d divides at least one of these numbers. We try to choose numbers such that one of them is as close as possible to n and divisible by d, and the other number is relatively small. As an example, consider the divisibility of 358 by 2. Instantly, we are sure that 2 divides 358 because 358 is even. We have been conditioned to look only at the units digit to determine if an integer is even. Why is that the case? Consider the following:

$$358 = 350 + 8$$
$$= 35(10) + 8$$

Now we know that $2|10$, so that $2|35(10)$, and that $2|8$, which tells us that $2|(35(10) + 8)$. Similarly, 2 divides any multiple of 10, so that to determine the divisibility of any integer by 2, we only consider whether the units digit is a multiple of 2. If it is, then the number is even and a multiple of 2. If not, then the number is odd and not divisible by 2.

We can develop similar tests for divisibility by 5 and 10. In general, we have the following divisibility rules.

DIVISIBILITY TEST FOR 2

An integer is divisible by 2 if, and only if, its units digit is divisible by 2.

DIVISIBILITY TEST FOR 5

An integer is divisible by 5 if, and only if, its units digit is divisible by 5, that is, if and only if, the units digit is 0 or 5.

DIVISIBILITY TEST FOR 10

An integer is divisible by 10 if, and only if, its units digit is divisible by 10, that is, if, and only if, the units digit is 0.

When we consider divisibility rules for 4 and 8, we see that $4 \nmid 10$ and $8 \nmid 10$, so it is not a matter of checking the units digit for divisibility by 4 and 8. However, 4 (which is 2^2) divides 10^2, and 8 (which is 2^3), divides 10^3.

We first develop a divisibility rule for 4. Consider any four-digit number n such that $n = a \cdot 10^3 + b \cdot 10^2 + c \cdot 10 + d$. Our *subgoal* is to *write the given number as a sum of two numbers,* one of which is as great as possible and divisible by 4. We know that $4|10^2$ because $10^2 = 4 \cdot 25$ and consequently, $4|10^3$. Because $4|10^2$, then $4|b \cdot 10^2$ and $4|a \cdot 10^3$. Finally, $4|a \cdot 10^3$ and $4|b \cdot 10^2$ imply $4|(a \cdot 10^3 + b \cdot 10^2)$. Now the divisibility of $a \cdot 10^3 + b \cdot 10^2 + c \cdot 10 + d$ by 4 depends on the divisibility of $(c \cdot 10 + d)$ by 4. Notice that $c \cdot 10 + d$ is the number represented by the last two digits in the given number n. We summarize this in the following test.

DIVISIBILITY TEST FOR 4

An integer is divisible by 4 if, and only if, the last two digits of the integer represent a number divisible by 4.

To investigate divisibility by 8, we note that the least positive power of 10 divisible by 8 is 10^3 since $10^3 = 8 \cdot 125$. Consequently, all integral powers of 10 greater than 10^3 also are divisible by 8. Hence, the following is a divisibility test for 8.

DIVISIBILITY TEST FOR 8

An integer is divisible by 8 if, and only if, the last three digits of the integer represent a number divisible by 8.

EXAMPLE 5-4   (a) Determine whether 97,128 is divisible by 2, 4, and 8.

(b) Determine whether 83,026 is divisible by 2, 4, and 8.

Solution   (a) $2|97,128$ because $2|8$.   (b) $2|83,026$ because $2|6$.
$4|97,128$ because $4|28$. $4\nmid 83,026$ because $4\nmid 26$.
$8|97,128$ because $8|128$. $8\nmid 83,026$ because $8\nmid 026$. ∎

REMARK In Example 5-4(a), it would have been sufficient to check that the given number is divisible by 8, because, if $8|a$, then $2|a$ and $4|a$. (Why?) However, if $8\nmid a$, we cannot conclude from this that $4\nmid a$ or $2\nmid a$. (Why?)

Next, we consider a divisibility test for 3. No power of 10 is divisible by 3, but the numbers 9, and 99, and 999, and others of this type are close to powers of 10 and are divisible by 3. For example, to determine whether 5721 is divisible by 3, we rewrite the number using 999, 99, and 9, as follows:

$$5721 = 5 \cdot 10^3 + 7 \cdot 10^2 + 2 \cdot 10 + 1$$
$$= 5(999 + 1) + 7(99 + 1) + 2(9 + 1) + 1$$
$$= 5 \cdot 999 + 5 \cdot 1 + 7 \cdot 99 + 7 \cdot 1 + 2 \cdot 9 + 2 + 1$$
$$= (5 \cdot 999 + 7 \cdot 99 + 2 \cdot 9) + (5 + 7 + 2 + 1)$$

The sum in the first set of parentheses is divisible by 3, so the divisibility of 5721 by 3 depends on the sum in the second set of parentheses. In this case, $5 + 7 + 2 + 1 = 15$ and $3|15$, so $3|5721$. Hence, to test 5721 for divisibility by 3, we test $5 + 7 + 2 + 1$ for divisibility by 3. Notice that $5 + 7 + 2 + 1$ is the sum of the digits of 5721. The example suggests the following test for divisibility by 3.

> **DIVISIBILITY TEST FOR 3**
>
> An integer is divisible by 3 if, and only if, the sum of its digits is divisible by 3.

We can use an argument similar to the one used to demonstrate that $3|5721$ to prove the test for divisibility by 3 on any integer and in particular for any four-digit number $n = a \cdot 10^3 + b \cdot 10^2 + c \cdot 10 + d$. Even though $a \cdot 10^3 + b \cdot 10^2 + c \cdot 10 + d$ is not necessarily divisible by 3, the number $a \cdot 999 + b \cdot 99 + c \cdot 9$ is close to n and *is* divisible by 3. We have the following:

$$\begin{aligned} a \cdot 10^3 + b \cdot 10^2 + c \cdot 10 + d &= a \cdot 1000 + b \cdot 100 + c \cdot 10 + d \\ &= a(999 + 1) + b(99 + 1) + c(9 + 1) + d \\ &= (a \cdot 999 + b \cdot 99 + c \cdot 9) + (a \cdot 1 + b \cdot 1 + c \cdot 1 + d) \\ &= (a \cdot 999 + b \cdot 99 + c \cdot 9) + (a + b + c + d) \end{aligned}$$

Because $3|9$, $3|99$, and $3|999$, it follows that $3|(a \cdot 999 + b \cdot 99 + c \cdot 9)$. If $3|(a + b + c + d)$, then $3|[(a \cdot 999 + b \cdot 99 + c \cdot 9) + (a + b + c + d)]$; that is, $3|n$. If, on the other hand, $3 \nmid (a + b + c + d)$, it follows from Theorem 5-2(b) that $3 \nmid n$.

Since $9|9$, $9|99$, $9|999$, and so on, a test similar to that for divisibility by 3 applies to divisibility by 9. (Why?)

> **DIVISIBILITY TEST FOR 9**
>
> An integer is divisible by 9 if, and only if, the sum of the digits of the integer is divisible by 9.

EXAMPLE 5-5 Use divisibility tests to determine whether each of the following numbers is divisible by 3 and divisible by 9.

(a) 1002   (b) 14,238

Solution   (a) Because $1 + 0 + 0 + 2 = 3$ and $3|3$, it follows that $3|1002$. Because $9 \nmid 3$, it follows that $9 \nmid 1002$.

(b) Because $1 + 4 + 2 + 3 + 8 = 18$ and $3|18$, it follows that $3|14{,}238$. Because $9|18$, it follows that $9|14{,}238$. ∎

Divisibility tests can be devised for 7 and 11. We state such tests but omit the proofs.

> **DIVISIBILITY TEST FOR 7**
>
> An integer is divisible by 7 if, and only if, the integer represented without its units digit, minus twice the units digit of the original integer, is divisible by 7.

DIVISIBILITY TEST FOR 11

An integer is divisible by 11 if, and only if, the sum of the digits in the places that are even powers of 10 minus the sum of the digits in the places that are odd powers of 10 is divisible by 11.

For example, to test whether 8,471,986 is divisible by 11, we check whether 11 divides the difference $(6 + 9 + 7 + 8) - (8 + 1 + 4)$, or 17. Because $11 \nmid 17$, it follows from the divisibility test for 11 that $11 \nmid 8,471,986$.

At this point, the only number less than 11 for which we have no divisibility test is 6. The divisibility test for 6 is related to the divisibility tests for 2 and 3. In Section 5-2, it will be shown that, if $2|n$ and $3|n$, then $2 \cdot 3|n$. Consequently, the following divisibility test is true.

DIVISIBILITY TEST FOR 6

An integer is divisible by 6 if, and only if, the integer is divisible by both 2 and 3.

EXAMPLE 5-6 Test each of the following numbers for divisibility by (i) 7, (ii) 11, and (iii) 6.

(a) 462   (b) 964,194

Solution   (a) (i) $7|(46 - 2 \cdot 2)$, so $7|462$.
  (ii) $11|(2 + 4 - 6)$, so $11|462$.
  (iii) $2|462$ and $3|462$, so $6|462$.

(b) (i) To determine whether or not 7 divides 964,194, we use the process several times.

$7|964,194$ if, and only if, $7|(96,419 - 2 \cdot 4)$, or $7|96,411$
$7|96,411$ if, and only if, $7|(9641 - 2 \cdot 1)$, or $7|9639$
$7|9639$ if, and only if, $7|(963 - 2 \cdot 9)$, or $7|945$
$7|945$ if, and only if, $7|(94 - 2 \cdot 5)$, or $7|84$

Because $7|84$ is true, it follows that $7|964,194$.
  (ii) $11|[(4 + 1 + 6) - (9 + 4 + 9)]$, so $11|964,194$.
  (iii) $2|964,194$ and $3|964,194$, so $6|964,194$. ∎

PROBLEM 1

A class from Washington School visited a neighborhood cannery warehouse. The warehouse manager told the class that there were 11,368 cans of juice in the inventory and that the cans were packed in boxes of 6 or 24, depending on the size of the can. One of the students, Sam, thought for a moment and announced that there was a mistake in the inventory. Is Sam's statement correct? Why or why not?

Understanding the Problem. The problem is to determine if the manager's inventory of 11,368 cans was correct. To solve the problem, we must assume that there are no partial boxes of cans; that is, a box must contain exactly 6 or exactly 24 cans of juice.

Devising a Plan. We know that the boxes contain either 6 cans or 24 cans, but we do not know how many boxes of each type there are. One strategy for solving this problem is to *find an equation* that involves the total number of cans in all the boxes.

The total number of cans, 11,368, equals the number of cans in all the 6-can boxes plus the number of cans in all the 24-can boxes. If there are n boxes containing 6 cans each, there are $6n$ cans altogether in those boxes. Similarly, if there are m boxes with 24 cans each, these boxes contain a total of $24m$ cans. Because the total was reported to be 11,368 cans, we have the equation $6n + 24m = 11,368$. Sam claimed that $6n + 24m \neq 11,368$.

One way to show that $6n + 24m \neq 11,368$ is to show that $6n + 24m$ and 11,368 do not have the same divisors. Both $6n$ and $24m$ are divisible by 6, which implies that $6n + 24m$ must be divisible by 6. If 11,368 is not divisible by 6, then Sam is correct.

Carrying Out the Plan. The divisibility test for 6 states that a number is divisible by 6 if, and only if, the number is divisible by both 2 and 3. Because 11,368 is an even number, it is divisible by 2. Is it divisible by 3?

The divisibility test for 3 states that a number is divisible by 3 if, and only if, the sum of the digits in the number is divisible by 3. We see that $1 + 1 + 3 + 6 + 8 = 19$, which is not divisible by 3, so 11,368 is not divisible by 3. Hence, Sam is correct.

Looking Back. Suppose 11,368 had been divisible by 6. Would that have implied that the manager was correct? The answer is no; it would have implied only that we would have to change our approach to the problem.

As a further Looking Back activity, suppose that, given different data, the manager is correct. Can we determine values for m and n? In fact, this can be done; if a computer is available, a program can be written to determine all possible natural-number values of m and n. ■

Equations involving only integers, as in Problem 1, are called *Diophantine equations* in honor of the Greek mathematician Diophantus, who lived in Alexandria in the third century A.D. The solution of various Diophantine equations has been the center of research in number theory from antiquity to the present. The equation $6n + 24m = 11,368$ of Problem 1 is a special case of the Diophantine equation $ax + by = c$, where a, b, and c are given integers and where integer solutions x and y are desired. As in Problem 1, if some integer divides both a and b but does not divide c, the equation has no solution.

HISTORICAL NOTE

A modern mathematician who worked in the area of number theory was American Julia Robinson (1919–1985). Robinson's work with the Russian mathematician Yuri Matijasevič on Diophantine equations led directly to the tenth of the famous set of 23 problems the German mathematician David Hilbert posed. Robinson was once described as "the most famous Robinson in the USSR after Robinson Crusoe." Robinson was the first woman mathematician to be elected to the National Academy of Sciences and the first woman president of the American Mathematical Society. She died of leukemia at the age of sixty-five.

PROBLEM SET 5-1

1. Classify each of the following as true or false:
   (a) 6 is a factor of 30.
   (b) 6 is a divisor of 30.
   (c) 6|30
   (d) 30 is divisible by 6.
   (e) 30 is a multiple of 6.
   (f) 6 is a multiple of 30.

2. Complete each of the following sentences. Simplify your answers, if possible.
   (a) If $7|14$ and $7|21$, then _____.
   (b) If $d|(213 - 57)$ and $d|57$, then _____.
   (c) If $d|(a - b)$ and $d|b$, then _____.
   (d) If $7|231$ and $7|14$, then _____.
   (e) If $d|(213 + 57)$ and $d|57$, then _____.
   (f) If $d|(a + b)$ and $d|b$, then _____.

3. There are 1379 children signed up to play Little League baseball. If exactly 9 players are assigned to each team, will any team be short of players?

4. A forester has 43,682 seedlings to be planted. Can these be planted in an equal number of rows with 11 seedlings in each row?

5. Justify each of the given statements, assuming that a, b, and c are integers. If a statement cannot be justified by one of the theorems in this section, answer "none."
   (a) $4|20$ implies $4|113 \cdot 20$.
   (b) $4|100$ and $4 \nmid 13$ imply $4 \nmid (100 + 13)$.
   (c) $4|100$ and $4 \nmid 13$ imply $4 \nmid 1300$.
   (d) $3|(a + b)$ and $3 \nmid c$ imply $3 \nmid (a + b + c)$.
   (e) $3|a$ implies $3|a^2$.

6. ▦ Determine each of the following without actually performing the division (justify your answers). Check your answers using the $\boxed{\text{INT}\div}$ button on your calculator.
   (a) Is 34,015 divisible by 17?
   (b) Is 34,051 divisible by 17?
   (c) Is 19,031 divisible by 19?
   (d) Is 19,031 divisible by 31?
   (e) Is $2^{64} + 1$ divisible by 2^{14}? (Will your calculator check this?)
   (f) Is $2 \cdot 3 \cdot 5 \cdot 7 \cdot 13 \cdot 17 + 1$ divisible by 2, 3, 5, 7, 13, or 17?

7. In a football game, a touchdown with an extra point is worth 7 points and a field goal is worth 3 points. Suppose that in a game the only scoring done by teams are touchdowns with extra points and field goals.
   (a) Which of the scores 1 to 25 are impossible for a team to score?
   (b) List all possible ways for a team to score 40 points.
   (c) A team scored 58 points with 6 touchdowns and extra points. How many field goals did the team score?

8. ▦(a) Use a calculator with an $\boxed{\text{INT}\div}$ button to determine the remainder when each of the following numbers is divided by 9:
   (i) 10,000   (ii) 2000   (iii) 30   (iv) 700
   (v) 12   (vi) 123   (vii) 1230   (viii) 4311
   (b) Make a conjecture about the remainder when $a \cdot 10^n$ is divided by 9, where $0 \leq a \leq 9$.
   (c) Make a conjecture about the remainder when a number, the sum of whose digits is less than 9, is divided by 9.

9. Complete the following table where n is the given integer.

n	Remainder when n is divided by 9	Sum of the digits of n	Remainder when the sum of the digits of n is divided by 9
(a) 31			
(b) 143			
(c) 345			
(d) 2987			
(e) 7652			

   (f) Make a conjecture about the remainder and the sum of the digits in a number when it is divided by 9.

10. A test for checking computations is called *casting out nines*. Consider the sum $193 + 24 + 786 = 1003$. The remainders when 193, 24, and 786 are divided by 9 are 4, 6, and 3, respectively. The sum of the remainders, 13, has a remainder of 4 when divided by 9, as does 1003. Checking the remainders in this manner provided a quasi-check for the computation. Find the following sums and use casting out nines to check your sums.
    (a) $12{,}343 + 4546 + 56$
    (b) $987 + 456 + 8765$
    (c) $10{,}034 + 3004 + 400 + 20$
    (d) ▶Will this check always work for addition? Give an example to illustrate your answer.
    (e) Try the check on the subtraction, $1003 - 46$.
    (f) Try the check on the multiplication, $345 \cdot 56$.
    (g) ▶Would it make sense to try the check on division? Why or why not?

11. Classify each of the following as true or false, assuming that a, b, c and d are integers. If a statement is false, give a counterexample.
    (a) If $d|(a + b)$, then $d|a$ and $d|b$.
    (b) If $d|(a + b)$, then $d|a$ or $d|b$.
    (c) If $d|a$ and $d|b$, then $d|b$.
    (d) If $d|ab$, then $d|a$ or $d|b$.
    (e) If $ab|c$, $a \neq 0$ and $b \neq 0$, then $a|c$ and $b|c$.
    (f) $1|a$.
    (g) $d|0$.

(h) If $a|b$ and $b|a$, then $a = b$.
(i) If $d|a$ and $d|b$, then $d|(ax + by)$, for any integers x and y.
(j) If $d \nmid a$ and $d \nmid b$, then $d \nmid (a + b)$.
(k) If $d|a^2$, then $d|a$.
(l) If $d \nmid a$, then $d \nmid a^2$.
(m) If $d \nmid a^2$, then $d \nmid a$.

12. Classify each of the following as true or false:
    (a) If every digit of a number is divisible by 3, the number itself is divisible by 3.
    (b) If a number is divisible by 3, then every digit of the number is divisible by 3.
    (c) A number is divisible by 3 if and only if every digit of the number is divisible by 3.
    (d) If a number is divisible by 6, then it is divisible by 2 and by 3.
    (e) If a number is divisible by 2 and 3, then it is divisible by 6.
    (f) If a number is divisible by 2 and 4, then it is divisible by 8.
    (g) If a number is divisible by 8, then it is divisible by 2 and 4.

13. Classify each of the statements in Problem 12 as "sometimes," "always," or "never" true.

14. Devise a test for divisibility by each given number.
    (a) 16   (b) 25

15. Jack owes $7812 on a new car. Can this be paid in 12 equal monthly installments?

16. A group of people ordered No-Cal candy bars. The bill was $2.09. If the original price of each was 12¢ but the price has been inflated, how much does each cost?

17. When the two missing digits in the given number are replaced, the number is divisible by 99. What is the number?
    85 _ _ 1

18. ▦ Test each of the following numbers for divisibility by 2, 3, 4, 5, 6, 7, 8, 9, 10, and 11. Use a calculator to check your answers.
    (a) 746,988   (b) 81,342   (c) 15,810
    (d) 183,324   (e) 901,815   (f) 4,201,012
    (g) 1001   (h) 10,001   (i) 30,860

19. Answer each of the following, and justify your answers.
    (a) If a number is not divisible by 5, can it be divisible by 10?
    (b) If a number is not divisible by 10, can it be divisible by 5?

20. Fill each blank with the greatest digit that makes the statement true.
    (a) $3|74\_$ (b) $9|83\_45$ (c) $11|6\_55$

21. ▶A number in which each digit except 0 appears exactly three times is divisible by 3. For example, 777,555,222 and 414,143,313 are divisible by 3. Explain why this statement is true.

22. Prove the following theorem: For any integers a, b, and c, with $a \neq 0$ and $b \neq 0$, if $a|b$ and $b|c$, then $a|c$.

23. Leap years occur in years that are divisible by 4. However, if the year ends in two zeros, in order for the year to be a leap year, it must be divisible by 400. Determine which of the following are leap years:
    (a) 1776   (b) 1986   (c) 2000   (d) 2024

24. A palindrome is a number that reads the same forward and backward.
    (a) Check the following four-digit palindromes for divisibility by 11.
        (i) 4554   (ii) 9339   (iii) 2002
        (iv) 2222
    ★(b) Prove that any four-digit palindrome is divisible by 11.
    (c) ▶Is every five-digit palindrome divisible by 11? Why or why not?
    (d) ▶Is every six-digit palindrome divisible by 11? Why or why not?

25. (a) Choose a two-digit number such that the number in the tens place is one greater than the number in the units place. Reverse the digits in your number, and subtract this number from your original number; for example, $87 - 78 = 9$. Make a conjecture concerning the results of performing these kinds of operations.
    (b) Choose any two-digit number such that the number in the tens place is two greater than the number in the units place. Reverse the digits in your number, and subtract this number from your original number; for example, $31 - 13 = 18$. Make a conjecture concerning the results of performing these kinds of operations.
    ★(c) Prove that, for any two-digit number, if the digits are reversed and the numbers subtracted, the difference is a multiple of 9.
    (d) Investigate what happens whenever two-digit numbers with equal digit sums are subtracted: for example, $62 - 35 = 27$.

26. A customer wants to mail a package. The postal clerk determines the cost of the package as $2.86, but only 6¢ and 15¢ stamps are available. Can the available stamps be used for the exact amount of postage for the package? Why or why not?

27. ▶Which of the following Diophantine equations can be shown not to have solutions? Explain your reasoning.
    (a) $18x + 27y = 3111$ (b) $2x + 6y = 113$
    (c) $10x + 25y = 1007$ (d) $4x + y = 108$
    (e) $8x + 108y = 4001$ (f) $5x + 12y = 606$

28. (a) Use the division algorithm from Chapter 3 to explain why, among any three consecutive integers, there is always one that is divisible by 3.
    (b) Generalize the statement in (a) to any n consecutive integers.

★29. Prove Theorem 5-2(b)

★30. Prove the test for divisibility by 9 for any five-digit number.

31. 🖩 Enter any three-digit number on the calculator; for example, enter 243. Repeat it: 243,243. Divide by 7. Divide by 11. Divide by 13. What is the answer? Try it again with any other three-digit number. Will this always work? Why?

32. A traveler wishes to purchase $610 worth of travelers checks. The checks are available only in denominations of $20 and $50. How many of each denomination should the traveler buy? Is the answer unique?

● COMPUTER CORNER

The following Logo program will determine if a positive integer N is divisible by another integer X. Type it into your computer.

```
TO TESTDIV :N:X
   IF INTEGER (:N/:X) = :N/:X PRINT [OKAY] ELSE PRINT
     [NOT DIVISIBLE]
END
```
(In LCSI, replace PRINT [OKAY] with [PRINT [OKAY]] and PRINT [NOT DIVISIBLE] with [PRINT [NOT DIVISIBLE]].)

1. Run this program using various values for :N and :X.
2. How does the program compare to using the $\boxed{\text{INT} \div}$ button on a calculator?

▼ BRAIN TEASER

Dee finds that she has an extraordinary social security number. Its nine digits contain all the numbers from 1 through 9. They also form a number with the following characteristics: when read from left to right, its first two digits form a number divisible by 2, its first three digits form a number divisible by 3, its first four digits form a number divisible by 4, and so on, until the complete number is divisible by 9. What is Dee's social security number?

Section 5-2 Prime and Composite Numbers

When we write $a|b$, we say that a is a divisor of b. One method (sometimes called the "candy bar method") used in elementary schools to determine the divisors of a number is to use squares of paper and to represent the number as a rectangle. This rectangle resembles a candy bar formed with small squares. The dimensions of the rectangle are divisors of the number. For example, Figure 5-3 shows rectangles to represent 12.

FIGURE 5-3

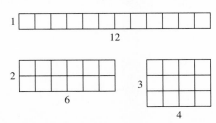

As the figure shows, the number 12 has six divisors: 1, 2, 3, 4, 6, and 12. If rectangles were used to find the divisors of 7, then we would find only 1 × 7 rectangle, as Figure 5-4 shows. Thus, 7 has exactly two divisors, 1 and 7.

FIGURE 5-4

To further illustrate the number of divisors of a number, we construct Table 5-1. Below each number listed across the top, we identify numbers less than or equal to 35 having that number of positive divisors. For example, 12 is in the 6 column because it has six divisors, and 7 is in the 2 column because it has only two divisors.

TABLE 5-1 Number of Factors

1	2	3	4	5	6	7	8
1	2	4	6	16	12		24
	3	9	8		18		30
	5	25	10		20		
	7		14		28		
	11		15		32		
	13		21				
	17		22				
	19		26				
	23		27				
	29		33				
	31		34				
	37		35				

prime
composite

Do you see any patterns forming in the table? Do you see why there will be no other entries in the 1 column? What is the next number in the 3 column? The numbers in the 2 column are of particular importance. Notice that they have exactly two divisors, namely, 1 and themselves. Any positive integer with exactly two distinct, positive divisors is called a *prime number,* or a **prime.** Any integer greater than 1 that has a positive factor other than 1 and itself is called a *composite number,* or a **composite.** For example, 4, 6, and 16 are composites because they have positive factors other than 1 and themselves. The number 1 has only one positive factor, so it is neither prime nor composite.

From the 2 column in Table 5-1, we see that the first twelve primes are 2, 3, 5, 7, 11, 13, 17, 19, 23, 29, 31, and 37. Other patterns in the table are explored in the problem set.

E X A M P L E 5 - 7   Show that the following numbers are composite:

(a) 1564   (b) 2781   (c) 1001

Solution   (a) Since 2|4, 1564 is divisible by 2.

(b) Since 3|(2 + 7 + 8 + 1), 2781 is divisible by 3.

(c) Since 11|[(1 + 0) − (0 + 1)], 1001 is divisible by 11.

Composite numbers can be expressed as products of two or more whole numbers greater than 1. For example, $18 = 2 \cdot 9$, $18 = 3 \cdot 6$, or $18 = 2 \cdot 3 \cdot 3$. Each expression of 18 as a product of factors is called a **factorization**.

factorization

PROBLEM 2

When his students asked Mr. Factor what his children's ages were, he answered, "I have three children. The product of their ages is 72 and the sum of their ages is the number of this room." The children asked for the door to be opened to verify the room number. Then, Sonja, the class math whiz, told the teacher that she needed more information to solve the problem. Mr. Factor said, "My oldest child is good at chess." Sonja then announced the correct ages for Mr. Factor's children. What are the ages of Mr. Factor's children?

Understanding the Problem. Mr. Factor has three children, and the product of their ages is 72. When Sonja was given the sum of the ages, she concluded that Mr. Factor did not provide enough information to determine the ages of the three children. After Mr. Factor announced that his oldest child is good at chess, Sonja was able to find the ages of the children. We are to determine the children's ages. From the given information, it seems that the fact that Mr. Factor has an oldest child is signficant.

Devising a Plan. To find the possible ages, we need to find three positive integers whose product is 72. We can do this systematically by *listing* the possible ages if there is a 1-year-old child in the family, then listing all the possible ages; if there is a 2-year-old in the family; and so on. Because $1 \cdot 2 \cdot 36 = 72$, the combination (1, 2, 36) is a possibility. However, because it does not matter in what order we list the ages, the combination (2, 1, 36) is the same as (1, 2, 36). Knowing that $72 = 2^3 \cdot 3^2$ can help us to list all the possible combinations, along with the corresponding sums, in a *table*. After examining the table, we hope to be able to determine how the additional information can be used to solve the problem.

Carrying Out the Plan. Table 5-2 shows all the possible ages whose product is 72, along with the corresponding sums. Notice that all the sums other than 14 appear only once in Table 5-2. Sonja knew the sum of the ages but could not determine the ages. The only

TABLE 5-2

Age	Age	Age	Sum of the ages
1	1	72	74
1	2	36	39
1	3	24	28
1	4	18	23
1	6	12	19
2	2	18	22
2	3	12	17
2	4	9	15
2	6	6	14
1	8	9	18
3	4	6	13
3	3	8	14

logical reason for this is that the classroom's number (the sum of the ages) must have been 14. There are two possible combinations that give the sum 14, (2, 6, 6) and (3, 3, 8). When Sonja was told that the oldest child was good at chess, she knew that (2, 6, 6) could not be a possible combination because, if the children were 2, 6, and 6 years old, there would not be an oldest among them. Thus, she concluded that the children's ages were 3, 3, and 8.

Looking Back. Is it possible to substitute another integer for 72 and solve the corresponding problem? If we choose the product of the ages to be 12, what similar problem can we pose? The possible triples are then (1, 1, 12), (1, 2, 6), (1, 3, 4), and (2, 2, 3), and the corresponding sums are 14, 9, 8, and 7. Given one of these numbers as a sum, we would be able to determine the triple, that is, the ages. But suppose that Mr. Factor said, "The youngest does not like spinach." We would know then that the first and the last triple are not possible as they do not determine a youngest child. To determine which of the triples (1, 2, 6), and (1, 3, 4) represents the ages of his children, Mr. Factor could say, "The middle child is a year older than the youngest." We would know then that the ages of his children are 1, 2, and 6. ∎

prime factorization   A factorization containing only prime numbers is called a **prime factorization.** To find a prime factorization of a given composite number, first rewrite the number as a product of two smaller numbers. Continue the process, factoring the lesser numbers until all factors are primes. For example, consider 260.

$$260 = 26 \cdot 10 = 2 \cdot 13 \cdot 2 \cdot 5 = 2 \cdot 2 \cdot 5 \cdot 13 = 2^2 \cdot 5 \cdot 13$$

factor tree   The procedure for finding a prime factorization of a number can be organized using a **factor tree,** as Figure 5-5(a) demonstrates. The last branches of the tree display the prime factors of 260.

A second way to factor 260 is shown in Figure 5-5(b). The two trees produce the same prime factorization, except for the order in which the primes appear in the products.

FIGURE 5-5

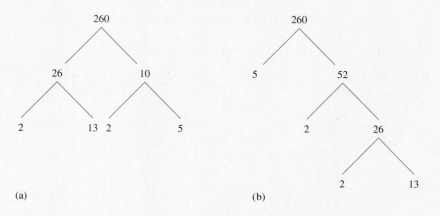

(a)   (b)

The *Fundamental Theorem of Arithmetic,* sometimes called the *Unique Factorization Theorem,* states that in general, if order is disregarded, the prime factorization of a number is unique.

> **THEOREM 5-3**
>
> **Fundamental Theorem of Arithmetic.** Each composite number can be written as a product of primes in one, and only one, way, aside from variation in the order of the prime factors.

The Fundamental Theorem of Arithmetic is a basis for an algorithmic approach to finding the prime factorization of a number. For example, consider 260. We start with the smallest prime, 2, and see if it divides 260. If not, we try the next greater prime and check for divisibility by this prime. Once we find a prime that divides the number in question, we must find the quotient of the number divided by the prime. This step in the prime factorization of 260 is shown in Figure 5-6(a). Next we check if the prime divides the quotient. If so, we repeat the process; if not, we try the next greater prime, 3, and check to see if it divides the quotient. We see that 260 divided by 2 yields 130, as shown in Figure 5-6(b). We continue the procedure, using greater primes, until a quotient of 1 is reached. The original number is the product of all the prime divisors used. The complete procedure for 260 is shown in Figure 5-6(c).

FIGURE 5-6

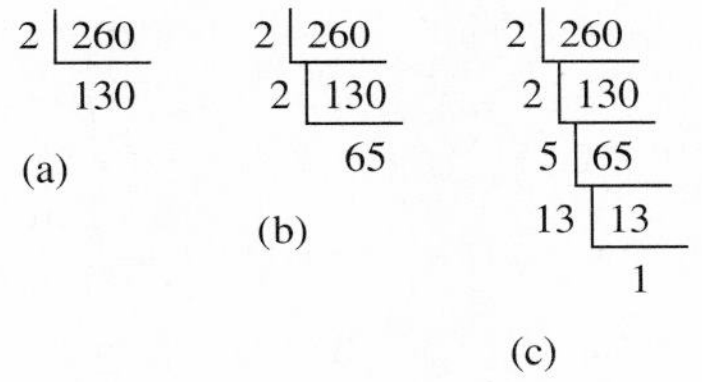

$$
\begin{array}{r|l}2 & 260 \\ \hline & 130\end{array} \quad \begin{array}{r|l}2 & 260 \\ \hline 2 & 130 \\ \hline & 65\end{array} \quad \begin{array}{r|l}2 & 260 \\ \hline 2 & 130 \\ \hline 5 & 65 \\ \hline 13 & 13 \\ \hline & 1\end{array}
$$

(a)   (b)   (c)

The primes in the prime factorization of a number are listed in increasing order from left to right and if a prime appears in a product more than once, exponential notation is used. Thus, the factorization of 260 is written as $2^2 \cdot 5 \cdot 13$.

We can use prime factorization to investigate the number of divisors a given number has. Consider the number 8, which has divisors 1, 2, 4, and 8. The prime factorization of 8 is 2^3, and its divisors may also be written as powers of 2: 2^0, 2^1, 2^2, and 2^3. We may generalize this notion for any prime p raised to a natural number power n as follows:

The divisors of p^n are $p^0, p^1, p^2, \ldots, p^n$.

As we can see, there are $n + 1$ divisors of p^n.

For a number such as 24 whose prime factorization is $2^3 \cdot 3^1$, we know that 2^3 has $3 + 1$, or 4, divisors and 3^1 has $1 + 1$, or 2, divisors. Since the products of all divisors of 2^3 and 3^1 are divisors of 24 (why?), then using the Fundamental Counting Principle (see Chapter 2), we know that 24 has $4 \cdot 2$, or 8, divisors. To check this, we *make an organized list* of divisors of 24 as shown:

Divisors of 2^3	2^0	2^1	2^2	2^3
Divisors of 3^1	3^0	3^1		
Divisors of 3^1 × divisors of 2^3 (Divisors of 24)	$3^0 \cdot 2^0 = 1$ $3^1 \cdot 2^0 = 3$	$3^0 \cdot 2^1 = 2$ $3^1 \cdot 2^1 = 6$	$3^0 \cdot 2^2 = 4$ $3^1 \cdot 2^2 = 12$	$3^0 \cdot 2^3 = 8$ $3^1 \cdot 2^3 = 24$

The divisors of 24 are seen grouped in pairs below:

1, 2, 4, 8, 3, 6, 12, 24

The process of determining the number of divisors of 24 can be generalized and is given in Theorem 5-4.

THEOREM 5-4

If the prime factorization of a number, n, is $n = p_1^{q_1} \cdot p_2^{q_2} \cdot p_3^{q_3} \cdot \ldots \cdot p_m^{q_m}$, then the number of divisors of n is $(q_1 + 1)(q_2 + 1)(q_3 + 1) \cdot \ldots \cdot (q_m + 1)$.

EXAMPLE 5-8 (a) Find all the divisors of 912.
(b) Find all the divisors of 324.

Solution (a) The prime factorization of 912 is $2^4 \cdot 3 \cdot 19$, and there are $5 \cdot 2 \cdot 2$, or 20, divisors. The divisors of 2^4 are 1, 2, 4, 8, and 16; the divisors of 3 are 1 and 3; and the divisors of 19 are 1 and 19. Thus, the divisors of 912 are as follows: 1, 2, 4, 8, 16, 3, 6, 12, 24, 48, 19, 38, 76, 152, 304, 57, 114, 228, 456, and 912.

(b) The prime factorization of 324 is $2^2 \cdot 3^4$, and there are $3 \cdot 5$, or 15, divisors. The divisors of 2^2 are 1, 2, and 4; the divisors of 3^4 are 1, 3, 9, 27, and 81. Thus, the divisors of 324 are 1, 2, 4, 3, 6, 12, 9, 18, 36, 27, 54, 108, 81, 162, and 324. ∎

In determining the factorization of a number such as 8127, observe that $9|8127$, or $8127 = 9k$, where k is an integer. Because $8127 = 9k$, then k is a factor of 8127 and $k = \dfrac{8127}{9}$. Theorem 5-5 states the general case.

THEOREM 5-5

If d is a non-zero factor of n, where $n \neq 0$, then $\dfrac{n}{d}$ is a factor of n.

Suppose that p is the *least* prime factor of the number n. Then by Theorem 5-5, n/p is a factor of n, and because p is the least factor of n, then $p \leq n/p$. If $p \leq n/p$, then $p^2 \leq n$. This idea is summarized in the following theorem.

THEOREM 5-6

If n is composite, then n has a prime factor p such that $p^2 \leq n$.

Theorem 5-6 can be used to help determine whether a given number is prime or composite. For example, consider the number 109. If 109 is composite, it must have a prime divisor p such that $p^2 \leq 109$. The primes whose squares do not exceed 109 are 2, 3, 5, and 7. Mentally, we can see that $2 \nmid 109$, $3 \nmid 109$, $5 \nmid 109$, and $7 \nmid 109$. Hence, 109 is prime. The argument used leads to the following theorem.

> ### THEOREM 5-7
>
> In n is an integer greater than 1 and not divisible by any prime p, where $p^2 \leq n$, then n is prime.

EXAMPLE 5-9 Is 397 composite or prime?

Solution The possible primes p such that $p^2 \leq 397$ are 2, 3, 5, 7, 11, 13, 17, and 19. Because $2 \nmid 397$, $3 \nmid 397$, $5 \nmid 397$, $7 \nmid 397$, $11 \nmid 397$, $13 \nmid 397$, $17 \nmid 397$, and $19 \nmid 937$, the number 397 is prime. ∎

PROBLEM 3

In an elaborate promotion for encouraging students to ride buses, Mountain Line Bus System (MLBS) took the first 1000 students to register at Kalispell College during Fall Quarter, wrote their names on pieces of paper numbered in sequence by ones from 1, and agreed to choose some pieces of paper for free passes. To choose the students, Jacques first placed the pieces of paper in numerical order face up, then flipped over all the pieces of paper that were labeled with even numbers. Then he started again and changed every third piece of paper beginning with the third one; that is, he turned the face-down pieces face up and turned the faced-up pieces face down. The process continued until Jacques completed his 1000th trip through the slips of paper. How many free passes were awarded with this scheme?

Understanding the Problem. The 1000 pieces of paper are numbered 1 through 1000. Jacques placed every piece of paper face up and in numerical order, returned to the start and flipped over every even-numbered piece of paper, returned and changed the state of every third piece of paper starting with the paper numbered 3, and so on. We must determine the number of pieces of paper that are facing up when the entire process is completed and Jacques has made 1000 passes through the slips of paper.

Devising a Plan. We use the strategy of *examining a simpler problem* in order to gain insight into the solution of the original problem. Suppose there were only 20 pieces of paper. If we denote a faced-up piece of paper with a u and a face-down piece of paper with a t, we can record the state of each slip changed by Jacques, as shown in Table 5-3. For example, on his fourth pass, Jacques puts slip 4 face up, puts slip 8 face up, puts slip 12 face down, puts slip 16 face up, and puts slip 20 face down.

Table 5-3 shows that after 20 passes, the only faced-up slips are 1, 4, 9, and 16. Each of these numbers is a perfect square. We must determine if this pattern continues. If it does, we must find out the number of perfect squares less than 1000.

Carrying Out the Plan. To determine if the pattern continues, consider slip 25. The slip is faced up in Jacques's first pass, faced down on his fifth pass, and faced up on his 25th pass. This suggests that the pattern is correct. (Note that 1, 5, and 25 are the only positive divisors of 25.) What happens with a slip such as 26, which is not a perfect square? Slip 26 is faced up on the first pass, faced down on the second pass, faced up on the 13th pass, and faced down on the 26th pass, and remains in that position. In general, we see that a slip is changed only on the passes whose order divide the slip number.

For the slip to be face up at the end, it must be faced up one more time than it is faced down; that is, the state must be changed an odd number of times. For this to happen, the number of the slip must have an odd number of divisors. We can show that the faced-up slips have numbers that are perfect squares by showing that only perfect squares have an odd number of divisors.

TABLE 5-3

	Numbers of Slips of Paper																			
	1	2	3	4	5	6	7	8	9	10	11	12	13	14	15	16	17	18	19	20
1	u	u	u	u	u	u	u	u	u	u	u	u	u	u	u	u	u	u	u	u
2		t		t		t		t		t		t		t		t		t		t
3			t			u			t			u			t			u		
4				u				u				t				u				u
5					t					u					u					t
6						t						u						t		
7							t							u						
8								t								t				
9									u									u		
10										t										u
11											t									
12												t								
13													t							
14														t						
15															t					
16																u				
17																	t			
18																		t		
19																			t	
20																				t

(Order of Jacques' Pass)

Recall that the divisors of a number appear in pairs. For example, the pairs of divisors of 80 and 81 are given by the following:

$$80 = 1 \cdot 80 = 2 \cdot 40 = 4 \cdot 20 = 5 \cdot 16 = 10 \cdot 8$$
$$81 = 1 \cdot 81 = 3 \cdot 27 = 9 \cdot 9$$

Thus, 80 has ten distinct divisors, or five pairs. On the other hand, the perfect square 81 has five distinct divisors: the pairs 1 and 81 and 3 and 27, and a single divisor, 9, which is paired with itself. We know that, if d is a divisor of n, then n/d is a divisor of n. Consequently, for all divisors d of n, if $d \neq n/d$, then each divisor can be paired with a different divisor, and n must have an even number of positive divisors. If for some divisor d, $d = n/d$, then $n = d^2$, and all the divisors of n, except d, are paired with a different divisor.

Hence, the number of divisors of n is odd. Because $d = n/d$ occurs only when $n = d^2$, it follows that n has an odd number of divisors if and only if n is a perfect square. As a result, the faced-up slips contain numbers that are perfect squares less than 1000, namely, 1^2, 2^2, 3^2, ..., 31^2. Therefore, 31 passes were awarded.

Looking Back. This problem suggests the following questions:

1. On which turns will Jacques touch only one piece of paper?
2. How many times will a slip with a prime number on it be touched?
3. Determine a method of finding the number of factors a number has, without actually listing all the factors. (*Hint:* Consider prime factorizations.) ∎

More About Primes

One way to find all the primes less than a given number is to use the Sieve of Eratosthenes, named after the Greek mathematician Eratosthenes (276–194 or 192 B.C.). If all the natural numbers greater than 1 are considered (or placed in the sieve), the numbers that are not prime are methodically crossed out (or drop through the holes of the sieve). The remaining numbers are prime. The following procedure illustrates this process:

1. In Table 5-4, we cross out 1 because 1 is not prime.
2. Circle 2 because 2 is prime.
3. Cross out other multiples of 2; they are not prime.
4. Circle 3 because 3 is prime.
5. Cross out other multiples of 3.
6. Circle 5 and 7 because they are primes; cross out their multiples.
7. In Table 5-4, we stop after step 6 because 7 is the greatest prime whose square, 49, is less than 100. All the numbers remaining in the list and not crossed out are prime.

TABLE 5-4

~~1~~	②	③	~~4~~	⑤	~~6~~	⑦	~~8~~	~~9~~	~~10~~
11	~~12~~	13	~~14~~	~~15~~	~~16~~	17	~~18~~	19	~~20~~
~~21~~	~~22~~	23	~~24~~	~~25~~	~~26~~	~~27~~	~~28~~	29	~~30~~
31	~~32~~	~~33~~	~~34~~	~~35~~	~~36~~	37	~~38~~	~~39~~	~~40~~
41	~~42~~	43	~~44~~	~~45~~	~~46~~	47	~~48~~	~~49~~	~~50~~
~~51~~	~~52~~	53	~~54~~	~~55~~	~~56~~	~~57~~	~~58~~	59	~~60~~
61	~~62~~	~~63~~	~~64~~	~~65~~	~~66~~	67	~~68~~	~~69~~	~~70~~
71	~~72~~	73	~~74~~	~~75~~	~~76~~	~~77~~	~~78~~	79	~~80~~
~~81~~	~~82~~	83	~~84~~	~~85~~	~~86~~	~~87~~	~~88~~	89	~~90~~
~~91~~	~~92~~	~~93~~	~~94~~	~~95~~	~~96~~	97	~~98~~	~~99~~	~~100~~

There are infinitely many whole numbers, infinitely many odd numbers, and infinitely many even numbers. Are there infinitely many primes? Because prime numbers do not appear in any known pattern, the answer to this question is not obvious. Euclid was the first to prove that there are infinitely many primes (see Problem Set 5-2, Problem 26).

Mathematicians have long looked for a formula that produces only primes, but no one has ever found one. One result was the expression $n^2 - n + 41$, where n is a whole number. Substituting 0, 1, 2, 3, . . . , 40 for n in the expression always results in a prime number. However, substituting 41 for n gives $41^2 - 41 + 41$, or 41^2, a composite number.

In 1971, the largest known prime was $2^{19,937} - 1$, found by Bryant Tuckerman of IBM. In 1978, two high school students (Laura Nickel and Curt Noll from Hayward, California) found a larger prime, $2^{23,209} - 1$, using 440 computer hours. Other larger primes have since been discovered, one of the latest being $2^{216,091} - 1$, which has 65,050 digits.

Goldbach's conjecture

There are many interesting problems concerning primes. For example, Christian Goldbach (1690–1764) asserted in a letter to Euler that every even integer greater than 2 is the sum of two primes. This statement is known as **Goldbach's conjecture.** For example, $4 = 2 + 2$, $6 = 3 + 3$, $8 = 3 + 5$, $10 = 3 + 7$, $12 = 5 + 7$, and $14 = 3 + 11$. In spite of the simplicity of the statement, no one knows for sure whether or not the statement is true.

PROBLEM 4

A woman with a basket of eggs finds that if she removes the eggs from the basket 2, 3, 4, 5, or 6 at a time, there is always 1 egg left. However, if she removes the eggs 7 at a time, there are no eggs left. If the basket holds up to 500 eggs, how many eggs does the woman have?

Understanding the Problem. When a woman removes eggs from the basket 2, 3, 4, 5, or 6 at a time, there is always 1 egg left. That means that if the number of eggs is divided by 2, 3, 4, 5, or 6, the remainder is always 1. We also know that when she removes the eggs 7 at a time, there are no eggs left; that is, the number of eggs is a multiple of 7. Finally, we know that the basket holds up to 500 eggs. We have to find the number of eggs in the basket.

Devising a Plan. One way to solve the problem is to *list* all the multiples of 7 between 7 and 500 and check which ones have a remainder of 1 when divided by 2, 3, 4, 5, or 6. Since this method is tedious, we look for a different approach. Let the number of eggs be n. Then, if n is divided by 2, the remainder is 1. Consequently, $n - 1$ will be divisible by 2. Similarly, 3, 4, 5, and 6 divide $n - 1$.

Since 2 and 3 divide $n - 1$, the primes 2 and 3 appear in the prime factorization of $n - 1$. Note that $4 | (n - 1)$ implies that $2 | (n - 1)$, and hence, from the information $2 | (n - 1)$ and $4 | (n - 1)$, we can conclude only that 2^2 appears in the prime factorization of $n - 1$. Since $5 | (n - 1)$, 5 appears in the prime factorization of $n - 1$. The fact that $6 | (n - 1)$ does not provide any new information, since it only implies that 2 and 3 are prime factors of $n - 1$, which we already know. Now, $n - 1$ may also have other prime factors. Denoting the product of these other prime factors by k, we have $n - 1 = 2^2 \cdot 3 \cdot 5 \cdot k = 60k$, where k is some natural number, and so $n = 60k + 1$. We now find all of the possible values for n in the form $60k + 1$ less than 500 and determine which ones are divisible by 7.

Carrying Out the Plan. Because $n = 60k + 1$ and k is any natural number, we substitute $k = 1, 2, 3, \ldots$ to obtain the following possible values for n that are less than 500:

$$61, 121, 181, 241, 301, 361, 421, 481.$$

Among these values, only 301 is divisible by 7; hence, 301 is the only possible answer to the problem.

Looking Back. In the preceding situation, we still had to test eight numbers for divisibility by 7. Is it possible to reduce the computations further? We know that $n = 60k + 1$ and that the possible values for k are $k = 1, 2, 3, 4, 5, 6, 7, 8$. We also know that $7 \mid n$; that is, $7 \mid (60k + 1)$. The problem is to find for which of the above values of k, $7 \mid (60k + 1)$. The question would have been easier to answer if, instead of $60k + 1$, we had a smaller number. We know that the least multiple of k closest to $60k$ that is divisible by 7 is $56k$. Since $7 \mid (60k + 1)$ and $7 \mid 56k$, we conclude that $7 \mid (60k + 1 - 56k)$; that is, $7 \mid (4k + 1)$. We now see that $7 \mid (60k + 1)$, if, and only if, $7 \mid (4k + 1)$. The only value of k between 1 and 8 that makes $4k + 1$ divisible by 7 is 5. Consequently, $7 \mid (60 \cdot 5 + 1)$, and 301 is the solution to the problem. ∎

HISTORICAL NOTE

During World War II, Alan Turing, Peter Hilton, and other British analysts helped crack the codes developed on the German Enigma cipher machine. In the 1970s, determining large prime numbers became extremely useful in coding and decoding secret messages. In all coding and decoding, the letters of an alphabet correspond in some way to nonnegative integers. A "safe" coding system, in which messages are unintelligible to everyone except the intended receiver, was devised by three Massachusetts Institute of Technology scientists (Ronald Rivast, Adi Shamir, and Leonard Adleman) and is referred to as the RSA (their initials) system. The secret deciphering key consists of two large prime numbers chosen by the user. The enciphering key is the product of these two primes. Because it is extremely difficult and time consuming to factor large numbers, it was practically impossible to recover the deciphering key from a known enciphering key. In 1982, new methods for factoring large numbers were invented, which resulted in the use of even greater primes to prevent the breaking of decoding keys.

PROBLEM SET 5-2

1. Use a factor tree to find the prime factorization for each of the following:
   (a) 504 (b) 2475 (c) 11,250
2. Which of the following numbers are primes?
   (a) 149 (b) 923 (c) 433
   (d) 101 (e) 463 (f) 897
3. What is the greatest prime you must consider to test whether or not 5669 is prime?
4. ▶In the Sieve of Eratosthenes in Table 5-4, explain why, after we cross out all the multiples of 2, 3, 5, and 7, the remaining numbers are primes.
5. Extend the Sieve of Eratosthenes to find all primes less than 200.
6. The factors of a locker number are 2, 5, and 9. If there are exactly nine additional factors, what is the locker number?

7. (a) When the U.S. flag had 48 stars, the stars formed a 6×8 rectangular array. In what other rectangular arrays could they have been arranged?
   (b) How many different rectangular arrays of stars could there be if there were only 47 states?
8. If the Spanish Armada had consisted of 177 galleons, could it have sailed in an equal number of small flotillas? If so, how many ships would have been in each?
9. Suppose that the 435 members of the House of Representatives are placed on committees consisting of more than 2 members but less than 30 members. Each committee is to have an equal number of members and each member is to be on only one committee.
   (a) What size committees are possible?
   (b) How many committees are there of each size?
10. Mr. Arboreta wants to set out fruit trees in a rectangular array. For each of the following numbers of trees, find all possible numbers of rows if each row is to have the same number of trees:
    (a) 36 (b) 28 (c) 17 (d) 144
11. ▶What is the smallest number that has exactly seven positive factors? Explain your answer.
12. The square root of a number n is a number q if $q^2 = n$. A number n is a perfect square if q is an integer. In the Farside cartoon below, answer the following:
    (a) What is the prime factorization of 5,248?
    (b) What is the number of divisors of 5,248?
    (c) Is 5,248 a perfect square?

"I asked you a question, buddy ... What's the square root of 5,248?"

THE FAR SIDE cartoon by Gary Larson is reprinted by permission of Chronicle Features, San Francisco, CA.

13. (a) Find a composite number different from 41^2 that is of the form $n^2 - n + 41$.
    ★(b) Prove that there are infinitely many composite numbers of the form $n^2 - n + 41$.
14. Find the least number divisible by each natural number less than or equal to 12.
15. ▶The primes 2 and 3 are consecutive integers. Is there another pair of consecutive integers both of which are prime? Justify your answer.
16. The prime numbers 11 and 13 are called **twin primes** because they differ by 2. Find all the twin primes less than 200. (The existence of infinitely many twin primes has not been proved.)
17. (a) Use the Fundamental Theorem of Arithmetic to justify that if $2|n$ and $3|n$, then $6|n$.
    (b) Is it always true that if $a|n$ and $b|n$, then $ab|n^2$? Either prove the statement or give a counter-example.
18. ▶In order to test for divisibility by 12, one student checked to determine divisibility by 3 and 4, while another checked for divisibility by 2 and 6. Are both students using a correct approach to divisibility by 12? Why or why not?
19. Show that, if 1 were considered a prime, every number would have more than one prime factorization.
20. If $42|n$, what other positive integers divide n?
21. Is it possible to find positive integers x, y, and z such that $2^x \cdot 3^y = 5^z$? Why or why not?
22. A prime such as 7331 is a superprime because any integers obtained by deleting digits from the right of 7331 are prime; for example, 733, 73, and 7.
    (a) In order for a prime to be a superprime, what digits cannot appear in the number?
    (b) Of the digits that can appear in a superprime, what digit cannot be the left-most digit of a superprime?
    (c) Find all of the two-digit superprimes.
    (d) Find a three-digit superprime.
23. Use Table 5-1 for each of the following:
    (a) Guess the next three numbers in the 3 column. Describe a pattern for forming the numbers.
    (b) Guess the next three numbers in the 5 column. Describe a pattern for forming the numbers.
    ★(c) Guess the next three numbers in the 4 column. Describe a pattern for forming the numbers.
24. ▶Explain why a prime number must have an odd number of prime divisors.
25. Find the greatest four-digit number that has exactly three factors.
★26. Complete the details for the following proof, which shows that there are infinitely many prime numbers.
    If the number of primes is finite, then there is a greatest prime denoted by p. Consider the product of all the primes, $2 \cdot 3 \cdot 5 \ldots p$, and let $N = (2 \cdot 3 \cdot 5 \ldots p) + 1$. Because $N > p$, where p is the greatest prime, N is com-

posite. Because N is composite, there is a prime q among the primes $2, 3, 5, \ldots, p$ such that $q|N$. However, none of the primes $2, 3, 5, \ldots, p$ divides N. (Why?)

Consequently, $q \nmid N$, which is a contradiction. Thus, the assumption that there are finitely many primes is false and the set of primes must be infinite.

27. ▶Explain why the product of any three consecutive numbers is divisible by 6.
28. ▶Explain why the product of any four consecutive numbers is divisible by 24.
29. It is not known whether there are infinitely many primes in the infinite sequence consisting only of ones: 1, 11, 111, 1111, ... Find infinitely many composite numbers in the sequence.
30. Find infinitely many composite numbers in the sequence whose nth term is $3n + 1$.
31. One formula yielding several primes is $n^2 + n + 17$. Substitute $n = 1, 2, 3, \ldots, 17$ in the formula and find which of the resulting numbers are primes and which are composites.

Review Problems

32. Classify each of the following as true or false:
    (a) 11 is a factor of 189.
    (b) 1001 is a multiple of 13.
    (c) $7|1001$ and $7 \nmid 12$ imply $7 \nmid (1001 - 12)$.
    (d) If a number is divisible by both 7 and 11, then its prime factorization contains 7 and 11.
33. Test each of the following for divisibility by 2, 3, 4, 5, 6, 7, 8, 9, 10, and 11:
    (a) 438,162   (b) 2,345,678,910
34. Prove that, if a number is divisible by 12, then it is divisible by 3.
35. Could $3376 be divided exactly among either 7 or 8 people?

● COMPUTER CORNER

The following Logo programs will determine whether or not a number is prime. Type the programs into your computer and use them to do Problem 2 in Problem Set 5-2. To execute the program, type PRIME with the number you wish to check as input.

```
TO PRIME :N
   CHECK :N INTEGER (SQRT :N)
END

TO CHECK :N :D
   IF :N = 1 PRINT [NEITHER PRIME NOR COMPOSITE] STOP
   IF :D = 1 PRINT "PRIME STOP
   IF REMAINDER :N :D = 0 PRINT "COMPOSITE STOP
   CHECK :N :D - 1
END
```

(In LCSI, place brackets around PRINT [NEITHER PRIME NOR COMPOSITE] STOP, PRINT "PRIME STOP, and PRINT "COMPOSITE STOP.)

LABORATORY ACTIVITY

In Figure 5-7 on the following page, a spiral starts with 41 at its center and continues in a counterclockwise direction. Primes are written in and squares representing composites are shaded. Continue the spiral until you reach the prime 439. Check the primes along the diagonal. Can you find each of the primes from the formula $n^2 + n + 41$ by substituting appropriate values for n?

244   CHAPTER 5   NUMBER THEORY

FIGURE 5-7

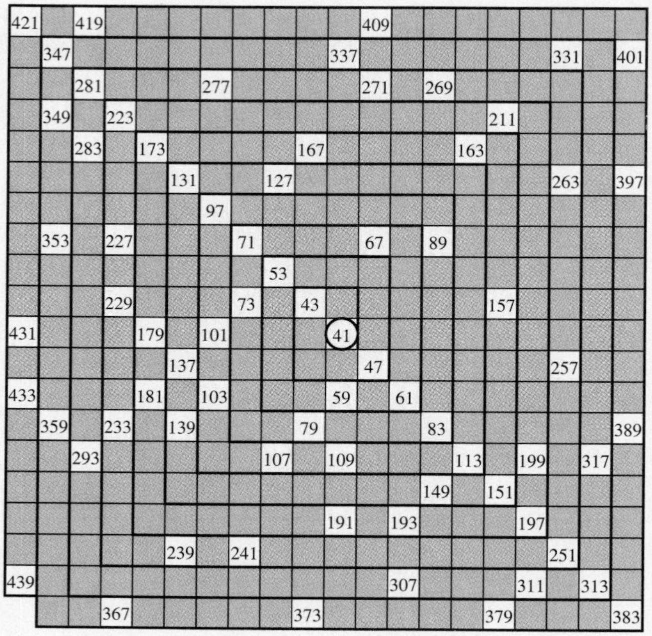

Section 5-3 Greatest Common Divisor and Least Common Multiple

Greatest Common Divisor

greatest common divisor (GCD)

The **greatest common divisor (GCD)** of two whole numbers is the greatest divisor or factor that the two numbers have in common. We can build a model of two or more numbers with Cuisenaire rods to determine the GCD of the numbers. For example, consider the 6 rod and the 8 rod in Figure 5-8(a).

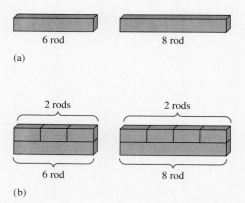

FIGURE 5-8

SECTION 5-3 GREATEST COMMON DIVISOR AND LEAST COMMON MULTIPLE

In order to find the GCD of 6 and 8, we must find the longest rod such that we can use multiples of that rod to build both the 6 rod and the 8 rod. The 1 rods and the 2 rods can be used to build both the 6 and 8 rods as shown in Figure 5-8(b); the 3 rods can be used to build the 6 rods but not the 8 rods; the 4 rods can be used to build the 8 rod but not the 6 rod; the 5 rods can be used to build neither; and the 6 rod cannot be used to build the 8 rod. Therefore, the GCD(6,8) = 2. In what follows, we present several other ways to find the GCD of two or more numbers.

The Intersection-of-Sets Method. In the intersection-of-sets method, we list all members of the set of positive divisors of the two numbers, then find the set of all *common divisors,* and, finally, pick the *greatest* element in that set. For example, to find the GCD of 20 and 32, denote the sets of divisors of 20 and 32 by D_{20} and D_{32}, respectively.

$$D_{20} = \{1, 2, 4, 5, 10, 20\}$$
$$D_{32} = \{1, 2, 4, 8, 16, 32\}$$

The set of all common positive divisors of 20 and 32 is

$$D_{20} \cap D_{32} = \{1, 2, 4\}.$$

Because the greatest number in the set of common positive divisors is 4, the GCD of 20 and 32 is 4, written GCD(20, 32) = 4.

The Prime Factorization Method. The intersection-of-sets method is rather time consuming and tedious if the numbers have many divisors. Another, more efficient, method is the *prime factorization method.* To find GCD(180, 168), first notice that

$$180 = 2 \cdot 2 \cdot 3 \cdot 3 \cdot 5$$
$$\text{and} \quad \updownarrow \quad \updownarrow \quad \quad \updownarrow$$
$$168 = 2 \cdot 2 \cdot 2 \cdot 3 \cdot 7.$$

We see that 180 and 168 have two factors of 2 and one of 3 in common. These common primes divide both 180 and 168. In fact, the only numbers other than 1 that divide both 180 and 168 must have no more than two 2s and one 3 and no other prime factors in their prime factorizations. The possible common divisors are 1, 2, 2^2, 3, $2 \cdot 3$, and $2^2 \cdot 3$. Hence, the greatest common divisor of 180 and 168 is $2^2 \cdot 3$. The procedure for finding the GCD of two or more numbers by using the prime factorization method is summarized as follows:

To find the GCD of two or more numbers, first find the prime factorizations of the given numbers, then take each common prime factor of the given numbers; the GCD is the product of the common factors, each raised to the lowest power of that prime that occurs in either of the prime factorizations.

relatively prime

If we apply the prime factorization technique to finding GCD(4, 9), we see that 4 and 9 have no common prime factors. Consequently, 1 is the only common divisor, so GCD(4, 9) = 1. Numbers such as 4 and 9, whose GCD is 1, are called **relatively prime.** Both the intersection-of-sets method and the prime factorization method are found in *Addison-Wesley Mathematics, Grade 7,* 1991, as seen on the student page given.

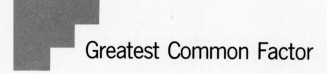

Greatest Common Factor

EXPLORE Analyze the Situation
The 7th graders are going to divide a 36 ft by 44 ft field into equal squares. The sides of the squares will be whole number lengths. Each square will be sold for $5. A goat will wander in the field for 30 minutes. The last square the goat eats from will be the winning square. What sizes could the squares be? What is the largest size the squares could be?

TALK ABOUT IT

1. What picture could you draw to help solve the problem?
2. How do you know that the squares cannot be 5 ft by 5 ft?
3. Can the field be divided into 6 ft by 6 ft squares? Explain.

The largest factor that two or more numbers have in common is called their **greatest common factor (GCF)**. If the GCF of two numbers is 1, the numbers are **relatively prime**.

Here are two methods for finding the GCF of 36 and 44.

Method 1 List the Factors	Method 2 Prime Factorization
Factors of 36: 1, 2, 3, 4, 6, 9, 12, 18, 36 Factors of 44: 1, 2, 4, 11, 22, 44 Common factors: 1, 2, 4 GCF = 4	$36 = 2 \cdot 2 \cdot 3 \cdot 3$ $44 = 2 \cdot 2 \cdot 11$ GCF $= 2 \cdot 2 = 4$ The GCF is the product of all the prime factors 36 and 44 have in common.

Examples Find the GCF of each pair of numbers.

A 4, 5
Factors of 4: 4, 2, 1
Factors of 5: 5, 1
GCF is 1; 4 and 5 are relatively prime.

B 45, 90
$45 = 3 \cdot 3 \cdot 5$
$90 = 2 \cdot 3 \cdot 3 \cdot 5$
GCF $= 3 \cdot 3 \cdot 5 = 45$

EXAMPLE 5-10   Find each of the following:

(a) GCD(108, 72)

(b) GCD(0, 13)

(c) GCD(x, y) if $x = 2^3 \cdot 7^2 \cdot 11 \cdot 13$ and $y = 2 \cdot 7^3 \cdot 13 \cdot 17$

(d) GCD(x, y, z) if $z = 2^2 \cdot 7$, using x and y from part (b)

Solution (a) Since $108 = 2^2 \cdot 3^3$ and $72 = 2^3 \cdot 3^2$, it follows that $GCD(108, 72) = 2^2 \cdot 3^2 = 36$.

(b) Since $13 \cdot 0 = 0$ and $13 \cdot 1 = 13$, then $GCD(0, 13) = 13$.

(c) $GCD(x, y) = 2 \cdot 7^2 \cdot 13 = 1274$.

(d) Because $x = 2^3 \cdot 7^2 \cdot 11 \cdot 13$, $y = 2 \cdot 7^3 \cdot 13 \cdot 17$, and $z = 2^2 \cdot 7$, then $GCD(x, y, z) = 2 \cdot 7 = 14$. Notice that $GCD(x, y, z)$ can also be obtained by finding the GCD of z and 1274, the answer from part (b). ∎

Using a Calculator. The TI Math Explorer™ calculator can be used to find the GCD of two numbers using the operations with fractions keys, $\boxed{/}$, $\boxed{\text{Simp}}$, and $\boxed{x \subset y}$. For example, to find the GCD(120, 180), use the following sequence of buttons to start: First, press $\boxed{1}$ $\boxed{2}$ $\boxed{0}$ $\boxed{/}$ $\boxed{1}$ $\boxed{8}$ $\boxed{0}$ $\boxed{\text{Simp}}$ $\boxed{=}$ to obtain the display $\boxed{\text{N/D} \to \text{n/d} \quad 60/90}$. By pressing the $\boxed{x \subset y}$ button, we see $\boxed{2}$ on the display as a common divisor of 120 and 180. By pressing the $\boxed{x \subset y}$ button again and pressing $\boxed{\text{Simp}}$ $\boxed{=}$ $\boxed{x \subset y}$, we see 2 again as a factor. The process is repeated to reveal 3, 5, and 1 as other common factors. When we see 1 as a factor, the process is complete. The GCD(120, 180) is the product of the common prime factors $2 \cdot 2 \cdot 3 \cdot 5$, or 60.

Euclidean Algorithm Method. Some numbers are hard to factor. For these numbers, another method is more efficient for finding the GCD. For example, suppose that we want to find GCD(676, 221). If we could find two smaller numbers whose GCD is the same as GCD(676, 221), our task would be easier. From Theorem 5-2, every divisor of 676 and 221 is also a divisor of $676 - 221$ and 221. Conversely, every divisor of $676 - 221$ and 221 is also a divisor of 676 and 221. Thus, the set of all the common divisors of 676 and 221 is the same as the set of all common divisors of $676 - 221$ and 221. Consequently, $GCD(676, 221) = GCD(676 - 221, 221)$. This process can be continued to subtract three 221's from 676 so that $GCD(676, 221) = GCD(676 - 3 \cdot 221, 221) = GCD(13, 221)$. To determine how many 221's can be subtracted from 676, we could have divided as follows:

$$\begin{array}{r} 3 \\ 221\overline{)676} \\ \underline{663} \\ 13 \end{array}$$

Continuing, we see that $GCD(13, 221) = GCD(0, 13)$ from the following division:

$$\begin{array}{r} 17 \\ 13\overline{)221} \\ \underline{13} \\ 91 \\ \underline{91} \\ 0 \end{array}$$

Because $GCD(0, 13) = 13$, the $GCD(676, 221) = 13$. Based on this illustration, we make the generalization outlined in the following theorem.

THEOREM 5-8

If a and b are any whole numbers and $a \geq b$, then GCD (a, b) = GCD(r, b), where r is the remainder when a is divided by b.

Euclidean algorithm

Finding the GCD of two numbers by repeatedly using Theorem 5-8 until the remainder 0 is reached is referred to as the **Euclidean algorithm.**

EXAMPLE 5-11 Use the Euclidean algorithm to find GCD(10,764, 2300).

Solution

$$\begin{array}{r} 4 \\ 2300\overline{)10{,}764} \\ 9{,}200 \\ \hline 1{,}564 \end{array}$$

Thus, GCD(10,764, 2300) = GCD(2300, 1564).

$$\begin{array}{r} 1 \\ 1564\overline{)2300} \\ 1564 \\ \hline 736 \end{array}$$

Thus, GCD(2300, 1564) = GCD(1564, 736).

$$\begin{array}{r} 2 \\ 736\overline{)1564} \\ 1472 \\ \hline 92 \end{array}$$

Thus, GCD(1564, 736) = GCD(736, 92).

$$\begin{array}{r} 8 \\ 92\overline{)736} \\ 736 \\ \hline 0 \end{array}$$

Thus, GCD(736, 92) = GCD(92, 0).

Because GCD(92, 0) = 92, it follows that GCD(10,764, 2300) = 92. ■

REMARK The procedure for finding the GCD by using the Euclidean algorithm can be stopped at any step at which the GCD is obvious.

Least Common Multiple

least common multiple (LCM)

Another useful concept in number theory is that of least common multiple. The **least common multiple (LCM)** of two natural numbers is the least positive multiple that the two numbers have in common. There are various methods for finding the LCM of two given natural numbers.

We can use a Cuisenaire rod model to determine the LCM of two numbers. For example, consider the 3 rod and 4 rod in Figure 5-9(a). We build trains of 3 rods and 4 rods until they are the same length, as shown in Figure 5-9(b). The LCM is the common length of train.

The Intersection-of-Sets Method. In the intersection-of-sets method, we first find the set of all positive *multiples* of both the first and second numbers, then find the set of all

FIGURE 5.9

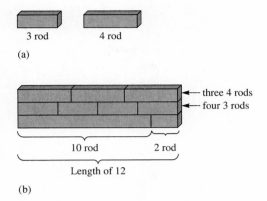

(a)

(b)

common multiples of both numbers, and finally pick the *least* element in that set. For example, to find the LCM of 8 and 12, denote the sets of positive multiples of 8 and 12 by M_8 and M_{12}, respectively.

$$M_8 = \{8, 16, 24, 32, 40, 48, 56, 64, 72, \ldots\}$$
$$M_{12} = \{12, 24, 36, 48, 60, 72, 84, 96, 108, \ldots\}$$

The set of common multiples is

$$M_8 \cap M_{12} = \{24, 48, 72, \ldots\}.$$

Because the least number in $M_8 \cap M_{12}$ is 24, the LCM of 8 and 12 is 24, written LCM(8, 12) = 24.

The Prime Factorization Method. The intersection-of-sets method for finding the LCM is often lengthy, especially when it is used to find the LCM of three or more natural numbers. Another, more efficient method for finding the LCM of several numbers is the *prime factorization method*. For example, to find LCM(40, 12), first find the prime factorizations of 40 and 12, namely, $2^3 \cdot 5$ and $2^2 \cdot 3$, respectively.

If $m = $ LCM(40, 12), then m is a multiple of 40, and must contain both 2^3 and 5 as factors. Also, m is a multiple of 12, and must contain 2^2 and 3 as factors. Since 2^3 is a multiple of 2^2, then $m = 2^3 \cdot 5 \cdot 3 = 120$. In general, we have the following:

To find the LCM of two natural numbers, first find the prime factorization of each number. Then take each of the primes that are factors of either of the given numbers. The LCM is the product of these primes, each raised to the greatest power of the prime that occurs in either of the prime factorizations.

EXAMPLE 5-12 Find the LCM of 2520 and 10,530.

Solution
$$2520 = 2^3 \cdot 3^2 \cdot 5 \cdot 7$$
$$10{,}530 = 2 \cdot 3^4 \cdot 5 \cdot 13$$
$$\text{LCM}(2520, 10{,}530) = 2^3 \cdot 3^4 \cdot 5 \cdot 7 \cdot 13$$

The Euclidean Algorithm Method. To see the connection between the GCD and LCM, consider the GCD and LCM of 6 and 9. Because $6 = 2 \cdot 3$ and $9 = 3^2$, it follows that GCD(6, 9) = 3 and LCM(6, 9) = 18. Notice that GCD(6, 9) $\cdot$ LCM(6, 9) = $3 \cdot 18 = 54$, and 54 is the product of the original numbers 6 and 9. In general, for any two natural numbers a and b, the connection between their GCD and LCM is given by Theorem 5-9.

> **THEOREM 5-9**
>
> For any two natural numbers a and b,
> $$\text{GCD}(a, b) \cdot \text{LCM}(a, b) = ab.$$

This result is useful for finding the LCM of two numbers a and b when their prime factorizations are not easy to find. GCD (a, b) can be found by the Euclidean algorithm, the product ab can be found by simple multiplication, and LCM(a, b) can be found by division.

EXAMPLE 5-13 Find LCM(731, 952).

Solution By the Euclidean algorithm, GCD(731, 952) = 17. By Theorem 5-9, $17 \cdot \text{LCM}(731, 952) = 731 \cdot 952$. Consequently,

$$\text{LCM}(731, 952) = \frac{731 \cdot 952}{17} = 40{,}936.$$

Although Theorem 5-9 cannot be used to find the LCM of more than two numbers, it is possible to find the LCM for three or more numbers. For example, to find LCM(12, 108, 120), we can use the prime factorization method.

$$12 = 2^2 \cdot 3$$
$$108 = 2^2 \cdot 3^3$$
$$120 = 2^3 \cdot 3 \cdot 5$$

Then, LCM(12, 108, 120) = $2^3 \cdot 3^3 \cdot 5 = 1080$.

The Division-by-Primes Method. Another procedure for finding the LCM of several natural numbers involves division by primes. For example, to find LCM(12, 75, 120), we start with the least prime that divides at least one of the given numbers and divide as follows:

$$\begin{array}{r|rrr} 2 & 12, & 75, & 120 \\ \hline & 6, & 75, & 60 \end{array}$$

Because 2 does not divide 75, simply bring down the 75. In order to obtain the LCM using this procedure, continue the division process until the row of answers consists of relatively prime numbers.

$$\begin{array}{r|rrr} 2 & 12, & 75, & 120 \\ 2 & 6, & 75, & 60 \\ 2 & 3, & 75, & 30 \\ 3 & 3, & 75, & 15 \\ 5 & 1, & 25, & 5 \\ \hline & 1, & 5, & 1 \end{array}$$

$\longrightarrow$ GCD is 3 (Why?)

Thus, LCM(12, 75, 120) = $2 \cdot 2 \cdot 2 \cdot 3 \cdot 5 \cdot 1 \cdot 5 \cdot 1 = 2^3 \cdot 3 \cdot 5^2 = 600$.

Two methods of finding the LCM of two numbers are given on the student page from *Addison-Wesley Mathematics,* Grade 7, 1991.

Least Common Multiple

LEARN ABOUT IT

EXPLORE Solve to Understand

Ralph picked two numbers, 6 and 8, out of a hat. Today every 6th customer will get a free sandwich and every 8th customer will get a free drink. Suppose 64 customers come into the store. Which ones will win both a free drink and a free sandwich?

TALK ABOUT IT

1. Which three customers were first to get free sandwiches? Free drinks?

2. Which customer was first to get both a free drink and a free sandwich?

To find the multiples of a number, multiply that number by 0, 1, 2, 3 and so on. The multiples of 12 are 0, 12, 24, 36, . . . The **least common multiple (LCM)** of two numbers is the smallest nonzero multiple which the numbers have in common.

Example Find the LCM of 12 and 18.

Method 1 List the nonzero multiples of each number until you reach a common multiple:
  multiples of 12: 12, 24, 36, . . .
  multiples of 18: 18, 36, . . .
The LCM of 12 and 18 is 36.

Method 2 List the prime factors of each number. Multiply the highest powers of each factor.
  $12 = 2 \cdot 2 \cdot 3 = 2^2 \cdot 3$
  $18 = 2 \cdot 3 \cdot 3 = 2 \cdot 3^2$
The LCM of 12 and 18 is $2^2 \cdot 3^2$ or 36.

PROBLEM SET 5-3

1. Find the GCD and the LCM for each of the following, using the intersection-of-sets method:
   (a) 18 and 10
   (b) 24 and 36
   (c) 8, 24, and 52

2. Find the GCD and the LCM for each of the following, using the prime factorization method:
   (a) 132 and 504
   (b) 65 and 1690
   (c) 900, 96, and 630
   (d) 108 and 360
   (e) 63 and 147
   (f) 625, 750, and 1000

3. Find the GCD for each of the following, using the Euclidean algorithm:
   (a) 220 and 2924
   (b) 14,595 and 10,856
   (c) 122,368 and 123,152

4. Find the LCM for each of the following, using any method:
   (a) 24 and 36
   (b) 72 and 90 and 96
   (c) 90 and 105 and 315

5. Find the LCM for each of the following pairs of numbers, using Theorem 5-9 and the answers from Problem 3.
   (a) 220 and 2924
   (b) 14,595 and 10,856
   (c) 122,368 and 123,152

6. Use Cuisenaire rods to find the GCD and the LCM of 6 and 10.

7. In Quinn's dormitory room, there are three snooze-alarm clocks, each of which is set at a different time. Clock A goes off every 15 min, clock B goes off every 40 min, and clock C goes off every 60 min. If all three clocks go off at 6:00 A.M., answer the following:
   (a) How long will it be before the clocks go off together again after 6:00 A.M.?
   (b) Would the answer to part (a) be different if clock B went off every 15 min and clock A went off every 40 min?

8. If a number is greater than the GCD(9, 12), less than the LCM(2, 3), and the number is odd, what is it?

9. At the Senior All-night Party, a money chest contained enough money so that from 1 to 6 winners could share the money equally. The winners were to be chosen from those still in attendance at 4:00 A.M., and no one who had left early could win.
   (a) What is the least amount of money that could be in the nonempty chest?
   (b) If there were actually five winners, how much would each receive?
   (c) If the prize money was to be given in $2 bills, how many bills were in the chest?

10. Midas has 120 gold coins and 144 silver coins. He wants to place his gold coins and his silver coins in stacks so that there are the same number of coins in each stack. What is the greatest number of coins that he can place in each stack?

11. Bill and Sue both work at night. Bill has every sixth night off and Sue has every eighth night off. If they are both off tonight, how many nights will it be before they are both off again?

12. By selling cookies at 24¢ each, José made enough money to buy several cans of pop costing 45¢ per can. If he had no money left over after buying the pop, what is the least number of cookies he could have sold?

13. Bijous I and II start their movies at 7:00 P.M. The movie at Bijou I takes 75 min, while the movie at Bijou II takes 90 min. If the shows run continuously, when will they start at the same time again?

14. Two bike riders ride around in a circular path. The first rider completes one round in 12 min and the second rider completes it in 18 min. If they both start at the same place and the same time and go in the same direction, after how many minutes will they meet again at the starting place?

15. Assume that a and b are any natural numbers, and answer each of the following:
    (a) If $GCD(a, b) = 1$, find $LCM(a, b)$.
    (b) Find $GCD(a, a)$ and $LCM(a, a)$.
    (c) Find $GCD(a^2, a)$ and $LCM(a^2, a)$.
    (d) If $a|b$, find $GCD(a, b)$ and $LCM(a, b)$.
    (e) If a and b are two different primes, find $GCD(a, b)$ and $LCM(a, b)$.
    (f) What is the relationship between a and b if $GCD(a, b) = a$?
    (g) What is the relationship between a and b if $LCM(a, b) = a$?

16. ▶Classify each of the following as true or false. Justify your answers.
    (a) If $GCD(a, b) = 1$, then a and b cannot both be even.
    (b) If $GCD(a, b) = 2$, then both a and b are even.
    (c) If a and b are even, then $GCD(a, b) = 2$.
    (d) For all natural numbers a and b, $LCM(a, b) | GCD(a, b)$.
    (e) For all natural numbers a and b, $LCM(a, b) | ab$.
    (f) $GCD(a, b) \leq a$
    (g) $LCM(a, b) \geq a$

17. In order to find $GCD(24, 20, 12)$, it is possible to find $GCD(24, 20)$, which is 4, and then find $GCD(4, 12)$, which is 4. Use this approach and the Euclidean algorithm to find $GCD(120, 75, 105)$.

18. ▶Is it true that $GCD(a, b, c) \cdot LCM(a, b, c) = abc$? Write an explanation of your answer.

19. (a) Show that 97,219,988,751 and 4 are relatively prime.
    (b) Show that 181,345,913 and 11 are relatively prime.

(c) Show that 181,345,913 and 33 are relatively prime.
20. Find all natural numbers x such that GCD(25, x) = 1 and $1 \leq x \leq 25$.
21. (a) A number is called *perfect* if it is equal to the sum of its proper divisors, that is, equal to the sum of all its divisors except the number itself. For example, because $6 = 1 + 2 + 3$, it is perfect. Find another perfect number less than 30.
   (b) Two numbers are said to be *amicable* if each is the sum of the proper divisors of the other. Show that 220 and 284 are amicable.
22. Is it always true that if $d | GCD(a, b)$, then $d | a$ and $d | b$? Why or why not?

Review Problems

23. Find two integers x and y such that $x \cdot y = 1,000,000$ and neither x nor y contains any zeros as digits.
24. Fill each blank space with a single digit that makes the corresponding statement true. Find all possible answers.
   (a) $3 | 83 \_ 51$ (b) $11 | 8 \_ 691$
   (c) $23 | 103 \_ 6$
25. Is 3111 a prime? Prove your answer.
26. Find a number that has exactly six prime factors.
27. Produce the least positive number that is divisible by 2, 3, 4, 5, 6, 7, 8, 9, 10, and 11.
28. What is the greatest prime that must be used to determine if 2089 is prime?
29. Given the set of numbers {61, 63, 65, 67, 70}, which one is a composite number between 62 and 72, has the sum of its digits as a prime number, and has more than 4 factors?
30. Refer to the drawing below and answer the following:
   (a) Which number is in all three shapes?
   (b) Which prime number is in the circle but not the square?
   (c) Which multiples of 3 are in the triangle?
   (d) Which composite number is neither in the square nor in the triangle?

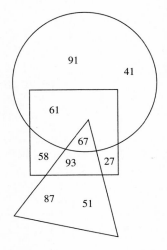

● COMPUTER CORNER

1. Type the following Logo procedure for finding the GCD of two positive integers into your computer, and then use the procedure to find the GCD of the given numbers.

```
TO GCD :A :B
  IF :B = 0 OUTPUT :A
  OUTPUT GCD :B (REMAINDER :A :B)
END
```

(*In LCSI, replace* IF :B = 0 OUTPUT :A *with* IF :B = 0 [OUTPUT :A].)

   (a) GCD (676, 221)
   (b) GCD (10,764, 2300)

2. Use Theorem 5-9 and the preceding GCD procedure to write a procedure LCM for finding the LCM of any two positive integers, :A and :B.

▼ BRAIN TEASER

For any $n \times m$ rectangle such that $GCD(n, m) = 1$, find a rule for determining the number of unit squares (1×1) that a diagonal passes through. For example, in the drawings below the diagonal passes through 8 and 6 squares, respectively.

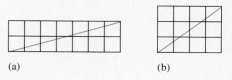

(a)   (b)

*Section 5-4 Clock and Modular Arithmetic

Clock Arithmetic

The book *Disquisitiones Arithmeticae* is among Karl Friedrich Gauss's great mathematical works. In this book, Gauss introduced a new topic, the theory of congruences, which very rapidly gained general acceptance and became a foundation for number theory.

One type of enrichment activity in elementary school involving congruences uses the arithmetic of a 12-hour clock. For example, if it is 9 o'clock, what time will it be 8 hours later? It is possible to use the clock in Figure 5-10 to determine that 8 hours after 9 o'clock is 5 o'clock. We record this as $9 \oplus 8 = 5$, where $\oplus$ denotes clock addition. The answer can also be obtained by performing the regular addition $9 + 8 = 17$, and by dividing 17 by 12 and taking the remainder. Thus, whenever the sum of two digits on a 12-hour clock under regular addition exceeds 12, add the numbers normally and then obtain the remainder when the sum is divided by 12.

To perform other operations on the clock such as $2 \ominus 9$, where $\ominus$ denotes clock subtraction, we could interpret it as the time 9 hours before 2 o'clock. Counting backward (counterclockwise) 9 units from 2 reveals that $2 \ominus 9 = 5$. If subtraction on the clock is defined in terms of addition, we have $2 \ominus 9 = x$, if and only if $2 = 9 \oplus x$. Consequently, $x = 5$.

FIGURE 5-10

E X A M P L E 5 - 1 4   Perform each of the following computations on a 12-hour clock:

(a) $8 \oplus 8$ (b) $4 \ominus 12$ (c) $4 \ominus 4$ (d) $4 \ominus 8$

Solution   (a) $(8 + 8) \div 12$ has remainder 4. Hence, $8 \oplus 8 = 4$.
(b) $4 \ominus 12 = 4$, since, by counting forward or backward 12 hours, you arrive at the original position.
(c) $4 \ominus 4 = 12$. This should be clear from looking at the clock, but it can also be found by using the definition of subtraction in terms of addition.
(d) $4 \ominus 8$ because $8 \oplus 8 = 4$

∎

Clock multiplication can be defined using repeated addition as with whole numbers. For example, $2 \otimes 8 = 8 \oplus 8 = 4$, where $\otimes$ denotes clock multiplication. Similarly, $3 \otimes 5 = (5 \oplus 5) \oplus 5 = 10 \oplus 5 = 3$.

Clock division can be defined in terms of multiplication. For example, $8 \oslash 5 = x$, where $\oslash$ denotes clock division, if and only if $8 = 5 \otimes x$, for a unique x in the set $\{1, 2, 3, \ldots, 12\}$. Because $5 \otimes 4 = 8$, then $8 \oslash 5 = 4$.

EXAMPLE 5-15 Perform the following operations on a 12-hour clock, if possible:

(a) $3 \otimes 11$ (b) $2 \oslash 7$ (c) $3 \oslash 2$ (d) $5 \oslash 12$

Solution (a) $3 \otimes 11 = (11 \oplus 11) \oplus 11 = 10 \oplus 11 = 9$
(b) $2 \oslash 7 = x$ if and only if $2 = 7 \otimes x$. Consequently, $x = 2$.
(c) $3 \oslash 2 = x$ if and only if $3 = 2 \otimes x$. Multiplying each of the numbers 1, 2, 3, 4, ..., 12 by 2 shows that none of the multiplications yields 3. Thus, the equation $3 = 2 \otimes x$ has no solution, and consequently, $3 \oslash 2$ is undefined.
(d) $5 \oslash 12 = x$ if and only if $5 = 12 \otimes x$. However, $12 \otimes x = 12$ for every x in the set $\{1, 3, 4, \ldots, 12\}$. Thus, $5 = 12 \otimes x$ has no solution on the clock; and therefore, $5 \oslash 12$ is undefined. ∎

Adding or subtracting 12 on a 12-hour clock gives the same result. Thus, 12 behaves as 0 does in a base-ten addition or subtraction and is the additive identity for addition on the 12-hour clock. Similarly, on a 5-hour clock, 5 behaves as 0 does.

Addition, subtraction, and multiplication on a 12-hour clock, can be performed for any two numbers, but as shown in Example 5-15, not all divisions can be performed. Division by 12, the additive identity, on a 12-hour clock either can never be performed or is not meaningful, since it does not yield a unique answer. However, there are clocks on which all divisions can be performed, except by the corresponding additive identities. One such clock is a 5-hour clock, shown in Figure 5-11.

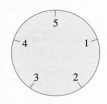

FIGURE 5-11

TABLE 5-5

(a)
$\oplus$	1	2	3	4	5
1	2	3	4	5	1
2	3	4	5	1	2
3	4	5	1	2	3
4	5	1	2	3	4
5	1	2	3	4	5

(b)
$\otimes$	1	2	3	4	5
1	1	2	3	4	5
2	2	4	1	3	5
3	3	1	4	2	5
4	4	3	2	1	5
5	5	5	5	5	5

On this clock, $3 \oplus 4 = 2$, $2 \ominus 3 = 4$, $2 \otimes 4 = 3$, and $3 \oslash 4 = 2$. Since adding 5 to any number yields the original number, 5 is the additive identity for this 5-hour clock, as seen in Table 5-5(a). Consequently, you might suspect that division by 5 is not possible on a 5-hour clock. To determine which divisions are possible, consider Table 5-5(b), a multiplication table for 5-hour clock arithmetic. To find $1 \oslash 2$, we write $1 \oslash 2 = x$, which is equivalent to $1 = 2 \otimes x$. The second row of part (b) of the table shows that $2 \otimes 1 = 2$, $2 \otimes 2 = 4$, $2 \otimes 3 = 1$, $2 \otimes 4 = 3$, and $2 \otimes 5 = 5$. The solution of $1 = 2 \otimes x$ is $x = 3$, so

$1 \oslash 2 = 3$. The information given in the second row of the table can be used to determine the following divisions:

$$2 \oslash 2 = 1 \text{ because } 2 = 2 \otimes 1$$
$$3 \oslash 2 = 4 \text{ because } 3 = 2 \otimes 4$$
$$4 \oslash 2 = 2 \text{ because } 4 = 2 \otimes 2$$
$$5 \oslash 2 = 5 \text{ because } 5 = 2 \otimes 5$$

Because every element occurs in the second row, division by 2 is always possible. Similarly, division by all other numbers, except 5, is always possible. In the problem set, you are asked to perform arithmetic on different clocks and to investigate for which clocks all computations, except division by the additive identity, can be performed.

Modular Arithmetic

Many of the concepts for clock arithmetic can be used to work problems involving a calendar. On the calendar in Figure 5-12, the five Sundays have dates 1, 8, 15, 22, and 29. Any two of these dates for Sunday differ by a multiple of 7. The same property is true for any other day of the week. For example, the second and thirtieth days fall on the same day, since $30 - 2 = 28$ and 28 is a multiple of 7. We say that 30 is congruent to 2, modulo 7, and we write $30 \equiv 2 \pmod{7}$. Similarly, because 18 and 6 differ by a multiple of 12, we write $18 \equiv 6 \pmod{12}$. This leads to the following definition.

APRIL						
S	M	T	W	T	F	S
1	2	3	4	5	6	7
8	9	10	11	12	13	14
15	16	17	18	19	20	21
22	23	24	25	26	27	28
29	30					

FIGURE 5-12

DEFINITION OF MODULAR CONGRUENCE

For integers a and b, **a is congruent to b modulo m**, written $a \equiv b \pmod{m}$, if and only if $a - b$ is a multiple of m, where m is a positive integer greater than 1.

EXAMPLE 5-16 Tell why each of the following is true.

(a) $23 \equiv 3 \pmod{10}$
(b) $23 \equiv 3 \pmod{4}$
(c) $23 \not\equiv 3 \pmod{7}$
(d) $10 \equiv {}^-1 \pmod{11}$
(e) $25 \equiv 5 \pmod{5}$

Solution
(a) $23 \equiv 3 \pmod{10}$, because $23 - 3$ is a multiple of 10.
(b) $23 \equiv 3 \pmod{4}$, because $23 - 3$ is a multiple of 4.
(c) $23 \not\equiv 3 \pmod{7}$, because $23 - 3$ is not a multiple of 7.
(d) $10 \equiv {}^-1 \pmod{11}$, because $10 - ({}^-1) = 11$ is a multiple of 11.
(e) $25 \equiv 5 \pmod{5}$, because $25 - 5 = 20$ is a multiple of 5. ∎

EXAMPLE 5-17 Find all integers x such that $x \equiv 1 \pmod{10}$.

Solution The solution is $x \equiv 1 \pmod{10}$ if, and only if, $x - 1 = 10k$, where k is any integer. Consequently, $x = 10k + 1$. Letting $k = 0, 1, 2, 3, \ldots$ yields the sequence 1, 11, 21, 31, 41, ... Likewise, letting $k = {}^-1, {}^-2, {}^-3, {}^-4, \ldots$ yields the negative integers

⁻9, ⁻19, ⁻29, ⁻39, ... The two sequences can be combined to give the solution set

$$\{\ldots, {}^-39, {}^-29, {}^-19, {}^-9, 1, 11, 21, 31, 41, 51, \ldots\}.$$ ∎

In Example 5-17, the positive integers obtained, 1, 11, 21, 31, 41, 51, ..., differ from each other by a multiple of 10; hence, they are congruent to each other modulo 10. Notice that each of the numbers 1, 11, 21, 31, 41, 51, ... has a remainder of 1 when divided by 10. In general, *two whole numbers are congruent modulo m if, and only if, their remainders, on division by m, are the same.*

The [INT÷] button on a calculator may be used to work with modular arithmetic. If we press the following sequence of buttons, we see that $4325 \equiv 5 \pmod 9$ because the remainder when 4325 is divided by 9 is 5.

[4] [3] [2] [5] [INT÷] [9] [=], and the display shows a remainder of 5.

Many properties of congruence are similar to properties for equality. Several of these are listed below.

PROPERTIES

For all integers a, b, and c:

1. $a \equiv a \pmod m$.
2. If $a \equiv b \pmod m$, then $b \equiv a \pmod m$.
3. If $a \equiv b \pmod m$ and $b \equiv c \pmod m$, then $a \equiv c \pmod m$.
4. If $a \equiv b \pmod m$, then $a + c \equiv b + c \pmod m$.
5. If $a \equiv b \pmod m$, then $ac \equiv bc \pmod m$.
6. If $a \equiv b \pmod m$ and $c \equiv d \pmod m$, then $ac \equiv bd \pmod m$.
7. If $a \equiv b \pmod m$ and k is a natural number, then $a^k \equiv b^k \pmod m$.

With these properties, it is possible to solve a variety of problems, such as the following.

PROBLEM 5

Find the remainder when 3^{100} is divided by 5.

Understanding the Problem. No calculator will accurately find 3^{100} so we cannot actually divide 3^{100} by 5 to find the remainder. The remainder should be 0, 1, 2, 3, or 4 when a number is divided by 5.

Devising a Plan. Since a calculator will not solve the problem, we look for an alternate plan. The use of modular arithmetic will help if we can find small integers that are equivalent to powers of 3, and use properties (5) and (7) to build up 3^{100} and find the mod 5 equivalent.

Carrying Out the Plan. We know that $3^2 \equiv 4 \pmod 5$
Thus,

$$3^3 \equiv 3 \cdot 4 \equiv 2 \pmod 5$$
$$3^4 \equiv 3 \cdot 2 \equiv 1 \pmod 5.$$

258   CHAPTER 5   NUMBER THEORY

Using property (7), we see that $(3^4)^{25} \equiv 1^{25}$ (mod 5) or $3^{100} \equiv 1$ (mod 5). It follows that 3^{100} and 1 have the same remainder when divided by 5. Thus, 3^{100} has remainder 1 when divided by 5.

Looking Back. This type of problem can be changed to find the remainders when dividing by different numbers or to find the units digit of numbers such as 2^{96}. ∎

EXAMPLE 5-18   (a) If it is now Monday, October 14, on what day of the week will October 14 fall next year, if next year is not a leap year?

(b) If Christmas falls on Thursday this year, on what day of the week will Christmas fall next year, if next year is a leap year?

Solution   (a) Because next year is not a leap year, we have 365 days in the year. Because $365 = 52 \cdot 7 + 1$, we have $365 \equiv 1$ (mod 7). Thus, 365 days after October 14 will be 52 weeks and one day later. Thus, October 14 will be on a Tuesday.

(b) Because there are 366 days in a leap year, we have $366 \equiv 2$ (mod 7). Thus, Christmas will be two days after Thursday, on Saturday. ∎

PROBLEM SET 5-4

1. Perform each of the following operations on a 12-hr clock, if possible:
   (a) $7 \oplus 8$ (b) $4 \oplus 10$ (c) $3 \ominus 9$
   (d) $4 \ominus 8$ (e) $3 \otimes 9$ (f) $4 \otimes 4$
   (g) $1 \oslash 3$ (h) $2 \oslash 5$

2. Perform each of the following operations on a 5-hr clock:
   (a) $3 \oplus 4$ (b) $3 \oplus 3$ (c) $3 \otimes 4$
   (d) $1 \otimes 4$ (e) $3 \otimes 4$ (f) $2 \otimes 3$
   (g) $3 \oslash 4$ (h) $1 \oslash 4$

3. (a) Construct an addition table for a 7-hr clock.
   (b) Using the addition table in (a), find $5 \ominus 6$ and $2 \ominus 5$.
   (c) Using the addition table in (a), show that subtraction can always be performed on a 7-hr clock.

4. (a) Construct a multiplication table for a 7-hr clock.
   (b) Use the multiplication table in (a) to find $3 \oslash 5$ and $4 \oslash 6$.
   (c) Use the multiplication table to find whether division by numbers different from 7 is always possible.

5. (a) Construct the multiplication tables for 3-, 4-, 6-, and 11-hr clocks.
   ▶(b) On which of the clocks in part (a) can divisions by numbers other than the additive identity always be performed? Explain your answer.

   (c) How do the multiplication tables of clocks for which division can always be performed (except by an additive identity) differ from the multiplication tables of clocks for which division is not always meaningful?

6. On a 12-hr clock, find each of the following:
   (a) Additive inverse of 2   (b) Additive inverse of 3
   (c) $(^-2) \oplus (^-3)$ (d) $^-(2 \oplus 3)$
   (e) $(^-2) \ominus (^-3)$ (f) $(^-2) \otimes (^-3)$

7. (a) If April 23 falls on Tuesday, what are the dates of the other Tuesdays in April?
   (b) If July 2 falls on Tuesday, list the dates of the Wednesdays in July.
   (c) If September 3 falls on Monday, on what day of the week will it fall next year, if next year is a leap year?

8. Fill in each blank so that the answer is nonnegative and the least possible number.
   (a) $29 \equiv$ _____ (mod 5)
   (b) $3498 \equiv$ _____ (mod 3)
   (c) $3498 \equiv$ _____ (mod 11)
   (d) $^-23 \equiv$ _____ (mod 10)

9. Show that each of the following statements is true:
   (a) $81 \equiv 1$ (mod 8)
   (b) $81 \equiv 1$ (mod 10)

(c) $1000 \equiv {}^-1 \pmod{13}$
(d) $10^{84} \equiv 1 \pmod 9$
(e) $10^{100} \equiv 1 \pmod{11}$
(f) $937 \equiv 37 \pmod{100}$

10. Show that $a \equiv 0 \pmod m$, if, and only if, $m|a$.
11. Translate each of the following statements into the language of congruences:
    (a) $8|24$ (b) $3|{}^-90$
    (c) Any integer n divides itself.
12. (a) Find all x such that $x \equiv 0 \pmod 2$.
    (b) Find all x such that $x \equiv 1 \pmod 2$.
    (c) Find all x such that $x \equiv 3 \pmod 5$.
13. Find the remainder for each of the following:
    (a) 5^{100} is divided by 6.
    (b) 5^{101} is divided by 6.
    (c) 10^{99} is divided by 11.
    (d) 10^{100} is divided by 11.
14. ▶Write an explanation of how the odometer on a car is an example of modular arithmetic. What is its mod?
★15. (a) Find a negative integer value for x such that $10^3 \equiv x \pmod{13}$ and $|x|$ is the least possible.
    (b) Find the remainder when 10^{99} is divided by 13.
★16. Use the fact that $100 \equiv 0 \pmod 4$ to find and prove a test for divisibility by 4.
★17. Show that, in general, the cancellation property for multiplication does not hold for congruences; that is, show that $ac \equiv bc \pmod m$ does not always imply $a \equiv b \pmod m$.

LABORATORY ACTIVITY

Most publishers include an International Standard Book Number (ISBN) on their books. If the ISBN number for a book is 0-8053-0390-1, what do the numbers mean? Why are they used?

▼ BRAIN TEASER

How many primes are in the following sequence?

9, 98, 987, 9876, . . . , 987654321, 9876543219, 98765432198, . . .

Solution to the Preliminary Problem

Understanding the Problem. In an essay contest, winners are ranked to receive $25 U.S. Savings Bonds. Winners are ranked without ties and the person with the highest ranking receives one more bond than the second-ranked person; the second-ranked person receives one more bond than the third-ranked person, and so on. We know that the total number of bonds to be awarded is 1000, the number of bonds each winner receives is a natural number, and if the highest-ranked person receives n bonds, then the second-ranked person receives $n - 1$ bonds, and so forth. The minimum number, k, of winners is an unknown to be determined.

Devising a Plan. Because we know the total number of bonds to be awarded to the winners, we may use the strategy of *writing an equation* to aid in solving the problem. Rank ordering the winners from greatest to least and using the strategy of *making a table* as in Table 5-6 reveals a pattern that will help in writing the desired equation.

TABLE 5-6

Rank of Winners	Number of Bonds Won
1	n
2	$n - 1$
3	$n - 2$
4	$n - 3$
5	$n - 4$
.	.
.	.
.	.
$k - 1$	$n - (k - 2)$
k	$n - (k - 1)$

Carrying Out the Plan. Table 5-6 reveals a pattern in the number of bonds received. The kth person receives $n - (k - 1)$ bonds. Because the total number of bonds is 1000, we have the following:

$$n + (n - 1) + (n - 2) + (n - 3) + \ldots + (n - (k - 1)) = 1000.$$

To simplify the equation, we sum the k consecutive numbers. We do this by recalling a *similar problem* (in Chapter 1) posed and solved by Gauss. This sum can be computed using the following:

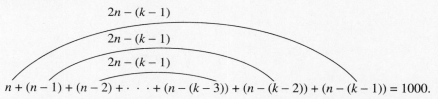

$$n + (n - 1) + (n - 2) + \cdots + (n - (k - 3)) + (n - (k - 2)) + (n - (k - 1)) = 1000.$$

We know there are k consecutive numbers, so there are $k/2$ pairs, each with the sum of $2n - (k - 1)$, and the sum is $k(2n - (k - 1))/2$. Therefore we have

$$k(2n - (k - 1))/2 = 1000$$
$$k(2n - (k - 1)) = 2000.$$

The last equation is a Diophantine equation in two unknowns and yields no apparent solution.

We do have one additional condition on the problem, which is that k is the minimum natural number greater than 1 that makes the awards feasible. Now k and $2n - (k - 1)$ are natural number factors of 2000, so they will be included in a list of the factors of 2000. We can use the prime factorization of 2000, or $2^4 \cdot 5^3$, to write the following list of factors:

1	2	4	8	16
5	10	20	40	80
25	50	100	200	400
125	250	500	1000	2000

If k is the least factor greater than 1 so that $2n - (k - 1)$ is a natural number, then k might be 2. If $k = 2$, then $2(2n - (2 - 1)) = 2000$, so that $n = 500 \ 1/2$ is not a natural number.

Hence, $k \neq 2$. If $k = 4$, then $4(2n - (4 - 1)) = 2000$, so that $n = 251\ 1/2$ is not a natural number. Therefore, $k \neq 4$. However, if $k = 5$, then $5(2n - (5 - 1)) = 2000$, and $n = 202$. If $k = 5$, there are five winners and the person with the highest rank receives 202 bonds, with the other winners receiving 201, 200, 199, and 198 bonds, respectively.

Looking Back. It is easy to check the solution because $202 + 201 + 200 + 199 + 198 = 1000$. Can we solve the problem if there is a different number of bonds? What solutions does the problem have if the number of bonds is not minimized? What is the maximum number of winners possible with this ranking scheme? Does the value of the bond matter?

QUESTIONS FROM THE CLASSROOM

1. A student claims that $a|a$ and $a|a$ implies $a|(a - a)$, and hence, $a|0$. Is the student correct?
2. A student writes, "If $d \nmid a$ and $d \nmid b$, then $d \nmid (a + b)$." How do you respond?
3. Your seventh-grade class has just completed a unit on divisibility rules. One of the better students asks why divisibility by numbers other than 3 and 9 cannot be tested by dividing the sum of the digits by the tested number. How should you respond?
4. A student claims that a number with an even number of digits is divisible by 7 if and only if each of the numbers formed by pairing the digits into groups of two is divisible by 7. For example, 49,562,107 is divisible by 7, since each of the numbers 49, 56, 21, and 07 is divisible by 7. Is this true?
5. A sixth-grade student argues that there are infinitely many primes because "there is no end to numbers." How do you respond?
6. A student claims that a number is divisible by 21 if, and only if, it is divisible by 3 and by 7, and, in general, a number is divisible by $a \cdot b$ if, and only if, it is divisible by a and by b. What is your response?
7. A student claims that, for any two integers a and b, GCD(a, b) divides LCM(a, b) and, hence, GCD$(a, b) <$ LCM(a, b). Is the student correct? Why or why not?
8. A student claims that there are infinitely many triples of positive integers x and y, z that make the equation $x^2 + y^2 = z^2$ true. How do you respond?
9. A student argues that 1 should be a prime because it has 1 and itself as divisors. How do you respond?
10. A student asks about the relation between least common multiple and least common denominator. How do you respond?

CHAPTER OUTLINE

I. Divisibility
  A. If a and b are any integers, then b **divides** a, denoted by $b|a$, if, and only if, there is an integer c such that $a = cb$.
  B. The following are basic divisibility theorems for integers a, b, and d:
    1. If $d|a$ and k is any integer, then $d|ka$.
    2. If $d|a$ and $d|b$, then $d|(a + b)$ and $d|(a - b)$.
    3. If $d|a$ and $d \nmid b$, then $d \nmid (a + b)$ and $d \nmid (a - b)$.
  C. Divisibility tests
    1. An integer is divisible by 2, 5, or 10 if and only if its units digit is divisible by 2, 5, or 10, respectively.
    2. An integer is divisible by 4 if, and only if, the last two digits of the integer represent a number divisible by 4.
    3. An integer is divisible by 8 if, and only if, the last three digits of the integer represent a number divisible by 8.
    4. An integer is divisible by 3 or by 9 if, and only if, the sum of its digits is divisible by 3 or 9, respectively.

5. An integer is divisible by 7 if, and only if, the integer represented without its units digit minus twice the units digit of the original number is divisible by 7.
6. An integer is divisible by 11 if, and only if, the sum of the digits in the places that are even powers of 10 minus the sum of the digits in the places that are odd powers of 10 is divisible by 11.
7. An integer is divisible by 6 if, and only if, the integer is divisible by both 2 and 3.

II. Prime and composite numbers
  A. Positive integers that have exactly two positive divisors are called **primes**. Integers greater than 1 and not primes are called **composites**.
  B. **Fundamental Theorem of Arithmetic:** Every composite number has one and only one prime factorization.
  C. Criterion for determining if a given number n is prime: *If n is not divisible by any prime p such that $p^2 \leq n$, then n is prime.*

III. Greatest common divisor and least common multiple
  A. The **greatest common divisor (GCD)** of two or more natural numbers is the greatest divisor, or factor, that the numbers have in common.
  B. **Euclidean algorithm:** If a and b are whole numbers and $a \geq b$, then $GCD(a, b) = GCD(b, r)$, where r is the remainder when a is divided by b. The procedure of finding the GCD of two numbers a and b by using the above result repeatedly is called the *Euclidean algorithm*.
  C. The **least common multiple (LCM)** of two or more natural numbers is the least positive multiple that the numbers have in common.
  D. $GCD(a, b) \cdot LCM(a, b) = ab$.
  E. If $GCD(a, b) = 1$, then a and b are **relatively prime**.

*IV. Modular arithmetic
  A. For any integers a and b, *a is congruent to b modulo m* if, and only if, $a - b$ is a multiple of m, where m is a positive integer greater than 1.
  B. Two integers are congruent modulo m if, and only if, their remainders upon division by m are the same.

CHAPTER TEST

1. Classify each of the following as true or false:
   (a) $8 \mid 4$ (b) $0 \mid 4$ (c) $4 \mid 0$
   (d) If a number is divisible by 4 and by 6, then it is divisible by 24.
   (e) If a number is not divisible by 12, then it is not divisible by 3.
2. Classify each of the following as true or false. If false, show a counterexample.
   (a) If $7 \mid x$ and $7 \nmid y$, then $7 \nmid xy$.
   (b) If $d \nmid (a + b)$, then $d \nmid a$ and $d \nmid b$.
   (c) If $16 \mid 10^4$, then $16 \mid 10^6$.
   (d) If $d \mid (a + b)$ and $d \nmid a$, then $d \nmid b$.
   (e) If $d \mid (x + y)$ and $d \mid x$, then $d \mid y$.
   (f) If $4 \nmid x$ and $4 \nmid y$, then $4 \nmid xy$.
3. Test each of the following numbers for divisibility by 2, 3, 4, 5, 6, 7, 8, 9, and 11:
   (a) 83,160   (b) 83,193
4. Assume that 10,007 is prime. Without actually dividing 10,024 by 17, prove that 10,024 is not divisible by 17.
5. Fill each blank with one digit to make each of the following true. (Find all the possible answers.)
   (a) $6 \mid 87\_4$
   (b) $24 \mid 4\_856$
   (c) $29 \mid 87\_\_4$
6. Determine whether each of the following numbers is prime or composite:
   (a) 143   (b) 223

7. ▶How can you tell if a number is divisible by 24? Check 4152 for divisibility by 24.
8. Find the GCD for each of the following:
   (a) 24 and 52
   (b) 5767 and 4453
9. Find the LCM for each of the following:
   (a) $2^3 \cdot 5^2 \cdot 7^3$, $2 \cdot 5^3 \cdot 7^2 \cdot 13$, and $2^4 \cdot 5 \cdot 7^4 \cdot 29$
   (b) 278 and 279
10. ▶Construct a number that has exactly five divisors. Explain your construction.
11. Find all divisors of 144.
12. Find the prime factorization of each of the following.
    (a) 172   (b) 288
    (c) 260   (d) 111
13. Jane and Ramon are running laps on a track. If they start at the same time and place and go in the same direction, with Jane running a lap in 5 min and Ramon running a lap in 3 min, how long will it take for them to be at the starting place at the same time if they continue to run at the same pace?
14. Candy bars priced at 50¢ each were not selling, so the price was reduced. Then they all sold in one day for a total of $31.93. What was the reduced price for each candy bar?
15. Two bells ring at 8:00 A.M. For the remainder of the day, one bell rings every half hour and the other bell rings every 45 min. What time will it be when the bells ring together again?

16. If we wanted each month of 12 months to have exactly the same number of days and there were exactly 365 days in a year, how many days are possible in each month?
17. If there were to be 9 boys and 6 girls at a party and the host wanted each to be given exactly the same number of candies that could be bought in packages containing 12 candies, what is the fewest number of packages that could be bought?
18. ▶Each place value of a natural number such as 2 in 4235 represents a multiple of a power of 10. Why does the entire number not represent a multiple of a power of 10?
★19. Prove the test for divisibility by 9 using a three-digit number n such that $n = a \cdot 10^2 + b \cdot 10 + c$.

*20. Find the remainder of each of the following:
   (a) 7^{100} is divided by 16.
   (b) 7^{100} is divided by 17.
   (c) 13^{1937} is divided by 10.
*21. The length of a week was probably inspired by the need for market days and religious holidays. The Romans, for example, once used an 8-day week. Assuming that April still had 30 days but was based on an 8-day week, if the first day of the month was on Sunday and the extra day after Saturday was called Venaday, on what day would the last day of the month fall?
*22. In measuring angles of rotation that a light on a small island lighthouse sweeps, what mod system would be used and why?

SELECTED BIBLIOGRAPHY

Bearing S. and B. Holtan. "Factors and Primes with a T Square." *Arithmetic Teacher* 34 (April 1987): 34.

Blocksma, M. *Reading the Numbers: A Survival Guide to the Measurements, Numbers, and Sizes Encountered in Everyday Life.* New York: Penguin Group, Viking Penguin, Inc., 1989.

Bunham, W. "Euclid and the Infinitude of Primes." *Mathematics Teacher* 80 (January 1987): 16–17.

Edwards, F. "Geometric Figures Make the LCM Obvious." *Arithmetic Teacher* 34 (March 1987): 17–18.

Ewbank, W. "LCM—Let's Put It in Its Place." *Arithmetic Teacher* 35 (November 1987): 45–47.

Hopkins, M. "Number Facts or Fantasy." *Arithmetic Teacher* 34 (March 1987): 38–42.

Lott, J., Ed. "Menu Madness." *Student Math Notes*. Reston, VA: National Council of Teachers of Mathematics (May 1991).

Olson, M. "On the Ball." *Student Math Notes*. Reston, VA: National Council of Teachers of Mathematics (September 1990).

Peterson, I. *The Mathematical Tourist: Snapshots of Modern Mathematics.* New York: W. H. Freeman and Company, 1988.

Tirman, A. "Pythagorean Triples." *Mathematics Teacher* 79 (November 1986): 652–655.

Wyatt, C. "Clock Beaters." *Arithmetic Teacher* 34 (September 1986): 20.

6 Rational Numbers as Fractions

Preliminary Problem

Daniel read that a grizzly bear can run 16 meters per second. Daniel found that he could run 8 meters per second. He also found that he could run from his well and be inside his cabin within 6 sec. If Daniel was at the well and was 50 m from a grizzly when it started to chase him, could Daniel safely make it to the cabin?

R ecall from Chapter 4 that because the equation $x + a = 0$ had no solution in the set of whole numbers, we devised a new number, denoted by ^-a, that is the unique solution of the equation. Similarly, a unique solution is needed to an equation like $6x = 5$. The unique solution is denoted by $\frac{5}{6}$. Thus $\frac{5}{6}$ can be thought of as a number such that $6 \cdot \frac{5}{6} = 5$. In the real world, we might think of $\frac{5}{6}$ as the part of the total pizza each person received when 5 equal-sized pizzas are to be divided equally among 6 people. In general, the unique solution for x in the equation $b \cdot x = a$, where $b \neq 0$, is denoted by $\frac{a}{b}$. Thus $\frac{a}{b}$ is a number such that

fraction $b \cdot \frac{a}{b} = a$. A **fraction** is a number of the form $\frac{a}{b}$, where a and b are *any numbers* ($b \neq 0$), not necessarily integers. In Chapter 7, we will see fractions in which a and b are not integers $\left(\text{for example, } \frac{\sqrt{2}}{2}\right)$.

The Set of Rational Numbers

rational numbers   Numbers such as $\frac{1}{3}, \frac{3}{5},$ and $\frac{2}{3}$ belong to the set of **rational numbers.** The set of rational numbers, denoted by Q, can be written as follows:

$$Q = \left\{ \frac{a}{b} \;\middle|\; a \text{ and } b \text{ are integers and } b \neq 0 \right\}.$$

numerator / denominator   In the rational number $\frac{a}{b}$, a is the **numerator** and b is the **denominator.** The rational number $\frac{a}{b}$ may also be represented as a/b or as $a \div b$. The word *fraction* is derived from the Latin word *fractus* meaning "to break." The word *numerator* comes from a Latin word meaning "numberer," and *denominator* comes from a Latin word meaning "namer." Hence, the numerator tells how many equal-sized parts there are, and the denominator tells what kind of parts there are.

The set of rational numbers is a subset of the set of fractions. Table 6-1 shows several different ways in which we use rational numbers.

Figure 6-1 illustrates the use of rational numbers as part of a whole and as part of a given set. For example, in the area model in Figure 6-1(a), one part out of three congruent parts, or $\frac{1}{3}$ of the largest rectangle, is shaded. In Figure 6-1(b), two parts out of three parts, or $\frac{2}{3}$ of the unit segment, are shaded. In Figure 6-1(c), three circles out of five circles, or $\frac{3}{5}$ of the circles, are shaded.

In this chapter, we discuss addition, subtraction, multiplication, and division of rational numbers, as well as solutions of equations and inequalities involving these numbers. The use of rationals as ratios and as negative exponents is discussed in the last two sections

(a)

(b)

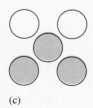
(c)

FIGURE 6-1

TABLE 6-1 Uses of Rational Numbers

Use	Example
Division problem or solution to a multiplication problem	The solution to $2x = 3$ is $\frac{3}{2}$.
Partition, or part, of a whole	Joe received one half of Mary's salary each month for alimony.
Ratio	The ratio of Republicans to Democrats in the Senate is three to five.
Probability	When you toss a fair coin, the probability of getting heads is $\frac{1}{2}$.

of this chapter. Throughout the chapter we emphasize estimation. According to the *Standards* (p. 36), *From children's earliest experiences with mathematics, estimation needs to be an ongoing part of their study of numbers, computation and measurement.*

HISTORICAL NOTE

The early Egyptian numeration system had symbols for fractions with numerators of 1. Most fractions with numerators other than 1 were expressed as a sum of different fractions with numerators of 1 $\left(\text{for example, } \frac{7}{12} = \frac{1}{3} + \frac{1}{4}\right)$.

Fractions with denominator 60 or powers of 60 were common in ancient Babylon about 2000 B.C., where 12,35 meant $12 + \frac{35}{60}$. The method was later adopted by the Greek astronomer Ptolemy (approximately A.D. 125). The same method was also used in Islamic and European countries and is presently used in the measurements of angles, where $13°19'47''$ means $13 + \frac{19}{60} + \frac{47}{60^2}$.

The modern notation for fractions with a bar between numerator and denominator is of Hindu origin. It came into general use in Europe in sixteenth-century books.

Section 6-1 Fractions and Rational Numbers

Our early exposure to fractions usually takes the form of oral descriptions rather than mathematical notations. We hear phrases such as "one half of a pizza," "one third of a cake," or "three fourths of a pie." The K-4 *Standards* (p. 58) contains the following with respect to introducing symbols for fractions: *Fraction symbols such as $\frac{1}{4}$ and $\frac{3}{2}$ should be introduced only after children have developed the concepts and oral language necessary for symbols to be meaningful and should be carefully connected to both the models and oral language.* We encounter such questions as "If two identical fruit bars are equally divided among three friends, how much does each get?" The answer is that each receives $\frac{2}{3}$ of a bar.

If we use the division representation of a rational number, then $a \div 1 = \frac{a}{1}$. Since $a \div 1 = a$, every integer a can be represented by $\frac{a}{1}$. This and the fact that not every rational number is an integer show that the set of integers is a proper subset of the set of rational numbers; that is, $I \subset Q$.

Rational numbers can be represented on a number line. Once the integers 0 and 1 are assigned to points on a line, every other rational number is assigned to a specific point. For example, to represent $\frac{3}{4}$ on the number line, we divide the segment from 0 to 1 into 4 segments of equal length. Then, starting from 0, we count 3 of these segments and stop at the mark corresponding to the right endpoint of the third segment the rational number $\frac{3}{4}$. Figure 6-2 shows the points that correspond to $\frac{3}{4}$, 1, $\frac{5}{4}$, 2, $\frac{-3}{4}$, $^-1$, $\frac{-5}{4}$, and $^-2$.

FIGURE 6-2

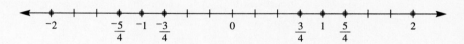

Equivalent Fractions

equivalent fractions

In the K-4 *Standards* (p. 32), we find: *If conceptual understandings are linked to procedures, children will not perceive of mathematics as an arbitrary set of rules; will not need to learn or memorize as many procedures; and will have the foundation to apply, recreate, and invent new ones when needed.* An example of this is the use of paper folding to generate **equivalent fractions**. In Figure 6-3(a), one of three congruent parts, or $\frac{1}{3}$ is shaded. In Figure 6-3(b), each of the thirds has been folded in half so that now we have six sections and two of six congruent parts, or $\frac{2}{6}$, are shaded. Thus both $\frac{1}{3}$ and $\frac{2}{6}$ represent exactly the same shaded portion. Although the symbols $\frac{1}{3}$ and $\frac{2}{6}$ do not look alike, they represent the same rational number. Strictly speaking, $\frac{1}{3}$ and $\frac{2}{6}$ are equivalent fractions. However, because they represent equal amounts, we write $\frac{1}{3} = \frac{2}{6}$, and say that $\frac{1}{3}$ equals $\frac{2}{6}$.

FIGURE 6-3

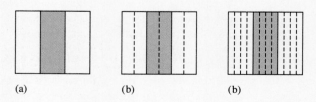

(a) (b) (b)

Figure 6-3(c) shows the rectangle with each of the original thirds folded into 4 equal parts with 4 parts of the 12 now shaded. Thus $\frac{1}{3}$ is equal to $\frac{4}{12}$ because the same portion

SECTION 6-1 FRACTIONS AND RATIONAL NUMBERS

of the model is shaded. Similarly, we could illustrate that $\frac{1}{3}, \frac{2}{6}, \frac{3}{9}, \frac{4}{12}, \frac{5}{15}, \ldots$ are equal. In other words, there are infinitely many ways of naming the rational number $\frac{1}{3}$. Similarly, there are infinitely many ways of naming any rational number.

This process of generating fractions equal to $\frac{1}{3}$ can be thought of as follows: If each of 3 equal-sized parts of a whole are halved, there must be twice as many of the smaller pieces. Hence, $\frac{1}{3} = \frac{2}{6}$. Similarly, $\frac{1}{3} = \frac{4}{12}$ because if each of 3 equal-sized parts of a whole are divided into 4 equal-sized parts, then there must be 4 times as many of the smaller pieces. In general, we have the following property for generating equivalent fractions, called the *Fundamental Law of Fractions*.

> **PROPERTY**
>
> **Fundamental Law of Fractions:** For any fraction $\frac{a}{b}$ and any number $c \neq 0$, $\frac{a}{b} = \frac{ac}{bc}$.

The Fundamental Law of Fractions may be stated as follows: *The value of a fraction does not change if its numerator and denominator are multiplied by the same nonzero number*. However, the Fundamental Law of Fractions does not imply that adding the same nonzero number to the numerator and the denominator results in an equivalent fraction. For example, $\frac{1}{2} \neq \frac{1+2}{2+2} = \frac{3}{4}$.

From the Fundamental Law of Fractions, $\frac{7}{-15} = \frac{-7}{15}$ because $\frac{7}{-15} = \frac{7 \cdot (-1)}{-15 \cdot (-1)} = \frac{-7}{15}$. Similarly, $\frac{a}{-b} = \frac{-a}{b}$. The form $\frac{-a}{b}$ is usually preferred. Equivalent fractions are illustrated on the student page from *Addison-Wesley Mathematics,* Grade 5, 1991, that appears on the following page.

EXAMPLE 6-1   Find a value for x so that $\frac{12}{42} = \frac{x}{210}$.

Solution   By the Fundamental Law of Fractions, $\frac{12}{42} = \frac{12 \cdot 5}{42 \cdot 5} = \frac{60}{210}$. Hence, $\frac{x}{210} = \frac{60}{210}$, and $x = 60$. ∎

In higher mathematics, rational numbers are often introduced as a collection of disjoint sets called equivalence classes. For example, the fractions in the set

$$\left\{ \ldots, \frac{-3}{-9}, \frac{-2}{-6}, \frac{-1}{-3}, \frac{1}{3}, \frac{2}{6}, \frac{3}{9}, \ldots \right\}$$

equivalent fractions   are **equivalent fractions,** and the set is an *equivalence class of fractions*. This particular class is typically represented by $\frac{1}{3}$.

Finding Equivalent Fractions

LEARN ABOUT IT

EXPLORE Solve to Understand
Keith bought paper for making origami figures. He bought 2 packages of orange paper and 3 packages of yellow paper. What fraction of the papers were orange?

TALK ABOUT IT

1. How many packages of each color did Keith buy?
2. How many papers are in each package?
3. How many papers did Keith buy altogether?

2 of the 5 packages were orange. $\frac{2}{5}$ of the packages were orange.
6 of the 15 papers were orange. $\frac{6}{15}$ of the papers were orange.

You can find a fraction that is equivalent to another fraction by multiplying or dividing the numerator and denominator by the same nonzero number.

 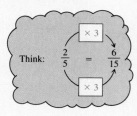 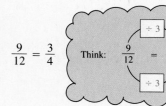

Simplifying Fractions

simplifying fractions The Fundamental Law of Fractions justifies a process called **simplifying fractions.** Consider the following:

$$\frac{60}{210} = \frac{6 \cdot 10}{21 \cdot 10} = \frac{6}{21}.$$

Also,

$$\frac{6}{21} = \frac{2 \cdot 3}{7 \cdot 3} = \frac{2}{7}.$$

We can simplify $\frac{60}{210}$ because the numerator and denominator have a common factor of 10. Also, we can simplify $\frac{6}{21}$ because 6 and 21 have a common factor of 3. However, we cannot simplify $\frac{2}{7}$ because 2 and 7 have no common factors other than 1. The fraction *simplest form* $\frac{2}{7}$ is called the **simplest form** of $\frac{60}{210}$.

Finding the simplest form of $\frac{60}{210}$ can be achieved with fewer steps by writing

$$\frac{60}{210} = \frac{2 \cdot 30}{7 \cdot 30} = \frac{2}{7}.$$

The number 30 is the GCD of 60 and 210. Simplifying $\frac{60}{210}$ amounts to dividing 60 and 210 by their greatest common divisor, 30. To write a fraction $\frac{a}{b}$ in simplest form, we divide both a and b by the GCD(a, b).

DEFINITION OF SIMPLEST FORM

A fraction $\frac{a}{b}$ is in simplest form if a and b have no common factor greater than 1, that is, if a and b are relatively prime.

We can use scientific/fraction calculators to simplify fractions. For example, to simplify $\frac{6}{12}$, we enter $\boxed{6}\boxed{/}\boxed{1}\boxed{2}$ and press $\boxed{\text{SIMP}}\boxed{=}$, and 3/6 appears on the screen. At this point, an indicator tells us that this is not in simplest form, and so we press $\boxed{\text{SIMP}}\boxed{=}$ again to obtain 1/2. At any time, we can view the factor that was removed by pressing $\boxed{x \supset y}$ key.

E X A M P L E   6 - 2   Write each of the following in simplest form:

(a) $\frac{45}{60}$ (b) $\frac{35}{17}$

Solution   (a) GCD(45, 60) = 15, so $\frac{45}{60} = \frac{3 \cdot 15}{4 \cdot 15} = \frac{3}{4}$.

(b) GCD(35, 17) = 1, so $\frac{35}{17}$ is in simplest form. ∎

Another method for writing the fraction in simplest form is to find a factorization of the numerator and denominator and then divide both numerator and denominator by all the common factors. For example,

$$\frac{a^2b}{ab^2} = \frac{a \cdot (ab)}{b \cdot (ab)} = \frac{a}{b}.$$

EXAMPLE 6-3 Write each of the following in simplest form:

(a) $\frac{28ab^2}{42a^2b^2}$ (b) $\frac{(a+b)^2}{3a+3b}$ (c) $\frac{x^2+x}{x+1}$ (d) $\frac{3+x^2}{3x^2}$ (e) $\frac{3+3x^2}{3x^2}$

Solution (a) $\frac{28ab^2}{42a^2b^2} = \frac{2(14ab^2)}{3a(14ab^2)} = \frac{2}{3a}$

(b) $\frac{(a+b)^2}{3a+3b} = \frac{(a+b) \cdot (a+b)}{3(a+b)} = \frac{a+b}{3}$

(c) $\frac{x^2+x}{x+1} = \frac{x(x+1)}{x+1} = \frac{x(x+1)}{1(x+1)} = \frac{x}{1}$

(d) $\frac{3+x^2}{3x^2}$ cannot be further reduced because $3 + x^2$ and $3x^2$ have no factors in common except 1.

(e) $\frac{3+3x^2}{3x^2} = \frac{3 \cdot (1+x^2)}{3 \cdot x^2} = \frac{1+x^2}{x^2}$ ∎

Equality of Fractions

We can use several methods to show that two fractions such as $\frac{12}{42}$ and $\frac{10}{35}$ are equal.

1. Reduce both fractions to the same simplest form.

$$\frac{12}{42} = \frac{2^2 \cdot 3}{2 \cdot 3 \cdot 7} = \frac{2}{7} \text{ and } \frac{10}{35} = \frac{5 \cdot 2}{5 \cdot 7} = \frac{2}{7}$$

Thus,

$$\frac{12}{42} = \frac{10}{35}.$$

2. Rewrite both fractions with the same least common denominator. Since LCM(42, 35) = 210, then

$$\frac{12}{42} = \frac{60}{210} \text{ and } \frac{10}{35} = \frac{60}{210}.$$

Thus,

$$\frac{12}{42} = \frac{10}{35}.$$

3. Rewrite both fractions with a common denominator (not necessarily the least). A common multiple of 42 and 35 may be found by finding the product $42 \cdot 35$ or 1470. Now,

$$\frac{12}{42} = \frac{420}{1470} \text{ and } \frac{10}{35} = \frac{420}{1470}.$$

Hence,
$$\frac{12}{42} = \frac{10}{35}.$$

The third method suggests a general algorithm for determining if two fractions $\frac{a}{b}$ and $\frac{c}{d}$ are equal. Rewrite both fractions with common denominator bd. That is,
$$\frac{a}{b} = \frac{ad}{bd} \text{ and } \frac{c}{d} = \frac{bc}{bd}.$$

Because the denominators are the same, $\frac{ad}{bd} = \frac{bc}{bd}$ if, and only if, $ad = bc$. For example, $\frac{24}{36} = \frac{6}{9}$ because $24 \cdot 9 = 216 = 36 \cdot 6$. In general, the following property results.

PROPERTY

Two fractions $\frac{a}{b}$ and $\frac{c}{d}$ are **equal** if, and only if, $ad = bc$.

Using a calculator, we may determine if two fractions are equal by using the property that $\frac{a}{b} = \frac{c}{d}$ if, and only if, $ad = bc$. We see that $\frac{2}{4} = \frac{1098}{2196}$, since both $\boxed{2}\boxed{\times}\boxed{2}\boxed{1}\boxed{9}\boxed{6}\boxed{=}$ and $\boxed{4}\boxed{\times}\boxed{1}\boxed{0}\boxed{9}\boxed{8}\boxed{=}$ yield a display of 4392.

PROBLEM SET 6-1

1. ▶Write a sentence illustrating the use of $\frac{7}{8}$ in each of the following ways:
   (a) As a division problem
   (b) As part of a whole
   (c) As a ratio

2. For each of the following, write a fraction to represent the shaded portion:

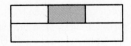
(a)

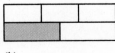
(b)

(c)

(d)

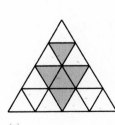
(e)

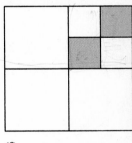
(f)

3. For each of the following four squares, write a fraction to represent the shaded portion. What property of fractions does the diagram illustrate?

(a)

(b)

(c)

(d)

4. Complete each figure so that it shows $\frac{3}{5}$.

(a)   (b)

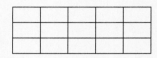
(c)

(d)

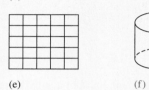
(e)   (f)

5. Refer to the figure below and represent each of the following as a fraction:
   (a) The dots inside the circle as a part of all the dots
   (b) The dots inside the rectangle as a part of all the dots
   (c) The dots in the intersection of the rectangle and the circle as a part of all the dots
   (d) The dots outside the circle but inside the rectangle as a part of all the dots

6. For each of the following, write three fractions equal to the given fraction:
   (a) $\frac{2}{9}$ (b) $\frac{-2}{5}$ (c) $\frac{0}{3}$ (d) $\frac{a}{2}$

7. Find the simplest form for each of the following fractions:
   (a) $\frac{156}{93}$ (b) $\frac{27}{45}$ (c) $\frac{-65}{91}$
   (d) $\frac{0}{68}$ (e) $\frac{84^2}{91^2}$ (f) $\frac{662}{703}$

8. ▶Mr. Gonzales and Ms. Price gave the same test to their fifth-grade classes. In Mr. Gonzales's class, 20 out of 25 students passed the test, and in Ms. Price's class, 24 out of 30 students passed the test. One of Ms. Price's students heard about the results of the tests and claimed that the classes did equally well. Is the student right? Explain.

9. Choose the expression in parentheses that equals or best describes the given fraction.
   (a) $\frac{0}{0}$ (1, undefined, 0)   (b) $\frac{5}{0}$ (undefined, 5, 0)
   (c) $\frac{0}{5}$ (undefined, 5, 0)
   (d) $\frac{2+a}{a}$ (2, 3, cannot be simplified)
   (e) $\frac{15+x}{3x}$ $\left(\frac{5+x}{x}, 5, \text{cannot be simplified}\right)$
   (f) $\frac{2^6 + 2^5}{2^4 + 2^7}$ $\left(1, \frac{2}{3}, \text{cannot be simplified}\right)$
   (g) $\frac{2^{100} + 2^{98}}{2^{100} - 2^{98}}$ $\left(2^{196}, \frac{5}{3}, \text{too large to simplify}\right)$

10. Find the simplest form for each of the following fractions:
    (a) $\frac{x}{x}$ (b) $\frac{14x^2y}{63xy^2}$ (c) $\frac{a^2 + ab}{a + b}$
    (d) $\frac{a^3 + 1}{a^3 b}$ (e) $\frac{a}{3a + ab}$ (f) $\frac{a}{3a + b}$

11. Determine if the following pairs are equal by writing each in simplest form:
    (a) $\frac{3}{8}$ and $\frac{375}{1000}$ (b) $\frac{18}{54}$ and $\frac{23}{69}$
    (c) $\frac{6}{10}$ and $\frac{600}{1000}$ (d) $\frac{17}{27}$ and $\frac{25}{45}$

12. Determine if the following pairs are equal by changing both to the same denominator:
    (a) $\frac{10}{16}$ and $\frac{12}{18}$ (b) $\frac{3}{12}$ and $\frac{41}{154}$
    (c) $\frac{3}{-12}$ and $\frac{-36}{144}$ (d) $\frac{-21}{86}$ and $\frac{-51}{215}$

13. A board is needed that is exactly $\frac{11}{32}$ in. wide to fill a hole. Can a board that is $\frac{3}{8}$ in. be shaved down to fit the hole? If so, how much must be shaved from the board?

14. Draw an area model to show that $\frac{3}{4} = \frac{6}{8}$.

15. If a fraction is equivalent to $\frac{3}{4}$ and the sum of the numerator and denominator is 84, what is the fraction?

16. Two parking meters are next to each other with the times left as shown. Which meter has more time left on it? How much more time is on it?

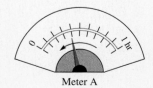
Meter A

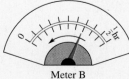
Meter B

17. Mr. Gomez filled his car's 16-gal gas tank. He took a short trip and used 6 gal of gas. Draw an arrow to show what his gas gauge looks like after the trip.

18. Read each measurement as shown on the ruler below.
    (a) A   (b) B   (c) C   (d) D

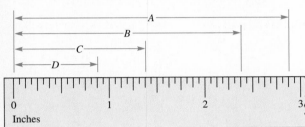

19. ▶When you multiply two positive fractions less than 1, how does the answer compare to the size of the fractions? Why?

20. ▶Is there a whole number that does not have a reciprocal? Explain. (For example, 2 and $\frac{1}{2}$ are reciprocals.)

21. ▶Should fractions always be reduced to their simplest form? Why or why not?

22. The numerator and denominator of a fraction are two-digit numbers, and the fraction is equivalent to $\frac{4}{7}$. The denominator is formed by reversing the order of the digits in the numerator (for example, $\frac{36}{63}$). Find all other possible solutions.

23. Solve for x in each of the following:
    (a) $\frac{2}{3} = \frac{x}{16}$ (b) $\frac{3}{4} = \frac{-27}{x}$ (c) $\frac{3}{x} = \frac{3x}{x^2}$

24. ▶(a) If $\frac{a}{c} = \frac{b}{c}$, what must be true?
    ▶(b) If $\frac{a}{b} = \frac{a}{c}$, what must be true?

25. Let W be the set of whole numbers, I be the set of integers, and Q be the set of rational numbers. Classify each of the following as true or false:
    (a) $W \subseteq Q$
    (b) $(I \cup W) \subset Q$
    (c) If Q is the universal set, $\overline{I} = W$.
    (d) $Q \cap I = W$
    (e) $Q \cap W = W$

26. 🖩 Use a calculator to check whether each of the following pairs of fractions are equal:
    (a) $\frac{24}{31}$ and $\frac{23}{30}$ (b) $\frac{86}{75}$ and $\frac{85}{74}$
    (c) $\frac{1513}{1691}$ and $\frac{1581}{1767}$

• COMPUTER CORNER

The following computer program will reduce fractions to their lowest terms.

```
10 PRINT "TYPE IN THE NUMERATOR AND THE DENOMINATOR"
20 PRINT "OF THE FRACTION SEPARATED BY A COMMA."
30 INPUT N,D
40 T = ABS (N) :B = ABS (D)
50 X = INT (T/B) :R = T - X*B
60 IF R = 0 THEN GOTO 80
70 T = B:B=R: GOTO 50
80 PRINT "THE SIMPLEST FORM IS " ;N/B; "/" ;D/B
90 END
```

Run the program for the following fractions:
(a) $\frac{124}{130}$ (b) $\frac{2002}{7007}$

Section 6-2 Addition and Subtraction of Rational Numbers

Addition of Rational Numbers with Like Denominators

Suppose that a pizza is divided into five parts of equal size. If one person ate one piece of the pizza and another person ate two pieces of the pizza, then they ate $\frac{1}{5}$ and $\frac{2}{5}$ of the pizza, respectively, as shown in the area model in Figure 6-4.

FIGURE 6-4

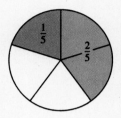

We see from the figure that $\frac{3}{5}$ of the pizza was eaten. That is,

$$\frac{1}{5} + \frac{2}{5} = \frac{1+2}{5} = \frac{3}{5}.$$

In general, we have the following definition for addition of rational numbers with like denominators.

DEFINITION OF RATIONAL NUMBER ADDITION

If $\frac{a}{b}$ and $\frac{c}{b}$ are rational numbers, then $\frac{a}{b} + \frac{c}{b} = \frac{a+c}{b}$.

Number-Line Model

The sum of two rational numbers can also be found using a number line. For example, to compute $\frac{1}{5} + \frac{3}{5}$, we use a number line with one unit divided into fifths, as shown in Figure 6-5.

FIGURE 6-5

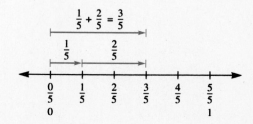

Addition of Rational Numbers with Unlike Denominators

To determine how to add fractions with unequal denominators, we use the *strategy of changing the problem into an equivalent problem* that we already know how to do. We know how to add fractions with the same denominators, so to add $\frac{2}{3} + \frac{1}{4}$, we rewrite $\frac{2}{3}$ and $\frac{1}{4}$ with the same denominators and add. The area model in Figure 6-6 demonstrates this.

FIGURE 6-6

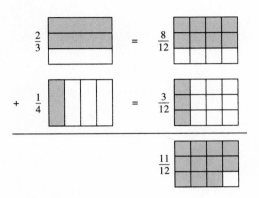

Therefore, $\frac{2}{3} + \frac{1}{4} = \frac{8}{12} + \frac{3}{12} = \frac{11}{12}$. The least common denominator of 3 and 4 is the LCM (3, 4). We add fractions with unlike denominators by finding the least common denominator. Then we rewrite each of the fractions using this denominator and add, for example,

$$\frac{2}{15} + \frac{5}{6} = \frac{4}{30} + \frac{25}{30} = \frac{29}{30}.$$

Another technique for adding fractions with unlike denominators is to write each of them over a common denominator (not necessarily the least). For example, to add $\frac{3}{8} + \frac{1}{6}$, a common denominator is $8 \cdot 6$. Hence,

$$\frac{3}{8} = \frac{3 \cdot 6}{8 \cdot 6} \quad \text{and} \quad \frac{1}{6} = \frac{8 \cdot 1}{8 \cdot 6}.$$

Thus,

$$\frac{3}{8} + \frac{1}{6} = \frac{3 \cdot 6}{8 \cdot 6} + \frac{8 \cdot 1}{8 \cdot 6} = \frac{3 \cdot 6 + 8 \cdot 1}{8 \cdot 6} = \frac{18 + 8}{48} = \frac{26}{48}, \text{ or } \frac{13}{24}.$$

In general, given two rational numbers $\frac{a}{b}$ and $\frac{c}{d}$, we may add the fractions as follows:

$$\frac{a}{b} + \frac{c}{d} = \frac{a \cdot d}{b \cdot d} + \frac{b \cdot c}{b \cdot d} = \frac{ad + bc}{bd}.$$

We summarize the result in the following.

PROPERTY

If $\frac{a}{b}$ and $\frac{c}{d}$ are any two rational numbers, then $\frac{a}{b} + \frac{c}{d} = \frac{ad + bc}{bd}$.

EXAMPLE 6-4 Find each of the following sums:

(a) $\frac{2}{15} + \frac{4}{21}$ (b) $\frac{2}{-3} + \frac{1}{5}$ (c) $\left(\frac{3}{4} + \frac{1}{5}\right) + \frac{1}{6}$ (d) $\frac{3}{x} + \frac{4}{y}$

Solution (a) $\frac{2}{15} + \frac{4}{21} = \frac{2 \cdot 7}{15 \cdot 7} + \frac{4 \cdot 5}{21 \cdot 5} = \frac{14}{105} + \frac{20}{105} = \frac{34}{105}$

(b) $\frac{2}{-3} + \frac{1}{5} = \frac{(2)(5) + (-3)(1)}{(-3)(5)} = \frac{10 + {-3}}{-15} = \frac{7}{-15}$

(c) $\frac{3}{4} + \frac{1}{5} = \frac{3 \cdot 5 + 4 \cdot 1}{4 \cdot 5} = \frac{19}{20}$. Hence, $\left(\frac{3}{4} + \frac{1}{5}\right) + \frac{1}{6} = \frac{19}{20} + \frac{1}{6} = \frac{19 \cdot 6 + 20 \cdot 1}{20 \cdot 6} = \frac{134}{120}$ or $\frac{67}{60}$.

(d) $\frac{3}{x} + \frac{4}{y} = \frac{3y}{xy} + \frac{4x}{xy} = \frac{3y + 4x}{xy}$ ∎

Mixed Numbers

mixed numbers In everyday life, we often use **mixed numbers,** that is, numbers that are made up of an integer and a fractional part of an integer. For example, Figure 6-7 shows that the nail is $2\frac{3}{4}$ in. long. The mixed number $2\frac{3}{4}$ means $2 + \frac{3}{4}$. It is sometimes inferred that $2\frac{3}{4}$ means 2 times $\frac{3}{4}$ since xy means $x \cdot y$, but this is not correct. Also, the number $-4\frac{3}{4}$ means $-(4 + \frac{3}{4})$, not $-4 + \frac{3}{4}$.

FIGURE 6-7

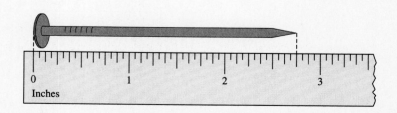

SECTION 6-2 ADDITION AND SUBTRACTION OF RATIONAL NUMBERS

A mixed number is a rational number, and therefore it can always be written in the form $\frac{a}{b}$. For example,

$$2\frac{3}{4} = 2 + \frac{3}{4} = \frac{2}{1} + \frac{3}{4} = \frac{2 \cdot 4 + 1 \cdot 3}{1 \cdot 4} = \frac{8 + 3}{4} = \frac{11}{4}.$$

proper fraction A fraction $\frac{a}{b}$, where $0 \leq |a| < |b|$, is called a **proper fraction.** For example, $\frac{4}{7}$ is a proper fraction, but $\frac{7}{4}, \frac{4}{4}$ and $\frac{-9}{7}$ are not proper fractions.

EXAMPLE 6-5 Change each of the following mixed numbers to the form $\frac{a}{b}$, where a and b are integers:

(a) $4\frac{1}{3}$ (b) $-3\frac{2}{5}$

Solution (a) $4\frac{1}{3} = 4 + \frac{1}{3} = \frac{4}{1} + \frac{1}{3} = \frac{4 \cdot 3 + 1 \cdot 1}{1 \cdot 3} = \frac{12 + 1}{3} = \frac{13}{3}$

(b) $-3\frac{2}{5} = -\left(3 + \frac{2}{5}\right) = -\left(\frac{3}{1} + \frac{2}{5}\right) = -\left(\frac{3 \cdot 5 + 1 \cdot 2}{1 \cdot 5}\right) = \frac{-17}{5}$

EXAMPLE 6-6 Change $\frac{29}{5}$ to a mixed number.

Solution $\frac{29}{5} = \frac{5 \cdot 5 + 4}{5} = \frac{5 \cdot 5}{5} + \frac{4}{5} = 5 + \frac{4}{5} = 5\frac{4}{5}$

REMARK In elementary schools, problems like Example 6-6 are usually solved using division.

$$\begin{array}{r} 5 \\ 5\overline{)29} \\ \underline{25} \\ 4 \end{array}$$

Hence, $\frac{29}{5} = 5 + \frac{4}{5} = 5\frac{4}{5}.$

Scientific/fraction calculators can be used to change improper fractions to mixed numbers. For example, if we enter $\boxed{2}\,\boxed{9}\,\boxed{/}\,\boxed{5}$ and press $\boxed{Ab/c}$, then 5⌴4/5 appears, which means $5\frac{4}{5}.$

EXAMPLE 6-7 Find $2\frac{4}{5} + 3\frac{5}{6}$.

Solution The problem is solved in two ways, for comparison.

Add the fractional parts and the integers of the mixed numbers separately.

$$2\frac{4}{5} = 2\frac{24}{30}$$
$$+3\frac{5}{6} = +3\frac{25}{30}$$
$$\overline{\phantom{+3\frac{5}{6}}} \quad \overline{5\frac{49}{30}}$$

But,

$$\frac{49}{30} = 1\frac{19}{30}$$

so

$$5\frac{49}{30} = 5 + \frac{49}{30} = 5 + 1\frac{19}{30} = 6\frac{19}{30}.$$

Change each mixed number into a rational number in the form $\frac{a}{b}$ and then add.

$$2\frac{4}{5} + 3\frac{5}{6} = \frac{14}{5} + \frac{23}{6}$$
$$= \frac{14 \cdot 6 + 5 \cdot 23}{5 \cdot 6}$$
$$= \frac{84 + 115}{30}$$
$$= \frac{199}{30}$$
$$= 6\frac{19}{30}$$

■

We can use scientific/fraction calculators to add mixed numbers. For example, to add $2\frac{4}{5} + 3\frac{5}{6}$, we enter $\boxed{2}$ $\boxed{\text{unit}}$ $\boxed{4}$ $\boxed{/}$ $\boxed{5}$ $\boxed{+}$ $\boxed{3}$ $\boxed{\text{unit}}$ $\boxed{5}$ $\boxed{/}$ $\boxed{6}$ $\boxed{=}$, and the display reads 5⊔49/30. We then press $\boxed{Ab/c}$ to obtain 6⊔19/30, which means $6\frac{19}{30}$.

Properties of Addition for Rational Numbers

Rational numbers have the following properties for addition: closure property, commutative property, associative property, additive identity property, and additive inverse property. To emphasize the additive inverse property of rational numbers, we state it explicitly.

PROPERTY

Additive Inverse Property of Rational Numbers: For any rational number $\frac{a}{b}$, there exists a unique rational number $-\frac{a}{b}$, called the additive inverse of $\frac{a}{b}$, such that $\frac{a}{b} + \left(-\frac{a}{b}\right) = 0 = \left(-\frac{a}{b}\right) + \frac{a}{b}$.

Another form of $-\frac{a}{b}$ can be found by considering the sum $\frac{a}{b} + \frac{-a}{b}$. Because

$$\frac{a}{b} + \frac{-a}{b} = \frac{a + -a}{b} = \frac{0}{b} = 0,$$

it follows that $-\frac{a}{b}$ and $\frac{-a}{b}$ are both additive inverses of $\frac{a}{b}$, so $-\frac{a}{b} = \frac{-a}{b}$.

EXAMPLE 6-8 Find the additive inverses for each of the following:

(a) $\frac{3}{5}$ (b) $\frac{-5}{11}$ (c) $4\frac{1}{2}$

Solution (a) $\frac{-3}{5}$ or $-\frac{3}{5}$ (b) $-\left(\frac{-5}{11}\right)$ or $\frac{5}{11}$ (c) $-4\frac{1}{2}$ ∎

Properties of the additive inverse for rational numbers are analogous to those of the additive inverse for integers, as shown in Table 6-2. As with the set of integers, the set of rational numbers also has the addition property of equality.

TABLE 6-2

Integers	Rational Numbers
1. $^-(^-a) = a$	1. $-\left(-\frac{a}{b}\right) = \frac{a}{b}$
2. $^-(a + b) = ^-a + ^-b$	2. $-\left(\frac{a}{b} + \frac{c}{d}\right) = \frac{-a}{b} + \frac{-c}{d}$

PROPERTY

Addition Property of Equality: If $\frac{a}{b}$ and $\frac{c}{d}$ are any rational numbers such that $\frac{a}{b} = \frac{c}{d}$, and if $\frac{e}{f}$ is any rational number, then $\frac{a}{b} + \frac{e}{f} = \frac{c}{d} + \frac{e}{f}$.

Subtraction of Rational Numbers

In elementary school, subtraction of rational numbers is usually introduced by using a take-away model. If we have $\frac{6}{7}$ of a pizza, and $\frac{2}{7}$ of the original pizza is taken away, $\frac{4}{7}$ of the pizza remains: that is, $\frac{6}{7} - \frac{2}{7} = \frac{6-2}{7} = \frac{4}{7}$. In general, subtraction of rational numbers with like denominators is determined as follows:

$$\frac{a}{b} - \frac{c}{b} = \frac{a-c}{b}.$$

In the following student page from *Heath Mathematics Connections,* Grade 5, 1992, we see that fraction bars provide another model for adding and subtracting fractions.

Fractions with Like Denominators

You add and subtract fractions with the same denominator, such as $\frac{3}{5}$ and $\frac{1}{5}$, just as you would numbers whose units of measure are the same.

3 meters + **1** meter = **4** meters

$3 + $1 = $4

$\frac{3}{5} + \frac{1}{5} = \frac{4}{5}$

▶ Add: $\frac{7}{8} + \frac{3}{8}$

- Check that the denominators are the same.

  $\frac{7}{8} + \frac{3}{8}$

- Add the numerators.

  $\frac{7}{8} + \frac{3}{8} = \frac{10}{8}$

- Simplify if you can.

  $\frac{10}{8} = 1\frac{2}{8}$ or $1\frac{1}{4}$

▶ Subtract: $\frac{7}{8} - \frac{3}{8}$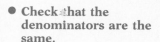

- Check that the denominators are the same.

  $\frac{7}{8} - \frac{3}{8}$

- Subtract the numerators.

  $\frac{7}{8} - \frac{3}{8} = \frac{4}{8}$

- Simplify if you can.

  $\frac{4}{8} = \frac{1}{2}$

The addition properties for whole numbers also apply to fractions.

Zero Property	$0 + \frac{3}{5} = \frac{3}{5}$	The sum of zero and one other addend is the other addend.
Commutative Property	$\frac{1}{6} + \frac{4}{6} = \frac{4}{6} + \frac{1}{6}$	The sum stays the same when the order of addends is changed.
Associative Property	$(\frac{3}{8} + \frac{4}{8}) + \frac{4}{8} = \frac{3}{8} + (\frac{4}{8} + \frac{4}{8})$	The sum stays the same when the grouping of addends is changed.

Subtraction of rational numbers, like subtraction of integers, can be defined in terms of addition as follows.

DEFINITION OF RATIONAL NUMBER SUBTRACTION

If $\frac{a}{b}$ and $\frac{c}{d}$ are any rational numbers, then $\frac{a}{b} - \frac{c}{d} = \frac{e}{f}$ if and only if $\frac{a}{b} = \frac{c}{d} + \frac{e}{f}$.

As with integers, we can see that subtraction of rational numbers can be performed by adding the additive inverses. The following theorem states this.

THEOREM 6-1

If $\frac{a}{b}$ and $\frac{c}{d}$ are any rational numbers, then $\frac{a}{b} - \frac{c}{d} = \frac{a}{b} + \frac{^-c}{d}$.

Now, using the definition of addition of rational numbers, we obtain the following:

$$\frac{a}{b} - \frac{c}{d} = \frac{a}{b} + \frac{^-c}{d}$$
$$= \frac{ad + b(^-c)}{bd}$$
$$= \frac{ad - bc}{bd}$$

We summarize this result in the following theorem.

THEOREM 6-2

If $\frac{a}{b}$ and $\frac{c}{d}$ are any rational numbers, then $\frac{a}{b} - \frac{c}{d} = \frac{ad - bc}{bd}$.

EXAMPLE 6-9 Find each difference.

(a) $\frac{5}{8} - \frac{1}{4}$ (b) $5\frac{1}{3} - 2\frac{3}{4}$

Solution (a) One approach is to find the least common denominator for the fractions. Because LCM(8,4) = 8, we have

$$\frac{5}{8} - \frac{1}{4} = \frac{5}{8} - \frac{2}{8} = \frac{3}{8}.$$

An alternate approach is given below.

$$\frac{5}{8} - \frac{1}{4} = \frac{5 \cdot 4 - 8 \cdot 1}{8 \cdot 4} = \frac{5 \cdot 4 - 8 \cdot 1}{32} = \frac{12}{32} \text{ or } \frac{3}{8}$$

(b) Two methods of solution are given.

$$5\frac{1}{3} = 5\frac{4}{12} = 4 + 1\frac{4}{12} = 4\frac{16}{12}$$
$$-2\frac{3}{4} = -2\frac{9}{12} = -2\frac{9}{12} = -2\frac{9}{12}$$
$$2\frac{7}{12}$$

$$5\frac{1}{3} - 2\frac{3}{4} = \frac{16}{3} - \frac{11}{4}$$
$$= \frac{16 \cdot 4 - 3 \cdot 11}{3 \cdot 4}$$
$$= \frac{64 - 33}{12}$$
$$= \frac{31}{12} \text{ or } 2\frac{7}{12} \quad \blacksquare$$

Estimation with Rational Numbers

In the 5–8 *Standards* (p. 97), we find: *Estimation is a powerful idea to be used both in solving problems and in checking the reasonableness of a result.* Consider, for example, a student who added $\frac{3}{4}$ and $\frac{1}{2}$ and obtained $\frac{4}{6}$ (most likely, the student confused the procedure with the procedure for multiplying fractions and added the numerators and then the denominators). An estimation of $\frac{3}{4} + \frac{1}{2}$ as a number greater than $\frac{1}{2} + \frac{1}{2}$ shows that the answer should be greater than 1 and that $\frac{4}{6}$ is unreasonable. Sometimes it is desirable to round fractions to a convenient fraction, such as $\frac{1}{2}, \frac{1}{3}, \frac{1}{4}, \frac{1}{5}, \frac{2}{3}, \frac{3}{4},$ or 1. For example, teachers should encourage an activity such as the one in Figure 6-8.

FIGURE 6-8

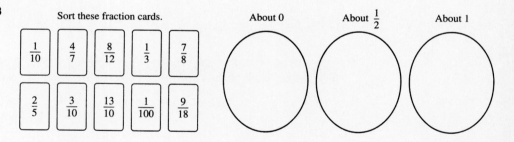

If a student had 59 correct answers out of 80 questions, the student answered $\frac{59}{80}$ of the questions correctly, which is approximately $\frac{60}{80}$, or $\frac{3}{4}$. Intuitively, $\frac{60}{80}$ is greater than $\frac{59}{80}$. On a number line, the greater fraction is to the right of the lesser. (The concepts of of greater than and less than are dealt with in more detail in Section 6-4.) The estimate $\frac{3}{4}$ for $\frac{59}{80}$ is a high estimate. In a similar way, we can estimate $\frac{31}{90}$ by $\frac{30}{90}$, or $\frac{1}{3}$. In this case, the actual answer is greater than the estimate of $\frac{1}{3}$. As the K–4 *Standards* (p. 36) point out, estimations *enhance the abilities of children to deal with everyday quantitative situations.*

EXAMPLE 6-10 A sixth-grade class is collecting cans to take to the recycling center. Becky's group brought the following amounts (in pounds). About how many pounds does her group have all together?

$$1\frac{1}{8}, \ 3\frac{4}{10}, \ 5\frac{7}{8}, \ \frac{6}{10}$$

Solution We can estimate the amount by using front-end estimation and then adjusting by using $0, \frac{1}{2},$ and 1 as reference points. The front-end estimate is $(1 + 3 + 5)$, or 9. The adjustment is $(0 + \frac{1}{2} + 1 + \frac{1}{2})$, or 2. An adjusted estimate would be 11 pounds. ■

EXAMPLE 6-11 Estimate each of the following:

(a) $\frac{27}{13} + \frac{10}{9}$ (b) $3\frac{9}{10} + 2\frac{7}{8} + \frac{11}{12}$

Solution (a) Because $\frac{27}{13}$ is more than 2 and $\frac{10}{9}$ is more than 1, an estimate is more than 3.

(b) We first add the front-end parts to obtain $3 + 2$, or 5. Because each of the fractions, $\frac{9}{10}, \frac{7}{8},$ and $\frac{11}{12}$, is close to but less than 1, their sum is less than 3. The approximate answer is a number close to but less than 8. ■

PROBLEM SET 6-2

1. Use a number line to find $\frac{1}{5} + \frac{2}{3}$.

2. In each case, perform the computation by using the least common denominator.
   (a) $\frac{3}{16} + \frac{7}{-8}$ (b) $\frac{4}{12} - \frac{2}{3}$
   (c) $\frac{5}{6} + \frac{-4}{9} + \frac{2}{3}$ (d) $\frac{2}{21} - \frac{3}{14}$

3. Use the definition of addition of rational numbers to find each of the following:
   (a) $\frac{6}{5} + \frac{-11}{4}$ (b) $\frac{4}{5} + \frac{6}{7}$
   (c) $\frac{-7}{8} + \frac{2}{5}$ (d) $\frac{5}{x} + \frac{-3}{y}$

4. Add the following rational numbers. Write your answers in simplest form.
   (a) $\frac{-3}{2x} + \frac{3}{2y} + \frac{-1}{4xy}$ (b) $\frac{-3}{2x^2y} + \frac{5}{6xy^2} + \frac{7}{x^2}$

5. Change each of the following fractions to mixed numbers:
   (a) $\frac{56}{3}$ (b) $\frac{14}{5}$
   (c) $-\frac{293}{100}$ (d) $-\frac{47}{8}$

6. Change each of the following mixed numbers to fractions in the form $\frac{a}{b}$, where a and b are integers:
   (a) $6\frac{3}{4}$ (b) $7\frac{1}{2}$
   (c) $-3\frac{5}{8}$ (d) $-4\frac{2}{3}$

7. Compute the following:
   (a) $\frac{5}{6} + 2\frac{1}{8}$ (b) $-4\frac{1}{2} - 3\frac{1}{6}$
   (c) $\frac{5}{2^4 \cdot 3^2} - \frac{1}{2^3 \cdot 3^4}$ (d) $11 - \left(\frac{3}{5} + \frac{-4}{45}\right)$

8. Place the numbers 2, 5, 6, and 8 in the boxes below to make the equation true.

   $$\frac{\square}{\square} + \frac{\square}{\square} = \frac{23}{24}$$

9. ▶Approximate each of the following situations with a convenient fraction, and explain your reasoning. Tell whether

your estimate is high or low.
(a) Giorgio had 15 base hits out of 46 times at bat.
(b) Ruth made 7 goals out of 41 shots.
(c) Laura answered 62 problems correctly out of 80.
(d) Jonathan made 9 baskets out of 19.

10. Use the information in the table to answer each of the following questions.

Team	Games Played	Games Won
Ducks	22	10
Beavers	19	10
Tigers	28	9
Bears	23	8
Lions	27	7
Wildcats	25	6
Badgers	21	5

(a) Which team won just over $\frac{1}{2}$ of its games?

(b) Which team won just under $\frac{1}{2}$ of its games?

(c) Which team won just over $\frac{1}{3}$ of its games?

(d) Which team won just under $\frac{1}{3}$ of its games?

(e) Which team won just over $\frac{1}{4}$ of its games?

(f) Which team won just under $\frac{1}{4}$ of its games?

11. Approximate each of the following fractions by 0, $\frac{1}{4}$, $\frac{1}{2}$, $\frac{3}{4}$, or 1. Tell whether your estimate is high or low.

(a) $\frac{19}{39}$ (b) $\frac{3}{197}$ (c) $\frac{150}{201}$

(d) $\frac{8}{9}$ (e) $\frac{113}{110}$ (f) $\frac{-2}{117}$

(g) $\frac{150}{198}$ (h) $\frac{999}{2000}$

12. Without actually finding the exact answer, state which of the numbers given in parentheses is the best approximation for the given sum or difference.

(a) $\frac{6}{13} + \frac{7}{15} + \frac{11}{23} + \frac{17}{35}$ $\left(1, 2, 3, 3\frac{1}{2}\right)$

(b) $\frac{30}{41} + \frac{1}{1000} + \frac{3}{2000}$ $\left(\frac{3}{8}, \frac{3}{4}, 1, 2\right)$

(c) $\frac{103}{300} + \frac{203}{601} - \frac{602}{897}$ $\left(1, \frac{1}{3}, \frac{2}{3}, 0\right)$

(d) $\frac{1}{100} - \frac{1}{101} + \frac{1}{102} - \frac{1}{103}$ $\left(\frac{1}{2}, 1, 0\right)$

13. Find the best estimate you can for each of the following:

(a) $4\frac{9}{10} + 1\frac{17}{18} + 3$

(b) $3\frac{1}{9} + 5\frac{1}{10} + 4\frac{3}{10} - 12\frac{1}{2}$

(c) $5\frac{4}{9} + 3\frac{5}{11} + 4\frac{1}{100}$ (d) $148\frac{3}{4} + 2\frac{1}{5}$

14. Use estimation to answer each of the following:

(a) Juan needs to make $11\frac{1}{4}$ lb of breakfast cereal. He bought $4\frac{7}{8}$ lb of oats, $3\frac{1}{4}$ lb cracked wheat, and $2\frac{15}{16}$ lb of triticale. Does Juan have enough grain?

(b) Jill expected to drive from Eugene to Seattle in less than 5 hr. It took her $1\frac{3}{4}$ hr to get from Eugene to Portland and $3\frac{5}{12}$ hr to get from Portland to Seattle. Did Jill make the trip in less than 5 hr?

15. Compute each of the following mentally:

(a) $1 - \frac{3}{4}$ (b) $6 - \frac{7}{8}$

(c) $3\frac{3}{8} + 2\frac{1}{4} - 5\frac{5}{8}$ (d) $2\frac{3}{5} + 4\frac{1}{10} + 3\frac{3}{10}$

16. The ruler shown below has regions marked M, A, T, H. Use mental mathematics and estimation to determine which region each of the following falls into (for example, $\frac{12}{5}$ in. falls in region A).

(a) $\frac{20}{8}$ in.   (b) $\frac{36}{8}$ in.   (c) $\frac{60}{16}$ in.   (d) $\frac{18}{4}$ in.

17. Perform the indicated operations, and write your answers in simplest form.
    (a) $\dfrac{d}{b} + \dfrac{a}{bc}$
    (b) $\dfrac{a}{a-b} + \dfrac{b}{a+b}$
    (c) $\dfrac{a}{a^2 - b^2} - \dfrac{b}{a - b}$

18. What, if anything, is wrong with each of the following?
    (a) $2 = \dfrac{6}{3} = \dfrac{3+3}{3} = \dfrac{3}{3} + 3 = 1 + 3 = 4$
    (b) $1 = \dfrac{4}{2+2} = \dfrac{4}{2} + \dfrac{4}{2} = 2 + 2 = 4$
    (c) $\dfrac{ab + c}{a} = \dfrac{\cancel{a}b + c}{\cancel{a}} = b + c$
    (d) $\dfrac{a^2 - b^2}{a - b} = \dfrac{a \cdot \cancel{a} - b \cdot \cancel{b}}{\cancel{a} - \cancel{b}} = a - b$
    (e) $\dfrac{a + c}{b + c} = \dfrac{a + \cancel{c}}{b + \cancel{c}} = \dfrac{a}{b}$

19. A class consists of $\dfrac{2}{5}$ freshmen, $\dfrac{1}{4}$ sophomores, and $\dfrac{1}{10}$ juniors; the rest are seniors. What fraction of the class is seniors?

20. The Naturals Company sells its products in many countries. The following two circle graphs show the fractions of the company's earnings for 1980 and 1990. Based on this information, answer the following questions:

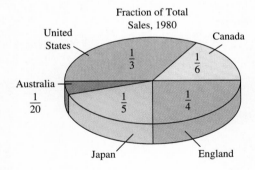

Fraction of Total Sales, 1980

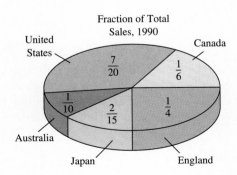

Fraction of Total Sales, 1990

(a) In 1980, how much greater was the fraction of sales for Japan than for Canada?

(b) In 1990, how much less was the fraction of sales for England than for the United States?
(c) How much greater was the fraction of total sales for the United States in 1990 than in 1980?
(d) Is it true that the amount of sales in dollars in Australia was less in 1990 than in 1980? Why?

21. Complete the following magic square:

	$\dfrac{1}{12}$	1
$\dfrac{1}{4}$	$\dfrac{11}{12}$	
$\dfrac{5}{6}$		$\dfrac{1}{6}$

22. A clerk sold three pieces of ribbon. One piece was $\dfrac{1}{3}$ yd long, another piece was $2\dfrac{3}{4}$ yd long, and the third was $3\dfrac{1}{2}$ yd long. What was the total length of ribbon sold?

23. ▶Sally claims that it is easier to add two fractions if she adds the numerators and then adds the denominators. How can you help her?

24. ▶Is any improper fraction equal to $\dfrac{4}{5}$? Why or why not?

25. A recipe requires $3\dfrac{1}{2}$ cups of milk. Ran put in $\dfrac{3}{4}$ cup and then another cup. How much more milk does he need to put in?

26. ▶Why should you consider the fractional parts of mixed numbers when you estimate?

27. Martine bought $8\dfrac{3}{4}$ yd of fabric. If she wants to make a skirt using $1\dfrac{7}{8}$ yd, pants using $2\dfrac{3}{8}$ yd, and a vest using $1\dfrac{2}{3}$ yd, how much fabric will be left over?

28. A plywood board $15\dfrac{3}{4}$ in. long is cut from a $38\dfrac{1}{4}$-in. board. The saw cut takes $\dfrac{3}{8}$ in. How long is the piece of board left after cutting?

29. Students from Rattlesnake School formed four teams to collect cans for recycling during the months of April and May. The students received 10¢ for each 5 lb of cans. A record of their efforts is given below.

NUMBER OF POUNDS COLLECTED

	Team 1	Team 2	Team 3	Team 4
April	$28\dfrac{3}{4}$	$32\dfrac{7}{8}$	$28\dfrac{1}{2}$	$35\dfrac{3}{16}$
May	$33\dfrac{1}{3}$	$28\dfrac{5}{12}$	$25\dfrac{3}{4}$	$41\dfrac{1}{2}$

(a) Which team collected the most for the two-month period? How much did they collect?

(b) What was the difference in the total amounts collected by the teams during the two months?

30. ▶What error pattern is occurring in each of the following?

(a) $\frac{13}{35} = \frac{1}{5}$, $\frac{27}{73} = \frac{2}{3}$, $\frac{16}{64} = \frac{1}{4}$

(b) $\frac{4}{5} + \frac{2}{3} = \frac{6}{8}$, $\frac{2}{5} + \frac{3}{4} = \frac{5}{9}$, $\frac{7}{8} + \frac{1}{3} = \frac{8}{11}$

(c) $8\frac{3}{4} - 6\frac{1}{8} = 2\frac{2}{4}$, $5\frac{3}{8} - 2\frac{2}{3} = 3\frac{1}{5}$, $2\frac{2}{7} - 1\frac{1}{3} = 1\frac{1}{4}$

(d) $\frac{2}{3} \cdot 3 = \frac{6}{9}$, $\frac{1}{4} \cdot 6 = \frac{6}{24}$, $\frac{4}{5} \cdot 2 = \frac{8}{10}$

31. Demonstrate by example that each of the following properties of rational numbers holds:
(a) Closure property of addition
(b) Commutative property of addition
(c) Associative property of addition

32. ▶Does each of the following properties hold for subtraction of rational numbers? Justify your answer.
(a) Closure (b) Commutative
(c) Associative (d) Identity
(e) Inverse

33. ▶For each of the following sequences, discover a pattern and write three more terms of the sequence if the pattern continues. Which of the sequences are arithmetic, and which are not? Justify your answers.

(a) $\frac{1}{4}, \frac{1}{2}, \frac{3}{4}, 1, \frac{5}{4}, \ldots$ (b) $\frac{1}{2}, \frac{2}{3}, \frac{3}{4}, \frac{4}{5}, \frac{5}{6}, \ldots$

(c) $\frac{2}{3}, \frac{5}{3}, \frac{8}{3}, \frac{11}{3}, \frac{14}{3}, \ldots$ (d) $\frac{5}{4}, \frac{3}{4}, \frac{1}{4}, \frac{-1}{4}, \frac{-3}{4}, \ldots$

34. Find the nth term in each of the sequences in Problem 33.

35. Insert five fractions between the numbers 1 and 2 so that the seven numbers (including 1 and 2) constitute an arithmetic sequence.

36. Let $f(x) = x + \frac{3}{4}$.
(a) Find the outputs if the inputs are:
(i) 0 (ii) $\frac{4}{3}$ (iii) $\frac{-3}{4}$
(b) For which inputs will the outputs be:
(i) 1 (ii) $^-1$ (iii) $\frac{1}{2}$

37. Let $f(x) = \frac{x+2}{x-1}$, and let the domain of the function be the set of all integers except 1. Find the following:
(a) $f(0)$ (b) $f(^-2)$ (c) $f(^-5)$ (d) $f(5)$

38. (a) Check that each of the following is true:
$\frac{1}{3} = \frac{1}{4} + \frac{1}{3 \cdot 4}$; $\frac{1}{4} = \frac{1}{5} + \frac{1}{4 \cdot 5}$
$\frac{1}{5} = \frac{1}{6} + \frac{1}{5 \cdot 6}$

(b) Based on the examples in (a), write $\frac{1}{n}$ as a sum of two unit fractions—that is, as a sum of fractions with numerator 1.

★(c) Prove your answer in (b).

Review Problems

39. Write each of the following fractions in simplest form:
(a) $\frac{14}{21}$ (b) $\frac{117}{153}$ (c) $\frac{5^2}{7^2}$ (d) $\frac{a^2+a}{1+a}$ (e) $\frac{a^2+1}{a+1}$

40. Determine if each of the following pairs of fractions is equal:
(a) $\frac{a^2}{b}$ and $\frac{a^2b^2}{b^3}$ (b) $\frac{377}{400}$ and $\frac{378}{401}$
(c) $\frac{0}{10}$ and $\frac{0}{-10}$ (d) $\frac{a}{b}$ and $\frac{a+1}{b+1}$, where $a \neq b$

▼ BRAIN TEASER

When Professor Sum was asked by Mr. Little how many students were in his classes, he answered, "All of them study either languages, physics, or not at all. One half of them study languages only, one fourth of them study French, one seventh of them study physics only, and there are 20 who do not study at all." How many students does Professor Sum have?

Section 6-3 Multiplication and Division of Rational Numbers

Multiplication of Rational Numbers

 In the 5–8 *Standards* (p. 67), we find a call for the greater use of visual models as well as a greater emphasis on concepts. Researchers have shown that area as well as length models are effective in the teaching of fractions.

To motivate the definition of multiplication of rational numbers, we use the interpretation of multiplication as repeated addition. Using repeated addition, we can interpret $3 \cdot \left(\frac{3}{4}\right)$ as follows:

$$3 \cdot \left(\frac{3}{4}\right) = \frac{3}{4} + \frac{3}{4} + \frac{3}{4} = \frac{9}{4} = 2\frac{1}{4}.$$

The area model in Figure 6-9 shows this.

FIGURE 6-9

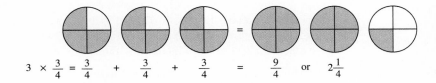

$3 \times \frac{3}{4} = \frac{3}{4} + \frac{3}{4} + \frac{3}{4} = \frac{9}{4}$ or $2\frac{1}{4}$

If the commutative property of multiplication of rational numbers is to be true, then $3 \cdot \left(\frac{3}{4}\right) = \left(\frac{3}{4}\right) \cdot 3 = \frac{9}{4}$.

Next, we consider what happens when both factors are fractions. If forests once covered about $\frac{3}{5}$ of the earth's land and only about $\frac{1}{2}$ of these forests remain, what fraction of the earth is covered with forests today? We can use an area model.

Figure 6-10(a) shows a one-unit rectangle separated into fifths, with $\frac{3}{5}$ shaded. Figure 6-10(b) shows the rectangle further separated into halves, with $\frac{1}{2}$ shaded. The green portion represents $\frac{1}{2}$ of $\frac{3}{5}$. In order to find $\frac{1}{2}$ of $\frac{3}{5}$, we could divide just the shaded portion of the rectangle in Figure 6-10(a) into 2 equal parts and take 1 of those parts. The result would be the green portion of Figure 6-10(b). However, the green portion represents 3 parts out of 10, or $\frac{3}{10}$ of the one-unit rectangle. Thus

$$\frac{1}{2} \cdot \frac{3}{5} = \frac{3}{10} = \frac{1 \cdot 3}{2 \cdot 5}.$$

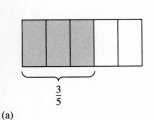

(a)

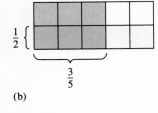

(b)

FIGURE 6-10

This discussion leads to the following definition of multiplication for rational numbers.

DEFINITION OF RATIONAL NUMBER MULTIPLICATION

If $\frac{a}{b}$ and $\frac{c}{d}$ are any rational numbers, then $\frac{a}{b} \cdot \frac{c}{d} = \frac{a \cdot c}{b \cdot d}$.

EXAMPLE 6-12   Find each of the following products:

(a) $\dfrac{5}{6} \cdot \dfrac{7}{11}$ (b) $6 \cdot \dfrac{1}{5}$ (c) $2\dfrac{1}{3} \cdot 3\dfrac{1}{5}$

Solution

(a) $\dfrac{5}{6} \cdot \dfrac{7}{11} = \dfrac{5 \cdot 7}{6 \cdot 11} = \dfrac{35}{66}$

(b) $6 \cdot \dfrac{1}{5} = \dfrac{6}{1} \cdot \dfrac{1}{5} = \dfrac{6 \cdot 1}{1 \cdot 5} = \dfrac{6}{5}$

(c) $2\dfrac{1}{3} \cdot 3\dfrac{1}{5} = \dfrac{7}{3} \cdot \dfrac{16}{5} = \dfrac{7 \cdot 16}{3 \cdot 5} = \dfrac{112}{15} = 7\dfrac{7}{15}$

HISTORICAL NOTE

In the Middle Ages, mathematical skill was admired and supported by the monarchs. Leonardo of Pisa (1170–1230), known as Fibonacci, was the most prominent of the medieval mathematicians. In 1225, Fibonacci participated in a mathematical tournament before the Roman Emperor Frederic II, who came to Pisa with a group of mathematicians to test Fibonacci's immense reputation. One of the questions was to find a rational number that is a square for example $\left(\dfrac{4}{9} = \left(\dfrac{2}{3}\right)^2\right)$, and that remains a square if it is decreased or increased by 5. Fibonacci found the number: $\dfrac{1681}{144}$, or $\left(\dfrac{41}{12}\right)^2$. When 5 is subtracted, it remains a square because $\dfrac{1681}{144} - 5 = \dfrac{961}{144} = \left(\dfrac{31}{12}\right)^2$, and when 5 is added, it remains a square because $\dfrac{1681}{144} + 5 = \dfrac{2401}{144} = \left(\dfrac{49}{12}\right)^2$.

Properties of Multiplication for Rational Numbers

Multiplication of rational numbers has properties analogous to the properties of addition of rational numbers. These include the following properties for multiplication: closure property, commutative property, associative property, multiplicative identity, and multiplicative inverse. For emphasis, we list the last two properties.

PROPERTIES

Multiplicative Identity of Rational Numbers: The number 1 is the unique number such that for every rational number $\dfrac{a}{b}$,

$$1 \cdot \left(\dfrac{a}{b}\right) = \dfrac{a}{b} = \left(\dfrac{a}{b}\right) \cdot 1.$$

Multiplicative Inverse of Rational Numbers: For any nonzero rational number $\dfrac{a}{b}$, $\dfrac{b}{a}$ is the unique rational number such that $\dfrac{a}{b} \cdot \dfrac{b}{a} = 1 = \dfrac{b}{a} \cdot \dfrac{a}{b}$. The multiplicative inverse of $\dfrac{a}{b}$ is also called the **reciprocal** of $\dfrac{a}{b}$.

reciprocal

SECTION 6-3 MULTIPLICATION AND DIVISION OF RATIONAL NUMBERS

> **REMARK** The multiplicative inverse property is a property we obtain when we expand from the set of integers to the set of rational numbers.

EXAMPLE 6-13 Find the multiplicative inverse of each of the following rational numbers:

(a) $\frac{2}{3}$ (b) $\frac{-2}{5}$ (c) 4 (d) 0 (e) $6\frac{1}{2}$

Solution (a) $\frac{3}{2}$

(b) $\frac{5}{-2}$, or $\frac{-5}{2}$

(c) Because $4 = \frac{4}{1}$, the multiplicative inverse of 4 is $\frac{1}{4}$.

(d) Even though $0 = \frac{0}{1}$, $\frac{1}{0}$ is undefined; there is no multiplicative inverse of 0.

(e) Because $6\frac{1}{2} = \frac{13}{2}$, the multiplicative inverse of $6\frac{1}{2}$ is $\frac{2}{13}$.

Multiplication and addition are connected through the distributive property of multiplication over addition.

PROPERTIES

Distributive Property of Multiplication over Addition for Rational Numbers: If $\frac{a}{b}$, $\frac{c}{d}$, and $\frac{e}{f}$ are any rational numbers, then

$$\frac{a}{b}\left(\frac{c}{d} + \frac{e}{f}\right) = \left(\frac{a}{b} \cdot \frac{c}{d}\right) + \left(\frac{a}{b} \cdot \frac{e}{f}\right).$$

Multiplication Property of Equality for Rational Numbers: If $\frac{a}{b}$ and $\frac{c}{d}$ are any rational numbers such that $\frac{a}{b} = \frac{c}{d}$, and $\frac{e}{f}$ is any rational number, then $\frac{a}{b} \cdot \frac{e}{f} = \frac{c}{d} \cdot \frac{e}{f}$.

Multiplication Property of Zero for Rational Numbers: If $\frac{a}{b}$ is any rational number, then $\frac{a}{b} \cdot 0 = 0 = 0 \cdot \frac{a}{b}$.

The properties of rational numbers are used to solve equations, as shown in the following examples.

EXAMPLE 6-14   Solve for x.

$$\frac{2}{3}x - \frac{1}{5} = \frac{3}{4}$$

Solution   To isolate x, we would like to eliminate $\frac{1}{5}$ from the left side of the equation. To achieve this goal, we add $\frac{1}{5}$ to both sides of the equation and proceed as shown.

$$\frac{2}{3}x - \frac{1}{5} = \frac{3}{4}$$
$$\frac{2}{3}x - \frac{1}{5} + \frac{1}{5} = \frac{3}{4} + \frac{1}{5}$$
$$\frac{2}{3}x = \frac{19}{20}$$
$$\frac{3}{2} \cdot \frac{2}{3}x = \frac{3}{2} \cdot \frac{19}{20}$$
$$x = \frac{57}{40}$$

REMARK   In Example 6-14, $\frac{2x}{3}$ can be treated as $\frac{2}{3} \cdot \frac{x}{1}$, or $\frac{2 \cdot x}{3 \cdot 1} = \frac{2x}{3}$. Hence, $\frac{2}{3}x = \frac{2x}{3}$.

EXAMPLE 6-15   A bicycle is on sale at $\frac{3}{4}$ of its original price. If the sale price is $330, what was the original price?

Solution   Let x be the original price. Then $\frac{3}{4}$ of the original price is $\frac{3}{4}x$. Because the sale price is $330, we have $\frac{3}{4}x = 330$. Solving for x gives

$$\frac{4}{3} \cdot \frac{3}{4}x = \frac{4}{3} \cdot 330$$
$$1 \cdot x = 440$$
$$x = 440.$$

Thus the original price was $440.

An alternate approach, which does not use algebra, follows. Because $\frac{3}{4}$ of the original price is $330, $\frac{1}{4}$ of the original price is $\frac{1}{3} \cdot 330$, or $110; thus, $4 \cdot \frac{1}{4}$ of the original price is $4 \cdot 110$, or $440.

EXAMPLE 6-16 At the end of the month, when Devora was paid for her paper route, she spent $50 on records and then $\frac{2}{5}$ of the remaining money on books. After that, with $\frac{1}{3}$ of the remaining amount, she bought presents; she was left with $48. How much was she paid at the end of the month?

Solution Let x denote Devora's earnings in dollars from her paper route. After spending $50, she had $x - 50$ dollars left. Since she spent $\frac{2}{5}$ of that on books, she was left with $\frac{3}{5}$ of $x - 50$, that is, $\frac{3}{5}(x - 50)$. Because she spent $\frac{1}{3}$ of this amount on presents, she was left with $\frac{2}{3}$ of the last amount: that is, $\frac{2}{3} \cdot \frac{3}{5} \cdot (x - 50)$. Because she was finally left with $48, we have

$$\frac{2}{3} \cdot \frac{3}{5}(x - 50) = 48$$
$$\frac{2}{5}(x - 50) = 48$$
$$\frac{5}{2} \cdot \frac{2}{5}(x - 50) = \frac{5}{2} \cdot 48$$
$$x - 50 = 120$$
$$x = 170.$$

Consequently, Devora earned $170 from her paper route. ∎

PROBLEM 1

Sonja wants to build a square deck with the floor made out of 1-in. by 6-in. boards. She wants the deck to be 30 boards wide. Boards come in lengths of 6, 8, 10, 12, 14, 16, 18, and 20 ft and sell for 32¢ per ft. How many boards of what length must Sonja order to make the floor? What is the minimum cost of the floor if she orders only complete boards?

Understanding the Problem. A square floor is to be made of 1-in. by 6-in. boards as described above. First we must understand that in fact a 1-in. by 6-in. board is in reality $5\frac{1}{2}$ in. wide and $\frac{3}{4}$ in. thick. Only complete boards can be used. We need to find the minimum cost.

Devising a Plan. Because the deck is a square and the deck is 30 boards wide, the length of the deck must be $30 \cdot (5\frac{1}{2}) = 165$ in. From this information, we can find the length of the boards needed and the cost.

Carrying Out the Plan. Since the length is 165 in., we divide by 12 to convert to feet and obtain $13\frac{3}{4}$ ft. Hence, we need to order 30 of the 14-ft boards. This gives $30 \cdot 14$, or 420 ft at 32¢ per ft. Thus, Sonja's bill would be $32¢ \times 420 = \$134.40$. This is the minimum cost because boards shorter than 14 ft will not work, boards longer than 14 ft cost more, and two lengths cannot be cut from any board.

Looking Back. We could vary this problem by changing the sizes of the deck or the boards. We could also work on related problems such as: Which size nail is needed to nail together three 1-in. by 6-in. boards so that the nail would go through two boards and go $\frac{1}{2}$ in. into the third board? (Nail sizes increase by $\frac{1}{4}$ in. For example, a 2-penny nail is 1 in. long; a 3-penny nail is $1\frac{1}{4}$ in. long; a 4-penny nail is $1\frac{1}{2}$ in. long, and so forth.) How many nails would be needed to nail down the deck? ∎

Division of Rational Numbers

Recall that $6 \div 3$ means "How many 3s are there in 6?". We found that $6 \div 3 = 2$ because $3 \cdot 2 = 6$. Consider $3 \div \left(\frac{1}{2}\right)$, which is equivalent to finding how many halves there are in 3. We see from the area model in Figure 6-11 that there are 6 half pieces in the 3 whole pieces. We record this as $3 \div \left(\frac{1}{2}\right) = 6$. Also note that $\left(\frac{1}{2}\right) \cdot 6 = 3$.

FIGURE 6-11

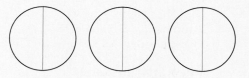

Next, consider $\left(\frac{3}{4}\right) \div \left(\frac{1}{8}\right)$. This means "how many $\frac{1}{8}$s are in $\frac{3}{4}$?". Figure 6-12 shows that there are six $\frac{1}{8}$s in the shaded portion, which represents $\frac{3}{4}$ of the whole. Therefore, $\left(\frac{3}{4}\right) \div \left(\frac{1}{8}\right) = 6$. Also, note that $\left(\frac{1}{8}\right) \cdot 6 = \frac{3}{4}$.

FIGURE 6-12

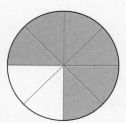

In the previous examples, we saw a relationship between division and multiplication of rational numbers. We can define division for rational numbers formally in terms of multiplication in the same way as we define division for integers.

> **DEFINITION OF RATIONAL NUMBER DIVISION**
>
> If $\frac{a}{b}$ and $\frac{c}{d}$ are any rational numbers and $\frac{c}{d}$ is not zero, then $\frac{a}{b} \div \frac{c}{d} = \frac{e}{f}$ if, and only if, $\frac{e}{f}$ is the unique rational number such that $\frac{c}{d} \cdot \frac{e}{f} = \frac{a}{b}$.

> **REMARK** In the definition of division, $\frac{c}{d}$ is not zero because division by zero is impossible. Also, $\frac{c}{d} \neq 0$ implies that $c \neq 0$.

Algorithm for Division of Rational Numbers

The student page from *Heath Mathematics Connection,* Grade 6, 1992, on the following page, motivates a division algorithm for rational numbers. Note that each *division sentence* on the student page is associated with a *multiplication sentence*. The two questions on the bottom of the page as well as Example 6-17 lead us to the familiar invert-and-multiply algorithm for division of rationals.

> **Algorithm for Division of Fractions**
>
> $$\frac{a}{b} \div \frac{c}{d} = \frac{a}{b} \cdot \frac{d}{c}, \text{ where } \frac{c}{d} \neq 0.$$

EXAMPLE 6-17 Compute the following:
(a) $1 \div \frac{2}{3}$ (b) $\frac{2}{3} \div \frac{5}{7}$

Solution (a) By definition, $1 \div \frac{2}{3} = x$ if, and only if, $\frac{2}{3} \cdot x = 1$. Since $\frac{2}{3}$ and x must be multiplicative inverses of each other, $x = \frac{3}{2}$. Thus $1 \div \frac{2}{3} = \frac{3}{2}$.

(b) Let $\frac{2}{3} \div \frac{5}{7} = x$. Then, $\frac{5}{7} \cdot x = \frac{2}{3}$. To solve for x, multiply both sides of the equation by the reciprocal of $\frac{5}{7}$, namely, $\frac{7}{5}$. Thus $\frac{7}{5}\left(\frac{5}{7}x\right) = \frac{7}{5} \cdot \frac{2}{3}$. Hence, $x = \frac{7}{5} \cdot \frac{2}{3} = \frac{14}{15}$. ∎

COOPERATIVE LEARNING
Reciprocals

Benny uses the scrap wood from his parents' woodworking shop to make coasters and bookends. He draws pictures and makes charts to help figure out how many pieces of wood he will end up with once he starts cutting.

Work with a partner. Use your recording sheet to help you answer the questions and complete each chart.

1. Fill in Chart 1. How many pieces of wood will Benny have if he cuts a 5-foot board into
   a. $\frac{1}{2}$-foot-long sections? **10**
   b. $\frac{1}{4}$-foot-long sections? **20**
   c. $\frac{1}{5}$-foot-long sections? **25**
   d. $\frac{1}{6}$-foot-long sections? **30**
   e. $\frac{1}{8}$-foot-long sections? **40**

CHART 1

	5-Foot Board	Division Sentence	Multiplication Sentence
a.	[board divided into 10 half-foot sections]	$5 \div \frac{1}{2} = \blacksquare$ 10	$5 \times \frac{2}{1} = \blacksquare$ 10
b.	[board with $\frac{1}{4}$ sections]	$5 \div \frac{1}{4} = \blacksquare$ 20	$5 \times \frac{4}{1} = 20$ 1
c.	[board with $\frac{1}{5}$ sections]	$5 \div \frac{1}{5} = \blacksquare$ 25	$5 \times \blacksquare = 25$ $\frac{5}{1}$
d.	[board]	$5 \div \frac{1}{6} = \blacksquare$ 30	$\frac{6}{1}$ $5 \times \blacksquare = \blacksquare$ 30
e.	[board]	$5 \div \frac{1}{8} = \blacksquare$ 40	$5 \blacksquare \times \blacksquare = \blacksquare$ 40 $\frac{8}{1}$

2. In each row of Chart 1, look at the fraction in the division sentence and the fraction in the multiplication sentence. What do you notice about the two fractions? **The numerators and denominators "switch" places.**

3. Find the product of each pair of fractions in Chart 1. What do you notice? **Each product is 1.**

SECTION 6-3 MULTIPLICATION AND DIVISION OF RATIONAL NUMBERS

The algorithm for division of fractions is usually justified in the middle grades by using the Fundamental Law of Fractions, $\frac{a}{b} = \frac{ac}{bc}$, where a, b, and c are all fractions. For example,

$$\frac{2}{3} \div \frac{5}{7} = \frac{\frac{2}{3}}{\frac{5}{7}} = \frac{\frac{2}{3} \cdot \frac{7}{5}}{\frac{5}{7} \cdot \frac{7}{5}} = \frac{\frac{2}{3} \cdot \frac{7}{5}}{\frac{5 \cdot 7}{7 \cdot 5}} = \frac{\frac{2}{3} \cdot \frac{7}{5}}{1} = \frac{2}{3} \cdot \frac{7}{5}.$$

Thus

$$\frac{2}{3} \div \frac{5}{7} = \frac{2}{3} \cdot \frac{7}{5}.$$

An alternative approach for developing an algorithm for division of fractions can be found by first dividing fractions with equal denominators. For example, $\frac{9}{10} \div \frac{3}{10} = 9 \div 3$ and $\frac{15}{23} \div \frac{5}{23} = 15 \div 5$. These examples suggest that, when two fractions with the same denominators are divided, the result can be obtained by dividing the numerator of the first fraction by the numerator of the second. To divide fractions with different denominators, we rename the fractions so that the denominators are equal. Thus

$$\frac{a}{b} \div \frac{c}{d} = \frac{ad}{bd} \div \frac{bc}{bd} = ad \div bc = \frac{ad}{bc}.$$

EXAMPLE 6-18 Perform each of the following divisions and write your answers in simplest form:

(a) $\dfrac{-5}{6} \div \dfrac{-3}{8}$ (b) $5\dfrac{1}{6} \div 4\dfrac{2}{3}$ (c) $\dfrac{\dfrac{1}{4} + \dfrac{-3}{2}}{\dfrac{5}{6} + \dfrac{7}{8}}$

Solution (a) $\dfrac{-5}{6} \div \dfrac{-3}{8} = \dfrac{-5}{6} \cdot \dfrac{8}{-3} = \dfrac{-40}{-18} = \dfrac{20}{9}$

(b) $5\dfrac{1}{6} \div 4\dfrac{2}{3} = \dfrac{31}{6} \div \dfrac{14}{3} = \dfrac{31}{6} \cdot \dfrac{3}{14} = \dfrac{93}{84} = \dfrac{31}{28}$, or $1\dfrac{3}{28}$

(c) We first perform the additions and then do the division.

$$\frac{1}{4} + \frac{-3}{2} = \frac{1}{4} + \frac{-6}{4} = \frac{-5}{4}$$

$$\frac{5}{6} + \frac{7}{8} = \frac{20}{24} + \frac{21}{24} = \frac{41}{24}$$

Hence,

$$\frac{\frac{1}{4} + \frac{-3}{2}}{\frac{5}{6} + \frac{7}{8}} = \frac{\frac{-5}{4}}{\frac{41}{24}} = \frac{-5}{4} \cdot \frac{24}{41} = \frac{-30}{41}.$$

Another method for dividing the two fractions in Example 6-18(c) is based on multiplying each fraction by the least common multiple (LCM) of the two fractions' denominators. Thus

$$\frac{\frac{-5}{4}}{\frac{41}{24}} = \frac{\frac{-5}{4} \cdot 24}{\frac{41}{24} \cdot 24} = \frac{-5 \cdot 6}{41} = \frac{-30}{41}.$$

EXAMPLE 6-19 A person has $35\frac{1}{2}$ yd of material available to make shirts. Each shirt requires $\frac{3}{8}$ of a yd of material.

(a) How many shirts can be made?
(b) How much material will be left over?

Solution (a) We need to find the integer part of the answer to $35\frac{1}{2} \div \frac{3}{8}$. The division follows.

$$35\frac{1}{2} \div \frac{3}{8} = \frac{71}{2} \cdot \frac{8}{3} = \frac{284}{3} = 94\frac{2}{3}$$

Thus we can make 94 shirts.

(b) Because the division in part (a) was by $\frac{3}{8}$, the amount of material left over is $\frac{2}{3}$ of $\frac{3}{8}$, or $\frac{2}{3} \cdot \frac{3}{8}$, or $\frac{1}{4}$ yd.

▼ BRAIN TEASER

A castle in the faraway land of Aluossim was surrounded by four moats. One day, the castle was attacked and captured by a fierce tribe from the north. Guards were stationed at each bridge. Prince Juanaricmo was allowed to take a number of bags of gold from the castle as he went into exile. However, the guard at the first bridge demanded half the bags of gold plus one more bag. Prince Juanaricmo met this demand and proceeded to the next bridge. The guards at the second, third, and fourth bridges made identical demands, all of which the prince met. When the prince finally crossed all the bridges, he had a single bag of gold left. With how many bags did he start?

PROBLEM SET 6-3

1. In the following figures, a unit rectangle is used to illustrate the product of two fractions. Name the fractions and their product.

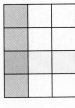
(a)

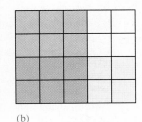
(b)

2. Use a rectangular region to illustrate each of the following products:
   (a) $\frac{3}{4} \cdot \frac{1}{3}$ (b) $\frac{1}{5} \cdot \frac{2}{3}$ (c) $\frac{2}{5} \cdot \frac{1}{3}$

3. ▶If the fractions represented by points C and D on the number line are multiplied, what point best represents the product? Explain why.

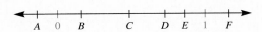

4. Find each product. Write your answers in simplest form.
   (a) $\frac{49}{65} \cdot \frac{26}{98}$ (b) $\frac{a}{b} \cdot \frac{b^2}{a^2}$ (c) $\frac{xy}{z} \cdot \frac{z^2 a}{x^3 y^2}$
   (d) $2\frac{1}{3} \cdot 3\frac{3}{4}$ (e) $\frac{22}{7} \cdot 4\frac{2}{3}$ (f) $\frac{-5}{2} \cdot 2\frac{1}{2}$

5. Use the distributive property to find each product.
   (a) $4\frac{1}{2} \cdot 2\frac{1}{3}$ [*Hint:* $\left(4 + \frac{1}{2}\right) \cdot \left(2 + \frac{1}{3}\right)$.]
   (b) $3\frac{1}{3} \cdot 2\frac{1}{2}$ (c) $248\frac{2}{5} \cdot 100\frac{1}{8}$

6. Find the multiplicative inverse for each of the following:
   (a) $\frac{-1}{3}$ (b) $3\frac{1}{3}$
   (c) $\frac{x}{y}$, if $x \neq 0$ and $y \neq 0$ (d) -7

7. ▶A plumber needed five sections of $2\frac{1}{8}$-ft pipe. Can this pipe be cut from a 12-ft section? If so, how much pipe would be left over? If not, explain why.

8. Express each of the following in simplest form:
   (a) $\dfrac{2\frac{3}{4}}{1\frac{1}{4}}$ (b) $2\frac{3}{4} \cdot 1\frac{4}{3}$
   (c) $\left(3\frac{2}{5} + 1\right)\left(4\frac{1}{3} - 2\frac{2}{3}\right)$ (d) $1\frac{1}{2} \cdot 1\frac{1}{3} \cdot 1\frac{1}{4}$
   (e) $\dfrac{\frac{1}{2} + \frac{1}{3}}{\frac{1}{2} - \frac{1}{3}}$ (f) $\dfrac{\frac{1}{4} + \frac{3}{2}}{\frac{5}{6} + \frac{-7}{8}}$
   (g) $\dfrac{1\frac{1}{2} - 2\frac{3}{4}}{\frac{1}{4} + \frac{-7}{8}}$ (h) $\frac{x}{y} \div \frac{x}{z}$
   (i) $\frac{x}{y} \cdot \frac{yz}{x}$ (j) $x \cdot \frac{5}{xy} \cdot \frac{y}{x}$ (k) $\frac{x^2 y^3}{z^3} \cdot \frac{z^2}{xy^2}$

9. Choose the number from among the numbers in parentheses that best approximates each of the following:
   (a) $3\frac{11}{12} \cdot 5\frac{3}{100}$ (8, 20, 15, 16)
   (b) $2\frac{1}{10} \cdot 7\frac{7}{8}$ (16, 14, 4, 3)
   (c) $20\frac{2}{3} \div 9\frac{7}{8}$ $\left(2, 180, \frac{1}{2}, 10\right)$
   (d) $\frac{1}{101} \div \frac{1}{103}$ $\left(0, 1, \frac{1}{2}, \frac{1}{4}\right)$

10. Estimate the following by rounding the fractions:
    (a) $5\frac{4}{5} \cdot 3\frac{1}{10}$ (b) $4\frac{10}{11} \cdot 5\frac{1}{8}$
    (c) $\dfrac{20\frac{8}{9}}{3\frac{1}{12}}$ (d) $\dfrac{12\frac{1}{3}}{1\frac{7}{8}}$

11. Without actually doing the computations, choose the number in parentheses that correctly describes each of the following:
    (a) $\frac{13}{14} \cdot \frac{17}{19}$ (greater than 1, less than 1)
    (b) $3\frac{2}{7} \div 5\frac{1}{9}$ (greater than 1, less than 1)
    (c) $4\frac{1}{3} \div 2\frac{3}{100}$ (greater than 2, less than 2)
    (d) $16 \div 4\frac{3}{18}$ (greater than 4, less than 4)
    (e) $16 \div 3\frac{8}{9}$ (greater than 4, less than 4)

12. A sewing project requires $6\frac{1}{8}$ yd of material that sells for 62¢ per yard and $3\frac{1}{4}$ yd that sells for 81¢ per yard. Choose the best estimate for the cost of the project.
    (a) Between $2 and $4
    (b) Between $4 and $6
    (c) Between $6 and $8
    (d) Between $8 and $10

13. Compute mentally. Find the exact answer.
    (a) $3\frac{1}{4} \cdot 8$
    (b) $7\frac{1}{4} \cdot 4$
    (c) $9\frac{1}{5} \cdot 10$
    (d) $8 \cdot 2\frac{1}{4}$
    (e) $3 \div \frac{1}{2}$
    (f) $3\frac{1}{2} \div \frac{1}{2}$
    (g) $3 \div \frac{1}{3}$
    (h) $4\frac{1}{2} \div 2$

14. ▶If the product of two numbers is 1, and one of the numbers is greater than 1, what do you know about the other number? Explain your answer.

15. Solve each of the following for x, and write your answers in simplest form:
    (a) $\frac{1}{3}x = \frac{7}{8}$
    (b) $\frac{1}{5} = \frac{7}{3}x$
    (c) $\frac{1}{2}x - 7 = \frac{3}{4}x$
    (d) $\frac{2}{3}\left(\frac{1}{2}x - 7\right) = \frac{3}{4}x$
    (e) $\frac{2}{5} \cdot \frac{3}{6} = x$
    (f) $x \div \frac{3}{4} = \frac{5}{8}$
    (g) $2\frac{1}{3}x + 7 = 3\frac{1}{4}$
    (h) $\frac{-2}{5}(10x + 1) = 1 - x$

16. ▶What are two reasonable estimates for $\frac{1}{7}$ of 39? Explain how you arrived at each estimate.

17. Di Paloma University had a faculty reduction and lost $\frac{1}{5}$ of its faculty. If 320 faculty members were left after the reduction, how many members were there originally?

18. Alberto owns $\frac{5}{9}$ of the stock in the North West Tofu Company. His sister, Renatta, owns half as much stock as Alberto. What part of the stock is owned by neither Alberto nor Renatta?

19. A person has $29\frac{1}{2}$ yd of material available to make doll uniforms. Each uniform requires $\frac{3}{4}$ yd of material.
    (a) How many uniforms can be made?
    (b) How much material will be left over?

20. ▶Suppose you divide a natural number, n, by a positive rational number less than 1. Will the answer always be less than n, sometimes less than n, or never less than n? Why?

21. Show that the following properties do *not* hold for division of rational numbers:
    (a) Commutative
    (b) Associative
    (c) Identity
    (d) Inverse

22. When you multiply a certain number by 3 and then subtract $\frac{7}{18}$, you get the same result as when you multiply the number by 2 and add $\frac{5}{12}$. What is the number?

23. Five eighths of the students at Salem State College live in dormitories. If 6000 students at the college live in the dormitories, how many students are there in the college?

24. A suit is on sale for $180. What was the original price of the suit if the discount was $\frac{1}{4}$ of the original price?

25. If every employee's salary at the Sunrise Software Company increases each year by $\frac{1}{10}$ of that person's salary the previous year, answer the following:
    (a) If Martha's present annual salary is $100,000, what will her salary be in 2 years?
    (b) If Aaron's present salary is $99,000, what was his salary 1 year ago?
    (c) If Juanita's present salary is $363,000, what was her salary 2 years ago?

26. At a certain company, three times as many men as women apply for work. If $\frac{1}{10}$ of the applicants are hired and $\frac{1}{20}$ of the men who apply are hired, what fraction of the women who apply are hired?

27. Jasmine is reading a book. She has finished $\frac{3}{4}$ of the book and has 82 pages left to read. How many pages has she read?

28. John took out all his money from his bank savings account. He spent $50 on a radio and $\frac{3}{5}$ of what remained on presents. Half of what was left he put back in his checking account, and the remaining $35 he donated to charity. How much money did John originally have in his savings account?

29. Peter, Paul, and Mary start at the same time walking around a circular track in the same direction. Peter takes $\frac{1}{2}$ hr to walk around the track, Paul takes $\frac{5}{12}$ hr, and Mary takes $\frac{1}{3}$ hr.
    (a) How many minutes does it take each person to walk around the track?
    (b) How many times will each person go around the track before all three meet again at the starting line?

30. A recipe calls for $2\frac{1}{3}$ packages of frozen beans. How many recipes can be made if 20 packages of beans are available?

31. The formula for converting degrees Celsius (C) to degrees Fahrenheit (F) is $F = \left(\frac{9}{5}\right) \cdot C + 32$.
    (a) If Samantha reads that the temperature is 32°C in Spain, what is the Fahrenheit temperature?
    (b) If the temperature dropped to −40°F in West Yellowstone, what is the temperature in degrees Celsius?

32. Glenn bought 175 shares of stock at $48\frac{1}{4}$ a share. A year later, he sold it at $35\frac{3}{8}$ a share. How much did Glenn lose on the transition?

33. Al gives $\frac{1}{2}$ of his marbles to Bev. Bev gives $\frac{1}{2}$ of these to Carl. Carl gives $\frac{1}{2}$ of these to Dani. If Dani has four marbles, how many did Al have originally?

34. ▶Detect the error pattern in the following:

$$\left(\frac{5}{8}\right)\cdot\left(\frac{2}{3}\right) = \left(\frac{8}{5}\right)\cdot\left(\frac{2}{3}\right) = \frac{16}{15}$$

$$\left(\frac{1}{2}\right)\cdot\left(\frac{1}{4}\right) = \left(\frac{2}{1}\right)\cdot\left(\frac{1}{4}\right) = \frac{2}{4}$$

How might you work with the student who did this work?

35. ▶How would you work the following as a mental math problem rather than a pencil and paper activity?

$$\left(\frac{1}{4}\right)\cdot 15 \cdot 12$$

36. Peppermint Patty is frustrated by a problem. Help her solve it.

PEANUTS reprinted by permission of UFS, Inc.

37. For each of the following sequences, (a) find a pattern and (b) write two more terms of the sequence, assuming that the pattern continues. Which of the sequences are geometric and which are not? Justify your answers.

(i) $1, \frac{1}{2}, \frac{1}{4}, \frac{1}{8}, \frac{1}{16}, \ldots$ (ii) $1, \frac{-1}{2}, \frac{1}{4}, \frac{-1}{8}, \frac{1}{16}, \ldots$

(iii) $\frac{4}{3}, 1, \frac{3}{4}, \frac{9}{16}, \frac{27}{64}, \ldots$ (iv) $\frac{1}{3}, \frac{2}{3^2}, \frac{3}{3^3}, \frac{4}{3^4}, \ldots$

38. There is a simple method for squaring any number that consists of a whole number and $\frac{1}{2}$. For example $\left(3\frac{1}{2}\right)^2 =$

$3\cdot 4 + \left(\frac{1}{2}\right)^2 = 12\frac{1}{4}$; $\left(4\frac{1}{2}\right)^2 = 4\cdot 5 + \left(\frac{1}{2}\right)^2 = 20\frac{1}{4}$; $\left(5\frac{1}{2}\right)^2$

$= 5\cdot 6 + \left(\frac{1}{2}\right)^2 = 30\frac{1}{4}$.

(a) Write a statement for $\left(n + \frac{1}{2}\right)^2$ that generalizes these examples, where n is a whole number.

★(b) Justify this procedure.

39. Let $f(x) = \frac{3x + 4}{4x - 5}$, where the domain is all rational numbers for which the function has a value.

(a) Find the outputs if the inputs are:
(i) 0 (ii) $\frac{2}{5}$ (iii) $\frac{-2}{5}$.

(b) For which imputs will the outputs be:
(i) 0 (ii) $\frac{2}{5}$ (iii) $\frac{-1}{2}$.

(c) What value for x is not in the domain of the function?

40. Consider these products.

First product: $\left(1 + \frac{1}{1}\right)\left(1 + \frac{1}{2}\right)$

Second product: $\left(1 + \frac{1}{1}\right)\left(1 + \frac{1}{2}\right)\left(1 + \frac{1}{3}\right)$

Third product: $\left(1 + \frac{1}{1}\right)\left(1 + \frac{1}{2}\right)\left(1 + \frac{1}{3}\right)\left(1 + \frac{1}{4}\right)$

(a) Calculate the value of each product. Based on the pattern in your answers, guess the value of the fourth product then check to determine if your guess is correct.

(b) Guess the value of the 100th product.

(c) Find as simple an expression as possible for the nth product.

★41. Investigate under what conditions, if any,

$$\frac{a}{b} = \frac{a + c}{b + c}.$$

★42. Let $S = \frac{1}{2} + \frac{1}{2^2} + \frac{1}{2^3} + \cdots + \frac{1}{2^{64}}$.

(a) Use the distributive property of multiplication over addition to find an expression for $2S$.

(b) Show that $2S - S = S = 1 - \left(\frac{1}{2}\right)^{64}$

(c) Find a simple expression for the sum

$$\frac{1}{2} + \frac{1}{2^2} + \frac{1}{2^3} + \cdots + \frac{1}{2^n}.$$

★43. In an arithmetic sequence, the first term is 1 and the 100th term is 2. Find the following:

(a) The 50th term
(b) The sum of the first 50 terms

Review Problems

44. Perform each of the following computations. Leave your answers in simplest form.

(a) $\frac{-3}{16} + \frac{7}{4}$ (b) $\frac{1}{6} + \frac{-4}{9} + \frac{5}{3}$

(c) $\dfrac{-5}{2^3 \cdot 3^2} - \dfrac{-5}{2 \cdot 3^3}$ (d) $3\dfrac{4}{5} + 4\dfrac{5}{6}$

(e) $5\dfrac{1}{6} - 3\dfrac{5}{8}$

(f) $-4\dfrac{1}{3} - 5\dfrac{5}{12}$

45. Each student at Sussex Elementary School takes one foreign language. Two thirds of the students take Spanish, $\dfrac{1}{9}$ take French, $\dfrac{1}{18}$ take German, and the rest take some other foreign language. If there are 720 students in the school, how many do not take Spanish, French, or German?

▼ BRAIN TEASER

A woman's will decreed that her cats be shared among her three daughters as follows: $\dfrac{1}{2}$ of the cats to the eldest daughter, $\dfrac{1}{3}$ of the cats to the middle daughter, and $\dfrac{1}{9}$ of the cats to the youngest daughter. Since the woman had 17 cats, the daughters decided that they could not carry out their mother's wishes. The judge who held the will agreed to lend the daughters a cat so they could share the cats as their mother wished. Now, $\dfrac{1}{2}$ of 18 is 9; $\dfrac{1}{3}$ of 18 is 6; and $\dfrac{1}{9}$ of 18 is 2. Since $9 + 6 + 2 = 17$, the daughters were able to divide the 17 cats and return the borrowed cat. They obviously did not need the extra cat to carry out their mother's bequest, but they could not divide 17 into halves, thirds, and ninths. Has the woman's will really been followed?

Section 6-4 Ordering Rational Numbers

Children know that $\dfrac{7}{8} > \dfrac{5}{8}$ because, if a pizza is divided into 8 parts, then 7 parts of a pizza is more than 5 parts. Similarly, $\dfrac{3}{7} < \dfrac{4}{7}$. Thus, given two fractions with common positive denominators, the one with the greater numerator is the greater fraction. This can be written as follows.

> **THEOREM 6-3**
>
> If a, b, and c are integers and $b > 0$, then $\dfrac{a}{b} > \dfrac{c}{b}$ if, and only if, $a > c$.

The condition $b > 0$ is essential in the property. If $b < 0$, the theorem is not necessarily true. Theorem 6-3 can be deduced from the definition that follows.

> **DEFINITION OF GREATER THAN**
>
> If $\dfrac{a}{b}$ and $\dfrac{c}{d}$ are rational numbers, then $\dfrac{a}{b} > \dfrac{c}{d}$ if, and only if, there is a positive rational number k such that $\dfrac{c}{d} + k = \dfrac{a}{b}$, or equivalently, if, and only if, $\dfrac{a}{b} - \dfrac{c}{d}$ is positive.

SECTION 6-4 ORDERING RATIONAL NUMBERS

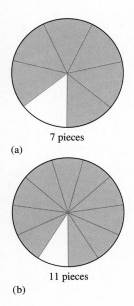

(a) 7 pieces

(b) 11 pieces

FIGURE 6-13

Suppose we have two pizzas of equal size, one cut into 7 pieces and the other cut into 11 pieces, as shown in Figure 6-13. Is there more pizza in 6 slices of the first than there is in 8 slices of the second? To answer this, we need to divide a whole into equal parts in such a way that $\frac{6}{7}$ and $\frac{8}{11}$ will be easily comparable. This can be accomplished by finding fractions equal to $\frac{6}{7}$ and $\frac{8}{11}$ with common denominators. We have $\frac{6}{7} = \frac{6 \cdot 11}{7 \cdot 11} = \frac{66}{77}$ and $\frac{8}{11} = \frac{7 \cdot 8}{7 \cdot 11} = \frac{56}{77}$. Because $66 > 56$, it follows that $\frac{66}{77} > \frac{56}{77}$, so $\frac{6}{7} > \frac{8}{11}$. Similarly, to compare $\frac{-5}{12}$ and $\frac{1}{-2}$, we first write them with common denominators, as follows:

$$\frac{1}{-2} = \frac{1 \cdot (-6)}{-2 \cdot (-6)} = \frac{-6}{12}$$

Because $-5 > -6$, $\frac{-5}{12} > \frac{-6}{12}$; therefore, $\frac{-5}{12} > \frac{1}{-2}$.

The technique of comparing fractions with unlike denominators using a common denominator is demonstrated on the following portion of a student page from *Addison-Wesley Mathematics*, Grade 5, 1991.

Look at the denominators.	Write equivalent fractions with a common denominator.	Compare the numerators.	The fractions compare the same way the numerators compare.
$\frac{3}{8}$ Not the same $\frac{2}{3}$	$\frac{3}{8} = \frac{9}{24}$ The same $\frac{2}{3} = \frac{16}{24}$	$9 < 16$	$\frac{9}{24} < \frac{16}{24}$ so $\frac{3}{8} < \frac{2}{3}$

A general criterion for the greater-than relation on rational numbers can be developed for the case in which the denominators are positive. Using the common denominator bd, the fractions $\frac{a}{b}$ and $\frac{c}{d}$ can be written as $\frac{ad}{bd}$ and $\frac{bc}{bd}$. Because $b > 0$ and $d > 0$, $bd > 0$, we apply Theorem 6-3 to realize that $\frac{ab}{bd} > \frac{bc}{bd}$ if, and only if, $ad > bc$.

EXAMPLE 6-20 Prove that the following order relations are true:

(a) $\frac{8}{9} > \frac{16}{19}$ (b) $\frac{1}{-4} < \frac{2}{11}$ (c) $\frac{1}{-4} > \frac{-1}{2}$

Solution (a) We write the fractions with the common denominator $9 \cdot 19$. We have $\frac{8}{9} > \frac{16}{19}$ is equivalent to $\frac{8 \cdot 19}{9 \cdot 19} > \frac{9 \cdot 16}{9 \cdot 19}$. The last inequality is true because $8 \cdot 19 > 9 \cdot 16$ (or $152 > 144$).

(b) $\frac{1}{-4} < \frac{2}{11}$ because a negative number is less than a positive number.

(c) We write the fractions with common positive denominators.

$$\frac{1}{-4} = \frac{-1}{4} \text{ and } \frac{-1}{2} = \frac{-1 \cdot 2}{2 \cdot 2} = \frac{-2}{4}$$

$$\frac{-1}{4} > \frac{-2}{4} \text{ because } -1 > -2.$$

As in Example 6-20(b), we can often compare the size of fractions by inspection and common sense. Such approaches should be strongly encouraged. Consider which is greater, $\frac{3}{4}$ or $\frac{4}{5}$. Three fourths of a pizza is a pizza with $\frac{1}{4}$ cut out, while $\frac{4}{5}$ of a pizza is a pizza with $\frac{1}{5}$ cut out. Can you find a similar way to determine which of the fractions $\frac{6}{5}$ and $\frac{7}{6}$ is greater? How about $\frac{134}{137}$ and $\frac{131}{130}$?

The proofs of the following theorems of the greater-than relation on rational numbers are similar to those involving integers and are left as exercises. Similar properties hold for $<$, $\leq$, and $\geq$.

THEOREM 6-4

Transitive Property of Greater Than For any rational numbers $\frac{a}{b}$, $\frac{c}{d}$, and $\frac{e}{f}$, if $\frac{a}{b} > \frac{c}{d}$ and $\frac{c}{d} > \frac{e}{f}$, then $\frac{a}{b} > \frac{e}{f}$.

THEOREM 6-5

Addition Property of Greater Than For any rational numbers, $\frac{a}{b}$, $\frac{c}{d}$, and $\frac{e}{f}$, if $\frac{a}{b} > \frac{c}{d}$, then $\frac{a}{b} + \frac{e}{f} > \frac{c}{d} + \frac{e}{f}$.

THEOREM 6-6

Multiplication Property of Greater Than For any rational numbers $\frac{a}{b}$, $\frac{c}{d}$, and $\frac{e}{f}$,

1. If $\frac{a}{b} > \frac{c}{d}$ and $\frac{e}{f} > 0$, then $\frac{a}{b} \cdot \frac{e}{f} > \frac{c}{d} \cdot \frac{e}{f}$.
2. If $\frac{a}{b} > \frac{c}{d}$ and $\frac{e}{f} < 0$, then $\frac{a}{b} \cdot \frac{e}{f} < \frac{c}{d} \cdot \frac{e}{f}$.

Solutions to Algebraic Inequalities

The preceding theorems can be used to aid in solving algebraic inequalities, as the following eample shows.

EXAMPLE 6-21 Solve for x, where x is a rational number.

(a) $\frac{3}{2}x < \frac{3}{4}$ (b) $\frac{1}{4}x + \frac{1}{5} \geq \frac{3}{8}x - \frac{1}{10}$

Solution (a) $\frac{3}{2}x < \frac{3}{4}$

$\left(\frac{2}{3}\right)\left(\frac{3}{2}x\right) < \left(\frac{2}{3}\right)\left(\frac{3}{4}\right)$

$x < \frac{6}{12}$, or $x < \frac{1}{2}$

(b) $\frac{1}{4}x + \frac{1}{5} \geq \frac{3}{8}x - \frac{1}{10}$

$\frac{1}{4}x + \frac{1}{5} + \frac{-1}{5} \geq \frac{3}{8}x - \frac{1}{10} + \frac{-1}{5}$

$\frac{1}{4}x \geq \frac{3}{8}x - \frac{3}{10}$

$\frac{-3}{8}x + \frac{1}{4}x \geq \frac{-3}{8}x + \frac{3}{8}x - \frac{3}{10}$

$\left(\frac{-3}{8} + \frac{1}{4}\right)x \geq \frac{-3}{10}$

$-\frac{1}{8}x \geq \frac{-3}{10}$

$-8\left(\frac{-1}{8}x\right) \leq -8\left(\frac{-3}{10}\right)$

$x \leq \frac{24}{10}$, or $2\frac{2}{5}$ ∎

Often, there is more than one way to solve an inequality. Two alternate methods for solving Example 6-21(b) follow.

1. First, add the fractions on each side of the inequality. Then, solve the resulting inequality.

$\frac{1}{4}x + \frac{1}{5} \geq \frac{3}{8}x - \frac{1}{10}$

$\frac{5x + 4}{20} \geq \frac{15x - 4}{40}$

$40(5x + 4) \geq 20(15x - 4)$

$200x + 160 \geq 300x - 80$

$-100x \geq -240$

$\left(\frac{-1}{100}\right)(-100x) \leq \frac{-1}{100}(-240)$

$x \leq \frac{240}{100}$

$x \leq \frac{12}{5}$, or $2\frac{2}{5}$

2. First, multiply both sides of the inequality by the least common multiple (LCM) of all the denominators. (This gives an inequality that does not involve fractions.) Then, solve the resulting inequality.

$\frac{1}{4}x + \frac{1}{5} \geq \frac{3}{8}x - \frac{1}{10}$

Since LCM(4, 5, 8, 10) = 40,

$$40\left(\frac{1}{4}x + \frac{1}{5}\right) \geq 40\left(\frac{3}{8}x - \frac{1}{10}\right)$$
$$10x + 8 \geq 15x - 4$$
$$^-5x \geq ^-12$$
$$x \leq \frac{^-12}{^-5}$$
$$x \leq \frac{12}{5}, \quad \text{or } 2\frac{2}{5}.$$

Estimations with Order

Properties of greater-than or less-than relations sometimes can be used to estimate the answers to problems. For example, to find the number of curtains requiring $5\frac{3}{8}$ yards of cloth that can be prepared from a bolt that contains $19\frac{3}{4}$ yards, we could compute $\dfrac{19\frac{3}{4}}{5\frac{3}{8}}$.

To estimate the quotient, we could use the *compatible number* strategy. Because $19\frac{3}{4}$ is approximately 20, and $5\frac{3}{8}$ is close to 5, we estimate that the answer should be approximately $\frac{20}{5}$, or 4. Is the actual answer greater than or less than 4? To answer that question, notice that, if two positive fractions have equal denominators, the one with the greater numerator is greater, and if two positive fractions have equal numerators, the one with the greater denominator is smaller. (Do you see intuitively why this is so?) Hence, $\dfrac{19\frac{3}{4}}{5\frac{3}{8}} < \dfrac{20}{5\frac{3}{8}} < \dfrac{20}{5}$. Thus (by Theorem 6-4), $\dfrac{19\frac{3}{4}}{5\frac{3}{8}} < \dfrac{20}{5}$. Similarly, it is possible to show that $\dfrac{19\frac{3}{8}}{5\frac{3}{8}} > \dfrac{18}{5\frac{3}{8}} > \dfrac{18}{6}$. Because $\dfrac{19\frac{3}{4}}{5\frac{3}{8}}$ is between 3 and 4, we can conclude that only 3 curtains can be prepared.

EXAMPLE 6-22 In each of the following, estimate the answer by finding two integers, one smaller than the answer and one greater than the answer.

(a) $9\frac{3}{4} \cdot 14\frac{5}{9}$ (b) $35\frac{1}{3} \div 6\frac{3}{5}$

Solution (a) $9 \cdot 14 < 9\frac{3}{4} \cdot 14\frac{5}{9} < 10 \cdot 15$

Hence,
$$126 < 9\frac{3}{4} \cdot 14\frac{5}{9} < 150.$$

(b) $\dfrac{35\frac{1}{3}}{6\frac{3}{5}} < \dfrac{36}{6\frac{3}{5}} < \dfrac{36}{6}.$ Hence, $\dfrac{35\frac{1}{3}}{6\frac{3}{5}} < 6.$

Similarly,   Thus,

$\dfrac{35\frac{1}{3}}{6\frac{3}{5}} > 5.$ $5 < \dfrac{35\frac{1}{3}}{6\frac{3}{5}} < 6.$ ∎

EXAMPLE 6-23 In each of the following, use estimation techniques to determine which is a correct description of the answer:

(a) $\dfrac{1}{2} + 1\dfrac{2}{3}$ (b) $\dfrac{8}{9} + \dfrac{17}{18} + \dfrac{29}{30}$ (c) $\dfrac{1}{2} \cdot \left(3\dfrac{9}{10}\right)$

between: (i) 1 and $1\dfrac{1}{2}$ between: (i) 2 and $2\dfrac{1}{2}$ between: (i) $\dfrac{1}{2}$ and 1

(ii) $1\dfrac{1}{2}$ and 2   (ii) $2\dfrac{1}{2}$ and 3   (ii) 1 and $1\dfrac{1}{2}$

(iii) 2 and $2\dfrac{1}{2}$ (iii) 3 and $3\dfrac{1}{2}$ (iii) $1\dfrac{1}{2}$ and 2

Solution We do each of the following mentally:

(a) $\dfrac{1}{2} + 1\dfrac{2}{3} > \dfrac{1}{2} + 1\dfrac{1}{2} = 2.$ Also $\dfrac{1}{2} + 1\dfrac{2}{3} < \dfrac{1}{2} + 2 = 2\dfrac{1}{2}.$ Hence, (iii) is correct.

(b) Each fraction is quite close to 1 but less than 1. Hence, (ii) is correct.

(c) This product is quite close to $\dfrac{1}{2} \cdot 4$ because $3\dfrac{9}{10}$ is close to 4. Hence, (iii) seems to be correct. A more precise approach is as follows:

$\dfrac{1}{2} \cdot 3 < \dfrac{1}{2} \cdot \left(3\dfrac{9}{10}\right) < \dfrac{1}{2} \cdot 4$ or $1\dfrac{1}{2} < \dfrac{1}{2} \cdot \left(3\dfrac{9}{10}\right) < 2.$ ∎

EXAMPLE 6-24 Without actually performing the computations, arrange the following in increasing order, from least to greatest:

(a) $\dfrac{41}{80}, \dfrac{29}{60}, \dfrac{3}{4}, \dfrac{19}{17}, \dfrac{9}{10}, 0, \dfrac{1}{10}, \dfrac{-3}{4}$

(b) $1\dfrac{5}{8} \cdot 6\dfrac{3}{5}, 2\dfrac{7}{19} \cdot 6\dfrac{3}{5}, \dfrac{1}{2} \cdot 5\dfrac{9}{19}, \dfrac{9}{10} \cdot 5\dfrac{1}{19}$

Solution (a) Because a negative number is less than 0 or a positive number, $\dfrac{-3}{4}$ is the least number. 0 is the next larger number. Notice that $\dfrac{41}{80}$ is $\dfrac{1}{80}$ more than $\dfrac{1}{2}$ and that $\dfrac{29}{60}$ is $\dfrac{1}{60}$ less than $\dfrac{1}{2}.$ Also, $\dfrac{9}{10}$ is close to 1 but less than 1, and $\dfrac{19}{17} > 1.$ The fraction $\dfrac{1}{10}$ is much less than $\dfrac{1}{2}.$ These

observations enable us to order the numbers as follows:
$$\frac{-3}{4} < 0 < \frac{1}{10} < \frac{29}{60} < \frac{41}{80} < \frac{3}{4} < \frac{9}{10} < \frac{19}{17}.$$

(b) Notice that each of the last two products is smaller than the first or the second product. The second product is greater than the first, and the fourth is greater than the third. Hence,
$$\frac{1}{2} \cdot 5\frac{9}{19} < \frac{9}{10} \cdot 5\frac{1}{19} < 1\frac{5}{8} \cdot 6\frac{3}{5} < 2\frac{7}{19} \cdot 6\frac{3}{5}.$$ ∎

Inequalities are useful in many applications of mathematics. Following are examples of problems that can be solved with inequalities.

PROBLEM 2

A plane increased its normal speed by 60 mph and traveled a distance of less than 2860 mi in $5\frac{1}{2}$ hr. When the plane decreased its normal speed by 100 mph, it covered a distance that was greater than 2400 mi in 8 hr. The pilot claimed that the plane's normal speed was less than 500 mph. Was the pilot correct?

Devising a Plan. Table 6-3 outlines the given information. We know that distance = rate × time, where rate is another name for speed. If we denote the normal speed by x miles per hour, we can write the distance traveled during the $5\frac{1}{2}$ hr and during the 8 hr in terms of x. From the information about the distances, we will have two inequalities involving x. The solution of these inequalities should give us more information about x.

TABLE 6-3

Speed (in miles per hour)	Time (in hours)	Distance (in miles)
Normal + 60	$5\frac{1}{2}$	Less than 2860
Normal − 100	8	Greater than 2400

Carrying Out the Plan. The outlined procedure for finding the distances is shown in Table 6-4.

TABLE 6-4

Normal Speed (in miles per hour)	New Speed (in miles per hour)	Time (in hours)	Distance (in miles, in terms of x)
x	$x + 60$	$5\frac{1}{2}$	$5\frac{1}{2}(x + 60)$
x	$x - 100$	8	$8(x - 100)$

Because we know that the first distance is less than 2860 and the second is greater than 2400, we have the following inequalities:

$$5\tfrac{1}{2}(x + 60) < 2860 \quad \text{and} \quad 8(x - 100) > 2400.$$

These inequalities are equivalent to each of the following pairs of inequalities:

$$5\tfrac{1}{2}(x + 60) < 2860 \quad \text{and} \quad 8(x - 100) > 2400$$

$$\tfrac{2}{11} \cdot \tfrac{11}{2}(x + 60) < \tfrac{2}{11} \cdot 2860 \quad \text{and} \quad \tfrac{1}{8} \cdot 8(x - 100) > \tfrac{1}{8} \cdot 2400$$

$$x + 60 < 520 \quad \text{and} \quad x - 100 > 300$$
$$x < 460 \quad \text{and} \quad x > 400$$

Consequently, $400 < x < 460$. Hence, the plane's normal speed is between 400 and 460 mph, and the pilot's claim was correct.

Looking Back. Even though we could not find the exact speed, the data in the question enabled us to find a range for the possible normal speeds.

Suppose we knew that the speed was a whole number. Could you make a change in numbers in the question (without changing any words) that would enable you to find the exact value of x? ∎

The set of rational numbers has a very special property that is not present for the set of whole numbers or for the set of integers. This property is given below.

DENSENESS PROPERTY

Given rational numbers $\tfrac{a}{b}$ and $\tfrac{c}{d}$, there is another rational number between these two numbers.

In the above property, between $\tfrac{a}{b}$ and the new rational number there is another rational number. Continuing this process, we see that between any two rational numbers $\tfrac{a}{b}$ and $\tfrac{c}{d}$, there are infinitely many rational numbers.

Consider $\tfrac{1}{2}$ and $\tfrac{2}{3}$. To find a rational number between $\tfrac{1}{2}$ and $\tfrac{2}{3}$, we first rewrite the fractions with a common denominator, as $\tfrac{3}{6}$ and $\tfrac{4}{6}$. Because there is no whole number between the numerators 3 and 4, we next find two fractions equivalent to $\tfrac{1}{2}$ and $\tfrac{2}{3}$ with greater denominators. For example, $\tfrac{1}{2} = \tfrac{6}{12}$ and $\tfrac{2}{3} = \tfrac{8}{12}$, and $\tfrac{7}{12}$ is between the two fractions $\tfrac{6}{12}$ and $\tfrac{8}{12}$. So $\tfrac{7}{12}$ is between $\tfrac{1}{2}$ and $\tfrac{2}{3}$.

Another way to find a rational number between two given rationals $\tfrac{a}{b}$ and $\tfrac{c}{d}$ is to find the *arithmetic mean* of the two numbers. To find the arithmetic mean of two numbers, we find the sum of the two numbers and divide by 2. For example, the arithmetic mean of $\frac{1}{2}$ and $\frac{2}{3}$ is $\frac{1}{2}\left(\frac{1}{2} + \frac{2}{3}\right)$, or $\frac{7}{12}$. The proof that the arithmetic mean of two given rational numbers always is between them is left as an exercise.

EXAMPLE 6-25 Find two fractions between $\frac{1}{2}$ and $\frac{7}{18}$.

Solution Because $\frac{1}{2} = \frac{1 \cdot 9}{2 \cdot 9} = \frac{9}{18}$, we see that $\frac{8}{18}$, or $\frac{4}{9}$, is between $\frac{7}{18}$ and $\frac{9}{18}$. To find another fraction between the given fractions, we find two fractions equivalent to $\frac{7}{18}$ and $\frac{9}{18}$, but with greater denominators. For example, $\frac{7}{18} = \frac{14}{36}$ and $\frac{9}{18} = \frac{18}{36}$. We now see that $\frac{15}{36}, \frac{16}{36},$ and $\frac{17}{36}$ are all between $\frac{14}{36}$ and $\frac{18}{36}$. ∎

▼ **BRAIN TEASER**

Two cyclists, David and Sara, started riding their bikes at 9:00 A.M. at City Hall. They followed the local bike trail and returned to City Hall at the same time. However, David rode three times as long as Sara rested on her trip and Sara rode four times as long as David rested on his trip. Assuming that each cyclist rode at a constant speed, who rode faster?

PROBLEM SET 6-4

1. For each of the following pairs of fractions, replace the comma with the correct symbol (<, =, >) to make a true statement.

   (a) $\frac{7}{8}, \frac{5}{6}$ (b) $2\frac{4}{5}, 2\frac{3}{6}$ (c) $\frac{-7}{8}, \frac{-4}{5}$

   (d) $\frac{1}{-7}, \frac{1}{-8}$ (e) $\frac{2}{5}, \frac{4}{10}$ (f) $\frac{0}{7}, \frac{0}{17}$

2. Illustrate each of the following fractions on a number line:

   $-2\frac{1}{4}, -1\frac{3}{8}, -\frac{1}{2}, 1\frac{1}{8}, 2\frac{5}{8}$

3. Arrange each of the following in decreasing order:

   (a) $\frac{11}{22}, \frac{11}{16}, \frac{11}{13}$ (b) $\frac{33}{16}, \frac{23}{16}, 3$

   (c) $\frac{-1}{5}, \frac{-19}{36}, \frac{-17}{30}$

4. Solve for x in each of the following:

   (a) $\frac{2}{3}x - \frac{7}{8} \leq \frac{1}{4}$

   (b) $x - \frac{1}{3} < \frac{2}{3}x + \frac{4}{5}$

   (c) $\frac{1}{5}x - 7 \geq \frac{2}{3}$

   (d) $5 - \frac{2}{3}x \leq \frac{1}{4}x - \frac{7}{8}$

5. ▶(a) If $b < 0$ and $d > 0$, is it true that $\frac{a}{b} > \frac{c}{d}$ if and only if $ad > bc$? Explain your answer.

   ▶(b) If $b < 0$ and $d < 0$, is it true that $\frac{a}{b} > \frac{c}{d}$ if, and only if, $ad > bc$? Explain your answer.

6. Estimate each of the following then perform the multiplications to see how good your estimates are.

   (a) $19\frac{8}{9} \cdot 20\frac{1}{9}$ (b) $19\frac{8}{9} \cdot 9\frac{1}{10}$ (c) $3\frac{9}{10} \cdot \frac{81}{82}$

7. In each of the following, choose the better estimate from the pair of estimates in parentheses:

   (a) $4\frac{5}{8} + 2\frac{5}{9}$ (under 7, over 7)

   (b) $7\frac{1}{10} + 5\frac{6}{11}$ (under 13, over 13)

   (c) $8\frac{1}{3} \div 8\frac{2}{3}$ (under 1, over 1)

(d) $6\frac{1}{10} \div \frac{11}{12}$ (under 6, over 6)

(e) $10 - 3\frac{4}{5}$ (under 6, over 6)

8. In Amy's algebra class, 6 out of 31 students received A's on a test. The same test was given to Bren's class and 5 out of 23 students received A's. Which class had the higher rate of A's?

9. In each of the following, estimate the answer:

   (a) $19\frac{8}{9} \cdot 9\frac{1}{10}$ (b) $80\frac{3}{4} \cdot 9\frac{1}{8}$ (c) $77\frac{3}{5} \cdot 6\frac{1}{4}$

   (d) $\dfrac{48\frac{2}{3}}{8\frac{4}{9}}$ (e) $\dfrac{5\frac{2}{3}}{2\frac{1}{17}}$

10. Estimate the number of $11\frac{3}{4}$-oz birdseed packages that can be produced from a supply of 21 lb of birdseed (16 oz = 1 lb).

11. (a) Choose several positive proper fractions. Square each of the fractions, and compare the size of the original fraction and its square. Make a conjecture concerning a fraction and its square.
    ★(b) Justify your conjecture in (a).
    (c) If a fraction is greater than 1, make a conjecture concerning which is greater: the fraction or its square.
    ★(d) Justify your conjecture in (c).

12. If $\frac{a}{b} < 1$ and $\frac{c}{d} > 0$, compare the size of $\frac{c}{d}$ with $\frac{a}{b} \cdot \frac{c}{d}$.

13. ▶If x and y are two rational numbers such that $x > 1$ and $y > 0$, which is greater: xy or y? Justify your answer.

14. Show that the sequence $\frac{1}{2}, \frac{2}{3}, \frac{3}{4}, \frac{4}{5}, \frac{5}{6}, \frac{6}{7}, \ldots$ is an increasing sequence; that is, show that each term in the sequence is greater than the preceding one.

15. Find an infinite, decreasing sequence (each term is smaller than the preceding one) of positive, rational numbers such that all the terms are greater than 1.

16. ▶(a) Explain why the system of whole numbers does not have the denseness property.
    ▶(b) Explain why the system of integers does not have the denseness property.

17. For each of the following, find two rational numbers between the given fractions:

    (a) $\frac{3}{7}$ and $\frac{4}{7}$ (b) $\frac{-7}{9}$ and $\frac{-8}{9}$
    (c) $\frac{5}{6}$ and $\frac{83}{100}$ (d) $\frac{-1}{3}$ and $\frac{3}{4}$

18. Find the greatest integer x, if one exists, satisfying each of the following:

    (a) $3x < 100$ (b) $\frac{3}{4}x < 100$
    (c) $\frac{3}{4}x < {}^-x + 1$ (d) $\frac{3}{8} < 2x - 13$

19. ▶How would you respond to each of the following students?
    (a) Iris claims that if we have two positive rational numbers, the one with the greatest numerator is the greatest.
    (b) Shirley claims that if we have two positive rational numbers, the one with the greatest denominator is the least.

20. Consider the number grid shown below. The circled numbers form a rhombus (that is, all sides are the same length).
    (a) If A is the sum of the four circled numbers and B is the sum of the four interior numbers, find A/B.
    (b) Form a rhombus by circling the numbers 6, 18, 25, and 37. Compute A and B as in (a); then find A/B.
    (c) How do the answers in (a) and (b) compare? Why does this happen?

1	2	3	4	5	6	7	8	9	10
11	12	13	14	15	16	17	18	19	20
21	22	23	24	25	26	27	28	29	30
31	32	33	34	35	36	37	38	39	40
41	42	43	44	45	46	47	48	49	50

21. The following are the first 4 terms of a sequence of fractions:

    $\dfrac{1+3}{5+7}, \dfrac{1+3+5}{7+9+11}, \dfrac{1+3+5+7}{9+11+13+15}, \dfrac{1+3+5+7+9}{11+13+15+17+19}$

    (a) Write each of the terms in simplest form.
    (b) Write the 100th term for this sequence and predict its value.
    ★(c) Justify your answer in part (b).

22. Show that the arithmetic mean of two rational numbers is between the two numbers; that is, for $0 < \frac{a}{b} < \frac{c}{d}$, prove that $0 < \frac{a}{b} < \frac{1}{2}\left(\frac{a}{b} + \frac{c}{d}\right) < \frac{c}{d}$.

★23. ▶If the same positive number is added to the numerator and denominator of a positive proper fraction, is the new fraction greater than, less than, or equal to the original fraction? Justify your answer.

Review Problems

24. Write each of the following in simplest form:

    (a) $3\frac{5}{8}$ (b) $3\frac{5}{8} \div 2\frac{5}{6}$
    (c) $\dfrac{-5}{12} \div \dfrac{-12}{5}$ (d) $\dfrac{(x-y)^2}{x^2-y^2} \cdot \dfrac{x+y}{x-y}$

25. The distance from Albertson to Florance is $28\frac{3}{4}$ mi. Roberto walks at the rate of $4\frac{1}{2}$ mph. How long will it take him to walk from Albertson to Florance?

26. Solve each of the following for x, and write your answer in simplest form:
    (a) $\frac{-3}{4}x = 1$
    (b) $^-x - \frac{3}{4} = \frac{5}{8}$
    (c) $\frac{3}{4}x = \frac{-2}{3}x + 2$
    (d) $\frac{3}{4}\left(1 - \frac{2}{3}x\right) = \frac{-3}{4}x$

27. A team practiced three times a week for 5 weeks. Each practice lasted $1\frac{3}{4}$ hr. How many hours did the team practice in the 5 weeks?

28. Estimate the following by finding two integers such that the answers are between the integers:
    (a) $\frac{8}{9} + 1\frac{19}{20} + \frac{14}{13}$
    (b) $\frac{7}{15} \cdot \left(16\frac{2}{9} - 6\frac{1}{10}\right)$

Section 6-5 Ratio and Proportion

ratio

One interpretation of rational numbers is as a ratio. For example, there may be a 2-to-3 ratio of Democrats to Republicans on a certain legislative committee, a friend may be given a speeding ticket for driving 63 miles per hour, or eggs may cost 98¢ a dozen. Each of these illustrates a **ratio**. A 1-to-2 ratio of males to females means that the number of males is $\frac{1}{2}$ the number of females, or that there is 1 male for every 2 females. The ratio 1 to 2 can be written as $\frac{1}{2}$ or 1:2. In general, a ratio is denoted by $\frac{a}{b}$ or $a:b$, where $b \neq 0$.

In the 5–8 *Standards* (p. 89), we find the following concerning ratios and proportion:

Ratios should be introduced gradually through discussing the many situations in which they occur naturally. It takes little effort to relate these situations to students' interests: 'If 245 of a company's 398 employees are women, how many of its 26 executives would you expect to be women?'

Through these practical exercises, students should come to recognize that ratios are not directly measurable but they contain two units and that the order of the items in the ratio pair in a proportion is critical. Thus, 23 persons per square mile is very different from 23 square miles per person.

EXAMPLE 6-26 There were 7 males and 12 females in the Dew Drop Inn on Monday evening. In the Game Room, next door, there were 14 males and 24 females.

(a) Express the number of males to females at the Inn as a ratio.
(b) Express the number of males to females at the Game Room as a ratio.

Solution (a) The ratio is $\frac{7}{12}$. (b) The ratio is $\frac{14}{24}$. ∎

proportional

In Example 6-26, the ratios $\frac{7}{12}$ and $\frac{14}{24}$ are equal and proportional to each other. In general, two ratios are **proportional** if, and only if, the fractions representing them are equal.

proportion Two equal ratios form a **proportion.** We know that, for rational numbers, $\frac{a}{b} = \frac{c}{d}$ if, and only if, $ad = bc$. Thus, $\frac{a}{b} = \frac{c}{d}$ is a proportion if, and only if, $ad = bc$. For example, $\frac{14}{24} = \frac{7}{12}$ is a proportion, because $14 \cdot 12 = 24 \cdot 7$.

Frequently, one term in a proportion is missing, as in

$$\frac{3}{8} = \frac{x}{16}.$$

We know that this equation is a proportion if, and only if,

$$3 \cdot 16 = 8 \cdot x$$
$$48 = 8 \cdot x$$
$$6 = x.$$

Another way to solve the equation is to multiply both sides by 16, as follows:

$$\frac{3}{8} \cdot 16 = \frac{x}{16} \cdot 16$$
$$3 \cdot 2 = x$$
$$x = 6$$

It is important to remember that in the ratio $a \div b$, a and b do not have to be integers. For example, if in Eugene, Oregon, $\frac{7}{10}$ of the population exercises regularly, then $\frac{3}{10}$ of the population does not exercise regularly, and the ratio of those who do exercise regularly to those who do not is $\frac{7}{10} \div \frac{3}{10}$, or $\frac{7}{3}$.

The following are examples of problems utilizing ratio and proportion.

E X A M P L E 6 - 2 7   If there should be 3 calculators for every 4 students in an elementary school class, how many calculators are needed for 44 students?

Solution   We use the strategy of *setting up a table,* as shown in Table 6-5:

TABLE 6-5

Number of calculators	3	x
Number of students	4	44

The ratio of calculators to students should always be the same.

$$\frac{\text{Calculators}}{\text{Students}} \longrightarrow \frac{3}{4} = \frac{x}{44}$$
$$3 \cdot 44 = 4 \cdot x$$
$$132 = 4x$$
$$33 = x$$

Thus, 33 calculators are needed.

It is important to notice units of measure when we work with proportions. For example, if a turtle travels 5 in. every 10 sec, how many feet does it travel in 50 sec? If units of measure are ignored, we might set up the following proportion:

$$\frac{5 \text{ in.}}{10 \text{ sec}} = \frac{x \text{ ft}}{50 \text{ sec}}$$

This statement is incorrect. A correct statement must involve the same units in each ratio. We may write the following:

$$\frac{5 \text{ in.}}{10 \text{ sec}} = \frac{x \text{ in.}}{50 \text{ sec}}$$

This implies that $x = 25$ in. Consequently, since 12 in. = 1 ft, the turtle travels $\frac{25}{12}$ ft, or $2\frac{1}{12}$ ft.

EXAMPLE 6-28 Kai, Paulus, and Judy made $2520 for painting a house. Kai worked 30 hr, Paulus worked 50 hr, and Judy worked 60 hr. They divided the money in proportion to the number of hours worked. How much did each earn?

Solution The ratio of hours worked is 30:50:60, or 3:5:6. If we denote the amount of money that Kai received by $3x$, then the amount of money that Paulus received must be $5x$, because then and only then will the ratios of the amounts $3x:5x$ be the same as $3:5$, as required. Similarly, Judy received $6x$. Because the total amount of money received is $3x + 5x + 6x$, we have

$$3x + 5x + 6x = 2520$$
$$14x = 2520$$
$$x = 180.$$

Hence,

Kai received $3x = 3 \cdot 180$, or $540
Paulus received $5x = 5 \cdot 180$, or $900
Judy received $6x = 6 \cdot 180$, or $1080.

To verify that the answers are correct, we find that

$$540 + 900 + 1080 = 2520$$

and that $540:900:1080$ is equivalent to $3:5:6$. ■

Properties of Proportions

Consider the proportion $\frac{15}{30} = \frac{3}{6}$. Because the ratios in the proportion are equal fractions and because equal nonzero fractions have equal reciprocals, it follows that $\frac{30}{15} = \frac{6}{3}$.

THEOREM 6-7

For any rational numbers $\frac{a}{b}$ and $\frac{c}{d}$, with $a \neq 0$ and $c \neq 0$, $\frac{a}{b} = \frac{c}{d}$ if, and only if, $\frac{b}{a} = \frac{d}{c}$.

Suppose Jaffa oranges sell at 7 for $1 in one store and 21 for $3 in another. Which store has a better buy? We see that the price of one orange is $1/7 in the first store and $3/21 in the second. Because $\frac{1}{7} = \frac{3}{21}$, each store charges the same price per orange. Another way to see this is to observe that if 7 oranges cost $1, then 3 times that many oranges should cost 3 times that much. Using ratios, we see that the ratio of the numbers of oranges is the same as the ratio of the prices; that is, $\frac{7}{21} = \frac{1}{3}$. This is true in general and is summarized in the following theorem, whose proof is left as an exercise.

THEOREM 6-8

For any rational numbers $\frac{a}{b}$ and $\frac{c}{d}$, with $c \neq 0$, $\frac{a}{b} = \frac{c}{d}$ if, and only if, $\frac{a}{c} = \frac{b}{d}$.

REMARK In the preceding theorem, it was not necessary to stipulate that $b \neq 0$ and $d \neq 0$, since these are inherent in the definition of rational numbers.

PROBLEM 3

In the Klysler auto factory, robots assemble cars. If 3 robots can assemble 17 cars in 10 min, how many cars can 14 robots assemble in 45 min if all robots work at the same rate all the time?

Understanding the Problem. We are to determine the number of cars that 14 robots can assemble in 45 minutes given that 3 robots can assemble 17 cars in 10 min. If we knew how many cars one robot could assemble in 45 min or how many cars one robot could assemble in 1 min, we could solve the problem.

Devising a Plan. Let x be the number of cars that 14 robots assemble in 45 min. Because the robots work at the same rate, we can express this rate by taking the information that 3 robots assemble 17 cars in 10 min and equating it with the information that 14 robots assemble x cars in 45 min. The rate would be the number of cars (or parts of a car) that 1 robot can assemble in 1 minute. We first need to find the number of cars that 1 robot can assemble in 1 minute. Then, we need to write and solve the desired equation to solve the problem.

Carrying Out the Plan. If 3 robots assemble 17 cars in 10 min, then the 3 robots assemble $\frac{17}{10}$ cars in 1 min. Consequently, 1 robot assembles $\frac{1}{3} \cdot \frac{17}{10}$, or $\frac{17}{30}$, of a car in 1 min. Similarly, if 14 robots assemble x cars in 45 min, then the 14 robots assemble $\frac{x}{45}$ cars in 1 min. Thus 1 robot assembles $\frac{1}{14} \cdot \frac{x}{45}$, or $\frac{x}{14 \cdot 45}$, of a car in 1 min. Because the rates are equal, we have the proportion $\frac{x}{14 \cdot 45} = \frac{17}{30}$. Solving this equation, we obtain $x = 357$, or 357, cars.

Looking Back. The problem can be solved without writing any equations, as follows. Because 1 robot assembles $\frac{17}{30}$ of a car in 1 min, 14 robots assemble $14 \cdot \frac{17}{30}$ cars in 1 min. Thus in 45 min, 14 robots assemble $45 \cdot 14 \cdot \frac{17}{30}$, or 357, cars.

The problem can be varied by changing the data or by considering two kinds of robots, each kind working at a different rate. Similar problems can be constructed concerning other jobs such as painting houses or washing cars. ∎

PROBLEM SET 6-5

1. There are 18 poodles and 12 cocker spaniels in a dog show.
   (a) What is the ratio of poodles to cocker spaniels?
   (b) What is the ratio of cocker spaniels to poodles?
2. In the English alphabet,
   (a) What is the ratio of vowels to consonants?
   (b) Write a word that has a ratio of 2:3 of vowels to consonants.
3. Solve for x in each proportion.
   (a) $\frac{12}{x} = \frac{18}{45}$ (b) $\frac{x}{7} = \frac{-10}{21}$
   (c) $\frac{5}{7} = \frac{3x}{98}$ (d) $3\frac{1}{2}$ is to 5 as x is to 15.
4. There are approximately 2 lb of muscle for every 5 lb of body weight. For a 90-lb child, how much of the weight is muscle?
5. There are five adult drivers to each teenage driver in Aluossim. If there are 12,345 adult drivers in Aluossim, how many teenage drivers are there?
6. If 4 grapefruits sell for 79¢, how much do 6 grapefruits cost?
7. On a map, $\frac{1}{3}$ in. represents 5 mi. If New York and Aluossim are 18 in. apart on the map, what is the actual distance between them?
8. David read 40 pages of a book in 50 min. How many pages should he be able to read in 80 min if he reads at a constant rate?
9. A candle is 30 in. long. After burning for 12 min, the candle is 25 in. long. How long would the whole candle take to burn?
10. Two numbers are in the ratio 3:4. Find the numbers if:
    (a) Their sum is 98 (b) Their product is 768

11. A rectangular yard has width-to-length ratio of 5:9. If the distance around the yard is 2800 ft, what are the dimensions of the yard?
12. Gary, Bill, and Carmella invested in a corporation in the ratio of 2:4:5, respectively. If they divide the profit of $82,000 proportionally to their investment, how much will each receive?
13. Sheila and Dora worked $3\frac{1}{2}$ hr and $4\frac{1}{2}$ hr, respectively, on a programming project. They were paid $176 for the project. How much did each earn?
14. Vonna scored 75 goals in her soccer practice. If her success-to-failure rate is 5:4, how many times did she attempt a goal?
15. The rise and span for a house roof are identified below. The pitch of a roof is the ratio of the rise to the half-span.
    (a) If the rise is 10 ft and the span is 28 ft, what is the pitch?
    (b) If the span is 16 ft and the pitch is $\frac{3}{4}$, what is the rise?

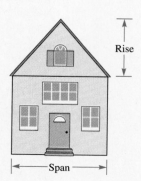

16. A grasshopper can jump 20 times its length. If jumping ability in humans were proportional, how far could a 6-ft-tall person jump?

17. Jim found out that after working for 9 mo he had earned 6 days of vacation time. How many days per year does he earn at this rate?

18. Gear ratios are used in industry. A gear ratio is the comparison of the number of teeth on two gears. When two gears are meshed, the revolutions per minute (rpm) are inversely proportional to the number of teeth; that is,

    $$\frac{\text{rpm of large gear}}{\text{rpm of small gear}} = \frac{\text{number of teeth on small gear}}{\text{number of teeth on large gear}}.$$

    (a) The rpm ratio of the large gear to the small gear is 4:6. If the small gear has 18 teeth, how many teeth does the large gear have?
    (b) The large gear revolves at 200 rpm and has 60 teeth. How many teeth are there on the small gear that has an rpm of 600?

19. A Boeing 747 jet is approximately 230 ft long and has a wingspan of 195 ft. If a scale model of the plane is about 40 cm long, what is the model's wingspan?

20. Jennifer weighs 160 lb on earth and 416 lb on Jupiter. Find Amy's weight on Jupiter if she weighs 120 lb on earth.

21. Iris has found some dinosaur bones and a fossil footprint. She found the length of the footprint to be 40 cm, the length of the thigh bone to be 100 cm, and the length of the body to be 700 cm.
    (a) What is the ratio of the footprint length to the length of the dinosaur?
    (b) Iris found a new track that she believes was made by the same species of dinosaur. If the footprint was 30 cm long and if the same ratio of foot length to body length holds, how long is the dinosaur?
    ▶ (c) In the same area, Iris also found a 50-cm thigh bone. Do you think this thigh bone belonged to the same dinosaur that made the 30-cm footprint that Iris found? Why or why not?

22. Three painters can paint 4 houses in 5 days. How long would it take 7 painters to paint 18 houses if all work was done at the same rate all the time?

23. Suppose that a 10-in. pizza costs $4. To find the price x of a 14-in. pizza, is it correct to set up the proportion $\frac{x}{4} = \frac{14}{10}$? Why or why not?

24. (a) If the ratio of boys to girls in a class is 2:3, what is the ratio of boys to all the students in the class? Why?
    (b) If the ratio of boys to girls in a class is $m:n$, what is the ratio of boys to all the students in the class?

25. If Sherwin can paint the house in 2 days working by himself and William can paint the house in 4 days working by himself, how many days should it take Sherwin and William working together?

★26. If Mary and Carter can paint a house in 5 hr and Mary alone can do the same job in 8 hr, how long would it take Carter working alone?

★27. Prove: For any rational numbers $\frac{a}{b}$ and $\frac{c}{d}$, if $\frac{a}{b} = \frac{c}{d}$, where $a \neq 0$ and $c \neq 0$, then $\frac{b}{a} = \frac{d}{c}$.

★28. Prove that the product of two proper fractions greater than 0 is less than either of the fractions.

29. (a) In Room A of the University Center there are one man and two women; in Room B there are two men and four women; and in Room C there are five men and ten women. If all the people in Rooms B and C go to Room A, what will be the ratio of men to women in Room A?
    ★(b) Prove the following generalization of the proportions used in (a):

    If $\frac{a}{b} = \frac{c}{d} = \frac{e}{f}$, then $\frac{a}{b} = \frac{c}{d} = \frac{e}{f} = \frac{a+c+e}{b+d+f}$.

★30. Prove that, if $\frac{a}{b} = \frac{c}{d}$, then:
    (a) $\frac{a+b}{b} = \frac{c+d}{d}$ $\quad \left(\text{Hint: } \frac{a}{b} + 1 = \frac{c}{d} + 1.\right)$
    (b) $\frac{a}{a+b} = \frac{c}{c+d}$
    (c) $\frac{a-b}{a+b} = \frac{c-d}{c+d}$

★31. Tom beat Dick by $\frac{1}{10}$ mi in a 5-mi race. Dick beat Harry by $\frac{1}{5}$ mi in the race. By how far did Tom beat Harry in the race?

Review Problems

32. Arrange each of the following in increasing order:
    (a) $\frac{-2}{5}, \frac{-3}{5}, 0, \frac{1}{5}, \frac{2}{5}$
    (b) $\frac{7}{12}, \frac{13}{18}, \frac{13}{24}$

33. Find the solution sets for each of the following:
    (a) $\frac{3}{4}x - \frac{5}{8} \geq \frac{1}{2}$
    (b) $\frac{-x}{5} + \frac{1}{10} < \frac{-1}{2}$
    (c) $\frac{-2}{5}(10x + 1) < 1 - x$
    (d) $\frac{2}{3}\left(\frac{1}{2}x - 7\right) \geq \frac{3}{4}x$

34. For each of the following, find three rational numbers between the given fractions:
    (a) $\frac{1}{3}$ and $\frac{2}{3}$ $\quad$ (b) $\frac{-5}{12}$ and $\frac{-1}{18}$

318   CHAPTER 6   RATIONAL NUMBERS AS FRACTIONS

▼ BRAIN TEASER

Janna walks from her home to a friend's house at the same speed every day. One day, without stopping, she walked to her friend's house and back to her own house and then another $\frac{3}{8}$ mi. She found that this took her exactly the same time as it took on another day to walk from her house to the friend's house and then another $\frac{5}{12}$ mi. What is the distance between the houses?

Section 6-6 Exponents Revisited

Recall that for whole numbers a, m, and n, with $a \neq 0$, the following properties hold:

1. $a^m = \underbrace{a \cdot a \cdot a \cdots a}_{m \text{ factors}}$

2. $a^m \cdot a^n = a^{m+n}$

3. $a^0 = 1$, where $a \neq 0$

Property 3 follows from property 2, if we assume that property 2 holds for $m = 0$. If $m = 0$, then $a^m \cdot a^n = a^{m+n}$ becomes $a^0 \cdot a^n = a^{0+n} = a^n$; and 1 is the only number that, on multiplying by a^n, gives a^n. The above notions can be extended for rational-number values of a. For example, consider the following:

$$\left(\frac{2}{3}\right)^4 = \frac{2}{3} \cdot \frac{2}{3} \cdot \frac{2}{3} \cdot \frac{2}{3}$$

$$\left(\frac{2}{3}\right)^2 \cdot \left(\frac{2}{3}\right)^3 = \left(\frac{2}{3} \cdot \frac{2}{3}\right) \cdot \left(\frac{2}{3} \cdot \frac{2}{3} \cdot \frac{2}{3}\right) = \left(\frac{2}{3}\right)^{2+3} = \left(\frac{2}{3}\right)^5$$

Also, $\left(\frac{2}{3}\right)^0 = 1$, and in general, for any nonzero rational number, we have $\left(\frac{a}{b}\right)^0 = 1$.

Exponents can also be extended to negative integers. Notice that as the exponents decrease by 1, the numbers on the right are divided by 10. Thus, the pattern might be continued.

$$10^3 = 10 \cdot 10 \cdot 10$$
$$10^2 = 10 \cdot 10$$
$$10^1 = 10$$
$$10^0 = 1$$
$$10^{-1} = \frac{1}{10} = \frac{1}{10^1}$$
$$10^{-2} = \frac{1}{10} \cdot \frac{1}{10} = \frac{1}{10^2}$$
$$10^{-3} = \frac{1}{10^2} \cdot \frac{1}{10} = \frac{1}{10^3}$$

If the pattern is extended, then we would predict that $10^{-n} = \frac{1}{10^n}$. This is true; and in general, for any nonzero number a, $a^{-n} = \dfrac{1}{a^n}$.

> **REMARK** Another explanation for the definition of a^{-n} is as follows. If the property $a^m \cdot a^n = a^{m+n}$ is to hold for all integer exponents, then $a^{-n} \cdot a^n = a^{-n+n} = a^0 = 1$. Thus a^{-n} is the multiplicative inverse of a^n, and consequently, $a^{-n} = \dfrac{1}{a^n}$.

Consider whether the property $a^m \cdot a^n = a^{m+n}$ can be extended to include all powers of a, where the exponents are integers. For example, is it true that $2^4 \cdot 2^{-3} = 2^{4+^-3} = 2^1$? The definitions of 2^{-3} and the properties of nonnegative exponents ensure that this is true, as shown next.

$$2^4 \cdot 2^{-3} = 2^4 \cdot \frac{1}{2^3} = \frac{2^4}{2^3} = \frac{2^1 \cdot 2^3}{2^3} = 2^1$$

Also, $2^{-4} \cdot 2^{-3} = 2^{-4+^-3} = 2^{-7}$ is true, because

$$2^{-4} \cdot 2^{-3} = \frac{1}{2^4} \cdot \frac{1}{2^3} = \frac{1 \cdot 1}{2^4 \cdot 2^3} = \frac{1}{2^{4+3}} = \frac{1}{2^7} = 2^{-7}.$$

In general, with integer exponents, the following property holds.

PROPERTY

For any nonzero rational number a and any integers m and n, $a^m \cdot a^n = a^{m+n}$.

Other properties of exponents can be developed by using the properties of rational numbers. For example,

$$\frac{2^5}{2^3} = \frac{2^3 \cdot 2^2}{2^3} = 2^2 = 2^{5-3} \qquad \frac{2^5}{2^8} = \frac{2^5}{2^5 \cdot 2^3} = \frac{1}{2^3} = 2^{-3} = 2^{5-8}.$$

With integer exponents, the following property holds.

PROPERTY

For any rational number a such that $a \neq 0$ and for any integers m and n, $\dfrac{a^m}{a^n} = a^{m-n}$.

Suppose a is a nonzero rational number and m and n are positive integers.

$$(a^m)^n = \underbrace{a^m \cdot a^m \cdot a^m \cdots a^m}_{n \text{ factors}} = \overbrace{a^{m+m+\cdots+m}}^{n \text{ terms}} = a^{nm} = a^{mn}$$

Thus $(a^m)^n = a^{mn}$. For example, $(2^3)^4 = 2^{3 \cdot 4} = 2^{12}$.

Does this property hold for negative-integer exponents? For example, does $(2^3)^{-4} = 2^{(3)(-4)} = 2^{-12}$? The answer is yes, because $(2^3)^{-4} = \frac{1}{(2^3)^4} = \frac{1}{2^{12}} = 2^{-12}$. Also, $(2^{-3})^4 = \left(\frac{1}{2^3}\right)^4 = \frac{1}{2^3} \cdot \frac{1}{2^3} \cdot \frac{1}{2^3} \cdot \frac{1}{2^3} = \frac{1^4}{(2^3)^4} = \frac{1}{2^{12}} = 2^{-12}$.

PROPERTY

For any rational number $a \neq 0$ and any integers m and n,
$$(a^m)^n = a^{mn}.$$

Using the definitions and properties developed, we can derive additional properties. Notice, for example, that
$$\left(\frac{2}{3}\right)^4 = \frac{2}{3} \cdot \frac{2}{3} \cdot \frac{2}{3} \cdot \frac{2}{3} = \frac{2 \cdot 2 \cdot 2 \cdot 2}{3 \cdot 3 \cdot 3 \cdot 3} = \frac{2^4}{3^4}.$$
This can be generalized as follows.

PROPERTY

For any nonzero rational number $\frac{a}{b}$ and any integer m,
$$\left(\frac{a}{b}\right)^m = \frac{a^m}{b^m}.$$

Note that, from the definition of negative exponents, the above property, and division of fractions, we have
$$\left(\frac{a}{b}\right)^{-m} = \frac{1}{\left(\frac{a}{b}\right)^m} = \frac{1}{\frac{a^m}{b^m}} = \frac{b^m}{a^m} = \left(\frac{b}{a}\right)^m.$$

Consequently, $\left(\frac{a}{b}\right)^{-m} = \left(\frac{b}{a}\right)^m$.

A property similar to this holds for multiplication. For example,
$$(2 \cdot 3)^{-3} = \frac{1}{(2 \cdot 3)^3} = \frac{1}{2^3 \cdot 3^3} = \left(\frac{1}{2^3}\right) \cdot \left(\frac{1}{3^3}\right) = 2^{-3} \cdot 3^{-3}$$
and in general, it is true that $(a \cdot b)^m = a^m \cdot b^m$ if a and b are rational numbers and m is an integer.

The definitions and properties of exponents are summarized in the following list.

PROPERTIES

Properties of Exponents

1. $a^m = \underbrace{a \cdot a \cdot a \cdots a}_{m \text{ factors}}$, where m is a positive integer
2. $a^0 = 1$, where $a \neq 0$
3. $a^{-m} = \dfrac{1}{a^m}$, where $a \neq 0$
4. $a^m \cdot a^n = a^{m+n}$
5. $\dfrac{a^m}{a^n} = a^{m-n}$, where $a \neq 0$
6. $(a^m)^n = a^{mn}$
7. $\left(\dfrac{a}{b}\right)^m = \dfrac{a^m}{b^m}$, where $b \neq 0$
8. $(ab)^m = a^m \cdot b^m$
9. $\left(\dfrac{a}{b}\right)^{-m} = \left(\dfrac{b}{a}\right)^m$

Observe that all the properties of exponents refer to powers with either the same base or the same exponent. Hence, to evaluate expressions using exponents where different bases or powers are used, perform all the computations or rewrite the expressions in either the same base or the same exponent if possible. For example, $\dfrac{27^4}{81^3}$ can be rewritten as $\dfrac{27^4}{81^3} = \dfrac{(3^3)^4}{(3^4)^3} = \dfrac{3^{12}}{3^{12}} = 1$.

EXAMPLE 6-29 Write each of the following in simplest form, using positive exponents in the final answer:

(a) $16^2 \cdot 8^{-3}$

(b) $20^2 \div 2^4$

(c) $(3x)^3 + 2y^2 x^0 + 5y^2 + x^2 \cdot x$, where $x \neq 0$

(d) $(a^{-3} + b^{-3})^{-1}$

Solution (a) $16^2 \cdot 8^{-3} = (2^4)^2 \cdot (2^3)^{-3} = 2^8 \cdot 2^{-9} = 2^{8+-9} = 2^{-1} = \dfrac{1}{2}$

(b) $\dfrac{20^2}{2^4} = \dfrac{(2^2 \cdot 5)^2}{2^4} = \dfrac{2^4 \cdot 5^2}{2^4} = 5^2$

(c) $(3x)^3 + 2y^2 x^0 + 5y^2 + x^2 \cdot x = 27x^3 + 2y^2 \cdot 1 + 5y^2 + x^3$
$= (27x^3 + x^3) + (2y^2 + 5y^2)$
$= 28x^3 + 7y^2$

(d) $(a^{-3} + b^{-3})^{-1} = \left(\dfrac{1}{a^3} + \dfrac{1}{b^3}\right)^{-1} = \left(\dfrac{b^3 + a^3}{a^3 b^3}\right)^{-1}$
$= \dfrac{1}{\left(\dfrac{a^3 + b^3}{a^3 b^3}\right)} = \dfrac{a^3 b^3}{a^3 + b^3}$

PROBLEM SET 6-6

1. Write each of the following in simplest form, with positive exponents in the final answer:
   (a) $3^{-7} \cdot 3^{-6}$
   (b) $3^7 \cdot 3^6$
   (c) $5^{15} \div 5^4$
   (d) $5^{15} \div 5^{-4}$
   (e) $(^-5)^{-2}$
   (f) $\dfrac{a^2}{a^{-3}}$, where $a \neq 0$
   (g) $\dfrac{a}{a^{-1}}$, where $a \neq 0$
   (h) $\dfrac{a^{-3}}{a^{-2}}$, where $a \neq 0$

2. Write each of the following in simplest form, using positive exponents in the final answer:
   (a) $\left(\dfrac{1}{2}\right)^3 \cdot \left(\dfrac{1}{2}\right)^7$
   (b) $\left(\dfrac{1}{2}\right)^9 \div \left(\dfrac{1}{2}\right)^6$
   (c) $\left(\dfrac{2}{3}\right)^5 \cdot \left(\dfrac{4}{9}\right)^2$
   (d) $\left(\dfrac{3}{5}\right)^7 \div \left(\dfrac{3}{5}\right)^7$
   (e) $\left(\dfrac{3}{5}\right)^{-7} \div \left(\dfrac{5}{3}\right)^4$
   (f) $\left[\left(\dfrac{5}{6}\right)^7\right]^3$

3. If a and b are rational numbers, with $a \neq 0$ and $b \neq 0$, and if m and n are integers, which of the following are true and which are false? Justify your answers.
   (a) $a^m \cdot b^n = (ab)^{m+n}$
   (b) $a^m \cdot b^n = (ab)^{mn}$
   (c) $a^m \cdot b^m = (ab)^{2m}$
   (d) $a^0 = 0$
   (e) $(a+b)^m = a^m + b^m$
   (f) $(a+b)^{-m} = \dfrac{1}{a^m} + \dfrac{1}{b^m}$
   (g) $a^{mn} = a^m \cdot a^n$
   (h) $\left(\dfrac{a}{b}\right)^{-1} = \dfrac{b}{a}$

4. Solve for the integer n in each of the following:
   (a) $2^n = 32$
   (b) $n^2 = 36$
   (c) $2^n \cdot 2^7 = 2^5$
   (d) $2^n \cdot 2^7 = 8$
   (e) $(2+n)^2 = 2^2 + n^2$
   (f) $3^n = 27^5$

5. A human being has approximately 25 trillion ($25 \cdot 10^{12}$) red blood cells, each with an average radius of $4 \cdot 10^{-3}$ mm (millimeters).
   (a) If these cells were placed end to end in a line, how long would the line be, in millimeters?
   (b) If 1 km is 10^6 mm, how long would the line be, in kilometers?

6. Solve each of the following inequalities for x, where x is an integer:
   (a) $3^x \leq 81$
   (b) $4^x < 8$
   (c) $3^{2x} > 27$
   (d) $2^x > 1$

7. Rewrite the following expressions, using positive exponents and expressing all fractions in simplest form:
   (a) $x^{-1} - x$
   (b) $x^2 - y^{-2}$
   (c) $2x^2 + (2x)^2 + 2^2 x$
   (d) $y^{-3} + y^3$
   (e) $\dfrac{3a - b}{(3a - b)^{-1}}$
   (f) $\dfrac{a^{-1}}{a^{-1} + b^{-1}}$

8. Which of the fractions in each pair is greater?
   (a) $\left(\dfrac{1}{2}\right)^3$ or $\left(\dfrac{1}{2}\right)^4$
   (b) $\left(\dfrac{3}{4}\right)^{10}$ or $\left(\dfrac{3}{4}\right)^8$
   (c) $\left(\dfrac{4}{3}\right)^{10}$ or $\left(\dfrac{4}{3}\right)^8$
   (d) $\left(\dfrac{3}{4}\right)^{10}$ or $\left(\dfrac{4}{5}\right)^{10}$
   (e) $\left(\dfrac{4}{3}\right)^{10}$ or $\left(\dfrac{5}{4}\right)^{10}$
   (f) $\left(\dfrac{3}{4}\right)^{100}$ or $\left(\dfrac{3}{4} \cdot \dfrac{9}{10}\right)^{100}$

9. Suppose that the amount of bacteria in a certain culture is given as a function of time by $Q(t) = 10^{10}\left(\dfrac{6}{5}\right)^t$, where t is the time in seconds and $Q(t)$ is the amount of bacteria after t seconds. Find the following:
   (a) The initial number of bacteria (that is, the number of bacteria at $t = 0$)
   (b) The number of bacteria after 2 sec

10. If the nth term of a sequence is given by $a_n = 3 \cdot 2^{-n}$, answer the following:
    (a) Find the first 5 terms.
    (b) Show that the first 5 terms are in a geometric sequence.
    (c) Find the first term that is less than $\dfrac{3}{1000}$.

11. If $f(n) = \dfrac{3}{4} \cdot 2^n$, find the following:
    (a) $f(0)$
    (b) $f(5)$
    (c) $f(^-5)$
    (d) The greatest integer value of n for which $f(n) < \dfrac{3}{400}$

12. (a) Which number is greater, 4^{300} or 3^{400}?
    ▶(b) Justify your answer to (a).
    (c) What happens when you try to evaluate these numbers on a calculator using the $\boxed{y^x}$ key?

13. Which number is greater?
    (a) 32^{50} or 4^{100}
    (b) $(^-27)^{-15}$ or $(^-3)^{-75}$

14. ▶Joe reported that $3^{20} = 3{,}486{,}784{,}406$. Is he correct? Why?

15. What are the last 3 digits of 5^{127}?

Review Problems

16. If a machine produces 6 items every 5 sec, how many items can the machine produce in 3 min?

17. If 3 out of every 80 items are defective, how many defective items are there among 720 items?

18. Find the simplest form for each of the following:
    (a) $\dfrac{24}{84}$
    (b) $\dfrac{12 \cdot 180}{18 \cdot 9}$
    (c) $\dfrac{8^4}{24^4}$
    (d) $\dfrac{13 \cdot 4}{40 \cdot 130}$
    (e) $\dfrac{4}{3} \cdot \dfrac{27}{16}$
    (f) $\dfrac{10^4 \cdot 7^8}{10^6 \cdot 7^6}$
    (g) $\dfrac{3}{4} \div \dfrac{4}{3}$
    (h) $\dfrac{x^3}{x^3 + x^2 y}$

19. Solve for x in each of the following:
    (a) $\dfrac{-3}{4} x = 1$
    (b) $\dfrac{2}{3} x = \dfrac{-3}{5}$

(c) $\frac{1}{3}x - 5 = \frac{-3}{4}x$ (d) $\frac{x}{3} = \frac{-3}{4}$

(e) $\frac{x}{3} = \frac{27}{x}$ (f) $\frac{x+1}{3} = \frac{3}{4}x$

20. If Rachel can paint $\frac{5}{6}$ of a house in 1 day, how long will it take her to paint the whole house?

21. If the ratio of boys to girls in a class is 3 to 8, will the ratio of boys to girls change if 2 new boys and 2 new girls join the class? Justify your answer.

22. Arrange the following in increasing order:

$$\frac{-2}{3}, \frac{-3}{4}, \frac{-6}{7}, \frac{-1}{2}, 0, \frac{4}{5}, \frac{6}{7}, \frac{7}{9}, \frac{9}{7}.$$

Solution to the Preliminary Problem

Understanding the Problem. Daniel runs 8 meters per second while a grizzly bear runs 16 meters per second. Daniel can run from his well to his cabin in 6 sec. We are to determine if the grizzly can catch Daniel before he gets to the cabin if Daniel has a 50-m head start.

Devising a Plan. We find the distance from the well to the cabin by multiplying the speed (8 meters per second) by the time (6 sec) to obtain 48 m. We must determine if Daniel can run 48 m before a grizzly can run (48 + 50), or 98 m.

Carrying Out the Plan. Since the bear can run 16 meters per second, it would take the bear $\frac{98}{16}$, or $6\frac{1}{8}$, seconds to get to the cabin. Therefore, we see that Daniel can beat the bear by $\frac{1}{8}$ sec to the cabin.

Looking Back. We could also think about the problem in the following way. Since the bear is twice as fast as Daniel, it can cover twice as much ground as Daniel in the same time period. Therefore, in the time it takes Daniel to cover 48 m, the bear can cover 96 m, which is not enough to catch Daniel because the bear needs 98 m to reach Daniel. We can alter the problem by changing the speeds of Daniel and/or the bear or by varying the distances.

QUESTIONS FROM THE CLASSROOM

1. A student wrote the solution set to the equation $\frac{x}{7} - 2 < -3$ as $\{-8, -9, -10, -11, \ldots\}$. Is the student correct?

2. A student simplified the fraction $\frac{m+n}{p+n}$ to $\frac{m}{p}$. Is that student correct?

3. Without thinking, one student argued that a pizza cut into 12 pieces was more than a pizza cut into 6 pieces. How would you respond?

4. When working on the problem of simplifying

$$\frac{3}{4} \cdot \frac{1}{2} \cdot \frac{2}{3},$$

a student did the following:

$$\frac{3}{4} \cdot \frac{1}{2} \cdot \frac{2}{3} = \left(\frac{3 \cdot 1}{4 \cdot 2}\right)\left(\frac{3 \cdot 2}{4 \cdot 3}\right) = \frac{3}{8} \cdot \frac{6}{12} = \frac{19}{96}.$$

What was the error?

5. A student asks, "If the ratio of boys to girls in the class is $\frac{2}{3}$ and 4 boys and 6 girls join the class, then the new ratio is $\frac{2+4}{3+6}$, or $\frac{6}{9}$. Since $\frac{2}{3} + \frac{4}{6} = \frac{2+4}{3+6}$, can all fractions be added in the same way?"

6. When the teacher asked the class to solve the equation $\frac{1}{4} + \frac{7}{4}\left(x + \frac{1}{5}\right) = x + \frac{6}{5}$, Nat wrote $\left(\frac{1}{4} + \frac{7}{4}\right)\left(x + \frac{1}{5}\right) = x + \frac{6}{5}$, solved the equation, and got the answer $x = \frac{4}{5}$, which is the correct answer to the original equation. The teacher told Nat that he had obtained the correct answer by using an incorrect method. Nat in turn responded that his method will also work for the equation $\frac{3}{8} + \frac{1}{4}(x - 1) = x - \frac{11}{8}$ and for the equation $1 + \frac{1}{2}(x - \frac{1}{4}) = x + \frac{1}{4}$. How would you respond now if you were the teacher?

7. Is $\frac{0}{6}$ in simplest form? Why or why not?

8. A student says that taking one half of a number is the same as dividing the number by one half. Is this correct?

9. A student writes $\frac{15}{53} < \frac{1}{3}$ because $3 \cdot 15 < 53 \cdot 1$. Another student writes $\frac{1\cancel{5}}{\cancel{5}3} = \frac{1}{3}$. Where is the fallacy?

10. On a test, a student wrote the following:
$$\frac{x}{7} - 2 < -3$$
$$\frac{x}{7} < -1$$
$$x > -7$$
What is the error?

11. A student claims that each of the following is an arithmetic sequence. Is the student right?

   (a) $\frac{1}{2}, \frac{2}{3}, \frac{3}{4}, \frac{4}{5}, \frac{5}{6}, \frac{6}{7}, \frac{7}{8}, \ldots$

   (b) $\frac{1}{2}, \left(\frac{1}{2}\right)^{-2}, \left(\frac{1}{2}\right)^{-5}, \left(\frac{1}{2}\right)^{-8}, \left(\frac{1}{2}\right)^{-11}, \ldots$

12. A student claims that she found a new way to obtain a fraction between two positive fractions: If $\frac{a}{b}$ and $\frac{c}{d}$ are two positive fractions, then $\frac{a+c}{b+d}$ is between these fractions. Is she right?

13. A student claims that if $\frac{a}{b} = \frac{c}{d}$, then $\frac{a+c}{b+d} = \frac{a}{b} = \frac{c}{d}$. Is he right?

14. A student claims that $\frac{1}{x} < \frac{1}{y}$ if, and only if, $x > y$. Is this correct?

15. A student claims that if x is positive, then $\frac{1}{x} > x$. What is your response?

CHAPTER OUTLINE

I. Fractions and rational numbers

   A. Numbers of the form $\frac{a}{b}$, where a and b are integers and $b \neq 0$, are called **rational numbers**.

   B. A rational number can be used as:
   1. A division problem or the solution to a multiplication problem
   2. A partition, or part, of a whole
   3. A ratio
   4. A probability

   C. **Fundamental Law of Fractions:** For any fractions $\frac{a}{b}$ and any number $c \neq 0$, $\frac{a}{b} = \frac{ac}{bc}$.

   D. Two fractions $\frac{a}{b}$ and $\frac{c}{d}$ are **equal** if, and only if, $ad = bc$.

   E. If $GCD(a, b) = 1$, then $\frac{a}{b}$ is said to be in **simplest form**.

   F. If $0 < |a| < |b|$, then $\frac{a}{b}$ is called a **proper fraction**.

II. Addition and subtraction of rational numbers

   A. $\frac{a}{b} + \frac{c}{b} = \frac{a+c}{b}$

   B. $\frac{a}{b} + \frac{c}{d} = \frac{ad+bc}{bd}$

   C. $\frac{a}{b} - \frac{c}{d} = \frac{ad-bc}{bd}$

   D. $\frac{a}{b} \cdot \frac{c}{d} = \frac{ac}{bd}$

   E. $\frac{a}{b} \div \frac{c}{d} = \frac{a}{b} \cdot \frac{d}{c} = \frac{ad}{bc}$, where $c \neq 0$

III. Properties of rational numbers
  A.

	Addition	Subtraction	Multiplication	Division
Closure	Yes	Yes	Yes	Yes, except for division by 0
Commutative	Yes	No	Yes	No
Associative	Yes	No	Yes	No
Identity	Yes	No	Yes	No
Inverse	Yes	No	Yes, except 0	No

  B. Distributive property of multiplication over addition for rational numbers x, y, and z:
  $$x(y + z) = xy + xz.$$
  C. Denseness property: Between any two rational numbers, there is another rational number.

IV. Ratio and proportion
  A. A quotient $a \div b$ is a **ratio**.
  B. A **proportion** is an equation of two ratios.
  C. Properties of proportions
  1. If $\frac{a}{b} = \frac{c}{d}$, then $\frac{b}{a} = \frac{d}{c}$, where $a \neq 0$ and $c \neq 0$.
  2. If $\frac{a}{b} = \frac{c}{d}$, then $\frac{a}{c} = \frac{b}{d}$, where $c \neq 0$.

V. Exponents
  A. $a^m = \underbrace{a \cdot a \cdot a \cdots a}_{m \text{ factors}}$, where m is a positive integer and a is a rational number.
  B. Properties of exponents involving rational numbers
  1. $a^0 = 1$, where $a \neq 0$
  2. $a^{-m} = \frac{1}{a^m}$, where $a \neq 0$
  3. $a^m \cdot a^n = a^{m+n}$
  4. $\frac{a^m}{a^n} = a^{m-n}$, where $a \neq 0$
  5. $(a^m)^n = a^{mn}$
  6. $\left(\frac{a}{b}\right)^m = \frac{a^m}{b^m}$, where $b \neq 0$
  7. $(ab)^m = a^m \cdot b^m$
  8. $\left(\frac{a}{b}\right)^{-m} = \left(\frac{b}{a}\right)^m$

CHAPTER TEST

1. For each of the following, draw a diagram illustrating the fraction:
   (a) $\frac{3}{4}$ (b) $\frac{2}{3}$ (c) $\frac{3}{4} \times \frac{2}{3}$

2. Write three rational numbers equal to $\frac{5}{6}$.

3. Reduce each of the following rational numbers to simplest form:
   (a) $\frac{24}{28}$ (b) $\frac{ax^2}{bx}$ (c) $\frac{0}{17}$
   (d) $\frac{45}{81}$ (e) $\frac{b^2 + bx}{b + x}$ (f) $\frac{16}{216}$

4. Replace the comma with $>$, $<$, or $=$ in each of the following pairs to make a true statement:
   (a) $\frac{6}{10}, \frac{120}{200}$ (b) $\frac{-3}{4}, \frac{-5}{6}$
   (c) $\left(\frac{4}{5}\right)^{10}, \left(\frac{4}{5}\right)^{20}$ (d) $\left(1 + \frac{1}{3}\right)^2, \left(1 + \frac{1}{3}\right)^3$

5. Perform each of the following computations:
   (a) $\frac{5}{6} + \frac{4}{15}$ (b) $\frac{4}{25} - \frac{3}{35}$
   (c) $\frac{5}{6} \cdot \frac{12}{13}$ (d) $\frac{5}{6} \div \frac{12}{15}$
   (e) $\left(5\frac{1}{6} + 7\frac{1}{3}\right) \div \frac{1}{4}$ (f) $\left(-5\frac{1}{6} + 7\frac{1}{3}\right) \div \frac{-9}{4}$

6. Find the additive and multiplicative inverses for each of the following:
   (a) 3 (b) $3\frac{1}{7}$ (c) $\frac{5}{6}$ (d) $-\frac{3}{4}$

7. Order the following numbers from least to greatest:
   $-1\frac{7}{8}$, 0, $-2\frac{1}{3}$, $\frac{69}{140}$, $\frac{71}{140}$, $\left(\frac{71}{140}\right)^{300}$, $\frac{1}{2}$, $\left(\frac{74}{73}\right)^{300}$

8. Simplify each of the following. Write your answer in the form $\frac{a}{b}$, where a and b are integers and $b \neq 0$.
   (a) $\dfrac{\frac{1}{2} - \frac{3}{4}}{\frac{5}{6} - \frac{7}{8}}$ (b) $\dfrac{\frac{3}{4} \cdot \frac{5}{6}}{\frac{1}{2}}$ (c) $\dfrac{\left(\frac{1}{2}\right)^2 - \left(\frac{3}{4}\right)^2}{\frac{1}{2} + \frac{3}{4}}$

9. Solve each of the following for x, where x is a rational number:
   (a) $\frac{1}{4}x - \frac{3}{5} \leq \frac{1}{2}(3 - 2x)$ (b) $\frac{x}{3} - \frac{x}{2} \geq \frac{-1}{4}$
   (c) $\frac{2}{3}\left(\frac{3}{4}x - 1\right) = \frac{2}{3} - x$ (d) $\frac{5}{6} = \frac{4 - x}{3}$

10. ▶Justify the invert-and-multiply algorithm for division of rational numbers.

11. If the ratio of boys to girls in Mr. Good's class is 3 to 5, the ratio of boys to girls in Ms. Garcia's is the same, and you

know that there are 15 girls in Ms. Garcia's class, how many boys are in her class?

12. Write each of the following in simplest form, with non-negative exponents in the final answer:
    (a) $\left(\frac{1}{2}\right)^4 \left(\frac{1}{2}\right)^7$
    (b) $5^{-16} \div 5^4$
    (c) $\left[\left(\frac{2}{3}\right)^7\right]^{-4}$
    (d) $3^{16} \cdot 3^2$

13. John has $54\frac{1}{4}$ yd of material. If he needs to cut the cloth into pieces that are $3\frac{1}{12}$ yd long, how many pieces can be cut? How much material will be left over?

14. Without actually performing the given operations, choose the most appropriate estimation (among the numbers in parentheses) for the given expression.
    (a) $\dfrac{30\frac{3}{8} \cdot 8\frac{1}{3}}{4\frac{1}{9} \cdot 3\frac{8}{9}}$ (15, 20, 8)
    (b) $\left(\dfrac{3}{800} + \dfrac{4}{5000} + \dfrac{15}{6}\right) \cdot 6$ (15, 0, 132)
    (c) $\dfrac{1}{407} \div \dfrac{1}{1609}$ $\left(\dfrac{1}{4}, 4, 0\right)$

15. Find two rational numbers between $\frac{3}{4}$ and $\frac{4}{5}$.

16. Rosa spent $10 of her savings on a record and $\frac{3}{5}$ of what remained on cloth. One fourth of what was left she put back in the bank account and was left with $18. What were Rosa's initial savings?

17. A fast food restaurant sells $4\frac{1}{2}$ times as many hamburgers as hot dogs and $\frac{3}{4}$ as many tacos as hot dogs. What is the ratio of hamburgers sold to tacos sold?

18. ▶Suppose that the ÷ button on your calculator is broken, but the 1/x button works. Explain how you could compute 504792/23.

19. Jim is starting a diet. When he arrived home, he ate $\frac{1}{3}$ of the half of pizza that was left from the previous night. The whole pizza contains approximately 2000 calories. How many calories did Jim consume?

20. Sid can type 1500 words in 20 min. At this rate, how long would it take him to type 1200 words?

21. Joyce has $\frac{1}{4}$ of her college credits in mathematics and $\frac{1}{3}$ of her credits in physics. She has 25 credits outside of math and physics. How many credits does she have in all?

SELECTED BIBLIOGRAPHY

Bennett, A., Jr. and P. Davidson. *Fraction Bars.* Palo Alto, CA: Creative Publications.

Bezuk, N. "Fractions in the Early Childhood Mathematics Curriculum." *Arithmetic Teacher* 35 (February, 1988):56–60.

Collyer, S. "Adding Fractions." *Mathematics Teaching* 116 (September 1986):9.

Edge, D. "Fractions and Panes." *Arithmetic Teacher* 34 (April 1987):13–17.

Ettline, J. "A Uniform Approach to Fractions." *Arithmetic Teacher* 32 (March 1985):42–43.

From the File. "Fractions." *Arithmetic Teacher* 32 (January 1985):43.

From the File. "Fractions Made Easy." *Arithmetic Teacher* 32 (September 1985):39.

Kalman, D. "Up Fractions! Up *n/m*!" *Arithmetic Teacher* 32 (April 1985):42–43.

Kennard, R. "Interpreting Fraction Form." *Mathematics Teaching* 112 (September 1985):46–47.

Lester, F. "Teacher Education: Preparing Teachers To Teach Rational Numbers." *Arithmetic Teacher* 31 (February 1984):54–56.

Malcolm, P. S. "Understanding Rational Numbers." *Mathematics Teacher* 80 (October 1987):518–521.

Ott, J. "A Unified Approach to Multiplying Fractions." *Arithmetic Teacher* 37 (March 1990):47–49.

Payne, J. and A. Towsley. "Implementing the Standards: Implications of NCTM's Standards for Teaching Fractions and Decimals." *Arithmetic Teacher* 37 (April 1990): 23–26.

Post, T. "Fractions and Other Rational Numbers." *Arithmetic Teacher* 37 (September 1989):3, 28.

Post, T. and K. Cramer. "Research into Practice: Children's Strategies in Ordering Rational Numbers." *Arithmetic Teacher* 35 (October 1987):33–35.

Quintero, A. "Helping Children Understand Ratios." *Arithmetic Teacher* 34 (April 1987):17–21.

Quintero, A. "Helping Children Understand Ratios." *Arithmetic Teacher* 34 (May 1987):17–21.

Steiner, E. "Division of Fractions: Developing Conceptual Sense with Dollars and Cents." *Arithmetic Teacher* 34 (May 1987):36–42.

Sweetland, R. "Understanding Multiplication of Fractions." *Arithmetic Teacher* 32 (September 1984):48–52.

Trafton, P., J. Zawojewki, R. Reys, and B. Reys. "Estimation with 'Nice' Fractions." *Mathematics Teacher* 79 (November 1986):629–630.

Van de Walle, J. and C. Thompson. "Fractions with Fraction Strips." *Arithmetic Teacher* 32 (December 1984): 48–52.

7 Decimals

Preliminary Problem

Diane received three discount coupons for 10%, 20%, and 30% off the retail price on certain sale items. On sale items, the store allows customers to combine the coupons in the following way: The customer selects a coupon and the discounted price is calculated. This discounted price is then discounted again when the customer selects a second discount coupon. The process is then repeated a third time. In what order should Diane apply the coupons to maximize her discount: 10%, 20%, and 30%; or 30%, 20%, and 10%; or in some other order? Assuming she uses the order of coupons that will maximize her discount, what will be her final discount off the original price of a sale item?

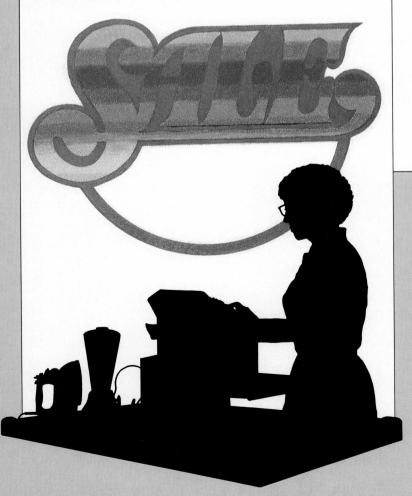

CHAPTER 7 DECIMAL

Although the Hindu-Arabic numeration system discussed in Chapter 3 was perfected around the sixth century, the extension of the system to decimals by the Dutch scientist Simon Stevin did not take place until about a thousand years later. The only significant improvement in the system since Stevin's time has been in notation. Even today there is no universally accepted form of writing a decimal point. For example in the United States, we write 6.75; in England, this number is written as 6 · 75; and in Germany and France, it is written 6,75.

Decimals are a natural extension of our base-ten system. Arithmetic operations on decimals are often much easier than on rational numbers. We emphasize the appropriateness of a given representation depending on the situation, as recommended by the *Standards* (p. 88):

Discussing the appropriateness of certain representations in a given situation, such as the fact that it is better to write "68/100 dollars" on a check than reduce to "17/25 dollars," helps students recognize that there is no single, uniform way to represent a fraction but that the "best" way depends largely on the situation. Students learn, for example, that $\frac{15}{100}$, $\frac{3}{20}$, 0.15, and 15% are all representations of the same number, appropriate for a fraction of a dollar on a bank check, the probability of winning a game, the tax on a purchase of $2.98, and a discount, respectively. Similarly, they learn that +8, $\frac{8}{1}$, and 8.0 are all appropriate representations of the same number, depending on whether they are subtracting integers, adding fractions, or labeling a coordinate axis with rational numbers.

The *Standards* (p. 87) also state that students should

- *understand, represent, and use numbers in a variety of equivalent forms (integer, fraction, decimal, percent, exponential, and scientific notation) in real-world and mathematical problem situations;*
- *understand and apply ratios, proportions, and percents in a wide variety of situations;*
- *investigate relationships among fractions, decimals, and percents.*

Section 7-1 Decimals and Decimal Operations

The word *decimal* comes from the Latin *decem*, meaning ten. Most people first see decimals when dealing with our notation for money. For example, a sign that says a bike costs $128.95 means that the cost is one-hundred twenty-eight whole dollars and some part of a dollar. The dot in $128.95 is called the **decimal point**. Because 95¢ is $\frac{95}{100}$ of a dollar, we have $128.95 = 128 + \frac{95}{100}$ dollars. Because 95¢ is 9 dimes and 5 cents and one dime is $\frac{1}{10}$ of a dollar, and 1 cent is $\frac{1}{100}$ of a dollar, 95¢ is $9 \cdot \frac{1}{10} + 5 \cdot \frac{1}{100}$ of a dollar.

decimal point

Consequently,

$$128.95 = 1 \cdot 10^2 + 2 \cdot 10 + 8 \cdot 1 + 9 \cdot \frac{1}{10} + 5 \cdot \frac{1}{10^2}.$$

The digits in 128.95 correspond to the place value groupings: 10^2, 10, 1, $\frac{1}{10}$, and $\frac{1}{10^2}$, respectively. Each group is $\frac{1}{10}$ of the group to the left. Similarly, 12.61843 represents

$$12 + \frac{6}{10^1} + \frac{1}{10^2} + \frac{8}{10^3} + \frac{4}{10^4} + \frac{3}{10^5} \quad \text{or} \quad 12\frac{61,843}{100,000}.$$

The decimal 12.61843 is read "twelve and sixty-one thousand eight hundred forty-three hundred-thousandths." (The decimal point is read as "and.") Each place to the right of a decimal point may be named by its power of 10. For example, the places of 12.61843 can be named as shown in Table 7-1.

TABLE 7-1

1	2	.	6	1	8	4	3
Tens	Units	And	Tenths	Hundredths	Thousandths	Ten-thousandths	Hundred-thousandths

Decimals can also be introduced with other concrete materials. If we use a set of base-ten blocks and decide that 1 flat will represent 1 unit, then 1 long represents $\frac{1}{10}$ and 1 cube represents $\frac{1}{100}$, as in Figure 7-1(a). In this model, Figure 7-1(b) represents 1.23.

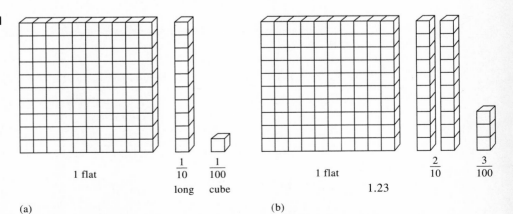

FIGURE 7-1

To represent a decimal such as 4.235, it is convenient to think of a block as a unit. Then a flat represents $\frac{1}{10}$ (one-tenth of a block), a long represents $\frac{1}{100}$, and a cube represents $\frac{1}{1000}$.

HISTORICAL NOTE

In 1584, Simon Stevin (1548–1620) wrote *La Disme*, a work that gave rules for computing with decimals. He not only stated the rules for decimal computations but also suggested practical applications for decimals and recommended that his government adopt the decimal system. Stevin's other achievements include being a quartermaster general for the Dutch army; making contributions in physics to statics and hydrostatics; working with military engineering; and inventing a carriage that carried 28 people and that was propelled by sails and ran along the seashore.

Table 7-2 shows other examples of decimals, their fractional notation, and their common fractional forms.

TABLE 7-2

Decimal	Fractional Notation	Common Fraction
5.3	$5 + \frac{3}{10}$	$5\frac{3}{10}$, or $\frac{53}{10}$
0.02	$0 + \frac{0}{10} + \frac{2}{100}$	$\frac{2}{100}$
2.0103	$2 + \frac{0}{10} + \frac{1}{100} + \frac{0}{1000} + \frac{3}{10,000}$	$2\frac{103}{10,000}$ or $\frac{20,103}{10,000}$
$^-3.6$	$-\left(3 + \frac{6}{10}\right)$	$^-3\frac{6}{10}$, or $-\frac{36}{10}$

Decimals can be written in expanded form using place value and negative exponents. Thus

$$12.61843 = 1 \cdot 10^1 + 2 \cdot 10^0 + 6 \cdot 10^{-1} + 1 \cdot 10^{-2} + 8 \cdot 10^{-3} + 4 \cdot 10^{-4} + 3 \cdot 10^{-5}.$$

To avoid negative exponents, most elementary school texts use fractional notation, as in Table 7-2.

Example 7-1 shows how to convert rational numbers, whose denominators are powers of 10, to decimals.

EXAMPLE 7-1 Convert each of the following to decimals:

(a) $\frac{56}{100}$ (b) $\frac{205}{10,000}$

Solution (a) $\frac{56}{100} = \frac{5 \cdot 10 + 6}{10^2} = \frac{5 \cdot 10}{10^2} + \frac{6}{10^2} = \frac{5}{10} + \frac{6}{10^2} = 0.56$

(b) $\frac{205}{10,000} = \frac{2 \cdot 10^2 + 0 \cdot 10 + 5}{10^4} = \frac{2 \cdot 10^2}{10^4} + \frac{0 \cdot 10}{10^4} + \frac{5}{10^4}$

$= \frac{2}{10^2} + \frac{0}{10^3} + \frac{5}{10^4} = \frac{0}{10^1} + \frac{2}{10^2} + \frac{0}{10^3} + \frac{5}{10^4} = 0.0205$ ∎

We reinforce the ideas in Example 7-1 through use of a calculator. For example, in part (a), press $\boxed{5}\,\boxed{6}\,\boxed{\div}\,\boxed{1}\,\boxed{0}\,\boxed{0}\,\boxed{=}$ and watch the display. Divide by 10 again and look at the new placement of the decimal point. Once more, divide by 10 (which amounts to dividing the original number, 56, by 10,000) and note the placement of the decimal point. This leads to the following general rule for dividing an integer by a power of 10:

To divide an integer by 10^n, count n digits from right to left, annexing zeros if necessary, and insert the decimal point to the left of the nth digit.

The fractions in Example 7-1 are easy to convert to decimals because the denominators are powers of 10. If the denominator of a fraction is not a power of 10, as in $\frac{3}{5}$, we use the problem-solving strategy of *converting the problem to one we already know how to do*. First, we change $\frac{3}{5}$ to a fraction in which the denominator is a power of 10, and then we convert the fraction to a decimal.

$$\frac{3}{5} = \frac{3 \cdot 2}{5 \cdot 2} = \frac{6}{10} = 0.6$$

The reason for multiplying the numerator and the denominator by 2 is apparent when we observe that $10 = 2 \cdot 5$. In general, because $10^n = (2 \cdot 5)^n = 2^n \cdot 5^n$, the prime factorization of the denominator must be $2^n \cdot 5^n$ in order for the denominator of a rational number to be 10^n. We use these ideas to write each fraction in Example 7-2 as a decimal.

EXAMPLE 7-2 Express each of the following as decimals:

(a) $\dfrac{7}{2^6}$ (b) $\dfrac{1}{2^3 \cdot 5^4}$ (c) $\dfrac{1}{125}$ (d) $\dfrac{7}{250}$

Solution (a) $\dfrac{7}{2^6} = \dfrac{7 \cdot 5^6}{2^6 \cdot 5^6} = \dfrac{7 \cdot 15{,}625}{(2 \cdot 5)^6} = \dfrac{109{,}375}{10^6} = 0.109375$

(b) $\dfrac{1}{2^3 \cdot 5^4} = \dfrac{1 \cdot 2^1}{2^3 \cdot 5^4 \cdot 2^1} = \dfrac{2}{2^4 \cdot 5^4} = \dfrac{2}{(2 \cdot 5)^4} = \dfrac{2}{10^4} = 0.0002$

(c) $\dfrac{1}{125} = \dfrac{1}{5^3} = \dfrac{1 \cdot 2^3}{5^3 \cdot 2^3} = \dfrac{8}{(5 \cdot 2)^3} = \dfrac{8}{10^3} = 0.008$

(d) $\dfrac{7}{250} = \dfrac{7}{2 \cdot 5^3} = \dfrac{7 \cdot 2^2}{(2 \cdot 5^3)2^2} = \dfrac{28}{(2 \cdot 5)^3} = \dfrac{28}{10^3} = 0.028$

A calculator can quickly convert fractions to decimals. For example, to find $\dfrac{56}{100}$, press $\boxed{5}\,\boxed{6}\,\boxed{\div}\,\boxed{1}\,\boxed{0}\,\boxed{0}\,\boxed{=}$. To find $\dfrac{7}{2^6}$, press $\boxed{7}\,\boxed{\div}\,\boxed{2}\,\boxed{y^x}\,\boxed{6}\,\boxed{=}$; to convert $\dfrac{1}{125}$ to a decimal, press $\boxed{1}\,\boxed{\div}\,\boxed{1}\,\boxed{2}\,\boxed{5}\,\boxed{=}$, or press $\boxed{1}\,\boxed{2}\,\boxed{5}\,\boxed{1/x}\,\boxed{=}$. The display on some calculators may show $\boxed{8.0 \quad -03}$, which is the calculator's notation for $\dfrac{8}{10^3}$, or $8 \cdot 10^{-3}$. This notation, called *scientific notation*, will be discussed in more detail later in this chapter.

terminating decimals

The answers in Example 7-2 are illustrations of **terminating decimals:** *decimals that can be written with only a finite number of places to the right of the decimal point.* If we attempt to rewrite $\frac{2}{11}$ as a terminating decimal using the method just developed, we first try to find a natural number b such that the following holds:

$$\frac{2}{11} = \frac{2b}{11b}, \quad \text{where } 11b \text{ is a power of } 10.$$

By the Fundamental Theorem of Arithmetic (discussed in Chapter 5), the only prime factors of a power of 10 are 2 and 5. Because $11b$ has 11 as a factor, we cannot write $11b$ as a power of 10, and therefore $\frac{2}{11}$ cannot be written as a terminating decimal. A similar argument using the Fundamental Theorem of Arithmetic holds in general, so we have the following result.

THEOREM 7-1

A rational number $\frac{a}{b}$ in simplest form can be written as a terminating decimal if, and only if, the prime factorization of the denominator contains no primes other than 2 or 5.

EXAMPLE 7-3 Which of the following fractions can be written as terminating decimals?

(a) $\frac{7}{8}$ (b) $\frac{11}{250}$ (c) $\frac{21}{28}$ (d) $\frac{37}{768}$

Solution (a) $\frac{7}{8} = \frac{7}{2^3}$. Because the denominator is 2^3, $\frac{7}{8}$ can be written as a terminating decimal.

(b) $\frac{11}{250} = \frac{11}{2 \cdot 5^3}$. The denominator is $2 \cdot 5^3$, so $\frac{11}{250}$ can be written as a terminating decimal.

(c) $\frac{21}{28} = \frac{21}{2^2 \cdot 7} = \frac{3}{2^2}$. The denominator of the fraction in simplest form is 2^2, so $\frac{21}{28}$ can be written as a terminating decimal.

(d) $\frac{37}{768} = \frac{37}{2^8 \cdot 3}$. This fraction is in simplest form and the denominator contains a factor of 3, so $\frac{37}{768}$ cannot be written as a terminating decimal. ∎

REMARK As Example 7-3(c) shows, to determine whether a rational number $\frac{a}{b}$ can be represented as a terminating decimal, we consider the prime factorization of the denominator *only* if the fraction is in simplest form.

Adding and Subtracting Decimals

 To develop an algorithm for addition of terminating decimals, consider the sum 2.16 + 1.73. We can compute the sum by *changing it to a problem we already know how to solve,* that is, to a sum involving fractions. We then use the commutative and associative properties of addition to complete the computation.

$$2.16 + 1.73 = \left(2 + \frac{1}{10} + \frac{6}{100}\right) + \left(1 + \frac{7}{10} + \frac{3}{100}\right)$$
$$= (2 + 1) + \left(\frac{1}{10} + \frac{7}{10}\right) + \left(\frac{6}{100} + \frac{3}{100}\right)$$
$$= 3 + \frac{8}{10} + \frac{9}{100}$$
$$= 3.89$$

This addition, using fractions, was accomplished by grouping the integers, the tenths, and the hundredths, and adding. Consequently, addition of decimals can be accomplished by lining up the decimal points and adding as with whole numbers.

In elementary school, base-ten blocks are recommended to demonstrate addition of decimals. Figure 7-2 shows how the above addition can be performed.

FIGURE 7-2

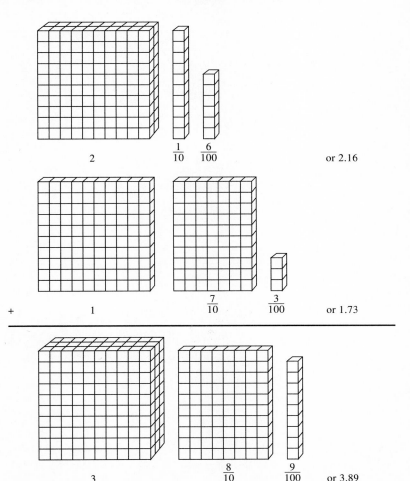

Trading or renaming can be used as in addition of whole numbers to find a sum such as 2.36 + 3.29.

Subtraction of terminating decimals also can be accomplished by lining up the decimal points and subtracting as with whole numbers. The student page from *Addison-Wesley Mathematics,* Grade 5, 1991, shows this.

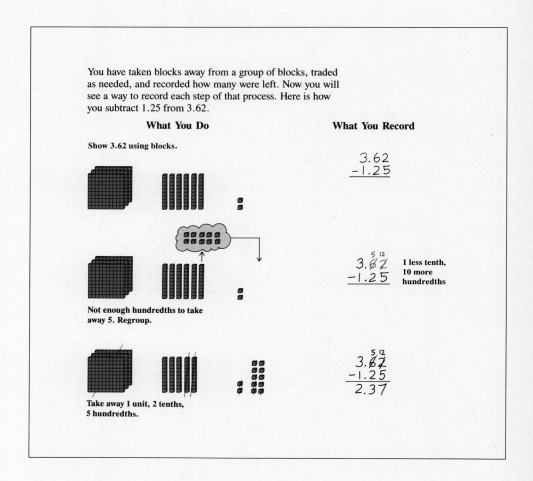

Multiplying Decimals

We can find algorithms for multiplication of terminating decimals by multiplying the corresponding fractions, each in the form $\frac{a}{b}$, where a and b are integers, $b \neq 0$. Consider the product $(4.62)(2.4)$.

$$(4.62)(2.4) = \frac{462}{100} \cdot \frac{24}{10} = \frac{462}{10^2} \cdot \frac{24}{10^1} = \frac{462 \cdot 24}{10^2 \cdot 10^1} = \frac{11{,}088}{10^3} = 11.088$$

Notice that the answer to this computation was obtained by multiplying the whole numbers 462 and 24 and then dividing the result by 10^3.

The algorithm for multiplying decimals can be stated as follows:

If there are n digits to the right of the decimal point in one number and m digits to the right of the decimal point in a second number, multiply the two numbers, ignoring the decimals, and then place the decimal point so that there are n + m digits to the right of the decimal point in the product.

> **REMARK** There are $n + m$ digits to the right of the decimal point in the product because $10^n \cdot 10^m = 10^{n+m}$.

EXAMPLE 7-4 Compute each of the following:

(a) (6.2)(1.43) (b) (0.02)(0.013) (c) (1000)(3.6)

Solution (a)
```
      1.4 3    (2 digits after the decimal point)
   ×  6.2     (1 digit after the decimal point)
      2 8 6
      8 5 8
      8.8 6 6  (3 digits after the decimal point)
```

(b)
```
      0.0 1 3
   ×    0.0 2
      0.0 0 0 2 6
```

(c)
```
         3.6
   ×   1 0 0 0
      3 6 0 0.0
```

> **REMARK** Example 7-4(c) suggests that multiplication by 10^n, where n is a positive integer, results in moving the decimal point in the multiplicand n places to the right.

Dividing Decimals

To develop an algorithm for dividing terminating decimals, first consider $0.96 \div 3$. We can approach this division by rewriting the decimal as a fraction and then dividing.

$$0.96 \div 3 = \frac{96}{100} \div \frac{3}{1} = \frac{96}{100} \cdot \frac{1}{3} = \frac{96 \cdot 1}{100 \cdot 3} = \frac{96}{3} \cdot \frac{1}{100}$$

$$= 32\left(\frac{1}{100}\right) = \frac{32}{100} = 0.32$$

We can also do the computation using the following procedure:

```
       0.32
   3)0.96
       9
       6
       6
       0
```

When the divisor is a whole number, we see that the division can be handled as with whole numbers and the decimal point can be placed directly over the decimal point in the dividend. When the divisor is not a whole number, as in $1.2032 \div 0.32$, we can obtain a whole-number divisor by expressing the quotient as a fraction and then multiplying the numerator and denominator of the fraction by 100.

$$\frac{1.2032}{0.32} = \frac{1.2032 \cdot 100}{0.32 \cdot 100} = \frac{120.32}{32}$$

This corresponds to rewriting the division problem in form (a) as an equivalent problem in form (b).

(a) $0.32 \overline{)1.2032}$ (b) $32 \overline{)120.32}$

In elementary school texts, this process is usually described as "moving" the decimal point two places to the right in both the dividend and the divisor. This process is usually indicated with arrows, as shown next.

```
              3.7 6
   0.3 2 )1.2 0 3 2      Multiply divisor and dividend by 100.
           9 6
           2 4 3
           2 2 4
             1 9 2
             1 9 2
                 0
```

EXAMPLE 7-5 Compute each of the following:

(a) $13.169 \div 0.13$ (b) $9 \div 0.75$

Solution (a)
```
            1 0 1.3              (b)         1 2
   0.1 3 )1 3.1 6 9                  0.7 5 )9.0 0
          1 3                                7 5
            1 6                              1 5 0
            1 3                              1 5 0
              3 9                                0
              3 9
                0
```

Notice that in Example 7-5(b), we annexed two zeros in the dividend because $\frac{9}{0.75} = \frac{9 \cdot 100}{0.75 \cdot 100} = \frac{900}{75}$.

EXAMPLE 7-6 An owner of a gasoline station must collect a gasoline tax of $0.11 on each gallon of gasoline sold. One week, the owner paid $1595 in gasoline taxes. The pump price of a gallon of gas that week was $1.35.

(a) How many gallons of gas were sold during the week?
(b) What was the revenue after taxes for the week?

Solution (a) To find the number of gallons of gas sold during the week, we must divide the total gas tax bill by the amount of the tax per gallon.

$$\frac{1595}{0.11} = 14{,}500$$

Thus 14,500 gallons were sold.

(b) To obtain the revenue after taxes, first determine the revenue before taxes. Then multiply the number of gallons sold by the cost per gallon.

$$(14{,}500)(\$1.35) = \$19{,}575$$

Next, subtract the cost remitted in gasoline taxes from the total revenue.

$$\$19{,}575 - \$1595 = \$17{,}980$$

Thus the revenue after gasoline taxes is $17,980. ∎

Mental Computation

Some of the tools used for mental computations with whole numbers can be used to perform mental computations with decimals as seen in the following:

1. *Breaking and bridging*

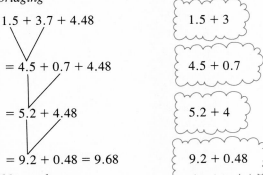

2. *Using compatible numbers*
   (Decimal numbers are compatible when they add to a whole number.)

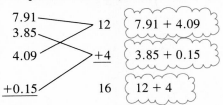

3. *Making compatible numbers*

$$\begin{aligned} 9.27 &= 9.25 + 0.02 \\ +3.79 &= 3.75 + 0.04 \\ &\ 13.00 + 0.06 = 13.06 \end{aligned}$$

4. *Balancing with decimals in subtraction*

$$\begin{array}{rcr} 4.63 = & 4.63 + 0.03 = & 4.66 \\ -1.97 = & -(1.97 + 0.03) = & -2.00 \\ \hline & & 2.66 \end{array}$$

5. *Balancing with decimals in division*

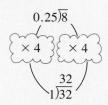

> **REMARK** Balancing with decimals in division uses the property $\dfrac{a}{b} = \dfrac{a \cdot c}{b \cdot c}$.

Calculator Computations

The *Standards* (p. 94), recommends that students *select and use the most appropriate tool*, and adds that they should be prepared to *select and use appropriate mental, paper-and-pencil, calculator, and computer methods*.

For example, if a car has been driven 462.8 miles and in the process it used 11.7 gallons of gas, then the number of miles per gallon is

$$11.7 \overline{)462.8}.$$

This division can be done faster with a calculator than with pencil and paper.

A different type of computation is necessary when a person needs to multiply a governmental budget of $14,329,846,537 by 6 to project the minimum budget for the coming 6 years. Most calculators have only an 8-digit display, but the budget contains 11 digits. If we desire accuracy, we can overcome the problem with a combination of calculator and pencil-and-paper computation. Recognizing that 14,329,846,537 = 14,329,000,000 + 846,537, we have the following:

$$\begin{aligned} 6 \times 14{,}329{,}846{,}537 &= 6(14{,}329{,}000{,}000 + 846{,}537) \\ &= 6 \cdot 14{,}329{,}000{,}000 + 6 \cdot 846{,}537 \\ &= 6 \cdot 14{,}329 \cdot 1{,}000{,}000 + 6 \cdot 846{,}537 \\ &= 85{,}974 \cdot 1{,}000{,}000 + 5{,}079{,}222 \end{aligned}$$

Therefore we could write the computation, using place value, as follows:

$$\begin{array}{r} 14{,}329{,}846{,}537 \\ \times\ 6 \\ \hline 5\ 079\ 222 \\ +85\ 974 \\ \hline 85{,}979{,}079{,}222 \end{array}$$

We see that the distributive property of multiplication over addition and the associative property of multiplication can be used for the multiplication of numbers with more digits than the calculator can accommodate. Similar computations are left as exercises.

PROBLEM SET 7-1

1. Write each of the following in expanded form:
   (a) 0.023
   (b) 206.06
   (c) 312.0103
   (d) 0.000132

2. Rewrite each of the following as decimals:
   (a) $4 \cdot 10^3 + 3 \cdot 10^2 + 5 \cdot 10 + 6 + 7 \cdot 10^{-1} + 8 \cdot 10^{-2}$
   (b) $4 \cdot 10^3 + 6 \cdot 10^{-1} + 8 \cdot 10^{-3}$
   (c) $4 \cdot 10^4 + 3 \cdot 10^{-2}$
   (d) $2 \cdot 10^{-1} + 4 \cdot 10^{-4} + 7 \cdot 10^{-7}$

3. Write each of the following as numerals:
   (a) Five hundred thirty-six and seventy-six ten-thousandths
   (b) Three and eight thousandths
   (c) Four hundred thirty-six millionths
   (d) Five million and two tenths

4. Write each of the following terminating decimals as fractions:
   (a) 0.436
   (b) 25.16
   (c) ⁻316.027
   (d) 28.1902
   (e) ⁻4.3
   (f) ⁻62.01

5. Without performing the actual divisions, determine which of the following represent terminating decimals:
   (a) $\frac{4}{5}$
   (b) $\frac{61}{2^2 \cdot 5}$
   (c) $\frac{3}{6}$
   (d) $\frac{1}{2^5}$
   (e) $\frac{36}{5^5}$
   (f) $\frac{133}{625}$
   (g) $\frac{1}{3}$
   (h) $\frac{2}{25}$
   (i) $\frac{1}{13}$

6. Where possible, write each of the numbers in Problem 5 as terminating decimals.

7. Compute each of the following:
   (a) $36.812 + 0.43 + 1.96$
   (b) $200.01 - 32.007$
   (c) $^-4.612 - 386.0193$
   (d) $(3.61)(0.413)$
   (e) $(^-2.6)(4)$
   (f) $10.7663 \div 2.3$

8. Calculate the following by converting each decimal to a fraction, performing the computation, and then converting the fraction answer to a decimal:
   (a) $13.62 + 4.082$
   (b) $12.62 - 4.082$
   (c) $(1.36)(0.02)$
   (d) $1.36 \div 0.02$

9. If Maura went to the store and bought a chair for $17.95, a lawn rake for $13.59, a spade for $14.86, a lawn mower for $179.98, and two six-packs of mineral water for $2.43 each, what was the bill?

10. ▶Explain why subtraction of terminating decimals can be accomplished by lining up the decimal points, subtracting as if the numbers were whole numbers, and then placing the decimal point in the difference.

11. Write each of the following as a decimal:
    (a) $4.63 \cdot 10^8$
    (b) $0.04 \cdot 10^8$
    (c) $46.3 \cdot 10^8$
    (d) $463.0 \cdot 10^8$
    (e) $0.00463 \cdot 10^8$
    (f) $0.0000000463 \cdot 10^8$
    (g) $4.63 \cdot 10^{-4}$
    (h) $0.04 \cdot 10^{-4}$
    (i) $46.3 \cdot 10^{-4}$
    (j) $0.0000463 \cdot 10^{-4}$
    (k) $4.63 \div 10^{-4}$
    (l) $0.04 \div 10^{-4}$
    (m) $46.3 \div 10^{-4}$
    (n) $0.0000463 \div 10^{-4}$

12. Which of the following divisions are equivalent to $18 \div 2$?
    (a) $20 \overline{) 180}$
    (b) $0.2 \overline{) 0.18}$
    (c) $0.002 \overline{) 0.018}$
    (d) $20 \overline{) 1800}$
    (e) $0.0002 \overline{) 0.00018}$
    (f) $0.2 \overline{) 1.8}$

13. ▶(a) Find the product of 0.22 and 0.35 on the calculator. How does the placement of the decimal point in the answer on the calculator compare with the placement of the decimal point using the rule in this chapter? Explain.
    ▶(b) In a similar manner, investigate placement of the decimal point in the quotient obtained by performing the division $0.2436 \div 0.0006$. Explain your solution.

14. The following are answers to various types of computations. Write an exercise for each answer.
    (a) 86.04 as an addition of two numbers
    (b) 353.76 as an addition of four numbers
    (c) 96.72 as a subtraction of two numbers
    (d) 0.0138 as a multiplication of two numbers
    (e) 0.12 as a subtraction of two numbers
    (f) 2.03 as a division of two numbers

15. At 60°F, 1 qt of water weighs 2.082 lb. One cu ft of water is 29.922 qt. What is the weight of a cubic foot of water, to the nearest thousandth of a pound?

16. Complete the following magic square; that is, make the sum of every row, column, and diagonal the same.

8.2		
3.7	5.5	
	9.1	2.8

17. Keith bought 30 lb of nuts at $3.00/lb. and 20 lb of nuts at $5.00/lb. If he wanted to buy 10 more pounds of a different kind of nut to make the average price per pound equal to $4.50, what price should he pay for the additional 10 lb?

18. A kilowatt hour means that 1000 watts of electricity are being used continuously for 1 hr. The electric utility company in Laura's town charges $0.03715 for each kilowatt hour used. Laura heats her house with three electric wall heaters that use 1200 watts per hour each.
    (a) How much does it cost to heat her house for one day?
    (b) How many hours would a 75-watt light bulb have to stay on to equal $1 in electricity charges?

19. Automobile engines used to be measured in cubic inches but are now usually measured in cubic centimeters. If 2.54 cm is equivalent to 1 in., answer the following:
    (a) Susan's 1963 Thunderbird has a 390-cu in. engine. Approximately how many cubic centimeters is this?
    (b) Dan's 1991 Taurus has a 3000-cm^3 engine. Approximately how many cubic inches is this?

20. Florence Griffith-Joyner set a world record for the women's 100-m dash at the 1988 Summer Olympics in Seoul, South Korea. She covered the distance in 10.49 sec. If 1 m is equivalent to 39.37 in., express Griffith-Joyner's speed in terms of miles per hour.

21. ▶Mary Kim invested $964 in 18 shares of stock. A month later, she sold the 18 shares at $61.48 per share. She also invested in 350 shares of stock for a total of $27,422.50. She sold this stock for $85.35 a share and paid $495 in total commissions. What was Mary Kim's profit or loss on the transactions to the nearest dollar? Explain your solution.

22. Luisa is traveling in Switzerland where the exchange rate is U.S. $1 = 1.56 Swiss francs for cash and U.S. $1 = 1.59 Swiss francs for traveler's checks.
    (a) Luisa is exchanging $235. How many Swiss francs will she get?
    (b) Luisa wants to buy a watch that costs 452.85 Swiss francs and hiking boots that cost 284.65 Swiss francs. What is the minimum number of dollars that Luisa needs to exchange to purchase both?
    ▶(c) Luisa wants to buy a suit that costs 687.75 Swiss francs. She has traveler's checks in U.S. dollars in denominations of $100 and $20. What is the least amount of dollars in traveler's checks she needs to exchange? Explain your solution.

23. Continue the decimal patterns shown below. (Assume each sequence is either arithmetic or geometric.)
    (a) 0.9, 1.8, 2.7, 3.6, 4.5
    (b) 0.3, 0.5, 0.7, 0.9, 1.1
    (c) 1, 0.5, 0.25, 0.125
    (d) 0.2, 1.5, 2.8, 4.1, 5.4

24. 🖩 In doing a calculator addition, a person pressed 42095 as an addend instead of 42.095. Using only one arithmetic operation, how could the person correct this error?

25. 🖩 Describe how the following might be computed on a calculator: If a satellite flies at the rate of 1565 mph, how long would it take to reach the surface of the sun, which is 93,000,000 mi away?

26. 🖩 Compute each of the following:
    (a) 123,456.7894 + 90,876,543.3212
    (b) 123,456.7894 − 90,876,543.3212
    (c) 4,234,567.891 × 36.56
    (d) 123,456,789.51432 ÷ 0.0012

27. 🖩 At a local bank, two different systems are available for charging for checking accounts. System A is a "dime-a-time" plan, as there is no monthly service charge and the charge is 10¢ per check written. System B is a plan with a service charge of 75¢ per month plus 7¢ per check written during that month.
    (a) Which plan is the more economical if an average of 12 checks per month is written?
    (b) Which system is the more economical if an average of 52 checks per month is written?
    ▶(c) What is the break-even point for the number of checks written, that is, the number of checks for which the costs of the two plans are as close as possible? Explain your solution.

28. 🖩 A bank statement from a local bank shows that a checking account has a balance of $83.62. The balance recorded in the checkbook shows only $21.69. After checking the canceled checks against the record of these checks, the customer finds that the bank has not yet recorded six checks in the amounts of $3.21, $14.56, $12.44, $6.98, $9.51, and $7.49. Is the bank record correct? (Assume the person's checkbook records *are* correct.)

29. 🖩 ▶The winner of the big sweepstakes has 15 min to decide whether to receive $1,000,000 cash immediately or to receive 1¢ on the first day of the month, 2¢ on the second day, 4¢ on the third, and so on, each day receiving double the previous day's amount, until the end of a 30-day month. However, only the amount received on that last day may be kept and all the rest of the month's "allowance" must be returned. Use a calculator to find which of these two options is more profitable, and determine how much more profitable one way is than the other. Explain your solution.

★30. ▶Given any reduced rational $\frac{a}{b}$ with $0 < a < b$, where b is of the form $2^m \cdot 5^n$ (m and n are whole numbers), determine a relationship between m and/or n and the number of digits in the terminating decimal. Justify your answer.

▼ B R A I N   T E A S E R

Arrange four 7s, using any operations and decimal points needed to obtain a value of 100.

Section 7-2 Decimals and Their Properties

The division process described in Section 7-1 can be used to develop a procedure for converting any rational number to a decimal. For example, 7/8 can be written as a terminating decimal as follows:

$$\frac{7}{8} = \frac{7}{2^3} = \frac{7 \cdot 5^3}{2^3 \cdot 5^3} = \frac{875}{1000} = 0.875.$$

The decimal for 7/8 can also be found by division, as follows:

```
     0.875
  8)7.000
    6 4
      60
      56
       40
       40
```

Similarly, nonterminating decimals can be obtained for other rational numbers. For example, to find a decimal representation for 2/11, consider the following division:

```
     0.18
  11)2.00
     1 1
       90
       88
        2
```

repeating decimal At this point, if the division is continued, the division pattern repeats, since the remainder 2 is the same as the original dividend. Thus the quotient is 0.181818.... A decimal of this type is called a **repeating decimal,** and the repeating block of digits is called the
repetend **repetend.** The repeating decimal is written as $0.\overline{18}$, where the bar indicates that the block of digits underneath is repeated infinitely.

EXAMPLE 7-7 Convert the following to decimals: (a) $\frac{1}{7}$ (b) $\frac{2}{13}$

Solution (a)
```
      0.142857
   7)1.000000
     7
     30
     28
      20
      14
       60
       56
        40
        35
         50
         49
          1
```
(b)
```
      0.153846
   13)2.000000
      1 3
      70
      65
       50
       39
       110
       104
         60
         52
          80
          78
           2
```

In (a), if the division process is continued at this point, the division pattern repeats; thus $\frac{1}{7} = 0.\overline{142857}$. Therefore in (b) $\frac{2}{13} = 0.\overline{153846}$. ∎

Notice in Example 7-7 that each repetend has 6 digits. Is it possible to predict how long the repetend will be, in general? In $\frac{1}{7}$, the remainders obtained in the division are 3, 2, 6, 4, 5, and 1. These are all the possible nonzero remainders that can be obtained when dividing by 7. (If we had obtained a remainder of 0, the decimal would terminate.) Consequently, the seventh division cannot produce a new remainder. Whenever a remainder recurs, the process repeats itself. Using similar reasoning, we could predict that the repetend for $\frac{2}{13}$ could not be longer than 12, as there are only 12 possible nonzero remainders. However, one of the remainders could repeat sooner than that, which is actually the case. In general, if $\frac{a}{b}$ is any rational number in simplest form with $b > a$ and it does not represent a terminating decimal, the repetend has at most $b - 1$ digits. Therefore, *a rational number may always be represented either as a terminating decimal or as a repeating decimal.*

EXAMPLE 7-8 Use a calculator to convert $\frac{1}{17}$ to a repeating decimal.

Solution In using a calculator, if we press $\boxed{1} \boxed{\div} \boxed{1} \boxed{7} \boxed{=}$, we obtain the following, shown as part of a division problem:

$$\begin{array}{r} 0.0588235 \\ 17\overline{)1} \end{array}$$

Without knowing whether or not the calculator has an internal round-off feature and with an 8-digit display, the greatest number of digits to be trusted in the quotient is 6 following the decimal. (Why?) If we use those 6 places and multiply 0.058823 times 17, we may continue the operation as follows:

$$\boxed{.}\boxed{0}\boxed{5}\boxed{8}\boxed{8}\boxed{2}\boxed{3}\boxed{\times}\boxed{1}\boxed{7}\boxed{=}$$

We then obtain 0.999991, which we may place in the preceding division:

$$\begin{array}{r} 0.058823 \\ 17\overline{)1.000000} \\ \underline{999991} \\ 9 \end{array}$$

We now divide 9 by 17 to obtain 0.5294118. Again ignoring the rightmost digit, we continue as before, completing the division as follows, where the repeating pattern is apparent:

$$\begin{array}{r} 0.05882352941176470588235 \\ 17\overline{)1.00000000000000000000000} \\ \underline{999991} \\ 9000000 \\ \underline{8999987} \\ 13000000 \\ \underline{12999985} \\ 15 \end{array}$$

Thus $\frac{1}{17} = 0.\overline{0588235294117647}$, and the repetend is 16 digits long.

SECTION 7-2 DECIMALS AND THEIR PROPERTIES

We have already considered how to write terminating decimals in the form $\frac{a}{b}$, where $a, b \in I$, $b \neq 0$. For example,

$$0.55 = \frac{55}{10^2} = \frac{55}{100}.$$

To write $0.\overline{5}$ in a similar way, we see that because the repeating decimal has infinitely many digits, there is no single power of 10 that can be placed in the denominator. To overcome this difficulty, we must somehow eliminate the infinitely repeating part of the decimal. Our *subgoal* is to write an equation for n without the repeating part. Suppose that $n = 0.\overline{5}$. It can be shown that $10(0.555\ldots) = 5.555\ldots = 5.\overline{5}$. Hence, $10n = 5.\overline{5}$. Using this information, we subtract to obtain an equation whose solution can be written as a rational number in the form a/b, where a and b are integers and $b \neq 0$.

$$\begin{aligned} 10n &= 5.\overline{5} \\ -n &= -0.\overline{5} \\ \hline 9n &= 5 \\ n &= \frac{5}{9} \end{aligned}$$

Thus $0.\overline{5} = \frac{5}{9}$. This result can be checked by performing the division $5 \div 9$. Notice that performing the subtraction gives an equation containing only integers. (The repeating blocks "cancel" each other.)

Suppose that a decimal has a repetend of more than one digit, such as $0.\overline{235}$. In order to write it in the form $\frac{a}{b}$, it is reasonable to multiply by 10^3, since there is a three-digit repetend. Let $n = 0.\overline{235}$. Our *subgoal* is again to write an equation for n without the repeating part of the decimal.

$$\begin{aligned} 1000n &= 235.\overline{235} \\ -n &= -0.\overline{235} \\ \hline 999n &= 235 \\ n &= \frac{235}{999} \end{aligned}$$

Hence, $0.\overline{235} = \frac{235}{999}$.

Notice that $0.\overline{5}$ repeats in blocks of one digit, and therefore, to write it in the form $\frac{a}{b}$, we first multiply by 10^1; $0.\overline{235}$ repeats in blocks of three digits, and therefore we first multiply by 10^3. In general, *if the repetend is immediately to the right of the decimal point, first multiply by 10^n, where n is the number of digits in the repetend, and then continue as in the preceding cases*.

Now, suppose that the repeating block does *not* occur immediately after the decimal point. For example, let $n = 2.3\overline{45}$. A strategy for solving this problem is to *change it to a related problem* we already know how to do; that is, change it to a problem where the repeating block immediately follows the decimal point. This is our new *subgoal*. To accomplish this, we multiply both sides by 10.

$$\begin{aligned} n &= 2.3\overline{45} \\ 10n &= 23.\overline{45} \end{aligned}$$

We now proceed as with previous problems. Because $10n = 23.\overline{45}$ and the number of digits in the repetend is 2, we multiply by 10^2 as follows:

$$100(10n) = 2345.\overline{45}$$

Thus

$$\begin{aligned} 1000n &= 2345.\overline{45} \\ -10n &= -23.\overline{45} \\ \hline 990n &= 2322 \\ n &= \frac{2322}{990}, \text{ or } \frac{387}{165} \end{aligned}$$

Hence, $2.3\overline{45} = \frac{2322}{990}$, or $\frac{387}{165}$.

To find the $\frac{a}{b}$ form of $0.\overline{9}$, we proceed as follows:

(1) $n = 0.\overline{9}$
(2) $10n = 9.\overline{9}$ (Multiply both sides of Eq. (1) by 10.)
(3) $9n = 9$ (Subtract Eq. (1) from Eq. (2).)
      $n = 9/9$, or 1   (Solve Eq. (3) for n.)

Hence, $0.\overline{9} = 1$. This approach to the problem may not be convincing. Another approach to show that $0.\overline{9}$ is really another name for 1 is shown next.

(1) $\frac{1}{3} = 0.33333333\ldots$

(2) $\frac{2}{3} = 0.66666666\ldots$

Adding Eqs. (1) and (2), we have $1 = 0.99999999\ldots$, or $0.\overline{9}$. Can you show that $4.\overline{9} = 5$? Deciding whether or not $0.\overline{9} = 1$ hinges on understanding the meaning of the decimal $0.\overline{9}$. This decimal represents the infinite sum $\frac{9}{10} + \frac{9}{10^2} + \frac{9}{10^3} + \ldots$. Such sums are defined as the limits of finite sums in more advanced mathematics courses.

Ordering Decimals

To find which of two given decimals is greater, we could convert each to rational numbers in the form $\frac{a}{b}$, where a and b are integers, and determine which is greater. For example, because $0.36 = \frac{36}{100}$ and $0.9 = 0.90 = \frac{90}{100}$ and $\frac{90}{100} > \frac{36}{100}$, it follows that $0.9 > 0.36$. One could also tell that $0.9 > 0.36$ because $\$0.90$ is 90¢ and $\$0.36$ is 36¢. This suggests how to order decimals without conversion to fractions. Another method is the following. Because $0.9 = 0.90$, we can line up the decimal points as follows:

$$\begin{aligned} 0.36 \\ 0.90 \end{aligned}$$

The digit in the tenths place of 0.36 is less than the tenths digit in 0.90, so 0.36 < 0.90. A similar procedure works for repeating decimals.

For example, to compare repeating decimals, such as $1.\overline{3478}$ and $1.3\overline{47821}$, we write the decimals one under the other, in their equivalent forms without the bars, and line up the decimal points.

$$1.34783478\ldots$$
$$1.34782178\ldots$$

The digits to the left of the decimal points and the first four digits after the decimal points are the same in each of the numbers. However, since the digit in the hundred-thousandths place of the top number, which is 3, is greater than the digit 2 in the hundred-thousandths place of the bottom number, $1.\overline{3478}$ is greater than $1.3\overline{47821}$.

It is easy to compare two fractions, such as $\frac{21}{43}$ and $\frac{37}{75}$, using a calculator. We convert each to a decimal and then compare the decimals.

$$\boxed{2}\,\boxed{1}\,\boxed{\div}\,\boxed{4}\,\boxed{3}\,\boxed{=} \longrightarrow 0.4883721$$
$$\boxed{3}\,\boxed{7}\,\boxed{\div}\,\boxed{7}\,\boxed{5}\,\boxed{=} \longrightarrow 0.4933333$$

Examining the digits in the hundredths place, we see that

$$\frac{37}{75} > \frac{21}{43}.$$

EXAMPLE 7-9 Find a rational number in decimal form between $0.\overline{35}$ and $0.\overline{351}$.

Solution First, line up the decimals.

$$0.353535\ldots$$
$$0.351351\ldots$$

To find a decimal between these two, observe that starting from the left, the first place at which the two numbers differ is the thousandths place. Clearly, one decimal between these two is 0.352. Others include 0.3514, $0.35\overline{15}$, and 0.35136. In fact, there are infinitely many others.

Rounding Decimals

Frequently, it is not necessary to know the exact numerical answer to a question. For example, if we want to know the distance to the moon or the population of New York City, the approximate answers of 239,000 miles and 7,800,000 people, respectively, may be adequate.

In order to approximate numbers, we adopt rules for rounding. The rounding rules given in the following flowchart are those used in most elementary schools and thus are the ones used in this text. As an example, directly below the flowchart, 0.867 is rounded to the nearest tenth.

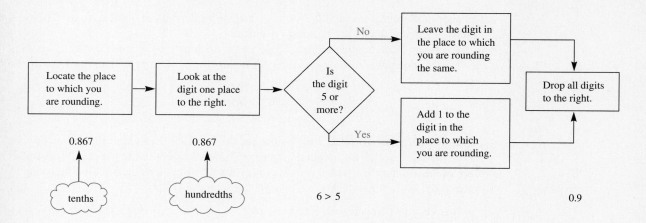

Therefore 0.867 rounded to the nearest tenth is 0.9, and we write $0.867 \doteq 0.9$ to symbolize the approximation. (Rounding rules often vary at the high school and college levels when the digit to the right of the last digit to be retained is 5.) The following rules, although *not* used in this text, are frequently used elsewhere. If the digit to the right of the last digit to be retained is 5, then increase by 1 the last digit to be retained when there is at last one nonzero digit to the right of the 5; when there is no nonzero digit to the right of the 5, increase by 1 the last digit to be retained only if the last digit is odd.

EXAMPLE 7-10 Round each of the following numbers:

(a) 7.456 to the nearest hundredth
(b) 7.456 to the nearest tenth
(c) 7.456 to the nearest unit
(d) 7456 to the nearest thousand
(e) 745 to the nearest ten
(f) 74.56 to the nearest ten

Solution
(a) $7.456 \doteq 7.46$ (b) $7.456 \doteq 7.5$
(c) $7.456 \doteq 7$ (d) $7456 \doteq 7000$
(e) $745 \doteq 750$ (f) $74.56 \doteq 70$

Estimating in Decimal Computations

Rounded numbers can be useful for estimating answers to computations. For example, consider each of the following:

1. Karly goes to the grocery store to buy items that cost the following amounts. She can estimate the total cost by rounding each amount to the nearest dollar and adding the

rounded numbers.

$$
\begin{aligned}
\$2.39 &\longrightarrow \$2 \\
0.89 &\longrightarrow 1 \\
6.13 &\longrightarrow 6 \\
4.75 &\longrightarrow 5 \\
+\ 5.05 &\longrightarrow \underline{\ 5\ } \\
& \quad\quad\ \$19
\end{aligned}
$$

Thus Karly's estimate for her grocery bill is $19.

2. Karly's bill for car repairs was $72.80, and she has a coupon for $17.50 off. She can estimate her total cost by rounding each amount to the nearest ten dollars and subtracting.

$$
\begin{array}{rr}
\$72.80 & \$70 \\
-\ 17.50 & -\ 20 \\ \hline
 & \$50
\end{array}
$$

Thus an estimate for the repair bill is $50.

3. Karly sees a flash of lightning and hears the thunder 3.2 sec later. She knows that sound travels at 0.33 km/sec. She may estimate the distance she is from the lightning by rounding the time to the nearest unit and the speed to the nearest tenth and multiplying.

$$
\begin{array}{rr}
0.33 \longrightarrow & 0.3 \\
\times\ 3.2 \longrightarrow & \times\ 3 \\ \hline
 & 0.9
\end{array}
$$

Thus Karly estimates that she is approximately 0.9 km from the lightning.

An alternate approach is to recognize that $0.33 \doteq \frac{1}{3}$ and 3.2 is close to 3.3, so an approximation using compatible numbers is $\left(\frac{1}{3}\right) \cdot 3.3$, or 1.1 km.

4. Karly wants to estimate the cost per kilogram of a frozen turkey that sells for $17.94 and weighs 6.42 kg. She rounds and divides as follows:

$$
6.42 \overline{)17.94} \longrightarrow 6\overline{)18.00}^{\,3.00}
$$

Thus the turkey sells for approximately $3.00/km.

When arithmetic operations are performed on rounded numbers, the results may be further off from the true situation. For example, suppose the distance, rounded to a tenth of a mile, along I-5 from Eugene to the first Albany exit is 42.6 mi, whereas the distance, rounded to the nearest tenth of a mile, from that exit to the first Salem exit is 22.4 mi. How far is it from Eugene to the first Salem exit? It seems that the answer is 42.6 + 22.4, or 65, mi. But how accurate is this answer? The distances might have been more accurately recorded as 42.55 and 22.35, when the sum would have been 64.9, or they may have been recorded as 42.64 and 22.44, when the sum would be 65.08, or 65.1 rounded to the nearest tenth. Thus the calculated sum of 65 miles could actually be 0.1 mi off in either direction. Similar errors may arise in other arithmetic operations.

Scientific Notation

scientific notation   When either very large or very small numbers are used, a special notation called **scientific notation** is common. For example, "the sun is 93,000,000 mi from Earth" is expressed as "the sun is $9.3 \cdot 10^7$ mi from Earth." A micrometer, a metric unit of measure that is 0.000001 m, is written as $1 \cdot 10^{-6}$ m.

> **DEFINITION OF SCIENTIFIC NOTATION**
>
> In **scientific notation,** a positive number is written as the product of a number greater than or equal to 1 and less than 10, and an integral power of 10.

EXAMPLE 7-11   Write each of the following in scientific notation:
(a) 413,682,000   (b) 0.0000231
(c) 83.7   (d) 10,000,000

Solution   (a) $413{,}682{,}000 = 4.13682 \cdot 10^8$ (b) $0.0000231 = 2.31 \cdot 10^{-5}$
(c) $8.37 \cdot 10^1$ (d) $1 \cdot 10^7$

EXAMPLE 7-12   Convert each of the following to standard numerals:
(a) $6.84 \cdot 10^{-5}$ (b) $3.12 \cdot 10^7$

Solution   (a) $6.84 \cdot 10^{-5} = 6.84 \cdot \left(\dfrac{1}{10^5}\right) = 0.0000684$

(b) $3.12 \cdot 10^7 = 31{,}200{,}000$

significant digits   In Example 7-12(a), the number $6.84 \cdot 10^{-5}$ is in scientific notation. The digits in 6.84 are called **significant digits.** Notice that if the number $6.84 \cdot 10^{-5}$ were written out in standard form, it would be 0.0000684. Even though there are eight digits in the number in standard form, only three are considered significant. Also consider the number 684,000. Written in scientific notation, this number is $6.84 \cdot 10^5$, and it too has only three significant digits. Because $604.3 = 6.043 \cdot 10^2$, it has four significant digits.

Numbers in scientific notation are easy to manipulate using the laws of exponents. For example, $(5.6 \cdot 10^5) \cdot (6 \cdot 10^4)$ can be rewritten as $(5.6 \cdot 6) \cdot (10^5 \cdot 10^4) = 33.6 \cdot 10^9$, which is $3.36 \cdot 10^{10}$ in scientific notation. Also,

$$(2.35 \cdot 10^{-15}) \cdot (2 \cdot 10^8) = (2.35 \cdot 2) \cdot (10^{-15} \cdot 10^8) = 4.7 \cdot 10^{-7}.$$

To investigate how your calculator handles scientific notation, consider the computation $41{,}368{,}200 \times 1000$. The answer is 41,368,200,000, or 4.13682×10^{10}. On a calculator, perform the following computation:

$$\boxed{4}\,\boxed{1}\,\boxed{3}\,\boxed{6}\,\boxed{8}\,\boxed{2}\,\boxed{0}\,\boxed{0}\,\boxed{\times}\,\boxed{1}\,\boxed{0}\,\boxed{0}\,\boxed{0}\,\boxed{=}$$

On many calculators, the display will read

$$\boxed{4.1368 \quad 10}$$

The number to the right of the space is the power to which 10 is raised when the number is written in scientific notation.

Calculators with an $\boxed{EE}$ key can be used to represent numbers in scientific notation.

For example, to find $5.2 \cdot 10^{16} \cdot 9.37 \cdot 10^4$, press

$\boxed{5}\boxed{.}\boxed{2}\boxed{EE}\boxed{1}\boxed{6}\boxed{\times}\boxed{9}\boxed{.}\boxed{3}\boxed{7}\boxed{EE}\boxed{4}\boxed{=}$

PROBLEM SET 7-2

1. Find the decimal representation for each of the following:
   (a) $\frac{4}{9}$ (b) $\frac{2}{7}$ (c) $\frac{3}{11}$ (d) $\frac{1}{15}$
   (e) $\frac{2}{75}$ (f) $\frac{1}{99}$ (g) $\frac{5}{6}$ (h) $\frac{1}{13}$

2. 🔲 This group of exercises concentrates on finding repeating decimals.
   (a) Use a calculator to find decimals for each of the following:
       (i) $\frac{1}{7}$ (ii) $\frac{2}{7}$
       (iii) $\frac{3}{7}$ (iv) $\frac{4}{7}$
       (v) $\frac{5}{7}$ (vi) $\frac{6}{7}$
   (b) How many places were used before each decimal repeated?
   ▶(c) Describe any relationship among your answers in (i)–(vi) of (a).

3. Using the answers obtained in Problem 2, add the numbers in the first half of each repetend to the numbers in the last half of each repetend. For example, $\frac{1}{7} = 0.\overline{142857}$, so we add $142 + 857$.
   (a) What do you discover?
   (b) Try the same experiment with $\frac{5}{13}$. Do you obtain the same type of result?
   ▶(c) Make a conjecture about the sums of numbers formed by the halves of the repetend of a repeating decimal.
   (d) Will this conjecture work on the repeating decimal for $\frac{1}{3}$?

4. 🔲 Find repeating decimals for each of the following:
   (a) $\frac{1}{13}$ (b) $\frac{1}{21}$ (c) $\frac{3}{19}$

5. Convert each of the following repeating decimals to fractions:
   (a) $2.4\overline{5}$ (b) $2.\overline{45}$
   (c) $2.4\overline{54}$ (d) $0.2\overline{45}$
   (e) $0.02\overline{45}$ (f) $-24.\overline{54}$
   (g) $0.\overline{4}$ (h) $0.\overline{6}$
   (i) $0.5\overline{5}$ (j) $0.\overline{34}$
   (k) $-2.\overline{34}$ (l) $-0.\overline{02}$

6. Order each of the following decimals from greatest to least:
   (a) $3.2, 3.\overline{22}, 3.\overline{23}, 3.2\overline{3}, 3.23$
   (b) $-1.454, -1.45\overline{4}, -1.45, -1,4\overline{54}, -1.\overline{454}$

7. Find a decimal between each of the following pairs of decimals:
   (a) 3.2 and 3.3   (b) 462.24 and 462.25
   (c) $462.2\overline{4}$ and $462.\overline{24}$ (d) 0.003 and 0.03

8. Find the decimal halfway between the two given decimals.
   (a) 3.2 and 3.3   (b) 462.24 and 462.25
   (c) 0.0003 and 0.03   (d) $462.2\overline{4}$ and $462.\overline{24}$

9. Some digits in the number shown below have been covered by squares. If each of the digits 1 through 9 is used exactly once in the number, what is the number in each of the following cases?

   $4\square3\square.\square8\square$

   (a) The number is as great as possible.
   (b) The number is the least possible.

10. Round each of the following numbers as specified:
    (a) 203.651 to the nearest hundred
    (b) 203.651 to the nearest ten
    (c) 203.651 to the nearest unit
    (d) 203.651 to the nearest tenth
    (e) 203.651 to the nearest hundredth

11. Jane's car travels 224 mi on 12 gal of gas. How many miles to the gallon does her car get, rounded to the nearest mile?

12. ▦ Sooner or later, most people are faced with the task of buying a number of items at the store, knowing that they have just barely enough money to cover the needed items. In order to avoid the embarrassment of coming up short and having to put some of the items back, they must use their estimating or rounding skills. For each of the following sets of items, estimate the cost. If you are short of funds, tell what must be returned to come just under the allotted amount to be spent. Use your calculator to check your answers.

    (a) Amount on hand: $2.98
    2 packs of gum at 24¢ each
    3 lollipops at 10¢ each
    1 licorice at 4 for 20¢
    1 soft drink at 75¢
    1 pack dental floss at 99¢

    (b) Amount on hand: $20.00
    7 gal of gas at $1.089/gal
    2 qt of oil at $1.05/qt
    A car wash at $1.99
    Flashlight batteries at $3.39
    Air in a tire at 0¢/lb
    Air freshener at 99¢
    Starter fluid at 99¢
    Parking ticket at $1.00
    Windshield wiper at $1.59
    Soft drink for your date at 75¢

13. Audrey wants to buy some camera equipment to take pictures on her daughter's birthday. To estimate the total cost, she rounds each price to the nearest dollar and adds the rounded prices. What is her estimate for the items listed?

    Camera $24.95
    Film $3.50
    Case $7.85

14. Estimate the sum or difference in each of the following by using (i) rounding and (ii) front-end estimation. Then perform the computations to see how close your estimates are to the actual answers.

    (a) 65.84   (b) 89.47
         24.29   −32.16
         12.18
        +19.75

    (c) 5.85   (d) 223.75
         6.13   − 87.60
         9.10
        +4.32

15. Express the following numbers in scientific notation:
    (a) 3325   (b) 46.32
    (c) 0.00013   (d) 930,146

16. Convert each of the following numbers to standard numerals:
    (a) $3.2 \cdot 10^{-9}$ (b) $3.2 \cdot 10^9$
    (c) $4.2 \cdot 10^{-1}$ (d) $6.2 \cdot 10^5$

17. Write the numerals in each of the following sentences in scientific notation:
    (a) The diameter of the Earth is about 12,700,000 m.
    (b) The distance from Pluto to the sun is 5,797,000 km.
    (c) Each year, about 50,000,000 cans are discarded in the United States.

18. Write the numerals in each sentence in standard form.
    (a) A computer requires $4.4 \cdot 10^{-6}$ sec to do an addition problem.
    (b) There are about $1.99 \cdot 10^4$ km of coastline in the United States.
    (c) The Earth has existed for approximately $3 \cdot 10^9$ yr.

19. Write the results of each of the following in scientific notation:
    (a) $(8 \cdot 10^{12}) \cdot (6 \cdot 10^{15})$
    (b) $(16 \cdot 10^{12}) \div (4 \cdot 10^5)$
    (c) $(5 \cdot 10^8) \cdot (6 \cdot 10^9) \div (15 \cdot 10^{15})$

20. ▶Which of the following numbers is the greatest: $100,000^3$; 1000^5; $100,000^2$? Justify your answer.

21. The speed of light is approximately 186,000 mi/sec. It takes light from the nearest star, Alpha Centauri, approximately 4 yr to reach the Earth. How many miles away is Alpha Centauri from the Earth? Express the answer in scientific notation.

22. ▦ ▶Use a calculator to find $\frac{26}{99}$ and $\frac{78}{99}$. Can you predict a decimal value for $\frac{51}{99}$? Will the technique used in your prediction always work? Explain why or why not?

23. Continue the following decimal patterns:
    (a) $0, 0.\overline{3}, 0.\overline{6}, 1, 1.\overline{3}$
    (b) $0, 0.5, 0.\overline{6}, 0.75, 0.8, 0.8\overline{3}$

★24. Suppose that $a = 0.\overline{32}$ and $b = 0.\overline{123}$.
    (a) Find $a + b$ by adding from left to right. How many digits are in the repetend of the sum?
    (b) Find $a + b$ if $a = 1.2\overline{34}$ and $b = 0.\overline{1234}$. Is the answer a rational number? How many digits are in the repetend?

Review Problems

25. (a) Human bones make up 0.18 of a person's total body weight. How much do the bones of a 120-lb person weigh?
    (b) Muscles make up about 0.4 of a person's body weight. How much do the muscles of a 120-lb person weigh?

26. John is a payroll clerk for a small company. Last month, the employees' gross earnings (earnings before deductions) totaled $27,849.50. John deducted $1520.63 for social security, $723.30 for unemployment insurance, and $2843.62 for federal income tax. What was the employees' net pay (their earnings after deductions)?

27. ▶How can you tell whether a fraction will represent a terminating decimal without performing the actual division?

28. Write each of the following decimals as fractions:
    (a) 16.72 (b) 0.003
    (c) ⁻5.07 (d) 0.123

● COMPUTER CORNER

The following BASIC program rounds decimals to a given number of places. See if it gives the same results as the rules identified in this chapter.

```
10 PRINT "THIS PROGRAM ROUNDS DECIMALS."
20 PRINT
30 PRINT "ENTER YOUR DECIMAL AND PRESS RETURN."
40 INPUT D
50 PRINT "HOW MANY DIGITS WOULD YOU LIKE TO THE "
55 PRINT "RIGHT OF THE DECIMAL POINT?"
60 INPUT N
70 LET S = INT (D * 10 ^ N + .5)
80 LET R = S / (10 ^ N)
90 PRINT
100 PRINT D ; " ROUNDS TO "; R
110 PRINT
120 PRINT "TO ENTER ANOTHER DECIMAL: TYPE RUN."
130 END
```

Section 7-3 Percents

percent

Percents are very useful in conveying information. People hear that there is a 60 percent chance of rain or that their savings accounts are drawing 6 percent interest. The word **percent** comes from the Latin phrase *per centum,* which means *per hundred*. For example, a bank that pays 6 percent simple interest on a savings account pays $6 for each $100 in the account for one year; that is, it pays $\frac{6}{100}$ of whatever amount is in the account for one year. We use the symbol % to indicate percent and, for example, write 6% for $\frac{6}{100}$.

In general, we have the following definition.

> **DEFINITION OF PERCENT**
>
> $$n\% = \frac{n}{100}$$

Thus $n\%$ of a quantity is $\frac{n}{100}$ of the quantity. Therefore 1% is one hundredth of a whole and 100% represents the entire quantity, whereas 200% represents $\frac{200}{100}$, or 2 times, the given quantity. Percents can be illustrated by using a hundreds grid. For example, what percent of the grid is shaded in Figure 7-3? Because 30 out of the 100, or $\frac{30}{100}$, of the squares are shaded, we say that 30% of the grid is shaded.

FIGURE 7-3

We can convert any number to a percent by first writing the number as a fraction with denominator 100. For instance, consider the example in the cartoon that follows. Obviously, the adult is incorrect.

BORN LOSER reprinted by permission of NEA, Inc.

The child in the cartoon missed 6 questions out of 10 and hence had 4 correct answers; therefore, $\frac{4}{10}$ of the answers were correct. Because $\frac{4}{10} = \frac{40}{100}$, the child had 40%, not 90%, correct answers.

Thus, we can convert a number to a percent by multiplying it by 100 and attaching the % symbol. For example,

$$0.0002 = 100 \cdot 0.0002\% = 0.02\%,$$
$$\frac{3}{4} = 100 \cdot \frac{3}{4}\% = \frac{300}{4}\% = 75\%.$$

EXAMPLE 7-13 Write each of the following as a percent:

(a) 0.03 (b) $0.\overline{3}$
(c) 1.2 (d) 0.00042
(e) 1 (f) $\frac{3}{5}$
(g) $\frac{2}{3}$ (h) $2\frac{1}{7}$

Solution
(a) $0.03 = 100 \cdot 0.03\% = 3\%$
(b) $0.\overline{3} = 100 \cdot 0.\overline{3}\% = 33.\overline{3}\%$
(c) $1.2 = 100 \cdot 1.2\% = 120\%$
(d) $0.00042 = 100 \cdot 0.00042\% = 0.042\%$
(e) $1 = 100 \cdot 1\% = 100\%$
(f) $\frac{3}{5} = 100 \cdot \frac{3}{5}\% = \frac{300}{5}\% = 60\%$
(g) $\frac{2}{3} = 100 \cdot \frac{2}{3}\% = \frac{200}{3}\% = 66.\overline{6}\%$
(h) $2\frac{1}{7} = 100 \cdot 2\frac{1}{7}\% = \frac{1500}{7}\% = 214\frac{2}{7}\%$

A number can also be converted to a percent by using a *proportion*. For example, to write $\frac{3}{5}$ as a percent, we need only find the value of n in the following proportion:

$$\frac{3}{5} = \frac{n}{100}.$$

Solving the proportion, we obtain $\left(\frac{3}{5}\right) \cdot 100 = n,$ or $n = 60,$ or $60\%.$

Still another way to convert a number to percent is to recall that $1 = 100\%$. Thus for example, $\frac{3}{4} = \frac{3}{4}$ of $1 = \frac{3}{4} \cdot 1 = \frac{3}{4} \cdot 100\% = 75\%.$

REMARK The % symbol is crucial in identifying the meaning of a number. For example, $\frac{1}{2}$ and $\frac{1}{2}\%$ are different numbers: $\frac{1}{2} = 50\%$, which is not equal to $\frac{1}{2}\%$. Similarly, 0.01 is different from 0.01%, which is 0.0001.

In doing computations, it is sometimes useful to convert percents to decimals. This can be done by writing the percent as a fraction and then converting the fraction to a decimal.

EXAMPLE 7-14 Write each percent as a decimal.

(a) 5%   (b) 6.3%   (c) 100%
(d) 250%   (e) $\frac{1}{3}$%   (f) $33\frac{1}{3}$%

Solution

(a) $5\% = \frac{5}{100} = 0.05$

(b) $6.3\% = \frac{6.3}{100} = 0.063$

(c) $100\% = \frac{100}{100} = 1$

(d) $250\% = \frac{250}{100} = 2.50$

(e) $\frac{1}{3}\% = \frac{\frac{1}{3}}{100} = \frac{0.\overline{3}}{100} = 0.00\overline{3}$

(f) $33\frac{1}{3}\% = \frac{33\frac{1}{3}}{100} = \frac{33.\overline{3}}{100} = 0.\overline{3}$

■

Another approach to writing percent as a decimal is to convert 1% to a decimal first. Because $1\% = \frac{1}{100} = 0.01$, we can conclude that $5\% = 5 \cdot 0.01 = 0.05$ and $6.3\% = 6.3 \cdot 0.01 = 0.063$. Some calculators have a percent key. For example, if the keys ③ ④ . ⑤ % are pressed in the order given, many calculators will display 0.345. You should investigate how your calculator handles percents.

Application problems involving percents usually take one of the following forms:

1. Finding a percent of a number.
2. Finding what percent one number is of another.
3. Finding a number when a percent of that number is known.

Before considering examples illustrating these forms, recall what it means to find a fraction "of" a number. For example, $\frac{2}{3}$ of 70 means $\frac{2}{3} \cdot 70$. Similarly, to find 40% of 70, we have $\frac{40}{100}$ of 70, which means $\frac{40}{100} \cdot 70$, or $0.40 \cdot 70 = 28$.

EXAMPLE 7-15 A house that sells for $72,000 requires a 20% down payment. What is the amount of the down payment?

Solution The down payment is 20% of $72,000, or $0.20 \cdot \$72,000 = \$14,400$. Hence, the amount of the down payment is $14,400.

■

SECTION 7-3 PERCENTS 355

EXAMPLE 7-16   If Alberto has 45 correct answers on an 80-question test, what percent of his answers are correct?

Solution   Alberto has $\frac{45}{80}$ of the answers correct. To find the percent of correct answers, we need to convert $\frac{45}{80}$ to a percent. We can do this by multiplying the fraction by 100 and attaching the % as follows:

$$\frac{45}{80} = 100 \cdot \frac{45}{80}\%$$
$$= 56.25\%.$$

Thus 56.25% of the answers are correct.

An alternate solution uses proportion. Let n be the percent of correct answers and proceed as follows:

$$\frac{45}{80} = \frac{n}{100}$$
$$\frac{45}{80} \cdot 100 = n$$
$$n = \frac{4500}{80} = 56.25$$

EXAMPLE 7-17   Forty-two percent of the parents of the school children in the Paxson School District are employed at Di Paloma University. If the number of parents employed by D.P.U. is 168, how many parents are in the school district?

Solution   Let n be the number of parents in the school district. Then 42% of n is 168. We *translate this information into an equation* and solve for n.

$$42\% \text{ of } n = 168$$
$$\frac{42}{100} \cdot n = 168$$
$$0.42 \cdot n = 168$$
$$n = \frac{168}{0.42} = 400$$

There are 400 parents in the school district.

The problem can be solved using a proportion. Forty-two percent, or $\frac{42}{100}$, of the parents are employed at D.P.U. If n is the total number of parents, then $168/n$ also represents the fraction of parents employed at D.P.U. Thus

$$\frac{42}{100} = \frac{168}{n}$$
$$42n = 100 \cdot 168$$
$$n = \frac{16{,}800}{42} = 400.$$

We can also solve the problem as follows:

$$42\% \text{ of } n \text{ is } 168$$
$$1\% \text{ of } n \text{ is } \frac{168}{42}$$
$$100\% \text{ of } n \text{ is } 100\left(\frac{168}{42}\right)$$
$$\text{Therefore } n \text{ is } 100\left(\frac{168}{42}\right), \text{ or } 400.$$

EXAMPLE 7-18 Mike bought a bicycle and then sold it for 20% more than he paid for it. If he sold the bike for $144, what did he pay for it?

Solution We are looking for the original price P that Mike paid for the bike. We know that he sold the bike for $144 and that this included a 20% profit. *Thus we can write the following equation:*

$$\$144 = P + \text{Mike's profit.}$$

Because Mike's profit is 20% of P, we proceed as follows:

$$144 = P + 20\% \cdot P$$
$$144 = P + 0.20 \cdot P$$
$$144 = (1 + 0.20)P$$
$$144 = 1.20P$$
$$\frac{144}{1.20} = P$$
$$120 = P$$

Thus Mike originally paid $120 for the bike.

EXAMPLE 7-19 Westerner's Clothing Store advertised a suit for 10% off, for a savings of $15. Later, the manager marked the suit at 30% off the original price. What is the amount of the current discount?

Solution A 10% discount amounts to a $15 savings. We could find the amount of the current discount if we knew the original price. Thus finding the original price becomes our *subgoal*. Because 10% of P is $15, we have the following:

$$10\% \cdot P = \$15$$
$$0.10 \cdot P = \$15$$
$$P = \$150$$

To find the current discount, we calculate 30% of $150. Because $0.30 \cdot \$150 = \$45$, the amount of the 30% discount is $45.

In the Looking Back stage of problem solving, we check the answer and look for other ways to solve the problem. A different approach leads to a more efficient solution and confirms the answer. If 10% of the price is $15, then 30% of the price is 3 times $15, or $45.

PROBLEM 1

A 100-lb watermelon was found to be 99% water. After it sat in the sunlight all day, some of the water evaporated, leaving the melon 98% water. How much did the melon weigh after the evaporation occurred?

Understanding the Problem. A watermelon weighing 100 lb was 99% water. After some evaporation occurred, the melon was only 98% water. We are to determine the weight of the melon when it was 98% water.

Devising a Plan. After finding the amount of water that evaporated, we can subtract that amount from 100 to find the new weight of the watermelon. Let w be the weight of the water that the melon lost due to evaporation. Use the strategy of *writing an equation* for w. The new weight of the melon after evaporation is $(100 - w)$ lb, and the new weight of the water content is 98% of $(100 - w)$ lb. The new weight of the water content can also be computed by subtracting the number of pounds of water lost from the original weight of the water, which is 99 lb. Thus the new weight of the water content is $(99 - w)$ lb. Hence, we have the following equation:

$$\begin{pmatrix} \text{New weight of} \\ \text{water content} \end{pmatrix} = 98\% \text{ of weight of melon after evaporation}$$

$$99 - w = 0.98 \times (100 - w)$$

Carrying Out the Plan. The equation given above is solved for w as follows:

$$99 - w = 0.98(100 - w)$$
$$99 - w = 98 - 0.98w$$
$$1 = 0.02w$$
$$50 = w$$

Thus the weight of the water that was lost due to evaporation is 50 lb, and hence the melon weighs $(100 - 50)$, or 50 lb after evaporation.

Looking Back. We can find an alternate solution using the fact that the amount of material other than water does not change after evaporation. Let x be the weight of the watermelon after evaporation. We know that in the original 100 lb of watermelon, there was 99% water and hence 1%, or 1 lb, was not water. After evaporation, the watermelon's weight was x lb and 98% of the weight was water; hence, 2% of it was not water. Because the amount of material that was not water did not change, we know that 2% of x equals 1, which implies that $x = 50$ lb.

Another way to solve the problem is to think of the question as, "2% of how much weight is equal to 1% of 100?" Because 2% of 50 is the same as 1% of 100, the answer is 50 lb. ∎

Mental Math with Percents

Mental math may be helpful when working with percents. Several techniques follow.

1. *Using fraction equivalents*

   Knowing fraction equivalents for some percents can make some computations easier. Table 7-3 gives several fraction equivalents. These equivalents can be used in such computations as the following:

$$50\% \text{ of } \$75 = \left(\frac{1}{2}\right)75 = \$37.50$$

$$66\frac{2}{3}\% \text{ of } 48 = \frac{2}{3}(48) = 32$$

TABLE 7-3

Percent	25%	50%	75%	$33\frac{1}{3}\%$	$66\frac{2}{3}\%$	10%	1%
Fraction Equivalent	$\frac{1}{4}$	$\frac{1}{2}$	$\frac{3}{4}$	$\frac{1}{3}$	$\frac{2}{3}$	$\frac{1}{10}$	$\frac{1}{100}$

2. *Using a known percent*

   Frequently, we may not know a percent of something, but we know a close percent of it. For example, to find 55% of 62, we might do the following:

   $$50\% \text{ of } 62 = \left(\frac{1}{2}\right)(62) = 31$$

   $$5\% \text{ of } 62 = \left(\frac{1}{2}\right)(10\%)(62) = \left(\frac{1}{2}\right)(6.2) = 3.1$$

   Adding, we see that 55% of 62 is $31 + 3.1 = 34.1$.

Estimations with Percents

Estimations with percents can be used to determine whether answers are reasonable. Following are some examples:

1. To estimate 27% of 598, we see that 27% of 598 is a little more than 25% of 598, but 25% of 598 is approximately the same as 25% of 600, or $\frac{1}{4}$ of 600, or 150. Here, we have adjusted 27% downward and 598 upward, so 150 should be a reasonable estimate. A better estimate might be obtained by estimating 30% of 600 and then subtracting 3% of 600 to obtain 27% of 600, giving $180 - 18$, or 162.

2. To estimate 148% of 500, we see that 148% of 500 should be slightly less than 150% of 500. 150% of 500 is $1.5(500) = 750$. Thus 148% of 500 should be a little less than 750.

EXAMPLE 7-20 Laura wants to buy a blouse originally priced at $26.50 but now on sale at 40% off. She has $17 in her wallet and wonders if she has enough cash. How can she mentally find out?

Solution It is easier to find 40% of $25 (versus $26.50) mentally. One way is to find 10% of $25, which is $2.50. Now, 40% is four times that much, that is, $4 \cdot \$2.50$, or $10. Thus, Laura estimates that the blouse will cost $\$26.50 - \$10$, or $16.50. Since the actual discount is greater than $10 (40% of 26.50 is greater than 40% of 25), Laura will have to pay less than $16.50 for the blouse and, hence, she has enough cash. ∎

Estimation is also used to check if an answer to a problem is reasonable, as seen on the student page from *Addison-Wesley Mathematics*, Grade 8, 1991.

Problem Solving
Determining Reasonable Answers

LEARN ABOUT IT

An important part of evaluating an answer to a problem is to check your work. This chart shows some ways you can do this.

Check Your Work
- Is the arithmetic correct?
- Did you use the strategies correctly?
- Is the answer reasonable?

Example

Do not solve the problem. Decide if the answer given is reasonable. If it is not reasonable, explain why.

Problem: The graph shows the percentage increase in house prices from 1987 to 1988 in 7 states. Kaki paid $209,500 for a home in Hawaii in 1987. By how much had the price increased in 1988?

Answer: The price had increased by $3,016.80.

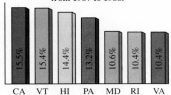

Increase in Home Prices from 1987 to 1988.

CA 15.5% | VT 15.4% | HI 14.4% | PA 13.2% | MD 10.6% | RI 10.4% | VA 10.4%

To check if the answer is reasonable, round 14.4% down to 14%. Then round $209,500 down to $200,000.

Multiply 200,000 by 14%.

The correct answer should be a little more than $28,000 because the numbers were rounded down.

$3,016.80 is not a reasonable answer.

```
14.4%  →  14%
$209,500 → $200,000
```

```
$200,000 · 0.14 = $28,000
```

TRY IT OUT

Do not solve the problem. Decide if the answer given is reasonable. If it is not reasonable, explain why.

1. Alan bought a home in Maryland in 1987. He paid $87,950. By how much had the price increased in 1988? Answer: The price had increased by $6,322.

2. In 1987, the median price of a home in Houston was $62,830. By 1988, it had dropped 9.6%. How much had it dropped? Answer: the median price dropped $6,031.68.

Use and Misuse of Percent

Sometimes it may not be clear which operations to perform with percent. The following example investigates this.

EXAMPLE 7-21 Which of the following statements are true and which are false? Explain your answers.

(a) Leonardo got a 10% raise at the end of his first year on the job and a 10% raise after another year. His total raise was 20% of his original salary.

(b) Jung and Dina paid 45% of their first department store bill of $620 and 48% of the second department store bill of $380. They paid 45% + 48% = 93% of the total bill of $1000.

(c) Bill spent 25% of his salary on food and 40% on housing. Bill spent 25% + 40% = 65% of his salary on food and housing.

(d) On their purchases, consumers pay a state tax of 7% and a city tax of 3%. Bonnie purchased a sofa for $420. She calculated her total tax by finding 10% of $420, that is, $42.

(e) In Bordertown, 65% of the adult population works in town, 25% works across the border, and 15% is unemployed.

(f) In Clean City, the fine for various polluting activities is a certain percentage of one's monthly income. The fine for smoking in public places is 40%, for driving a polluting car is 50%, and for littering is 30% of one's monthly income. Mr. Schmutzig committed all three polluting crimes in one day and paid a fine of 120% of his monthly salary.

Solution

(a) Percent is always of a quantity. For example, 10% of a quantity and another 10% of the same quantity is 20% of that quantity. In Leonardo's case, the first 10% raise was calculated based on his original salary and the second 10% raise was calculated on his new salary. Consequently, the percentages cannot be added, and the statement is false.

(b) The answer does not make sense. Jung and Dina paid less than $\frac{1}{2}$ of each bill, so they could not have paid 93% (almost all) of the total. In fact, $\frac{1}{2}$ of one bill plus $\frac{1}{2}$ of another bill is not $\frac{1}{2} + \frac{1}{2}$ or 1, the full amount of the total bill.

(c) Because the percentages are of the same quantity, the statement is true.

(d) Again, because the percentages are of the same quantity ($420), the statement is true.

(e) Because the percentages are of the same quantity, that is, the number of adults, we can add them: 65% + 25% + 25% = 105%. But 105% of the population accounts for more (5% more) than the town's population, which is impossible. Hence, the statement is false.

(f) Again, the percentages are of the same quantity, that is, the individual's monthly income. Hence, we can add them: 120% of one's monthly income is a stiff fine, but possible. ∎

PROBLEM SET 7-3

1. Express each of the following as percents:
   (a) 7.89 (b) 0.032 (c) 193.1 (d) 0.2
   (e) $\frac{5}{6}$ (f) $\frac{3}{20}$ (g) $\frac{1}{8}$ (h) $\frac{3}{8}$
   (i) $\frac{5}{8}$ (j) $\frac{1}{6}$ (k) $\frac{4}{5}$ (l) $\frac{1}{40}$

2. Convert each of the following percents to decimals:
   (a) 16% (b) $4\frac{1}{2}$% (c) $\frac{1}{5}$% (d) $\frac{2}{7}$%
   (e) $13\frac{2}{3}$% (f) 125% (g) $\frac{1}{3}$% (h) $\frac{1}{4}$%

3. Fill in the blanks to find other expressions for 4%.
   (a) _____ for every 100 (b) _____ for every 50
   (c) 1 for every _____ (d) 8 for every _____
   (e) 0.5 for every _____

4. 🖩 Different calculators compute percents in various ways. To investigate this, consider 5 · 6%.
   (a) If the following sequence of keys is pressed, is the correct answer of 0.3 displayed on your calculator?
   [5] [×] [6] [%] [=]
   (b) Press [6] [%] [×] [5] [=]. Is the answer 0.3?

5. Answer each of the following:
   (a) Find 6% of 34.
   (b) 17 is what percent of 34?
   (c) 18 is 30% of what number?
   (d) Find 7% of 49.
   (e) 61.5 is what percent of 20.5?
   (f) 16 is 40% of what number?

6. Marc had 84 boxes of candy to sell. He sold 75% of the boxes. How many did he sell?

7. Gail made $16,000 last year and received a 6% raise. How much does she make now?

8. Gail received a 7% raise last year. If her salary is now $15,515, what was her salary last year?

9. ▶A company bought a used typewriter for $350, which was 80% of the original cost. What was the original cost of the typewriter? Explain your solution.

10. Joe sold 180 newspapers out of 200. Bill sold 85% of his 260 newspapers. Ron sold 212 newspapers, 80% of those he had.
    (a) Who sold the most newspapers? How many?
    (b) Who sold the greatest percent of his newspapers? What percent?
    (c) Who started with the greatest number of newspapers? How many?

11. If a dress that normally sells for $35 is on sale for $28, what is the "percent off"? (This could be called a *percent of decrease,* or a *discount.*)

12. A car originally cost $8000. One year later, it was worth $6800. What is the percent of depreciation?

13. On a certain day in Glacier Park, 728 eagles were counted. Five years later, 594 were counted. What was the percent of decrease in the number of eagles counted?

14. Mort bought his house in 1975 for $29,000. It was recently appraised at $55,000. What is the *percent of increase* in value?

15. ▶Xuan weighed 9 lb when he was born. At 6 mo, he weighed 18 lb. What was the percent of increase in Xuan's weight? Explain your solution.

16. Sally bought a dress marked at 20% off. If the regular price was $28.00, what was the sale price?

17. What is the sale price of a softball if the regular price is $6.80 and there is a 25% discount?

18. If a $\frac{1}{4}$-c serving of Crunchies breakfast food has 0.5% of the minimum daily requirement of Vitamin C, how many cups would you have to eat in order to obtain the minimum daily requirement of Vitamin C?

19. An airline ticket costs $320 without the tax. If the tax rate is 5%, what is the total bill for the airline ticket?

20. Bill got 52 correct answers on an 80-question test. What percent of the questions did he not answer correctly?

21. A real estate broker receives 4% of an $80,000 sale. How much does the broker receive?

22. A survey reported that $66\frac{2}{3}$% of 1800 employees favored a new insurance program. How many employees favored the new program?

23. A family has a monthly income of $2400 and makes a monthly house payment of $400. What percent of the income is the house payment?

24. A plumber's wage this year is $19.80/hr. This is a 110% increase over last year's hourly wage. What is the dollar increase in the hourly wage over last year?

25. ▶Ms. Price has received a 10% raise in salary in each of the last 2 yrs. If her annual salary this year is $100,000, what was her salary 2 yrs ago, rounded to the nearest penny? Explain your reasoning.

26. Soda is advertised at 45¢ a can or $2.40 a six-pack. If 6 cans are to be purchased, what percent is saved by purchasing the six-pack?

27. John paid $330 for a new mountain bicycle to sell in his shop. He wants to price it so that he can offer a 10% discount and still make 20% of the price he paid for it. At what price should the bike be marked?

28. ▶The price of a suit that sold for $100 was reduced by 25%. By what percent must the price of the suit be increased to bring the price back to $100? Explain.

29. Howard entered a store and said to the owner, "Give me as much money as I have with me and I will spend 80% of the total." After this was done, Howard repeated the operation at a second store and at a third store and was finally left with $12. How much money did he have at first?

30. 🖩 The car Elsie bought 1 yr ago has depreciated by $1116.88, which is 12.13% of the price she paid for it. How much did she pay for the car, to the nearest cent?

31. Solve each of the following, using mental mathematics:
    (a) 15% of $22
    (b) 20% of $120
    (c) 5% of $38
    (d) 25% of $98

32. If we build a 10 × 10 model with blocks, as shown in the figure, and paint the entire model, what percent of the cubes will have each of the following?
    (a) Four faces painted
    (b) Three faces painted
    (c) Two faces painted

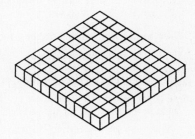

33. 🖩 Answer the questions in Problem 32 for models of the following sizes:
    (a) 9 × 9
    (b) 8 × 8
    (c) 7 × 7
    (d) 12 × 12

34. ▶Which of the following statements is true and which is false? If a statement is false, explain why and find an appropriate correct answer.
    (a) The price of a house increased 10% in 1 yr and a year later it dropped by 10%. Hence, the total change in price was 10% − 10%, or 0%, that is, the price after 2 yr was the same as the original price.
    (b) The price of a house decreased 10% in 1 yr and a year later it increased by 10%. Hence, the total change in price was 0 dollars.

35. To be safe but to still achieve a cardiovascular training effect, people should monitor their heart rates while exercising. The maximum heart rate can be approximated by subtracting your age from 220. You can safely achieve a training effect if you maintain your heart rate between 60% and 80% of that number for at least 20 min three times a week.
    (a) Determine the range for your age.
    (b) At the top of a long hill, Jeannie slows her bike and takes her pulse. She counts 41 beats in 15 sec.
        (i) Express in decimal form the amount of time in seconds between successive beats.
        (ii) Express the amount in terms of minutes.

36. (a) By what percent is 50 larger than 40?
    (b) By what percent is 40 smaller than 50?

    (c) If you get a 20% raise but the cost of living remains steady, by what percent does your purchasing power increase?
    (d) ▶If your income stays the same, but the cost of living drops by 20%, by what percent does your purchasing power increase? Justify your answer.

37. A crew consists of one apprentice, one journeyman, and one master carpenter. The crew receives a check for $4200 for a job they just finished. A journeyman makes 200% of what an apprentice makes, and a master makes 150% of what a journeyman makes. How much does each person in the crew earn?

38. (a) In an incoming freshman class of 500 students, only 20 claimed to be math majors. What percent of the freshman class is this?
    (b) When the survey was repeated the next year, 5% of the nonmath majors had decided to switch and become math majors.
        (i) How many math majors were there now?
        (ii) What percent of the freshman class is this?

39. ▶Nick and Nora are seated in a restaurant that does not accept credit cards when they discover they have only $35 in cash. They want to leave a 15% tip. How much can they spend on dinner and still have enough for the tip? Explain your solution.

40. Use the pie charts from an information page from a 1990 state income tax instruction booklet to answer the following questions:

WHERE YOUR TAX DOLLAR GOES

Source of General Fund revenues 1989–1991

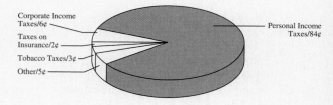

Expenditure of General Fund revenues 1989–1991

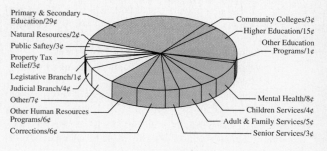

(a) What percent of revenue comes from personal and corporate income taxes?
(b) What percent of revenue from income taxes comes from corporate income taxes?
(c) What percent of revenue is spent on education?
(d) What percent of the education budget is devoted to primary and secondary education?
(e) Find a combination of programs that spends exactly 50% of the human resources budget.
(f) Many of the percents may have been rounded in the preparation of the pie charts. Does each pie chart total 100%?

41. Use the 1990 state income tax rate chart in Problem 40 to answer the following questions:
    (a) Marie, a single person, had a 1990 taxable income of $16,250. How much income tax did she owe?
    (b) Peter and Holly are married. Peter's 1990 taxable income was $12,321 whereas Holly's was $6,532. Should they file separately or jointly to minimize their tax bill?

Review Problems

42. A state charged a company $63.27/day for overdue taxes. The total bill for the overdue taxes was $6137.19. How many days were the taxes overdue?
43. Write 33.21 as a fraction in simplest form.
44. Write $\frac{2}{9}$ as a decimal.
45. Write $31.0\overline{5}$ as a fraction in simplest form.
46. Write each of the following in scientific notation:
    (a) 3,250,000   (b) 0.00012
47. Round 32.015 to the indicated place.
    (a) The nearest tenth   (b) The nearest ten

Page 2-Form 40 1990   **TAX RATE CHARTS**

Tax Rate Chart A: For persons filing SINGLE, or MARRIED FILING SEPERATELY		Tax Rate Chart B: For persons filing JOINTLY, HEAD OF HOUSEHOLD or QUALIFYING WIDOW(ER) WITH DEPENDENT CHILD	
If your taxable income is:	Your tax is	If your taxable income is:	Your tax is
Not over $2,000	5% of taxable income	Not over $4,000	5% of taxable income
Over $2,000 but not over $5,000	$100 plus 7% of excess over $2,000	Over $4,000 but not over $10,000	$200 plus 7% of excess over $4,000
Over $5,000	$310 plus 9% of excess over $5,000	Over $10,000	$620 plus 9% of excess over $10,000

▼ BRAIN TEASER

The crust of a certain pumpkin pie is 25% of the pie. By what percent should the amount of crust be reduced in order to make it constitute 20% of the pie?

---

 *Section 7-4 **Computing Interest**

When a bank advertises a $5\frac{1}{2}$% interest rate on a savings account, the amount of money

interest the bank will pay for using that money is called **interest.** The original amount deposited or
principal borrowed is the **principal.** The percent used to determine the interest is the **interest rate.**
interest rate Interest rates are given for specific periods of time, such as years, months, or days. Interest
simple interest computed on the original principal is **simple interest.** For example, suppose we borrow $5000 from a company at a simple interest rate of 12% for 1 yr. The interest we owe on the loan for 1 yr is 12% of $5000, or $5000 · 0.12. In general, if a principal P is invested at an

annual interest rate of r, then the simple interest after 1 yr is $Pr \cdot 1$; after t years, it is Prt, or Prt. If I represents simple interest, we have

$$I = Prt.$$

The amount needed to pay off a $5000 loan at 12% simple interest is the $5000 borrowed plus the interest on the $5000, that is, $5000 + 5000 \cdot 0.12$, or $5600. In general,

amount / balance   an **amount** (*or* **balance**) A is equal to the principal P plus the interest I, that is,

$$A = P + I = P + Prt = P(1 + rt).$$

EXAMPLE 7-22   Vera opens a savings account that pays simple interest at the rate of $5\frac{1}{4}\%$ per year. If she deposits $2000 and makes no other deposits, find the interest and the final amount for the following periods of time:

(a) 1 yr   (b) 90 days

Solution   (a) To find the interest for 1 yr, we proceed as follows:

$$I = \$2000 \cdot 5\frac{1}{4}\% \cdot 1 = \$2000 \cdot 0.0525 = \$105.$$

Thus her amount at the end of 1 yr is

$$\$2000 + \$105 = \$2105.$$

(b) When the interest rate is annual and the interest period is given in days, we represent the time as a fractional part of a year by dividing the number of days by 365. Thus

$$I = \$2000 \cdot 5\frac{1}{4}\% \cdot \frac{90}{365}$$

$$= \$2000 \cdot 0.0525 \cdot \frac{90}{365} \doteq \$25.89.$$

Hence, $A \doteq \$2000 + \$25.89 = \$2025.89.$ ∎

EXAMPLE 7-23   Find the annual interest rate if a principal of $10,000 increased to $10,900 at the end of 1 yr.

Solution   Let the annual interest be $x\%$. We know that $x\%$ of $10,000 is the increase. Because the increase is $10,900 - \$10,000 = \$900$, we use the strategy of *writing an equation* for x as follows:

$$x\% \text{ of } 10{,}000 = 900$$

$$\frac{x}{100} \cdot 10{,}000 = 900$$

$$x = 9$$

Thus the interest is 9%. We can solve this problem mentally by asking, "What percent of 10,000 is 900?" Because 1% of 10,000 is 100, to obtain 900, take 9% of 10,000.   ∎

Compound Interest

compound interest

In business transactions, interest is usually calculated daily (365 times a year). In the case of savings, the earned interest is added daily to the principal, and each day the interest is earned on a different amount; that is, it is earned on the previous interest as well as the principal. When interest is computed in this way, it is called **compound interest.** Compounding usually is done annually (once a year), semiannually (twice a year), quarterly (4 times a year), or monthly (12 times a year). However, even if the interest is compounded, it is given as an annual rate. For example, if the annual rate is 6% compounded monthly, the interest per month is $\frac{6}{12}$%, or 0.5%. If it is compounded daily, the interest per day is $\frac{6}{365}$%. In general, *the interest rate per period is the annual interest rate divided by the number of periods in a year.*

EXAMPLE 7-24 If you invest $100 at 8% compounded quarterly, how much will you have in the account after 1 year?

Solution The quarterly interest rate is $\frac{1}{4} \cdot 8$%, or 2%. It seems that we would have to calculate the interest 4 times. But we can also reason as follows. If at the beginning of any of the 4 periods there are x dollars in the account, at the end of that period there will be

$$x + 2\% \text{ of } x$$
$$= x + 0.02x$$
$$= x(1 + 0.02)$$
$$= x \cdot 1.02 \text{ dollars.}$$

Hence, to find the amount at the end of any period, we need only multiply the amount at the beginning of the period by 1.02. From Table 7-4, we see that the amount at the end of the fourth period is $100 \cdot 1.02^4$. On a scientific calculator, we can find the amount using $\boxed{1}\boxed{0}\boxed{0}\boxed{\times}\boxed{1}\boxed{.}\boxed{0}\boxed{2}\boxed{y^x}\boxed{4}\boxed{=}$. The calculator displays 108.24322. Thus the amount at the end of 1 yr is approximately $108.24.

TABLE 7-4

Period	Initial Amount	Final Amount
1	100	$100 \cdot 1.02$
2	$100 \cdot 1.02$	$(100 \cdot 1.02) \cdot 1.02$ or $100 \cdot 1.02^2$
3	$100 \cdot 1.02^2$	$(100 \cdot 1.02)^2 \cdot 1.02$ or $100 \cdot 1.02^3$
4	$100 \cdot 1.02^3$	$(100 \cdot 1.02^3) \cdot 1.02$ or $100 \cdot 1.02^4$

REMARK Because the amount at the beginning of any period in Example 7-24 is multiplied by the fixed number 1.02 to obtain the amount at the end of this period, the final amounts at the end of the periods form a geometric sequence. Hence, finding the final amount at the end of the nth period amounts to finding the nth term of a geometric sequence whose first term is $100 \cdot 1.02$ (amount at the end of the first period) and whose ratio is 1.02.

We can generalize the discussion in Example 7-24. If the interest rate per period is i (a fraction or a decimal), and the principal is P, then the amount A after n periods can be found as follows. If, at the beginning of any period, there are x dollars in the account, at the end of that period there will be x plus the interest I for that period. That is, $x + I = x + x \cdot i = x(1 + i)$. Hence, to find the amount at the end of any period, we need only to multiply the amount at the beginning of the period by $1 + i$. This is shown in Table 7-5. Therefore the amount at the end of the nth period is $P(1 + i)^n$, and we have the formula $A = P(1 + i)^n$.

TABLE 7-5

Period	Initial Amount	Final Amount
1	P	$P(1 + i)$
2	$P(1 + i)$	$[P(1 + i)](1 + i)$, or $P(1 + i)^2$
3	$P(1 + i)^2$	$[P(1 + i)^2](1 + i)$, or $P(1 + i)^3$
4	$P(1 + i)^3$	$[P(1 + i)^3](1 + i)$, or $P(1 + i)^4$
$\vdots$	$\vdots$	$\vdots$
n	$p(1 + i)^{n-1}$	$[P(1 + i)^{n-1}](1 + i)$, or $P(1 + i)^n$

EXAMPLE 7-25 Suppose we deposit $1000 in a savings account that pays 6% interest compounded quarterly.

(a) What is the balance at the end of 1 year?
(b) What is the *effective annual yield* on this investment; that is, what is the rate that would have been paid if the amount had been invested using simple interest?

Solution (a) An annual interest rate of 6% earns $\frac{1}{4}$ of 6%, or an interest rate of $\frac{0.06}{4}$, in one quarter. Because there are four periods, we have the following:

$$A = 1000\left(1 + \frac{0.06}{4}\right)^4 \doteq \$1061.36.$$

Thus the balance at the end of 1 year is $1061.36.

(b) Because the interest earned is $1061.36 - \$1000.00 = \$61.36$, the effective annual yield can be computed by using the simple interest formula, $I = Prt$.

$$61.36 = 1000 \cdot r \cdot 1$$
$$\frac{61.36}{1000} = r$$
$$0.06136 = r$$
$$6.136\% = r$$

Hence, the effective annual yield is 6.136%.

To save for their child's college education, a couple deposits $3000 into an account that pays 11% annual interest compounded daily. Find the amount in this account after 12 yr.

Solution The principal in the problem is $3000, the daily rate i is $0.11/365$, and the number of compounding periods is $12 \cdot 365$, or 4380. Thus we have

$$A = \$3000\left(1 + \frac{0.11}{365}\right)^{4380} \doteq \$11{,}228.$$

Hence the amount in the account is $11,228. ∎

PROBLEM SET 7-4

You will need a calculator for most of these problems.

1. Complete the following compound-interest chart:

Compounding Period	Principal	Annual Rate	Length of Time (Years)	Interest Rate Per Period	Number of Periods	Amount of Interest Paid
(a) Semiannual	$1000	6%	2			
(b) Quarterly	$1000	8%	3			
(c) Monthly	$1000	10%	5			
(d) Daily	$1000	12%	4			

2. Ms. Jackson borrowed $42,000 at 13% annual simple interest to buy her house. If she won the Irish Sweepstakes exactly 1 yr later and was able to repay the loan without penalty, how much interest would she owe?

3. Carolyn went on a shopping spree with her Bankamount card and made purchases totaling $125. If the interest rate is 1.5% per month on the unpaid balance and she does not pay this debt for 1 yr, how much interest will she owe at the end of the year?

4. A man collected $28,500 on a loan of $25,000 he made 4 yrs ago. If he charged simple interest, what was the rate he charged?

5. Burger Queen will need $50,000 in 5 yr for a new addition. To meet this goal, money is deposited today in an account that pays 9% annual interest compounded quarterly. Find the amount that should be invested to total $50,000 in 5 yr.

6. A company is expanding its line to include more products. To do so, it borrows $320,000 at 13.5% annual simple interest for a period of 18 mo. How much interest must the company pay?

7. ▶An amount of $3000 was deposited in a bank at a rate of 5% compounded quarterly for 3 yr; the rate then increased to 8% and was compounded quarterly for the next 3 yr. If no money was withdrawn, what was the balance at the end of this time period? Explain your reasoning.

8. To save for their retirement, a couple deposits $4000 in an account that pays 9% interest compounded quarterly. What will be the value of their investment after 20 yr?

9. ▶A money-market fund pays 14% annual interest compounded daily. What is the value of $10,000 invested in this fund after 15 yr? Explain your solution.

10. A car company is offering car loans at a simple-interest rate of 9%. Find the interest charged to a customer who finances a car loan of $7200 for 3 yr.

11. ▶The New Age Savings Bank advertises 9% interest rates compounded daily, while the Pay More Bank pays 10.5% interest compounded annually. Which bank offers a better rate for a customer if she plans to leave her money in for exactly 1 yr? Justify your answer.

12. Johnny and Carolyn have three different savings plans, which accumulated the following amounts of interest for 1 yr:
    (i) A passbook savings account that accumulated $53.90 on a principal of $980.
    (ii) A certificate of deposit that accumulated $55.20 on a principal of $600.
    (iii) A money market certificate that accumulated $158.40 on a principal of $1200.
    Which of these accounts paid the best interest rate for the year?

13. If a hamburger costs $1.35 and if the price continues to rise at a rate of 11% a year for the next 6 yr, what will the price of a hamburger be at the end of 6 yr?
14. If college tuition is $10,000 this year, what will it be 10 yr from now, if we assume a constant inflation rate of 9% a year?
15. Sara invested money at a bank that paid 6.5% compounded quarterly. If she had $4650 at the end of 4 yr, what was her initial investment?
16. ▶A car is purchased for $15,000. If each year the car depreciates by 10% of its value the preceding year, what will its value be at the end of 3 yr? Explain your reasoning.
17. If Adrien and Jarrell deposit $300 on January 1 in a holiday savings account that pays 1.1% per month interest and they withdraw the money on December 1 of the same year, what is the effective annual yield?
18. ▶Because of a recession, the value of a new house depreciated 10% each year for 3 yr in a row. Then, for the next 3 yr, the value of the house increased 10% each year.
    (a) Did the value of the house increase or decrease after 6 yr? Explain.
    (b) By what percent of the original cost?
19. The number of trees in a rainforest decreases each month by 0.5%. If the forest has approximately $2.34 \cdot 10^9$ trees, how many trees will be left after 20 yr?
20. ▶Determine the number of years (to the nearest tenth) that it would take for any amount of money to double deposited at a 10% interest rate compounded annually. Explain your reasoning.

Section 7-5 Real Numbers

Every rational number can be expressed either as a repeating decimal or as a terminating decimal. The ancient Greeks discovered numbers that are not rational. Such numbers must have a decimal representation that neither terminates nor repeats. To find such decimals, we focus on the characteristics they must have:

1. There must be an infinite number of nonzero digits to the right of the decimal point.
2. There cannot be a repeating block of digits (a repetend).

One way to construct a nonterminating, nonrepeating decimal is to devise a pattern of infinite digits in such a way that there will definitely be no repeated block. Consider the number 0.1010010001... . If the pattern shown continues, the next groups of digits are four zeros followed by 1, five zeros followed by 1, and so on. It is possible to describe a pattern for this decimal, but there is no repeating block of digits. Because this decimal is nonterminating and nonrepeating, it cannot represent a rational number. Numbers that are not rational numbers are called **irrational numbers.**

irrational numbers

π (pi)

In the mid-eighteenth century, it was proved that the number that is the ratio of the circumference of a circle to its diameter, symbolized by **π (pi)**, is an irrational number. The numbers $\frac{22}{7}$, 3.14, or 3.14159 are rational-number approximations of π. The value of π has been computed to thousands of decimal places with no apparent pattern.

Irrational numbers occur in the study of area. For example, to find the area of a square, we use the formula $A = s^2$, where A is the area and s is the length of a side of the square. If a side of a square is 3 cm long, then the area of the square is 9 cm² (square centimeters). Conversely, we can use the formula to find the length of a side of a square, given its area. If the area of a square is 25 cm², then $s^2 = 25$, so $s = 5$ or $^-5$. Each of these solutions is called a **square root** of 25. However, because lengths are always nonnegative, 5 is the only possible solution. The positive solution of $s^2 = 25$ (namely, 5) is called the **principal square root** of 25 and is denoted by $\sqrt{25}$. Similarly, the principal square root of 2 is denoted by $\sqrt{2}$. Note that $\sqrt{16} \neq {}^-4$, because $^-4$ is not the principal square root of 16.

square root
principal square root

SECTION 7-5 REAL NUMBERS

> **DEFINITION OF THE PRINCIPAL SQUARE ROOT**
>
> If a is any whole number, the **principal square root** of a is the nonnegative number b such that $b^2 = a$.

radical sign / radicand

> **REMARK** The principal square root of a is denoted by $\sqrt{a}$, where the symbol $\sqrt{}$ is called a **radical sign** and a is called the **radicand.**

HISTORICAL NOTE

The discovery of irrational numbers by members of the Pythagorean Society (founded by Pythagoras) is one of the greatest events in the history of mathematics. This discovery was very disturbing to the Pythagoreans, who believed that everything depended on whole numbers, so they decided to keep the matter secret. One legend has it that Hippasus, a society member, was drowned because he relayed the secret to persons outside the society.

Christoff Rudolff, a German mathematician, was the first to use the symbol $\sqrt{}$, in 1525.

EXAMPLE 7-26   Find:

(a) The square roots of 144

(b) The principal square root of 144

(c) $\sqrt{\dfrac{4}{9}}$

Solution (a) The square roots of 144 are 12 and $^-12$.

(b) The principal square root of 144 is 12.

(c) $\sqrt{\dfrac{4}{9}} = \dfrac{2}{3}$

Some square roots are rational numbers. Other square roots, like $\sqrt{2}$, are irrational numbers. To see this, note that $1^2 = 1$ and $2^2 = 4$, and that there is no whole number s such that $s^2 = 2$. Is there a rational number $\dfrac{a}{b}$ such that $\left(\dfrac{a}{b}\right)^2 = 2$? We use the strategy of *indirect reasoning*. If we assume there is such a rational number, then the following must be true:

$$\left(\dfrac{a}{b}\right)^2 = 2$$
$$\dfrac{a^2}{b^2} = 2$$
$$a^2 = 2b^2$$

Because $a^2 = 2b^2$, the Fundamental Theorem of Arithmetic says that the prime factorizations of a^2 and $2b^2$ are the same. In particular, the prime 2 appears the same number of times in the prime factorization of a^2 as it does in the factorization of $2b^2$. Because $b^2 = b \cdot b$, then no matter how many times 2 appears in the prime factorization of b, it appears twice as many times in $b \cdot b$. Also, a^2 has an even number of 2s for the same reason that b^2 does. In $2b^2$, another factor of 2 is introduced, resulting in an odd number of 2s in the prime factorization of $2b^2$ and, hence, of a^2. But 2 cannot appear both an odd number of times and an even number of times in the same prime factorization of a^2. We have a contradiction. This contradiction could have been caused only by the assumption that $\sqrt{2}$ is a rational number. Consequently, $\sqrt{2}$ must be an irrational number. We can use a similar argument to show that $\sqrt{3}$ is irrational or $\sqrt{n}$ is irrational, where n is a whole number but not the square of another whole number.

EXAMPLE 7-27 Prove that $2 + \sqrt{2}$ is an irrational number.

Solution We use the strategy of *indirect reasoning*. Suppose $2 + \sqrt{2} = \dfrac{a}{b}$, where $\dfrac{a}{b}$ is a rational number. Then

$$\sqrt{2} = \frac{a}{b} - 2$$

$$\sqrt{2} = \frac{a - 2b}{b}.$$

But $\dfrac{a - 2b}{b}$ is a rational number (why?), and this is a contradiction because $\sqrt{2}$ is an irrational number. Thus $2 + \sqrt{2}$ is an irrational number. ∎

> **REMARK** In a similar manner, we could prove $m + n\sqrt{2}$ is an irrational number for all rational numbers m and n except $n = 0$.

Pythagorean Theorem

Many irrational numbers can be interpreted geometrically. For example, a point on a number line can be found to represent $\sqrt{2}$ by using the **Pythagorean Theorem** (see Chapter 13). That is, if a and b are the lengths of the shorter sides (legs) of a right triangle and c is the length of the longer side (hypotenuse), then $a^2 + b^2 = c^2$, as shown in Figure 7-4.

FIGURE 7-4

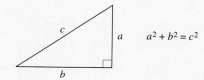

Figure 7-5 shows a segment one unit long constructed perpendicular to a number line at point P. Thus two sides of the triangle shown are one unit long. If $a = b = 1$, then $c^2 = 2$ and $c = \sqrt{2}$. To find a point on the number line that corresponds to $\sqrt{2}$, we need to find a point Q on the number line such that the distance from 0 to Q is $\sqrt{2}$. Because $\sqrt{2}$ is the length of the hypotenuse, the point Q can be found by marking an arc with center 0 and radius c. The intersection of the positive number line with the arc is Q.

FIGURE 7-5

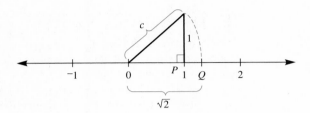

Similarly, other square roots can be constructed, as shown in Figure 7-6.

FIGURE 7-6

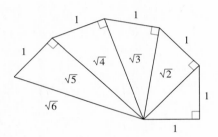

From Figure 7-5, we see that $\sqrt{2}$ must have a value between 1 and 2; that is, $1 < \sqrt{2} < 2$. To obtain a closer approximation of $\sqrt{2}$, we attempt to "squeeze" $\sqrt{2}$ between two numbers that are between 1 and 2. Because $(1.5)^2 = 2.25$ and $(1.4)^2 = 1.96$, it follows that $1.4 < \sqrt{2} < 1.5$. Because a^2 can be interpreted as the area of a square with side of length a, this discussion can be pictured geometrically, as in Figure 7-7.

FIGURE 7-7

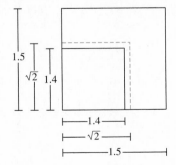

If we desire a more accurate approximation for $\sqrt{2}$, we can continue this squeezing process. We see that $(1.4)^2$, or 1.96, is closer to 2 than is $(1.5)^2$, or 2.25, so we choose numbers closer to 1.4 in order to find the next approximation. We find the following:

$$(1.42)^2 = 2.0164$$

$$(1.41)^2 = 1.9981$$

Thus $1.41 < \sqrt{2} < 1.42$. We can continue this process until we obtain the desired approximation. Note that if the calculator has a square-root key, we can obtain the approximation directly.

The System of Real Numbers

real numbers The set of **real numbers** R is the union of the set of rational numbers and the set of irrational numbers. Real numbers represented as decimals can be terminating, repeating, or nonterminating and nonrepeating.

Every integer is a rational number as well as a real number. Every rational number is a real number, but not every real number is rational, as has been shown with $\sqrt{2}$. The relationships among these sets of numbers is summarized in the Venn diagram in Figure 7-8, where the universe is the set of real numbers and the complement of the set of rationals is the set of irrational numbers. The concept of fractions can now be extended to include all numbers of the form $\frac{a}{b}$, where a and b are real numbers with $b \neq 0$, such as $\frac{\sqrt{3}}{5}$. Addition, subtraction, multiplication, and division are defined on the set of real numbers in such a way that all the properties of these operations on rationals still hold. The properties are summarized next.

FIGURE 7-8

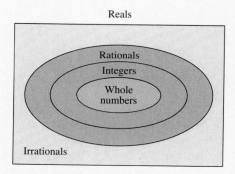

PROPERTIES

Closure Properties: For real numbers a and b, $a + b$ and $a \cdot b$ are unique real numbers.

Commutative Properties: For real numbers a and b, $a + b = b + a$ and $a \cdot b = b \cdot a$.

Associative Properties: For real numbers a, b, and c, $a + (b + c) = (a + b) + c$ and $a \cdot (b \cdot c) = (a \cdot b) \cdot c$.

Identity Properties: The number 0 is the unique additive identity and 1 is the unique multiplicative identity such that, for any real number a, $0 + a = a = a + 0$ and $1 \cdot a = a = a \cdot 1$.

Inverse Properties: (1) For every real number a, ^-a is its unique additive inverse; that is, $a + {}^-a = 0 = {}^-a + a$. (2) For every nonzero real number a, $\frac{1}{a}$ is its unique multiplicative inverse; that is, $a \cdot \left(\frac{1}{a}\right) = 1 = \left(\frac{1}{a}\right) \cdot a$.

Distributive Property of Multiplication over Addition: For real numbers a, b, and c, $a \cdot (b + c) = a \cdot b + a \cdot c$.

Denseness Property: For real numbers a and b, there exists a real number c such that $a < c < b$.

Properties of equality and inequality similar to those for rational numbers hold for real numbers. Using these properties, we can solve real-number equations and inequalities. The number line can be used to picture real numbers, because every point on a number line corresponds to a real number and every real number corresponds to a point on the number line. Because such a one-to-one correspondence is possible, solution sets of real-number equations and inequalities can be graphed on a number line.

EXAMPLE 7-28 Solve each of the following and show the solution on a number line:

(a) $x - 3 \leq \sqrt{2} + {}^-2$ (b) $\dfrac{3x^2}{2} - 4 = 5$ (c) $|x| \geq 2$

Solution (a) $x - 3 \leq \sqrt{2} + {}^-2$
$x \leq \sqrt{2} + 1$

Thus the solution is $x \leq \sqrt{2} + 1$, where x is a real number, is shown in Figure 7-9.

FIGURE 7-9

(b) $\dfrac{3x^2}{2} - 4 = 5$

$\dfrac{3x^2}{2} = 9$

$3x^2 = 18$

$x^2 = 6$

$x = \sqrt{6}$, or $x = {}^-\sqrt{6}$

The solution is shown on the number line in Figure 7-10.

FIGURE 7-10

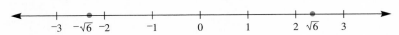

(c) To solve $|x| \geq 2$, we look for all the points on the number line whose distance from the origin is greater than or equal to 2. All such points are shown on the number line in Figure 7-11. The answer can be written as $x \leq {}^-2$ or $x \geq 2$. Note that the answer could not be written as $2 \leq x \leq {}^-2$. Why not?

FIGURE 7-11

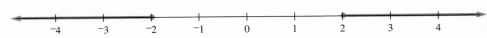

REMARK If the inequality in Example 7-28(a) had been $x - 3 < \sqrt{2} + {}^-2$, then the solution would be $x < \sqrt{2} + 1$, as seen in Figure 7-12, with a hollow dot to indicate that $\sqrt{2} + 1$ is not included.

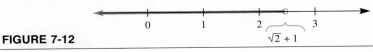

FIGURE 7-12

PROBLEM SET 7-5

1. Without using a radical sign, write an irrational number all of whose digits are 2s and 3s.
2. Arrange the following real numbers in order from least to greatest:
   $0.78, 0.\overline{7}, 0.\overline{78}, 0.7\overline{88}, 0.7\overline{8}, 0.7\overline{88}, 0.77, 0.787787778\ldots$
3. Arrange the following real numbers in order from greatest to least:
   $0.9, 0.\overline{9}, 0.\overline{98}, 0.9\overline{88}, 0.9\overline{98}, 0.\overline{898}$
4. Which of the following represent irrational numbers?
   (a) $\sqrt{51}$ (b) $\sqrt{64}$ (c) $\sqrt{324}$
   (d) $\sqrt{325}$ (e) $2 + 3\sqrt{2}$ (f) $\sqrt{2} \div 5$
5. Find the square roots, correct to tenths, for each of the following without a calculator, if possible:
   (a) 225   (b) 251   (c) 169
   (d) 512   (e) $^-81$ (f) 625
6. Find the approximate square roots for each of the following, rounded to hundredths. Use the squeezing method.
   (a) 17   (b) 7   (c) 21
   (d) 0.0120   (e) 20.3   (f) 1.64
7. Classify each of the following as true or false. If false, give a counterexample.
   (a) The sum of any rational number and any irrational number is a rational number.
   (b) The sum of any two irrational numbers is an irrational number.
   (c) The product of any two irrational numbers is an irrational number.
   (d) The difference of any two irrational numbers is an irrational number.
8. ▶Is it true that $\sqrt{a+b} = \sqrt{a} + \sqrt{b}$? Explain.
9. Find three irrational numbers between 1 and 3.
10. Find an irrational number between $0.\overline{53}$ and $0.\overline{54}$.
11. ▶Pi (π) is an irrational number. Could $\pi = \frac{22}{7}$? Why or why not?
12. ▶Without using a calculator or doing any computation, determine if $\sqrt{13} = 3.60\overline{5}$. Explain why or why not.
13. If R is the set of real numbers, Q is the set of rational numbers, I is the set of integers, W is the set of whole numbers, and S is the set of irrational numbers, find each of the following:
    (a) $Q \cup S$ (b) $Q \cap S$ (c) $Q \cap R$
    (d) $S \cap W$ (e) $W \cup R$ (f) $Q \cup R$
14. Solve each of the following for real numbers and graph the solution:
    (a) $5x - 1 \le \frac{7}{2}x + 3$ (b) $4 + 3x \ge \sqrt{5} - 7x$
    (c) $\frac{2}{3}x + \sqrt{3} \le {}^-5x$ (d) $(2x - 1)^2 = 4$
    (e) $|x| \ge 7$ (f) $|x| \le 3$

15. If the following letters correspond to the sets listed in Problem 13, put a check mark under each set of numbers for which a solution to the problem exists. (N is the set of natural numbers.)

	N	I	Q	R	S
(a) $x^2 + 1 = 5$					
(b) $2x - 1 = 32$					
(c) $x^2 = 3$					
(d) $x^2 = 4$					
(e) $\sqrt{x} = {}^-1$					
(f) $\frac{3}{4}x = 4$					

16. For what real values of x, if any, is each statement true?
    (a) $\sqrt{x} = 8$ (b) $\sqrt{x} = {}^-8$
    (c) $\sqrt{{}^-x} = 8$ (d) $\sqrt{{}^-x} = {}^-8$
    (e) $\sqrt{x} > 0$ (f) $\sqrt{x} < 0$
17. A diagonal brace is placed in a 4- by 5-ft rectangular gate. What is the length of the brace to the nearest tenth of a foot? (*Hint:* Use the Pythagorean Theorem.)
18. For a simple pendulum of length l, given in centimeters, the time of the period T in seconds is given by $T = 2\pi\sqrt{\frac{l}{g}}$, where $g = 9.8$ cm/sec^2. Find the time T rounded to hundredths if
    (a) $l = 20$ cm;   (b) $l = 100$ cm.
19. ▶The sequence $0.13, 0.1313, 0.131313, 0.13131313, \ldots$ is an increasing sequence, that is, each term is greater than the preceding one. Find the smallest possible rational number $\frac{a}{b}$, where a and b are integers, such that all terms of the above sequence are less than $\frac{a}{b}$. Justify your answer.
★20. Prove: $\sqrt{3}$ is irrational.
★21. Prove: If p is a prime number, then $\sqrt{p}$ is an irrational number.
22. (a) For what whole numbers m is $\sqrt{m}$ a rational number?
    ★(b) Prove your answer in (a).
23. (a) Show that $0.5 + \frac{1}{0.5} \ge 2$.
    ★(b) Prove that any positive real number x plus its reciprocal $\frac{1}{x}$ is greater than or equal to 2.

Review Problems

24. Write a repeating decimal equal to each of the following without using zeros:
    (a) 5   (b) 5.1   (c) $\frac{1}{2}$

25. Write 0.00024 as a fraction in simplest form.
26. Arrange the following from least to greatest:
    4.09, 4.099, 4.0$\overline{9}$, 4.09$\overline{1}$
27. Write 0.$\overline{24}$ as a fraction in simplest form.
28. Write each of the following as a standard numeral:
    (a) $2.08 \cdot 10^5$ (b) $3.8 \cdot 10^{-4}$
29. Joan's salary this year is $18,600. If she receives a 9% raise, what will be her salary next year?
30. If 800 of the 2000 students at the university are males, what percent of the university students are females?

---

*Section 7-6 Radicals and Rational Exponents

nth root
index

The positive solution to $x^2 = 5$ is $\sqrt{5}$. Similarly, the positive solution to $x^4 = 5$ can be denoted as $\sqrt[4]{5}$. In general, the positive solution to $x^n = 5$ is $\sqrt[n]{5}$ and is called the **nth root** of 5. The numeral n is called the **index**. Note that in the expression $\sqrt{5}$, the index 2 is understood and is not expressed.

In general, the positive solution to $x^n = b$, where b is nonnegative, is $\sqrt[n]{b}$. Substituting $\sqrt[n]{b}$ for x in the equation $x^n = b$ gives the following:

$$(\sqrt[n]{b})^n = b.$$

If b is negative, $\sqrt[n]{b}$ cannot always be defined. For example, consider $\sqrt[4]{-16}$. If $\sqrt[4]{-16} = x$, then $x^4 = {}^-16$. Because any nonzero real number raised to the fourth power is positive, there is no real-number solution to $x^4 = {}^-16$ and therefore $\sqrt[4]{-16}$ is not a real number. Similarly, it is not possible to find *any* even root of a negative number. However, the value $^-2$ satisfies the equation $x^3 = {}^-8$. Hence, $\sqrt[3]{-8} = {}^-2$. In general, *the odd root of a negative number is a negative number*.

Because $\sqrt{a}$, if it exists, is positive by definition, $\sqrt{(-3)^2} = \sqrt{9} = 3$ and not $^-3$. Many students think that $\sqrt{a^2}$ always equals a. This is true if $a \geq 0$, but false if $a < 0$. In general, $\sqrt{a^2} = |a|$. Similarly, $\sqrt[4]{a^4} = |a|$ and $\sqrt[6]{a^6} = |a|$, but $\sqrt[3]{a^3} = a$ (why?).

Now, consider an expression such as $4^{1/2}$. What does it mean? By extending the properties of exponents previously developed for integer exponents, we have $4^{1/2} \cdot 4^{1/2} = 4^{1/2 + 1/2} = 4^1$. This implies that $(4^{1/2})^2 = 4$, or $4^{1/2}$, is a square root of 4. The number $4^{1/2}$ is assumed to be the principal square root of 4, that is, $4^{1/2} = \sqrt{4}$. In general, if x is a nonnegative real number, then $x^{1/2} = \sqrt{x}$. Similarly, $(x^{1/3})^3 = x^{(1/3) \cdot 3} = x^1$, and $x^{1/3} = \sqrt[3]{x}$. This discussion leads to the following definition.

DEFINITION

For any real number x and any positive integer n, $x^{1/n} = \sqrt[n]{x}$, where $\sqrt[n]{x}$ is meaningful.

Also, because $(x^m)^{1/n} = \sqrt[n]{x^m}$, and if $(x^m)^{1/n} = x^{m/n}$, it follows that $x^{m/n} = \sqrt[n]{x^m}$.

EXAMPLE 7-29   Simplify each of the following:
(a) $16^{1/4}$ (b) $32^{1/6}$
(c) $(^-8)^{1/3}$ (d) $64^{3/2}$

Solution   (a) $16^{1/4} = \sqrt[4]{16} = 2$ (b) $32^{1/6} = \sqrt[6]{32}$
(c) $(^-8)^{1/3} = \sqrt[3]{-8} = {}^-2$ (d) $64^{3/2} = (2^6)^{3/2} = 2^9 = 512$ ∎

The properties of integer exponents also hold for rational exponents. These properties are equivalent to the corresponding properties of radicals if the expressions involving radicals are meaningful.

PROPERTIES

Let r and s be any rational numbers, x and y be any real numbers, and n be any nonzero integer.

(a) $(xy)^r = x^r \cdot y^r$ implies $(xy)^{1/n} = x^{1/n} y^{1/n}$ and $\sqrt[n]{xy} = \sqrt[n]{x} \sqrt[n]{y}$.

(b) $\left(\dfrac{x}{y}\right)^r = \dfrac{x^r}{y^r}$ implies $\left(\dfrac{x}{y}\right)^{1/n} = \dfrac{x^{1/n}}{y^{1/n}}$ and $\sqrt[n]{\dfrac{x}{y}} = \dfrac{\sqrt[n]{x}}{\sqrt[n]{y}}$.

(c) $(x^r)^s = x^{rs}$ implies $(x^{1/n})^s = x^{s/n}$ and hence, $(\sqrt[n]{x})^s = \sqrt[n]{x^s}$.

The preceding properties can be used to simplify the square roots of many numbers. For example, $\sqrt{96} = \sqrt{16 \cdot 6} = \sqrt{16}\sqrt{6} = 4\sqrt{6}$. When $\sqrt{96}$ is written as $4\sqrt{6}$, it is said to be in *simplest form*. In general, *to write an nth root in simplest form, factor out as many nth powers as possible*. Note that $\sqrt{32} = \sqrt{4 \cdot 8} = 2\sqrt{8}$. Hence, $2\sqrt{8}$ is a simplified form, but not the simplest form. The simplest form of $\sqrt{32}$ is $\sqrt{16 \cdot 2} = \sqrt{16} \cdot \sqrt{2} = 4\sqrt{2}$.

EXAMPLE 7-30 Write each of the following in simplest form:

(a) $\sqrt{200}$ (b) $\sqrt{75}$ (c) $\sqrt[3]{240}$
(d) $\sqrt{3} \cdot \sqrt{15}$ (e) $\sqrt[3]{81} \cdot \sqrt[3]{32}$

Solution
(a) $\sqrt{200} = \sqrt{100 \cdot 2} = \sqrt{100}\sqrt{2} = 10\sqrt{2}$
(b) $\sqrt{75} = \sqrt{25 \cdot 3} = \sqrt{25}\sqrt{3} = 5\sqrt{3}$
(c) $\sqrt[3]{240} = \sqrt[3]{8 \cdot 30} = \sqrt[3]{8}\sqrt[3]{30} = 2\sqrt[3]{30}$
(d) $\sqrt{3} \cdot \sqrt{15} = \sqrt{3 \cdot 15} = \sqrt{45} = \sqrt{9 \cdot 5} = \sqrt{9} \cdot \sqrt{5} = 3\sqrt{5}$
(e) $\sqrt[3]{81} \cdot \sqrt[3]{32} = \sqrt[3]{81 \cdot 32} = \sqrt[3]{3^4 \cdot 2^5} = \sqrt[3]{3^3 \cdot 2^3 \cdot 3 \cdot 2^2} = \sqrt[3]{3^3 \cdot 2^3} \cdot \sqrt[3]{3 \cdot 2^2} = 6\sqrt[3]{12}$

Some expressions in the form $\sqrt{x} + \sqrt{y}$ can be simplified. For example,

$$\sqrt{24} + \sqrt{54} = \sqrt{4 \cdot 6} + \sqrt{9 \cdot 6}$$
$$= \sqrt{4}\sqrt{6} + \sqrt{9}\sqrt{6}$$
$$= 2\sqrt{6} + 3\sqrt{6}$$
$$= (2 + 3)\sqrt{6}$$
$$= 5\sqrt{6}.$$

Be careful! Notice that $\sqrt{9} + \sqrt{4} = 3 + 2 = 5$, but $\sqrt{9 + 4} = \sqrt{13}$. Thus $\sqrt{9} + \sqrt{4} \neq \sqrt{9 + 4}$ and in general, $\sqrt{x} + \sqrt{y} \neq \sqrt{x + y}$.

EXAMPLE 7-31 Write each expression in simplest form.

(a) $\sqrt{20} + \sqrt{45} - \sqrt{80}$ (b) $\sqrt{12} + \sqrt{13}$
(c) $\sqrt{49x} + \sqrt{4x}$

Solution (a) $\sqrt{20} + \sqrt{45} - \sqrt{80} = 2\sqrt{5} + 3\sqrt{5} - 4\sqrt{5} = \sqrt{5}$
(b) $\sqrt{12} + \sqrt{13} = 2\sqrt{3} + \sqrt{13}$
(c) $\sqrt{49x} + \sqrt{4x} = 7\sqrt{x} + 2\sqrt{x} = 9\sqrt{x}$

PROBLEM SET 7-6

1. Write each of the following square roots in simplest form:
   (a) $\sqrt{180}$ (b) $\sqrt{529}$ (c) $\sqrt{363}$
   (d) $\sqrt{252}$ (e) $\sqrt{\dfrac{169}{196}}$ (f) $\sqrt{\dfrac{49}{196}}$

2. Write each of the following in simplest form:
   (a) $\sqrt[3]{-27}$ (b) $\sqrt[5]{96}$ (c) $\sqrt[5]{32}$
   (d) $\sqrt[3]{250}$ (e) $\sqrt[5]{-243}$ (f) $\sqrt[4]{64}$

3. Write each of the following expressions in simplest form:
   (a) $2\sqrt{3} + 3\sqrt{2} + \sqrt{180}$
   (b) $\sqrt[3]{4} \cdot \sqrt[3]{10}$
   (c) $(2\sqrt{3} + 3\sqrt{2})^2$
   (d) $\sqrt{6} \div \sqrt{12}$
   (e) $5\sqrt{72} + 2\sqrt{50} - \sqrt{288} - \sqrt{242}$
   (f) $\sqrt{8/7} \div \sqrt{4/21}$

4. Rewrite each of the following in simplest form:
   (a) $16^{1/2}$ (b) $27^{2/3}$ (c) $64^{5/6}$
   (d) $3^{1/2} \cdot 3^{3/2}$ (e) $(32)^{-2/5}$ (f) $9^{2/3} \cdot 27^{2/9}$
   $\left(\dfrac{1}{16}\right)^{-3/2}$
   (g) $16^{-1/2}$ (h) $27^{-2/3}$ (i) $32^{2/5}$
   (j) $8^{3/2} \cdot 4^{1/4}$ (k) $(10^{1/3} \cdot 10^{-1/6})^6$ (l) $16^{5/12} \cdot 16^{1/3}$
   (m) $64^{2/3}$ (n) $64^{-1/3}$ (o) $(2 \cdot 64^{1/2})^{1/2}$

5. ▶Is $\sqrt{x^2 + y^2} = x + y$ for all values of x and y? Explain your reasoning.

6. ▶Answer the following as being true "sometimes," "always," or "never." Justify your answers.
   (a) $\sqrt{a^2} = a$ (b) $\sqrt{(^-x)^2} = {^-x}$
   (c) $\sqrt{(^-x)^2} = |x|$ (d) $\sqrt{(a+b)^2} = a + b$
   (e) $\sqrt[4]{a^2} = \sqrt{a}$

7. The numbers $3^{3/4}, 3, 3^{5/4}, \ldots, 729$ are in a geometric sequence. Find the number of terms in the sequence.

8. The following exponential function approximates the number of bacteria after t hours: $E(t) = 2^{10} \cdot 16^t$.
   (a) What is the initial number of bacteria, that is, the number when $t = 0$?
   (b) After $\dfrac{1}{4}$ hr, how many bacteria are there?
   (c) After $\dfrac{1}{2}$ hr, how many bacteria are there?

9. ▶Without using a calculator, arrange the following in increasing order. Explain your reasoning.
   $(4/25)^{-1/3}, (25/4)^{1/3}, (4/25)^{-1/4}$

10. ▶Without using a calculator, determine which is greater in each of the following. Explain your reasoning.
    (a) $\sqrt{3}$ or $\sqrt[3]{4}$
    (b) $\sqrt[3]{3}$ or $\sqrt{2}$
    ★(c) $\sqrt{12} + \sqrt{14}$ or $\sqrt{11} + \sqrt{15}$

11. Write $\sqrt{2\sqrt{2\sqrt{2}}}$ in the form $\sqrt[n]{2^m}$, where n and m are positive integers.

12. Solve for x, where x is a rational number.
    (a) $3^x = 81$ (b) $4^x = 8$
    (c) $128^{-x} = 16$ (d) $\left(\dfrac{4}{9}\right)^{3x} = \dfrac{32}{243}$

13. What is the value of $\sqrt[3]{(x-2)^{-2}}$ when $x = 6$?

14. (a) For what values of n is $\sqrt[n]{a}$ meaningful when $a < 0$?
    (b) For what values of m and n is $\sqrt[n]{a^m}$ meaningful when $a < 0$?

15. Classify each of the following numbers as rational or irrational:
    (a) $\sqrt{2} - \dfrac{2}{\sqrt{2}}$ (b) $(\sqrt{2})^{-4}$
    (c) $\dfrac{1}{1 + \sqrt{2}}$ (d) $\dfrac{1}{1 + \sqrt{2}} + 1 - \sqrt{2}$

Solution to the Preliminary Problem

Understanding the Problem. Diane has three discount coupons for 10%, 20%, and 30%, respectively. On certain sale items, she is allowed to combine the coupons to increase her discount. The coupons are applied successively after Diane chooses the order in which she wishes them to be applied. We are asked to determine in which order Diane should ask to have the discounts calculated and what her final discount will be.

Devising a Plan. Part of the problem is one of reference. When the first discount coupon is applied, this discount is calculated on the original price. The second discount coupon is then applied to the discounted price, not the original price. Likewise, the third coupon is then applied to the twice-discounted price. There are a total of six different orders in which these three discount coupons can be applied. We *could* perform the calculations in all six cases to determine which order is best. But suppose that Diane had a coupon for 40% in addition to the other three. We would then have 24 different orders and the task of checking all of these would be tedious. We will use the strategy of *examining a simpler case*. Suppose there were only two coupons, one for 10% and one for 20%. Now there are only two possible orders to consider. We will calculate each of these, examine our results, and see if we can generalize. We will further *simplify the problem* by assuming that the original price of Diane's purchase is $100.

Carrying Out the Plan. First, we consider the case of using the 10% coupon first, then the 20% one. The new price after applying the 10% coupon is $100 - 10\%$ of $100 = 100 - 0.10 \cdot 100 = 90$. The new price after the second discount is applied is $90 - 20\%$ of $90 = 90 - 0.20 \cdot 90 = 72$. Now we consider the other case. The new price after applying the 20% coupon is $100 - 20\%$ of $100 = 100 - 0.20 \cdot 100 = 80$. The new price after the second discount is applied is $80 - 10\%$ of $80 = 80 - 0.10 \cdot 80 = 72$. In both cases, we got $72 for a final discounted price. Notice that 10% "off" a price is the same as 90% of the original price. Likewise, 20% "off" is the same as 80% of the original price, and so forth. Thus we could have calculated the first case as $[100 \cdot 0.90] \cdot 0.80$. Because the original problem also involves a 30% coupon, the final price will be $100 \cdot 0.90 \cdot 0.80 \cdot 0.70 = \$50.40$. So the final discount is $100 - 50.4 = \$49.6$, or a 49.6% discount. Since multiplication is associative and commutative, the order of applying the coupons makes no difference.

Looking Back. We could have considered 1 of the 6 cases in the following manner. The discounted price of a $100 item would have been 70% of 80% of 90% of $100. (The word "of" usually indicates the operation of multiplication.) Since multiplication is commutative and associative, the order would make no difference.

QUESTIONS FROM THE CLASSROOM

1. A student says that $3\frac{1}{4}\% = 0.03 + 0.25 = 0.28$. What is the error, if any?
2. Why is $\sqrt{25} \neq {}^-5$?
3. A student claims that $\sqrt{(-5)^2} = {}^-5$ because $\sqrt{a^2} = a$. Is this correct?
4. Another student says that $\sqrt{(-5)^2} = [(^-5)^2]^{1/2} = (^-5)^{2/2} = (^-5)^1 = {}^-5$. Is this correct?
5. A student claims that the equation $\sqrt{-x} = 3$ has no solution since the square root of a negative number does not exist. Why is this argument wrong?
6. On a test, a student wrote the following:

$$\frac{x^2}{7} - 2 \geq {}^-3$$

$$\frac{x^2}{7} \geq {}^-1$$

$$x^2 \geq {}^-7$$

Because $\sqrt{-7}$ does not exist, there is no solution. What is the error?

7. A student reports that it is impossible to mark a product up 150% because 100% of something is all there is. What is your response?
8. A student reports that $-438{,}340{,}000$ cannot be written in scientific notation. How do you respond?
9. A student multiplies $(6.5)(8.5)$ to obtain:

$$\begin{array}{r} 8.5 \\ \times\ 6.5 \\ \hline 4\ 2\ 5 \\ 5\ 1\ 0 \\ \hline 5\ 5.2\ 5 \end{array}$$

However, when the student multiplies $8\tfrac{1}{2} \cdot 6\tfrac{1}{2}$, the following is obtained:

$$\begin{array}{r} 8\tfrac{1}{2} \\ \times\ 6\tfrac{1}{2} \\ \hline 4\tfrac{1}{4}\quad \left(\tfrac{1}{2}\cdot 8\tfrac{1}{2}\right) \\ 48\phantom{\tfrac{1}{4}}\quad (6\cdot 8) \\ \hline 52\tfrac{1}{4} \end{array}$$

How is this possible?

10. A student claims that 0.36 is greater than 0.9 because 36 is greater than 9. How do you respond?
11. Respond to the following:
    (a) A student claims that $\dfrac{9443}{9444}$ and $\dfrac{9444}{9445}$ are equal because both display 0.9998941 on his scientific calculator when the divisions are performed.
    (b) Another student claims that the fractions are not equal and wants to know if there is any way the same calculator can determine which is greater.
12. A student says that the solution of $x^2 = 5$ is written as $x = \pm\sqrt{5}$ and therefore $\sqrt{5}$ has two values, one positive and one negative. How do you respond?
13. A student argues that a $p\%$ increase in salary followed by a $q\%$ decrease is equivalent to a $q\%$ decrease followed by a $p\%$ increase because of the commutative property of multiplication. How do you respond?
14. A student argues that $0.01\% = 0.01$ because in 0.01%, the percent is already written as a decimal. How do you respond?

CHAPTER OUTLINE

I. Decimals
  A. Every rational number can be represented as a terminating or repeating decimal.
  B. A rational number $\dfrac{a}{b}$, whose denominator is of the form $2^m \cdot 5^n$, where m and n are whole numbers, can be expressed as a **terminating decimal.**
  C. A **repeating decimal** is a decimal with a block of digits, called the **repetend**, repeated infinitely many times.
  D. A number is in **scientific notation** if it is written as the product of a number n that is greater than or equal to 1 and an integral power of 10. The number of digits in n is called the number of **significant digits** of n.
  E. An **irrational number** is represented by a nonterminating, nonrepeating decimal.
  F. **Percent** means *per hundred*. Percent is written using the % symbol; $x\% = \dfrac{x}{100}$.

* II. Interest
  A. **Simple interest** is computed using the formula $I = Prt$, where I is the interest, P is the principal, r is the annual interest rate, and t is the time in years.
  B. When **compound interest** is involved, we use the formula $A = P(1 + i)^n$, where A is the balance, P is the principal, i is the interest rate per period, and n is the number of periods.

III. Real numbers
  A. The set of **real numbers** is the set of all decimals, namely, the union of the set of rational numbers and the set of irrational numbers.
  B. If a is any whole number, then the **principal square root** of a, denoted by $\sqrt{a}$, is the nonnegative number b such that $b \cdot b = b^2 = a$.
  C. Square roots can be found using the **squeezing method.**

*IV. Radicals and rational exponents
  A. $\sqrt[n]{x}$, or $x^{1/n}$, is called **nth root** of x and n is called the **index.**
  B. The following properties hold for radicals if the expressions involving radicals are meaningful:
    (a) $\sqrt[n]{xy} = \sqrt[n]{x} \cdot \sqrt[n]{y}$
    (b) $\sqrt[n]{\dfrac{x}{y}} = \dfrac{\sqrt[n]{x}}{\sqrt[n]{y}}$
    (c) $(\sqrt[n]{x})^m = \sqrt[n]{x^m}$

CHAPTER TEST

1. Perform the following operations:
   (a) $3.6 + 2.007 - 6.3$
   (b) $(5.2)(6.07)$
   (c) $(5.1 + 6.32)0.02$
   (d) $0.12032 \div 3.76$
   (e) $0.012 - 0.109$
   (f) $(0.02)^4$

2. Write each of the following in expanded form:
   (a) 32.012
   (b) 0.00103

3. ▶Give a test to determine if a fraction can be written as a terminating decimal, without actually performing the division. Explain why this test is valid.

4. A board is 442.4 cm long. How many shelves can be cut from it if each shelf is 55.3 cm long? (Disregard the width of the cuts.)

5. Write each of the following as a decimal:
   (a) $\frac{4}{7}$
   (b) $\frac{1}{8}$
   (c) $\frac{2}{3}$
   (d) $\frac{5}{8}$

6. Write each of the following as a fraction in simplest form:
   (a) 0.28
   (b) $0.\overline{3}$
   (c) $2.0\overline{8}$

7. Round each of the following numbers as specified:
   (a) 307.625 to the nearest hundredth
   (b) 307.625 to the nearest tenth
   (c) 307.625 to the nearest unit
   (d) 307.625 to the nearest hundred

8. Solve each of the following for x, where x is a real number:
   (a) $0.2x - 0.75 \geq \frac{1}{2}(x - 3.5)$
   (b) $0.\overline{9} + x = 1$
   (c) $23\%(x) = 4600$
   (d) 10 is x percent of 50
   (e) 17 is 50% of x
   (f) $0.\overline{3} + x = 1$

9. Answer each of the following:
   (a) 6 is what percent of 24?
   (b) What is 320% of 60?
   (c) 17 is 30% of what number?
   (d) 0.2 is what percent of 1?

10. Change each of the following to percents:
    (a) $\frac{1}{8}$
    (b) $\frac{3}{40}$
    (c) 6.27
    (d) 0.0123
    (e) $\frac{3}{2}$

11. Change each of the following percents to decimals:
    (a) 60%
    (b) $\left(\frac{2}{3}\right)\%$
    (c) 100%

12. ▶Answer each of the following and explain your answers:
    (a) Is the set of irrational numbers closed under addition?
    (b) Is the set of irrational numbers closed under subtraction?
    (c) Is the set of irrational numbers closed under multiplication?
    (d) Is the set of irrational numbers closed under division?

13. Find an approximation for $\sqrt{23}$ correct to three decimal places.

14. Rewrite each of the following in scientific notation:
    (a) 426,000
    (b) 32
    (c) 0.00000237
    (d) 0.325

15. What is the number of significant digits in each of the numbers in Problem 14?

16. Classify each of the following as rational or irrational. (Assume the patterns shown continue.)
    (a) 2.191199119999119999119 . . .
    (b) $\frac{1}{\sqrt{2}}$
    (c) $\frac{4}{9}$
    (d) 0.0011001100110011 . . .
    (e) 0.001100011000011 . . .

17. Sandy received a dividend that equals 11% of the value of her investment. If her dividend was $1020.80, how much was her investment?

18. Five computers in a shipment of 150 were found to be defective. What percent of the computers were defective?

19. On a mathematics examination, a student missed 8 of 70 questions. What percent of the questions, rounded to the nearest tenth, did the student do correctly?

20. ▶A microcomputer system costs $3450 at present. This is 60% of the cost 4 yr ago. What was the cost of the system 4 yr ago? Explain your reasoning.

21. If, on a purchase of one new suit, you are offered successive discounts of 5%, 10%, or 20%, in any order you wish, what order should you choose?

22. Jane bought a bicycle and sold it for 30% more than she paid for it. She sold it for $104. How much did she pay for it?

*23. A company was offered a $30,000 loan at a 12.5% annual interest rate for 4 yr. Find the simple interest due on the loan at the end of 4 yr.

*24. A money-market fund pays 14% annual interest compounded quarterly. What is the value of a $10,000 investment after 3 yr?

*25. Find the simplest form for each of the following:
(a) $\sqrt{242}$ (b) $\sqrt{288}$
(c) $\sqrt{360}$ (d) $\sqrt[3]{162}$

*26. Write each of the following in simplest form, with non-negative exponents in the final answer:
(a) $\left(\frac{1}{2}\right)^4 \left(\frac{1}{2}\right)^7$ (b) $5^{-16} \div 5^4$
(c) $\left[\left(\frac{2}{3}\right)^7\right]^{-4}$ (d) $3^{16} \cdot 3^2$

SELECTED BIBLIOGRAPHY

Barson, A. and L. Barson "Ideas." *Arithmetic Teacher* 35 (January 1988): 19–24.

Boling, B. "A Different Method for Solving Percentage Problems." *Mathematics Teacher* 78 (October 1985): 523–524.

Chow, P. and T. Lin. "Extracting Square Roots Made Easy." *Arithmetic Teacher* 29 (November 1981): 48–50.

Coburn, T. "Percentage and the Hand Calculator." *Mathematics Teacher* 79 (May 1986): 361–367.

Lester, F. "Teacher Education: Preparing Teachers to Teach Rational Numbers." *Arithmetic Teacher* 31 (February 1984): 54–56.

Malcom, P. S. "Understanding Rational Numbers." *Mathematics Teacher* 80 (October 1987): 518–521.

Payne, J. "Curricular Issues: Teaching Rational Numbers." *Arithmetic Teacher* 31 (February 1984): 14–17.

Prevost, F. "Teaching Rational Numbers—Junior High School." *Arithmetic Teacher* 31 (February 1984): 43–46.

Quintero, A. "Helping Children Understand Ratios." *Arithmetic Teacher* 34 (April 1987): 17–21.

Rossini, B. "Using Percent Problems to Promote Critical Thinking." *Mathematics Teacher* 81 (January 1988): 31–34.

Sherzer, L. "Expanding the Limits of the Calculator Display." *Mathematics Teacher* 79 (January 1986): 20–21.

Skypek, D. "Special Characteristics of Rational Numbers." *Arithmetic Teacher* 31 (February 1984): 10–12.

Soler, F. and R. Schuster. "Compound Growth and Related Situations: A Problem-Solving Approach." *Mathematics Teacher* 75 (November 1982): 640–643.

Sullivan, K. "Money—A Key to Mathematical Success." *Arithmetic Teacher* 29 (November 1981): 34–35.

Teahan, T. "How I Learned to Do Percents." *Arithmetic Teacher* 27 (January 1979): 16–17.

Wiebe, J. "Manipulating Percentages." *Mathematics Teacher* 79 (January 1986): 21, 23–26.

Zawojewski, J. "Initial Decimal Concepts: Are They Really So Easy?" *Arithmetic Teacher* 30 (March 1983): 52–56.

8 Probability

Preliminary Problem

Stephen placed three letters in envelopes while he was having a telephone conversation. He addressed the envelopes and sealed them without checking to see if each letter was in the correct envelope. What is the probability that each of the letters was inserted correctly?

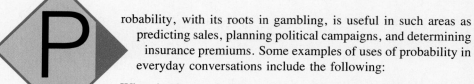

Probability, with its roots in gambling, is useful in such areas as predicting sales, planning political campaigns, and determining insurance premiums. Some examples of uses of probability in everyday conversations include the following:

What is the *probability* that the Braves will win the World Series?

The *odds* are 2 to 1 that Millie will win the dog show.

There is *no chance* you will get a raise.

There is a 50% *chance* of rain today.

Probability plays an important role in both the K–4 and the 5–8 *Standards*. The following quote from the 5–8 *Standards* (p. 110) shows how important the study of probability is in solving problems.

To see how the predictions we hear and see every day are based on probability, students must use their knowledge of probability to solve problems. In modeling problems, conducting simulations, and collecting, graphing, and studying data, students will come to understand how predictions can be based on data. Mathematically derived probabilities can be determined by building a table or tree diagram, creating an area model, making a list, or using simple counting procedures. Students develop an appreciation of the power of simulation and experimentation by comparing experimental results to the mathematically derived probabilities.

In this chapter, we use tree diagrams and geometric probabilities (area models) to solve problems and to analyze games involving spinners, cards, and dice. Counting techniques are introduced and the role of simulations in probability is discussed. Many of the ideas in this chapter are adapted from the materials developed in the Comprehensive School Mathematics Project (CSMP).

HISTORICAL NOTE

The originator of probability theory is not known. However, Blaise Pascal (1623–1662), the French philosopher and mathematician, solved two gambling problems posed by the Chevalier de Meré, a professional gambler, and with Pierre de Fermat (1601–1665), is considered one of the founders of probability theory. One of de Meré's questions was how to divide the stakes if two players start, but fail to complete, a game in which the winner is the one who wins three matches out of five. A Dutch mathematician, Christian Huygens (1629–1695) also worked on the problem and as a result wrote *De ratiociniis in ludo alea,* the first treatise on probability.

Section 8-1 How Probabilities Are Determined

experiment
outcome

Probabilities are ratios, expressed as fractions, decimals, or percents, determined by considering results or outcomes of experiments. An **experiment** is an activity such as tossing a coin. Each of the possible results of an experiment is an **outcome.** If we toss a coin then, assuming that the coin cannot land on its edge, there are two distinct possible outcomes: heads (*H*) and tails (*T*).

sample space   A set of all possible outcomes for an experiment is called a **sample space.** In a single coin toss, the sample space *S* is given by $S = \{H, T\}$. The sample space can be modeled by a tree diagram, as shown in Figure 8-1. Each outcome of the experiment is designated by a separate branch in the tree diagram.

FIGURE 8-1

The sample space *S* for rolling the standard die in Figure 8-2(a) is $S = \{1, 2, 3, 4, 5, 6\}$. A tree diagram for the sample space is given in Figure 8-2(b).

FIGURE 8-2

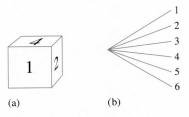

(a)   (b)

event   Any subset of a sample space is called an **event.** For example, the set of all even-numbered rolls $\{2, 4, 6\}$ is a subset of all possible rolls of a die $\{1, 2, 3, 4, 5, 6\}$ and is an event.

E X A M P L E 8 - 1   Suppose that an experiment consists of drawing 1 slip of paper from a jar containing 12 slips of paper, each with a different month of the year written on it. Find each of the following:

(a) The sample space *S* for the experiment
(b) The event *A* consisting of outcomes having a month beginning with J
(c) The event *B* consisting of outcomes having the name of a month that has exactly four letters
(d) The event *C* consisting of outcomes having a month that begins with M or N

Solution   (a) $S = \{$January, February, March, April, May, June, July, August, September, October, November, December$\}$
(b) $A = \{$January, June, July$\}$
(c) $B = \{$June, July$\}$
(d) $C = \{$March, May, November$\}$

Determining Probabilities

Around 1900, the English statistician Karl Pearson tossed a coin 24,000 times and recorded 12,012 heads. During World War II, the Dane, John Kerrich, a prisoner of war, tossed a coin 10,000 times. A subset of his results is given in Table 8-1. The *relative frequency* column on the right is obtained by dividing the number of heads by the number of tosses of the coin.

TABLE 8-1

Number of Tosses	Number of Heads	Relative Frequency
10	4	0.400
50	25	0.500
100	44	0.440
500	255	0.510
1,000	502	0.502
5,000	2,533	0.507
8,000	4,034	0.504
10,000	5,067	0.507

After 10 tosses, the data in Table 8-1 suggest that heads might occur $\frac{4}{10}$ of the time. After 50 tosses, the data in Table 8-1 suggest that we could expect heads about $\frac{25}{50}$ of the time. Sometimes Kerrich tossed heads a little less than half the time and sometimes a little more than half the time; but as the number of tosses increased, he obtained heads close to half the time. The relative frequency for Pearson's 24,000 tosses gives a similar result of 12,012/24,000 or 0.5005.

experimentally
empirically

When a probability is determined by observing outcomes of experiments, it is said to be determined **experimentally,** or **empirically.** The exact number of heads that occurs when a fair coin is tossed a few times cannot be accurately predicted. Probabilities only suggest what will happen in the "long run." When a fair coin is tossed many times and the fraction (or proportion) of heads is near $\frac{1}{2}$, we say that the probability of heads occurring

equally likely

is $\frac{1}{2}$ and write $P(H) = \frac{1}{2}$. We also say that the two outcomes, H and T, are **equally likely** because one outcome is just as likely to occur as the other.

> **REMARK** Technically, we should write $P(\{H\})$ because the event of tossing a head is a set; H is an element of the set. However, common usage is to write $P(H)$.

We could also argue that each side of a fair coin should appear roughly $\frac{1}{2}$ of the time in long strings of tosses because the coin is symmetric and has two sides. Hence, we would again conclude that

$$P(H) = P(T) = \frac{1}{2}.$$

SECTION 8-1 HOW PROBABILITIES ARE DETERMINED

theoretically In this case, assuming ideal conditions, we determine the probability **theoretically.** If an experiment is repeated a great number of times, we expect the experimental probability to approach the theoretical probability.

If a *fair* die (one that is just as likely to land on any of the numerals 1 through 6) is rolled many times, each outcome will appear about $\frac{1}{6}$ of the time and each outcome has a probability of $\frac{1}{6}$. We write, for example, $P(4) = \frac{1}{6}$, for the probability of tossing a 4. Similarly, any other face should also be assigned probability $\frac{1}{6}$.

For a fair die, $S = \{1, 2, 3, 4, 5, 6\}$ and $P(1) = P(2) = P(3) = P(4) = P(5) = P(6) = \frac{1}{6}$. In each case, the probability for each outcome is a number between 0 and 1 and the sum of the probabilities for the distinct outcomes in the sample space is equal to 1. For a sample space with equally likely outcomes, the probability of an event A can be defined as follows.

> **DEFINITION OF PROBABILITY OF AN EVENT WITH EQUALLY LIKELY OUTCOMES**
>
> For an experiment with sample space S and equally likely outcomes, the **probability of an event A** is given by
> $$P(A) = \frac{\text{Number of elements of } A}{\text{Number of elements of } S} = \frac{n(A)}{n(S)}.$$

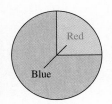

FIGURE 8-3

This definition applies only to sample spaces with equally likely outcomes. Applying the definition to outcomes that are not equally likely leads to incorrect conclusions. For example, the sample space for spinning the spinner in Figure 8-3 is given by $S = \{\text{Red}, \text{Blue}\}$, but the outcome Blue is more likely to occur than is the outcome Red, and $P(\text{Red})$ is not equal to $\frac{1}{2}$ but to $\frac{90}{360}$, or $\frac{1}{4}$. If the spinner were spun 100 times, it would seem reasonable to expect that about $\frac{1}{4}$, or 25, of the outcomes would be Red, whereas about $\frac{3}{4}$, or 75, of the outcomes would be Blue.

at random **EXAMPLE 8-2** If each of the 25 numbers in set $S = \{1, 2, 3, 4, 5, \ldots, 25\}$ is being chosen **at random**, that is, with an equal chance of being drawn, calculate each probability below:

(a) The event A that an even number is drawn
(b) The event B that a number less than 10 and greater than 20 is drawn
(c) The event C that a number less than 26 is drawn
(d) The event D that a prime number is drawn
(e) The event E that a number both even and prime is drawn

Solution Each of the 25 numbers in set S has an equal chance of being drawn.

(a) $A = \{2, 4, 6, 8, 10, 12, 14, 16, 18, 20, 22, 24\}$, so $n(A) = 12$. Thus,
$$P(A) = \frac{n(A)}{n(S)} = \frac{12}{25}.$$

(b) $B = \emptyset$, so $n(B) = 0$. Thus, $P(B) = \frac{0}{25} = 0$.

(c) $C = S$ and $n(C) = 25$. Thus, $P(C) = \frac{25}{25} = 1$.

(d) $D = \{2, 3, 5, 7, 11, 13, 17, 19, 23\}$, so $n(D) = 9$. Thus,
$$P(D) = \frac{n(D)}{n(S)} = \frac{9}{25}.$$

(e) $E = \{2\}$, so $n(E) = 1$. Thus, $P(E) = \frac{1}{25}$. ∎

impossible event

In Example 8-2(b), event B is the empty set. An event such as B that has no outcomes in it is called an **impossible event** and *has probability* 0. If the word *and* were replaced by *or* in Example 8-2(b), then event B would no longer be the empty set. In Example 8-2(c), event C consists of drawing a number less than 26 on a single draw. Because every number in S is less than 26, $P(C) = \frac{25}{25} = 1$. An event that has probability 1 is called a **certain event**.

certain event

An event is a subset of a sample space. The event can have no more outcomes than in the sample space. In addition, an event can have no fewer than 0 outcomes. Thus, if A is any event, it occurs between 0% and 100% of the time, and we have the following:

$$0\% \leq P(A) \leq 100\% \quad \text{or} \quad 0 \leq P(A) \leq 1.$$

Mutually Exclusive Events

Consider one spin of the wheel shown in Figure 8-4. For this experiment, we have $S = \{0, 1, 2, 3, 4, 5, 6, 7, 8, 9\}$. If $A = \{0, 1, 2, 3, 4\}$ and $B = \{5, 7\}$, then $A \cap B = \emptyset$. Two such events are called **mutually exclusive** events. If event A occurs, then event B cannot occur, and we have the following definition.

mutually exclusive

FIGURE 8-4

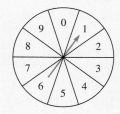

DEFINITION OF MUTUALLY EXCLUSIVE EVENTS

Events A and B are **mutually exclusive** if $A \cap B = \emptyset$.

Each outcome in the preceding sample space S is equally likely, with probability $\frac{1}{10}$. Thus, if we write the probability of A or B as $P(A \cup B)$, we have the following:

$$P(A \cup B) = \frac{n(A \cup B)}{n(S)} = \frac{7}{10} = \frac{5+2}{10} = \frac{5}{10} + \frac{2}{10}$$
$$= \frac{n(A)}{n(S)} + \frac{n(B)}{n(S)} = P(A) + P(B)$$

The result developed in this example is true for any mutually exclusive events. In general, we have the following property.

PROPERTY

If events A and B are mutually exclusive, then $P(A \cup B) = P(A) + P(B)$.

This property follows immediately from the fact that, if $A \cap B = \emptyset$, then $n(A \cup B) = n(A) + n(B)$.

The probability of an event with mutually exclusive outcomes but without equally likely outcomes may be computed in the same manner. Consider a loaded die in which $P(1) = P(2) = \frac{3}{10}$, and $P(3) = P(4) = P(5) = P(6) = \frac{1}{10}$. If event A consists of tossing a 2 and event B consists of tossing a 4, then events A and B are mutually exclusive, and the following represents the probability of tossing a 2 or a 4:

$$P(A \cup B) = P(A) + P(B) = P(2) + P(4) = \frac{3}{10} + \frac{1}{10} = \frac{4}{10}.$$

We find the probability of tossing a 2 or a 4 by adding the probabilities of events A and B. Likewise, the probability of an event C of tossing an even number $\{2, 4, 6\}$ is given by finding the sum of $P(A \cup B)$ and $P(6)$ as shown below:

$$P(C) = [P(A) + P(B)] + P(6) = \left(\frac{3}{10} + \frac{1}{10}\right) + \frac{1}{10} = \frac{5}{10}.$$

This leads to the following definition, which holds for probabilities with or without equally likely outcomes.

DEFINITION OF PROBABILITY OF AN EVENT

The **probability of an event** is equal to the sum of the probabilities of all the outcomes in the event.

EXAMPLE 8-3 If we draw a card at random from an ordinary deck of playing cards, what is the probability that the card is an ace?

Solution There are 52 cards in a deck, and four are aces. If event A is drawing an ace, then $A = \left\{ \boxed{A\spadesuit}, \boxed{A\clubsuit}, \boxed{A\diamondsuit}, \boxed{A\heartsuit} \right\}$. We first use the definition of probability for equally likely outcomes to compute the following:

$$P(A) = \frac{n(A)}{n(S)} = \frac{4}{52}.$$

An alternate approach is to find the sum of each of the probabilities of the outcomes in the event, where the probability of drawing any single ace from the deck is $\frac{1}{52}$:

$$P(A) = \frac{1}{52} + \frac{1}{52} + \frac{1}{52} + \frac{1}{52} = \frac{4}{52}.$$

∎

Complementary Events

complement of A

Two mutually exclusive events whose union is the sample space are called *complementary events*. If A is an event, the **complement of A**, written $\overline{A}$, where A fails to happen is also an event.

For example, consider the event, $A = \{2, 4\}$, of tossing a 2 or a 4 using a standard die. The complement of A is the set $\overline{A} = \{1, 3, 5, 6\}$. Because the sample space S is $\{1, 2, 3, 4, 5, 6\}$, we have $P(A) = \frac{2}{6}$, and $P(\overline{A}) = \frac{4}{6}$. In this case, $P(\overline{A}) = 1 - P(A)$. This is true in general for any set A and its complement. Thus, we have the following:

$$A \cup \overline{A} = S$$
$$n(A \cup \overline{A}) = n(A) + n(\overline{A})$$
$$P(A \cup \overline{A}) = P(A) + P(\overline{A}) = P(S)$$

However, S is a certain event and has probability 1. Therefore,

$$P(A) + P(\overline{A}) = 1$$
$$P(A) = 1 - P(\overline{A}), \text{ or } P(\overline{A}) = 1 - P(A)$$

Non–Mutually Exclusive Events

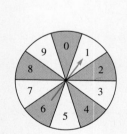

FIGURE 8-5

Consider the spinner in Figure 8-5. Let E be the event of spinning an even number and T be the event of spinning a number divisible by 3.

$$E = \{0, 2, 4, 6, 8\}$$
$$T = \{0, 3, 6, 9\}$$

The event of spinning an even number and a number divisible by 3, denoted by $E \cap T$, is $\{0, 6\}$. Because $E \cap T = \{0, 6\}$, E and T are not mutually exclusive. However, we can compute the probability of E or T as follows:

$$P(E \cup T) = \frac{n(E \cup T)}{n(S)}.$$

Because $E \cup T = \{0, 2, 4, 6, 8, 3, 9\}$, $n(E \cup T) = 7$. Also, $n(S) = 10$, and the $P(E \cup T) = \frac{7}{10}$.

In general, we can also compute the probability of E or T by using notions of sets from Chapter 2. We know

$$n(E \cup T) = n(E) + n(T) - n(E \cap T).$$

Therefore,

$$P(E \cup T) = \frac{n(E \cup T)}{n(S)}$$
$$= \frac{n(E) + n(T) - n(E \cap T)}{n(S)}$$
$$= \frac{n(E)}{n(S)} + \frac{n(T)}{n(S)} - \frac{n(E \cap T)}{n(S)}$$
$$= P(E) + P(T) - P(E \cap T)$$

This result is true in general, and the properties of probability are summarized below.

PROPERTIES OF PROBABILITY

1. $P(\emptyset) = 0$ (impossible event).
2. $P(S) = 1$, where S is the sample space (certain event).
3. For any event A, $0 \leq P(A) \leq 1$.
4. If A and B are events and $A \cap B = \emptyset$, then $P(A \cup B) = P(A) + P(B)$.
5. If A and B are events and $A \cap B \neq \emptyset$, then $P(A \cup B) = P(A) + P(B) - P(A \cap B)$.
6. If A is an event, then $P(\overline{A}) = 1 - P(A)$.

EXAMPLE 8-4 A golf bag contains two red tees, four blue tees, and five white tees.

(a) What is the probability of the event A that a tee drawn at random is red?
(b) What is the probability of the event "not A," that is, that a tee drawn at random is not red?
(c) What is the probability of the event that a tee drawn at random is either red or blue?

Solution

(a) Because the bag contains a total of $2 + 4 + 5$, or 11, tees and 2 tees are red, $P(A) = \frac{2}{11}$.

(b) The bag contains 11 tees and 9 are not red, so the probability of "not A" is $\frac{9}{11}$. Also, notice that $P(\overline{A}) = 1 - P(A) = 1 - \frac{2}{11} = \frac{9}{11}$.

(c) The bag contains 2 red tees and 4 blue tees and $R \cap B = \emptyset$, so $P(R \cup B) = \frac{2}{11} + \frac{4}{11}$, or $\frac{6}{11}$. ∎

EXAMPLE 8-5 Find the probability of rolling a sum of 7 or 11 when rolling a fair pair of dice.

Solution To solve this problem, we use the strategy of *making a table,* as in Figure 8-6(a), to show all possible outcomes of tossing the dice. We know that there are six possible results from tossing the first die and six from tossing the second die, so by the Fundamental Counting Principle, there are $6 \cdot 6$, or 36, entries in the chart. It may be easier to read the results when they are recorded as ordered pairs, as in Figure 8-6(b), where the first component represents the number on the first die and the second component represents the number on the second die. We find the possible sums in rolling the pair of dice {2, 3, 4, 5, 6, 7, 8, 9, 10, 11, 12} by adding the components of the ordered pairs, as shown in Figure 8-6(c).

FIGURE 8-6

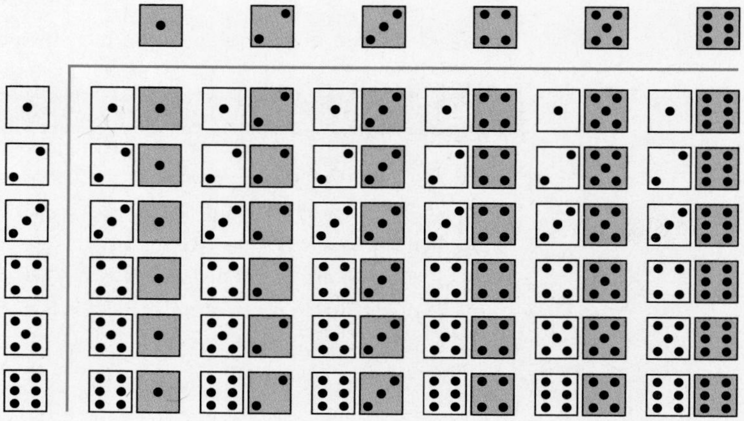

(a)

(b)

	Number on Second Die					
Number on First Die	1	2	3	4	5	6
1	(1,1)	(1, 2)	(1, 3)	(1, 4)	(1, 5)	(1, 6)
2	(2, 1)	(2, 2)	(2, 3)	(2, 4)	(2, 5)	(2, 6)
3	(3, 1)	(3, 2)	(3, 3)	(3, 4)	(3, 5)	(3, 6)
4	(4, 1)	(4, 2)	(4, 3)	(4, 4)	(4, 5)	(4, 6)
5	(5, 1)	(5, 2)	(5, 3)	(5, 4)	(5, 5)	(5, 6)
6	(6, 1)	(6, 2)	(6, 3)	(6, 4)	(6, 5)	(6, 6)

(c)

	Number on Second Die					
Number on First Die	1	2	3	4	5	6
1	2	3	4	5	6	7
2	3	4	5	6	7	8
3	4	5	6	7	8	9
4	5	6	7	8	9	10
5	6	7	8	9	10	11
6	7	8	9	10	11	12

From Figure 8-6(c), we see that a sum of 7 can be obtained in six different ways. Similarly, a sum of 11 can be obtained in two different ways. Because there are 36 different elements in the sample space, we have $P(7) = \frac{6}{36}$ and $P(11) = \frac{2}{36}$ and $P(7 \text{ or } 11) = P(7) + P(11) = \frac{6}{36} + \frac{2}{36} = \frac{8}{36}$.

PROBLEM SET 8-1

1. Write the sample space for each of the following experiments:
   (a) Selecting at random one of the ten most recent presidents of the United States
   (b) Choosing at random one of your classmates in this mathematics class
   (c) Choosing at random one of the members of the U.S. House of Representatives from your home state

2. An experiment consists of selecting the last digit of a telephone number. Assume that each of the ten digits is equally likely to appear as a last digit. List each of the following:
   (a) The sample space
   (b) The event consisting of outcomes that the digit is less than 5
   (c) The event consisting of outcomes that the digit is odd
   (d) The event consisting of outcomes that the digit is not 2
   (e) Find the probability of each of the events (b)–(d).

3. If the spinner shown is spun, find the probabilities of obtaining each of the following:

   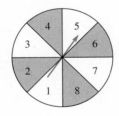

   (a) P(factor of 35)
   (b) P(multiple of 3)
   (c) P(even number)
   (d) P(6 or 2)
   (e) P(11)
   (f) P(composite number)
   (g) P(neither a prime nor a composite)

4. A card is selected from an ordinary bridge deck consisting of 52 cards. Find the probabilities for each of the following:
   (a) A red card
   (b) A face card
   (c) A red card or a ten
   (d) A queen
   (e) Not a queen
   (f) A face card or a club
   (g) A face card and a club
   (h) Not a face card and not a club

5. A drawer contains six black socks, four brown socks, and two green socks. Suppose that one sock is drawn from the drawer and that it is equally likely that any one of the socks is drawn. Find the probabilities for each of the following:
   (a) The sock is brown.
   (b) The sock is either black or green.
   (c) The sock is red.
   (d) The sock is not black.

6. Each letter of the alphabet is written on a separate piece of paper and placed in a box and then one piece of paper is drawn at random.
   (a) What is the probability that the paper has a vowel written on it?
   (b) What is the probability that the paper has a consonant written on it?

7. If the probability of being able to board a connecting flight to Boston is 0.2, what is the probability of missing the connecting flight?

8. Riena has six unmarked computer disks in a box, where each is dedicated to her homework areas of English, mathematics, French, American history, chemistry, and computer science. Answer the following questions:
   (a) If she chooses a computer disk at random, what is the probability she chooses the English disk?
   (b) What is the probability that Riena chooses a humanities disk?

9. The questions below refer to a very popular dice game, craps, in which a player rolls two dice.
   (a) Rolling a sum of 7 or 11 on the first roll of the dice is a win. What is the probability of winning on the first roll?
   (b) Rolling a sum of 2, 3, or 12 on the first roll of the dice is a loss. What is the probability of losing on the first roll?
   (c) Rolling a sum of 4, 5, 6, 8, 9, or 10 on the first roll is neither a win nor a loss. What is the probability of neither winning nor losing on the first roll?
   (d) After rolling a sum of 4, 5, 6, 8, 9, or 10, a player must roll the same sum again before rolling a sum of 7. Which sum, 4, 5, 6, 8, 9, or 10, has the highest probability of occurring again?
   (e) What is the probability of rolling a sum of 1 on any roll of the dice?
   (f) What is the probability of rolling a sum less than 13 on any roll of the dice?
   (g) If the two dice are rolled 60 times, predict about how many times a sum of 7 will be rolled.

10. ▶According to a weather report, there is a 30% chance that it will rain tomorrow. What is the probability that it will not rain tomorrow? Explain your answer.

11. A roulette wheel has 38 slots around the rim. The first 36 slots are numbered from 1 to 36. Half of these 36 slots are red and the other half are black. The remaining two slots are numbered 0 and 00 and are colored green. As the rou-

lette wheel is spun in one direction, a small ivory ball is rolled along the rim in the opposite direction. The ball has an equally likely chance of falling into any one of the 38 slots. Find each of the following:

(a) The probability the ball lands in a black slot
(b) The probability the ball lands on 0 or 00
(c) The probability the ball does not land on a number from 1 through 12
(d) The probability the ball lands on an odd number or on a green slot

12. If the roulette wheel in problem 11 is spun 190 times, predict about how many times the ball will land on 0 or 00.

13. Determine if each player has an equal probability of winning each of the following games.
    (a) Toss a fair coin. If heads appears, I win; if tails appears, you lose.
    (b) Toss a fair coin. If heads appears, I win; otherwise, you win.
    (c) Toss a fair die numbered 1 through 6. If 1 appears, I win; if 6 appears, you win.
    (d) Toss a fair die numbered 1 through 6. If an even number appears, I win; if an odd number appears, you win.
    (e) Toss a fair die numbered 1 through 6. If a number greater than or equal to 3 appears, I win; otherwise, you win.
    (f) Toss two fair dice numbered 1 through 6. If a 1 appears on each die, I win; if a 6 appears on each die, you win.
    (g) Toss two fair dice numbered 1 through 6. If the sum is 3, I win; if the sum is 2, you win.
    (h) Toss two dice numbered 1 through 6; one die is red and one die is white. If the number on the red die is greater than the number on the white die, I win; otherwise, you win.

14. A bowler has made 45 strikes in the last 150 frames she has bowled. What is the estimated probability that she will get a strike in the next frame she bowls?

15. Suppose that two coins are tossed. Find the probability for each of the following:
    (a) Exactly one head   (b) At least one head
    (c) At most one head

16. If A and B are mutually exclusive and if $P(A) = 0.3$ and $P(B) = 0.4$, what is $P(A \cup B)$?

17. ▶ If $P(A) = 0.8$ and $P(B) = 0.9$, can events A and B be mutually exclusive? Explain your answer.

18. A calculus class is composed of 35 men and 45 women. There are 20 business majors, 30 biology majors, 10 computer science majors, and 20 mathematics majors. No person has a double major. If a single student is chosen from the class, what is the probability that the student is:
    (a) Female
    (b) A computer science major
    (c) Not a mathematics major
    (d) A computer science major or a mathematics major

LABORATORY ACTIVITY

1. Suppose that a paper cup is tossed in the air. The different ways it can land are shown below.

Top   Bottom   Side

Toss a cup 100 times and record each result. From this information, calculate the experimental probability of each outcome. Do the outcomes appear to be equally likely? Based on experimental probabilities, how many times would you predict the cup will land on its side if tossed 100 times?

2. Toss a coin 100 times and record the results. From this information, calculate the experimental probability of getting heads on a particular toss. Does the experimental result agree with the expected theoretical probability of $\frac{1}{2}$?

SECTION 8-1 HOW PROBABILITIES ARE DETERMINED

3. Hold a coin upright on its edge under your forefinger on a hard surface and then spin it with your other finger so that it spins before landing. Repeat this experiment 100 times, and compare your experimental probabilities with those in activity 2.

▼ BRAIN TEASER

A Stanford University statistician, Bradley Efron, designed the following set of nonstandard dice. They are designed in such a way that if you choose a die and roll it, I can choose one and know that two-thirds of the time, my roll will be greater than yours. Which die should I choose if you choose A, B, C, or D?

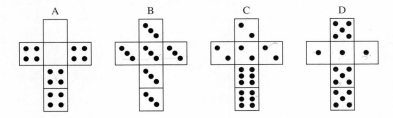

(In 1970, Martin Gardner discussed this type of dice in *Scientific American* in an article, "The Paradox of the Nontransitive Dice and the Elusive Principle of Indifference.")

● COMPUTER CORNER

The Logo procedures below will simulate flipping a coin and output the result:

```
TO TOSS
    OUTPUT PICK.ONE [HEADS TAILS]
END
TO PICK.ONE :FLIP
    OUTPUT ITEM (1 + RANDOM COUNT :FLIP) :FLIP
END
```

Enter the procedures in your computer and then complete the following:

1. Execute PRINT TOSS.
2. To simulate tossing a coin 25 times, execute the following:
   REPEAT 25 [PRINT TOSS]

3. Based upon the results in Number 2, what is the experimental probability for obtaining HEADS? For obtaining TAILS?

4. Edit the Logo procedure TOSS to make the list be [HEADS HEADS TAILS TAILS TAILS]. Now execute the following:

   REPEAT 25 [PRINT TOSS]

   What is the experimental probability for obtaining HEADS now? For obtaining TAILS now?

Section 8-2 Multistage Experiments with Tree Diagrams and Geometric Probabilities

In Section 8-1, we considered one-stage experiments: that is, experiments that are over after one step. Next, we consider several multistage experiments. For example, the box in Figure 8-7 contains one black and two white balls. A ball is drawn at random and its color is recorded. The ball is then *replaced* and a second ball is drawn and its color is recorded.

FIGURE 8-7

The sample space for this two-stage experiment may be recorded using ordered pairs as {(●, ●), (●, ○), (○, ●), (○, ○)} or, more commonly, as {●●, ●○, ○●, ○○}, as shown in the tree diagram in Figure 8-8.

FIGURE 8-8

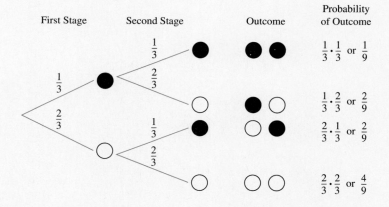

To assign the probability of the outcomes in this experiment, consider, for example, the path for the outcome ● ○. In the first stage, the probability of obtaining a black ball is $\frac{1}{3}$. Then, the probability of obtaining a white ball in the second stage (second draw) is $\frac{2}{3}$. Thus, we expect to obtain a black ball on the first draw $\frac{1}{3}$ of the time and then to draw a white ball $\frac{2}{3}$ of those times that we obtained a black ball, that is, $\frac{2}{3}$ of $\frac{1}{3}$, or $\frac{2}{3} \cdot \frac{1}{3}$. Observe that this product may be obtained by multiplying the probabilities along the branches used for the path leading to ●○, that is, $\frac{1}{3} \cdot \frac{2}{3}$, or $\frac{2}{9}$. The probabilities shown in Figure 8-8 are obtained by following the paths leading to each of the four outcomes and multiplying the probabilities along the paths. This discussion yields the following property for tree diagrams.

> **PROPERTY: MULTIPLICATION RULE FOR PROBABILITIES**
>
> For all multistage experiments, the probability of the outcome along any path is equal to the product of all the probabilities along the path.

> **REMARK** The sum of the probabilities on branches from any point always equals 1, and the sum of the probabilities for the possible outcomes must also be 1.

Look at the box pictured in Figure 8-7 again. This time, suppose two balls are drawn one by one *without replacement*. A tree diagram for this experiment, along with the set of possible outcomes, is shown in Figure 8-9.

FIGURE 8-9

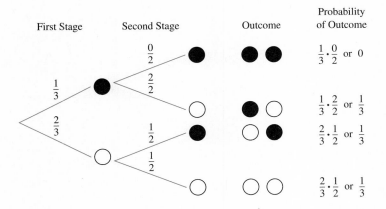

The denominators of the fractions along the second stage are all 2. Since the draws are made without replacement, there are only two balls remaining for the second draw.

Consider event A, consisting of the outcomes for drawing exactly one black ball in the two draws without replacement. This event is given by $A = \{●○, ○●\}$. Since the outcome ●○ appears $\frac{1}{3}$ of the time, and the outcome ○● appears $\frac{1}{3}$ of the time, then either ●○ or ○● will appear $\frac{2}{3}$ of the time. Thus, $P(A) = \frac{1}{3} + \frac{1}{3} = \frac{2}{3}$.

398   CHAPTER 8   PROBABILITY

Event B, consisting of outcomes for drawing *at least* one black ball, could be recorded as $B = \{● ○, ○ ●, ● ●\}$. Because $P(● ○) = \frac{1}{3}$, $P(○ ●) = \frac{1}{3}$, and $P(● ●) = 0$, then $P(B) = \frac{1}{3} + \frac{1}{3} + 0 = \frac{2}{3}$. Because $\overline{B} = \{○ ○\}$ and $P(\overline{B}) = \frac{1}{3}$, the probability of B could have been computed as follows: $P(B) = 1 - P(\overline{B}) = 1 - \frac{1}{3} = \frac{2}{3}$.

EXAMPLE 8-6 Figure 8-10 shows a box with 11 letters. Suppose four letters are drawn at random from the box one by one without replacement. What is the probability of the outcome BABY, with the letters chosen in exactly the order given?

FIGURE 8-10

PROBABILITY

Solution We do not need the entire tree diagram to find this probability because we are interested in only the branch leading to the outcome BABY. The portion needed is shown in Figure 8-11.

FIGURE 8-11

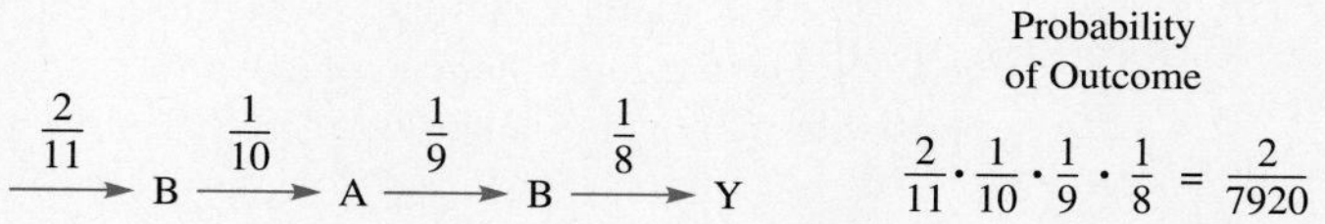

$\xrightarrow{\frac{2}{11}} B \xrightarrow{\frac{1}{10}} A \xrightarrow{\frac{1}{9}} B \xrightarrow{\frac{1}{8}} Y$ Probability of Outcome $\frac{2}{11} \cdot \frac{1}{10} \cdot \frac{1}{9} \cdot \frac{1}{8} = \frac{2}{7920}$

The probability of the first B is $\frac{2}{11}$ because there are two B's out of 11 letters. The probability of the second B is $\frac{1}{9}$ because there are nine letters left after one B and one A have been chosen. Then, $P(\text{BABY})$ is $\frac{2}{7920}$, as shown. ∎

In Example 8-6, suppose four letters are drawn one by one from the box and the letters are replaced after each drawing. In this case, the branch needed to find $P(\text{BABY})$ is pictured in Figure 8-12. Then, $P(\text{BABY}) = \left(\frac{2}{11}\right) \cdot \left(\frac{1}{11}\right) \cdot \left(\frac{2}{11}\right) \cdot \left(\frac{1}{11}\right)$, or $\frac{4}{14,641}$.

FIGURE 8-12

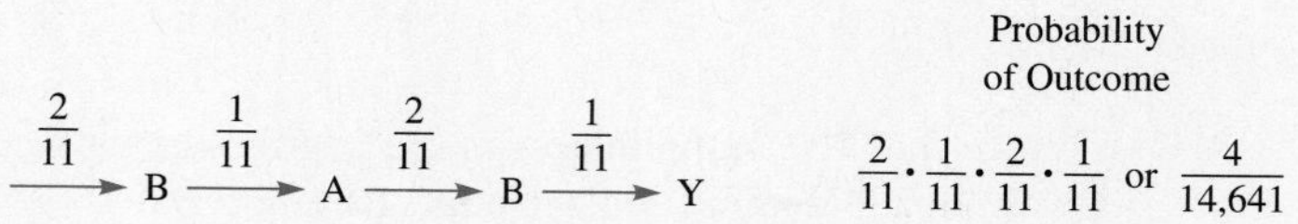

$\xrightarrow{\frac{2}{11}} B \xrightarrow{\frac{1}{11}} A \xrightarrow{\frac{2}{11}} B \xrightarrow{\frac{1}{11}} Y$ Probability of Outcome $\frac{2}{11} \cdot \frac{1}{11} \cdot \frac{2}{11} \cdot \frac{1}{11}$ or $\frac{4}{14,641}$

EXAMPLE 8-7 Consider the three boxes in Figure 8-13. A letter is drawn from box 1 and placed in box 2. Then, a letter is drawn from box 2 and placed in box 3. Finally, a letter is drawn from box 3. What is the probability that the letter drawn from box 3 is B?

SECTION 8-2 MULTISTAGE EXPERIMENTS

FIGURE 8-13

Solution A tree diagram for this experiment is given in Figure 8-14. Notice that the denominators in the second stage are 3 rather than 2 because in this stage there are now three letters in box 2. The denominators in the third stage are 5 because in this stage there are five letters in box 3. To find the probability that a *B* is drawn from box 3, add the probabilities for the outcomes *AAB*, *ABB*, *BAB*, and *BBB* that make up this event.

FIGURE 8-14

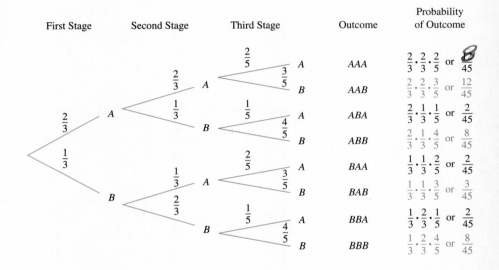

Thus, the probability of obtaining a *B* on the draw from box 3 in this experiment is $\frac{12}{45} + \frac{8}{45} + \frac{3}{45} + \frac{8}{45} = \frac{31}{45}$.

Modeling Games

The *Teaching Standards* contain statements that teachers need common experiences to build and extend their knowledge of mathematics. These standards also state (p. 135),

In the process of constructing and developing these experiences, appropriate attention to, and use of, mathematical modeling and technology should be included to enhance the teaching and learning of the mathematical ideas.

We can use models to analyze games involving probability. Consider the following game, which Arthur and Guinevere play.

There are two black marbles and one white marble in a box. Guinevere mixes the marbles, and Arthur draws two marbles at random without replacement. If they match,

Arthur wins; otherwise, Guinevere wins. Does each player have an equal chance of winning? We *develop a model* for analyzing the game. One possible model is a tree diagram, as shown in Figure 8-15.

FIGURE 8-15

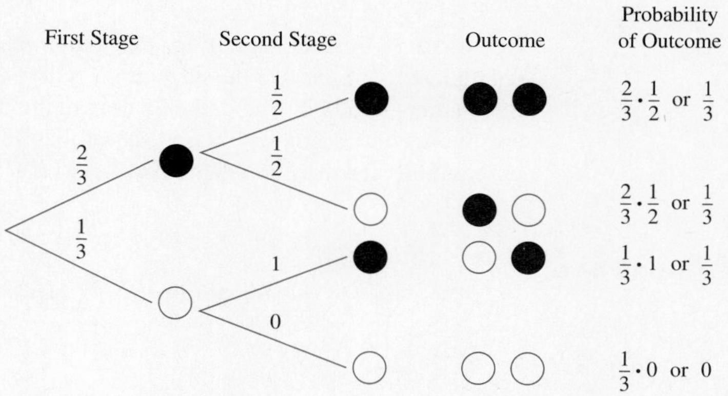

First Stage	Second Stage	Outcome	Probability of Outcome
$\frac{2}{3}$	$\frac{1}{2}$ ●	●●	$\frac{2}{3} \cdot \frac{1}{2}$ or $\frac{1}{3}$
	$\frac{1}{2}$ ○	●○	$\frac{2}{3} \cdot \frac{1}{2}$ or $\frac{1}{3}$
$\frac{1}{3}$	1 ●	○●	$\frac{1}{3} \cdot 1$ or $\frac{1}{3}$
	0 ○	○○	$\frac{1}{3} \cdot 0$ or 0

The probability that the marbles are the same color is $\frac{1}{3} + 0$, or $\frac{1}{3}$, and the probability that they are not the same color is $\frac{1}{3} + \frac{1}{3}$, or $\frac{2}{3}$. Because $\frac{1}{3} \neq \frac{2}{3}$, the players do not have the same chance of winning.

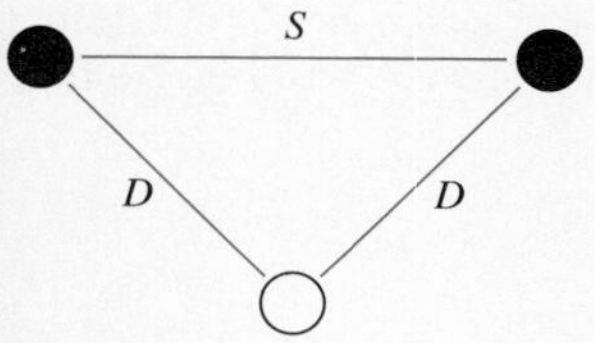

An alternate model for analyzing this game is given in Figure 8-16, where the black and white marbles are shown along with the possible ways of drawing two marbles. Each line segment in the diagram represents one pair of marbles that could be drawn; S indicates that the marbles in the pair are the same color, and D indicates that the marbles are different colors. Because there are two D's in Figure 8-16, we see that the probability of drawing two different-colored marbles is $\frac{2}{3}$. Likewise, the probability of drawing two marbles of the same color is $\frac{1}{3}$. Because $\frac{2}{3} \neq \frac{1}{3}$, the players do not have an equal chance of winning. Will adding another white marble give each player an equal chance of winning? With two white and two black marbles, we have the model in Figure 8-17. Therefore, $P(D) = \frac{4}{6}$, or $\frac{2}{3}$, and $P(S) = \frac{2}{6}$, or $\frac{1}{3}$. We see that adding another white marble does not change the probabilities.

FIGURE 8-16

FIGURE 8-17

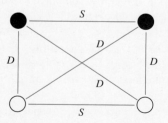

Next, consider a game with the same rules but using three black marbles and one white marble. A model for this situation is shown in Figure 8-18.

FIGURE 8-18

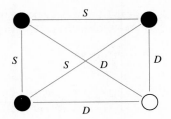

Thus, the probability of drawing two marbles of the same color is $\frac{3}{6}$, and the probability of drawing two marbles of different colors is $\frac{3}{6}$. Finally, we have a game in which each player has an equal chance of winning.

Does each player have an equal chance of winning if only one white marble and one black marble are used and the ball is replaced after the first draw? Can you find additional fair games involving different numbers of marbles? Can you find a pattern for the numbers of black and white marbles that allow each player to have an equal chance of winning?

PROBLEM 1

In a party game, a child is handed six strings, as shown in Figure 8-19(a). Another child ties the top ends two at a time, forming three separate knots, and the bottom ends, forming three separate knots, as in Figure 8-19(b). If the strings form one closed ring, as in Figure 8-19(c), the child wins a prize. What is the probability that the child wins a prize on the first try?

FIGURE 8-19

 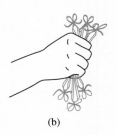 

(a)   (b)   (c)

Understanding the Problem. The problem is to determine the probability that one closed ring will be formed. One closed ring means that all six pieces are joined end to end to form one, and only one, ring, as shown in Figure 8-19(c).

Devising a Plan. Figure 8-20(a) shows what happens when the ends of the strings of one set are tied in pairs at the top. Notice that no matter in what order those ends are tied, the result appears as in Figure 8-20(a).

FIGURE 8-20

 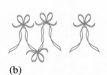

(a)   (b)   (c)

Then, the other ends are tied in a three-stage experiment. If we pick any string in the first stage, then there are five choices for its mate. Four of these choices are favorable choices for forming a ring. Thus, the probability of forming a favorable first tie is $\frac{4}{5}$. Figure 8-20(b) shows a favorable tie at the first stage.

For any one of the remaining four strings, there are three choices for its mate. Two of these choices are favorable ones. Thus, the probability of forming a favorable second tie is $\frac{2}{3}$. Figure 8-20(c) shows a favorable tie at the second stage.

Now, two ends remain. Since nothing can go wrong at the third stage, the probability of making a favorable tie is 1. If we use the probabilities completed at each stage and a single branch of a tree diagram, we can calculate the probability of performing three successful ties in a row and hence the probability of forming one closed ring.

Carrying Out the Plan. If we let S represent a successful tie at each stage, then the branch of the tree with which we are concerned is the one shown in Figure 8-21.

FIGURE 8-21

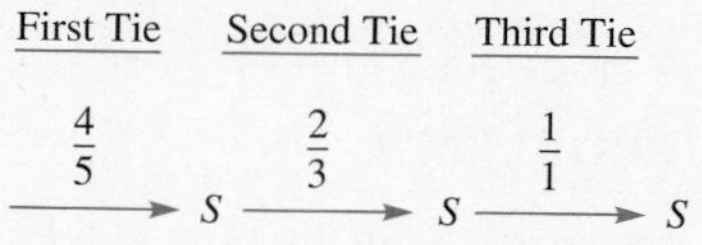

$$\xrightarrow{\frac{4}{5}} S \xrightarrow{\frac{2}{3}} S \xrightarrow{\frac{1}{1}} S$$

First Tie   Second Tie   Third Tie

Thus, the probability of forming one ring is $P(\text{ring}) = \frac{4}{5} \cdot \frac{2}{3} \cdot \frac{1}{1} = \frac{8}{15} = 0.5\overline{3}$.

Looking Back. The probability that a child will form a ring on the first try is $\frac{8}{15}$. A class might simulate this problem several times with strings to see how the fraction of successes compares with the theoretical probability of $\frac{8}{15}$.

Related problems that could be attempted include the following:

1. If a child fails to get a ring 10 times in a row, the child may not play again. What is the probability of such a streak of bad luck?
2. If the number of strings is reduced to three and the rule is that an upper end must be tied to a lower end, what is the probability of a single ring?
3. If the number of strings is three, but an upper end can be tied to either an upper or a lower end, what is the probability of a single ring?
4. What is the probability of forming three rings in the original problem?
5. What is the probability of forming two rings in the original problem?

Geometric Probability

A different model used in probability employs geometric shapes and is called an *area model* in the *Standards*. When using area models to determine probabilities geometrically, outcomes are associated with points chosen at random in a geometric region representing the sample space. For example, suppose we throw darts at a square target two units long on a side and divided into four congruent triangles, as shown in Figure 8-22. If the dart must hit the target somewhere and if all spots can be hit with equal probability, what is the probability that the dart will land in the shaded region? The entire target, which has an area of four square units, represents the sample space. The shaded area is the event of a successful toss. The area of the shaded part is $\frac{1}{4}$ of the sample space. Thus, the probability of the dart's landing in the shaded region is the ratio of the area of the event to the area of the sample space, or $\frac{1}{4}$.

FIGURE 8-22

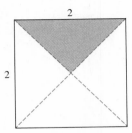

Other geometric probability problems are investigated in the problem set. Problem 2 is solved by using both tree diagrams and a geometric approach.

PROBLEM 2

On a quiz show, a contestant stands at the entrance to a maze that opens into two rooms, as shown in Figure 8-23. The master of ceremonies' assistant is to place a new car in one room and a donkey in the other. The contestant must walk through the maze into one of the rooms and wins whatever is in that room. If the contestant makes each decision in the maze at random, in which room should the assistant place the car to give the contestant the best chance to win?

FIGURE 8-23

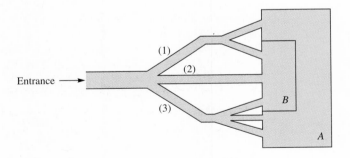

Understanding the Problem. The contestant must first choose one of the paths marked 1, 2, or 3 and then choose other paths as she proceeds through the maze. In order to determine the room most likely to be chosen by the contestant, the assistant must be able to determine the probability of the contestant's reaching each room. To compute the probability, the assistant must understand what it means for something to be chosen at random and how to compute multistage probabilities.

 Devising a Plan. One way to determine where the car should be placed is to *model the choices* with a tree diagram and to compute the probabilities along the branches of the tree.

Carrying Out the Plan. A tree diagram for the maze is shown in Figure 8-24, along with the possible outcomes and the probabilities of each branch. Thus, room *B* has the greater probability of being chosen. This is where the car should be placed for the contestant to have the best chance of winning it.

FIGURE 8-24

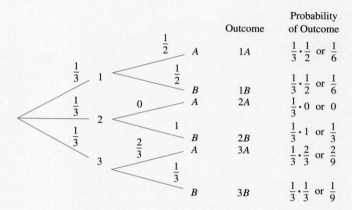

Looking Back. An alternate model for this problem and for many probability problems is an area model. The rectangle in Figure 8-25(a) represents the first three choices that the contestant can make. Because each choice is equally likely, each is represented by an equal area. If the upper path is chosen, then rooms *A* and *B* have an equal change of being chosen. If the middle path is chosen, then only room *B* can be entered. If the lower path is chosen, then room *A* is entered $\frac{2}{3}$ of the time. This can be expressed in terms of the area model as shown in Figure 8-25(b). Dividing the rectangle into pieces of equal area, we obtain the model in Figure 8-25(c), in which the area representing room *B* is shaded. Because the area representing room *B* is greater than the area representing room *A*, room *B* has the greater probability of being chosen. If desired, Figure 8-25(c) enables us to find the probability of choosing room *B*. Because the shaded area consists of 11 rectangles out of a total of 18 rectangles, the probability of choosing room *B* is $\frac{11}{18}$. We can vary the problem by changing the maze or by changing the locations of the rooms.

FIGURE 8-25

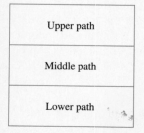

(a)

(b)

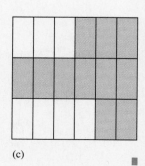

(c)

PROBLEM SET 8-2

1. (a) Use a tree diagram to develop the sample space for tossing a fair coin twice.
   (b) Use a tree diagram to develop the sample space for an experiment consisting of tossing a fair coin and then rolling a die.

2. Suppose that an experiment consists of spinning X and then spinning Y, as given below. Find each of the following:

   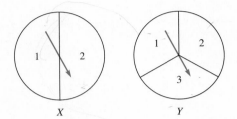

   (a) The sample space S for the experiment
   (b) The event A consisting of outcomes from spinning an even number followed by an even number
   (c) The event B consisting of outcomes from spinning at least one 2
   (d) The event C consisting of outcomes from spinning exactly one 2

3. A box contains six letters, as shown. What is the probability of the outcome DAN if three letters are drawn one by one (a) with replacement? (b) without replacement?

   | RANDOM |

4. Three boxes containing letters are shown.

   | MATH | | AND | | HISTORY |
   | 1 | | 2 | | 3 |

   Answer each of the following questions about the boxes:
   (a) From box 1, three letters are drawn one by one without replacement and recorded in order. What is the probability that the outcome is HAT?
   (b) From box 1, three letters are drawn one by one with replacement and recorded in order. What is the probability that the outcome is HAT?
   (c) One letter is drawn at random from box 1, then another from box 2, and then another from box 3, with the results recorded in order. What is the probability that the outcome is HAT?
   (d) If a box is chosen at random and then a letter is drawn at random from the box, what is the probability that the outcome is A?

5. ▶An executive committee consisted of ten members: four women and six men. Three members were selected at random to be sent to a meeting in Hawaii. A blindfolded woman drew three of the ten names from a hat. All three names drawn were women. What was the probability of such luck? Explain your reasoning.

6. Two boxes with letters follow. You are to choose a box and draw three letters at random, one by one, without replacement. If the outcome is SOS, you win a prize.

   | SOS | | SOSSOS |
   | 1 | | 2 |

   (a) Which box should you choose?
   (b) Which box would you choose if the letters are to be drawn with replacement?

7. Three boxes containing balls are shown. Draw a ball from box 1 and place it in box 2. Then draw a ball from box 2 and place it in box 3. Finally, draw a ball from box 3.
   (a) What is the probability that the last ball, drawn from box 3, is white?
   (b) What is the probability that the last ball drawn is black?

8. ▶Carolyn will win a large prize if she wins two tennis games in a row out of three games. She is to play alternately against Billie and Bobby. She may choose to play Billie-Bobby-Billie or Bobby-Billie-Bobby. She wins against Billie 50% of the time and against Bobby 80% of the time. Which alternative should she choose, and why?

9. Two boxes with black and white balls are shown. A ball is drawn at random from box 1, and then a ball is drawn at random from box 2, and the colors are recorded in order.

   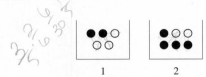

   Find each of the following:
   (a) The probability of two white balls
   (b) The probability of at least one black ball
   (c) The probability of at most one black ball
   (d) The probability of ●○ or ○●

10. A penny, a nickel, a dime, and a quarter are tossed. What is the probability of at least three heads?

11. Assume that the probability is $\frac{1}{2}$ that a child born is a boy. What is the probability that, if a family is going to have four children, they will all be boys?

12. Brittany is going to ascend a four-step staircase. At any time, she is just as likely to stride up one step or two steps.

Find the probabilities that she will ascend the four steps in:
(a) two strides
(b) three strides
(c) four strides

13. ▶A witness to a crime observed that the criminal had blond hair and blue eyes and drove a red car. When the police look for a suspect, is the probability greater that they will find someone with blond hair and blue eyes, or that they will find someone with blond hair and blue eyes who drives a red car? Explain your answer.

14. The numbers of symbols on each of the three dials of a standard slot machine are shown in the table.

Symbol	Dial 1	Dial 2	Dial 3
Bar	1	3	1
Bell	1	3	3
Plum	5	1	5
Orange	3	6	7
Cherry	7	7	0
Lemon	3	0	4
Total	20	20	20

Find the probability for each of the following:
(a) Three plums   (b) Three oranges
(c) Three lemons   (d) No plums

15. If a person takes a five-question true-false test, what is the probability that the score is 100% correct if the person guesses on every question?

16. In a drawer containing only black and blue socks, there are 10 blue socks and 12 black socks. Suppose that it is dark and you choose three socks at random. What is the probability that you will choose a matching pair?

17. Rattlesnake and Paxson Colleges play four games against each other in a chess tournament. Rob Fisher, the chess whiz from Paxson, withdrew from the tournament, so the probabilities that Rattlesnake and Paxson will win each game are $\frac{2}{3}$ and $\frac{1}{3}$, respectively. What are the following probabilities?
(a) Paxson loses all four games.
(b) The match is a draw with each school winning two games.

18. The combinations on the lockers at the high school consist of three numbers, each ranging from 0 to 39. If a combination is chosen at random, what is the probability that the first two numbers are multiples of nine and the third number is a multiple of four?

19. A box contains the 11 letters shown. The letters are drawn one by one without replacement, and the results are recorded in order. Find the probability of the outcome MISSISSIPPI.

| MIIIIPPSSSS |

20. Consider the following dart board. If a dart may hit any point on the board with equal probability, what is the probability that it will land in:
(a) Section A
(b) Section B
(c) Section C

(Assume that all quadrilaterals are squares and that the x's represent equal measures.)

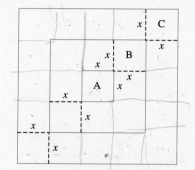

21. In the square dart board pictured below, suppose a dart is equally likely to land in any region of the board. Points are given as follows:

Region	Points
A	10
B	8
C	6
D	4
E	2

Answer the following:
(a) What is the total area of the board?
(b) What is the probability of a dart's landing in each of the regions of the board?
(c) If two darts are tossed, what is the probability of scoring 20 points?
(d) What is the probability that the dart will land in neither regions D nor E?

22. The land area of Earth is approximately 57,500,000 sq mi. The water area of Earth is approximately 139,600,000 sq mi. If a meteor landing at random hits Earth, what is the probability, to the nearest tenth, that it will hit water?

23. An electric clock is stopped by a power failure. What is the probability that the second hand is stopped between the 3 and the 4?

24. A husband and wife discover that there is a 10% probability of passing on a hereditary disease to any one of their children. If they plan to have three children, what is the probability that at least one child will inherit the disease?

25. Let $A = \{x | -1 < x < 1\}$ and $B = \{x | -3 < x < 2\}$. If a real number is picked at random from set B, what is the probability that it will be in A?

26. Suppose a sock drawer holds only red and black socks. What is the least number of each color of socks in the drawer if the probability of drawing exactly one red sock on two draws is exactly $\frac{1}{2}$?

27. ▶Use graph paper to design a dart board such that the probability of hitting a certain part of the board is $\frac{3}{5}$. Explain your reasoning.

28. ▶An experiment consists of tossing a coin twice. A student reasons that there are three possible outcomes: two heads, one head and one tail, or two tails. Thus, $P(HH) = \frac{1}{3}$. How would you answer the student, and why?

29. At a certain hospital, 40 patients have lung cancer, 30 patients smoke, and 25 have lung cancer and smoke. Suppose that the hospital contains 200 patients. If a patient chosen at random is known to smoke, what is the probability that the patient has lung cancer?

30. There are 40 employees in a certain firm. We know that 28 of these employees are males, two of these males are secretaries, and there are ten secretaries employed by the firm. What is the probability that an employee chosen at random is a secretary, given that the person is a male?

31. ▶On August 18, 1913, the roulette wheel at a casino in Monte Carlo came up black 26 times in a row. Many people in the crowd during the streak placed large bets on the red because they were convinced that the "law of averages" would catch up with the wheel. If the wheel was fair, explain how you think that the wheel would have shown on the 17th spin and why.

32. An assembly line has two inspectors. The probability that the first inspector will miss a defective item is 0.05. If the defective item passes the first inspector, the probability that the second inspector will miss it is 0.01. What is the probability that a defective item will pass by both inspectors?

33. In a certain population of caribou, the probability of an animal's being sickly is $\frac{1}{20}$. If a caribou is sickly, the probability of its being eaten by wolves is $\frac{1}{3}$. If a caribou is not sickly, the probability of its being eaten by wolves is $\frac{1}{150}$. If a caribou is chosen at random from the herd, what is the probability that it will be eaten by wolves?

★34. Consider the three spinners A, B, and C shown in the figure.

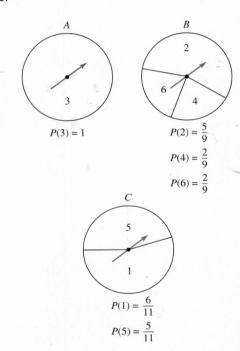

(a) Suppose that you choose a spinner and then a friend chooses a spinner, and each person spins the chosen spinner. If the person spinning the higher number wins, which spinner should you choose if you want to win?

(b) This time you are to play the same game as in (a), but two friends play. If each player chooses a spinner, would you make the same choice that you did in (a)? Why?

★35. Abe proposes the following game. He lets you choose one of the four equally likely outcomes obtained by tossing a coin twice: {HH, HT, TH, TT}. Abe then chooses one of the other outcomes. A coin is flipped until either your choice or Abe's choice appears. For example, suppose that you choose *TT* and Abe chooses *HT*. If the first two flips yield *TH*, then no one wins and the game continues. If, after five flips, the string TH<u>HHT</u> appears, then Abe is the winner because the sequence *HT* finally appeared. Is this game fair? Why?

★36. Jane has two tennis serves, a hard serve and a soft serve. Her hard serve is in (a good serve) 50% of the time, and her soft serve is in (good) 75% of the time. If her hard serve is in, she wins 75% of her points. If her soft serve is in, she wins 50% of her points. Since she is allowed to re-serve one time if her first serve is out, what should her serving strategy be? That is, should she serve hard followed by soft; both hard; soft followed by hard; or both soft?

Review Problems

37. Match the phrase to the probability that describes it.
    (a) A certain event   (i) $\frac{1}{1000}$
    (b) An impossible event   (ii) $\frac{999}{1000}$
    (c) A very likely event   (iii) 0
    (d) An unlikely event   (iv) $\frac{1}{2}$
    (e) A 50% chance   (v) 1

38. A date in the month of April is chosen at random. Find the probabilities of the date's being each of the following:
    (a) April 7
    (b) April 31
    (c) Before April 20

BRAIN TEASER

Suppose that *n* people are in a room. Two people bet on whether at least two of the people in the room have a birthday on the same date during the year (for example, October 14). Assume that a person is as likely to be born on one day as another, and ignore leap years. How many people must be in the room before the bet is even? (If *n* = 366, it is a sure bet.) Poll your class to see if two people have the same birthday. Use this information to find an experimental answer. Then find a theoretical solution. A calculator is very helpful for the computations.

Section 8-3 Using Simulations in Probability

Students can use simulations to study phenomena too complex to analyze by other means. Using simulations, students can estimate probabilities rather than determine probabilities analytically. The *Teaching Standards* (p. 136) assert:

Students should have opportunities to explore empirical probability from simulations and from data they have collected and to analyze theoretical probability on the basis of a description of the underlying sample space. . . . The power of simulation as a problem-solving technique for making decisions under uncertainty should be a prominent experience.

Suppose we want to simulate the results of tossing a coin 100 times. We could do this using random digits, as Table 8-2 shows. Random-digit tables are lists of digits selected at random, often by a computer. To simulate the coin toss, pick a number at random to start and then read across the table, letting an even digit represent heads and an odd number represent tails. Continue this process for 100 digits. The experimental probability of heads is the ratio of the number of even digits found (heads) to 100.

Similarly, to determine the probability of a couple having two girls (GG) in an expected family, we could use the random-digit table, with an even digit representing a girl and an odd digit representing a boy. Because there are two children, we need to consider pairs of digits. If we examine 100 pairs, then the experimental probability of GG will be the number of pairs of even digits divided by 100, the total number of pairs considered.

TABLE 8-2   Random Digits

36422	93239	76046	81114	77412	86557	19549	98473	15221	87856
78496	47197	37961	67568	14861	61077	85210	51264	49975	71785
95384	59596	05081	39968	80495	00192	94679	18307	16265	48888
37957	89199	10816	24260	52302	69592	55019	94127	71721	70673
31422	27529	95051	83157	96377	33723	52902	51302	86370	50452
07443	15346	40653	84238	24430	88834	77318	07486	33950	61598
41348	86255	92715	96656	49693	99286	83447	20215	16040	41085
12398	95111	45663	55020	57159	58010	43162	98878	73337	35571
77229	92095	44305	09285	73256	02968	31129	66588	48126	52700
61175	53014	60304	13976	96312	42442	96713	43940	92516	81421
16825	27482	97858	05642	88047	68960	52991	67703	29805	42701
84656	03089	05166	67571	25545	26603	40243	55482	38341	97782
03872	31767	23729	89523	73654	24626	78393	77172	41328	95633
40488	70426	04034	46618	55102	93408	10965	69744	80766	14889
98322	25528	43808	05935	78338	77881	90139	72375	50624	91385
13366	52764	02407	14202	74172	58770	65348	24115	44277	96735
86711	27764	86789	43800	87582	09298	17880	75507	35217	08352
53886	50358	62738	91783	71944	90221	79403	75139	09102	77826
99348	21186	42266	01531	44325	61042	13453	61917	90426	12437
49985	08787	59448	82680	52929	19077	98518	06251	58451	91140
49807	32863	69984	20102	09523	47827	08374	79849	19352	62726
46569	00365	23591	44317	55054	99835	20633	66215	46668	53587
09988	44203	43532	54538	16619	45444	11957	69184	98398	96508
32916	00567	82881	59753	54761	39404	90756	91760	18698	42852
93285	32297	27254	27198	99093	97821	46277	10439	30389	45372
03222	39951	12738	50303	25017	84207	52123	88637	19369	58289
87002	61789	96250	99337	14144	00027	43542	87030	14773	73087
68840	94259	01961	42552	91843	33855	00824	48733	81297	80411
88323	28828	64765	08244	53077	50897	91937	08871	91517	19668
55170	71062	64159	79364	53088	21536	39451	95649	65256	23950

The following page from *Addison-Wesley Mathematics*, Grade 7, 1991, shows how middle school students learn to use simulations to solve probability problems.

ENRICHMENT
Simulating a Probability Problem

Robin Hood is returning from Nottingham to Sherwood Forest. The map shows the different routes he can take. The Sheriff of Nottingham is trying to catch Robin at one of the 4 bridges. The probability that Robin will find an open bridge is $\frac{1}{2}$. What is the probability that Robin will find an open route to Sherwood?

You can **simulate** or **model** the problem this way.

- Use a **random digit generator** to get a list of digits 0 to 9. Let even digits = open bridge and odd digits = closed bridge.
- Keep a record of open and closed bridges. Then after each trial use the map to decide if there is an open route for Robin Hood.

Trial 1: 2, 5, 1, 3
Trial 2: 7, 4, 4, 6

Trial	Bridge 1	Bridge 2	Bridge 3	Bridge 4	Open Route?
1	Open	Blocked	Blocked	Blocked	No
2	Blocked	Open	Open	Open	Yes

1. Try the simulation of the problem for 20 trials. How many times was there an open route?

2. What is the experimental probability of an open route?
   Exp. $P(\text{Open}) = \dfrac{\text{No. of open routes}}{\text{No. of trials}}$

3. Combine your results for 20 trials with those of your classmates. What is the probability of an open route using the combined trials?

4. Is Robin Hood more likely to find an open route to Sherwood Forest or is he more likely to find a blocked route?

The Peanuts cartoon also suggests a simulation problem concerning chocolate chip cookies.

PEANUTS reprinted by permission of UFS, Inc.

EXAMPLE 8-8 Suppose that Lucy makes enough batter for exactly 100 chocolate chip cookies and mixes 100 chocolate chips into the batter. If the chips are distributed at random and Charlie chooses a cookie at random from the 100 cookies, estimate the probability that it will contain exactly one chocolate chip.

Solution A simulation can be used to estimate the probability of choosing a cookie with exactly one chocolate chip. We construct a 10×10 grid, as shown in Figure 8-26(a), to represent the 100 cookies Lucy made. Each square (cookie) can be associated with some ordered pair, where the first component is for the horizontal scale and the second component is for the vertical scale. For example, the squares (0, 2) and (5, 3) are pictured in Figure 8-26(a). Using the table of random digits, close your eyes, take a pencil, and point to one number to start. Look at the number and the number immediately following it. Consider these numbers as an ordered pair and continue on until 100 ordered pairs are obtained. For example, suppose we start at a 3 and the numbers following 3 are as shown:

39968   80495   00192 . . .

FIGURE 8-26

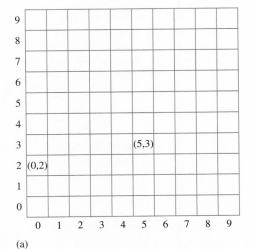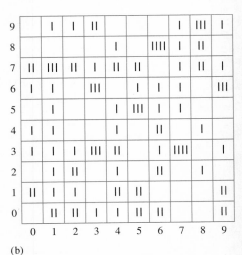

(a)   (b)

Then the ordered pairs would be given as (3, 9), (9, 6), (8, 8), (0, 4), and so on. Use each pair of numbers as the coordinates for the square (cookie) and place a tally on the grid to represent each chip, as shown in Figure 8-26(b). We estimate the probability that a cookie has exactly 1 chip by counting the number of squares with exactly 1 tally and dividing by 100. Table 8-3 shows the results of one simulation. Thus, the estimate for the probability of Charlie's receiving a cookie with exactly one chip is $\frac{34}{100}$.

TABLE 8-3

Number of Chips	Number of Cookies
0	38
1	34
2	20
3	6
≥4	2

Try a simulation on your own, and compare your results with the preceding ones and with the results given in Table 8-4, obtained by theoretical methods.

TABLE 8-4

Number of Chips	Number of Cookies
0	36.8
1	36.8
2	18.4
3	6.1
≥4	1.9

EXAMPLE 8-9 A baseball player, Reggie, has a batting average of 0.400; that is, his probability of getting a hit on any particular time at bat is 0.400. Estimate the probability that he will get at least one hit in his next three times at bat.

Solution We use a random-digit table to simulate this example. We choose a starting point and place the random digits in groups of three. Because Reggie's probability of getting a hit on any particular time at bat is 0.400, we could use the occurrence of four particular numbers from 0 through 9 to represent a hit. Suppose a hit is represented by the digits 0, 1, 2, and 3. At least one hit is obtained in three times at bat if, in any sequence of 3 digits, a 0, 1, 2, or 3 appears. Data for 50 trials are given below.

```
780  862  760  580  783  720  590  506  021  366
848  118  073  077  042  254  063  667  374  153
377  883  573  683  780  115  662  591  685  274
279  652  754  909  754  892  310  673  964  351
803  034  799  915  059  006  774  640  298  961
```

We see that a 0, 1, 2, or 3 appears in 42 out of the 50 trials; thus, an estimate for the probability of at least one hit in Reggie's next three times at bat is $\frac{42}{50}$. Try to determine the theoretical probability for this experiment.

From a random sample, we can deduce information about the population from which the sample was taken. To see how this can be done, consider Example 8-10.

EXAMPLE 8-10 To determine the number of fish in a certain pond, suppose we capture 300 fish, mark them, and throw them back into the pond. Suppose that, the next day, 200 fish are caught and 20 of these are already marked. Then these 200 fish are thrown back into the pond. Estimate how many fish are in the pond.

Solution Because 20 of the 200 fish are marked, we assume that $\frac{20}{200}$, or $\frac{1}{10}$, of the fish are marked. Thus, $\frac{1}{10}$ of the population is marked. If n represents the population, then $\frac{1}{10}n = 300$, and $n = 300 \cdot 10 = 3000$. Hence, an estimate for the fish population of the pond is 3000 fish. ∎

PROBLEM 3

A publishing company hires two proofreaders, Al and Betsy, to read a manuscript. Al finds 48 errors and Betsy finds 42 errors. The editor finds that 30 common errors were found by the proofreaders; that is, 30 errors were found by both Al and Betsy. What is your estimate of the number of errors not yet found?

Understanding the Problem. Of the 48 errors and 42 errors found by Al and Betsy in a manuscript, 30 errors were listed by both of them. From this information, we are to estimate the number of undetected errors in the manuscript. If we could estimate how efficient either Al or Betsy was at finding errors, then we could estimate the fraction of errors that he or she could be expected to find and thus be able to estimate the number of errors not yet found. This amounts to finding the probability that either Al or Betsy will find a given error.

Devising a Plan. To find the probability that either Al or Betsy will find a given error, we *draw a Venn diagram* for the given information. From Figure 8-27, we see that the total number of errors found is 18 + 30 + 12, or 60.

To compute the probability that Al will find a given error, we use the set of errors found by Betsy as a sample space. Thus, Al found $\frac{30}{42}$ of the errors in the set of 42 errors. From this information, we can estimate the number of errors in the manuscript.

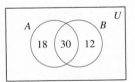

FIGURE 8-27

Carrying Out the Plan. We have estimated that Al finds $\frac{30}{42}$ of the errors in a given sample. If we let n be the total number of errors in the manuscript, then we could *set up the proportion* $30/42 = 48/n$. Therefore, $30n = 42 \cdot 48$ and $n = (42 \cdot 48)/30$, which is approximately 67. Because 60 errors were found, the estimate is that 67 − 60, or 7, errors remain undetected.

Looking Back. We might wonder what result is obtained if we use Betsy's probability of finding an error instead of Al's probability. If we use Al's 48 errors as a sample space, then Betsy found $\frac{30}{48}$ of these errors. Solving as before, we have $30/48 = 42/n$, and $n \doteq 67$. The estimate for the number of undetected errors is 67 − 60, or 7. We see that it makes no difference which proofreader is chosen to make the estimates. Do you think it would be worthwhile to hire a third proofreader? Investigate this probability by trying various inputs for a third proofreader.

PROBLEM 4

Assume that Carmen Smith, a basketball player, makes free throws with 80% probability of success and is placed in a one-and-one situation where she is given a second foul shot only if the first shot goes through the basket. How can we simulate 25 attempts from the foul line in one-and-one situations to determine how many times we would expect Carmen to score 0 points, 1 point, and 2 points?

Understanding the Problem. The probability that Carmen makes any given free throw is 80%. She is shooting in a one-and-one situation: If she misses the first shot, she receives 0 points; if she makes the first shot, she receives 1 point and is allowed to shoot *one* more time. Each basket made counts as 1 point. Thus, Carmen has the opportunity to score 0, 1, or 2 points. We are to determine by simulation how many times Carmen can be expected to score 0 points, 1 point, and 2 points in 25 attempts at one-and-one situations.

Devising a Plan. One way to simulate the problem is to use a random-digit table. Because Carmen's probability of making any basket is 80%, we could use the occurrence of a 0, 1, 2, 3, 4, 5, 6, or 7 to simulate making the basket and the occurrence of an 8 or a 9 to simulate missing the basket. Another way to simulate the problem is to construct a spinner with 80% of the spinner devoted to making a basket and 20% of the spinner devoted to missing the basket. This could be done by constructing the spinner with 80% of the 360 degrees (that is, 288 degrees) devoted to making the basket and 72 degrees devoted to missing the basket. A spinner for this simulation is shown in Figure 8-28.

FIGURE 8-28

Carrying Out the Plan. We spin the spinner in Figure 8-28 to simulate 25 sets of one-and-one situations. If the spinner lands on "miss," 0 points are recorded for the set. If the spinner lands on "make," a second spin is taken. If the spinner shows another "make," then 2 points are recorded; otherwise, 1 point is recorded. This is repeated for 25 sets and the results are recorded. Four simulations of 25 sets are given in Table 8-5. We used four trials to obtain a better estimate than we would get from only a single trial.

TABLE 8-5

Number of Points	Trial 1	Trial 2	Trial 3	Trial 4	Total	Estimated Probability
0	4	6	5	5	20	$\frac{20}{100}$
1	2	4	5	4	15	$\frac{15}{100}$
2	19	15	15	16	65	$\frac{65}{100}$

To solve the problem, we use the estimated probability and multiply by 25 to obtain the results in Table 8-6.

SECTION 8-3 USING SIMULATIONS IN PROBABILITY   415

TABLE 8-6

Number of Points	Expected Number of Times Points Are Scored in 25 Attempts
0	5
1	3.75
2	16.25

 Looking Back. We can compute the theoretical probability for this experiment by using a *tree diagram,* as shown in Figure 8-29.

FIGURE 8-29

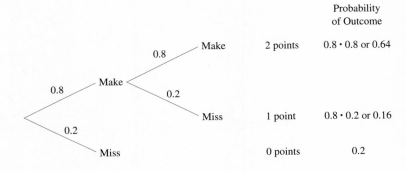

Thus, theoretical estimates for the number of points scored in 25 attempts can be computed. These estimates are given in Table 8-7. Compare these results with the experimental probability obtained previously.

TABLE 8-7

Number of Points	Expected Number of Times Points Are Scored in 25 Attempts
0	5
1	4
2	16

Geometric probability could also be used to solve this problem. We represent the sample space with the 10 × 10 grid shown in Figure 8-30(a), separate it into two parts, with a vertical line to represent the dividing line between making 80% on the first shot and missing 20% on it. That is, 80 of the 100 squares are devoted to making the first shot, and 20 of the 100 squares are devoted to missing the first shot. If the first shot is made, a second shot is taken with an 80% probability of success. Therefore, we subdivide the "make" area from the first shot into two portions of 80% and 20%. This is done by marking off 8 of the 10 rows in the "make" area. We then assign the appropriate number of points to each of the three areas, as shown in Figure 8-30(b). Now we see that 64 of the 100 squares (or

64% of the sample space) are devoted to scoring 2 points, 16 of the 100 squares (or 16%) are devoted to scoring 1 point, and 20 of the 100 squares (20%) are devoted to scoring 0 points. These results are consistent with the results obtained by using the tree diagram. ∎

FIGURE 8-30

(a)

(b)

PROBLEM SET 8-3

1. Try the simulation of the problem on the student page of this section for 20 trials. How many times was there an open route?

2. ▶Describe how you might use a deck of cards to simulate the chance of the birth of a girl.

3. The weather forecast in Pelican, Alaska, is for a 90% chance of rain on any given day.
   ▶(a) Describe how you might use the ten playing cards to simulate the expectation of rain on any given day.
   (b) Use the cards to find an experimental probability of not having rain for seven days in a row.
   (c) What is the theoretical probability of not having rain for seven days in a row?

4. How might you use a table of random digits to simulate each of the following?
   (a) Tossing a single die
   (b) Choosing three people at random from a group of 20 people
   (c) Spinning the spinner, where the probability of each color is as shown

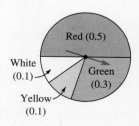

Red (0.5)
White (0.1)
Green (0.3)
Yellow (0.1)

5. To estimate the fish population of a certain pond, 200 fish were caught, marked, and returned to the pond. The next day, 300 fish were caught, of which 50 had been marked the previous day. Estimate the fish population of the pond.

6. Pick a block of two digits from the random-digit table. What is the probability that the number picked is less than 30?

7. ▶A cereal company places a coupon bearing a number from 1 to 9 in each box of cereal. If the numbers are distributed at random in the boxes of cereal, estimate the number of boxes of cereal, on the average, that would have to be purchased in order to obtain all nine numbers. Explain how the table of random digits could be used to estimate the number of coupons.

8. ▶A school has 500 students. The principal is to pick 30 students at random from the school to go to the Rose Bowl. How can this be done by using a random-digit table?

9. In a certain city, the probability that it will rain on a certain day is 0.8 if it rained the day before. The probability that it will be dry on a certain day is 0.3 if it was dry the day before. It is now Sunday and it is raining. Use the table of random digits to simulate the weather for the rest of the week.

10. How many cards would you expect to have to turn over on the average in an ordinary bridge deck before an ace appears? Try this experiment with cards and see how close your guess is.

11. Suppose that, in the World Series, the two teams are even-

ly matched. The two teams play until one team wins four games, and no ties are possible.
  (a) What is the maximum number of games that could be played?
  (b) Use simulation to approximate the probabilities that the series will end in four games and in seven games.
12. ▶The probability of the home team's winning a basketball game is 80%. Describe a simulation of the probability that the home team will win three home games in a row.
13. It is reported that 15% of people who came into contact with a person infected with strep throat contracted the disease. Use the random-digit table to simulate the probability that at least one child in a three-child family will catch the disease, given that each of the children has come into contact with the infected person.
★14. Montana duck hunters are all perfect shots. Ten Montana hunters are in a duck blind when ten ducks fly over. All ten hunters pick a duck at random to shoot at, and all ten hunters fire at the same time. How many ducks could be expected to escape, on the average, if this experiment were repeated a large number of times? How could this problem be simulated?

Review Problems Don't Do

15. A single card is drawn from an ordinary bridge deck. What is the probability of obtaining each of the following?
    (a) A club
    (b) A queen and a spade
    (c) Not a queen
    (d) Not a heart
    (e) A spade or a heart
    (f) The six of diamonds
    (g) A queen or a spade
    (h) Either red or black
16. From a sack containing seven red marbles, eight blue marbles, and four white marbles, marbles are drawn at random for several experiments. What is the probability of each of the following events?
    (a) One marble drawn at random is either red or blue.
    (b) The first draw is red and the second is blue, where one marble is drawn at random, its color is recorded, the marble is replaced, and another marble is drawn.
    (c) The event in (b), where the first marble is not replaced.

● COMPUTER CORNER

Enter the Logo procedure below into the computer to simulate the rolling of a die.

```
TO ROLL
    OUTPUT (1 + RANDOM 6)
END
```

Execute the procedure with the following line:

```
REPEAT 100 [PRINT ROLL]
```

Count the number of times each digit is printed to estimate the probabilities of obtaining a 1, 2, 3, 4, 5, or 6 when a die is tossed. (a) How close is this approximation to the theoretical probability for each of those events? (b) How would you change the procedure to simulate Problem 3(a) in the Problem Set?

Section 8-4 Odds and Expected Value

Computing Odds

odds

People talk about the *odds in favor of* and the *odds against* a particular team in an athletic contest. When the **odds** in favor of the Falcons' winning a particular football game are 4 to 1, this refers to how likely the Falcons are to win relative to how likely they are to lose. The probability of their winning is four times the probability of their losing. If W represents the event Falcons win and L represents the event Falcons lose, then $P(W) = 4P(L)$ or as a proportion, we have

$$\frac{P(W)}{P(L)} = \frac{4}{1}, \text{ or } 4:1.$$

Because W and L are complements of each other, we have

$$\frac{P(W)}{P(\overline{W})} = \frac{P(W)}{1 - P(W)} = \frac{4}{1}, \text{ or } 4:1.$$

The odds against the Falcons' winning are how likely the Falcons are to lose relative to how likely they are to win. Using the information above, we have the following:

$$\frac{P(L)}{P(W)} = \frac{1}{4}, \text{ or } 1:4.$$

Because $L = \overline{W}$, we have

$$\frac{P(\overline{W})}{P(W)} = \frac{1 - P(W)}{P(W)} = \frac{1}{4}, \text{ or } 1:4.$$

Formally, odds are defined as follows.

DEFINITION OF ODDS

Let $P(A)$ be the probability that A occurs, and let $P(\overline{A})$ be the probability that A does not occur. Then the **odds in favor** of an event A are

$$\frac{P(A)}{P(\overline{A})} \quad \text{or} \quad \frac{P(A)}{1 - P(A)}.$$

The **odds against** an event A are

$$\frac{P(\overline{A})}{P(A)} \quad \text{or} \quad \frac{1 - P(A)}{P(A)}.$$

When odds are calculated, denominators of the probabilities divide out. Thus, alternate definitions for odds in case of *equally likely* outcomes are as follows:

$$\text{Odds in favor} = \frac{\text{Number of favorable outcomes}}{\text{Number of unfavorable outcomes}}$$

$$\text{Odds against} = \frac{\text{Number of unfavorable outcomes}}{\text{Number of favorable outcomes}}$$

When you roll a die, the number of favorable ways of rolling a 4 in one throw of a die is 1, and the number of unfavorable ways is 5. Thus the odds in favor of rolling a 4 are 1 to 5.

EXAMPLE 8-11 For each of the following, find the odds in favor of the events occurring:

(a) Rolling a number less than 5 on a die
(b) Tossing heads on a fair coin
(c) Drawing an ace from an ordinary 52-card deck
(d) Drawing a heart from an ordinary 52-card deck

Solution (a) The probability of rolling a number less than 5 is $\frac{4}{6}$; the probability of rolling a number not less than 5 is $\frac{2}{6}$. The odds in favor of rolling a number less than 5 are $\left(\frac{4}{6}\right) \div \left(\frac{2}{6}\right)$, or $4:2$, or $2:1$.

(b) $P(H) = \frac{1}{2}$ and $P(\overline{H}) = \frac{1}{2}$; the odds in favor of getting heads are $\left(\frac{1}{2}\right) \div \left(\frac{1}{2}\right)$, or $1:1$.

(c) The probability of drawing an ace is $\frac{4}{52}$, and the probability of not drawing an ace is $\frac{48}{52}$; the odds in favor of drawing an ace are $\left(\frac{4}{52}\right) \div \left(\frac{48}{52}\right)$, or $4:48$, or $1:12$.

(d) The probability of drawing a heart is $\frac{13}{52}$, or $\frac{1}{4}$, and the probability of not drawing a heart is $\frac{39}{52}$, or $\frac{3}{4}$; the odds in favor of drawing a heart are $\left(\frac{13}{52}\right) \div \left(\frac{39}{52}\right) = \frac{13}{39}$, or $13:39$, or $1:3$. ∎

REMARK The preceding examples could have been worked just as easily by using the alternate definition for odds. In Example 8-11(c), we could reason that, since there are four ways of drawing an ace (favorable outcomes) and 48 ways of not drawing an ace (unfavorable outcomes), the odds in favor of drawing an ace are $4:48$, or $1:12$. Rework the other three parts of this example, using the alternate definition for odds.

Given the probability of an event, it is possible to find the odds in favor of (or against) the event and vice versa. For example, if the odds in favor of an event A are $5:1$, then the following proportion holds:

$$\frac{P(A)}{1 - P(A)} = \frac{5}{1}$$
$$P(A) = 5[1 - P(A)]$$
$$6P(A) = 5$$
$$P(A) = \frac{5}{6}$$

The probability $\frac{5}{6}$ is a ratio. The exact number of favorable outcomes and the exact total of all outcomes are not necessarily known.

EXAMPLE 8-12 In the cartoon, find the probability of making totally black copies if the odds of 3 to 1 are against making totally black copies.

TODAY'S ODDS

Makes totally black copies	3-1
Makes copies with wavy black lines	4-1
Misfeeds	2-1
Gives no change	3-1
Gives double change	8-1
Mystery light appears and 2 or more of the above occur	5-1

COPIES 20¢

Cable

Solution If the odds against making totally black copies are 3 to 1, B represents the event of making a totally black copy, then $\overline{B}$ represents not making a totally black copy, and we have

$$\frac{P(\overline{B})}{1 - P(\overline{B})} = \frac{3}{1}$$

$$P(\overline{B}) = 3(1 - P(\overline{B}))$$

$$P(\overline{B}) = \frac{3}{4}$$

$$P(B) = 1 - P(\overline{B}), \text{ or } 1 - \frac{3}{4}$$

$$P(B) = \frac{1}{4}$$

Expected Value

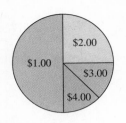

FIGURE 8-31

Consider the spinner in Figure 8-31, with the payoff in each sector of the circle. If the spinner is fair, then the following probabilities can be assigned to each region:

$$P(\$1.00) = \frac{1}{2}$$

$$P(\$2.00) = \frac{1}{4}$$

$$P(\$3.00) = \frac{1}{8}$$

$$P(\$4.00) = \frac{1}{8}$$

Should the owner of this spinner expect to make money over an extended period of time if the charge is $2.00 per spin?

To determine the average payoff over the long run, we multiply the probability of landing on the payoff by the payoff and then find the sum of the products. This computation is given by $E = \left(\frac{1}{2}\right)1 + \left(\frac{1}{4}\right)2 + \left(\frac{1}{8}\right)3 + \left(\frac{1}{8}\right)4 = 1.875$, or about 1.88. The owner can expect to pay out about $1.88 per spin, and $1.88 is less than the $2.00 charge, so the owner should make a profit on the spinner if it is used many times.

expected value   The approximate total of the products in this example, $1.88, is called the **expected value,** or *mathematical expectation,* of the experiment.

DEFINITION OF EXPECTED VALUE

If, in an experiment, the possible outcomes are numbers $a_1, a_2, \ldots, a_n$, occurring with probabilities $p_1, p_2, \ldots, p_n$, respectively, then the **expected value** (mathematical expectation) E is given by the equation

$$E = a_1 \cdot p_1 + a_2 \cdot p_2 + a_3 \cdot p_3 + \cdots + a_n \cdot p_n.$$

The expected value is an average of winnings for the long run. Expected value can be used to predict the average result of an experiment when it is repeated many times, but an expected value cannot be used to determine the outcome of any single experiment.

Suppose that Mega-Mouth Toothpaste Company is giving away $20,000 in a contest. To win the contest, a person must send in a postcard with his or her name on it (no purchase of toothpaste is necessary). Suppose the company expects to receive one million postcards. Is this contest fair? *A game is considered fair if the net winnings are* $0; that is, the expected value of the game must equal the price of playing the game. The expected value is one-millionth of $20,000; that is, $E = (1/1{,}000{,}000) \cdot (20{,}000) = \frac{2}{100} = 0.02$. Because the cost of the postcard and postage exceeds $0.02, the contest is not fair.

EXAMPLE 8-13   Suppose that you pay $5.00 to play the following game. Two coins are tossed. You receive $10 if two heads occur, $5 if exactly one head occurs, and nothing if no heads appear. Is this a fair game?

Solution Before we determine the average payoff, recall that $P(HH) = \frac{1}{4}$, $P(HT$ or $TH) = \frac{1}{2}$, and $P(TT) = \frac{1}{4}$. To find the expected value we perform the following computation:

$$E = \left(\frac{1}{4}\right) \cdot (\$10) + \left(\frac{1}{2}\right) \cdot (\$5) + \left(\frac{1}{4}\right) \cdot (0) = \$5.$$

Because the price of playing is equal to the average payoff, the net winnings are $0, and this is a fair game. ∎

PROBLEM 5

Al and Betsy played a coin-tossing game in which a fair coin was tossed until a total of either three heads or three tails occurred. Al was to win when a total of three heads were tossed, and Betsy was to win when a total of three tails were tossed. Each bet $50 on the game. If the coin was lost when Al had two heads and Betsy had one tail, how should the stakes be fairly split if the game is not continued?

Understanding the Problem. Al and Betsy each bet $50 on a coin-tossing game in which a fair coin was to be tossed five times. Al was to win when a total of three heads was obtained; Betsy was to win when a total of three tails was obtained. When Al had two heads and Betsy had one tail, the coin was lost. The problem is how to split the stakes fairly.

If the stakes of the game are to be split fairly, then there could be many interpretations. Possibly, though, the best is to split the pot in proportion to the probabilities of each player's winning the game when play was halted. We must calculate the expected value for each player and split the pot accordingly.

Devising a Plan. A third head would make Al the winner, whereas Betsy needs two more tails to win. A *tree diagram* that simulates the completion of the game allows us to find the probability of each player's winning the game. Once the probabilities are found, all that is necessary is to multiply the probabilities by the amount of the pot, $100, to determine each player's fair share.

Carrying Out the Plan. The tree diagram in Figure 8-32 shows the possibilities for game winners if the game is completed. We can find the probabilities of each player's winning as follows:

$$P(\text{Betsy wins}) = \frac{1}{2} \cdot \frac{1}{2} = \frac{1}{4}$$

$$P(\text{Al wins}) = 1 - \frac{1}{4} = \frac{3}{4}$$

FIGURE 8-32

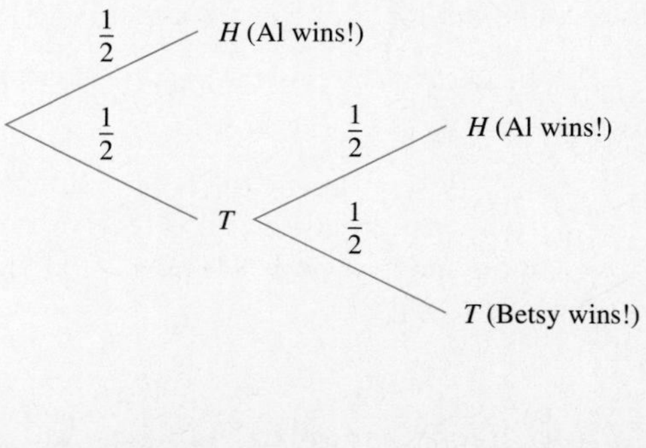

Hence, the fair way to split the stakes is for Al to receive $\frac{3}{4}$ of $100, or $75, whereas Betsy should receive $\frac{1}{4}$ of $100, or $25.

Looking Back. The problem could be made even more interesting by assuming that the coin is not fair so that the probability is not $\frac{1}{2}$ for each branch in the tree diagram. Other possibilities arise if the players have unequal amounts of money in the pot or if more tosses are required in order to win. ∎

PROBLEM SET 8-4

1. (a) What are the odds in favor of drawing a face card from an ordinary deck of playing cards?
   (b) What are the odds against drawing a face card?
2. On a single roll of a pair of dice, what are the odds against rolling a sum of 7?
3. If the probability of a boy's being born is $\frac{1}{2}$, and a family plans to have four children, what are the odds against having all boys?
4. Diane tossed a coin nine times and got nine tails. Assume that Diane's coin is fair and answer each of the following questions:
   (a) What is the probability of tossing a tail on the tenth toss?
   (b) What is the probability of tossing ten more tails in a row?
   (c) What are the odds against tossing ten more tails in a row?
5. If the odds against Sam's winning his first prize fight are 3 to 5, then what is the probability that he will win the fight?
6. What are the odds in favor of tossing at least two heads if a fair coin is tossed three times?
7. If the probability of rain for the day is 60%, what are the odds against its raining?
8. On an American roulette wheel, half of the slots numbered 1–36 are red and half are black. Two slots, numbered 0 and 00, are green. What are the odds against a red slot's coming up on any spin of the wheel?
9. ▶A game involves tossing two coins. A player wins $1.00 if both tosses result in heads. What should you pay to play this game in order to make it a fair game? Explain your answer.
10. Suppose that a player rolls a fair die and receives the number of dollars equal to the number of spots showing on the die. What is the expected value?
11. A punch-out card contains 500 spaces. One particular space pays $1000, five other spaces each pay $100, and the remaining spaces pay nothing. If a player chooses one space, what is the expected value of the game?

12. The chart below shows the probabilities assigned by Stu to the number of hours spent on homework on a given night:

Hours	Probability
1	0.15
2	0.20
3	0.40
4	0.10
5	0.05
6	0.10

If Stu's friend Stella calls and asks how long his homework will take, what would you expect his answer to be, based on this table?

13. In the cartoon, Snoopy is told that the odds are 1000 to 1 that he will end up with a broken arm if he touches Linus's blanket. What is the probability of this event E?

PEANUTS reprinted by permission of UFS, Inc.

14. Suppose five quarters, five dimes, five nickels, and ten pennies are in a box. One coin is selected at random. What is the expected value of this experiment?

15. If the odds in favor of Fast Leg's winning a horse race are 5 to 2 and the first prize is $14,000, what is the expected value of Fast Leg's winning?

16. Al and Betsy are playing a coin-tossing game in which a fair coin is tossed. Al wins when a total of ten heads are tossed, and Betsy wins when a total of ten tails are tossed.
    (a) If nine heads and eight tails have been tossed and the game is stopped, how should a pot of $100 be fairly divided?
    (b) What are the odds against Betsy's winning at the time the game was stopped in (a)?
    ★(c) Suppose eight heads and five tails have been tossed when the game is stopped. How should a pot of $100 be fairly divided?
    (d) What are the odds in favor of Al's winning at the time the game was stopped in (c)?

17. Suppose you pay $5.00 to play a game in which two coins are tossed. You receive $10 if two heads occur, $5 if exactly one head occurs, and $0 if no heads appear. Is this a fair game?

18. Lori spends $1.00 for 1 ticket in a raffle with a $100 prize. If 200 tickets are sold, is $1.00 a fair price to pay for the ticket?

19. The odds against the University of Montana basketball team winning the 1991 Men's NCAA Tournament were listed as one billion to one. Assuming these odds, what was the probability that the University of Montana would win the tournament?

20. ▶A prominent newspaper reported that the odds of getting AIDS in June, 1991 were 68,000 to 1. Explain why you believe or disbelieve this report.

21. On a tote board at a race track, the odds for Gameylegs are listed as 26:1. Tote boards list the odds that the horse will lose the race, not win the race. If this is the case, what is the probability of Gameylegs winning the race?

Review Exercises

22. Write the sample space for each of the following experiments:
    (a) Spin spinner 1 once.
    (b) Spin spinner 2 once.
    (c) Spin spinner 1 once and then spin spinner 2 once.
    (d) Spin spinner 2 once and then roll a die.
    (e) Spin spinner 1 twice.
    (f) Spin spinner 2 twice.

23. Draw a spinner with two sections, red and blue, such that the probability of getting (Blue, Blue) on two spins is $\frac{25}{36}$.

24. When drawing two letters from the alphabet with replacement, find the probability of getting two vowels.

Spinner 1   Spinner 2

---

*Section 8-5 Methods of Counting

Permutations of Unlike Objects

Consider how many ways the owner of an ice cream parlor can display 10 ice cream flavors in a row along the front of the display case. The first position can be filled in ten ways, the second position in nine ways, the third position in eight ways, and so on. By the Fundamental Counting Principle, there are $10 \cdot 9 \cdot 8 \cdot 7 \cdot 6 \cdot 5 \cdot 4 \cdot 3 \cdot 2 \cdot 1$, or 3,628,800, ways to display the flavors. If there were 16 flavors, there would be $16 \cdot 15 \cdot 14 \cdot 13 \cdots 3 \cdot 2 \cdot 1$ ways to arrange them. In general, *if there are n objects, then the number of possible ways to arrange the objects in a row is the product of all the natural numbers from n to 1, inclusive.* This expression is called ***n* factorial** and is denoted by ***n*!**, as shown next.

n factorial / n!

$$n! = n \cdot (n-1) \cdot (n-2) \cdots 3 \cdot 2 \cdot 1$$

For example, $5! = 5 \cdot 4 \cdot 3 \cdot 2 \cdot 1$, $3! = 3 \cdot 2 \cdot 1$, and $1! = 1$. Using factorial notation is helpful in counting and probability problems.

SECTION 8-5 METHODS OF COUNTING

permutation   An arrangement of things in a definite order with no repetitions is called a **permutation**. For example, RAT, RTA, TAR, TRA, ART, and ATR are all different arrangements of the letters R, A, and T.

Consider the set of people in a small club, {Al, Betty, Carl, Dan}. To elect a president and a secretary, order is important and no repetitions are possible. Counting the number of possibilities is a permutation problem. Since there are four ways of choosing a president and then three ways of choosing a secretary, by the Fundamental Counting Principle, there are $4 \cdot 3$, or 12, ways of choosing a president and a secretary. Choosing two officers from a club of four is a permutation of four people chosen two at a time. The number of possible permutations of four objects taken two at a time, denoted by $_4P_2$, may be counted using the Fundamental Counting Principle, as seen in Figure 8-33. Therefore, we have $_4P_2 = 4 \cdot 3$, or 12.

FIGURE 8-33

| 4 choices | 3 choices |
| President | Secretary |

In general, *if n objects are chosen r at a time, then the number of possible permutations, denoted by $_nP_r$, is*

$$_nP_r = n \cdot (n - 1) \cdot (n - 2) \cdots [n - (r - 1)],$$

or

$$_nP_r = n \cdot (n - 1) \cdot (n - 2) \cdots (n - r + 1).$$

The formula for permutations can be written in terms of factorials. Consider a permutation of 20 objects three at a time, $_{20}P_3$:

$$_{20}P_3 = 20 \cdot 19 \cdot 18$$
$$= \frac{20 \cdot 19 \cdot 18 \cdot 17 \cdots \cdot 3 \cdot 2 \cdot 1}{17 \cdot 16 \cdots \cdot 3 \cdot 2 \cdot 1}$$
$$_{20}P_3 = \frac{20!}{17!}$$
$$_{20}P_3 = \frac{20!}{(20 - 3)!}$$

The above can be generalized as follows:

$$_nP_r = \frac{n!}{(n - r)!}$$

$_nP_n$ is the number of permutations of n objects chosen n at a time—that is, the number of ways of rearranging n objects in a row. We have seen that this number is $n!$. If we use the formula for $_nP_r$ to compute $_nP_n$, we obtain

$$_nP_r = \frac{n!}{(n - n)!} = \frac{n!}{0!}$$

Consequently, $n! = n!/0!$. It is to make this equation true, that we define 0! to be 1.

EXAMPLE 8-14   (a) A baseball team has nine players. Find the number of ways the manager can arrange the batting order.

(b) Find the number of ways of choosing three initials from the alphabet if none of the letters can be repeated.

Solution (a) Because there are nine ways to choose the first batter, eight ways to choose the second batter, and so on, there are $9 \cdot 8 \cdot 7 \cdots 2 \cdot 1 = 9!$, or 362,880, ways of arranging the batting order. Using the formula for permutations, we have $_9P_9 = 9! = 362,880$.

(b) There are 26 ways of choosing the first letter, 25 ways of choosing the second letter, and 24 ways of choosing the third letter; hence, there are $26 \cdot 25 \cdot 24$, or 15,600, ways of choosing the three letters. Using the formula for permutations we have $_{26}P_3 = 26 \cdot 25 \cdot 24 = 15,600$. ∎

Permutations Involving Like Objects

In the previous counting examples, each individual object to be counted was distinct. Suppose we wanted to rearrange the letters in the word ZOO. How many choices would we have? A tree diagram, as in Figure 8-34, suggests that there might be $3 \cdot 2 \cdot 1 = 3!$, or 6, possibilities. However, looking at the list of possibilities shows that each of ZOO, OZO, and OOZ appears twice because the O's are not different. We need to determine how to remove the duplication in arrangements such as this where some objects are the same. To eliminate the duplication, we divide the number of arrangements shown by the number of ways the two O's can be rearranged, which is 2!. Consequently, there are $\frac{3!}{2!}$, or 3, ways of arranging the letters in ZOO. The arrangements are ZOO, OZO, and OOZ.

FIGURE 8-34

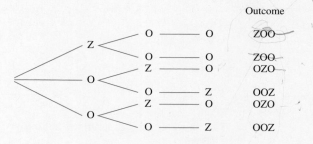

PERMUTATIONS OF LIKE OBJECTS

If a set contains n elements, of which r_1 are of one kind, r_2 are of another kind, and so on through r_k, then the number of different arrangements of all n elements is equal to
$$\frac{n!}{r_1! \cdot r_2! \cdot r_3! \cdots r_k!}$$

EXAMPLE 8-15 Find the number of rearrangements of the letters in each of the following words:

(a) Bubble (b) Statistics

Solution (a) There are six letters with *b* repeated three times. Hence, the number of arrangements is

$$\frac{6!}{3!} = 6 \cdot 5 \cdot 4 = 120.$$

(b) There are 10 letters in the word *statistics*, with 3 *s*'s, 3 *t*'s, and 2 *i*'s duplicated in the word. Hence the number of arrangements is

$$\frac{10!}{3! \cdot 3! \cdot 2!} = \frac{10 \cdot 9 \cdot 8 \cdot 7 \cdot 6 \cdot 5 \cdot 4 \cdot 3 \cdot 2 \cdot 1}{3 \cdot 2 \cdot 1 \cdot 3 \cdot 2 \cdot 1 \cdot 2 \cdot 1} = 50,400.$$

■

Combinations

combination

Reconsider the club {Al, Betty, Carl, Dan}. Suppose that a two-person committee is selected with no chair. In this case, order is not important, and an Al-Betty choice is the same as a Betty-Al choice. An arrangement of objects in which the order does not make any difference is called a **combination**. A comparison of the results of electing a president and secretary for the club and the results of simply selecting a two-person committee are shown in Figure 8-35. We see that the number of combinations is the number of permutations divided by 2, or

$$\frac{4 \cdot 3}{2} = 6.$$

Because each two-person choice can be arranged in 2!, or 2, ways, we divide the number of permutations by 2.

In how many ways can a committee of three people be selected from the club {Al, Betty, Carl, Dan}? To solve this problem, we first solve the problem of finding the number of three-person committees, assuming that a president, vice president, and secretary are chosen. A partial list of possibilities for both problems is shown in Figure 8-36.

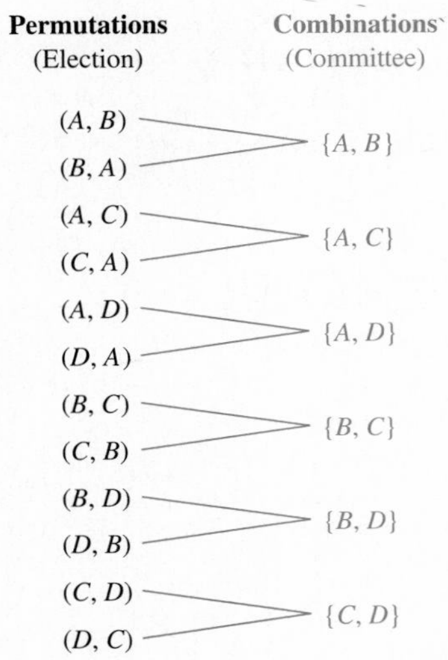

Permutations (Election)	Combinations (Committee)	Permutations (Election)	Combinations (Committee)
(A, B) (B, A)	{A, B}	(A, B, C) (A, C, B) (B, A, C) (B, C, A) (C, A, B) (C, B, A)	{A, B, C}
(A, C) (C, A)	{A, C}		
(A, D) (D, A)	{A, D}		
(B, C) (C, B)	{B, C}	(A, B, D) (A, D, B) (B, A, D) (B, D, A) (D, A, B) (D, B, A)	{A, B, D}
(B, D) (D, B)	{B, D}		
(C, D) (D, C)	{C, D}	⋮	⋮

FIGURE 8-35 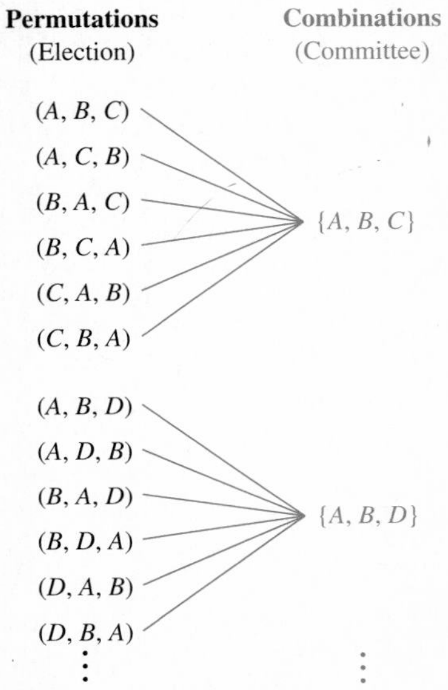   **FIGURE 8-36**

By the Fundamental Counting Principle, the number of ways to choose three people from the list of 4 is $4 \cdot 3 \cdot 2$, or 24. However, with each triple chosen, there are 3! ways to rearrange the triple. Hence, there are 3!, or 6, times as many permutations as there are combinations. In general, we use the following rule to count combinations: *To find the number of combinations possible in a counting problem, first use the Fundamental Counting Principle to find the number of permutations, and then divide by the number of ways in which each choice can be arranged.*

Symbolically, the number of combinations of n objects taken r at a time is denoted by ${}_nC_r$. From the preceding rule, we develop the following formula:

$$ {}_nC_r = \frac{{}_nP_r}{{}_rP_r} = \frac{\frac{n!}{(n-r)!}}{r!} = \frac{n!}{r!(n-r)!}. $$

It is not necessary to memorize this formula, since we can always find the number of combinations by using the reasoning developed in the committee example. Such reasoning is used on the student page from *Addison-Wesley Mathematics*, Grade 8, 1989.

Counting Selections: Combinations

How many selections of 2 topics from the 4 topics are possible?

First we find the number of permutations of 2 topics from 4 topics.

Permutations = **4 × 3 = 12**

The **order** of the topics is not important. Choosing A, then B, is the same as choosing B, then A. We need to divide by the number of permutations of 2 topics from 2 topics.

American History
Write a report on any 2 topics.
A. Treaty of Paris
B. Boston Tea Party
C. The Stamp Act
D. The Loyalists

Permutations of
2 topics from 4 topics → $\frac{4 \times 3}{2 \times 1} = \frac{12}{2} = 6$
Permutations of
2 topics from 2 topics

List of Permutations

A,B	A,C	A,D	B,C	B,D	C,D
B,A	C,A	D,A	C,B	D,B	D,C

There are 6 possible selections.

A selection of a number of objects from a set of objects, *without regard to order*, is called a **combination** of the objects.

Selections

A and B	A and D	B and D
A and C	B and C	C and D

Other Examples

How many combinations or selections of 3 topics from 5 topics are possible?

Combinations = $\frac{5 \times 4 \times 3}{3 \times 2 \times 1} = \frac{60}{6} = 10$

Topics	Combinations of 3 topics from 5 topics	
A	ABC	ADE
B	ABD	BCD
C	ABE	BCE
D	ACD	BDE
E	ACE	CDE

SECTION 8-5 METHODS OF COUNTING

EXAMPLE 8-16 The Library of Science Book Club offers three free books from a list of 42. If you circle three choices from a list of 42 numbers on a postcard, how many possible choices are there?

Solution By the Fundamental Counting Principle, there are $42 \cdot 41 \cdot 40$ ways to choose the three free books. Because each set of three circled numbers could be rearranged $3 \cdot 2 \cdot 1$ different ways, there is an extra factor of $3!$ in the original $42 \cdot 41 \cdot 40$ ways. Therefore, the number of combinations possible for 3 books is

$$\frac{42 \cdot 41 \cdot 40}{3!} = 11{,}480.$$

EXAMPLE 8-17 At the beginning of the second quarter of a mathematics class for elementary school teachers, each of the 25 students shook hands with each of the other students exactly once. How many handshakes took place?

Solution Since the handshake between persons A and B is the same as that between persons B and A, this is a problem of choosing combinations of 25 people two at a time. There are

$$\frac{25 \cdot 24}{2!} = 300$$

different handshakes.

PROBLEM 6

In the cartoon, suppose that Peppermint Patty took a six-question true-false test. If she answered each question true or false at random, what is the probability that she answered 50% of the questions correctly?

PEANUTS reprinted by permission of UFS, Inc.

Understanding the Problem. A score of 50% indicates that Peppermint Patty answered $\frac{1}{2}$ of the six questions, or three questions, correctly. She answered the questions true or false at random, so the probability that she answered a given question correctly is $\frac{1}{2}$.

We are asked to determine the probability that Patty answered exactly three of the questions correctly.

Devising a Plan. We do not know which three questions Patty missed. She could have missed any three questions out of six on the test. Suppose that she answered questions 2, 4, and 5 incorrectly. In this case, she would have answered questions 1, 3, and 6 correctly. We can compute the probability of this set of answers by *using the branch of a tree diagram* in Figure 8-37, where C represents a correct answer and I represents an incorrect answer. Multiplying the probabilities along the branches, we obtain $\left(\frac{1}{2}\right)^6$ as the probability of answering questions 1–6 in the following way: C I C I I C. There are other ways to answer exactly three questions correctly: for example, C C C I I I. The probability of answering questions 1–6 in this way is also $\left(\frac{1}{2}\right)^6$. The number of ways to answer the questions is simply the number of ways of arranging three C's and three I's in a row, which is also the number of ways of choosing three correct questions out of six, that is, $_6C_3$. Because all these arrangements give Patty a score of 50%, the desired probability is the sum of the probabilities for each arrangement.

FIGURE 8-37

Carrying Out the Plan. There are $_6C_3$, or 20 sets of answers similar to the one in Figure 8-37, with three correct and three incorrect answers. The product of the probabilities for each of these sets of answers is $\left(\frac{1}{2}\right)^6$, so the sum of the probabilities for all 20 sets is $20 \cdot \left(\frac{1}{2}\right)^6$, or approximately 0.3125. Thus, Peppermint Patty has a probability of 0.3125 of obtaining a score of exactly 50% on the test.

Looking Back. It seems paradoxical to learn that the probability of obtaining a score of 50% on a six-question true-false test is not close to $\frac{1}{2}$. As an extension of the problem, suppose that a passing score is a score of at least 70%. Now what is the probability that Peppermint Patty will pass? What is the probability of her obtaining a score of at least 50% on the test? If the test is a six-question multiple-choice test with five alternative answers for each question, what is the probability of obtaining a score of at least 50% by random guessing?

PROBLEM SET 8-5

1. ▶The Fundamental Counting Principle, permutations, and combinations are used in counting problems. In your own words, compare these terms.
2. (a) In how many ways can the letters in each of the following words be arranged?
    (i) GOOD (ii) MATHEMATICS
   (b) In a car race there are six Chevrolets, four Fords, and two Pontiacs. In how many ways can the 12 cars finish if we consider only the makes of the cars?
3. Suppose the Department of Motor Vehicles uses only six numbers to create its license plates. Answer the following:
   (a) How many license plates are possible?
   (b) Based upon the 1990 census, are there any states

where the answer in (a) might provide enough licenses?

▶(c) If you were in charge of constructing license plates for the state of California, describe the method you would use to ensure you would have enough license plates available.

4. The eighth-grade class at a grade school has 16 girls and 14 boys. How many different possible boy-girl dates can be arranged?

5. If a coin is tossed five times, in how many different ways can the sequence of heads and tails appear?

6. The telephone prefix for a university is 243. The prefix is followed by four digits. How many telephones are possible before a new prefix is needed?

7. Radio stations in the United States have call letters that begin with either K or W. Some have a total of three letters, whereas others have four letters. How many sets of three-letter call letters are possible? How many sets of four-letter call letters are possible?

8. Carlin's Pizza House offers three kinds of salads, 15 kinds of pizza, and four kinds of desserts. How many different three-course meals can be ordered?

9. Decide whether each of the following is true or false:

   (a) $6! = 6 \cdot 5!$

   (b) $3! + 3! = 6!$

   (c) $\dfrac{6!}{3!} = 2!$

   (d) $\dfrac{6!}{3} = 2!$

   (e) $\dfrac{6!}{5!} = 6$

   (f) $\dfrac{6!}{4!2!} = 15$

   (g) $n!(n+1) = (n+1)!$

10. In how many ways can the letters in the word SCRAMBLE be rearranged?

11. How many two-person committees can be formed from a group of six people?

12. Find the number of ways to rearrange the letters in the following words:

    (a) OHIO
    (b) ALABAMA
    (c) ILLINOIS
    (d) MISSISSIPPI
    (e) TENNESSEE

13. Assume a class has 30 members.

    (a) In how many ways can a president, vice president, and secretary be selected?

    (b) How many committees of three can be chosen?

14. A basketball coach was criticized in the newspaper for not trying out every combination of players. If the team roster has 12 players, how many five-player combinations are possible?

15. Solve the problem posed by the following cartoon. (AAUGHH! is not an acceptable answer.)

PEANUTS reprinted by permission of UFS, Inc.

16. A five-volume numbered set of books is placed randomly on a shelf. What is the probability that the books will be numbered in the correct order from left to right?

17. Take ten points in a plane, no three of them on a line. How many straight lines can be drawn if each line is drawn through a pair of points?

18. A bicycle lock has 3 reels, each of which contains the numbers 0–9. To open the lock, you must have the right combination of numbers, such as 355 or 962, where one number is chosen from each reel. How many different combinations are possible for the lock?

19. Sally has four red flags, three green flags, and two white flags. How many nine-flag signals can she run up a flagpole?

20. Find the number of shortest paths from point A to point B along the edges of the cubes in each of the following. (For example, in (a) one shortest path is A-C-D-B.)

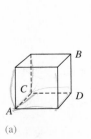
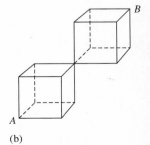

(a)   (b)

CHAPTER 8 PROBABILITY

21. At a party, 28 handshakes took place. Each person shook hands exactly once with each of the others present. How many people were at the party?
22. A committee of three people is selected at random from a set consisting of seven Americans, five French people, and three English people.
    (a) What is the probability that the committee consists of all Americans?
    (b) What is the probability that the committee has no Americans?
23. The triangular array of numbers pictured is called **Pascal's triangle.** Notice the first and last numbers in each row are 1's. Every other number is the sum of the two numbers immediately above it. The rows are counted starting at 0.

    ```
                    1              Row
                  1   1             (0)
                1   2   1           (1)
              1   3   3   1         (2)
            1   4   6   4   1       (3)
          1   5  10  10   5   1     (4)
        1   6  15  20  15   6   1   (5)
                                    (6)
    ```

It can be shown that the entries in Pascal's triangle are the numbers of combinations of n objects taken r at a time, where n is the number of the row and r is the number of the item in the row. (Note that r could be 0.) For example, in the third row we have $_3C_0 = 1$, $_3C_1 = 3$, $_3C_2 = 3$, and $_3C_3 = 1$. Use the triangle to determine the following:
(a) $_5C_3$
(b) $_5C_5$
(c) $_6C_0$
(d) $_3C_2$

24. The probability of a basketball player's making a free throw successfully at any time in a game is $\frac{2}{3}$. If the player attempts ten free throws in a game, what is the probability that exactly six free throws are made?
★25. In how many ways can five couples be seated in a row of ten chairs if no couple is separated?

▼ **BRAIN TEASER**

An airplane can complete its flight if at least $\frac{1}{2}$ of its engines are working. If the probability that an engine fails is 0.01 and all engine failures are independent events, what is the probability of a successful flight if the plane has the given number of engines?
(a) 2 engines (b) 4 engines

Solution To The Preliminary Problem

Understanding the Problem. Stephen sealed three letters in addressed envelopes without checking to see if each was in the correct envelope. We are to determine the probability that each of the three letters was placed correctly. This probability could be found if we knew the sample space, or at least how many elements are in the sample space.

Devising a Plan. To aid in solving the problem, we represent the respective letters as a, b, and c and the respective envelopes as A, B, and C. For example, a correctly placed letter a would be in envelope A. To construct the sample space, we use the strategy of *making a table*. The table should show all the possible arrangements of letters in envelopes. Once the table is completed, we can determine the probability that each letter is correctly placed.

Carrying Out the Plan. Table 8-8 is constructed by using the envelope labels A, B, and C as headings and listing all possibilities of letters a, b, and c below the headings. Case 1 is the only case out of six in which each of the envelopes is labeled correctly, so the probability that each envelope is labeled correctly is $\frac{1}{6}$.

TABLE 8-8

	A	B	C
1	a	b	c
2	a	c	b
3	b	a	c
4	b	c	a
5	c	a	b
6	c	b	a

(Addresses across top; Letters down side)

Looking Back. Is the probability of having each letter placed incorrectly the same as the probability of having each letter placed correctly? A first guess might be that the probabilities are the same, but that is not true. Why?

We also could have used a counting argument to solve the problem. Given an envelope, there is only one correct letter to place in the envelope. Thus, there is one correct way to place the letters in the envelopes. By the Fundamental Counting Principle, there are $3 \cdot 2 \cdot 1$ ways of choosing the letters to place in the envelopes, so the probability of having the letters correctly placed is $\frac{1}{6}$.

QUESTIONS FROM THE CLASSROOM

1. A student claims that, if a fair coin is tossed and comes up heads five times in a row, then, according to the law of averages, the probability of tails on the next toss is greater than the probability of heads. What is your reply?

2. A student observes the spinner and claims that the color red has the highest probability of appearing since there are two red areas on the spinner. What is your reply?

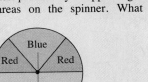

3. A student tosses a coin three times, and tails appears each time. The student concludes that the coin is not fair. What is your response?

4. An experiment consists of tossing a coin twice. The student reasons that there are three possible outcomes: two heads, one head and one tail, or two tails. Thus, $P(HH) = \frac{1}{3}$. What is your reply?

5. In response to the question, "If a fair die is rolled twice, what is the probability of rolling a pair of 5s?" a student replies, "One third, because $\frac{1}{6} + \frac{1}{6} = \frac{1}{3}$." How do you respond?

6. A student wonders why probabilities cannot be negative. What is your response?

7. A student claims that "if the probability of an event is $\frac{3}{5}$, then there are three ways the event can occur and only five elements in the sample space." How do you respond?

8. A student does not understand the meaning of $_4P_0$. The student wants to know how we can choose four objects 0 at a time. How do you respond?

9. A student wants to know why, if we can define 0! as 1, we cannot define $\frac{1}{0}$ as 1. How do you respond?

CHAPTER OUTLINE

I. Probability
   A. Probabilities can be determined **experimentally (empirically)** or **theoretically**.
   B. A **sample space** is the set of all possible outcomes of an **experiment**.
   C. An **event** is a subset of a sample space.
   D. Outcomes are **equally likely** if each outcome is as likely to occur as another.
   E. If all outcomes of an experiment are *equally likely*, the **probability of an event A** from sample space S is given by
   $$P(A) = \frac{n(A)}{n(S)}.$$
   F. An **impossible event** is an event with a probability of zero. An impossible event can never occur.
   G. A **certain event** is an event with a probability of one. A certain event is sure to happen.
   H. Two events are **mutually exclusive** if and only if exactly one of the events can occur at any given time—that is, if and only if the events are disjoint.
   I. The probability of the **complement of an event** is given by $P(\overline{A}) = 1 - P(A)$, where A is the event and $\overline{A}$ is its complement.
   J. **Multiplication Rule for Probabilities** For all **multistage experiments**, the probability of the outcome along any path of a tree diagram is equal to the product of all the probabilities along the path.
   K. **Simulations** can play an important part in probability. Fair coins, dice, spinners, and random-digit tables are useful in performing simulations.

II. Odds and expected value
   A. The **odds in favor** of an event A are given by
   $$\frac{P(A)}{P(\overline{A})} = \frac{P(A)}{1 - P(A)}.$$
   B. The **odds against** an event A are given by
   $$\frac{P(\overline{A})}{P(A)} = \frac{1 - P(A)}{P(A)}.$$
   C. If, in an experiment, the possible outcomes are numbers $a_1, a_2, \ldots, a_n$, occurring with probabilities $p_1, p_2, \ldots, p_n$, respectively, then the **expected value** E is defined as
   $$E = a_1 \cdot p_1 + a_2 \cdot p_2 + a_3 \cdot p_3 + \cdots + a_n \cdot p_n.$$

III. Counting principles
   A. **Fundamental Counting Principle** If an event M can occur in m ways and, after it has occurred, event N can occur in n ways, then event M followed by event N can occur in $m \cdot n$ ways.
   B. **Permutations** are arrangements in which order is important.
   $$_nP_r = \frac{n!}{(n-r)!}$$
   C. The expression $n!$, called n **factorial**, represents the product of all the natural numbers less than or equal to n. 0! is defined as 1.
   D. **Permutations of like objects** If a set contains n elements, of which r_1 are of one kind, r_2 are of another kind, and so on through r_k, then the number of different arrangements of all n elements is equal to
   $$\frac{n!}{r_1! \cdot r_2! \cdot r_3! \cdots r_k!}.$$
   E. **Combinations** are arrangements in which order is *not* important. To find the number of combinations possible, first use the Fundamental Counting Principle to find the number of permutations, and then divide by the number of ways in which each choice can be arranged.
   $$_nC_r = \frac{_nP_r}{_rP_r}$$

CHAPTER TEST

1. Suppose that the names of the days of the week are placed in a box and one name is drawn at random.
   (a) List the sample space for this experiment.
   (b) List the event consisting of outcomes that the day drawn starts with the letter T.
   (c) What is the probability of drawing a day that starts with T?
2. If you have a jar of 1000 jelly beans, and you know that $P(\text{Blue}) = \frac{4}{5}$ and $P(\text{Red}) = \frac{1}{8}$, what can you say about the beans in the jar?
3. In the 1960 presidential election, John F. Kennedy received 34,226,731 votes, and Richard M. Nixon received 34,108,157. If a voter is chosen at random, answer the following.
   (a) What is the probability that the person voted for Kennedy?
   (b) What is the probability that the person voted for Nixon?
   (c) What are the odds that a person chosen at random did not vote for Nixon?
4. A box contains three red balls, five black balls, and four white balls. Suppose that one ball is drawn at random. Find the probability of each of the following events:
   (a) A black ball is drawn.
   (b) A black or a white ball is drawn.
   (c) Neither a red nor a white ball is drawn.
   (d) A red ball is not drawn.
   (e) A black ball and a white ball are drawn.
   (f) A black or white or red ball is drawn.
5. One card is selected at random from an ordinary set of 52 cards. Find the probability of each of the following events:
   (a) A club is drawn.
   (b) A spade and a 5 are drawn.
   (c) A heart or a face card is drawn.
   (d) A jack is not drawn.
6. A box contains five black balls and four white balls. If three balls are drawn one by one, find the probability that they are all white if the draws are made as follows:
   (a) With replacement
   (b) Without replacement
7. Consider the two boxes in the figure. If a letter is drawn from box 1 and placed into box 2, and then a letter is drawn from box 2, what is the probability that the letter is an L?

8. Use the following boxes for a two-stage experiment. First select a box at random, and then select a letter at random from the box. What is the probability of drawing an A?

9. Consider the boxes shown. Draw a ball from box 1, and put it into box 2. Then draw a ball from box 2, and put it into box 3. Finally, draw a ball from box 3. Construct a tree diagram for this experiment, and calculate the probability that the last ball chosen is black.

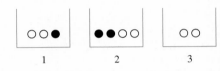

10. What are the odds in favor of drawing a jack when one card is drawn from an ordinary deck of playing cards?
11. A die is rolled once. What are the odds against rolling a prime number?
12. If the odds in favor of a certain event are 3 to 5, what is the probability that the event will occur?
13. A game consists of rolling two dice. Rolling double 1s pays $7.20. Rolling double 6s pays $3.60. Any other roll pays nothing. What is the expected value for this game?
14. A total of 3000 tickets have been sold for a drawing. If one ticket is drawn for a single prize of $1000, what is a fair price for a ticket?
*15. How many four-digit numbers can be formed if the first digit cannot be zero and the last digit must be two?
*16. A club consists of ten members. In how many different ways can a group of three people be selected to go on a European trip?
*17. Find the number of different ways that four flags can be displayed on a flagpole, one above the other, if ten different flags are available.
*18. Five women live together in an apartment. Two of the women have blue eyes. If two of the women are chosen at random, what is the probability that they both have blue eyes?
*19. Five horses (Deadbeat, Applefarm, Bandy, Cash, and Egglegs) run in a race.
   (a) In how many ways can the first-, second-, and third-place horses be determined?
   (b) Find the probability that Deadbeat finishes first and Bandy finishes second in the race.
   (c) Find the probability that the first-, second-, and third-place horses are Deadbeat, Egglegs, and Cash, in that order.

20. Al and Ruby each roll an ordinary die once. What is the probability that the number of Ruby's roll is greater than the number of Al's roll?

*21. Amy has a quiz on which she is to answer any three of the five questions. If she is equally well versed on all questions and chooses three questions at random, what is the probability that question one is not chosen?

22. On a certain street there are three traffic lights. At any given time, the probability that a light is green is 0.3. What is the probability that a person will hit all three lights when they are green?

23. A three-stage rocket has the following probabilities for failure. The probability for failure at stage one is $\frac{1}{6}$; at stage two, it is $\frac{1}{8}$; and at stage three, it is $\frac{1}{10}$. What is the probability of a successful flight, given that the first stage was successful?

24. ▶How could each of the following be simulated by using a random-digit table?
    (a) Tossing a fair die
    (b) Picking three months at random from the 12 months of the year
    (c) Spinning the spinner shown

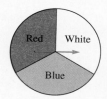

25. If a dart is thrown at the following tangram dart board and we assume that the dart lands at random on the board, what are the probabilities of its landing in each of the following areas?
    (a) Area A
    (b) Area B
    (c) Area C

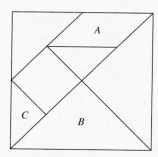

26. The points M, N, O, P, and Q represent exits on a highway. An accident occurs at random between points M and Q. What is the probability that it has occurred between N and O?

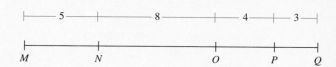

SELECTED BIBLIOGRAPHY

Bright, G. "Teaching Mathematics with Technology: Probability Simulations." *Arithmetic Teacher* 36 (May 1989): 16–18.

Engel, A. *A Short Course in Probability*. St. Louis, Mo.: Comprehensive School Mathematics Project, 1970.

Erickson, D., M. Frank, and R. Kelley. "WITPO (What Is the Probability Of)." *Mathematics Teacher* 84 (April 1991): 258–264.

Fennell, F. "Implementing the *Standards:* Probability." *Arithmetic Teacher* 38 (December 1990): 18–22.

Lappan, G., et al. "Area Models for Probability." *Mathematics Teacher* 80 (November 1987): 650–654.

Lappan, G. and M. Winter. "Probability Simulation in Middle School." *Mathematics Teacher* 73 (September 1980): 446–449.

Milton, J. "Probability in Ancient Times; or, Shall I Go Down after the Philistines?" *Mathematics Teacher* 82 (March 1989): 211–213.

Shaughnessy, J., and T. Dick. "Monty's Dilemma: Should You Stick or Switch?" *Mathematics Teacher* 84 (April 1991): 252–256.

Shulte, A. "Learning Probability Concepts in Elementary School Mathematics." *Arithmetic Teacher* 34 (January 1987): 32–33.

Shultz, H., and B. Leonard. "Probability and Intuition." *Mathematics Teacher* 82 (January 1989): 52–53.

Walton, K. "Probability, Computer Simulation, and Mathematics." *Mathematics Teacher* 83 (January 1990): 22–25.

9 Statistics: An Introduction

Preliminary Problem

A racetrack measures 1 mile around. If you drive around the track once and average 30 mph, what speed must you average on your second trip around the track in order to average 60 mph for the two laps?

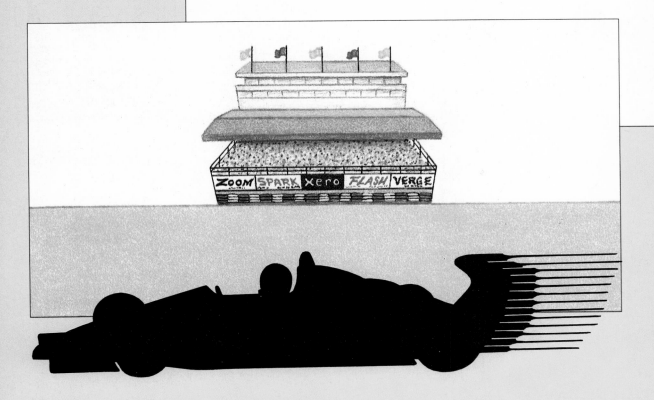

For a long time, the word *statistics* referred to numerical information about state or political territories. The word itself comes from the Latin *statisticus*, meaning "of the state." Statistics as we know it today took several centuries and many great minds to develop. John Graunt was one of the first people to record his work in the area of statistics.

Statistics are both used and abused. Sometimes the abuse is of little consequence and entirely unintentional. This may or may not be true in Sally's case, below.

PEANUTS reprinted by permission of UFS, Inc.

 Statistics plays an important role in the *Standards* at both the K–4 level and the 5–8 level. Following is an excerpt from the 5–8 *Standards* (p. 105):

In grades K–4, students begin to explore basic ideas of statistics by gathering data appropriate to their grade level, organizing them in charts or graphs, and reading information from displays of data. These concepts should be expanded in the middle grades. Students in grades 5–8 have a keen interest in trends in music, movies, fashion, and sports. An investigation of how such trends are developed and communicated is an excellent motivator for the study of statistics. Students need to be actively involved in each of the steps that comprise statistics, from gathering information to communicating results.

 In the *Teaching Standards* (p. 136) we find the following concerning what all teachers of grades K–4 should know:

Teachers should have a variety of experiences in the collection, organization, representation, analysis, and interpretation of data. Key statistical concepts for all teachers include measures of central tendency, measures of variation (range, standard deviation, interquartile range, and outliers), and general distributions. Representations of data should include various types of graphs, including bar, line, circle, and pictographs as well as line plots, stem-and-leaf plots, box plots, histograms, and scatter plots.

Additional statistical topics, including misuses of statistics, are listed for teachers of grades 5–8. In this chapter, we cover these topics listed in the *Teaching Standards* and include a section on possible abuses of statistics.

HISTORICAL NOTE

John Graunt (1620–1674) was an English haberdasher who studied birth and death records and discovered that more boys were born than girls. He also found that, because men were more subject to death from occupational accidents, diseases, and war, the numbers of men and women of marriageable age were about equal.

In 1662, Graunt's book *Natural and Political Observations upon the Bills of Mortality* was published. His work was the first to analyze statistics and to draw conclusions on the basis of such analysis. Graunt's work led to the development of actuarial science, which is used by life insurance companies.

Section 9-1 Statistical Graphs

Visual illustrations are an important part of statistics. Such illustrations or graphs take many forms: pictographs, circle graphs, pie charts, line plots, stem-and-leaf plots, frequency tables, histograms, bar graphs and frequency polygons or line graphs. A *graph* is a picture that displays numerical facts called *data*. Data can be presented in various forms once they are collected. The K–4 *Standards* point out (p. 55):

Children should learn that data can be displayed in different ways and that depending on the question being asked, one type of display might be more appropriate than another. A variety of experiences helps children build a foundation for creating conventional graphs.

Pictographs

pictograph One type of graph that children often encounter and that we often see in newspapers and magazines is a **pictograph.** In a pictograph, a symbol or an icon is used to represent a quantity of items. A *key* is usually presented, which tells what the symbol represents. Pictographs are frequently used to show comparisons of outputs. Figure 9-1 shows examples; a major disadvantage of pictographs is evident in Figure 9-1(a). The month of September contains a partial bundle of newspapers. It is impossible to tell from the graph the weight of that bundle with any accuracy.

FIGURE 9-1

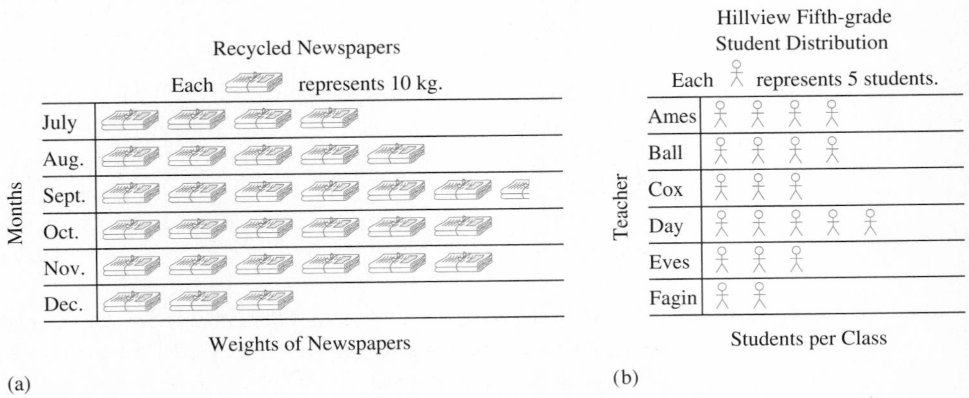

(a) Recycled Newspapers — Weights of Newspapers (Months: July, Aug., Sept., Oct., Nov., Dec.; Each bundle represents 10 kg.)

(b) Hillview Fifth-grade Student Distribution — Students per Class (Teachers: Ames, Ball, Cox, Day, Eves, Fagin; Each figure represents 5 students.)

Line Plots

line plot

Next we examine a type of graph called a **line plot.** A line plot is somewhat like a pictograph but no numerical values are lost in the graph. Line plots provide a quick, simple way of organizing data. Typically, we use them with fewer than 50 values, and especially when there is only one group of data.

Suppose that the 30 students in Mr. Abel's class received the following test scores:

$$82 \quad 97 \quad 70 \quad 72 \quad 83 \quad 75 \quad 76 \quad 84 \quad 76 \quad 88 \quad 80 \quad 81 \quad 81 \quad 52 \quad 82$$
$$82 \quad 73 \quad 98 \quad 83 \quad 72 \quad 84 \quad 84 \quad 76 \quad 85 \quad 86 \quad 78 \quad 97 \quad 97 \quad 82 \quad 77$$

A line plot for Mr. Abel's class consists of a horizontal number line, on which each score is denoted by an *x* above the corresponding value on the number line, as shown in Figure 9-2. The number of *x*'s above each score indicates how many times each score occurred.

FIGURE 9-2

Scores on Mr. Abel's Test

Figure 9-2 yields information about Mr. Abel's exam. For example, three students scored 76 and four students scored greater than 90. We also see that the low score was 52, the high score was 98, and the most frequent score was 82. Several features of the data become more obvious when line plots are used. For example, outliers, clusters, and gaps are apparent. **Outliers** are data points whose values are significantly larger or smaller than other values, such as the score of 52 in Figure 9-2. Outliers will be discussed in greater detail in the next section. **Clusters** are isolated groups of points, such as the one located at scores of 97 and 98. **Gaps** are large spaces between points, such as the one between 88 and 97.

outliers

clusters

gaps

Stem-and-Leaf Plots

stem-and-leaf plot

Another graph mentioned in the *Teaching Standards* is the **stem-and-leaf plot.** The stem-and-leaf plot is closely related to the line plot except that the number line is usually vertical and digits are used rather than *x*'s. A stem-and-leaf plot of test scores for Mr. Abel's class is shown in Figure 9-3.

FIGURE 9-3

Scores on Mr. Abel's Test

```
5 | 2
6 |
7 | 0223566678
8 | 0112222334444568     9 | 7 represents 97
9 | 7778
```

stems

leaves

The numbers on the left side of the vertical line are the **stems,** and the numbers on the right side are the **leaves.** In Figure 9-3, the stems are the tens digits of the scores on the test, and the leaves are the unit digits. In this case, 5|2 represents a score of 52, and 9|8 represents a score of 98. We now construct a stem-and-leaf plot using Table 9-1, which lists the presidents of the United States and their ages at death.

TABLE 9-1

President	Age at Death	President	Age at Death	President	Age at Death
George Washington	67	Millard Fillmore	74	Theodore Roosevelt	60
John Adams	90	Franklin Pierce	64	William Taft	72
Thomas Jefferson	83	James Buchanan	77	Woodrow Wilson	67
James Madison	85	Abraham Lincoln	56	Warren Harding	57
James Monroe	73	Andrew Johnson	66	Calvin Coolidge	60
John Q. Adams	80	Ulysses Grant	63	Herbert Hoover	90
Andrew Jackson	78	Rutherford Hayes	70	Franklin Roosevelt	63
Martin Van Buren	79	James Garfield	49	Harry Truman	88
William H. Harrison	68	Chester Arthur	57	Dwight Eisenhower	78
John Tyler	71	Grover Cleveland	71	John Kennedy	46
James K. Polk	53	Benjamin Harrison	67	Lyndon Johnson	64
Zachary Taylor	65	William McKinley	58		

The presidents died in their 40s, 50s, 60s, 70s, 80s, or 90s. Thus, we concentrate on numbers from 40 to 99. We choose the tens digits of the numbers as the stems. The leaves are the units digits. The plot is formed by placing the stem digits in a column from least to greatest on the left side of a vertical line, as shown in Figure 9-4(a). The leaves (which represent the units digits of the ages) are given on the right side of the vertical line, in whichever row contains their stem, as shown in Figure 9-4(b).

FIGURE 9-4

```
Stem | Leaf        Ages of Presidents at Death
  4  |              4 | 96
  5  |              5 | 36787
  6  |              6 | 785463707034
  7  |              7 | 38914701128       4 | 9 represents 49
  8  |              8 | 3508                  years old
  9  |              9 | 00
 (a)                 (b)
```

In Figure 9-4(b), the top row has 4 as a stem and 9 and 6 as leaves. These numbers represent the ages 49 and 46, the ages at death of James Garfield and John Kennedy, respectively. Normally, the graph is titled and is accompanied by a legend telling how to interpret the symbols used in it.

In some sense, the data in Figure 9-4(b) are still not orderly, because the numbers within each leaf are not in order from least to greatest on a given row. To make an **ordered stem-and-leaf plot,** we arrange the leaves on their given rows from least to greatest, starting at the left, as in Figure 9-5.

ordered stem-and-leaf plot

FIGURE 9-5

```
Ages of Presidents at Death
4 | 69
5 | 36778
6 | 003344567778
7 | 01112347889       4 | 9 represents 49
8 | 0358                  years old
9 | 00
```

There is no unique way to construct stem-and-leaf plots. Smaller numbers are usually placed at the top so that, when the plot is turned counterclockwise 90°, it resembles a histogram (discussed later in this section). Important advantages of stem-and-leaf plots are that they can be created by hand rather easily and that they do not become unmanageable when the number of values becomes large. Moreover, no original values are lost in a stem-and-leaf plot. For example, we can still tell that the youngest age at death was 46 and that exactly two presidents died when they were 90. A disadvantage of stem-and-leaf plots is that we do lose some information; for example, we know from the plot that some president died at age 88, but we do not know which one.

Following is a summary of how to construct a stem-and-leaf plot:

---

1. Find the high and low values of the data.
2. Decide on the stems.
3. List the stems in a column from least to greatest.
4. Use each piece of data to create leaves to the right of the stems on the appropriate rows.
5. If the plot is to be ordered, list the leaves in order from least to greatest.
6. Add a legend identifying the values represented by the stems and leaves. For example, 5|6 represents 56.
7. Add a title explaining what the graph is about.

The student page, on page 443, from *Exploring Mathematics,* Grade 6 (Scott, Foresman and Company, 1991), shows another example of the construction of a stem-and-leaf plot.

If two sets of data are to be compared, a *back-to-back stem-and-leaf plot* can be used. Two plots are made: one with leaves to the right, and one with leaves to the left. For example, if Mr. Abel gave the same test to two classes, he might prepare a back-to-back stem-and-leaf plot, as shown in Figure 9-6. Which class do you think did better on the test? Why?

FIGURE 9-6

Mr. Abel's Test Scores

Second Period Class			Fifth Period Class
0 | 5 | represents a score of 50	20	5 | 2	| 5 | 2 represents a score of 52
	531	6 | 24	
	99987542	7 | 1257	
	875420	8 | 4456999	
	1	9 | 2457	
		10 | 0	

REMARK When back-to-back stem-and-leaf plots are used, both sets of data should be about the same size.

SCIENCE CONNECTION

Stem-and-Leaf Plots

Build Understanding

As part of a science project, Julio tested 20 size-AA batteries to see how long each would last. He followed these steps to make a *stem-and-leaf plot* of the data in his table.

Lifetime of Batteries To the Nearest Hour

56	67	63	70
46	57	71	67
58	60	72	67
57	60	90	63
88	78	49	65

a. Find the least and greatest values in the table. The least value, 46, has a 4 in the tens place, and the greatest value, 90, has a 9 in the tens place. The stems are the tens digits 4 through 9.

b. Write the stems in a column. Draw a vertical line to the right.

```
4 |
5 |
6 |
7 |
8 |
9 |
```

c. The leaves are the ones digits. Write each leaf to the right of its stem.

```
4 | 6     Battery Age: 46
5 |       Stem = 4, leaf = 6
6 |
7 |
8 |
9 | 0     Battery Age: 90
          Stem = 9, leaf = 0
```

d. Enter each battery lifetime from the chart.

```
4 | 6, 9
5 | 6, 7, 8, 7
6 | 7, 3, 7, 0, 7, 0, 3, 5
7 | 0, 1, 2, 8
8 | 8
9 | 0
```

e. Rewrite each leaf in increasing order.

```
4 | 6, 9
5 | 6, 7, 7, 8
6 | 0, 0, 3, 3, 5, 7, 7, 7
7 | 0, 1, 2, 8
8 | 8
9 | 0
```

■ **Write About Math** Using the information from the stem-and-leaf plot, write a paragraph describing the results of Julio's test.

CHAPTER 9   STATISTICS: AN INTRODUCTION

EXAMPLE 9-1   Group the presidents in Table 9-1 into two groups, the first consisting of George Washington to Ulysses Grant, and the second consisting of Rutherford Hayes to Lyndon Johnson.

(a) Create back-to-back stem-and-leaf plots of the two groups and see if there appears to be a difference in ages at death between the two groups.

(b) Which group of presidents seems to have lived longer?

Solution   (a) Because the ages at death vary from 46 to 90, the stems vary from 4 to 9. In Figure 9-7, the first 18 presidents are listed on the left and the remaining 17 presidents are listed on the right.

FIGURE 9-7

Ages of President's at Death

Early Presidents		Later Presidents
	4	96
63	5	787
364587	6	707034
741983	7	0128
053	8	8
0	9	0

3 | 8 | represents 83 years old

| 6 | 7   represents 67 years old

(b) The early presidents seem, on average, to have lived longer because the ages at the high end, especially in the 70s and 80s, come more often from the early presidents. The ages at the lower end come more often from the later presidents. For the stems in the 50s and 60s, the numbers of leaves are about equal. ∎

The stem-and-leaf plot is easier to manage than a line plot when the number of pieces of data is great. When there are 50 or fewer values, the choice of a line plot or a stem-and-leaf plot is a matter of personal preference. When the number of pieces of data is between 50 and 250, the stem-and-leaf plot is more appropriate than a line plot. A stem-and-leaf plot shows how wide a range of values the data cover, where the values are concentrated, whether the data has any symmetry, where gaps in the data are, and whether any data points are decidedly different from the rest of the data.

Frequency Tables

frequency table   A slightly different way to display data is to use a frequency table. A **frequency table** shows how many times a certain piece of data occurs. For example, suppose that Dan offers the following deal. He rolls a die. If any number other than 6 appears, he pays $5. If a 6 appears, you pay him $5. With a fair die, the probability of Dan's winning is $\frac{1}{6}$.

Thus, Dan will not win unless the die is loaded or unless 6 appears considerably more often than could normally be expected. The data in Table 9-2 show the results of 60 rolls with Dan's die.

The results of Dan's die tosses may be summarized, as shown in the frequency table in Table 9-3.

TABLE 9-2

Results of Dan's Die Tosses

1	6	6	2	6	3	6	6	4	6
6	2	6	6	4	5	6	6	1	6
1	6	6	5	6	6	4	6	5	6
6	5	6	2	4	2	5	6	3	4
3	6	1	6	3	6	6	1	6	6
6	4	6	3	6	3	6	4	6	5

According to the frequency table, 6 appears many more times than could be expected from a fair die. (If the die were fair, the number of 6s should be closer to $\frac{1}{6} \cdot 60$, or 10.)

TABLE 9-3

Number	Tally	Frequency
1	⁄⁄⁄⁄⁄	5
2	⁄⁄⁄⁄	4
3	⁄⁄⁄⁄⁄ ⁄	6
4	⁄⁄⁄⁄⁄ ⁄⁄	7
5	⁄⁄⁄⁄⁄ ⁄	6
6	⁄⁄⁄⁄⁄ ⁄⁄⁄⁄⁄ ⁄⁄⁄⁄⁄ ⁄⁄⁄⁄⁄ ⁄⁄⁄⁄⁄ ⁄⁄⁄⁄⁄ ⁄⁄	32
	Total	60

Histograms and Bar Graphs

histogram   The data from Table 9-3 may be pictured graphically using a **histogram,** a graph that is closely related to a stem-and-leaf plot. Figure 9-8 shows a histogram of the frequencies in Table 9-3.

FIGURE 9-8

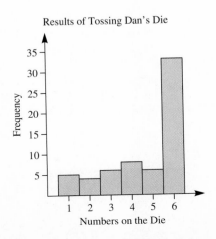

Results of Tossing Dan's Die

446 CHAPTER 9 STATISTICS: AN INTRODUCTION

A histogram is made up of adjoining vertical rectangles, or bars. In this case, the numbers on the die are shown on the horizontal axis. The numbers along the vertical axis give the scale for the frequency. The frequencies of the numbers on the die are shown by the bars, which are all the same width. The higher the bar, the greater the frequency. The scale on the vertical axis must also be of uniform interval size. In addition, all histograms should have the axes labeled and should include a title identifying the graph's content.

Histograms can easily be made from single-sided stem-and-leaf plots. For example, if we take the stem-and-leaf plot in Figure 9-4(b) and enclose each row (set of leaves) in a bar, as in Figure 9-9, we have what looks like a histogram. We can make Figure 9-9 resemble Figure 9-8 by rotating the graph 90° counterclockwise. Histograms show gaps and clusters just as stem-and-leaf plots do. However, with a histogram we cannot retrieve data as we can in a stem-and-leaf plot. Another disadvantage of a histogram is that it is often necessary to estimate the heights of the bars.

FIGURE 9-9

Ages of President's at Death

```
4 | 96
5 | 36787
6 | 785463707034      4 | 9 represents
7 | 38914701128           49 years old
8 | 3508
9 | 00
```

bar graph   A **bar graph** is like a histogram except that a bar graph has spaces between the bars. A typical bar graph showing the heights in centimeters of five students is given in Figure 9-10.

FIGURE 9-10

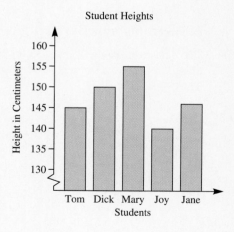

Student Heights

The break in the vertical axis, denoted by a squiggle, indicates that part of the scale has been omitted. Therefore, the scale is not accurate from 0 to 130. The height of each bar represents the height in centimeters of each student named on the horizontal axis. Each space between bars is usually one-half the width of the bars. Graphs using a break in the vertical axis can be used to create different visual impressions, which are sometimes misleading. For example, consider the two graphs in Figure 9-11, which represent the number of girls trying out for basketball at each of three middle schools. As we can see, the graph in Figure 9-11(a) portrays a different picture than the one in Figure 9-11(b).

FIGURE 9-11

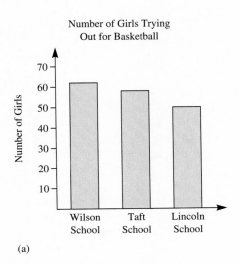

(a)

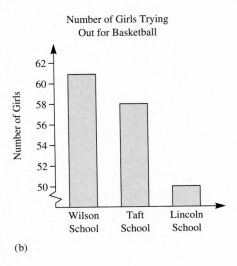

(b)

double bar graphs   **Double bar graphs** can be used to make comparisons in data. The student page from *Addison-Wesley Mathematics,* Grade 6, 1991, on the following page shows a double bar graph. Note that bar graphs can be drawn either vertically or horizontally.

Reading Graphs

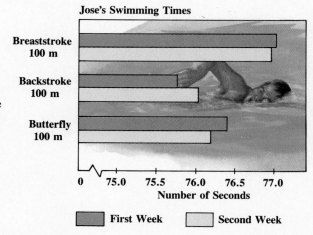

Jose's Swimming Times

LEARN ABOUT IT

EXPLORE Study the Data

Every week, Jose tries to improve his swimming times. This **double bar graph** shows his best times for two different weeks and three different events. The numbers along the bottom of the graph are called the **scale** of the graph.

TALK ABOUT IT

1. Why do you think the graph above is called a double bar graph?
2. What numbers are missing along the scale? Why do you think they were left out?
3. What do the two bars for each event compare?

- A squiggle at the beginning of a scale means that part of the scale has been omitted.
- On most graphs you must estimate to find the approximate number represented by the bar or point on a graph.

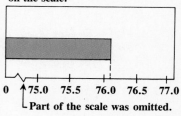

The bar represents about 76.2 on the scale.

Part of the scale was omitted.

TRY IT OUT

1. In how many seconds did Jose swim the breaststroke the first week? The second week?

2. For which events did he improve his time the second week? Which showed the greatest improvement?

3. Approximately what was his best time for the backstroke?

Frequency Polygons (Line Graphs)

frequency polygon / line graph

Another graphic form used in presenting the data from a frequency table is a **frequency polygon,** or **line graph.** A frequency polygon can be plotted from a frequency table by using line segments to connect the points representing the data, or it can be constructed from a histogram by using line segments to connect the midpoints of the tops of each of the rectangular bars. Figure 9-12(a) shows the frequency polygon (line graph) for the data from Table 9-3. Figure 9-12(b) shows how to obtain the same frequency polygon from the histogram in Figure 9-8.

FIGURE 9-12

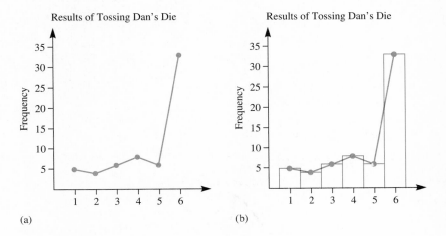

Line graphs are used primarily to show an amount of change and a direction of change over a period of time. In Figure 9-12, the histogram is more appropriate because the data fall into distinct categories, which we want to compare. We are not interested in change over time. Figure 9-13 shows a more appropriate use of a line graph.

FIGURE 9-13

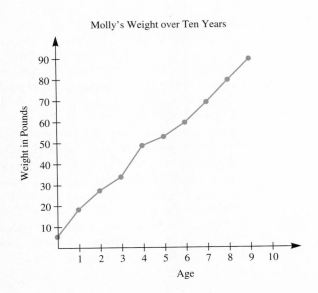

Grouped Data

The greater the amount of data, the more difficult it becomes to construct a frequency table for individual items. In such cases, the data may be grouped. For example, consider the scores in Table 9-4. A stem-and-leaf plot for this data is shown in Figure 9-14.

TABLE 9-4

50 Student Scores									
52	56	25	56	68	73	66	64	56	100
20	39	9	50	98	54	54	40	50	96
36	44	18	97	109	65	21	60	44	54
92	49	37	94	72	88	89	35	59	34
48	32	15	53	84	72	88	16	52	60

FIGURE 9-14

Student Scores

```
 0 | 9
 1 | 856
 2 | 051
 3 | 692754
 4 | 84904
 5 | 266034460924
 6 | 856400
 7 | 232
 8 | 4898
 9 | 27486           10 | 9 represents
10 | 09                  a score of 109
```

classes

grouped frequency table

class mark

The construction of the stem-and-leaf plot leads in a natural way to a grouping of scores in intervals. The intervals are called **classes;** for the data in Figure 9-14, the following classes are used: 0–9, 10–19, 20–29, 30–39, 40–49, 50–59, 60–69, 70–79, 80–89, 90–99, 100–109. Each class has an interval size of 10; that is, ten different scores can fall within the interval 0 through 9. (Students often incorrectly report the interval size as 9 because $9 - 0 = 9$.) The **grouped frequency table** for the data in Table 9-4 with intervals of length 10 is given in Table 9-5. Figure 9-14 contains more information than does Table 9-5, because the raw scores themselves are not available in Table 9-5. Although Table 9-5 shows that 12 scores fall in the interval 50–59, it does not show the particular scores in the interval. The greater the size of the interval, the greater the amount of information lost. The choice of interval size may vary. Classes should be chosen to accommodate *all* the data, and each item should fit into only one class; that is, the classes should not overlap.

A bar graph or histogram can be used to display the data from a grouped frequency table. A bar graph of the data in Table 9-5 is shown in Figure 9-15(a). Rather than placing the intervals below the bars as in Figure 9-15, we could represent each interval by a single score known as a **class mark.** To find the class mark, we find the midpoint of each class; for example, $(0 + 9)/2 = 4.5$, $(10 + 19)/2 = 14.5$, and so on. Class marks are used in the graph in Figure 9-15(b).

TABLE 9-5

Classes	Tally	Frequency
0–9	I	1
10–19	III	3
20–29	III	3
30–39	IHI I	6
40–49	IHI	5
50–59	IHI IHI II	12
60–69	IHI I	6
70–79	III	3
80–89	IIII	4
90–99	IHI	5
100–109	II	2

FIGURE 9-15

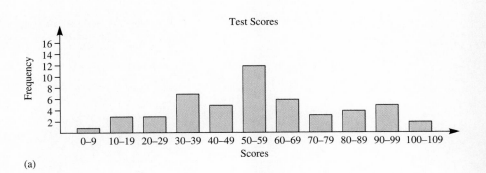

(a)

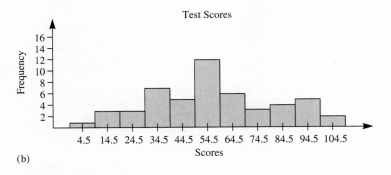

(b)

Circle Graphs (Pie Charts)

circle graph
pie chart

Another type of graph used to represent data is the circle graph. A **circle graph,** or **pie chart,** consists of a circular region partitioned into disjoint sections, with each section representing a part or percentage of the whole. A circle graph shows how parts are related to the whole. An example of a circle graph is given in Figure 9-16.

FIGURE 9-16

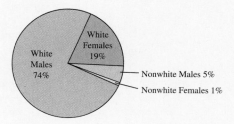

Example 9-2 shows how a circle graph can be constructed from given data.

EXAMPLE 9-2 Construct a circle graph for the information in Table 9-6, which is based on information taken from a U.S. Bureau of the Census Report, 1986.

TABLE 9-6

Age	Number of People (to the nearest million)
Under 5	18
5–17	45
18–24	28
25–34	43
35–44	33
45–64	45
Over 65	29

Solution The whole circle of 360° represents the 241 million people we obtain by adding all the numbers for the different age groups. The area of each sector of the graph is proportional to the fraction or percentage of the population the section represents. For example, the sector for the under-5 group is $\frac{18}{241}$, or approximately 7%, of the circle. Because the whole circle is 360 degrees, then 7% of 360, or approximately 25°, should be devoted to the under-5 group. Similarly, we can compute the number of degrees for each age group, as shown in Table 9-7.

TABLE 9-7

Age	Ratio	Approximate Percent	Approximate Degrees
Under 5	18/241	7	25
5–17	45/241	19	68
18–24	28/241	12	43
25–34	43/241	18	65
35–44	33/241	14	50
45–64	45/241	19	68
Over 65	29/241	12	43

The percentages and degrees are only approximate. We can then use a protractor to draw the circle graph, as shown in Figure 9-17.

FIGURE 9-17

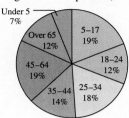

Ages of U.S. Population, 1986

PROBLEM SET 9-1

1. Make a pictograph to represent the data, using 🥛 to represent ten glasses of lemonade sold.

   Glasses of Lemonade Sold

	Tally	Frequency
Monday	|||| |||| ||||	15
Tuesday	|||| |||| |||| ||||	20
Wednesday	|||| |||| |||| |||| |||| ||||	30
Thursday	||||	5
Friday	|||| ||||	10

2. The pictograph below shows the approximate number of people who speak the six most common languages.
   (a) About how many people speak Spanish?
   (b) About how many people speak English?
   (c) About how many more people speak Mandarin than Arabic?

   Number of People Speaking the Six Most Common Languages

Arabic	● ◐
English	● ● ● ◐
Hindi	● ● ◐
Mandarin	● ● ● ● ● ◐
Russian	● ● ◐
Spanish	● ● ◐

   Each ● represents 100 million people.

3. The ages of the 30 students from Washington School who participated in the city track meet are listed below. Draw a line plot to represent these data.

10	10	11	10	13	8	10	13	14	9
14	13	10	14	11	9	13	10	11	12
11	12	14	13	12	8	13	14	9	14

4. The following stem-and-leaf plot gives the weight in pounds of all 15 students in the Algebra 1 class at East Junior High.
   (a) Write the weights of the 15 students.
   (b) What is the weight of the lightest student in the class?
   (c) What is the weight of the heaviest student in the class?

   Weights of Students in East Junior High Algebra 1 Class

7	24
8	112578
9	2478
10	3
11	
12	35

   10|3 represents 103 lb

5. Draw a histogram based on the stem-and-leaf plot in problem 4.

6. Toss a coin 30 times.
   (a) Construct a line plot for the data.
   (b) Draw a histogram for the data.

7. The figure shows a bar graph of the rainfall in centimeters during the last school year. Answer each of the following questions:
   (a) Which month had the most rainfall, and how much did it have?
   (b) How much total rain fell in October, December, and January?

10. Draw a histogram to represent the data given below.

Distances from the Sun

Planet	Distance (in millions of miles)
Earth	93
Mars	142
Mercury	36
Venus	67

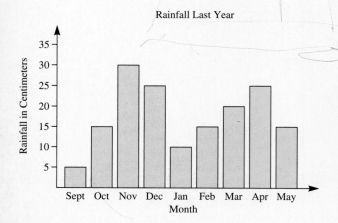

Rainfall Last Year

11. The given data represent total car sales for Johnson's car lot from January through June. Draw a bar graph for the data.

Month	Jan.	Feb.	Mar.	Apr.	May	June
Number of Cars Sold	90	86	92	96	90	100

12. Five coins are tossed 64 times. A distribution for the number of heads obtained is shown below. Draw a histogram for the data.

Number of Heads	0	1	2	3	4	5
Frequency	2	10	20	20	10	2

8. HKM Company employs 40 people of ages as shown:

   34 58 21 63 48 52 24 52 37 23
   23 34 45 46 23 26 21 18 41 27
   23 45 32 63 20 19 21 23 54 62
   41 32 26 41 25 18 23 34 29 26

   (a) Draw a stem-and-leaf plot for the data.
   (b) Are more employees in their 40s or in their 50s?
   (c) How many employees are less than 30 years old?
   (d) What percent of the people are 50 years or older?

9. Given the following bar graph, estimate the length of the following rivers:
   (a) Mississippi   (b) Columbia

13. The grade distribution for the final examination in the mathematics course for elementary teachers is shown.

Grade	Frequency
A	4
B	10
C	37
D	8
F	1

(a) Draw a bar graph of the data.
(b) Draw a circle graph of the data.

14. The following are the amounts (rounded to the nearest dollar) paid by 25 students for textbooks during the fall term:

   35 42 37 60 50
   42 50 16 58 39
   33 39 23 53 51
   48 41 49 62 40
   45 37 62 30 23

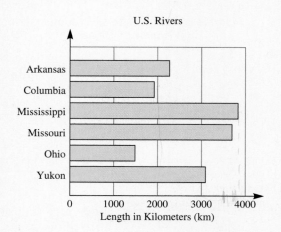

U.S. Rivers

(a) Draw an ordered stem-and-leaf plot to illustrate the data.
(b) Construct a grouped frequency table for the data, starting the first class at $15.00 with intervals of $5.00 each.
(c) Draw a histogram of the data.
(d) Draw a frequency polygon of the data.

15. ▶The following graphs give the temperatures for a certain day. Which graph is more helpful for guessing the actual temperature at 10:00 A.M.? Why?

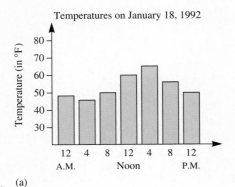
(a)

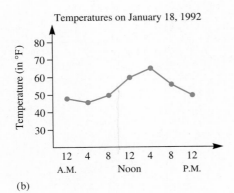
(b)

16. Tell whether it would be more appropriate to use a bar graph or a line graph for each of the following. Then draw the appropriate graph in each case.

(a) **U.S. Population**

Year	Population
1920	105,710,620
1930	122,775,046
1940	131,669,275
1950	150,697,361
1960	179,323,175
1970	203,302,031
1980	226,545,805

(b) **Continents of the World**

Continent	Area in Square Miles (mi²)
Africa	11,694,000
Antarctica	5,100,000
Asia	16,968,000
Australia	2,966,000
Europe	4,066,000
North America	9,363,000
South America	6,886,000

17. The following horizontal bar graph gives the top speeds of several animals.
(a) Which is the slowest animal shown?
(b) How fast can a chicken run?
(c) Which animal can run twice as fast as a rabbit?
(d) Can a lion outrun a zebra?

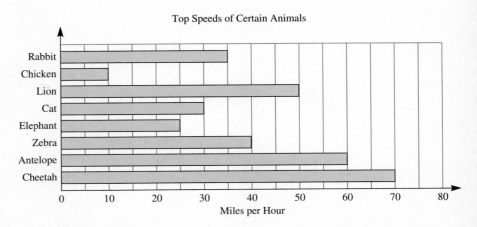

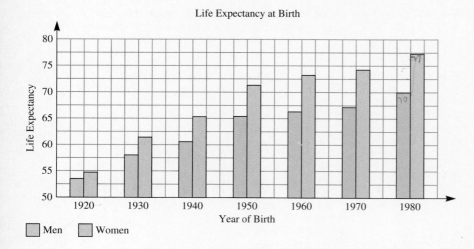

Life Expectancy at Birth

18. Consider the bar graphs above showing the life expectancies for men and women.
    (a) Whose life expectancy has changed the most since 1920?
    (b) In 1920, about how much longer was a woman expected to live than a man?
    (c) In 1980, about how much longer was a woman expected to live than a man?
19. ▶Discuss an example of when a circle graph would be preferable to a bar graph or a line graph.
20. ▶Discuss an example of when a line graph would be preferable to a bar graph.
21. ▶Give an example of a set of data for which a stem-and-leaf plot is more informative than a histogram.
22. The graph below shows how the value of a car depreciates each year. This graph will allow us to find the trade-in value of a car for each of 5 yr. The percentages given in the graph are based on the selling price of the new car.

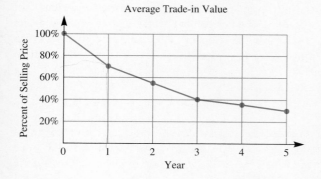

(a) What is the approximate trade-in value of a $12,000 car after 1 yr?
(b) How much has a $20,000 car depreciated after 5 yr?
(c) What is the approximate trade-in value of a $20,000 car after 4 yr?
(d) Dani wants to trade in her car before it loses half its value. When should she do this?

23. Given the circle graph below, answer the following questions:
    (a) Which is the largest continent?
    (b) Which continent is about twice the size of Antarctica?
    (c) How does Africa compare in size to Asia?
    (d) Which two continents make up about half of the Earth's surface?
    (e) What is the ratio of the size of Australia to North America?
    (f) If Europe has approximately 4.1 million square miles of land, what is the total area of the land on Earth?

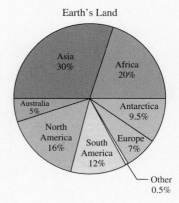

24. Following are the names and records of the home run leaders in the American and National Leagues from 1976 to 1990. Construct an ordered back-to-back stem-and-leaf

plot for the data. Are any patterns of difference evident in comparisons of the two groups of data?

Year	National League	HR	American League	HR
1976	Mike Schmidt	38	Graig Nettles	32
1977	George Foster	52	Jim Rice	39
1978	George Foster	40	Jim Rice	46
1979	Dave Kingman	48	Gorman Thomas	45
1980	Mike Schmidt	48	Reggie Jackson, Ben Oglivie	41
1981	Mike Schmidt	31	Bobby Grich, Eddie Murray, Tony Armas, Dwight Evans	22
1982	Dave Kingman	37	Gorman Thomas, Reggie Jackson	32
1983	Mike Schmidt	40	Jim Rice	39
1984	Mike Schmidt, Dale Murphy	36	Tony Armas	43
1985	Dale Murphy	37	Darrell Evans	40
1986	Mike Schmidt	37	Jesse Barfield	40
1987	Andre Dawson	49	Mark McGuire	49
1988	Darryl Strawberry	39	José Canseco	42
1989	Kevin Mitchel	47	Fred McGriff	36
1990	Ryne Sandberg	40	Cecil Fielder	51

25. A list of presidents, with the number of children for each, follows:

1. Washington, 0
2. J. Adams, 5
3. Jefferson, 6
4. Madison, 0
5. Monroe, 2
6. J. Q. Adams, 4
7. Jackson, 0
8. Van Buren, 4
9. W. H. Harrison, 10
10. Tyler, 14
11. Polk, 0
12. Taylor, 6
13. Fillmore, 2
14. Pierce, 3
15. Buchanan, 0
16. Lincoln, 4
17. A. Johnson, 5
18. Grant, 4
19. Hayes, 8
20. Garfield, 7
21. Arthur, 3
22. Cleveland, 5
23. B. Harrison, 3
24. McKinley, 2
25. T. Roosevelt, 6
26. Taft, 3
27. Wilson, 3
28. Harding, 0
29. Coolidge, 2
30. Hoover, 2
31. F. D. Roosevelt, 6
32. Truman, 1
33. Eisenhower, 2
34. Kennedy, 3
35. L. B. Johnson, 2
36. Nixon, 2
37. Ford, 4
38. Carter, 3
39. Reagan, 4
40. Bush, 5

(a) Construct a line plot for these data.
(b) Make a frequency table for these data.
(c) What is the most frequent number of children?

Section 9-2 Measures of Central Tendency and Variation

The media present us with a variety of data and statistics. For example, we find in the *World Almanac* that the average person's lifetime includes 6 yr of eating, 4 yr of cleaning, 2 yr of trying to return telephone calls to people who never seem to be in, 6 mo waiting at stop lights, 1 yr looking for misplaced objects, and 8 mo opening junk mail. In the previous section, we examined data by looking at graphs to display the overall distribution of values. In this section, we find that we can describe specific aspects of data by a few carefully chosen numbers that tell us much about the data. Two important aspects of data are its *center* and its *spread*. The mean, median, and mode are three numbers that describe where data are centered, and these are called **measures of central tendency**. Each of these measures is a single number that describes the data but each does it in a slightly different way. The *range, variance,* and *standard deviation* introduced later also describe the spread of the data.

measures of central tendency

A word that is often used in statistics is *average*. Most of the time, average means *typical*. To explore more about averages, examine the following set of data for three teachers, each of whom claims that his or her class scored better *on the average* than the other two classes did.

Mr. Smith: 62, 94, 95, 98, 98
Mr. Jones: 62, 62, 98, 99, 100
Ms. Leed: 40, 62, 85, 99, 99

All of these teachers are correct in their assertions, because each one has used a different number to characterize the scores in his or her particular class. In the following, we examine how each teacher can justify the claim.

Computing Means

arithmetic mean
average / mean

The number most commonly used to characterize a set of data is the **arithmetic mean,** frequently called the **average,** or the **mean.** To find the mean of scores for each of the teachers given above, we find the sum of the scores in each case and divide by 5, the number of scores.

$$\text{Mean (Smith):} \quad \frac{62 + 94 + 95 + 98 + 98}{5} = \frac{447}{5} = 89.4$$

$$\text{Mean (Jones):} \quad \frac{62 + 62 + 98 + 99 + 100}{5} = \frac{421}{5} = 84.2$$

$$\text{Mean (Leed):} \quad \frac{40 + 62 + 85 + 99 + 99}{5} = \frac{385}{5} = 77$$

Thus, in terms of the mean, Mr. Smith's class scored better than the other two classes. In general, we define the *arithmetic mean* as follows.

DEFINITION OF MEAN

The **arithmetic mean** of the numbers $x_1, x_2, \ldots, x_n$, denoted by $\bar{x}$ and read "x bar," is given by

$$\bar{x} = \frac{x_1 + x_2 + x_3 + \cdots + x_n}{n}.$$

Understanding the Mean as a Balance Point

Because the mean is the most widely used measure of central tendency, we provide a model for thinking about it. Suppose Jill, a student at a rural school, reports that the mean number of pets for the six students in her group is 5. Do we know anything about the distribution of these pets? One way that we could have a mean of 5 is that all six students have exactly 5 pets, as shown in the line plot in Figure 9-18(a).

FIGURE 9-18

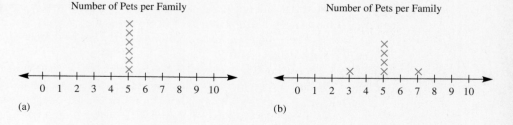

In this graph, all the pets are centered at the mean of 5. Although this distribution is possible, it is not very likely. If we change the line plot as shown in Figure 9-18(b), the mean is still 5. Notice that the new line plot could be obtained by moving one value from Figure 9-18(a) two units to the right and then balancing this by moving one value two units to the left. We can think of the mean as a *balance point,* where the total distance on one side of the mean (fulcrum) is the same as the total distance on the other side.

FIGURE 9-19

Number of Children per Family

Consider Figure 9-19, which shows the number of children for each family in Jill's group. The mean of 5 is the balance point where the sum of the total distances above the mean equals the sum of the total distances below the mean. The sum of the distances below the mean is $1 + 2 + 2 + 3$, or 8. The sum of the distances above the mean is $3 + 5$, or 8. In this case, we see that the data are centered about the mean, but the mean does not belong to the set of data.

Computing Medians

median   The value exactly in the middle of an ordered set of numbers is called the **median.** To find the median for the teachers' scores, we arrange each of their scores in increasing order and pick the middle score. Intuitively, we know that half the scores are greater than the median and half are less.

$$\text{Median (Smith):} \quad 62, 94, \;\widehat{95},\; 98, 98 \qquad \text{median} = 95$$

$$\text{Median (Jones):} \quad 62, 62, \;\widehat{98},\; 99, 100 \qquad \text{median} = 98$$

$$\text{Median (Leed):} \quad 40, 62, \;\widehat{85},\; 99, 99 \qquad \text{median} = 85$$

Thus, in terms of the median, Mr. Jones's class scored better than the other two classes.

With an odd number of scores, as in the present example, the median is the middle score. With an even number of scores, however, the median is defined as the mean of the middle two scores; thus, to find the median, we add the middle two scores and divide by 2. For example, the median of the scores

$$64, 68, \boxed{70, 74}\; 82, 90$$

is given by

$$\frac{70 + 74}{2}, \text{ or } 72.$$

In general, to find the median for a set of n numbers, proceed as follows:

---

1. Arrange the numbers in order from least to greatest.
2. (a) If n is odd, the median is the middle number.
   (b) If n is even, the median is the mean of the two middle numbers.

Finding Modes

mode   The **mode** of a set of data is the number that appears most frequently, if there is one. In some distributions no number appears more than once, and in other distributions there may be more than one mode. For example, the set of scores 64, 79, 80, 82, 90 has no mode (or *bimodal*   five modes). The set of scores 64, 75, 75, 82, 90, 90, 98 is **bimodal** (two modes), because both 75 and 90 are modes. It is possible for a set of data to have too many modes for this type of number to be useful in describing the data.

For the three classes listed earlier, if the mode is used as a criterion, Ms. Leed's class scored better than the other two classes.

$$\begin{array}{lll} \text{Mode (Smith):} & 62, 94, 95, 98, 98 & \text{mode} = 98 \\ \text{Mode (Jones):} & 62, 62, 98, 99, 100 & \text{mode} = 62 \\ \text{Mode (Leed):} & 40, 62, 85, 99, 99 & \text{mode} = 99 \end{array}$$

EXAMPLE 9-3   Find (a) the mean, (b) the median, and (c) the mode for the following collection of data:

$$60 \quad 60 \quad 70 \quad 95 \quad 95 \quad 100$$

Solution   (a) $\bar{x} = \dfrac{60 + 60 + 70 + 95 + 95 + 100}{6} = \dfrac{480}{6} = 80$

(b) The median is $\dfrac{70 + 95}{2}$, or 82.5.

(c) The set of data is bimodal and has both 60 and 95 as modes. ∎

Choosing the Most Appropriate Average

Although the *mean* is the number most commonly used to describe a set of data, it may not always be the best number to use. Suppose, for example, that a company employs 20 people. The president of the company earns $200,000, the vice president earns $75,000, and 18 employees earn $10,000 each. The mean salary for this company is

$$\frac{\$200{,}000 + \$75{,}000 + 18(\$10{,}000)}{20} = \frac{\$455{,}000}{20} = \$22{,}750.$$

In this case, the mean salary of $22,750 is not representative, and either the median or mode, both of which are $10,000, would better describe the typical salary. Notice that *the mean is affected by extreme values*.

In most cases, the *median* is not affected by extreme values. The median, however, can be misleading. For example, suppose that nine students make the following scores on a test: 30, 35, 40, 40, 92, 92, 93, 98, 99. From the median score of 92, one might infer that the individuals all scored very well, yet 92 is certainly not a typical score.

The *mode*, too, can be misleading in describing a set of data with very few items or many frequently occurring items. For example, the scores 40, 42, 50, 62, 63, 65, 98, 98 have a mode of 98, which is not a typical value.

The choice of which number to use to represent a particular set of data is not always easy. In the example involving the three teachers, each teacher chose the number that best suited his or her claim. The type of number used should always be specified.

SECTION 9-2   MEASURES OF CENTRAL TENDENCY AND VARIATION

PROBLEM 1

Dr. Van Gruff asked his students to keep track of their own grades. One day, he asked the students to report their grades. One of the students, Eddy, had lost his papers but remembered the grades on four of six assignments: 100, 82, 74, and 60. Also, according to Eddy, the mean of all six papers was 69, and the other two papers had identical grades. What were the grades on Eddy's other two homework papers?

Understanding the Problem. Eddy had scores of 100, 82, 74, and 60 on four of six papers. The mean of all six papers was 69, and two identical scores were missing. We must find the two missing grades.

Devising a Plan.   To find the missing grades, we use the strategy of *writing an equation* for x. Because the mean is obtained by finding the sum of the scores and then dividing by the number of scores, which is 6, if we let x stand for each of the two missing grades, we have

$$69 = \frac{100 + 82 + 74 + 60 + x + x}{6}.$$

Carrying Out the Plan.   We now solve the equation as follows:

$$69 = \frac{100 + 82 + 74 + 60 + x + x}{6}$$

$$69 = \frac{316 + 2x}{6}$$

$$49 = x$$

Since the solution to the equation is $x = 49$, we conclude that each of the two missing scores was 49.

Looking Back.   The answer of 49 seems reasonable, since the mean of 69 is less than three of the four given scores. We can easily check this by computing the mean of the scores 100, 82, 74, 60, 49, 49 and showing that it is indeed 69.   ∎

Measures of Dispersion

The mean, median, and mode provide information about where the central portion of a distribution is located, but limited information about the whole distribution. If you sit in the sauna for 30 min and then in a freezer for 30 min, the average temperature of your surroundings for this hour might sound comfortable. In this case, we need at least one more statistic to better understand the data. We need a measure to tell how much the data are "scattered." Such measures are called *measures of dispersion*. The range is the simplest measure of dispersion, but it is of little use when extreme values are present. The variance and the standard deviation are used when the mean is chosen to be the measure of central tendency. The need for a measure of spread is shown in the following discussion.

Suppose that Professors Abel and Babel each taught a section of a statistics course and each had six students. Both professors gave the same final exam. The results, along with the means for each group of scores, are given in Table 9-8, with stem-and-leaf plots in Figure 9-20(a) and (b), respectively. As the stem-and-leaf plots show, the sets of data are very different. The first set of scores is more spread out, or varies more, than the second. However, each set of scores has 60 as the mean. Each median also equals 60. Although the mean and the median for these two groups are the same, the two distributions of scores are very different.

TABLE 9-8

Abel	Babel
100	70
80	70
70	60
50	60
50	60
10	40
$\bar{x} = \dfrac{360}{6} = 60$	$\bar{x} = \dfrac{360}{6} = 60$

FIGURE 9-20

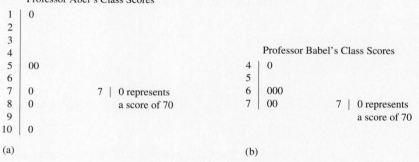

There are several ways to measure the spread (variation) of data. The simplest way is to subtract the least number from the greatest number. This difference is called the **range**. The range for Professor Abel's class is 100 − 10, or 90. The range for Professor Babel's class is 70 − 40, or 30. Although the range is easy to calculate, it has the disadvantage of being determined by two scores. For example, the sets of scores 10, 20, 25, 30, 100 and 10, 80, 85, 90, 90, 100 both have a range of 90.

range

Variance and Standard Deviation

variance
standard deviation

The two most commonly used measures of dispersion are the **variance** and the **standard deviation**. These measures are based on how far the scores are from the mean. To find out how far each value deviates from the mean, we first find the mean and then subtract each value in the data from the mean to obtain the deviation of each data point from the mean. Some of these values may be positive, and others may be negative. Recall that the mean is the balance point and the total of the deviations above the mean will equal the total of the deviations below the mean. At this point, the mean of the deviation is 0 because the sum of the deviations is 0. Squaring the deviations will make them all positive. The mean of the squared deviations is called the *variance*. Because the variance involves squaring the deviations, it does not have the same units of measurement as the original observations. For example, lengths measured in feet have a variance measured in square feet. To obtain the same units as the original observations, we take the square root of the variance and obtain the *standard deviation*.

The steps involved in calculating the variance v and standard deviation s of n numbers are as follows:

---

1. Find the mean of the numbers.
2. Subtract the mean from each number.
3. Square each difference found in step 2.
4. Find the sum of the squares in step 3.
5. Divide by n to obtain the variance, v.
6. Find the square root of v to obtain the standard deviation, s.

These six steps can be summarized for the numbers $x_1, x_2, x_3, \ldots, x_n$ as follows, where $\bar{x}$ is the mean of these numbers.

$$s = \sqrt{v} = \sqrt{\frac{(x_1 - \bar{x})^2 + (x_2 - \bar{x})^2 + (x_3 - \bar{x})^2 + \cdots + (x_n - \bar{x})^2}{n}}$$

REMARK In some textbooks, this formula involves division by $n - 1$ instead of by n. Division by $n - 1$ is more useful for advanced work in statistics.

The variances and standard deviations for the final exam data from the classes of Professors Abel and Babel are calculated by using Tables 9-9 and 9-10, respectively.

TABLE 9-9 Abel's Scores

x	$x - \bar{x}$	$(x - \bar{x})^2$
100	40	1600
80	20	400
70	10	100
50	-10	100
50	-10	100
10	-50	2500
Totals 360	0	4800

$\bar{x} = \dfrac{360}{6} = 60$

$v = \dfrac{4800}{6} = 800$

$s = \sqrt{800} \doteq 28.3$

TABLE 9-10 Babel's Scores

x	$x - \bar{x}$	$(x - \bar{x})^2$
70	10	100
70	10	100
60	0	0
60	0	0
60	0	0
40	-20	400
Totals 360	0	600

$\bar{x} = \dfrac{360}{6} = 60$

$v = \dfrac{600}{6} = 100$

$s = \sqrt{100} = 10$

Values far from the mean on either side will have large positive squared deviations whereas values close to the mean will have small positive squared deviations. Therefore, the standard deviation is a large number when the values from a set of data are widely spread. The standard deviation is a small number (close to 0) when the data values are close together. This is further illustrated in Example 9-4.

EXAMPLE 9-4   Professor Boone gave two exams. Exam A had grades of 0, 0, 0, 100, 100, and 100, and exam B had grades of 50, 50, 50, 50, 50, and 50. Find the following for each exam:

(a) The mean   (b) The median   (c) The standard deviation

Solution   (a) The means for exams A and B are each 50.

(b) The medians for the exams are each 50.

(c) The standard deviations for exams A and B are as follows:

$$s_A = \sqrt{\frac{3(0-50)^2 + 3(100-50)^2}{6}} = 50$$

$$s_B = \sqrt{\frac{6(50-50)^2}{6}} = 0$$

∎

EXAMPLE 9-5   Given the data 32, 41, 47, 53, 57, find each of the following:

(a) The range   (b) The variance   (c) The standard deviation

Solution   (a) The range is $57 - 32$, or 25.

(b) The variance v is computed by using the information in Table 9-11.

(c) $s = \sqrt{78.4} \doteq 8.9$

TABLE 9-11

x	$x - \bar{x}$	$(x - \bar{x})^2$
32	−14	196
41	−5	25
47	1	1
53	7	49
57	11	121
Totals 230	0	392

$\bar{x} = \dfrac{230}{5} = 46$

$v = \dfrac{392}{5} = 78.4$

∎

Box Plots

box plot

Line plots or stem-and-leaf plots become unwieldy when a large amount of data is involved. A **box plot,** sometimes called a *box-and-whisker plot,* is a display that is especially useful for handling many data values. When there is only one variable, box plots make it easy to see patterns of the data. When two or more variables are present, box plots allow us to explore the data and draw informal conclusions. Box plots show only certain statistics rather than all the data. A box plot is a visual representation of what is called the *five-number summary,* which consists of the median, the quartiles, and the smallest and greatest values in the distribution. The center, the spread, and the overall range of distribution are immediately evident by looking at a box plot.

SECTION 9-2 MEASURES OF CENTRAL TENDENCY AND VARIATION

To construct a box plot, we must find the *lower quartile* and the *upper quartile* of a given set of data. If scores are arranged from lowest to highest, the lower quartile, the median, and the upper quartile divide the data into four groups that are approximately the same size. Consider the following set of test scores:

$$20 \quad 25 \quad 40 \quad 50 \quad 50 \quad 60 \quad 70 \quad 75 \quad 80 \quad 80 \quad 90 \quad 100 \quad 100$$

We first find the median, which is 70, and draw a line segment through it.

$$20 \quad 25 \quad 40 \quad 50 \quad 50 \quad 60 \quad 7|0 \quad 75 \quad 80 \quad 80 \quad 90 \quad 100 \quad 100$$

Next, we consider only the values to the left of the segment and draw a line segment where the median of those values is located.

$$20 \quad 25 \quad 40 \mid 50 \quad 50 \quad 60$$

lower quartile
first quartile (Q_1)
upper quartile (Q_3)

The score of $45 = (40 + 50)/2$ is the median of the scores less than the median of all scores and therefore is the **lower quartile**. The lower quartile is often called the **first quartile** and is denoted by Q_1. Similarly, we can find the upper (or third) quartile (Q_3), which is $(80 + 90)/2$, or 85. The **upper quartile (Q_3)** is the median of the scores greater than the median of all scores. Thus, we have divided the scores into four groups of three scores each:

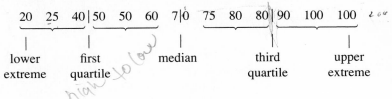

interquartile range (IQR)

The **interquartile range (IQR)** is the difference between the upper quartile and the lower quartile. In this case, $IQR = 85 - 45 = 40$. The IQR is itself another useful measure of variation because it is less influenced by extreme values. With the IQR, we have limited the range to the middle 50% of the values.

Next we draw short horizontal lines at the median and the two quartiles, and we connect them to form a *box*. We then draw segments from each end of the box to the extreme values. These segments are called *whiskers*. The result is the upright box plot shown on the right-hand side of Figure 9-21. Similarly, a box plot can be drawn lengthwise by using a horizontal scale.

FIGURE 9-21

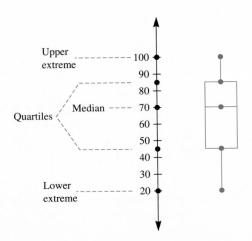

The box plot gives a fairly clear picture of the spread of the data. If we look at the graph in Figure 9-21, we see that the median is 70, the maximum value is 100, the minimum value is 20, and the upper and lower quartiles are 45 and 85.

In the box plot in Figure 9-21, the median is above the center of the box and so there are more scores above than below it. Another example of the construction of a box plot is given on the facing page, on the student page from *Addison-Wesley Mathematics,* Grade 8, 1991.

EXAMPLE 9-6 What are the minimum and maximum values, the median, and the lower and upper quartiles of the box plot shown in Figure 9-22?

FIGURE 9-22

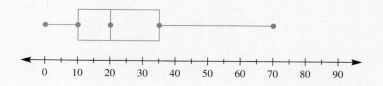

Solution The minimum value is 0, the maximum value is 70, the median is 20, the lower quartile is 10, and the upper quartile is 35. ∎

Outliers

Recall that an *outlier* is a value that is widely separated from the rest of a group of data. For example, in a set of scores such as

$$91 \quad 92 \quad 92 \quad 93 \quad 93 \quad 93 \quad 94$$

all data are grouped close together and no values are widely separated. However, in a set of scores such as

$$21 \quad 92 \quad 92 \quad 93 \quad 93 \quad 93 \quad 95 \quad 150$$

both 21 and 150 are widely separated from the rest of the data. These values are potential outliers. The upper and lower extreme values are not necessarily outliers. In data such as

$$75 \quad 90 \quad 91 \quad 92 \quad 92 \quad 93 \quad 93$$

it is not easy to decide, so we develop a rule. The rule we use to determine outliers is: *An* **outlier** *is any value that is more than 1.5 interquartile ranges above the upper quartile or more than 1.5 interquartile ranges below the lower quartile.* Statisticians sometimes use values different from 1.5 to determine outliers.

It is common practice to indicate outliers with asterisks. Whiskers are then drawn to the extreme points that are *not* outliers. To investigate how this works, we use the data in Table 9-12.

Box and Whisker Graphs

LEARN ABOUT IT

Graphic images help us interpret data. A **box and whisker graph** is one way to provide a picture of the central tendency of data.

EXPLORE Study the Table

You can draw a box and whisker graph to display the data given in the stem and leaf plot at the right.

TALK ABOUT IT

1. How many items of data are listed in the stem and leaf plot?

2. List the data in order. Which item has the highest value? Which item has the lowest value?

3. What is the median of the data?

To complete a box and whisker graph of the above data follow the steps below.

- Find the median of the upper half of the data and label it Q_U to represent the **upper quartile**.
- Find the median of the lower half of the data and label it Q_L to represent the **lower quartile**.
- Mark an appropriate vertical scale and draw a box that connects the upper quartile to the lower quartile. A line across the box indicates the median. Label the median "MD." For this data, MD = 82.
- Draw lines, "whiskers," from the box to the **highest** (H) and **lowest** (L) data items.

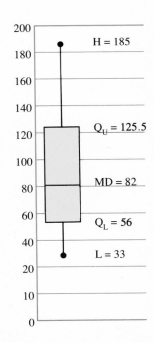

TRY IT OUT

Draw a box and whisker graph for each of these sets of data.

1. 12, 23, 24, 24, 28, 37, 49, 51, 53, 54, 54, 63, 65, 67, 92, 98
2. 23, 45, 46, 46, 49, 25, 72, 48, 63, 18, 29, 53

TABLE 9-12 Final Medal Standings for the Top 20 Countries—1988 Olympics

Country	Number of Medals
USSR	132
East Germany	102
United States	94
West Germany	40
Bulgaria	35
South Korea	33
China	28
Romania	24
Great Britain	24
Hungary	23
France	16
Poland	16
Italy	14
Japan	14
Australia	14
New Zealand	13
Yugoslavia	12
Sweden	11
Canada	10
Kenya	9

The extreme scores are 132 and 9, the median is 19.5, and the quartiles are 34 and 13.5, with IQR = 20.5. Outliers are scores that are greater than 34 + 1.5(20.5), or 64.75, or less than 13.5 − 1.5(20.5), or ⁻17.25. Therefore, in this data set, there are three outliers: 94, 102, and 132. A box plot is given in Figure 9-23. Notice that the whisker stops at the extreme point, 40, and the three outliers are indicated by asterisks.

FIGURE 9-23

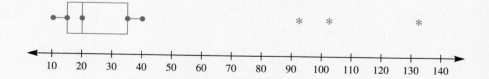

Comparing Sets of Data

Box plots are used primarily for large sets of data or for comparing several distributions. The stem-and-leaf plot is usually a much clearer display for a single distribution. Several box plots drawn below the same number line give us the easiest comparison of medians, extreme scores, and the quartiles for the sets of data. As an example, we construct box plots comparing the data in Table 9-13.

SECTION 9-2 MEASURES OF CENTRAL TENDENCY AND VARIATION

TABLE 9-13 Batting Average Champions 1978–1988

Year	National League		American League	
1978	Parker	.334	Carew	.333
1979	Hernandez	.344	Lynn	.333
1980	Buckner	.324	Brett	.390
1981	Madlock	.341	Lansford	.336
1982	Oliver	.331	Wilson	.332
1983	Madlock	.323	Boggs	.361
1984	Gwynn	.351	Mattingly	.343
1985	McGee	.353	Boggs	.368
1986	Raines	.334	Boggs	.357
1987	Gwynn	.369	Boggs	.363
1988	Gwynn	.313	Boggs	.366

Before constructing horizontal box plots, we must find the five important values for each class. These are given in Table 9-14. Next we draw the horizontal scale and construct the box plots for the American and National Leagues, as shown in Figure 9-24.

TABLE 9-14

Value	National League	American League
Maximum	.369	.390
Upper quartile	.351	.366
Median	.334	.357
Lower quartile	.324	.333
Minimum	.313	.332

FIGURE 9-24

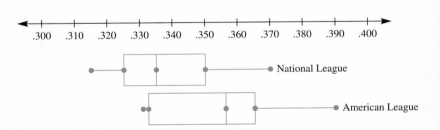

From Figure 9-24, we can see that the length of the box (IQR) for the American League is longer than the box (IQR) for the National League. This implies that the top batting averages in the American League vary more from year to year than do those in the National League. The top batting averages in the American League tend to be higher than those in the National League, since the extreme scores, median, and quartiles for the American League are all higher than those for the National League. Also, almost 75% of the top batting averages in the American League are higher than the median of the top averages in the National League.

In this example, the IQR for the National League is .351 − .324, or .027. The upper cutoff point for identifying outliers is .351 + 1.5(.027), or .3915. The lower cutoff point for outliers is .324 − 1.5(.027), or .2835. Because no top batting averages in the National League are greater than .3915 or less than .2835, there are no outliers in the National League data. For the American League, the IQR is .366 − .333 = .033. Values greater than .366 + 1.5(.033), or .4155, or less than .333 − 1.5(.033), or .2835, are outliers. Therefore, there are no outliers in the American League data, either.

Using Box Plots

Although we cannot spot clusters or gaps in box plots (as we can with stem-and-leaf or line plots), we can more easily compare data from different sets. With box plots, we do not need to have sets of data that are approximately the same size, as we did for stem-and-leaf plots. To compare data from two or more sets using their box plots, we first study the boxes to see if they are located in approximately the same places. Next, we consider the lengths of the boxes to see if the variability of the data is about the same. We also check whether the median, the quartiles, and the extreme values in one set are greater than those in another set. If they are, the data in the first set are greater than those of the other set, no matter how we compare them. If they are not, we can continue to study the data for other similarities and differences.

▼ **BRAIN TEASER**

The speeds of racing cars were timed after 3 mi, $4\frac{1}{2}$ mi, and 6 mi. Freddy averaged 140 miles per hour (mph) for the first 3 mi, 168 mph for the next $1\frac{1}{2}$ mi, and 210 mph for the last $1\frac{1}{2}$ mi. What was his mean speed for the total 6-mi run?

PROBLEM SET 9-2

1. Calculate the mean, the median, and the mode for each of the following collections of data:
   (a) 2, 8, 7, 8, 5, 8, 10, 5
   (b) 10, 12, 12, 14, 20, 16, 12, 14, 11
   (c) 18, 22, 22, 17, 30, 18, 12
   (d) 82, 80, 63, 75, 92, 80, 92, 90, 80, 80
   (e) 5, 5, 5, 5, 5, 10

2. (a) If each of six students scored 80 on a test, find each of the following for the set of six scores:
   (i) Mean
   (ii) Median
   (iii) Mode
   (b) Make up another set of six scores that are not all the same but in which the mean, median, and mode are all 80.

3. The mean score on a set of 20 tests is 75. What is the sum of the 20 test scores?

4. The tram at a ski area has a capacity of 50 people with a load limit of 7500 lb. What is the mean weight of the passengers if the tram is loaded to capacity?

5. The mean for a set of 28 scores is 80. Suppose two more students take the test and score 60 and 50. What is the new mean?

6. The names and ages for each person in a family of five follow:

Name	Dick	Jane	Kirk	Jean	Scott
Age	40	36	8	6	2

(a) What is the mean age?
(b) Find the mean of the ages 5 yr from now.
(c) Find the mean 10 yr from now.
(d) Describe the relationships among the means found in (a), (b), and (c).

7. ▶Suppose that you own a hat shop and decide to order hats in only *one* size for the coming season. To decide which size to order, you look at last year's sales figures, which are itemized according to size. Should you find the mean, median, or mode for the data? Why?

8. A table showing Jon's fall quarter grades follows. Find his grade point average for the term (A = 4, B = 3, C = 2, D = 1, F = 0).

Course	Credits	Grades
Math	5	B
English	3	A
Physics	5	C
German	3	D
Handball	1	A

9. If the mean weight of seven linemen on a football team is 230 lb and the mean weight of the four backfield members is 190 lb, what is the mean weight of the eleven-man team?

10. If 99 people had a mean income of $12,000, how much is the mean income increased by the addition of a single income of $200,000?

11. The following table gives the annual salaries of the 40 players of a certain professional football team.
    (a) Find the mean annual salary for the team.
    (b) Find the median annual salary.
    (c) Find the mode.

Salary	Number of Players
$ 18,000	2
22,000	4
26,000	4
35,000	3
38,000	12
44,000	8
50,000	4
80,000	2
150,000	1

12. In a gymnastics competition, each competitor receives six scores. The highest and lowest scores are eliminated, and the official score is the mean of the four remaining scores.
    (a) If the only events in the competition are the balance beam, the uneven bars, and the floor exercise, find the winner of each event.
    (b) Find the overall winner of the competition if the overall winner is the person with the highest combined official scores.

Gymnast	Scores
Balance Beam	
Meta	9.2 9.2 9.1 9.3 9.8 9.6
Lisa	9.3 9.1 9.4 9.6 9.9 9.4
Olga	9.4 9.5 9.6 9.6 9.9 9.6
Uneven Bars	
Meta	9.2 9.1 9.3 9.2 9.4 9.5
Lisa	10.0 9.8 9.9 9.7 9.9 9.8
Olga	9.4 9.6 9.5 9.4 9.4 9.4
Floor Exercises	
Meta	9.7 9.8 9.4 9.8 9.8 9.7
Lisa	10.0 9.9 9.8 10.0 9.7 10.0
Olga	9.4 9.3 9.6 9.4 9.5 9.4

13. Maria needed 8 gal of gas to fill her car's gas tank. The mileage odometer read 42,800 mi. When the odometer read 43,030, Maria filled the tank with 12 gal. At the end of the trip, she filled the tank with 18 gal and the odometer read 43,390 mi. How many miles per gallon (mpg) did she get for the entire trip?

14. If Janet traveled 45 mi in 90 min, what was her mean speed?

15. Emily worked the following hours for the week. How many hours did she average (mean) per day for the week?

    Monday: $5\frac{1}{2}$ hr Tuesday: $3\frac{1}{2}$ hr Wednesday: $5\frac{1}{4}$ hr

    Thursday: $6\frac{3}{4}$ hr Friday: 8 hr

16. The youngest person in the company is 24 years old. The range of ages is 34 yr. How old is the oldest person in the company?

17. Choose the set(s) of numbers that fits the descriptions given in each of the following:
    (a) The mean is 6.
        The range is 6.
        Set A: 3, 5, 7, 9
        Set B: 2, 4, 6, 8
        Set C: 2, 3, 4, 15
    (b) The mean is 11.
        The median is 11.
        The mode is 11.
        Set A: 9, 10, 10, 11, 12, 12, 13
        Set B: 11, 11, 11, 11, 11, 11, 11
        Set C: 9, 11, 11, 11, 11, 12, 12

(c) The mean is 3.
 The median is 3.
 It has no mode.

 Set A: 0, $2\frac{1}{2}$, $6\frac{1}{2}$
 Set B: 3, 3, 3, 3
 Set C: 1, 2, 4, 5

18. Carl had scores of 90, 95, 85, and 90 on his first four tests.
    (a) Find the median, mean, and mode.
    ▶(b) Carl scored a 20 on his fifth exam. Which of the three averages would Carl want the instructor to use to compute his average? Why?
    (c) Which measure is affected the most by an extreme score?

19. ▶The mean of the five numbers given below is 50:

    20   35   50   60   85

    (a) Add four numbers to the list so that the mean of the nine numbers is still 50.
    (b) Explain how you could choose the four numbers to add to the list so that the mean did not change.
    (c) How does the mean of the four numbers you added to the list compare to the original mean of 50? Why?

20. What is the standard deviation of the heights of seven trapeze artists if their heights are 175 cm, 182 cm, 190 cm, 180 cm, 192 cm, 172 cm, and 190 cm?

21. ▶What happens to the mean and to the standard deviation of a set of data when the same number is added to each value in the data?

22. (a) If all the numbers in a set are equal, what is the standard deviation?
    (b) If the standard deviation of a set of numbers is zero, must all the numbers in the set be equal?

23. In a Math 131 class at DiPaloma University, the grades on the first exam were as follows:

    | 96 | 71 | 43 | 77 | 75 | 76 | 61 |
    | 83 | 71 | 58 | 97 | 76 | 74 | 91 |
    | 74 | 71 | 77 | 83 | 87 | 93 | 79 |

    (a) Find the mean.
    (b) Find the median.
    (c) Find the mode.
    (d) Find the variance of the scores.
    (e) Find the standard deviation of the scores.

24. To receive an A in a class, Willie needs at least a mean of 90 on five exams. Willie's grades on the first four exams were 84, 95, 86, and 94. What minimum score does he need on the fifth exam to receive an A in the class?

25. Ginny's median score on three tests was 90. Her mean score was 92 and her range was 6. What were her three test scores?

26. The mean of five numbers is 6. If one of the five numbers is removed, the mean becomes 7. What is the value of the number that was removed?

27. ▶Sue drives 5 mi at 30 mph and then 5 mi at 50 mph. Is the mean speed for the trip 40 mph? Why or why not?

28. Construct a box plot for the gas mileages (mpg) of the various company cars given below:

    22 18 14 28 30 12 38 22
    30 39 20 18 14 16 10

29. Box plots comparing the ticket prices of two performing arts theaters are given below.
    (a) What is the median ticket price for each theater?
    (b) Which theater has the greatest range of prices?
    (c) What is the highest ticket price at either theater?
    ▶(d) Make some statements comparing the ticket prices at the two theaters.

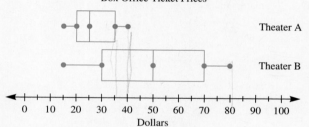

Box Office Ticket Prices

30. Construct a box plot for the following set of test scores. Indicate outliers, if any, with asterisks.

    20 95 40 70 90 70 80 80 90 95

31. Following are the heights in feet of the tallest ten buildings in Los Angeles and in Minneapolis.

Los Angeles	Minneapolis
858	950
750	775
735	668
699	579
625	561
620	447
578	440
571	416
534	403
516	366

    (a) Draw horizontal box plots to compare the data.
    (b) Are there any outliers in this data? If so, which values are they?

32. In a school system, teachers start at a salary of $15,200 and have a top salary of $31,800. The teachers' union is

bargaining with the school district for next year's salary increment.
  (a) If every teacher is given a $1000 raise, what happens to each of the following?
      (i) Mean   (ii) Median   (iii) Extremes
      (iv) Quartiles   (v) Standard deviation
  (b) If every teacher received a 5% raise, what does this do to the following?
      (i) Mean   (ii) Standard deviation

33. (a) Find the mean and the median of the following arithmetic sequences:
      (i) 1, 3, 5, 7, 9
      (ii) 1, 3, 5, 7, 9, . . . , 199
      (iii) 7, 10, 13, 16, . . . , 607
  (b) Based on your answers in (a), make a conjecture about the mean and the median of any arithmetic sequence.

34. Show that the following formula for variance is equivalent to the one given in the text:
$$v = \frac{x_1^2 + x_2^2 + \cdots + x_n^2}{n} - \bar{x}^2$$

Review Problems

35. Given the following bar graph, answer the following:
  (a) Which mountain is the highest? Approximately how high is it?
  (b) Which mountains are higher than 6000 m?

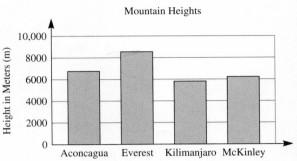

Mountain Heights

36. Raw test scores from a history test are as follows:

86	85	87	96	55
90	94	82	68	77
88	89	85	74	90
72	80	76	88	73
64	79	73	85	93

  (a) Construct an ordered stem-and-leaf plot for the given data.
  (b) Construct a grouped frequency table for these scores with intervals of 5, starting the first class at 55.
  (c) Draw a histogram of the data.
  (d) Draw a frequency polygon of the data.
  (e) If a circle graph of the grouped data in (b) were drawn, how many degrees would be in the section representing the 85–89 interval?

▼ **BRAIN TEASER**

The mean age of the first seven people to arrive at Grandpa Elmer's birthday party was 21. When Jeff, who is 29, arrived at the party, the mean age increased to 22. Mary, who is also 29, arrived next. Did the mean age increase to 23 with Mary's arrival? The tenth and last person to arrive was Grandpa Elmer, and the mean age increased to 30 yr. How old is Grandpa Elmer on this birthday?

LABORATORY ACTIVITY

We can model the mean as a measure of central tendency using a strip of cardboard that is 1 in. wide and 1 ft long with holes 1 in. apart and $\frac{1}{8}$ in. from the edge, as shown below.

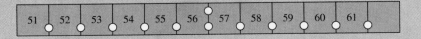

Use string and tape to suspend the strip from a desk with the string tied through a hole punched between 56 and 57. Paper clips of equal size are then used, as shown below, to investigate means.

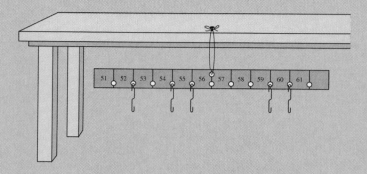

1. (a) If paper clips are hung in the holes at 51, 53, and 60, where should an additional clip be hung in order to achieve a balance?
   (b) If paper clips are hung at 51, 54, and 60, where should two additional paper clips be hung to achieve a balance?
   (c) If paper clips are hung at 51, 53, 54, and 55, where should four additional clips be hung to achieve a balance?
2. Find the mean of the data of all the numbers in each part in (1) and compare this answer with the number in the center of the strip.
3. Would any of the means in (2) change if we hung an additional paper clip in the center of the strip?
4. Find the median and mode for 51, 53, 54, 56, 58, 58, and 59.
5. Hang paper clips in each hole in (4). If a number appears more than once, hang that number of paper clips in the hole. Is there a balance around the median? Is there a balance around the mode?
6. Under what conditions do you think there would be a balance around the (a) mean, (b) median, and (c) mode? Test your conjecture using the cardboard strip.

*Section 9-3 Normal Distributions

To better understand how standard deviations are used as measures of dispersion, we next consider normal distributions. The graphs of normal distributions are the bell-shaped curves called normal curves. These curves often describe distributions such as IQ scores for the population of the United States.

normal curve

A **normal curve** is a smooth, bell-shaped curve that depicts frequency values distributed symmetrically about the mean. (Also, the mean, median, and mode all have the same value.) The normal curve is a theoretical distribution that extends infinitely in both directions. It gets closer and closer to the *x*-axis but never reaches it. On a normal curve, about 68% of the values lie within one standard deviation of the mean, about 95% lie within two standard deviations, and about 99.8% are within three standard deviations. The percentages represent approximations of the total percent of area under the curve. The curve and the percentages are illustrated in Figure 9-25.

FIGURE 9-25

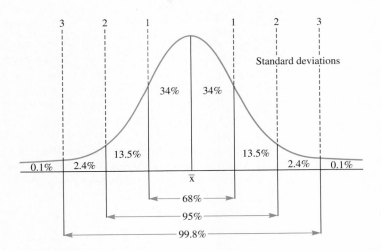

Suppose the area under the curve represents the population of the United States. Psychologists claim that the mean IQ is 100 and the standard deviation is 15. They also claim that an IQ score of over 130 represents a superior score. Because 130 is equal to the mean plus two standard deviations, we see from Figure 9-25 that only 2.5% of the population fall into this category.

EXAMPLE 9-7   When a standardized test was scored, there was a mean of 500 and a standard deviation of 100. Suppose that 10,000 students took the test and their scores had a mound-shaped distribution, making it possible to use a normal curve to approximate the distribution.

(a) How many scored between 400 and 600?
(b) How many scored between 300 and 700?
(c) How many scored between 200 and 800?

Solution   (a) Since one standard deviation on either side of the mean is from 400 to 600, about 68% of the scores fall in this interval. Thus, 0.68(10,000), or 6800, students scored between 400 and 600.

(b) About 95% of 10,000, or 9500, students scored between 300 and 700.

(c) About 99.8% of 10,000, or 9980, students scored between 200 and 800.

REMARK About 0.2%, or 20, students' scores in Example 9-7 fall outside three standard deviations. About 10 of these students did very well on the test, and about 10 students did very poorly.

HISTORICAL NOTE

Abraham De Moivre (1667–1754), a French Huguenot, was the first to develop and study the normal curve. He was one of the first to study actuarial information, in his book *Annuities upon Lives*. He also worked in trigonometry and complex numbers. There is an interesting fable about the death of De Moivre. It is reported that he noticed that each day he required one-quarter of an hour more sleep than he had on the previous day; when the arithmetic progression for sleep reached 24 hr, he died. De Moivre's work with the normal curve went essentially unnoticed. Later, the normal curve was developed independently by Pierre Laplace (1749–1827) and Karl Friedrich Gauss (1777–1855). Gauss found so many applications for the normal curve that it is sometimes referred to as the *Gaussian curve*.

Suppose that a group of students asked their teacher to grade "on a curve." If the teacher gave a test to 200 students and the mean on the test was 71, with a standard deviation of 7, the graph in Figure 9-26 shows how the grades could be assigned. In Figure 9-26, the teacher has used the normal curve in grading. (The use of the normal curve presupposes that the teacher had a mound-shaped distribution of scores and also that the teacher arbitrarily decided to use the lines marking standard deviations to determine the boundaries of the A's, B's, C's, D's, and F's). Thus, based on the normal curve in Figure 9-26, Table 9-15 shows the range of grades that the teacher might assign if the grades are rounded. Students who ask their teachers to grade on the curve may wish to reconsider if the normal curve is to be used.

FIGURE 9-26

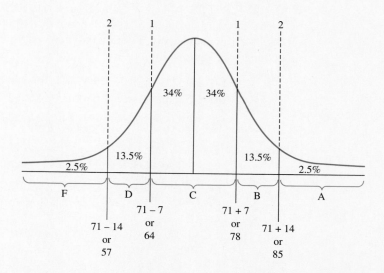

TABLE 9-15

Test Score	Grade	Number of People per Grade	Percentage Receiving Grade
85 and above	A	5	2.5%
78–84	B	27	13.5%
64–77	C	136	68%
57–63	D	27	13.5%
Below 57	F	5	2.5%

percentiles

When students take a standardized test such as the ACT or SAT, their scores are often reported in **percentiles.** A percentile shows a person's score relative to other scores. For example, if a student's score is at the 82nd percentile, this means that approximately 82% of those taking the test scored lower than the student and approximately 18% had higher scores.

Comparing Scores from Different Tests

Scores by themselves may have very little meaning. For example, Wayne's receiving a score of 20 on his first math quiz does not mean much unless we know more about the possible set of points on the test and how the other students scored. However, if we knew that the quiz on which Wayne received a score of 20 had a mean of 16 and a standard deviation of 2.2, we would have enough information to determine how he scored in relation to the rest of the class. Now suppose that Wayne received a score of 52 on his second quiz, where the mean was 48 and the standard deviation was 3.1. Did Wayne do better on the first quiz or on the second quiz? We need a way to compare the scores in these two cases.

z-score

One way to deal with this problem is to translate all scores into *standard scores*. One such standard score is the *z*-score. A **z-score** gives the number of standard deviations by which the score differs from the mean as well as whether the score is above or below the mean.

> **DEFINITION OF *z*-SCORE**
>
> The **z-score** of any number x in a normal distribution is given by
> $$z = \frac{x - \bar{x}}{s},$$
> where x is the score, $\bar{x}$ is the mean, and s is the standard deviation.

For Wayne's two quizzes, we have the following:

$$z = \frac{20 - 16}{2.2} \doteq 1.81 \quad \text{and} \quad z = \frac{52 - 48}{3.1} \doteq 1.29$$

Therefore we can say that Wayne did better in relation to the rest of his class on the first quiz than on the second quiz. To see why we can make this claim and this comparison, we solve the equation

$$z = \frac{x - \bar{x}}{s}$$

for x to obtain $x = \bar{x} + sz$. Therefore, on Wayne's first quiz, his raw score was $\bar{x} + 1.81s$; that is, his score was 1.81 standard deviations above the mean. On his second quiz, his score was $\bar{x} + 1.29s$, or his score was 1.29 standard deviations above the mean.

Consider three student scores on a test with a mean of 71 and a standard deviation of 7. If the student scores are 71, 64, and 85, then the respective z-scores are as follows:

$$z = \frac{71 - 71}{7} = 0 \quad \text{(Student 1)}$$

$$z = \frac{64 - 71}{7} = {}^-1 \quad \text{(Student 2)}$$

$$z = \frac{85 - 71}{7} = 2 \quad \text{(Student 3)}$$

Figure 9-27 can be used to interpret the z-scores *if the scores in the class have a mound-shaped distribution.* A z-score of 0 indicates that the score of 71 is the mean. A z-score of $^-1$ indicates that the score of 64 is one standard deviation below the mean. Similarly, a z-score of 2 indicates that the score of 85 is two standard deviations above the mean. Using Figure 9-25, we see that 97.5% of the students taking the test scored lower than the third student.

FIGURE 9-27

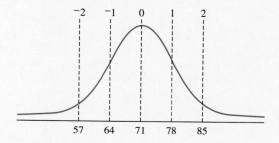

In comparisons of results from various tests taken by the same reference group, z-scores are useful. Suppose that a group of students took both an English test and a mathematics test; then a comparison of the z-scores would be reasonable. If a college class and a fourth-grade class took a mathematics test, a comparison of z-scores would not be reasonable, because the reference groups are different.

EXAMPLE 9-8 For a certain group of people, the mean height is 182 cm, with a standard deviation of 11 cm. Juanita's height has a z-score of 1.4. What is her height?

Solution We use the formula for z-scores, $z = \frac{x - \bar{x}}{s}$. We know that $z = 1.4$, $\bar{x} = 182$, and $s = 11$.

$$1.4 = \frac{x - 182}{11}$$
$$15.4 = x - 182$$
$$x = 197.4$$

Therefore, Juanita's height is 197.4 cm.

In a large group of scores, such as may be obtained on various standardized tests, we generally expect the scores to approximate a normal curve. If all scores are translated to z-scores, then with any given z-score we should be able to determine the approximate percentage of people who scored either above or below this z-score. For example, suppose a z-score on a test is 1.5. In determining the percentage of people who scored below this, we know from the graph in Figure 9-25 that the percentage is more than 84% and less than 97.5%. Table 9-16 can be used to find that percentage.

TABLE 9-16

z-score	Percentage Below	z-score	Percentage Below	z-score	Percentage Below
⁻3.0	0.13	⁻1.0	15.87	1.0	84.13
⁻2.9	0.19	⁻0.9	18.41	1.1	86.43
⁻2.8	0.26	⁻0.8	21.19	1.2	88.49
⁻2.7	0.35	⁻0.7	24.20	1.3	90.32
⁻2.6	0.47	⁻0.6	27.42	1.4	91.92
⁻2.5	0.62	⁻0.5	30.85	1.5	93.32
⁻2.4	0.82	⁻0.4	34.46	1.6	94.52
⁻2.3	1.07	⁻0.3	38.21	1.7	95.54
⁻2.2	1.39	⁻0.2	42.07	1.8	96.41
⁻2.1	1.79	⁻0.1	46.02	1.9	97.13
⁻2.0	2.27	0.0	50.00	2.0	97.73
⁻1.9	2.87	0.1	53.98	2.1	98.21
⁻1.8	3.59	0.2	57.93	2.2	98.61
⁻1.7	4.46	0.3	61.79	2.3	98.93
⁻1.6	5.48	0.4	65.54	2.4	99.18
⁻1.5	6.68	0.5	69.15	2.5	99.38
⁻1.4	8.08	0.6	72.58	2.6	99.53
⁻1.3	9.68	0.7	75.80	2.7	99.65
⁻1.2	11.51	0.8	78.81	2.8	99.74
⁻1.1	13.57	0.9	81.59	2.9	99.81
				3.0	99.87

Table 9-16 gives the percentage of scores falling below a given z-score for normal distributions. The percentage corresponding to any observation from a normal distribution can be found by converting the observation to a corresponding z-score and then looking in the table. In our example, we see from Table 9-16 that approximately 93.32% of the population had z-scores below 1.5. More detailed tables show z-scores carried out to more decimal places. Figure 9-28 provides a graphic representation of the type of information given in Table 9-16.

FIGURE 9-28

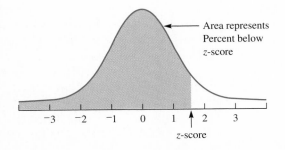

Suppose that the grades on a certain test are normally distributed and we want to grade an exam so that a grade of C is given to all students scoring within 1.2 standard deviations of the mean. What percentage of the students should receive Cs? Table 9-16 shows that 88.49% of the population scored less than 1.2 standard deviations above the mean. We know that 50% of the population scored below the mean. Therefore, 88.49% − 50% = 38.49% of the population lies between the mean and 1.2 standard deviations. Because the normal curve is symmetric, we have the same percent on the other side of the mean, for a total of 2 · (38.49%) = 76.98%. Therefore, approximately 77% of the students should receive C's.

EXAMPLE 9-9 The heights of American young women are approximately normally distributed with a mean of 65.5 in. and a standard deviation of 2.5 in.

(a) What is the z-score for a woman who is 69 in. tall?

(b) What is the z-score for a woman who is 60 in. tall?

(c) What percentage of the population of American young women fall below each of these heights?

Solution (a) A woman 69 in. tall has a z-score of
$$\frac{69 - 65.5}{2.5} = 1.4.$$

(b) A woman 60 in. tall has a z-score of
$$\frac{60 - 65.5}{2.5} = {}^-2.2.$$

(c) If $z = 1.4$, then from Table 9-16 the percentage of the population below this height is 91.92%. If $z = {}^-2.2$, the percentage of the population below this height is 1.39%. ∎

Scattergrams

We often wonder if there is a relationship between two sets of data. For example, at the college level, mathematics placement exams are given to determine readiness for calculus. If students who score better on the exam do better in calculus than students who do poorly, we say that there is a positive correlation between the placement scores and achievement in the course. We can analyze relationships between two sets of data by using a **scattergram**, such as the one in Figure 9-29(a). In Figure 9-29(a), the points are not connected like those in a line graph, and there may be more than one point for a given number on either scale. Figure 9-29(a) shows that the highest score was 10 and the lowest was 1. We see that three students studied 4 hr for the test and that the mode was 5 hr.

All of the points on a scattergram usually do not fall on a particular line, but on some scattergrams the points fall near a line called the **trend line**. The trend line is used to make predictions. If the trend line slopes up from left to right as in Figure 9-29(b), then we say there is a *positive correlation*. From the trend line in Figure 9-29(b), we see that students who studied 7 hr typically scored six correct answers.

FIGURE 9-29

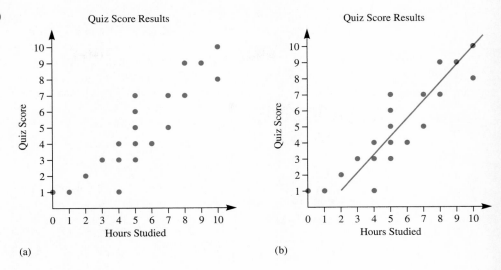

(a) (b)

If the trend line slopes downward to the right, we can also make predictions; we say there is a *negative correlation*. If the points do not approximately fall on any line, we say there is *no correlation*. Scattergrams also show clusters of points and outliers. Examples of various correlations are given in Figure 9-30.

FIGURE 9-30

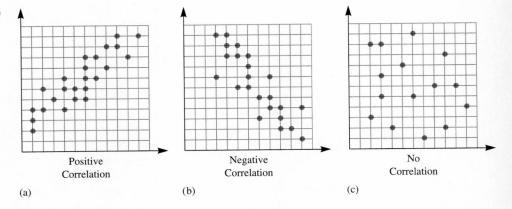

(a) Positive Correlation   (b) Negative Correlation   (c) No Correlation

The student page from *Addison-Wesley Mathematics*, Grade 7, 1991, on the following page gives an example of a scattergram. Read through the student page and answer the questions at the bottom of the page.

Scattergrams

LEARN ABOUT IT

EXPLORE Examine the graph

Ordered pairs of data called **data points** can be shown in a **scattergram.** Here, each data point stands for the cost and rating of one brand of windsurfer board.

Brands	
A Freestyle Surfer	E Wild Board
B Tahiti Board	F Shark Board
C Caribbi 270	G XJ300 Wind
D Surfer Mate	H Sleek Surf
	J Breeze Board

Consumer Ratings: Windsurfer Boards

TALK ABOUT IT

1. Which board is the most expensive? the least expensive?
2. Which board has the highest rating? the lowest rating?
3. Which boards would you consider to be good buys? Why?

One relationship between measured quantities is the **correlation.**

Positive Correlation
Both sets of data increase together.

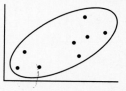

Negative Correlation
One set of data increases as the other decreases.

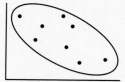

If data points are scattered over the graph, we say there is **no correlation.**

TRY IT OUT

Use the Windsurfer Boards scattergram above to answer these questions.

1. What is the approximate cost and rating of the Breeze Board?
2. Does the scattergram show positive, negative, or no correlation?

PROBLEM SET 9-3

1. The mean IQ score for 1500 students is 100, with a standard deviation of 15. Assuming the scores have a normal curve,
   (a) How many have an IQ between 85 and 115?
   (b) How many have an IQ between 70 and 130?
   (c) How many have an IQ over 145?

2. Sugar Plops boxes say they hold 16 oz. To make sure they do, the manufacturer fills the box to a mean weight of 16.1 oz, with a standard deviation of 0.05 oz. If the weights have a normal curve, what percentage of the boxes actually contains 16 oz or more?

3. For certain workers, the mean wage is $5.00/hr, with a standard deviation of $0.50. If a worker is chosen at random, what is the probability that the worker's wage is between $4.50 and $5.50? Assume a normal distribution of wages.

4. A job-applicant test consisted of three parts: verbal, quantitative, and logical reasoning. The mean and standard deviation for each part are given below.

	Verbal	Quantitative	Logical Reasoning
$\bar{x}$	84	118	14
s	10	18	4

   (a) Holly's scores were 90 on verbal, 133 on quantitative, and 18 on logical reasoning. Determine her z-score for each part.
   (b) Use the answers in (a) to determine each of the following:
     (i) On which part did she perform relatively the highest?
     (ii) On which part did she perform relatively the lowest?
     (iii) To determine an overall composite score, we find the mean of the z-scores. What is Holly's composite score?

5. Assume a normal distribution and that the average phone call in a certain town lasted 4 min with a standard deviation of 2 min. What percentage of the calls lasted fewer than 2 min?

6. ▶ In a normal distribution, how are the mean and the median related? Why?

7. According to psychologists, IQs are normally distributed with a mean of 100 and a standard deviation of 15.
   (a) What percentage of the population have IQs between 100 and 130?
   (b) What percentage of the population have IQs lower than 85?

8. Use Table 9-16 to find the percentages of scores below each of the following z-scores:
   (a) $^-2.3$ (b) 1.7   (c) $^-2.0$

9. Use Table 9-16 to find the percentage of scores between z-scores of 1.4 and 1.5.

10. The weights of newborn babies are distributed normally with a mean of approximately 105 oz and a standard deviation of 20 oz. If a newborn is selected at random, what is the probability that the baby weighs less than 125 oz?

11. If the mean is 63 and a score of 53 corresponds to a z-score of $^-1.25$, what is the standard deviation?

12. On a certain exam, the mean is 72 and the standard deviation is 9. If a grade of A is given to any student who scores at least two standard deviations above the mean, what is the lowest score that a person could receive and still get an A?

13. Assume that the heights of American women are approximately normally distributed, with a mean of 65.5 in. and a standard deviation of 2.5 in. Within what range are the heights of 95% of American women?

14. A tire company tested a particular model of tire and found the tires to be normally distributed with respect to wear. The mean was 28,000 mi, and the standard deviation was 2500 mi. If 2000 tires are tested, about how many are likely to wear out before 23,000 mi?

15. A standardized mathematics test was given to 10,000 students, and the scores were normally distributed. The mean was 500, and the standard deviation was 60. If a student scored below 440 points, the student was considered deficient in mathematics. About how many students were rated deficient?

16. Coach Lewis kept track of his team's high-jump records for a 10-yr period.

Year	1983	1984	1985	1986	1987	1988
Record (nearest in.)	65	67	67	68	70	74

Year	1989	1990	1991	1992
Record (nearest in.)	77	78	80	81

    (a) Draw a scattergram for the data.
    (b) What kind of correlation is there for these data?

17. Given the scattergram shown, answer the following questions:
    (a) What type of correlation exists for these data?
    (b) About how many movies does an average 25-year-old attend?

(c) Based on the scattergram, how old do you think a person is who attends 16 movies a year?

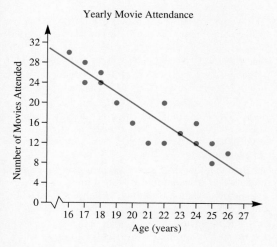

Review Problems

18. On the English 100 exam, the scores were as follows:

43	91	73	65
56	77	84	91
82	65	98	65

(a) Find the mean.
(b) Find the median.
(c) Find the mode.
(d) Find the variance.
(e) Find the standard deviation.

19. If the mean of a set of 36 scores is 27 and two additional scores of 40 and 42 are added, what is the new mean?

20. On a certain exam, Tony corrected 10 papers and found the mean for his group to be 70. Alice corrected the remaining 20 papers and found that the mean for her group was 80. What is the mean of the combined group of 30 students?

21. Following are the men's gold-medal times for the 100-m run in the Olympic games from 1896 to 1964. Construct an ordered stem-and-leaf plot for the data.

Year	Time (in seconds)
1896	12.0
1900	11.0
1904	11.0
1908	10.8
1912	10.8
1920	10.8
1924	10.6
1928	10.8
1932	10.3
1936	10.3
1948	10.3
1952	10.4
1956	10.5
1960	10.2
1964	10.0

*Section 9-4 Abuses of Statistics

Not only are statistics frequently used, they are also frequently abused. Benjamin Disraeli (1804–1881), an English prime minister, once remarked, "There are three kinds of lies: lies, damned lies, and statistics." People sometimes deliberately use statistics to mislead others. In the past, this has been seen in advertising. More often, the misuse of statistics is the result of misinterpreting what the statistics actually mean. For example, if we were told that the average depth of water in a lily pond was 2 ft, most of us would presume that a heron could stand up in any part of the pond. That this is not necessarily the case is seen in the accompanying Far Side cartoon.

SECTION 9-4 ABUSES OF STATISTICS 485

THE FAR SIDE © 1985 Far Works, Inc. Reprinted with permission of Universal Press Syndicate. All rights reserved.

Now consider an advertisement in which it is reported that, of the people responding to a recent survey, 98% said that Buffepain is the most effective pain reliever of headaches and arthritis of all those tested. To certify that the statistics are not being misused, the following information also should have been reported:

1. The number of people surveyed
2. The number of people responding
3. How the people participating in the survey were chosen
4. The number and type of pain relievers tested

Without the information listed, the following situations are possible, all of which could cause the advertisement to be misleading:

1. Suppose that 1,000,000 people nationwide were sent the survey, and only 50 responded. This would mean that there was only a 0.005% response, which would certainly cause us to mistrust the ad.
2. Of the 50 responding in (1), suppose that 49 responses were affirmative. The 98% claim is true, but 999,950 people did not respond at all.
3. Suppose that all the people sent the survey were chosen from a town in which the major industry was the manufacture of Buffepain. It is very doubtful that the survey would represent an unbiased sample.
4. Suppose that only two pain relievers were tested: Buffepain, whose active ingredient is 100% aspirin, and a placebo containing only powdered sugar.

This is not to say that advertisements of this type are all misleading or dishonest but simply that statistics are only as honest as their users. The next time you hear an advertisement such as "After using Ultraguard toothpaste, Joseph has 40% fewer cavities," you might ask whether or not Joseph has 40% fewer teeth than an average person.

Another example of the misuse of statistics is a headline from a student newspaper in Texas that claimed that "$33\frac{1}{3}\%$ of the Female Mathematics Faculty Marry Their Students." The headline was in fact true since there were three female mathematics faculty members, and one of them *did* marry one of her students. As we can see, statistics can be used to distort facts without really lying.

A different type of misuse of statistics involves graphs. Among the things to look for in a graph are the following; if they are not there, then the graph may be misleading:

1. Title
2. Labels on both axes of a line or bar chart and on all sections of a pie chart
3. Source of the data
4. Key in a pictograph
5. Uniform size of symbols in a pictograph
6. Scale: Does it start with zero? If not, is there a break shown?
7. Scale: Are the numbers evenly spaced?

As an example of a misleading use of graphs, consider how they can be used to distort data or exaggerate certain pieces of information. A frequency polygon, histogram, or bar graph can be altered by changing the scale of the graph. For example, consider the data in Table 9-17 for the number of graduates from a community college for the years 1988 to 1992.

TABLE 9-17

Year	1988	1989	1990	1991	1992
Number of graduates	140	180	200	210	160

The two graphs in Figure 9-31(a) and (b) represent the same data, but different scales are used in each. The statistics presented are the same, but these two graphs do not convey the same psychological message. Notice that, in Figure 9-31(b), the years on the horizontal axis of the graph are spread out and the numbers on the vertical axis are condensed. Both of these changes minimize the variability of the data. A college administrator might use the graph in Figure 9-31(b) to convince people that the college was not in serious enrollment trouble.

Other graphs can also be misleading. Suppose, for example, that the number of boxes of cereal sold by Sugar Plops last year was 2 million and the number of boxes of cereal sold by Korn Krisps was 8 million. The Korn Krisps executives prepared the graph in Figure 9-32 to demonstrate the data. The Sugar Plops people objected. Do you see why?

The graph in Figure 9-32 clearly distorts the data, since the figure for Korn Krisps is both four times as high and four times as wide as the bar for Sugar Plops. Thus, the area representing Korn Krisps is 16 times the area representing Sugar Plops, rather than four times the area, as would be justified by the original data.

Other ways to distort bar graphs include omitting a scale, as in Figure 9-33(a). The scale is given in Figure 9-33(b).

SECTION 9-4 ABUSES OF STATISTICS 487

FIGURE 9-31

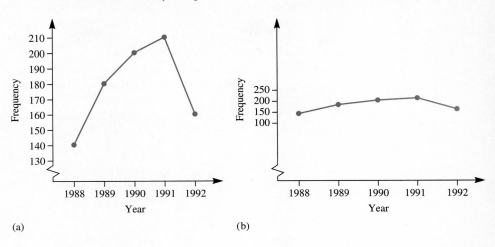

No. of Graduates of Community College (a)

No. of Graduates of Community College (b)

FIGURE 9-32

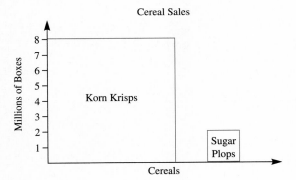

Cereal Sales

FIGURE 9-33

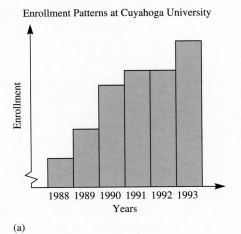

Enrollment Patterns at Cuyahoga University (a)

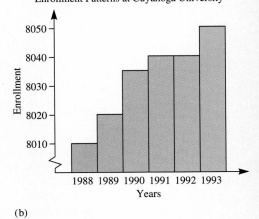

Enrollment Patterns at Cuyahoga University (b)

Circle graphs easily become distorted when attempts are made to depict them as three-dimensional. Many graphs of this type do not acknowledge either the variable thickness of the depiction or the distortion due to perspective. Observe that the 27% sector pictured in Figure 9-34 looks far greater than the 23% sector, although they should be very nearly the same size.

FIGURE 9-34

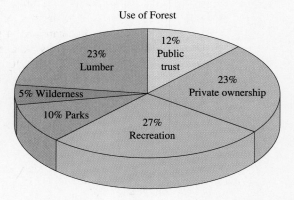

Figure 9-35 shows how the comparison of Sugar Plops and Korn Krisps cereals from Figure 9-32 might look if the figures were made three-dimensional. The figure for Korn Krisps has a volume 64 times the volume of the Sugar Plops figure.

FIGURE 9-35

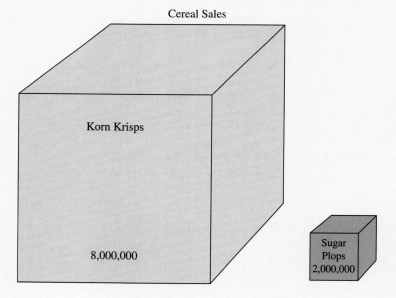

The final examples we discuss of the misuses of statistics involve misleading uses of mean, median, and mode. All these are "averages" and may be used to suit a person's purposes. As discussed in Section 9-2 in the example involving the teachers Smith, Jones, and Leed, each teacher had reported that his or her class had done better than the other two. Each of the teachers was using a different number to represent the test scores.

The use of statistics in this way is often misleading. For example, college administrators wishing to portray to prospective employees a rosy salary picture may find a mean salary of $38,000 for ranked professors along with deans, vice-presidents, and presidents in the schedule of salaries. At the same time, a faculty union or teachers' group that is bargaining for faculty salaries may include part-time employees and instructors along with ranked professors and may exclude all administration personnel in order to present a mean salary of $29,000 at the bargaining table. The important thing to watch for when a mean is reported is disparate cases in the reference group. If the sample is small, then a few extremely high or low scores can have a great influence on the mean.

If the median is being used as the average, then suppose that Figure 9-36 shows the salaries of both administrators and faculty members at the college. The median in this case might be $33,500, which is a representative of neither major group of employees. The bimodal distribution allows the median to be nonrepresentative of the distribution.

FIGURE 9-36

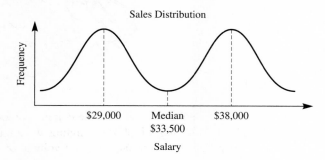

To conclude the comments on the misuse of statistics, consider this quote from Darrell Huff's book *How to Lie with Statistics* (p. 8):

So it is with much that you read and hear. Averages and relationships and trends and graphs are not always what they seem. There may be more in them than meets the eye, and there may be a great deal less.

The secret language of statistics, so appealing in a fact-minded culture, is employed to sensationalize, inflate, confuse, and oversimplify. Statistical methods and statistical terms are necessary in reporting the mass data of social and economic trends, business conditions, "opinion" polls, census. But without writers who use the words with honesty and understanding and readers who know what they mean, the result can be semantic "nonsense."

PROBLEM SET 9-4

1. ▶Discuss whether the following claims could be misleading. Explain why.
    (a) A car manufacturer claims that its car is quieter than a glider.
    (b) A motorcycle manufacturer claims that more than 95% of its cycles sold in the United States in the last 15 yr are still on the road.
    (c) A company claims its fruit juice has 10% more fruit solids than is required by U.S. government standards. (The government requires 10% fruit solids.)
    (d) A brand of bread claims to be 40% fresher.
    (e) A used car dealer claims that a car he is trying to sell will get up to 30 mpg.
    (f) Sudso claims that its detergent will leave your clothes brighter.
    (g) A sugarless gum company claims that 8 of every 10 dentists responding to the survey recommend sugarless gum.

(h) Most accidents occur in the home. Therefore, to be safer, you should stay out of your house as much as possible.

(i) Over 95% of the people who fly to a certain city do so on Airline A. Therefore, most people prefer Airline A to other airlines.

2. ▶The city of Podunk advertised that its temperature was the ideal temperature in the country because its mean temperature was 25°C. What possible misconceptions could people draw from this advertisement?

3. ▶Jenny averaged 70 on her quizzes during the first part of the quarter and 80 on her quizzes during the second part of the quarter. When she found out that her final average for the quarter was not 75, she went to argue with her teacher. Give a possible explanation for Jenny's misunderstanding.

4. ▶Suppose that the following circle graphs are used to illustrate the fact that the number of elementary teaching majors at teachers' colleges has doubled between 1980 and 1990, while the percentage of male elementary teaching majors has stayed the same. What is misleading about the way the graphs are constructed?

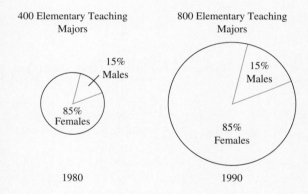

5. ▶What is wrong with the line graph shown?

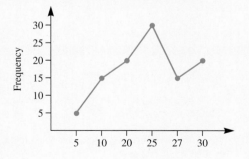

6. ▶Can you draw any valid conclusions about a set of data in which the mean is less than the median?

7. ▶A student read that nine out of ten pickup trucks sold in the last 10 yr are still on the road. She concluded that the average life of a pickup is around 10 yr. Is she correct?

8. ▶General Cooster once asked a person by the side of a river if it was too deep to ride his horse across. The person responded that the average depth was 2 ft. If General Cooster rode out across the river, what assumptions did he make on the basis of the person's information?

9. Doug's Dog Food Company wanted to impress the public with the magnitude of the company's growth. Sales of Doug's Dog Food had doubled from 1989 to 1990, so the company displayed the following graph, in which the radius of the base and the height of the 1990 can are double those of the 1989 can. What does the graph really show with respect to the growth of the company? (*Hint:* The volume of a cylinder is given by $V = \pi r^2 h$, where r is the radius of the base and h is the height.)

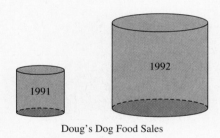

Doug's Dog Food Sales

10. ▶Explain what is wrong with the following graph.

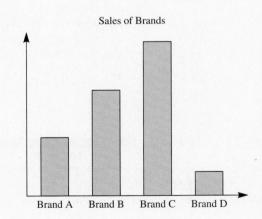

11. ▶Consider the following pictograph. Ms. Lead claims that on the basis of this information, we can conclude that men are worse drivers than women. Discuss whether you can reach that conclusion from this graph or whether you

would need more information. If more information is needed, what would you like to know?

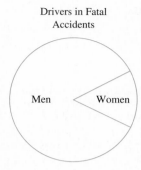

12. ▶The graph below was prepared to compare prices of camcorders at three different stores. Tell which of the statements below is true and explain why or why not.

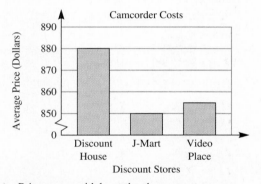

(a) Prices vary widely at the three stores.
(b) The price of Discount House is four times as great as that at J-Mart.
(c) The prices at J-Mart and Video Place differ by less than $10.

13. ▶The number of accidents per year on a certain highway for a 5-yr period is given below.

Year	1988	1989	1990	1991	1992
Number of accidents	24	26	30	32	38

(a) Draw a bar graph to convince people that the number of accidents is on the rise and that something should be done about it.
(b) Draw a bar graph to show that the rate of accidents is constant and that nothing needs to be done.

14. Write a list of scores for which the mean and median are not representative of the list.

15. ▶Graphs should be easily interpretable by the reader. Consider the following graph, which appeared in McDonald's *1990 Annual Report*. Interpret the graph and tell whether or not you think the graph is clear.

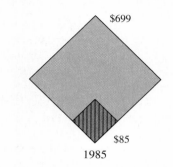

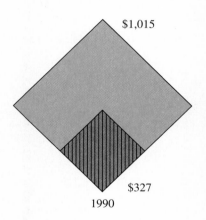

Franchised Restaurant Margins
(in millions of dollars)

■ United States   ▨ Outside of the United States

Solution to the Preliminary Problem

Understanding the Problem. A racetrack has a perimeter of 1 mi and one lap is completed at an average (mean) speed of 30 mph. We must determine the average speed for a second lap in order that the two-lap average is 60 mph.

Devising a Plan. We know that distance *(d)*, speed *(s)*, and time *(t)* are related by the formula $d = s \cdot t$. Therefore, $t = d/s$ and the time it took for the first lap is equal to the distance of 1 mi divided by the average speed of 30 mph, that is, $\frac{1}{30}$ hr, or 2 min. Because we want the average speed for two laps to be 60 mph, we determine the time it would take to complete two laps at 60 mph. From this time, we could compute the necessary speed for the second lap.

Carrying Out the Plan. To average 60 mph for 2 mi, it would take $\frac{2}{60}$ hr, or 2 min. However, we have used the entire 2 min to drive the first lap. Therefore, no matter how fast we drive the second lap, we can never reach an average speed of 60 mph. The greater the speed on the second lap, the closer we can come to an average speed of 60 mph, but we can never reach it.

Looking Back. Many people will answer this problem by reasoning that if we drive 90 mph on the second lap, then we can average 60 mph for the two laps since the average of 30 and 90 is 60, but this is not correct. Consider a straight 2-mi track. If we travel at 30 mph for the first mile, then it takes 2 min to complete the first half of the track. If we now travel at 90 mph for the other mile, we use $\frac{1}{90}$ hr, or $\frac{2}{3}$ min. For the total 2-mi track, it takes $2\frac{2}{3}$ min traveling 90 mph for the second mile. To compute the average speed, we divide the distance of 2 mi by the time of $2\frac{2}{3}$ min to obtain $\frac{6}{8}$ mi/min, or 45 mph—not 60 mph. ∎

QUESTIONS FROM THE CLASSROOM

1. A student asks, "If the average income of each of 10 people is $10,000 and one person gets a raise of $10,000, is the median, the mean, or the mode changed and, if so, by how much?"
2. A student asks for an example of when the mode is the best average. What is your response?
3. A student says that a stem-and-leaf plot is always the best way to present data. How do you respond?
4. Suppose that the class takes a test and the following averages are obtained: mean, 80; median, 90; mode, 70. Tom, who scored 80, would like to know if he did better than half the class. What is your response?
5. A student wants to know the advantages of presenting data in graphic form rather than in tabular form. What is your response? What are the disadvantages?
6. A student asks if it is possible to find the mode for data in a grouped frequency table. What is your response?
7. A student asks if she can draw any conclusions about a set of data if she knows that the mean for the data is less than the median. How do you answer?
8. A student asks if it is possible to have a standard deviation of ⁻5. How do you respond?
9. Mel's mean on 10 tests for the quarter was 89. He complained to the teacher that he should be given an A because he missed the cutoff of 90 by only a single point. Did he really miss an A by only a single point?
10. A student claims that bar graphs can be used to give the same information as line graphs so she should not have to learn how to do line graphs. What is your response?

CHAPTER OUTLINE

I. **Descriptive statistics**
   Information can be summarized in each of the following forms:
   1. **Pictographs**
   2. **Line plots**
   3. **Stem-and-leaf plots**
   4. **Frequency tables**
   5. **Histograms** or **bar graphs**
   6. **Frequency polygons** or **line graphs**
   7. **Circle graphs** or **pie charts**
   8. **Box plots**

II. **Measures of central tendency**
   A. The **mean** of n given numbers is the sum of the numbers divided by n.
   B. The **median** of a set of numbers is the middle number if the numbers are arranged in numerical order; if there is no middle number, the median is the mean of the two middle numbers.
   C. The **mode** of a set of numbers is the number or numbers that occur most frequently in the set.

III. **Measures of variation**
   A. The **range** is the difference between the greatest and least numbers in the set.
   B. The **variance** is found by subtracting the mean from each value, squaring each of these differences, finding the sum of these squares, and dividing by n, where n is the number of observations.
   C. The **standard deviation** is equal to the square root of the variance.

D. **Box plots** focus attention on the median, the quartiles, and the extremes and invite comparisons among them.
   1. The **lower quartile** is the median of the subset of data less than the median of all the values in the data set.
   2. The **upper quartile** is the median of the subset of data greater than the median of all the values in the data set.
   3. The **interquartile range (IQR)** is calculated as the difference between the upper quartile and the lower quartile.
   4. An **outlier** is any value more than 1.5 IQR above the upper quartile or more than 1.5 IQR below the lower quartile.

*E. In a **normal curve,** 68% of the values are within one standard deviation of the mean, 95% are within two standard deviations of the mean, and 99.8% are within three standard deviations of the mean.

*F. A **z-score** gives the position of a score in relation to the remainder of the distribution, using the standard deviation as the unit of measure.

$$z = \frac{x - \bar{x}}{s}$$

*G. **Scattergrams** are graphs of ordered pairs that allow us to examine the relationship (correlation) between two sets of data.

CHAPTER TEST

1. ▶Suppose that you read that "the average family in Rattlesnake Gulch has 2.41 children." What average is being used to describe the data? Explain your answer. Suppose that the sentence had said 2.5; then what are the possibilities?

2. At Bug's Bar-B-Q restaurant, the average weekly wage for full-time workers is $150. If there are ten part-time employees whose average weekly salary is $50 and the total weekly payroll is $3950, how many full-time employees are there?

3. Find the mean, the median, and the mode for each of the following groups of data:
   (a) 10, 50, 30, 40, 10, 60, 10
   (b) 5, 8, 6, 3, 5, 4, 3, 6, 1, 9

4. Find the range, variance, and standard deviation for each set of scores in Problem 3.

5. The masses, in kilograms, of children in Ms. Rider's class follow:

40	49	43	48	46	42	439	39
42	41	42	39	41	40	45	43

   (a) Make a line plot for the data.
   (b) Make an ordered stem-and-leaf plot for the data.
   (c) Make a frequency table for the data.
   (d) Make a bar graph of the data.

6. The grades on a test for 30 students follow:

96	73	61	76	77	84
78	98	98	80	67	82
61	75	79	90	73	80
85	63	86	100	94	77
86	84	91	62	77	64

   (a) Make a grouped frequency table for these scores, using four classes and starting the first class at 61.
   (b) Draw a histogram of the grouped data.
   (c) Draw a frequency polygon of the data.

7. The budget for the Wegetem Crime Co. is $2,000,000. If $600,000 is spent on bribes, $400,000 is spent for legal fees, $300,000 for bail money, $300,000 for contracts, and $400,000 for public relations, draw a circle graph to indicate how the company spends its money.

8. ▶What, if anything, is wrong with the following bar graph?

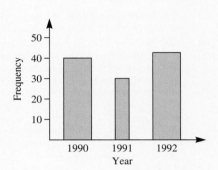

9. The mean salary of 24 people is $9000. How much will one additional salary of $80,000 increase the mean salary?

10. A cheetah can run 70 mph, a lion can run 50 mph, and a human can run 28 mph. Draw a bar graph to represent these data.

11. The life expectancies at birth for males and females are given in the following table.
    (a) Draw back-to-back ordered stem-and-leaf plots to compare the data.
    (b) Draw box plots to compare the data.

Year	Male	Female
1970	67.1	74.7
1971	67.4	75.0
1972	67.4	75.1
1973	67.6	75.3
1974	68.2	75.9
1975	68.8	76.6
1976	69.1	76.8
1977	69.5	77.2
1978	69.6	77.3
1979	70.0	77.8
1980	70.0	77.5
1981	70.4	77.8
1982	70.9	78.1
1983	71.0	78.1
1984	71.2	78.2
1985	71.2	78.2
1986	71.3	78.3
1987	71.5	78.4
1988	71.4	78.3
1989	71.8	78.5

12. Larry and Moe both took the same courses last quarter. Each had bet that he would receive the better grades. Their courses and grades are as follows:

Courses	Larry's Grades	Moe's Grades
Math (4 credits)	A	C
Chemistry (4 credits)	A	C
English (3 credits)	B	B
Psychology (3 credits)	C	A
Tennis (1 credit)	C	A

Moe claimed that the results constituted a tie, since both received 2 A's, 1 B, and 2 C's. Larry said that he won the bet because he had the higher grade-point average for the quarter. Who is correct? (Allow 4 points for an A, 3 points for a B, 2 points for a C, 1 point for a D, and 0 points for an F.)

13. Following are the lengths in yards of the 9 holes of the University Golf Course.

160	360	330
350	180	460
480	450	380

Find each of the following measures with respect to the lengths of the holes:
(a) Median (b) Mode
(c) Mean (d) Standard deviation

14. The speeds in miles per hour of 30 cars were checked by radar. The data are as follows:

62 67 69 72 75 60 58 86 74 68
56 67 82 88 90 54 67 65 64 68
74 65 58 75 67 65 66 64 45 64

(a) Find the median.
(b) Find the upper and lower quartiles.
(c) Draw a box plot for the data, and indicate outliers (if any exist) with asterisks.
(d) What percent of the scores is in the interquartile range?
(e) If every person driving faster than 70 mph received a ticket, what percentage of the drivers received speeding tickets?
(f) Is the median in the center of the box? Why or why not?

*15. The heights of 1000 girls at East High School were measured, and the mean was found to be 64 in., with a standard deviation of 2 in. If the heights are approximately normally distributed, about how many of the girls are
(a) Over 68 in. tall?
(b) Between 60 and 64 in. tall?
(c) If a girl is selected at random at East High School, what is the probability that she will be over 66 in. tall?

*16. A standardized test has a mean of 600 and a standard deviation of 75. If 1000 students took the test and their scores approximated a normal curve, how many scored between 600 and 750?

*17. If a student scored 725 on the test in Problem 16, what is his or her *z*-score?

*18. ▶Two different companies tested their products to determine the average (mean) life of their products. Company A had an average life of 150 hr with a standard deviation of 10 hr. Company B had an average life of 145 with a standard deviation of 2 hr. Which company would you rather buy from? Explain why.

*19. The scattergram given below was obtained from the girls trying out for the high school basketball team.

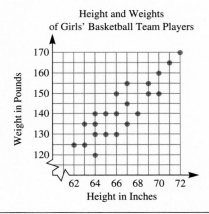

Height and Weights of Girls' Basketball Team Players

(a) What kind of correlation exists between the heights and weights that are listed?
(b) What is the weight of the girl who is 72 in. tall?
(c) How tall is the girl who weighs 145 lb?
(d) What is the mode of the heights?
(e) What is the range of the weights?

*20. ▶The Nielsen Television Index rating of 30 means that an estimated 30% of American televisions are tuned to the show with that rating. The ratings are based on the preferences of a scientifically selected sample of 1200 homes.

(a) Discuss possible ways in which viewers could bias this sample.
(b) How could networks attempt to bias the results?

*21. ▶List and give examples of several ways to misuse statistics graphically.

SELECTED BIBLIOGRAPHY

Barbella, P. "Realistic Examples in Elementary Statistics." *Mathematics Teacher* 80 (December 1987): 740–743.

Bryan E. "Exploring Data with Box Plots." *Mathematics Teacher* 81 (November 1988): 658–663.

Burrill, G. "Statistics and Probability." *Mathematics Teacher* 83 (February 1990): 113–118.

Corwin, R., and S. Friel. *Statistics: Prediction and Sampling.* Palo Alto, Calif: Dale Seymour Publishing, 1990.

Davis, G. "Using Data Analysis to Explore Class Enrollment." *Mathematics Teacher* 83 (February 1990): 104–106.

Fennell, F. "Ya Gotta Play to Win: A Probability and Statistics Unit for the Middle Grades." *Arithmetic Teacher* 31 (March 1983): 26–30.

Friel, S., and R. Corwin. "The Statistics Standards in K–8 Mathematics." *Arithmetic Teacher* 38 (October 1990): 35–39.

Goldman, P. "Teaching Arithmetic Averaging: An Activity Approach." *Arithmetic Teacher* 37 (March 1990): 38–43.

Huff, D. *How to Lie with Statistics.* New York: Norton, 1954.

Kelly, I., and J. Beamer. "Central Tendency and Dispersion: The Essential Union." *Mathematics Teacher* 79 (January 1986): 59–65.

Kimberling, C. "Mean, Standard Deviation, and Stopping the Stars." *Mathematics Teacher* 77 (November 1984): 633–636.

Landwehr, J., and A. Watkins. *Exploring Data.* Palo Alto, Calif.: Dale Seymour Publishing, 1987.

Landwehr, J., and A. Watkins. "Stem-and-Leaf Plots." *Mathematics Teacher* 78 (October 1985): 528–532, 537–538.

MacDonald, A. "A Stem-Leaf Plot: An Approach to Statistics." *Mathematics Teacher* 75 (January 1982): 25, 27, 28.

Mitchem, J. "Paradoxes in Averages." *Mathematics Teacher* 82 (April 1989): 250–253.

Mullenex, J. "Box Plots: Basic and Advanced." *Mathematics Teacher* 83 (February 1990): 108–112.

Olson, A. "Exploring Baseball Data." *Mathematics Teacher* 80 (October 1987): 565–569, 584.

Schaffer, R. "Why Data Analysis?" *Mathematics Teacher* 83 (February 1990): 90–93.

Scheinok, P. "A Summer Program in Probability and Statistics for Inner-City Seventh Graders." *Mathematics Teacher* 81 (April 1988): 310–314.

Schulte, A., ed. *Teaching Statistics and Probability*. Reston, Va.: National Council of Teachers of Mathematics, 1981.

Zawojewski, J. "Research into Practice: Teaching Statistics: Mean, Median, and Mode." *Arithmetic Teacher* 35 (March 1988): 25.

10 Introductory Geometry

Preliminary Problem

Mario was studying right angles and wondered if during his math class, the minute and hour hands of the clock formed a right angle. If his class meets from 2:00 to 3:00 P.M., determine whether a right angle is formed. If it is, figure out to the nearest minute when the hands form the right angle.

The word *geometry* comes from two Greek words, *ge* and *metria*, which might literally be translated as "earth measuring." The approach to geometry developed by Euclid of Alexandria in *The Elements* has been used for over 2000 yr as the basis of secondary-school geometry. Elementary-school geometry has not been as standardized as secondary-school geometry. According to the 5–8 *Standards* p. 113, *"Students should learn to use correct vocabulary, including common words like and, or, all, some, always, never, and if . . . then to reason, as well as words like parallel, perpendicular, and similar to describe."* The *Standards* (p. 113) further suggests that such vocabulary should be learned in an investigative manner. In this chaper, we consider some basic notions of geometry and some elementary network theory.

HISTORICAL NOTE

Euclid of Alexandria (ca. 300 B.C.) was a teacher at the Museum, a school formed by Ptolemy I of Egypt. Little is known of Euclid's life, although legend has it that he was a kindly gentleman who studied geometry for its beauty and logic. When asked by a student what use there was in the study of geometry, he requested that the student be given some money, "since he needs make gain of what he learns." Euclid is best known for *The Elements,* a work so systematic and encompassing that many earlier mathematical works were simply discarded and thus lost to all future generations. *The Elements,* composed of 13 books, included not only geometry but arithmetic and topics in algebra. American high school geometry classes cover material found in six of these books. Euclid started out with a set of statements that he assumed to be true and showed that geometric discoveries followed logically from his assumptions. In doing this, Euclid set up a *deductive system*.

Research suggests that children may learn geometry along the lines of a structure for reasoning developed by Dina and Pierre van Hiele of the Netherlands in the 1950s. Soviet educators became interested in the van Hiele research, and, as a result, changed their geometry curriculum. The *Standards* brought the van Hiele theory of phases of learning (see the Historical Note) closer to implementation in the United States by stressing the importance of sequential learning and an activity approach. The geometry chapters' Laboratory Activities demonstrate the van Hiele levels of learning. Developing geometry totally along the lines of the van Hiele structure is beyond the scope of this text.

HISTORICAL NOTE

Dina van Hiele-Geldof and Pierre van Hiele, high school teachers in the Netherlands, were disturbed by the way their students learned geometry. In their dissertations at the University of Utrecht, they devised a structure for helping students develop insight into reasoning in geometry. The following levels were developed by Dina, who died shortly after completing her dissertation; the levels were further developed and written about by Pierre, and they were modified by Alan Hoffer in 1981.

Level 0: Students recognize figures by their global appearance. They say such words as *triangle* and *square* but do not recognize properties of these figures.

Level 1: Students analyze component parts of figures but do not interrelate figures and properties. They may know that all sides of a square are congruent and that the diagonals of a rhombus are perpendicular bisectors of each other.

Level 2: Students may relate figures and their properties, but they do not organize sequences of statements to justify their observations. They may know that all squares are rhombuses but may not be able to state why in an organized way.

Level 3: Students at this level can reason deductively within the mathematical system to justify their observations.

Level 4: Students at this level can compare different axiom systems with a high degree of rigor, even without concrete models.

It is equally important to recognize the sequence of phases the van Hieles specified to help students move from one level of learning to another. These phases are as follows:

Phase 1. Inquiry: Teachers and students engage in dialogue about the topic to be studied. Use of vocabulary is extremely important at this stage.

Phase 2. Directed orientation: Teachers sequence activities to be explored by students in such a way that students become familiar with the structures involved.

Phase 3. Expliciting: With little help, students build on experiences and refine their vocabulary to discuss the relations of the structures.

Phase 4. Free orientation: Students encounter multistep tasks to be completed in different ways. They gain experience in resolving tasks on their own and make explicit many relations among the objects of the structures being studied.

Phase 5. Integration: Students are able to internalize and unify relations into a new body of thought. Teachers help by giving global surveys of what students already know.

Section 10-1 Basic Notions

The *Teaching Standards* (p. 137) point out:

Geometry should focus on intuitive "common-sense" investigations of geometric concepts in such a way that general properties emerge and are used as the basis for conjectures and

deductions. Later, observations and deductions can be studied more formally as part of a mathematical system.

The first "common-sense" notions that we discuss are the three building blocks of geometry: *points*, *lines*, and *planes*. We present these as *undefined* terms. The use of undefined terms is necessary in mathematics to avoid circular definitions. For example, the definition of *line* in one dictionary is "the path of a moving point." However, the definition of path is "a line of movement." Another example showing the frustration involved in starting out without some basic undefined notions is shown in the cartoon.

Reprinted by permission of Johnny Hart and Creators Syndicate, Inc.

An intuitive description of points, lines, and planes is given next. Other geometric concepts are then developed from these notions.

Points

point   A **point,** the basic unit in geometry, represents a location in space. It has no dimension. We use physical models such as those in Figure 10-1 to demonstrate points. Points are usually represented by dots and a capital letter as in the Figure 10-1(d).

FIGURE 10-1

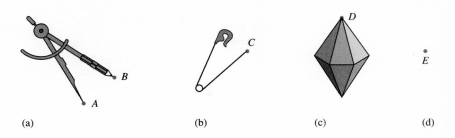

Lines

line A **line** consists of infinitely many points in a straight arrangement. A line has no thickness, and it extends forever in two directions. A "straight" highway centerline as in Figure 10-2(a) might be considered a line in the everyday world, but it could represent only a part of a line in mathematics. We represent a line as shown in Figure 10-2(c) where the arrowheads indicate the directions of the line. We name a line with either a single lower-case letter or by using two points on the line. The line in Figure 10-2(c) can be written as ℓ or $\overleftrightarrow{AB}$.

FIGURE 10-2

A line is determined by any two points. Thus points A and B are elements of $\overleftrightarrow{AB}$, and we write $A \in \overleftrightarrow{AB}$ and $B \in \overleftrightarrow{AB}$.

When discussing situations involving points and lines, we rely on such undefined relations as "contains," "belongs to," "is on," and "is between." In Figure 10-3, for example, line ℓ contains points A, B, and C but does not contain point D. Also, points A, B, and C belong to line ℓ, but point D does not. Similarly, points A, B, and C are on line ℓ, *collinear* or are **collinear**, but points B, C, and D are *noncollinear*. If three collinear points A, B, and *between* C are arranged as in Figure 10-3, B is **between** A and C, written A-B-C. Point D is not between B and C, because B, D, and C are noncollinear.

FIGURE 10-3

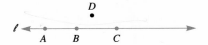

Using points and lines, we define other geometric figures. A point separates the line into two disjoint pieces called *half-lines* and the point itself. When the point is joined to either half-line, a *ray* is formed. These subsets of a line along with a *line segment* are defined in Table 10-1 with symbols shown below the appropriate illustration.

TABLE 10-1

Definition	Illustration
A **line segment**, or **segment**, is a subset of a line that contains two points of the line and all points between those two points.	$\overline{AB}$ or $\overline{BA}$
A **half-line** is a subset of a line that contains all points on the line on one side of a given point (excluding the point itself).	$\overset{\circ\!\!\longrightarrow}{AB}$
A **ray** is a subset of a line that contains a point and all points on the line on one side of the point.	$\overrightarrow{AB}$

REMARK $\overrightarrow{AB} \neq \overrightarrow{BA}$ because they do not have the same endpoint. They have only $\overline{AB}$ in common. Likewise, $\overset{\circ\!\!\longrightarrow}{AB} \neq \overset{\circ\!\!\longrightarrow}{BA}$. However, $\overline{AB} = \overline{BA}$ and $\overleftrightarrow{AB} = \overleftrightarrow{BA}$.

It is sometimes convenient to refer to other subsets of a line. A *half-open* (or *half-closed*) segment is a segment without one endpoint, symbolized, for example, by $\overset{\circ\!\!\rule{0.4cm}{0.4pt}}{AB}$; an open segment is a segment without either endpoint, symbolized, for example, by $\overset{\circ\!\!\rule{0.4cm}{0.4pt}\!\!\circ}{AB}$.

Planes

A tabletop, a floor, a ceiling, a wall, or any other smooth level surface is commonly thought of as a plane. However, each of these is only part of a plane because a plane extends endlessly in two dimensions. A **plane** is usually represented by a four-sided figure, as pictured in Figure 10-4(c), and it is commonly denoted by a lowercase Greek letter, such as alpha (α), beta (β), or gamma (γ) or by capital letters representing three noncollinear

FIGURE 10-4

(a)

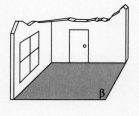
(b)

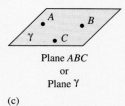
Plane *ABC*
or
Plane γ
(c)

points in the plane, such as *ABC*. Three noncollinear points uniquely determine a plane, just as two points uniquely determine a line. That is why $\overleftrightarrow{AB}$ is used for a line and *ABC* is used for a plane.

coplanar points

Points that belong to the same plane are **coplanar points.** Figure 10-5 shows coplanar points *D, G,* and *E; D, G, E,* and *F* are noncoplanar points, because no single plane contains all four points. Points are *noncoplanar* if, and only if, no one plane can contain them.

FIGURE 10-5

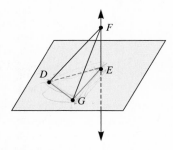

Similarly, we can define *coplanar* and *noncoplanar lines*. In Figure 10-5, lines $\overleftrightarrow{DE}$ and $\overleftrightarrow{GE}$ are coplanar, whereas lines $\overleftrightarrow{GF}$ and $\overleftrightarrow{DE}$ are noncoplanar.

Definitions and illustrations of various types of lines are given in Table 10-2. Exercises using these definitions are given in the problem set.

TABLE 10-2

Definition	Illustration
Two coplanar lines *m* and *n* are called **intersecting lines** if, and only if, they have exactly one point in common.	
Concurrent lines are lines that contain the same point. Concurrent lines may be either coplanar or noncoplanar.	
Two distinct coplanar lines *m* and *n* that have no points in common are called **parallel lines.**	
Two lines that cannot be contained in the same plane are called **skew lines.** $\overleftrightarrow{AB}$ and $\overleftrightarrow{CD}$ are skew lines.	

REMARK Two segments, two rays, or a ray and a segment are parallel if they lie on parallel lines. Skew lines cannot intersect. However, skew lines are not parallel because no single plane can contain them.

Other examples of some of the basic geometric figures just discussed are given in the student page from *Mathematics in Action,* Macmillan/McGraw-Hill, Grade 5, 1991.

Other Relations among Points, Lines, and Planes

The model of two walls intersecting in a line suggests that if two distinct planes have any points in common, then that set of points is a line, as in Figure 10-6(a). If we consider the spine of a book to be a line and each page to be a plane, as in Figure 10-6(b), we see that infinitely many planes can contain a given line. However, it is surprising to many students that three distinct planes may have a line in common, no common points, or just a single common point. In Figure 10-6(a), we see the walls and the floor (three distinct planes) intersecting at a single point, the corner. As we consider the drawings in Figure 10-6, other questions about the relations among points and lines may come to mind.

1. How many different lines in a plane can be drawn through two distinct points?
2. If lines were defined as lines of longitude on a globe, would the answer to Question 1 be different? Would it matter where the two points were located?
3. Why do surveyors and photographers include a tripod in their equipment?
4. What is the minimum number of points needed to determine a level surface or a plane?

FIGURE 10-6

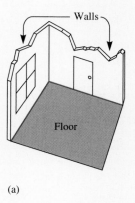

(a)

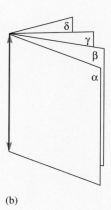

(b)

The set of properties on page 506 summarizes some common notions about points, lines, and planes. (These are sometimes referred to as axioms.)

DEVELOPING A CONCEPT
Geometry Around Us

Some basic geometric figures are suggested in this picture.

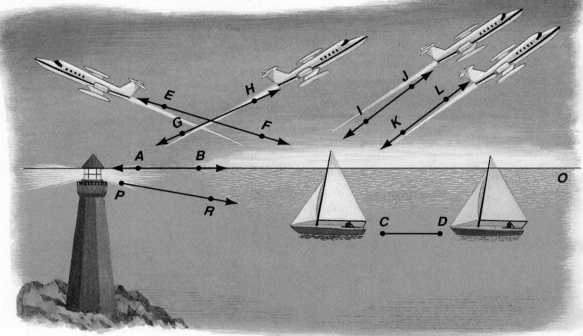

	Read	**Symbol**
The source of the light from the lighthouse suggests a point. A **point** is an exact location in space.	point P	P
A light beam coming from the lighthouse suggests a ray. A **ray** is part of a line.	ray PR	$\overrightarrow{PR}$
The horizon suggests a line. A **line** is made up of points and goes on and on in both directions.	line AB or line BA	$\overleftrightarrow{AB}$ or $\overleftrightarrow{BA}$
The distance between the ships suggests a line segment. A **line segment** is part of a line.	line segment CD or line segment DC	$\overline{CD}$ or $\overline{DC}$
The surface of the ocean suggests a plane. A **plane** is a flat surface that goes on and on in all directions.	plane O	plane O
The pair of jet trails at the top left suggest intersecting lines. **Intersecting lines** cross each other. The pair of jet trails at the top right suggest parallel lines. **Parallel lines** are lines in the same plane that never intersect.	$\overleftrightarrow{EF}$ intersects $\overleftrightarrow{GH}$ $\overleftrightarrow{IJ}$ is parallel to $\overleftrightarrow{KL}$ $\overleftrightarrow{IJ} \parallel \overleftrightarrow{KL}$	

PROPERTIES OF POINTS, LINES, AND PLANES

1. There is exactly one line that contains any two distinct points.
2. There is exactly one plane that contains any three distinct noncollinear points.
3. If two points lie in a plane, then the line containing the points lies in the plane.
4. If two distinct planes intersect, then their intersection is a line.

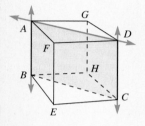

FIGURE 10-7

Property 2 gives one method of determining a plane, and implicit in this property are several other ways of determining a plane. Some of these are listed next and are illustrated in terms of the cube in Figure 10-7.

1. A line and a point not on the line determine a plane. (*Example:* $\overleftrightarrow{AB}$ and point C determine plane ABC.)
2. Two parallel lines determine a plane. (*Example:* $\overleftrightarrow{AB}$ and $\overleftrightarrow{CD}$ determine plane ABC.)
3. Two intersecting lines determine a plane. (*Example:* $\overleftrightarrow{AB}$ and $\overleftrightarrow{DA}$ determine plane ABD.)

REMARK Unless they are parallel, two nonintersecting lines do not determine a plane. For example, in Figure 10-7, $\overleftrightarrow{AB}$ and $\overleftrightarrow{GD}$ cannot be placed in the same plane. They are skew lines.

parallel planes

Two distinct planes either intersect in a line or are parallel. In Figure 10-8(a), the planes are **parallel** ($\alpha \| \beta$); that is, they have no points in common. Figure 10-8(b) shows two planes that intersect in $\overleftrightarrow{AB}$.

FIGURE 10-8

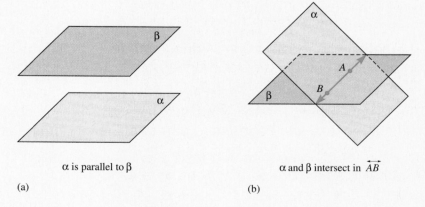

α is parallel to β

(a)

α and β intersect in $\overleftrightarrow{AB}$

(b)

A line and a plane can be related in one of three possible ways. If a line and a plane have no points in common, we define the line to be parallel to the plane, as in Figure 10-9(a). If two points of a line are in the plane, then the entire line containing the points is contained in the plane, as in Figure 10-9(b). If a line intersects a plane but is not contained in the plane, it intersects the plane at only one point, as in Figure 10-9(c).

FIGURE 10-9

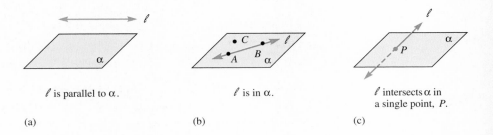

ℓ is parallel to α.

(a)

ℓ is in α.

(b)

ℓ intersects α in a single point, P.

(c)

Just as a point separates a line into two half-lines and the point itself, a line separates a plane into two half-planes and the line itself. In Figure 10-9(b), line ℓ separates plane ABC into two half-planes. A half-plane may be specified by the line determining the half-plane and one point in the half-plane, such as the half-plane determined by $\overleftrightarrow{AB}$ and containing point C in Figure 10-9(b). A line and the two half-planes determined by the line are three disjoint subsets of a plane. Points, lines, and planes are all subsets of space. In three dimensions, **space** is the set of all points. A plane separates space into two half-spaces. A plane and the two half-spaces determined by the plane are three disjoint subsets of space. The notion of a half-plane might be modeled by one side of a road on a plane. Can you think of a model for a half-space?

space

Angles

angle
sides
vertex

When two rays share a common endpoint, an **angle** is formed. This is illustrated in Figure 10-10(a). The rays of an angle are called the **sides** of the angle, and the common endpoint is called the **vertex** of the angle. An angle can be named by three different points: the vertex and a point on each ray, with the vertex always listed between the other two points. Thus the angle in Figure 10-10(a) may be named $\angle CBA$ or $\angle ABC$. The latter is read "angle ABC." When there is no risk of confusion, it is customary simply to name an angle by its vertex, by a number, or (occasionally) by a lowercase Greek letter. The angle in Figure 10-10(a) can therefore be named $\angle B$ or $\angle 1$. In Figure 10-10(b), however, more than one angle has vertex P, namely, $\angle QPR$, $\angle RPS$, and $\angle QPS$. Thus the notation $\angle P$ is inadequate for naming any one of the angles α, β, or $\angle QPS$.

In Figure 10-10(c), $\angle B$ separates the plane into three disjoint sets: the interior of the angle, the angle itself, and the exterior of the angle. Using the concept of the interior of an

FIGURE 10-10

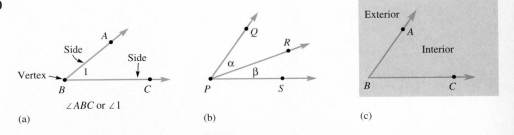

$\angle ABC$ or $\angle 1$

(a)

(b)

(c)

adjacent angles angle, we define adjacent angles, such as ∠QPR and ∠RPS in Figure 10-10(b), as follows: **Adjacent angles** are angles that share a common vertex and a common side and have nonoverlapping interiors.

Angle Measurement

degree

Angles are measured according to the amount of "opening" between two rays. A unit commonly used for measuring angles is the **degree.** A complete rotation about a point is an opening of 360°. One degree is then $\frac{1}{360}$ of a complete rotation. Figure 10-11 shows that ∠BAC has a measure of 30 degrees, written $m(\angle BAC) = 30°$. The measuring device pictured is a **protractor.** A degree is subdivided into 60 equal parts called **minutes,** and each minute is further subdivided into 60 equal parts called **seconds.** The measurement 29 degrees, 47 minutes, 13 seconds is written 29°47′13″.

protractor / minutes seconds

FIGURE 10-11

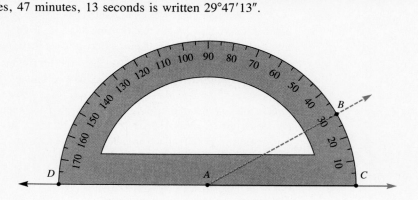

HISTORICAL NOTE

In 1634, Pierre Herigone used a symbol for an angle. It was not until 1923 that the Mathematical Association of America recommended ∠ as the standard symbol for angle in the United States. The use of 360° to measure angles seems to date to the Babylonian culture (4000–3000 B.C.), with minutes and seconds coming from Latin translations of Arabic translations of Babylonian sexagesimal (base 60) fractions.

EXAMPLE 10-1 (a) In Figure 10-12, find the measure of ∠BAC if $m(\angle 1) = 47°45′$ and $m(\angle 2) = 29°58′$.

(b) Express 47°45′ as a number of degrees.

Solution (a) $m(\angle BAC) = 47°45′ - 29°58′$
$= 46°(60 + 45)′ - 29°58′$
$= 46°105′ - 29°58′$
$= (46 - 29)° + (105 - 58)′$
$= 17°47′$

(b) $47°45′ = 47\frac{45°}{60} = 47.75°$

FIGURE 10-12

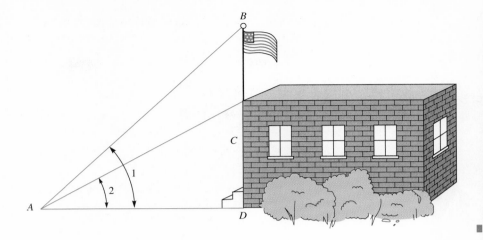

REMARK In the solution for Example 10-1(a), we used the fact that $m(\angle BAC) = m(\angle 1) - m(\angle 2)$. In general, if C is in the interior of $\angle BAD$, then $m(\angle BAD) = m(\angle BAC) + m(\angle DAC)$. This implies $m(\angle BAD) - m(\angle BAC) = m(\angle DAC)$.

Types of Angles

We can create different types of angles by paperfolding, especially with wax paper. Consider the folds shown in Figure 10-13(a) and (b). A piece of paper is folded in half and then reopened. If any point on the fold line labeled ℓ is chosen as the vertex, then the measure of the angle pictured is 180°. If the paper is refolded and folded once more, as in Figure 10-13(c), and then is reopened, as in Figure 10-13(d), four angles of the same size are created. Each angle has measure 90°.

FIGURE 10-13

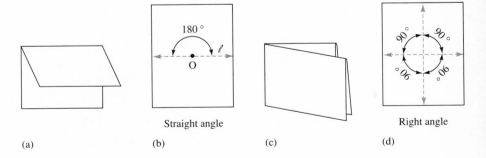

If the paper is folded as shown in Figure 10-14 and reopened, then angles α and β are formed, with measures that are less than 90° and greater than 90°, respectively. (Note that β has measure less than 180°.)

FIGURE 10-14

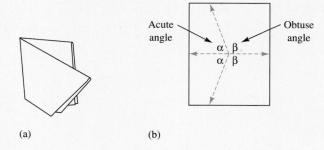

(a)   (b)

The different types of planar angles we have just discovered are shown in Figure 10-15, along with their definitions.

FIGURE 10-15

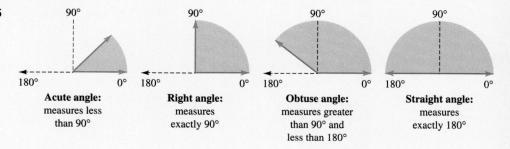

Acute angle: measures less than 90°

Right angle: measures exactly 90°

Obtuse angle: measures greater than 90° and less than 180°

Straight angle: measures exactly 180°

Perpendicular Lines

perpendicular lines

When two lines intersect so that the angles formed are right angles, as in Figure 10-16, the lines are **perpendicular lines.** In Figure 10-16, lines m and n are perpendicular, and we write $m \perp n$. (Also, in Figure 10-16, the symbol ⊐ is used to indicate a right angle.) Two intersecting segments, two intersecting rays, or a segment and a ray that intersect are called perpendicular if they lie on perpendicular lines. For example, in Figure 10-16, $\overline{AB} \perp \overline{BC}$, $\overrightarrow{BA} \perp \overrightarrow{BC}$, and $\overrightarrow{AB} \perp \overrightarrow{BC}$.

FIGURE 10-16

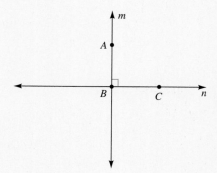

If a line and a plane intersect, they can be perpendicular. For example, consider Figure 10-17, where planes β and γ represent two walls intersecting along $\overleftrightarrow{AB}$. The edge $\overleftrightarrow{AB}$ is perpendicular to the floor. Also, every line in the plane of the floor (plane α) passing through point A is perpendicular to $\overleftrightarrow{AB}$. This discussion leads to the following definition.

FIGURE 10-17

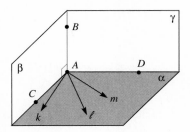

> **DEFINITION OF A LINE PERPENDICULAR TO A PLANE**
>
> **A line and a plane are perpendicular** if, and only if, they intersect and the line is perpendicular to every line in the plane that passes through the point of intersection.

> **REMARK** It is possible to prove that a line is perpendicular to a plane if, and only if, it is perpendicular to and concurrent with two intersecting lines forming the plane.

The plane containing one wall and the plane containing the floor of a typical room, such as α and β in Figure 10-17, are perpendicular planes. Notice that β and γ contain $\overleftrightarrow{AB}$, which is perpendicular to α. In fact, any plane containing $\overleftrightarrow{AB}$ is perpendicular to plane α. Thus β ⊥ α and γ ⊥ α.

> **DEFINITION OF PERPENDICULAR PLANES**
>
> **Two planes are perpendicular** if, and only if, one plane contains a line perpendicular to the other plane.

Dihedral Angles

dihedral angle

An alternate method for determining whether two planes are perpendicular is to determine the measure of an angle formed by the intersecting planes. If the angle formed by the planes is a right angle, then the planes are perpendicular. A **dihedral angle** is the union of two half-planes and the common line defining the half-planes. The half-planes are usually referred to as the *faces* of the dihedral angle, and the common line is referred to as the *edge* of the dihedral angle. In Figure 10-18, dihedral angle O-AC-D is formed by intersecting planes α and β. Note that point O is in plane α, $\overleftrightarrow{AC}$ is the edge of the dihedral angle, and point D is in plane β. In what other ways could the dihedral angle be named?

FIGURE 10-18

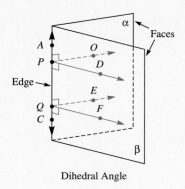

Dihedral Angle

With a given dihedral angle, we can associate a planar angle by choosing a point P on the edge of the dihedral angle and drawing two rays, one in each face, perpendicular to the edge. In Figure 10-18, $\angle OPD$ and $\angle EQF$ are planar angles associated with the dihedral angle. It can be shown that $\angle OPD$ and $\angle EQF$ are congruent and, in general, that all planar angles associated with a given dihedral angle are congruent. This property enables us to measure a dihedral angle. *We define the measure of a dihedral angle as the measure of any of the associated planar angles.*

EXAMPLE 10-2 Given Figure 10-19, answer each of the following:

(a) Name two pairs of skew lines.
(b) Are $\overleftrightarrow{BD}$ and $\overleftrightarrow{FH}$ parallel, skew, or intersecting lines?
(c) Are $\overleftrightarrow{BD}$ and $\overleftrightarrow{GH}$ parallel?
(d) Find the intersection of $\overleftrightarrow{BD}$ and plane EFG.
(e) Find the intersection of $\overleftrightarrow{BH}$ and plane DCG.
(f) Name two pairs of perpendicular planes.
(g) Name two lines that are perpendicular to plane EFH.
(h) Name a planar angle that could be used to measure dihedral angle E-FH-B.
(i) What is the measure of dihedral angle D-HG-F?

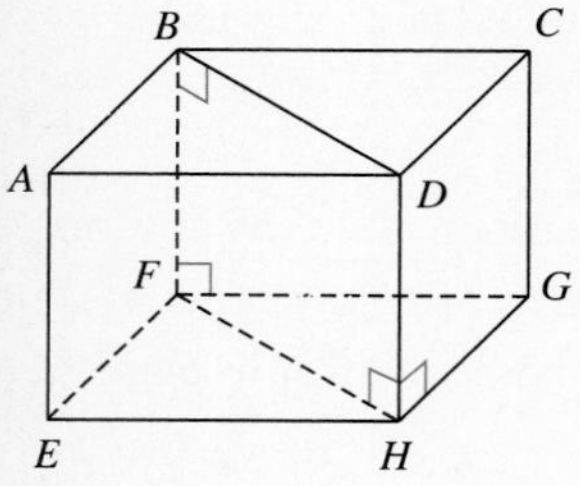

FIGURE 10-19

Solution
(a) $\overleftrightarrow{BC}$ and $\overleftrightarrow{DH}$, or $\overleftrightarrow{AE}$ and $\overleftrightarrow{BD}$. Others are possible.
(b) $\overleftrightarrow{BD}$ and $\overleftrightarrow{FH}$ are parallel.
(c) No, $\overleftrightarrow{BD}$ and $\overleftrightarrow{GH}$ are skew lines.
(d) The intersection is the empty set, because $\overleftrightarrow{BD}$ and plane EFG have no points in common.
(e) The intersection of $\overleftrightarrow{BH}$ and plane DCG is point H.
(f) Planes BFC and EFG are perpendicular, as are planes BFD and EFG.
(g) $\overleftrightarrow{BF}$ and $\overleftrightarrow{DH}$ are perpendicular to plane EFH.
(h) One possibility is $\angle EFB$.
(i) 90°

PROBLEM SET 10-1

1. Write each of the following in symbolic form:
   (a) Line AB
   (b) Line segment AB
   (c) Ray AB
   (d) Half-line AB
   (e) Line AB is parallel to line CD
   (f) Open line segment AB
   (g) Line AB is perpendicular to line CD.
   (h) The measure of angle ABC is 30 degrees.

2. A line, a line segment, or a ray can be considered as a set of points. Given the following figure, find a more concise name for each of the following:

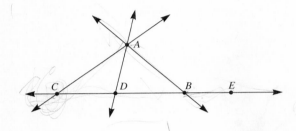

   (a) $\overleftrightarrow{AC} \cap \overleftrightarrow{BE}$
   (b) $\overleftrightarrow{AC} \cap \overline{BE}$
   (c) $\overrightarrow{CA} \cap \overrightarrow{EB}$
   (d) $\overrightarrow{CA} \cap \overrightarrow{BC}$
   (e) $\overline{CB} \cup \overline{BE}$
   (f) $\overrightarrow{AB} \cup \overrightarrow{AB}$
   (g) $\overrightarrow{AB} \cup \overrightarrow{BA}$
   (h) $\overrightarrow{AD} \cup \overrightarrow{DA}$

3. ▶Letters on a computer monitor are formed when a series of pixels (picture elements) are lighted. The more lights used, the more distinct the letter becomes. Consider the following magnified symbol for the number 1. Is the symbol a true geometric segment? Why or why not?

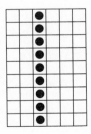

4. Use the accompanying drawing of one of the Great Pyramids of Egypt to find the following:
   (a) The intersection of $\overline{AD}$ and $\overline{CE}$
   (b) The dihedral angle formed by planes BDE and BDA
   (c) The intersection of planes ABC, ACE, and BCE
   (d) The intersection of $\overleftrightarrow{AD}$ and $\overleftrightarrow{CA}$
   (e) The intersection of $\overrightarrow{AD}$ and $\overrightarrow{CA}$
   (f) The intersection of $\overline{AD}$ and $\overline{CA}$
   (g) A pair of skew lines
   (h) A pair of parallel lines
   (i) A plane that is not determined by one of the triangular faces or by the base

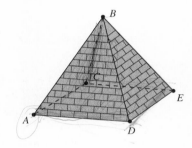

5. Determine whether each of the following is true or false:
   (a) Two distinct planes either intersect in a line or are parallel.
   (b) If two points are common to a line and to a plane, then the entire line is in the plane.
   (c) It is always possible to find a plane through four given points in space.
   (d) If two distinct lines do not intersect, they are parallel.
   (e) The intersection of three planes may be a single point.
   (f) If two distinct lines intersect, one, and only one, plane contains the lines.
   (g) Infinitely many planes contain both of two skew lines.
   (h) If each of two parallel lines is parallel to a plane α, then the plane determined by the two parallel lines is parallel to α.
   (i) If three points are coplanar, then they must be collinear.
   (j) If two distinct lines are parallel to a third line in space, then the two lines are parallel to each other.
   (k) If a plane α contains one line ℓ, but not another line m, and ℓ is parallel to m, then α is parallel to m.
   (l) A line parallel to each of two intersecting planes is parallel to the line of intersection of the planes.

6. For every false statement in Problem 5, give an example of a physical situation that demonstrates its falsity.

514 CHAPTER 10 INTRODUCTORY GEOMETRY

7. How many pairs of adjacent angles are in the following figure:

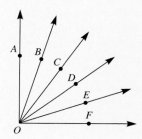

8. Identify a possible physical model for each of the following:
   (a) Perpendicular lines
   (b) An acute angle
   (c) An obtuse angle
   (d) An obtuse dihedral angle
   (e) An acute dihedral angle
   (f) A line containing two distinct points
   (g) Four noncoplanar points

9. ▶(a) If two parallel planes α and β intersect a third plane γ in two lines ℓ and m, are ℓ and m necessarily parallel? Explain your answer.
   ▶(b) If two planes α and β intersect a third plane in two parallel lines, are α and β always parallel? Why?
   ▶(c) Suppose that two intersecting lines are both parallel to a plane α. Is the plane determined by these intersecting lines parallel to α? Why?

10. Find the measure of each of the following angles:
    (a) ∠EAB
    (b) ∠EAD
    (c) ∠GAF
    (d) ∠CAF

11. Use a protractor to find the measure of each of the following pictured angles:

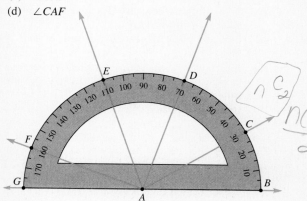

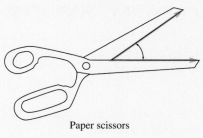

(a) Paper scissors

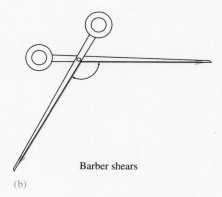

(b) Barber shears

12. (a) Perform each of the following operations, leaving your answers in simplest form:
    (i) 18°35'29" + 22°55'41"
    (ii) 93°38'14" − 13°49'27"
    (b) Express each of the following in degrees, minutes, and seconds, without decimals:
    (i) 0.9°
    (ii) 15.13°

13. How many rays are determined by each of the following?
    (a) Three collinear points
    (b) Four collinear points
    (c) Five collinear points
    (d) n collinear points

14. (a) How many lines are determined by three noncollinear points?
    (b) How many lines are determined by four points, no three of which are collinear?
    (c) How many lines are determined by five points, no three of which are collinear?
    (d) How many lines are determined by n points, no three of which are collinear?

15. (a) In the following table, sketch the possible intersections of the given number of lines. (Using dry spaghetti may help.) Two sketches are given for you.

	Number of Intersection Points						
Number of Lines		0	1	2	3	4	5 ...
	2		✕	Not possible	Not possible	Not possible	Not possible
	3					Not possible	Not possible
	4				※		A
	5						
	6						

(b) Find a formula for determining the greatest possible number of intersection points, given n lines.

16. ▶Explain mathematically why a three-legged stool is always stable and a four-legged stool sometimes rocks.

17. In the following figure, points A and B are in plane α:

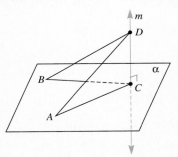

▶(a) Is $\angle BDC$ a right angle? Explain your answer.
▶(b) Is it possible to find a point P in plane α such that $\angle DPC$ is obtuse? Justify your answer.
▶(c) Is the plane determined by the points A, D, and C perpendicular to plane α? Why?

18. ▶(a) Is it possible for a line to be perpendicular to one line in a plane but not to be perpendicular to the plane? Explain.
▶(b) Is it possible for a line to be perpendicular to two distinct lines in a plane and yet not be perpendicular to the plane? Explain.
▶(c) If a line not in a given plane is perpendicular to two distinct lines in the plane, is the line necessarily perpendicular to the plane? Explain.

19. Trace each of the following drawings. In your tracing, use dashed lines for segments that would not be seen, and use solid lines for segments that would be seen. (Different people may see different perspectives.)

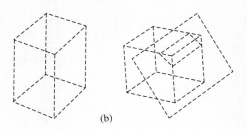
(a)   (b)

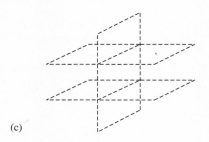
(c)

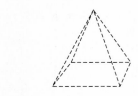
(d)

20. Into how many regions is the plane separated by each of the following. (Do not count the lines.)
(a) Two parallel lines
(b) Two intersecting lines
(c) Three parallel lines
(d) Three concurrent lines

21. Given the three sets below, list all regions that must be empty.

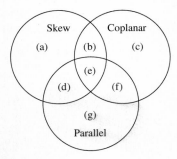

22. ▶Explain why $\overrightarrow{AB}$ and $\overleftarrow{BA}$ do not represent the same figure.

23. ▶(a) Can two skew lines be perpendicular? Explain why or why not.
▶(b) Can the measures of two obtuse angles sum to 180°? Explain why or why not.

▶(c) Can the measures of two acute angles sum to 180°? Explain why or why not.

24. ▶A student asks if there are such things as skew planes. How do you respond?

25. ▶A student claims that if line segments $\overline{AB}$ and $\overline{CD}$ are in the same plane and do not intersect, then they are parallel. How do you respond?

26. ▶Maggie claims that to make the measure of an angle greater, you just extend the rays further. How do you respond?

27. ▶Henry claims that a line segment has a finite number of points because it has two endpoints. How do you respond?

28. ▶Is it possible to locate four points in a plane such that the number of lines determined by the points is neither 1, 4, nor 6? Explain.

★29. Prove that if two parallel planes are intersected by a third plane, the lines of intersection are parallel.

30. Write Logo procedures to draw each of the following:
    (a) A procedure called ANGLE with input :SIZE to draw a variable-sized angle
    (b) A procedure called SEGMENT with input :LENGTH to draw a variable-sized segment
    (c) A procedure called PERPENDICULAR with inputs :LENGTH1 and :LENGTH2 to draw two variable-sized perpendicular segments
    (d) A procedure called PARALLEL with inputs :LENGTH 1 and :LENGTH2 to draw two variable-sized parallel segments

▼ BRAIN TEASER

In a connect-the-dots drawing of points around a circle, as shown, Eli decided to draw all possible segments connecting the 25 distinct points. How many segments did he draw?

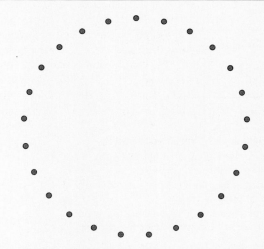

LABORATORY ACTIVITY

The following activities involve Level 0 of the van Hiele structure for learning geometry.

1. Within the classroom, identify a physical object with the following shapes:
   (a) Parallel lines   (b) Parallel planes
   (c) Skew lines   (d) Dihedral angle
   (e) Right angles

> **2.** On a sheet of dot paper or on a geoboard, as pictured, create the following shapes:
>
>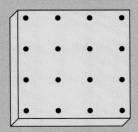
>
> (a) Right angle   (b) Acute angle
> (c) Obtuse angle   (d) Adjacent angles
> (e) Parallel lines (f) Intersecting lines

Section 10-2 Polygonal Curves

Suppose we were to take a pencil and draw a path on a piece of paper without lifting the pencil and without retracing any part of the path except for single points. The fact that we are drawing on a sheet of paper restricts us to a plane and the fact that we do not lift the pencil implies that we have no breaks in our drawing, or that the curve is "connected." Although each curve in Figure 10-20 is in the plane and is connected, there are some obvious differences. For example, the curves in (b), (c), (f), (g), (h), (i), (j), and (k) could be drawn by starting and stopping at the same point. These are called **closed curves.** The closed curves in (b) and (f) cross themselves whereas those in (c), (g), (h), (i), (j), and (k) do not. A curve is **simple** if it does not cross itself. The curves in (c), (g), (h), (i), (j), and (k) are simple, closed curves. The curves in (d), (e), (f), (g), (h), and (j) are similar in that

closed curves

simple curves

FIGURE 10-20

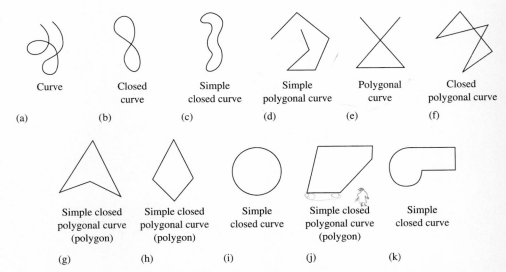

polygonal curves
polygons

they are made up entirely of line segments. These are called **polygonal curves**. **Polygons** are polygonal curves that are both simple and closed, as in Figure 10-20 (g), (h), and (j).

> **REMARK** Not all mathematicians agree on how a curve should be defined. Our examples imply that curves have a beginning and an endpoint, and we will make this assumption. An alternate definition might not require this assumption, and so lines, rays, open segments, and so on would also be considered curves.

The line segments forming a polygon are the *sides* of the polygon. A point where two sides meet is a *vertex* of the polygon. Every simple closed polygon separates the plane into three disjoint subsets: the interior of the polygon, the exterior of the polygon, and the polygon itself. This is illustrated in Figure 10-21(b). Together, a polygon and its interior form a **polygonal region.**

polygonal region

FIGURE 10-21

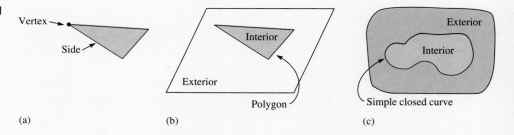

Jordan Curve Theorem

Figure 10-21(b) shows a special case of the **Jordan Curve Theorem,** which states that *any simple closed curve separates a plane into three disjoint sets: the exterior, the interior, and the curve itself.* Another example is shown in Figure 10-21(c). Deciding whether a point is inside or outside a curve is investigated in Problem 1.

PROBLEM 1

Determine whether point X is inside or outside the simple closed curve of Figure 10-22.

FIGURE 10-22

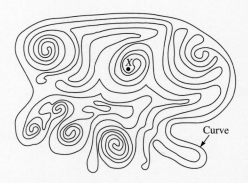

Understanding the Problem. We must determine whether point X in Figure 10-22 belongs to the exterior or to the interior of the curve.

Devising a Plan. One approach to this problem is to start shading the area surrounding point X. If we stay between the lines, we should be able to decide whether the shaded area is inside or outside the curve.

Carrying Out the Plan. If we shade the region around point X as described, we obtain Figure 10-23. The shaded part of Figure 10-23 indicates that point X is located outside the curve.

FIGURE 10-23

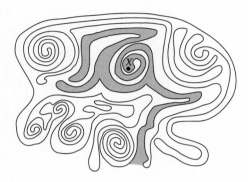

Looking Back. Another strategy is explored in Figure 10-24(a) and (b), in which point X is connected with any point Y, where point Y is definitely outside the curve. How many times does the dashed segment cross the curve in each of these cases? Try different locations for point Y outside the curve in Figure 10-24(a) and (b). What is your conjecture? ∎

FIGURE 10-24

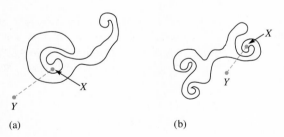

(a)   (b)

More about Polygons

How are the polygons in Figure 10-25 different? Intuitively, we might say that the polygon in (a) has no indentations whereas the one in Figure 10-25(b) is indented. Mathematically, if the segment connecting any two points of the polygonal region is a subset of the polygonal region, then the polygon is a **convex polygon.** For example, in Figure 10-25(a), no matter what two points of the hexagonal region are chosen as the endpoints of a segment, the entire segment lies in the hexagonal region. This is not the case in Figure 10-25(b), where it is possible to draw a segment between two points of the polygonal region such that part of the segment lies outside the region. Figure 10-25(b) is called a *concave polygon*.

convex polygon

FIGURE 10-25

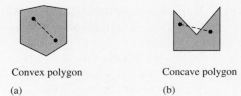

Convex polygon
(a)

Concave polygon
(b)

Polygons are classified according to the number of sides or vertices they have. For example, consider the polygons listed in Table 10-3.

TABLE 10-3

Polygon	Number of Sides or Vertices
Triangle	3
Quadrilateral	4
Pentagon	5
Hexagon	6
Heptagon	7
Octagon	8
Nonagon	9
Decagon	10
n-gon	n

interior angle, or angle, of a polygon
exterior angle of a polygon

Polygons are referred to by listing the capital letters representing consecutive vertices. For example, the polygon in Figure 10-26(a) can be referred to as *ABCD* or *CDAB*. The vertices *A* and *B* are said to be *consecutive* vertices, whereas vertices such as *A* and *C* are *nonconsecutive* vertices. Any two sides of a polygon having a common vertex determine an **interior angle,** or **angle, of the polygon,** such as ∠1 of polygon *ABCD* in Figure 10-26(a). An **exterior angle of a polygon** is determined by a side of the polygon and the extension of a contiguous side of the polygon. An example is ∠2 in Figure 10-26(b).

FIGURE 10-26

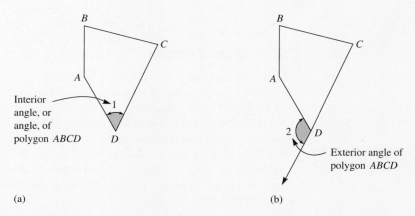

(a) (b)

diagonal   Any line segment connecting nonconsecutive vertices of a polygon is a **diagonal.** In Figure 10-27(a), segments $\overline{AC}$, $\overline{AD}$, $\overline{BE}$, $\overline{BD}$, and $\overline{CE}$ are diagonals of the given pentagon. In Figure 10-27(b), segments $\overline{QS}$ and $\overline{PR}$ are the diagonals of the quadrilateral *PQRS*. In Figure 10-27(a), the diagonals (except for their endpoints) of the pentagon lie in the interior of the pentagon. In contrast, the quadrilateral in Figure 10-27(b) has a diagonal that, except for its endpoints, lies in the exterior of the quadrilateral. What type of polygon must have a diagonal such that a part of the diagonal falls outside the polygon?

FIGURE 10-27

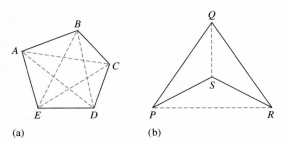

String art is often constructed on the basis of polygons and their diagonals. Problem 2 investigates the number of diagonals in such a polygon.

PROBLEM 2

How many diagonals does the 24-gon pictured in Figure 10-28 have?

FIGURE 10-28

Understanding the Problem. We have a polygon with 24 sides and are to determine how many different diagonals can be drawn.

 Devising a Plan. We use the strategy of *examining related simpler cases* of the problem in order to develop a pattern for the original problem. Figure 10-29 shows that a triangle has no diagonals, a square has two diagonals, a pentagon has five diagonals, and a hexagon has nine diagonals.

FIGURE 10-29

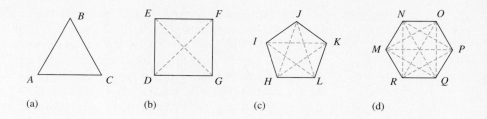

(a) (b) (c) (d)

Examining Figure 10-29(d), we see that from any vertex, we can draw only three diagonals from that vertex. In general, we cannot draw a diagonal from a chosen vertex to itself or to either of the two adjacent vertices. In Figure 10-29(d), the number of diagonals that can be drawn from any one vertex is three less than the total number of vertices, that is, $6 - 3$, or 3. Similarly, in a polygon with 24 sides, the number of diagonals that can be drawn from any one vertex is three less than the number of vertices, that is, $24 - 3$, or 21. Because 21 diagonals emanate from each of the 24 vertices, we should be able to use this information to determine the total number of diagonals.

Carrying Out the Plan. It might appear that we have 24(21), or 504, diagonals in a 24-gon. However, based on this notion, we should also have $6(6 - 3)$, or 18, diagonals in a hexagon. This result does not agree with the actual number of 9. This is because each diagonal is determined by two vertices and when we counted the number of diagonals from each vertex, we counted each diagonal twice. Hence, in a 24-gon, there must be 24(21)/2, or 252, diagonals.

Looking Back. Based on this reasoning, the number of diagonals in an n-gon is $n(n - 3)/2$. This formula gives results consistent with the number of diagonals pictured in Figure 10-29. An alternate solution to this problem uses the notion of combinations developed in Chapter 8. The number of ways that all the vertices in an n-gon can be connected two at a time is the number of combinations of n vertices chosen two at a time, that is, $_nC_2$, or $\dfrac{n(n-1)}{2}$. This number of segments includes both the number of diagonals and the number of sides. If we subtract the number of sides n from $\dfrac{n(n-1)}{2}$, the number of diagonals is $\dfrac{n(n-1)}{2} - n$. This expression can be simplified to $n(n-3)/2$. ∎

Congruent Segments and Angles

congruent parts   Most modern industries operate on the notion of creating **congruent parts,** parts that are of the same size and shape. For example, the specifications for all cars of a particular model are the same, and all parts produced for that model are basically "the same." Most frequently, when we discuss congruent figures, we are discussing figures in a plane. For
congruent segments   example, two line **segments** are described as **congruent** if and only if they have the same length. (Length is discussed in Chapter 13.) In the case of line segments, having the same length means that a tracing of one line segment can be fitted exactly on top of the other. If $\overline{AB}$ is congruent to $\overline{CD}$, we write $\overline{AB} \cong \overline{CD}$. The symbol $\cong$ is read "is congruent to." In a
congruent angles   similar manner, we say that two **angles** are **congruent** to each other if, and only if, they have the same measure. Congruent segments and congruent angles are shown in Figure 10-30(a) and (b), respectively.

SECTION 10-2 POLYGONAL CURVES 523

FIGURE 10-30

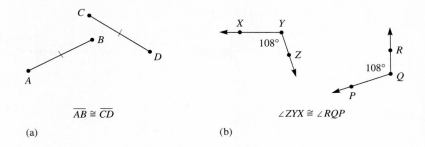

(a) $\overline{AB} \cong \overline{CD}$

(b) $\angle ZYX \cong \angle RQP$

Regular Polygons

regular polygons

Polygons in which all the angles are congruent and all the sides are congruent are called **regular polygons.** We say that a regular polygon is both *equiangular* and *equilateral*. A regular triangle is an equilateral triangle. A regular pentagon and a regular hexagon are illustrated in Figure 10-31. The congruent sides and congruent angles are marked.

FIGURE 10-31

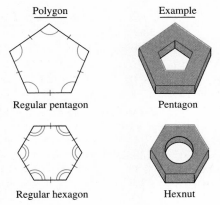

Polygon — Regular pentagon / Regular hexagon

Example — Pentagon / Hexnut

Triangles and Quadrilaterals

Triangles may be classified according to their angle measures, as shown in Table 10-4. Triangles and quadrilaterals may also be classified as shown in Table 10-5.

TABLE 10-4

Definition	Illustration	Example
A triangle containing a right angle is a **right triangle.**	right triangle	antenna tower
A triangle in which all the angles are acute is an **acute triangle.**	acute triangle	YIELD sign
A triangle containing an obtuse angle is an **obtuse triangle.**	obtuse triangle	jet aircraft

TABLE 10-5

Definition	Illustration	Example
A triangle with no sides congruent is a **scalene triangle**.		
A triangle with at least two sides congruent is an **isosceles triangle**.		
A triangle with three sides congruent is an **equilateral triangle**.		
A **trapezoid** is a quadrilateral with at least one pair of parallel sides.		
A **kite** is a quadrilateral with two distinct pairs of consecutive sides congruent.		
An **isosceles trapezoid** is a trapezoid with a pair of base angles congruent.		
A **parallelogram** is a quadrilateral in which each pair of opposite sides is parallel.		
A **rectangle** is a parallelogram with a right angle.		
A **rhombus** is a parallelogram with all sides congruent.		
A **square** is a rectangle with all sides congruent.		

Some texts give different definitions for a trapezoid and an isosceles trapezoid. Many elementary texts define a trapezoid as a quadrilateral with *exactly* one pair of parallel sides. Note the definition of a trapezoid on the student page from *Addison-Wesley Mathematics*, Grade 4, 1991. The student page makes use of an excellent teaching aid called a *tangram*. Try to work through the explorations.

Classifying Quadrilaterals

LEARN ABOUT IT

EXPLORE Use a Tangram Puzzle

- How many quadrilaterals of different shapes can you make using any combination of pieces A and B? Draw each one.
- How many different quadrilaterals can you make with pieces C, D, and E? Draw each one.

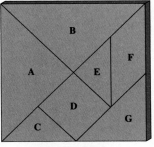

Tangram Puzzle

TALK ABOUT IT

1. Which of your quadrilaterals have at least one right angle?
2. Which have one pair of parallel sides?
3. Which have two pairs of sides that are the same length?
4. Which have all sides the same length?

Here are some types of quadrilaterals.

Square

All sides the same length
All angles right angles

Rectangle

Two pairs of same-length sides
All angles right angles

Trapezoid

Exactly one pair of parallel sides

Parallelogram

Two pairs of same-length sides
Two pairs of parallel sides

Notice from these definitions that every triangle is a polygon and every equilateral triangle is also isosceles. However, not every isosceles triangle is equilateral. Using set concepts, we can say that the set of all triangles is a proper subset of the set of all polygons; also, the set of all equilateral triangles is a proper subset of the set of all isosceles triangles. This hierarchy is shown in Figure 10-32, where more general terms appear above more specific ones.

FIGURE 10-32

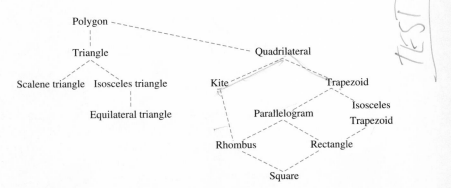

Using the definitions in Tables 10-4 and 10-5, we can prove many properties of both triangles and quadrilaterals. Among these properties are the following:

1. An equilateral triangle is isosceles.
2. A square is a regular quadrilateral.
3. A square is a rhombus with a right angle.
4. A rectangle is an isosceles trapezoid.
5. Some isosceles trapezoids are kites.

Circles

circle
center

In Figure 10-33, the regular 24-gon resembles a circle. The more sides a regular n-gon has, the closer it resembles a circle. (This type of thinking has led to a procedure in Logo for drawing a turtle-type circle.) A **circle** is defined as the set of all points in a plane that lie the same distance from a given point, called the **center**. Circles are simple closed curves and are discussed in more detail in Chapter 11.

FIGURE 10-33

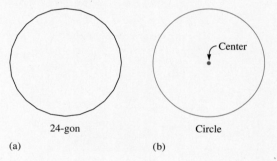

(a) 24-gon   (b) Circle

PROBLEM SET 10-2

1. For each of the following, which figures labeled (1)–(12) can be classified under the given term?
   (a) Polygonal curve
   (b) Simple polygonal curve
   (c) Closed polygonal curve
   (d) Polygon
   (e) Convex polygon
   (f) Concave polygon

(9)   (10)   (11)   (12)

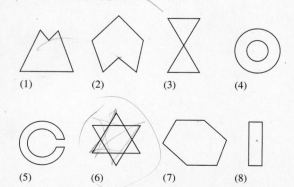

(1) (2) (3) (4) (5) (6) (7) (8)

2. Which of the printed capital letters of the English alphabet are simple, closed curves?

3. In each of the following diagrams, determine whether point X is inside or outside the curve:

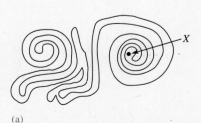

(a)

(b)

4. ▶For each of the following figures, determine if it is possible to connect like numerals by curves that do not cross each other or any other curves in the figure. Explain your answer.

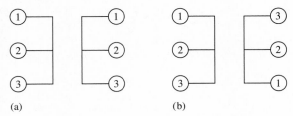

(a)   (b)

5. What is the maximum number of intersection points between a quadrilateral and a triangle (where no sides of the polygons are on the same line)?

6. Which of the following figures are convex, and which are concave?

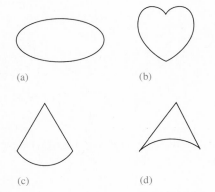

7. On a geoboard, construct each of the following:
   (a) A scalene triangle   (b) A square
   (c) A trapezoid   (d) A convex hexagon
   (e) A concave quadrilateral (f) A parallelogram

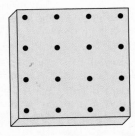

8. If possible, draw the following triangles. If it is not possible to do so, state why.
   (a) An obtuse scalene triangle
   (b) An acute scalene triangle
   (c) A right scalene triangle
   (d) An obtuse equilateral triangle
   (e) A right equilateral triangle
   (f) An obtuse isosceles triangle
   (g) An acute isosceles triangle
   (h) A right isosceles triangle

9. How many diagonals does each of the following have?
   (a) Decagon
   (b) 20-gon
   (c) 100-gon

10. Identify each of the following triangles as scalene, isosceles, or equilateral:

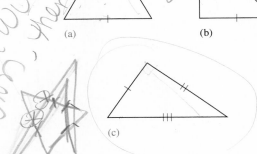

11. ▶Tell whether each of the following is true or false. If the statement is false, explain why.
    (a) Every isosceles triangle is equilateral.
    (b) All equilateral triangles are isosceles.
    (c) All squares are rectangles.
    (d) Some rectangles are rhombuses.
    (e) All parallelograms are quadrilaterals.
    (f) Every rhombus is a regular quadrilateral.
    (g) Every parallelogram is a trapezoid.
    (h) Every equilateral triangle is a scalene triangle.
    (i) Every square is a kite.
    (j) Some rectangles are squares.
    (k) No square is a rectangle.
    (l) No trapezoid is a parallelogram.
    (m) Some right triangles are isosceles.
    (n) An isosceles trapezoid may not be a kite.
    (o) No parallelogram is an isosceles trapezoid.

12. ▶Jodi identifies the following figure in (a) as a rectangle and the figure in (b) as a square. She claims that the figure

in (b) is not a rectangle because it is a square. How do you respond?

(a)   (b)

13. ▶Millie claims that a rhombus is regular because all the sides are congruent. How do you respond?
14. Describe regions (a) and (b) in the following Venn diagram:

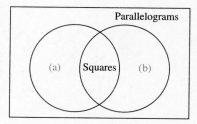

15. Use the labeled points in the drawing below to answer the following:

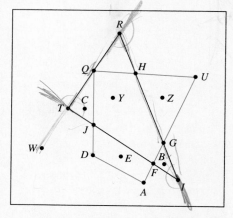

(a) Which points belong to triangle *TRI*?
(b) Which points belong to the interior of the quadrilateral *QUAD*?
(c) Which points belong to the exterior of triangle *TRI*?
(d) Which points belong to triangle *TRI* and quadrilateral *QUAD*?
(e) Which points belong to the intersection of the interiors of triangle *TRI* and quadrilateral *QUAD*?

16. If three toothpicks are used, an equilateral triangle can be formed (recorded as a 1-1-1 triangle). No triangle can be formed using four toothpicks. If seven toothpicks are used, two isosceles triangles (as shown below) can be formed. Complete the following table if toothpicks must be placed end to end in the same plane:

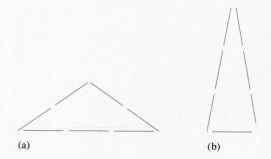

(a)   (b)

Number of Toothpicks	Possible Triangles	Type of Triangle
3	1-1-1	Equilateral
4	none	N.A.
5	2-2-1	Isosceles
6	2-2-2	Equilateral
7	3-2-2 or 3-3-1	Isosceles
8		
9		
10		
11		
12		

17. Write Logo procedures to draw each of the following:
    (a) A simple polygonal curve
    (b) A closed polygonal curve
    (c) A nonsimple nonclosed polygonal curve
    (d) A simple closed polygonal curve

18. Write a Logo program to draw each of the following:
    (a) A square
    (b) A rectangle

Review Problems

19. If three distinct rays with the same vertex are drawn as shown next, then three different angles are formed: $\angle AOB$, $\angle AOC$, and $\angle BOC$.
    (a) How many different angles are formed by using 10 distinct noncollinear rays with the same vertex?
    (b) How many different angles are formed by using n distinct noncollinear rays with the same vertex?

SECTION 10-2   EXERCISES   529

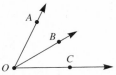

20. What are the possible intersection sets of a line and an angle?
21. Find each of the following in the given figure:

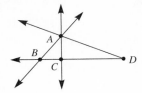

(a) $\overleftrightarrow{AC} \cap \overline{BD}$
(b) $\overline{BD} \cup \overline{CD}$
(c) Three line segments containing point A
(d) $\overleftrightarrow{DC} \cap \overrightarrow{DA}$

22. Classify the following as true or false. If false, tell why.
   (a) A ray has two endpoints.
   (b) For any points M and N, $\overleftrightarrow{MN} = \overleftrightarrow{NM}$.
   (c) Skew lines are coplanar.
   (d) $\overrightarrow{MN} = \overrightarrow{NM}$
   (e) A line segment contains an infinite number of points.
   (f) If two distinct planes intersect, their intersection is a line segment.

LABORATORY ACTIVITY

1. The following van Hiele Level 1 activity consists of using cutouts of different quadrilaterals. Sort the shapes according to the following attributes:
   (a) Number of parallel sides
   (b) Number of right angles
   (c) Number of congruent sides
   (d) Polygons with congruent diagonals

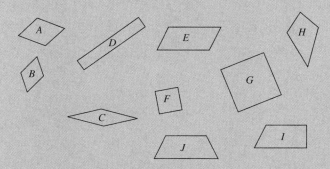

2. Use the cutouts to identify properties characteristic of different classes of figures. For example, "Congruent opposite sides describe a parallelogram."

3. For the following van Hiele Level 2 activity, work with a partner. One of you should construct a figure on a geoboard. Do not show the figure to your partner, but tell your partner the properties of the figure you constructed. Have your partner try to identify the figure you constructed. Try this with each of the figures named.
   (a) Scalene triangle   (b) Isosceles triangle   (c) Square
   (d) Parallelogram   (e) Trapezoid   (f) Rectangle
   (g) Kite   (h) Rhombus

> **BRAIN TEASER**
>
> Given three buildings A, B, and C, as shown, and three utility centers for electricity (E), gas (G), and water (W), is it possible to connect each of the three buildings to each of the three utility centers without crossing lines?
>
>

Section 10-3 More about Angles

Two intersecting lines form four nonstraight angles in the plane. In Figure 10-34, the four angles are ∠1, ∠2, ∠3, and ∠4.

FIGURE 10-34

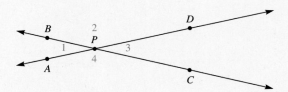

vertical angles
supplementary angles

Angles formed by two intersecting lines, such as ∠1 and ∠3 in Figure 10-34, are **vertical angles.** Another pair of vertical angles in Figure 10-34 is ∠2 and ∠4.

Figure 10-34 also illustrates pairs of supplementary angles. Two angles are **supplementary angles** if the sum of their measures is 180°. Each is a *supplement* of the other. Angles 1 and 2, 1 and 4, 2 and 3, and 3 and 4 are pairs of supplementary angles in Figure 10-34.

complementary angles

Two angles are **complementary angles** if the sum of their measures is 90°. Each is a *complement* of the other. Figure 10-35 shows examples of supplementary and complementary angles.

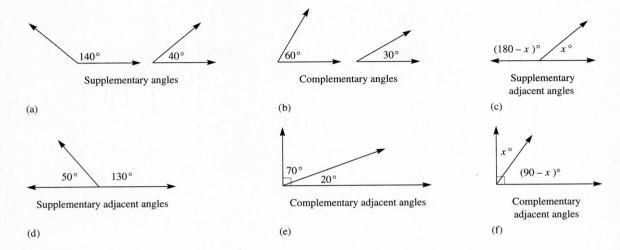

FIGURE 10-35

Using the notion of congruence of angles, we can derive the following theorems involving supplementary and complementary angles.

THEOREM 10-1

(a) Supplements of the same angle, or of congruent angles, are congruent.
(b) Complements of the same angle, or of congruent angles, are congruent.

Using Theorem 10-1(a), we can easily prove that vertical angles are congruent. Look at Figure 10-36.

FIGURE 10-36

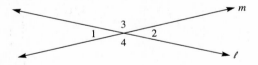

Because ℓ is a straight line, $\angle 1$ is a supplement of $\angle 4$. Because m is a straight line, $\angle 2$ is a supplement of $\angle 4$. As $\angle 1$ and $\angle 2$ are the supplements of the same angle, $\angle 4$, they are congruent and equal in measure. Similarly, $\angle 3$ and $\angle 4$ are supplements of $\angle 1$ and therefore are congruent. Thus vertical angles are congruent. We summarize the preceding result in the following theorem.

THEOREM 10-2

Vertical angles are congruent.

Theorem 10-1(b) can also be used to deduce congruence relationships among certain angles, as shown in Example 10-3.

EXAMPLE 10-3 In Figure 10-37, suppose ∠APC and ∠BPD are right angles. Why are ∠1 and ∠3 congruent?

FIGURE 10-37

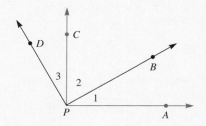

Solution Because ∠APC is a right angle, ∠1 is a complement of ∠2. Because ∠BPD is a right angle, ∠3 is also a complement of ∠2. Thus ∠1 and ∠3 are complements of the same angle and hence are congruent. ∎

transversal

Vertical angles are formed by two intersecting lines. Angles are also formed when a line intersects two distinct lines. Any line that intersects a pair of lines is called a **transversal** of those lines. In Figure 10-38(a), line p is a transversal of lines m and n. Angles formed by these lines are named according to their placement in relation to the transversal and the two given lines. Various types of angles, together with examples of each in Figure 10-38, are listed next.

FIGURE 10-38

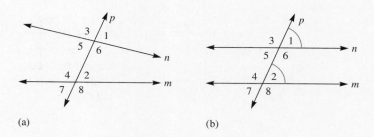

(a)   (b)

interior angles **Interior angles:** ∠2, ∠4, ∠5, ∠6
exterior angles **Exterior angles:** ∠3, ∠1, ∠7, ∠8
alternate interior angles **Alternate interior angles:** ∠5 and ∠2, ∠4 and ∠6
alternate exterior angles **Alternate exterior angles:** ∠1 and ∠7, ∠3 and ∠8
corresponding angles **Corresponding angles:** ∠3 and ∠4, ∠5 and ∠7, ∠1 and ∠2, ∠6 and ∠8

If corresponding angles, such as ∠1 and ∠2, are congruent, as in Figure 10-38(b), it can be shown that each pair of corresponding angles, alternate interior angles, and alternate exterior angles are congruent.

If we further examine Figure 10-38(b), we see that lines m and n appear to be parallel when ∠1 is congruent to ∠2. Conversely, if the lines are parallel, the sets of angles mentioned previously are congruent. The preceding discussion leads to the following theorem, which we state without proof.

THEOREM 10-3

If any two distinct lines are cut by a transversal, then a pair of corresponding angles, alternate interior angles, or alternate exterior angles are congruent if, and only if, the lines are parallel.

The Sum of the Measures of the Angles of a Triangle

The sum of the measures of the angles in a particular triangle can intuitively be shown to be 180° by using a torn triangle, as in Figure 10-39. Angles 1, 2, and 3 of triangle *ABC* in Figure 10-39(a) are torn as pictured and then placed along line ℓ, as shown in Figure 10-39(b). Because the sum of the measures of the angles equals the measure of a straight angle, the sum of the measures of the angles of triangle *ABC* is 180°.

FIGURE 10-39

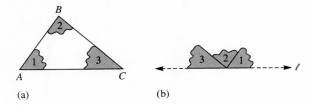

An alternate technique for showing that the sum of the measures of the angles in a triangle is 180° is shown in Figure 10-40. Suppose we start at vertex *A* in Figure 10-40(a) facing *B*, walk all the way around the triangle, and stop at the same position and pointed in the same direction as when we started; then we turned 360°. The 360° we turned is the sum of the shaded exterior angles of the triangle. Since an exterior angle and its adjacent interior angle are supplementary, the sum of all the interior and exterior angles is 3 · 180°, or 540°. By performing the subtraction 540° − 360°, we see that the sum of the interior angles of the triangle is 180°.

FIGURE 10-40

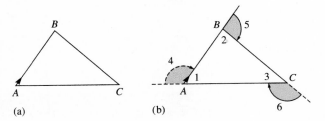

The amount of turning that takes place in walking around the triangle in Figure 10-40(b) and returning to the original position and heading is equal to the amount of turning in a complete circle. This concept can be expressed in generalized form in a statement often referred to in Logo as the Total Turtle Trip Theorem.

TOTAL TURTLE TRIP THEOREM

Any convex polygon can be drawn with a total turning of 360°.

The Total Turtle Trip Theorem leads to the following theorem concerning the sum of the exterior angles of any convex polygon.

THEOREM 10-4

The sum of the exterior angles of any convex polygon is 360°.

A more general case of the Total Turtle Trip Theorem is the Closed Path Theorem.

CLOSED PATH THEOREM

The total turning around any closed path is a multiple of 360°.

For a more formal proof that the sum of the measures of the angles of a triangle is 180°, consider triangle ABC in Figure 10-41(a). We can prove that $m(\angle 1) + m(\angle 2) + m(\angle 3) = 180°$ by showing that the sum of the measures of the three angles of the triangle equals the measure of a straight angle. Because a straight angle is formed by a line, our *subgoal* is to draw a line through one of the vertices of $\triangle ABC$, which will create angles congruent to $\angle 1$, $\angle 2$, and $\angle 3$. We do this by drawing line ℓ parallel to $\overleftrightarrow{BC}$ through vertex A, as shown in Figure 10-41(b). Because ℓ and $\overleftrightarrow{BC}$ are parallel, with transversals $\overleftrightarrow{AB}$ and $\overleftrightarrow{AC}$, it follows that alternate interior angles are congruent. Consequently, $m(\angle 1) = m(\angle 4)$ and $m(\angle 3) = m(\angle 5)$. Thus, $m(\angle 1) + m(\angle 2) + m(\angle 3) = m(\angle 4) + m(\angle 2) + m(\angle 5) = 180°$. So, $m(\angle 1) + m(\angle 2) + m(\angle 3) = 180°$. From this, we obtain the following theorem.

FIGURE 10-41

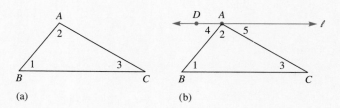

THEOREM 10-5

The sum of the measures of the interior angles of a triangle is 180°.

EXAMPLE 10-4 (a) If one angle of a parallelogram is 50°, what are the measures of the other angles?

(b) Find x, the measure of $\angle Q$ in Figure 10-42, where $\overrightarrow{AB} \parallel \overrightarrow{CD}$.

FIGURE 10-42

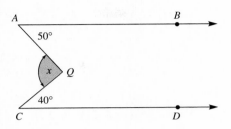

Solution (a) In Figure 10-43, we draw the lines containing the sides of parallelogram *ABCD*. Since $\angle 4$ and the angle with measure $50°$ are corresponding angles formed by parallel lines $\overleftrightarrow{AB}$ and $\overleftrightarrow{CD}$ cut by transversal $\overleftrightarrow{AD}$, then $m(\angle 4) = 50°$. Because $\angle 1$ and $\angle 4$ are supplementary, $m(\angle 1) = 180° - 50° = 130°$. Using similar reasoning, we find that $m(\angle 2) = 50°$ and $m(\angle 3) = 130°$.

FIGURE 10-43

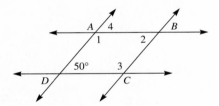

(b) We know how to *solve a related problem* when angles are formed by parallel lines and a transversal. To obtain a transversal, we extend $\overrightarrow{AQ}$ as shown in Figure 10-44. Now, using alternate interior angles formed by transversal $\overleftrightarrow{AQ}$ cutting parallel rays $\overrightarrow{AB}$ and $\overrightarrow{CD}$, we see that $m(\angle AEC) = 50°$. Thus $m(\angle CQE) = 180 - (50 + 40) = 90°$. Because $\angle x$ is supplementary to $\angle CQE$, $m(\angle x) = 180 - 90 = 90°$.

FIGURE 10-44

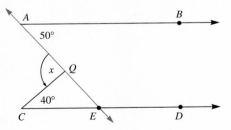

EXAMPLE 10-5 (a) In Figure 10-45(a), $m(\angle D) = 90°$ and $m(\angle E) = 25°$. Find $m(\angle A)$.

(b) In Figure 10-45(b), the measures of $\angle A$ and $\angle B$ are twice the measure of $\angle C$. Find the measures of each of the angles in the triangle.

(c) In Figure 10-45(c), $m(\angle A) = 70°$ and $m(\angle B) = 30°$. Find $m(\angle 1)$.

FIGURE 10-45

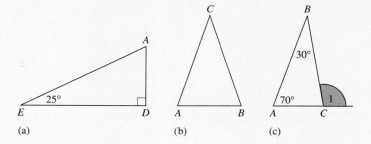

Solution (a) The sum of the measures of the angles in a triangle is 180°. Thus, $m(\angle A) + 90° + 25° = 180°$, and, consequently, $m(\angle A) = 65°$.

(b) Suppose that $m(\angle C) = x$. Then $m(\angle A) = 2x$ and $m(\angle B) = 2x$. Thus, $x + 2x + 2x = 180°$. Consequently, $5x = 180°$ and $x = 36°$. Because $2x = 72°$, we see that $m(\angle C) = 36°$, $m(\angle A) = 72°$, and $m(\angle B) = 72°$.

(c) The sum of the measures of the angles of a triangle is 180°. Thus, $m(\angle BCA) = 180° - (30° + 70°)$, or 80°. Now, $m(\angle BCA) + m(\angle 1) = 180°$, so $m(\angle 1) = 180° - 80°$, or 100°. ∎

Any n-gon has n interior angles and n exterior angles. Since the sum of every interior angle and its adjacent exterior angle is 180°, the sum of all the interior and exterior angles is $180n$. Because the sum of the exterior angles of any convex n-gon is 360°, the sum of the interior angles is $180n - 360°$, or $(n - 2)180°$. Two other methods of developing this result used in the elementary school are explored in the problem set.

In a regular n-gon, all n angles are congruent, and the sum of their measures is $(n - 2)180°$, so the measure of a single angle is $(n - 2)180°/n$.

The results from this discussion are summarized in the following theorem.

THEOREM 10-6

(a) The sum of the measures of the interior angles of any convex polygon with n sides is $(n - 2)180°$.

(b) The measure of a single interior angle of a regular n-gon is $(n - 2)180°/n$.

EXAMPLE 10-6 (a) Find the measure of each angle of a regular decagon.

(b) Find the number of sides of a regular polygon, each of whose angles has a measure of 175°.

Solution (a) Because a decagon has 10 sides, the sum of the measures of the angles of a decagon is $(10 - 2)180°$, or 1440°. A regular decagon has 10 angles, all of which are congruent, so each one has a measure of $\frac{1440°}{10}$, or 144°. As an alternate solution, each exterior angle is $\frac{360°}{10}$, or 36°. Hence, each interior angle is $180 - 36$, or 144°.

(b) Each interior angle of the regular polygon is 175°. Thus the measure of each exterior angle of the polygon is 180° − 175°, or 5°. Because the sum of the measures of all exterior angles of a convex polygon is 360°, the number of exterior angles is $\frac{360}{5}$, or 72. Hence, the number of sides is 72. ∎

▼ BRAIN TEASER

Find the sum of the measures of ∠1, ∠2, ∠3, ∠4, and ∠5 in any five-pointed star like the one in the accompanying figure. What is the sum of the measures of the angles in any seven-pointed star? What is the sum of the measures of the angles in any odd-pointed star, no vertex of which is in the interior of any angle of the star?

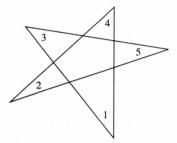

PROBLEM SET 10-3

1. For each of the following figures, which pairs of angles marked are adjacent and which are vertical?

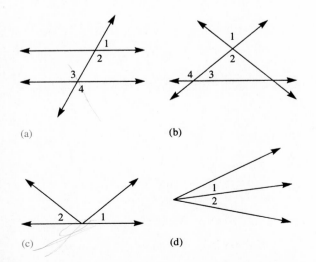

2. ▶Explain how you might find the measure of the angle of inclination of a staircase.

3. For each of the following, sketch a pair of angles whose intersection is given:
   (a) The empty set
   (b) Exactly two points
   (c) Exactly three points
   (d) Exactly four points
   (e) More than four points

4. If five lines all meet in a single point, how many pairs of vertical angles are formed?

5. Find the measure of the third angle in each of the following triangles:

(a)

(b)

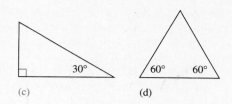

(c)   (d)

6. In each of the following pictured cases, are *m* and *n* parallel lines? Justify your answer.

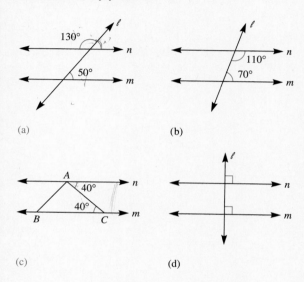

7. ▶(a) If one of the angles in a triangle is obtuse, can another angle be obtuse? Why?
   ▶(b) If one of the angles in a triangle is acute, can the other two angles be acute? Why?
   ▶(c) Can a triangle have two right angles? Why?
   ▶(d) If a triangle has one acute angle, is the triangle necessarily acute? Why?

8. In the following figure, $\overleftrightarrow{DE} \parallel \overleftrightarrow{BC}$, $\overleftrightarrow{EF} \parallel \overleftrightarrow{AB}$, and $\overleftrightarrow{DF} \parallel \overleftrightarrow{AC}$.

   Also, $m(\angle 1) = 45°$ and $m(\angle 2) = 65°$. Find each of the following values.
   (a) $m(\angle 3)$ (b) $m(\angle D)$
   (c) $m(\angle E)$ (d) $m(\angle F)$

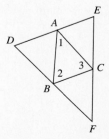

9. Find the measures of the angles marked *x* and *y*.

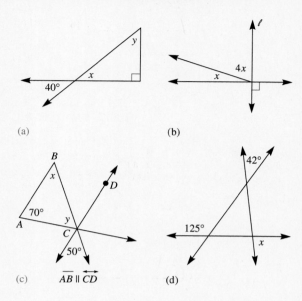

10. (a) What is the measure of an angle whose measure is twice that of its complement?
    (b) If two angles of a triangle are complementary, what is the measure of the third angle?

11. Two angles are complementary and the ratio of their measures is 7:2. What are the angle measures?

12. If the measures of the three angles of a triangle are $(3x + 15)°$, $(5x - 15)°$, and $(2x + 30)°$, what is the measure of each angle?

13. Find the sum of the measures of the marked angles in each of the following:

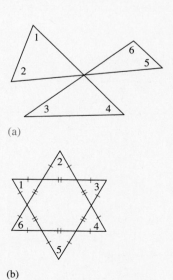

SECTION 10-3   EXERCISES   539

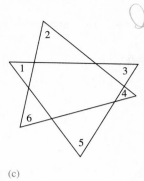

(c)

14. ▶In the following figure, A is a point not on line ℓ. Why is it impossible to have two distinct perpendicular segments from A to ℓ?

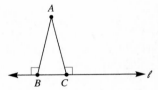

15. (a) In a regular polygon, the measure of each angle is 162°. How many sides does the polygon have?
    (b) Find the measure of each of the angles of a regular dodecagon.

16. (a) Show how to find the sum of the measures of the angles of any convex pentagon by choosing any point P in the interior and constructing triangles, as shown in the figure.

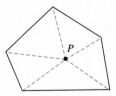

    (b) Using the method suggested by the diagram in (a), find the sum of the measures of the angles of any convex n-gon. Is your answer the same as the one already obtained in this section, that is, $(n - 2)180°$?

17. ▶(a) Explain how to find the sum of the measures of the angles of any convex pentagon by drawing all diagonals of the pentagon from one vertex, as shown in the following figure:

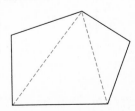

(b) Using the method suggested by the diagram in (a), explain how to find the sum of the measures of the angles of any convex n-gon.

18. Classify each of the following as true or false:
    (a) All angles in a regular polygon are congruent.
    (b) Any polygon with congruent sides must have congruent angles.
    (c) Any polygon with congruent angles must have congruent sides.
    (d) The sum of the measures of the interior angles of any obtuse triangle is 180°.
    (e) The interior angles of a triangle are supplementary, since the sum of their measures is 180°.

19. (a) In the figure, what is the relationship between $m(\angle 4)$ and $[m(\angle 1) + m(\angle 2)]$?

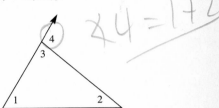

    (b) Justify your conjecture in (a).

20. Calculate the measure of each angle of a pentagon, where the measures of the angles form an arithmetic sequence and the least measure is 60°.

21. ▶A student claims that angles 1 and 2 are congruent because they are vertical angles. How do you respond?

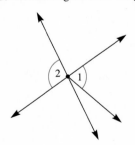

22. ▶Regular hexagons have been used to tile floors. Can a floor by tiled using only regular pentagons? Why or why not?

23. Two sides of a regular octagon are extended as shown in the following figure. Find the measure of angle 1.

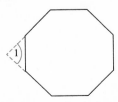

24. ▶If two angles of one triangle are congruent to two angles

in another triangle, must the third angle be congruent? Why or why not?

25. If lines ℓ and m are parallel and angles 1 and 2 are two interior angles on the same side of the transversal, what must be true about the angles?

26. Find the measure of angle x in the following:

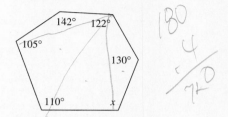

27. Find the measures of angles 1, 2, and 3 given that TRAP is a trapezoid.

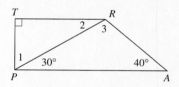

28. Home plate on a baseball field has three right angles and two congruent angles. Find the measures of each of these two congruent angles.

★29. Prove Theorem 10-1(a) and (b).

★30. Prove that two distinct coplanar lines perpendicular to the same line are parallel.

★31. Suppose that the following polygon ABCD is a parallelogram. Prove each of the following:

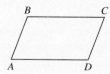

(a) $m(\angle A) + m(\angle B) = 180°$
(b) $m(\angle A) = m(\angle C)$ and $m(\angle B) = m(\angle D)$

32. What is the measure of the angle between the hands of the clock at exactly 4:37 P.M.?

33. (a) Study the accompanying figure and notice that a parallelogram can be determined by knowing the length of two sides :L and :W and the measure of one angle :A.

Write a procedure called PARALLELOGRAM with inputs :L, :A, and :W that draws such a parallelogram.

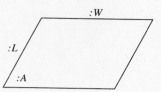

(b) Write a procedure called RECTANGLE that calls the PARALLELOGRAM procedure to draw a rectangle.
(c) Write a procedure called RHOMBUS that calls the PARALLELOGRAM procedure to draw a rhombus.
(d) How could a square of size 50 be generated by the PARALLELOGRAM procedure?
(e) How could a square of size 50 be generated by the RHOMBUS procedure?

Review Problems

34. If four distinct lines lie in a plane, what is the maximum number of intersection points of the four lines?
35. ▶Is it possible for the union of two rays to be a line segment? Explain your answer.
36. Draw a polygonal curve that is closed but not simple.
37. Describe how you might determine the dihedral angle A-BC-D formed on the hip roof of a house.

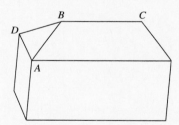

38. If a cube, as pictured next, intersects a plane, what possible figures can be obtained by the intersection. Sketch the planes and figures obtained in each case.

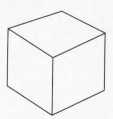

39. Name the geometric figure suggested by each of the following:

(a) A stop sign
(b) A school crossing sign
(c) A railroad crossing sign
(d) A speed limit sign
(e) A deer crossing sign

40. ▶In each of the following, find the required properties. If it is not possible to find the indicated properties, explain why.
    (a) Two properties that hold true for all rectangles but not for all rhombuses
    (b) Two properties that hold true for all squares but not for all isosceles trapezoids
    (c) Two properties that hold true for all parallelograms but not for all squares

---

LABORATORY ACTIVITY

As a van Hiele Level 2 activity, prepare a set of cards labeled with the following names of quadrilaterals: rectangle, parallelogram, square, trapezoid, rhombus, and quadrilateral. Use colored strings and develop a Venn diagram arranging the cards in their proper places as subsets of the Venn diagram, if possible.

Section 10-4 Geometry in Three Dimensions

Simple Closed Surfaces

A visit to the grocery store exposes us to many three-dimensional objects that have simple closed surfaces. Examples are shown in Figure 10-46.

FIGURE 10-46

  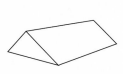 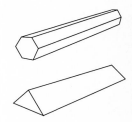

simple closed surfaces
sphere
center

solid

Simple closed surfaces have exactly one interior and no holes and are hollow. An example is a sphere. A **sphere** is defined as the set of all points at a given distance from a given point, the **center**. As with the relation of a polygon to the plane containing it, a simple closed surface partitions space into three disjoint sets: points outside the surface, points belonging to the surface, and points inside the surface. The union of all points on a simple closed surface and all interior points is referred to as a **solid**. Figures 10-47(a), (b), (c), and

FIGURE 10-47

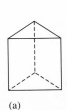 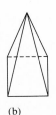 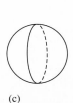  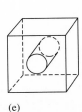 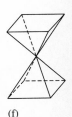

(a) (b) (c) (d) (e) (f)

polyhedron (d) are examples of simple closed surfaces; (e) and (f) are not. A **polyhedron** (polyhedra is the plural) is a simple closed surface made up of polygonal regions. (Recall that a polygonal region is formed by a polygon and its interior.) The word *polyhedron* is self-explanatory: *poly* means "many," and *hedron* means "flat surfaces." Figures 10-47(a) and (b) are examples of polyhedra, but (c), (d), (e), and (f) are not.

face
vertices
edges

Each of the polygonal regions of a polyhedron is a **face**. The vertices of the polygonal regions are the **vertices** of the polyhedron, and the sides of each polygonal region are the **edges** of the polyhedron.

prism

A **prism** is a polyhedron in which two congruent polygonal faces lie in parallel planes, and the other faces are bounded by parallelograms. Figure 10-48 shows four different prisms. The upper and lower parallel faces of a prism, like the faces *ABC* and *DEF* at the bottom and top of the prism in Figure 10-48(a), are the **bases** of the prism. A prism usually is named after its bases, as the figure suggests.

bases

FIGURE 10-48

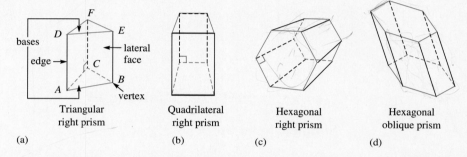

Triangular right prism (a)   Quadrilateral right prism (b)   Hexagonal right prism (c)   Hexagonal oblique prism (d)

lateral faces
right prism
oblique prism

The **lateral faces** of a prism, the faces other than the bases, are bounded by parallelograms. If the lateral faces of a prism are all bounded by rectangles, the prism is a **right prism.** The first three prisms in Figure 10-48 are right prisms. Figure 10-48(d) is an **oblique prism** because all its lateral edges are *not* perpendicular to the bases, and therefore its faces are *not* bounded by rectangles.

Students often have trouble drawing three-dimensional figures. Figure 10-49 gives an example of how to draw a pentagonal prism.

FIGURE 10-49

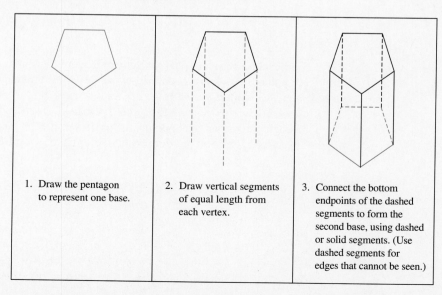

1. Draw the pentagon to represent one base.

2. Draw vertical segments of equal length from each vertex.

3. Connect the bottom endpoints of the dashed segments to form the second base, using dashed or solid segments. (Use dashed segments for edges that cannot be seen.)

An example of students working with spatial visualization and drawing three-dimensional figures is given on the student page from *Heath Mathematics Connections*, Grade 6, 1992.

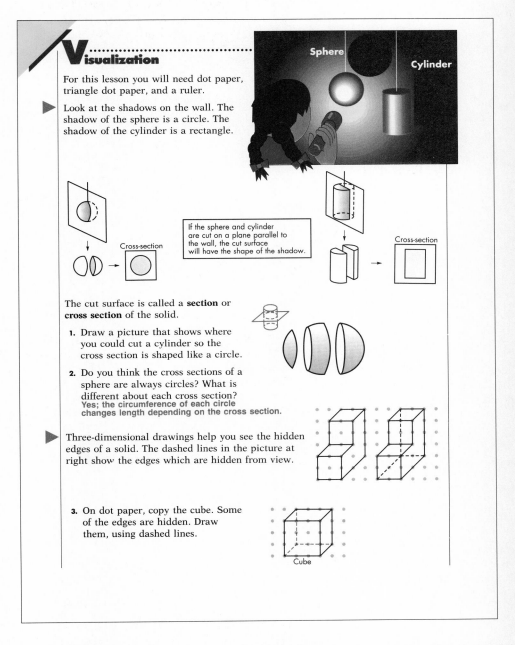

A **pyramid** is a polyhedron determined by a simple closed polygonal region, a point not in the plane of the region, and triangular regions determined by the point and each pair of consecutive vertices of the polygonal region. The polygonal region is the **base** of the pyramid, and the point is the **apex**. The faces other than the base are **lateral faces**. Pyramids are classified according to their bases, as shown in Figure 10-50.

base
apex / lateral faces

FIGURE 10-50

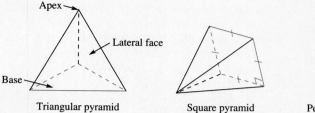

Triangular pyramid   Square pyramid   Pentagonal pyramid

To draw a pyramid, follow the steps in Figure 10-51.

FIGURE 10-51

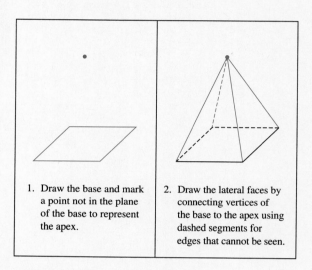

1. Draw the base and mark a point not in the plane of the base to represent the apex.
2. Draw the lateral faces by connecting vertices of the base to the apex using dashed segments for edges that cannot be seen.

convex polyhedron   A polyhedron is a **convex polyhedron** if, and only if, the segment connecting any two points in the interior of the polyhedron is itself in the interior. Figure 10-52 shows a concave polyhedron (that is, one that is caved in).

FIGURE 10-52

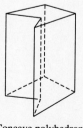

Concave polyhedron

Regular Polyhedra

regular polyhedron   A **regular polyhedron** is a convex polyhedron whose faces are congruent regular polygonal regions such that the number of edges that meet at each vertex is the same for all the vertices of the polyhedron.

HISTORICAL NOTE

The regular solid polyhedra are called the **Platonic solids,** after the Greek philosopher Plato (fourth century B.C.). Plato attached a mystical significance to the five regular polyhedra, associating them with what he believed were the four elements (earth, air, fire, water) and the universe. Plato suggested that the smallest particles of earth have the form of a cube, those of air an octahedron, those of fire a tetrahedral, those of water an icosahedron, and those of the universe a dodecahedron.

Regular polyhedra have fascinated mathematicians for centuries. At least three of them were identified by the Pythagoreans (ca. 500 B.C.). Two others were known to the followers of Plato (ca. 350 B.C.). Three of the five polyhedra occur in nature in the form of crystals of sodium sulphantimoniate, sodium chloride (common salt), and chrome alum, respectively, as seen in Figure 10-53. The other two do not occur in crystalline form but have been observed as skeletons of microscopic sea animals called radiolaria.

FIGURE 10-53

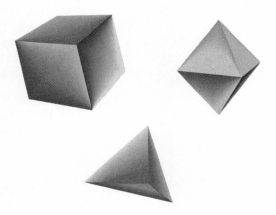

PROBLEM 3

How many regular polyhedra are there?

Understanding the Problem. Each face of a regular polyhedron is congruent to each of the other faces of that polyhedron, and each face is a regular polygon. We are to find the number of different regular polyhedra.

Devising a Plan. The sum of the measures of all the angles at a vertex of a regular polyhedron must be less than 360°. (Do you see intuitively why this is true?) We next examine the measures of the interior angles of regular polygons to determine which of the polygons could be faces of a regular polyhedron and then try to determine how many types of polyhedra there are.

Carrying Out the Plan. We determine the size of an angle of some regular polygons as shown in Table 10-6. Could a regular heptagon be a face of a regular polyhedron? At least three figures must fit together at a vertex to make a polyhedron. (Why?) If three angles of a regular heptagon were together at one vertex, then the sum of the measures of these angles would be $\frac{3 \cdot 900°}{7}$, or $\frac{2700°}{7}$, which is greater than 360°. Similarly, more than three angles cannot be used at a vertex. Thus, a heptagon cannot be used to make a regular polyhedron.

TABLE 10-6

Polygon	Measure of an Interior Angle
Triangle	60°
Square	90°
Pentagon	108°
Hexagon	120°
Heptagon	$\left(\frac{900}{7}\right)°$

Because the measure of an interior angle of a regular polygon increases as the number of sides of the polygon increases (why?), any polygon with more than six sides will have an interior angle greater than 120°. Hence, if three angles were to fit together at a vertex, the sum of the measures of the angles would be greater than 360°. Thus, the only polygons that might be used to make regular polyhedra are equilateral triangles, squares, regular pentagons, and regular hexagons. Consider the possibilities in Table 10-7 on the facing page.

Notice that we were not able to use six equilateral triangles to make a polyhedron, because $6(60°) = 360°$ and the triangles would lie in a plane. Similarly, we could not use four squares or any hexagons. We also could not use more than three pentagons, because if we did the sum of the angles would be more than 360°.

semiregular polyhedra   **Looking Back.** Interested readers may want to investigate **semiregular polyhedra.** These are also formed by using regular polygons as faces, but the regular polygons used need not have the same number of sides. For example, a semiregular polyhedron might have squares and regular octagons as its faces.

The patterns in Figure 10-54, called *nets,* may be used to construct the five regular polyhedra. It is left as an exercise to determine other patterns for constructing the regular polyhedra.

FIGURE 10-54

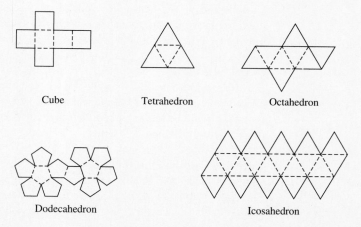

Cube   Tetrahedron   Octahedron

Dodecahedron   Icosahedron

TABLE 10-7

Polygon	Measure of an Interior Angle	Number of Polygons at a Vertex	Sum of the Angles at the Vertex	Polyhedron Formed	Model
Triangle	60°	3	180°	**Tetrahedron**	
Triangle	60°	4	240°	**Octahedron**	
Triangle	60°	5	300°	**Icosahedron**	
Square	90°	3	270°	**Cube**	
Pentagon	108°	3	324°	**Dodecahedron**	

Euler's Formula

Euler's formula

A simple relationship among the number of faces, the number of edges, and the number of vertices of any polyhedron was discovered by the French mathematician and philosopher René Descartes (1596–1650) and rediscovered by the Swiss mathematician Leonhard Euler (1707–1783). Table 10-8 suggests a relationship among the numbers of vertices (V), edges (E), and faces (F). In each case, $V + F - E = 2$. This result is known as **Euler's formula.** Prisms, pyramids, and Euler's formula are investigated in the problem set.

TABLE 10-8

Name	V	F	E
Tetrahedron	4	4	6
Cube	8	6	12
Octahedron	6	8	12
Dodecahedron	20	12	30
Icosahedron	12	20	30

Cylinders and Cones

cylinder
bases

A cylinder is an example of a simple closed surface that is not a polyhedron. Consider a line segment $\overline{AB}$ and a line ℓ as shown in Figure 10-55. When $\overline{AB}$ moves so that it always remains parallel to a given line ℓ and points A and B trace simple closed planar curves other than polygons, the surface generated by $\overline{AB}$, along with the simple closed curves and their interiors, forms a **cylinder.** The simple closed curves traced by A and B along with their interiors are the **bases** of the cylinder and the remaining points constitute the *lateral surface of the cylinder*. Three different cylinders are pictured in Figure 10-55.

FIGURE 10-55

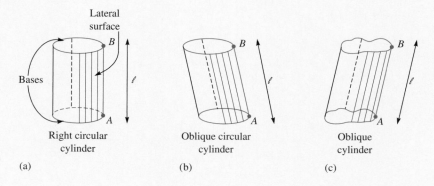

circular cylinder
right cylinder
oblique cylinder

If a base of a cylinder is a circular region, the cylinder is a **circular cylinder.** If the line segment forming a cylinder is perpendicular to a base, the cylinder is a **right cylinder.** Cylinders that are not right cylinders are **oblique cylinders.** The cylinder in Figure 10-55(a) is a right cylinder; those in Figures 10-55(b) and (c) are oblique cylinders.

cone / vertex

Suppose we have a simple closed curve, other than a polygon, in a plane and a point P not in the plane of the curve. The union of line segments connecting point P to each point of a simple closed curve and the simple closed curve and the interior of the curve is a **cone.** Cones are pictured in Figure 10-56. Point P is the **vertex** of the cone. The points of the cone that are not in the base constitute the *lateral surface of the cone*. A line segment from the vertex P perpendicular to the plane of the base is the **altitude.** A **right circular cone,** such as the one in Figure 10-56(a), is a cone whose altitude intersects the base (a circular region) at the center of the circle. Figure 10-56(b) illustrates an oblique cone, and Figure 10-56(c) illustrates an **oblique circular cone.**

altitude
right circular cone

oblique circular cone

FIGURE 10-56

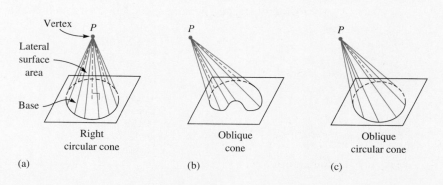

HISTORICAL NOTE

Leonhard Euler went blind in 1766 and for the remaining 17 years of his life continued to do mathematics by dictating to a secretary and by writing formulas in chalk on a slate for his secretary to copy down. He published 530 papers in his lifetime and left enough work to supply the *Proceedings of the St. Petersburg Academy* for the next 47 years.

PROBLEM SET 10-4

1. Identify each of the following polyhedra. If a polyhedron can be described in more than one way, give as many names as possible.

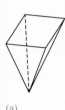
(a)

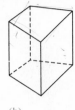
(b)

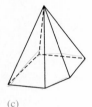
(c)

2. Given the tetrahedron below, name the following:
   (a) Vertices (b) Edges (c) Faces
   (d) Intersection of face *DRW* and edge $\overline{RA}$

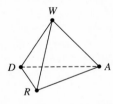

3. Identify five different shapes of containers that can be found in the grocery store.
4. For each of the following, what is the minimum number of faces possible?
   (a) Prism   (b) Pyramid   (c) Polyhedron
5. Classify each of the following as true or false:
   (a) If the lateral faces of a prism are rectangles, it is a right prism.
   (b) Every pyramid is a prism.
   (c) Every pyramid is a polyhedron.
   (d) The bases of a prism lie in perpendicular planes.
   (e) The bases of all cones are circles.
   (f) A cylinder has only one base.
   (g) All lateral faces of an oblique prism are rectangular regions.
   (h) All regular polyhedra are convex.

6. ▶How many possible pairs of bases does a rectangular prism have? Explain.
7. For each of the following, draw a prism and a pyramid that have the given region as a base:
   (a) Triangle   (b) Pentagon
   (c) Regular hexagon
8. Two prisms are sketched on the following dot paper. Complete the drawings by using dashed segments for the hidden edges.

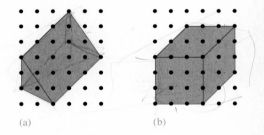

(a)   (b)

9. Name each polyhedron that can be constructed using the following nets:

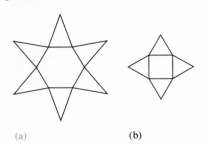

(a)   (b)

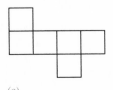

(c)

550 CHAPTER 10   INTRODUCTORY GEOMETRY

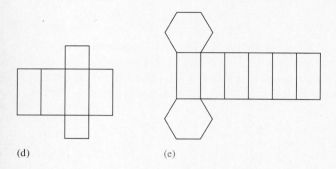

(d)   (e)

10. Match each 3-d figure sliced by a plane with its cross-section.

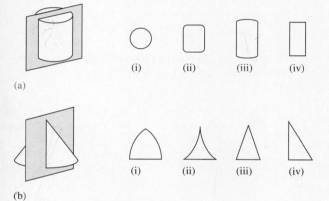

11. The figure on the left in each of the following represents a card attached to a wire as shown. Match each figure on the left with what it would look like if it were revolved by spinning the wire between your fingers.

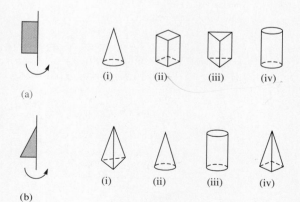

12. Which of the three-dimensional figure(s) below could be used to make the shadow shown in (a)? In (b)?

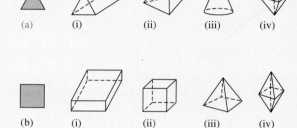

13. A diagonal of a prism is any segment determined by two vertices that do not lie in the same face as shown next. Complete the following table showing the total number of diagonals for various prisms:

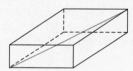

Prism	Vertices per Base	Diagonals per Vertex	Total Number of Diagonals
Quadrilateral	4	1	4
Pentagonal	5		
Hexagonal			
Heptagonal			
Octagonal			
$\vdots$			
n-gonal			

14. ▶Which of the following could be drawings of a quadrilateral pyramid? If yes, where would you be standing in each case? Explain why.

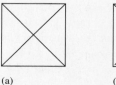

(a)   (b)

15. If possible, sketch each of the following:
    (a) An oblique square prism
    (b) An oblique square pyramid
    (c) A noncircular right cone
    (d) A noncircular cone that is not right
16. Consider a jar with a lid, as illustrated in the following figure. The jar is half filled with water. In each of the marked drawings, sketch the water.

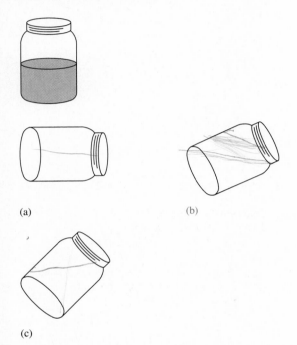

17. On the left of each of the following figures is a net for a three-dimensional object. On the right are several objects. Which object will the net fold to make?

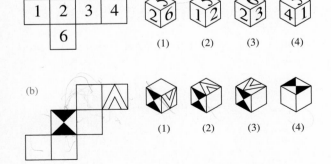

18. Sketch the intersection of each of the following:

(a) Cube (b) Remainder of unseen figure completes the cube

 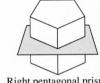

(c) Sphere (d) Right pentagonal prism

(e) Right circular cone (plane parallel to base) (f) Right circular cylinder (plane not parallel to base)

19. Verify Euler's formula for each of the polyhedra in Problem 1.
20. Answer each of the following questions about a pyramid and a prism, each having an n-gon as a base:
    (a) How many faces does each have?
    (b) How many vertices does each have?
    (c) How many edges does each have?
    (d) Use your answers to (a), (b), and (c) to verify Euler's formula for all pyramids and all prisms.
21. Complete the table for each of the polyhedra described in the table.

Polyhedron	Vertices	Faces	Edges
(a)		8	12
(b)	20	30	
(c)	6		15

22. Check whether Euler's formula holds for each of the following figures.

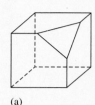

(a)

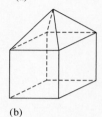

(b)

23. A circle may be considered a "many-sided" polygon. Use this notion to describe the relationship between each of the following:
    (a) A pyramid and a cone
    (b) A prism and a cylinder
24. ▶(a) Can a prism have exactly 33 edges? Explain why or why not.
    ▶(b) Can a pyramid have exactly 33 edges? Explain why or why not.

Review Problems

25. If a toilet-tissue roll is cut along the seams, what is the shape obtained?

26. Triangles *ABC* and *CDE* are equilateral triangles. Find the measure of ∠*BCD*.

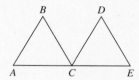

27. What is the measure of each angle in a regular nonagon?
28. Classify the following as true or false. If false, tell why.
    (a) Every rhombus is a parallelogram.
    (b) Every polygon has at least three sides.
    (c) Triangles can have at most two acute angles.
29. (a) If two angles of a triangle are complementary, what type of triangle is it?
    ▶(b) Justify your answer.

▼ **B R A I N   T E A S E R**

A rectangular region can be rolled to form the lateral surface of a right circular cylinder. What shape of paper is needed to make an oblique circular cylinder? (See "Making a Better Beer Glass" by A. Hoffer.)

LABORATORY ACTIVITY

1. As a van Hiele Level 0 activity, use isometric dot paper, as shown, to construct two-dimensional representations of three-dimensional figures. In this case, the letter *O* is pictured.

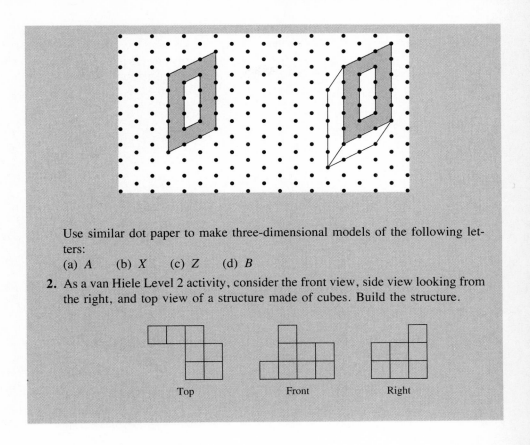

Use similar dot paper to make three-dimensional models of the following letters:
(a) A (b) X (c) Z (d) B

2. As a van Hiele Level 2 activity, consider the front view, side view looking from the right, and top view of a structure made of cubes. Build the structure.

*Section 10-5 Networks

In the 1700s, the people of Königsberg, Germany, used to enjoy walking over the bridges of the Pregel River. There were two islands in the river and seven bridges, as shown in Figure 10-57. These walks eventually led to the following problem:

Is it possible to walk across all the bridges so that each bridge is crossed exactly once on the same walk?

FIGURE 10-57

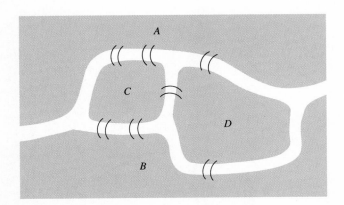

There is no restriction on where to start the walk or where to finish. Try to find a path before reading further. This problem has become known as the *Königsberg bridge problem*. Leonhard Euler became interested in this problem and solved it in 1736. He represented the problem in much simpler form by representing the land masses, islands, and bridges in what he called a **network,** as shown in the colored portion of Figure 10-58.

network

FIGURE 10-58

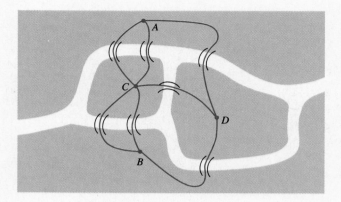

vertices / arcs

The points in a network are **vertices,** and the curves are **arcs.** Using a network diagram, we can restate the Königsberg bridge problem as follows: *Is there a path through the network beginning at some vertex and ending at the same or another vertex such that each arc is traversed exactly once?* A network having such a path is **traversable;** that is, each arc is passed through exactly once.

traversable

We can walk around an ordinary city block, as in Figure 10-59(a). It is not necessary to start at any particular point, and, in general, we can traverse any simple closed curve. Now consider walking around two city blocks and down the street that runs between them, as shown in Figure 10-59(b). To traverse this network, it is necessary to start at vertex B or C. Starting at points other than B or C might suggest that the figure is not traversable, but this is not the case, as shown in Figure 10.55(b). If we start at B, we end at C and vice versa. Note that it is permissible to pass through a vertex more than once but an arc may be traversed only once. Vertices B and C are endpoints of three line segments, and each of the other vertices are endpoints of two segments.

FIGURE 10-59

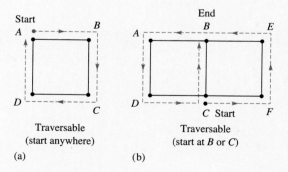

Traversable
(start anywhere)
(a)

Traversable
(start at B or C)
(b)

A traversable network is the type of network, or route, that a highway inspector would like to have if given the responsibility of checking out all the roads in a highway system. The inspector needs to traverse each road (arc) in the system but would save time by not having to make any repeat journeys during any inspection tour. It would be feasible for the inspector to go through any town (vertex) more than once on the route. Consider the networks in Figure 10-60. Is it possible for the highway inspector to do the job with these networks without traversing any road twice?

FIGURE 10-60

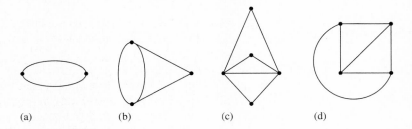

(a) (b) (c) (d)

The first three networks in Figure 10-60 are traversable; the fourth network, (d), is not. Notice that the number of arcs meeting at each vertex in networks (a) and (c) is even. Any such vertex is an **even vertex.** If the number of arcs meeting at a vertex is odd, it is an **odd vertex.** In network (b), only the odd vertices will work as starting or stopping points. In network (d), which is not traversable, all the vertices are odd. If a network is traversable, each arrival at a vertex other than a starting or a stopping point requires a departure. Thus, each vertex that is not a starting or stopping point must be even. The starting and stopping vertices in a traversable network may be even or odd, as seen in Figure 10-60(a) and (b), respectively.

even vertex
odd vertex

In general, networks have the following properties:

1. *If a network has all even vertices, it is traversable. Any vertex can be a starting point, and the same vertex must be the stopping point.*
2. *If a network has two odd vertices, it is traversable. One odd vertex must be the starting point, and the other odd vertex must be the stopping point.*
3. *If a network has more than two odd vertices, it is not traversable.*
4. *There is no network with exactly one odd vertex.*

EXAMPLE 10-7 Which of the networks in Figure 10-61 are traversable?

FIGURE 10-61

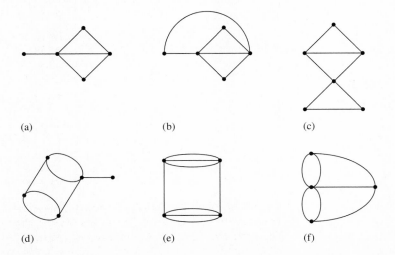

(a) (b) (c)

(d) (e) (f)

Solution Networks in (b) and (e) have all even vertices and therefore are traversable.

Networks in (a) and (c) have exactly two odd vertices and are traversable.

Networks in (d) and (f) have four odd vertices and are not traversable. ∎

The network in Figure 10-61(f) represents the Königsberg bridge problem. The network has four odd vertices, and consequently is not traversable; hence, no walk is possible to complete the problem.

A problem similar to the highway inspector problem involves a traveling salesperson. Such a person might have to travel networks comparable to those of the highway inspector; however, the salesperson is interested only in visiting each town (vertex) once—and not necessarily in following each road. It is not known for which networks this can be accomplished. Can you find a route for the traveling salesperson for each network in Figure 10-61?

A different type of application of network problems is discussed in Example 10-8.

EXAMPLE 10-8 Look at the floor plan of the house shown in Figure 10-62. Is it possible for a security guard to go through all the rooms of the house and pass through each door exactly once?

FIGURE 10-62

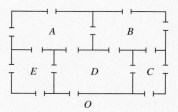

Solution Represent the floor plan as a network, as in Figure 10-63. Designate the rooms and the outside as vertices, and the paths through the doors as arcs. The network has more than two odd vertices, namely, A, B, D, and O. Thus the network is not traversable, and it is impossible to go through all the rooms and pass through each door exactly once.

FIGURE 10-63

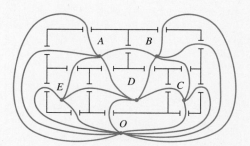

PROBLEM SET 10-5

1. Which of the following networks are traversable? If the network is traversable, draw an appropriate path through it, labeling the starting and stopping vertices.

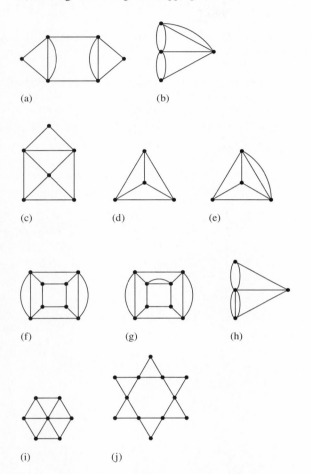

2. Which of the networks in Problem 1 can be efficiently traveled by the traveling salesperson, with no vertex visited more than once?

3. A city contains a river, three islands, and 10 bridges, as shown in the accompanying figure. Is it possible to take a walk around the city by starting at any land area and returning after visiting every part of the city and crossing each bridge exactly once? If so, show such a path both on the original figure and on the corresponding network.

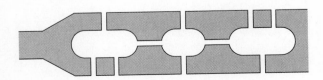

4. Use the accompanying floor plans for each of the following:
   (a) Draw a network that corresponds to each floor plan.
   (b) Determine if it is possible to pass through each room of each house by passing through each door exactly once. If it is possible, draw such a trip.

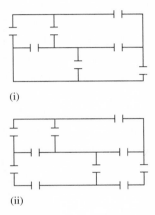

5. Can a person walk through each door once and only once and also go through both of the following houses in a single path? If it is possible, draw such a path.

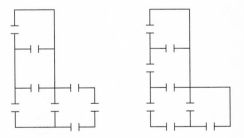

6. The following drawing represents the floor plan of an art museum. All tours begin and end at the entry. If possible, design a tour route that will allow a person to see every room but not go through any room twice.

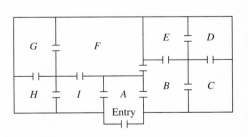

7. Each of the networks in Problem 1 separate the plane into several subsets. If R is the number of interior and exterior regions of the plane, V is the number of vertices, and A is the number of arcs, complete the following chart using each of the networks. (The first one is done for you.)

Network	R	V	A	R+V−A
(a)	6	6	10	2

8. Molly is making her first trip to the United States and would like to tour the eight states pictured. She would like to plan her trip so that she can cross each border between neighboring states exactly once—that is, the Washington-Oregon border, the Washington-Idaho border, and so on. Is such a trip possible? If so, does it make any difference in which state she starts her trip?

9. Draw a network that is not traversable; use as few vertices and arcs as possible.

10. ▶If you were commissioned to build an eighth bridge to make the Königsberg bridge problem traversable, where would you build your bridge? Is there more than one location? Explain why.

LABORATORY ACTIVITY

1. Take a strip of paper like the one shown in the figure. Give one end a half-twist and join the ends by taping them. The surface obtained is called a Möbius strip.

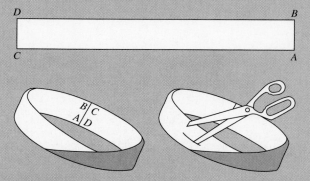

(a) Use a pencil to shade one side of a Möbius strip. What do you discover?

(b) Imagine cutting a Möbius strip all around midway between the edges. What do you predict will happen? Now do the actual cutting. What is the result?

(c) Imagine cutting a Möbius strip one-third of the way from an edge and parallel to the edge all the way through until you return to the starting point. Predict the result. Then actually do the cutting. Was your prediction correct?

(d) Imagine cutting around a Möbius strip one-fourth of the way from an edge. Predict the result. Then actually do the cutting. How does the result compare with the result of experiment (c)?

2. (a) Take a strip of paper and give it two half-twists (one full twist). Then join the ends together. Answer the questions in part 1.
   (b) Repeat the experiment in (a), using three half-twists.
   (c) Repeat the experiment in (a), using four half-twists. What do you find for odd-numbered twists? Even-numbered twists?

3. Take two strips of paper and tape each of them in a circular shape. Join them as shown below.

   (a) What happens if you cut completely around the middle of each strip as shown?
   (b) Repeat part (a) if both strips are Möbius strips. Does it make any difference if the half twists are in opposite directions?

*Section 10-6 Introducing Logo as a Tool in Geometry

Appendix II on Logo should be completed before this section is begun. In this section, we show some examples of how Logo might be used as a problem-solving tool in helping teach and learn geometric topics.

Logo Quadrilaterals

In Appendix II, you will find the following procedures for drawing variable-sized squares and rectangles:

```
TO SQUARE :SIDE
  REPEAT 4 [FORWARD :SIDE RIGHT 90]
END

TO RECTANGLE :HEIGHT :WIDTH
 REPEAT 2 [FORWARD :HEIGHT RIGHT 90 FORWARD
      :WIDTH RIGHT 90]
END
```

The procedures for drawing a square and rectangle just presented are natural procedures to develop and execute. However, because a square is a rectangle whose length and width are the same, we should be able to use the RECTANGLE procedure to draw a square as follows:

```
TO SQUARE :SIDE
 RECTANGLE :SIDE :SIDE
END
```

With a view toward designing other procedures for quadrilaterals, consider the diagram in Figure 10-64. Since rhombuses, rectangles, and squares are parallelograms, we first write a PARALLELOGRAM procedure so that it can be used to write the RHOMBUS and RECTANGLE procedures. Consider Figure 10-65, which depicts two parallelograms made out of tagboard and brads. The lengths of the sides are the same, but the parallelograms are very different.

FIGURE 10-64

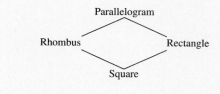

FIGURE 10-65

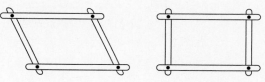

Therefore it is necessary to specify lengths of sides and the measure of an angle when a parallelogram is to be constructed. If we specify the measure of one angle of the parallelogram, then the measures of the other angles follow, as seen in Figure 10-66.

FIGURE 10-66

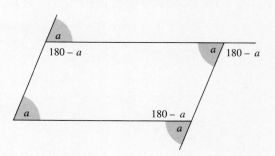

A possible procedure for drawing a parallelogram is as follows:

```
TO PARALLELOGRAM :SIDE1 :SIDE2 :ANGLE
 REPEAT 2 [FORWARD :SIDE1 RIGHT :ANGLE FORWARD
    :SIDE2 RIGHT 180-:ANGLE]
END
```

It is left as an exercise to use the PARALLELOGRAM procedure to write procedures for drawing a rectangle and a rhombus. Use of the developed rhombus procedure can also yield a different procedure for drawing a square.

Other Logo Geometry and Problem-solving Notions

The Total Turtle Trip Theorem can be used to write procedures to draw regular polygons. For example, drawing a regular pentagon (five sides) requires each of the five angles to have a measure that is $\frac{1}{5}$ of the total turning. Thus each turn angle must have a measure of 360°/5, or 72°.

To draw a five-pointed star, we recall a theorem called the Closed Path Theorem, which states that the total turning around any closed path must be a multiple of 360°. Thus we could write the following STARS procedure, where :N is the number of sides, :SIZE is the length of each side, and :MULT is an integer we multiply times 360 to give multiples of 360.

```
TO STARS :N :SIZE :MULT
 REPEAT :N [FD :SIZE RT (360*:MULT)/:N]
END
```

STARS 5 50 1 gives a pentagon; STARS 5 50 2 gives the desired star, and hence the turning angle is (360° · 2)/5, or 144°.

What results can you obtain with other inputs? Could you draw a seven-pointed star by using the STARS procedure? Is only one seven-pointed star possible? How about a six-pointed star?

PROBLEM SET 10-6

1. Use the PARALLELOGRAM procedure of this section to write procedures to draw quadrilaterals for the figures named by the following procedures:
   (a) RECTANGLE   (b) RHOMBUS
2. Use the RHOMBUS procedure developed in Problem 1 to write a new SQUARE procedure.
3. Use the RHOMBUS procedure from Problem 1 to write a CUBE procedure to draw the following. (Do you see a cube?)

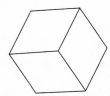

4. How many degrees should the turn be at each angle to produce a drawing of each of the following?
   (a) A regular hexagon (6 sides)
   (b) A regular heptagon (7 sides)
   (c) A regular octagon (8 sides)

5. Predict the results of executing each of the following and then use the computer to check your predictions:
   (a) STARS 5 50 3
   (b) STARS 5 50 4
   (c) STARS 5 50 5
   (d) STARS 5 50 6
   (e) STARS 5 50 7
6. ▶Is it possible to draw a six-pointed star by using the methods developed in this section? Why or why not?
7. Write procedures for drawing each of the following:

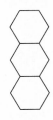

(a)

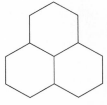

(b)

8. Write a procedure called THIRTY that draws a 30° angle.
9. (a) Write a procedure called SEG that draws a segment of length 100 steps, with midpoint at home and heading of 45.
   (b) Write a procedure called PAR to draw a segment of length 100 steps that is parallel to the segment drawn in (a).
10. Describe what the following Logo procedure will draw:
    TO GUESS
     REPEAT 30[FD 4 RT 12]
    END
11. Write a procedure called FILL.RECT that will fill in any rectangle that it draws. Do not use the FILL command that is available on many versions of Logo.
12. Complete the following POLYGON procedure so that it will draw regular polygons where :NUM is the number of sides in the polygon and :LEN is the length of any given side.

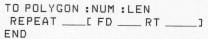

TO POLYGON :NUM :LEN
  REPEAT ___[FD ___ RT ___]
END

13. Write a procedure called COUNT.ANGLES to count the number of angles formed by n concurrent rays with the same endpoint. An example is pictured next, where six angles are formed by four rays.

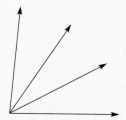

LABORATORY ACTIVITY

Find a set of attribute blocks and/or tangram pieces and write Logo procedures to draw each of the pieces. Use the pieces to form some type of figure. Then write a superprocedure that makes use of the other procedures to draw the figure.

Solution to the Preliminary Problem

Understanding the Problem. We are to determine if clock hands form a right angle between 2:00 and 3:00 P.M.. Also, if a right angle can be formed, then we are to determine to the nearest minute the time that the hands form this angle. A right angle has a measure of 90 degrees.

Devising a Plan. We can quickly determine that a right angle can be formed by the hands of the clock between 2:00 and 3:00 P.M. by noticing that at approximately 2:20 P.M., we have an acute angle and at 2:40 P.M. we have an obtuse angle. As the minute hand sweeps around, it must form all angles between the acute angle and the obtuse angle; thus, a right angle must be formed. To determine the time when the right angle is formed, we note that there are 360° in one complete revolution of a hand on a clock. Because there are 60 minutes in one evolution of the minute hand, each time the hand moves 1 minute, it moves $\frac{360}{60}$, or 6°. Each time the hour hand moves one hour, it moves $\frac{360}{12}$, or 30°. Therefore, the sum of the measures of angles 2 and 3 as shown in Figure 10-67(a) is 60°. Because the hour hand must be somewhere between 2 and 3 to form a right angle, the minute hand must be somewhere between 5 and 6 and the sum of angles 1 and 4 in Figure 10-67(b) must be 30°. (Why?) One strategy for determining the time when the sum of angles 1 and 4 is 30° is *guess and check.*

FIGURE 10-67

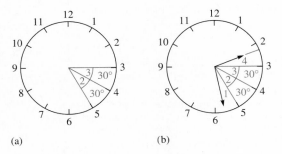

Carrying Out the Plan. Suppose we guess that the time when we obtain a right angle is 2:30 P.M.. Because the minute hand is on the 6, the measure of angle 1 is 30°. At 2:30 P.M., the hour hand is halfway between the 2 and the 3 so that the measure of angle 4 is 15°. Therefore, the measure of the angle between the hands at 2:30 P.M. is 30 + 30 + 30 + 15, or 105°, which is too large. In a similar manner, we can see that the angle formed at 2:25 P.M. is too small. We now know that the correct time is somewhere between 2:25 and 2:30 P.M.. Suppose we choose 2:27 P.M.. Because the minute hand has moved 2 minutes past the 5 and every time it moves 1 minute, it moves 6°, the measure of angle 1 is 12°. The measure of angle 4 is harder to determine this time. We know that the hour hand is $\frac{27}{60}$ of the way between the 2 and the 3 on the clock and has $\frac{33}{60}$ of the way to go to reach the 3. (Why?) Therefore, the measure of angle 4 is $\frac{33}{60}$ of 30° or 16.5°. Adding, we find that the measure of the angle at 2:27 P.M. is 12 + 30 + 30 + 16.5, or 88.5°. A similar calculation will show that the measure of the angle at 2:28 P.M. is 94°. Now we know that the time for a right angle is between 2:27 and 2:28 P.M. and that it is closer to 2:27 P.M.. (Why?) Therefore, the closest the hands come to forming a right angle to the nearest minute in Maria's class is at 2:27 P.M..

Looking Back. It is possible to find the time when the hands are closer to forming a right angle. For example, suppose we add 0.3 of a minute to 2:27 P.M. and see how this affects the angle between the hands. The measure of angle 1 is now 2.3 × 6, or 13.8°. The measure of angle 4 is now $\left(\frac{32.7}{60}\right) \times 30$, or 16.35°. Therefore, the new angle has measure 13.8 + 30 + 30 + 16.35, or 90.15°. We could continue in this manner to get closer and closer to a right angle. We could also use rate equations and find the time exactly. Another possible problem is, "How many times during a 12-hour period are right angles formed?"

QUESTIONS FROM THE CLASSROOM

1. A student claims that if any two planes that do not intersect are parallel, then any two lines that do not intersect should also be parallel. How do you respond?
2. A student says that it is actually impossible to measure an angle, since each angle is the union of two rays that extend infinitely and therefore continue forever. What is your response?
3. A student asks whether a polygon whose sides are congruent is necessarily a regular polygon and whether a polygon with all angles congruent is necessarily a regular polygon. How do you answer?
4. A student thinks that a square is the only regular polygon with all right angles. The student asks if this is true and if so, why. How do you answer?
5. A student says that a line is parallel to itself. How do you reply?
6. A students says that a line in the plane of the classroom ceiling cannot be parallel to a line in the plane of the classroom floor because the lines are not in the same plane. Is this student correct? Why?
7. "If a line were a great circle, as on a globe, would there be any parallel lines?" asks one student. What is your answer?
8. One student says, "My sister's high school geometry book talked about equal angles. Why don't we use equal angles instead of congruent angles?" How do you reply?
9. A student says there can be only 360 different rays emanating from a point, since there are only 360° in a circle. How do you respond?

CHAPTER OUTLINE

I. Basic geometric notions
   A. Points, lines, and planes
   1. **Points, lines,** and **planes** are basic, but undefined terms.
   2. **Collinear points** are points that belong to the same line.
   3. Important subsets of lines are **segments, half-lines,** and **rays**.
   4. **Coplanar points** are points that lie in the same plane. **Coplanar lines** are defined similarly.
   5. Two lines with exactly one point in common are **intersecting lines.**
   6. **Concurrent lines** are lines that contain a common point.
   7. Two distinct coplanar lines with no points in common are **parallel.**
   8. **Skew lines** are lines that cannot be contained in the same plane.
   9. **Parallel planes** are planes with no points in common.
   10. **Space** is the set of all points.
   11. An **angle** is the union of two rays with a common endpoint.
   12. Angles are classified according to size as **acute, obtuse, right,** or **straight.**
   13. Two lines that meet to form a right angle are **perpendicular.**
   14. A **dihedral angle** is the union of two half-planes and the common line defining the half-planes.
   B. Plane figures
   1. A **closed curve** is a curve that, when traced, has the same starting and stopping points and may cross itself at individual points.
   2. A **simple curve** is a curve that does not cross itself when traced, except that it is possible for the starting and stopping points to be the same.
   3. A **polygonal curve** is a curve made up of line segments.
   4. A **polygon** is a simple closed polygonal curve.
      (a) A **diagonal** is any line segment connecting two nonconsecutive vertices of a polygon.
      (b) A **convex polygon** is one such that, if any two points of the polygonal region are connected by a segment, the segment is a subset of the polygonal region.
      (c) A **concave polygon** is a nonconvex polygon.
      (d) A **regular polygon** is a polygon in which all the angles are congruent and all the sides are congruent.

5. A **polygonal region** is the union of a polygon and its interior.
6. **Jordan Curve theorem:** A simple closed curve separates the plane into three disjoint subsets: the interior of the curve, the exterior of the curve, and the curve itself.
7. **Congruent figures** are figures with the same size and shape.
8. Triangles are classified according to the lengths of their sides as **scalene, isosceles,** or **equilateral,** and according to the measures of their angles as **acute, obtuse,** or **right.**
9. Quadrilaterals with special properties are **trapezoids, parallelograms, rectangles, kites, isosceles trapezoids, rhombuses,** and **squares.**
10. A **circle** is a set of points in a plane that lie at the same distance from a given point, called the **center.**

II. Theorems involving angles
  A. **Supplements** of the same angle, or of congruent angles, are congruent.
  B. **Complements** of the same angle, or of congruent angles, are congruent.
  C. **Vertical angles** formed by intersecting lines are congruent.
  D. If any two distinct lines are cut by a transversal, then a pair of **corresponding angles, alternate interior angles,** or **alternate exterior angles** are congruent if, and only if, the lines are parallel.
  E. The sum of the measures of the angles of a triangle is 180°.
  F. The sum of the measures of the interior angles of any convex polygon with n sides is $(n-2)180°$. One angle of a regular polygon measures $(n-2)180°/n$.
  G. The sum of the measures of the exterior angles of any convex polygon is 360°.
  H. **Total Turtle Trip Theorem:** Any convex polygon can be drawn with a total turning of 360°.
  I. **Closed Path Theorem:** The total turning around any closed path is a multiple of 360°.

III. Three-dimensional figures
  A. A **polyhedron** is a simple closed surface formed by polygonal regions.
  B. Three-dimensional figures with special properties are **prisms, pyramids, regular polyhedra, cylinders, cones,** and **spheres.**
  C. **Euler's formula,** $V + F - E = 2$, holds for polyhedra, where V, E, and F represent the number of vertices, the number of edges, and the number of faces of a polyhedron, respectively.

*IV. Networks
  A. A **network** is a collection of points called **vertices** and a collection of curves called **arcs.**
  B. A vertex of a network is called an **even vertex** if the number of arcs meeting at the vertex is even. A vertex is called an **odd vertex** if the number of arcs meeting at a vertex is odd.
  C. A network is called **traversable** if there is a path through the network such that each arc is passed through exactly once.
      1. If all the vertices of a network are even, then the network is traversable. Any vertex can be a starting point and the same vertex must be the stopping point.
      2. If a network has two odd vertices, it is traversable. One odd vertex must be the starting point and the other odd vertex must be the stopping point.
      3. If a network has more than two odd vertices, it is not traversable.
      4. No network has exactly one odd vertex.

CHAPTER TEST

1. Sketch diagrams such that each of the following is true:
   (a) The intersection of two segments is a segment.
   (b) The intersection of two rays is a ray.
   (c) The intersection of two rays is a segment.
   (d) The intersection of two angles is an angle.
   (e) The intersection of two angles is a segment.
2. (a) List three different names for line m.

   (b) Name two different rays on m that have endpoint B.
   (c) Find a simpler name for $\overrightarrow{AB} \cap \overrightarrow{BA}$.
   (d) Find a simpler name for $\overleftrightarrow{AB} \cap \overrightarrow{BC}$.
   (e) Find a simpler name for $\overrightarrow{BA} \cap \overrightarrow{AC}$.
3. In the following figure, $\overleftrightarrow{PQ}$ is perpendicular to α:

   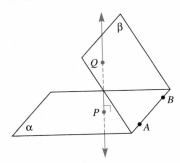

   (a) Name a pair of skew lines.

(b) Using only the letters in the figure, name as many planes as possible that are each perpendicular to α.
(c) What is the intersection of planes *APQ* and β?
▶(d) Is there a single plane containing *A, B, P,* and *Q*? Explain your answer.

4. For each of the following, sketch two parallelograms, if possible, that satisfy the given conditions:
   (a) Their intersection is a single point.
   (b) Their intersection is exactly two points.
   (c) Their intersection is exactly three points.
   (d) Their intersection is exactly one line segment.

5. Draw each of the following curves:
   (a) A simple closed curve
   (b) A closed curve that is not simple
   (c) A concave hexagon
   (d) A convex decagon

6. ▶(a) Can a triangle have two obtuse angles? Justify your answer.
   ▶(b) Can a parallelogram have four acute angles? Justify your answer.

7. In a certain triangle, the measure of one angle is twice the measure of the smallest angle. The measure of the third angle is seven times greater than the measure of the smallest angle. Find the measures of each of the angles in the triangle.

8. ▶(a) Explain how to derive an expression for the sum of the measures of the angles in a convex *n*-gon.
   (b) In a certain regular polygon, the measure of each angle is 176°. How many sides does the polygon have?

9. (a) Sketch a convex polyhedron with at least 10 vertices.
   (b) Count the number of vertices, edges, and faces for the polyhedron in (a) and determine if Euler's formula holds for this polyhedron.

10. Sketch each of the following:
    (a) Three planes that intersect in a point
    (b) A plane and a cone that intersect in a circle
    (c) A plane and a cylinder that intersect in a segment
    (d) Two pyramids that intersect in a triangle

11. Sketch drawings to illustrate different possible intersections of a square pyramid and a plane.

12. If $3x°$ and $(6x - 18)°$ are measures of corresponding angles formed by two parallel lines and a transversal, what is the value of *x*?

13. Find 6°48′59″ + 28°19′36″. Write your answer in simplest terms.

14. In the figure, ℓ is parallel to *m*, and $m(\angle 1) = 60°$. Find each of the following:

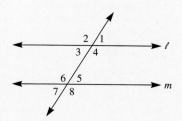

   (a) $m(\angle 3)$ (b) $m(\angle 6)$ (c) $m(\angle 8)$

15. If a pyramid has an octagon for a base, how many lateral faces does it have?

16. If *ABC* is a right triangle and $m(\angle A) = 42°$, what is the measure of the other acute angle?

*17. (a) Which of the following networks are traversable?
     (b) Find a corresponding path for the networks that are traversable.

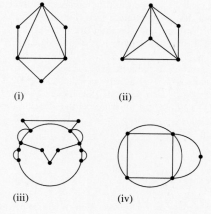

(i)   (ii)

(iii)   (iv)

*18. 🖥 Write a Logo program to construct two perpendicular segments.

*19. 🖥 Write a Logo program called ISOS to draw a variable-sized isosceles triangle with inputs :LEG for the length of one of the congruent sides and :ANGLE for the angle between the congruent sides.

SELECTED BIBLIOGRAPHY

Adele, G. "When Did Euclid Live? An Answer Plus a Short History of Geometry." *Mathematics Teacher* 82 (September 1989): 460–463.

Billstein, R., and J. Lott. "The Turtle Deserves a Star." *Arithmetic Teacher* 33 (March 1986): 14–16.

Bledsoe, G. "Guessing Geometric Shapes." *Mathematics Teacher* 80 (March 1987): 178–180.

Bright, G., and J. Harvey. "Games, Geometry, and Teaching." *Mathematics Teacher* 81 (April 1988): 250–259.

Bright, G., and J. Harvey. "Learning and Fun with Geometry Games." *Arithmetic Teacher* 35 (April 1988): 22–26.

Carroll, W. "Cross Sections of Clay Solids." *Arithmetic Teacher* 35 (March 1988): 6–11.

Fuys, D., D. Geddes, and R. Tischler. *The van Hiele Model of Thinking in Geometry among Adolescents*. Reston, Va.: National Council of Teachers of Mathematics, 1988.

Hoffer, A. "Making a Better Beer Glass." *Mathematics Teacher* 75 (May 1982): 378–379.

Hoffer, A. *Van Hiele-based Research*. In R. Lesh and M. Landau, *Acquisition of Mathematics Concepts and Processes*. New York: Academic Press, 1983.

Kriegler, S. "The Tangram—It's More Than an Ancient Puzzle." *Arithmetic Teacher* 38 (May 1991): 38–43.

Morrow, L. "Geometry through the Standards." *Arithmetic Teacher* 38 (April 1991): 21–25.

Posamentier, A. "Geometry: A Remedy for the Malaise of Middle School Mathematics." *Mathematics Teacher* 82 (December 1989): 678–680.

Souza, R. "Golfing with a Protractor." *Arithmetic Teacher* 35 (April 1988): 52–56.

Teppo, A. "Van Hiele Levels of Geometric Thought Revisited." *Mathematics Teacher* 84 (March 1991): 210–221.

Thiessen, D., and M. Matthias. "Selected Children's Books for Geometry." *Arithmetic Teacher* 37 (December 1989): 47–51.

Van Hiele. *Structure and Insight*. New York: Academic Press, 1986.

Wilson, M. "Measuring a van Hiele Geometry Sequence: A Reanalysis." *Journal for Research in Mathematics Education* 21 (May 1990): 230–237.

Winter, M., et al. *Middle Grades Mathematics Project Spatial Visualization*. Menlo Park, Calif.: Addison-Wesley, 1986.

11 Constructions, Congruence, and Similarity

Preliminary Problem

Mary Rose needed a wooden triangle for her art project with one side 18 in. long. The angles of the triangle needed to be 36°, 84°, and 60°. Mary Rose had a carpenter make such a triangle. A few months later, she needed another triangle with the same specifications. The first carpenter was unavailable so she contacted a different one. When the second triangle arrived, she was surprised that it was not congruent to the first one. Explain how this is possible.

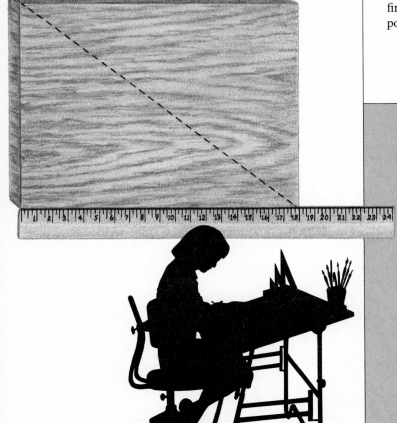

570 CHAPTER 11 CONSTRUCTIONS, CONGRUENCE, AND SIMILARITY

In the *Standards* (p. 112), it is recommended that in grades 5–8 *students examine and discover relationships and develop spatial sense by constructing, drawing, measuring, visualizing, comparing, transforming, and classifying geometric figures.* In this chapter, we introduce the concepts of congruence and similarity and investigate properties of congruent triangles through compass-and-straightedge constructions. The straight line and circle were considered the basic geometric figures by the Greeks and the straightedge and compass are their physical analogues. It is also believed that the Greek philosopher Plato (427–347 B.C.) rejected the use of other mechanical devices for geometric constructions because they emphasized practicality rather than "ideas," which he regarded as more important. Constructions are also done in this chapter by means of paper folding and through use of a plastic device called a Mira.

Throughout the chapter, we use linear measurement and the notion of length, although a formal discussion of measurement is postponed until Chapter 13.

Section 11-1   Congruence Through Constructions

congruent   In mathematics, the word **congruent** is used to describe objects that have the same size and shape. In elementary schools, tracing is a method for determining congruence. For example, the squares in Figure 11-1 are congruent because a tracing of one square can be made to match the other. We say that square $ABCD$ is congruent to square $EFGH$, and we write $ABCD \cong EFGH$.

FIGURE 11-1
Congruent squares

Any two line segments have the same shape, so *two line segments are congruent if they have the same size (length)*. The length of the line segment $\overline{AB}$ is denoted by AB. In a similar way, two angles are congruent if, and only if, their measures are the same.

$$\overline{AB} \cong \overline{CD} \text{ if, and only if, } AB = CD$$
$$\angle ABC \cong \angle DEF \text{ if, and only if, } m(\angle ABC) = m(\angle DEF)$$

The congruence relation has the reflexive, symmetric, and transitive properties.

HISTORICAL NOTE

The three most famous compass-and-ruler construction problems of antiquity are:

1. Constructing a square equal in area to a given circle (often referred to as "squaring the circle").
2. Constructing the edge of a cube whose volume is double that of a cube of a given edge.

3. Trisecting any angle.

For generations, mathematicians endeavored to solve these problems, but their efforts were unsuccessful. In the nineteenth century, as a result of the work of the French mathematician Evariste Galois (1811–1832), it was proved that these constructions cannot be done using only a compass and a straightedge. Although Galois died in a duel at age 21, his discoveries had a lasting influence on modern mathematics.

Geometric Constructions

A geometric construction is a task in which, given some geometric elements such as points, segments, angles, or circles, other elements are derived by using certain well-defined instruments. Geometers were for generations interested in knowing which problems could be solved with the given instruments.

Ancient Greek mathematicians constructed geometric figures with a straightedge (no markings on it) and a collapsible compass. Figure 11-2 shows a modern compass. It can be used to mark off and duplicate lengths and to construct circles or arcs with a radius of a given measure. To draw a circle when given the radius PQ of a circle, we follow the steps illustrated in Figure 11-2.

FIGURE 11-2 Construct a circle given its radius.

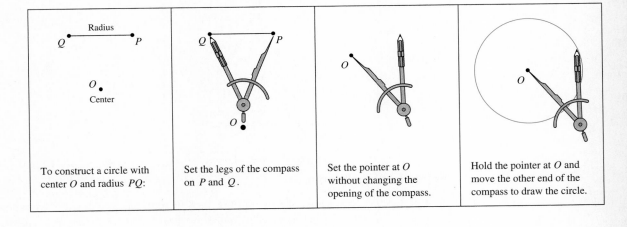

The figure formed in Figure 11-2 is a circle with center O and is called circle O. Any other circle is congruent to circle O if the radii of the two circles are congruent. In general, *two circles are congruent if their radii are congruent.*

arc
center of an arc

An **arc** of a circle can be thought of as any part of the circle that can be drawn without lifting a pencil. An arc is either a part of a circle or the entire circle. The **center of an arc** is the center of the circle containing the arc.

To avoid the ambiguity of two points on a circle determining two different arcs, an arc is normally named by three letters, such as arc ACB, denoted by $\widehat{ACB}$ in Figure 11-3. In this notation, the first and last letters indicate the endpoints of the arc, and the middle letter indicates which of two possible arcs is intended. If there is no danger of ambiguity in a

minor arc

discussion, two letters are used to name the arc formed. In Figure 11-3, $\widehat{ACB}$, is a **minor**

major arc **arc,** and $\widehat{ADB}$ is a **major arc.** If the major arc and the minor arc of a circle are the same
semicircle size, each is a **semicircle.**

FIGURE 11-3

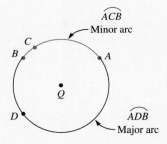

Constructing Segments

There are many ways to construct a segment congruent to a given segment $\overline{AB}$. A natural approach is to use a ruler, measure $\overline{AB}$, and then draw a congruent segment. A different way is to trace $\overline{AB}$ onto a piece of paper. A third method is to use a straightedge and a compass (as in Figure 11-4).

FIGURE 11-4 Construct a line segment congruent to a given segment.

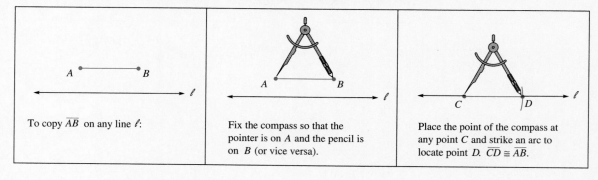

Triangle Congruence

Informally, two figures are congruent if it is possible to fit one figure onto the other so that all matching parts are congruent. If we were to trace $\triangle ABC$ in Figure 11-5 and put the tracing over $\triangle A'B'C'$ so that the tracing of A is over A', the tracing of B is over B', and the tracing of C is over C', we would see that the tracing of $\triangle ABC$ is identical to the tracing of $\triangle A'B'C'$. This suggests the following definition of congruent triangles.

> **DEFINITION OF CONGRUENT TRIANGLES**
>
> $\triangle ABC$ is congruent to $\triangle A'B'C'$, written $\triangle ABC \cong \triangle A'B'C'$, if $\angle A \cong \angle A'$, $\angle B \cong \angle B'$, $\angle C \cong \angle C'$, $\overline{AB} \cong \overline{A'B'}$, $\overline{BC} \cong \overline{B'C'}$, and $\overline{AC} \cong \overline{A'C'}$.

To show that two triangles are congruent using the above definition, we need to find a one-to-one correspondence between the vertices of one triangle and the vertices of the other such that each pair of corresponding angles and each pair of corresponding sides are congruent. The statement, "Corresponding parts of congruent triangles are congruent," is sometimes abbreviated CPCTC.

If $\triangle ABC \cong \triangle A'B'C'$, as in Figure 11-5, then any rearrangement of the letters ABC and the corresponding rearrangement of $A'B'C'$ results in another symbolic representation of the same congruence. For example, $\triangle BCA \cong \triangle B'C'A'$.

FIGURE 11-5

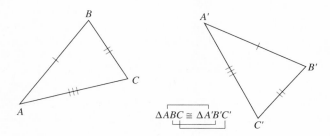

Because there are $3 \cdot 2 \cdot 1$, or 6, ways to rearrange the letters A, B, and C, each pair of congruent triangles can be symbolized in six ways. However, each of the six symbolic representations gives the same information about the triangles.

E X A M P L E 1 1 - 1 Assume that each of the pairs of triangles in Figure 11-6 is congruent, and write an appropriate symbolic congruence in each case.

FIGURE 11-6

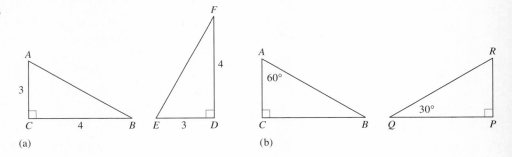

Solution (a) Vertex C corresponds to D because the angles at C and D are right angles. Also, $\overline{CB} \cong \overline{DF}$ and C corresponds to D, so B corresponds to F. Consequently, the remaining vertices must correspond; that is, A corresponds to E. Thus, one possible symbolic congruence is $\triangle ABC \cong \triangle EFD$.

(b) Vertex C corresponds to P because the angles at C and P are both right angles. To establish the other correspondences, we first find the missing angles in the triangles. We see that $m(\angle B) = 90° - 60° = 30°$ and $m(\angle R) = 90° - 30° = 60°$. Consequently, A corresponds to R because $m(\angle A) = m(\angle R) = 60°$, and B corresponds to Q because $m(\angle B) = m(\angle Q) = 30°$. Thus, one possible symbolic congruence is $\triangle ABC \cong \triangle RQP$.

Side, Side, Side Property (SSS)

In an automotive assembly line, the entire production process is designed in such a way that the same model cars are congruent to each other. (Outside paint and different colors of upholstery fabric and interiors provide individual differences.) Calibration experts work to ensure that car parts are interchangeable, so the same part fits on all basic models of the same car. To ensure that the cars are congruent, the parts must be congruent. In considering congruence of figures in geometry, we apply the same process. In the assembly line of automotive production, decisions have to be made about the minimal set of items to consider for eventual congruency.

If three sides and three angles of one triangle are congruent to the corresponding six items of another triangle, then we can conclude from the definition of congruent triangles that the triangles are congruent. Do we need to know that all six parts of one triangle are congruent to the corresponding parts of the second triangle to conclude that the triangles are congruent?

Consider the triangle formed by attaching three segments, as in Figure 11-7. Such a triangle is *rigid*. Its size and shape cannot be changed. Because of this property, a manufacturer can make duplicates if the lengths of the sides are known. Many bridges or other structures that have exposed frameworks demonstrate the practical use of rigidity of triangles. Using drinking straws and straight pins, you can experiment with different triangles to verify that all triangles are rigid.

Because a triangle is completely determined by its three sides, we have the following property.

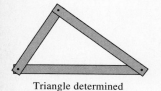

Triangle determined by its 3 sides

FIGURE 11-7

PROPERTY

Side, Side, Side (SSS) If the three sides of one triangle are congruent, respectively, to the three sides of a second triangle, then the triangles are congruent.

EXAMPLE 11-2 For each of the parts in Figure 11-8, use SSS to explain why the given triangles are congruent.

FIGURE 11-8

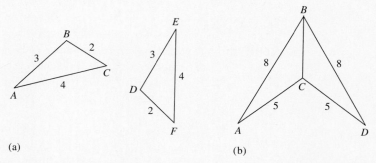

(a) (b)

Solution (a) $\triangle ABC \cong \triangle EDF$ by SSS because $\overline{AB} \cong \overline{ED}$, $\overline{BC} \cong \overline{DF}$, and $\overline{AC} \cong \overline{EF}$.

(b) $\triangle ABC \cong \triangle DBC$ by SSS because $\overline{AB} \cong \overline{DB}$, $\overline{AC} \cong \overline{DC}$, and $\overline{BC} \cong \overline{BC}$.

Constructing a Triangle Given Three Sides

Using the SSS property, we can construct a duplicate triangle if given a triangle or construct a triangle if given the lengths of the three sides. In Figure 11-9, $\triangle ABC$ is given, and we are to construct $\triangle A'B'C'$ congruent to $\triangle ABC$.

FIGURE 11-9

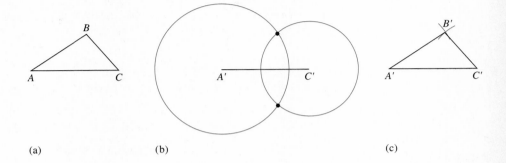

First, we construct a segment congruent to one of the three segments. For example, we may construct $\overline{A'C'}$ so that it is congruent to $\overline{AC}$. To complete the triangle construction, the other vertex, B', must be located. The distance from A' to B' is AB. All points at a distance AB from A' are on a circle with center at A' and radius of length AB. Similarly, B' must be on a circle with center C' and radius of length BC. The only possible locations for B' are at the points where the two circles intersect. Either point is acceptable. Usually, a picture of the construction shows only one possibility, and the construction uses only arcs, as pictured in Figure 11-9(c).

REMARK Starting the construction with a segment $\overline{A'B'}$ congruent to $\overline{AB}$ or with $\overline{B'C'}$ congruent to $\overline{BC}$ would also result in triangles congruent to $\triangle ABC$.

From the preceding construction, it may seem that, given any three segments, it is possible to construct a triangle whose sides are congruent to the given segments. However, this is not the case. For example, consider the segments in Figure 11-10(a), whose measures are p, q, and r. If we choose the base of the triangle to be a side of length p and attempt to find the third vertex by intersecting arcs, as in Figure 11-10(b), we find that the arcs do not intersect. Because no intersection occurs, a triangle is not determined.

FIGURE 11-10

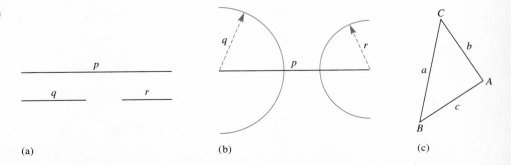

In Figure 11-10(c), we see that the path from B to A along $\overline{BC}$ and then $\overline{CA}$ is longer than the path from B to A along $\overline{AB}$. Thus, $a + b > c$. Consequently, we have the following property.

> **PROPERTY**
>
> **Triangle Inequality** The sum of the measures of any two sides of a triangle must be greater than the measure of the third side.

The above property tells us that segments of length 3 cm, 5 cm, and 9 cm do not determine a triangle, because $3 + 5$ is not greater than 9.

Constructing Congruent Angles

We use the SSS notion of congruent triangles to construct an angle congruent to a given angle $\angle B$ by making $\angle B$ a part of an isosceles triangle and then reproducing this triangle, as in Figure 11-11.

FIGURE 11-11
Copy an angle.

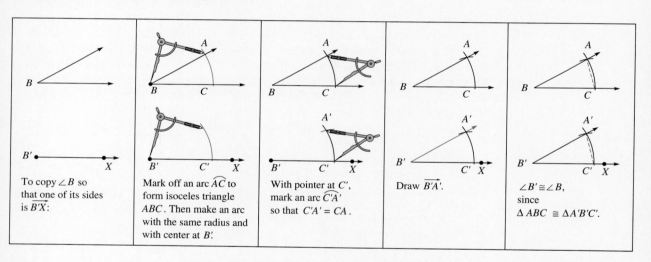

We have seen that, given three segments, no more than one triangle can be constructed. Could more than one triangle be constructed from only two segments? Consider Figure 11-12(b), which shows three different triangles with sides congruent to the segments given in Figure 11-12(a). The length of the third side depends on the measure of the angle between the other two sides, the **included angle.**

included angle

FIGURE 11-12

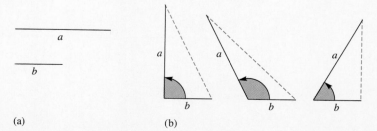

SECTION 11-1 CONGRUENCE THROUGH CONSTRUCTIONS

Side, Angle, Side Property (SAS)

It appears that, if we knew the lengths of two sides and the measure of the angle included between these two sides, we could construct a unique triangle. This is true, and we can express the rule as the **Side, Angle, Side (SAS)** property.

Side, Angle, Side (SAS)

> **PROPERTY**
>
> **Side, Angle, Side (SAS)** If two sides and the included angle of one triangle are congruent to two sides and the included angle of another triangle, respectively, then the two triangles are congruent.

EXAMPLE 11-3 For each part of Figure 11-13, use SAS to show that the given pair of triangles are congruent.

FIGURE 11-13

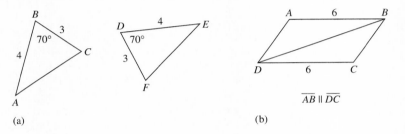

Solution (a) $\triangle ABC \cong \triangle EDF$ by SAS because $\overline{AB} \cong \overline{ED}$, $\angle B \cong \angle D$, and $\overline{BC} \cong \overline{DF}$.

(b) Because $\overline{AB} \cong \overline{CD}$ and $\overline{DB} \cong \overline{BD}$, we need either another side or another angle to show that the triangles are congruent. We know nothing about the sides except that $\overline{AB} \parallel \overline{DC}$. Since parallel segments $\overline{AB}$ and $\overline{DC}$ are cut by transversal $\overline{BD}$, we have alternate interior angles $\angle ABD$ and $\angle BDC$ congruent. Now $\triangle ABD \cong \triangle CDB$ by SAS. ∎

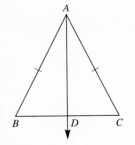

EXAMPLE 11-4 Given isosceles triangle ABC with $\overline{AB} \cong \overline{AC}$ and $\overrightarrow{AD}$ the bisector of $\angle A$, as shown in Figure 11-14, show that $\angle B \cong \angle C$.

Solution Because $\overrightarrow{AD}$ is the bisector of $\angle A$, then $\angle BAD \cong \angle CAD$. Also, $\overline{AD} \cong \overline{AD}$ and $\overline{AB} \cong \overline{AC}$, so $\triangle BAD \cong \triangle CAD$ by SAS. Therefore, $\angle B \cong \angle C$ because the angles are corresponding parts of congruent triangles. ∎

FIGURE 11-14

Example 11-4 proves the following theorem.

> **THEOREM 11-1**
>
> If two sides of a triangle are congruent, then the angles opposite these sides are congruent.

Constructions Involving Two Sides and an Angle of a Triangle

Figure 11-15 shows how to construct a triangle congruent to △ABC by using two sides $\overline{AB}$ and $\overline{AC}$ and the *included angle,* ∠A, formed by these sides. First, a ray with an arbitrary endpoint A' is drawn, and $\overline{A'C'}$ is constructed congruent to $\overline{AC}$. Then, ∠A' is constructed so that ∠A' ≅ ∠A, and B' is marked on the side of ∠A' not containing C' so that $\overline{A'B'}$ ≅ $\overline{AB}$. Connecting B' and C' completes △A'B'C' so that △A'B'C' ≅ △ABC.

FIGURE 11-15

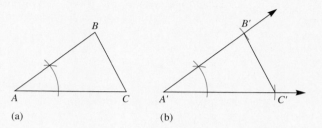

(a)　　　(b)

If, in two triangles, two sides and an angle not included between these sides are congruent, respectively, the information is not sufficient to guarantee congruent triangles. For example, consider △ABC in Figure 11-16. We locate point B' on $\overline{BC}$ so that $\overline{AB'}$ ≅ $\overline{AB}$. (How?) Then, in △ABC and △AB'C, $\overline{AB}$ ≅ $\overline{AB'}$, $\overline{AC}$ ≅ $\overline{AC}$, and ∠C ≅ ∠C. Thus, in the two triangles, two sides and an angle not included between these sides are congruent, but the triangles are not congruent. In certain special cases, it is impossible to locate a point B' on $\overline{BC}$ so that $\overline{AB'}$ ≅ $\overline{AB}$.

FIGURE 11-16

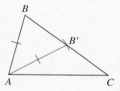

PROBLEM SET 11-1

1. (a) Draw any triangle ABC in which the measure ∠A is greater than the measure of ∠B. Compare BC and AC. What did you find?
   (b) Draw any triangle ABC in which BC is greater than AC. Measure the angles opposite $\overline{BC}$ and $\overline{AC}$. Compare the angle measures. What did you find?
   ▶ (c) Based on your findings in (a) and (b), make a conjecture concerning the lengths of sides and the measures of angles of a triangle.
2. Using a ruler, protractor, compass, or tracing paper, construct each of the following, if possible:
   (a) A segment congruent to $\overline{AB}$ and an angle congruent to ∠CAB

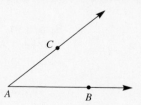

   (b) A triangle with sides of lengths 2 cm, 3 cm, and 4 cm
   (c) A triangle with sides of lengths 4 cm, 3 cm, and 5 cm (What kind of triangle is it?)
   (d) A triangle with sides 4 cm, 5 cm, and 10 cm

(e) An equilateral triangle with sides 5 cm
(f) A triangle with sides 6 cm and 7 cm and an included angle of measure 75°
(g) A triangle with sides 6 cm and 7 cm and a nonincluded angle of measure 40°
(h) A triangle with sides 6 cm and 6 cm and a nonincluded angle of measure 40°
(i) A right triangle with legs 4 cm and 8 cm (The legs include the right angle.)

3. ▶For each of the conditions in Problem 2(b)–(h), does the given information determine a unique triangle? Explain why or why not.

4. How many different triangles can be constructed with toothpicks by connecting the toothpicks only at their ends if each side of a triangle can contain at most five toothpicks?

5. For each of the following, determine whether the given conditions are sufficient to prove that $\triangle PQR \cong \triangle MNO$. Justify your answers.
   (a) $\overline{PQ} \cong \overline{MN}$, $\overline{PR} \cong \overline{MO}$, $\angle P \cong \angle M$
   (b) $\overline{PQ} \cong \overline{MN}$, $\overline{PR} \cong \overline{MO}$, $\overline{QR} \cong \overline{NO}$
   (c) $\overline{PQ} \cong \overline{MN}$, $\overline{PR} \cong \overline{MO}$, $\angle Q \cong \angle N$

6. ▶A rancher designed a wooden gate as shown. Explain the purpose of the diagonal boards on the gate.

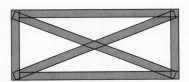

7. ▶A rural homeowner had his television antenna held in place by three guy wires, as shown below. If the distances to each of the stakes from the base of the antenna are the same, what is true about the lengths of the wires? Why?

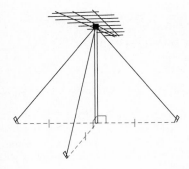

8. ▶A group of students on a hiking trip wants to find the distance AB across a pond. One student suggests choosing any point C, connecting it with B, and then finding point D such that $\angle DCB \cong \angle ACB$ and $\overline{DC} \cong \overline{AC}$. How and why does this help in find the distance AB?

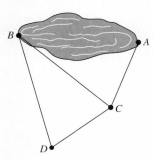

9. Using only a compass and a straightedge, perform each of the following:
   (a) Reproduce $\angle A$.

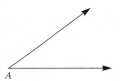

   (b) Construct an equilateral triangle with side $\overline{AB}$.

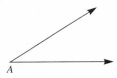

   (c) Construct a 60° angle.
   (d) Construct an isosceles triangle with $\angle A$ as the angle included between the two congruent sides.

10. Using only a compass and a straightedge, perform each of the following:
    (a) Construct $\angle C$ so that $m(\angle C) = m(\angle A) + m(\angle B)$.

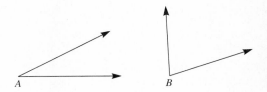

    (b) Using the angles in (a), construct $\angle C$ so that $m(\angle C) = m(\angle B) - m(\angle A)$.

11. An equilateral triangle ABC is congruent to itself.
    (a) Write all possible true correspondences between the triangle and itself.

(b) Use one of your answers in (a) to show that an equilateral triangular is also equiangular.

12. In the following drawing, $\overrightarrow{BD}$ bisects $\angle ABC$ of isosceles triangle ABC with $\overline{AB} \cong \overline{CB}$.

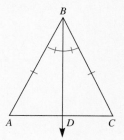

(a) Make a conjecture about a relation between $\overline{AD}$ and $\overline{CD}$.
▶(b) Justify your conjecture in (a).
(c) What are the measures of $\angle ADB$ and $\angle CDB$?
▶(d) Justify your answer in (c).

13. We have seen that, if in two triangles, two sides and an angle not included between these sides are congruent, the triangles are not necessarily congruent. For what kinds of angles would it be possible to conclude that the triangles are congruent?

14. Suppose that the following polygon $ABCD$ is any square:
   (a) What is the relationship between point F and the diagonals $\overline{BD}$ and $\overline{AC}$?
   ▶(b) Justify your answer in (a).
   (c) What are the measures of $\angle BFA$ and $\angle AFD$?
   ▶(d) Justify your answer in (c).

15. The diagonals of a quadrilateral bisect each other (that is, each diagonal is divided by the point of intersection into two congruent segments).
   (a) What kind of quadrilateral must it be?
   ▶(b) Justify your answer in (a).

16. Construct several noncongruent rhombuses and several noncongruent parallelograms that are not rhombuses. In each case, construct the diagonals.
   (a) Based on your observations, what is true about the angles formed by the diagonals of a rhombus that is not necessarily true about the angles formed by the diagonals of a parallelogram that is not a rhombus?
   ▶(b) Justify your conjecture in (a).

17. (a) What kind of figure is a quadrilateral in which both pairs of opposite sides are congruent?
   ▶(b) Justify your answer in (a).

★18. ▶The theorem stating that the base angles of an isosceles triangle are congruent followed from Example 11-4, where the angle bisector $\overrightarrow{AD}$ was drawn. It is also possible to justify this theorem without drawing an angle bisector. Show that $\triangle BAC \cong \triangle CAB$ by using one of the congruence properties, and hence conclude the theorem.

19. 🖥 Write a Logo procedure to draw a variable-sized equilateral triangle.

20. 🖥 Logo programs can be used to construct triangles based on parts of a triangle. Type the following program into your computer and then run the following:
   (a) SAS 50 75 83
   (b) SAS 60 120 60

```
TO SAS :SIDE1 :ANGLE :SIDE2
  DRAW
  FORWARD :SIDE1
  RIGHT 180 - :ANGLE
  FORWARD :SIDE2
  HOME
END
```

(In LCSI, replace DRAW with CLEARSCREEN (CS).)

21. 🖥 Following is a different procedure for constructing a triangle when given two sides and an included angle. Compare this procedure to the one in Problem 20.

```
TO SAS1 :SIDE1 :ANGLE :SIDE2
  DRAW
  BACK :SIDE1
  RIGHT :ANGLE
  FORWARD :SIDE2
  HOME
END
```

(In LCSI, replace DRAW with CLEARSCREEN (CS).)

22. (a) In Problem 21, if SAS 50 190 60 is executed, what is the result?
   (b) Is it possible in reality to draw a triangle with sides of 50 and 60 units and an included angle of 190°? Why?
   (c) What line could be added to the procedure in Problem 21 to correct the "bug" you encountered in (b)?

LABORATORY ACTIVITY

In this van Hiele Level 3 activity, you are given the name of a shape. (i) List sufficient properties to define that shape. For example, if given the words *isosceles triangle*, you might say that it is a triangle with at least two sides congruent. (ii) Now derive other properties of the shape, based upon your answer in (i). For example, you could show that the base angles are congruent in the isosceles triangle.

(a) Parallelogram

(b) Rectangle

(c) Kite

(d) Rhombus

(e) Square

Section 11-2 Other Congruence Properties

Angle, Side, Angle (ASA)

We have seen that triangles can be determined to be congruent by SSS and SAS. Can a triangle be constructed congruent to a given triangle by using two angles and a side? Figure 11-17 shows the construction of a triangle $A'B'C'$ such that $\overline{A'C'} \cong \overline{AC}$, $\angle A' \cong \angle A$, and $\angle C' \cong \angle C$. It seems that $\triangle A'B'C' \cong \triangle ABC$. This construction illustrates a property of congruence called **Angle, Side, Angle**, abbreviated **ASA**.

Angle, Side, Angle (ASA)

FIGURE 11-17

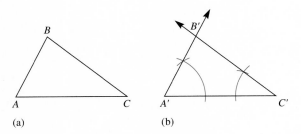

(a)   (b)

PROPERTY

Angle, Side, Angle (ASA) If two angles and the included side of one triangle are congruent to two angles and the included side of another triangle, respectively, then the triangles are congruent.

In Figure 11-18, △ABC and △DEF have two pairs of angles congruent and a pair of sides congruent. If ∠A ≅ ∠D and ∠B ≅ ∠E, we can deduce that ∠C ≅ ∠F because both are equal to 180° − (70 + 40)°. Then, since $\overline{AC} \cong \overline{DF}$, we have △ABC ≅ △DEF by ASA.

FIGURE 11-18

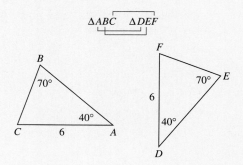

In general, by using the ASA property, we can justify the following property.

PROPERTY

Angle, Angle, Side (AAS) If two angles and a corresponding side of one triangle are congruent to two angles and a corresponding side of another triangle, respectively, then the two triangles are congruent.

EXAMPLE 11-5 Show that each of the given pairs of triangles in Figure 11-19 is congruent.

FIGURE 11-19

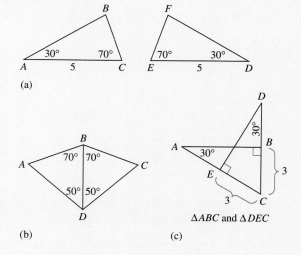

Solution (a) ∠A ≅ ∠D, $\overline{AC} \cong \overline{DE}$, and ∠C ≅ ∠E. Consequently, by ASA, △ABC ≅ △DFE.

(b) $\angle ABD \cong \angle CBD$, $\overline{BD} \cong \overline{BD}$, and $\angle ADB \cong \angle CDB$. Consequently, by ASA, $\triangle ABD \cong \triangle CBD$.

(c) $\angle A \cong \angle D$, $\angle ABC \cong \angle DEC$, and $\overline{BC} \cong \overline{EC}$. Consequently, by AAS, $\triangle ABC \cong \triangle DEC$. ∎

Is it possible to have two angles and a side of one triangle congruent to two angles and a side of another triangle and yet not have two congruent triangles? This question is explored in the problem set.

In Figure 11-20, the angles of one triangle are congruent to corresponding angles in another triangle, and the triangles are not congruent. Thus, an AAA property for congruency does not exist. (The triangles are *similar,* a concept discussed later in this chapter.)

FIGURE 11-20

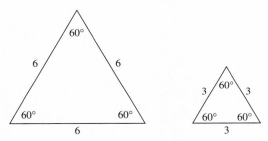

Using properties of congruent triangles, we can deduce various properties of quadrilaterals. Table 11-1 summarizes the definitions and lists some properties of five quadrilaterals. These and other properties of quadrilaterals are further investigated in the problem set.

Notice that the kite that is pictured in Table 11-1 is convex. Can you sketch a concave kite?

Congruent Polygons

Determining congruency conditions for polygons other than triangles is not an easy task. For example, the SSS property for congruent triangles has no analogy for quadrilaterals. The quadrilaterals in Figure 11-21 do not have the same shape, despite having four congruent sides. *One way to be sure that two polygons are congruent is to know that all corresponding sides and angles of the polygons are congruent.* This may be done by "moving" one figure to see if it "fits" exactly on top of the other figure.

FIGURE 11-21

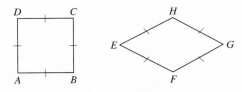

TABLE 11-1

Quadrilateral and Its Definition	Properties of the Quadrilateral
Parallelogram: A quadrilateral in which each pair of opposite sides is parallel	(a) Opposite sides are congruent. (b) Opposite angles are congruent. (c) Diagonals bisect each other.
Rectangle: A parallelogram with a right angle	(a) A rectangle has all the properties of a parallelogram. (b) All the angles of a rectangle are right angles. (c) A quadrilateral in which all the angles are right angles is a rectangle.
Rhombus: A parallelogram with all sides congruent	(a) A rhombus has all the properties of a parallelogram. (b) A quadrilateral in which all the sides are congruent is a rhombus. (c) The diagonals of a rhombus are perpendicular to each other. (d) Diagonals bisect opposite angles.
Square: A rectangle with all sides congruent	A square has all the properties of a parallelogram, a rectangle, and a rhombus.
Kite: A quadrilateral with two distinct pairs of consecutive sides congruent	(a) Lines containing the diagonals are perpendicular to each other. (b) A line containing one diagonal is a bisector of the other. (c) A line containing one diagonal bisects nonconsecutive angles.

PROBLEM SET 11-2

1. Use a ruler, protractor, and compass to construct each of the following, if possible:
   (a) A triangle with angles of 60° and 70° and an included side of 8 in.
   (b) A triangle with angles of 60° and 70° and a nonincluded side of 8 cm on a side of the 60° angle.
   (c) A right triangle with one acute angle of 75° and a leg of 5 cm on a side of the 75° angle.
   (d) A triangle with angles of 30°, 70°, and 80°.

2. ▶For each of the conditions in Problem 1(a)–(d), is it possible to construct two noncongruent triangles? Explain why or why not.

3. ▶For each of the following, determine whether the given conditions are sufficient to prove that $\triangle PQR \cong \triangle MNO$. Justify your answers.
   (a) $\angle Q \cong \angle N, \angle P \cong \angle M, \overline{PQ} \cong \overline{MN}$
   (b) $\angle R \cong \angle O, \angle P \cong \angle M, \overline{QR} \cong \overline{NO}$
   (c) $\overline{PQ} \cong \overline{MN}, \overline{PR} \cong \overline{MO}, \angle N \cong \angle Q$
   (d) $\angle P \cong \angle M, \angle Q \cong \angle N, \angle R \cong \angle O$

4. ▶A parallel ruler, as shown below, can be used to draw parallel lines. The distance between the parallel segments $\overline{AB}$ and $\overline{DC}$ can vary. The ruler is constructed so that the distance between A and B equals the distance between D and C. The distance between A and C is the same as the distance between B and D. How do you know that $\overline{AB}$ and $\overline{DC}$ are always parallel? Explain.

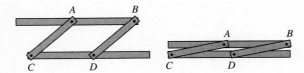

5. ▶Most ironing boards are collapsible for storage and can be adjusted to fit the height of the person using them. The surface of the board, though, remains parallel to the floor, regardless of the height. Explain how to construct the legs of an ironing board to ensure that the surface is always parallel to the floor.

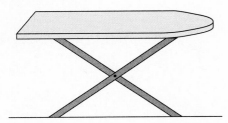

6. In each of the following, if possible, fill in the blank by choosing one of the words *parallelogram, rectangle, rhombus,* or *square* so that the resulting sentence is true. If none of the words makes the sentence true, answer "none" and justify your answer.
   (a) A quadrilateral is a _____ if, and only if, its diagonals bisect each other.
   (b) A quadrilateral is a _____ if, and only if, its diagonals are congruent.
   (c) A quadrilateral is a _____ if, and only if, its diagonals are perpendicular to each other.
   (d) A quadrilateral is a _____ if, and only if, its diagonals are congruent and bisect each other.
   (e) A quadrilateral is a _____ if, and only if, its diagonals are perpendicular to each other and bisect each other.
   (f) A quadrilateral is a _____ if, and only if, its diagonals are congruent, perpendicular to each other, and bisect each other.
   (g) A quadrilateral is a _____ if, and only if, a pair of opposite sides is parallel and congruent.

7. Classify each of the following statements as either true or false. If the statement is false, provide a counterexample.
   (a) The diagonals of a square are perpendicular bisectors of each other.
   (b) If all sides of a quadrilateral are congruent, the quadrilateral is a rhombus.
   (c) If a rhombus is a square, it must also be a rectangle.
   (d) An isosceles trapezoid can be a rectangle.
   (e) A square is a trapezoid.
   (f) A trapezoid is a parallelogram.
   (g) A parallelogram is a trapezoid.
   (h) No rectangle is a rhombus.
   (i) No trapezoid is a square.
   (j) Some squares are trapezoids.

8. (a) Construct quadrilaterals having exactly one, two, and four right angles.
   ▶(b) Why can a quadrilateral not have exactly three right angles?
   ▶(c) Can a parallelogram have exactly two right angles?

9. Each fourth grader is given a protractor, two 30-cm sticks and two 20-cm sticks and is asked to form a quadrilateral with a 75° angle. Sketch all possibilities.

10. ▶Stan is standing on the bank of a river wearing a baseball cap. Standing erect and looking directly at the other bank, he pulls the bill of his cap down until it just obscures his vision of the opposite bank. He then turns around, being careful not to disturb the cap and picks out a spot that is just obscured by the bill of his cap. He then paces off the distance to this spot and claims that the distance across the river is approximately equal to the distance he paced. What is the justification for Stan's claim?

11. In parallelogram *ABCD*, suppose that we connect *P*, a point on $\overline{DC}$, to *O* and extend $\overline{PO}$ until it intersects $\overline{AB}$ at *Q*.
    (a) How are $\overline{OP}$ and $\overline{QO}$ related?
    ▶(b) Justify your answer in (a).

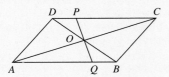

12. Figure *ABCD* is a kite.

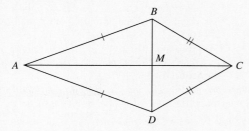

    ▶(a) Argue that $\overline{AC}$ bisects ∠*A* and ∠*C*.
    ▶(b) Let *M* be the point where the diagonals of kite *ABCD* intersect. Measure ∠*AMD*, and make a conjecture concerning the angle between the diagonals of a kite. Justify your conjecture.
    (c) Show that $\overline{BM} \cong \overline{MD}$.

13. (a) In an isosceles trapezoid, make a conjecture concerning the lengths of the sides opposite the congruent angles.
    (b) Make a conjecture concerning the diagonals of an isosceles trapezoid.
    ▶(c) Justify your conjectures in (a) and (b).

14. Using a straightedge and a compass, construct any convex kite. Then construct a second kite that is not congruent to the first but whose sides are congruent to the corresponding sides of the first kite.

15. (a) What type of figure is formed by joining the midpoints of a rectangle?

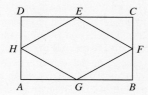

    ★(b) Prove your answer in (a).
    (c) What type of figure is formed by joining the midpoints of the sides of a parallelogram?
    ★(d) Prove your answer in (c).

    (e) Make a conjecture concerning the type of figure that is formed by joining the midpoints of any quadrilateral.

16. What information is needed to determine congruency for each of the following?
    (a) Two squares
    (b) Two rectangles
    (c) Two parallelograms

★17. Suppose that polygon *ABCD* is any parallelogram. Use congruent triangles to justify each of the following:
    (a) ∠*A* ≅ ∠*C* and ∠*B* ≅ ∠*D* (opposite angles are congruent).
    (b) $\overline{BC} \cong \overline{AD}$ and $\overline{AB} \cong \overline{CD}$ (opposite sides are congruent).
    (c) $\overline{BF} \cong \overline{DF}$ and $\overline{AF} \cong \overline{CF}$ (the diagonals bisect each other).
    (d) ∠*DAB* and ∠*ABC* are supplementary.

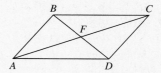

18. (a) Logo programs can be used to construct triangles, using the ASA property. Type the given procedures into your computer and then run the following:
    (i) ASA 60 50 70
    (ii) ASA 80 50 60
    (iii) AAS 60 50 70
    (iv) AAS 130 20 50

    ```
    TO ASA :ANGLE1 :SIDE :ANGLE2
      DRAW
      FORWARD 120
      BACK 120
      LEFT :ANGLE1
      FORWARD :SIDE
      RIGHT (180 - :ANGLE2)
      FORWARD 120
    END
    TO AAS :ANGLE1 :ANGLE2 :SIDE
      ASA :ANGLE1 :SIDE  180 -
        (:ANGLE1 + :ANGLE2)
    END
    ```

    (In LCSI, replace DRAW with CLEARSCREEN (CS).)
    (b) Predict the outcomes when the following are executed:
    (i) AAS 130 50 100   (ii) AAS 90 45 120
    (c) Add a line to the ASA procedure so that inputs for angles that are impossible in a triangle are not allowed.

19. (a) Write a Logo procedure called RHOMBUS with inputs :SIDE and :ANGLE in which the first variable

is the length of the side of the rhombus and the second is the measure of an interior angle of the rhombus.
▶(b) Execute RHOMBUS 80 50 and RHOMBUS 80 130. What is the relationship between the two figures? Why?
(c) Write a Logo procedure called SQ.RHOM with input :SIDE that will draw a square by calling on the RHOMBUS procedure.
20. Write a Logo procedure for starting at home and drawing an isosceles triangle given the length of two congruent sides and the measure of two congruent angles.

Review Problems

21. If possible, construct a triangle having the following three segments a, b, and c as its sides.

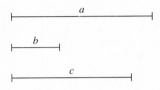

22. Construct an equilateral triangle whose sides are congruent to the following segment:

23. For each pair of triangles shown, determine whether the given conditions are sufficient to show that the triangles are congruent. If the triangles are congruent, tell which property can be used to verify this fact.

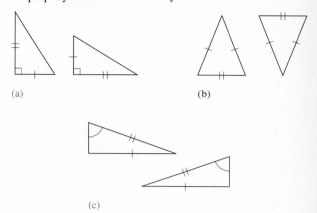

(a) (b)

(c)

LABORATORY ACTIVITY

As a van Hiele Level 3 activity, consider the statements "If a quadrilateral has opposite sides congruent, then it is a parallelogram," and "If one pair of sides of a quadrilateral is congruent and parallel, then the quadrilateral is a parallelogram." Show that, if one statement is true, then the other must also be true.

Section 11-3 Other Constructions

We use the definition of the rhombus and the following properties (given in Section 11-2) to accomplish basic compass-and-straightedge constructions.

1. A rhombus is a parallelogram in which all the sides are congruent.
2. The diagonals of a rhombus are perpendicular to each other.
3. The diagonals of a rhombus bisect the opposite angles.
4. The diagonals of a rhombus are bisectors of each other.

Constructing Parallel Lines

To construct a line parallel to a given line ℓ through a point P not on ℓ, as in the leftmost panel of Figure 11-22, our strategy is to construct a rhombus with one of its vertices at P

and one of its sides on line ℓ. Because the opposite sides of a rhombus are parallel, one of the sides through P will be parallel to ℓ. This construction is shown in Figure 11-22.

FIGURE 11-22

Constructing parallel lines (rhombus method).

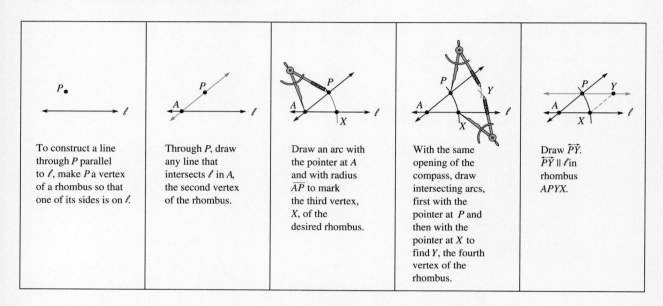

Figure 11-23 shows another way to do the construction. If congruent corresponding angles are formed by a transversal cutting two lines, then the lines are parallel. Thus, the first step is to draw a transversal through P that intersects ℓ. The angle marked α is formed by the transversal and line ℓ. By constructing an angle with a vertex at P congruent to α, we create congruent corresponding angles; therefore, $\overleftrightarrow{PQ} \parallel \ell$.

FIGURE 11-23

Constructing parallel lines (corresponding angle method).

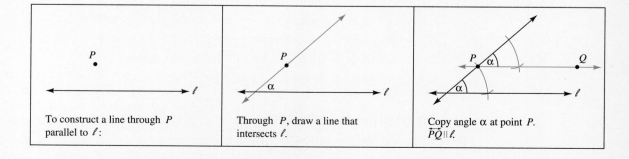

Parallel lines are frequently constructed using a ruler and a triangle or two triangles. If a ruler and a triangle are used, the ruler is left fixed and the triangle is slid so that one side of the triangle touches the ruler at all times. In Figure 11-24, the hypotenuses of the right triangles are all parallel (also the legs not on the ruler are all parallel). How can this method be used to accomplish the construction in Figure 11-23?

FIGURE 11-24

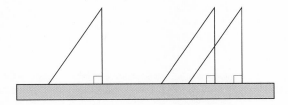

Constructing Angle Bisectors

angle bisector   Another construction based on a property of a rhombus is the construction of an **angle bisector,** a ray that separates an angle into two congruent angles. The diagonal of a rhombus with vertex A bisects $\angle A$, as shown in Figure 11-25.

FIGURE 11-25   Bisecting an angle.

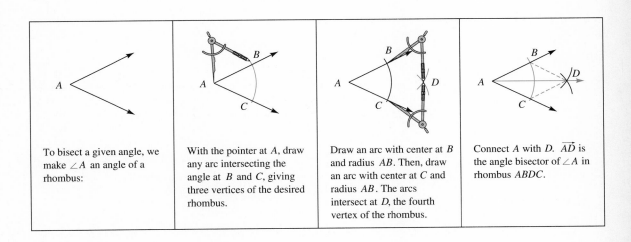

| To bisect a given angle, we make $\angle A$ an angle of a rhombus: | With the pointer at A, draw any arc intersecting the angle at B and C, giving three vertices of the desired rhombus. | Draw an arc with center at B and radius AB. Then, draw an arc with center at C and radius AB. The arcs intersect at D, the fourth vertex of the rhombus. | Connect A with D. $\overrightarrow{AD}$ is the angle bisector of $\angle A$ in rhombus $ABDC$. |

Constructing Perpendicular Lines

In order to construct a line through P perpendicular to line ℓ, where P is not a point on ℓ, as in Figure 11-26, recall that the diagonals of a rhombus are perpendicular to each other. If we construct a rhombus with a vertex at P and two vertices A and B on ℓ, as in Figure 11-26, the segment connecting the fourth vertex Q to P is perpendicular to ℓ because $\overline{AB}$ and $\overline{PQ}$ are diagonals of the rhombus.

FIGURE 11-26

Constructing a perpendicular to a line from a point not on a line.

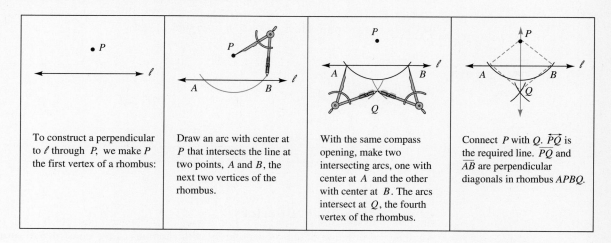

perpendicular bisector

The line perpendicular to a segment at its midpoint is the **perpendicular bisector** of the segment. To construct the perpendicular bisector of a line segment, as in Figure 11-27, we use the fact that the diagonals of a rhombus are perpendicular bisectors of each other. The construction yields a rhombus such that the original segment is one of the diagonals of the rhombus and the other diagonal is the perpendicular bisector, as in Figure 11-27.

FIGURE 11-27

Bisecting a line segment.

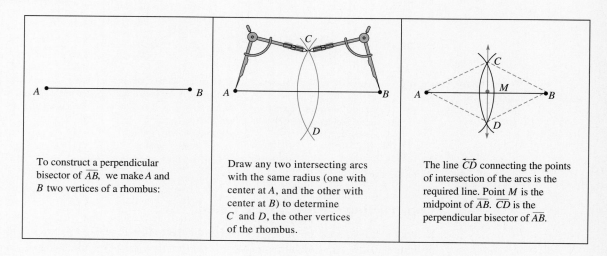

Constructing a perpendicular to a line ℓ at a point M on ℓ is based on the same property of a rhombus just used. That is, the diagonals of a rhombus are perpendicular bisectors of each other. Observe in Figure 11-27 that $\overline{CD}$ is a perpendicular to $\overline{AB}$ through M. Thus, we construct a rhombus whose diagonals intersect at point M, as in Figure 11-28.

FIGURE 11-28

Constructing a perpendicular to a line from a point on the line.

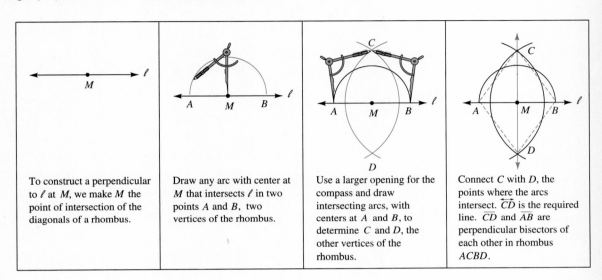

| To construct a perpendicular to ℓ at M, we make M the point of intersection of the diagonals of a rhombus. | Draw any arc with center at M that intersects ℓ in two points A and B, two vertices of the rhombus. | Use a larger opening for the compass and draw intersecting arcs, with centers at A and B, to determine C and D, the other vertices of the rhombus. | Connect C with D, the points where the arcs intersect. $\overline{CD}$ is the required line. $\overline{CD}$ and $\overline{AB}$ are perpendicular bisectors of each other in rhombus $ACBD$. |

altitude Constructing perpendiculars is useful in locating altitudes of a triangle. An **altitude** of a triangle is the perpendicular segment from a vertex of the triangle to the line containing the opposite side of the triangle. The construction of altitudes is described in Example 11-6.

EXAMPLE 11-6 Given triangle ABC, construct an altitude from vertex A in each part of Figure 11-29.

FIGURE 11-29

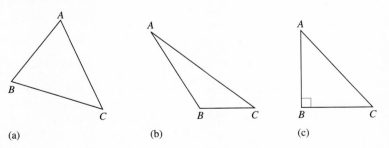

Solution (a) Since an altitude is the perpendicular from a vertex to the line containing the opposite side of a triangle, we need to construct a perpendicular from point A to the line containing $\overline{BC}$. Such a construction is shown in Figure 11-30. $\overline{AD}$ is the required altitude.

FIGURE 11-30

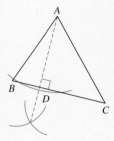

(b) The construction of the altitude from vertex A is shown in Figure 11-31. Notice that the required altitude $\overline{AD}$ does not intersect the interior of △ABC.

FIGURE 11-31

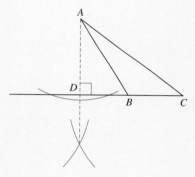

(c) Triangle ABC is a right triangle. The altitude from vertex A is the side $\overline{AB}$. No construction is required. ∎

Paper Folding and Mira Constructions

Perpendicularity constructions can also be completed by means of paper folding or by using a Mira. A Mira is a plastic device that acts as a reflector so that the image of an object can be seen behind the Mira. The drawing edge of the Mira acts as a folding line on paper. Any construction demonstrated in this text using paper folding can also be done with a Mira.

To use paper folding to construct a perpendicular to a given line ℓ at a point P on the line, we fold the line onto itself, as in Figure 11-32(a). The fold line is perpendicular to ℓ. To perform the construction with a Mira, we place the Mira with the drawing edge on P, as in Figure 11-32(b), so that ℓ is reflected onto itself. The line along the drawing edge is the required perpendicular.

FIGURE 11-32

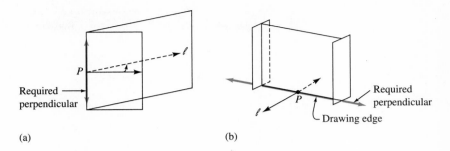

(a)   (b)

To construct the bisector of an angle with a Mira, we place the drawing edge of the Mira on the vertex of the angle and reflect one side of the angle onto the other. Similarly, we can bisect an angle by folding a line through the vertex so that one side of the angle folds onto the other side. For example, in Figure 11-33, to bisect ∠ABC, we fold and crease the paper through the vertex, B, so that $\overrightarrow{BA}$ coincides with $\overrightarrow{BC}$.

FIGURE 11-33

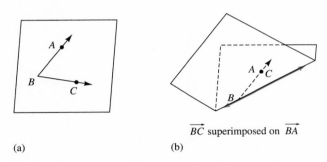

(a)   (b) $\overrightarrow{BC}$ superimposed on $\overrightarrow{BA}$

Properties of Angle Bisectors and Perpendicular Bisectors

Consider the angle bisector in Figure 11-34. It seems that any point P on the angle bisector is equidistant from the sides of the angle; that is, PD = PE. (The distance from a point to a line is the length of the perpendicular from the point to the line.)

FIGURE 11-34

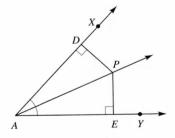

To justify this, we find two congruent triangles that have these segments as corresponding sides. The only triangles pictured are △ADP and △AEP. Because $\overrightarrow{AP}$ is the angle bisector, ∠DAP ≅ ∠EAP. Also, ∠PDA and ∠PEA are right angles and are thus congruent. Because $\overline{AP}$ is congruent to itself, △PDA ≅ △PEA by AAS. Thus, $\overline{PD} ≅ \overline{PE}$ because they are corresponding parts of congruent triangles PDA and PEA. Consequently, we have the following theorem.

THEOREM 11-2

Any point P on an angle bisector is equidistant from the sides of the angle.

REMARK It can also be shown that, if a point is in the interior of the angle and is equidistant from the sides of the angle, it must be on the angle bisector of that angle.

Next, consider the perpendicular bisector of a segment. In Figure 11-35(a), ℓ is the perpendicular bisector of $\overline{AB}$. Consider some point P on ℓ. It appears that P is equidistant from points A and B; that is, $\overline{PA} \cong \overline{PB}$. This can be justified using congruent triangles and is stated as a theorem.

FIGURE 11-35

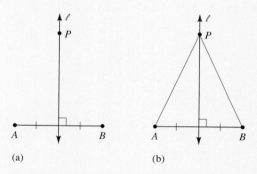

THEOREM 11-3

Any point on the perpendicular bisector of a line segment is equidistant from the endpoints of the bisected segment.

REMARK The converse of Theorem 11-3 is also true: That is, a point equidistant from the endpoints of a segment must be on the perpendicular bisector of the segment.

PROBLEM SET 11-3

1. Use a compass and a straightedge to construct a line m through P parallel to ℓ, using each of the following:
   (a) Alternate interior angles
   (b) Alternate exterior angles

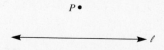

2. Construct each of the following, using paper folding.
   (a) Bisector of ∠A

   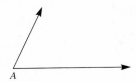

   (b) Perpendicular bisector of $\overline{AB}$

   (c) Perpendicular from P to ℓ

3. Complete the constructions in Problem 2, using a compass and a straightedge.
4. Complete the constructions in Problem 2, using a Mira, if one is available.
5. ▶Given an angle and a roll of tape, describe how you might construct the bisector of the angle.
6. Construct the three altitudes of each of the following types of triangles, using any method.
   (a) Acute triangle
   (b) Right triangle   (c) Obtuse triangle

   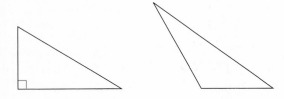

   (d) Make a conjecture about the lines containing the three altitudes of an acute triangle.
   (e) Make a conjecture about the lines containing the three altitudes of a right triangle.
   (f) Make a conjecture about the lines containing the three altitudes of an obtuse triangle.
7. Construct the perpendicular bisectors of each of the triangles in Problem 6.
   (a) Make a conjecture about the perpendicular bisectors of an acute triangle.
   (b) Make a conjecture about the perpendicular bisectors of a right triangle.
   (c) Make a conjecture about the perpendicular bisectors of an obtuse triangle.
8. A **chord** of a circle is a segment with endpoints on the circle.
   (a) Construct a circle, several chords, and a perpendicular bisector of each chord. Make a conjecture concerning the perpendicular bisector of a chord and the center of the circle.
   ▶(b) Justify your conjecture in (a).
   (c) Given a circle with an unmarked center, find the center of the circle.
9. A **median** of a triangle is a segment from a vertex of the triangle to the midpoint of the opposite side. Construct the three medians of each triangle in Problem 6. Their point of intersection is the **centroid,** the center of gravity of the triangle.
10. Construct a square with $\overline{AB}$ as a side.

11. ▶Suppose that you are "charged" 10¢ each time you use your straightedge to draw a line segment and 10¢ each time you use your compass to draw an arc. Using only a compass and a straightedge, what is the cheapest way to construct a square? Explain your reasoning.
12. Using a compass and a straightedge, construct a parallelogram with A, B, and C as vertices.

    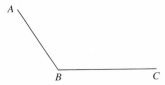

13. ▶In the concave quadrilateral APBQ, $\overline{PQ}$ and $\overline{AB}$ are the diagonals; $\overline{AP} \cong \overline{BP}$, $\overline{AQ} \cong \overline{BQ}$, and $\overline{PQ}$ has been extended until it intersects $\overline{AB}$ at C.
    (a) Make a conjecture concerning $\overrightarrow{PQ}$ and $\overline{AB}$.
    ▶(b) Justify your conjecture in (a).
    (c) Make conjectures concerning the relationships between $\overrightarrow{PQ}$ and ∠APB and between $\overrightarrow{QC}$ and ∠AQB.
    ▶(d) Justify your conjectures in (c).

    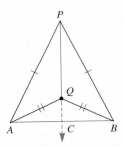

14. ▶Construct each of the following, if possible. If the construction is not possible, explain why.
    (a) A square, given one side
    (b) A square, given one diagonal
    (c) A rectangle, given one diagonal
    (d) A parallelogram, given two of its adjacent sides
    (e) A rhombus, given two of its diagonals
    (f) A triangle with two obtuse angles
    (g) A parallelogram with exactly three right angles
    (h) A kite with two right angles
    (i) A kite with three right angles
    ★(j) An isosceles triangle, given its base and the angle opposite the base
    ★(k) A trapezoid, given four of its sides

15. Using only a compass and a straightedge, construct angles with each of the following measures:
    (a) 30°
    (b) 15°
    (c) 45°
    (d) 75°
    (e) 105°

16. Given $\overline{AB}$ in the following figure, use a compass and a straightedge to construct the perpendicular bisector of $\overline{AB}$. You are not allowed to put any marks below $\overline{AB}$.

17. ▶(a) Explain why the hypotenuses in Figure 11-24 are all parallel.
    (b) Use the "sliding triangle" method described in Figure 11-24 to construct a line through a point P parallel to a line ℓ.

18. Draw a line ℓ and point P, not on the line and use the sliding triangle method described in Figure 11-24 to construct a perpendicular to ℓ through P using a straightedge and a right triangle.

19. ▶Explain how to construct a line through a given point parallel to a given line using alternate interior angles.

20. 💻 Write a Logo procedure to draw an equilateral triangle and three segments containing the altitudes of the triangle, as shown in the figure.

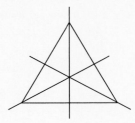

21. 💻 Use variables to write a Logo procedure to draw each of the following:
    (a) The angle bisector of a variable-sized angle
    (b) The perpendicular bisector of a variable-sized segment
    (c) A set of two variable-sized parallel segments

Review Problems

22. Given that $\overleftrightarrow{AB} \parallel \overleftrightarrow{ED}$ and $\overline{BC} \cong \overline{CE}$, why is $\overline{AC} \cong \overline{CD}$?

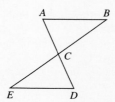

23. Draw $\triangle ABC$; then construct $\triangle PQR$ congruent to $\triangle ABC$, using each of the following combinations:
    (a) Two sides of $\triangle ABC$ and an angle included between these sides
    (b) The three sides of $\triangle ABC$
    (c) Two angles and a side included between these angles

24. ▶(a) LUCY is a trapezoid with diagonals intersecting at O. Is this enough information to conclude that LUCY contains one or more pairs of congruent triangles? Which ones are congruent, if any? Justify your answer.
    (b) Suppose that LUCY is an isosceles trapezoid, with $\overline{LY} \cong \overline{UC}$. Answer the question in (a).

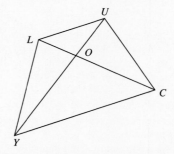

LABORATORY ACTIVITY

Consider the following drawing. The circles have congruent radii. (a) As a van Hiele Level 1 activity, use a marked ruler and protractor to find a relationship between segments $\overline{AB}$ and $\overline{CD}$. (b) As a van Hiele Level 2 activity, explain why the relationship you discovered is true.

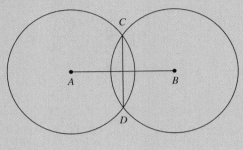

Section 11-4 Circles and Spheres

A circle is a set of points in a plane that are equidistant from a given point in the plane called the center, as shown in Figure 11-36. The radius is the length of any segment connecting the center with a point on the circle. Any segment with both endpoints on the circle is a *chord*. A line that contains a chord is a **secant**. A chord that passes through the center of the circle is a **diameter**. A diameter is the longest chord of the circle, and its length equals twice the length of the radius.

FIGURE 11-36

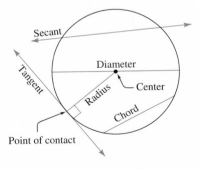

point of contact
point of tangency

A line that intersects the circle at exactly one point and is perpendicular to a radius at that point of intersection is a **tangent**. The point of intersection is the **point of contact,** or **point of tangency.**

The symbol for a fallout shelter, pictured in Figure 11-37(a), contains three equilateral triangles with a vertex at the center of the circle. It suggests that congruent chords in a circle intersect the circle to form congruent arcs. In Figure 11-37(b), we have congruent chords $\overline{AB}$ and $\overline{CD}$. Triangles AOB and COD are formed by connecting the endpoints of the chords to the center of the circle, O. These triangles are congruent, giving us congruent

central angles

angles $\angle AOB$ and $\angle COD$. (Why?) These angles are **central angles** because their shared vertex is the center of the circle. In general, *congruent chords determine congruent central* *angles and congruent arcs of the circle.* Conversely, it can be shown that *congruent central angles determine congruent chords and congruent arcs of a circle.*

FIGURE 11-37

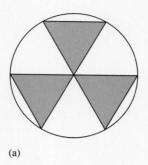 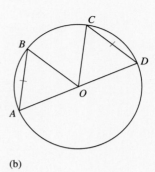

(a) (b)

Inscribing Polygons in a Circle

inscribed polygon

When all the vertices of a polygon are points of a given circle, the polygon is called an **inscribed polygon.** A regular hexagon inscribed in a circle is shown in Figure 11-38(a). All sides of a regular hexagon are congruent, so the corresponding arcs are congruent and the six corresponding central angles are congruent. Because the sum of the measures of these angles is 360°, the measure of each central angle is 60°. This fact is sufficient to inscribe a hexagon in a given circle, using a protractor. A compass-and-straightedge construction can also be accomplished. Look at $\triangle AOB$. Because $\overline{OA} \cong \overline{OB}$, the triangle is isosceles, and the base angles $\angle BAO$ and $\angle ABO$ are congruent. The central angle is 60°, so $m(\angle BAO) + m(\angle ABO) = 120°$. Consequently, $m(\angle BAO) = m(\angle ABO) = 60°$, and the triangle is equiangular and equilateral. Thus, $\overline{AB}$ is congruent to a radius of the circle. As a result, to inscribe a regular hexagon in a circle, we pick any point P on the circle and mark off chords congruent to the radius. Figure 11-38(b) shows such a construction.

FIGURE 11-38

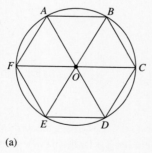 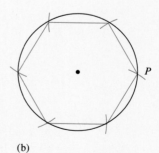

(a) (b)

To inscribe a square in a circle, we determine four congruent central angles. The central angles must be right angles because the sum of their measures is 360°. Hence, we need only construct two perpendicular diameters of the circle. Figure 11-39 shows the construction. First, we draw any diameter $\overline{PQ}$. Then, we construct a perpendicular to $\overline{PQ}$ at O and thus determine points R and S. Quadrilateral *PRQS* is the required square.

FIGURE 11-39

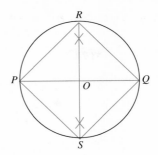

HISTORICAL NOTE

Determining which polygons can and which cannot be inscribed in a circle, using only a compass and a straightedge, has intrigued mathematicians for centuries. At age 19, Carl Friedrich Gauss proved that a regular 17-gon could be inscribed in a circle with these tools. He considered it one of his master achievements and wanted a replica of his construction placed on his tombstone. Later, he proved that a regular n-gon can be constructed with compass and straightedge if, and only if, either (1) n is a prime number of the form $2^{2k} + 1$, or a product of distinct primes of this form, (2) n is a power of 2, or (3) n is a product of numbers satisfying conditions (1) and (2). Thus, a regular septagon cannot be inscribed in a circle with a compass and a straightedge because 7 is not of the form $2^{2k} + 1$. However, because the only factors of 17 are 1 and 17 and $2^{2^2} + 1 = 2^4 + 1 = 17$, it follows from Gauss's theorem that a 17-gon can be inscribed in a circle with the aid of only a compass and a straightedge. On the basis of these notions, it has been shown that regular polygons with 257 and 65,537 sides can also be constructed. In 1832, F. J. Richelot published a study of the regular polygon of 257 sides, and, reportedly, a professor from Lingen, Germany, named Hermes spent 10 years of his life trying to construct a regular polygon of 65,537 sides.

Circumscribing Circles about Triangles

circumscribing

We can inscribe a triangle in a circle by connecting any three points of the circle with line segments. Conversely, given three vertices of any triangle, we can draw a circle that contains the vertices. This process is called **circumscribing** a circle about a triangle. For example, in Figure 11-40, circle O is circumscribed about $\triangle ABC$. Such a circle must contain $\overline{AB}, \overline{BC},$ and $\overline{AC}$ as chords. Also, it must have $\overline{OA} \cong \overline{OB} \cong \overline{OC}$, since they are all radii. Hence, O must be equidistant from A and B and, consequently, O must be on the perpendicular bisector of $\overline{AB}$. Similarly, $\overline{O}$ is on the perpendicular bisectors of $\overline{BC}$ and $\overline{AC}$. Hence, to find the center of the circle, we construct perpendicular bisectors of any two chords. The point of intersection of the chords is the center of the circle. Segments connecting O with A, B, and C are the radii of the circle.

FIGURE 11-40

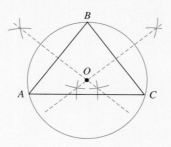

Inscribing Circles in Triangles

Recall that a tangent was defined as being perpendicular to a radius drawn to the point of contact. Thus, to construct a tangent to a given circle at any point on the circle, we construct a perpendicular to the radius at that point.

inscribed   A circle is **inscribed** in a triangle if it is tangent to the three sides of the triangle. For example, in Figure 11-41(a), circle O is inscribed in $\triangle DEF$ and A, B, and C are the points of contact. Because $\overline{OA}$, $\overline{OB}$, and $\overline{OC}$ are radii, they all have the same length, and they are perpendicular to the three sides of the triangle they each intersect. Thus, O is equidistant from the sides, so it lies on the bisectors of $\angle 1$, $\angle 2$, and $\angle 3$ (see Section 11-3).

To inscribe a circle in a triangle, we first construct the bisectors of two of the angles. Their intersection point O is the center of the inscribed circle. The radius of the circle can be determined by constructing a perpendicular from O to a side of the triangle. Figure 11-41(b) shows the construction. The circle with center O and radius $\overline{OC}$ is the required circle.

FIGURE 11-41

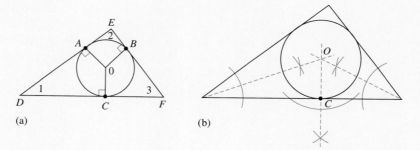

(a)   (b)

Spheres

A sphere is the three-dimensional analogue of a circle. The definitions of chord, secant, and tangent apply to spheres as well as to circles. A plane is tangent to a sphere if it intersects the sphere at exactly one point.

If a plane intersects a sphere at more than one point, then the intersection is a circle, as shown in Figure 11-42(a).

great circle   The largest of all such circles, a circle that contains a diameter of a sphere, is called a **great circle.** Any plane containing the center of the sphere intersects the sphere in a great circle. On the globe, the equator is a great circle. If the globe is cut by a plane parallel to the equator, the circles obtained are called **circles of latitude.** They become smaller the farther

circles of latitude
circle of longitude

they are from the equator, and they become very small near the North and South Poles, as shown in Figure 11-42(b). Any circle passing through the North and South Poles is called a **circle of longitude.** Each circle of longitude is a great circle. (Why?)

FIGURE 11-42

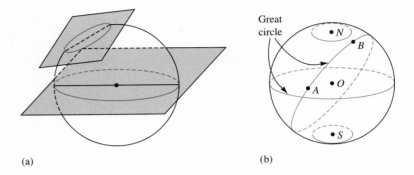

(a) (b)

In a plane, the shortest route between two points is the segment connecting the points. What is the shortest distance between two points A and B on a sphere? It is possible to prove that the shortest route between two points on a sphere is the minor arc $\overset{\frown}{AB}$ of the great circle obtained by cutting the sphere with a plane through A, B, and the center of the sphere. For that reason, if you take a nonstop flight between Seattle and London, it is likely that you will pass over Greenland.

REMARK Note that, when the two points A and B are relatively close, the distance along the minor arc $\overset{\frown}{AB}$ is close to the length of the segment $\overline{AB}$.

PROBLEM SET 11-4

1. Draw a circle on a piece of paper, and use paper folding (or a Mira) to determine the center of the circle.
2. ▶**Concentric circles** are circles that have the same center. Can you construct a common tangent to two distinct concentric circles? Why or why not?
3. In the following figure, $\overline{AB}$ is a diameter of the circle. What is $m(\angle ACB)$ for each value of $m(\angle CAB)$?
    (a) 30°
    (b) 50°
    (c) 71°
    (d) Based on your answers in (a)–(c), make a conjecture.
    ▶(e) Justify your conjecture.

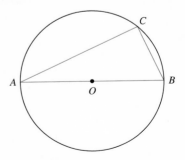

4. Inscribe a regular dodecagon (12-gon) in a circle by first inscribing a regular hexagon in the circle.

5. Inscribe a regular octagon in a circle.
6. Draw a circle.
   (a) Inscribe several quadrilaterals.
   (b) Measure the angles of the quadrilaterals from (a), and find the sums of the measures of pairs of opposite angles.
   (c) What seems to be true about the relationships among the angles?
7. Inscribe a circle in the given square, using only a compass and a straightedge.

8. ▶Is it possible to inscribe a circle in every quadrilateral? Explain.
9. Construct a circle with center O that is tangent to ℓ, using only a compass and a straightedge.

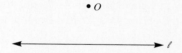

10. Construct a circle that is tangent to lines ℓ, m, and n, where $\ell \parallel m$.

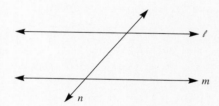

11. In the accompanying figure, $\overline{AC}$ is a diameter of circle O and $\overline{CB}$ is a chord of the circle.
    (a) What type of triangle is $\triangle OCB$?
    (b) Find a relationship between $m(\angle 1) + m(\angle 2)$ and $m(\angle 3)$.
    (c) Find a relationship between $m(\angle 1)$ and $m(\angle 3)$.

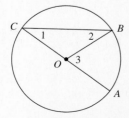

(d) Use your answer in (c) to find a relationship between angle measures α and β in each of the accompanying figures. (O is the center of the circle in each figure.)

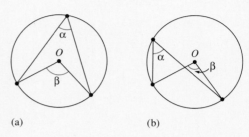

(a)   (b)

(e) Find the relationships among the measures of angles 1, 2, and 3, where the points P_1, P_2, and P_3 are arbitrary points on the major arc $\widehat{AP_1B}$.

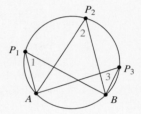

12. (a) In how many points can a line intersect a sphere?
    (b) In how many points can a plane intersect a sphere?
13. (a) Suppose you are standing on the equator, facing north. You walk until you reach the North Pole, turn right 90°, walk until you reach the equator, turn right 90°, walk until you reach your starting point and turn right 90°. What is the angle sum in your "triangle"?
    (b) Now start again at the equator facing north and walk through the North Pole until you reach the equator on the opposite side. Turn right 90°, and walk along the equator until you reach your starting point and turn 90°. What is the angle sum in the figure made by the walk?
★14. Show that congruent chords in a circle are the same distance from the center.
★15. Construct a circle that contains point P and is tangent to the two given parallel lines ℓ and m.

SECTION 11-5   SIMILAR TRIANGLES AND SIMILAR FIGURES   603

16. 🖥 Write a Logo procedure called FILL.CIRCLE with variable :RADIUS to shade the interior of a circle of a given radius.

17. 🖥 Write a Logo procedure to draw a circle and a segment containing one of its diameters.

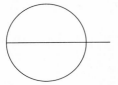

Review Problems

18. In $\triangle ABC$, name the *included side* between angles $\angle ABC$ and $\angle CAB$.

19. In $\triangle ABC$, name the *included angle* between sides $\overline{AB}$ and $\overline{BC}$.

20. In two right triangles, $\triangle ABC$ and $\triangle DEF$, if $\angle A$ and $\angle D$ are congruent and $\overline{AC}$ and $\overline{DF}$ are congruent, what do we know about the two triangles? Why?

21. ▶Use a compass and a straightedge to draw a pair of obtuse vertical angles and the angle bisector of one of these angles. Extend the angle bisector. Does the extended angle bisector bisect the other vertical angle? Justify your answer.

LABORATORY ACTIVITY

As a van Hiele Level 0 activity on a geoboard, construct right triangles, squares, rectangles, and other polygons. Which of the following, if any, can you construct?

(a) Pentagon
(b) Hexagon
(c) Equilateral triangle
(d) Circle

Section 11-5   Similar Triangles and Similar Figures

When a germ is examined under a microscope, when a slide is projected on a screen, or when a wet wool sweater shrinks in a clothes dryer, the shapes in each case remain the same, but the sizes are altered. In mathematics, we say that *two figures that have the same shape but not necessarily the same size are* **similar.**

similar

   For example, on the following page, a reproduction of the student page of Silver, Burdett, and Ginn, 1991, *Mathematics: Exploring Your World*, Grade 7, shows a photograph enlarged to poster size. The ratio of the corresponding sides, $\frac{24}{8}$ or 3, is called the

scale factor

scale factor. In particular, the ratio of the length of the larger picture to the length of the smaller picture is 3. In fact, it seems that, in any such enlargement, (or reduction for that matter), the results will be similar; that is, the corresponding angle measures remain the same, and the corresponding sides are proportional.

   These observations suggest the following definition of similar triangles.

DEFINITION OF SIMILAR TRIANGLES

$\triangle ABC$ is similar to $\triangle DEF$, written $\triangle ABC \sim \triangle DEF$, if, and only if, $\angle A \cong \angle D$, $\angle B \cong \angle E$, $\angle C \cong \angle F$, and $\frac{AB}{DE} = \frac{AC}{DF} = \frac{BC}{EF}$.

GETTING STARTED

Many copy machines reduce or enlarge pictures. How are the copy and the original alike? How are they different?

Similar Figures

The nature photographer often interprets nature's power and beauty. This dramatic photograph was enlarged to poster size. Find the length of the poster.

These rectangles are **similar** because they have the same shape.

▶ When two figures are similar, the ratios of the lengths of corresponding sides are equal. That is, the corresponding sides are in proportion.

$$\frac{\text{length} \longrightarrow 10}{\text{width} \longrightarrow 8} = \frac{n}{24}$$

$$10 \times 24 = 8 \times n$$

$$240 = 8n$$

$$30 = n$$

The length of the poster is 30 in.

△ABC is similar to △DEF. Find the length of $\overline{AC}$.

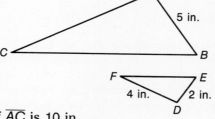

$$\frac{AB}{DE} = \frac{AC}{DF}$$

$$\frac{5}{2} = \frac{n}{4}$$

$$5 \times 4 = 2 \times n$$

$$20 = 2n$$

$$10 = n$$

The length of $\overline{AC}$ is 10 in.

$\overline{AB}$ corresponds to $\overline{DE}$.
$\overline{BC}$ corresponds to $\overline{EF}$.
$\overline{AC}$ corresponds to $\overline{DF}$.

SECTION 11-5 SIMILAR TRIANGLES AND SIMILAR FIGURES

> **REMARK** In the one-to-one correspondence between similar triangles, the angles at the corresponding vertices are congruent and the ratios of the corresponding sides are the same.

EXAMPLE 11-7 Given the pairs of similar triangles in Figure 11-43, find a one-to-one correspondence among the vertices of the triangles such that the corresponding angles are congruent. Then write the proportion for the corresponding sides that follows from the definition.

FIGURE 11-43

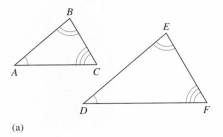

(a)

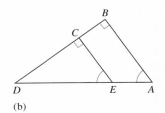

(b)

Solution (a) $\triangle ABC \sim \triangle DEF$

$$\frac{AB}{DE} = \frac{BC}{EF} = \frac{AC}{DF}$$

(b) $\triangle ABD \sim \triangle ECD$

$$\frac{AB}{EC} = \frac{BD}{CD} = \frac{AD}{ED}$$

Angle, Angle, Angle Property (AAA)

As with congruent triangles, minimal conditions may be used to determine when two triangles are similar. For example, suppose that two triangles each have angles with measures of 50°, 30°, and 100°, but the side opposite the 100° angle is 5 units long in one of the triangles and 1 unit long in the other.

The triangles appear to have the same shape, as shown in Figure 11-44. The figure suggests that if the angles of the two triangles are congruent, then the sides are proportional and the triangles are similar. This statement is true in general. It is called the **Angle, Angle, Angle** property of similarity for triangles, abbreviated **AAA**.

Angle, Angle, Angle (AAA)

FIGURE 11-44

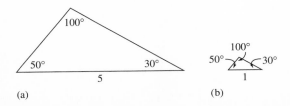

(a)   (b)

PROPERTY OF SIMILAR TRIANGLES—ANGLE, ANGLE, ANGLE (AAA)

Angle, Angle, Angle (AAA) If three angles of one triangle are congruent, respectively, to the three angles of a second triangle, then the triangles are similar.

Angle, Angle (AA)

Given the measures of any two angles of a triangle, the measure of the third angle can be found. Hence, if two angles in one triangle are congruent to two angles in another triangle, then the third angles must also be congruent. Consequently, the AAA condition may be reduced to **Angle, Angle (AA)**.

EXAMPLE 11-8 For each part of Figure 11-45, find a pair of similar triangles.

FIGURE 11-45

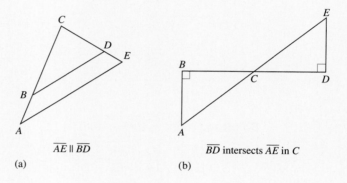

Solution (a) Because $\overline{AE} \parallel \overline{BD}$, congruent corresponding angles are formed by a transversal cutting the parallel segments. Thus, $\angle CBD \cong \angle CAE$, and $\angle CDB \cong \angle CEA$. Also, $\angle C \cong \angle C$, so $\triangle CBD \sim \triangle CAE$ by AAA.
(b) $\angle B \cong \angle D$ because both are right triangles. Also, $\angle ACB \cong \angle ECD$ because they are vertical angles. Thus, $\triangle ACB \sim \triangle ECD$ by AA. ∎

In general, knowing that the corresponding angles are congruent is not sufficient to determine similarity for any two polygons. For example, in a square of side 4 cm and a rectangle 2 cm by 4 cm, all of the angles are congruent, but the two figures are not similar. In fact, *two* **polygons** are **similar** *if and only if the corresponding angles are congruent and the corresponding sides are proportional.*

similar polygons

EXAMPLE 11-9 In Figure 11-46, find x.

FIGURE 11-46

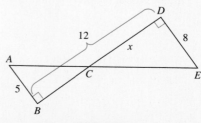

$\triangle ABC \sim \triangle EDC$

SECTION 11-5 SIMILAR TRIANGLES AND SIMILAR FIGURES

Solution $\triangle ABC \sim \triangle EDC$, so

$$\frac{AB}{ED} = \frac{AC}{EC} = \frac{BC}{DC}.$$

Now, $AB = 5$, $ED = 8$, and $CD = x$, so that $BC = 12 - x$. Thus,

$$\frac{5}{8} = \frac{12 - x}{x}$$
$$5x = 8(12 - x)$$
$$5x = 96 - 8x$$
$$13x = 96$$
$$x = \frac{96}{13}.$$

Properties of Proportion

Similar triangles give rise to various properties involving proportions. For example, in Figure 11-47, if $\overline{BC} \parallel \overline{DE}$, then $\frac{AB}{BD} = \frac{AC}{CE}$. This can be justified as follows: $\overline{BC} \parallel \overline{DE}$, so $\triangle ADE \sim \triangle ABC$. (Why?) Consequently, $\frac{AD}{AB} = \frac{AE}{AC}$, which may be written as follows:

$$\frac{x + y}{x} = \frac{z + w}{z}$$
$$\frac{x}{x} + \frac{y}{x} = \frac{z}{z} + \frac{w}{z}$$
$$1 + \frac{y}{x} = 1 + \frac{w}{z}$$
$$\frac{y}{x} = \frac{w}{z}$$
$$\frac{x}{y} = \frac{z}{w}$$

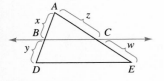

FIGURE 11-47

This result is summarized in the following theorem.

THEOREM 11-4

If a line parallel to one side of a triangle intersects the other sides, then it divides those sides into proportional segments.

The converse of Theorem 11-4 is also true: That is, if in Figure 11-47, we know that $\frac{AB}{BD} = \frac{AC}{CE}$, then we can conclude that $\overline{BC} \parallel \overline{DE}$. We summarize this result in the following theorem.

THEOREM 11-5

If a line divides two sides of a triangle into proportional segments, then the line is parallel to the third side.

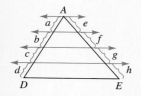

FIGURE 11-48

Similarly, if lines parallel to $\overline{DE}$ intersect $\triangle ADE$, as shown in Figure 11-48, so that $a = b = c = d$, it can be shown that $e = f = g = h$. This result is stated in the following theorem.

THEOREM 11-6

If parallel lines cut off congruent segments on one transversal, then they cut off congruent segments on any transversal.

Theorem 11-6 can be used to divide a given segment into any number of congruent parts. For example, using only a compass and a straightedge, we can divide segment $\overline{AB}$ in Figure 11-49 into three congruent parts by making the construction resemble Figure 11-48.

FIGURE 11-49

Separation of a Segment into Congruent Parts

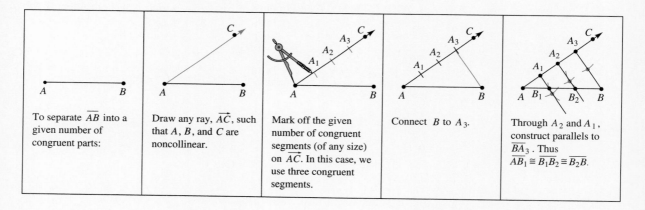

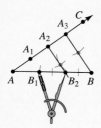

FIGURE 11-50

REMARK It is necessary to construct only $\overline{A_2B_2}$. We can then use a compass to mark off point B_1, making $B_1B_2 = BB_2$, as in Figure 11-50.

Indirect Measurements

Similar triangles have also been used to make indirect measurements since the time of Thales of Miletus (ca. 600 B.C.), who is believed to have determined the height of the Great Pyramid of Egypt. Most likely, he used ratios involving shadows, similar to those in Figure 11-51. The sun is so far away that it should make approximately congruent angles at B and B'. Because the angles at C and C' are right angles, $\triangle ABC \sim \triangle A'B'C'$. Hence,

$$\frac{AC}{A'C'} = \frac{BC}{B'C'}.$$

SECTION 11-5 SIMILAR TRIANGLES AND SIMILAR FIGURES

FIGURE 11-51

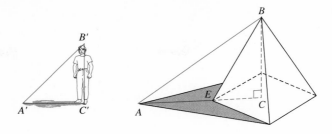

And because $AC = AE + EC$, the following proportion is obtained:

$$\frac{AE + EC}{A'C'} = \frac{BC}{B'C'}$$

The person's height and shadow can be measured. Also, the length of the shadow of the pyramid AE can be measured, and EC can be found because the base of the pyramid is a square. Each term of the proportion except the height of the pyramid is known. Thus, the height BC of the pyramid can be found by solving the proportion.

EXAMPLE 11-10 On a sunny day, a tall tree casts a 40-m shadow. At the same time, a meter stick held vertically casts a 2.5-m shadow. How tall is the tree?

SOLUTION Look at Figure 11-52. The pictured triangles are similar by AA because the tree and the stick both meet the ground at right angles, and the angles formed by the sun's rays are congruent (because the shadows are measured at the same time).

$$\frac{x}{40} = \frac{1}{2.5}$$
$$2.5x = 40$$
$$x = 16$$

The tree is 16 m tall.

FIGURE 11-52

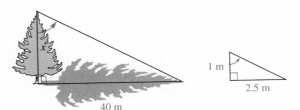

PROBLEM 1

Two neighbors, Smith and Wesson, planned to erect flagpoles in their yards. Smith wanted a 10-ft pole, and Wesson wanted a 15-ft pole. In order to keep the poles straight while the concrete bases hardened, they agreed to tie guy wires from the tops of the flagpoles to a fence post on the property line and from the fence post to the bases of the flagpoles, as shown in Figure 11-53. How high should the fence post be and how far apart should they erect flagpoles for this scheme to work?

FIGURE 11-53

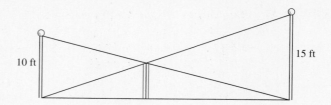

Understanding the Problem. In Figure 11-54, $AB = 10$ ft and $DC = 15$ ft; we need to find EF and AC. We know that $\overline{AB}$, $\overline{EF}$, and $\overline{DC}$ are perpendicular to $\overline{AC}$ and are therefore parallel.

FIGURE 11-54

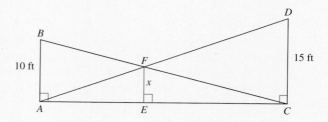

Devising a Plan. $\overline{EF}$ is a side in $\triangle EFC$. Because $\triangle EFC$ and $\triangle ABC$ are right triangles and share $\angle BCA$, it follows by AA that $\triangle EFC \sim \triangle ABC$. $\overline{EF}$ is also a side in $\triangle AFE$, and we have $\triangle AFE \sim \triangle ADC$. (Why?) From the two pairs of similar triangles, we will have proportions containing the unknown FE. The distance AC between the poles will appear in those proportions as well. We will then attempt to find FE and AC by solving the equations.

Carrying Out the Plan. Let $FE = x$. Then, from $\triangle EFC \sim \triangle ABC$ and $\triangle AFE \sim \triangle ADC$, we have:

$$\frac{x}{10} = \frac{EC}{AC}$$

$$\frac{x}{15} = \frac{AE}{AC}.$$

 To find x, our *subgoal* is to obtain an equation in which x is the only unknown. We know that $EC + AE = AC$. This suggests adding the proportions:

$$\frac{x}{10} + \frac{x}{15} = \frac{EC}{AC} + \frac{AE}{AC}$$

or

$$\frac{x}{10} + \frac{x}{15} = \frac{EC + AE}{AC} = \frac{AC}{AC} = 1.$$

Thus, $\frac{x}{10} + \frac{x}{15} = 1$. We may solve this equation as follows:

$$\left(\frac{1}{10} + \frac{1}{15}\right)x = 1$$

$$\frac{5}{30}x = 1$$

$$\frac{1}{6}x = 1$$

$$x = 6$$

Thus the fence post should be 6 ft high. Notice that, when finding the height of the fence post, we did not have to know how far apart the poles are. This means that, if the poles are placed farther apart, we would have the same height for the fence post. Thus, the poles could be any distance apart.

Looking Back. We could solve the problem for flagpoles of any length. If $AB = a$ and $CD = b$, then we would have $\frac{x}{a} + \frac{x}{b} = 1$ (why?) and, hence,

$$\left(\frac{1}{a} + \frac{1}{b}\right)x = 1 \quad \text{or} \quad \frac{a+b}{ab} \cdot x = 1 \quad \text{or} \quad x = \frac{ab}{a+b}.$$

▼ **BRAIN TEASER**

A building was to be built on a triangular piece of property. The architect was given the measurements of the angles of the triangular lot as approximately 54°, 39°, and 87° and the lengths of two of the sides as 100 m and 80 m. When the architect began the design on drafting paper, she drew a triangle to scale with the corresponding measures and found that the lot was considerably smaller than she had been led to believe. It appeared that the proposed building would not fit. The surveyor was called; he confirmed each of the measurements and could not see any problem with the size. Neither could understand the reason for the other's opinion. Determine the reason for the miscommunication, and suggest a way to provide an accurate description of the lot.

LABORATORY ACTIVITY

The device pictured below is called a pantograph. It is used to draw enlarged versions of figures. Well-made adjustable pantographs are available from drafting or art supply stores, but you can make a crude one from wooden lath or other material. A pointer at D is used to trace along an original figure, which causes the pencil at F to draw an enlarged version of the figure. Make or obtain a pantograph and experiment with enlarging figures. Explain how and why this works.

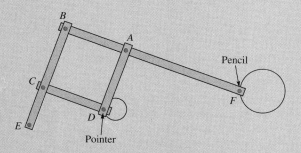

PROBLEM SET 11-5

1. ▶Which of the following are always similar? Why?
   (a) Any two equilateral triangles
   (b) Any two squares
   (c) Any two rectangles
   (d) Any two rhombuses
   (e) Any two circles
   (f) Any two regular polygons
   (g) Any two regular polygons with the same number of sides

2. Use grid paper to draw figures that have sides three times as large as the given ones, as shown in the following:

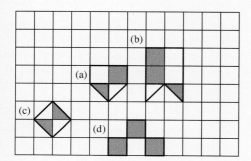

3. ▶Are congruent triangles similar? Why?

4. (a) Construct a triangle with sides of lengths 4 cm, 6 cm, and 8 cm.
   (b) Construct another triangle with sides of lengths 2 cm, 3 cm, and 4 cm.
   (c) Make a conjecture about the similarity of triangles that have proportional sides only.

5. (a) Construct a triangle with sides of lengths 4 cm and 6 cm and an included angle of 60°.
   (b) Construct a triangle with sides of lengths 2 cm and 3 cm and an included angle of 60°.
   (c) Make a conjecture about the similarity of triangles that have two sides proportional and congruent included angles.

6. (a) Sketch two nonsimilar polygons for which corresponding angles are congruent.
   (b) Sketch two nonsimilar polygons for which corresponding sides are proportional.

7. Examine several examples of similar polygons, and make a conjecture concerning the ratio of their perimeters.

8. ▶(a) Which of the following pairs of triangles are similar? If they are similar, explain why.
   (b) For each pair of similar triangles, find the scale factor of the sides of the triangles.

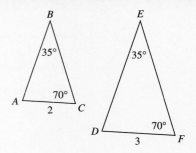
(i)

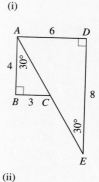
(ii)

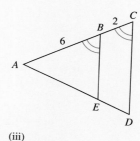
(iii)

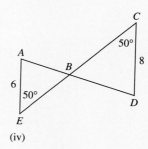
(iv)

9. Assume that in the following figures the triangles in each part are similar and find the measures of the unknown sides.

SECTION 11-5   EXERCISES   613

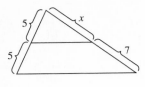
(a)

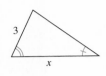 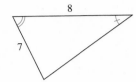
(b)

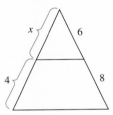
(c)

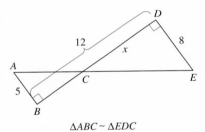
△ABC ~ △EDC
(d)

10. In the following figure, use a compass and a straightedge to separate $\overline{AB}$ into five congruent pieces.

11. ▶Assuming that the lines on an ordinary piece of notebook paper are parallel and equidistant, describe a method for using the paper to divide a piece of licorice evenly among 2, 3, 4, or 5 children. Explain why it works.

12. In the following right triangle ABC, we have $\overline{CD} \perp \overline{AB}$.

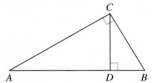

(a) Find three pairs of similar triangles. Justify your answers.
(b) Write the corresponding proportions for each set of similar triangles.

13. ▶In the cartoon, at the bottom of this page, if a smaller map were obtained and its scale was half the size of the original map, would the distance to be traveled be any different? Why?

14. Find the distance AB across the pond using the similar triangles shown.

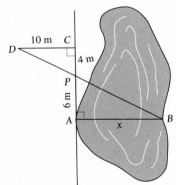

Reprinted with special permission of King Features Syndicate, Inc.

15. To find the height of a tree, a group of Girl Scouts devised the following method. A girl walks away from the tree along its shadow until the shadow of the top of her head coincides with the shadow of the top of the tree. If the girl is 150 cm tall, her distance to the foot of the tree is 1500 cm, and the length of her shadow is 300 cm, how tall is the tree?

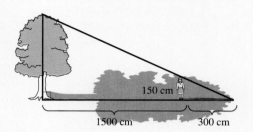

16. The angle bisector of one of the angles in an isosceles triangle is constructed. This angle bisector partitions the original triangle into two isosceles triangles.
    (a) What are the angle measures of the original triangle? (There are two possibilities.)
    ▶(b) Which, if any, triangles are congruent? Similar? Explain your answers.

17. Samantha wants to know how far above the ground the top of a leaning flagpole is. At high noon, when the sun is directly overhead, the shadow cast by the pole is 7 ft long. Samantha holds a plumb bob with a string 3 ft long up to the flagpole and determines that the point of the plumb bob touches the ground 13 in. from the base of the flagpole. How far above the ground is the top of the pole?

18. For her Earth Week information booth, Dian has prepared a report on the gorilla, an endangered species. For her backdrop, she wants to project a life-sized image of a gorilla on a screen. She has a slide showing a gorilla standing erect. In the slide, the image of the gorilla is $\frac{3}{4}$ in. tall. Dian's research shows that adult gorillas often reach 6 ft in height. If the bulb in the slide projector is 3 in. from the slide, where should the projector be placed so that the gorilla appears life-size on the screen?

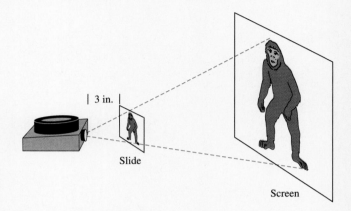

★19. Poles $\overline{AB}$, $\overline{CF}$, and $\overline{DE}$ are constructed perpendicular to the ground in such a way that, when B, C, D, and F, are connected by wires, $BCDF$ is a rectangle. If $\overline{AB}$ is 9 m long and $\overline{DE}$ is 4 m long, how long is $\overline{CF}$? How far apart are $\overline{AB}$ and $\overline{DE}$?

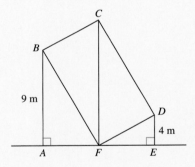

20. (a) Write a procedure called RECTANGLE that draws a rectangle of variable size with inputs :LEN and :WID; then write a procedure called SIM.RECT that draws a rectangle whose sides measure twice as long as those of the rectangle drawn by RECTANGLE when the same inputs are used for :LEN and :WID.
    (b) Write a procedure called SIM.RECTANGLE that draws a rectangle similar to the one drawn by REC-TANGLE, with a scale factor called :SCALE that affects the size of the rectangle.
    (c) Write a procedure called PARALLELOGRAM that draws a parallelogram of variable size with inputs :LEN, :WID, and :ANGLE; then write a procedure called SIM.PAR that generates similar parallelograms.

21. (a) Write a Logo procedure called TRISECT that draws a line segment of length determined by input :LEN and has the turtle divide it into three congruent parts.
    (b) Write a procedure called PARTITION that draws a line segment of length determined by input :LEN and has the turtle divide it into :NUM parts, where :NUM is also an input.

Review Problems

22. ▶If a person holds a mirror at arm's length and looks into it, is the image seen congruent to the original? Why or why not?

23. Given the length of the base of an isosceles triangle and the length of the altitude to that base, construct the triangle.

   _____
              Base

   _____
              Altitude

▼ **BRAIN TEASER**

A toy maker wants to cut the following plastic rectangle *EFGH* into four right triangles and a rectangle. Given the measurements shown, how long is $\overline{CE}$?

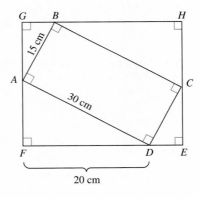

*Section 11-6 More on Geometry and Logo

In the exercises of this chapter, we have investigated several Logo procedures for constructing triangles. In this section we consider the construction of isosceles triangles. In the general case, an isosceles triangle can be assumed to have only two sides congruent. In Figure 11-55, we see an isosceles triangle with sides $S1$ and $S2$ and base angles with measure A. This implies that the measure of the other angle of the triangle is $180 - 2A$. Thus, if the length of any side and the measure of one base angle is given, we should be able to construct the triangle by ASA.

FIGURE 11-55

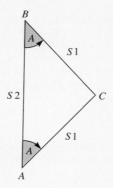

In the process of writing a Logo procedure for drawing isosceles triangles, we exhibit the use of recursive procedures for geometric purposes. It is likely that the following procedures require a deeper understanding than most elementary students have, but they may be simply given to the students as a piece of useful software.

We begin our Logo procedures by assuming that the length of the two congruent sides and the measure of one of the two base angles are given, as in Figure 11-56. If we start the turtle at home with a heading of 0, we might first draw the length of the known side :S1, turn the turtle right 2*:A (why?), move forward the length of the known side :S1, and then send the turtle home. This procedure follows.

```
TO ISOSCELES1  :S1  :A
    DRAW
    FORWARD  :S1
    RIGHT 2*:A
    FORWARD  :S1
    HOME
END
```

(In LCSI, replace DRAW with CLEARSCREEN.)

FIGURE 11-56

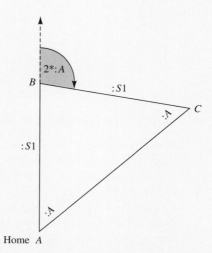

A different procedure for drawing △ABC (Figure 11-57) is required if we are given the length of the base, :S2, and the measure of a base angle, :A. Again suppose that the turtle starts at home with a heading of 0 and that we draw the length of the known side, :S2, first.

FIGURE 11-57

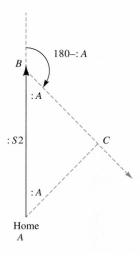

In Figure 11-57, this would leave the turtle at point B with a heading of 0. We might then turn the turtle $180 - :A$ in order to draw $\overrightarrow{BC}$. We do not know how far the turtle should move to reach point C, but we want the turtle to move along $\overrightarrow{BC}$ until point C is located, and then we want it to return HOME to complete the triangle. To have the turtle do this, we might consider programming it to check many points along $\overrightarrow{BC}$ until it has the desired heading when pointed in the direction of A. In Figure 11-58(a), if the turtle were at point A (HOME) and pointed toward point C, it would have a heading of :A. However, if it were at point C and pointed toward point A, as in Figure 11-58(b), what would its heading be?

FIGURE 11-58

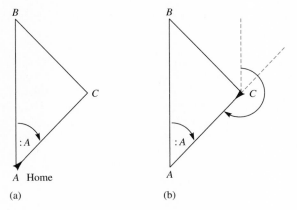

Since the two directions are on the same line but in opposite directions, the headings differ by 180°. Thus, the heading at point C toward point A is $180 + :A$, as shown in Figure 11-58(b).

With this in mind, we move the turtle along $\overrightarrow{BC}$, stopping it at intervals to turn toward point A to check whether it has the desired heading. Suppose that a procedure called CHECKHEAD is written to accomplish this. If it were, then the ISOSCELES2 procedure might be the following:

```
TO ISOSCELES2 :S2 :A
   DRAW
   FORWARD :S2
   RIGHT 180 - :A
   CHECKHEAD :A
   HOME
END
```

(In LCSI, replace DRAW with CLEARSCREEN.)

In the CHECKHEAD procedure, we want the turtle to move a bit along $\overrightarrow{BC}$, stop, turn toward point A, see if its heading is 180 + :A, and, if it is, stop checking. If the heading is not 180 + :A, then we want the turtle to turn back along $\overrightarrow{BC}$ with its original heading, move forward a bit more, and repeat the process. Because we do not know how many times the process has to be repeated, we use a recursive procedure. A first draft of the procedure might be the following:

```
DRAFT
TO CHECKHEAD :A
   FORWARD 1
   SETHEADING TOWARDS 0 0
   IF HEADING = 180 + :A THEN STOP ELSE
      SETHEADING 180 - :A CHECKHEAD :A
END
```

(In LCSI, replace SETHEADING TOWARDS 0 0 with SETHEADING TOWARDS [0 0]. Also replace IF HEADING = 180 + :A THEN STOP ELSE SETHEADING 180 − :A CHECKHEAD :A with IF HEADING = 180 + :A [STOP] [SETHEADING 180 − :A CHECKHEAD :A].)

This draft is a good idea, but if you try it, it may not do what you want. One reason for this is that all measurements on the computer are approximations, and the exact heading may never be reached. If it cannot be reached, you have to decide how close an approximation to demand. This issue is entirely up to you. Suppose that you decide that being within 2° is acceptable. If so, you will want to know whether or not the absolute value of the difference of the heading and 180 + :A is less than 2°. As was mentioned earlier, Logo does not have an absolute value procedure. Here is the one written earlier.

```
TO ABS :VALUE
   IF :VALUE < 0 OUTPUT - :VALUE ELSE OUTPUT :VALUE
END
```

(In LCSI, replace IF :VALUE < 0 OUTPUT − :VALUE ELSE OUTPUT :VALUE with IF :VALUE < 0 [OUTPUT − :VALUE] [OUTPUT :VALUE].)

With this discussion in mind, the CHECKHEAD procedure might be as follows:

```
TO CHECKHEAD :A
    FORWARD 1
    SETHEADING TOWARDS 0 0
    IF ABS (HEADING - ( 180 + :A)) < 2 THEN STOP ELSE
        SETHEADING 180 - :A CHECKHEAD :A
END
```

(In LCSI, replace SETHEADING TOWARDS 0 0 with SETHEADING TOWARDS [0 0]. Also replace IF ABS (HEADING - (180 + :A)) < 2 THEN STOP ELSE SETHEADING 180 - :A CHECKHEAD :A with IF ABS (HEADING - (180 + :A)) < 2 [STOP] [SETHEADING 180 - :A CHECKHEAD :A].)

Other Logo geometry procedures are investigated in the exercises.

PROBLEM SET 11-6

1. Write a procedure called SIMSQS to draw two similar squares when given the length of a side :SIDE of one of the squares and the scale factor :K.
2. Write a procedure called RTISOS to draw an isosceles right triangle when given the length of the hypotenuse, :HYPOT.
3. A 30°-60°-90° triangle is formed by drawing an altitude of an equilateral triangle. Use this hint to write a procedure called TRI30 to draw a 30°-60°-90° triangle if given the length of the hypotenuse, :HYPOT.
4. Write a procedure called STAR to draw a regular hexagon for which, on each side of the hexagon, there is an equilateral triangle with congruent sides of length :SIDE, where :SIDE is the length of a side of the hexagon.
★ 5. Write a procedure called ISOSCELES3 to draw an isosceles triangle when given the measure of one base angle, :ANGLE, and the height, :HEIGHT.

Solution to the Preliminary Problem

Understanding the Problem. Mary Rose ordered a wooden triangle with one side 18 in. long and angles of 36°, 84°, and 60°. Later, she ordered another triangle with the same specifications, but to her surprise it was not congruent to the first triangle. We are to explain how this is possible.

Devising a Plan. To explain the situation we try to construct two noncongruent triangles with Mary Rose's specifications. First, we need to construct a triangle with the given side and angles. Then, because the second triangle has the same angles, it must be similar to the first by AAA. Consequently, we try to construct a triangle similar to the first but not congruent to it.

Carrying Out the Plan. In Figure 11-59, $\overline{AB}$ represents the 18-in.-long side. We then use a protractor to construct $\angle A$ measuring 60°, and $\angle B$ measuring 36°. Hence, $m(\angle C) = 84°$. (Why?) Next, we are to construct a noncongruent triangle similar to $\triangle ABC$ with one side as long as $\overline{AB}$. Because this may seem difficult, we consider *a simpler but related problem*. We drop the condition that the triangle must have a side of length AB and consider constructing a triangle similar to $\triangle ABC$. This can be conveniently achieved by using one of the existing vertices of $\triangle ABC$ and drawing a side parallel to the opposite side. For example, in Figure 11-59 $\overline{C_1B_1} \parallel \overline{CB}$ and $\triangle ACB \sim \triangle AC_1B_1$. (Why?) There are infinitely many such triangles AC_1B_1. We imagine sliding $\overrightarrow{BC}$ along $\overrightarrow{AB}$ and $\overrightarrow{AC}$ so that $\overline{B_1C_1} \parallel \overline{BC}$ and until $AC_1 = AB$. Consequently, $\triangle ACB$ and $\triangle AC_1B_1$ are two noncongruent similar triangles such that $AC_1 = AB$ and therefore there are two noncongruent triangles with Mary Rose's specifications.

FIGURE 11-59

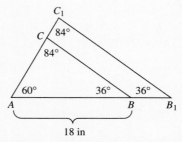

Looking Back. If Mary Rose gave each carpenter the measurements of two sides of the desired triangles and all the angles, could she then be sure to obtain a unique triangle?

QUESTIONS FROM THE CLASSROOM

1. On a test, a student wrote $AB \cong CD$ instead of $\overline{AB} \cong \overline{CD}$. Is this answer correct?
2. A student asks if there are any constructions that cannot be done with a compass and a straightedge. How do you answer?
3. A student asks for a mathematical definition of congruence that holds for all figures. How do you respond? Is your response the same for similarity?
4. One student claims that by trisecting $\overrightarrow{AB}$ and drawing $\overrightarrow{CD}$ and $\overrightarrow{CE}$ as shown, she has trisected $\angle ACB$.

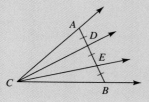

How do you convince her that her construction is wrong?

5. In the following drawing, a student claims that polygon $ABCD$ is a parallelogram if $\angle 1 \cong \angle 2$. Is he correct?

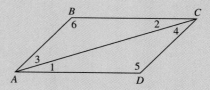

6. A student claims that, when the midpoints of the sides of any polygon are connected, a polygon similar to the original results. Is this true?
7. A student asks why $\cong$ rather than $=$ is used to discuss triangles that have the same size and shape. What do you say?

8. A student draws the following figure and claims that, because every triangle is congruent to itself, we can write △ABC ≅ △BCA. What is your response?

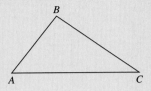

9. A student says that, since every circle can have a quadrilateral inscribed in it, we should be able to inscribe a circle in every quadrilateral. How do you respond?
10. A student claims that on a very large circle it is possible to find three points that are collinear. How do you respond?

CHAPTER OUTLINE

I. Congruence
   A. Two geometric figures are **congruent** if, and only if, they have the same size and shape.
   B. Two triangles are congruent if they satisfy any of the following properties:
      1. **Side, Side, Side (SSS)**
      2. **Side, Angle, Side (SAS)**
      3. **Angle, Side, Angle (ASA)**
      4. **Angle, Angle, Side (AAS)**
   C. **Triangle Inequality:** The sum of the measures of any two sides of a triangle must be greater than the measure of the third side.

II. Circles and spheres
   A. A **circle** is a set of all points in a plane that are the same distance (radius) from a given point (center).
   B. An **arc** of a circle is any part of the circle that can be drawn without lifting a pencil. The **center of the arc** is the center of the circle containing the arc.
   C. A **chord** is a segment whose endpoints lie on a circle.
   D. A **secant** is a line that contains a chord of a circle.
   E. A **tangent** is a line that intersects a circle at exactly one point; it is perpendicular to a radius drawn at that point.
   F. A **sphere** is a set of all points in space that are the same distance (radius) from a given point (center).
   G. A **central angle** is an angle whose vertex is at the center of the circle in which it is defined.

III. Similar figures
   A. Two polygons are **similar** if and only if their corresponding angles are congruent and their corresponding sides are proportional.
   B. **AAA or AA:** If three (two) angles of one triangle are congruent to three (two) angles of a second triangle, the triangles are **similar**.

IV. Proportion
   A. If a line parallel to one side of a triangle intersects the other sides, it divides those sides into proportional segments.
   B. If parallel lines cut off congruent segments on one transversal, they cut off congruent segments on any transversal.

V. Constructions that can be accomplished using a compass and a straightedge
   A. Copy a line segment.
   B. Copy a circle.
   C. Copy an angle.
   D. Bisect a segment.
   E. Bisect an angle.
   F. Construct a perpendicular from a point to a line.
   G. Construct a perpendicular bisector of a segment.
   H. Construct a perpendicular to a line through a point on the line.
   I. Construct a parallel to a line through a point not on the line.
   J. Divide a segment into congruent parts.
   K. Inscribe some regular polygons in a circle.
   L. Circumscribe a circle about a triangle.
   M. Inscribe a circle in a triangle.

CHAPTER TEST

1. Each of the following figures contains at least one pair of congruent triangles. Identify them and tell why they are congruent.

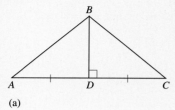
(a)

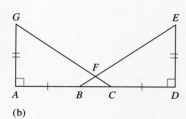
(b)

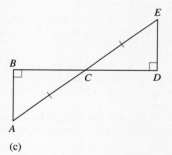
(c)

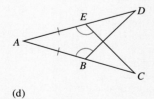
(d)

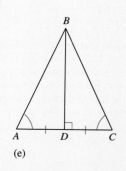
(e)

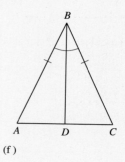
(f)

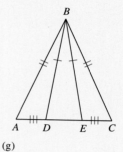
(g)

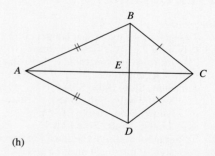
(h)

2. ▶In the figure, $ABCD$ is a square and $\overline{DE} \cong \overline{BF}$. What kind of figure is $AECF$? Justify your answer.

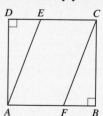

3. Construct each of the following by using (1) compass and straightedge and (2) paper folding:

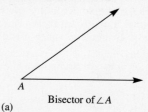
(a) Bisector of ∠A

CHAPTER TEST   623

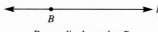
(b) Perpendicular to *l* at *B*

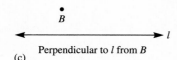
(c) Perpendicular to *l* from *B*

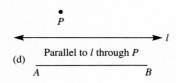
(d) Parallel to *l* through *P*

4. For each of the following pairs of similar triangles, find the missing measures:

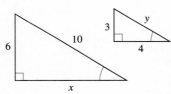
(a)

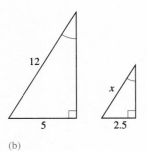
(b)

5. Divide the given segment into five congruent parts.

6. ▶If *ABCD* is a trapezoid, $\overline{EF} \parallel \overline{AD}$, and $\overline{AC}$ is a diagonal, what is the relationship between $\frac{a}{b}$ and $\frac{c}{d}$? Why?

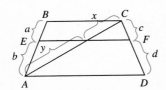

7. Construct a circle that contains *A* and *B* and has its center on ℓ.

8. For each of the following, show that appropriate triangles are similar, and find *x* and *y*:

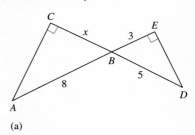
(a)

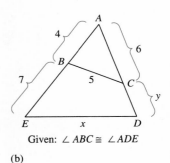
Given: ∠*ABC* ≅ ∠*ADE*
(b)

9. ▶Determine whether each of the following is true or false. If false, explain why.
  (a) A radius of a circle is a chord of the circle.
  (b) A diameter of a circle may be a tangent of the circle.
  (c) If a radius bisects a chord of a circle, then it is perpendicular to the chord.
  (d) Two spheres may intersect at exactly one point.
  (e) Two spheres may intersect in a circle.

10. A person 2 m tall casts a shadow 1 m long when a building has a 6-m shadow. How high is the building?

11. (a) Which of the polygons on the following page can be inscribed in a circle? Assume that all sides of each polygon are congruent and that all the angles of polygons (iii) and (iv) are congruent.
   (b) Based on your answer in (a), make a conjecture about what kinds of polygons can be inscribed in a circle.

624   CHAPTER 11   CONSTRUCTIONS, CONGRUENCE, AND SIMILARITY

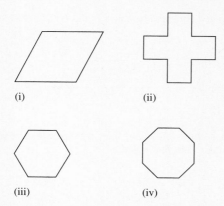

(i)   (ii)

(iii)   (iv)

12. What is the vertical height of the playground slide shown?

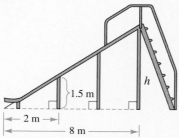

13. Find the distance d across the river sketched below.

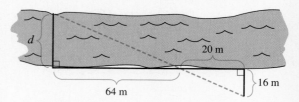

14. ▶Are the following statements always true, always false, or true in some cases and false in others? Explain your answers.
    (a) A quadrilateral whose diagonals are congruent and perpendicular is a square.
    (b) A circle can be circumscribed about every isosceles trapezoid.

SELECTED BIBLIOGRAPHY

Battista, M. "MATHSTUFF Logo Procedures: Bridging the Gap between Logo and School Geometry." *Arithmetic Teacher* 35 (September 1987): 7–11.

Billstein, R., S. Libeskind, and J. Lott. *Apple Logo, Programming, and Problem Solving.* Menlo Park, Calif.: Benjamin Cummings, 1986.

Billstein, R., S. Libeskind, and J. Lott. *Logo, MIT Logo for the Apple.* Menlo Park, Calif.: Benjamin/Cummings, 1985.

Burger, W. "Geometry." *Arithmetic Teacher* 32 (February 1985): 52–56.

Friedlander, A., and G. Lappan. "Similarity: Investigations at the Middle Grade Level." In *Learning and Teaching Geometry, K–12.* Reston, Va.: National Council of Teachers of Mathematics, 1987.

Hurd, S. "An Application of the Criteria ASASA for Quadrilaterals." *Mathematics Teacher* 81 (February 1988): 124–126.

Lappan, G., and R. Even. "Similarity in the Middle Grades." *Arithmetic Teacher* 35 (May 1988): 32–35.

Lappan, G., and E. Phillips. "Spatial Visualization." *Mathematics Teacher* 79 (November 1984): 618–623.

Lennie, J. "A Lab Approach for Teaching Basic Geometry." *Mathematics Teacher* 79 (October 1986): 523–524.

Mathematics Resource Project. *Geometry and Visualization.* Palo Alto, Calif.: Creative Publications, 1985.

Newton, J. "From Pattern-Block Play to Logo Programming." *Arithmetic Teacher* 35 (May 1988): 6–9.

Robertson, J. "Geometric Constructions Using Hinged Mirrors." *Mathematics Teacher* 79 (May 1986): 380–386.

Senk, S. L., and D. B. Hirschorn. "Multiple Approaches to Geometry: Teaching Similarity." *Mathematics Teacher* 83 (April 1990): 274–280.

Tarte, L. "Dropping Perpendiculars the Easy Way." *Mathematics Teacher* 80 (January 1987): 30–31.

Van de Walle, J., and C. Thompson. "Promoting Mathematical Thinking." *Arithmetic Teacher* 32 (February 1985): 7–13.

Walter, M. *Boxes, Squares, and Other Things.* Reston, Va.: National Council of Teachers of Mathematics, 1970.

12 Motion Geometry and Tessellations

Preliminary Problem

The people of Climate County want to build a road connecting Sunny Street and Shady Lane. They want the road to be parallel to and the same length as the west side of Rainbow Park. Show where the new road should be built.

CHAPTER 12   MOTION GEOMETRY AND TESSELLATIONS

Euclid envisioned moving one geometric figure in a plane and placing it on top of another to determine if the two figures were congruent. Intuitively, we know this can be done by making a tracing of one figure, shifting, turning, or flipping the tracing, and placing it back down atop the other figure. Elementary school students seem to be able to identify congruences by this type of pre-deductive activity. The 5–8 *Standards* (p. 114) supports this approach: *Explorations of flips, slides, turns, stretchers, and shrinkers will illuminate the concepts of congruence and similarity.*

The notions of motion geometry are introduced in this chapter, along with *tessellations* of a plane, that is, the filling of a plane with repetitions of a figure in such a way that no figures overlap and there are no gaps.

Section 12-1   Translations and Rotations

Translations

Plane figures can be moved from one position to another by different motions. One such motion is described in Figure 12-1, where a child moves down a slide without any accompanying twisting or turning. This type of motion is called a **translation,** or a **slide.**

translation / slide

FIGURE 12-1

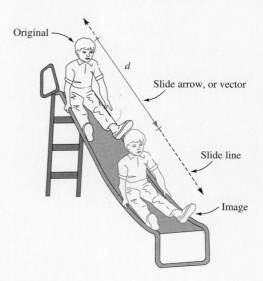

slide arrow / vector
slide line / image

Notice that in Figure 12-1, the child has moved a certain distance in a certain direction along a line. The distance and direction the original figure is moved are marked with a **slide arrow,** or **vector,** along a **slide line** to obtain the **image.** As another example of this type of motion, consider Figure 12-2(a), where figures in plane α are traced on plane α'. The tracing is then slid without twisting or turning along the slide arrow so that point P falls on point Q. Once the sliding is completed, we punch holes in α' to mark points A', B', and C', find the image of the figure determined by points A, B, and C, and complete the drawing, as in Figure 12-2(b). In Figure 12-2(b), $AA' = d$ and $\overline{AA'}$ is parallel to $\overline{PQ}$. Similarly, $BB' = d$ and $\overline{BB'}$ is parallel to $\overline{PQ}$. In this example, we have a one-to-one correspondence between the plane and itself. Any one-to-one correspondence between a plane and itself is a **transformation** of the plane. A transformation is determined if it is possible to find the image of every point in the plane. A translation is a special type of transformation, as defined next.

transformation

FIGURE 12-2

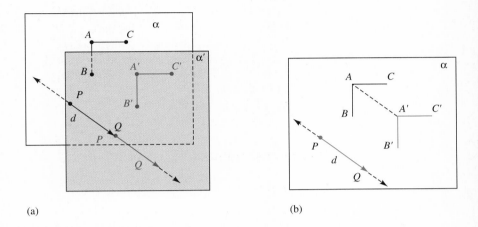

(a)   (b)

DEFINITION OF A TRANSLATION

A **translation** is a transformation of a plane that moves every point of the plane a specified distance in a specified direction along a straight line.

Figure 12-3 shows a translation that takes $\triangle ABC$ to $\triangle A'B'C'$. The translation is determined by the slide arrow, or vector, from M to N. The vector determines the image of any point in the plane in the following way: The image of a point A in the plane is the point A' obtained by sliding A along a line parallel to $\overleftrightarrow{MN}$ in the direction from M to N by the distance MN. (MN is denoted by d in Figure 12-3.) Dashed segments have been used to connect the vertices of $\triangle ABC$ with their respective images under the translation. Notice the $AA' = BB' = CC' = d$ and $\overline{AA'} \parallel \overline{BB'} \parallel \overline{CC'}$. It appears that under the translation, figures do not change their shapes or sizes. This is true in general, and we say that a translation

FIGURE 12-3

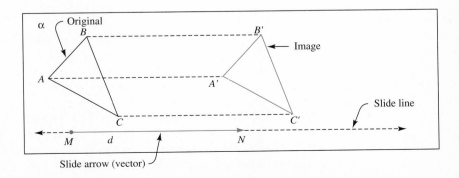

isometry
rigid motion

preserves both length and angle size (and hence, the congruence) of figures. Any transformation that preserves distance is called an **isometry** (derived from Greek and meaning "equal measure"), or a **rigid motion.** Thus a translation is an isometry.

> ### HISTORICAL NOTE
>
> In 1872, at age 23, Felix Klein (1849–1925) was appointed to a chair at the University of Erlangen, Germany. His inaugural address, referred to as the *Erlanger Programm,* described geometry as the study of properties of figures that do not change under a particular set of transformations. Specifically, Euclidean geometry was described as the study of such properties of figures as area and lengths, which remain unchanged under a set of transformations called *isometries*. It was not until 1975 that a secondary-school text was introduced in the United States that utilized Klein's ideas in a major way.

More Constructions of Translations

The image of a figure *under a translation* can be constructed with tracing paper. We can also construct a translation image of a figure by using only a compass and a straightedge. To construct such an image, we first need to know how to find the image of a given point, since geometric figures are made of points. Given a translation determined by the slide arrow from M to N in Figure 12-4(a), A', the image of A under this translation, must be such that $\overline{AA'}$ is parallel to $\overline{MN}$ and $AA' = MN$. This implies that $AA'NM$ is a parallelogram. (Why?) Hence, to find A', we construct parallelogram $AA'NM$, where A, N, and M are given. For an efficient construction of the parallelogram, recall that a quadrilateral in which each pair of opposite sides is congruent is a parallelogram. Thus we construct vertex A' by making $AA' = MN$ and $NA' = MA$. This is shown in Figure 12-4(b).

FIGURE 12-4

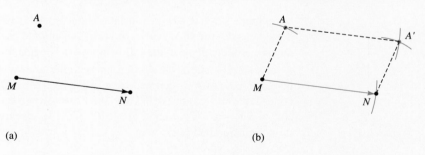

(a)   (b)

> **REMARK** There exist two parallelograms with vertices A, M, and N, but only one, as indicated in Figure 12-4(b), takes A to A' in the direction of the arrow $\overrightarrow{MN}$.

To find the image of a triangle under a translation, we find the images of the three vertices by a process similar to that used in Figure 12-4 and connect these images with segments to form the triangle's image.

On a geoboard or a grid, it is often possible to find an image of a point, as the following example shows.

EXAMPLE 12-1 Find the image of $\overline{AB}$ under the translation from X to X' pictured on the dot paper in Figure 12-5.

FIGURE 12-5

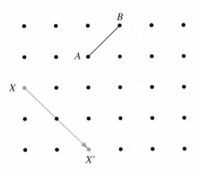

Solution To find the image of $\overline{AB}$, we first find the images of A and B by using tracing paper. The result is shown in Figure 12-6.

FIGURE 12-6

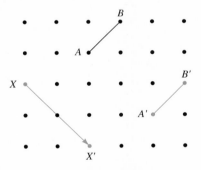

REMARK The image could have been found without tracing by noticing that the image of each point can be determined by shifting each point down two units and then shifting it right two units.

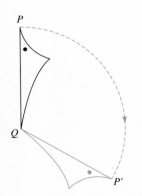

FIGURE 12-7

Rotations

rotation / turn

A **rotation,** or **turn,** is another kind of isometry. Figure 12-7 illustrates congruent figures resulting from a rotation about point Q. A rotation can be constructed by using tracing paper, as in Figure 12-8. In Figure 12-8(a), $\triangle ABC$ and point O are traced on tracing paper. Holding point O fixed, we can turn the tracing paper to obtain an image, $\triangle A'B'C'$, as

turn center / turn angle

shown in Figure 12-8(b). Point O is the **turn center,** and $\angle COC'$ is the **turn angle.**

FIGURE 12-8

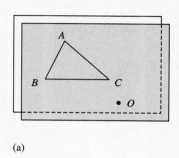
(a)

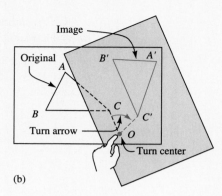
(b)

In order to determine a rotation, we must be given three things: the turn center; the direction of the turn, either clockwise or counterclockwise; and the amount of the turn. The amount and the direction of the turn can be illustrated by a **turn arrow,** or it can be specified as a number of degrees.

turn arrow

Figure 12-9 shows an example of a rotation about point O through 30° in a counterclockwise direction. The image of the letter **F** is shown in green.

FIGURE 12-9

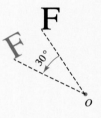

This discussion leads us to the following definition.

> **DEFINITION OF ROTATION**
>
> A **rotation** is a transformation of the plane determined by holding one point—the center—fixed and rotating the plane about this point by a certain amount in a certain direction.

Constructions of Rotations

The definition of a rotation can be used to find the image of any point of the plane under a rotation with center O and through a specified amount, as pictured in Figure 12-10(a). Consider what must be done to find the image of any point P in the plane. We need to draw a circle with center O and radius OP, as in Figure 12-10(b), and then, starting at P, we need to move along the circle—counterclockwise if $\alpha > 0$, and clockwise if $\alpha < 0$—until we locate the point P' such that $m(\angle POP') = \alpha$. Using a compass and a straightedge, we can find the location of P' by constructing $\angle POP'$ congruent to the angle of rotation so that P'

will be drawn in the desired direction indicated by the amount of the rotation, or the turn angle. This is illustrated in Figure 12-10(b).

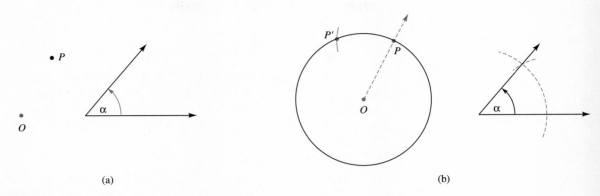

FIGURE 12-10

Because a rotation is an isometry, the image of a figure under a rotation is congruent to the original figure. Thus, under a rotation as under a translation, the image of a line is a line, the image of a circle is a circle, and the images of parallel lines are parallel.

Rotations may be constructed on dot paper, as demonstrated in Example 12-2.

EXAMPLE 12-22 Find the image of $\triangle ABC$ under the rotation with center O in Figure 12-11.

FIGURE 12-11

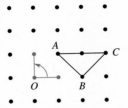

Solution We use tracing paper to find $\triangle A'B'C'$, the image of $\triangle ABC$, as shown in Figure 12-12.

FIGURE 12-12

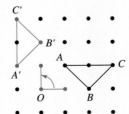

half-turn

A rotation of 360° will transform any figure onto itself. A rotation of 180° about a point is also of particular interest. Such a rotation is called a **half-turn.** Because a half-turn is a rotation, it has all the properties of rotations. Figure 12-13 shows some shapes and their images under a half-turn about point O.

FIGURE 12-13

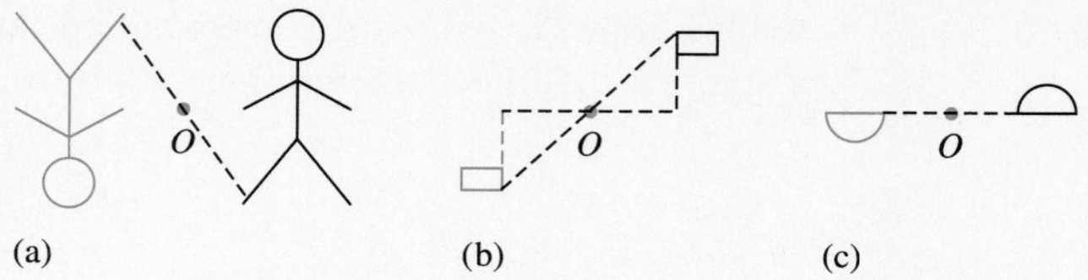

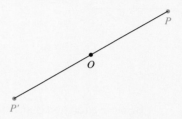

(a)   (b)   (c)

Figure 12-14 shows a point P and its image P' under a half turn about O. Because $\angle POP'$ measures $180°$, points P, O, and P' are collinear.

FIGURE 12-14

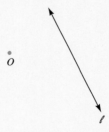

These observations make it easy to construct an image of a figure under a half-turn using a straightedge and a compass, as demonstrated in the following example.

E X A M P L E 1 2 - 3   Use a compass and a straightedge to find the image of a line ℓ under a half-turn about point O in Figure 12-15.

FIGURE 12-15

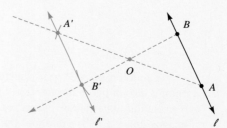

Solution   Because a line is determined by two points and the image of a line is a line, it is sufficient for us to pick any two points on ℓ and find the images of these points under the half-turn. In Figure 12-16, we pick two arbitrary points A and B on ℓ and find their images A' and B' by drawing $\overrightarrow{AO}$ and marking off $OA' = OA$ so that O is the midpoint of $\overline{AA'}$. We find B' similarly. The line connecting A' and B' is the image of ℓ. The result is illustrated in Figure 12-16.

FIGURE 12-16

PROBLEM SET 12-1

1. What type of motion is involved in each of the following?
   (a) A skier skiing straight down a slope
   (b) A leaf floating down a stream

2. For each of the following, find the image of the given quadrilateral under a translation from A to B:

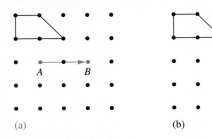

(a) (b)

3. Find the figure whose image is given in each of the following under a translation from X to X':

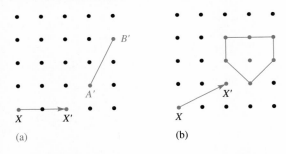

(a) (b)

4. Construct the image of $\overline{BC}$ under the translation pictured, using the following:
   (a) Tracing paper
   (b) Compass and straightedge

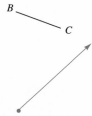

5. Name three everyday examples of rotations.

6. Find the image of the given quadrilateral in a 90° counterclockwise rotation about O.

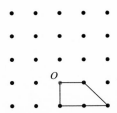

7. The images of segment $\overline{AB}$ under various rotations are given in the accompanying drawings. Find $\overline{AB}$ in each case.

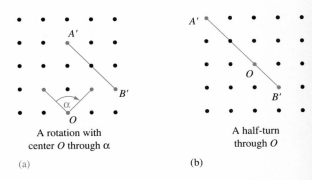

A rotation with center O through α A half-turn through O

(a) (b)

8. (a) The image of NOON is still NOON after a special half-turn. List some other such words. What letters can such words contain?
   (b) 1961 is the image of 1961 after a special half-turn. What other natural numbers less than 10,000 have this property?

9. In each of the following, find the image of the circle M under a 120° counterclockwise rotation about O:

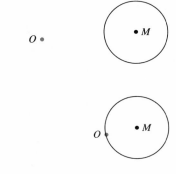

(a)

(b)

10. In each of the following figures, find the image of the figure under a half-turn about O:

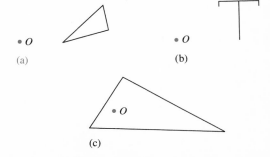

(a) (b)

(c)

11. Draw three arbitrary points P, P', and Q. If P' is the image of P under a half-turn, construct the image of Q under that half-turn.
12. (a) In succession, perform the two rotations, each with center O, in the figure.
    (b) What is the result of the two rotations?
    (c) Is the order of the rotations important?
    (d) Could the result have been accomplished in one rotation?

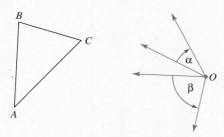

13. When $\triangle ABC$ is rotated about a point O by $360°$, each of the vertices traces a path.
    (a) What geometric figure does each vertex trace?
    ▶(b) Identify all points O for which two vertices trace an identical path. Justify your answer.
    ▶(c) Given any $\triangle ABC$, is there a point O such that the three vertices trace an identical path? If so, describe how to find such a point. Justify your answer.

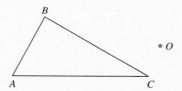

14. ▶In each of the following, trace the given figure on tracing paper, rotate it by $180°$ about the given point O, sketch the image and make a conjecture about the kind of figure that is formed by the union of the original figure and its image. In each case explain why you think your conjecture is true.

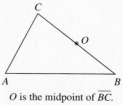

O is the midpoint of $\overline{BC}$.
(a)

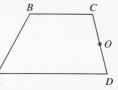

$ABCD$ is a trapezoid and O is the midpoint of $\overline{CD}$.
(b)

$ABCD$ is a square and O is the intersection of its diagonals
(c)

15. ▶Draw any line and label it ℓ. Use tracing paper to find ℓ', the image of ℓ under the given rotation. In each case, describe in words how ℓ' is related to ℓ.
    (a) Half-turn about point O on ℓ.
    (b) Half-turn about a point O, not on ℓ.
    (c) A $90°$ turn counterclockwise about point O, not on ℓ.
    (d) A $60°$ turn counterclockwise about point O, not on ℓ.

★16. (a) Through point P, construct a segment $\overline{AB}$ that is bisected by point P, where point A is on line ℓ and point B is on line m.
    (b) Is the construction always possible if lines ℓ and m are parallel?

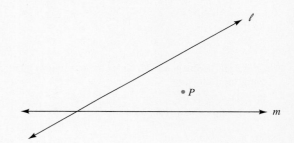

17. (a) Translations may be explored in Logo by using a figure called an EE. Type the given programs into your computer and then run the following with the turtle starting at home with heading 0:
    (i) SLIDE 40 45
    (ii) SLIDE 200 57
    (iii) SLIDE (−50) (−75)
    ```
    TO SLIDE :DIRECTION :DISTANCE
      EE
      PENUP
      SETHEADING :DIRECTION
      FORWARD :DISTANCE
      PENDOWN
      SETHEADING 0
      EE
    END
    ```

```
TO EE
  FORWARD 50 RIGHT 90
  FORWARD 25 BACK 25
  LEFT 90 BACK 25
  RIGHT 90 FORWARD 10
  BACK 10 LEFT 90
  BACK 25 RIGHT 90
  FORWARD 25 BACK 25
  LEFT 90
END
```

(b) Edit the SLIDE procedure in (a) so that it will slide an equilateral triangle.

18. Write a Logo procedure called ROTATE that will draw a square and produce the image of the square when the square is rotated by an arbitrary angle :A about one of its vertices.

19. Write a Logo procedure called TURN.CIRCLE that will draw a circle passing through the home of the turtle and produce the image of the circle under the following transformations:
   (a) A half-turn about the turtle's home
   (b) A 90° counterclockwise rotation about the turtle's home

▼ BRAIN TEASER

In the drawing on the left, a coin is shown atop another coin. Suppose the top coin is rotated around the circumference of the bottom coin until it rests directly below the bottom coin. Will the head be straight up or upside down? Explain why.

LABORATORY ACTIVITY

1. As a van Hiele Level 1 activity, look around your classroom and find and list the following:
   (a) Congruent objects such that a translation will take one object to another
   (b) Congruent objects such that a rotation will take one object to another

2. As a van Hiele Level 3 activity, use a compass and a straightedge in the following drawing to construct two chords of equal length through points P and Q that are perpendicular to each other.

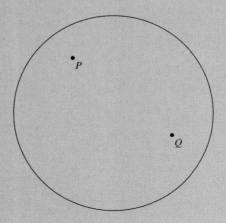

636　CHAPTER 12　MOTION GEOMETRY AND TESSELLATIONS

Section 12-2　Reflections and Glide Reflections

Reflections

reflection / flip　Another isometry is a **reflection,** or a **flip.** One example of a reflection often encountered in our daily lives is a mirror image. Figure 12-17 shows a figure with its mirror image.

FIGURE 12-17

Another reflection is shown in the B.C. cartoon.

Reprinted by permission of Johnny Hart and Creators Syndicate, Inc.

reflecting line　In a plane, we can simulate reflections in various ways. Consider the half tree shown in Figure 12-18(a). Folding the paper along the **reflecting line** and drawing the image gives *mirror image*　the **mirror image,** or *image,* of the half tree. In Figure 12-18(b), the paper is unfolded. The total figure obtained is symmetric about the fold line in much the same way that a mirror gives symmetry in space. Another way to simulate a reflection in a line involves using a Mira and is illustrated in Figure 12-18(c).

FIGURE 12-18

 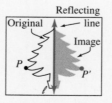 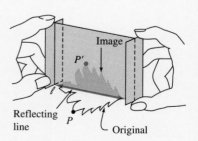

(a)　　　　　　　(b)　　　　　　　(c)

In Figure 12-19(a), the image of P under a reflection in line ℓ is P'. $\overline{PP'}$ is both perpendicular to and bisected by ℓ, or equivalently, ℓ is the perpendicular bisector of $\overline{PP'}$. In Figure 12-19(b), P is its own image under the reflection in line ℓ. If ℓ were a mirror, then P' would be the mirror image of P. This leads us to the following definition of a reflection.

FIGURE 12-19

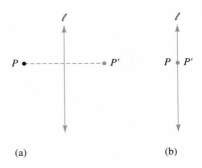

DEFINITION OF REFLECTION

A **reflection** in a line ℓ is a transformation of a plane that pairs each point P of the plane with a point P' in such a way that ℓ is the perpendicular bisector of $\overline{PP'}$, as long as P is not on ℓ. If P is on ℓ, then $P = P'$.

In Figure 12-20, we see another property of a reflection. In the original triangle, if we walk clockwise around the vertices, starting at vertex A, we see the vertices in the order A-B-C. However, in the reflection image of triangle ABC, if we start at A' (the image of A) and walk clockwise, we see the vertices in the following order: A'-C'-B'. Thus a reflection does something that neither a translation nor a rotation does: It reverses the orientation of the original figure. There are many methods of constructing a reflection image. We illustrate such constructions with paper folding, tracing paper, a Mira, compass, and straightedge.

FIGURE 12-20

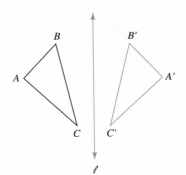

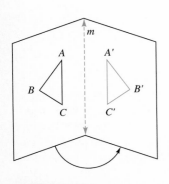

FIGURE 12-21

Constructing a Reflection by Paper Folding. To find the image of $\triangle ABC$ in Figure 12-21, we fold the paper along reflecting line m and mark the image of the triangle. To do this may require punching holes in the paper or indenting the vertices heavily with a pencil so that they are visible.

Paper folding and mirrors are often used in elementary school to explore reflection in a line, as shown on the student page from *Addison-Wesley Mathematics*, Grade 7, 1991.

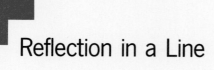

Reflection in a Line

LEARN ABOUT IT

EXPLORE Complete the Activity

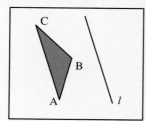

Draw △ABC and line *l*. Fold along *l* keeping △ABC on the outside of the paper.

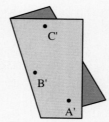

Mark points A', B', and C' that coincide with the vertices of △ABC.

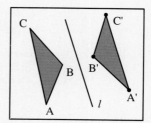

Unfold the paper and draw △A'B'C'.

TALK ABOUT IT

1. Set a mirror on line *l* perpendicular to the paper. What do you see?

2. How is the fold line *l* related to $\overline{BB'}$?

When you look at △ABC, in the mirror on the reflection line *l*, △ABC appears to be △A'B'C'. So, △A'B'C' is called the mirror or **reflection image** of △ABC in line *l*.

The reflection image of ABCD in *l* is EFGH.

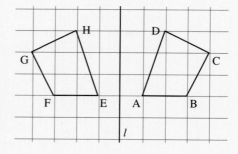

Constructing a Reflection Using Tracing Paper. Another method of constructing the image of an original under a reflection involves using tracing paper or an acetate sheet. Figure 12-22(a) shows the use of tracing paper. We trace the original figure, the reflecting line, and a point on the reflecting line, which we use as a reference point. When we flip the tracing paper over to perform the reflection, we align the reflecting line and the reference point, as in Figure 12-22(b). Aligning the reference point ensures that no translating occurs along the reflecting line when the reflection is performed. If we wish the image to be on the paper with the original, we may indent the tracing paper or acetate sheet to mark the images of the original vertices.

FIGURE 12-22

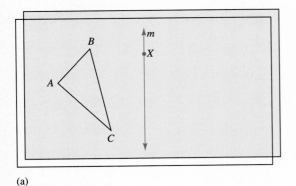
(a)

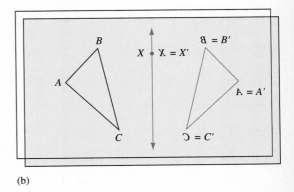
(b)

Constructing a Reflection with a Mira. To construct the image of an original under a reflection in line m by using a Mira, as in Figure 12-23, we align the drawing edge of the Mira along line m and mark the image of the original.

FIGURE 12-23

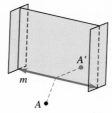

Constructing a Reflection with Compass and Straightedge. To construct the image of a figure under a reflection with a compass and a straightedge, first recall that the reflecting line m is the perpendicular bisector of the segment connecting any point P and its image P'. Thus, given point P and reflecting line m, as in Figure 12-24(a), we construct the image of P in two stages. First we construct a perpendicular ray from P to m, as in Figure 12-24(b), that intersects m in point X. Then we find P' on $\overrightarrow{PX}$ such that $XP' = PX$. If the compass setting is not changed in the construction, then P' is found automatically, as in Figure 12-24(b). (Why?)

FIGURE 12-24

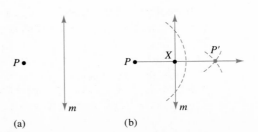

640   CHAPTER 12   MOTION GEOMETRY AND TESSELLATIONS

EXAMPLE 12-4   Use a compass and a straightedge to construct the image of $\overleftrightarrow{AB}$ under a reflection in line m in Figure 12-25.

FIGURE 12-25

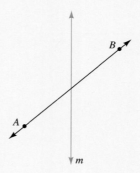

Solution   Under a reflection, the image of a line is a line. Thus, to find the image of $\overleftrightarrow{AB}$, it is sufficient to choose any two points on the line and find their images. The images determine the line that is the image of $\overleftrightarrow{AB}$. We choose two points whose images are easy to find. Point X, the intersection of $\overleftrightarrow{AB}$ and m, is its own image. If we choose point A and use the compass-and-straightedge method explained earlier, we produce the construction shown in Figure 12-26.

FIGURE 12-26

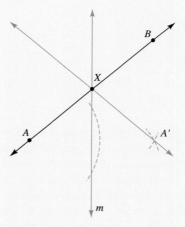

REMARK   We could have used the compass and straightedge to find the images of both A and B in order to complete the construction.

Tracing paper and dot paper or geoboards may be used to find the images of figures under a reflection, as described in Example 12-5.

EXAMPLE 12-5   Find the image of △ABC under a reflection in line m, as in Figure 12-27.

FIGURE 12-27

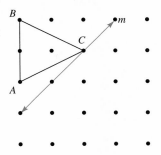

Solution   We can use tracing paper to find the image of the reflection. The result is shown in Figure 12-28. Note that C is the image of itself and that the images of vertices A and B are A' and B' such that m is the perpendicular bisector of $\overline{AA'}$ and $\overline{BB'}$.

FIGURE 12-28

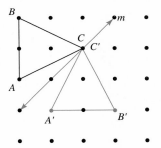

It is possible to find the reflecting line, given an original figure and its reflection image. An example of this is provided in Example 12-6.

EXAMPLE 12-6   Given the △ABC and its reflection image △A'B'C', as shown in Figure 12-29, find the line of reflection.

FIGURE 12-29

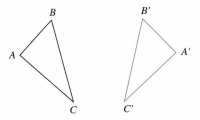

Solution   Because the reflecting line m is the perpendicular bisector of all segments' connecting points and their images, it is sufficient to find the perpendicular bisector of $\overline{AA'}$. This is shown in Figure 12-30.

FIGURE 12-30

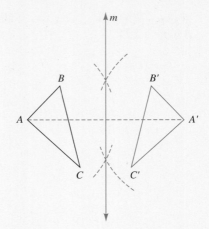

> **REMARK** Any vertex other than A could also have been used. Another technique is to fold $\triangle ABC$ onto $\triangle A'B'C'$ and crease the fold line. The fold line is the reflecting line.

PROBLEM 1

The hiker in Figure 12-31, carrying a bucket, sees that his tent is on fire. To what point on the bank of the river should the hiker run to fill his bucket in order to make his trip to the tent as short as possible?

FIGURE 12-31

 Understanding the Problem. To understand the problem better, we first *draw a diagram*, as seen in Figure 12-32. We label the hiker H, the tent T, and the river r. The hiker needs to find point P on the bank of the river so that the distance $HP + PT$ is as short as possible.

FIGURE 12-32

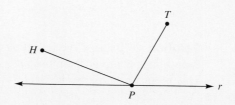

Devising a Plan. From the properties of reflections, we know that if T' is the reflection of the tent T in line r, then r is the perpendicular bisector of $\overline{TT'}$, as shown in Figure 12-33(a). Hence, any point on r is equidistant from T and T'. Thus $PT = PT'$. Therefore the hiker may solve the problem by solving a *related problem* of finding a point P on r such that the path from H to P and then to T' is as short as possible. The shortest path connecting H and T' is a segment. The intersection of $\overline{HT'}$ and r determines the point on the river toward which the hiker should run.

FIGURE 12-33

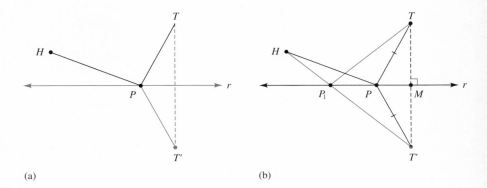

(a)   (b)

Carrying Out the Plan. Connect H and T', as shown in Figure 12-33(b). The point of intersection P_1 is the required point. The hiker should run to P_1 and then from P_1 to T.

Looking Back. To prove that the path from H to P_1 to T is the shortest possible, we need to prove that $HP_1 + P_1T < HP + PT$, where P is any point on r different from P_1. Because $HP_1 + P_1T = HP_1 + P_1T' = HT'$ (why?) and $HP + PT = HP + PT'$ (why?), the inequality that we needed to prove is equivalent to $HT' < HP + PT'$. This last inequality follows from the Triangle Inequality. ∎

Glide Reflections

glide reflection   Another basic isometry is a **glide reflection.** An example of a glide reflection is shown in the footprints of Figure 12-34. We consider the footprint labeled F_1 to have been translated to footprint F_2 and then reflected over line m to yield F_3, the image of F_1. The illustration in Figure 12-34 leads us to the following definition.

FIGURE 12-34

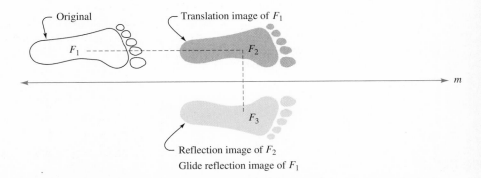

DEFINITION OF GLIDE REFLECTION

A **glide reflection** is a translation followed by a reflection in a line parallel to the slide arrow of the translation.

Another illustration of a glide reflection is shown in Figure 12-35, where $\triangle A_1 B_1 C_1$ is the image of $\triangle ABC$ under a translation taking M to N, and then $\triangle A'B'C'$ is the image of $\triangle A_1 B_1 C_1$ under a reflection in m, where m is parallel to $\overleftrightarrow{MN}$. Hence, the glide reflection takes $\triangle ABC$ to $\triangle A'B'C'$. Because constructing a glide reflection involves constructing a translation and a reflection, which we have already seen how to perform, the task of constructing a glide reflection is not a new problem. Exercises involving the construction of images of figures under glide reflections are given in the problem set.

FIGURE 12-35

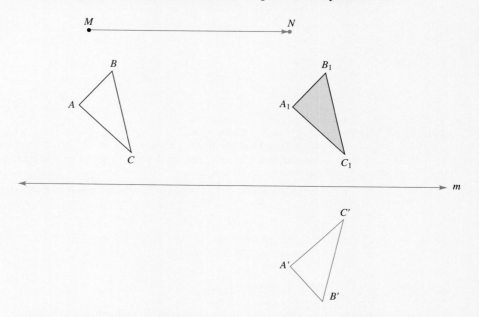

We have seen that under an isometry, the image of a figure is a congruent figure. Also, given two congruent figures, it is possible to show that one figure can be transformed to the other by a succession of isometries. The following example shows how this can be done in such cases.

EXAMPLE 12-7 Given that $ABCD$ in Figure 12-36 is a rectangle, describe a sequence of isometries to show: (a) $\triangle ADC \cong \triangle CBA$; (b) $\triangle ADC \cong \triangle BCD$; and (c) $\triangle ADC \cong \triangle DAB$.

Solution
(a) A half-turn with center E will cause the desired transformation.
(b) A reflection in a line passing through E and parallel to $\overline{AD}$ will produce the desired transformation.
(c) A reflection in a line passing through E and parallel to $\overline{DC}$ will produce the desired transformation.

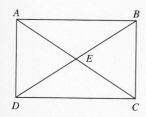

FIGURE 12-36

PROBLEM SET 12-2

1. For each of the following, find the image of the given quadrilateral under a reflection in ℓ:

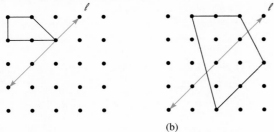

(b)

2. (a) Find the image of △ABC under a reflection in line ℓ.

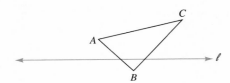

▶(b) Is it possible to find the image of △ABC in (a) without first finding the image of B? If so, explain how.

3. Draw a line and a circle whose center is not on the line. Find the image of the circle under a reflection in the line.

4. Which of the following figures have a reflecting line such that the image of the figure under the reflecting line is the figure itself? In each case, find as many such reflecting lines as possible, sketching appropriate drawings.
   (a) Circle
   (b) Segment
   (c) Ray
   (d) Square
   (e) Rectangle
   (f) Scalene triangle
   (g) Isosceles triangle
   (h) Equilateral triangle
   (i) Trapezoid whose base angles are not congruent
   (j) Isosceles trapezoid
   (k) Arc
   (l) Kite
   (m) Rhombus
   (n) Regular hexagon
   (o) Regular n-gon

5. Justify your answers in Problem 4 by paper folding.

6. What is the result of performing two successive reflections in line ℓ in the figure?

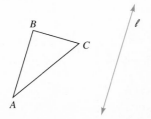

7. Suppose lines ℓ and m are parallel and △ABC is reflected in ℓ to obtain △A'B'C' and then △A'B'C' is reflected in m to obtain △A"B"C". Determine whether the same final image is obtained if △ABC is reflected first in m and then in ℓ.

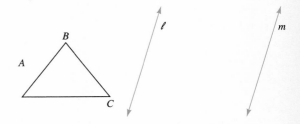

8. In Problem 7, △ABC is reflected in ℓ and then its image is reflected in m.
   (a) Use the drawing in Problem 7 to determine whether the same final image is obtained if △ABC is first reflected in m and then its image is reflected in ℓ.
   (b) Answer the question in (a) for the case when ℓ and m are perpendicular.

9. For the following, use any construction methods to find the image of △ABC if △ABC is reflected in ℓ to obtain △A'B'C' and then △A'B'C' is reflected in m to obtain △A"B"C" (ℓ and m intersect at O):

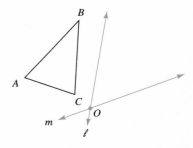

10. (a) The word TOT is its own image when it is reflected through a vertical line through O. List some other such words.

(a)

(b) The image of BOOK is still BOOK when it is reflected through a horizontal line. List some other such words. Which uppercase letters can you use?

(c) The image of 1881 is 1881 after reflection in either a horizontal or vertical line as shown. What other natural numbers less than 2000 have this property?

11. ▶(a) Draw an isosceles triangle *ABC* and then construct a line such that the image of △*ABC* when reflected in the line is △*ABC*. Explain why the line you constructed has the required property.
   ▶(b) For what kind of triangles is it possible to find more than one line with the property in (a)? Justify your answer.
   ▶(c) Given a scalene triangle, *ABC*, is it possible to find a line ℓ such that when △*ABC* is reflected in ℓ, its image is △*ABC*? Explain your answer.
   ▶(d) Draw a circle with center *O* and a line with the property that the image of the circle, when reflected in the line, is the original circle. Identify all such lines. Justify your answer.

12. Find the image of the footprint in the glide reflection that is the result of a translation from *M* to *N* followed by a reflection in ℓ.

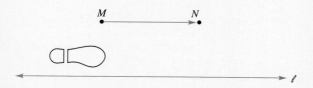

13. A glide reflection was defined as a translation followed by a reflection in appropriate lines.
   (a) Use the drawing in Problem 12 to determine whether the same final image is obtained if the reflection is followed by the translation.
   (b) Based on your answer in (a), are the reflection and translation involved in the glide reflection commutative?

14. For the given figure numbered 1, decide whether a reflection, a translation, a rotation, or a glide reflection will transform the figure into each of the other numbered figures. (There may be more than one possible answer.)

*15. If a Mira is available, use it to investigate Problems 2, 7, 8, and 9.

★16. When a billiard ball bounces off a side of a pool table, the angle of incidence is usually congruent to the angle of reflection; that is, $\angle 1 \cong \angle 2$. If a cue ball is at point *A*, show how the player should aim to hit three sides of the table and then the ball at *B*.

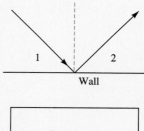

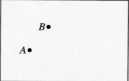

*17. ▶Two cities, represented by points A and B, are located near two perpendicular roads as shown. The cities' mayors found it necessary to build another road connecting A with a point P on road 1, then connecting P with a point Q on road 2, and finally connecting Q with B. How should the road $APQB$ be constructed if it is to be as short as possible? Copy the figure shown and use a straightedge and a compass, a Mira, or paper folding to construct the shortest possible path. Explain why the path you found is the shortest.

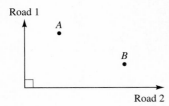

18. ▶Use the following drawing to explain how a periscope works:

19. 🖥 Enter the given procedures into your computer and then describe the transformations illustrated in each of the following:

```
TO SQ
  REPEAT 4 [FD 40 RT 90]
END
TO FSQ
  REPEAT 4 [FD 40 LT 90]
END
```

(a) ```
TO MOVE1
  SQ
  RT 150
  SQ
END
```
(b) ```
TO MOVE2
  SQ
  FSQ
END
```
(c) ```
TO MOVE3
  SQ
  PU RT 45 FD 60 PD
  SQ
END
```

20. (a) 🖥 Write a simple Logo procedure called FIG1 using FD, BK, RT, and LT several times. Run your procedure.
   (b) Write a new procedure, FIG2, in which you replace every FD in your FIG1 procedure with BK and every BK with FD. How does the drawing made with FIG2 compare to the drawing made with FIG1?
   (c) Write a new procedure, FIG3, in which you replace every RT in your FIG1 procedure with LT and every LT with RT. How does the drawing made with FIG3 compare with the drawings made with FIG1 and FIG2?

21. (a) 🖥 Write a Logo procedure called EQTRI to draw a variable-sized equilateral triangle.
   (b) Now write a procedure called EQTRI2 in which all the RIGHTs in EQTRI have been replaced with LEFTs, and vice versa.
   (c) Execute EQTRI and EQTRI2 on the same screen without erasing anything. What single transformation could produce the same result on the original triangle drawn by EQTRI?
   (d) Execute EQTRI 50 and then EQTRI −50. What single transformation could produce the same result on the original triangle drawn by EQTRI 50?

**Review Exercises**

22. Which capital printed letters of the English alphabet are their own images under a rotation?
23. Which capital letters of the English alphabet are their own images under a half-turn?
24. **MOW** is an example of a word that could be transformed into itself by which isometry?
25. (a) Find all possible rotations that transform a circle into itself.
   (b) By what other kinds of transformations can a circle be transformed into itself?
```

BRAIN TEASER

Two cities are on opposite sides of a river, as shown. The cities' engineers want to build a bridge across the river that is perpendicular to the banks of the river and access roads to the bridge so that the total distance between the cities is as short as possible. Where should the bridge and the roads be built?

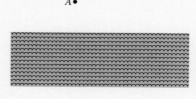

LABORATORY ACTIVITY

As a van Hiele Level 1 activity, take a 1-by-1-ft square of linoleum tile and carve a pattern in the tile comparable to the one shown in the following figure.

Now spread ink over the uncarved surface and press a piece of paper onto the ink, being careful not to let it slide across the tile. Next, remove the paper and consider the printed impression made on the paper. How are the printed paper and the original carved tile related to one another?

Section 12-3 Size Transformations

The transformations we have investigated so far preserved distance; consequently, the image of a figure under one of these transformations was a figure congruent to the original. A different type of transformation happens when a slide is projected on a screen. All objects on the slide are enlarged on the screen by the same factor. Figure 12-37 is another example of such a transformation. Each side of △A'B'C' is twice as long as the corresponding side of △ABC. The point O is the *center* of the *size transformation* and 2 is the *scale factor*. Notice that O, A, and A' are collinear and $OA' = 2 \cdot OA$.

FIGURE 12-37

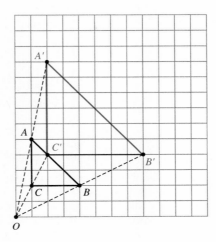

In general, we have the following definition.

DEFINITION OF SIZE TRANSFORMATION

A **size transformation** from the plane to the plane with center O and scale factor r ($r > 0$) is a transformation that assigns to each point A in the plane a point A', such that O, A, and A' are collinear and $OA' = r \cdot OA$ and so that O is not between A and A'.

Figure 12-38 shows the image of a point A under a size transformation with scale factor $r > 1$.

FIGURE 12-38

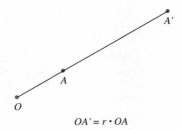

$OA' = r \cdot OA$

EXAMPLE 12-8

(a) In Figure 12-39(a), find the image of point P under a size transformation with center O and scale factor $\frac{2}{3}$.

(b) Find the image of the quadrilateral *ABCD* in Figure 12-39(b) under the size transformation with center *O* and scale factor $\frac{2}{3}$.

FIGURE 12-39

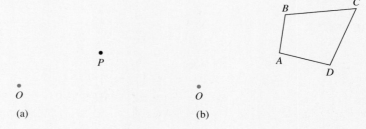

(a) (b)

Solution

(a) In Figure 12-40(a), we connect *O* with *P* and divide $\overline{OP}$ into three equal parts. The point *P'* is the image of *P* because $OP' = \frac{2}{3} OP$.

(b) We find the image of each of the vertices and connect the images to obtain the quadrilateral *A'B'C'D'* shown in Figure 12-40(b).

FIGURE 12-40

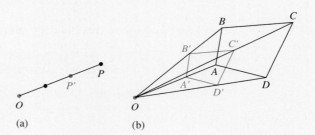

(a) (b)

In Figure 12-40(b), the sides of the quadrilateral *A'B'C'D'* are all parallel to the corresponding sides of the original quadrilateral and the angles of the quadrilateral *A'B'C'D'* are congruent to the corresponding angles of quadrilateral *ABCD*. Also, each side in the quadrilateral *A'B'C'D'* is $\frac{2}{3}$ as long as the corresponding side of quadrilateral *ABCD*. These properties are true for any size transformation and are summarized in the following theorem.

> **THEOREM 12-1**
>
> A size transformation with center *O* and scale factor r $(r > 0)$ has the following properties:
>
> 1. The image of a line segment is a line segment parallel to the original segment and *r* times as long.
> 2. The image of an angle is an angle congruent to the original angle.

From Theorem 12-1, it follows that the image of a polygon under a size transformation is a similar polygon. (Why?) However, given any two similar polygons it is not always possible to find a size transformation so that the image of one polygon under the transformation is the other polygon. But, given two similar polygons, we can "move" one of the polygons to a place so that it will be the image of the other polygon under a size transformation. The following examples show such instances.

EXAMPLE 12-9   Show that $\triangle ABC$ in Figure 12-41 is the image of $\triangle ADE$ under a size transformation. Identify the center of the size transformation and the scale factor.

FIGURE 12-41

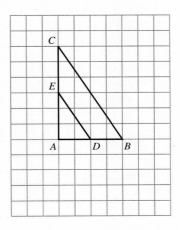

Solution   Because $\dfrac{AB}{AD} = \dfrac{AC}{AE} = 2$, we choose A as the center of the size transformation and 2 as the scale factor. Notice that under this transformation, the image of A is A itself. The image of D is B and the image of E is C. ∎

EXAMPLE 12-10   Show that $\triangle ABC$ in Figure 12-42(a) is the image of $\triangle APQ$ under a succession of isometries with a size transformation.

FIGURE 12-42

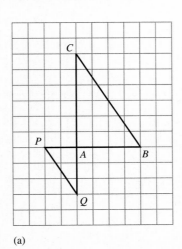

(a)

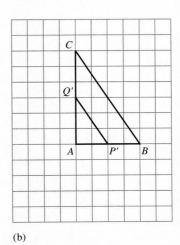

(b)

Solution We use the strategy of *looking at a related problem*. In Example 12-9, the common vertex served as the center of the size transformation, which was possible because the corresponding sides of the triangles were parallel. To achieve a similar situation, we first transform △APQ by a half-turn in A and obtain △AP'Q', shown in Figure 12-42(b). Now C is the image of Q' under a size transformation with center at A and scale factor 2. B is the image of P', and A is the image of itself under this transformation. Thus △ABC can be obtained from △APQ by first finding the image of △APQ under a half-turn in A and then applying a size transformation with center A and a scale factor 2 to that image. ∎

The above examples are a basis for an alternate definition of similar figures.

DEFINITION OF SIMILAR FIGURES

Two figures are similar if it is possible to transform one onto the other by a sequence of isometries followed by a size transformation.

PROBLEM SET 12-3

1. In each of the following drawings, reflect the original triangle in line ℓ and then reflect the image found in line m. After that, start again—this time first reflecting the original triangle in line m and then reflecting the image found in line ℓ. Is the final image in both cases the same?

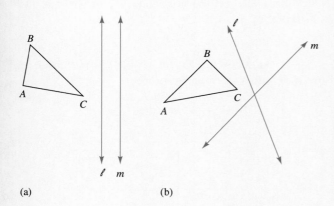

(a)   (b)

2. Describe a transformation that would "undo" each of the following:
   (a) A translation from M to N
   (b) A rotation of 75° with center O in a clockwise direction
   (c) A rotation of 45° with center A in a counterclockwise direction
   (d) A glide reflection that is the composition of a reflection in line m and a translation that takes A to B
   (e) A reflection in line n

3. In the following coordinate plane, find the images of each of the given points in the transformation that is the composition of a reflection in line m followed by a reflection in line n:
   (a) (4,3)   (b) (0,1)
   (c) (⁻1,0)   (d) (0,0)

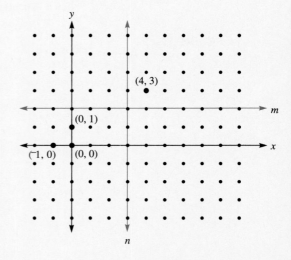

SECTION 12-3   EXERCISES   653

4. ►Copy the following figure onto grid paper and determine the center and the scale factor of the size transformation. Explain why there is only one possibility for the center.

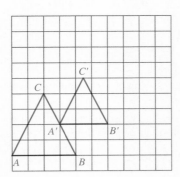

5. In the following figures, describe a sequence of isometries followed by a size transformation so that the larger triangle is the final image of the smaller one.

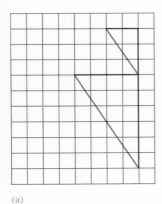
(a)

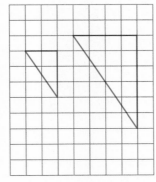
(b)

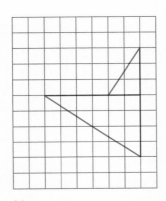
(c)

6. In the following drawing, find the image of △ABC under the size transformation with center O and scale factor $\frac{1}{2}$:

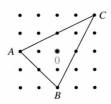

7. In each of the following drawings, find transformations that will take △ABC to its image, △A'B'C', which is similar:

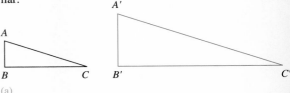
(a)

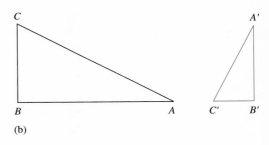
(b)

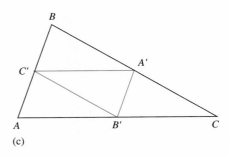
(c)

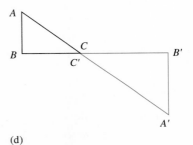
(d)

654   CHAPTER 12   MOTION GEOMETRY AND TESSELLATIONS

LABORATORY ACTIVITY

As a van Hiele Level 3 activity, consider △ABC and its image after being reflected in lines *m*, *n*, and *p* in order. Find a single line *q* that could be used to reflect the original △ABC onto the final image. Explain why it is always possible to find such a line.

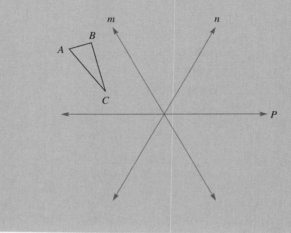

Section 12-4 Symmetries

Line Symmetries

The concept of a reflection can be used to identify line symmetries of a figure. All the drawings in Figure 12-43 have symmetries about the dashed lines.

FIGURE 12-43

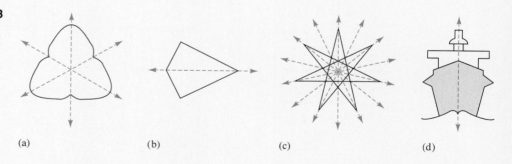

(a)   (b)   (c)   (d)

line of symmetry   Mathematically, a geometric figure is said to have a **line of symmetry** if it is its own image under a reflection in line ℓ. A method of creating a symmetrical figure is seen in Example 12-11.

EXAMPLE 12-11 In Figure 12-44, we are given a figure and a line *m*. Do the minimum amount of drawing to create a figure from the given figure so that the result is symmetric about line *m*.

FIGURE 12-44

Solution In order for the resulting figure both to be symmetric about line *m* and to incorporate the existing figure, we need to reflect the existing figure about line *m*. The desired result of doing that is the combination of the original and the image. The resulting figure is shown in Figure 12-45.

FIGURE 12-45

> **REMARK** The type of problem shown in Example 12-11 is much easier to do with a Mira.

EXAMPLE 12-12 How many lines of symmetry does each of the drawings in Figure 12-46 have?

FIGURE 12-46

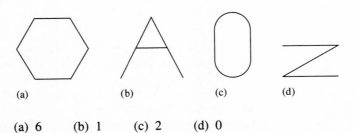

(a)   (b)   (c)   (d)

Solution   (a) 6   (b) 1   (c) 2   (d) 0

PROBLEM 2

At the site of an ancient settlement, archaeologists found a fragment of a saucer, as shown in Figure 12-47. To restore the saucer, the archaeologists had to determine the radius of the original saucer. How can they find the radius?

FIGURE 12-47

Understanding the Problem. The border of the shard shown in Figure 12-47 was part of a circle. In order to reconstruct the saucer, we are to determine the radius of the circle of which the shard is a part.

Devising a Plan. A *model* can be used to determine the radius. We trace an outline of the circular edge of the three-dimensional shard on a piece of paper. The result is an arc of a circle, as shown in Figure 12-48. To determine the radius, we find the center O. A circle has infinitely many lines of symmetry and each line passes through the center of the circle, where all the lines of symmetry intersect.

FIGURE 12-48

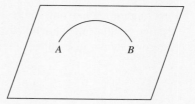

Carrying Out the Plan. To find a line of symmetry, fold the paper containing $\overarc{AB}$ so that a portion of the arc is folded onto itself. Then unfold the paper and draw the line of symmetry on the fold mark, as shown in Figure 12-49(a). By refolding the paper in Figure 12-49(a) so that a different portion of the arc $\overarc{AB}$ is folded onto itself, determine a second line of symmetry, as shown in Figure 12-49(b). The two dotted lines of symmetry intersect at O, the center of the circle of which $\overarc{AB}$ is an arc. To complete the problem, measure the length of either $\overline{OB}$ or $\overline{OA}$. (They should be the same.)

FIGURE 12-49

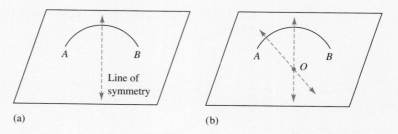

(a)   (b)

Looking Back. In the first fold, endpoint B of the arc was folded onto another point of the arc. Label this other point X. The result is shown in Figure 12-50. Because the center of the circle lies on the perpendicular bisector of a chord (why?) and the fold line ℓ is a line of symmetry of the circle containing $\overarc{AB}$, the fold line must be the perpendicular bisector of $\overline{XB}$ and it must contain the center of the circle. We could have used this property to determine the center of the circle by choosing two chords on the arc and finding the point where the perpendicular bisectors of the chords intersect. Alternatively, we could have used a compass and a straightedge.

FIGURE 12-50

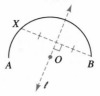

A related problem is: What would happen if the piece of pottery had been part of a sphere? Would the same ideas still work? ∎

Rotational (Turn) Symmetries

rotational symmetry / turn symmetry

A figure has **rotational symmetry,** or **turn symmetry,** when the traced figure can be rotated less than 360° about some point so that it matches the original figure. Note that the condition "less than 360°" is necessary because any figure will coincide with itself after being rotated 360°. In Figure 12-51, the equilateral triangle coincides with itself after a rotation of 120° about point O. Hence, we say that the triangle has 120° rotational symmetry. In addition, in Figure 12-51, if we rotated the triangle another 120°, we would find that it again matches the original, so we can say that the triangle also has 240° rotational symmetry.

FIGURE 12-51

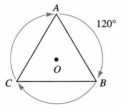

Other examples of figures that have rotational symmetry are shown in Figure 12-52. In Figure 12-52, (a), (b), (c), and (d) have 72°, 90°, 180°, and 180° rotational symmetries, respectively [parts (a) and (b) also have other rotational symmetries].

FIGURE 12-52

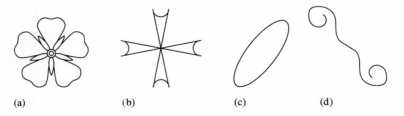

(a)   (b)   (c)   (d)

In general, we can determine whether a figure has rotational symmetry by tracing it and turning the tracing about a point (the center of the figure) to see if it aligns on the figure before the tracing has turned in a complete circle, or 360°. The amount of the rotation can be determined by measuring the angle ∠POP' through which a point P is rotated around a point O to match another point P' when the figures align. Such an angle, ∠POP', is labeled with points P, O, and P' of Figure 12-53 and has measure 120°. Point O, the point held fixed when the tracing is turned, is the turn center.

FIGURE 12-53

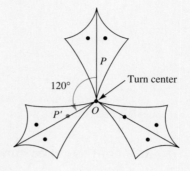

EXAMPLE 12-13 Determine the amount of the turn for the rotational symmetries of each part of Figure 12-54.

FIGURE 12-54

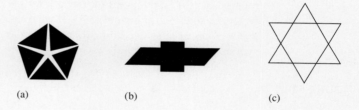

(a)   (b)   (c)

Solution (a) The amounts of the turns are $\frac{360°}{5}$ or 72°, 144°, 216°, and 288°.

(b) The amount of the turn is 180°.

(c) The amounts of the turns are 60°, 120°, 180°, 240°, and 300°. ∎

The rotation in Figure 12-54(b) exemplifies yet another type of symmetry, namely, point symmetry.

Point Symmetry

point symmetry   Any figure that has 180° rotational symmetry is said to have **point symmetry** about the turn center. Figures with point symmetry are shown in Figure 12-55.

FIGURE 12-55

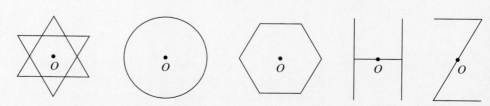

Suppose P is any point of a figure with point symmetry, such as in Figure 12-56(a). If the figure is rotated 180° about its center, point O, there is a corresponding point P', as shown in part (b) of the figure. Points P, O, and P' are on the same line, and O divides the segment connecting points P and P' into two parts of equal length; that is, O is the midpoint of $\overline{PP'}$.

FIGURE 12-56

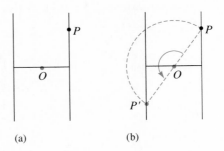

(a)   (b)

Plane Symmetry

plane of symmetry   A three-dimensional figure has a **plane of symmetry** when every point of the figure on one side of the plane has a mirror image on the other side of the plane. Examples of figures with plane symmetry are shown in Figure 12-57. Solids can also have point symmetry, line symmetry, and turn symmetry. These symmetries are analogous to the two-dimensional symmetries and are investigated in the problem set.

FIGURE 12-57

  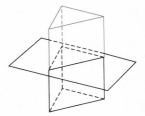

Applications of Symmetries in Geometry

Geometric figures in a plane can be classified according to the number of symmetries they have. Consider a triangle described as having exactly one line of symmetry and no turn symmetries at all. What could the triangle look like? The only possibility is a triangle in which two sides are congruent, that is, an isosceles triangle. The line of symmetry passes through a vertex, as shown in Figure 12-58.

FIGURE 12-58

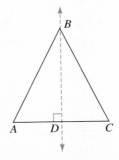

Just as we used the number of lines of symmetry to describe an isosceles triangle, we can describe equilateral and scalene triangles in terms of the number of lines of symmetry they have. This is left as an exercise.

A square, as in Figure 12-59, can be defined as a four-sided figure with four lines of symmetry—d_1, d_2, h, and v—and three turn symmetries about point O. In fact, we can use lines of symmetry and turn symmetries to define various types of quadrilaterals normally used in geometry. It is left as an exercise to see how these definitions differ from those in Chapter 10.

FIGURE 12-59

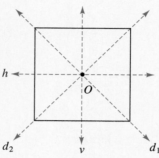

PROBLEM SET 12-4

1. Various international signs have symmetries. Determine which of the following have (i) line symmetry, (ii) rotational symmetry, and (iii) point symmetry:

Rendezvous point
(a)

Light switch
(b)

Bar
(c)

Observation deck
(d)

2. Design symbols that have each of the following symmetries, if possible:
   (a) Line symmetry but not rotational symmetry
   (b) Rotational symmetry but not point symmetry
   (c) Rotational symmetry but not line symmetry

3. In each of the following, complete the sketches so that they have line symmetry about ℓ:

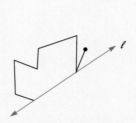

Line symmetry about ℓ
(a)

Line symmetry about ℓ
(b)

4. (a) Determine the number of lines of symmetry in each of the following flags.
   (b) Sketch the lines of symmetry for each flag.

Switzerland
(i)

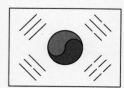

South Korea
(ii)

Israel
(iii)

Barbados
(iv)

(c) Can a figure have point, line, and rotational symmetry? If so, sketch a figure with these properties.

▶(d) If a figure has point symmetry, must it have line symmetry? Is the converse true?

▶(e) If a figure has both point and line symmetry, must it have rotational symmetry? Why?

9. In each of the following, complete the sketches so that they have the indicated symmetry:

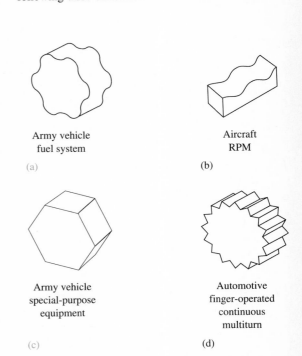

Point symmetry about O 60° rotational symmetry about O
(a)   (b)

5. Find the lines of symmetry, if any, for each of the following trademarks:

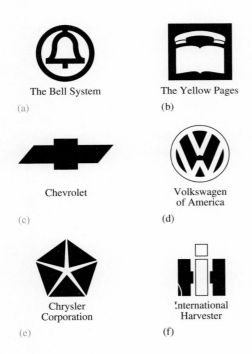

The Bell System (a)   The Yellow Pages (b)

Chevrolet (c)   Volkswagen of America (d)

Chrysler Corporation (e)   International Harvester (f)

10. How many planes of symmetry, if any, does each of the following three-dimensional vehicle controls have?

Army vehicle fuel system
(a)

Aircraft RPM
(b)

Army vehicle special-purpose equipment
(c)

Automotive finger-operated continuous multiturn
(d)

6. If possible, sketch a triangle that satisfies each of the following:
   (a) It has no lines of symmetry.
   (b) It has exactly one line of symmetry.
   (c) It has exactly two lines of symmetry.
   (d) It has exactly three lines of symmetry.

7. Sketch a figure that has point symmetry but no line symmetry.

8. Answer each of the following. If your answer is no, provide a counterexample.
   ▶(a) If a figure has point symmetry, must it have rotational symmetry? Why?
   ▶(b) If a figure has rotational symmetry, must it have point symmetry? Why?

11. Write a Logo procedure that draws a square and produces a figure with rotational symmetry of
    (a) 60°, (b) 120°, (c) 180°, (d) 240°, and (e) 300°.

12. Write a Logo procedure that draws an equilateral triangle and produces a figure with rotational symmetry of
    (a) 60°, (b) 120°, (c) 240°, and (d) 300°.

Review Problems

13. For each case, find the image of the given figure using paper folding.

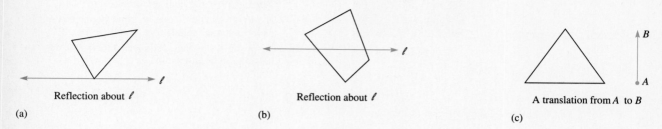

(a) Reflection about ℓ (b) Reflection about ℓ (c) A translation from A to B

14. Construct each of the images in Problem 13 using a compass and a straightedge.

● COMPUTER CORNER

The following Logo INSPI procedure produces drawings with various symmetries depending upon the inputs for :ANGLE. Some of these drawings are shown in the following figure. Determine what kinds of symmetries each of the figures has. Try some of your own inputs and see what drawings they produce and what kinds of symmetries the resulting figures have.

```
TO INSPI :SIDE :ANGLE
  FD :SIDE RT :ANGLE
  INSPI :SIDE :ANGLE + 5
END
```

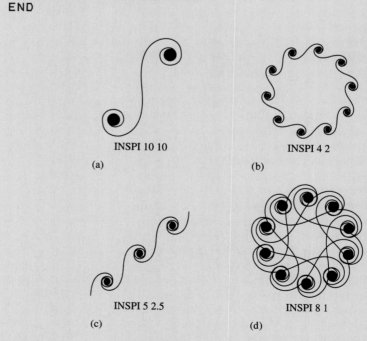

(a) INSPI 10 10
(b) INSPI 4 2
(c) INSPI 5 2.5
(d) INSPI 8 1

SECTION 12-5 TESSELLATIONS OF THE PLANE   663

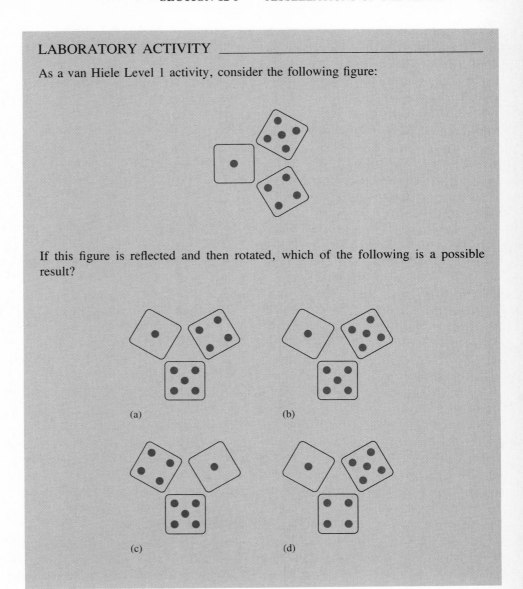

LABORATORY ACTIVITY

As a van Hiele Level 1 activity, consider the following figure:

If this figure is reflected and then rotated, which of the following is a possible result?

(a) (b) (c) (d)

*Section 12-5   Tessellations of the Plane

tessellation   A **tessellation** of a plane is the filling of the plane with repetitions of figures in such a way that no figures overlap and there are no gaps. (Similarly, one can tessellate space.) The tiling of a floor and various mosaics are examples of tessellations. Maurits C. Escher, born in the Netherlands in 1902, was a master of tessellations. Many of his drawings have fascinated mathematicians for decades. An example of his work, *Study of Regular Division of the Plane with Reptiles,* pen, ink, and watercolor, 1939, contains an exhibit of a tessellation of the plane by a lizardlike shape, as shown in Figure 12-60.

FIGURE 12-60

© 1939 M.C. Escher/Cordon Art–Baarn–Holland.

At the heart of the tessellation in Figure 12-60, we see a regular hexagon, but perhaps the simplest tessellation of the plane can be achieved with squares. Figure 12-61 shows two different tessellations of the plane with squares.

FIGURE 12-61

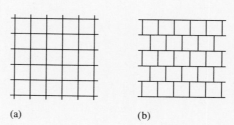

(a)   (b)

Regular Tessellations

Tessellations with regular polygons are appealing and interesting because of their simplicity. Figure 12-62 shows portions of tessellations with equilateral triangles and with regular hexagons. To determine other regular polygons that tessellate the plane, we investigate the possible size of the interior angle of a tessellating polygon. If n is the number of sides of a regular polygon, then because the sum of the measures of the exterior angles is 360°, the measure of an exterior angle is $360°/n$. Hence, the measure of an interior angle is $180° - 360°/n$. Table 12-1 gives some values of n, the type of regular polygon related to each, and the angle measure of an interior angle found by using the expression $180° - 360°/n$. If a regular polygon tessellates the plane, the sum of the congruent angles of the polygons around every vertex must be 360°. Thus 360 divided by the angle measure gives the number of angles around a vertex and hence must be an integer. If we divide 360 by each of the angle measures in the table, we find that of these measures only 60, 90, and 120 divide 360; hence, only an equilateral triangle, a square, and a regular hexagon can tessellate the plane. Can other regular polygons tessellate the plane? Notice that $\frac{360}{120} = 3$, and hence

360 divided by a number greater than 120 is smaller than 3; however, the number of sides of a polygon cannot be less than 3. Because a polygon with more than six sides has an interior angle greater than 120°, it actually is not necessary to consider polygons with more than six sides.

FIGURE 12-62

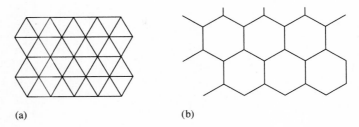

(a) (b)

TABLE 12-1

Number of Sides	Regular Polygon	Measure of Interior Angle
3	Triangle	60°
4	Square	90°
5	Pentagon	108°
6	Hexagon	120°
7	Heptagon	900/7°
8	Octagon	135°
9	Nonagon	140°
10	Decagon	144°

Next we consider tessellating the plane with arbitrary convex quadrilaterals. Before reading on, you may wish to investigate the problem yourself, with the help of cardboard quadrilaterals. Figure 12-63 shows an arbitrary convex quadrilateral and a way to tessellate the plane with the quadrilateral. Successive 180° turns of the quadrilateral about the midpoints P, Q, R, and S of its sides will produce four congruent quadrilaterals around a common vertex. Notice that the sum of the measures of the angles around vertex A is $a + b + c + d$, which is the sum of the measures of the interior angles of the quadrilateral, or 360°. As we have seen earlier, a regular pentagon does not tessellate the plane. However,

FIGURE 12-63

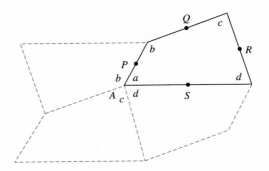

some nonregular pentagons do tessellate the plane. One such pentagon, along with a tessellation of the plane by the pentagon, is shown in Figure 12-64.

FIGURE 12-64

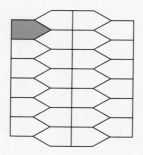

The study of which irregular pentagons tessellate is a surprisingly rich problem. Mathematicians thought that they had solved the problem and had classified eight types of pentagons that would tessellate. They believed they had all of them. But then in 1975, Marjorie Rice, a woman with no formal training in mathematics discovered a ninth type of tessellating pentagon. She went on to discover four more types of tessellating pentagons by 1977. Her interest was piqued by reading an article in *Scientific American* by Martin Gardner. Two of these are shown in Figure 12-65. The problem of how many different types of pentagons tessellate remains unsolved.

FIGURE 12-65

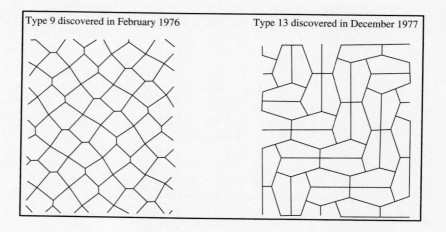

Tessellations are commonly studied in elementary school, as shown on the student page from *Addison-Wesley Mathematics,* Grade 8, 1991, on page 667. Other tessellations involving irregular figures are possible by using rotations and reflections. Consider the *Study of Regular Division of the Plane with Birds,* india ink and watercolor, 1955, by M. C. Escher in Figure 12-66. There are various ways to construct the basic figure in Figure 12-66, shown on page 668. One is suggested in *Creating Escher-type Drawings* by Ranucci and Teeters. Another is presented in Figure 12-67, also on page 668, which utilizes graph paper, semicircles, translations, and rotations. This figure can be used to form the basis of a tessellation drawn with Logo procedures, as seen in "Escher-like Logo-type Tessellations" by Lott.

Tessellations

LEARN ABOUT IT

A collection of polygons form a **tessellation** if they cover the entire plane with no overlapping. The tessellation at right is based on a sketch from one of Leonardo da Vinci's notebooks. It shows ways in which both squares and triangles can tessellate the plane. Would any quadrilateral or triangle tessellate the plane?

EXPLORE Use Dot Paper

Work in groups. Arrange about 12 of each polygon on dot paper to form a tessellation of that polygon.

TALK ABOUT IT

1. In your tessellation of triangles do your vertices of the triangles touch other triangles only at vertices? How many triangles surround each vertex?

2. In your tessellation of quadrilaterals how many of them surround each vertex?

Did your tessellations look something like these? How are we using the relationships about the sum of the measures of a polygon?

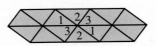 

$m\angle 1 + m\angle 2 + m\angle 3 = 180°$ $m\angle 1 + m\angle 2 + m\angle 3 + m\angle 4 = 360°$

All triangles tessellate the plane. All quadrilaterals tessellate the plane.

668 CHAPTER 12 MOTION GEOMETRY AND TESSELLATIONS

FIGURE 12-66

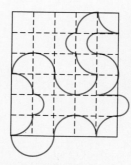

FIGURE 12-67

Another category of interesting tessellations of the plane are the Penrose tilings, named after the British mathematical physicist, Roger Penrose. The simplest of these tessellations involves two rhombuses, one with angles of 72° and 108°, the other with angles of 36° and 144°. The sides of the two rhombuses are congruent to each other. Surprisingly, these shapes do not form pentagons when put together but rather five-sided stars or other pentagonal shapes. There is an infinite number of different ways that Penrose tilings can be constructed. One such tessellation is shown in Figure 12-68.

FIGURE 12-68

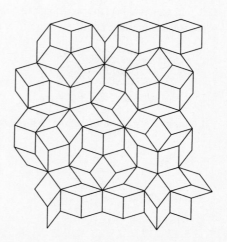

PROBLEM SET 12-5

1. On dot paper, draw a tessellation of the plane using the given figures.

   (a)

   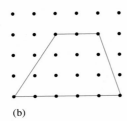
   (b)

2. (a) Tessellate the plane with the quadrilateral shown.

   ▶ (b) Is it possible to tessellate the plane with any quadrilateral?

3. On square-dot paper, use each of the following four pentominoes, one at a time, to make a tessellation of the plane, if possible. (A pentomino is a polygon composed of five congruent, nonoverlapping squares.) Which of the pentominoes tessellate the plane?

   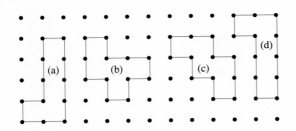

4. The **dual of a tessellation** is the tessellation obtained by connecting the centers of the polygons in the original tessellation that share a common side. The dual of the tessellation of equilateral triangles is the tessellation of regular hexagons (shown in color).

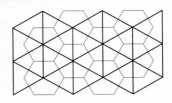

   Describe and show the dual of each of the following:
   (a) The regular tessellation of squares shown in Figure 12-61(a)
   (b) The tessellation of squares in Figure 12-61(b)
   (c) A tessellation of regular hexagons

5. We have seen that equilateral triangles, squares, and regular hexagons are the only regular polygons that will tessellate the plane by themselves. However, there are many ways to tessellate the plane by using combinations of these and other regular polygons, as shown in the figure. Try to produce other such tessellations, using the following:
   (a) Only equilateral triangles, squares, and regular hexagons
   (b) Regular octagons (8-gons) and squares

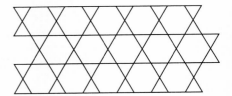

6. 🖳 Write Logo procedures to draw tessellations with the following figures. Have each tessellation appear on the screen in the form of two vertical strips.
   (a) Squares
   (b) Equilateral triangles
   (c) Regular hexagons

7. 🖳 A sidewalk is made of tiles of the type shown in the figure. Each tile is made of three regular hexagons from which three sides have been removed. Write a Logo procedure to draw a tessellation composed of four such figures.

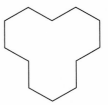

LABORATORY ACTIVITY

As a van Hiele Level 0 activity, use pattern blocks to construct tessellations using each of the following types of pieces:

1. Squares
2. Equilateral triangles
3. Octagons and squares
4. Rhombuses

*Section 12-6 Escher-like Logo-type Tessellations

Logo, with recursion and turtle graphics, provides a natural environment in which to create Escher-like tessellations. A monitor screen can be tessellated with an equilateral triangle, a square, and a regular hexagon, as mentioned in Section 12-5. Specifically, the set of procedures given here will tessellate the screen with squares using $^{-}120$ and 120 as boundaries for x-coordinates of the squares and $^{-}100$ and 100 as boundaries for the y-coordinates of the squares. The main procedure is WALL, which requires inputs for the coordinates of the point at which the first square is to have a lower left vertex, as well as an input for the length of a side of that square. First, the WALL procedure calls the SETUP procedure, which moves the turtle to the proper position at the lower left of the screen to begin the drawing. It then calls WALLPAPER, which does the drawing of the tessellation. The plan is to draw a strip of vertical squares, move over to the right side of the first square drawn, and repeat the process. This should continue until the screen is filled with squares. Figure 12-69(a) shows the start of the procedure; Figure 12-69(b) shows what happens when WALL -60 (-60) 30 is executed.

FIGURE 12-69

(a)

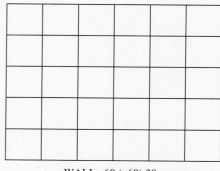

WALL –60 (–60) 30

(b)

SECTION 12-6 ESCHER-LIKE LOGO-TYPE TESSELLATIONS

The procedures for tessellating with a square are given next.

```
TO WALL :XPT :YPT :SIDE          TO SQUARE :SIDE
  DRAW                             REPEAT 4[FORWARD
  SETUP :XPT :YPT                    :SIDE RIGHT 90]
  WALLPAPER :YPT :SIDE           END
END

TO SETUP :XPT :YPT
  PENUP
  SETXY :XPT :YPT
  PENDOWN
END
```

(In LCSI, replace DRAW with CLEARSCREEN and SETXY :XPT :YPT with SETPOS (LIST :XPT :YPT).)

```
TO WALLPAPER :YPT :SIDE
  SQUARESTRIP :SIDE
  PENUP
  SETUP (XCOR + :SIDE) :YPT
  PENDOWN
  WALLPAPER :YPT :SIDE
END

TO SQUARESTRIP :SIDE
  IF XCOR + :SIDE > 120 TOPLEVEL
  IF (ANYOF (XCOR - :SIDE < -120)(XCOR + :SIDE >
    120) (YCOR - :SIDE < -100)(YCOR + :SIDE > 100)) STOP
  SQUARE :SIDE
  FORWARD :SIDE
  SQUARESTRIP :SIDE
END
```

(In LCSI, replace IF XCOR + :SIDE > 120 TOPLEVEL with IF XCOR + :SIDE > 120 [THROW "TOPLEVEL]. Also replace IF (ANYOF (XCOR - :SIDE < -120) (XCOR + :SIDE > 120)(YCOR - :SIDE < -100) (YCOR + :SIDE > 100)) STOP with IF (OR (XCOR - :SIDE < -120) (XCOR + :SIDE > 120)(YCOR - :SIDE < -100) (YCOR + :SIDE > 100)) [STOP].)

To create Escher-like tessellations using Logo, one possibility is to create a tessellation using deformations of sides of a regular polygon and a translation to complete the drawing. We simply determine how to deform the sides of the squares in such a way that they will fit when translated. Consider Figure 12-70(a), where a square is drawn; Figure 12-70(b), where one side of the square is deformed; and Figure 12-70(c), where the deformation is translated to the side of the square parallel to the one where the original deformation took place and is attached to that side, as shown.

FIGURE 12-70

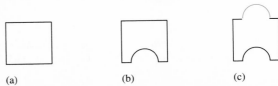

(a)   (b)   (c)

In order to write a set of procedures for a tessellation based on the drawing in Figure 12-70(c), we use the SETUP procedure, edit other procedures just presented, and use arc procedures listed next from the ARCS file of the Logo Utilities disk.

```
TO WALL1 :XPT :YPT :SIDE
  DRAW
  SETUP :XPT :YPT
  WALLPAPER1 :YPT :SIDE
END
```

(In LCSI, replace DRAW with CLEARSCREEN.)

```
TO WALLPAPER1 :YPT :SIDE
  SQUARARCSTRIP :SIDE
  PENUP
  SETUP (XCOR + :SIDE) :YPT
  PENDOWN
  WALLPAPER1 :YPT :SIDE
END

TO SQUARARCSTRIP :SIDE
  IF XCOR + :SIDE > 120 TOPLEVEL
  IF (ANYOF (XCOR - :SIDE < -120) (XCOR + :SIDE >
    120) (YCOR - :SIDE < -100) (YCOR + :SIDE > 100))
    STOP
  SQARC :SIDE
  FORWARD :SIDE
  SQUARARCSTRIP :SIDE
END
```

(In LCSI, replace IF XCOR + :SIDE > 120 TOPLEVEL with IF XCOR + :SIDE > 120 [THROW "TOPLEVEL] and replace IF (ANYOF (XCOR − :SIDE < −120) (XCOR + :SIDE > 120) (YCOR − :SIDE < −100) (YCOR + :SIDE > 100)) STOP with IF (OR (XCOR − :SIDE < −120) (XCOR + :SIDE > 120) (YCOR − :SIDE < −100) (YCOR + :SIDE > 100)) [STOP].)

```
TO SQARC :SIDE
  FORWARD :SIDE RIGHT 90
  FORWARD :SIDE/4 LEFT 90
  RARC :SIDE/4 180
  LEFT 90 FORWARD :SIDE/4
  RIGHT 90 FORWARD :SIDE
  RIGHT 90 FORWARD :SIDE/4 RIGHT 90 LARC :SIDE/4 180
  RIGHT 90 FORWARD :SIDE/4
  RIGHT 90
END
```

(In LCSI, replace RARC with ARCR and replace LARC with ARCL. Both ARCR and ARCL are loaded into the computer when you boot LCSI without inserting your own disk.)

SECTION 12-6 ESCHER-LIKE LOGO-TYPE TESSELLATIONS

```
TO RARC :RADIUS :DEGREES
  RIGHT 2.5
  RARC1 RADIUS*0.0174532 :DEGREES
  LEFT 2.5
END

TO RARC1 :SIZE :DEGREES
  REPEAT QUOTIENT :DEGREES 5 [FORWARD :SIZE *5 RIGHT 5]
    CORRECTARCR :SIZE (REMAINDER :DEGREES 10)
END

TO CORRECTARCR :SIZE :AMOUNT
  FORWARD :SIZE * :AMOUNT
  RIGHT :AMOUNT
END

TO LARC :RADIUS :DEGREES
  LEFT 2.5
  LARC1 :RADIUS*0.0174532 :DEGREES
  RIGHT 2.5
END

TO LARC1 :SIZE :DEGREES
  REPEAT QUOTIENT :DEGREES 5 [FORWARD :SIZE*5 LEFT 5]
    CORRECTARCL :SIZE (REMAINDER :DEGREES 10)
END

TO CORRECTARCL :SIZE :AMOUNT
  FORWARD :SIZE * :AMOUNT
  LEFT :AMOUNT
END
```

Using these procedures, we obtain a tessellation similar to the one shown in Figure 12-71.

FIGURE 12-71

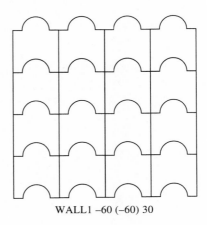

WALL1 –60 (–60) 30

One method of using an equilateral triangle to create a different tessellation is to change the triangle's shape by altering a side and rotating the changed side about a vertex to form the third side and a new figure with which to tessellate. Figure 12-72 shows how this might be done.

The set of procedures for drawing a tessellation with this shape are more like a tessellation drawn with a regular hexagon. Depending on where the starting and stopping points are, there will appear to be gaps in the tessellation. The interested reader should try these procedures to establish that this is not the case. The entire set of procedures for creating the tessellation are given next. SETUP is used, as before.

FIGURE 12-72

```
TO WALL2 :XPT :YPT :SIDE
 DRAW
 SETUP :XPT :YPT
 WALLPAPER2 :YPT :SIDE
END
```
(In LCSI, replace DRAW with CLEARSCREEN.)

```
TO WALLPAPER2 :YPT :SIDE
 MAKE "X XCOR
 RTRISTRIP :SIDE
 PENUP
 SETUP (:X + :SIDE*SQRT 3) :YPT
 PENDOWN
 WALLPAPER2 :YPT :SIDE
END
```

```
TO RTRISTRIP :SIDE
 IF XCOR + :SIDE > 120 TOPLEVEL
 IF (ANYOF (XCOR < -120)(XCOR + :SIDE *SQRT 3 >
  120)(YCOR < -100)(YCOR + :SIDE *3 > 100)) STOP
 REPEAT 3 [RTRIANGLE :SIDE RIGHT 120]
 PENUP
 FORWARD :SIDE
 RIGHT 60
 FORWARD :SIDE LEFT 60
 PENDOWN
 RTRISTRIP :SIDE
END
```

*(In LCSI, replace IF XCOR + :SIDE > 120 TOPLEVEL with IF XCOR + :SIDE > 120 [THROW "TOPLEVEL]. Also replace IF (ANYOF (XCOR<-120) (XCOR + :SIDE*SQRT3>120) (YCOR<-100) (YCOR + :SIDE*3 > 100)) STOP with IF (OR (XCOR < -120) (XCOR + :SIDE*SQRT 3 > 120) (YCOR < -100)(YCOR + :SIDE *3 > 100)) [STOP].)*

```
TO RTRIANGLE :SIDE
 FORWARD :SIDE RIGHT 150
 LARC :SIDE 60
 RIGHT 120
 RARC :SIDE 60
 RIGHT 90
END
```

(In LCSI, replace LARC and RARC with ARCL and ARCR, respectively.)

It is left as an exercise to determine why the conditions are as they are in the inputs to ANYOF. *(Use OR if in LCSI.)* A tessellation drawn using these procedures is given in Figure 12-73. The tessellation in Figure 12-73 makes use of the altered triangle to draw many figures and also uses what might be called the negative space to complete the picture. This means that not every altered triangle in the final picture is traced around. This technique is sometimes helpful, especially if the figure to be used to tessellate is complicated and takes some time to draw.

FIGURE 12-73

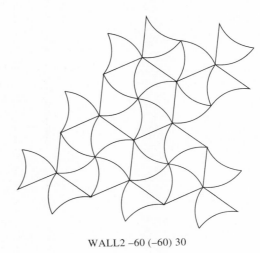

WALL2 –60 (–60) 30

The tessellations given here are but a start and should be used only as ideas for creating your own. It is a good idea to use graph paper to help in the creation of design.

PROBLEM SET 12-6

1. Write a set of procedures to tessellate the screen with an equilateral triangle.
2. Why are the inputs to ANYOF *(OR in LCSI)* in the RTRISTRIP procedure as they are?
3. Write a set of procedures to tessellate the screen with a regular hexagon. (*Hint:* The "width" of a hexagon is $\sqrt{3}$ times the length of a side.)
4. (a) Draw any figure you wish.
   (b) Write a procedure to draw a vertical strip of these figures. Will they necessarily fit one atop the other?
5. ▶Will any set of figures that will fit between two parallel lines tessellate the plane? Explain your answer.
6. Write a set of procedures to tessellate the screen with a rectangle.
7. Edit the first set of procedures in this section to tessellate the screen with the following figure:

CHAPTER 12 MOTION GEOMETRY AND TESSELLATIONS

● COMPUTER CORNER

Fractal geometry was introduced at IBM in 1977 by mathematician Benoit Mandelbrot for the purpose of modeling natural phenomena such as irregular coastlines, arteries and veins, the branching structure of plants, the thermal agitation of molecules in a fluid, and sponges. In 1906, Helge von Koch came up with a curve that has infinite perimeter. To visualize this curve, we construct in Figure 12-74 a sequence of polygons S_1, S_2, S_3, . . . as follows.

(a) S_1 is an equilateral triangle.

(b) S_2 has an equilateral triangle constructed on each side of S_1 but with the base removed.

(c) S_3 is obtained from S_2 like S_2 was obtained from S_1.

We continue in a similar way to obtain the other polygons of the sequence in Figure 12-74(d) and (e). These polygons come closer and closer to a curve, called the *snowflake curve*, which is an example of a fractal. Type the following SNOWFLAKE procedure into your computer and display the polygons shown in Figure 12-74.

FIGURE 12-74

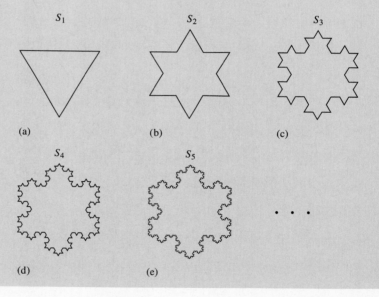

● COMPUTER CORNER (cont'd)

```
TO SNOWFLAKE :LEVEL
  CS PU SETXY 90 52
  MAKE "SIZE 180
  MAKE "N 5 - :LEVEL
  MAKE "LEVEL 1
  REPEAT :N[MAKE "LEVEL :LEVEL *3]
  REPEAT 3[DRAW.IT :SIZE :LEVEL RT 120]
END
```
(In LCSI replace SETXY (90) 52 with SETPOS [90 52].)

```
TO DRAW.IT :SIZE :LEVEL
  IF :SIZE < :LEVEL THEN FD :SIZE STOP
  DRAW.IT :SIZE/3 :LEVEL LT 60
  DRAW.IT :SIZE/3 :LEVEL RT 120
  DRAW.IT :SIZE/3 :LEVEL LT 60
  DRAW.IT :SIZE/3 :LEVEL
END
```
(In LCSI replace IF :SIZE < :LEVEL THEN FD :SIZE STOP with IF :SIZE < :LEVEL [FD :SIZE STOP].)

Figure 12-75 shows another example of a fractal—a computer-generated picture of the Mandelbrot set known as the "Tail of the Seahorse."

FIGURE 12-75

Solution to the Preliminary Problem

Understanding the Problem. Figure 12-76 is a drawing for the problem. $\overline{AB}$ is the west side of the park. Line ℓ represents Shady Lane and line s represents Sunny Street. $\overline{PQ}$ represents the road to be built parallel to and the same length as $\overline{AB}$. We need to find points P and Q.

FIGURE 12-76

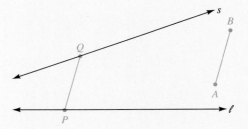

Devising a Plan. Because $PQ = AB$ and $\overline{PQ} \parallel \overline{AB}$; Q is the image of P under translation from A to B. We do not know the location of either P or Q, but we know that Q is on line s and that Q is the image of point P on ℓ under translation from A to B. We do not know where P lies on ℓ, but we could find the images of all the points on ℓ under translation from A to B and see which ones (if any) lie on s. One of those points will be Q. Finding the image of ℓ is our *subgoal*; it can be accomplished by translating ℓ from A to B.

Carrying Out the Plan. Because a line is determined by two points, to find the image of ℓ under translation from A to B choose any two points X and Y on ℓ and find their images X' and Y'. The *line* connecting X' and Y' is ℓ'. In Figure 12-77 Q is the intersection of ℓ' with s and P is the image of Q under translation from B to A (why?) and $\overline{PQ}$ represents the road to be built.

FIGURE 12-77

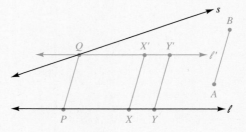

Looking Back. The point P could have also been found in a way similar to the way Q was found, namely, by finding the image s' of s under translation from B to A and finding the intersection of s' with ℓ.

The problem could be varied by requiring that the road has some other characteristics, for example, the same length as $\overline{AB}$ and perpendicular to $\overline{AB}$.

QUESTIONS FROM THE CLASSROOM

1. A student asks, "If I have a point and its image, is that enough to determine whether the image was found using a translation, reflection, rotation, or glide reflection?" How do you respond?
2. Another student asks a question similar to Question 1 but is concerned about a segment and its image. How do you respond to this student?
3. A student claims that a kite has no lines of symmetry. How do you respond?
4. A student says that every three-dimensional figure that has plane symmetry automatically has line symmetry. Do you agree?
5. A student claims that anything that can be accomplished by a rotation can also be accomplished by a reflection in a line. She claims that if A' is the image of A under a rotation about point O by some amount, then A' can also be obtained from A by a reflection in the perpendicular bisector of $\overline{AA'}$. Hence, a reflection and a rotation are the same. How do you respond?
6. A student says that in a size transformation where the scale factor is 0, we do not have a transformation. Is that true?
7. A student asks if every translation on a grid can be accomplished by a translation along a vertical direction followed by a translation along a horizontal direction. How do you respond?

CHAPTER OUTLINE

I. Transformations
  A. **Isometries** are transformations that preserve distance.
    1. A **translation**, or **slide**, is a transformation of a plane that moves every point a specified distance in a specified direction along a straight line.
    2. A **rotation** is a transformation of the plane determined by holding one point (the center) fixed and rotating the plane about this point by a certain amount in a certain direction.
    3. A **half-turn** is a rotation of 180°.
    4. A **reflection** in a line m is a transformation among points of the plane that pairs each point P of the plane with a point P' in such a way that m is the perpendicular bisector of $\overline{PP'}$, as long as P is not on m. If P is on m, then $P = P'$.
    5. A **glide reflection** is the composition of a translation and a reflection in a line parallel to the slide arrow of the translation.
  B. A **size transformation** S from the plane to the plane has the following properties: Some point O, the center of the size transformation, is its own image. For any other point Q of the plane, its image Q' is such that $OQ'/OQ = r$, where r is a positive real number, and O, Q, and Q' are collinear.
  C. A **similarity** is the composition of an isometry and a size transformation.

II. Symmetries
  A. A figure has **line symmetry** if it is its own image under a reflection.
  B. A figure has **rotational symmetry** if it is its own image under a rotation of less than 360° about its center.
  C. A figure that has 180° rotational symmetry is said to have **point symmetry**.
  D. A three-dimensional figure has a **plane of symmetry** when every point of the figure on one side of the plane has a mirror image on the other side of the plane.

*III. Tessellations
  A **tessellation** of a plane is the filling of the plane with repetitions of figures in such a way that no figures overlap and there are no gaps.

CHAPTER TEST

1. Complete each of the following motions:

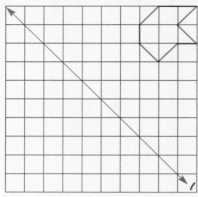

A reflection in ℓ
(a)

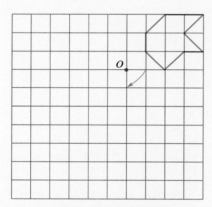

A rotation in O through the given arc
(b)

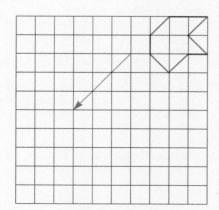

A translation, as pictured
(c)

2. For each of the following, construct the image of $\triangle ABC$:

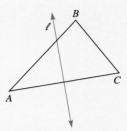

Through a reflection in ℓ
(a)

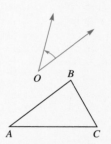

Through the given rotation in O
(b)

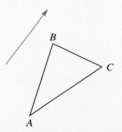

Through the translation pictured
(c)

3. How many lines of symmetry, if any, does each of the following figures have?

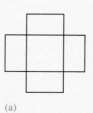

(a)   (b)

CHAPTER TEST 681

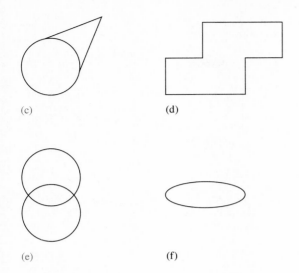

(c) (d)

(e) (f)

4. For each of the following, identify the types of symmetry (line, rotational, or point) possessed by the given figure:

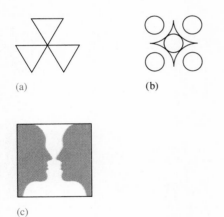

(a) (b)

(c)

5. How many planes of symmetry does each of the following have?
   (a) A ball
   (b) A right cylindrical water pipe
   (c) A box that is a right rectangular prism but not a cube
   (d) A cube

6. What type of symmetries (line, rotational, or point) does each of the lowercase letters of the printed English alphabet have?

7. ▶Explain why a regular octagon cannot tessellate the plane.

8. Given points A and B and circle O, find point C on circle O such that $\triangle ABC$ is isosceles.

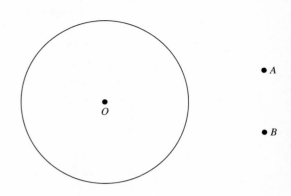

9. Given $\triangle A'B'C'$, the image of $\triangle ABC$ under a size transformation, locate points A, B, and C such that A' is the center of the size transformation and $BC = \frac{1}{2}B'C'$.

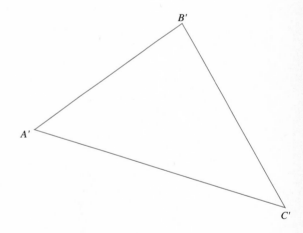

10. Given that STAR is a parallelogram, describe a sequence of isometries to show the following:
    (a) $\triangle STA \cong \triangle ARS$
    (b) $\triangle TSR \cong \triangle RAT$

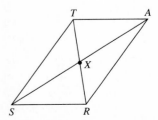

682 CHAPTER 12 MOTION GEOMETRY AND TESSELLATIONS

11. Given that BEAUTY is a regular hexagon, describe a sequence of isometries that will transform the following:
    (a) BEAU into AUTY
    (b) BEAU into YTUA

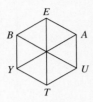

12. Given that △SNO ≅ △SWO, describe an isometry or isometries that will transform △SNO into △SWO.

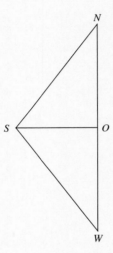

13. Show that △SER is the image of △HOR under a succession of isometries with a size transformation.

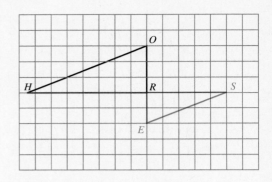

14. Show that △BAT is the image of △PIG under a succession of isometries with a size transformation.

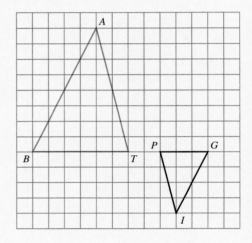

★15. Write a Logo procedure called RHOMSTRIP that will draw a vertical strip of rhombuses.

SELECTED BIBLIOGRAPHY

Bidwell, J. "Using Reflections to Find Symmetric and Asymmetric Patterns." *Arithmetic Teacher* 34 (March 1987): 10–15.

Billstein, R., S. Libeskind, and J. Lott. *Logo: MIT Logo for the Apple*. Menlo Park, Calif.: Benjamin/Cummings, 1985.

Byrne, D. "The Bank Shot." *Mathematics Teacher* 79 (September 1986): 429–430, 487.

DeTemple, D. "Reflection Borders for Patchwork Quilts." *Mathematics Teacher* 80 (February 1986): 138–143.

Lappan, G., and R. Even. "Research into Practice: Similarity in the Middle Grades." *Arithmetic Teacher* 35 (May 1988): 32–35.

Lott, J. "Escher-like Logo-type Tessellations." *Logo Exchange* 6 (November 1987): 7–11.

May, B. "Reflections on Miniature Golf." *Mathematics Teacher* 78 (May 1985): 351–353.

Ranucci, E., and J. Teeters. *Creating Escher-type Drawings*. Palo Alto, Calif.: Creative Publications, 1977.

Reesink, C. "Crystals: Through the Looking Glass with Planes, Points, and Rotational Symmetry." *Mathematics Teacher* 80 (May 1987): 377–388.

Renshaw, B. "Symmetry the Trademark Way." *Arithmetic Teacher* 34 (September 1986): 6–12.

Sawada, D. "Symmetry and Tessellations from Rotational

Transformations on Transparencies." *Arithmetic Teacher* 33 (December 1985): 12–13.

Senk, S. L., and D. B. Hirschhorn. "Multiple Approaches to Geometry: Teaching Similarity." *Mathematics Teacher* 83 (April 1990): 274–280.

Shyers, J. "Reflective Paths to Minimum-Distance Solutions." *Mathematics Teacher* 79 (March 1986): 174–177, 203.

Sicklick, F., B. Turkel, and F. R. Curcio. "The Transformation Game." *Arithmetic Teacher* 36 (October 1988): 37–41.

Thompson, P. "A Piagetian Approach to Transformation Geometry via Microworlds." *Mathematics Teacher* 78 (September 1985): 465–471.

Willcutt, B. "Triangular Tiles for Your Patio." *Arithmetic Teacher* 34 (May 1987): 43–45.

Woods, J. "Let the Computer Draw the Tessellations That You Design." *Mathematics Teacher* 81 (February 1988): 138–141.

Zurstadt, B. "Tessellations and the Art of M. C. Escher." *Arithmetic Teacher* 31 (January 1984): 54–55.

13 Concepts of Measurement

Preliminary Problem

Three planes, in a holding pattern, fly in a circle centered over the Memphis airport in an airshow. The planes are equally spaced in a circle of $\frac{1}{2}$ mi radius at an altitude of 1 mi. What is the shortest distance between any two planes, and how far is each from the airport?

In the *Teaching Standards* (p. 136), we find the following:

> *The attributes of what we measure include length, area, volume, capacity, time, temperature, angles, weight, and mass. Teachers should understand that the units to record measure are different from the process of measurement itself. These ideas should be reinforced through varied experiences, using both standard and nonstandard units where students learn to estimate lengths, areas, and so on. Of particular importance should be an understanding of the Systeme International d'Unites (the metric system).*

In this chapter, we develop the metric and English systems of measurement for length, area, volume, mass, and temperature with the philosophy that students should learn to think within a measurement system. Consequently, conversions among units of measure in the metric and the English systems are not considered. In measuring geometric objects, our goal is to describe the size of the object relative to a given standard object of fixed size and shape. The standard figure is referred to as a *unit of measure*. Once a unit of measure is chosen, we can determine the measure of a given object by finding how many congruent unit objects are necessary to cover the object.

We develop formulas for the areas of plane figures and for surface areas and volumes of solids. We also use the concept of area in discussing the Pythagorean theorem. The *Standards* (p. 116) for grades 5–8 states: *The curriculum should focus on the development of understanding, not the rote memorization of formulas.* In that regard, we attempt to show how area and volume formulas can be developed through exercises that show different developmental techniques, as well as applications of the formulas.

Section 13-1 Units of Length

Three fundamental quantities of measure are length, mass, and time. Early attempts at measurement lacked a standard unit object and used hands, arms, and feet as units of measure. These early crude measurements were eventually refined and standardized by the English into a very complicated system. At this time, the United States is the only major industrial nation in the world that continues to use the English system.

The English System

Originally, in the English system, a yard was the distance from the tip of the nose to the end of an outstretched arm of an adult person and a foot was the length of a human foot. In 1893, the United States defined the yard and other units in terms of metric units. Some

TABLE 13-1

Unit	Equivalent in Other Units
yard (yd)	3 ft
foot (ft)	12 in.
mile (mi)	1760 yd, or 5280 ft

units of length in the English system and relationships among them are summarized in Table 13-1.

EXAMPLE 13-1 Convert each of the following:

(a) 218 ft = _____ yd   (b) 8432 yd = _____ mi
(c) 0.2 mi = _____ ft   (d) 64 in. = _____ yd

Solution

(a) Because 1 ft = $\frac{1}{3}$ yd, 218 ft = $218 \cdot \frac{1}{3}$ yd $\doteq$ 72.67 yd.

(b) Because 1 yd = $\frac{1}{1760}$ mi, 8432 yd = $8432 \cdot \frac{1}{1760}$ mi = 4.79 mi.

(c) 1 mi = 5280 ft. Hence, 0.2 mi = $0.2 \cdot 5280$ ft = 1056 ft.

(d) We first find a connection between yards and inches. We have 1 yd = 3 ft and 1 ft = 12 in. Hence, 1 yd = 3 ft = $3 \cdot 12$ in. = 36 in. Hence, 1 in. = $\frac{1}{36}$ yd; therefore 64 in. = $64 \cdot \frac{1}{36}$ yd $\doteq$ 1.78 yd. ∎

The Metric System

The metric system, a decimal system, was proposed in France in 1670 by Gabriel Mouton. However, not until the French Revolution in 1790 did the French Academy of Sciences bring various groups together to develop the system. The Academy recognized the need for a standard base unit of linear measurement. The members chose $\frac{1}{10{,}000{,}000}$ of the distance from the equator to the North Pole on a meridian through Paris as the base unit of length and called it the **meter (m)**. In 1960, the meter was redefined in terms of krypton 86 wavelengths and still later as the distance traveled by light in a vacuum during $\frac{1}{299{,}792{,}458}$ sec. Since 1893, the yard in the United States has been defined as $\frac{3600}{3937}$ of a meter, whatever the definition of a meter.

meter

The scientific definition of meter may not be very meaningful in everyday life, but if you turn your head away from your outstretched arm, the distance from your nose to your fingertip is about 1 m. Also, 1 m is about the distance from a doorknob to the floor; and 1 m is about 39 in., slightly longer than 1 yd.

Different units of length in the metric system are obtained by multiplying a power of 10 times the base unit. The prefixes for these units, the multiplication factors, and their symbols are given in Table 13-2.

TABLE 13-2

Prefix	Symbol	Factor	
kilo	k	1000	(one thousand)
hecto	h	100	(one hundred)
deka	da	10	(ten)
deci	d	0.1	(one-tenth)
centi	c	0.01	(one-hundredth)
milli	m	0.001	(one-thousandth)

REMARK Hecto, deka, and deci are not common prefixes and have limited use. These should not be stressed when teaching the metric system. Kilo, hecto, and deka are Greek prefixes, whereas deci, centi, and milli are Latin prefixes.

The metric prefixes with the base unit meter names the different units of length. Table 13-3 gives these units, their relationship to the meter, and the symbol for each.

TABLE 13-3

Unit	Symbol	Relationship to Base Unit
kilometer	km	1000 meters
*hectometer	hm	100 meters
*dekameter	dam	10 meters
meter	**m**	**base unit**
*decimeter	dm	0.1 meter
centimeter	cm	0.01 meter
millimeter	mm	0.001 meter

*Not commonly used.

REMARK Two other prefixes, mega (1,000,000) and micro (0.000001), are used, respectively, for very large and very small units.

For easy references for metric measures of length, estimations for a meter, a decimeter, a centimeter, and a millimeter are shown in Figure 13-1. The kilometer is commonly used for measuring longer distances. Because "kilo" stands for 1000, 1 km = 1000 m. Nine football fields, including end zones, laid end to end are approximately 1 km long.

FIGURE 13-1

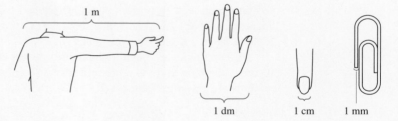

Conversions among metric lengths are accomplished by multiplying or dividing by powers of 10. As with money, we simply move the decimal point to the left or right, depending on the units. For example,

0.123 km = 1.23 hm = 12.3 dam = 123 m = 1230 dm = 12,300 cm = 123,000 mm.

It is possible to convert units by using the chart in Figure 13-2. We count the number of steps from one unit to the other and move the decimal point that many steps in the same direction.

SECTION 13-1 UNITS OF LENGTH

FIGURE 13-2

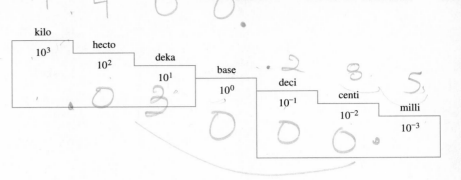

EXAMPLE 13-2 Convert each of the following:

(a) 1.4 km = _____ m   (b) 285 mm = _____ m   (c) 0.03 km = _____ cm

Solution   (a) To change kilometers to meters, we must multiply by 1000, because 1 km = 1000 m, or move the decimal point three places to the right. Hence, 1.4 km = 1400 m.

(b) To change from millimeters to meters, we must multiply by 0.001, because 1 mm = 0.001 m, or move the decimal point three places to the left. Thus 285 mm = 0.285 m.

(c) To change kilometers to centimeters, we must first multiply by 1000 to convert kilometers to meters, and then multiply by 100 to convert meters to centimeters. Therefore we move the decimal five places to the right to obtain 0.03 km = 3000 cm. ■

Units of length are referred to as *linear measures* and are commonly measured with rulers. Figure 13-3 shows part of a centimeter ruler. Rulers can be used to measure distance. Following are three basic properties of distance.

FIGURE 13-3

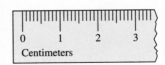

PROPERTIES

1. The distance between any two points A and B is greater than or equal to 0, written $AB \geq 0$.
2. The distance between any two points A and B is the same as the distance between B and A, written $AB = BA$.
3. *Triangle Inequality:* For any three points A, B, and C, the distance between A and B plus the distance between B and C is greater than or equal to the distance between A and C, written $AB + BC \geq AC$.

Property 3 is illustrated in Figure 13-4(a), where $AB + BC = AC$ if, and only if, A, B, and C are collinear and B is between A and C. If A, B, and C are not collinear, as in Figure 13-4(b), then $AB + BC > AC$.

FIGURE 13-4

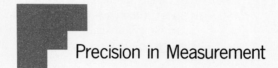

(a)   (b)

Measuring distances in the real world frequently results in errors. Because of this, many industrial plants using parts from a variety of sources rely on portable calibration units that are taken from plant to plant to test measuring instruments used in constructing the parts. This is done so that the final assembly plant can fit all the parts together to make the product. To calibrate the measuring instruments, technicians must establish the greatest possible error (GPE) allowable in order to obtain the final fit. The *Addison-Wesley Mathematics,* Grade 7, 1991, shows one approach to GPE for middle-school students.

Precision in Measurement

LEARN ABOUT IT

EXPLORE Study the Chart
The chart shows the measurements of coins from five different countries. The coins were first measured using centimeters (A). They were measured again using millimeters (B). Place the coins in order from largest to smallest using each set of measurements.

Coin	A Diameter (nearest cm)	B Diameter (nearest mm)
Haiti 10 centimes	2 cm	2.1 cm
Panama $\frac{1}{4}$ balboa	2 cm	2.4 cm
Peru 10 centavos	2 cm	2.1 cm
Greenland kroner	3 cm	3.3 cm
Liberia 50 cent	3 cm	2.9 cm

TALK ABOUT IT

1. Which coin has the largest diameter?

2. Why do you think the Panamanian and the Peruvian coins have the same A measurements, but different B measurements?

3. How could you make an even more precise measurement of the Haitian coin?

The **greatest possible error (GPE)** of a measurement is half (0.5) the measurement unit used. For example, if the diameter of a coin is measured at 3 cm to the nearest cm, the actual length of the diameter must be between 2.5 cm and 3.5 cm. In this case the GPE is 0.5 cm.

Example Which measurement is more precise, 23 m to the nearest meter or 23.40 m to the nearest centimeter?

The GPE of the first measurement is 0.5 m. The GPE of the second is 0.5 cm. This means 23.40 m to the nearest centimeter is a more precise measurement than 23 m to the nearest meter.

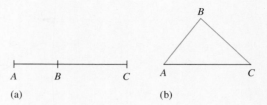

When drawings are given, we assume that the measures listed are accurate. When actually measuring figures in the real world, we find that such accuracy is usually impossible.

Distance Around a Plane Figure

perimeter The **perimeter** of a simple closed curve is the length of the curve, that is, the distance around the figure. If a figure is a polygon, its perimeter is the sum of the lengths of the sides. Perimeter is always expressed in linear measure.

EXAMPLE 13-3 Find the perimeter of each of the shapes in Figure 13-5.

FIGURE 13-5

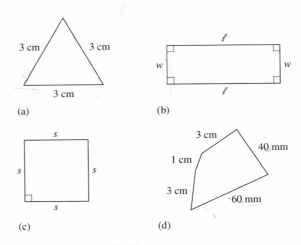

(e) a regular n-gon with side s.

(f) a kite with two sides of lengths ℓ and w.

Solution
(a) The perimeter is $3(3) = 9$, or 9 cm.
(b) The perimeter is $2w + 2\ell$.
(c) The perimeter is $4s$.
(d) Because 40 mm = 4 cm and 60 mm = 6 cm, the perimeter is $1 + 3 + 4 + 6 + 3 = 17$, or 17 cm.
(e) Because all sides of a regular n-gon are congruent, the perimeter is ns.
(f) Because two sets of consecutive sides of a kite are congruent, the lengths are $\ell, \ell, w,$ and w. Therefore the perimeter is $2\ell + 2w$. ∎

Circumference of a Circle

circumference The perimeter of a circle is called its **circumference**. The ancient Greeks discovered that if they divided the circumference of any circle by the length of its diameter, they always obtained approximately the same number. The number is approximately 3.14 (see the

692 CHAPTER 13 CONCEPTS OF MEASUREMENT

pi Laboratory Activity at the end of this section). Today, the ratio of circumference C to diameter d is symbolized as π **(pi)**. *For most practical purposes, π is approximated by $\frac{22}{7}$, $3\frac{1}{7}$, or 3.14. These values are only approximations and are not exact values of π.* If you are asked for the exact circumference of a circle with diameter 6 cm, the answer is 6π cm. Circumference is always expressed in linear measure. In the late eighteenth century, mathematicians proved that the ratio $\frac{C}{d}$, or π, is not a terminating or repeating decimal but an irrational number.

The relationship $\frac{C}{d} = \pi$ is a formula for finding the circumference of a circle and normally is written as $C = \pi d$ or $C = 2\pi r$ because the length of diameter d is twice the radius (r) of the circle.

HISTORICAL NOTE

$\pi = 3.14159$
26535
89793
23846
26433
83279
50288
41971
69399
37510
58209
.
.
.

Archimedes (b. 287 B.C.) found an approximation for π given by the inequality $3\frac{10}{71} < \pi < 3\frac{10}{70}$. A Chinese astronomer thought that $\pi = \frac{355}{113}$. Ludolph van Ceulen (1540–1610), a German mathematician, calculated π to 35 decimal places. The approximation was engraved on his tombstone. Leonhard Euler adopted the symbol π in 1737 and caused its wide usage. In 1761, Johann Lambert, an Alsatian mathematician, proved that π is an irrational number. In 1989, using U. S.-made computers, Columbia University mathematicians and Soviet émigré brothers, David and Gregory Chudnovsky, established 480 million digits of π. If these digits were printed along a line, the line would extend 600 mi.

Arc Length

The length of an arc depends on the radius of the circle and the central angle determining the arc. If the central angle has a measure of 180°, as in Figure 13-6(a), the arc is a *semicircle* **semicircle.** The length of a semicircle is $\frac{1}{2} \cdot 2\pi r$, or πr. The length of an arc whose cen-

FIGURE 13-6

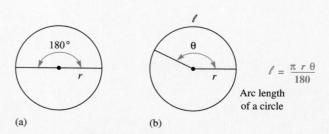

(a)   (b)

$\ell = \frac{\pi r \theta}{180}$

Arc length of a circle

tral angle is $\theta°$ can be developed as in Figure 13-6(b). Since a circle has $360°$, an angle of θ degrees determines $\theta/360$ of a circle. Because the circumference of a circle is $2\pi r$, an arc of $\theta°$ has length $\dfrac{\theta}{360} \cdot 2\pi r$ or $\dfrac{\pi r \theta}{180}$.

EXAMPLE 13-4 Find each of the following:

(a) The circumference of a circle if the radius is 2 m
(b) The radius of a circle if the circumference is 15π m
(c) The length of a $25°$ arc of a circle of radius 10 cm
(d) The radius of an arc whose central angle is $87°$ and whose length is 154 cm

Solution (a) $C = 2\pi(2) = 4\pi$; thus the circumference is 4π m.

(b) $C = 2\pi r$ implies $15\pi = 2\pi r$. Hence, $r = \dfrac{15}{2}$. Thus the radius is $\dfrac{15}{2}$ m.

(c) The arc length is $\dfrac{\pi r \theta}{180} = \dfrac{\pi \cdot 10 \cdot 25}{180}$ cm, or $\dfrac{25\pi}{18}$ cm, or approximately 4.36 cm.

(d) The arc length ℓ is $\dfrac{\pi r \theta}{180}$, so that $154 = \dfrac{\pi r \cdot 87}{180}$. Therefore $r \doteq 101.4$ cm. ∎

LABORATORY ACTIVITY

To approximate the value of π, you need several different-sized round tin cans or jars, string, and a marked ruler. Pick a can and wrap the string tightly around the can. Use a pen to mark a point on the string where the beginning of the string meets the string again. Unwrap the string and measure its length. Next, determine the diameter of the can by tracing the bottom of the can on a piece of paper. Fold the circle onto itself to find a line of symmetry. The chord determined by the line is a diameter of the circle. Measure the diameter and determine the ratio of the circumference to the diameter. (Use the same units in all of your measurements.) Repeat the experiment with at least three cans and find the average of the corresponding ratios.

PROBLEM SET 13-1

1. Use the ruler pictured to find each of the following lengths:
   (a) AB   (b) OE   (c) CJ
   (d) EF   (e) IJ   (f) AF
   (g) IC   (h) GB

2. Convert each of the following:
   (a) 100 in. = _____ yd
   (b) 400 yd = _____ in.
   (c) 300 ft = _____ yd
   (d) 372 in. = _____ ft

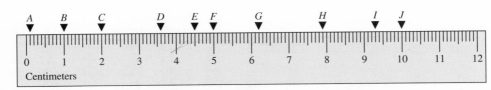

3. Draw segments that you estimate to be of the following lengths. Use a metric ruler to check the estimates.
   (a) 10 mm   (b) 100 mm   (c) 1 cm
   (d) 10 cm   (e) 0.01 m   (f) 15 cm
   (g) 0.1 m   (h) 27 mm

4. Estimate the length of the following segment and then measure it. Express the measurement in each of the following units:

   |—————————————————————————|

   (a) Millimeters   (b) Centimeters

5. Choose an appropriate metric unit and estimate each of the following measures:
   (a) The length of a pencil
   (b) The diameter of a nickel
   (c) The width of the top of a desk
   (d) The thickness of the top of a desk
   (e) The length of this sheet of paper
   (f) The height of a door
   (g) Your height
   (h) Your hand span

6. Redo Problem 5 using English measures.

7. Complete the following table:

Item	m	cm	mm
(a) Length of a piece of paper		35	
(b) Height of a woman	1.63		
(c) Width of a filmstrip			35
(d) Length of a cigarette			100
(e) Length of two meter sticks laid end to end	2		

8. For each of the following, place a decimal point in the number to make the sentence reasonable:
   (a) A stack of 10 dimes is 1000 mm high.
   (b) The desk is 770 m high.
   (c) It is 100 m across the street.
   (d) A dollar bill is 155 cm long.
   (e) The basketball player is 1950 cm tall.
   (f) A new piece of chalk is about 8100 cm long.
   (g) The speed limit in town is 400 km/hr.

9. List the following in decreasing order:
   8 cm, 5218 mm, 245 cm, 91 mm, 6 m, 700 mm

10. Draw each of the following as accurately as possible:
    (a) A regular polygon whose perimeter is 12 cm
    (b) A circle whose circumference is 4 in.
    (c) A triangle whose perimeter is 4 in.
    (d) A nonconvex quadrilateral whose perimeter is 8 cm

11. Guess the perimeter of each of the following figures in centimeters and then check the estimates using a ruler:

(a)

(b)

(c)

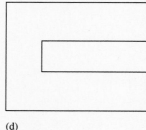
(d)

12. Complete each of the following:
    (a) 10 mm = _____ cm
    (b) 17 cm = _____ m
    (c) 262 m = _____ km
    (d) 3 km = _____ m
    (e) 30 mm = _____ m
    (f) 0.17 km = _____ m
    (g) 35 m = _____ cm
    (h) 359 mm = _____ m
    (i) 1 mm = _____ cm
    (j) 647 mm = _____ cm
    (k) 0.1 cm = _____ mm
    (l) 5 km = _____ m
    (m) 51.3 m = _____ cm

13. Draw a triangle ABC. Measure the length of each of its sides in millimeters. For each of the following, tell which is greater and by how much:
    (a) $AB + BC$ or AC (b) $BC + CA$ or AB
    (c) $AB + CA$ or BC

14. Which of the following cannot be the lengths of the sides of a triangle?
    (a) 23 cm, 50 cm, 60 cm
    (b) 10 cm, 40 cm, 50 cm
    (c) 410 mm, 260 mm, 14 cm

15. ▶(a) Is it possible to draw a rhombus whose perimeter is exactly four times the length of a diagonal? Explain your answer.
    (b) Do you think it is possible to draw a square whose perimeter is exactly four times the length of one of its diagonals? Explain your answer.

16. ▶Explain how proportions can be used to find the length of an arc.
17. Take an $8\frac{1}{2}$-×-11-in. piece of typing paper, fold it as shown below, and then cut the folded paper along the diagonal segment.

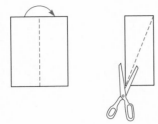

(a) Put the pieces back together to find the maximum perimeter.
(b) Rearrange the pieces to find the minimum perimeter.

18. The following figure made of six unit squares has a perimeter of 12 units. The figure is made in such a way that any two squares share a common side or a common vertex or have no points in common and each square shares an edge with another.

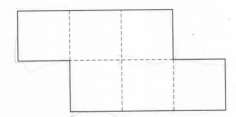

(a) Add more squares to the figure so that the perimeter of the new figure is 18.
(b) What is the minimum number of squares required to make a figure of perimeter 18?
(c) What is the maximum number of squares that can be used to make a figure of perimeter 18?

19. For each of the following circumferences, find the length of the radius of the circle:
 (a) 12π cm (b) 6 m (c) 0.67 m (d) 92π cm

20. For each of the following, if a circle has the dimensions given, what is its circumference?
 (a) 6 cm diameter (b) 3 cm radius
 (c) $\frac{2}{\pi}$ cm radius (d) 6π cm diameter

21. ▶What happens to the circumference of a circle if the length of the radius is doubled?

22. The following figure is a circle whose radius is r units. The diameters of the two semicircular regions inside the large circle are both r units as well. Compute the length of the curve that separates the shaded and white regions.

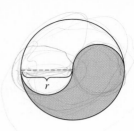

23. ▶A student has a tennis can containing three tennis balls. To the student's surprise, the perimeter of the top of the can is longer than the height of the can. The student wants to know if this fact can be explained without performing any measurements. Can you help?

24. (a) If in two similar triangles, the ratio between the lengths of the corresponding sides is 2:1, what is the ratio between their perimeters?
 (b) Make a conjecture concerning the relationship between the ratio of the perimeters of two similar triangles and the ratio of the corresponding sides.
 ▶(c) Justify your conjecture in (b).

25. (a) Astronomers use a unit of distance called a light year, which is the distance that light travels in one year. If the speed of light is 300,000 km/sec, how long is one light year in kilometers?
 (b) The nearest star (other than the sun), Alpha Centauri, is 4.34 light years away from Earth. How far is that in kilometers?
 (c) How long will it take a rocket traveling 60,000 km/hr to reach Alpha Centauri?
 (d) How long will it take the rocket in (c) to travel to the sun, if it takes approximately 8 min 19 sec for light from the sun to reach Earth?

26. Since jet planes can exceed the speed of sound, a new measurement called *Mach number* was invented to measure the speed of such planes. Mach 2 is twice the speed of sound. (Mach number is a number indicating the ratio of the speed of an object through a medium to the speed of sound in the medium.) The speed of sound in air is approximately 344 m/sec.
 (a) Express Mach 2.5 in kilometers per hour.
 (b) Express Mach 3 in meters per second.
 (c) Express the speed of 5000 km/hr as a Mach number.

27. Refer to the following figure and determine the perimeter of the paint lane and the semicircle determined by the free-throw line on a basketball court.

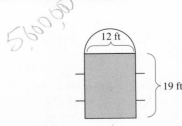

B.C.   by johnny hart

Reprinted by permission of Johnny Hart and Creators Syndicate, Inc.

28. In the cartoon above, Clumsy describes a foot long hot dog as being 6 in. long. If that is the case, what would you expect the lengths of each of the following to be in "footlongs"?

(a) Yard   (b) Mile

▶In track, the second lane from the inside is longer than the inside lane. Use this information to explain why runners are lined up as they are in running events that require a complete lap of the track.

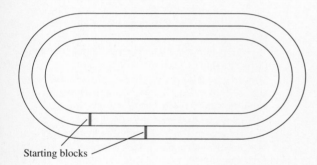

Starting blocks

30. ▶If polygons are constructed as shown below, what can you say about the numbers that can be the perimeters of the polygons constructed? Explain your answer.

Allowed arrangements

Disallowed arrangements

★31. Using squares one unit on a side to form polygons as in Problem 30, complete the table below to try to determine a formula for the maximum and minimum areas of polygons with fixed perimeters. (*Hint:* Count the squares to determine the areas.)

Perimeter	Minimum Area	Maximum Area
4	1	1
6	2	2
8	3	4
10	4	
12		9
14		
16		
18		
20		
22		
24		
26		
$2n$		

BRAIN TEASER

Suppose a wire is stretched tightly around Earth. (The radius of Earth is approximately 6400 km.) If the wire is cut, its length is increased by 20 m, and the wire is then placed back around the earth so that the wire is the same distance from Earth at every point, could you walk under the wire?

LABORATORY ACTIVITY

1. As a van Hiele Level 1 activity, use a metric ruler to find the perimeter, in millimeters, of each of the following figures:

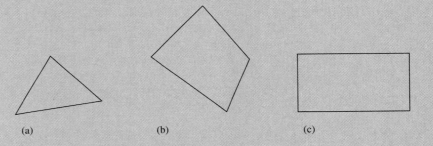

(a) (b) (c)

2. As a van Hiele Level 1 activity, use an English ruler to measure the indicated part of each of the following figures in inches:

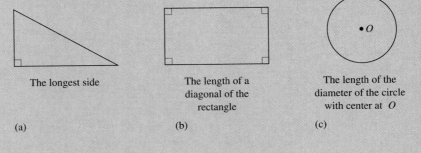

(a) The longest side

(b) The length of a diagonal of the rectangle

(c) The length of the diameter of the circle with center at O

● COMPUTER CORNER

Write Logo procedures to construct each of the following:

(a) An equilateral triangle with a perimeter of 150 turtle steps
(b) An isosceles triangle with a perimeter of 150 turtle steps

(continued)

(c) A regular septagon whose perimeter is 100 turtle steps

(d) A square whose diagonal is 30 turtle steps. (Approximately what is the perimeter of the square?)

Section 13-2 Areas of Polygons and Circles

area **Area** is a number assigned to the amount of surface in the interior region of a figure. Using a square as the basic unit, we see that the area of a region is the number of such squares required to tessellate the region. *The measure of area is always expressed in square units.*

A square measuring 1 in. on each side has area of 1 square inch denoted by 1 in^2. A square measuring 1 cm on each side has an area of 1 square centimeter, denoted by 1 cm^2. A square measuring 1 m on each side has an area of 1 square meter, denoted by 1 m^2.

To determine how many 1-cm squares are in a square meter, look at Figure 13-7(a). There are 100 cm in 1 m, so each side of the square meter has a measure of 100 cm. Thus it takes 100 rows of 100-cm squares each to fill a square meter, that is, $100 \cdot 100$, or 10,000-cm squares. Because the area of each centimeter square is $1 \text{ cm} \cdot 1 \text{ cm}$, or 1 cm^2, there are 10,000 cm^2 in 1 m^2. In general, the area A of a square that is s units on a side is s^2, as shown in Figure 13-7(b).

FIGURE 13-7

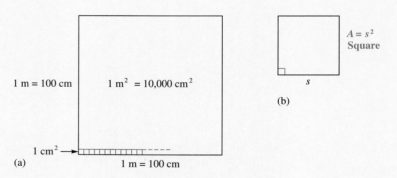

Other metric conversions of area measure can be developed by using the formula for the area of a square. For example, Figure 13-8(a) shows that 1 m^2 = 10,000 cm^2 = 1,000,000 mm^2. Likewise, Figure 13-8(b) shows that 1 m^2 = 0.000001 km^2. Similarly, 1 cm^2 = 100 mm^2 and 1 km^2 = 1,000,000 m^2.

FIGURE 13-8

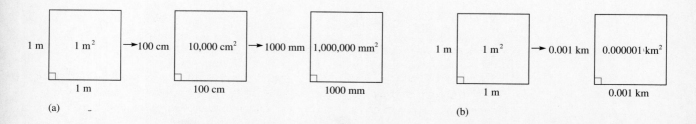

SECTION 13-2 AREAS OF POLYGONS AND CIRCLES

Table 13-4 shows the symbols for metric units of area and their relationship to the square meter.

TABLE 13-4

Unit	Symbol	Relationship to Square Meter
square kilometer	km²	1,000,000 m²
*square hectometer	hm²	10,000 m²
*square dekameter	dam²	100 m²
square meter	**m²**	1 m²
*square decimeter	dm²	0.01 m²
square centimeter	cm²	0.0001 m²
square millimeter	mm²	0.000001 m²

*Not commonly used.

EXAMPLE 13-5   Convert each of the following:

(a) 5 cm² = _____ mm²   (b) 124,000,000 m² = _____ km²

Solution (a) 1 cm² = 100 mm² implies 5 cm² = 5 · 1 cm² = 5 · 100 mm² = 500 mm².

(b) 1 m² = 0.000001 km² implies 124,000,000 m² = 124,000,000 · 1 m² = 124,000,000 · 0.000001 km² = 124 km². ∎

REMARK Students sometimes confuse the area of 5 cm² with the area of a square 5 cm on each side. The area of a square 5 cm on each side is (5 cm)², or 25 cm². Five squares each 1 cm by 1 cm have the area of 5 cm². Thus 5 cm² ≠ (5 cm)².

Based on the relationship among units of length in the English system, it is possible to convert among English units of area. For example, because 1 yd = 3 ft, it follows that (1 yd)² = 1 yd · 1 yd = 3 ft · 3 ft = 9 ft². Similarly, because 1 ft = 12 in., (1 ft)² = 1 ft · 1 ft = 12 in. · 12 in. = 144 in². Table 13-5 summarizes various relationships among units of area in the English system.

TABLE 13-5

Unit of Area	Equivalent of Other Units
1 ft²	$\frac{1}{9}$ yd², or 144 in.²
1 yd²	9 ft²
1 mi²	3,097,600 yd², or 27,878,400 ft²

Land Measure

One of the most common applications of area today is in land measure. Old deeds in the United States include land measures in terms of chains, poles, rods, acres, sections, lots, *acre*   and townships. The common unit of land measure in the English system is the **acre**. There *square mile* are 4840 yd² in 1 acre. For very large land measures in the English system, the **square mile** (**mi²**), or 640 acres, is used.

In the metric system, small land areas are measured in terms of a square unit 10 m on a side, called an **are** (pronounced "air") and denoted by **a**. Thus 1 a = 10 m · 10 m, or 100 m². Larger land areas are measured in **hectares**. A hectare is 100 a. A hectare, denoted by **ha**, is the amount of land whose area is 100 m · 100 m, or 10,000 m². It follows that one hectare is the area of a square that is 100 m on a side. For very large land measures, the **square kilometer**, denoted by km², is used. One square kilometer is the area of a square with a side 1 km, or 1000 m, long. Land area measures are summarized in Table 13-6.

are / a
hectares / ha

square kilometer

TABLE 13-6

Unit of Area	Equivalent in Other Units
1 a	100 m²
1 ha	100 a, or 10,000 m²
1 km²	1,000,000 m²
1 acre	4840 yd²
1 mi²	640 acres

EXAMPLE 13-6 (a) A square field has a side of 400 m. Find the area of the field in hectares.

(b) A square field has a side of 400 yds. Find the area of the field in acres.

Solution (a) $A = (400 \text{ m})^2 = 160{,}000 \text{ m}^2 = \frac{160{,}000}{10{,}000} \text{ ha} = 16 \text{ ha}$

(b) $A = (400 \text{ yd})^2 = 160{,}000 \text{ yd}^2 = \frac{160{,}000}{4840} \text{ acre} \doteq 33.1 \text{ acre}$ ∎

Areas on a Geoboard

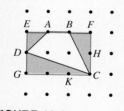

FIGURE 13-9

Activities that foster an intuitive grasp of the concept of area should precede the teaching of formulas for finding the areas of different figures. Such activities can be accomplished on a geoboard. In Figure 13-9, the distance between two adjacent nails in a row or column is one unit. To find the area of quadrilateral *ABCD*, we construct the colored rectangle *EFCG* around the quadrilateral and then subtract the areas of triangles *EAD*, *BFC*, and *DGC*. The area of the rectangle *EFCG* is 6 square units. The area of △*EAD* is $\frac{1}{2}$ square unit, and the area of △*BFC* is half the area of rectangle *BFCK*, or $\frac{1}{2}$ of 2, or 1, square unit. Similarly, the area of △*DGC* is half the area of rectangle *DHCG*, that is, $\frac{1}{2} \cdot 3$, or $\frac{3}{2}$, square units. Consequently, the area of *ABCD* is $6 - \left(\frac{1}{2} + 1 + \frac{3}{2}\right)$, or 3, square units.

EXAMPLE 13-7 Using a geoboard, find the areas of each of the shaded figures in Figure 13-10.

FIGURE 13-10

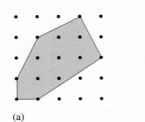
(a)

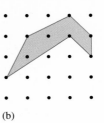
(b)

Solution (a) We construct a square around the hexagon, and then subtract the areas of regions (a), (b), (c), (d), and (e) from the area of this square, as shown in Figure 13-11. Therefore the area of the hexagon is $16 - (3 + 1 + 1 + 1 + 1)$, or 9, square units.

FIGURE 13-11

(b) The area of the hexagon equals the area of the surrounding rectangle shown in Figure 13-12 minus the sum of the area of figures (a), (b), (c), (d), (e), (f), and (g). Thus the area of the hexagon is $12 - (3 + 1 + \frac{1}{2} + \frac{1}{2} + 1 + 1 + 1)$, or 4, square units. ∎

FIGURE 13-12

Area of a Rectangle

To measure area, we may count the number of units of area contained in any given region. For example, suppose that the square in Figure 13-13(a) represents 1 square unit. Then, the rectangle $ABCD$ in Figure 13-13(b) contains $3 \cdot 4$, or 12, square units.

FIGURE 13-13

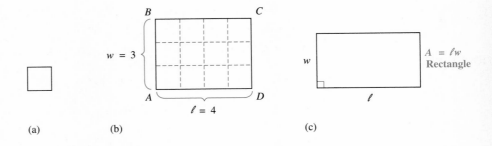

If the unit in Figure 13-13(a) is 1 cm², then the area of rectangle $ABCD$ is 12 cm². In general, the area A of any rectangle may be found by multiplying the lengths of two adjacent sides ℓ and w, or $A = \ell w$, as given in Figure 13-13(c).

EXAMPLE 13-8 Find the area of each rectangle in Figure 13-14.

FIGURE 13-14

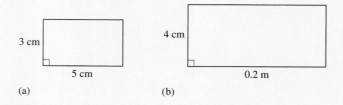

(a)   (b)

Solution (a) $A = (3 \text{ cm})(5 \text{ cm}) = 15 \text{ cm}^2$

(b) First, write the lengths of the sides in the same unit of length. Because 0.2 m = 20 cm, $A = (4 \text{ cm})(20 \text{ cm}) = 80 \text{ cm}^2$. Alternatively, 4 cm = 0.04 m, so $A = (0.04 \text{ m})(0.2 \text{ m}) = 0.008 \text{ m}^2$. ∎

Area of a Parallelogram

A formula for the area of a parallelogram is developed by *reducing the problem to one that we already know how to do,* that of finding the area of a rectangle. Consider the parallelogram *ABCD* in Figure 13-15(a). The parallelogram can be separated into two parts. The shaded triangle is translated to the right of the parallelogram, as in Figure 13-15(b), to obtain a rectangle with length *b* and width *h*. The translation image of *D* is *C*. The parallelogram and the rectangle have the same area. (Why?) Because the area of the rectangle is *bh*, the area of the original parallelogram *ABCD* is also *bh*.

FIGURE 13-15

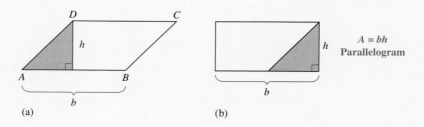

(a)   (b)

base In general, any side of a parallelogram can be designated as a **base,** with measure *b*.
height The **height** *(h)* is the distance between the bases and is always the length of a segment perpendicular to the lines containing the bases. In Figure 13-16, *EB*, or *g*, is the height that

FIGURE 13-16

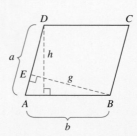

corresponds to the bases $\overline{AD}$ and $\overline{BC}$, each of which has measure a. Consequently, the area of the parallelogram $ABCD$ is ag. Similarly, its area can be expressed as bh. Therefore $A = ag = bh$.

Area of a Triangle

A formula for the area of a triangle is *developed from a known problem* for finding the area of a parallelogram using the strategy of *drawing a diagram*. In Figure 13-17(a), $\triangle BAC$ has base b and height h. Let $\triangle BAC'$ be the image of $\triangle BAC$ when $\triangle BAC$ is rotated 180° about M, the midpoint of $\overline{AB}$, as in Figure 13-17(b). Proving that quadrilateral $BCAC'$ is a parallelogram is left as an exercise. Parallelogram $BCAC'$ has area bh and is constructed of congruent triangles BAC and BAC'. So the area of $\triangle ABC$ is $\frac{1}{2}bh$. Thus the area of a triangle is equal to half the product of the length of a side and the altitude to that side.

FIGURE 13-17

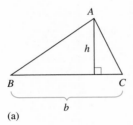

(a)

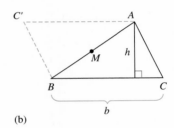

(b)

$A = \frac{1}{2}bh$
Triangle

In Figure 13-18, $\overline{BC}$ is a base of $\triangle ABC$, and the corresponding height h_1, or AE, is the distance from the opposite vertex A to the line containing $\overline{BC}$. Similarly, $\overline{AC}$ can be chosen as a base. Then h_2, or BG, the distance from the opposite vertex B to the line containing $\overline{AC}$, is the corresponding height. If AB is chosen as a base, then the corresponding height is h_3, or FC. Thus the area A of $\triangle ABC$ is

$$A = \frac{bh_1}{2} = \frac{ah_2}{2} = \frac{ch_3}{2}.$$

FIGURE 13-18

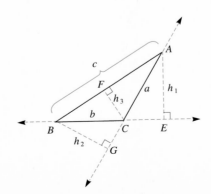

CHAPTER 13 CONCEPTS OF MEASUREMENT

EXAMPLE 13-9 Find the areas in Figure 13-19. Assume that the quadrilaterals in (a) and (b) are parallelograms.

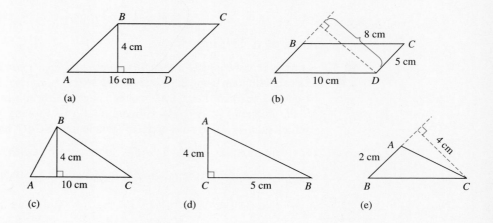

FIGURE 13-19

Solution
(a) $A = bh = (16 \text{ cm})(4 \text{ cm}) = 64 \text{ cm}^2$
(b) $A = bh = (5 \text{ cm})(8 \text{ cm}) = 40 \text{ cm}^2$
(c) $A = \frac{1}{2}bh = \frac{1}{2}(10 \text{ cm})(4 \text{ cm}) = 20 \text{ cm}^2$
(d) $A = \frac{1}{2}bh = \frac{1}{2}(5 \text{ cm})(4 \text{ cm}) = 10 \text{ cm}^2$
(e) $A = \frac{1}{2}bh = \frac{1}{2}(2 \text{ cm})(4 \text{ cm}) = 4 \text{ cm}^2$ ∎

Area of a Trapezoid

Areas of general polygons can be found by partitioning the polygons into triangles, finding the areas of the triangles, and summing those areas. In Figure 13-20(a), trapezoid $ABCD$ has bases b_1 and b_2 and height h. By continuing to *draw in the diagram*, connecting points B and D, as in Figure 13-20(b), we create two triangles: one with base $\overline{AB}$ and height DE and the other with base $\overline{CD}$ and height BF. Because $\overline{DE} \cong \overline{BF}$, each has height h. Thus the areas of triangles ADB and DCB are $\frac{1}{2}(b_1 h)$ and $\frac{1}{2}(b_2 h)$, respectively. Hence, the area of trapezoid $ABCD$ is $\frac{1}{2}(b_1 h) + \frac{1}{2}(b_2 h)$, or $\frac{1}{2}h(b_1 + b_2)$. That is, the area of a trapezoid is equal to half the height times the sum of the lengths of the bases. Other ways to examine this area are found in the problem set.

FIGURE 13-20

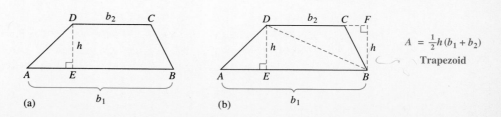

EXAMPLE 13-10   Find the areas of the trapezoids in Figure 13-21.

FIGURE 13-21

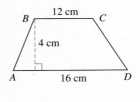
(a)

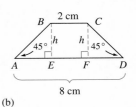
(b)

Solution (a) $A = \frac{1}{2}h(b_1 + b_2) = \frac{1}{2}(4 \text{ cm})(12 \text{ cm} + 16 \text{ cm}) = 56 \text{ cm}^2$

(b) To find the area of trapezoid *ABCD*, we *use a subgoal* to find the height, h. In Figure 13-21(b), $BE = CF = h$. Also, $\overline{BE}$ is a side of $\triangle ABE$, which has angles with measures of 45° and 90°. Consequently, the third angle in triangle *ABE* is $180 - (45 + 90)$, or 45°. Therefore $\triangle ABE$ is isosceles and $AE = BE = h$. Similarly, it follows that $FD = h$. Because $AD = 8 = h + EF + h$, we could find h if we knew the value of EF. From Figure 13-21(b), $EF = BC = 2$ because *BCFE* is a rectangle (why?) and opposite sides of a rectangle are congruent. Now $h + EF + h = h + 2 + h = 8$. Thus $h = 3$ cm and the area of the trapezoid is $A = \frac{1}{2}(3 \text{ cm})(2 \text{ cm} + 8 \text{ cm})$, or 15 cm².

PROBLEM 1

Larry purchased a plot of land surrounded by a fence. The former owner had subdivided the land into 13 equal-sized square plots, as shown in Figure 13-22. To reapportion the property into two plots of equal area, Larry wishes to build a single, straight fence beginning at the far left corner (point *P* on the drawing). Is such a fence possible? If so, where should the other end be?

FIGURE 13-22

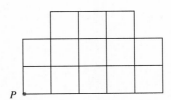

Understanding the Problem. To divide the land in Figure 13-22 into two plots of equal area by means of a straight fence starting at point *P*, *a subgoal* becomes to find the other endpoint. Because the area of the entire plot is 13 square units, the area of each part formed by the fence must be $\frac{1}{2} \times 13$, or $6\frac{1}{2}$ square units.

Devising a Plan. To find an approximate location for the fence, consider a fence connecting P with point A, as shown in Figure 13-23. The area of the land below fence $\overline{PA}$ is the sum of the areas of $\triangle APD$ and the rectangle $DAFE$. The area of $\triangle APD$ is 4 and the area of rectangle $DAFE$ is 2, so that the area below the fence $\overline{PA}$ is $4 + 2$, or 6, square units. We want an area of $6\frac{1}{2}$ square units; consequently, the other end of the fence should be above point A.

FIGURE 13-23

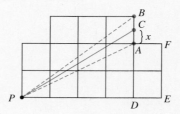

A similar argument shows that the area below $\overline{PB}$ is 8 square units and the end of the fence should be below B. Thus the other end of the fence should be at a point C between A and B. To find the exact location of point C, we designate CA by x, *write an equation* for x by finding the area below $\overline{PC}$ in terms of x, make it equal to $6\frac{1}{2}$, and solve for x.

Carrying Out the Plan. The area below $\overline{PC}$ equals the area of $\triangle PCD$ plus the area of the rectangle $DAFE$. The area of $\triangle PCD$ is

$$\frac{PD \cdot DC}{2} = \frac{4(2 + x)}{2} = 2(2 + x).$$

The area of rectangle $DAFE$ is 2 so that $2(2 + x) + 2$ should equal half the area of the plot. Consequently, we have the following:

$$2(2 + x) + 2 = 6\frac{1}{2}$$

$$4 + 2x + 2 = \frac{13}{2}$$

$$2x = \frac{1}{2}$$

$$x = \frac{1}{4}$$

Therefore the fence should be built along the line connecting point P to the point C, which is $\frac{1}{4}$ unit directly above point A. Point C can be found by dividing $\overline{AB}$ into four congruent parts.

Looking Back. We check that the solution is correct by finding the area above $\overline{PC}$. The problem can be varied by changing the shape of the plot, by determining if the problem could be solved if Larry wanted the plot divided into thirds, fourths, and so on. We could also approach the problem as if Larry wanted to subdivide the land but wanted to use the existing lines that mark the squares. ∎

Area of a Regular Polygon

The area of a triangle can be used to find the area of any regular polygon, as illustrated *using a simpler case* involving a regular hexagon (see Figure 13-24a). The hexagon can be separated into six congruent triangles, each with a vertex at the center, with side s and height a. (The height of such a triangle of a regular polygon is called the *apothem* and is denoted by a.) The area of each triangle is $\frac{1}{2}as$. Because six triangles make up the hexagon, the area of the hexagon is $6(\frac{1}{2}as)$, or $\frac{1}{2}a(6s)$. However, $6s$ is the perimeter p of the hexagon, so the area of the hexagon is $\frac{1}{2}ap$. The same process can be used to develop the formula for the area of any regular polygon. That is, the area of any regular polygon is $\frac{1}{2}ap$, where a is the height of one of the triangles involved and p is the perimeter of the polygon, as shown in Figure 13-24(b).

FIGURE 13-24

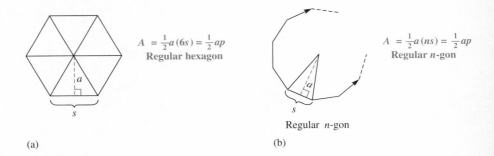

(a) $A = \frac{1}{2}a(6s) = \frac{1}{2}ap$
Regular hexagon

(b) $A = \frac{1}{2}a(ns) = \frac{1}{2}ap$
Regular n-gon

Area of a Circle

We use the strategy of *examining a related problem* to find a formula for the area of a circle. The area of a regular polygon inscribed in a circle, as in Figure 13-25, approximates the area of the circle, and we know that the area of any regular n-gon is $\frac{1}{2}ap$, where a is the height of a triangle of the n-gon and p is the perimeter. If the number of sides n is made very large, then the perimeter and the area of the n-gon are close to those of the circle. Also, the apothem a is approximately equal to the radius r of the circle, and the

FIGURE 13-25

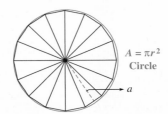

$A = \pi r^2$
Circle

perimeter p approximates the circumference $2\pi r$. Because the area of the circle is approximately equal to the area of the n-gon, $\frac{1}{2}ap \doteq \frac{1}{2}r \cdot 2\pi r = \pi r^2$. In fact, the area of the circle is precisely πr^2.

A similar approach to the preceding derivation of the area of a circle was given in 1609 by the astronomer Johann Kepler (1571–1630). The approach shown on the student page from *Addison-Wesley Mathematics,* Grade 8, 1991, is similar to Kepler's.

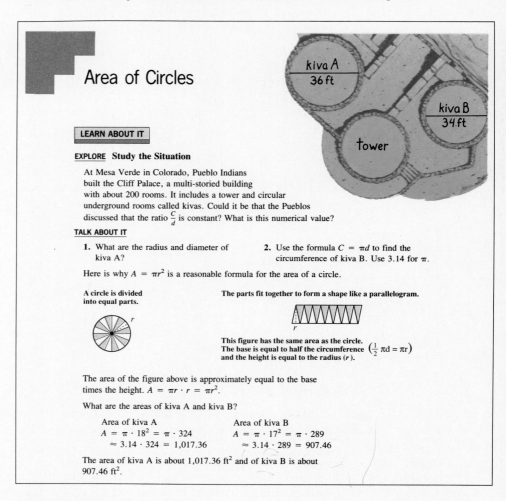

Area of a Sector

sector   A **sector** of a circle is a pie-shaped region of the circle determined by a central angle of the circle. The area of a sector depends on the radius of the circle and the measure of the central angle determining the sector. If the angle has a measure of 90°, as in Figure 13-26(a), the area of the sector is one-fourth the area of the circle, or $\frac{90}{360}\pi r^2$. In any circle, there are 360°, so the area of a sector whose central angle has measure θ degrees is $\frac{\theta}{360}(\pi r^2)$, as shown in Figure 13-26(b).

FIGURE 13-26

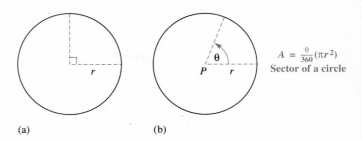

(a) (b)

PROBLEM SET 13-2

1. Choose the most appropriate metric units (cm², m², or km²) and English units (in.², yd², mi²) for measuring each of the following:
   (a) Area of a sheet of notebook paper
   (b) Area of a quarter
   (c) Area of a desktop
   (d) Area of a classroom floor
   (e) Area of a parallel parking space
   (f) Area of an airport runway

2. Estimate and then measure each of the following using cm², m², or km²:
   (a) Area of a door   (b) Area of a chair seat
   (c) Area of a desktop (d) Area of a chalkboard

3. Complete the following conversion table:

Item	m²	cm²	mm²
Area of a sheet of paper		588	
Area of a cross section of a crayon			192
Area of a desktop	1.5		
Area of a dollar bill		100	
Area of a postage stamp		5	

4. ▶Explain the difference between a 2-m square and 2 m².

5. Complete the following conversions using a calculator:
   (a) 4000 ft² = _____ yd²
   (b) 10⁶ yd² = _____ mi²
   (c) 10 mi² = _____ A
   (d) 3 A = _____ ft²

6. Complete each of the following:
   (a) A football field is about 49 m by 100 m or _____ m².
   (b) About _____ ares are in two football fields.
   (c) About _____ hectares are in two football fields.

7. Find the areas of each of the following figures if the distance between two adjacent nails in a row or a column is one unit:

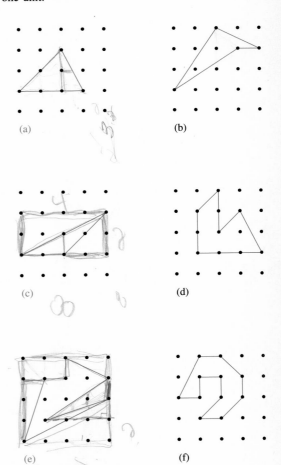

8. If all vertices of a polygon are points on square-dot paper, the polygon is called a **lattice polygon.** In 1899, G. Pick discovered a surprising theorem involving I, the number of dots *inside* the polygon, and B, the number of dots that lie *on* the polygon. The theorem states that the

area of any lattice polygon is $I + \frac{1}{2}B - 1$. Check that this is true for the polygons in Problem 7.

9. Find the area of △ABC in each of the following triangles:

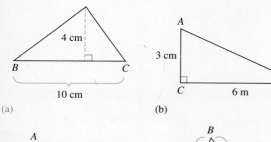

(a) (b)

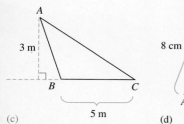

(c)

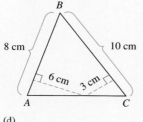

(d)

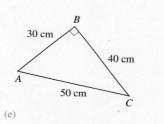

(e)

10. ▶In the following figure, $\ell \parallel \overleftrightarrow{AB}$. If the area of △ABP is 10 cm², what are the areas of △ABQ, △ABR, △ABS, △ABT, and △ABU? Explain your answers.

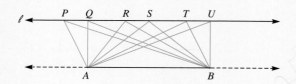

11. Find the area of each of the following quadrilaterals:

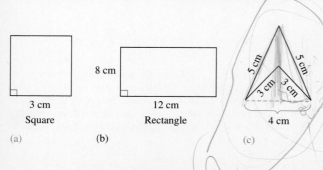

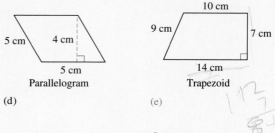

(d) Parallelogram (e) Trapezoid

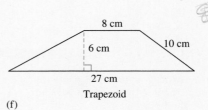

(f) Trapezoid

12. (a) A rectangular piece of land is 1300 by 1500 m.
   (i) What is the area in square kilometers?
   (ii) What is the area in hectares?
   (b) A rectangular piece of land is 1300 by 1500 yd.
   (i) What is the area in square miles?
   (ii) What is the area in acres?
   ▶(c) Explain which measuring system you would rather use to solve (a) and (b) above.

13. For a parallelogram whose sides are 6 cm and 10 cm, which of the following is true?
   (a) The data are insufficient to enable us to determine the area.
   (b) The area equals 60 cm².
   (c) The area is greater than 60 cm².
   (d) The area is less than 60 cm².

14. (a) Find the area of the trapezoid shown.

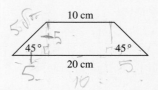

★(b) Show that the area of a trapezoid similar to the one in (a), with bases a and b and with base angles of 45° each, is $\frac{a^2 - b^2}{4}$.

15. If the diagonals of a rhombus are a and b units long, find the area of the rhombus in terms of a and b.

16. Find the cost of carpeting the following rectangular rooms:
   (a) Dimensions: 6.5 by 4.5 m; cost = $13.85/m²
   (b) Dimensions: 15 by 11 ft; cost = $30/yd²

17. Find the area of each of the following. Leave your answers in terms of π.

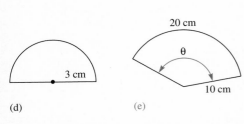

(a) (b) (c)

(d) (e)

18. Joe uses stick-on square carpet tiles to cover his 3-by-4-m bathroom. If each tile is 10 cm on a side, how many tiles does he need?

19. A rectangular plot of land is to be seeded with grass. If the plot is 22 by 28 m and if a 1-kg bag of seed is needed for 85 m² of land, how many bags of seed will it take?

20. Consider the following figures made of five congruent squares constructed in such a way that any two squares share a common side or a common vertex or have no points in common and each square shares an edge with another. Each square is considered a unit square.

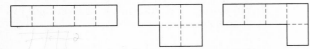

(a) What are the greatest and least perimeters possible of a figure made of unit squares whose area is
   (i) 5?   (ii) 6?   (iii) 30?
   Draw the appropriate figures.
(b) What are the greatest and least possible areas of a figure made of unit squares whose perimeter is
   (i) 12 units?
   (ii) 26 units?
   Draw the appropriate figures.
(c) What is the greatest possible perimeter of a figure made of unit squares whose area is n square units?
★(d) What is the least possible area of a figure made of unit squares whose perimeter is $2n$ units long?

21. Find the area of each regular polygon.

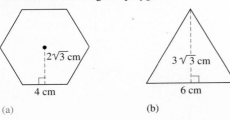

(a) (b)

22. (a) If a circle has a circumference of 8π cm, what is its area?
    (b) If a circle with radius r and a square with a side of length s have the same area, express r in terms of s.

23. Find the area of each of the following shaded parts. Assume all arcs are circular.

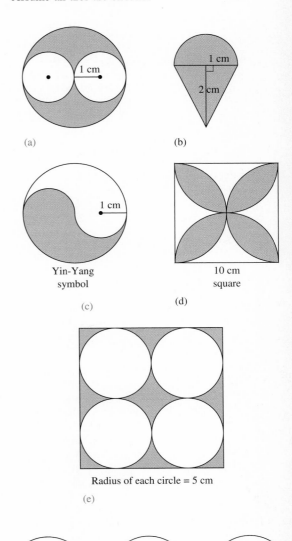

24. A circular flower bed is 6 m in diameter and has a circular sidewalk around it 1 m wide. Find the area of the sidewalk in square meters.

25. (a) To make lids, congruent circles are cut out of a rectangular piece of tin, as shown in the following figure. Find what percentage of the tin is wasted.

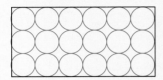

(b) Suppose the rectangular piece of tin is the same size as in (a), but smaller congruent circles are cut out. Also suppose that the circles are still tangent to each other and to the sides of the rectangle. What percent of the tin is wasted if the radius of each circle is as follows?
  (i) Half the radius of the circles in (a)
  (ii) One-third the radius of the circles in (a)

26. ▶(a) Explain how the following drawing can be used to determine a formula for the area of △ABC.

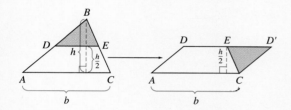

(b) Use paper cutting to reassemble △ABC in (a) into parallelogram ADD′C.

27. ▶(a) If the ratio of the sides of two squares is 2:3, what is the ratio of the areas? Justify your answer.
(b) If the ratio between the diagonals of two squares is 2:3, what is the ratio of their areas? Justify your answer.

28. (a) If, in two similar triangles, the ratio of the lengths of the corresponding sides is 2:1, what is the ratio of their areas?
(b) Make a conjecture concerning the relationship between the ratio of the areas of two similar triangles and the ratio of the corresponding sides.
▶(c) Justify your conjecture in (b).

29. ▶(a) The screens of two television sets are similar rectangles. The 20-in. set (the length of the diagonal is 20 in.) costs $400, whereas the 27-in. set with similar features costs $600. If a customer is concerned about the size of the viewing area and is willing to pay the same amount per square foot, which is a better buy? Why?
(b) What should the length of the diagonal of a TV set be in order for the viewing area to be twice the viewing area of the 20-in. set?

30. ▶The area of a parallelogram can be found by using the concept of a half-turn (a turn by 180°). Consider the parallelogram ABCD and let M and N be the midpoints of $\overline{AB}$ and $\overline{CD}$, respectively. Rotate the shaded triangle with vertex M about M by 180° clockwise, and rotate the shaded triangle with vertex N about N by 180° counterclockwise. What kind of figure do you obtain? Now complete the argument to find the area of the parallelogram.

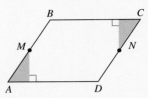

31. Consider the trapezoid ABCD. Rotate the trapezoid 180° clockwise about the midpoint M of $\overline{BC}$. Use the figure obtained from the union of the original trapezoid and its image to derive the formula for the area of a trapezoid.

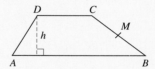

32. The following figure consists of five congruent squares. Find a line through point P that divides the figure into two parts of equal area.

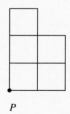

33. ▶A landscape architect wants to divide a circular garden into four congruent regions, each bounded only by arcs. How can this be done?

34. Find the shaded area enclosed by two semicircles and two tangents to the semicircles as shown.

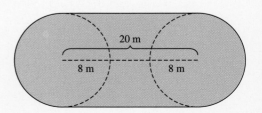

35. An aircraft company starts with a rectangular piece of metal measuring 10 by 10 in. and wants to remove a strip x in. wide from all sides to form another rectangle with an area of 64 in.^{2}. Find x.

★36. In the drawing, quadrilateral $ABCD$ is a parallelogram and P is any point on $\overline{AC}$. Prove that the area of $\triangle BCP$ is equal to the area of $\triangle DPC$.

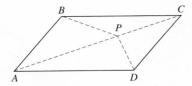

Review Problems

37. Complete each of the following:
    (a) 100 mm = _____ cm
    (b) 10.4 cm = _____ mm
    (c) 350 mm = _____ m
    (d) 0.04 m = _____ mm
    (e) 8 km = _____ m
    (f) 6504 m = _____ km

38. Find the perimeters for each of the following if all arcs shown are semicircles:

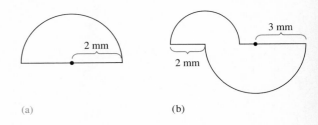

(a)   (b)

39. ▶Explain why $BCAC'$ in Figure 13-17(b) is a parallelogram.

LABORATORY ACTIVITY

As van Hiele Level 2 activities, answer the following without using any area formulas:

1. On a 5-by-5 geoboard, make $\triangle DEF$. Keep the rubber band around D and E fixed and move the vertex F to all the possible locations so that the triangles formed will have the same area as the area of $\triangle DEF$. How do the locations for the third vertex relate to D and E?

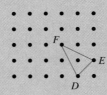

2. On a 5-by-5 geoboard, construct, if possible, squares of areas 1, 2, 3, 4, 5, 6, and 7 square units.

3. On a 5-by-5 geoboard, construct triangles that have areas $\frac{1}{2}$, 1, $1\frac{1}{2}$, 2, . . . , until the maximum-sized triangle is reached.

▼ BRAIN TEASER

The accompanying rectangle was apparently formed by cutting the square shown along the dotted lines and reassembling the pieces as pictured.

1. What is the area of the square?
2. What is the area of the rectangle?
3. How do you explain the discrepancy?

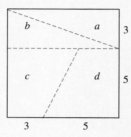
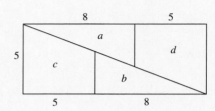

Section 13-3 The Pythagorean Relationship

One of the most remarkable and useful discoveries in geometry, known as the Pythagorean theorem, involves right triangles. This theorem was illustrated on a Greek stamp in 1955, as shown in Figure 13-27, to honor the 2500th anniversary of the founding of the Pythagorean School.

FIGURE 13-27

hypotenuse
legs

In the triangle on the stamp, the side opposite the right angle is the **hypotenuse.** The other two sides are **legs.** Interpreted in terms of area, the Pythagorean theorem states that the area of a square with the hypotenuse of a right triangle as a side is equal to the sum of the areas of the squares with the legs as sides.

Because the Pythagoreans affirmed geometric results on the basis of special cases, mathematical historians believe it is possible they may have discovered the theorem by looking at a floor tiling like the one in Figure 13-28. Each square can be divided by its diagonal into two congruent isosceles right triangles, so we see that the shaded square constructed with $\overline{AB}$ as a side consists of four triangles, each congruent to $\triangle ABC$. Similarly, each of the shaded squares with legs $\overline{BC}$ and $\overline{AC}$ as sides consists of two triangles congruent to $\triangle ABC$. Thus the area of the larger square is equal to the sum of the areas of the two smaller squares. The theorem is true in general and is stated below using Figure 13-29.

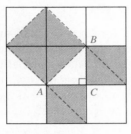

FIGURE 13-28

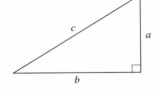

FIGURE 13-29

THEOREM 13-1

Pythagorean theorem: If a right triangle has legs of lengths a and b and hypotenuse of length c, then $c^2 = a^2 + b^2$.

There are hundreds of known proofs for the Pythagorean theorem today. The classic book *The Pythagorean Proposition*, by E. Loomis, contains many of these proofs. Some proofs involve the strategy of *drawing diagrams* with a square area c^2 equal to the areas a^2 and b^2 of two other squares. One such proof is given in Figure 13-30; others are discussed in Problem Set 13-3. In Figure 13-30(a), let the measures of the legs of a right triangle ABC be a and b and let c be the measure of the hypotenuse. We draw a square with sides of length $a + b$ and subdivide it, as shown in Figure 13-30(b). In Figure 13-30(c), another square with side of length $a + b$ is drawn and each of its sides is divided into two segments of length a and b, as shown.

FIGURE 13-30

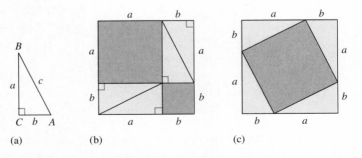

(a)   (b)   (c)

Each yellow triangle is congruent to $\triangle ABC$. (Why?) Consequently, each of the triangles has hypotenuse c and the same area, $\frac{1}{2}ab$. Thus the length of each side of the inside gray quadrilateral in Figure 13-30(c) is c and hence the figure is a rhombus. In fact, it is possible to show that the figure is a square whose area is c^2. To complete the proof, consider the four triangles in Figure 13-30(b) and (c). Because the areas of the sets of four triangles in both Figure 13-30 (b) and (c) are equal, the sum of the areas of the two shaded squares in Figure 13-30(b) equals the area of the shaded square in Figure 13-30(c), that is, $a^2 + b^2 = c^2$.

■ HISTORICAL NOTE

Pythagoras (ca. 582–507 B.C.), a Greek philosopher and mathematician, was head of a group known as the Pythagoreans. Members of the group regarded Pythagoras as a demigod and attributed all their discoveries to him. The Pythagoreans believed in the transmigration of the soul from one body to another. One of Pythagoras's most unusual discoveries was the dependence of the musical intervals on the ratio of the length of strings at the same tension, the ratio 2:1 giving the octave, 3:2, the fifth, and 4:3, the fourth.

EXAMPLE 13-11 For each drawing in Figure 13-31, find x using the Pythagorean theorem.

FIGURE 13-31

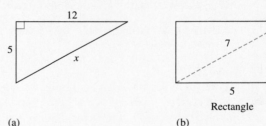

(a) (b)

Solution (a) By the Pythagorean theorem,

$$5^2 + 12^2 = x^2$$
$$25 + 144 = x^2$$
$$169 = x^2$$
$$13 = x.$$

(b) In the rectangle, the diagonal partitions the rectangle into two right triangles, each with lengths 5 units and width x units. Thus we have the following:

$$5^2 + x^2 = 7^2$$
$$25 + x^2 = 49$$
$$x^2 = 24$$
$$x = \sqrt{24}, \text{ or approximately } 4.9$$ ■

The Pythagorean theorem is used in solving many real-life problems, as shown on the following student page from *Addison-Wesley Mathematics,* Grade 8, 1989.

SECTION 13-3   THE PYTHAGOREAN RELATIONSHIP   717

Finding the Length of the Hypotenuse

The size of a rectangular display screen is given as the length of the diagonal of the screen. The length of the screen is 24 cm and the width is 18 cm. What is the length of the diagonal?

The diagonal is the hypotenuse of a right triangle whose legs measure 18 cm and 24 cm.

We can use the Pythagorean formula to find the length.

$$a^2 + b^2 = c^2$$
$$18^2 + 24^2 = c^2$$
$$900 = c^2$$
$$\sqrt{900} = \sqrt{c^2}$$
$$30 = c$$

To solve this equation, find the square root of the numbers on each side.

The diagonal is 30 cm long.

When using the Pythagorean theorem, we must work with a right triangle. At times, though, the segment whose length we want to find may not be a side of any known right triangle. The following examples deal with such situations.

EXAMPLE 13-12   A pole $\overline{BD}$, 28 ft high, is perpendicular to the ground. Two wires $\overline{BC}$ and $\overline{BA}$, each 35 ft long, are attached to the top of the pole and to stakes A and C on the ground, as in Figure 13-32. If points A, D, and C are collinear, how far are the stakes A and C from each other?

Solution $\overline{AC}$ is not a side in any known right triangle, but we want to find AC. Because a point equidistant from the endpoints of a segment must be on a perpendicular bisector of the segment, then $AD = DC$. Therefore AC is twice as long as DC. We may

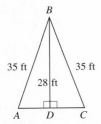

FIGURE 13-32

find DC by applying the Pythagorean theorem in triangle BDC. This results in the following:

$$28^2 + (DC)^2 = 35^2$$
$$(DC)^2 = 35^2 - 28^2$$
$$DC = \sqrt{441}, \text{ or } 21 \text{ ft}$$
$$AC = 2 \cdot DC = 42 \text{ ft}$$

EXAMPLE 13-13 How tall is the Great Pyramid of Cheops, a right regular square pyramid, if the base has a side 775 ft and the slant height is 608 ft?

FIGURE 13-33

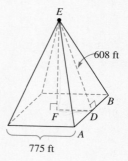

Solution In Figure 13-33, $\overline{EF}$ is a leg of a right triangle formed by $\overline{FD}$, $\overline{EF}$, and $\overline{ED}$. Because the pyramid is a right regular pyramid, $\overline{EF}$ intersects the base at its center. Thus $DF = \left(\frac{1}{2}\right)AB$, or $\left(\frac{1}{2}\right)775$, or 387.5 ft. Now ED, the slant height, has length 608 ft, and we can apply the Pythagorean theorem as follows:

$$(EF)^2 + (DF)^2 = (ED)^2$$
$$(EF)^2 + (387.5)^2 = (608)^2$$
$$(EF)^2 = 219{,}507.75$$
$$EF \doteq 468.5 \text{ ft}$$

Thus the Great Pyramid is 468.5 ft tall.

The student page from *Houghton Mifflin Mathematics*, Grade 8, 1991, illustrates two special right triangles. In the isosceles right triangle pictured, each leg is 1 unit long and the hypotenuse is $\sqrt{2}$ units long. This property is generalized when the isosceles right triangle has a leg of length a as follows.

PROPERTY

Property of 45°-45°-90° Triangle: In an isosceles right triangle, if the length of each leg is a, then the hypotenuse has length $a\sqrt{2}$.

Similarly, the student page shows that a 30°-60°-90° triangle is half of an equilateral triangle. When the equilateral triangle has side 2 units long, then in the 30°-60°-90° triangle, the leg opposite the 30° angle is 1 unit long and the leg opposite the 60° angle has a

SECTION 13-3 THE PYTHAGOREAN RELATIONSHIP

SPECIAL RIGHT TRIANGLES

In an isosceles right triangle, the two legs are equal in length. The measures of the angles opposite the legs are equal.

Suppose each leg of an isosceles right triangle is 1 unit long. What is the length of the hypotenuse?

$$c^2 = a^2 + b^2$$
$$c^2 = 1^2 + 1^2$$
$$c^2 = 2$$
$$c = \sqrt{2}$$

The length of the hypotenuse is $\sqrt{2}$

In an isosceles right triangle, if the length of each leg is a, you can use $a\sqrt{2}$ to find the length of the hypotenuse.

$\triangle ABC$ is a 30°–60° right triangle. It is half of the equilateral $\triangle ABD$. Suppose $\overline{BC}$ is 1 unit long. Then $\overline{AB}$ is 2 units long. What is AC?

$(AC)^2 + (BC)^2 = (AB)^2$ ← Use the Pythagorean theorem.
$(AC)^2 + 1^2 = 2^2$
$(AC)^2 = 3$
$AC = \sqrt{3}$ So, AC is $\sqrt{3}$.

In a 30°–60° right triangle, if the shorter leg has a length a, then the longer leg has a length $a\sqrt{3}$, and the hypotenuse has a length $2a$.

FIGURE 13-34

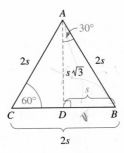

length of $\sqrt{3}$. This example may also be generalized using the similar triangle in Figure 13-34. When the side of the equilateral triangle ABC is $2s$, then in triangle ABD, the side opposite the 30° angle, $\overline{BD}$, is s units long, and AD may be found using the Pythagorean theorem to have a length of $s\sqrt{3}$ units. This discussion is generalized in the following property.

PROPERTY

Property of 30°-60°-90° Triangle: In a 30°-60°-90° triangle, the length of the hypotenuse is two times as long as the leg opposite the 30° angle and the leg opposite the 60° angle is $\sqrt{3}$ times the shorter leg.

Converse of the Pythagorean Theorem

The converse of the Pythagorean theorem is also true and provided a useful way for early surveyors, sometimes called Egyptian rope stretchers, to determine right angles. Figure 13-35(a) shows a knotted rope with 12 equally-spaced knots. Figure 13-35(b) shows how the rope might be held to form a triangle with sides of lengths 3, 4, and 5. The triangle formed is a right triangle and contains a 90° angle.

FIGURE 13-35

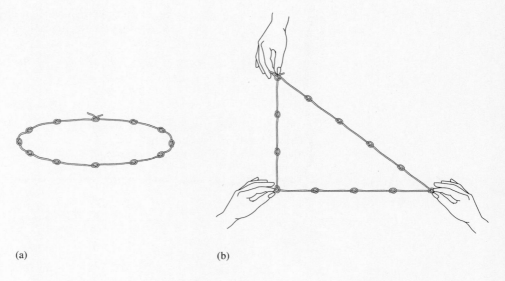

(a)   (b)

Given a triangle with sides of lengths a, b, and c such that $a^2 + b^2 = c^2$, must the triangle be a right triangle? This is the case, and we state the following without proof.

THEOREM 13-2

Converse of the Pythagorean theorem: If $\triangle ABC$ is a triangle with sides of lengths a, b, and c such that $a^2 + b^2 = c^2$, then $\triangle ABC$ is a right triangle with the right angle opposite the side of length c.

EXAMPLE 13-14 Determine whether the following can be the lengths of the sides of a right triangle:

(a) 51, 68, 85   (b) 2, 3, $\sqrt{13}$ (c) 3, 4, 7

Solution (a) $51^2 + 68^2 = 7225 = 85^2$, so 51, 68, and 85 can be the lengths of the sides of a right triangle.
(b) $2^2 + 3^2 = 4 + 9 = 13 = (\sqrt{13})^2$, so 2, 3, and $\sqrt{13}$ can be the lengths of the sides of a right triangle.
(c) $3^2 + 4^2 \neq 7^2$, so the measures cannot be the lengths of the sides of a right triangle. ∎

PROBLEM SET 13-3

1. Use the Pythagorean theorem to find x in each of the following:

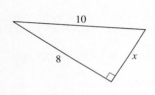
(a)

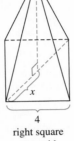
square
(b)

right square pyramid
(i)

right circular cone
(j)

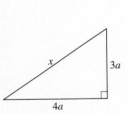
(c)

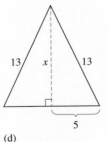
(d)

cube
(k)

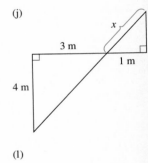
(l)

equilateral triangle
(e)

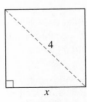
square
(f)

2. If the hypotenuse of a right triangle is 30 cm long and one of the legs is twice as long as the other, how long are the legs of the triangle?

3. For each of the following, can the given numbers represent lengths of sides of a right triangle?
  (a) 10, 24, 16
  (b) 16, 34, 30
  (c) $\sqrt{2}, \sqrt{2}, 2$
  (d) 2, $\sqrt{3}$, 1
  (e) $\sqrt{2}, \sqrt{3}, \sqrt{5}$
  (f) $\dfrac{3}{2}, \dfrac{4}{2}, \dfrac{5}{2}$

4. What is the longest line segment that can be drawn in a right rectangular prism that is 12 cm wide, 15 cm long, and 9 cm high?

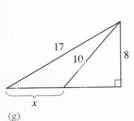
(g)

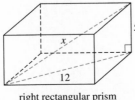
right rectangular prism
(h)

5. For each of the following, solve for the unknowns:

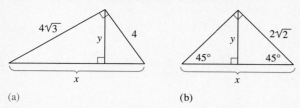

(a)   (b)

6. Two cars leave a house at the same time. One car travels 60 km/hr north, while the other car travels 40 km/hr east. After 1 hr, how far apart are the cars?

7. Two airplanes depart from the same place at 2:00 P.M. One plane flies south at a speed of 376 km/hr, and the other flies west at a speed of 648 km/hr. How far apart are the airplanes at 5:30 P.M.?

8. Starting from point A, a boat sails due south for 6 mi, then due east for 5 mi, and then due south for 4 mi. How far is the boat from A?

9. ▶If the hypotenuse and a leg of one right triangle are congruent to the hypotenuse and a leg of another right triangle, respectively, must the triangles be congruent? Explain your answer.

10. A 15-ft ladder is leaning against a wall. The base of the ladder is 3 ft from the wall. How high above the ground is the top of the ladder?

11. Find the area of a regular hexagon inscribed in a circle whose radius is as follows:
    (a) 5 cm
    (b) r cm

12. In the following figure, two poles are 25 m and 15 m high. A cable 14 m long joins the tops of the poles. Find the distance between the poles.

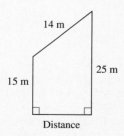

13. Find the area of each of the following:

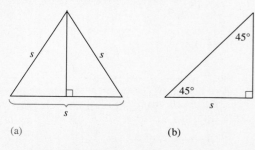

(a)   (b)

14. A builder needs to calculate the dimensions of a regular hexagonal window. Assuming the altitude CD of the window is 1.3 m high, find the width AB (O is the midpoint of $\overline{AB}$) in the following figure:

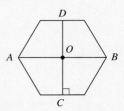

15. The length of the diagonal $\overline{AC}$ of a rhombus $ABCD$ is 20 cm. The distance between $\overline{AB}$ and $\overline{DC}$ is 12 cm. Find the length of the sides of the rhombus and the length of the other diagonal.

16. ▶If the hypotenuse in a 30°-60°-90° triangle is $c/2$ units long, what is the length of the side opposite the 60° angle? Explain your answer.

17. If the length of the hypotenuse in a 45°-45°-90° triangle is c, find the length of a leg.

18. ▶Given the following square, describe how to use a compass and a straightedge to construct a square whose area is as specified:
    (a) Twice the area of the given square
    (b) Half the area of the given square

19. If possible, draw a square with the given number of square units on a geoboard grid. (You will have to draw your own geoboard grid.)
    (a) 5   (b) 7   (c) 8   (d) 14   (e) 15

20. Use the following drawing to prove the Pythagorean theorem by using corresponding parts of similar triangles $\triangle ACD$, $\triangle CBD$, and $\triangle ABC$. Lengths of sides are indicated by a, b, c, x, and y. (*Hint:* Show that $b^2 = cx$ and $a^2 = cy$.)

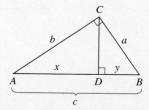

21. Before he was elected president of the United States, James Garfield discovered a proof of the Pythagorean theorem. He formed a trapezoid like the one that follows and found the area of the trapezoid in two different ways. Can you discover his proof?

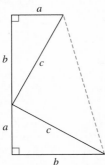

22. Construct semicircles on right triangle ABC with $\overline{AB}$, $\overline{BC}$, and $\overline{AC}$ as diameters. Is the area of the semicircle on the hypotenuse equal to the sum of the areas of the semicircles on the legs?

23. ▶On each side of a right triangle, construct an equilateral triangle.
    (a) Is the area of the triangle constructed on the hypotenuse always equal to the sum of the areas of the triangles constructed on the legs?
    (b) Justify your answer.

24. Use the given figure to prove the Pythagorean theorem by first proving that the quadrilateral with side c is a square; then, computing the area of the square with side $a + b$ in two different ways: (a) as $(a + b)^2$; and (b) as the sum of the areas of the four triangles and the square with side c.

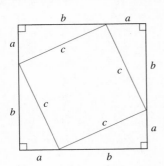

25. ▶What does it mean if a sign marking an incline on a highway says 6% grade?

26. If $\overline{AB}$, a diameter of circle O, has length 10 cm, point C is on circle O, and AC is 6 cm, how long is $\overline{BC}$?

27. Georgette wants to put a diagonal brace on a gate that is 3 ft wide and 5 ft high. If she uses a board that is 6 in. wide and 8 ft long, how much will she have left?

28. If a third baseman on the base throws to first base, how far is the ball thrown? (*Hint:* The distance from home plate to first base is 90 ft.)

29. What is the longest piece of straight spaghetti that will fit in a cylindrical can that has a radius of 2 in. and height of 10 in.?

Review Problems

30. Arrange the following in decreasing order: 3.2 m, 322 cm, 0.032 km, 3.020 mm.

31. Find the area of each figure.

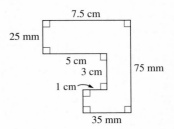
(a)

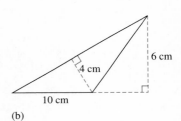
(b)

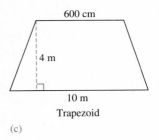
(c)

32. Complete the following table concerning circles:

	Radius	Diameter	Circumference	Area
(a)	5 cm			
(b)		24 cm		
(c)				17π m²
(d)			20π cm	

33. A wire 10 m long is wrapped around a circular region. If the wire fits exactly, what is the area of the region?

BRAIN TEASER

A spider sitting at A, the midpoint of the edge of the ceiling in the room shown in the following figure, spies a fly on the floor at C, the midpoint of the edge of the floor. If the spider must walk along the wall, ceiling, or floor, what is the length of the shortest path the spider can travel to reach the fly?

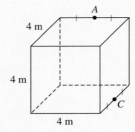

LABORATORY ACTIVITY

As a van Hiele Level 1 activity, consider the following drawings. Without using formulas, check whether the area of the square constructed on the hypotenuse of the shaded right triangle equals the sum of the areas of the squares constructed on the legs of the triangle.

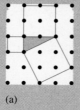 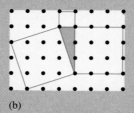 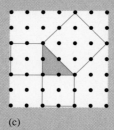

(a)   (b)   (c)

Section 13-4 Applications Involving Areas

Painting houses, buying roofing, seal-coating driveways, and buying carpet are among the common applications that involve computing areas. In many real-world problems, we must find the surface areas of such three-dimensional figures as prisms, cylinders, pyramids, cones, and spheres. Formulas for finding these areas are usually based on finding the area of two-dimensional pieces of the three-dimensional figures. In this section, we use the notion of **nets,** two-dimensional patterns that can be used to construct three-dimensional figures, to aid in determining surface areas of the figures.

Surface Area of Right Prisms

Consider the cereal box from the student page from *Addison-Wesley Mathematics*, Grade 8, 1991. To find the amount of cardboard necessary to make the box, the box is cut along the edges and made to lie flat. We obtain a net and see that the box is composed of a series of rectangles. We find the area of each rectangle and sum those areas to find the surface area of the box.

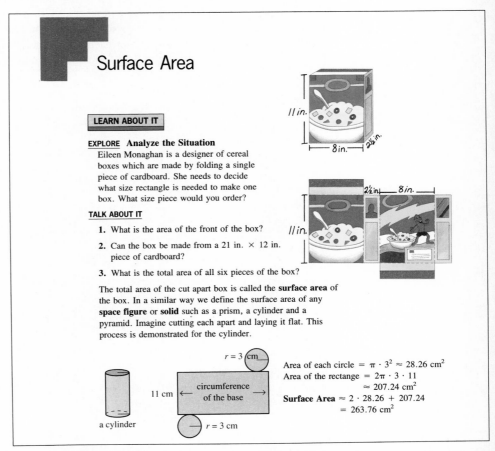

A similar process can be used for many three-dimensional figures. For example, cubes are the simplest polyhedra. The surface area of the cube in Figure 13-36(a) is the sum of the areas of the faces of the cube. Because each of the six faces is a square of area 16 cm^2, the surface area is $6 \cdot (16 \text{ cm}^2)$, or 96 cm^2, or in general, for a cube whose edge is e units, the surface area is $6e^2$.

FIGURE 13-36

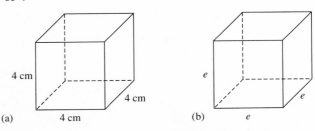

726   CHAPTER 13   CONCEPTS OF MEASUREMENT

lateral surface area / surface area

To find the surface area of right prisms, we find the sum of the areas of the rectangles that constitute the lateral faces and the areas of the top and bottom. The sum of the areas of the lateral faces is called the **lateral surface area.** Thus the **surface area** is the sum of the lateral surface area and the area of the bases.

Figure 13-37(a) shows a right pentagonal prism, and Figure 13-37(b) shows a net for the figure. The section formed by the lateral faces is stretched out flat. It forms a rectangle whose length is $s_1 + s_2 + s_3 + s_4 + s_5$ and whose width is h. Because $s_1 + s_2 + s_3 + s_4 + s_5$ is the perimeter p of the base of the prism, the lateral surface area is $(s_1 + s_2 + s_3 + s_4 + s_5) \cdot h$, or ph. If B stands for the area of each of the prism's bases, then the surface area, S.A., of the right prism is given by the following formula:

$$S.A. = ph + 2B.$$

This formula holds for any right prism, regardless of the shape of its bases.

FIGURE 13-37

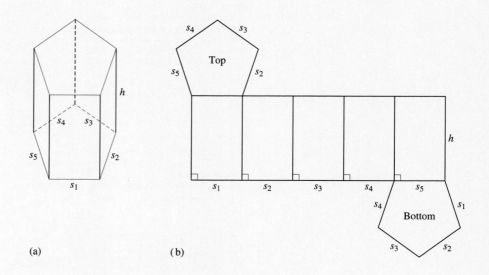

(a)   (b)

EXAMPLE 13-15   Find the surface area of each of the right prisms in Figure 13-38.

FIGURE 13-38

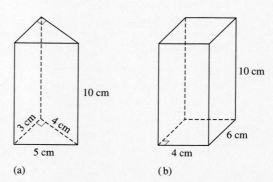

(a)   (b)

Solution (a) Each base is a right triangle. The area of the bases is $2\left(\frac{1}{2} \cdot 3 \text{ cm} \cdot 4 \text{ cm}\right)$, or 12 cm². The perimeter of a base is 3 cm + 4 cm + 5 cm, or 12 cm. Hence, the lateral surface area is (12 cm)(10 cm), or 120 cm², and the surface area is 120 cm² + 12 cm², or 132 cm².

(b) The area of the bases is 2(4 cm)(6 cm), or 48 cm². The lateral surface area is 2(4 cm + 6 cm) · 10 cm, or 200 cm², so the surface area of the right prism is 248 cm². ∎

Surface Area of a Cylinder

To find the surface area of the right circular cylinder shown in Figure 13-39, we cut off the bases and slice the lateral surface open by cutting along any line perpendicular to the bases. Such a slice is shown as a dotted segment in Figure 13-39(a); then we unroll the cylinder to form a rectangle, as shown in Figure 13-39(b). To find the total surface area, we find the area of the rectangle and the areas of the top and bottom circles. The length of the rectangle is the circumference of the circular base $2\pi r$ and its width is the height of the cylinder h. Hence, the area of the rectangle is $2\pi rh$. The area of each base is πr^2. Because the surface area is the sum of the areas of the two circular bases and the lateral surface area, we have

$$S.A. = 2\pi r^2 + 2\pi rh.$$

FIGURE 13-39

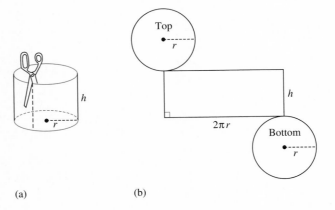

(a)   (b)

Surface Area of a Pyramid

The surface area of a pyramid is the sum of the lateral surface area of the pyramid and the area of the base. A *right regular pyramid* is a pyramid such that the segments connecting the apex to each vertex of the base are congruent and the base is a regular polygon. The lateral faces of the right regular pyramid pictured in Figure 13-40 are congruent triangles. Each triangle has an altitude of length ℓ called the *slant height*. Because the pyramid is right regular, each side of the base has the same length b. Thus, to find the lateral surface area of a right regular pyramid, we need to find the area of one face $\frac{1}{2}b\ell$ and multiply it by n, the number of faces. Adding the lateral surface area $n\left(\frac{1}{2}b\ell\right)$ to the area of the base B gives the surface area. Because nb is the perimeter of the base, the formula reduces to the following:

$$S.A. = B + \frac{1}{2}p\ell.$$

FIGURE 13-40

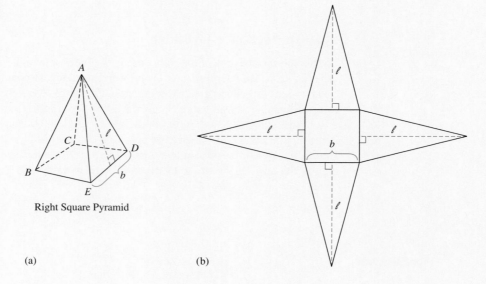

(a) Right Square Pyramid

(b)

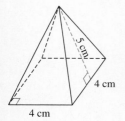

FIGURE 13-41

EXAMPLE 13-16 Find the surface area of the right regular pyramid in Figure 13-41.

Solution The surface area consists of the area of the square base plus the area of the four triangular faces. Hence, the surface area is

$$4 \text{ cm} \cdot 4 \text{ cm} + 4 \cdot \left(\frac{1}{2} \cdot 4 \text{ cm} \cdot 5 \text{ cm}\right) = 16 \text{ cm}^2 + 40 \text{ cm}^2$$
$$= 56 \text{ cm}^2.$$

Surface Area of a Cone

It is possible to find a formula for the surface area of a cone by approximating the cone with a pyramid. As shown in Figure 13-42, we inscribe in the circular base of the cone a regular polygon with many sides. The polygon can be used as the base of a regular right pyramid. The lateral surface area of the pyramid is close to the lateral surface area of the cone; and the greater the number of faces of the pyramid, the closer the surface area of the pyramid is to that of the cone. We know that the lateral surface of the pyramid is $\frac{1}{2}p \cdot h$, where p is the perimeter of the base and h is the height of each triangle. With many sides in the pyramid, the perimeter of its base is close to the perimeter of the circle, $2\pi r$. The height of each triangle of the pyramid is close to the slant height ℓ, a segment connecting the vertex of the cone with a point on the circular base, as shown in Figure 13-42(b). Consequently, it is reasonable that the lateral surface of the cone becomes $\frac{1}{2} \cdot 2\pi r \cdot \ell$, or $\pi r \ell$. To find the total surface area of the cone, we add πr^2, the area of the base. Thus $S.A. = \pi r^2 + \pi r \ell$.

FIGURE 13-42

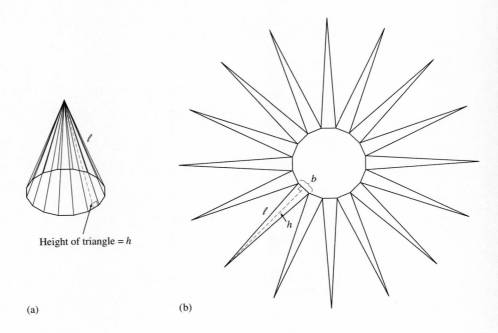

(a) Height of triangle = h

(b)

PROBLEM 2

If a right cirular cone as in Figure 13-43(a) is cut apart as in Figure 13-43(b) to form a sector of a circle and a circular base as in Figure 13-43(c), determine the angle, θ, of the sector in terms of the radius, r, of the base and the slant height, ℓ, of the cone.

FIGURE 13-43

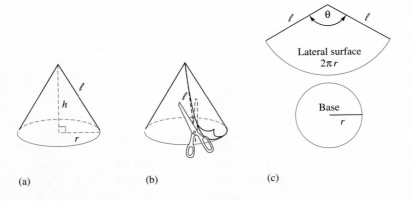

(a) (b) (c)

Understanding the Problem. It may not be apparent that a cone can be separated into the circular base and a sector as in Figure 13-43(c). (Readers should try this for themselves.) We are to find the measure of the angle θ forming the sector of the circle in terms of r and ℓ.

Devising a Plan. We have *solved related problems* when we found the arc length of a sector in Section 13-1 to be $\dfrac{\theta}{360}(\pi r)$ given the radius, r, of the sector and when we found the circumference of a circle of radius r to be $2\pi r$. To find θ in Figure 13-43(c), we use the strategy of *writing an equation* by considering the arc length of the circle in two different ways. We set these lengths equal and solve for θ.

Carrying Out the Plan. The length of the arc of the sector can be written in terms of θ and ℓ as $\left(\dfrac{\theta}{360}\right)(2\pi\ell)$. (Why?) Also, the length of the arc is $2\pi r$ because it is the circumference of the circular base of the cone. Hence,

$$\frac{\theta}{360}(2\pi\ell) = 2\pi r$$
$$\theta(2\pi\ell) = 720\pi r$$
$$\theta = \frac{360r}{\ell}.$$

Consequently, the measure of the angle forming the sector of a circle that becomes the lateral surface of a cone can be written as $\dfrac{360r}{\ell}$.

Looking Back. We can use the above information to find the surface area of a cone in a different way. The area of the sector is found by determining what proportion of a circle the sector is. If the angle determining a sector is $\dfrac{360r}{\ell}$, then that sector has area $\left(\dfrac{\frac{360r}{\ell}}{360}\right)(\pi\ell^2)$, or $\pi r\ell$. Thus the surface area of the cone is $\pi r\ell$ plus the area of the base, πr^2; that is, $S.A. = \pi r^2 + \pi r\ell$.

EXAMPLE 13-17 Given the cone in Figure 13-44, find the surface area of that cone.

Solution The base of the cone is a circle with radius 3 cm and area $\pi(3 \text{ cm})^2$, or 9π cm². The lateral surface has area $\pi(3 \text{ cm})(5 \text{ cm})$, or 15π cm². Thus we have the following surface area:

$$S.A. = \pi(3 \text{ cm})^2 + \pi(3 \text{ cm})(5 \text{ cm})$$
$$= 9\pi \text{ cm}^2 + 15\pi \text{ cm}^2$$
$$= 24\pi \text{ cm}^2$$

Right circular cone

FIGURE 13-44

Surface Area of a Sphere

Finding a formula for the surface area of a sphere is a simple task using calculus, but it is not easy in elementary mathematics. The area of a sphere is four times the area of a great circle of the sphere. Therefore the formula is $S.A. = 4\pi r^2$, as pictured in Figure 13-45.

FIGURE 13-45

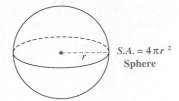

LABORATORY ACTIVITY

Create different cones from sectors of a circle. Use a compass to draw a sector of a circle whose diameter is almost as large as the width of a page of paper. Draw two such sectors with the same radii but with different central angles. In one sector, make the central angle smaller than 180°, and in the other, make it greater than 180°. Then make a cone from each sector by gluing the edges of each sector together. Can you predict which cone will be taller? Without performing the experiment, can you explain which cone will be taller?

PROBLEM SET 13-4

1. Find the surface area of each of the following:

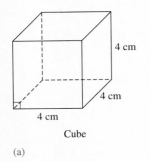
Cube
(a)

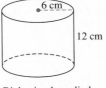
Right circular cylinder
(b)

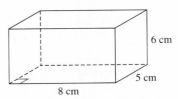
Right rectangular prism
(c)

Sphere
(d)

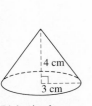
Right circular cone
(e)

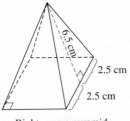
Right square pyramid
(f)

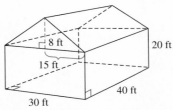
(g)

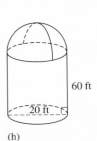
(h)

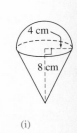
(i)

2. How many liters of paint are needed to paint the walls of a room that is 6 m long, 4 m wide, and 2.5 m tall if 1 L of paint covers 20 m²? (Assume there are no doors or windows.)

3. The napkin ring pictured is to be resilvered. How many square millimeters of surface area must be covered?

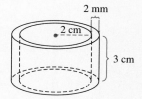

4. Assume the radius of Earth is 6370 km and Earth is a sphere. What is its surface area?

5. Two cubes have sides of length 4 cm and 6 cm, respectively. What is the ratio of their surface areas?

6. Suppose one cylinder has radius 2 m and height 6 m and another has radius 6 m and height 2 m.
   (a) Which cylinder has the greater lateral surface area?
   (b) Which cylinder has the greater total surface area?

7. The base of a right pyramid is a regular hexagon with sides of length 12 m. The altitude of the pyramid is 9 m. Find the total surface area of the pyramid.

8. ▶(a) What happens to the surface area of a cube if the length of each edge is tripled? Explain your answer.
   (b) What is the effect on the lateral surface area of a cylinder if the height is doubled? Explain your answer.

9. How does the surface area of a box (including top and bottom) change if
   (a) Each dimension is doubled?
   (b) Each dimension is tripled?
   (c) Each dimension is multiplied by a factor of k?

10. How does the lateral surface area of a cone change if
    (a) The slant height is tripled but the radius of the base remains the same?
    (b) The radius of the base is tripled but the slant height remains the same?
    (c) The slant height and the radius of the base are tripled?

11. ▶What happens to the surface area of a sphere if the radius is
    (a) Doubled?
    (b) Tripled?

12. Suppose a structure is composed of cubes with at least one face of each cube connected to the face of another cube, as shown below.

   (a) If one cube is added, what is the maximum surface area the structure can have?
   (b) What is the minimum surface area the structure can have?
   ▶(c) Is it possible to design a structure so that one can add a cube and add nothing to the surface area of the structure? (*Hint:* Cubes might have to be glued together.) Explain your answer.
   (d) Given 12 cubes, sketch the structure with the least surface area. What is the surface area of this structure?

13. ▶Design a net for a polyhedron in such a way that the surface area of the polyhedron is 10 cm². Explain what polyhedron the net will form and why its surface area is 10 cm².

14. ▶If a square pyramid is formed with the same base and height as a cube, what is the range of possibilities for the surface area of the pyramid?

15. ▶Describe how to find the surface area of a doughnut.

16. The sector shown is rolled into a cone so that the dotted edges just touch. Find the following:
    (a) The lateral surface area of the cone
    (b) The total surface area of the cone

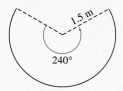

17. The sector shown is rolled into a cone so that the dotted edges just touch. If the area of the sector is 270 cm², find the following:
    (a) The area of the base of the cone
    (b) The height of the cone

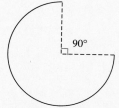

18. ▶A sphere is inscribed in a right cylinder, as shown. Compare the surface area of the sphere with the lateral surface area of the cylinder.

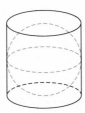

19. Karen is using a lawn mower to cut a rectangular field of grass. She follows the boundary of the field and finds that it takes her 12 rounds to cut one-half of the field and an additional 15 rounds to cut the remainder. If the lawn mower cuts a swath 3 ft wide, what are the dimensions of the field?

20. Find the surface area of a square pyramid if the area of the base is 100 cm² and the height of the pyramid is 20 cm.

21. Each of the regions shown revolves about the indicated axis. In each case sketch the three-dimensional figure obtained and find its surface area.

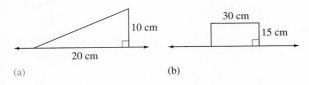

(a) (b)

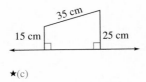

★(c)

22. 🖩 The total surface area of a cube is 10,648 cm³. What is the length of each of the following?
    (a) One of the sides
    (b) A diagonal that is not a diagonal of a face
★23. The surface area of a regular tetrahedron is 400 cm². Find the length of the side of the tetrahedron.
★24. Find the total surface area of the following stand, which was cut from a right circular cone:

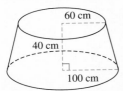

★25. A cylinder is inscribed in a cone, as shown. Find the lateral surface area of the cylinder if the height of the cone is 40 cm, the height of the cylinder 30 cm, and the radius of the base of the cone is 25 cm.

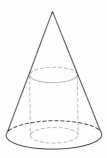

Review Problems

26. Complete each of the following:
    (a) $10 \text{ m}^2 = $ _____ cm^2
    (b) $13{,}680 \text{ cm}^2 = $ _____ m^2
    (c) $5 \text{ cm}^2 = $ _____ mm^2
    (d) $2 \text{ km}^2 = $ _____ m^2
    (e) $10^6 \text{ m}^2 = $ _____ km^2
    (f) $10^{12} \text{ mm}^2 = $ _____ m^2

27. The sides of a rectangle are 10 cm and 20 cm. Find the length of a diagonal of the rectangle.

28. The length of the side of a rhombus is 30 cm. If the length of one of the diagonals is 40 cm, find the length of the other diagonal.

29. Find the perimeters and the areas of the following figures:

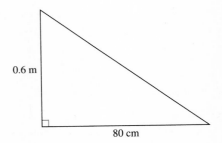

(a)

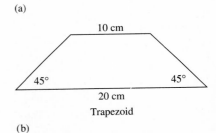

(b) Trapezoid

30. In the following figure, the length of the longer diagonal $\overline{AC}$ of rhombus $ABCD$ is 40 cm; $AE = 24$ cm. Find the length of a side of the rhombus and the length of the other diagonal.

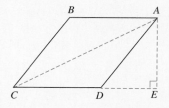

▼ BRAIN TEASER

A manufacturer of paper cups wants to produce paper cups in the form of truncated cones 16 cm high, with one circular base of radius 11 cm and the other of radius 7 cm, as shown. When the base of such a cup is removed and the cup is slit and flattened, the flattened region looks like a part of a circular ring. To design a pattern to make the cup, the manufacturer needs the data required to construct the flattened region. Find these data.

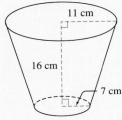

LABORATORY ACTIVITY

1. As a van Hiele Level 1 activity, use four cubes of the same size to build shapes such as the ones shown in the following figure. In each case, find the surface area using a square face of the cube as a unit of area.

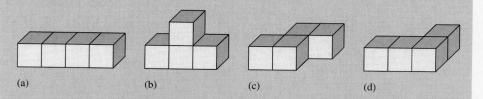

(a)　　　(b)　　　(c)　　　(d)

2. As a van Hiele Level 2 activity, consider the various figures that can be built by using four congruent cubes in such a way that any two cubes in your figure (a) have no points in common, (b) have an edge in common, or (c) have a whole face in common.
   (i) What figure will have the greatest surface area?
   (ii) What figure will have the least surface area?

Section 13-5 Volume Measure and Volumes

Metric Measure of Volume

Volume describes how much space a three-dimensional figure occupies. The unit of measure for volume must be a shape that will tessellate space. Cubes can tessellate space, that is, they can be closely stacked so that they leave no gaps and fill space. Standard units of volume are based on cubes and are called *cubic units*. The volume of a rectangular right prism can be measured by determining how many cubes are needed to build it. To find the volume, count how many cubes cover the base and then how many layers of these cubes are used to fill the prism, as shown in Figure 13-46(a). There are $8 \cdot 4$, or 32, cubes required to cover the base and there are five such layers. Hence, the volume of the rectangular prism is $8 \cdot 4 \cdot 5$, or 160 cubic units. For any rectangular right prism with dimensions ℓ, w, and h measured in the same linear units, the volume of the prism is given by the area of the base, ℓw, times the height, h, or $V = \ell w h$, as shown in Figure 13-46(b).

FIGURE 13-46

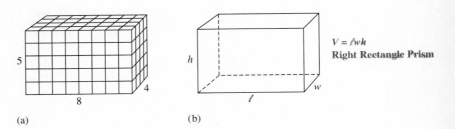

(a)   (b)

cubic centimeter
cubic meter

The most commonly used metric units of volume are the **cubic centimeter** and the **cubic meter.** A cubic centimeter is the volume of a cube whose length, width, and height are each one centimeter. One cubic centimeter is denoted by 1 cm³. Similarly, a cubic meter is the volume of a cube whose length, width, and height are each 1 m. One cubic meter is denoted by 1 m³. Other metric units of volume are symbolized similarly.

Figure 13-47 shows that since 1 dm = 10 cm, 1 dm³ = (10 cm) · (10 cm) · (10 cm) = 1000 cm³. Figure 13-48 shows that 1 m³ = 1,000,000 cm³ and that 1 dm³ = 0.001 m³.

FIGURE 13-47

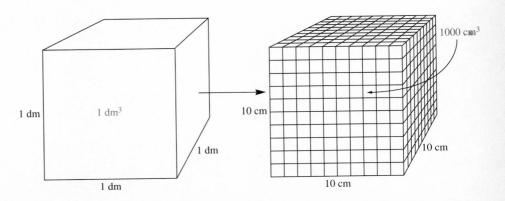

FIGURE 13-48

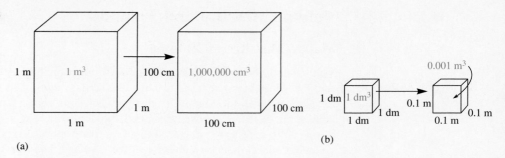

(a)   (b)

Each metric unit of length is 10 *times as great as the next smaller unit. Each metric unit of area is* 100 *times as great as the next smaller unit. Each metric unit of volume is* 1000 *times as great as the next smaller unit.* For example, because 1 cm = 0.01 m, then 1 cm^3 = (0.01 × 0.01 × 0.01) m^3, or 0.000001 m^3. Thus to convert from cubic centimeters to cubic meters, all that is required is to move the decimal point six places to the left.

EXAMPLE 13-18   Convert each of the following:

(a) 5 m^3 = _____ cm^3 (b) 12,300 mm^3 = _____ cm^3

Solution   (a) 1 m = 100 cm, so 1 m^3 = (100 cm)(100 cm)(100 cm), or 1,000,000 cm^3. Thus 5 m^3 = (5)(1,000,000 cm^3) = 5,000,000 cm^3.

(b) 1 mm = 0.1 cm, so 1 mm^3 = (0.1 cm)(0.1 cm)(0.1 cm), or 0.001 cm^3. Thus 12,300 mm^3 = 12,300(0.001 cm^3) = 12.3 cm^3. ∎

In the metric system, cubic units may be used for either dry or liquid measure, although units such as liters and milliliters are usually used for liquid measures. By definition, a **liter,** symbolized by L, equals, or is the capacity of, a cubic decimeter; that is, 1 L = 1 dm^3. (In the United States, L is the symbol for liter, but this is not universally accepted.)

liter

Because 1 L = 1 dm^3 and 1 dm^3 = 1000 cm^3, it follows that 1 L = 1000 cm^3 and 1 cm^3 = 0.001 L. Prefixes can be used with all base units in the metric system, so 0.001 L = 1 milliliter = 1 mL. Hence, 1 cm^3 = 1 mL. These relationships are summarized in Figure 13-49.

FIGURE 13-49

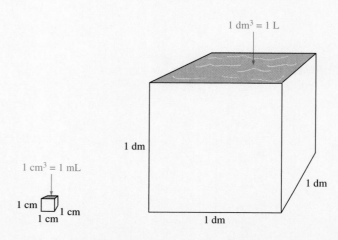

SECTION 13-5 VOLUME MEASURE AND VOLUMES

Table 13-7 shows how metric units involving the liter are related.

TABLE 13-7

Unit	Symbol	Relation to Liter
kiloliter	kL	1000 liters
*hectoliter	hL	100 liters
*dekaliter	daL	10 liters
liter	**L**	**1 liter**
*deciliter	dL	0.1 liter
centiliter	cL	0.01 liter
milliliter	mL	0.001 liter

*Not commonly used.

EXAMPLE 13-19 Convert each of the following as indicated:

(a) 27 L = _____ mL   (b) 362 mL = _____ L
(c) 3 mL = _____ cm^3 (d) 3 m^3 = _____ L

Solution (a) 1 L = 1000 mL, so 27 L = 27 · 1000 mL = 27,000 mL.
(b) 1 mL = 0.001 L, so 362 mL = 362(0.001 L) = 0.362 L.
(c) 1 mL = 1 cm^3, so 3 mL = 3 cm^3.
(d) 1 m^3 = 1000 dm^3 and 1 dm^3 = 1 L, so 1 m^3 = 100 L and 3 m^3 = 3000 L. ■

English Measure of Volume

Basic units of volume in the English system are the cubic foot (1 ft^3), the cubic yard (1 yd^3), and the cubic inch (1 $in.^3$). In the United States, 1 gal = 231 $in.^3$, which is about 3.7853 L, and 1 qt = $\frac{1}{4}$ gal or 57.749 $in.^3$.

Relationships among the one-dimensional units enable us to convert from one unit of volume to another, as shown in the following example.

EXAMPLE 13-20 Convert each of the following, as indicated:

(a) 45 yd^3 = _____ ft^3
(b) 4320 $in.^3$ = _____ yd^3
(c) 10 gal = _____ ft^3
(d) 3 ft^3 = _____ yd^3

Solution (a) Because 1 yd^3 = $(3\ ft)^3$ = 27 ft^3, we have 45 yd^3 = 45 · 27 ft^3, or 1215 ft^3.

(b) Because 1 in. = $\frac{1}{36}$ yd, we have 1 $in.^3$ = $\left(\frac{1}{36}\right)^3$ yd^3. Consequently, 4320 $in.^3$ = 4320 · $\left(\frac{1}{36}\right)^3$ $yd^3 \doteq$ 0.0926 yd^3, or approximately 0.1 yd^3.

(c) Because 1 gal = 231 in.^{3} and 1 in.^{3} = $\left(\frac{1}{12}\right)^3$ ft^3, we have 10 gal = 2310 in.^{3} = $2310\left(\frac{1}{12}\right)^3$ ft$^3 \doteq 1.337$ ft^3, or approximately 1.3 ft^3.

(d) As seen in part (a), 1 ft^3 = $\frac{1}{27}$ yd^3. Hence 3 ft^3 = $3 \cdot \frac{1}{27}$ yd^3 = $\frac{1}{9}$ yd^3. ∎

Volume of Right Prisms and Right Cylinders

The volume of a right rectangular prism is the area of the base times the height of the prism. If we denote the area of the base by B and the height by h, then we have $V = Bh$ as the formula for the volume of any right prism. Similarly, the volume of a right cylinder is the area of the base πr^2 times the height of the cylinder, $V = \pi r^2 h$. The same type of formula holds for any three-dimensional figure *if* all cross sections parallel to the base are congruent to the base. These formulas can be intuitively justified by considering three-dimensional figures made of clay, as in Figure 13-50. For example, consider the clay prism in Figure 13-50(a). Because all the parallel cross sections of the prism are congruent, we should be able to deform the clay model into the right rectangular prism shown in Figure 13-50(b) without changing its altitude or the area of the base. Because the amount of material does not change, the volume of the rectangular prism should be the same as the volume of the original prism.

FIGURE 13-50

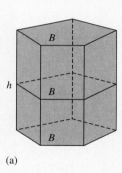
(a)

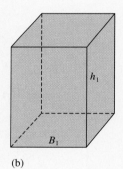
(b)

Cavalieri's Principle

Formulas for the volumes of many three-dimensional figures can be derived by using a principle based on a phenomenon described above. In Figure 13-51(a), a rectangular box

FIGURE 13-51

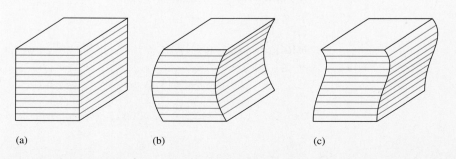

(a)   (b)   (c)

has been sliced into thin layers. If the layers are shifted to form the solids in Figure 13-51(b) and (c), the volume of each of the three solids is the same as the volume of the original rectangular box. The principle described is referred to as *Cavalieri's Principle*.

CAVALIERI'S PRINCIPLE

Two solids with bases in the same plane have equal volumes if every plane parallel to the bases intersects the solids in cross sections of equal area.

EXAMPLE 13-21 Find the volume of each of the figures in Figure 13-52.

FIGURE 13-52

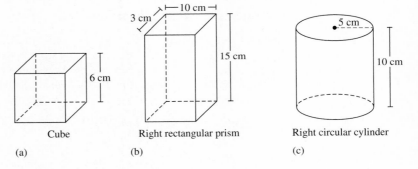

(a) Cube   (b) Right rectangular prism   (c) Right circular cylinder

Solution
(a) $V = Bh = (6 \text{ cm} \cdot 6 \text{ cm}) \cdot 6 \text{ cm} = 216 \text{ cm}^3$
(b) $V = Bh = (10 \text{ cm} \cdot 3 \text{ cm}) \cdot 15 \text{ cm} = 450 \text{ cm}^3$
(c) $V = \pi r^2 h = \pi (5 \text{ cm})^2 \cdot 10 \text{ cm} = 250\pi \text{ cm}^3$ ∎

HISTORICAL NOTE

Bonaventura Cavalieri (1598–1647), an Italian mathematician and disciple of Galileo, contributed to the development of geometry, trigonometry, and algebra in the Renaissance. He became a Jesuit at an early age and later, after reading Euclid's *Elements*, was inspired to study mathematics. In 1629, Cavalieri became a professor at Bologna and held that post until his death. Cavalieri is best known for his principle concerning the volumes of solids.

Using Cavalieri's Principle, it can be shown that two pyramids with the same height and the same base areas also have the same volume. Consequently, we have the following:

The *volume of any pyramid is* $\frac{1}{3} Bh$, where B is the area of the base and h is the height.

Figure 13-53 shows a right triangular prism that has an equilateral triangle as a base. This prism can be separated into three pyramids with bases of equal area and with heights the same as that of the prism. In this special case, the volume of the right triangular pyramid is one-third the volume of the triangular prism, that is, $V = \frac{1}{3}Bh$. This can be demonstrated by building three paper models of the pyramids and fitting them together into a prism.

FIGURE 13-53

Volume of a Right Circular Cone

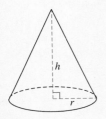

FIGURE 13-54

To find the volume of a right cone with a circular base, as shown in Figure 13-54, consider the polygonal base of a pyramid with many sides. The base approximates a circle, and the volume of the pyramid is approximately the volume of the cone with the circle as a base and the same height as the pyramid. The area of the base is approximately πr^2, where r is the apothem of the polygon. Hence, the formula for the volume of a right circular cone with a circular base is $V = \frac{1}{3}\pi r^2 h$ or $V = \frac{1}{3}Bh$.

EXAMPLE 13-22 Find the volume of each of the figures in Figure 13-55.

FIGURE 13-55

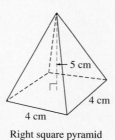

Right square pyramid
(a)

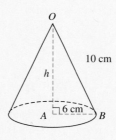

Right circular cone
(b)

Solution (a) The figure is a pyramid with a square base whose area is 4 cm · 4 cm and whose height is 5 cm. Hence, $V = \frac{1}{3}Bh = \frac{1}{3}(4\text{ cm} \cdot 4\text{ cm})(5\text{ cm}) = \frac{80}{3}\text{ cm}^3$.

(b) The base of the cone is a circle of radius 6 cm. Because the volume of

the cone is given by $V = \frac{1}{3}\pi r^2 h$, we need to know the height. In the right triangle OAB, $OA = h$ and by the Pythagorean theorem, $h^2 + 6^2 = 10^2$. Hence, $h^2 = 100 - 36$, or 64, and $h = 8$ cm. Thus
$$V = \frac{1}{3}\pi r^2 h = \frac{1}{3}\pi(6 \text{ cm})^2(8 \text{ cm}) = 96\pi \text{ cm}^3.$$ ∎

EXAMPLE 13-23 Figure 13-56 is a net for a pyramid. If each triangle is equilateral, find its volume.

FIGURE 13-56

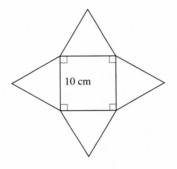

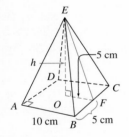

FIGURE 13-57

Solution The folded-up pyramid obtained is shown in Figure 13-57. The volume of the pyramid is $V = \frac{1}{3}Bh = \frac{1}{3} \cdot 10^2 h$. We must find h. Notice that h is a leg in the right triangle EOF, where F is the midpoint of CB. We know that $OF = 5$ cm. If we knew $\overline{EF}$, we could find h by applying the Pythagorean theorem to $\triangle EOF$. To find the length of $\overline{EF}$, notice that $\overline{EF}$ is a leg in the right triangle EBF. ($\overline{EF}$ is the perpendicular bisector of $\overline{BC}$ in the equilateral triangle BEC.) In the right triangle EBF, we have $(EB)^2 = (BF)^2 + (EF)^2$. Because $EB = 10$ cm and $BF = 5$ cm, it follows that $10^2 = 5^2 + (EF)^2$, or $EF = \sqrt{75}$ cm $\doteq 8.66$ cm. In $\triangle EOF$, we have $h^2 + 5^2 = (EF)^2$, or $h^2 + 25 = 75$. Thus $h = \sqrt{50}$ cm $\doteq 7.07$ cm, and $V = \frac{1}{3} \cdot 10^2 \cdot 7.07 \doteq 235.7$ cm^3. ∎

Volume of a Sphere

To find the volume of a sphere, imagine that a sphere is composed of a great number of congruent pyramids with apexes at the center of the sphere and that the vertices of the base touch the sphere, as shown in Figure 13-58. If the pyramids have very small bases, then the

FIGURE 13-58

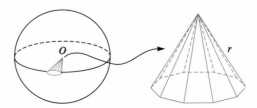

height of each pyramid is nearly the radius r. Hence, the volume of each pyramid is $\frac{1}{3}Bh$ or $\frac{1}{3}Br$, where B is the area of the base. If there are n pyramids each with base area B, then the total volume of the pyramids is $V = \frac{1}{3}nBr$. Because nB is the total surface area of all the bases of the pyramids and because the sum of the areas of all the bases of the pyramids is very close to the surface area of the sphere, $4\pi r^2$, the volume of the sphere is given by $V = \frac{1}{3}(4\pi r^2)r = \frac{4}{3}\pi r^3$.

EXAMPLE 13-24   Find the volume of a sphere whose radius is 6 cm.

Solution $V = \frac{4}{3}\pi(6 \text{ cm})^3 = \frac{4}{3}\pi(216 \text{ cm}^3) = 288\pi \text{ cm}^3$ ∎

PROBLEM 3

A manufacturer of metal cans has a large quantity of rectangular metal sheets 20 by 30 cm. Without cutting the sheets, the manager wants to make cylindrical pipes with circular cross sections from some of the sheets and box-shaped pipes with square cross sections from the other sheets so that the volume of the box-shaped pipes is greater than the volume of the cylindrical pipes. Is this possible? If so, how should the pipes be made and what are their volumes?

Understanding the Problem. We use 20-by-30-cm rectangular sheets of metal to make some cylindrical pipes and some box-shaped pipes with square cross sections that have a greater volume than the cylindrical pipes. Is this possible, and if so, how should the pipes be designed and what are their volumes?

Figure 13-59 shows a sheet of metal and two sections of pipe made from it, one cylindrical and the other box-shaped. A model for such pipes can be designed from a piece of paper by bending it into a cylinder or by folding it into a right rectangular prism, as shown.

FIGURE 13-59

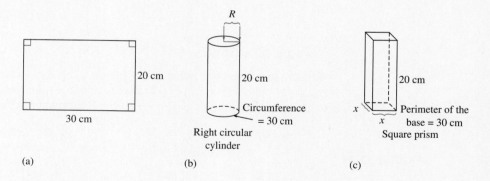

Devising a Plan. If we compute the volume of the cylinder in Figure 13-59(b) and the volume of the prism in Figure 13-59(c), we can determine which is greater. If the prism has a greater volume, the solution of the problem will be complete, otherwise, we look for other ways to design the pipes before concluding that a solution is impossible.

To compute the volume of the cylinder, we find the area of the base. The area of the circular base is πr^2. To find r, we note that the circumference of the circle $2\pi r$ is 30 cm. Thus $r = \dfrac{30}{2\pi} \doteq 4.77$ cm, and the area of the circle is $\pi r^2 = \pi(4.77)^2 \doteq 71.48$ cm^2. With the given information, we can also find the area of the base of the rectangular box.

Similarly, because the perimeter of the base of the prism is $4x$, we have $4x = 30$, or $x = 7.5$ cm. Thus the area of the square base is $x^2 = (7.5)^2$, or 56.25 cm^2.

Carrying Out the Plan. Denoting the volume of the cylindrical pipe by V_1 and the volume of the box-shaped pipe by V_2, we have $V_1 \doteq 71.48 \cdot 20$, or approximately 1429.6 cm^3. For the volume of the box-shaped pipe, we have $V_2 \doteq 56.25 \cdot 20$, or 1125 cm^2. We see that in the first design for the pipes, the volume of the cylindrical pipe is greater than the volume of the box-shaped pipe, which is not the required outcome.

Rather than bending the rectangular sheet of metal along the 30-cm side, we could bend it along the 20-cm side to obtain either pipe, as shown in Figure 13-60. Denoting the radius of the cylindrical pipe by r, the side of the box-shaped pipe by y, and their volumes by V_3 and V_4, respectively, we have $V_3 = \pi r^2 \cdot 30 = \pi(20/2\pi)^2 \cdot 30 = (10^2 \cdot 30)/\pi$, or approximately 954.9 cm^3. Also, $V_4 = y^2 \cdot 30 = \left(\dfrac{20}{4}\right)^2 \cdot 30 = 25 \cdot 30$, or 750 cm^3. Because $V_2 = 1125$ cm^3 and $V_3 = 945.9$ cm^3, we see that the volume of the box-shaped pipe with an altitude of 20 cm is greater than the volume of the cylindrical pipe with an altitude of 30 cm.

FIGURE 13-60

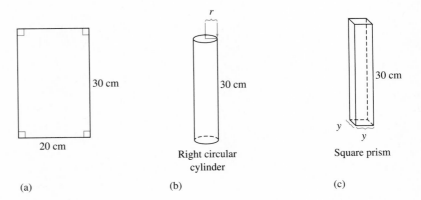

(a)   (b) Right circular cylinder   (c) Square prism

Looking Back. We could ask for the volumes of other three-dimensional objects that can be obtained by bending the rectangular sheets of metal. Also, because the lateral surface areas of the four types of pipes were the same but their volumes were different, we might want to investigate whether there are other cylinders and prisms that have the same lateral surface area and the same volume. Is it possible to find a circular cylinder with lateral surface area of 600 cm^2 and smallest possible volume? Similarly, is there a circular cylinder with the given surface area and greatest possible volume?

PROBLEM SET 13-5

1. Complete each of the following:
   (a) $8 \text{ m}^3 = $ _____ dm^3
   (b) $500 \text{ cm}^3 = $ _____ m^3
   (c) $675{,}000 \text{ m}^3 = $ _____ km^3
   (d) $3 \text{ m}^3 = $ _____ cm^3
   (e) $7000 \text{ mm}^3 = $ _____ cm^3
   (f) $0.002 \text{ m}^3 = $ _____ cm^3
   (g) $400 \text{ in.}^3 = $ _____ yd^3
   (h) $25 \text{ yd}^3 = $ _____ ft^3
   (i) $0.2 \text{ ft}^3 = $ _____ in.^3
   (j) $1200 \text{ in.}^3 = $ _____ ft^3

2. ▶ Why is a unit sphere, a sphere with radius 1 cm, not a "good" unit of volume measure, even for measuring the volume of another sphere?

3. Find the volume of each of the following:

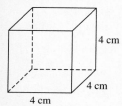
(a) Right rectangular prism

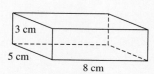
(b) Right rectangular prism

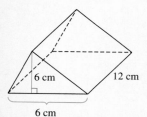
(c) Right triangular prism

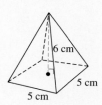
(d) Square pyramid

(e) Right circular cone

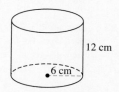
(f) Right circular cylinder

(g) Sphere

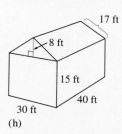
(h)

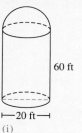
(i)

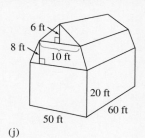
(j)

(k)

4. Complete the following chart:

	(a)	(b)	(c)	(d)	(e)	(f)
cm^3		500			750	4800
dm^3	2					
L			1.5			
mL				5000		

5. Place a decimal point in each of the following to make it an accurate sentence:
   (a) A paper cup holds about 2000 mL.
   (b) A regular soft drink bottle holds about 320 L.
   (c) A quart milk container holds about 10 L.
   (d) A teaspoonful of cough syrup is about 500 mL.

6. What volume of silver is needed to make the napkin ring in the following figure out of solid silver? Give your answer in cubic millimeters.

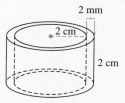

7. Two cubes have sides of lengths 4 cm and 6 cm, respectively. What is the ratio of their volumes?

8. ▶What happens to the volume of a sphere if the radius is doubled?
9. Complete the following chart for right rectangular prisms with the given dimensions:

	(a)	(b)	(c)	(d)
Length	20 cm	10 cm	2 dm	15 cm
Width	10 cm	2 dm	1 dm	2 dm
Height	10 cm	3 dm		
Volume (cm³)				
Volume (dm³)				7.5 dm³
Volume (L)			4 L	

10. A right cylindrical tank holds how many liters if it is 6 m long and 13 m in diameter?
11. If the length of Earth's diameter is approximately four times the length of the diameter of the moon and both bodies are spheres, what is the ratio of their volumes?
12. A bread pan is 18 cm × 18 cm × 5 cm. How many liters does it hold?
13. An Olympic-sized pool in the shape of a right rectangular prism is 50 m long and 25 m wide. If it is 2 m deep throughout, how many liters of water does it hold?
14. If a faucet is dripping at the rate of 15 drops/min and there are 20 drops/mL, how many liters of water are wasted in a 30-day month?
15. A standard straw is 25 cm long and 4 mm in diameter. How much liquid can be held in the straw at one time?
16. A cross section of a water pipe is a ring whose large circle has a radius of 14.5 cm and whose smaller circle has a radius of 5 cm.
    (a) Find the number of liters of water the pipe can hold if it is 10 m long.
    (b) How long should the pipe be if it is to hold 1,000,000 L of water?
17. ▶A theater decides to change the shape of its popcorn container from a regular box to a right regular pyramid and charge only half as much. If the containers, shown in the following figure, are the same height and the tops are the same size, is this a bargain for the customer?

18. ▶Which is the better buy, a grapefruit 5 cm in radius that costs 22¢ or a grapefruit 6 cm in radius that costs 31¢?

19. Two spherical cantaloupes of the same kind are sold at a fruit and vegetable stand. The circumference of one of the melons is 60 cm and that of the other is 50 cm. The larger melon is $1\frac{1}{2}$ times as expensive as the smaller. Which melon is the better buy and why?
20. A regular square pyramid is 3 m high and the perimeter of its base is 16 m. Find the volume of the pyramid.
21. A right rectangular prism with base $ABCD$ as the bottom is shown in the following figure. Suppose that X is drawn so that $AX = 3 \cdot AP$, where AP is the height of the prism and X is connected to A, B, C, and D, forming a pyramid. How do the volumes of the pyramid and the prism compare?

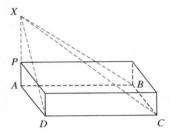

22. A right cylindrical can is to hold exactly 1 L of water. What should the height of the can be if the radius is 12 cm?
23. A box is packed with six pop cans, as shown. What percentage of the volume of the interior of the box is not occupied by the pop cans?

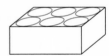

24. An engineer is to design a square-based pyramid whose volume is to be 100 m³.
    (a) Find the dimensions (the length of a side of the square and the altitude) of one such pyramid.
    (b) How many (noncongruent) such pyramids are possible? Why?
25. A square pyramid has a lateral surface area of 2 m². Its height equals the length of the side of the square base. Find the volume of the pyramid.
26. A square sheet of cardboard measuring y cm on a side is to be used to produce an open-top box when the maker cuts off a small square x cm by x cm from each corner and bends up the sides.
    (a) Find the volume of the box if $y = 200$ cm and $x = 20$ cm.
    ★(b) Assume that y is known; find the expression for the volume V as a function of x.

★27. If each edge and each side of the base of a right regular pyramid is increased or decreased by the same factor, the new right regular pyramid is said to be similar to the original one. Show that the ratio of the volumes of two square right pyramids equals the cube of the ratio of their corresponding heights.

28. ▶A grocer stacked oranges starting with 10 on the bottom, 6 on the next level, 3 on the third level, and 1 on the top to form a pyramidal shape. Assume each orange has a 3-in. diameter.
    (a) Find the volume of the oranges.
    (b) Find the approximate volume of the pyramidal shape. Explain your answer.

29. In the 1950s, 45 rpm records had 3.5-in. radii and a hole in the middle with a 1.5-in. diameter. If each record was $\frac{1}{16}$ in. thick, approximately what was the volume of a stack of 20 records?

30. ▶Explain how you would find the volume of an irregular shape.

31. Approximately what is the volume of a cord of wood?

32. If a scoop of ice cream is a perfect sphere with radius 5 cm, it fits exactly along one of its great circles in a sugar cone as shown below. Suppose the ice cream melts and the cone does not absorb any of the ice cream. If the cone is 10 cm tall, will it hold the melted ice cream? If not, how tall would the cone have to be to hold the melted ice cream?

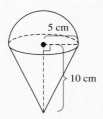

★33. Half of the air is let out of a spherical balloon. If the balloon remains in the shape of a sphere, how does the radius of the smaller balloon compare to the original radius?

Review Problems

34. Find the surface areas of the following figures:

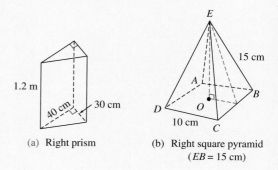

(a) Right prism   (b) Right square pyramid ($EB = 15$ cm)

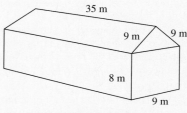

(c) Barn (include floor)

35. The diagonal of a rectangle has measure 1.3 m, and a side has measure 120 cm. Find each of the following:
    (a) Perimeter of the rectangle
    (b) Area of the rectangle

36. Find the area of a triangle that has sides of 3 m, 3 m, and 2 m.

37. A poster is to contain 0.25 m² of printed matter, with margins of 12 cm at top and bottom and 6 cm at each side. Find the width of the poster if its height is 74 cm.

LABORATORY ACTIVITY

1. As a van Hiele Level 1 activity, work with 30 congruent cubes. Build the cube with greatest volume possible by stacking the cubes. How many of the given cubes did you need to build your cube?

2. As a van Hiele Level 2 activity, use 24 congruent cubes to build solids with the following characteristics:
    (a) The greatest possible surface area
    (b) The least possible surface area

*Section 13-6   Mass and Temperature

Three centuries ago, Isaac Newton pointed out that in everyday life, *weight* is used for what is really mass. *Mass* is a quantity of matter as opposed to *weight,* which is a force exerted by gravitational pull. When an astronaut is in orbit above Earth, his weight has changed even though his mass remains the same. In common parlance on Earth, weight and mass are still used interchangeably. In the English system, weight is measured in such units as tons, pounds, and ounces. We use the avoirdupois units here. One pound (lb) equals 16 ounces (oz) and 2000 lb equals 1 English ton.

gram In the metric system, the base unit for mass is the **gram,** denoted by g. A gram is the mass of 1 cm³ of water. An ordinary paper clip or a thumbtack each has a mass of about 1 g.

As with other base metric units, prefixes are added to gram to obtain other units. For example, a kilogram (kg) is 1000 g. Because 1 cm³ of water has a mass of 1 g, the mass of 1 L of water is 1 kg. Two standard loaves of bread have a mass of about 1 kg. A person's mass also is measured in kilograms; a newborn baby has a mass of about 4 kg. Another unit of mass in the metric system is the metric ton (t), which is equal to 1000 kg. The metric ton is used to record the masses of objects such as cars and trucks. A small foreign car has a mass of about 1 t. Mega and micro are other prefixes used with the base unit.

Table 13-8 lists metric units of mass. Conversions involving metric units of mass are handled in the same way as conversions involving metric units of length.

TABLE 13-8

Unit	Symbol	Relationship to Gram
ton (metric)	t	1,000,000 grams
kilogram	kg	1000 grams
*hectogram	hg	100 grams
*dekagram	dag	10 grams
gram	**g**	**1 gram**
*decigram	dg	0.1 gram
*centigram	cg	0.01 gram
milligram	mg	0.001 gram

*Not commonly used.

EXAMPLE 13-25   Complete each of the following:

(a) 34 g = _____ kg   (b) 6836 kg = _____ t

Solution   (a) 34 g = 34(0.001 kg) = 0.034 kg
(b) 6836 kg = 6836(0.001 t) = 6.836 t   ∎

The relationship among the units of volume and mass in the metric system is illustrated in Figure 13-61.

FIGURE 13-61

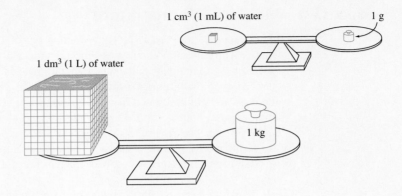

EXAMPLE 13-26 A waterbed is 180 cm wide, 210 cm long, and 20 cm thick.

(a) How many liters of water can it hold?

(b) What is its mass in kilograms when it is full of water?

Solution (a) The volume of the waterbed is found by multiplying the length times the width times the height.

$$V = \ell w h$$
$$= 180 \text{ cm} \cdot 210 \text{ cm} \cdot 20 \text{ cm}$$
$$= 756{,}000 \text{ cm}^3, \text{ or } 756{,}000 \text{ mL}$$

Because 1 mL = 0.001 L, the volume is 756 L.

(b) Because 1 L of water has a mass of 1 kg, 756 L of water has a mass of 756 kg.

> **REMARK** To see one advantage of the metric system, suppose that the bed is 6 ft by 7 ft by 9 in. Try to find the volume in gallons and the weight of the water in pounds.

Temperature

degree Kelvin

The base unit of temperature for the metric system, the **degree Kelvin,** is used only for scientific measurements and is an absolute temperature. The freezing point of water is 273° on this scale. For normal temperature measurements in the metric system, the base unit is

degree Celsius

the **degree Celsius,** named for Anders Celsius, the Swedish scientist who invented the system. The Celsius scale has 100 equal divisions between 0 degrees Celsius (0°C), the freezing point of water, and 100 degrees Celsius (100°C), the boiling point of water, as seen in Figure 13-62. In the English system, the Fahrenheit scale has 180 equal divisions between 32°F, the freezing point of water, and 212°F, the boiling point of water.

Figure 13-62 gives other temperature comparisons of the two scales and further illustrates the relationship between them. Because the Celsius scale has 100 divisions between the freezing point and the boiling point of water, whereas the Fahrenheit scale has 180 divisions, the relationship between the two scales is 100 to 180, or 5 to 9. For every 5 degrees on the Celsius scale, there are 9 degrees on the Fahrenheit scale, and for each

SECTION 13-6 MASS AND TEMPERATURE   749

FIGURE 13-62

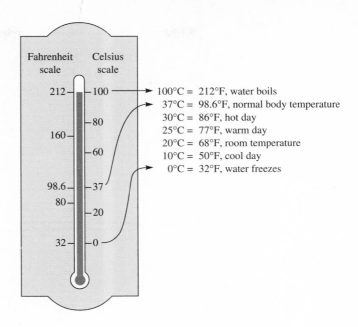

degree on the Fahrenheit scale, there is $\frac{5}{9}$ degree on the Celsius scale. Because the ratio between the number of degrees above freezing on the Celsius scale and the number of degrees above freezing on the Fahrenheit scale remains the same and equals $\frac{5}{9}$, we may convert temperatures from one system to the other. For example, suppose that we want to convert 50° on the Fahrenheit scale to the corresponding number on the Celsius scale. On the Fahrenheit scale, 50° is 50 − 32, or 18°, above freezing, but on the Celsius scale, it is $\frac{5}{9} \cdot 18$, or 10°, above freezing. Because the freezing temperature on the Celsius scale is 0°, 10° above freezing is 10° Celsius. Thus 50°F = 10°C. In general, F degrees is $F - 32$ above freezing on the Fahrenheit scale, but only $\frac{5}{9}(F - 32)$ above freezing on the Celsius scale. Thus we have the relation $C = \frac{5}{9}(F - 32)$. If we solve the equation for F, we obtain $F = \frac{9}{5}C + 32$. Rather than memorize these formulas, we encourage you to reason as shown to convert from one scale to the other.

EXAMPLE 13-27   Convert 20°C to degrees Fahrenheit.

Solution   For 100 divisions on the Celsius scale, we have 212 − 32, or 180, divisions on the Fahrenheit scale. Hence, for every 1 degree on the Celsius scale, there are $\frac{180}{100}$, or $\frac{9}{5}$, degrees on the Fahrenheit scale. Because 20°C is 20° above freezing, on the Fahrenheit scale it would be $\frac{9}{5} \cdot 20$, or 36, degrees above freezing, or 32 + 36, or 68 degrees. Thus 20°C = 68°F. ∎

> **HISTORICAL NOTE**
>
> Gabriel Daniel Fahrenheit (1686–1736) was a German physicist who contributed to the construction of improved thermometers and introduced the scale bearing his name. He lived most of his life in England and Holland and made a living manufacturing meteorological instruments.
>
> Anders Celsius (1701–1744) was a Swedish astronomer. He was a professor at the University of Uppsala. In 1742, he described the centigrade scale in a paper before the Swedish Academy of Sciences.

PROBLEM SET 13-6

1. For each of the following, select the appropriate metric unit of measure (gram, kilogram, or metric ton):
   (a) Car
   (b) Adult
   (c) Can of frozen orange juice
   (d) Elephant
   (e) Jar of mustard
   (f) Bag of peanuts
   (g) Army tank
   (h) Cat
   (i) Dictionary

2. For each of the following, choose the correct unit (milligram, gram, or kilogram) to make each sentence reasonable:
   (a) A staple has a mass of about 340 _____.
   (b) A professional football player has a mass of about 110 _____.
   (c) A vitamin tablet has a mass of 1100 _____.
   (d) A dime has a mass of 2 _____.
   (e) The recipe said to add 4 _____ of salt.
   (f) One strand of hair has a mass of 2 _____.

3. Complete each of the following:
   (a) 15,000 g = _____ kg
   (b) 8000 kg = _____ t
   (c) 0.036 kg = _____ g
   (d) 72 g = _____ kg
   (e) 4230 mg = _____ g
   (f) 3 g 7 mg = _____ g
   (g) 5 kg 750 g = _____ g
   (h) 5 kg 750 g = _____ kg
   (i) 0.03 t = _____ kg
   (j) 2.6 lb = _____ oz
   (k) 25 oz = _____ lb
   (l) 50 oz = _____ lb
   (m) 3.8 lb = _____ oz

4. If a paper dollar has a mass of approximately 1 g, is it possible to lift $1,000,000 in the following denominations?
   (a) $1 bills   (b) $10 bills   (c) $100 bills
   (d) $1000 bills   (e) $10,000 bills

5. A fish tank, which is a right rectangular prism, is 40-by-20-by-20 cm. If it is filled with water, what is the mass of the water?

6. In a grocery store, one kind of meat costs $5.80/kg. How much does 400 g of this meat cost?

7. If a certain spice costs $20/kg, how much does 1 g cost?

8. ▶Abel bought a kilogram of Moxwill coffee for $9 and Babel bought 400 g of the same brand of coffee for $4.60. Who made the better buy? Why?

9. Convert each of the following from degrees Fahrenheit to the nearest integer degree Celsius:
   (a) 10°F   (b) 0°F   (c) 30°F
   (d) 100°F   (e) 212°F   (f) ⁻40°F

10. Answer each of the following:
    (a) The thermometer reads 20°C. Can you go snow skiing?
    (b) The thermometer reads 26°C. Will the outdoor ice rink be open?
    (c) Your temperature is 37°C. Do you have a fever?
    (d) If your body temperature is 39°C, are you ill?
    (e) It is 40°C. Will you need a sweater at the outdoor concert?
    (f) The temperature reads 35°C. Should you go water skiing?
    (g) The temperature reads ⁻10°C. Is it appropriate to go ice fishing?
    (h) Your bath water is 16°C. Will you have a hot, warm, or chilly bath?
    (i) It's 30°C in the room. Are you comfortably hot or cold?

11. Convert each of the following from degrees Celsius to the nearest integer degree Fahrenheit:
    (a) 10°C   (b) 0°C   (c) 30°C
    (d) 100°C   (e) 212°C   (f) ⁻40°C

Review Problems

12. Find the perimeter and the area of the following figures:

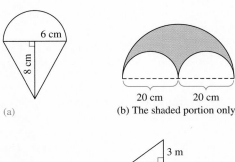

(a)

(b) The shaded portion only

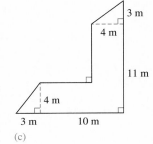

(c)

13. Complete the following:
    (a) 350 mm = _____ cm
    (b) 1600 cm² = _____ m²
    (c) 0.4 m² = _____ mm²
    (d) 5.2 m³ = _____ cm³
    (e) 5.2 m³ = _____ L
    (f) 3500 cm³ = _____ m³

14. Determine whether each of the following is a right triangle:

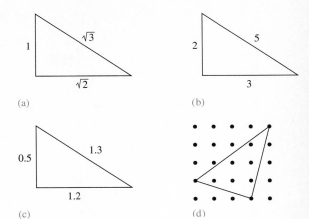

(a)

(b)

(c)

(d)

15. A person walks 5 km north, 3 km east, 1 km north, and then 2 km east. How far is the person from the starting point?

16. Find the volume and the surface area of each of the following solids:

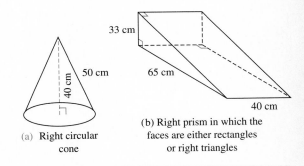

(a) Right circular cone

(b) Right prism in which the faces are either rectangles or right triangles

---

LABORATORY ACTIVITY

1. Record the mass of each U.S. coin. Which coin has the greatest mass? Which of the following sets have the same mass?

   (a) A half-dollar vs. two quarters
   (b) A quarter vs. two dimes and a nickel
   (c) A dime vs. two nickels
   (d) A dime vs. 10 pennies
   (e) A nickel vs. five pennies

2. Record the temperature of the room on a Celsius thermometer. Pour 200 mL of water into a liter container. Record the temperature of the water. Add 100 mL of ice to the water. Wait 1 min. and record the temperature of the ice water.

*Section 13-7 Using Logo to Draw Circles

The following POLYGON procedure was developed in Chapter 10 for drawing regular polygons with :N and :SIZE representing the number of sides and the length of a side, respectively:

```
TO POLYGON :N :SIZE
  REPEAT :N [FD :SIZE RT 360/:N]
END
```

The POLYGON procedure can be used to demonstrate the mathematical notion that as the number of sides in a regular polygon increases, the figure approaches a circle. When the turtle is at home with heading 0, the following procedure, called POLYGONS, can be used to draw an equilateral triangle, followed by a square, followed by a regular pentagon, followed by a regular hexagon, and so on, until a regular 20-gon is drawn. The result of executing POLYGONS 3 is shown in Figure 13-63.

FIGURE 13-63

```
TO POLYGONS :N
  FULLSCREEN
  POLYGON :N 20
  IF :N > 20 STOP
  POLYGONS :N + 1
END
```

POLYGONS 3

(In LCSI, change IF:N>20 STOP to IF :N>20 [STOP].)

> **REMARK** The command FULLSCREEN gives full graphics capability for drawing.

Drawing Turtle-type Circles

To instruct the turtle to draw a figure that looks like a circle (*turtle-type circle*), we must tell the turtle to move forward a little and turn a little and then repeat this sequence of motions until it comes back to its original position. If the turtle is told to go forward 1 unit and then to turn right 1°, then this sequence of motions repeated 360 times yields a turtle-type circle. The CIRCLE1 procedure from Appendix II along with its output is given in Figure 13-64.

FIGURE 13-64

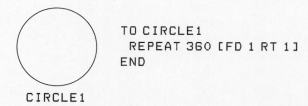

SECTION 13-7 USING LOGO TO DRAW CIRCLES

The turtle-type circle determined by CIRCLE1 is actually a regular 360-gon. One way to vary the size of the circle is to vary the amount the turtle moves forward. For example, suppose the turtle moves forward 0.5 unit rather than 1 unit, or, in general, any number of units S. We define a new procedure called VCIRCLE that accepts a variable input :S. Figure 13-65 shows various circles drawn by the VCIRCLE procedure.

FIGURE 13-65

VCIRCLE 0.8   VCIRCLE 0.5   VCIRCLE 0.1

```
TO VCIRCLE :S
  REPEAT 360 [FD :S RT 1]
END
```

Sometimes, it is more useful to draw a circle of a given radius. To write a procedure for drawing a circle of radius R, we use the procedure for VCIRCLE and the formula $2\pi r$ for the circumference of a circle. To understand the general case better, we first *solve a special case* of the problem. Suppose we want to draw a circle with radius 50. What should the side S of the approximating polygon be? Because the perimeter of the polygon is $360 \cdot S$ and the circumference of a circle with radius 50 is $2\pi \cdot 50$, it follows that $360 \cdot S \doteq 2\pi \cdot 50$. Now, $S \doteq (2\pi \cdot 50)/360$, or 0.873. Thus VCIRCLE 0.873 will draw a circle with radius approximately 50 units long.

In general, to find a procedure to draw a circle with radius R as an input, we need to find S in terms of R. This can be done as follows:

$$360 \cdot S \doteq 2\pi R$$
$$S \doteq \frac{2\pi R}{360}$$

Hence, a procedure called CIRCLE for drawing a circle with radius R can be written as follows, where π is approximated as 3.1416:

```
TO CIRCLE :R
  VCIRCLE (2*3.1416*:R) / 360
END
```

We can check that CIRCLE 50 draws a turtle-type circle of radius 50 by turning the turtle toward the center (after the circle has been completed) and instructing the turtle to go forward 100 units, the diameter of a circle with radius 50. If the turtle draws a diameter of the circle, the circle has a radius of 50 units. This check can be accomplished by executing the line

```
RT 90 FD 100 HT
```

It should be noted that this check is not totally accurate because our turtle-type circle is not a true circle.

Drawing Circles Faster

One way to draw circles faster when executing the CIRCLE procedure is to hide the turtle so that the turtle does not have to be drawn on the screen for each move. We can also speed up construction of the circle by varying the number of repeats. To close the figure, the turtle must complete a 360° trip. To ensure that this happens, the number of times the moves are repeated, multiplied by the number of degrees for each turn, must equal 360. For example, if we decide that the turtle should turn 15° each time, the process should be completed $\frac{360}{15}$, or 24, times because $15° \cdot 24 = 360°$.

In addition, if we wish the size of the faster circle to be approximately the same as that of the original circle, we need to make the perimeter of the polygon approximating the faster circle be the same as the perimeter of the polygon that drew the original approximation. This can be done by adapting the length of each side of the polygon. For example, if the original figure has perimeter 360 units and the new figure is to have 24 sides and perimeter 360 units, then each side must have length $\frac{360}{24}$, or 15, turtle steps. Consequently, we obtain the following FCIRCLE procedure:

```
TO FCIRCLE
  REPEAT 24 [FD 15 RT 15]
END
```

It is left as an exercise to write a procedure for drawing circles using variables for both the length of a side and the amount of the turn.

Procedures can also be designed for drawing arcs, that is, continuous parts of a circle. For example, if we wished to draw a semicircle, we could write the following procedure:

```
TO SEMICIRCLE
  REPEAT 180 [FD 1 RT 1]
END
```

As an exercise, develop a procedure called RARC for drawing arcs of variable sizes.

PROBLEM SET 13-7

1. (a) In the CIRCLE1 procedure in this section, suppose the instruction RT 1 is changed to RT 5. How does the new figure differ from the original one?
   (b) How does the figure differ if RT 1 is changed to LT 1?

2. Generalize the FCIRCLE procedure to include variable-sized sides and a variable number of repeats.

3. In the VCIRCLE procedure, if the turtle repeats the sequence FD :S RT 1 fewer than 360 times, it will draw an arc of a circle.
   (a) Write a procedure for drawing an arc for which the inputs are the length of a side of an approximating polygon and the number of degrees in the arc.
   (b) Write a procedure for drawing an arc, given the radius of the arc and the number of degrees in the arc.

4. Create your own designs using the arc procedure developed in Problem 3(b).

5. Write "fast" procedures to draw figures similar to each of the following:

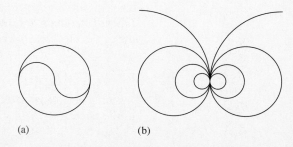

(a)   (b)

SECTION 13-7 EXERCISES 755

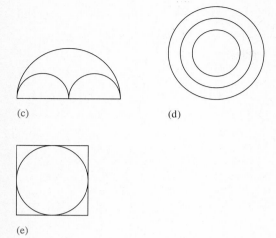

(c)   (d)

(e)

6. Write a procedure called SYMBOL for drawing the Olympic symbol, shown in the following figure, in variable sizes.

7. Use the procedures developed in Problem 3 to write a procedure to draw variable-sized flowers similar to the one shown.

8. Write a procedure called DIAMCIRC, with radius :R of a circle as an input, that draws the circle and its diameter.

9. Write a procedure to draw the following "Mastercard-like" figure:

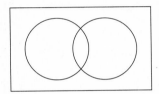

10. What should be the length of a side of a polygon to draw a turtle-type circle approximating a circle of circumference 150?

▼ BRAIN TEASER

Each student in a class is given a rectangular sheet of paper 10 in. by 17 in. and some tape. The students are to roll the paper into a circular cylinder, tape the edges together, and make the cylinder stand on the desk so that it will hold popcorn. How should students roll the sheets of paper in order for the resulting cylinders to hold the maximum amount of popcorn?

Solution to the Preliminary Problem

Understanding the Problem. Three planes are in a circular holding pattern over the Memphis airport. The planes are equally spaced in a circle of $\frac{1}{2}$ mi radius and are each at an altitude of 1 mi. We are asked how far apart the planes are and how far each is from the airport. Two distinct problems are presented. *Drawing a diagram* may aid in understanding the problems. In Figure 13-66, the planes are on the base of a right circular cone with the airport at the apex of the cone. One of the problems is to determine the slant height, that is, the distance from the airport to each plane. To solve the other problem, that

756 CHAPTER 13 CONCEPTS OF MEASUREMENT

is, determine how far apart the planes are, we are concerned only with the circular base and the distance between any two of the planes.

FIGURE 13-66

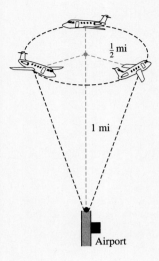

Devising a Plan. Using one of the planes as a vertex, the airport as a vertex, and the center of the base of the circle as the third vertex of a triangle, as shown in Figure 13-67, we can use the Pythagorean theorem to find the slant height of the cone.

FIGURE 13-67

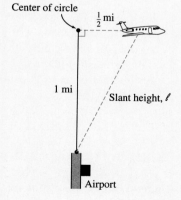

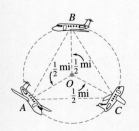

FIGURE 13-68

To find the distance between two planes, we consider the circular base of the cone, as in Figure 13-68. Because the planes are equally spaced on the circle, the planes themselves form the vertices of an equilateral triangle inscribed in a circle. We know the radius of the circle and must determine the length of a side of the triangle.

Carrying Out the Plan. To find the slant height, ℓ, in Figure 13-67, we have the following:

$$\left(\frac{1}{2}\right)^2 + (1)^2 = \ell^2$$

$$\frac{1}{4} + 1 = \ell^2$$

$$\frac{5}{4} = \ell^2$$

$$\frac{\sqrt{5}}{2} = \ell, \text{ or the plane is approximately 1.1 mi from the airport}$$

To find the distance, BC, that the planes are apart, we examine triangle ABC in the circle of radius OC, as in Figure 13-68. We know that triangle ABC is an equilateral triangle with each angle having measure 60°. We also know that $m(\angle BCO) = 30°$. (Why?) We know a relation among the sides of a 30°-60°-90° triangle. If $\overline{OC}$ were extended until it intersected $\overline{AB}$ in D, as in Figure 13-69, then triangle BCD would be a 30°-60°-90° triangle (why?). Thus *a subgoal* is to find a side of triangle BDC. We do not know the length of any side of triangle BDC, but completing this triangle reveals another 30°-60°-90° triangle, that is, triangle BDO. Now, the hypotenuse of triangle BDO is $\frac{1}{2}$ mi, so using the fact that the sides of this triangle are in the ratio $1:\sqrt{3}:2$, we know that DO is $\frac{1}{4}$ mi and BD is $\left(\frac{1}{4}\right)\sqrt{3}$, or $\frac{\sqrt{3}}{4}$, mi. Thus BC is $2\left(\frac{\sqrt{3}}{4}\right)$, or 0.9, mi. Hence, the planes are approximately 0.9 mi apart.

FIGURE 13-69

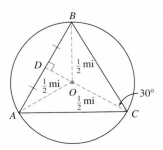

Looking Back. A different method of solving the problem involves drawing a diameter having one endpoint B in Figure 13-69. What type of triangle is formed by the diameter and point C? We could devise different types of problems, such as what is the arc length between planes or the volume of the airspace in the cone formed by the circling airplanes at the base and the airport at the apex. Also, what is the volume of the triangular pyramid formed by the three airplanes and the airport? We might create similar problems using more planes. Consider if there were six planes. How does the perimeter of the hexagon formed by the six planes compare to the perimeter of the triangle formed by three planes?

QUESTIONS FROM THE CLASSROOM

1. A student asks if the units of measure must be the same for each term in order to use the formulas for volumes. How do you respond?
2. In the discussion of the Pythagorean theorem, squares were constructed on each side of a right triangle. A student asks, "If different similar figures are constructed on each side of the triangle, does the same type of relationship still hold?" How do you reply?
3. A student asks, "Can I find the area of an angle?" How do you respond?
4. A student argues that a square has no area because its interior can be thought of as the union of infinitely many line segments, each of which has no area. How do you react?
5. A student asks whether the volume of a prism can ever be the same number as its surface area. How do you answer?
6. A student asks, "Why should the United States switch to the metric system?" How do you reply?
7. A student claims that in a triangle with 20° and 40° angles, the side opposite the 40° angle is twice as long as the side opposite the 20° angle. How do you reply?
8. A student interpreted 5 cm³ as shown in the following drawing. What is wrong with this interpretation?

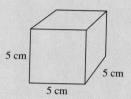

9. To approximate the value of π, a student measures the circumference C of a circle and the diameter d of the circle and finds the ratio $C:d$. The student wonders whether the ratio will change if she measures in English units rather than in metric units. How do you respond?
10. A student claims that because are and hectare are measures of area, we should say "square are" and "square hectare." How do you respond?
11. A student claims that because she can measure the circumference of a circle and its diameter and write the ratio as a fraction involving decimals, then the ratio is π. Therefore π must be a rational number. How do you respond?
12. A student claims that the area of his hand does not exist because it cannot be found by any formula. How do you respond?

CHAPTER OUTLINE

I. The English system
   A. Linear measure
      1 ft = 12 in.
      1 yd = 3 ft
      1 mi = 5280 ft = 1760 yd
   B. Area measure
      1. Units commonly used are the **square inch** (in.²), **square yard** (yd²), and **square foot** (ft²).
      2. Land can be measured in **acres**.
   C. Volume measure
      Units commonly used are the **cubic inch** (in.³), **cubic foot** (ft³), **cubic yard** (yd³), and **gallon**.
   *D. Mass
      Units of mass commonly used are **pound** (lb), **ounce** (oz) (1 oz = $\frac{1}{16}$ lb), and **ton** (1 ton = 2000 lb).

II. The metric system
   A. A summary of relationships among prefixes and the base unit of linear measure follows:

Prefix	Unit	Relationship to Base Unit		Symbol
kilo	kilometer	1000	meters	km
*hecto	hectometer	100	meters	hm
*deka	dekameter	10	meters	dam
	meter	1	meter	m
*deci	decimeter	0.1	meter	dm
centi	centimeter	0.01	meter	cm
milli	millimeter	0.001	meter	mm

*Not commonly used.

B. Area measure
  1. Units commonly used are the **square kilometer** (km^2), **square meter** (m^2), **square centimeter** (cm^2), and **square millimeter** (mm^2).
  2. Land can be measured using the **are** (100 m^2) and the **hectare** (10,000 m^2).
C. Volume measure
  1. Units commonly used are the **cubic meter** (m^3), **cubic decimeter** (dm^3), and **cubic centimeter** (cm^3).
  2. 1 dm^3 = 1 L and 1 cm^3 = 1 mL.
*D. Mass
  1. Units of mass commonly used are the **milligram** (mg), **gram** (g), **kilogram** (kg), and **metric ton** (t).
  2. 1 L and 1 mL of water have masses of approximately 1 kg and 1 g, respectively.
*E. Temperature
  1. The official unit of metric temperature is the **degree Kelvin**, but the unit commonly used is the **degree Celsius**. (In the English system, the unit of temperature is the **degree Fahrenheit**.)
  2. Basic temperature reference points are the following:
     100°C—boiling point of water
     37°C—normal body temperature
     20°C—comfortable room temperature
     0°C—freezing point of water
  3. $C = \frac{5}{9}(F - 32)$ and $F = \frac{9}{5}C + 32$

III. Distance
  A. **Distance properties.** Given points A, B, and C:
    1. $AB \geq 0$
    2. $AB = BA$
    3. $AB + BC \geq AC$
  B. The distance around a two-dimensional figure is called the **perimeter.** The distance C around a circle is called the **circumference.** $C = 2\pi r = \pi d$, where r is the radius of the circle and d is the diameter.

IV. Areas
  A. Formulas for areas
    1. **Square:** $A = s^2$, where s is a side.
    2. **Rectangle:** $A = \ell w$, where ℓ is the length and w is the width.
    3. **Parallelogram:** $A = bh$, where b is the base and h is the height.
    4. **Triangle:** $A = \frac{1}{2}bh$, where b is the base and h is the altitude to that base.
    5. **Trapezoid:** $A = \frac{1}{2}h(b_1 + b_2)$, where b_1 and b_2 are the bases and h is the height.
    6. **Regular polygon:** $A = \frac{1}{2}ap$, where a is the apothem and p is the perimeter.
    7. **Circle:** $A = \pi r^2$, where r is the radius.
    8. **Sector:** $A = \theta\pi r^2/360$, where θ is the measure of the central angle forming the sector and r is the radius of the circle containing the sector.
  B. **The Pythagorean theorem:** In any right triangle, the square of the length of the hypotenuse is equal to the sum of the squares of the lengths of the legs.
  C. Triangle Relations
    1. Property of 30°-60°-90° Triangle: The hypotenuse in a 30°-60°-90° triangle is two times the length of the leg opposite the 30° angle, and the length of the leg opposite the 60° angle is $\sqrt{3}$ times the short leg.
    2. Property of 45°-45°-90° Triangle: The hypotenuse of a 45°-45°-90° triangle is $\sqrt{2}$ times a leg.
  D. **Converse of the Pythagorean theorem:** In any triangle ABC with sides of lengths a, b, and c such that $a^2 + b^2 = c^2$, $\triangle ABC$ is a right triangle with the right angle opposite the side of length c.

V. Surface areas and volumes
  A. Formulas for areas
    1. **Right prism:** $S.A. = 2B + ph$, where B is the area of a base, p is the perimeter of the base, and h is the height of the prism.
    2. **Right circular cylinder:** $S.A. = 2\pi r^2 + 2\pi rh$, where r is the radius of the circular base and h is the height of the cylinder.
    3. **Right circular cone:** $S.A. = \pi r^2 + \pi r\ell$, where r is the radius of the circular base and ℓ is the slant height.
    4. **Right regular pyramid:** $S.A. = B + \frac{1}{2}p\ell$, where B is the area of the base, p is the perimeter of the base, and ℓ is the slant height.
    5. **Sphere:** $S.A. = 4\pi r^2$, where r is the radius of the sphere.
  B. Formulas for volumes
    1. **Right prism:** $V = Bh$, where B is the area of the base and h is the height.
       (a) **Right rectangular prism:** $V = \ell wh$, where ℓ is the length, w is the width, and h is the height.
       (b) **Cube:** $V = e^3$, where e is an edge.
    2. **Right circular cylinder:** $V = \pi r^2 h$, where r is the radius of the base and h is the height of the cylinder.
    3. **Pyramid:** $V = \frac{1}{3}Bh$, where B is the area of the base and h is the height of the pyramid.
    4. **Circular cone:** $V = \frac{1}{3}\pi r^2 h$, where r is the radius of the circular base and h is the height.
    5. **Sphere:** $V = \frac{4}{3}\pi r^3$, where r is the radius of the sphere.

CHAPTER TEST

1. Complete the following chart for converting metric measures:

	mm	cm	m	km
(a)				0.05
(b)		320		
(c)	260,000,000			
(d)			190	

2. For each of the following, choose an appropriate metric unit—millimeter, centimeter, meter, or kilometer:
   (a) The thickness of a penny
   (b) The length of a new lead pencil
   (c) The diameter of a dime
   (d) The distance the winner travels in the Indianapolis 500
   (e) The height of a doorknob
   (f) The length of a soccer field

3. ▶For each of the following, describe how you can find the area of the parallelogram
   (a) Using DE: (b) Using BF:

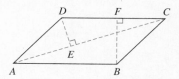

4. What is the area of the shaded region in the figure?

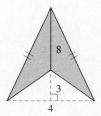

5. What is the area of the shaded region on each geoboard if the unit of measure is 1 cm²?

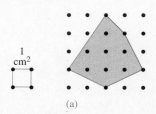
(a)

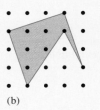
(b)

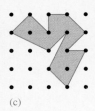
(c)

6. Find the area of the kite shown in the following figure:

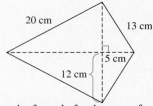

7. ▶Explain how the formula for the area of a trapezoid can be found by using the given pictures.

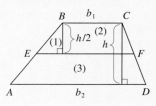

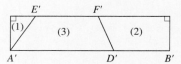

8. Use the figure to find each of the following:
   (a) The area of the hexagon
   (b) The area of the circle

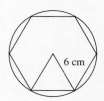

9. Find the area of each shaded region in the following figures:

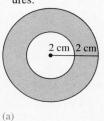
(a)

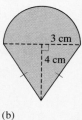
(b)

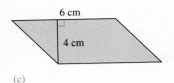
(c)

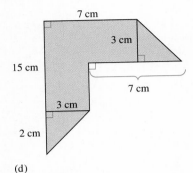
(d)

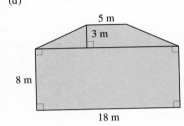
(e)

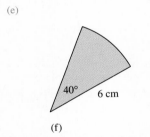
(f)

10. ▶For each of the following, can the measures represent sides of a right triangle? Explain your answers.
    (a) 5 cm, 12 cm, 13 cm
    (b) 40 cm, 60 cm, 104 cm

11. Find the surface area and volume of each of the following figures:

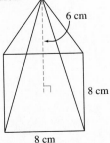
Right square pyramid
(a)

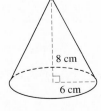
Right circular cone
(b)

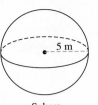
Sphere
(c)

Right circular cone
(d)

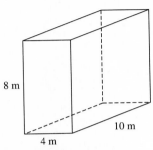
Right rectangular prism
(e)

12. Find the lateral surface area of the following cone:

13. ▶Doug's Dog Food Company wanted to impress the public with the magnitude of the company's growth. Sales of Doug's Dog Food had doubled from 1991 to 1992, so the company displayed the following graph, which shows the radius of the base and the height of the 1992 can to be double those of the 1991 dog food can. What does the graph really show with respect to the company's growth? Explain your answer.

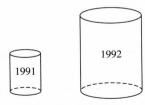
Doug's Dog Food Sales

14. Complete each of the following:
    *(a) Very heavy objects have mass that is measured in _____.
    (b) A cube whose length, width, and height are each 1 cm has a volume of _____.

- *(c) If the cube in (b) is filled with water, the mass of the water is _____.
- *(d) Which has a larger volume, 1 L or 1 dm³?
- (e) If a car uses 1 L of gas to go 12 km, the amount of gas needed to go 300 km is _____ L.
- (f) 20 ha = _____ a
- (g) 51.8 L = _____ cm³
- (h) 10 km² = _____ m²
- (i) 50 L = _____ mL
- (j) 5830 mL = _____ L
- (k) 25 m³ = _____ dm³
- (l) 75 dm³ = _____ mL
- *(m) 52,813 g = _____ kg
- *(n) 4800 kg = _____ t

15. Complete the following. (Use a calculator whenever convenient.)
    - (a) 50 ft = _____ yd
    - (b) 947 yd = _____ mi
    - (c) 9800 ft² = _____ yd²
    - (d) 3.4 mi² = _____ acres
    - (e) 18 yd³ = _____ ft³
    - (f) 0.8 ft³ = _____ in.³
    - (g) 3.8 lb = _____ oz
    - (h) 49 oz = _____ lb
    - *(i) 28°C = _____ °F
    - *(j) 95°F = _____ °C

16. Two cones are defined to be similar if the ratio between their heights equals the ratio between their radii. If two similar cones have heights h_1 and h_2, find the ratio between their volumes in terms of h_1 and h_2.

17. (a) A tank that is a right rectangular prism is 1 m by 2 m by 3 m. If the tank is filled with water, what is the mass of the water?
    - (b) Suppose the tank is exactly half full of water and then a heavy metal sphere of radius 30 cm is put into the tank. How high is the water now if the height of the tank is 3 m?
    - (c) What is the radius of the largest possible metal sphere that can be put into the half-full tank without causing the water to overflow?

*18. For each of the following, fill in the correct unit to make the sentence reasonable:
    - (a) Anna filled the gas tank with 80 _____.
    - (b) A man has a mass of about 82 _____.
    - (c) The textbook has a mass of 978 _____.
    - (d) A nickel has a mass of 5 _____.
    - (e) A typical adult cat has a mass of about 4 _____.
    - (f) A compact car has a mass of about 1.5 _____.
    - (g) The amount of coffee in the cup is 180 _____.

*19. For each of the following, decide if the situation is likely or unlikely:
    - (a) Carrie's bath water has a temperature of 15°C.
    - (b) She found 26°C too warm and lowered the thermostat to 21°C.
    - (c) Jim is drinking water that has a temperature of ⁻5°C.
    - (d) The water in the teakettle has a temperature of 120°C.
    - (e) The outside temperature dropped to 5°C, and ice appeared on the lake.

*20. Complete each of the following:
    - (a) 2 dm³ of water has a mass of _____ g.
    - (b) 1 L of water has a mass of _____ g.
    - (c) 3 cm³ of water has a mass of _____ g.
    - (d) 4.2 mL of water has a mass of _____ kg.
    - (e) 0.2 L of water has a volume of _____ m³.

SELECTED BIBLIOGRAPHY

Binswanger, R. "Discovering Perimeter and Area with Logo." *Arithmetic Teacher* 36 (September 1988): 18–24.

Bright, G., and J. Harvey. "Games, Geometry, and Teaching." *Mathematics Teacher* 81 (April 1988): 250–259.

Clopton, E. "Sharing Teaching Ideas: Area and Perimeter Are Independent." *Mathematics Teacher* 84 (January 1991): 33–35.

Cohen, D. "Estimating the Volumes of Solid Figures with Curved Surfaces." *Mathematics Teacher* 84 (May 1991): 392–395.

Gerver, R. "Discovering Pi—Two Approaches." *Arithmetic Teacher* 37 (April 1990): 18–22.

Kilmer, J. "Triangles of Equal Area and Perimeter and Inscribed Circles." *Mathematics Teacher* 81 (January 1988): 65–69.

Loomis, E. *The Pythagorean Proposition*. Washington, D. C.: National Council of Teachers of Mathematics. Reston, Va.: NCTM, 1976.

Pudelka, P. "Sharing Teaching Ideas: Formulas and Sugar Cubes." *Mathematics Teacher* 83 (February 1990): 119–120.

Smith, L. "Areas and Perimeters of Geoboard Polygons." *Mathematics Teacher* 83 (May 1990): 392–398.

Stover, D. "Sharing Teaching Ideas: Area of a Triangle." *Mathematics Teacher* 83 (February 1990): 120.

14 Coordinate Geometry

Preliminary Problem

McNairy County is planning to build a new regional airport and wants it to be equally close to each of the Triangular Cities: Adamsville, Selmer, and Leapwood. If Adamsville and Selmer are 50 mi apart, Leapwood and Adamsville are 30 mi apart, and Leapwood and Selmer are 40 mi apart, where should the airport be located?

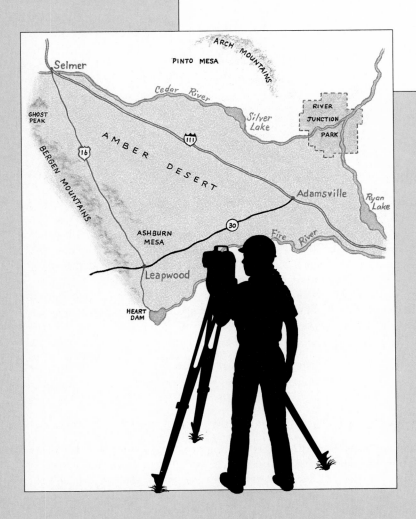

The Cartesian coordinate system (named for René Descartes) enables us to study geometry using algebra and to interpret algebraic phenomena geometrically.

The 5–8 *Standards* (p. 102) state that the mathematics curriculum should include explorations so that students can *develop confidence in solving linear equations using concrete, informal, and formal methods; investigate inequalities and nonlinear equations informally; apply algebraic methods to solve a variety of real-world and mathematical problems.*

In addition, the *Teaching Standards* (p. 136) includes the following:

Synthetic, coordinate, and transformational geometry should be used to provide opportunities for teachers to solve problems and to hone their skills in building justifications and coherent arguments for the plausibility of conjectures.

In this chapter, we review algebra skills and provide a framework for a study of beginning algebra involving coordinates and lines.

Cartesian coordinate systems are typically introduced using coordinates on a map, as seen on the student page from *Addison-Wesley Mathematics,* Grade 7, 1991, on page 765.

HISTORICAL NOTE

René Descartes (1596–1650), a French philosopher and mathematician and the inventor of coordinate geometry, studied law but never practiced it. Instead, he made lasting contributions to philosophy and science. His greatest contribution to mathematics was published in the book *La géométrie* in 1637. In 1649, Queen Christina of Sweden invited Descartes to come to her court and instruct her. He was apprehensive about going "to live in the land of bears among rocks and ice" but accepted the invitation. The queen fixed 5 o'clock in the morning as the time for her lessons. Descartes, returning one chilly morning from instructing the queen, caught a severe cold and died within a few weeks.

Section 14-1 Coordinate System in a Plane

origin / x-axis
y-axis

A Cartesian coordinate system is constructed by placing two number lines perpendicular to each other at the point where both have coordinate 0, as in Figure 14-1(a). The intersection point of the two lines is the **origin,** the horizontal line is the **x-axis,** and the vertical line is the **y-axis.**

SECTION 14-1 COORDINATE SYSTEM IN A PLANE

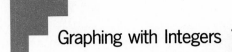
Graphing with Integers

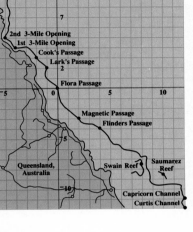

LEARN ABOUT IT

EXPLORE Study the Map

The Great Barrier Reef, the longest reef in the world, is 1,250 miles long. It has many passages and channels. Glenn and Susan were planning to go scuba diving at the reef. They marked some of the many openings on their map to help guide them. They also drew a grid over the map to assist them in estimating distances. They chose the Flora Passage as the origin (0, 0) because they were familiar with that area of the reef.

TALK ABOUT IT

1. The ordered pair for Magnetic Passage is (2, ⁻3). Describe how to find the Magnetic Passage on the map above.

2. What ordered pairs describe the locations of the Great N.E. Channel and Capricorn Channel?

On the **coordinate plane** above, the horizontal line through (0, 0) is called the **x-axis**. The vertical line through (0, 0) is called the **y-axis**.

Examples Draw a coordinate plane and plot the points for these locations.

A Flinders Passage: (4, ⁻4)
Start at (0, 0). Go 4 units right, then down 4 units.

B Lark's Passage: (⁻1, 2)
Start at (0, 0). Go 1 unit left, then up 2 units.

quadrants Together, the *x-axis* and the *y-axis* separate the plane into four parts called **quadrants**. Figure 14-1(a) shows the numbered quadrants. The quadrants do not include points on the axes.

FIGURE 14-1

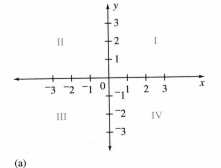

(a)

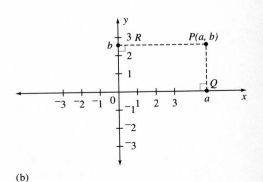

(b)

766 CHAPTER 14 COORDINATE GEOMETRY

The location of any point P can be described by an ordered pair of numbers, as shown in Figure 14-1(b). If a perpendicular from P to the x-axis intersects at a point with coordinate a and a perpendicular from P to the y-axis intersects at a point with coordinate b, point P has coordinates (a, b). The first component in the ordered pair (a, b) is the **abscissa, or x-coordinate,** of P. The second component is the **ordinate, or y-coordinate,** of P. To each point in the plane, there corresponds an ordered pair (a, b) and vice versa. Hence, there is a one-to-one correspondence between all the points in the plane and all the ordered pairs of real numbers.

abscissa / x-coordinate
ordinate / y-coordinate

For example, in Figure 14-2, the x-coordinate of P is $^-3$ and the y-coordinate of P is 2, so P has coordinate $(^-3, 2)$. Similarly, R has a coordinate $(^-4, ^-3)$, written as $R(^-4, ^-3)$.

FIGURE 14-2

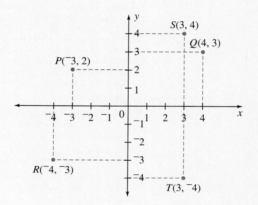

The Distance Formula

One way to *approximate* the distance between two points in a coordinate plane is to measure it with a ruler that has the same scale as the coordinate axes. Using algebraic techniques, we can calculate the *exact distance* between two points in the plane. First, suppose the two points are on one of the axes. For example, in Figure 14-3(a), $A(2, 0)$ and $B(5, 0)$ are on the x-axis. The distance between these two points is three units.

$$AB = OB - OA = 5 - 2 = 3$$

FIGURE 14-3

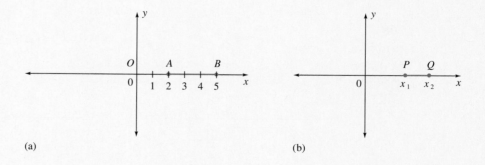

(a)   (b)

In general, if two points P and Q are on the x-axis, as in Figure 14-3(b), with x-coordinates x_1 and x_2, respectively, and $x_2 > x_1$, then $PQ = x_2 - x_1$. In fact, *the distance between two points on the x-axis is always the absolute value of the difference between the x-coordinates of the points* (Why?) A similar result holds for any two points on the y-axis.

Figure 14-4 shows two points in the plane, $C(2, 5)$ and $D(6, 8)$. The distance between C and D can be found by *drawing a picture*. Perpendiculars from the points to the x-axis and to the y-axis, respectively, are drawn defining triangle CDE. The lengths of the legs of triangle CDE are found by using horizontal and vertical distances and properties of rectangles.

$$CE = |6 - 2| = 4$$
$$DE = |8 - 5| = 3$$

FIGURE 14-4

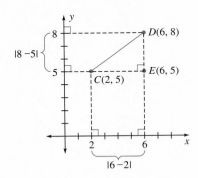

The distance between C and D can be found by applying the Pythagorean theorem to the triangle.

$$CD^2 = DE^2 + CE^2$$
$$= 3^2 + 4^2$$
$$= 25$$
$$CD = \sqrt{25}, \text{ or } 5$$

The method can be generalized to find a formula for the distance between any two points $A(x_1, y_1)$ and $B(x_2, y_2)$. Construct a right triangle with $\overline{AB}$ as one of its sides by drawing a segment through A parallel to the x-axis and a segment through B parallel to the y-axis, as shown in Figure 14-5. The lines containing the segments intersect at point C, forming right triangle ABC. Now, apply the Pythagorean theorem.

FIGURE 14-5

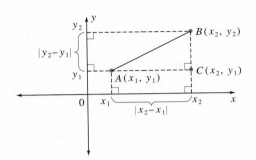

In Figure 14-5 we show that $AC = |x_2 - x_1|$ and $BC = |y_2 - y_1|$. By the Pythagorean theorem, $(AB)^2 = |x_2 - x_1|^2 + |y_2 - y_1|^2$, and consequently *distance formula*

$AB = \sqrt{|x_2 - x_1|^2 + |y_2 - y_1|^2}$. Because $|x_2 - x_1|^2 = (x_2 - x_1)^2$ and $|y_2 - y_1|^2 = (y_2 - y_1)^2$, $AB = \sqrt{(x_2 - x_1)^2 + (y_2 - y_1)^2}$. This result is known as the **distance formula.**

DISTANCE FORMULA

The distance between the points $A(x_1, y_1)$ and $B(x_2, y_2)$ is given by
$$AB = \sqrt{(x_2 - x_1)^2 + (y_2 - y_1)^2}.$$

REMARK It makes no difference whether $x_2 - x_1$ or $x_1 - x_2$ is used in the distance formula because $(x_2 - x_1)^2 = (x_1 - x_2)^2$. The same is true for the y-values.

EXAMPLE 14-1 For each of the following, determine the distance between P and Q:

(a) $P(2, 7), Q(3, 5)$

(b) $P(0, 0), Q(3, ^-4)$

Solution (a) $PQ = \sqrt{(3 - 2)^2 + (5 - 7)^2} = \sqrt{1 + 4} = \sqrt{5}$

(b) $PQ = \sqrt{(0 - 3)^2 + [0 - (^-4)]^2} = \sqrt{9 + 16} = \sqrt{25} = 5$ ∎

EXAMPLE 14-2 (a) Show that $A(7, 4)$, $B(^-2, 1)$, and $C(10, ^-5)$ are the vertices of an isosceles triangle.

(b) Show that $\triangle ABC$ in (a) is a right triangle.

Solution (a) Using the distance formula, we find the lengths of the sides.

$AB = \sqrt{(^-2 - 7)^2 + (1 - 4)^2} = \sqrt{(^-9)^2 + (^-3)^2} = \sqrt{90}$
$BC = \sqrt{[10 - (^-2)]^2 + (^-5 - 1)^2} = \sqrt{12^2 + (^-6)^2} = \sqrt{180}$
$AC = \sqrt{(10 - 7)^2 + (^-5 - 4)^2} = \sqrt{3^2 + (^-9)^2} = \sqrt{90}$

Thus $AB = AC$, so the triangle is isosceles.

(b) Because $(\sqrt{90})^2 + (\sqrt{90})^2 = (\sqrt{180})^2$, $\triangle ABC$ is a right triangle with $\overline{BC}$ as hypotenuse and $\overline{AB}$ and $\overline{AC}$ as legs. ∎

Equation of a Circle

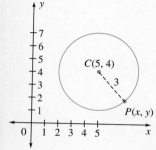

FIGURE 14-6

Because a circle is determined by its center and the radius, we can use the distance formula and a general point with coordinates (x, y) to write an equation for the circle. For example, Figure 14-6 shows the circle with center $C(5, 4)$ and radius 3 units long.

To find the equation of this circle, we first consider a point $P(x, y)$ on the circle. The distance from P to the center of the circle is 3 units; that is, $CP = 3$. By the distance formula, $\sqrt{(x - 5)^2 + (y - 4)^2} = 3$. Squaring both sides of the equation, we obtain $(x - 5)^2 + (y - 4)^2 = 9$. Thus the coordinates of any point of the circle satisfy the equation $(x - 5)^2 + (y - 4)^2 = 9$.

This process can be generalized to find the equation of any circle with center $C(a, b)$ and radius r. In Figure 14-7, P is on the circle if, and only if, $\sqrt{(x - a)^2 + (y - b)^2} = r$. Squaring both sides results in the *equation of the circle*.

FIGURE 14-7

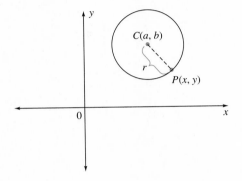

EQUATION OF A CIRCLE

The equation of a circle with center (a, b) and radius r is $(x - a)^2 + (y - b)^2 = r^2$.

EXAMPLE 14-3 (a) Find the equation of the circle with the center at the origin and radius 4.

(b) Find the equation of the circle with center at $C(^-4, 3)$ and radius 5.

(c) Sketch the graph of $(x + 2)^2 + (y - 3)^2 = 9$.

(d) Write a condition for the set of points in the interior of the circle given in (c).

Solution (a) The center is at $(0, 0)$ and $r = 4$, so the equation $(x - a)^2 + (y - b)^2 = r^2$ becomes $(x - 0)^2 + (y - 0)^2 = 4^2$, or $x^2 + y^2 = 16$.

(b) The equation is $[x - (^-4)]^2 + (y - 3)^2 = 5^2$, or $(x + 4)^2 + (y - 3)^2 = 25$.

(c) The equation $(x + 2)^2 + (y - 3)^2 = 9$ is in the form $(x - a)^2 + (y - b)^2 = r^2$ if and only if $x - a = x + 2$, $y - b = y - 3$, and $r^2 = 9$. Thus $a = ^-2$, $b = 3$, and $r = 3$. Hence, the equation is that of a circle with center at $(^-2, 3)$ and $r = 3$. Figure 14-8 shows the graph of the circle.

FIGURE 14-8

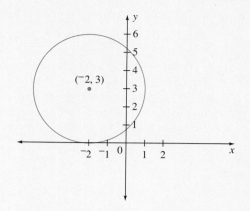

(d) A point (x, y) is in the interior of the circle if, and only if, the distance between the point and the center of the circle is less than the radius. In this case, $\sqrt{[x - (^-2)]^2 + (y - 3)^2} < 3$, or $(x + 2)^2 + (y - 3)^2 < 9$. ∎

EXAMPLE 14-4 Find all points on the y-axis that are 10 units away from $C(8, 3)$.

Solution Suppose P is a point on the y-axis satisfying the given requirements. Then, P has 0 as its x-coordinate. Thus the coordinates of P are $(0, y)$. We find y so that $PC = 10$.

$$\sqrt{(0 - 8)^2 + (y - 3)^2} = 10$$

Square both sides of the equation and solve for y.

$$(0 - 8)^2 + (y - 3)^2 = 100$$
$$64 + (y - 3)^2 = 100$$
$$(y - 3)^2 = 36$$

Thus $y - 3 = 6$ or $y - 3 = ^-6$, so $y = 9$ or $y = ^-3$. Hence, two points, $P_1(0, 9)$ and $P_2(0, ^-3)$, satisfy the condition. ∎

The Midpoint Formula

In Chapter 11, we discovered how to find the midpoint of a segment using a compass and straightedge. We can also find the coordinates of the midpoint of a segment given its endpoints. With two points $A(x_1, y_1)$ and $B(x_2, y_2)$, we can find the coordinates of the midpoint M of the segment $\overline{AB}$ as follows. First, *consider a simpler problem* using the two points $A(x_1, 0)$ and $B(x_2, 0)$ on the x-axis, as shown in Figure 14-9 with $x_1 < x_2$.

To find the x-coordinate of the midpoint M, use the given information to write an equation for x in terms of x_1 and x_2, and then solve for x. Because M is the midpoint of $\overline{AB}$, $AM = MB$, which implies that $x - x_1 = x_2 - x$; therefore

$$2x - x_1 = x_2$$
$$2x = x_1 + x_2$$
$$x = \frac{x_1 + x_2}{2}.$$

FIGURE 14-9

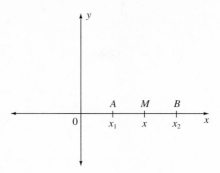

Similarly, for the case in which two points lie on the y-axis, the two points are $A(0, y_1)$ and $B(0, y_2)$, and the y-coordinate of the midpoint is $\frac{y_1 + y_2}{2}$.

Now, consider the general case. Let $A(x_1, y_1)$ and $B(x_2, y_2)$ be the endpoints of segment $\overline{AB}$ whose midpoint $M(x, y)$ is shown in Figure 14-10. Because $\overline{AA_1}$, $\overline{MM_1}$, and $\overline{BB_1}$ are parallel and M is the midpoint of $\overline{AB}$, M_1 is the midpoint of $\overline{A_1B_1}$. (Why?) Hence, $x = \frac{x_1 + x_2}{2}$. By an analogous argument, $y = \frac{y_1 + y_2}{2}$. We summarize this result below.

FIGURE 14-10

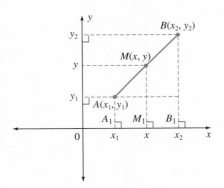

MIDPOINT FORMULA

Given $A(x_1, y_1)$ and $B(x_2, y_2)$, the midpoint M of $\overline{AB}$ is

$$M\left(\frac{x_1 + x_2}{2}, \frac{y_1 + y_2}{2}\right).$$

REMARK To find the midpoint of a line segment, simply find the arithmetic mean of the respective coordinates of the two endpoints.

EXAMPLE 14-5 (a) Find the coordinates of the midpoint of $\overline{AB}$ if A has coordinates $(^-3, 2)$ and B has coordinates $(3, ^-5)$.

(b) Suppose M is the midpoint of $\overline{AB}$, A has coordinates $(2, ^-3)$, and M has coordinates $(^-2, 1)$. Find the coordinates of B.

Solution (a) Let (x, y) be the coordinates of the midpoint of $\overline{AB}$. Then we use the midpoint formula.

$$x = \frac{x_1 + x_2}{2} = \frac{^-3 + 3}{2} = 0$$

$$y = \frac{y_1 + y_2}{2} = \frac{2 + (^-5)}{2} = \frac{^-3}{2}$$

Hence, the midpoint has coordinates $\left(0, \frac{^-3}{2}\right)$.

(b) Let the coordinates of B be (x, y), as shown in Figure 14-11. The coordinates of the midpoint M are the arithmetic means of the respective coordinates of the endpoints of $\overline{AB}$.

$$^-2 = \frac{x + 2}{2} \qquad 1 = \frac{y + (^-3)}{2}$$
$$^-4 = x + 2 \qquad 2 = y + (^-3)$$
$$x = ^-6 \qquad y = 5$$

FIGURE 14-11

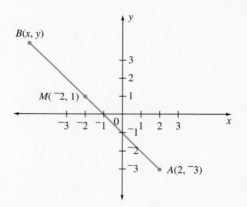

Consequently, $B(^-6, 5)$ is the required point.

PROBLEM SET 14-1

1. ▶(a) Describe the application of a grid system in developing a map for a new flat, midwestern town with two major perpendicular streets.

(b) Draw such a grid and locate each of the following:
   (1) A school in quadrant I
   (2) A church in quadrant II
   (3) City hall at the origin
   (4) A museum in quadrant IV
   (5) A college in quadrant III

2. (a) Give the coordinates of each of the points A, B, C, D, E, F, G, and H of the following figure:

SECTION 14-1 EXERCISES

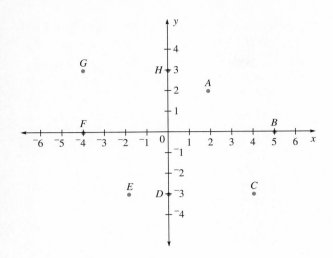

8. Use the graph to answer the following questions:

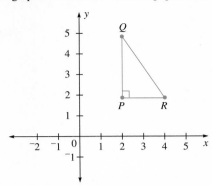

(a) Give the coordinates of the images of points P, Q, and R if $\triangle PQR$ is reflected in the x-axis.
(b) Give the coordinates of the images of points P, Q, and R if $\triangle PQR$ is rotated 90° counterclockwise, with the origin as the center of the rotation.
(c) Find the image of triangle PQR under a translation that takes P to R.

(b) Find the coordinates of another point (not drawn) that is collinear with E, D, and C.

3. Name the quadrant in which each of the following ordered pairs is located:
   (a) (3, 7)   (b) (⁻5, ⁻8)   (c) (⁻10, 32)
   (d) (10, ⁻40) (e) (0, 7)

4. Describe the set of points in each quadrant using set-builder notation.

5. Plot each of the points $A(^-3, ^-2)$, $B(^-3, 6)$, and $C(4, 6)$, and then find the coordinates of a point D such that quadrilateral $ABCD$ is a rectangle.

6. Plot at least six points, each having the sum of its coordinates equal to 4.

7. Use the figure to answer the following questions:

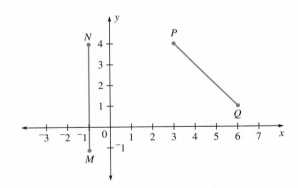

(a) Give the coordinates of the endpoints of $\overline{PQ}$.
(b) Give the coordinates of the endpoints of $\overline{MN}$.

9. Find the coordinates of the images of each of the points (0, 1), (1, 0), (2, 4), (⁻2, 4), (⁻2, ⁻4), and (2, ⁻4) under the following transformations:
   (a) A reflection in the x-axis
   (b) A reflection in the y-axis
   (c) A rotation by 90° counterclockwise about (0, 0)
   (d) A half-turn whose center is (0, 0)
   (e) A translation along the segment $\overline{OA}$ from $O(0, 0)$ to $A(0, ^-4)$

10. Consider $\triangle ABC$, whose vertices are at $A(^-2, 5)$, $B(2, 6)$, and $C(5, 1)$. Find the coordinates of the vertices of the image of the triangle if $\triangle ABC$ is transformed by the following:
    (a) A reflection in the x-axis
    (b) A reflection in the y-axis
    (c) A reflection in the y-axis followed by a reflection in the x-axis

11. (a) Reflect the point $P(2, 4)$ in the y-axis, and then reflect its image in the x-axis. What are the coordinates of the final image point?
    (b) If the point $P(a, b)$ is in the first quadrant and is reflected in the y-axis, and then its image is reflected in the x-axis, what are the coordinates of the final image point?

12. Find the images of each point $A(1, 0)$, $B(2, 2)$, $C(3, 1)$, $D(3, ^-1)$, and $E(a, b)$ when it is reflected in the line $\overleftrightarrow{OP}$, where P has the coordinates (5, 5) and O is the origin.

13. Find the coordinates of the image of point $P(a, b)$ when it is rotated 90° counterclockwise about the following points:
    (a) $(a, 0)$ (b) The origin

14. For each of the following, find the length of $\overline{AB}$:
    (a) $A(0, 3), B(0, 7)$ (b) $A(0, ^-3), B(0, ^-7)$
    (c) $A(0, 3), B(4, 0)$ (d) $A(0, ^-3), B(^-4, 0)$
    (e) $A(^-1, 2), B(3, ^-4)$ (f) $A(0, 0), B(^-4, 3)$
    (g) $A(4, ^-5), B\left(\frac{1}{2}, \frac{^-7}{4}\right)$ (h) $A(4, 0), B(5.2, ^-3.7)$
    (i) $A(5, 3), B(5, ^-2)$ (j) $A(5, 2), B(^-3, 4)$

15. Find the perimeter of the triangle with vertices at $A(0, 0)$, $B(^-4, ^-3)$, and $C(^-5, 0)$.

16. Show that $(0, 6), (^-3, 0)$, and $(9, ^-6)$ are the vertices of a right triangle.

17. Show that the triangle whose vertices are $A(^-2, ^-5), B(1, ^-1)$, and $C(5, 2)$ is isosceles.

18. Find x if the distance between $P(1, 3)$ and $Q(x, 9)$ is 10 units.

19. For each of the following, find the midpoint of the line segment whose endpoints have the given coordinates:
    (a) $(^-3, 1)$ and $(3, 9)$
    (b) $(4, ^-3)$ and $(5, ^-1)$
    (c) $(1.8, ^-3.7)$ and $(2.2, 1.3)$
    (d) $(1 + a, a - b)$ and $(1 - a, b - a)$

20. One endpoint of a diameter of a circle with center $C(^-2, 5)$ is given by $(3, ^-1)$. Find the coordinates of the other endpoint.

21. (a) Find the midpoints of the sides of a triangle whose vertices have the coordinates $(0, 0), (^-4, 6)$, and $(4, 2)$.
    (b) Find the lengths of the medians of the triangle in (a). (A *median* is a segment connecting the vertex of a triangle to the midpoint of the opposite side.)

22. For each of the following, write the equation of the circle with center C and radius r:
    (a) $C(3, ^-2)$ and $r = 2$
    (b) $C(^-3, ^-4)$ and $r = 5$
    (c) $C(^-1, 0)$ and $r = 2$
    (d) $C(0, 0)$ and $r = 3$

23. Given the circle whose equation is $x^2 + y^2 = 9$, which of the following points are in its interior, which are in its exterior, and which are on the circle?
    (a) $(3, ^-3)$ (b) $(2, ^-2)$
    (c) $(1, 8)$ (d) $(3, 1982)$
    (e) $(5.1234, ^-3.7894)$ (f) $\left(\frac{1}{387}, \frac{1}{1983}\right)$
    (g) $\left(\frac{^-1}{2}, \frac{35}{2}\right)$ (h) $(0, 3)$

24. Find the equation of the circle that has its center at the origin and that contains the point with coordinates $(^-3, 5)$.

25. For each of the following, find the equation of the circle that has its center at $C(4, ^-3)$ and that passes through the following point:
    (a) The origin (b) $(5, ^-2)$

26. Find the equation of the circle that has a diameter with endpoints at $(^-8, 2)$ and $(4, ^-6)$.

27. Graph each of the following, if possible:
    (a) $x^2 + y^2 > 4$ (b) $x^2 + y^2 \leq 4$
    (c) $x^2 + y^2 = 9$ (d) $x^2 + y^2 - 4 = 0$
    (e) $x^2 + y^2 + 4 = 0$ (f) $x^2 + y^2 = 4$

28. Find the equation of the circle that passes through the origin and the point $(5, 2)$ and has its center on the x-axis.

29. ▶Is $2x^2 + 2y^2 = 1$ an equation of a circle? If it is, find its center and radius; if it is not, explain why not.

★30. Use the distance formula to show that the points with coordinates $(^-1, 5), (0, 2)$, and $(1, ^-1)$ are collinear.

31. Use coordinates to prove that the midpoint M of the hypotenuse of a right triangle is equidistant from the vertices. (*Hint:* Use the coordinate system shown in the accompanying figure.)

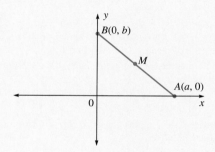

★32. One day, Linda left home H for school S. Rather than stopping at school, she went on to the corner A, which is twice as far from home as the school and on the same street as her home and the school. Then she headed for the ice cream parlor I. Passing the ice cream parlor, she headed straight for the next corner B, which is on the same street as A and I and twice as far from A as I. Walking toward the park P, she continued beyond it to the next corner C, so that C, P, and B are on the same street and C is twice as far from B as P. At this point, she again headed for the school but continued walking in a straight line twice as far, reaching point D. Then, she headed for the ice cream parlor, but continued in a straight line twice as far to point E. From E, Linda headed for the park but continued in a straight line twice as far to F, where she stopped. The following drawing shows the first part of Linda's walk. What is the location of Linda's final stop?

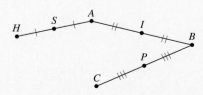

★**33.** In the Larson cartoon, if the professors are using a simple 5 × 5 grid to obtain their lattice point coordinates and time is ignored, what is the probability of their choosing identical coordinates?

Tempers flare when Professor Carlson and Lazzell, working independently, ironically set their time machines to identical coordinates.

THE FAR SIDE © 1985 Far Works, Inc. Reprinted with permission of Universal Press Syndicate. All rights reserved.

LABORATORY ACTIVITY

As van Hiele Level 1 activities, draw each of the following on the grid and in each case, find the coordinates of the vertices:

(a) A rectangle (b) A parallelogram (c) A kite

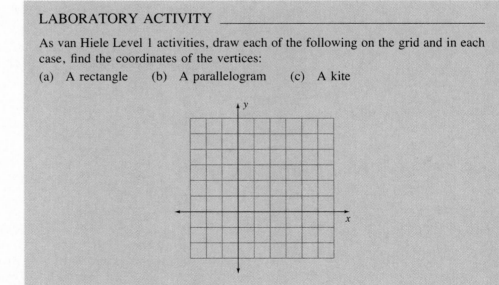

Section 14-2 Equations of Lines

Graphing Lines

Every point on the x-axis has a y-coordinate of zero. Thus the x-axis can be described as the set of all points (x, y) such that $y = 0$. This set of points on the x-axis has equation $y = 0$. Similarly, the y-axis can be described as the set of all points (x, y) for which $x = 0$ and y is an arbitrary real number. Thus $x = 0$ is the equation of the y-axis. If we plot the set of all points that satisfy a given condition, the resulting picture on the Cartesian coordinate system is called the **graph** of the set.

EXAMPLE 14-6 Sketch the graph for each of the following:

(a) $x = 2$ (b) $y = 3$

(c) $x < 2$ and $y = 3$

Solution (a) The equation $x = 2$ represents the set of all points (x, y) for which $x = 2$ and y is any real number. This set is the line perpendicular to the x-axis at $(2, 0)$, as in Figure 14-12.

FIGURE 14-12

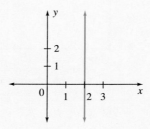

(b) The equation $y = 3$ represents the set of all points (x, y) for which $y = 3$ and x is any real number. This set is the line perpendicular to the y-axis at $(0, 3)$, as in Figure 14-13.

FIGURE 14-13

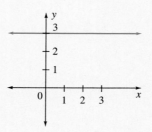

(c) Together, the statements represent the set of all points (x, y) for which $x < 2$, but y is always 3. The set describes a half-line, as shown in Figure 14-14. Note that the hollow dot at $(2, 3)$ indicates that this point is not included in the solution.

FIGURE 14-14

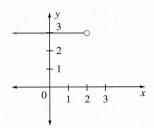

In Example 14-6, we found the graphs of the equations $x = 2$ and $y = 3$. In general, the graph of the equation $x = a$, where a is some real number, is a line perpendicular to the x-axis through the point with coordinates $(a, 0)$, as shown in Figure 14-15. Similarly, the graph of the equation $y = b$ is a line perpendicular to the y-axis through the point with coordinates $(0, b)$.

FIGURE 14-15

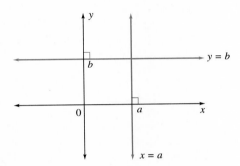

TABLE 14-1

Number of Term	Term
1	4
2	7
3	10
4	13
.	.
.	.
.	.
n	$3n + 1$

In Example 14-6, we examined sets of points resulting in graphs of lines parallel to the x- and y-axes, respectively. Now we consider the set of points in Table 14-1. The general formula for the nth term of this arithmetic sequence is $3n + 1$. If the number of the term is the x-coordinate and the corresponding term is the y-coordinate, we see that the set of points appear to lie on a line, as in Figure 14-16, which is not parallel to either the x- or y-axes.

FIGURE 14-16

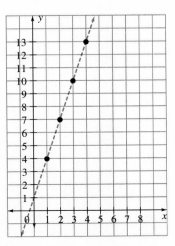

CHAPTER 14   COORDINATE GEOMETRY

Do all arithmetic sequences lie along lines? To help answer this question, we consider related simpler sequences, the multiples of selected numbers as in Tables 14-2(a)–(e).

TABLE 14-2

Multiples of 1		Multiples of 2		Multiples of $\frac{1}{2}$		Multiples of $^-1$		Multiples of $^-2$	
1	1	1	2	1	$\frac{1}{2}$	1	$^-1$	1	$^-2$
2	2	2	4	2	1	2	$^-2$	2	$^-4$
3	3	3	6	3	$\frac{3}{2}$	3	$^-3$	3	$^-6$
4	4	4	8	4	2	4	$^-4$	4	$^-8$
5	5	5	10	.	.	.	.	.	.
6	6	6	12	.	.	.	.	.	.
7	7	.	.	.	.	.	.	.	.
8	8	.	.	.	.	n	^-n	n	^-2n
9	9	.	.	n	$\frac{1}{2}n$		(d)		(e)
10	10	n	$2n$		(c)				
.	.		(b)						
.	.								
n	n								
(a)									

If the sets of ordered pairs in the tables are plotted on a graph as in Figure 14-17, then we see that the pairs of each table appear to determine lines. Those lines and their corresponding equations are given in Figure 14-17.

FIGURE 14-17

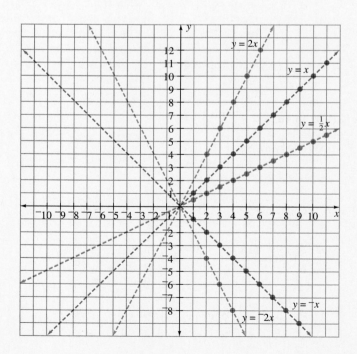

All five lines in Figure 14-17 have equations of the form $y = mx$, where m takes the values 2, 1, $\frac{1}{2}$, ⁻1, and ⁻2. If all points along a given dashed line are connected, then all the points on that line satisfy the corresponding equation. The number m is a measure of steepness and is called the **slope** of the line whose equation is $y = mx$. The graph goes up from left to right (increases) if m is positive, and it goes down from left to right (decreases) if m is negative. If m is 0, what happens to the line? What happens when m is very large?

slope

All lines in Figure 14-17 pass through the origin. This is true for any line whose equation is $y = mx$. If $x = 0$, then $y = m \cdot 0 = 0$, and $(0, 0)$ is a point on the graph of $y = mx$. Conversely, it is possible to show that any nonvertical line passing through the origin has an equation of the form $y = mx$, for some value of m.

EXAMPLE 14-7 Find the equation of the line that contains $(0, 0)$ and $(2, 3)$.

Solution The line goes through the origin; therefore its equation has the form $y = mx$. To find the equation of the line, we must find the value of m. The line contains $(2, 3)$, so we substitute 2 for x and 3 for y in the equation $y = mx$ to obtain $3 = m \cdot 2$, and thus $m = \frac{3}{2}$. Hence, the required equation is $y = \frac{3}{2}x$. ∎

Next, we consider equations of the form $y = mx + b$, where b is a real number. To do this, we examine the graphs of $y = x + 2$ and $y = x$. Given the graph of $y = x$, we can obtain the graph of $y = x + 2$ by "raising" each point on the first graph by two units because, for a certain value of x, the corresponding y value is two units greater. This is shown in Figure 14-18(a). Similarly, to sketch the graph of $y = x - 2$, we first draw the graph of $y = x$ and then lower each point vertically by two units, as shown in Figure 14-18(b).

FIGURE 14-18

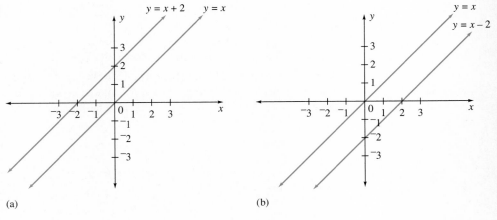

(a)   (b)

The graphs of $y = x + 2$ and $y = x - 2$ are straight lines. Moreover, the lines whose equations are $y = x$, $y = x + 2$, and $y = x - 2$ are parallel. In general, for a given value of

m, the graph of $y = mx + b$ is a straight line through $(0, b)$ and parallel to the line whose equation is $y = mx$.

Further, the graph of the line $y = mx + b$, where $b > 0$, can be obtained from the graph of $y = mx$ by sliding $y = mx$ up b units, as shown in Figure 14-19. If $b < 0$, $y = mx$ must be slid down $|b|$ units.

y-intercept
slope-intercept form
x-intercept

The graph of $y = mx + b$ in Figure 14-19 crosses the y-axis at point $P(0, b)$. The value of y at the point of intersection of any line with the y-axis is the **y-intercept.** Thus b is the y-intercept of $y = mx + b$, and this form of the equation of a straight line is the **slope-intercept form.** Similarly, the value of x at the point of intersection of a line with the x-axis is the **x-intercept.**

FIGURE 14-19

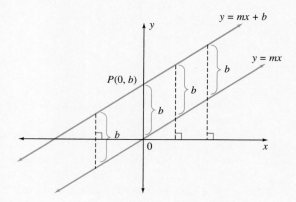

EXAMPLE 14-8

Given the equation $y - 3x = {}^-6$:

(a) Find the slope of the line.
(b) Find the y-intercept.
(c) Find the x-intercept.
(d) Sketch the graph of the equation.

Solution

(a) To write the equation in the form $y = mx + b$, we add $3x$ to both sides of the given equation to obtain $y = 3x + ({}^-6)$. Hence, the slope is 3.

(b) The form $y = 3x + ({}^-6)$ shows that $b = {}^-6$, which is the y-intercept. (The y-intercept can also be found directly by substituting $x = 0$ in the equation and finding the corresponding value of y.)

(c) The x-intercept is the x-coordinate of the point where the graph intersects the x-axis. At that point, $y = 0$. Substituting 0 for y in $y = 3x - 6$ gives 2 as the x-intercept.

(d) The y-intercept and the x-intercept are located at $(0, {}^-6)$ and $(2, 0)$, respectively, on the line. We plot these points and draw the line through them to obtain the desired graph in Figure 14-20. Note that any two points of the line can be used to sketch the graph because any two points determine a line.

FIGURE 14-20

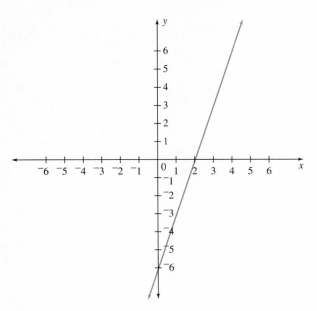

The equation $y = b$ can be written in slope-intercept form as $y = 0 \cdot x + b$. Consequently, its slope is 0 and its y-intercept is b. This should not be surprising; because the line is parallel to the x-axis, its steepness, or slope, should be 0. Any vertical line has equation $x = a$ for some real number a. This equation cannot be written in slope-intercept form. The slope of a vertical line is undefined and will be discussed later in the chapter. In general, *every straight line has an equation of either the form $y = mx + b$ or $x = a$*. Any equation that can be put in one of these forms is a **linear equation.**

linear equation

> **EQUATION OF A LINE**
>
> Every line has an equation of either the form $y = mx + b$ or $x = a$.

The Slope as Rise/Run

Because a line is determined by any two of its points, it is possible, given the coordinates of two points on a line, to find the equation of the line. For example, given $A(4, 2)$ and $B(1, 6)$, we can find the equation of $\overleftrightarrow{AB}$. Because the line is not perpendicular to the x-axis (why?), it must be of the form $y = mx + b$. Substituting the coordinates of A and B in $y = mx + b$ results in the following equations:

$$2 = m \cdot 4 + b \quad \text{or} \quad 2 = 4m + b$$
$$6 = m \cdot 1 + b \quad \text{or} \quad 6 = m + b$$

To find the equation of the line, we must find the values of m and b. If we solve for b in each of these equations, we obtain $b = 2 - 4m$ and $b = 6 - m$, respectively. Consequently, $2 - 4m = 6 - m$, so $m = \dfrac{-4}{3}$. Substituting this value of m in either of the equations gives $b = \frac{22}{3}$. As a consequence, the equation of the line through A and B is $y = -\frac{4}{3}x + \frac{22}{3}$. The correctness of this equation can be checked by substituting the coordinates of the two given points, $A(4, 2)$ and $B(1, 6)$, in the equation.

Using an analogous approach, we can find a general formula for the slope of a line, given two points on the line, $A(x_1, y_1)$ and $B(x_2, y_2)$. If the line is not a vertical line, its equation is given by $y = mx + b$. Substituting the coordinates of A and B into this equation gives the following:

$$y_1 = mx_1 + b \quad \text{and therefore} \quad y_1 - mx_1 = b$$
$$y_2 = mx_2 + b \quad \text{and therefore} \quad y_2 - mx_2 = b$$

By equating these two values for b and solving for m, we identify the formula for slope.

$$y_1 - mx_1 = y_2 - mx_2$$
$$mx_2 - mx_1 = y_2 - y_1$$
$$m(x_2 - x_1) = y_2 - y_1$$
$$m = \frac{y_2 - y_1}{x_2 - x_1}$$

run / rise

The slope of the line $\overleftrightarrow{AB}$ is the change in y-coordinates divided by the corresponding change in x-coordinates of any two points on $\overleftrightarrow{AB}$. The difference $x_2 - x_1$ is the **run**, and the difference $y_2 - y_1$ is the **rise**. Thus the slope is often defined as "rise over run," or $\frac{\text{rise}}{\text{run}}$. The slope formula can be interpreted geometrically, as shown in Figure 14-21. The ratio $\frac{y_2 - y_1}{x_2 - x_1}$ is always the same, regardless of which two points on a given nonvertical line are chosen. This fact is illustrated in Figure 14-22, where right triangles have been constructed and shaded on each line. In each triangle, the horizontal side is the run and the vertical side is the rise.

FIGURE 14-21

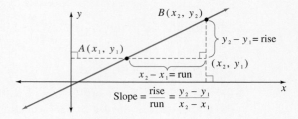

The slope of each line can be calculated as the rise over the run in any of the shaded triangles, with hypotenuse along the given line. To test this fact, notice that

$$\text{for } y = \frac{1}{2}x, \quad m = \frac{\text{rise}}{\text{run}} = \frac{1\frac{1}{2}}{3} = \frac{2}{4} = \frac{1}{2}$$

$$\text{for } y = x, \quad m = \frac{\text{rise}}{\text{run}} = \frac{1}{1} = \frac{2}{2} = \frac{3}{3}$$

$$\text{for } y = 3x, \quad m = \frac{3}{1} = \frac{6}{2}.$$

FIGURE 14-22

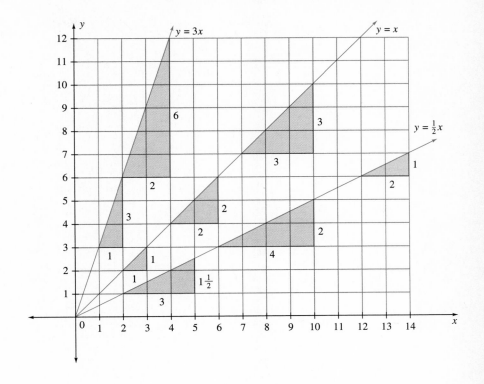

The discussion of slope is summarized in the following.

SLOPE FORMULA

Given two points $A(x_1, y_1)$ and $B(x_2, y_2)$ with $x_1 \neq x_2$, the slope m of the line $\overleftrightarrow{AB}$ is given by

$$m = \frac{y_2 - y_1}{x_2 - x_1} = \frac{\text{rise}}{\text{run}}.$$

By multiplying both the numerator and the denominator on the right side of the slope formula by $^-1$, we obtain

$$m = \frac{y_2 - y_1}{x_2 - x_1} = \frac{(y_2 - y_1)(^-1)}{(x_2 - x_1)(^-1)} = \frac{y_1 - y_2}{x_1 - x_2}.$$

This shows that while it does not matter which point is named (x_1, y_1) and which is named (x_2, y_2), *the order of the coordinates in the subtraction must be consistent.*

When a line is inclined from the left downward to the right, the slope is negative. This is illustrated in Figure 14-23, where the graph of the line $y = {}^-2x$ is shown. The slope of line $y = {}^-2x$ can be calculated as $\dfrac{\text{rise}}{\text{run}} = \dfrac{^-4}{2} = \dfrac{^-2}{1}$.

FIGURE 14-23

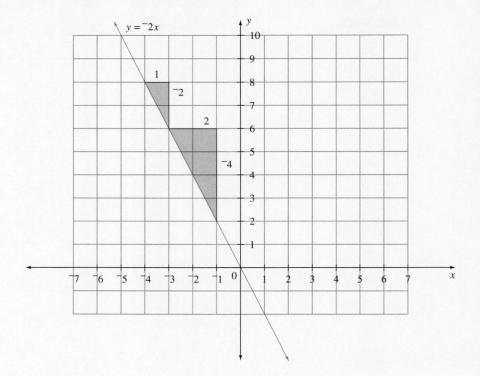

EXAMPLE 14-9 (a) Find the slope of $\overleftrightarrow{AB}$, given $A(3, 1)$ and $B(5, 4)$.

(b) Find the slope and the equation of the line passing through the points $A(^-3, 4)$ and $B(^-1, 0)$.

Solution (a) $m = \dfrac{4-1}{5-3} = \dfrac{3}{2}$, or $\dfrac{1-4}{3-5} = \dfrac{^-3}{^-2} = \dfrac{3}{2}$

(b) $m = \dfrac{4-0}{^-3-(^-1)} = \dfrac{4}{^-2} = {}^-2$ ∎

The equation of the line can be written in the form $y = mx + b$. Because $m = {}^-2$, it follows that $y = {}^-2x + b$ and the value of b must be found. The required line contains each of the given points, so the coordinates of each point must satisfy the equation. We substitute the coordinates of $B(^-1, 0)$ into $y = {}^-2x + b$ and proceed as follows:

$$y = {}^-2x + b$$
$$0 = {}^-2(^-1) + b$$
$$0 = 2 + b$$
$${}^-2 = b$$

The required equation is $y = {}^-2x + {}^-2$, or $y = {}^-2x - 2$. To check that $y = {}^-2x - 2$ is the required equation, we would have to verify that the coordinates of both A and B satisfy the equation. The verification is left to the reader.

We use the slope formula to find the equation of a line, given any point on the line and the slope of the line. In Figure 14-24, line ℓ has slope m and contains a given point (x_1, y_1). Point (x, y) represents any other point on line ℓ if, and only if, the slope determined by

FIGURE 14-24

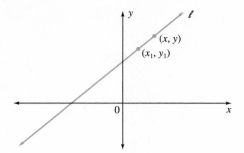

points (x_1, y_1) and (x, y) is m. We use the slope formula and proceed as follows:

$$\frac{y - y_1}{x - x_1} = m$$

$$y - y_1 = m(x - x_1)$$

point-slope form The result is the **point-slope form** of a line.

> **POINT-SLOPE FORM OF A LINE**
>
> The equation of a line with slope m through a given point (x_1, y_1) is $y - y_1 = m(x - x_1)$.

In Chapter 13, we developed a formula for conversion between Fahrenheit and Celsius temperatures. This formula can be developed using the notion of slope. Figure 14-25(a) shows a dual-scale thermometer, and Figure 14-25(b) shows the corresponding points plotted on a graph. The points appear to lie along a line. We use the points (0, 32) and

FIGURE 14-25

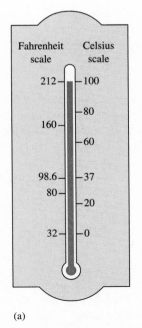

(a)

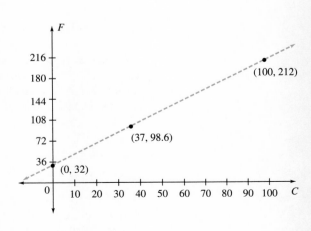

(b)

(100, 212) to find the slope as follows:

$$m = \frac{212 - 32}{100 - 0} = \frac{180}{100} = \frac{9}{5}.$$

Using the point-slope form of a line, we write the equation as follows:

$$F - 32 = \left(\frac{9}{5}\right)(C - 0), \text{ or } F = \left(\frac{9}{5}\right)C + 32.$$

We could check corresponding points from the thermometer to see that the equation holds in those cases.

To examine the slope of a vertical line, we pick any two points on the line, (x_1, y_1) and (x_2, y_2). Since the line is vertical, $x_1 = x_2$. Consequently,

$$m = \frac{y_2 - y_1}{x_2 - x_1} = \frac{y_2 - y_1}{0},$$

which is not meaningful. Thus *the slope of a vertical line is undefined.* Because two nonvertical lines with the same slope are parallel and vertical lines are parallel, consequently, we have the following.

PROPERTY OF SLOPES OF PARALLEL LINES

Any two lines are parallel if they both have the same slope or if both lines have undefined slope.

Perpendicular Lines

What is the relationship between the slopes of two perpendicular lines that are not vertical? We first consider a special case when the lines go through the origin. Suppose the slopes of the lines ℓ_1 and ℓ_2 shown in Figure 14-26 are m_1 and m_2, respectively. Because the slope of a line is equal to rise over run, the slope of ℓ_1 can also be determined from $\triangle OAB$, in which we choose $OA = 1$. We have $m_1 = \dfrac{\text{rise}}{\text{run}} = \dfrac{BA}{1} = BA.$

We now rotate the plane 90° counterclockwise about center O. The image of ℓ_1 is ℓ_2. To determine the image of $\triangle OBA$, we need only determine the images of each of the vertices of the triangle. The image of O is O itself. The image of A is A' on the y-axis (why?), and the image of B is B' on ℓ_2 (why?). Because rotation preserves congruence, $\triangle OB'A' \cong \triangle OBA$; consequently, $\angle A'$ is a right angle, $A'B' = m_1$, and $OA' = 1$. Thus as shown in Figure 14-26, point B' is at $(^-m_1, 1)$. We can use the slope formula to find the slope of ℓ_2 as follows:

$$m_2 = \frac{1 - 0}{^-m_1 - 0} = \frac{1}{^-m_1} = \frac{^-1}{m_1}.$$

Thus $m_2 = {^-1}/m_1$, or $m_1 m_2 = {^-1}$.

The relationship between the slopes m_1 and m_2 of two perpendicular lines (neither of which is vertical) that do not intersect at the origin can always be found using two lines parallel to the original lines but that pass through the origin. Because parallel lines have equal slopes, the relationship between the slopes of the perpendicular lines is the same as

FIGURE 14-26

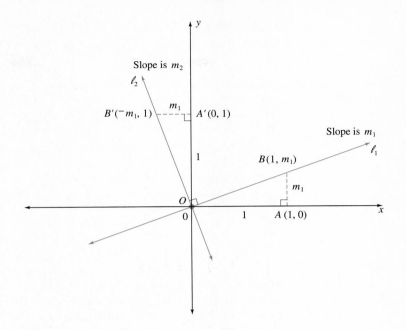

the relationship between the slopes of the perpendicular lines through the origin, that is, $m_1 m_2 = {}^-1$.

It is also possible to prove the converse statement, that is, if the slopes of two lines satisfy the condition $m_1 m_2 = {}^-1$, then the lines are perpendicular. We summarize these results in the following property.

> **PROPERTY OF SLOPES OF PERPENDICULAR LINES**
>
> Two lines, neither of which is vertical, are perpendicular if, and only if, their slopes m_1 and m_2 satisfy the condition $m_1 m_2 = {}^-1$. Any vertical line is perpendicular to a line with slope 0.

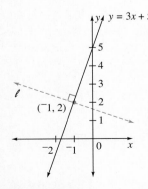

FIGURE 14-27

EXAMPLE 14-10   Find the equation of line ℓ through point $({}^-1, 2)$ and perpendicular to the line $y = 3x + 5$.

Solution   To find the equation of the dashed line in Figure 14-27, we need only know a point on the line and the slope. We have such a point, $({}^-1, 2)$, on line ℓ. Suppose m is the slope of ℓ. Then, because the line $y = 3x + 5$ has slope 3 and is perpendicular to ℓ, we have $m \cdot 3 = {}^-1$; therefore $m = -\dfrac{1}{3}$. We find the equation of ℓ by using the point-slope form of a line:

$$y - 2 = -\frac{1}{3}(x + 1).$$

This equation can also be written as

$$y = -\frac{1}{3}x + \frac{5}{3}.$$

EXAMPLE 14-11 In Figure 14-28, $\overline{AB}$ is a diameter and C is any other point on the circle. For $\triangle ABC$, prove that $\angle C$ is a right angle.

FIGURE 14-28

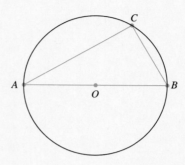

Solution We draw a coordinate system with origin through the center O of the circle, as in Figure 14-29. If the radius of the circle is r, then the coordinates of B are $(r, 0)$ and the coordinates of A are $(^-r, 0)$. Further, because the point C is an arbitrary point on the circle, we designate its coordinates by (x, y). To prove that $\angle C$ is a right angle, we establish a *subgoal* that $\overline{AC}$ is perpendicular to $\overline{BC}$ by showing that the product of the slopes of $\overleftrightarrow{AC}$ and $\overleftrightarrow{BC}$ is $^-1$. If m_{AC} and m_{BC} are the slopes of $\overleftrightarrow{AC}$ and $\overleftrightarrow{BC}$, respectively, we have

$$m_{AC} \cdot m_{BC} = \frac{y-0}{x-(^-r)} \cdot \frac{y-0}{x-r} = \frac{y^2}{(x+r)(x-r)} = \frac{y^2}{x^2-r^2}.$$

FIGURE 14-29

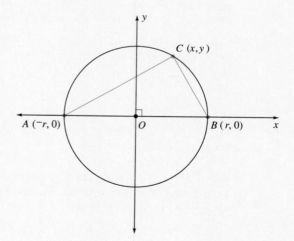

We need to show now that $\frac{y^2}{x^2 - r^2} = ^-1$. So far, we have not used the fact that C is on the circle. Because the distance from $C(x, y)$ to $(0, 0)$ is r, we have $\sqrt{(x-0)^2 + (y-0)^2} = r$, or $x^2 + y^2 = r^2$, which is the equation of the circle. Consequently, $x^2 - r^2 = {}^-y^2$. Substituting ${}^-y^2$ for $x^2 - r^2$, we have

$$m_{AC} \cdot m_{BC} = \frac{y^2}{x^2 - r^2} = \frac{y^2}{-y^2} = {}^-1.$$

Consequently, $\overleftrightarrow{AC}$ and $\overleftrightarrow{BC}$ are perpendicular, and therefore $\angle C$ is a right angle. ∎

Graphing Inequalities

In mathematics, we often need to graph inequalities as well as equations. For example, consider the inequality $x < 2$. Mathematicians agree that in a plane, the statement $x < 2$ indicates all points (x, y) in the plane for which $x < 2$ and y is any real number. Because there are no restrictions on the y-coordinates, the coordinates of any point to the left of the line with equation $x = 2$, but not on the line, satisfy the condition. The fact that the line is not included is indicated using a dashed line. The half-plane in Figure 14-30(a) is the graph.

Now consider the inequality $y - 3x > {}^-6$. This inequality is equivalent to $y > 3x + ({}^-6)$. A point whose coordinates satisfy $y > 3x + ({}^-6)$ is above the line represented by $y = 3x + ({}^-6)$. Consequently, the graph of the inequality is the half-plane above the line given by $y = 3x + ({}^-6)$. The graph is sketched in Figure 14-30(b).

FIGURE 14-30

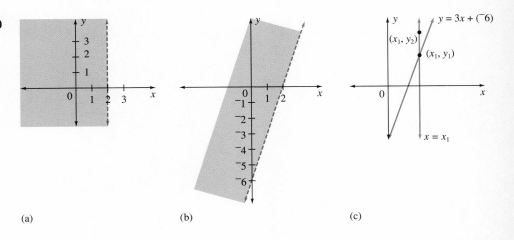

(a)   (b)   (c)

The fact that the graph of $y > 3x + ({}^-6)$ is the region above the line can be explained by using Figure 14-30(c) as follows. For every point (x_1, y_1) on the line $y = 3x + ({}^-6)$, we have $y_1 = 3x_1 + ({}^-6)$. If $y_2 > y_1$, then (x_1, y_2) is on the line $x = x_1$ above the point (x_1, y_1).

The graph of any inequality in one of the forms $y > mx + b$ or $y < mx + b$ is a half-plane either above or below the line $y = mx + b$. Thus in order to graph an inequality like $y > mx + b$, we first graph the corresponding straight line. Then we check some point not on the line to see if it satisfies the inequality. If it does, the half-plane containing the point is the graph, and if not, the half-plane not including the point is the graph. For example, checking $(0, 0)$ in $y - 3x > {}^-6$ gives $0 - 3 \cdot 0 > {}^-6$, which is a true statement. Thus the half-plane determined by $y - 3x = {}^-6$ and containing the origin is the graph of the inequality, as pictured in Figure 14-30(b).

EXAMPLE 14-12 Graph the following inequalities on the same coordinate system to determine all points that satisfy both inequalities:

(1) $2x + 3y > 6$

(2) $x - y \leq 0$

Solution First, we graph the lines represented by the equations $2x + 3y = 6$ and $x - y = 0$. To determine the half-planes to be shaded, we check point $(0, 0)$ in inequality (1) and obtain $0 > 6$, a false statement, so the required half-plane determined by the first inequality does not contain the point $(0, 0)$. In Figure 14-31, the graph of this inequality is marked with magenta. Substituting $x = 0$ and $y = 0$ in inequality (2) gives $0 \leq 0$, a true statement. Thus $(0, 0)$ is part of the required solution. However, $(0, 0)$ is on the line $x - y = 0$, so it is in neither of the half-planes determined by this line. Another point must be checked. Consider, for example, $(1, 2)$. Substituting $x = 1$ and $y = 2$ in inequality (2) gives $^-1 \leq 0$, a true statement; hence, $(1, 2)$ is in the half-plane determined by equation (2). In Figure 14-31, the graph of this inequality is marked with yellow. Hence, the set of common points is the brown portion of Figure 14-31.

FIGURE 14-31

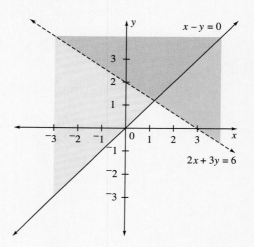

PROBLEM SET 14-2

1. Sketch the graphs of the equations $y = ^-x$ and $y = ^-x + 3$ on the same coordinate system.

2. ▶The graph of $y = mx$ is given in the figure to the right. Sketch the graphs for each of the following on the same figure. Explain your answers.
   (a) $y = mx + 3$ (b) $y = mx - 3$

3. Sketch the graphs for each of the following equations or inequalities:
   (a) $y = \dfrac{^-3}{4}x + 3$ (b) $y = ^-3$
   (c) $y \geq 15x - 30$ (d) $x = ^-2$
   (e) $y = 3x - 1$ (f) $y \leq \dfrac{1}{20}x$

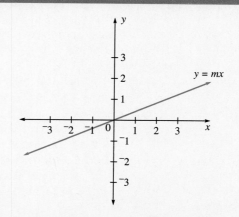

4. Find the x-intercept and y-intercept for the equations in Problem 3, if they exist.

5. In Chapter 13, a relationship between the Fahrenheit and Celsius scales for measuring temperature was discussed. Sketch the graphs of (a) and (b).
   (a) The temperature y in degrees Fahrenheit as a function of the temperature x in degrees Celsius using the formula $y = \frac{9}{5}x + 32$
   (b) The temperature y in degrees Celsius as a function of the temperature x in degrees Fahrenheit using the formula $y = \frac{5}{9}(x - 32)$
   ▶(c) If the graphs in (a) and (b) are drawn in the same coordinate system, describe their relationship to each other.

6. Write each of the following equations in slope-intercept form:
   (a) $3y - x = 0$
   (b) $x + y = 3$
   (c) $\frac{x}{3} + \frac{y}{4} = 1$
   (d) $3x - 4y + 7 = 0$
   (e) $x = 3y$
   (f) $x - y = 4(x - y)$

7. For each of the following, find the slope, if it exists, of the line determined by the given pair of points:
   (a) $(4, 3)$ and $(^-5, 0)$
   (b) $(^-4, 1)$ and $(5, 2)$
   (c) $(\sqrt{5}, 2)$ and $(1, 2)$
   (d) $(^-3, 81)$ and $(^-3, 198)$
   (e) $(1.0001, 12)$ and $(1, 10)$
   (f) (a, a) and (b, b)

8. For each of the following, write the equation of the line determined by the given pair of points in slope-intercept form or in the form $x = a$:
   (a) $(^-4, 3)$ and $(1, ^-2)$
   (b) $(0, 0)$ and $(2, 1)$
   (c) $(0, 1)$ and $(2, 1)$
   (d) $(2, 1)$ and $(2, ^-1)$
   (e) $\left(0, \frac{-1}{2}\right)$ and $\left(\frac{1}{2}, 0\right)$
   (f) $(^-a, 0)$ and $(a, 0)$, $a \neq 0$

9. For each of the following, find the equation of the line that passes through the given point and has the given slope:
   (a) $(^-3, 0)$ with slope $\frac{-1}{2}$
   (b) $(1, ^-3)$ with slope $\frac{2}{3}$
   (c) $(2, ^-3)$ with slope 0
   (d) $(^-1, ^-5)$ with slope $\frac{-5}{7}$

10. Use slopes to determine which of the following pairs of lines are parallel:
    (a) $y = 2x - 1$ and $y = 2x + 7$
    (b) $4y - 3x + 4 = 0$ and $8y - 6x + 1 = 0$

(c) $y - 2x = 0$ and $4x - 2y = 3$
(d) $\frac{x}{3} + \frac{y}{4} = 1$ and $y = \frac{4}{3}x$

11. For each of the following, find the equation of a line through $P(^-2, 3)$ and parallel to the line represented by the given equation:
    (a) $y = ^-2x$
    (b) $3y + 2x + 1 = 0$
    (c) $x = 0$
    (d) $y = ^-1$
    (e) $x = 3$
    (f) $y = ^-4$
    (g) $x + y = 2$
    (h) $\frac{x}{2} + \frac{y}{3} = 1$

12. Classify each of the following pairs of lines as parallel, perpendicular, or neither:
    (a) $3y - x = 1$, $2y + 6x - 1 = 0$
    (b) $2y - x - 3 = 0$, $2x - 4y + 16 = 0$
    (c) $x = 3$, $y = 0$
    (d) $x + y = 3$, $x - 2y = 1$

13. The door on a house was 4 ft above ground level. To allow handicap access, a ramp with a slope of $\frac{1}{10}$ was placed from the ground to the door. How long was the ramp?

14. Use slopes to show that the vertices $A(2, 1)$, $B(3, 5)$, $C(^-5, 1)$, and $D(^-6, ^-3)$ form a parallelogram.

15. Use slopes to show that the points represented by $(0, ^-1)$, $(1, 2)$, and $(^-1, ^-4)$ are collinear.

16. Find the x-intercept and y-intercept of the line whose equation is $\frac{x}{a} + \frac{y}{b} = 1$, where $a \neq 0$ and $b \neq 0$.

17. Find the equation of the reflection image of the line $y = 3x + 1$ in each of the following:
    (a) The x-axis
    (b) The y-axis
    (c) The line $y = x$

18. Given the points $A(^-1, 1)$, $B(1, 2)$ and $C(^-3, ^-2)$, write an equation for each of the following:
    (a) The line through A perpendicular to $\overleftrightarrow{BC}$
    (b) The line through B perpendicular to $\overleftrightarrow{AC}$

19. Use slopes to determine whether the triangle with the given vertices is a right triangle or not.
    (a) $A(^-2, 3)$, $B(^-3, 5)$, $C(4, 6)$
    (b) $A(3, 1)$, $B(2, 0)$, $C(^-1, 1)$

20. Verify that the diagonals of the following square $OABC$ are perpendicular bisectors of each other:

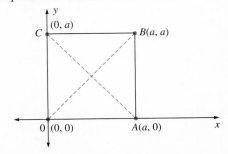

21. Find the coordinates of two other points collinear with each of the following pairs of given points:
    (a) $P(2, 2), Q(4, 2)$
    (b) $P(^-1, 0), Q(^-1, 2)$
    (c) $P(^-3, 0), Q(3, 0)$
    (d) $P(0, ^-2), Q(0, 3)$
    (e) $P(0, 0), Q(0, 1)$
    (f) $P(0, 0), Q(1, 1)$

22. For each of the following, give as much information as possible about x and y:
    (a) The ordered pairs $(^-2, 0), (^-2, 1)$, and (x, y) represent collinear points.
    (b) The ordered pairs $(^-2, 1), (0, 1)$, and (x, y) represent collinear points.
    (c) The ordered pair (x, y) is in the fourth quadrant.

23. Consider the lines through $P(2, 4)$ and perpendicular to the x- and y-axes, respectively. Find both the area and the perimeter of the rectangle formed by these lines and the axes.

24. Sketch the graphs for each of the following:
    (a) $x = ^-3$
    (b) $y = ^-1$
    (c) $x > ^-3$
    (d) $y > ^-1$
    (e) $x > ^-3$ and $y = 2$
    (f) $y \geq ^-1$ and $x = 0$

25. Find the equations for each of the following:
    (a) The line containing $P(3, 0)$ and perpendicular to the x-axis
    (b) The line containing $P(0, ^-2)$ and parallel to the x-axis
    (c) The line containing $P(^-4, 5)$ and parallel to the x-axis
    (d) The line containing $P(^-4, 5)$ and parallel to the y-axis

26. A *lattice point* is a point whose coordinates are integers. Find the lattice points in the graph of each of the following:
    (a) $x + y = 5$, and the points are in the first quadrant or on the coordinate axes.
    (b) $x - y = 5$, and the points are in the fourth quadrant or on the coordinate axes.
    (c) $|x + y| = 5$, and the points are in the first or third quadrant or on the coordinate axes.

27. Graph all the lattice points satisfying each of the following:
    (a) $y = x + 2$, x and y are integers, and $0 \leq x \leq 4$
    (b) $y = ^-x + 1$, x is an integer, and $^-1 \leq x \leq 3$
    (c) $x^2 + y^2 = 9$, and either x or y is an integer
    (d) $y = x^2$, and $|x| \leq 4$ and x is an integer
    (e) $xy = 5$, and $|x| \leq 4$ and x is an integer

28. Describe algebraically all the points in the following regions:

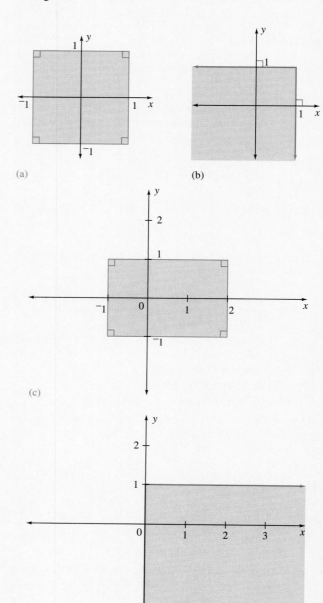

29. Find the equation of the image of the line $x = 3$ in the following circumstances:
    (a) The line is reflected in the x-axis.
    (b) The line is reflected in the y-axis.
    (c) The line is rotated 90° counterclockwise about the origin.

30. Graph each of the following inequalities:
    (a) $x - y + 3 > 0$
    (b) $2x > 3y$
    (c) $x - 2y + 1 \leq 0$
    (d) $x - 2y + 1 \geq 0$
31. Graph all points that satisfy the following:
    (a) $x - y + 3 > 0$ and $y - 2x - 1 < 0$
    (b) $x > 3$, $y < {}^-4$ and $x - y - 3 > 0$
    (c) $x + y + 3 = 0$ and $x - y > 3$
32. Find the equation of the image of the line $y = x$ under the following transformations:
    (a) Reflection in the x-axis
    (b) Reflection in the y-axis
    (c) Rotation of 45° counterclockwise about the origin
    (d) Rotation of 90° counterclockwise about the origin
    (e) Half-turn with the center at the origin
    (f) Translation in the direction of the x-axis, three units to the right
    (g) Translation in the direction of the y-axis, three units up
    (h) Sketch the graphs of $y = 2x + 3$ and $y = {}^-2x - 3$ on the same coordinate system. How are the graphs related?
    (i) How are the graphs $y = mx + b$ and $y = {}^-mx - b$ related?
33. Find the equation of the image of the line $x - y = 1$ under the following transformations:
    (a) Reflection in the x-axis
    (b) Reflection in the y-axis
    (c) Reflection in the line $y = x$
    (d) Half-turn with the center at the origin

★34. Graph each of the following equations:
    (a) $y = |x|$
    (b) $|y| = x$
    (c) $|y| = |x|$
★35. ▶What is an empty graph? Write an equation whose graph is an empty set. Explain your answer.

Review Problems

36. Find the equations for each of the following:
    (a) The line containing $({}^-7, {}^-8)$ and parallel to the x-axis
    (b) The line containing $({}^-7, {}^-8)$ and perpendicular to the x-axis
37. Find the coordinates of two other points collinear with the given points.
    (a) $P({}^-2, {}^-2)$, $Q({}^-4, {}^-2)$
    (b) $P({}^-7, {}^-8)$, $Q(3, 4)$
38. Plot each of the points $A(2, 2)$, $B(6, 6)$, and $C(8, {}^-4)$, and then find the coordinates of a point D such that quadrilateral $ABCD$ is a parallelogram.
39. Find the area of the triangle whose vertices are the following:
    (a) $(0, 0), (3, 0), (1, 1)$
    (b) $(0, 0), (6, 3), (10, 0)$
    (c) $(0, 0), (0, {}^-5), (3, 3)$
    (d) $(1, 2), (5, 2), ({}^-8, 8)$
    (e) $(0, 0), (1, 1), (7, 1)$

LABORATORY ACTIVITY

As van Hiele Level 2 activities, answer the following:

1. Find the slopes of each of the segments drawn on the following geoboard:

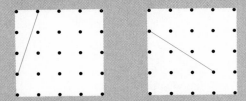

2. Use the concept of slope to draw two segments perpendicular to each of the segments in Activity 1.

Section 14-3 Systems of Linear Equations

The mathematical descriptions of many problems involve more than one equation, each involving more than one unknown. To solve such problems, we must find a common solution to the equations, if it exists. An example is given on the student page from *Addison-Wesley Mathematics*, Grade 8, 1991.

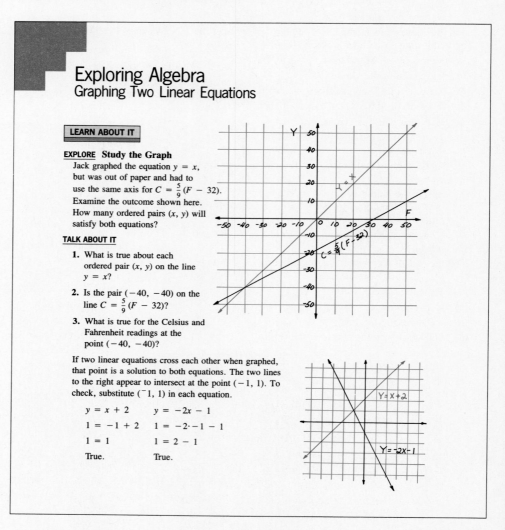

Any solution to a system of linear equations is an ordered pair (x, y) that satisfies both equations. Systems of linear equations arise in many story problems. Consider the following.

EXAMPLE 14-13 May Chin ordered lunch for herself and several friends by phone without checking prices. Once, she paid $7.00 for 5 soyburgers and 4 orders of fries, and another time she paid $6.00 for 4 of each. Set up a system of equations with two unknowns representing the prices of a soyburger and an order of fries, respectively.

Solution Let x be the price in dollars of a soyburger, and let y be the price of an order of fries. Five soyburgers cost $5x$ dollars, and 4 orders of fries cost $4y$ dollars. Because May paid $7.00 for the order, we have $5x + 4y = 7$. Similarly, $4x + 4y = 6$, or $2x + 2y = 3$. ∎

Systems with One Solution

An ordered pair satisfying both equations is a point that belongs to each of the lines. Figure 14-32 shows the graphs of $5x + 2y = 6$ and $x - 4y = -1$. The two lines appear to intersect at $\left(1, \frac{1}{2}\right)$. Thus $\left(1, \frac{1}{2}\right)$ appears to be the solution of the given system of equations. This solution can be checked by substituting 1 for x and $\frac{1}{2}$ for y in each equation. Because two distinct lines intersect in only one point, $\left(1, \frac{1}{2}\right)$ is the only solution to the system.

FIGURE 14-32

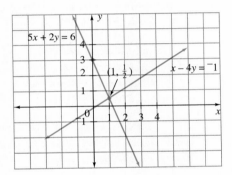

Drawbacks to estimating a solution to a system of equations graphically include an inability to read noninteger real-number coordinates of points. However, there are algebraic methods for solving systems of linear equations. Consider, for example, the system $y = x + 3$ and $y = 2x - 1$. By substitution, $x + 3 = 2x - 1$, an equation with one unknown. Solving for x gives $3 + 1 = 2x - x$, and, hence, $4 = x$. Substituting 4 for x in either equation gives $y = 7$. Thus $(4, 7)$ is the solution to the given system. As before, this solution can be checked by substituting the obtained values for x and y in the original equation. This method for solving a system of linear equations is called the **substitution method**.

substitution method

EXAMPLE 14-14 Solve the following system:

$$3x - 4y = 5$$
$$2x + 5y = 1$$

Solution First, rewrite each equation, expressing y in terms of x.

$$y = \frac{3x - 5}{4} \quad \text{and} \quad y = \frac{1 - 2x}{5}$$

Then equate the expressions for y and solve the resulting equation for x.

$$\frac{3x-5}{4} = \frac{1-2x}{5}$$
$$5(3x-5) = 4(1-2x)$$
$$15x - 25 = 4 - 8x$$
$$23x = 29$$
$$x = \frac{29}{23}$$

Substituting $\frac{29}{23}$ for x in $y = \frac{3x-5}{4}$ gives $y = \frac{^-7}{23}$. Hence, $x = \frac{29}{23}$ and $y = \frac{^-7}{23}$. This can be checked by substituting the values for x and y in the original equations. ∎

> **REMARK** Sometimes it is more convenient to solve a system of equations by expressing x in terms of y in one of the equations and substituting the obtained expression for x in the other equation.

elimination method

The **elimination method** for solving two equations with two unknowns is based on eliminating one of the variables by adding or subtracting the original or equivalent equations. For example, consider the following system:

$$x - y = ^-3$$
$$x + y = 7$$

By adding the two equations, we can eliminate the variable y. The resulting equation can then be solved for x.

$$\begin{array}{rl} x - y = & ^-3 \\ x + y = & 7 \\ \hline 2x = & 4 \\ x = & 2 \end{array}$$

Substituting 2 for x in the first equation (either equation may be used) gives $y = 5$. Checking this result shows that $x = 2$ and $y = 5$, or $(2, 5)$, is the solution to the system.

Often, another operation is required before equations are added so that an unknown can be eliminated. For example, consider the following system:

$$3x + 2y = 5$$
$$5x - 4y = 3$$

Adding the equations does not eliminate either unknown. However, if the first equation contained $4y$ rather than $2y$, the variable y could be eliminated by adding. To obtain $4y$ in the first equation, we multiply both sides of the equation by 2 to obtain the equivalent equation $6x + 4y = 10$. Adding the equations in the equivalent system gives the following:

$$\begin{array}{rl} 6x + 4y = & 10 \\ 5x - 4y = & 3 \\ \hline 11x = & 13 \\ x = & \frac{13}{11} \end{array}$$

To find the corresponding value of y, we substitute $\frac{13}{11}$ for x in either of the original equations and solve for y, or we use the elimination method again and solve for y.

An alternate method is to eliminate the x-values from the original system by multiplying the first equation by 5 and the second by $^-3$ (or the first by $^-5$ and the second by 3). Then we add the two equations and solve for y.

$$15x + 10y = 25$$
$$^-15x + 12y = {}^-9$$
$$\overline{22y = 16}$$
$$y = \frac{16}{22}, \quad \text{or} \quad \frac{8}{11}$$

Consequently, $\left(\frac{13}{11}, \frac{8}{11}\right)$ is the solution of the original system. This solution, as always, should be checked by substitution in the *original* equations.

Solutions to Other Systems

All examples thus far have had unique solutions. However, other situations may arise. Geometrically, a system of two linear equations can be characterized as follows:

1. *The system has a unique solution if, and only if, the graphs of the equations intersect in a single point.*
2. *The system has no solution if, and only if, the equations represent parallel lines.*
3. *The system has infinitely many solutions if, and only if, the equations represent the same line.*

Consider the following system:

$$2x - 3y = 1$$
$$^-4x + 6y = 5$$

In an attempt to solve for x, we multiply the first equation by 2 and then add as follows:

$$4x - 6y = 2$$
$$^-4x + 6y = 5$$
$$\overline{0 = 7}$$

A false statement results. Logically, a false result can occur only on the basis of a false assumption or an incorrect procedure. Because our procedure is correct in this case, there must be a false assumption. We assumed that the system has a solution. That assumption caused a false statement; therefore the assumption itself must be false. Hence, the system has no solution. In other words, the solution set is $\emptyset$. This situation arises if, and only if, the corresponding lines are parallel.

Next, consider the following system:

$$2x - 3y = 1$$
$$^-4x + 6y = {}^-2$$

To solve this system, we multiply the first equation by 2 and add as follows:

$$4x - 6y = 2$$
$$^-4x + 6y = {}^-2$$
$$\overline{0 = 0}$$

The resulting statement, $0 = 0$, is always true. Rewriting the equation as $0 \cdot x + 0 \cdot y = 0$ shows that all values of x and y satisfy this equation. The values of x and y that satisfy both $0 \cdot x + 0 \cdot y = 0$ and $2x - 3y = 1$ are those that satisfy $2x - 3y = 1$. Infinitely many such pairs x and y correspond to points on the line $2x - 3y = 1$ and hence to $^-4x + 6y = ^-2$.

One way to check whether a system has infinitely many solutions is to see if each of the original equations represents the same line. In the preceding system, both equations may be written as $y = \frac{2}{3}x - \frac{1}{3}$. Another way to check whether a system has infinitely many solutions is to observe if one equation can be multiplied by some number to obtain the second equation. For example, multiplying the equation $2x - 3y = 1$ by $^-2$ yields the second equation, $^-4x + 6y = ^-2$.

EXAMPLE 14-15 Identify each of the following systems as having a unique solution, no solution, or infinitely many solutions:

(a) $2x - 3y = 5$
$\frac{1}{2}x - y = 1$

(b) $\frac{x}{3} - \frac{y}{4} = 1$
$3y - 4x + 12 = 0$

(c) $6x - 9y = 5$
$^-8x + 12y = 7$

Solution One approach is to attempt to solve each of the systems. Another approach is to write each of the equations in slope-intercept form and interpret the system geometrically.

(a) *First method.* To eliminate x, multiply the second equation by $^-4$ and add the equations.

$$2x - 3y = 5$$
$$^-2x + 4y = ^-4$$
$$y = 1$$

Substituting 1 for y in either equation gives $x = 4$. Thus $(4, 1)$ is the unique solution of the system.

Second method. In slope-intercept form, the first equation is $y = \frac{2}{3}x - \frac{5}{3}$. The second equation is $y = \frac{1}{2}x - 1$. The slopes of the corresponding lines are $\frac{2}{3}$ and $\frac{1}{2}$, respectively. Consequently, the lines are distinct and are not parallel; therefore they intersect in a single point whose coordinates are the unique solution to the original system.

(b) *First method.* Multiply the first equation by 12 and rewrite the second equation as $^-4x + 3y = ^-12$. Adding the resulting equations gives the following:

$$4x - 3y = 12$$
$$^-4x + 3y = ^-12$$
$$0 = 0$$

Because every pair (x, y) satisfies $0 \cdot x + 0 \cdot y = 0$, the original system has infinitely many solutions.

Second method. In slope-intercept form, both equations have the form
$$y = \frac{4}{3}x - 4.$$
Thus the two lines are identical and the system has infinitely many solutions.

(c) *First method.* To eliminate y, multiply the first equation by 4 and the second by 3, then add the resulting equations.
$$24x - 36y = 20$$
$$\underline{^-24x + 36y = 21}$$
$$0 = 41$$

No pair of numbers satisfies $0 \cdot x + 0 \cdot y = 41$, so this equation has no solutions; consequently, the original system has no solutions.

Second method. In slope-intercept form, the first equation is $y = \frac{2}{3}x - \frac{5}{9}$. The second equation is $y = \frac{2}{3}x + \frac{7}{12}$. The corresponding lines have the same slope, $\frac{2}{3}$, but different y-intercepts. Consequently, the lines are parallel and the original system has no solution. ∎

EXAMPLE 14-16 The Park Street Restaurant offers two eggs with sausage for $1.80 and one egg with sausage for $1.35. If there is no break in price for quantity, what is the cost of one egg?

Solution Let x be the cost of one egg and y be the cost of the sausage. The cost of two eggs and sausage is $2x + y$, or $1.80. The cost of one egg and sausage is $x + y$, or $1.35. Thus the following system is obtained:
$$2x + y = 1.80$$
$$x + y = 1.35$$
Subtracting the equations gives $x = 0.45$. Thus the cost of one egg is $0.45. ∎

PROBLEM SET 14-3

1. Use the equation $2x - 3y = 5$ for each of the following:
   (a) Find four solutions of the equation.
   (b) Graph all the solutions for which $^-2 \leq x \leq 2$.
   (c) Graph all the solutions for which $0 \leq y \leq 2$.

2. Solve each of the following systems, if possible. Indicate whether the system has a unique solution, infinitely many solutions, or no solution.
   (a) $y = 3x - 1$
   $y = x + 3$
   (b) $2x - 6y = 7$
   $3x - 9y = 10$
   (c) $3x + 4y = ^-17$
   $2x + 3y = ^-13$
   (d) $8y - 6x = 78$
   $9x - 12y = 12$
   (e) $5x - 18y = 0$
   $x - 24y = 0$
   (f) $2x + 3y = 1$
   $3x - y = 1$

3. ▶Solve each of the following systems, if possible. If a solution is not possible, explain why not.
   (a) $y = x + 3$
   $3x - 4y + 1 = 0$
   (b) $\frac{x}{3} - \frac{y}{4} = 1$
   $\frac{x}{5} - \frac{y}{3} = 2$
   (c) $3x - 4y = ^-x + 3$
   $x - 2 = 4(y - 3)$
   (d) $x - y = \frac{2}{3}(x + y)$
   $x + y = \frac{2}{3}(x - y)$
   (e) $\sqrt{2}x - y = 3$
   $x - \sqrt{2}y = 1$
   (f) $x - y = \frac{x}{3}$
   $y - x = \frac{3}{4}y + 1$

4. ▶Consider the following systems of equations:
   - (i) $2x + 3y = 7$
     $x - y = 3$
   - (ii) $x - y = 13$
     $5x - 5y = 7$
   - (iii) $8x + 2y = 6$
     $4x + y = 3$

   (a) Find the slopes of each of the lines in the systems.
   (b) Determine the number of solutions to each of the systems.
   (c) Make a conjecture about a relationship between the slopes of the lines in the systems and the number of solutions of the systems.

5. Using the concept of slope in Problem 4, identify whether each of the following systems has a unique solution, infinitely many solutions, or no solution:
   - (a) $3x - 4y = 5$
     $\frac{x}{3} - \frac{y}{5} = 1$
   - (b) $4y - 3x + 4 = 0$
     $8y - 6x + 40 = 0$
   - (c) $3y - 2x = 15$
     $\frac{2}{3}x - y + 5 = 0$
   - (d) $x - y = x + y$
     $x = 0$

6. The vertices of a triangle are given by (0, 0), (10, 0), and (6, 8). Show that the segments connecting (5, 0) and (6, 8), (10, 0) and (3, 4), and (0, 0) and (8, 4) intersect at a common point.

7. Two adjacent sides of a parallelogram are on lines with equations $x - 3y + 3 = 0$ and $x + 2y - 2 = 0$. One vertex is at (0, ⁻4). Write the equations of the lines containing the other two sides.

8. Find the area of the triangle bounded by the lines $y = 2x + 3$, $x + y = 5$, and $x = 3$.

9. Triangle ABC has vertices A(0, 0), B(2, 5), C(6, 1).
   (a) Find the point of intersection P of the lines containing the altitude from A to $\overline{BC}$ and the altitude from B to $\overline{AC}$.
   (b) Show that the third altitude of △ABC goes through the point P.

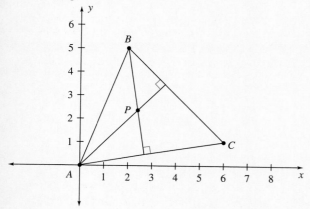

10. The sum of two numbers is $\frac{3}{4}$ and their difference is $\frac{7}{9}$. Find the numbers.

11. The owner of a 5000-gal oil truck loads the truck with gasoline and kerosene. The profit on each gallon of gasoline is 13¢ and on each gallon of kerosene is 12¢. Find how many gallons of each fuel the owner loaded if the profit was $640.

12. A health-food store has two different kinds of granola: cashew nut granola selling at $1.80/lb and golden granola selling at $1.20/lb. How much of each kind should be mixed to produce a 200-lb mixture selling at $1.60/lb?

13. A laboratory carries two different solutions of the same acid: a 60% solution and a 90% solution. How many liters of each solution should be mixed in order to produce 150 L of 80% solution?

14. A physician invests $80,000 in two stocks. At the end of the year, the physician sells the stocks: the first at a 15% profit and the second at a 20% profit. How much did the physician invest in each stock if the total profit was $15,000?

15. At the end of 10 mo, the balance of an account earning simple interest is $2100.
   (a) If, at the end of 18 mo, the balance is $2180, how much money was originally in the account?
   (b) What is the rate of interest?

16. If five times the width of a waterbed equals four times its length and its perimeter is 270 in., what are the length and width of the bed?

17. Josephine's bank contains 27 coins. If all the coins are either dimes or quarters and the value of the coins is $5.25, how many of each kind of coin are there?

18. ▶(a) Solve each of the following systems of equations. What do you notice about the answers?
   - (i) $x + 2y = 3$
     $4x + 5y = 6$
   - (ii) $2x + 3y = 4$
     $5x + 6y = 7$
   - (iii) $31x + 32y = 33$
     $34x + 35y = 36$

   (b) Write another system similar to those in (a). What solution did you expect? Check your guess.
   (c) Write a general system similar to those in (a). What solution does this system have? Why?

★19. Consider the following figure. Show that the three altitudes of a triangle are concurrent, that is, that they intersect in a single point.

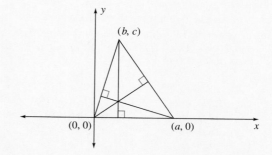

Review Problems

20. Which of the following are not equations of lines?
    (a) $x + y = 3$
    (b) $xy = 7$
    (c) $2x + 3y \leq 4$
    (d) $y = 5$

21. Find the slope and y-intercept of each of the following:
    (a) $6y + 5x = 7$
    (b) $\frac{2}{3}x + \frac{1}{2}y = \frac{1}{5}$
    (c) $0.2y - 0.75x - 0.37 = 0$
    (d) $y = 4$

22. Write the equations of each of the following:
    (a) A line through $(4, 7)$ and $(^-6, ^-2)$
    (b) A line through $(4, 7)$ parallel to the line $y = \frac{5}{3}x + 6$
    (c) A line through $(^-6, ^-8)$ perpendicular to the y-axis

23. Graph each of the following:
    (a) $3x + 2y = 14$
    (b) $^-x - y = ^-8$
    (c) $2x + 3y \leq 4$
    (d) $x - 16 \geq ^-8y$

24. Determine the slope of ℓ in each of the following:

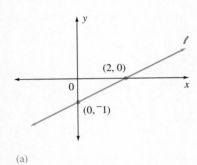

(a)

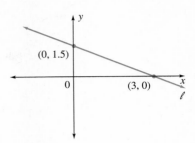

(b)

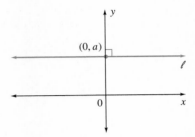

(c)

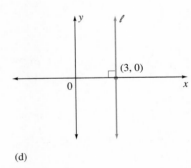

(d)

25. Find the equation of the line parallel to the x-axis that contains the point of intersection of the lines with equations $3x + 5y = 7$ and $2x - y = ^-4$.

▼ BRAIN TEASER

A school committee meeting began between 3:00 and 4:00 P.M. and ended between 6:00 and 7:00 P.M. The positions of the minute hand and the hour hand of the clock were reversed at the end of the meeting from what they were at the beginning of the meeting. When did the meeting start and end?

LABORATORY ACTIVITY

1. As a van Hiele Level 1 activity, approximate the area of $\triangle OAB$ in the following figure.

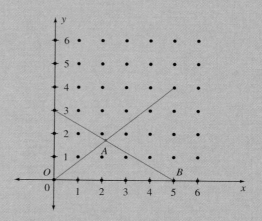

2. As a van Hiele Level 2 activity, find the exact area of $\triangle OAB$.
3. As a van Hiele Level 1 activity, estimate the area of the following $\triangle ABC$.

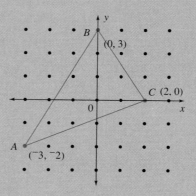

4. As van Hiele Level 2 activities, do the following:
   (a) Find the exact area of $\triangle ABC$ in the figure for Activity 3. (i) by using the distance formula, and (ii) without using the distance formula.
   (b) Draw a triangle that has the same area as $\triangle ABC$ in part (a) but a different perimeter.

*Section 14-4 Coordinate Geometry and Logo

In Logo, the screen is treated as a Cartesian coordinate system in which each point of the screen is associated with an ordered pair of numbers (x, y). The origin, $(0, 0)$, is at the turtle's home. Figure 14-33 pictures the point $(-30, 20)$.

FIGURE 14-33

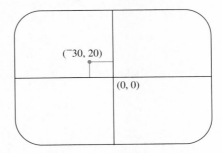

SETX

SETY

The primitive **SETX** is used with a number input to move the turtle in a horizontal direction to the position whose x-coordinate is the given input. This command changes neither the turtle's heading nor its y-coordinate. **SETY** works in the same way with the y-coordinate. For example, if the turtle is at home, SETY 50 moves the turtle vertically up to the point where it has 50 as its y-coordinate. Now SETY -50 moves the turtle down to the point where the y-coordinate is $^-50$. SETY does not affect the turtle's heading. Explore at which inputs to SETX and SETY the turtle begins to wrap around the screen. As an example of SETX and SETY, suppose we start the turtle at home and type the following:

```
SETX 50 SETY 70 SETX -30 SETY -50
```

The turtle's final position has coordinates $(^-30, ^-50)$, and the path drawn is shown in Figure 14-34.

FIGURE 14-34

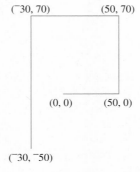

To place the turtle at the point with coordinates $(^-30, ^-50)$, we could type SETX -30 and then SETY -50. With the SETXY command, the operation can be done in one step.

SETXY **SETXY** (*SETPOS in LCSI*) takes two inputs: The first is the x-coordinate and the second is the y-coordinate. (*In LCSI, the inputs to SETPOS must be made in the form of a list, that is, they must be enclosed in brackets.*) For example, typing SETXY 30 50 (*SETPOS [30 50] in LCSI*) moves the turtle from its present position to the point with coordinates (30, 50) without changing the heading. Do you think that typing SETXY −30 −50 (*SETPOS [¯30 ¯50] in LCSI*) will move the turtle to the point with coordinates (¯30, ¯50)? Try it. If the experiment did not work, Logo probably treated −30 −50 as a subtraction and thought that there was only one input, namely, ¯80. Parentheses can be used to overcome this, as in SETXY −30 (−50). (*The parentheses are not necessary in LCSI.*)

EXAMPLE 14-17   Write a procedure, using coordinate commands, to draw the parallelogram in Figure 14-35.

FIGURE 14-35

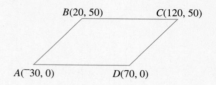

Solution   The solution is given in the following P.GRAM procedure:

```
TO P.GRAM
  PENUP
  SETXY -30 0
  PENDOWN
  SETXY 20 50
  SETXY 120 50
  SETXY 70 0
  SETXY -30 0
END
```

(*In LCSI, SETXY is replaced with SETPOS and the inputs are enclosed in brackets.*)

REMARK   Note that in P.GRAM, we could have used SETX 120 in place of SETXY 120 50 and SETX −30 in place of SETXY −30 0 (*or in place of SETPOS [120 50] and SETPOS [−30 0] in LCSI*).

Logo coordinate commands can be used for drawing geometric figures. Compare the following two procedures:

```
TO RECTANGLE :WIDTH :LENGTH
  REPEAT 2 [FD :WIDTH RT 90 FD :LENGTH RT 90]
END
```

```
TO RECTANGLE1 :WIDTH :LENGTH
  SETXY :WIDTH 0
  SETXY :WIDTH :LENGTH
  SETXY 0 :LENGTH
  SETXY 0 0
END
```

(In LCSI, when variables are used as inputs, SETXY is replaced with SETPOS LIST; for example, SETPOS LIST :WIDTH :LENGTH.)

If the turtle starts at home with heading 0, the RECTANGLE and RECTANGLE1 procedures produce identical figures. If the turtle is not at home, the figures drawn by the two procedures are different. Experiment with these procedures to see that this is true, and explain the phenomenon.

We use the coordinate system to investigate several other geometric concepts. Consider how we might write a procedure to draw a line segment connecting two points, given their coordinates (:X1, :Y1) and (:X2, :Y2). To draw the segment, we must pick up the pen, move the turtle to the coordinates of the first point, put down the pen, and have the turtle move to the second point. The procedure called SEGMENT is given in Figure 14-36, along with a figure produced by SEGMENT.

FIGURE 14-36

```
TO SEGMENT :X1 :Y1 :X2 :Y2
  PENUP
  SETXY :X1 :Y1
  PENDOWN
  SETXY :X2 :Y2
END
```
(In LCSI, replace SETXY with SETPOS LIST.)

SEGMENT 0 0 50 50

Now, consider how to write a procedure called COOR.TRI to draw a triangle, given the coordinates of its three vertices. The SEGMENT procedure can be used to draw the three sides. The procedure is given in Figure 14-37, along with an execution of it.

FIGURE 14-37

```
TO COOR.TRI :X1 :Y1 :X2 :Y2 :X3 :Y3
  SEGMENT :X1 :Y1 :X2 :Y2
  SEGMENT :X2 :Y2 :X3 :Y3
  SEGMENT :X3 :Y3 :X1 :Y1
END
```

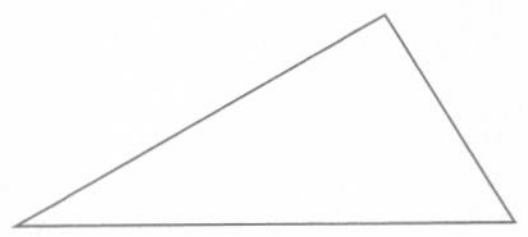

COOR.TRI 0 0 80 50 110 0

Next, we consider how to construct the three medians of the triangles generated by COOR.TRI. Recall that medians are line segments drawn from a vertex of a triangle to the midpoint of the opposite side. The midpoint is determined by using the midpoint formula as seen in Figure 14-38.

FIGURE 14-38
```
TO MIDPOINT :X1 :Y1 :X2 :Y2
   SETXY (:X1+:X2)/2 (:Y1+:Y2)/2
END
```
(In LCSI, replace SETXY with SETPOS LIST.)

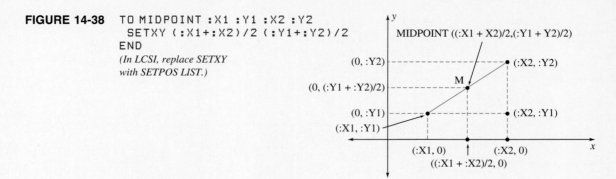

Once the sides of the triangle are drawn and the midpoints determined, the medians can be drawn by using the TRI.MEDIANS procedure given in Figure 14-39.

FIGURE 14-39
```
TO TRI.MEDIANS :X1 :Y1 :X2 :Y2 :X3 :Y3
   COOR.TRI :X1 :Y1 :X2 :Y2 :X3 :Y3
   SEGMENT :X1 :Y1 (:X2+:X3)/2 (:Y2+:Y3)/2
   SEGMENT :X2 :Y2 (:X1+:X3)/2 (:Y1+:Y3)/2
   SEGMENT :X3 :Y3 (:X1+:X2)/2 (:Y1+:Y2)/2
END
```

TRI.MEDIANS 0 0 80 50 110 0

medial triangle   A **medial triangle** is a triangle formed by connecting the midpoints of its three sides. A procedure called MEDIAL.TRI for drawing a medial triangle is given in Figure 14-40.

FIGURE 14-40
```
TO MEDIAL.TRI :X1 :Y1 :X2 :Y2 :X3 :Y3
   COOR.TRI :X1 :Y1 :X2 :Y2 :X3 :Y3
   MIDPOINT :X1 :Y1 :X2 :Y2
   MIDPOINT :X2 :Y2 :X3 :Y3
   MIDPOINT :X3 :Y3 :X1 :Y1
   MIDPOINT :X1 :Y1 :X2 :Y2
END
```

MEDIAL.TRI 0 0 80 50 110 0

Finding Lengths of Line Segments

It is possible to find the lengths of a line segment by using coordinate commands. Recall that we can find the length of a segment with the distance formula $d = \sqrt{(x_2 - x_1)^2 + (y_2 - y_1)^2}$. We can write a Logo procedure for computing the distance between any two points by using the distance formula and the SQRT primitive. **SQRT** takes one numerical input and outputs the square root of the number. The primitive OUTPUT (OP) is useful here. **OUTPUT (OP)** takes one input and causes the current procedure to stop and output the result to a calling procedure. For example, SQRT 2 gives RESULT: 1.41421. A DISTANCE procedure that prints out the distance between any two points

SQRT

OUTPUT (OP)

follows:

```
TO DISTANCE :X1 :Y1 :X2 :Y2
  OUTPUT SQRT (:X2 - :X1)*(:X2 - :X1) + (:Y2 - :Y1)
    *(:Y2 - :Y1)
END
```

If we execute DISTANCE $-10\ 20\ 20\ 40$, where $(-10, 20)$ and $(20, 40)$ are the coordinates of the two points, then RESULT: 36.0555 is displayed on the screen.

The DISTANCE procedure can also be used to compute the distance from a point with given coordinates even if we do not know the exact coordinates of the turtle. Logo has two primitives, XCOR and YCOR, that can be used to determine the coordinates of the turtle.

XCOR   If we type PRINT **XCOR** and press RETURN, the *x*-coordinate of the turtle is displayed.
YCOR   Similarly, PRINT **YCOR** yields the *y*-coordinate of the turtle. With this in mind, we can determine the distance from any point where the turtle is located to the point (5, 10) by executing DISTANCE 5 10 XCOR YCOR.

Constructing Triangles with SAS

MAKE   The **MAKE** command can be used to assign a value to a variable. For example, if we execute MAKE "X 5 and then type PRINT :X, which means "Print the value associated with X," then the computer will print 5. The MAKE command can be used inside or outside a procedure. In general, the MAKE command accepts two inputs: The first, preceded by a quotation mark, is the name of the variable and the second is the value of the variable we are defining. The MAKE command can be used with the other commands introduced in this chapter. For example, we could write a procedure called SAS.TRI to draw a triangle with two sides of length 80 and 50 and an included angle of 30°. To write a SAS.TRI procedure to accomplish this, we could tell the turtle to go forward 80 units, the length of the first side, and learn the *x*- and *y*-coordinates of its position. We could next reverse the turtle 80 units, turn it right 30°, and move it forward 50 units, the length of the second side. To complete the triangle, we could then send the turtle to the coordinates it learned when it drew the first side. The SAS.TRI procedure follows, and the result is shown in Figure 14-41.

FIGURE 14-41

```
TO SAS.TRI
  FD 80
  MAKE "X XCOR
  MAKE "Y YCOR
  BK 80
  RT 30
  FD 50
  SETXY :X :Y
END
```

SAS.TRI

(In LCSI, replace SETXY with SETPOS LIST.)

An alternate method of writing this procedure utilizes the primitive TOWARDS.
TOWARDS   **TOWARDS** takes two inputs, which are interpreted as *x*- and *y*-coordinates of a point, and outputs the heading from the turtle to that point. For example, if the turtle is at home and

pointed up, then typing TOWARDS 50 0 yields RESULT: 90. (*In LCSI, 50 0 must be placed in brackets.*) With SETHEADING, TOWARDS will turn the turtle in the direction of a point. For example, SETHEADING TOWARDS 50 0 will turn the turtle in place until it points toward the point with coordinates (50, 0). (*In LCSI, use SETHEADING TOWARDS [50 0].*) Thus the SAS.TRI procedure could be rewritten as follows:

```
TO SAS.TRI
    FD 80
    MAKE "X XCOR
    MAKE "Y YCOR
    BK 80
    RT 30
    FD 50
    SETHEADING TOWARDS :X :Y
    SETXY :X :Y
END
```

(*In LCSI, replace TOWARDS with TOWARDS LIST, and SETXY with SETPOS LIST.*)

It is left as an exercise to write a variable procedure called SAS with inputs :S1, :A, :S2 for drawing a triangle, given two sides and the included angle.

SUMMARY OF COMMANDS

Command	Description
SETX	Takes one number input and moves the turtle horizontally to the point with that *x*-coordinate. Draws a trail if the pen is down.
SETY	Takes one number input and moves the turtle vertically to the point with that *y*-coordinate. Draws a trail if the pen is down.
SETXY*	Takes two number inputs *A* and *B* and moves the turtle to the point with the given coordinates (*A*, *B*). Draws a trail if the pen is down.
SQRT	Takes one nonnegative number input and yields the square root of that input.
XCOR	Takes no inputs; outputs the turtle's current *x*-coordinate.
YCOR	Takes no inputs; outputs the turtle's current *y*-coordinate.
TOWARDS†	Takes two number inputs, which are interpreted as *x*- and *y*-coordinates of a point, and outputs the heading from the turtle to that point.
OUTPUT (OP)	Takes one input and causes the current procedure to stop and output the result to the calling procedure.
MAKE	Takes two inputs. The first, preceded by a quotation mark, is the name of a variable. The second is the value of the variable.

*In LCSI, SETPOS is used with the list of coordinates as input, for example, SETPOS [20, 30].

†In LCSI, TOWARDS is used with the list of coordinates as input, for example, TOWARDS [20, 30].

PROBLEM SET 14-4

1. Use SETX, SETY, and SETXY (*in LCSI, SETPOS*) to draw the largest rectangle possible on the monitor screen.
2. How are the triangles generated by the MEDIAL.TRI procedure related?
3. Write a procedure called AXES that uses the SETXY (*in LCSI, SETPOS*) primitive to draw the *x*- and *y*-axes with the origin placed at home.
4. How many different squares can be drawn if given two vertices of a square?

5. Write a procedure called FILL.RECT that uses the SETXY *(in LCSI, SETPOS)* command to fill (color in) a rectangle that has length 50 and width 30. Assume the turtle starts at home with heading 0.
6. Write a procedure called CCIRCLE to draw a circle in which the coordinates of the center of the circle and the radius are inputs. For example, CCIRCLE 10 20 30 should draw a circle that has center (10, 20) and radius 30.
7. Write a procedure called QUAD to draw a quadrilateral if the coordinates of the vertices are input.
8. (a) Write a procedure called MEDIAL.QUAD to draw a medial figure for a quadrilateral when the coordinates of the vertices are inputs.
   (b) Execute your procedure in (a) for several cases, and make a conjecture concerning the quadrilateral formed by the midpoints.
9. Write a procedure called MEDIAL.QUADS with inputs :X1, :Y1, :X2, :Y2, :X3, :Y3, :X4, and :Y4, which are the coordinates of the vertices, and :NUM, which gives the number of medial figures drawn.
10. Write a variable procedure called SAS to draw a triangle, given two sides :S1 and :S2 and the included angle :A.
11. Write a procedure called R.ISOS.TRI to draw a right isosceles triangle of side length given by the input :LEN.
12. Write a procedure to draw a circle that has radius 50 and passes through the point (⁻20, ⁻40).
13. Write a procedure to draw a circle in which the coordinates of the center of the circle and the radius are inputs.
★14. Write a procedure called MEDIAL.TRIS that uses recursion such that :NUM, the number of medial triangles that are to be drawn, is included as input.

Solution to the Preliminary Problem

Understanding the Problem. A county wants to build a regional airport that is equally close to three cities that determine a triangle. The cities are situated as shown in Figure 14-42. It is apparent that the distances shown among the cities fit the Pythagorean theorem, $30^2 + 40^2 = 50^2$ and thus determine a right triangle. We use the fact that the cities are vertices of a right triangle to determine the location for the airport.

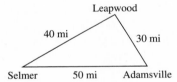

FIGURE 14-42

Devising a Plan. The distance between Adamsville and Selmer is the greatest, so it must be the length of the hypotenuse of the triangle, and Leapwood must be the vertex of the right angle. We could place Leapwood at the origin, with Selmer and Adamsville on the *x*- and *y*-axes, respectively. The coordinate system with the cities and the unknown location of the airport are shown in Figure 14-43. We could then use the distance formula to determine the required location (*x*, *y*) of the airport.

FIGURE 14-43

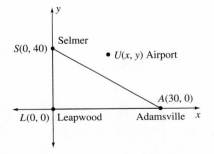

Carrying Out the Plan. Using the information of Figure 14-43, we obtain the following distances where A represents Adamsville, S represents Selmer, L represents Leapwood, and U represents the airport.

$$US = \sqrt{(x-0)^2 + (y-40)^2}$$
$$UL = \sqrt{(x-0)^2 + (y-0)^2}$$
$$UA = \sqrt{(x-30)^2 + (y-0)^2}$$

Because each of the distances must be equal, we have the following:

$$US = UL$$
$$\sqrt{(x-0)^2 + (y-40)^2} = \sqrt{(x-0)^2 + (y-0)^2}$$
$$x^2 + y^2 - 80y + 1600 = x^2 + y^2$$
$$^-80y + 1600 = 0$$
$$^-80y = ^-1600$$
$$y = 20$$

Similarly, we have

$$UL = UA$$
$$\sqrt{(x-0)^2 + (y-0)^2} = \sqrt{(x-30)^2 + (y-0)^2}$$
$$x^2 + y^2 = x^2 - 60x + 900 + y^2$$
$$0 = ^-60x + 900$$
$$60x = 900$$
$$x = 15.$$

Therefore the airport should be located on the plot of Figure 14-43 at the point with coordinates (15, 20), that is, 15 mi east and 20 mi north of Leapwood on the grid.

Looking Back. We can check that the location of the airport with coordinates (15, 20) is correct by considering whether $US = UA$, as follows:

$$US = \sqrt{(15-0)^2 + (20-40)^2} = \sqrt{225 + 400} = \sqrt{625} = 25$$
$$UA = \sqrt{(15-30)^2 + (20-0)^2} = \sqrt{225 + 400} = \sqrt{625} = 25$$

Thus the location is correct.

We also observe that because $AS = 50$, $UA = 25$, and $US = 25$, U is halfway between A and S, or U is the midpoint of $\overline{AS}$. Another way to solve the problem is to realize that the point equidistant from S, L, and A is the center of the circumscribing circle of triangle SLA. The center can be found by writing the equations of the perpendicular bisectors of two sides and finding the point of intersection of those equations. ∎

Q UESTIONS FROM THE CLASSROOM

1. A student asks, "If slope is such an important concept, why is slope of a vertical line undefined?" What is your response?

2. A student argues, "Since 0 is nothing and a horizontal line has slope 0, a horizontal line also has no slope." How do you reply?

3. Trying to find the midpoint M of $\overline{AB}$, where A and B have coordinates $(x_1, 0)$ and $(x_2, 0)$ and $x_2 > x_1$, a student argues as follows: "Since $AB = x_2 - x_1$, half that distance is $\frac{x_2 - x_1}{2}$, and since the midpoint M is halfway between A and B, the x-coordinate of M is $\frac{x_2 - x_1}{2}$." How do you respond?

4. A student does not understand why, when solving a system of linear equations, it is necessary to check the solution in the original equations rather than in some simpler equivalent equations. How do you respond?

5. A student claims that the graph of every inequality of the form $ax + by + c > 0$ is the half-plane above the line $ax + by + c = 0$, while the graph of $ax + by + c < 0$ is always below the line. How do you respond?

6. A student claims that the distance formula can be further simplified as follows:
$$d = \sqrt{(x_1 - x_2)^2 + (y_1 - y_2)^2}$$
$$= \sqrt{(x_1 - x_2)^2} + \sqrt{(y_1 - y_2)^2}$$
$$= |x_1 - x_2| + |y_1 - y_2|$$
How do you respond?

7. A student claims that if lines k and ℓ have slopes m_1 and m_2, respectively, and $m_1 m_2 = 1$, then the line k' (the reflection of k in the x-axis) is perpendicular to line l. She wants to know if this is always true, and why. What is your response?

*8. A student asks why when using Logo and the turtle is not at home, the rectangle procedures of Section 14-4 do not draw the same figure. How do you respond?

CHAPTER OUTLINE

I. Coordinate system in a plane
   A. Any point in the plane can be described by an ordered pair of real numbers, the first of which is the ***x*-coordinate** and the second of which is the ***y*-coordinate**.
   B. Together, the x-axis and y-axis divide the plane into four **quadrants**.

II. Distance concepts
   A. **Distance Formula:** The **distance** between the points (x_1, y_1) and (x_2, y_2) is given by $d = \sqrt{(x_2 - x_1)^2 + (y_2 - y_1)^2}$.
   B. **Midpoint Formula:** Given $A(x_1, y_1)$ and $B(x_2, y_2)$, the coordinates of the **midpoint** M of $\overline{AB}$ are $\left(\frac{x_1 + x_2}{2}, \frac{y_1 + y_2}{2}\right)$.
   C. The **equation of the circle** with center at (a, b) and radius r is $(x - a)^2 + (y - b)^2 = r^2$.

III. Equations of sets of points and notions of slope
   A. The slope of a line is a measure of its steepness.
      1. Given (x_1, y_1) and (x_2, y_2) with $x_2 \neq x_1$, the **slope** m of the line through the two points is given by $m = \frac{y_2 - y_1}{x_2 - x_1}$.
      2. The slope of a vertical line is not defined.
   B. The equation of any nonvertical line can be written in the following forms:
      1. **Slope intercept form:** $y = mx + b$, where m is the slope and b the y-intercept.
      2. **Point-slope form:** $y - y_1 = m(x - x_1)$, where (x_1, y_1) is a point on the line and m is the slope.
   C. The equation of any vertical line can be written in the form $x = a$.
   D. Two nonvertical lines are **parallel** if, and only if, they have equal slopes. Vertical lines have no slope and are parallel.
   E. Two lines, neither of which is vertical, are **perpendicular** if, and only if, the product of their slopes is $^-1$. If one line is vertical and the other has slope 0, the lines are perpendicular.

IV. Systems of linear equations
   A. A system of **linear equations** can be solved graphically by drawing the graphs of the equations.
      1. If the equations represent two intersecting lines, the system has a unique solution, that is, the ordered pair corresponding to the point of intersection.
      2. If the equations represent two different parallel lines, the system has no solutions.
      3. If the two equations represent the same line, the system has infinitely many solutions.
   B. A system of linear equations can be solved algebraically by either the **substitution method** or the **elimination method**.

*V. Coordinate geometry and Logo
The following primitives are used:
```
SETX        SETY
SETXY       SQRT
XCOR        YCOR
TOWARDS     MAKE
OUTPUT
```

CHAPTER TEST

1. Find the perimeter of the triangle with vertices at $A(0, 0)$, $B(^-4, 3)$, and $C(0, 6)$.

2. Show algebraically in at least two different ways that $(4, 2)$, $(0, ^-1)$, and $(^-4, ^-4)$ are collinear points.

3. Sketch the graphs for each of the following:
   (a) $3x - y = 1$
   (b) $3x - y \leq 1$
   (c) $2x + 3y + 1 = 0$

4. For each of the following, write the equation of the line determined by the given pair of points:
   (a) $(2, ^-3)$ and $(^-1, 1)$
   (b) $(^-3, 0)$ and $(^-3, 2)$
   (c) $(^-2, 3)$ and $(2, 3)$

5. The vertices of $\triangle ABC$ are $A(^-3, 0)$, $B(0, 4)$, and $C(2, 5)$. Find each of the following:
   (a) The equation of the line through C and parallel to $\overleftrightarrow{AB}$.
   (b) The equation of the line through C and parallel to the x-axis.
   (c) The point where the line found in (b) intersects $\overleftrightarrow{AB}$.

6. Find the equation of the line passing through the y-intercept of the line $2x + 3y + 1 = 0$ and perpendicular to that line.

7. ▶Solve each of the following systems, if possible. Indicate whether the system has a unique solution, infinitely many solutions, or no solution. Explain your answer.
   (a) $x + 2y = 3$
       $2x - y = 9$
   (b) $\dfrac{x}{2} + \dfrac{y}{3} = 1$
       $4y - 3x = 2$
   (c) $x - 2y = 1$
       $4y - 2x = 0$

8. A store sells nuts in two types of containers: regular and deluxe. Each regular container contains 1 lb of cashews and 2 lb of peanuts. Each deluxe container contains 3 lb of cashews and 1.5 lb of peanuts. The store sold 170 lb of cashews and 205 lb of peanuts. How many containers of each kind were sold?

9. (a) Find the midpoint of the line segment whose endpoints are given by $(^-4, 2)$ and $(6, ^-3)$.
   (b) The midpoint of a segment is given by $(^-5, 4)$ and one of its endpoints is given by $(^-3, 5)$. Find the coordinates of the other endpoint.

10. Graph each of the following:
    (a) $x^2 + y^2 = 16$
    (b) $(x + 1)^2 + (y - 2)^2 = 9$
    (c) $x^2 + y^2 \leq 16$

11. Find the equation of the circle that has its center at $C(^-3, 4)$ and that passes through the origin.

12. Graph each of the following systems:
    (a) $x \geq ^-7$
        $y \leq 4$
    (b) $y \leq 2x + 3$
        $y \geq x + ^-7$

13. The sum of the numbers of red and black jelly beans on Ronnie's desk is 12. Also, the sum is twice the difference of the numbers of the two colors. How many jelly beans of each color does he have if there are more red than black jelly beans?

14. The freshman class at the university has 225 fewer students enrolled than are in the sophomore class. The number in the sophomore class is only 50 students short of being twice as great as the number in the freshman class. How many students are in each class?

15. In the presidential election of 1932, Franklin D. Roosevelt received 6,563,988 more votes than Herbert Hoover. If one-fifth of Roosevelt's votes had been won by Herbert Hoover, then Hoover would have won the election by 2,444,622 votes. How many votes did each receive?

16. ▶(a) Find the coordinates of the point of intersection of any two perpendicular bisectors of two sides of $\triangle OBC$, if O is the origin and the other two vertices are $B(1, 4)$ and $C(5, 0)$.
    (b) Show that the third perpendicular bisector goes through the point you found in (a).
    (c) Use the distance formula to show that the point you found in (a) is equidistant from the vertices of $\triangle OBC$.
    (d) Find the equation of the circle that circumscribes $\triangle OBC$.

17. Find the point on the line $x + y = 3$ that is nearest to the origin.

18. Find the distance from the point $(1, 0)$ to the line $y = 2x$.

19. Find the distance between the line $y = ^-2x + 1$ and $y = ^-2x + 3$. The distance between two parallel lines is the distance from any point of one to the other.

20. A point P in space can be determined by giving its location relative to the x-, y-, and z-axes, as shown.

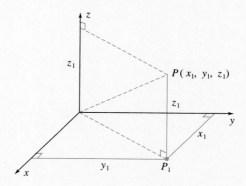

For $A(5, 3, 8)$ and $B(7, 6, 4)$, find the following:
(a) The point of intersection between the line through A parallel to the z-axis and the xy-plane (the plane determined by the x- and y-axes)

(b) The point of intersection between the line through A parallel to the x-axis and the yz-plane

(c) The point of intersection between the line through A parallel to the y-axis and the xy-plane

★ (d) The distance between the points (5, 3, 0) and (7, 6, 0)

★ (e) The distance between A and (5, 3, 1)

★ (f) The distance between A and (⁻5, 3, 8)

SELECTED BIBLIOGRAPHY

Dion, G. "The Graphics Calculator: A Tool for Critical Thinking." *Mathematics Teacher* 83 (October 1990): 564–571.

Graham, K., and J. Ferrini-Mundy. "Activities: Functions and Their Representations." *Mathematics Teacher* 83 (March 1990): 209–216.

Grocki, J. "Sharing Teaching Ideas: The Gymnasium—A Dynamic Classroom for Teaching Slope." *Mathematics Teacher* 83 (November 1990): 636–637.

Hopkinson, R. "Interpreting and Applying the Distance Formula." *Mathematics Teacher* 80 (October 1987): 572–579.

Morrell, L. "Tips for Beginners: Graphing Inequalities with the y-intercept." *Mathematics Teacher* 84 (April 1991): 294–295.

Taback, S. "Coordinate Geometry: A Powerful Tool for Solving Problems." *Mathematics Teacher* 83 (April 1990): 264–268.

APPENDIX I

An Introduction to BASIC

 The *Standards* recommends that students have access to a computer for individual or group work. Our goal in this appendix is to familiarize you with the rudiments of the BASIC language and to provide practice in using some BASIC commands. The word BASIC stands for "Beginner's All-Purpose Symbolic Instruction Code." BASIC was developed in 1964 at Dartmouth College by John Kemeny and Thomas Kurtz.

Section AI-1   BASIC: Variables and Operations

program

To communicate with a computer, we usually write a set of instructions called a **program** or we use progams that other people have written. We can also use the computer as a calculator. This is called the *immediate-execution mode*. To calculate in this mode, we use the PRINT command and the symbols shown in Table AI-1.

TABLE AI-1

Operation	Math Symbol	Math Example	BASIC Symbol	BASIC Example
Addition	$+$	$2 + 3$	$+$	$2 + 3$
Subtraction	$-$	$5 - 2$	$-$	$5 - 2$
Multiplication	$\times$ or $\cdot$	5×3 or $5 \cdot 3$	$*$	$5 * 3$
Division	$\div$	$16 \div 4$	$/$	$16/4$
Exponentiation		2^3	$\wedge$ or $\uparrow$ or $**$	$2 \wedge 3$ or $2 \uparrow 3$ or $2 ** 3$
Square Root	$\sqrt{}$	$\sqrt{4}$	SQR	SQR(4)
Absolute Value	$||$	$|-2|$	ABS	ABS(-2)

Note that symbols used in BASIC are sometimes different from those used in math. For example, to calculate 8^5, we type

```
PRINT 8 ∧ 5
```

and press the RETURN key to see the display read 32768. To shorten the typing, it is possible to replace PRINT with a question mark, as in ?8 ∧ 5. Parentheses may be used to group numbers, as in PRINT SQR(3 ∧ 2 + 4 ∧ 2) to find $\sqrt{3^2 + 4^2}$.

We can also use the computer in *delayed-execution mode*. In this mode, statements can be put together using line numbers and commands to form a program. Figure AI-1 illustrates two examples of very simple BASIC programs: the first for printing a message and the second for calculating the value of Y for a given value of X, where

$$Y = 5X^4 + \frac{4}{X} - 3.$$

In BASIC, numeric variables may be denoted by single letters, by two letters placed next to each other with no spaces between them, by a letter followed by a single digit, or by a letter followed by a number in parentheses.

FIGURE AI-1

(a) ```
10 PRINT "HERE WE GO"
20 PRINT "LET'S LEARN TO PROGRAM"
30 END
```

(b) ```
10 INPUT X
20 LET Y = 5 * X ∧ 4 + 4 / X - 3
30 PRINT Y
40 END
```

string variables

In Figure AI-1(b), the variables X and Y are used. A different type of variable, a **string variable,** is denoted by a letter followed by a dollar sign, for example, A$. A string variable may be used for nonnumeric values such as words or strings of symbols.

System Commands

line number
RUN

Each line in the programs in Figure AI-1 is a statement that gives the computer an instruction. Notice that each statement begins with a **line number.** A system command such as RUN does not require a line number. When **RUN** is entered with a program in the internal memory, the computer executes the program's instructions in the order of the line numbers. Line numbers must be positive integers. In most cases, the numbers 10, 20, 30, . . . or 100, 200, 300, . . . are used for line numbers to leave room for forgotten statements, which may be added later. For example, suppose that after typing line 30 in Figure AI-1(b) (PRINT Y), you realize that you should have had a statement between lines 10 and 20. All that is necessary is to type the new statement preceded by a line number between 10 and 20, such as 15. The new line can be typed at any time when the program is still active. The computer will rearrange all the lines in the correct order and will execute the program in the correct order. To delete errors in a given line, you may retype the line correctly. To delete a complete line, type the line number and then press RETURN . It should be noted that some computers allow full-screen editing, in which parts of a line may be changed without retyping an entire line. When the system command **LIST** is entered, the program will be displayed with all the line numbers in proper numerical order and all corrections included. LIST can be used at any time to see the complete program that is currently in the computer's memory.

LIST

SECTION AI-1 BASIC: VARIABLES AND OPERATIONS

NEW   Another useful system command if more than one program is being written is **NEW**. If we type NEW and press the RETURN key, the computer's memory is cleared, and a new program can be written without the danger of having the lines of the new program merged with a program that is already in memory. However, remember that when NEW is typed and the RETURN key is pressed, the old program in the internal memory is lost. If we want the old program saved, it may be stored in memory. The computer manual may be consulted to see how to accomplish this.

Programming Commands

The four BASIC key words that were used in Figure AI-1(b) are INPUT, LET, PRINT, and END. These are called programming commands. In line 10, the key word **INPUT** is followed by the variable X. The INPUT statement causes the computer to stop during the run and print "?". The user must assign a specific numerical value to X in this particular program by typing in that value and pressing the RETURN key. After the value is entered, the program will continue. An INPUT statement always consists of the word INPUT and a variable or list of variables separated by commas.

INPUT

LET   In line 20, we see a statement that contains the key word **LET**. A LET statement is used to assign a value to a variable. The LET statement places data in a memory location. In Figure AI-1(b), the variable Y is assigned the value

```
5 * X ∧ 4 + 4 / X - 3
```

and this value is stored in the memory location labeled Y. The statement is composed of a line number, followed by the word LET, followed by a variable that equals a mathematical expression. The command LET is frequently optional. For example, LET $X = 17$ is simply written as $X = 17$ on many machines.

PRINT   In line 30, we see the key word **PRINT**. PRINT Y is a direction to the computer to print the value of the variable Y when the program is run. (Other uses of the key word PRINT will be discussed later.)

END   The **END** statement indicates the end of a program. Some systems do not require an END statement, but it is good programming style to use it.

When the program in Figure AI-1(b) is run, the computer acts as the function machine shown in Figure AI-2. When the value of the variable X is input, the machine assigns a value for Y according to the rule $Y = 5X^4 + \frac{4}{X} - 3$. Thus, if the input is 2 (that is, $X = 2$), the output will be $Y = 5 \cdot 2^4 + \frac{4}{2} - 3 = 79$.

FIGURE AI-2

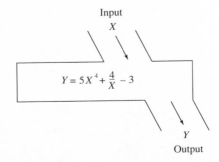

To use the program in Figure AI-1(b) to find the value of Y for a given X, such as $X = 2$, type RUN and then press the [RETURN] key. (*For the remainder of this appendix, it is assumed that the* [RETURN] *key is pressed as needed*.) The computer will print a question mark, which indicates that the user should input a value, in our case, 2. The computer then prints the corresponding value of Y, which is 79. The interaction with the computer will look like the following (*the boldface characters are typed by the computer*):

RUN
? 2
79

On many computers, the word READY is printed when the run is completed. Other machines may indicate this by the word OK or by a flashing cursor or some other device. These are the computer's way of saying that it is ready for your next instruction. If you wish to input a new number for X into the program, type RUN again and press the [RETURN] key. Another question mark will appear, and the new number may be input. If we input 180 for X and run the program, the printout appears as follows:

RUN
? 180
5.2488E + 09

E-notation   The answer, 5.2488E + 09, is represented using **E-notation**. This notation is like scientific notation in that the letter E means "exponent" and indicates the power or exponent on the base 10. The use of E-notation may vary from one brand of computer to another.

1.876E + 07 means $1.876 \cdot 10^7$, or 18,760,000

7.9325E − 04 means $7.9325 \cdot 10^{-4}$, or 0.00079325

The order of operations that a computer uses is the same as those discussed earlier in this book; that is, exponentiations (if any) are done first; multiplications and divisions are done next, in order from left to right; and additions and subtractions are done last, also in order from left to right unless parentheses indicate otherwise. *When you are in doubt as to how the computer will execute the operations, use parentheses.*

EXAMPLE AI-1   Write (a) through (c) in BASIC notation and (d) and (e) in base-10 notation.

(a) $b^2 - 4ac$ (b) $\dfrac{3(4 + b)}{c}$

(c) $3 + 4c^3$ (d) 3.4798E + 08

(e) 4.5935E − 05

Solution   (a) B $\wedge$ 2 − 4 ∗ A ∗ C   (b) 3 ∗ (4 + B) / C
(c) 3 + 4 ∗ C $\wedge$ 3   (d) $3.4798 \cdot 10^8 = 347{,}980{,}000$
(e) $4.5935 \cdot 10^{-5} = 0.000045935$

Several BASIC programming commands are used in the following example.

EXAMPLE AI-2   Fair State University is increasing the salaries of its faculty based on performance. A department secretary needs a computer program to check the controller's figures. The program must calculate the next year's salary for each faculty member, given the present salary and the percent of increase of the salary of that faculty member. Write a program for the secretary to calculate each new salary. Try your program for an old salary of $38,479 and a 17% increase.

Solution If the old salary and the percent of increase are denoted by S and P, respectively, then the new salary N is given by $N = S + S \cdot \dfrac{P}{100}$, or $N = S \cdot \left(1 + \dfrac{P}{100}\right)$. A program for this computation follows:

```
10 REM UNIVERSITY SALARY INCREASE
20 PRINT "ENTER THE OLD SALARY AND THE PERCENT"
30 PRINT "OF INCREASE SEPARATED BY A COMMA."
40 INPUT S, P
50 LET N = S * (1 + P / 100)
60 PRINT "THE NEW SALARY IS $"; N
70 END
```

When the program is run for the given salary and percent of increase, the printout will be as follows:

**ENTER THE OLD SALARY AND THE PERCENT
OF INCREASE SEPARATED BY A COMMA.**
?38479,17
THE NEW SALARY IS $45020.43

In lines 20 and 30 of Example AI-2, we have used the programming command PRINT. When the program is run, the computer prints everything inside the quotation marks. When a semicolon is placed at the end of a PRINT statement, the computer continues the next PRINT statement on the same line, one space to the right of the last printed character. (This placement may vary on different machines.) In line 60, PRINT is used with a phrase in quotation marks followed by a semicolon and the letter N. When the semicolon is used, the value of N is printed immediately following the phrase. (You should investigate what happens when a comma is used in place of a semicolon.) Execute the program to determine the output.

REM   A programmer may wish to make a remark or title the program on the very first line. This can be done by using the programming command PRINT or a new programming command, **REM** (for REMARK). If a REM statement is used as in line 10 in the program above, the computer will save the line as an explanation of the program for anybody who desires to list the program. The REM statement is ignored during execution and is not printed during a run.

Section AI-2   Branching

IF-THEN Statements

branching   One possible use of the computer in education is in computer-assisted instruction, for example, in designing programs to create drill problems. Suppose we want to design a program that can be used repeatedly to allow students to practice adding two numbers. We want the students to keep working on a problem until they get it correct; then, they will receive a new problem. This can be accomplished through **branching.** One way branching occurs is when decisions are made in a program through the use of the programming

IF-THEN statement **IF-THEN.** (This type of branching is called *conditional branching*.) The general form of an IF-THEN statement is IF (*condition*) THEN (*line number or command*). For example,

```
30 IF X < 2 THEN 60
```

When the computer executes line 30, it evaluates the condition $X < 2$. If the value of X is less than 2, then the computer branches to the line number following the word THEN, in this case, line 60. If the value of X is not less than 2, the computer executes the next statement listed in numerical order. The condition following the IF part always includes an equality or inequality symbol. However, in BASIC, some of these symbols differ from their mathematical counterparts. The symbols are compared in Table AI-2.

We now attempt to write the two-number drill program mentioned previously. In order for a program to be meaningful to another user, it should contain instructions telling the user exactly what to do. These instructions can be given by using PRINT statements. The numbers to be added must be either input by the user or generated by the computer's random-number generator. For this program, we choose the second option. However, before doing this, we investigate the computer's random-number generator.

TABLE AI-2

Mathematical Symbol	BASIC Symbol	Meaning
$=$	$=$	Equals
$<$	$<$	Is less than
$\leq$	$<=$	Is less than or equal to
$>$	$>$	Is greater than
$\geq$	$>=$	Is greater than or equal to
$\neq$	$<>$	Is not equal to

RND The general form of the function is **RND(X),** where **RND** is the name of the function and X is the argument of the function. On some computers, the value of the argument is important; on others, it is not. On still others, the argument is not needed at all. The reader should consult the user's manual for specifics. For this text, we use RND(1).

The RND function causes some computers to select a random (unpredictable) six-digit number between 0 and .999999, such as .893352, .158723, or .931562. Because many programs require the use of random numbers in ranges other than from 0 to 1, we must find a way to select numbers in other ranges. For example, suppose we want the computer to select a nonnegative integer less than 100. If we multiply the RND function by 100, it will generate values between 0 and 99.9999. If we want random-integer values between 0 and 99.9999, we use the INT function, as follows:

```
10 LET N = INT(100 * RND(1))
```

Now, N takes on nonnegative integer values less than 100.

INT **INT,** the greatest-integer function, is a BASIC command used in the form of INT(X), where X is the argument. When used in a program, INT(X) will produce the greatest integer less than or equal to X. For example, executing PRINT INT (3.5) produces the computer output **3.**

If we want our two-number addition program to give us addition facts involving numbers between 1 and 10, we use the expression INT(10 * RND(1)) + 1 in our program.

Once the numbers are chosen and added, we must inform the user whether the sum is correct. We must also determine whether the user wishes to continue the practice. IF-THEN and GOTO statements are used to accomplish this. **GOTO** statements allow the program to branch to another statement instead of following the statements in the order given by the line numbers. (Observe that the GOTO command allows unconditional branching to another line of the program.) These ideas are incorporated in the following program:

GOTO

```
10 PRINT "THIS PROGRAM PROVIDES ADDITION PRACTICE."
20 LET A = INT (10 * RND(1)) + 1
30 LET B = INT (10 * RND(1)) + 1
40 PRINT "AFTER THE QUESTION MARK, TYPE THE SUM."
50 PRINT A; "+"; B; "=";
60 INPUT C
70 IF A + B = C THEN 100
80 PRINT "SORRY, TRY AGAIN."
90 GOTO 40
100 PRINT "VERY GOOD, DO YOU WANT TO ADD"
110 PRINT "OTHER NUMBERS?"
120 INPUT D$
130 IF D$ = "YES" THEN 20
140 END
```

There are two IF-THEN statements in the preceding program. When the program is run and line 70 is reached and if $A + B = C$ is true, then the computer branches to line 100 and continues the run. In line 70, if $A + B = C$ is not true, then the computer automatically goes to the next line, line 80, and continues the run.

When the program is run and the computer reaches line 90, the program automatically loops back to line 40, and the run continues. It should be noted that if a student continually misses the addition, the run will never get past line 90. (The program could have been written to ask the student if another try is desired.) On different machines, the procedure to get out of such a loop varies. On many machines, the keys $\boxed{\text{STOP}}$ or $\boxed{\text{BREAK}}$ are used. On some other machines, the user should press the $\boxed{\text{CONTROL}}$ key and the $\boxed{\text{C}}$ key at the same time (called a CONTROL-C) or press the $\boxed{\text{RESET}}$ key or $\boxed{\text{CONTROL}}$ $\boxed{\text{RESET}}$. This may have to be done several times to stop the run, at which time the computer will indicate that it is ready for you to proceed.

Counters

Suppose we want to keep a record of how many drill exercises an individual attempts. This can be done with a *counter,* which uses either an IF-THEN or a GOTO statement along with a LET statement. A LET statement allows us to set the value of a variable; for example, LET $X = 10$. BASIC also allows a variable to be defined in terms of itself; for example, LET $X = X + 1$. The computer does not interpret $X = X + 1$ as an equation, but rather as "replace X with $X + 1$." This is useful in a program that counts how many exercises were correctly answered or, in general, how many times a certain section of a program has been executed. Consider the following program:

```
10 LET X = 0
20 LET X = X + 1
30 PRINT X
40 IF X < 10 THEN 20
50 END
```

initialized In line 10, X is **initialized,** or set to 0. Line 20 is the counter. When the program is run, the output appears as follows:

1
2
3
.
.
.
10

To understand this output, notice that in line 10, X equals 0. In line 20, the new value of X becomes $0 + 1$, or 1. Hence, at line 30, the value 1 is printed. Next, the computer executes the instruction in line 40. Because $1 < 10$, it branches to line 20. As X now has value 1, the new value of X in line 20 becomes $1 + 1$, or 2. Thus 2 is printed, and so on. After 9 is printed, we are again at line 40. Because $9 < 10$, the computer again branches to line 20, and the new value of X becomes $9 + 1$, or 10. This value is printed; because $10 < 10$ is false, the program continues to line 50 and ends.

We now return to our original addition-drill program. To keep a record of the number of drill exercises attempted, a counter is inserted between lines 30 and 40. A PRINT statement in line 140 outputs the desired record. The revised program is as follows:

```
10 PRINT "THIS PROGRAM PROVIDES ADDITION PRACTICE."
15 LET X = 0
20 LET A = INT (10 * RND(1)) + 1
30 LET B = INT (10 * RND(1)) + 1
35 LET X = X + 1
40 PRINT "AFTER THE QUESTION MARK, TYPE THE SUM."
50 PRINT A;"+";B;"=";
60 INPUT C
70 IF A + B = C THEN 100
80 PRINT "SORRY, TRY AGAIN."
90 GOTO 40
100 PRINT "VERY GOOD, DO YOU WANT TO ADD"
110 PRINT "ANOTHER PAIR OF NUMBERS?"
120 INPUT D$
130 IF D$ = "YES" THEN 20
140 PRINT "THE NUMBER OF EXERCISES WAS";X
150 END
```

The program could be edited to allow the computer to tell not only the number of exercises attempted but also the number done correctly.

FOR-NEXT Statements

FOR-NEXT Another method for having the computer count how many times an action is performed is with the use of the key words **FOR-NEXT**. In a BASIC program, FOR-NEXT statements always occur in pairs, with the FOR statement preceding the associated NEXT statement. For example, we could utilize FOR-NEXT statements as shown.

```
20 FOR X = 1 TO 10
 .
 .
 .
120 NEXT X
```

If a program containing the preceding lines is run and line 20 is reached, the computer initializes a counter by assigning the value of 1 to X. At line 120, the computer loops back to line 20, assigns the value 2 to X, again continues to line 120, loops back, and so on, until $X = 10$. When X is 10 and line 120 is reached, the computer continues on to the next line of the program following line 120. With no other directions in line 20, the computer automatically increments X by 1 each time the loop is passed through. X can be incremented for *STEP* other values by using the word **STEP** and the desired value. For example, replace line 20 with the following:

```
20 FOR X = 1 TO 10 STEP 0.5
```

This line will initialize X with a value of 1 and increase this value by 0.5 each time line 20 is reached. The values of X will be 1, 1.5, 2, 2.5, 3, . . . , 9, 9.5, 10.

STEP can also be used to decrement, rather than increment, the FOR-NEXT loop. For example, consider the following line:

```
20 FOR X = 10 TO 1 STEP -0.5
```

Here, the value of X will be initialized at 10 and will decrease in steps of 0.5 until it reaches $X = 1$. Thus the values of X will be 10, 9.5, 9, 8.5, 8, . . . , 1.

FOR-NEXT commands are very useful when you know how many times to go through the loop, as seen in the following program, which will output the first 20 powers of 2:

```
10 REM POWERS OF 2
20 FOR E = 1 TO 20
30 LET T = 2 ∧ E
40 PRINT T
50 NEXT E
60 END
```

PROBLEM 1

Professor Anna Litik was asked by the university athletic director to devise a chart for converting the heights of visiting basketball players from inches to centimeters. Help Dr. Litik by writing a program that will print a table giving the conversions incremented by half inches from 65 in. to 84 in.

Understanding the Problem. The problem is to write a program that will output the number of centimeters that corresponds to a certain number of inches. Furthermore, the numbers of inches are from 65 to 84, inclusive, and the conversions made must be for each half inch. To solve this problem, the conversion factor 1 in. = 2.54 cm is needed. Thus 2 in. = 2(2.54), or 5.08, cm, and, in general, N in. = N(2.54) cm.

Devising a Plan. We can write a program using FOR-NEXT statements to produce a table for converting heights from inches to centimeters. We let X represent the number of inches starting at 65, increment by 0.5, and end at 84. Between the FOR and NEXT statements, we must print the value of X to represent the equivalent number of centimeters. Because it would be inappropriate to report that a player's height is 196.62 cm, we need to report the heights as integers. We do this by using INT. In our program, we can use INT $(X * 2.54)$ to assign an integer number of centimeters to X in. However, INT does not round numbers. If we would like to build in a rounding factor, we might consider using INT$(X * 2.54 + 0.5)$. (Why?)

To write the program, we set the first value of X equal to 65 in. Next, we convert 65 in. to centimeters by multiplying by 2.54 and then choose the greatest integer less than or equal to $65 \cdot 2.54 + 0.5$. We then print X and the obtained integer and loop back for the next value of X, which is 0.5 greater than the last value converted. This procedure continues until all values have been converted. The complete program follows. In line 50, the PRINT statement is used to provide the desired output in a more readable format. When a PRINT statement is typed with no inputs, a blank line is left between the outputs.

```
10 REM THIS PROGRAM WILL PRINT A
20 REM TABLE FOR CONVERTING INCHES
30 REM TO CENTIMETERS
40 PRINT "INCHES", "CENTIMETERS"
50 PRINT
60 FOR X = 65 TO 84 STEP 0.5
70 LET Y = INT(X * 2.54 + 0.5)
80 PRINT X, Y
90 NEXT X
100 END
```

Carrying Out the Plan. A shortened version of the printout obtained when the program is run follows. Notice that the space between the headings and the data was obtained from line 50 in the program.

INCHES	CENTIMETERS
65	165
65.5	166
66	168
66.5	169
67	170
67.5	171
68	173
68.5	174
69	175
69.5	177
.	.
.	.
83.5	212
84	213

Looking Back. This program could be revised to generate other conversion charts. All that would be necessary would be to find the required conversion factors and edit the program accordingly.

Read and Data Statements

READ / DATA When large blocks of data are used, the key words **READ** and **DATA** are very useful. The formats for the READ and DATA statements are as follows:

READ Variable(s)

DATA Constant(s)

The variables in a READ statement are separated by commas. Similarly, the constants in a DATA statement are separated by commas. A comma should not be used at the end of the data. An example of statements containing the commands READ and DATA follow:

```
10 READ A1, B, B2, Z
50 DATA 9, 6, 2, .5
```

When a program containing READ-DATA statements is run, the computer assigns the constants in the DATA statement to the variables in the READ statement in the order they are listed. A branching command, such as GOTO, is normally used to have the computer loop back to the READ statement so that it will use all DATA when the program is run. It should be noted that several READ statements can be used with one DATA statement as long as there are sufficient data for each of the variables in the READ statement; or, one READ statement may be used with several DATA statements. An example of a program using READ-DATA commands is given below. This program computes and prints the numbers 1 through 20 with their square roots and their squares.

```
10 REM THIS PROGRAM COMPUTES SQUARES AND SQUARE ROOTS
20 PRINT "NUMBER", "SQUARE", "SQUARE ROOT"
30 PRINT
40 READ N
50 LET S = N * N
60 LET R = SQR(N)
70 PRINT N, S, R
80 GOTO 40
90 DATA 1, 2, 3, 4, 5, 6, 7, 8, 9, 10
100 DATA 11, 12, 13, 14, 15, 16, 17, 18, 19, 20
110 END
```

PROBLEM 2

Dr. Jubal, an English professor, had five test scores for each of his five graduate students in English. The grades for the students are shown in Table AI-3 on page 826. Determine averages for Dr. Jubal by writing a program to find the mean for each student.

Understanding the Problem. We are to write a program to calculate the mean for each student in Dr. Jubal's class. We need to find the sum of the five scores for each student and divide by 5.

TABLE AI-3

Student # (S)	Grammar (G)	Writing (W)	Drama (D)	Poetry (P)	Novel (N)
1	72	93	86	82	86
2	83	86	95	74	80
3	91	82	76	60	89
4	88	94	72	76	84
5	75	84	98	92	79

Devising a Plan. To find the mean for each student, we use a READ statement with the variable S for the student number and variables G, W, D, P, and N for grammar, writing, drama, poetry, and novel scores, respectively. For example, for student 1, we want the computer to calculate $M = (G + W + D + P + N)/5$, or $(72 + 93 + 86 + 82 + 86)/5$. After the mean is computed for each student, we need a printout that lists students' numbers and corresponding means. A sample program to compute the means is as follows:

```
10 REM THIS PROGRAM PRINTS MEANS FOR DR. JUBAL.
20 PRINT "STUDENT #","MEAN"
30 READ S,G,W,D,P,N
40 LET M = (G + W + D + P + N) / 5
50 PRINT
60 PRINT S,M
70 GOTO 30
80 DATA 1,72,93,86,82,86,2,83,86,95,74,80,3,91,82,76,60,89
90 DATA 4,88,94,72,76,84,5,75,84,98,92,79
100 END
```

> **REMARK** Note that not all the data could be listed in one DATA line. The computer will accept data from as many DATA lines as you wish, but the number of DATA items should be a multiple of the number of variables in a READ statement. The DATA statements may be anywhere in the program, but they are usually at the end.

Carrying Out the Plan. A run for the program would appear as follows:

STUDENT #	MEAN
1	85.8
2	83.6
3	79.6
4	82.8
5	85.6

?OUT OF DATA ERROR IN 30

Now we have the averages of each of Dr. Jubal's students. Note the message at the end of the output. This message tells us that all the data have been used.

Looking Back. Various computers may print different messages or no message when all the data are used. To eliminate the message at the end of the printout, a method called *flagging* **flagging** can be used. See a BASIC computer manual for more information about this method.

A FOR-NEXT loop could also be used to signal that all the data for Dr. Jubal's class have been used, as seen in the following modified program:

```
10 REM THIS PROGRAM PRINTS MEANS FOR DR. JUBAL.
20 PRINT "STUDENT #", "MEAN"
30 FOR S = 1 TO 5
40 READ S, G, W, D, P, N
50 LET M = (G + W + D + P + N) / 5
60 PRINT
70 PRINT S, M
80 NEXT S
90 DATA 1, 72, 93, 86, 82, 86, 2, 83, 86, 95, 74, 80, 3, 91, 82, 76, 60, 89
100 DATA 4, 88, 94, 72, 76, 84, 5, 75, 84, 98, 92, 79
110 END
```

We now have three methods of getting data into the computer: LET statements, INPUT statements, and READ-DATA statements. A primary advantage of READ-DATA statements over INPUT statements is the speed with which the computer is able to accept and use the information. Another advantage is not having to enter information every time the program is run. There is no lag time while the computer waits for the user to input information during the run of the program. However, when using READ-DATA statements, the program itself must be changed to input new data; the use of INPUT allows the user to input new data without altering the program.

PROBLEM 3

In 1980, the number of cars using the Chicago Loop and Inter-Mountain Bypass highways was 220,819,000 and 66,944,000, respectively. If the annual usage growth rates for these two highways were 1.5% and 2.9%, respectively, and if these usage rates were to remain constant, in what year would the number of cars using the Inter-Mountain Bypass equal or surpass the number of cars using the Chicago Loop? How many cars would use each highway in that year?

Understanding the Problem. A useful strategy in problem solving is to compare the given problem to any *related problems* that have previously been solved. Each number of cars in the current problem plays the role of the principal in a compound-interest problem. Thus the number of cars using the Chicago Loop is growing according to the formula $220{,}819{,}000(1 + .015)^N$, where N is the number of years involved. Similarly, the number of cars using the Inter-Mountain Bypass is growing according to the formula

$$66{,}944{,}000(1 + .029)^N.$$

Devising a Plan. The program must be written so that the number of cars on each road is computed for different years; in addition, the first time the number of Inter-Mountain cars is greater than or equal to the number of Chicago Loop cars, the computer should print the value of N added to 1980, giving the desired years. The total number of cars for both highways should also be output at this time. The program must have a counter, and it must contain a loop to compute values for different Ns. A program based on this discussion follows:

```
10 REM A TRAFFIC PROBLEM INVOLVING THE CHICAGO LOOP AND IM BYPASS
15 LET N = 0
20 LET N = N + 1
30 LET C = 220819000 * 1.015 ∧ N
40 LET I = 66944000 * 1.029 ∧ N
50 IF I >= C THEN 70
60 GOTO 20
70 PRINT "IN ";N + 1980;", THE NUMBER OF CHICAGO"
80 PRINT "LOOP CARS WILL BE "; C
90 PRINT "IN ";N + 1980;", THE NUMBER OF IM BYPASS"
100 PRINT "CARS WILL BE ";I
110 END
```

Carrying Out the Plan. A run of the preceding program produces the following:

**IN 2068, THE NUMBER OF CHICAGO
LOOP CARS WILL BE 818555517
IN 2068, THE NUMBER OF IM BYPASS
CARS WILL BE 828446826**

Looking Back. An alternative activity might be to modify the program to find the number of cars of both the Chicago Loop and the Inter-Mountain Bypass in the year 2000, assuming the usage rates in the problem remain constant. In what year will the number of cars on the Chicago Loop exceed one-half billion?

SUMMARY OF COMMANDS

Command	Description
RUN	Causes the computer to execute a program.
LIST	Causes the computer to print a listing of a program.
NEW	Clears the computer's memory.
INPUT	Takes one input and causes the computer to pause during execution and output "?" as a prompt for data.
LET	Assigns a value to a variable.
PRINT	Causes the computer to print an input.
END	Indicates the end of a program.
REM	Used to include a remark or reminder.
IF-THEN	Used in the form IF (condition) THEN (line number of condition).
RND(X)	Causes the computer to select a random six-digit number between 0 and .999999.
INT(X)	Outputs the greatest integer less than or equal to X.
GOTO	Used in the form GOTO (line number) for branching.
FOR-NEXT	FOR-NEXT statements occur in pairs in a program, with the FOR statement preceding the NEXT statement.
STEP	Increments values of a variable.
READ-DATA	Used to enter large blocks of data in a program.

PROBLEM SET A1

1. Which of the following are valid variables in BASIC?
   - (a) P
   - (b) M4
   - (c) 3Z
   - (d) M
   - (e) AB
   - (f) BA(3)
   - (g) A4

2. Write each of the following using BASIC notation:
   - (a) $X^2 + Y^2 - 3Z$
   - (b) $\left(\dfrac{24 \cdot 34}{2}\right)^3$
   - (c) $a + b - \dfrac{c^2}{d}$
   - (d) $\dfrac{a+b}{c+d}$
   - (e) $\dfrac{15}{a(2b^2 + 5)}$

3. Perform the following calculations in the same order a computer would:
   - (a) $3 * 5 - 2 * 6 + 4$
   - (b) $7 * (6 - 2) / 4$
   - (c) $3 * (5 + 7) / (3 + 3)$
   - (d) $3 \wedge 3 / 9$
   - (e) $(2 * (3 - 5)) \wedge 2$
   - (f) $9 \wedge 2 - 5 \wedge 3$

4. Write each of the following as a base-10 number in standard form:
   - (a) $3.52E + 07$
   - (b) $1.93E - 05$
   - (c) $-1.233E - 06$
   - (d) $-7.402E + 03$

5. Predict the output, if any, for each of the following programs. If possible, check your answers on a computer.
   - (a)
     ```
     10 A = 5
     20 B = 10
     30 PRINT A,B
     40 END
     ```
   - (b)
     ```
     10 A = 5
     20 B = 10
     30 PRINT "A,B"
     40 END
     ```
   - (c)
     ```
     10 A = 5
     20 B = 10
     30 PRINT A, ,B
     40 END
     ```
   - (d)
     ```
     10 A = 5
     20 B = 10
     30 PRINT A;B
     40 END
     ```
   - (e)
     ```
     10 A = 5
     20 B = 10
        PRINT A,B
     40 END
     ```
   - (f)
     ```
     10 A = 5
     20 B = 10
     30 PRINT A
     40 PRINT B
     50 END
     ```
   - (g)
     ```
     10 A = 5
     20 B = 10
     30 PRINT A;
     40 PRINT B
     50 END
     ```
   - (h)
     ```
     10 A = 5
     20 B = 10
     30 PRINT "THE VALUE OF A"
     35 PRINT " IS ";A;"."; 
     40 PRINT " THE VALUE OF B"
     45 PRINT "IS ";B;"."
     50 END
     ```
   - (i)
     ```
     10 A = 5
     20 B = 10
     30 PRINT "THE VALUE OF A"
     35 PRINT "IS ",A,".";
     40 PRINT " THE VALUE OF B"
     45 PRINT "IS ";B;"."
     50 END
     ```

6. (a) Write a computer program for calculating the value of Y for a given X, where
   $$Y = 13X^5 - \dfrac{27}{X} + 3.$$
   (b) Use the program in (a) to find the value of Y for: (i) $X = 1.873$; (ii) $X = 7$.

7. (a) Write a program (including a title and directions) for converting degrees Fahrenheit into degrees Celsius, using the formula
   $$C = \dfrac{5}{9}(F - 32).$$
   (b) Use the program from (a) to convert the following Fahrenheit temperatures to Celsius:
   - (i) 212°F
   - (ii) 98.6°F
   - (iii) 68°F
   - (iv) 32°F
   - (v) $-40°$F
   - (vi) $-273°$F

8. Write a program for computing the volume of a right rectangular prism given its three dimensions L, W, and H, where $V = LWH$. Try your program for $L = 8$, $W = 5$, and $H = 3$.

9. $100 is invested in a bank for 25 yr at 18% interest compounded annually.
   - (a) Find the balance by using a single PRINT statement.
   - (b) Find the balance by writing a program with an input N, where N is the number of years the money is invested.

10. Determine the output for the following programs and then run the programs on the computer to check your answers:
    - (a)
      ```
      10 FOR I = 1 TO 15
      20 PRINT I,
      30 NEXT I
      40 END
      ```

(b) 10 FOR I = 1 TO 15
    20 PRINT I * 10,
    30 NEXT I
    40 END

(c) 10 FOR X = 1 TO 4 STEP 0.2
    20 PRINT X,
    30 NEXT X
    40 END

(d) 10 FOR X = 15 TO 1 STEP - 1
    20 PRINT X,
    30 NEXT X
    40 END

(e) 10 LET X = 100
    20 PRINT X,
    30 LET X = X - 1
    40 IF X < 0 THEN 60
    50 GOTO 20
    60 END

(f) 10 LET X = 10
    20 LET X = X + 1
    30 PRINT X
    40 IF X <= 10 THEN 20
    50 END

(g) 10 READ X, Y
    20 DATA 2, 3
    30 IF X > Y THEN 50
    40 PRINT X; " IS LESS THAN ";Y
    50 END

(h) 10 FOR I = 2 TO 5
    20 FOR J = 1 TO 3
    30 PRINT I; J
    40 NEXT I
    50 NEXT J
    60 END

11. Determine the outputs of each of the following programs. Why are the outputs different?

    (a) 10 PRINT "HEY YOU OUT THERE"
        15 LET K = 0
        20 LET K = K + 1
        30 IF K > 5 THEN 50
        40 GOTO 20
        50 END

    (b) 10 LET K = K + 1
        20 PRINT "HEY YOU OUT THERE"
        30 IF K > 5 THEN 50
        40 GOTO 10
        50 END

12. Rewrite the following program using a FOR-NEXT loops so that the outputs in (a) and (b) are obtained:
    10 LET N = 1
    20 IF N > 20 THEN 60
    30 PRINT 2 * N
    40 LET N = N + 1
    50 GOTO 20
    60 END

    (a) 2
        4
        6
        .
        .
        .

    (b) 2 4 6 8 . . .

13. Write a program to find the sum of the squares of the first 100 positive integers.

14. Modify the two-number addition program of this section to do two-number multiplications.

15. Examine the following program and determine what the computer would print. Then run the program to see if you were correct.

    10 FOR A = 1 TO 3
    20 FOR B = 4 TO 5
    30 LET C = A + B
    40 PRINT A, B, C
    50 NEXT B
    60 NEXT A
    70 END

16. Write a program to determine if any positive integer N is a prime number.

17. Write a program to convert degrees Fahrenheit to degrees Celsius, using the formula $C = \frac{(F - 32)5}{9}$. Make the program print Fahrenheit temperatures and corresponding Celsius temperatures from $^-40°F$ to $220°F$ in intervals of $10°F$.

18. Write a program to calculate the area A and the circumference C of a circle for any radius R that is input. Use $A = \pi R^2$ and $C = 2\pi R$, with $\pi \doteq 3.14159$.

19. Write a program to compute N! where N is any natural number. $(N! = 1 \cdot 2 \cdot 3 \cdot 4 \cdot \ldots \cdot (N - 1) \cdot N.)$

20. Laura won a lottery prize of $50,000. She put the money into a Certificate of Deposit at 15%/yr paid annually. Suppose that just after the interest is paid at the end of the first year, she withdraws $10,000 for a trip. The remainder is reinvested with the same terms. If she repeats this act year after year, write a program to find out how long her money will last.

21. Write a program to verify that
$$1^3 + 2^3 + 3^3 + \cdots + N^3 = (1 + 2 + 3 + \cdots + N)^2$$
is true for the first 10 natural numbers.

22. Write a program to compute the sum of the first 1000 odd natural numbers.

23. Write a program to find the sum
$$1 + \frac{1}{2} + \frac{1}{3} + \frac{1}{4} + \frac{1}{5} + \frac{1}{6} + \cdots + \frac{1}{N}.$$
Run the program for $N = 100$.

24. Plastic Card Company computes its bills on the last day of the month. If a customer pays the bill within the first 10 days of the next month, a 5% discount is given. If the bill

is paid within the next 10 days, the face value of the bill is paid. If the bill is paid after the 20th day, a penalty of 2% is added to the original bill computed at the end of the previous month. Write a program to take the customer I.D. number and the amount of the bill and print the amounts to be paid for each of the three options.

25. Write a program to compute and print out the value of the cube and the cube root of the first 20 natural numbers.

26. If you have $1000 invested in a bank at 5% annual interest compounded daily, how long would you have to leave it in the bank to have a balance of $5000?

27. Write a program to generate the first N Fibonacci numbers where N is any natural number. The Fibonacci sequence 1, 1, 2, 3, 5, 8, 13, 21, . . . is obtained by starting with 1, 1 and generating each successive term by summing the two previous terms. Check your program by finding the first 10 Fibonacci numbers.

28. In baseball, the number of hits H a player gets divided by the number of times at bat B is called the "batting average." Write a program to have the computer print a player's number and batting average. (The decimal need not be rounded.)

29. Imagine that you are paid 1¢ on the first day, 2¢ on the second day, 4¢ on the third day, 8¢ on the fourth day, and, in general, 2^{n-1} cents on the nth day. Each day's salary is double that of the previous day. Compare the salary on the 15th day with the sum of the salaries for the first 14 days.

30. Write and run a program to find all numbers less than 40 that can be written as a sum of two square numbers.

APPENDIX II

An Introduction to Logo Turtle Graphics

turtle

Turtle graphics, implemented using the computer language Logo, are especially suited for studying geometry. Students give instructions to a **turtle,** a triangular figure on the display screen.

The following discussion refers to MIT Logo (available from Terrapin Software, Inc.). Most commands also work in the LCSI (Logo Computer Systems, Inc.) version of Logowriter. If you are using Logowriter, you will need to consult your manual on how to clear the screen, how to use the "flipside" of the page to write procedures, and how to deal with various other differences when working with files. If commands differ, suggestions for LCSI Logo are given in parentheses. If an abbreviation can be used in place of a Logo command, then it will be given in parentheses immediately following the command when it is introduced.

Section AII-1 Introducing the Turtle

nodraw mode

draw mode

DRAW

In Logo, the computer accepts instructions in the nodraw, draw, and edit modes. After Logo is loaded into the computer, the first mode that appears on the screen is the **nodraw mode.** A question mark, called a *prompt,* and a flashing *cursor* appear on the screen as the computer awaits instructions. To execute turtle graphic commands, we enter the **draw mode** by typing **DRAW** *(in LCSI CLEARSCREEN (CS))* and pressing the RETURN

834 APPENDIX II AN INTRODUCTION TO LOGO TURTLE GRAPHICS

NODRAW (ND) key. After the DRAW command is executed, we can return to the nodraw mode by executing **NODRAW (ND)** *(in LCSI, TEXTSCREEN)*. In the draw mode, the turtle appears in the center of the screen, as shown in Figure AII-1. The turtle's position in the center of the screen is called "home." These built-in words that the computer understands, such as *primitives* DRAW and NODRAW, are called **primitives**.

FIGURE AII-1

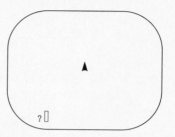

In Figure AII-1, the prompt at the bottom left of the screen indicates that the computer is ready to accept instructions, and the rectangular cursor shows where the next character that is typed will appear. Logo reserves the lines at the bottom of the screen for the user's input and the computer's responses. The remainder of the screen is used for drawing. This division of the screen into a drawing portion and a text portion is referred to as the *split-screen mode* **split-screen mode**.

Moving the Turtle

FORWARD (FD) To make the turtle change position, we use the primitives **FORWARD (FD)** and **BACK**
BACK (BK) **(BK)**, *followed by a space* and a numerical input. The numerical input tells the turtle how far to move. For example, after the DRAW command is executed, typing FORWARD 100 or FD 100 and pressing RETURN causes the turtle to move 100 "turtle units" in the direction it is pointing, as shown in Figure AII-2(a). Similarly, the BACK command may be used with a numerical input. For example, BACK 75 or BK 75 causes the turtle to move backwards 75 units. Giving the turtle too great an input causes the turtle to "wrap around" the screen. To explore how the turtle wraps, try FD 250 and observe what happens.

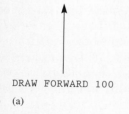

DRAW FORWARD 100
(a)

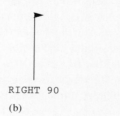

RIGHT 90
(b)

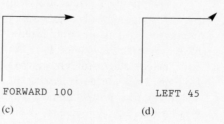

FORWARD 100
(c)

LEFT 45
(d)

BACK 75
(e)

FIGURE AII-2

Turning the Turtle

RIGHT (RT) To make the turtle change direction, we used the commands **RIGHT (RT)** and **LEFT**
LEFT (LT) **(LT)**. The RIGHT and LEFT commands, along with numerical inputs, cause the turtle to turn in place the specified number of degrees. For example, typing RIGHT 90 or RT 90 and pressing RETURN causes the turtle to turn 90° to the right of the direction it previously pointed. A sequence of moves illustrating these commands is given in Figure AII-2.

SECTION AII-1 INTRODUCING THE TURTLE

Logo accepts a sequence of commands written on one line. For example, Figure AII-2(e) could be drawn by typing the following and pressing RETURN :

DRAW FD 100 RT 90 FD 100 LT 45 BK 75

(In LCSI, replace DRAW with CLEARSCREEN.)

PENUP (PU)
PENDOWN (PD)
HIDETURTLE (HT)
SHOWTURTLE (ST)

To move the turtle without leaving a trail, we use the command **PENUP (PU)**. To make the turtle leave a trail again, type **PENDOWN (PD)**. It is possible to hide the turtle by typing **HIDETURTLE (HT)**. To make the turtle reappear, type **SHOWTURTLE (ST)**.

HOME

To return the turtle to the center of the screen with heading 0, type the command **HOME**. However, a trail to the center of the screen will be drawn from the position the turtle occupied before HOME was typed unless the command PENUP is used before HOME.

To start a new drawing with a clear screen, we type DRAW *(in LCSI, replace DRAW with CLEARSCREEN (CS))*. This returns the turtle to its initial position and direction in the center of the screen and clears the screen. Any time the turtle points straight north (up), we say it has heading 0. A heading of 90 is directly east, 180 is directly south, and 270 is directly west. The screen could be marked as shown in Figure AII-3.

FIGURE AII-3

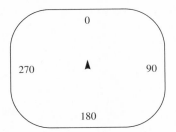

HEADING

To learn the turtle's heading, we use HEADING. **HEADING** needs no inputs; typing HEADING in the draw mode and pressing RETURN causes the computer to output the turtle's heading. To have the computer print only the value of the heading, we use the primitive **PRINT (PR)** along with HEADING, as in PRINT HEADING. For example, if the turtle is at home with heading 0 and we type RT 45 PR HEADING, then 45 will be displayed. If we execute RT 45 PR HEADING again, then 90 will be displayed.

PRINT (PR)

SETHEADING (SETH)

The command **SETHEADING (SETH)** can be used to turn the turtle in a direction from the 0 heading. For example, SETH 100 turns the turtle so that it has a heading of 100. This command can be used no matter where the turtle is located or what its heading is at the time. Figure AII-4 gives an example of the use of the SETH and HOME commands.

FIGURE AII-4

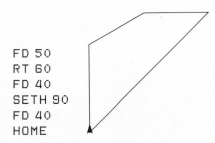

> **REMARK** PRINT (PR) normally takes one input, causes the input to be printed on the screen, and moves the cursor to the next line. PRINT (PR) may take a word or a list of words as input. For example, PR "TURTLE and PR [TURTLE POWER] are both acceptable.

A summary of commands introduced thus far is shown in Table AII-1.

TABLE AII-1

Command	Abbreviation	Example
DRAW*		
NODRAW†	ND	
FORWARD	FD	FD 50
BACK	BK	BK 60
RIGHT	RT	RT 90
LEFT	LT	LT 45
PENUP	PU	
PENDOWN	PD	
HIDETURTLE	HT	
SHOWTURTLE	ST	
HEADING		PR HEADING
PRINT	PR	PR "LOGO
HOME		
SETHEADING	SETH	SETH 270

*The command in *LCSI* is CLEARSCREEN (CS).
†The command in *LCSI* is TEXTSCREEN (CTRL-T).

■ HISTORICAL NOTE

The computer language Logo was developed in 1967 at Bolt, Beranek, and Newman, Inc. of Cambridge, Massachusetts, and the Massachusetts Institute of Technology (MIT) by Daniel Bobrow, Wallace Feurzeig, and Seymour Papert. The name "Logo" is derived from the Greek word for "thought." The developers of Logo were influenced by the field of artificial intelligence, the computer language LISP, and the theories of Jean Piaget. The tradition of calling the display creature a turtle can be traced to early experiments involving robot-like creatures referred to as "tortoises." When computer graphics were implemented, the screen creature inherited the turtle terminology.

People studying Logo are encouraged to "play turtle" and act out their commands. For example, to act out drawing a square, we may walk around the square by moving forward 50 units, turning right 90°, moving forward 50 units, turning right 90°, moving forward 50 units, turning right 90°, and finally moving forward 50 units. The sequence of commands for these moves is summarized in Figure AII-5(a), with the resulting square and final position of the turtle shown in Figure AII-5(b).

SECTION AII-1 INTRODUCING THE TURTLE   837

FIGURE AII-5

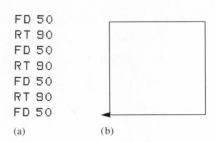

```
FD 50
RT 90
FD 50
RT 90
FD 50
RT 90
FD 50
```
(a)   (b)

The turtle's final position in Figure AII-5(b) is the same as its initial position, but its heading is different. When drawing a figure, it is often convenient to have the turtle's final state be the same as its initial state. When this happens, we say that the set of commands is *state transparent*.

state transparent

To return the turtle to its initial state in Figure AII-5(b), we turn it right 90° by adding the line RT 90 at the end of the sequence of commands in Figure AII-5(a). The new sequence of commands is given in Figure AII-6(a), with the resulting square and turtle shown in Figure AII-6(b).

FIGURE AII-6

```
FD 50
RT 90
FD 50
RT 90
FD 50
RT 90
FD 50
RT 90
```
(a)   (b)

The sequence of commands in Figure AII-6(a) contains the instructions FD 50 and RT 90 repeated four times. Logo allows us to use the REPEAT command to repeat a list of instructions. For example, to draw the square in Figure AII-6(b), we could type the following:

REPEAT 4 [FD 50 RT 90]

REPEAT In general, **REPEAT** takes two inputs: a number and a list of commands. The commands in the brackets are repeated the designated number of times.

E X A M P L E   A II - 1   Predict the results of each of the following, indicating the initial and final turtle states. Then, check your answers with a computer. In each case, assume the turtle starts at home with heading 0.

(a) FD 100
    RT 135
    FD 100
    RT 45
    FD 100
    RT 135
    FD 100
    RT 45

(b) REPEAT 2 [FD 100 RT 135 FD 100 RT 45]
(c) REPEAT 8 [FD 50 RT 45]
(d) BK 100 SETH 270 FD 100 HOME
(e) SETH 90 REPEAT 5[FD 10 PU FD 5 PD]

Solution Results are depicted in Figure AII-7.

FIGURE AII-7

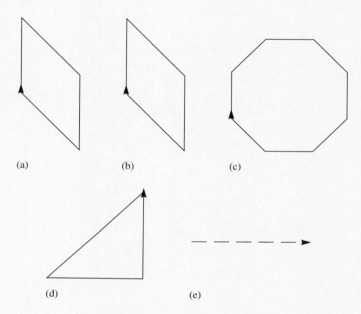

Defining Procedures

The sequence of commands in Figure AII-6(a) instructed the turtle to draw a square. If the screen is cleared, the figure is lost. To redraw the square, the entire sequence of commands must be retyped. Fortunately, with Logo, it is possible to store instructions in the computer's memory by creating a procedure. A **procedure** is a group of one or more instructions to the computer that the computer can store to be used later.

procedure

TO

To create a procedure in MIT Logo, we type **TO**, followed by the name we wish to call the procedure and press RETURN. (*In LCSI, we type* EDIT (ED), *followed by quotation marks and the procedure name, for example,* ED "TRIANGLE.) When RETURN is pressed, the computer enters the **edit mode,** or the teaching mode. In this mode, the lines that follow are not executed but may be stored in memory under the given name. The name must be a sequence of symbols with no spaces, and it may not be the name of a primitive. For example, to create a procedure to draw a square, the following is entered. To signify the end of a procedure, it is good practice to type **END** as the last line of the procedure. It is necessary to type END at the end of a procedure if another procedure is to be defined without your leaving the edit mode.

edit mode

END

```
TO SQUARE1
  REPEAT 4 [FD 50 RT 90]
END
```

SECTION AII-1 INTRODUCING THE TURTLE

CONTROL C / (CTRL-C)

To define and store a procedure and exit the edit mode, we press the [C] key while holding down the [CONTROL] key. This is called a **CONTROL-C (CTRL-C).** After the procedure has been defined, typing the name of the procedure and pressing [RETURN] causes the computer to immediately execute the instructions in the procedure. In the remainder of this section, we assume that the SQUARE1 procedure just given and all subsequent procedures are stored in the computer's memory and can be reused.

In Logo, it is possible for one procedure to call on another procedure, as shown in Example AII-2. If the SQUARE1 procedure has not been defined on your computer, please define it before working the example.

EXAMPLE AII-2 Predict what figures will be drawn by defining and executing each of the following procedures. Assume the turtle starts at home with heading 0.

(a) TO SQUARE2
    RT 90
    SQUARE1
    END

(b) TO SQUARESTACK
    SQUARE1
    RT 90
    SQUARE1
    END

(c) TO STAIR
    SQUARE1
    RT 180
    SQUARE1
    END

(d) TO TURNSQUARE
    SQUARE1
    RT 45
    SQUARE1
    END

Solution Results are depicted in Figure AII-8.

FIGURE AII-8

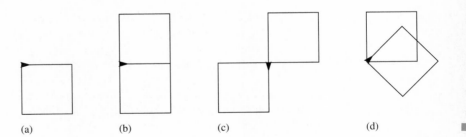

(a)　　(b)　　(c)　　(d)

PROBLEM 1

Write a procedure for drawing a triangle whose sides each have length of 50 turtle steps and whose angles each have measure of 60°. This type of triangle is called an *equilateral triangle*.

Understanding the Problem. We are to write a procedure to draw a triangle with all sides of length 50 turtle units and all angles of measure 60°. We can start at any position with any heading.

Devising a Plan. It is helpful to sketch the triangle to determine what angle the turtle needs to turn at each vertex. Suppose the turtle starts at point A with heading 0 and moves 50 turtle steps to point B, as in Figure AII-9. This can be done by telling the turtle to move FD 50. At point B, the turtle still has heading 0. To walk on $\overrightarrow{BC}$, the turtle may turn 120° to the right. Thus the next command should be RT 120. The triangle has three sides of equal length, so three turns are necessary to achieve the turtle's initial heading. If we repeat the sequence FD 50 RT 120 three times, this should cause the turtle to walk around the triangle and finish in its original position with its original heading.

FIGURE AII-9

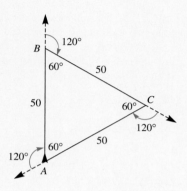

Carrying Out the Plan. A procedure called TRIANGLE1 based on the preceding discussion follows:

```
TO TRIANGLE1
 REPEAT 3 [FD 50 RT 120]
END
```

Looking Back. Executing the TRIANGLE1 procedure yields the desired figure. Additional investigations include writing a procedure to draw the same type of triangle by turning left instead of right or writing a procedure to draw a triangle with one horizontal side. Procedures for drawing other polygons could also be explored. ∎

One of the great advantages of Logo is its ability to use procedures to define new procedures. Consider the following problem.

PROBLEM 2

Write a procedure to draw the "house" in Figure AII-10.

FIGURE AII-10

Understanding the Problem. The top of the house appears to be a triangle similar to the one in Problem 1, and the bottom appears to be a square. We must write a super procedure in which procedures for drawing a square and a triangle will be incorporated to draw the house.

top-down programming

Devising a Plan. One way to solve the problem is to break down the problem of drawing a house into *simpler problems,* that of drawing the bottom of the house (the square) and that of drawing the roof (the triangle). The type of programming that starts with a general idea and breaks down the problem into smaller parts is called **top-down programming.** We have a procedure SQUARE1 for drawing a square of length 50 units, and a procedure TRIANGLE1 for drawing a triangle of length 50 units. If we use these two procedures, then we should be able to draw the house.

Carrying Out the Plan. If the turtle has heading 0, it may seem that typing SQUARE1 followed by TRIANGLE1 would draw the desired house. The result of this effort is shown in Figure AII-11(a). Why did this not produce the desired figure? The turtle first drew the square in Figure AII-11(a) and returned to its initial position at point *A* with heading 0. Then, typing TRIANGLE1 caused the turtle to draw triangle *ABC* as in Figure AII-9. To draw the roof in proper position, we need the turtle to be at the upper left vertex of the square. This can be achieved by typing FD 50. But, if TRIANGLE1 is typed now, we obtain the shape in Figure AII-11(b), which is still not the desired house.

FIGURE AII-11

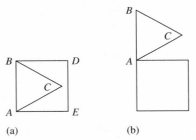

(a)   (b)

After typing SQUARE1 and FD 50, the turtle is at point *A* with heading 0. To form the roof as in Figure AII-12(a) and to walk on $\overline{AB}$, the turtle needs to turn right by 90° − 60°, or 30°. With the turtle having this heading, typing TRIANGLE1 should cause the turtle to draw the desired roof. The complete procedure, called HOUSE, is shown in Figure AII-12(b).

FIGURE AII-12

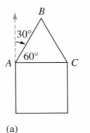

```
TO HOUSE
   SQUARE1
   FD 50
   RT 30
   TRIANGLE1
   HIDETURTLE
END
```

(a)   (b)

Looking Back. If the HOUSE procedure is executed, the desired figure is obtained. Alternate techniques for drawing the figure could also be explored. Houses of other sizes could be drawn and windows and doors could be added. ∎

As Problem 2 shows, trial and error helps the user to get acquainted with the problem and eventually to find the correct solution. This process of rewriting a program that does not do what we want it to do is called *debugging*.

To help you understand how Logo works when a procedure calls another procedure, a telescoping model of the HOUSE procedure in Problem 2 is given in Figure AII-13.

FIGURE AII-13

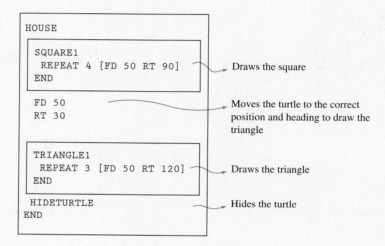

When HOUSE is run, it encounters the call for SQUARE1. At this point, all the lines of SQUARE1 are inserted. After SQUARE1 has been completed, control is returned to the procedure that called it, namely, HOUSE. Now, HOUSE continues where it left off and executes FD 50 RT 30, then calls on the TRIANGLE1 procedure. After TRIANGLE1 has been executed, it returns control to HOUSE, which hides the turtle and encounters its own END statement.

In working through the HOUSE procedure, we went through several steps. A summary of these steps follows. These steps might be useful in solving a variety of problems presented in this text.

1. Sketch your drawing on paper (preferably graph paper) to get an idea of the scale to be used and of how the final picture should look.

2. Divide the drawing into parts that are repeated, that you already know how to draw, or that are smaller parts of the whole. Separate procedures for drawing each individual part are easier to check than a single procedure for the whole drawing.

3. Decide how your individual procedures are going to fit together to form the complete picture. Some procedures might be necessary just to move the turtle to the right position for drawing the individual parts.

4. Write your procedures. One approach is to write individual procedures, make sure they work, and then try to put them all together to form the complete picture. Another approach is to fit the procedures together as they are completed. Either approach is an

SECTION AII-1 INTRODUCING THE TURTLE   843

acceptable problem-solving strategy, and each has advantages in different situations.

We demonstrate how these steps can be used in a problem-solving format in Problem 3.

PROBLEM 3

Write a procedure to draw the figure sketched on the graph paper in Figure AII-14.

FIGURE AII-14

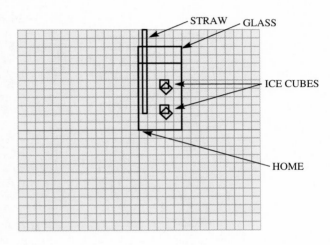

Understanding the Problem. We are to write a procedure to draw a figure similar to the one shown in Figure AII-14. Each length of a square on the grid paper represents 10 turtle steps.

Devising a Plan. The figure can be broken into three separate parts: the glass, the straw, and the ice cubes. Using top-down programming, a procedure called DRINK to draw this figure might appear as follows:

```
TO DRINK
 GLASS
 ICE.CUBES
 ICE.CUBES
 STRAW
 HT
END
```

To complete the problem, we must write procedures for each portion of the DRINK procedure.

Carrying Out the Plan. First, we design a procedure called GLASS for drawing the glass. If the turtle starts at home with heading 0, one possible procedure and its output are given in Figure AII-15.

FIGURE AII-15

```
TO GLASS
  REPEAT 2 [FD 100 RT 90 FD 50 RT 90]
  FD 80 RT 90
  FD 50 BK 50
  LT 90 BK 80
END
```

Likewise, procedures called STRAW and ICE.CUBES can be designed to draw the other two parts, as shown in Figure AII-16(a) and (b).

FIGURE AII-16

```
TO STRAW
  REPEAT 2 [FD 100 RT 90 FD 5 RT 90]
END

TO ICE.CUBES
  SQUARE3
  RT 45
  SQUARE3
  LT 45
END

TO SQUARE3
  REPEAT 4 [FD 10 RT 90]
END
```

(a)

(b)

If we now execute DRAW *(CS in LSC1)* and attempt to execute the DRINK procedure as defined, the result is as shown in Figure AII-17.

FIGURE AII-17

Here, we encounter a bug. To correct the DRINK procedure so it will draw the desired figure, we must keep track of the turtle's position and heading. Sometimes, it is convenient to move the turtle to the required positions and headings by using a set of procedures. The following procedures—called SETUP.CUBES1, SETUP.CUBES2, and SETUP.STRAW—move the turtle to the correct position and heading to draw each part:

```
TO SETUP.CUBES1
  PU FD 50 RT 90 FD 25 LT 90 PD
END

TO SETUP.CUBES2
  PU BK 30 PD
END

TO SETUP.STRAW
  PU LT 90 FD 20 RT 90 PD
END
```

FIGURE AII-18

```
TO DRINK
  GLASS
  SETUP.CUBES1
  ICE.CUBES
  SETUP.CUBES2
  ICE.CUBES
  SETUP.STRAW
  STRAW
  HT
END
```

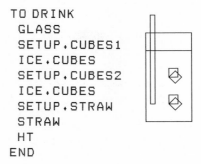

Looking Back. When the DRINK procedure is executed, it yields the desired figure. The procedure could have been written in many different ways. Although various strategies could be used to develop the procedure, we see that the top-down strategy can be very useful. One advantage of the top-down strategy is that it is easier to work with and makes it easier to debug smaller portions of the figure rather than to try to do the complete figure all at one time. ∎

Writing Procedures with Variables

The SQUARE1 procedure in this section allowed us to draw only squares of side 50. If we want to draw smaller or larger squares, we must write a new procedure. It would be more convenient if we could write one procedure that would work for a square of any size. This can be accomplished in Logo by using a variable as input, rather than a fixed number such as 50 in FD 50. When we use variable input in Logo, we warn the computer that the "thing" we are going to type is a variable by using a colon before the variable name. For example, a variable input to the SQUARE procedure might be called :SIDE, where :SIDE stands for the length of a side of the square. Note that there is no space between the colon and the word SIDE. We define a new SQUARE procedure with variable input :SIDE, and we place the name of the variable in the title line.

```
TO SQUARE :SIDE
  REPEAT 4 [FD :SIDE RT 90]
END
```

If we want the turtle to draw a square of size 40, we type SQUARE 40. Notice that we do not type SQUARE :40 because 40 is not a variable. (In fact, the computer will not understand the instruction SQUARE :40.) Investigate what happens if SQUARE is typed with no inputs.

> **REMARK** We call the new variable square procedure SQUARE instead of SQUARE1. If we attempt to enter the edit mode to define a new SQUARE1 procedure and the old procedure has not been erased, the computer will display the old SQUARE1 procedure on the screen for us to edit. You should consult your Logo manual for directions on how to edit procedures.

A procedure may have more than one input. For example, consider the following equivalent procedures for drawing a rectangle. Two variables are used so that two inputs can be accepted.

```
TO RECTANGLE :HEIGHT :WIDTH
  FD :HEIGHT RT 90
  FD :WIDTH RT 90
  FD :HEIGHT RT 90
  FD :WIDTH RT 90
END

TO RECTANGLE :HEIGHT :WIDTH
  REPEAT 2 [FD :HEIGHT RT 90 FD :WIDTH RT 90]
END
```

Figure AII-19 shows rectangles drawn by either of the RECTANGLE procedures with different inputs for the sides.

FIGURE AII-19

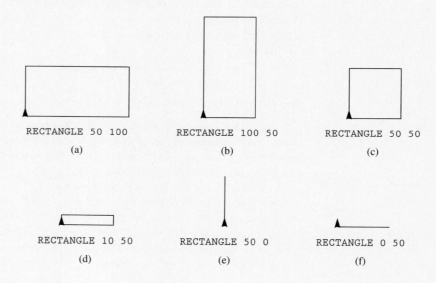

BRAIN TEASER

Write a procedure for drawing the following figure, using a continuous path and without retracing any segment. (Single points may be retraced.)

SUMMARY OF COMMANDS

DRAW*	Needs no input. It sends the turtle home and clears the graphics screen.
NODRAW (ND)†	Needs no input. It exits the graphics mode, clears the screen, and places the cursor in the upper left corner.
FORWARD (FD)	Takes one input. If the input is positive, it moves the turtle forward (in the direction the turtle is facing) the number of turtle units that are input. For example, FD 20 moves the turtle forward 20 units.
BACK (BK)	Takes one input. If the input is positive, it moves the turtle backwards the number of turtle units that are input. For example, BK 40 moves the turtle backwards 40 units.
RIGHT (RT)	Takes one input. If the input is positive, it turns the turtle right from its present heading the number of degrees that are input. For example, RT 90 turns the turtle right 90°.
LEFT (LT)	Takes one input. If the input is positive, it turns the turtle left from its present heading the number of degrees that are input. For example, LT 90 turns the turtle left 90°.
PENUP (PU)	Needs no input. In the graphics mode, it enables the turtle to move without leaving a track.
PENDOWN (PD)	Needs no input. In the graphics mode, it causes the turtle to leave a track.
HIDETURTLE (HT)	Needs no input. It causes the turtle to disappear.
SHOWTURTLE (ST)	Needs no input. It causes the turtle to reappear.
PRINT (PR)	Takes one input. It causes the input to be printed on the screen and moves the cursor to the next line.
HEADING	Needs no input. In the draw mode, it outputs the turtle's heading.
SETHEADING (SETH)	Takes one input. It turns the turtle to the heading indicated by the input.
END	Used at the end of a procedure. Tells the computer that there are no more instructions to be given in the procedure.
HOME	Needs no input. It returns the turtle to the center of the screen and sets its heading to 0. If the pen is down, it leaves a track from the turtle's present location to the home position.
REPEAT	Takes a number and a list as input. It executes the instructions in the list the designated number of times.
TO‡	Takes the name of a procedure as input and causes Logo to enter the edit mode.

*In *LCSI*, this command is CLEARSCREEN (SC).
†In *LCSI*, this command is TEXTSCREEN (CTRL-T).
‡In *LCSI*, this command is EDIT (ED).

APPENDIX II AN INTRODUCTION TO LOGO TURTLE GRAPHICS

PROBLEM SET AII-1

1. Sketch figures drawn by the turtle using each of the following sets of instructions. Check your sketches by executing the instructions on a computer. Type DRAW *(CS in LCSI)* after each lettered part.

   (a) FD 50
   RT 90
   FD 50
   RT 45
   FD 50
   RT 135
   FD 50

   (b) FD 50
   RT 90
   BK 50
   RT 60
   FD 50

   (c) FD -50 FD 50
   (d) LT -90 BK -50 RT 40 PR HEADING
   (e) RT 360 PR HEADING
   (f) SETH 30 REPEAT 3[FD 50 RT 120]
   (g) FD 100/2 RT 5*6 BK 100+20

2. Experiment with the turtle to find the dimensions of the screen.

3. Predict what the turtle will draw with the following sets of instructions. Check your answers by executing the instructions on the computer.
   SQUARE1 and TRIANGLE1 are defined in the text.
   (a) REPEAT 8 [SQUARE1 RT 45]
   (b) REPEAT 6 [TRIANGLE1 RT 60]
   (c) REPEAT 36 [SQUARE1 RT 10]

4. Write procedures to draw figures similar to each of the following:

Rectangle that is not a square
(a)

Flag
(b)

A hat
(c)

The letter *T*
(d)

A rhombus that is not a square
(e)

A square with a smaller square inside
(f)

5. Write a procedure to draw the following:

6. Use any procedures in this section to write new procedures that will draw each of the following figures:

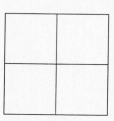

(a)

(b)

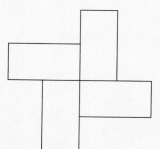

(c)

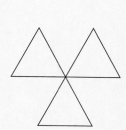
(d)

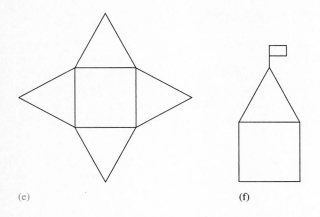

(e)   (f)

7. Write procedures to draw figures similar to each of the following:

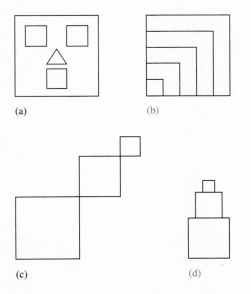

(a)   (b)

(c)   (d)

8. Use top-down programming to write a procedure called DOG to draw a figure similar to the following:

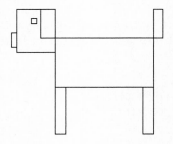

9. Use top-down programming to write a procedure called KITE to draw a figure similar to the following:

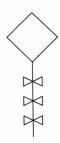

10. Write procedures to draw figures similar to those in Problem 4, but of variable size.

11. Write a procedure called BLADES to draw a figure similar to the following, but of variable size:

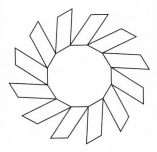

12. Write a procedure called RECTANGLES to draw a figure similar to the following, but of variable size:

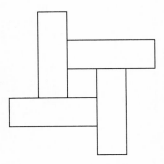

Section AII-2 Using Recursion as a Problem-solving Tool

recursion

In Section AII-1, we considered procedures that called on other procedures. **Recursion** is the process of a procedure calling on a copy of itself. As a first example of recursion, we write a procedure called CIRC for drawing a "turtle-type" circle. This could be done by having the turtle move forward "a little," then turn right "a little," and continuing this process until a closed figure is obtained. Thus we could start the procedure with FD 1 RT 1 and then have the turtle start the procedure anew each time the instruction is executed. Such a procedure follows:

```
TO CIRC
 FD 1 RT 1
 CIRC
END
```

To understand how CIRC works and, in general, what happens when a procedure calls itself, we use the telescoping model in Figure AII-20.

FIGURE AII-20

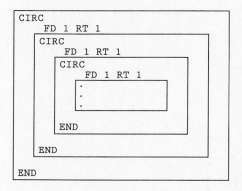

When CIRC is executed, FD 1 RT 1 causes the turtle to move forward one unit and then turn right 1°. CIRC then calls a copy of the CIRC procedure, which again executes FD 1 RT 1 and in turn calls another copy of CIRC, and so on. The process continues because we have made CIRC one of the instructions in the CIRC procedure. The END statement is never reached, and the instruction FD 1 RT 1 is executed indefinitely. You may stop the execution of the procedure at any time by pressing **CTRL-G**.

CTRL-G

tail-end recursion

embedded recursion

The repetitive process shown in the CIRC procedure occurs in the type of recursion called **tail-end recursion.** In tail-end recursion, only one recursive call is made within the body of the procedure, and it is the final step before the END statement. Later, we introduce another type of recursion that is sometimes called **embedded recursion.**

> **REMARK** When drawing a circle with right turns of 1°, the turtle only has to turn through 360° to complete the circle. A CIRCLE1 procedure could be written using the REPEAT command as follows.

SECTION AII-2 USING RECURSION AS A PROBLEM-SOLVING TOOL

```
TO CIRCLE1
  REPEAT 360 [FD 1 RT 1]
END
```

The CIRCLE1 procedure stops, whereas the CIRC procedure does not.

Recursion is particularly valuable when we do not know how many times to repeat a set of instructions to accomplish some goal. For example, consider the shapes that can be drawn by repeating the instruction "Go forward some fixed distance and turn right some fixed angle." A recursive procedure called POLY that does this is as follows:

```
TO POLY :SIDE :ANGLE
  FD :SIDE RT :ANGLE
  POLY :SIDE :ANGLE
END
```

To execute the POLY procedure, we need two numerical inputs, one for :SIDE and the other for :ANGLE. Figure AII-21 shows shapes drawn by POLY with eight different inputs. The drawings were stopped using CTRL-G.

The POLY procedure draws regular polygons (polygons that have congruent sides and congruent angles), as in Figure AII-21(a), (b), and (c), and also star shapes as in Figure AII-21(d), (e), and (f).

FIGURE AII-21

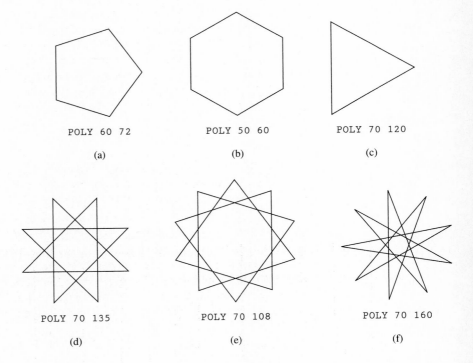

POLY 60 72   POLY 50 60   POLY 70 120
   (a)   (b)   (c)

POLY 70 135   POLY 70 108   POLY 70 160
   (d)   (e)   (f)

Try other executions of POLY, such as POLY 50 180, POLY 50 181, POLY 60 288, POLY 6000 300, and POLY 7000 135. Try to predict which inputs produce regular polygons and which produce star shapes.

All the figures drawn by the POLY procedure in Figure AII-21 are closed; that is, they can be drawn by starting and stopping at the same point. Will all figures drawn by POLY be closed? We can also ask the following questions:

1. Given the value of :ANGLE in the POLY procedure, is it possible to predict (before the figure is drawn) how many vertices the figure will have?
2. If we wish the POLY procedure to draw a figure with a given number of vertices, can we determine what the correct angle input should be?

With the help of recursion, we can accomplish tasks that cannot be easily done with just the REPEAT command, especially if we do not know how many times to repeat a sequence of instructions. Consider drawing a square-type spiral as shown in Figure AII-22.

FIGURE AII-22

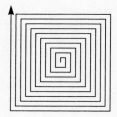

Suppose each side of the figure is five units longer than the preceding side. If the turtle starts at home, the figure can be drawn by telling the turtle to move forward a certain length :LEN, turn right 90°, move forward a distance five units greater than the previous value of :LEN, and so on. A recursive procedure called SQSPI shows how this can be done.

```
TO SQSPI :LEN
 FD :LEN RT 90
 SQSPI :LEN + 5
END
```

Each time SQSPI calls itself, the length of :LEN is increased by five units. When SQSPI is run, the sides grow too large to fit on the screen. Rather than stopping SQSPI with CTRL-G, it is possible to write a "stop" instruction in the procedure. This can be done *IF* with the IF and STOP primitives. **IF** is a primitive that tests one of three conditions: equal (=), less than (<), or greater than (>). The IF primitive is used in the following form:

IF *(Condition)* *(Action to be taken if condition is true)*

The parentheses should not be typed. *(In LCSI, the format is IF (Condition) [Action to be taken if condition is true], where the square brackets must be typed.)*

For example, if we do not want the turtle to draw any segment longer than 100 units, we insert the following instruction:

IF :LEN > 100 STOP

STOP *(In LSCI, IF :LEN > 100 [STOP].)* When this line is inserted into a procedure, the IF statement causes the computer to check whether the value of :LEN is greater than 100. If it is, the procedure stops; if not, the next line is executed. The **STOP** primitive causes the current procedure to stop and returns control to the calling procedure, if there is one. An edited form of the SQSPI procedure is as follows:

```
TO SQSPI :LEN
  IF :LEN > 100 STOP
  FD :LEN RT 90
  SQSPI :LEN + 5
END
```

(In LCSI, replace STOP with [STOP].)

The SQSPI procedure can be generalized to draw other spiral-type figures. Investigate the following POLYSPI procedure for various inputs:

```
TO POLYSPI :SIDE :ANGLE
  IF :SIDE > 100 STOP
  FD :SIDE RT :ANGLE
  POLYSPI :SIDE + 5 :ANGLE
END
```

(In LCSI, replace STOP with [STOP].)

What inputs should be given to POLYSPI in order to achieve the same effect as SQSPI does? Also, investigate what happens when :ANGLE, rather than :SIDE, is incremented each time the recursive call is made.

Embedded Recursion

embedded recursion

Tail-end recursion involves only one recursive call within the body of the procedure, and it is the final step before the END statement. Recursive calls also can be **embedded;** that is, the recursive call is not the last line before the END statement. For example, consider the tail-end recursion in the T.SQUARE procedure and the embedded recursion in the E.SQUARE procedure that follow. Predict the results of executing each of these procedures with input 30; then, execute them to see if you were correct.

```
TO T.SQUARE :SIDE
  IF :SIDE < 5 STOP
  REPEAT 4 [FD :SIDE RT 90]
  T.SQUARE :SIDE - 10
END

TO E.SQUARE :SIDE
  IF :SIDE < 5 STOP
  REPEAT 4 [FD :SIDE RT 90]
  E.SQUARE :SIDE - 10
  LT 45 FD :SIDE
END
```

(In LCSI, remember to replace STOP with [STOP].)

The T.SQUARE procedure probably did exactly what you expected; however, the E.SQUARE procedure may have surprised you. To see why E.SQUARE behaved the way it did, we use the telescoping model in Figure AII-23.

FIGURE AII-23

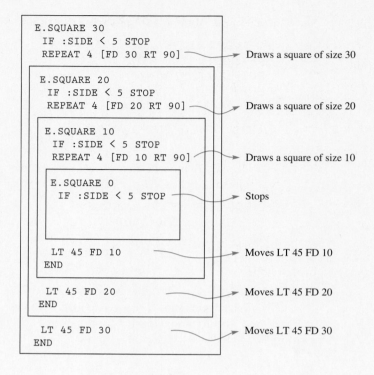

Notice that when :SIDE < 5 in E.SQUARE 0, the STOP command is finally reached. However, STOP stops only the procedure it is in, not the calling procedure. Control is then returned to the previous procedure, and so on.

From our model, we see that recursion works according to the following rules:

1. Executions in Logo programs proceed line by line. When a procedure calls itself, it puts on hold any instructions that are written after the call and inserts a copy of itself at the point where the call occurs. If the called procedure stops, control is returned to the calling procedure at the point where the call occurred. The remainder of the lines in the calling procedure are then executed.

2. The process in (1) applies to any successive calls.

SUMMARY OF COMMANDS

IF*	Takes two inputs. The first input must be either true or false. The second input contains instructions that are carried out if, and only if, the first input is true.
STOP	Takes no inputs. It causes the current production to stop and returns control to the calling procedure.

*In *LCSI*, the second input must be enclosed in brackets, for example, IF :SIDE > 100 [STOP].

PROBLEM SET AII-2

1. Predict the shapes that will be drawn by the following procedures and then check your predictions on the computer. The SQUARE and TRIANGLE procedures are defined as follows. *(In LCSI, replace STOP with [STOP] in (b), (d), (e), and (f).)*

   ```
   TO TRIANGLE :SIDE
     REPEAT 3 [FD :SIDE RT 120]
   END
   TO SQUARE :SIDE
     REPEAT 4 [FD :SIDE RT 90]
   END
   ```

   (a) ```
       TO FIGURE :SIDE
         TRIANGLE :SIDE
         RT 10
         FIGURE :SIDE
       END
       ```
   (b) ```
       TO FIGURE1 :SIDE
         IF :SIDE < 5 STOP
         TRIANGLE :SIDE
         RT 10
         FIGURE1 :SIDE - 5
       END
       ```
   (c) ```
       TO TOWER :SIDE
         SQUARE :SIDE
         FD :SIDE
         TOWER :SIDE * 0.5
       END
       ```
   (d) ```
       TO TOWER1 :SIDE
         IF :SIDE < 2 STOP
         SQUARE :SIDE
         FD :SIDE
         TOWER1 :SIDE * 0.5
       END
       ```
   (e) ```
       TO SQ :SIDE
         IF :SIDE < 2 STOP
         SQUARE :SIDE
         SQ :SIDE - 5
       END
       ```
   (f) ```
       TO SPIRAL :SIDE
         IF :SIDE > 50 STOP
         FD :SIDE
         RT 30
         SPIRAL :SIDE + 3
       END
       ```

2. Given the following NEWPOLY, POLYSPIRAL, and INSPI procedures, predict the shapes that will be drawn by each and then check your predictions on the computer:

   ```
   TO NEWPOLY :SIDE :ANGLE
     FD :SIDE RT :ANGLE
     FD :SIDE RT :ANGLE * 2
     NEWPOLY :SIDE :ANGLE
   END
   TO POLYSPIRAL :SIDE :ANGLE :INC
     FD :SIDE RT :ANGLE
     POLYSPIRAL (:SIDE + :INC) :ANGLE :INC
   END
   TO INSPI :SIDE :ANGLE :INC
     FD :SIDE RT :ANGLE
     INSPI :SIDE (:ANGLE + :INC) :INC
   END
   ```

   (a) NEWPOLY 50 30
   (b) NEWPOLY 50 144
   (c) NEWPOLY 50 125
   (d) POLYSPIRAL 2 85 3
   (e) POLYSPIRAL 1 119 2
   (f) POLYSPIRAL 1 100 5
   (g) INSPI 10 2 20
   (h) INSPI 2 0 10
   (i) INSPI 10 5 10

3. Write recursive procedures to draw figures similar to the following six figures. Use the STOP command in your procedures.

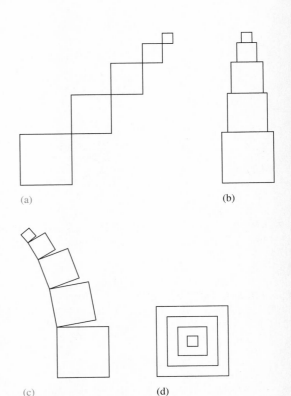

(a) (b) (c) (d)

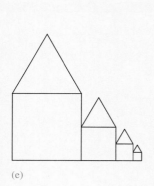
(e)

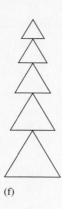
(f)

4. Write a recursive procedure with a STOP command to draw a figure similar to the following:

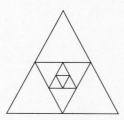

5. (a) Execute each of the following to see how embedded recursion can be used to "unwind" the turtle. *(Use [STOP] in LCSI)*.

```
TO WOW :X
  IF :X < 5 STOP
  FD :X RT 87
  WOW :X - 2
  LT 87 BK :X
END
TO WOW1 :X
  IF :X < 5 STOP
  FD :X RT 118
  WOW1 :X - 2
  LT 118 BK :X
END
```

(b) Write a procedure, using embedded recursion, that "unwinds" like those in (a).

6. (a) Predict what happens when REFLECT 6 is executed. Check your predictions by executing the procedure. *(Use [STOP] in LCSI)*.

```
TO REFLECT :N
  IF :N < 3 STOP
  REPEAT :N [FD 30 RT 360/:N]
  REFLECT :N - 1
  REPEAT :N [FD 30 RT 360/:N]
END
```

(b) Try REFLECT with various other values of :N.

7. Write a procedure called SPIN.SQ that uses recursion and a STOP command to spin a variable-sized square while "shrinking" its size, as shown in the following figure:

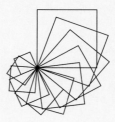

8. Write a recursive procedure with a STOP statement that draws the following variable-sized figure made of squares:

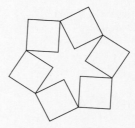

9. Write a procedure utilizing embedded recursion to draw a variable-sized figure similar to the following:

Answers to Selected Problems

CHAPTER 1

Problem Set 1-1
2. (a) $5 \times 6, 6 \times 7, 7 \times 8$ (d) 26, 37, 50 (g) 34, 55, 89 (k) $2^{32}, 2^{64}, 2^{128}$ (o) 0, 22, 44, 66, 88, 110
3. (c) 96, 192, 384; geometric (f) 66, 77, 88; arithmetic (i) $6^3, 7^3, 8^3$; neither
5. (b) 10,100
7. (a) 41 (b) $4n + 1$
9. 15 liters
12. 19
13. 23rd year
15. (a) 10,000 (b) n^2
16. (a) 42
17. (c) 15, 17, 21, 27, 35, 45
19. (a) 101 (d) 87 (f) 21
20. (a) Yes. The difference between terms in the new sequence is the same as in the old sequence.
23. If the sequences have the same ratio, the resulting sequence is geometric. If the sequences are

$$a, ar, ar^2, ar^3, \ldots$$

and

$$b, br, br^2, br^3, \ldots$$

then the resulting sequence is

$$a + b, (a + b)r, (a + b)r^2, (a + b)r^3, \ldots$$

This is a geometric sequence with ratio r and first term $a + b$.

25. 48, 72, 108
26. (a) 6, 12
27. (a) 3, 6, 11, 18, 27 (b) 4, 9, 14, 19, 24 . . .
28. (a) 1, 1, 2, 3, 5, 8, 13, 21, 34, 55, 89, 144, (b) Yes; the sum of the first 4 terms equals the (6th term) $- 1$. The sum of the first 5 terms equals the (7th term) $- 1$. The sum of the first 6 terms equals the (8th term) $- 1$, equals 20 (c) 143 (d) The sum of the 1st n terms equals the $(n + 2)$th term $- 1$
32. (a) 300, 500, 700, 900, 1100, 1300, . . . , $(2n + 1)100$ (b) 2, 4, 8, 16, 32, 64, . . . , 2^n The sequence in (b) becomes greater than the sequence in (a) on the 12th term.

Problem Set 1-2
1. (a) 4950
4. 18
6. (a) 204 squares
8. 12
11. $2.45
12. 16 days
16. David had 78 marbles, Judy had 42, and Jacobo had 24.
19. (a) 11 (b) 63
23. (a) If both numbers were less than or equal to 9, then their product would be less than or equal to $9 \times 9 = 81$, which is not greater than 82.
24. (c) 20,503
27. $13,500
28. $78, $42, and $24. This is the same problem mathematically as problem 16.
30. 35 moves
31. (a) 21, 24, 27 (b) 243, 2, 729
32. $22 + (n - 1)10$ or $10n + 12$
33. 21 terms
34. 903

Problem Set 1-3
1. (a) (i) 541×72
3. $3.99 + 5.87 + 6.47 = \$16.33$
6. 275,000,000
11. $5,256,000 per year
13. 625
18. $5459 = 53 \times 103$. Divide by primes up to 73, which is close to the square root of 5459.
23. Play second; at your turn, make the calculator display a multiple of 4.
26. (a) 35, 42, 49 (b) 1, 16, 1
27. $20n - 8$
28. 21 terms
29. 9 ways

857

Chapter 1 Test
1. (a) 15, 21, 28 (c) 400, 200, 100 (e) 17, 20, 23
3. (a) $3n + 2$ (b) n^3 (c) 3^n
4. (b) 2, 6, 12, 20, 30, . . .
8. 89 years
9. The worm will climb out on the 10th day.
12. 21 posts
15. 44,000,000 rotations
17. 39 boxes
19. (b) $(10^{n+1} - 1)/9$
20. 9 hours

Problem Set 2-1
2. (a) {m, a, t, h, e, i, c, s} (c) {January, June, July} (e) {x|x is a state in the United States} (g) {Alaska, Hawaii, Washington, California, Oregon}
3. (a) B = {x, y, z, w} (c) {1, 2} ⊂ {1, 2, 3, 4} (e) A ⊄ B (g) {0} ≠ ∅
5. (a) Yes (c) Yes (e) No
7. (a) 24 (c) n!
10. $\overline{A}$ is the set of all college students without a straight A average.
11. (a) 7
12. (b) They are equal.
14. (a) ∈ (c) ∉ (e) ∉
17. (a) True (c) True
18. (a) A = {1, 2}; n(A) = 2.
    B = {1, 2, 3, 4}; n(B) = 4.
    A is a proper subset of B.
   (c) A = ∅; n(A) = 0.
    B = {1, 2, 3}; n(B) = 3
    A is a proper subset of B.
20. (a) 63

Problem Set 2-2
1. (a) Yes (c) Yes (e) No
2. (a) True (c) True (e) True (g) True
3. (a) B
4.

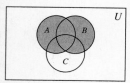

(a)

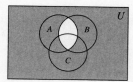

(c)

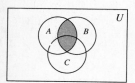

(e)

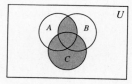

(g)

5. (a) U (c) S (e) S (g) ∅ (i) $\overline{S}$ (k) ∅
7. (a) A (c) ∅
8. (a) $B \cap \overline{A}$ (c) $A \cap B \cap C$ (e) $(A \cap C) \cap \overline{B}$
9.

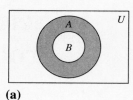

(a)

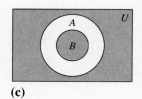

(c)

10. (a) False (c) False (e) False
11. (a) (i) 5
       (ii) 2
13. (a) The set of all natural numbers (c) E, the set of all even natural numbers
14.

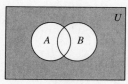

(a) $\overline{A \cup B} = \overline{A} \cap \overline{B}$
(c) (a) A = {1, 2, 3}; B = {5}; U = {x|x is a natural number} $\overline{A \cup B}$ = {4, 6, 7, 8. . .} = {4, 5, 6, 7. . .} ∩ {1, 2, 3, 4, 6. . .} = $\overline{A} \cap \overline{B}$
16. (a) The set of college basketball players/students who are more than 200 cm tall. (c) The set of all college basketball players or college students who are more than 200 cm tall. (e) The set of all college students more than 200 cm tall who are not basketball players.
18. 18
20.

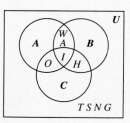

21. (a) 20 (c) 10
22. (a) False If A = {apples, oranges}
             B = {1, 2}
             n(A) = n(B) but A ≠ B
   (c) False If A = {1, 2, 3, 5}
              B = {1, 2, 3, 4, 5}
              A − B = ∅ but A ≠ B
   (e) True; unless A and B are infinite sets; in that case, n(A) needs to be defined.

24. (a) {(x, a), (x, b), (x, c), (y, a), (y, b), (y, c)}
(c) ∅ **(e)** ∅ **(g)** {(x, 0), (y, 0), (a, 0), (b, 0), (c, 0)}
26. (a) C = {a}
    D = {b, c, d, e}
(c) C = {0, 1}
    D = {1, 0}
27. (a) 20 **(c)** (mn)p
28. (a) 0 **(c)** 0
29. 5
32. 60
33. 93
36. {a, b, c}, {a, b}, {a, c}, {b, c}, {a}, {b}, {c}, ∅
37. Yes
39. (a) {Maine, Minnesota, Michigan, Mississippi, Missouri, Maryland, Massachusetts, Montana}
(b) Z = {x|x is a state in the United States that begins with the letter M}

Problem Set 2-3
1. (a) "is the square root of"; (5, 25) (10, 100) **(c)** "is the lowercase form of"; (q, Q) (p, P)
3. Answers vary. An example is (Garfield, Jon Arbuckle).
5. (a) Not reflexive, not symmetric, not transitive **(c)** Reflexive, symmetric, transitive. Equivalence relation. **(e)** Not reflexive, symmetric, not transitive **(g)** Not reflexive, not symmetric, transitive
6. (a) Reflexive, symmetric, transitive; equivalence relation **(c)** Not reflexive, symmetric, not transitive
8. (a) f(x) = 2x **(c)** f(x) = x + 6
9. (a) Not a function since 1 is paired with a and d **(c)** Function
13. (a) 5 **(c)** 35
14. (b) 32
15. (a)

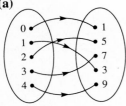

(c)

x	f(x)
0	1
1	3
2	5
3	7
4	9

16. (a) ⁻5 **(c)** 65
17. (b) 57¢
18. (a) 2 chirps/second
19. (b) f(n) = 2n + $1.50
21. (a) 51 **(c)** 2

22. (a) 5n − 2 **(c)** 2n
23. (a) Yes
24.

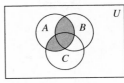 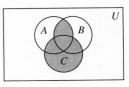

(a) $A \cap (B \cup C)$ $(A \cap B) \cup C$

26. (a) {x|x = 2n; n > 6; n ∈ N}
27. (a) {a, b, c, d} **(c)** ∅ **(e)** ∅
30. (a) 6

Problem Set 2-4
1. (a) False statement **(c)** False statement **(e)** Not a statement **(g)** True statement **(i)** Not a statement
2. (a) There exists x = 3 such that x + 8 = 11. **(c)** There exists a natural number x such that $x^2 = 4$.
4. (a) The book does not have 500 pages. **(c)** $3 \cdot 5 \neq 15$ **(e)** There exists a dog which does not have 4 legs. **(g)** There exists a square which is not a rectangle. **(i)** There exists a natural number x such that $x + 3 \neq 3 + x$. **(k)** There exists a counting number not divisible by itself and 1. **(m)** There exists a natural number x such that $5x + 4x \neq 9x$.
5. (a)

~p	~(~p)
F	T
T	F

(c) Yes
6. (a) q ∧ r **(c)** ~(q ∧ r)
7. (a) F **(c)** T **(e)** F **(g)** F **(i)** F
9. (a) No **(c)** No
11.

~p	~q	~p ∨ q
F	F	T
F	T	F
T	F	T
T	T	T

12. (a) Today's not Wednesday or this is not the month of June. **(c)** It's not true that it's both raining and the month is July.

Problem Set 2-5
1. (a) p → q **(c)** p → ~q **(e)** ~q → ~p
2. (a) Converse: If you're good in sports, then you eat Meaties. Inverse: If you don't eat Meaties, then you're not good in sports. Contrapositive: If you're not good in sports, then you don't eat Meaties.
(c) Converse: If you have cavities, then you don't use Ultra Brush toothpaste. Inverse: If you use Ultra Brush toothpaste, then you don't have cavities. Contrapositive: If you don't have cavities, then you use Ultra Brush toothpaste.

3. (a)

p	q	p ∨ q	p → (p ∨ q)
T	T	T	T
T	F	T	T
F	T	T	T
F	F	F	T

(c)

p	~p	~(~p)	p → ~(~p)
T	F	T	T
F	T	F	T

4. (a) T **(c)** F **(e)** T
7. No
8. (a) No **(c)** No
9. If a number is not a multiple of 4, it is not a multiple of eight.
10. (a)

p	q	r	(p → q)	(p ∧ r)	[(p ∧ r) → q]	(p → q) → [(p ∧ r) → q]
T	T	T	T	T	T	T
T	T	F	T	F	T	T
T	F	T	F	T	F	T
T	F	F	F	F	T	T
F	T	T	T	F	T	T
F	T	F	T	F	T	T
F	F	T	T	F	T	T
F	F	F	T	F	T	T

(c)

p	q	~p	~q	p → q	[(p → q) ∧ (~q)]	[(p → q) ∧ (~q)] → (~p)
T	T	F	F	T	F	T
T	F	F	T	F	F	T
F	T	T	F	T	F	T
F	F	T	T	T	T	T

12. (a) It's false. **(c)** Yes, if q is true, then p is false, and p → q can still be true and all conditions are met.
14. (a) Valid **(c)** Valid
15. (a) Helen is poor. **(c)** If I study for the final, I will look for a teaching job.
16. (a) If a figure is a square, then it's a rectangle. **(c)** If a figure has exactly 3 sides, then it may be a triangle.

Chapter 2 Test
1. {x|x is a letter in the Greek alphabet}
4. (a) {r, a, v, e} **(c)** {u, n, i, v, r} **(e)** {u, v, s} **(g)** {i, n} **(i)** 5
5.

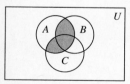

(a) A ∩ (B ∪ C)

6. (a) {(i, s), (i, e), (i, t), (d, s), (d, e), (d, t), (e, s), (e, e), (e, t), (a, s), (a, e), (a, t)} **(c)** 0
7. 7! or 5040
10. (a) B ∪ (A ∩ C)
12. (a) False; if, for example, they have no members in common, neither is a subset of the other. **(c)** False; {1, 2, 3} ~ {3, 4, 5} but the two sets are not equal. **(e)** False; Infinite sets are equivalent to proper subsets of themselves. **(g)** True
13. (a) 17 **(c)** 0
14. 7 students
16. (a) 36 students **(c)** 5
17. (a) Yes **(c)** Yes
18. (a) 7 **(c)** 37
19. (a) Answers vary, but a function assigns exactly one Zipcode to each post office.
20. (a) range = {3, 4, 5, 6} **(c)** range = {0, 1, 4, 9, 16}
21. (a) Reflexive, symmetric, transitive **(c)** Symmetric
22. (a) Yes **(c)** No
23. (a) No women smoke. **(c)** Some heavy metal rock is not loud.

24. (a)

p	q	~q	p ∨ ~q	[p ∨ (~q)] ∧ p
T	T	F	T	T
T	F	T	T	T
F	T	F	F	F
F	F	T	T	F

(c)

p	q	~q	p → (~q)	(~q) → p	[p → (~q)] ∧ [(~q) → p]
T	T	F	F	T	F
T	F	T	T	T	T
F	T	F	T	T	T
F	F	T	T	F	F

25. (a) Equivalent
27. (a) Joe Czernyu loves Mom and apple pie.
(c) Albertina will pass Math 100.
29. (a) Valid (c) Valid

CHAPTER 3

Problem Set 3-1
1. (a) M̄CDXXIV (b) 46,032 (c) < ▼▼ (d) 𓀀 ∩ | (e) 𓁹

2. (a) MCML; MCMXLVIII (c) M; CMXCVIII (d) << <▼▼; << <
6. (a) CXXI (c) LXXXIX
7. (a) ∩∩∩∩∩ || (c) 𓂀 |||

8. (a) ▼ <▼▼; ∩∩∩∩∩∩||; LXXII; ⋮⋮
(b) 602; 999 ||; DCII; ⋮⋮
(c) 1223; << << ▼▼▼; MCCXXIII; ⋮⋮⋮

13. (a) Hundreds (c) Thousands
14. (a) 3,004,005 (c) 3,560
16. (a) 86
19. 4,782,969
20. Assume an eight-digit display without scientific notation. (a) 98,765,432 (c) 99,999,999
21. (a) Answers vary, e.g., subtract 2020

Problem Set 3-2
1. (a) $k = 2$
2. No. If $k = 0$, we would have $k = 0 + k$, implying $k > k$.
4. (a)

```
       6 + 3 = 9
    ┌───────────┐
    ┌──── 6 ───┬─ 3 ─┐
    ├─┼─┼─┼─┼─┼─┼─┼─┼─▶
    0          6     9
```

5. (a) 5 (c) 0, 1, 2,

6. (a) 3 (c) a (e) 3, 4, 5, 6, 7, 8, 9
7. (a) Yes (c) Yes (e) Yes
8. (b) $213 = x + 119$
9. (a) Commutative Property for Addition (b) Associative Property for Addition (c) Commutative Property for Addition
10. (a) 3820, 3802, 8023
11. (a) 33, 38, 43 (b) 56, 49, 42
14. (a) For example, $5 - 3 \neq 3 - 5$ (c) $4 - 0 \neq 0 - 4$ and $0 - 4 \neq 4$
15. (a) 9 (c) 3 (e) 5
16. (a) 1 (c) 8 or 9
17. 0
18. (a)

8	1	6
3	5	7
4	9	2

19.

8	3
4	12

21.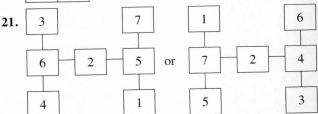

26. 45 points
27. 400
32. (a) 70 (c) 1100 (e) 3470

35. 26
36. (a) CMLIX **(b)** XXXVIII
37. There are fewer symbols to remember and place value is used.
38. $5 \cdot 10^3 + 2 \cdot 10^2 + 8 \cdot 10^1 + 6 \cdot 1$

Problem Set 3-3
2. $35.00
3. (a) Yes **(c)** Yes **(e)** Yes
4. (a) No, $2 + 3 = 5$ **(b)** Yes
6. $8 \cdot 3 = (6 + 2) \cdot 3 = 6 \cdot 3 + 2 \cdot 3 = 18 + 6 = 24$
7. (a) Commutative Property for Multiplication **(c)** Commutative Property of Addition **(e)** Identity Property of Multiplication **(g)** Distributive Property of Multiplication over Addition **(h)** Distributive Property of Multiplication over Addition
8. (b) 4
9. (a) $ac + ad + bc + bd$ **(d)** $x^2 + xy + xz + yx + y^2 + yz$, or $x^2 + 2xy + xz + y^2 + yz$
10. (a) 11 **(c)** 16
11. (a) $(4 + 3) \times 2 = 14$ **(c)** $(5 + 4 + 9) \div 3 = 6$
15. (a) 6 **(c)** 4
16. (b)

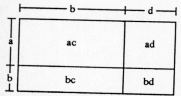

$(a + b)(c + d) = ac + ad + bc + bd$

17. (a) $40 = 8 \cdot 5$ **(c)** $48 = x \cdot 16$
19. (a) $(5 \times 2) + 6 = 16$ **(c)** $(15 \div 3) - 4 = 1$
20. (a) $2 \div 1 \neq 1 \div 2$ **(c)** $8 \div (2 + 2) \neq (8 \div 2) + (8 \div 2)$
21. $32
23. 9 minutes
24. (a)

□	△
0	34
1	26
2	18
3	10
4	2

25. 1 and 36, 2 and 18, 3 and 12, 4 and 9, 6 and 6
27. 30
31. (a) Yes **(b)** Yes **(c)** Yes, a **(d)** Yes
32. (a) 3 **(c)** 2 **(e)** 4
33. The answers depend upon the keys available on your calculator.

(a)
$3 = 1 + 9 - 7$ $11 = 7 + 1 + \sqrt{9}$
$4 = 1^7 + \sqrt{9}$ $12 = 19 - 7$
$5 = 7 - \sqrt{9} + 1$ $13 = 91 \div 7$
$6 = 7 - 1^9$ $14 = 7(\sqrt{9} - 1)$
$7 = 7 \cdot 1^9$ $15 = 7 + 9 - 1$
$8 = 7 + 1^9$ $16 = (7 + 9) \cdot 1$
$9 = 1^7 \cdot 9$ $17 = 7 + 9 + 1$
$10 = 1^7 + 9$ $18 = \sqrt{9}(7 - 1)$
     $19 = ?$
     $20 = 7\sqrt{9} - 1$
(c) For example, $22 + 2$.
34. (i) ∩∩∩∩∩∩∩||||| **(ii)** LXXV
(iii) ▼<▼▼▼▼▼▼▼
35. $3 \cdot 10^4 + 5 \cdot 10^3 + 2 \cdot 10^2 + 0 \cdot 10^1 + 6$
36. For example, {0, 1}.
37. No. For example, $5 - 2 \neq 2 - 5$.
38.

```
       11 - 3      3
     |--------|--------|
              11
     |-----------------|

  +--+--+--+--+--+--+--+--+--+--+--+
  0           5         10 11
```

Problem Set 3-4
1. (b)
```
   ¹5²2 4
   3 2 8́
   5 6 7²
 + 1⁴3²5₄
   1 5 5 4
```

4. (a)   981   **(c)**   1,069
            +421   2,094
            ————   9,546
            1402   9,003
                                 +7,064
                                 ———————
                                  28,776

5. (a) 87693   **(c)**   383
           −46414   −159
           ————   ———
           41279   224

6. (a) One possibility: 863
                              +752
                              ————
                              1615

7. If only positive numbers are used: **(a)**   876
                                                   −235
                                                   ————
                                                    641

8. 15,782
9. (a) 34, 39, 44
11. No, not all at dinner. He can have either the steak or the salad.
12. Molly, 55 lbs; Karly, 50 lbs; Samantha, 65 lbs.
15. (a) (i) No, not clustered **(ii)** Yes, clustered around 500
17. (a) About 121 weeks **(c)** Answers vary
27. $8 + 8 + 8 + 88 + 888$
29. It is doubling the second number in the operation
30. (a) 34; 34; 34 **(b)** 34 **(c)** 34 **(d)** Yes **(e)** Yes
31. $5280 = 5 \cdot 10^3 + 2 \cdot 10^2 + 8 \cdot 10 + 0 \cdot 1$

32. For example, $2 + (3 + 4) = (2 + 3) + 4$
$2 + 7 = 5 + 4$
$9 = 9$

33.

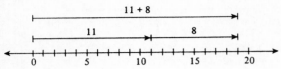

34. 1,000,410
35. (a) $a \cdot (x + 1)$ (b) $(3 + a) \cdot (x + y)$
36. 15

Problem Set 3-5
1. (a)

	7	2	8	
6	6/3	1/8	7/2	9
8	2/8	8/8	3/2	4
	4	3	2	

3. (a) 426
   × 783
   ─────
   1278
   3408
   2982
   ──────
   333558

5. (b) 6^{15} (d) 10^{12}
6. (a) 2^{100}
8. (c)   363
   × 84
   ─────
   2904
   1452
   ──────
   30492

9. → 17 × 63   63
       8   126   +1008
       4   252   ─────
       2   504   1071
    → 1   1008

10. (a) 21 (c) 304
11. (a) 22 (c) 7
12. (a) $15 \cdot (10 + 2) = 150 + 30 = 180$ (c) $30 \cdot 99 = 30(100 - 1) = 3000 - 30 = 2970$
14. (a) 1332 (b) Jane, 330 more calories (c) Maurice, 96 more calories
16. $60
18. (a) $3\overline{)876}$ (b) $8\overline{)367}$
19. (a) Monthly payments are more expensive. (b) $3,700
21. 3
22. 8 cars (remember the match)

23.

2	11
4	15
0	7
6	19
12	31

25. (b) If $ab \cdot cd = ba \cdot dc$ then $ac = bd$
28. 3 hrs.
31. 10 seconds
33. 58 buses needed, not all full
34. 11 km/L
36. (a) (i) $70 (b) on the 12th trip
37. (a) (i) $27 \times 198 = 5346$ (ii) $48 \times 159 = 7632$ (iii) $39 \times 186 = 7254$ (c) 1
38. (a) 763   (b) 678
       × 8   × 3
       ────   ────
       6104   2034

39. (a) 762
       × 83
       ──────
       63,246
40. 7,500,000 cows
41. (a)   37
       ×43
       ────
       111
       1480
       ────
       1591
42. (a) 1; 121; 12,321; 1,234,321; your calculator may not produce the pattern after this term, but it continues through $111,111,111 \times 111,111,111$.
43. $60; $3600; $86,400; $604,800; $2,592,000 (30 days); $31,536,000 (365 days); $630,720,000 or $631,152,000 (with leap years).
45. 999999∩∩∩∩∩∩∩IIII
46. 300,260
47. For example, $3 + 0 = 3 = 0 + 3$.
48. (a) $x \cdot (a + b + 2)$ (b) $(3 + x)(a + b)$
49. 6979 miles
50. 724

Problem Set 3-6
1. (a) (1, 10, 11, 100, 101, 110, 111, 1000, 1001, 1010, 1011, 1100, 1101, 1110, 1111)$_{two}$
2. 20
3. $2032_{four} = (2 \cdot 10^3 + 0 \cdot 10^2 + 3 \cdot 10 + 2)_{four}$
4. (a) 111_{two} (c) 999_{ten}
5. (a) ETE_{twelve}; $EE1_{twelve}$ (c) 554_{six}; 1000_{six} (e) 444_{five}; 1001_{five}
6. (a) There is no numeral 4 in base four. (c) There is no numeral T in base three.
7. (a) 3212_{five} (c) 12110_{four}
8. 100010_{two}
9. (a) 117 (c) 1331 (e) 157
11. 1 prize of $625, 2 prizes of $125, and 1 of $25.

864 ANSWERS TO SELECTED PROBLEMS

13. (a) 8 weeks, 2 days (c) 1 day, 5 hours
14. $E66_{twelve}$; 1662
15. (a) 6 (c) nine
17. 4; 1, 2, 4, 8; 1, 2, 4, 8, 16
19. (a) 121_{five} (c) 1010_{five} (e) 1001_{two}
21. (b) 1 hour 39 minutes 40 seconds
22. (a) 2 quarts, 1 pint, 0 cups, or 1 half-gallon, 0 quarts, 1 pint, 0 cups (c) 2 quarts, 1 pint, 1 cup
24.
$$\begin{array}{r} {}^{3}3\,2 \\ 1^{1}3 \\ 2\ 2^{0} \\ 4\ 3 \\ 2^{3}3^{0} \\ 1^{0}2_{0} \\ \hline 3\ 1\ 0_{five} \end{array}$$
25. (a) 3 gross 10 dozen 9 ones
26. (a) 22 students on Tuesday; (b) 1 gal., 1 half-gallon, 1 qt., 1 pint, and 1 cup
27. (a) 70 (b) 87
28. There is no numeral 5 in base five; $2_{five} + 3_{five} = 10_{five}$.
29. (a)
$$\begin{array}{r} 230_{five} \\ -\ 22_{five} \\ \hline 203_{five} \end{array}$$
30. (a) 233_{five} (c) 2144_{five} (e) 67_{eight} (g) 110_{two}
31. (a) Nine (c) Six
32. 30221_{five}
33. $a = 5, b = 7$

Chapter 3 Test
1. (a) 400,044 (c) 1704 (e) 1448
2. (a) CMXCIX (c) ••• (e) 1000_{twelve} (g) 1241_{nine}
3. (a) 3^{17} (c) 3^5
4. (a) Distributive Property of Multiplication over Addition (c) Identity Property for Multiplication (e) Commutative Property of Multiplication
5. (a) $3 < 13$, because $3 + 10 = 13$
6. $1000 \cdot 438 = 10^3(4 \cdot 10^2 + 8 \cdot 10 + 3)$
$= 4 \cdot 10^5 + 8 \cdot 10^4 + 3 \cdot 10^3$
$= 4 \cdot 10^5 + 8 \cdot 10^4 + 3 \cdot 10^3 + 0 \cdot 10^2 + 0 \cdot 10^1 + 0 \cdot 1$
$= 483,000$
7. (a) 1119 (b) $173E_{twelve}$
8. (a) 60,074 (b) 14150_{eight}
9. (a) 5 remainder 243 (b) 91 remainder 10 (c) 120_{five} remainder 2_{five} (d) 11_{two} remainder 10_{two}
10. (a) $5 \cdot 912 + 243 = 4803$ (b) $91 \cdot 11 + 10 = 1011$ (c) $23_{five} \cdot 120_{five} + 2_{five} = 3312_{five}$ (d) $11_{two} \cdot 11_{two} + 10_{two} = 1011_{two}$
11. (a) tens
12. (a) 10, 11, 12, 13, 14, 15 (c) All whole numbers

13. (a)
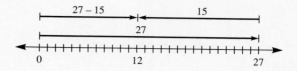

(c)

14. (a) $15a$ (d) $(x + 5)(3 + y)$
16. $395
18. 2600
20. (a) Several answers are possible. For example,
$$\begin{array}{r} 296 \\ +\ 541 \\ \hline 837 \end{array} \qquad \begin{array}{r} 569 \\ +\ 214 \\ \hline 783 \end{array}$$
21. 69 miles
22. 40 cans
23. 12 outfits
25. 26
26. $2.16
28. There are 36 bikes and 18 trikes
30. $400

CHAPTER 4

Problem Set 4-1
1. (b) 5 (f) $^-a + ^-b$ or $-(a + b)$
3. (a) 5 (d) $^-5$
4. (a)

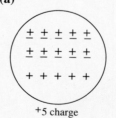
+5 charge

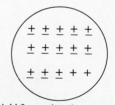
Add 3 negative charges; net result 2 positive charges

(d)

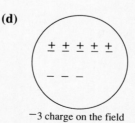
$^-3$ charge on the field

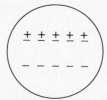
Add 2 negative charges; net result 5 negative charges

7. (f) 2 (g) 2
8. (a) −7 (d) −$150
10. (c)

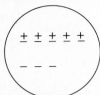
−3 charge on the field

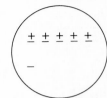
Take away 2 negative charges; net result 1 negative charge

12. (b) 3 − 1 = 2; 2 − 1 = 1; 1 − 1 = 0; 0 − 1 = −1; −1 − 1 = −2; −2 − 1 = −3
14. (c) 13
15. (b) 3 (e) −13
17. (a) 1 + 4x (b) 2x + y
18. (b) All positive integers (e) There are none. (g) There are none.

21.
2	−13	8
5	−1	−7
−10	11	−4

Other answers are possible.

23. 33 points
24. (c) 192°C (d) 56 or 80
26. −4 pounds
28. (a) 0 (c) 1
29. (c) 0 or 2.
31. (b) 19 (c) 19
32. Greatest possible value: $a − (b − c) − d$, or 8. Least possible value: $a − b − (c − d)$, or −6.
34. (b) 3775
37. (b) 516 (c) 10,894
39. (a) $(a + b) + (−a + −b) = a + b + −a + −b = a + −a + b + −b = (a + −a) + (b + −b) = 0 + 0 = 0$
41. (b) True (c) True (e) False; let $x = −1$

Problem Set 4-2

3. If you are now at 0 moving west at 4 km/h, you will be at 8 km west of 0 two hours from now.

4. (a) $−20 \times 4$ (b) 20×4 (c) $−20n$ (d) $20n$

5. (g) 0 (h) 16
7. (d) −10 (e) a; if $b \neq 0$ (j) −4 (k) Impossible (n) −2
8. (b) $32°C − 30 \cdot (3°C)$ (d) $25°C − 20 \cdot (3°C)$
9. (b) −66 divided by 11 = −6; He lost 6 yards.
12. (c) −1000 (d) 81
13. (c) −5 (f) −9 (g) −13 (h) −8
14. (b), (c), (g), (h) are always positive; (a), (f) are always negative
16. (a) Commutative Property of Multiplication (d) Distributive Property of Multiplication over Addition
18. (b) $2xy$ (e) $x + 2y$ (f) b
19. (g) All integers except 0 (l) All integers except 0 (n) All integers
20. (a) $−2x + 2$ (e) $−2x − 2y + 2z$ (f) $−x^2 + xy + 3x$ (i) $−x^4 + 3x^2 − 2$
21. (b) $25 − 10,000 = −9975$ (e) $x^2 − 1$
23. (a) $8x$ (e) $x(x + y)$ (h) $x(3x + y − 1)$ (j) $(a + b)c$ (m) $(2x + 5y)(2x − 5y)$
26. (a) False (b) True (c) True (d) True
28. (b) $^-8, ^-11, d = −3$, nth term is $−3n + 13$ (d) $−128, 256, r = ^-2$, nth term is $(−2)^n$ (f) $−8 \cdot 2^7, 9 \cdot 2^8$
30. (a) $−9, −6, −1, 6, 15$ (e) $−1, 4, −9, 16, −25$ (f) $2, −8, 24, −64, 160$
31. 7, 2
34. $−(a + b) = (−1)(a + b)$ by first part of problem 33.
   $= (−1)a + (−1)b$ by Distributive Property.
   $= −a + −b$ by first part of problem 33.
36. (c) 13 (d) 3
40. 400 pounds

Problem Set 4-3

1. (c) $−100, −20, −15, −13, 0$
2. (a) $−5 + 2 = −3$
3. (c) 18 (i) −2 (j) $x \geq −2$ and x is an integer (n) $x < −3$ and x is an integer
4. (a) True (b) False except when $x = 3$ (c) True (d) True (e) True (f) False
6. (b) 0
7. (c) $4n − 7$ (f) $4n$ (h) $n − 4$
8. (a) $60t$ (d) $175d$ cents (g) $3m$ (j) $40°F − (3°F)t$
10. Tom is older than 11.
14. Factory A produces 2800 cars per day
    Factory B produces 1400 cars per day
    Factory C produces 3100 cars per day
16. 524 student tickets
18. 78, 80, 82
20. Eldest, $30,000; middle, $24,000; youngest, $10,000
22. 200 student tickets, 800 adult tickets
25. 48 miles
26. (a) $−4 < x < 4$

28. Proof: $0 < a < b$ implies that $0 < a$ and $0 < b$. Now $a < b$ and $0 < a$ imply $a^2 < ab$. Also $a < b$ and $0 < b$ imply $ab < b^2$. Hence $a^2 < b^2$.
31. (a) $-3, -2, -1, 0, 1$
33. (a) -10 (c) -4 (g) -4 (j) -4 (p) -1

Chapter 4 Test
1. (e) $x - y$ (g) 32
2. (d) 0 (e) 8
3. (c) Any integer except 0 (f) Any integer
6. (a) $(x - y)(x + y) = (x - y)x + (x - y)y$
$= x^2 - yx + xy - y^2$
$= x^2 - xy + xy - y^2$
$= x^2 - y^2$
(b) $4 - x^2$
7. (c) $3x - 1$ (f) $-9 - 6x - x^2$
8. (b) $x(x + 1)$ (e) $5(1 + x)$ (f) $(x - y)x$
9. (a) -2 (c) $1, 2, 3, 4, \ldots$
10. (a) False (c) False
11. (b) $3 - (4 - 5) \neq (3 - 4) - 5$ (d) $8/(4 - 2) \neq 8/4 - 8/2$
13. $-7°$ C
15. 7 nickels, 17 dimes
17. 42 gallons
18. 4000 pounds of Spanish peanuts, 4000 pounds of cashews, 2000 pounds of pecans

CHAPTER 5

Problem Set 5-1
1. (a) True (c) True (e) True
2. (a) Answers will vary. $7|35$ (c) $d|a$ (e) $d|213$
4. No.
5. (a) True by Theorem 5-1 (c) None (e) True by Theorem 5-1
6. (a) No; $17|34,000$ and $17\nmid15$ imply $17\nmid(34,000 + 15)$ (c) No; $19|19,000$ and $19\nmid 31$ imply $19\nmid(19,000 + 31)$ (e) No; $2^{14}|2^{64}$ and $2^{14}\nmid 1$ imply $2^{14}\nmid(2^{64} + 1)$
7. (a) $1, 2, 4, 5, 8, 11$ (c) Impossible if one scores an extra point with every touchdown. If not, then 6 or 7.
8. (a) (i) 1 (iii) 3 (v) 3 (vii) 6
(c) The remainder is the sum of the digits.

9. (a)

4	4	4
(c) 3	12	3
(e) 2	20	2

10. (a) $12,343 + 4546 + 56 = 16,945$
$4 + 1 + 2 = 7$
(c) $10,034 + 3004 + 400 + 20 = 13,458$
$8 + 7 + 4 + 2 = 21$ has remainder 3 when divided by 9 as does $1 + 3 + 4 + 5 + 8$.

(e) $1003 - 46 = 957$
$4 - 1 = 3$ has remainder 3 when divided by 9 as does $9 + 5 + 7 = 21$. (g) Yes; the divisions could be done by repeated subtractions, and if the check works for subtractions, it should work for divisions.
11. (a) False; $2|4$, but $2\nmid 1$ and $2\nmid 3$ (c) True (e) True (g) True (i) True (k) False; $50|10^2$, but $50\nmid 10$ (m) True
12. (a) True (c) False (e) True (g) True
14. (a) A number is divisible by 16 if and only if the last four digits form a number divisible by 16.
16. Each candy bar costs $.19.
18. (a) $2, 3, 4, 6, 11$ (c) $2, 3, 5, 6, 10$ (e) $3, 5$ (g) $7, 11$ (i) $2, 4, 5, 10$.
19. (a) No. If $5\nmid d$ for any integer d, then there is no integer m such that $5m = d$. If we assume $10|d$, this means there exists n such that $10n = d$ or $5(2n) = d$. This contradicts the original assumption that d is not divisible by 5.
20. (a) 7 (c) 6
23. (a) Yes (c) Yes
24. (a) (i) Yes; (ii) Yes; (iii) Yes; (iv) Yes
25. (a) The result is always 9. (c) Let the number be $a \cdot 10 + b$. The number with the digits reversed is $b \cdot 10 + a$.
Now, $a \cdot 10 + b - (b \cdot 10 + a)$
$= a \cdot 10 + b - b \cdot 10 - a$
$= a \cdot 10 - a + b - b \cdot 10$
$= 9a - 9b$
$= 9(a - b)$
Thus, the difference is a multiple of 9.
26. No. Both 6 and 15 are multiples of 3; 286 is not.
28. (a) Consider any sequence $a, a + 1, a + 2$. According to the division algorithm, there exists a unique quotient and remainder for $a, a + 1$, and $a + 2$ when we divide by three. Since there are only 3 possible remainders when dividing by 3, namely 1, 2, and 0, when we divide by 3, we get a quotient g with a remainder of 1, 2, or 0. If the quotient is g and r = 1, then $a + 1$ divided by 3 has a quotient of g and r = 2. Also, $a + 2$ divided by 3 yields a quotient of g with r = 0. In fact, the remainders for the sequence $a, a + 1, a + 2$ when dividing by 3 will follow the sequence $0, 1, 2, 0, 1, 2, 0, 1, 2 \ldots$ with 0 a member of any sequential triple.
30. Let $n = a \cdot 10^4 + b \cdot 10^3 + c \cdot 10^2 + d \cdot 10 + e$.
$a \cdot 10^4 = a(10,000) = a(9999 + 1) = a \cdot 9999 + a$
$b \cdot 10^3 = b(1000) = b(999 + 1) = b \cdot 999 + b$
$c \cdot 10^2 = c(100) = c(99 + 1) = c \cdot 99 + c$
$d \cdot 10^1 = d(10) = d(9 + 1) = d \cdot 9 + d$
Thus, $n = (a \cdot 9999 + b \cdot 999 + c \cdot 99 + d \cdot 9) + (a + b + c + d + e)$. Because $9|9, 9|99, 9|999$ and $9|9999$, it follows that $9|[(a \cdot 9999 + b \cdot 999 + c \cdot 99 + d \cdot 9) + (a + b + c + d + e)]$; that is, $9|n$. If, on the other hand, $9\nmid(a + b + c + d + e)$, it follows that $9\nmid n$.
32. $(3 \times \$20) + (11 \times \$50)$, or $(8 \times \$20) + (9 \times \$50)$, or $(13 \times \$20) + (7 \times \$50)$, or $(18 \times \$20) + (5 \times \$50)$, or $(23 \times \$20) + (3 \times \$50)$, or $(28 \times \$20) + (1 \times \$50)$. The answer is not unique.

Problem Set 5-2

1. (a) 504

```
      504
     /   \
    2    252
        /   \
       2    126
           /   \
          2    63
              /  \
             3   21
                /  \
               3   7
```
$504 = 2^3 \cdot 3^2 \cdot 7$

(c) 11250

```
      11250
     /    \
    2    5625
         /   \
        3    875
            /   \
           3    625
               /   \
              5    125
                  /   \
                 5    25
                     /  \
                    5    5
```
$11{,}250 = 2 \cdot 3^2 \cdot 5^4$

2. (a) 149 is prime. **(c)** 433 is prime. **(e)** 463 is prime.
3. 73
5. 101, 103, 107, 109, 113, 127, 131, 137, 139, 149, 151, 157, 163, 167, 173, 179, 181, 191, 193, 197, 199
7. (a) 1×48; 2×24; 3×16; 4×12
(b) Only one; 1×47.
9. (a) 3, 5, 15, 29, people **(b)** 145 committees of 3; 87 committees of 5; 29 committees of 15; 15 committees of 29.
10. (a) 1, 2, 3, 4, 6, 9, 12, 18, or 36 **(c)** 1 or 17
12. (a) $2^7 \cdot 41$ **(c)** No
13. (b) Let $n = 41a$ where $a \in N$. Then $n^2 - n + 41 = (41a)^2 - 41a + 41 = 41(41a^2 - a + 1)$
14. 27,720
16. 3, 5; 5, 7; 11, 13; 17, 19; 29, 31; 41, 43; 59, 61; 71, 73; 101, 103; 107, 109; 137, 139; 149, 151; 179, 181; 191, 193; 197, 199
17. (a) The Fundamental Theorem of Arithmetic says that n can be written as a product of primes in one and only one way. Since $2|n$ and $3|n$ and 2 and 3 are both prime, they must be included in the unique factorization.
That is, $2 \cdot 3 \cdot p_1 \cdot p_2 \cdot \ldots \cdot p_m = n$
Therefore, $(2 \cdot 3)(p_1 \cdot p_2 \cdot \ldots \cdot p_m) = n$
Thus, $6|n$.
19. Every number would have its "usual" factorization $1(p_1 \cdot p_2 \cdot p_3 \cdot \ldots \cdot p_n)$, along with infinitely many such other factorizations because $1^n = 1$; n may be any natural number.
21. No, because 5^2 has no factors of either 2 or 3.
22. (a) 4, 6, 8, 0 **(c)** 23, 29, 31, 37, 53, 59, 71, 73, 79
23. (a) 49, 121, 169; Squares of prime numbers (column 2) **(c)** 38, 39, 46; Products of 2 primes or cubes of primes.
26. None of the primes 2, 3, 5 ..., p divides N because if any one of the primes divided N, then it must also divide 1 which is impossible.
29. There are infinitely many composites of the form 1, 11, 111, 1111, 11111, 111111... since every 3rd member of this sequence will be divisible by 3.
31. Composites are obtained when $n = 16$ or 17.
32. (a) False **(c)** True
33. (a) 2, 3, 6
34. If $12|n$, there exists an integer a such that
$12a = n$.
$(3 \cdot 4)a = n$
$3(4a) = n$
Thus, $3|n$.

Problem Set 5-3

1. (a) $D_{18} = \{1, 2, 3, 6, 9, 18\}$
$D_{10} = \{1, 2, 5, 10\}$
$GCD(18, 10) = 2$
$M_{18} = \{18, 36, 54, 72, 90 \ldots\}$
$M_{10} = \{10, 20, 30, 40, 50, 60, 70, 80, 90 \ldots\}$
$LCM(18, 10) = 90$
(c) $D_8 = \{1, 2, 4, 8\}$
$D_{24} = \{1, 2, 3, 4, 6, 8, 12, 24\}$
$D_{52} = \{1, 2, 4, 13, 26, 52\}$
$GCD(8, 24, 52) = 4$
$M_8 = \{8, 16, 24, 32, 40, 48, 56, 64, 72, 80, 88, 96 \ldots\}$
$M_{24} = \{24, 48, 72, 96, 120, 144, 168, 192, 216, 240, 264, 288, 312 \ldots\}$
$M_{52} = \{52, 104, 156, 208, 160, 312 \ldots\}$
$LCM(8, 24, 52) = 312$
2. (a) $132 = 2^2 \cdot 3 \cdot 11$
$504 = 2^3 \cdot 3^2 \cdot 7$
$GCD(132, 504) = 2^2 \cdot 3 = 12$
$LCM(132, 504) = 2^3 \cdot 3^2 \cdot 7 \cdot 11 = 5544$
(c) $96 = 2^5 \cdot 3$
$900 = 2^2 \cdot 3^2 \cdot 5^2$
$630 = 2 \cdot 3^2 \cdot 5 \cdot 7$
$GCD(96, 900, 630) = 2 \cdot 3 = 6$
$LCM(96, 900, 630) = 2^5 \cdot 3^2 \cdot 5^2 \cdot 7 = 50{,}400$
(e) $63 = 3^2 \cdot 7$
$147 = 3 \cdot 7^2$
$GCD(63, 147) = 3 \cdot 7 = 21$
$LCM(63, 147) = 3^2 \cdot 7^2 = 441$
3. (a) $GCD(2924, 220) = GCD(220, 64) = GCD(64, 28) = GCD(28, 8) = GCD(8, 4) = GCD(4, 0) = 4$ **(c)** $GCD(123, 152, 122{,}368) = GCD(122{,}368{,}784) = GCD(784, 64) = GCD(64, 16) = GCD(16, 0) = 16$
4. (a) 72 **(c)** 630

5. (a) 220 · 2924/4 or 160,820 **(c)** 123,152 · 122,368/16 or 941,866,496
7. (a) LCM (15, 40, 60) = 120 minutes = 2 hours So the clocks alarm again together at 8:00 A.M.
9. (a) $60 **(c)** 30
11. 24 nights
13. After 7 1/2 hours, or 2:30 A.M.
15. (a) ab **(c)** GCD (a^2, a) = a; LCM (a^2, a) = a^2 **(e)** GCD (a, b) = 1; LCM (a,b) = ab **(g)** b|a
17. GCD (120, 75) = 15
GCD (15, 105) = 15
GCD (120, 75, 105) = 15
19. (a) $4 = 2^2$. Since 97,219,988,751 is odd, it has no prime factors of two. Consequently, 1 is their only common divisor and they are relatively prime. **(c)** 33 has prime factors 3 and 11. 181,345,913 is divisible by neither 3 nor 11. Therefore, they're relatively prime.
20. {1, 2, 3, 4, 6, 7, 8, 9, 11, 12, 13, 14, 16, 17, 18, 19, 21, 22, 23, 24}
21. (a) 28
23. x = 15,625; y = 64
24. (a) 83,151; 83,451; 83,751 **(c)** 10,396
26. Answers may vary. 30,030 = 2 · 3 · 5 · 7 · 11 · 13
28. 43
30. (a) 67 **(c)** 93, 87, 51

Problem Set 5-4

1. (a) 3 **(c)** 6 **(e)** 3 **(g)** Does not exist
2. (a) 2 **(c)** 2 **(e)** 2 **(g)** 2
3. (a)

⊕	1	2	3	4	5	6	7
1	2	3	4	5	6	7	1
2	3	4	5	6	7	1	2
3	4	5	6	7	1	2	3
4	5	6	7	1	2	3	4
5	6	7	1	2	3	4	5
6	7	1	2	3	4	5	6
7	1	2	3	4	5	6	7

4. (a)

⊗	1	2	3	4	5	6	7
1	1	2	3	4	5	6	7
2	2	4	6	1	3	5	7
3	3	6	2	5	1	4	7
4	4	1	5	2	6	3	7
5	5	3	1	6	4	2	7
6	6	5	4	3	2	1	7
7	7	7	7	7	7	7	7

6. (a) 10 **(c)** 7 **(e)** 1
7. (a) 2, 9, 16, 30 **(c)** 366 ≡ 2(mod 7); Wednesday
8. (a) 4 **(c)** 0
9. (a) 8|(81 − 1) **(c)** 13|(1000 − ⁻1) **(e)** $10^2 \equiv 1$ (mod 11) implies $(10^2)^{50} \equiv 1^{50}$ (mod 11)
11. (a) 24 ≡ 0(mod 8) **(c)** n ≡ 0(mod n)
12. (a) x = 2k, k is an integer. x ∈ {. . . , ⁻4, ⁻2, 0, 2, 4, . . .} **(c)** x − 3 = 5k implies x = 3 + 5k where k is an integer. x ∈ {. . . , ⁻7, ⁻2, 3, 8, 13, . . .}

13. (a) 1 **(c)** 10
15. (a) ⁻1
17. For example, 2 · 11 ≡ 1 · 11 (mod 11), but 2 ≡ 1 (mod 11) is false.

Chapter 5 Test

1. (a) False **(c)** True **(e)** False; 9, for example
2. (a) False; 7|7 and 7∤3 yet 7|3 · 7 **(c)** True **(e)** True
3. (a) Divisible by 2, 3, 4, 5, 6, 7, 8, 9, 11.
5. (a) 87<u>2</u>4; 86<u>5</u>4; 87<u>8</u>4 **(c)** 87,1<u>7</u>4; 87,<u>4</u>64; 87,<u>7</u>54
6. (a) Composite
8. (a) 4
9. (a) $2^4 \cdot 5^3 \cdot 7^4 \cdot 13 \cdot 29$
11. 1, 2, 3, 4, 6, 8, 9, 12, 16, 18, 24, 36, 48, 72, 144
12. (a) $2^2 \cdot 43$ **(c)** $2^2 \cdot 5 \cdot 13$
13. 15 minutes
15. 9:30 A.M.
17. 15
19. $n = a \cdot 10^2 + b \cdot 10 + c$
$n = a(99 + 1) + b(9 + 1) + c$
$n = 99a + 9b + (a + b + c)$
Since 9|99a and 9|9b, 9|[99a + 9b + (a + b + c)] if and only if 9|(a + b + c)
20. (a) 1 **(c)** 3
22. mod 360. It would cover all the area encircling the lighthouse.

CHAPTER 6

Problem Set 6-1

2. (a) 1/6 **(c)** 2/6 **(e)** 5/16
3. (a) 2/3 **(b)** 4/6 **(c)** 6/9 **(d)** 8/12. The diagram illustrates the Fundamental Law of Fractions
4.

(a) 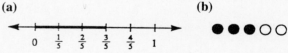 **(b)**

(d) **(f)**

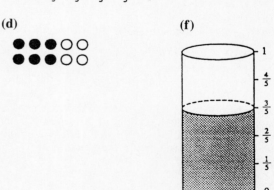

5. (a) 9/24 or 3/8 **(b)** 12/24 or 1/2 **(c)** 4/24 or 1/6 **(d)** 8/24 or 1/3

7. (a) 52/31 (b) 3/5 (c) −5/7
9. (a) undefined (c) 0 (e) cannot be simplified (g) 5/3
10. (a) 1 (c) a/1 (e) 1/(3 + b)
11. (a) equal (c) equal
12. (a) not equal (c) equal
15. $\dfrac{36}{48}$
16. Meter A, 12 minutes
18. (a) $2\dfrac{7}{8}$ (c) $1\dfrac{3}{8}$
22. $\dfrac{12}{21}, \dfrac{24}{42}, \dfrac{48}{84}$
23. (a) 32/3 (c) x is any rational number except 0.
24. (a) $a = b, c \neq 0$.
25. (a) T (b) T (c) F (d) F (e) T
26. (a) Not equal (c) Equal

Problem Set 6-2

2. (a) $\dfrac{-11}{16}$ (c) $\dfrac{19}{18}$
3. (a) $\dfrac{-31}{20}$ (c) $\dfrac{-19}{40}$
4. (a) $\dfrac{-6y + 6x - 1}{4xy}$
5. (a) $18\dfrac{2}{3}$ (c) $-2\dfrac{93}{100}$
6. (a) $\dfrac{27}{4}$ (c) $\dfrac{-29}{8}$
7. (a) $\dfrac{71}{24}$ (c) $\dfrac{43}{2^4 \cdot 3^4}$
8. $\dfrac{2}{6} + \dfrac{5}{8}$
9. (a) 1/3, high (c) 3/4, low
10. (a) Beavers (c) Bears (e) Lions
11. (a) 1/2, high (c) 3/4, high (e) 1, low (g) 3/4, low
12. (a) 2 (b) 3/4 (c) 0 (d) 0
13. (a) 10 (c) 13
14. (a) No
15. (a) 1/4 (c) 0
16. (a) A (c) T
17. (a) $\dfrac{dc + a}{bc}$ (c) $\dfrac{a - ab - b^2}{a^2 - b^2}$
18. (a) $\dfrac{3 + 3}{3} \neq \dfrac{3}{3} + 3$ (c) $\dfrac{ab + c}{a} \neq \dfrac{\cancel{a}b + c}{\cancel{a}}$ (e) $\dfrac{a + c}{b + c} \neq \dfrac{a + \cancel{c}}{b + \cancel{c}}$
20. (a) $\dfrac{1}{30}$ (c) $\dfrac{1}{60}$

21.

$\dfrac{5}{3}$	$\dfrac{1}{12}$	1
$\dfrac{1}{4}$	$\dfrac{11}{12}$	$\dfrac{19}{12}$
$\dfrac{5}{6}$	$\dfrac{7}{4}$	$\dfrac{1}{6}$

22. $6\dfrac{7}{12}$ yards
25. $1\dfrac{3}{4}$ cups
27. $2\dfrac{5}{6}$ yards
28. $22\dfrac{1}{8}$ inches
29. (a) Team 4, $76\dfrac{11}{16}$ pounds
31. (a) $\dfrac{1}{2} + \dfrac{3}{4} = \dfrac{5}{4} \in Q$ (c) $\left(\dfrac{1}{2} + \dfrac{1}{3}\right) + \dfrac{1}{4} = \dfrac{1}{2} + \left(\dfrac{1}{3} + \dfrac{1}{4}\right)$
33. (a) $\dfrac{6}{4}, \dfrac{7}{4}, 2$; arithmetic, $\dfrac{1}{2} - \dfrac{1}{4} = \dfrac{3}{4} - \dfrac{1}{2} = 1 - \dfrac{3}{4} = \dfrac{5}{4} - 1$ (c) $\dfrac{17}{3}, \dfrac{20}{3}, \dfrac{23}{3}$; arithmetic; $\dfrac{5}{3} - \dfrac{2}{3} = \dfrac{8}{3} - \dfrac{5}{3} = \dfrac{11}{3} - \dfrac{8}{3} = \dfrac{14}{3} - \dfrac{11}{3}$
36. (a) (i) $\dfrac{3}{4}$ (ii) $\dfrac{25}{12}$ (iii) 0
37. (a) $f(0) = -2$ (c) $f(-5) = \dfrac{1}{2}$
38. (b) $\dfrac{1}{n} = \dfrac{1}{n + 1} + \dfrac{1}{n(n + 1)}$
39. (a) $\dfrac{2}{3}$ (b) $\dfrac{13}{17}$ (c) $\dfrac{25}{49}$ (d) $\dfrac{a}{1}$, or a (e) Reduced
40. (a) Equal (b) Not equal (c) Equal (d) Not equal

Problem Set 6-3

1. (a) $\dfrac{1}{4} \cdot \dfrac{1}{3} = \dfrac{1}{12}$ (b) $\dfrac{2}{4} \cdot \dfrac{3}{5} = \dfrac{6}{20}$
2. (a)

4. (a) $\dfrac{1}{5}$ (c) $\dfrac{za}{x^2 y}$ (e) $\dfrac{44}{3}$
5. (a) $10\dfrac{1}{2}$
6. (a) -3 (c) $\dfrac{y}{x}$

870   ANSWERS TO SELECTED PROBLEMS

8. (a) $\frac{11}{5}$ **(c)** $\frac{22}{3}$ **(e)** 5 **(g)** 2 **(i)** z **(k)** $\frac{xy}{z}$
9. (a) 20 **(c)** 2
10. (a) 18 **(c)** 7
11. (a) Less than 1 **(c)** Greater than 2 **(e)** Greater than 4
13. (a) 26 **(c)** 92 **(e)** 6 **(g)** 9
15. (a) $\frac{21}{8}$ **(c)** -28 **(e)** $\frac{1}{5}$ **(g)** $\frac{-45}{28}$
17. 400
19. (a) 39 uniforms **(b)** $\frac{1}{4}$ yards left
22. 29/36
24. 240
25. (b) $90,000
26. 1/4
28. $225
29. (a) Peter, 30 min.; Paul, 25 min.; Mary, 20 min.
30. 8
32. $2253\frac{1}{8}$
33. 32
35. $\left(\frac{1}{4} \cdot 12\right) \cdot 15 = 3 \cdot 15 = 45$
38. (a) $n(n+1) + \left(\frac{1}{2}\right)^2$
39. (a) (i) $\frac{-4}{5}$ (ii) $\frac{-26}{17}$ (iii) $\frac{-14}{33}$ **(b)** (i) $\frac{-4}{3}$
(ii) $\frac{-30}{7}$ (iii) $\frac{-3}{10}$ **(c)** $\frac{5}{4}$
41. $c = 0$ or $a = b$, where $b \neq 0$.
43. (a) $1\frac{49}{99}$
44. (a) $\frac{25}{16}$ **(b)** $\frac{25}{18}$ **(c)** $\frac{5}{216}$ **(d)** $\frac{259}{30}$ **(e)** $\frac{37}{24}$
(f) $\frac{-39}{4}$
45. 120 students

Problem Set 6-4
1. (a) > **(c)** < **(e)** =
2.

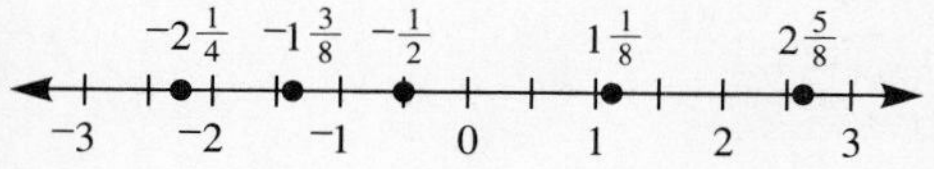

(number line showing $-2\frac{1}{4}$, $-1\frac{3}{8}$, $-\frac{1}{2}$, $1\frac{1}{8}$, $2\frac{5}{8}$)

3. (b) 3, $\frac{33}{16}$, $\frac{23}{12}$
4. (a) $x \leq \frac{27}{16}$ **(c)** $x \geq \frac{115}{3}$

5. (a) No. Multiplication by bd, which is negative, reverses the order.
6. (a) $399\frac{80}{81}$ **(c)** $3\frac{699}{820}$
7. (a) over 7 **(c)** under 1 **(e)** over 6
9. (a) about 180 **(c)** about 468 **(e)** about 3
10. Every 3 pounds of birdseed yields about 4 packages. Thus, there are about 28 packages.
11. (a) A proper fraction is greater than its square. **(c)** If a fraction is greater than 1, it is less than its square.
18. (a) 33 **(c)** $x < 4/7$ so $x = 0$
20. (a) 1 **(b)** 1
21. (a) They are equal $\frac{1}{3}$
(b) $\dfrac{1 + 3 + 5 + 7 + \ldots + 201}{203 + 205 + 207 + \ldots + 403} = \dfrac{1}{3}$
23. We are considering $\dfrac{a}{b}$ and $\dfrac{a+x}{b+x}$ when $a < b$.
$\dfrac{a}{b} < \dfrac{a+x}{b+x}$ because $ab + ax < ab + bx$.
24. (a) $\frac{29}{8}$ **(b)** $\frac{87}{68}$ **(c)** $\frac{25}{144}$ **(d)** 1 (provided that $|x| \neq |y|$)
25. $6\frac{7}{18}$ hours
26. (a) $\frac{-4}{3}$ **(b)** $\frac{-11}{8}$ **(c)** $\frac{24}{17}$ **(d)** -3
27. $26\frac{1}{4}$ hours
28. (a) 3 and 4 **(b)** (4 and 5) or (3 and 6), depending on estimation used.

Problem Set 6-5
1. (a) 3:2 **(b)** 2:3
2. (a) 5:21
3. (a) 30 **(c)** $23\frac{1}{3}$
4. 36 pounds
6. $1.19
8. 64
10. (a) 42, 56
12. $14,909.09, $29,818.18, $37,272.73
14. 135
15. (a) $\frac{5}{7}$
17. 8 days
18. (a) 27
20. 312 pounds
21. (a) $\frac{40}{700}$ or $\frac{4}{70}$ **(c)** Probably not, $\frac{2}{5} \neq \frac{3}{5}$
22. $9\frac{9}{14}$ days

24. (a) 2:5. Because the ratio is 2:3, there are $2x$ boys and $3x$ girls, hence the ratio of boys to all the students is $2x/(2x + 3x) = 2/5$. (b) $m : (m + n)$

25. $1\frac{1}{3}$ days

27. $\frac{a}{b} = \frac{c}{d}$ implies $ad = bc$, which is equivalent to $d = \frac{bc}{a}$; then, $\frac{d}{c} = \frac{b}{a}$.

29. (a) $\frac{1}{2}$ (b) Let $\frac{a}{b} = \frac{c}{d} = \frac{e}{f} = r$.
Then $a = br$
$c = dr$
$e = fr$
So, $a + c + e = br + dr + fr$,
$a + c + e = r(b + d + f)$,
$\frac{a + c + e}{b + d + f} = r$

31. $\frac{37}{125}$ of a mile

32. (a) $\frac{-3}{5}, \frac{-2}{5}, 0, \frac{1}{5}, \frac{2}{5}$ (b) $\frac{13}{24}, \frac{7}{12}, \frac{13}{18}$

33. (a) $x \geq \frac{3}{2}$ (b) $x > 3$ (c) $x > \frac{-7}{15}$ (d) $x \leq \frac{-56}{5}$

34. Answers vary. Examples are: (a) $\frac{11}{30}, \frac{12}{30}, \frac{13}{30}$
(b) $\frac{-1}{12}, \frac{-1}{9}, \frac{-5}{36}$

Problem Set 6-6

1. (a) $\frac{1}{3^{13}}$ (c) 5^{11} (e) $\frac{1}{(-5)^2}$, or $\frac{1}{5^2}$ (g) a^2

2. (a) $\left(\frac{1}{2}\right)^{10}$ (c) $\left(\frac{2}{3}\right)^9$ (e) $\left(\frac{5}{3}\right)^3$

3. (a) False $2^3 \cdot 2^4 \neq (2 \cdot 2)^{3+4}$ (c) False $2^3 \cdot 2^3 \neq (2 \cdot 2)^{2 \cdot 3}$ (e) False $(2 + 3)^2 \neq 2^2 + 3^2$ (g) False $a^{mn} = (a^m)^n \neq a^m \cdot a^n$

4. (a) 5 (c) -2 (e) 0

5. (a) $2 \cdot 10^{11}$ (b) $2 \cdot 10^5$

6. (a) $x \leq 4$ (c) $x \geq 2$

7. (a) $\frac{1 - x^2}{x}$ (c) $6x^2 + 4x$ (e) $(3a - b)^2$

8. (a) $\left(\frac{1}{2}\right)^3$ (c) $\left(\frac{4}{3}\right)^{10}$ (e) $\left(\frac{4}{3}\right)^{10}$

9. (a) 10^{10} (b) $10^{10} \cdot (6/5)^2$

10. (a) 3/2, 3/4, 3/8, 3/16, 3/32 (c) 3/1024

11. (a) 3/4 (c) 3/128

12. (a) 3^{400} (b) $4^{300} = (4^3)^{100} = 64^{100}$, $3^{400} = (3^4)^{100} = 81^{100}$ and $81^{100} > 64^{100}$

13. (a) 32^{50} because $32^{50} = (2^5)^{50} = 2^{250}$ and $4^{100} = 2^{200}$

14. No, it must end in 1, 3, 9, or 7

16. 216

17. 27

18. (a) $\frac{2}{7}$ (b) $\frac{40}{3}$ (c) $\frac{1}{3^4}$ (d) $\frac{1}{100}$ (e) $\frac{9}{4}$ (f) $\frac{49}{100}$ (g) $\frac{9}{16}$ (h) $\frac{x}{x+y}$

19. (a) $\frac{-4}{3}$ (b) $\frac{-9}{10}$ (c) $\frac{60}{13}$ (d) $\frac{-9}{4}$ (e) 9 or -9 (f) $\frac{4}{5}$

20. $1\frac{1}{5}$ days

21. It will become greater because $\frac{3x}{8x} < \frac{3x + 2}{8x + 2}$.

22. $\frac{-6}{7}, \frac{-3}{4}, \frac{-2}{3}, \frac{-1}{2}, 0, \frac{7}{9}, \frac{4}{5}, \frac{6}{7}, \frac{9}{7}$

Chapter 6 Test

1. (a) (c)

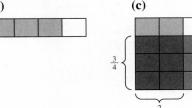

3. (a) $\frac{6}{7}$ (c) $\frac{0}{1}$ (e) $\frac{b}{1}$

4. (a) $=$ (c) $>$

5. (a) $\frac{11}{10}$ (c) $\frac{10}{13}$ (e) $\frac{50}{9}$

6. (a) $-3, \frac{1}{3}$ (c) $\frac{-5}{6}, \frac{6}{5}$

7. $-2\frac{1}{3}, -1\frac{7}{8}, 0, (71/140)^{300}, 69/140, 1/2, 71/140, (74/73)^{300}$

8. (a) 6 (c) $\frac{-1}{4}$

9. (a) $x \leq \frac{42}{25}$ (c) $x = \frac{8}{9}$

11. 9

12. (a) $\frac{1}{2^{11}}$ (c) $\left(\frac{3}{2}\right)^{28}$, or $\frac{3^{28}}{2^{28}}$

13. 17 pieces, $\frac{11}{6}$ yards left.

14. (a) 15 (c) 4

16. $70

17. 6:1

19. $333\frac{1}{3}$ calories

20. 16 minutes

21. 60 credits

CHAPTER 7

Problem Set 7-1
1. (a) $0 \cdot 10^0 + 0 \cdot 10^{-1} + 2 \cdot 10^{-2} + 3 \cdot 10^{-3}$
(b) $2 \cdot 10^2 + 0 \cdot 10 + 6 \cdot 10^0 + 0 \cdot 10^{-1} + 6 \cdot 10^{-2}$
(c) $3 \cdot 10^2 + 1 \cdot 10 + 2 \cdot 10^0 + 0 \cdot 10^{-1} + 1 \cdot 10^{-2} + 0 \cdot 10^{-3} + 3 \cdot 10^{-4}$ (d) $0 \cdot 10^0 + 0 \cdot 10^{-1} + 0 \cdot 10^{-2} + 0 \cdot 10^{-3} + 1 \cdot 10^{-4} + 3 \cdot 10^{-5} + 2 \cdot 10^{-6}$
2. (b) 4000.608 (d) 0.2004007
4. (b) 2516/100 = 629/250 (d) 281,902/10,000 = 140,951/5000
5. (a), (b), (c), (d), (e), (f), and (h) can be represented by terminating decimals.
9. $231.24
11. (f) 4.63 (h) 0.000004 (n) 0.463
15. 62.298 lb
17. $8.00
19. (a) 6390.95 cubic cm (b) 183.07 cubic in
22. (a) $235 cash = 366.60 Swiss francs; $235 traveler's checks = 373.65 Swiss francs (b) Exchange at least $472.76 cash or $463.84 in traveler's checks

Problem Set 7-2
5. (a) 221/90 (e) 243/9900 = 27/1100 (i) 5/9
6. (a) $3.2\overline{3}, 3.\overline{23}, 3.23, 3.\overline{22}, 3.20$
7. Answers may vary: (b) 462.245
8. (b) 462.245
9. (a) 49,736.5281
10. (a) 200 (d) 203.7
13. $37
15. (b) $4.632 \cdot 10$ (c) $1.3 \cdot 10^{-4}$
17. (a) $1.27 \cdot 10^7$
19. (b) $4 \cdot 10^7$
21. $2.3462784 \cdot 10^{13}$
23. (a) $1.\overline{6}, 2, 2.\overline{3}, 2.\overline{6}, 3, 3.\overline{3}, 3.\overline{6}, 4, \ldots$
24. (a) $0.\overline{446355}$; 6
25. (a) 21.6 pounds (b) 48 pounds

Problem Set 7-3
2. (c) 0.002 (f) 1.25 (g) $0.00\overline{3}$
3. (b) 2 (d) 200
5. (b) 50% (d) 3.43 (f) 40
7. $16,960
8. $14,500
10. (a) Bill sold 221 (c) Ron started with 265
12. 15%
14. $\frac{26}{29} \cdot 100\%$ or approximately 89.7%
16. $22.40
19. $336
21. $3200
24. $10.37/hr
27. $440
29. $187.50
31. (a) $3.30 (d) $24.50
35. (a) Answers may vary. (b) (i) 0.366 seconds between beats (ii) 0.0061 minutes between beats
37. Apprentice makes $700, Journeyman makes $1400, Master makes $2100
39. $30.43
41. (a) $1322.50 (b) Filing separately: $1416.77. Filing jointly: $1416.77.
It doesn't make a difference whether they file separately or jointly.
42. 97 days
45. 2795/90 = 559/18
47. (a) 32.0

Problem Set 7-4
2. $5,460.00
4. 3.5%
6. $64,800
8. $23,720.58
10. $1944
12. (iii) 13.2%
14. Approximately $23,673.64
17. Approximately 12.79%
19. Approximately $7.026762 \cdot 10^8$

Problem Set 7-5
1. Answers may vary. One answer is 0.232233222333 . . .
2. $0.77, 0.\overline{7}, 0.78, 0.787787778\ldots, 0.\overline{78}, 0.788$, $0.\overline{78} = 0.\overline{788}$
7. (a) False, $\sqrt{2} + 0$ (b) False, $-\sqrt{2} + \sqrt{2}$
(c) False, $\sqrt{2} \cdot \sqrt{2}$ (d) False, $\sqrt{2} - \sqrt{2}$
10. Answers may vary. For example, 0.54544544454444 . . . assuming the pattern continues.
14.

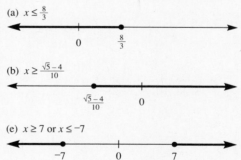

(a) $x \leq \frac{8}{3}$

(b) $x \geq \frac{\sqrt{5} - 4}{10}$

(e) $x \geq 7$ or $x \leq -7$

16. (a) 64 (d) none (f) none
18. (a) 8.98 seconds (b) 20.07 seconds
23. (a) $0.5 + 1/0.5 = 0.5 + 2 = 2.5 \geq 2$
(b) Suppose $x + 1/x < 2$. Since $x > 0$, $x^2 + 1 < 2x$ so that $x^2 - 2x + 1 < 0$, or $(x - 1)^2 < 0$, which is false. Therefore, $x + 1/x \geq 2$.
25. 3/12,500
27. 8/33
29. $20,274

Problem Set 7-6
1. (a) $6\sqrt{5}$ (e) 13/14 (f) 1/2

2. (a) −3 (c) 2 (e) −3
3. (a) $2\sqrt{3} + 3\sqrt{2} + 6\sqrt{5}$ (d) $\sqrt{2}/2$ or $\sqrt{1/2}$ (f) $\sqrt{6}$
4. (a) 4 (e) 1/256 (f) 9 (k) 10 (m) 16 (o) 4
7. 22
11. $\sqrt[8]{2^7}$
12. (a) 4 (d) 5/6
13. $\sqrt[3]{1/16}$
14. (a) n is odd (b) When m is even, then n can be any number except 0. When m is odd, then n must also be odd.

Chapter 7 Test
4. 8
7. (a) 307.63 (c) 308
8. (a) $x \leq 3.\overline{3}$ (d) 20% (f) $0.\overline{6}$
9. (a) 25% (d) 20%
10. (d) 1.23% (e) 150%
11. (b) 0.006
15. (a) 3 (b) 3 (c) 2 (d) 3
16. (a) Irrational (b) Irrational (c) Rational (d) Rational (e) Irrational
18. $3.\overline{3}\%$
21. It makes no difference.
22. $80
24. $15,110.69
25. (d) $3\sqrt[3]{6}$
26. (a) $1/2^{11}$ (d) 3^{18}

CHAPTER 8

Problem Set 8-1
1. (a) {Bush, Reagan, Carter, Ford, Nixon, Johnson, Kennedy, Eisenhower, Truman, F. Roosevelt}
2. (a) {0, 1, 2, 3, 4, 5, 6, 7, 8, 9} (c) {1, 3, 5, 7, 9}
3. (a) 3/8 (c) 4/8, or 1/2 (e) 0 (g) 1/8
4. (a) 26/52, or 1/2 (c) 28/52, or 7/13 (e) 48/52, or 12/13 (g) 3/52
5. (a) 4/12, or 1/3 (c) 0
8. (a) 1/6
9. (a) 8/36, or 2/9 (c) 24/36, or 2/3 (e) 0 (g) 10 times
11. (a) 18/38, or 9/19 (c) 26/38, or 13/19
13. (a) No (c) Yes (e) No (g) No
15. (a) 2/4, or 1/2 (c) 3/4
16. 0.7
18. (a) 45/80, or 9/16 (c) 60/80, or 3/4

Problem Set 8-2
1. (a)

```
        H
    H <
   /    T
  <
   \    H
    T <
        T
```

2. (a) {(1, 1), (1, 2), (1, 3), (2, 1), (2, 1), (2, 3)} (c) {(1, 2), (2, 1), (2, 2), (2, 3)}
3. (a) 1/216
4. (a) 1/24 (c) 1/84
6. (a) Box 1, with probability 1/3 (Box 2 has probability 1/5)
7. (a) 64/75
9. (a) 1/5 (c) 11/15
11. 1/16
12. (a) 1/4 (c) 1/16
14. (a) 1/320 (c) 0
15. 1/32
16. 1
18. 1/256
20. (a) 1/25 (c) 16/25
21. (a) 100 square units (c) 1/625
22. 0.7
23. 1/12
25. 2/5
26. 3 Reds, 1 Black
29. 25/30, or 5/6
32. 0.0005
34. (a) Spinner A
36. She should serve the first one hard; it does not matter with the next one.
37. (a) v (c) ii (e) iv
38. (a) 1/30 (c) 19/30

Problem Set 8-3
3. (c) 0.0000001
4. (a) Let 1, 2, 3, 4, 5, and 6 represent the numbers of the die and ignore the numbers 0, 7, 8, 9.
(c) Represent Red by the numbers 0, 1, 2, 3, 4; Green by the number 5, 6, 7; Yellow by the number 8; and White by the number 9.
5. 1200 fish
7. Pick a starting spot in the table and count the number of digits it takes before all the numbers 1 through 9 are obtained. Repeat this experiment many times and find the average number of coupons.
11. (a) 7
12. Answers may vary, e.g., use a random digit table. Let the digits 1–8 represent a win and the digits 0 and 9 represent losses. Mark off blocks of 3. If only the digits 1–8 appear, then this represents 3 wins in a row.
14. Let the 10 ducks be represented by the digits 0, 1, 2, 3, . . . , 8, 9. Then pick a starting point in the table and mark off 10 digits to simulate which ducks the hunters shoot at. Count how many of the digits 0 through 9 are not in the 10 digits and this represents the ducks that escaped. Do this experiment many times and take the average to determine an answer. See how close your simulation comes to 3.49 ducks.
15. (a) 1/4 (c) 48/52 or 12/13 (e) 1/2 (g) 16/52 or 4/13
16. (a) 15/19 (c) 28/171

874 ANSWERS TO SELECTED PROBLEMS

Problem Set 8-4
1. (a) 12 to 40, or 3 to 10
3. 15 to 1
4. (a) 1/2 (c) 1023 to 1
6. 1 to 1
7. 4 to 6, or 2 to 3
10. $3.50
12. 3 hours
13. 1000/1001
15. $10,000
16. (a) Since Al's probability of winning at this point was 3/4 and Betsy's was 1/4, Al should get $75 and Betsy $25. (b) 3 to 1
18. No.
21. 1/27
22. (a) {1, 2, 3, 4} (c) {(1, Red), (1, Blue), (2, Red), (2, Blue), (3, Red), (3, Blue), (4, Red), (4, Blue)} (e) {(1, 1), (1, 2), (1, 3), (1, 4), (2, 1), (2, 2), (2, 3), (2, 4), (3, 1), (3, 2), (3, 3), (3, 4), (4, 1), (4, 2), (4, 3), (4, 4)}
24. 25/676

Problem Set 8-5
2. (a) (i) 12
3. (a) 1,000,000
4. 224
5. 32
7. 1352 with 3 letter call letters; 35,152 with 4 letter call letters.
9. (a) True (c) False (e) True (g) True
11. 15
12. (a) 12 (c) 3360 (e) 3780
13. (a) 24,360
15. 362,880
16. 1/120
18. 1000
20. (a) 6
22. (a) 1/13
23. (a) 10 (c) 1
25. 3840

Chapter 8 Test
1. (a) {Monday, Tuesday, Wednesday, Thursday, Friday, Saturday, Sunday} (c) 2/7
3. (a) Approximately 0.501 (c) 34,226,731 to 34,108,157
4. (a) 5/12 (c) 5/12 (e) 0
5. (a) 13/52, or 1/4 (c) 22/52, or 11/26
6. (a) 64/729
7. 6/25
9. 7/45
10. 4 to 48, or 1 to 12
12. 3/8
13. $.30
15. 900
17. 5040

18. 2/20, or 1/10
19. (a) 5 · 4 · 3, or 60 (c) 1/60
22. 0.027
23. 63/80
25. (a) 1/8 (c) 1/16
26. 8/20, or 2/5

CHAPTER 9

Problem Set 9-1
2. (a) 225 million (c) 550 million
3.

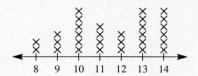

Student Ages at Washington School

4. (a) 72, 74, 81, 82, 85, 87, 88, 92, 94, 97, 98, 103, 123, 125 (b) 72 lbs. (c) 125 lbs.
5.

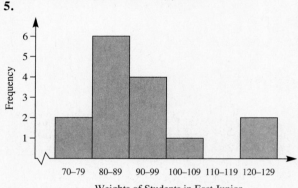

Weights of Students in East Junior High Algebra I Class

7. (a) November, 30 cm (b) 50 cm
8. (a)

Ages of HKM Employees

```
6 | 332
5 | 8224
4 | 8561511
3 | 474224          3 | 4 represents
2 | 14333617301365396    34 years old
1 | 898
```

(c) 20
9. (a) Approx. 3800 km (b) Approx. 1900 km

10.

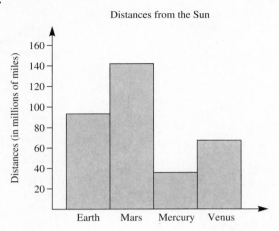

11.

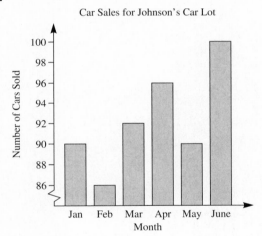

13. (a)

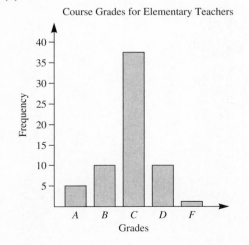

13. (b) Course Grades For Elementary Teachers

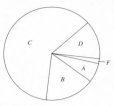

14. (a) Fall Text Books Costs

```
1 | 6
2 | 3 3
3 | 0 3 5 7 7 9 9
4 | 0 1 2 2 5 8 9
5 | 0 0 1 3 8
6 | 0 2 2           2|3 represents $23
```

(b) Fall Textbook Costs

Classes	Tally	Frequency	Classes	Tally	Frequency
$15–19	I	1	$45–49	III	3
$20–24	II	2	$50–54	IIII	4
$25–29		0	$55–59	I	1
$30–34	II	2	$60–64	III	3
$35–39	IIII	5			25
$40–44	IIII	4			

(c) (d) Frequency polygon and histogram on same graph.

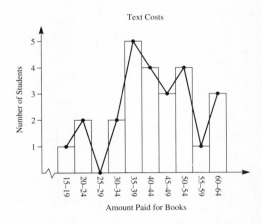

15. The line graph is more helpful because we can approximate the point midway between 8:00 and 12:00 noon and then draw a vertical line upward until it hits the line graph. An approximation for the 10:00 temperature can then be obtained from the vertical axis.

17. (a) Chicken **(c)** Cheetah
18. (a) Women **(c)** Approx. 7.5 years
22. (a) Approx. $8400 **(c)** Approx. $7,200
23. (a) Asia **(c)** It is about $\frac{2}{3}$ as large. **(e)** 5:16

24.

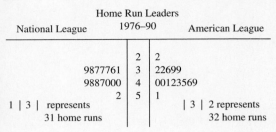

25. (a)

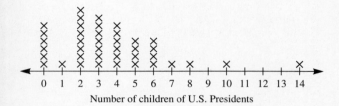

Problem Set 9-2
1. (a) Mean = 6.625, median = 7.5, mode = 8
(c) Mean $\doteq$ 19.9, median = 18, modes = 18 and 22
(e) Mean = 5.8$\overline{3}$, median = 5, mode = 5
2. (a) The mean, median, and mode are all 80.
4. 150 pounds
6. (a) $\bar{x}$ = 18.4 years (c) 28.4 years
8. Approximately 2.59
9. Approximately 215.45 pounds
11. (a) $41,275 (b) $38,000 (c) $38,000
12. (a) Balance Beam–Olga (9.575); Uneven Bars–Lisa (9.85); Floor–Lisa (9.925)
14. 30 mph
16. 58 years old
17. (a) A (c) C
18. (a) Mean–90; Median–90; Mode–90. (c) Mean
20. $s \doteq 7.3$ cm
22. (a) $s = 0$ (b) Yes
23. (a) Approximately 76.81 (c) 71
25. 96, 90, and 90
26. 2
29. (a) A–$25, B–$50 (b) B (c) $80 at B
30.

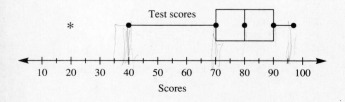

31. (a)

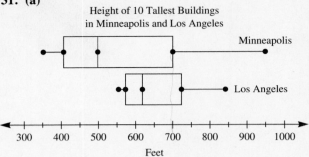

(b) There are no outliers.
33. (a) (i) $\bar{x} = 5$, $m = 5$ (ii) $\bar{x} = 100$, $m = 100$ (iii) $\bar{x} = 307$, $m = 307$ (b) The mean and median of an arithmetic sequence are the same.
35. (a) Everest, approx. 8500 m (b) Aconcagua, Everest, McKinley
36. (a)

History Test Scores

5	5
6	48
7	2334679
8	0255567889
9	00346

7 | 2 represents a score of 72

(b)

History Test Scores

Classes	Tallies	Frequency
55–59	I	1
60–64	I	1
65–69	I	1
70–74	IIII	4
75–79	III	3
80–84	II	2
85–89	IIII III	8
90–94	IIII	4
95–99	I	1

(c)

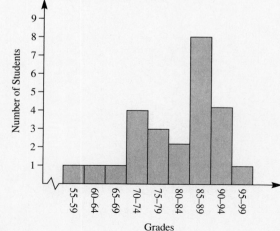

(d)

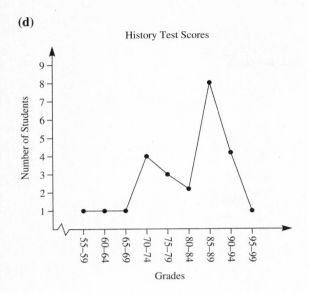

(e) Approximately 115°

Problem Set 9-3
1. **(a)** 1020   **(c)** 1.5, so 1 or 2 people
2. 97.5%
3. 0.68
4. **(a)** verbal; 0.6; quantitative, $0.8\overline{3}$; logical reasoning, 1
6. They are equal.
7. **(a)** 47.5%
8. **(a)** 1.07%   **(c)** 2.27%
9. 1.4%
10. 0.84
11. 8
12. 90
14. 50
16. **(a)**

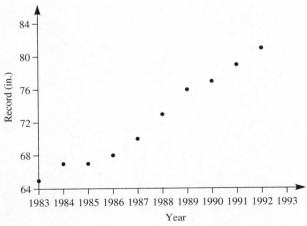

16. **(b)** Positive
17. **(a)** Negative   **(b)** Approx. 10   **(c)** 22 years old
18. **(a)** 74.17   **(b)** 75   **(c)** 65   **(d)** 237.97   **(e)** 15.43
19. 27.74
20. $76.\overline{6}$
21.   Men's Olympic
         100 meter Run Times
               1896–1964
    1 0 | 0 2 3 3 3 4 5 6 8 8 8 8
    1 1 | 0 0
    1 2 | 0   10|0 represents
                          10.0 seconds

Problem Set 9-4
3. She could have taken a different number of quizzes during the first part of the quarter than the second part.
4. When the radius of a circle is doubled, the area is quadrupled, which is misleading since the population has only doubled.
5. The horizontal axis does not have uniformly-sized intervals and both the horizontal axis and the graph are not labeled.
6. There were more scores above the mean than below, but the mean was affected more by low scores.
8. Answers vary; however, he is assuming that there are no deep holes in the river where he crosses.
9. The three-dimensional drawing distorts the graph. The result of doubling the radius and the height of the can are to increase the volume by a factor of 8.
10. No labels so we can compare actual sales.
12. **(a)** False, prices vary only by $30.   **(b)** False, the bar has 4 times the area but this is not true of prices.   **(c)** True

Chapter 9 Test
2. $2\overline{3}$
3. **(a)** Mean = 30, median = 30, mode = 10
5. **(a)**

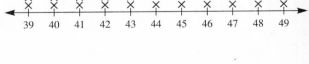

(b)

       Miss Rider's Class
       Masses in Kilograms

    3 | 9 9
    4 | 0 0 1 1 2 2 2 2 3 3 4 5 6 7 8 9 9 9   4 | 0 represents
                                                     40 kg

(c)

Miss Rider's Class
Masses in Kilograms

Mass	Tally	Frequency
39	II	2
40	II	2
41	II	2
42	IIII	4
43	II	2
44	I	1
45	I	1
46	I	1
47	I	1
48	I	1
49	III	3
		20

(d)

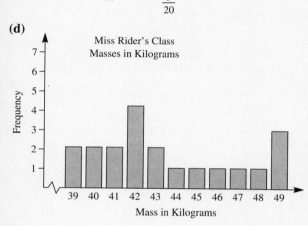

6. (a)

Test Grades

Classes	Tally	Frequency
61–70	LHT I	6
71–80	LHT LHT I	11
81–90	LHT II	7
91–100	LHT I	6
		30

(b) and **(c)** are shown on the same graph.

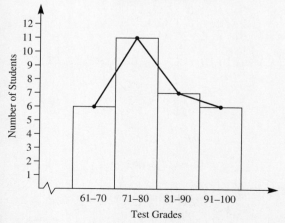

7.

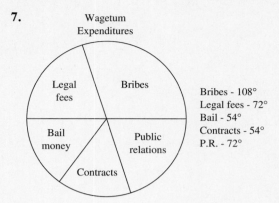

Bribes - 108°
Legal fees - 72°
Bail - 54°
Contracts - 54°
P.R. - 72°

8. The widths of the bars are not uniform and the graph has no title.

9. $2840

10.

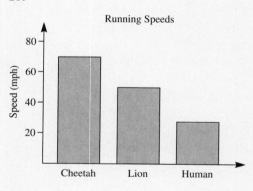

11. (a)

Life Expectancy
for Males and Females

Females		Males
	67	1446
	68	28
	69	156
	70	0049
	71	0223458
	72	
	73	
7	74	
9310	75	
86	76	
88532	77	
54332211	78	
7	79	

7 | 74 | represents
74.7 years old

| 67 | 1 represents
67.1 years old

(b)

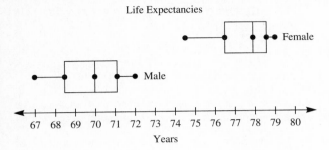

13. (a) 360 **(c)** 350
14. (a) 67
(c)

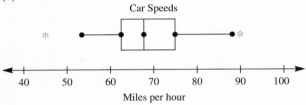

(e) 30%
15. (a) 25 **(c)** 0.16
17. $1.\overline{6}$
19. (a) Positive **(c)** 67 in. **(e)** 50 lb.

CHAPTER 10

Problem Set 10-1

1. (a) $\overleftrightarrow{AB}$ **(b)** $\overline{AB}$ **(c)** $\overrightarrow{AB}$ **(d)** $\overrightarrow{AB}$
(e) $\overleftrightarrow{AB} \parallel \overleftrightarrow{CD}$ **(f)** $\overline{AB}$ **(g)** $\overleftrightarrow{AB} \perp \overleftrightarrow{CD}$
(h) $m(\angle ABC) = 30°$
2. (a) $\{C\}$ **(c)** $\{C\}$ **(e)** $\overline{CE}$ **(g)** $\overrightarrow{BA}$
4. (a) $\varnothing$ **(c)** $\{C\}$ **(e)** $\{A\}$ **(g)** Answers vary. **(i)** Plane BCD or Plane BEA
5. (a) True **(c)** False **(e)** True **(g)** False **(i)** False **(k)** True
9. (a) Yes; otherwise a point of intersection of the two lines would also be a point of intersection of the two parallel planes. **(c)** Yes; if the planes were not parallel, they would intersect in a line m and at least one of the given lines would intersect m and would then intersect plane α, which contradicts the fact that the given lines are parallel to α.
10. (a) 110° **(c)** 20°
11. (a) Approximately 36° **(b)** Approximately 120°
12. (a) (i) 43°31′10″ **(ii)** 79°48′47″ **(b) (i)** 54′ **(ii)** 15°7′48″
13. (a) 4 **(c)** 8
14. (a) 3 **(c)** 10

15. (a)

	Number of Intersection Points					
	0	1	2	3	4	5
2	⇄	✕	Not Possible	Not Possible	Not Possible	Not Possible
3	⇄	✱	⇄	✕	Not Possible	Not Possible
4	⇄	✱	Not Possible	⇄	⇄	✕
5	⇄	✱	Not Possible	Not Possible	⇄	⇄
6	⇄	✱	Not Possible	Not Possible	Not Possible	⇄

(Number of lines on the vertical axis.)

17. (a) No; if $\angle BDC$ were a right angle, then both $\overleftrightarrow{BD}$ and $\overleftrightarrow{BC}$ would be perpendicular to $\overleftrightarrow{DC}$ and thus be parallel. **(c)** Yes; use the definition of perpendicular planes.
20. (a) 3 **(c)** 4
23. (a) No, skew lines do not intersect so they cannot be perpendicular **(c)** No, if each angle is less than 90°, the sum of the 2 angles will be less than 180°
25. For line segments in the same plane to be parallel, they must lie on parallel lines.
28. No, if the 4 points are collinear, there is 1 line; if 3 are collinear, there are 4 lines; and if no 3 are collinear, there are 6 lines.
30. (a)
```
TO ANGLE :SIZE
  FD 100 BK 100
  RT :SIZE FD 100
  BK 100 LT :SIZE
END
```
(b)
```
TO SEGMENT :LENGTH
  FD :LENGTH BK :LENGTH
END
```
(c)
```
TO PERPENDICULAR :LENGTH1
    :LENGTH2
  FD :LENGTH BK :LENGTH1/2
  RT 90 FD :LENGTH2
  BK :LENGTH2 LT 90
  BK :LENGTH1/2
END
```
(d)
```
TO PARALLEL :LENGTH1 :LENGTH2
  DRAW
  FD :LENGTH1 PENUP
  RT 90 FD 10 RT 90
  PENDOWN FD :LENGTH2
  PENUP HOME
  RT 180 PENDOWN
END
```
(In LCSI, replace DRAW with CLEARSCREEN).

Problem Set 10-2

1. (a) 1, 2, 3, 6, 7, 8, 9, 11, 12 **(c)** 1, 2, 3, 6, 7, 8, 9, 11 **(e)** 7, 8

3. (a) Outside
5. 8 (with nonconvex quadrilateral)
6. (a) and (c) are convex; (b) and (d) are concave.
7. Answers may vary.
9. (a) 35 (c) 4850
10. (a) Isosceles and equilateral (c) Scalene
11. (a) False. To be isosceles, the triangle may have only 2 congruent sides, not necessarily
3. (c) True (e) True (g) True (i) True
(k) False. All squares are rectangles. (m) True
(o) False. See (n).
13. To be regular all sides must be congruent but also all angles must be congruent. All angles are not congruent unless the rhombus is a square.
15. (a) T, Q, R, H, G, I, F, J (c) W, D, A, Z, U, E (e) Y
19. (a) 45 (b) $n(n-1)/2$
20. $\emptyset$, 1 point, 2 points, ray
21. (a) $\{C\}$ (c) $\overline{AB}$, $\overline{AC}$, and $\overline{AD}$
22. (a) False. A ray has only one endpoint.
(c) False. Skew lines cannot be contained in the single plane. (e) True

Problem Set 10-3
1. (a) $\angle 1$ and $\angle 2$ are adjacent angles; $\angle 3$ and $\angle 4$ are vertical angles. (c) $\angle 1$ and $\angle 2$ are neither vertical nor adjacent angles.
5. (a) 60° (c) 60°
6. (a) Yes; a pair of corresponding angles are 50° each. (c) Yes; a pair of alternate interior angles are 40° each.
7. (a) No. Two or more obtuse angles will produce a sum of more than 180°. (c) No. The sum of the measures of the three angles would be more than 180°.
8. (a) 70° (c) 65°
9. (a) $x = 40°$ and $y = 50°$ (c) $x = 50°$ and $y = 60°$
10. (a) 60°
11. 70° and 20°
12. 60°
13. (a) 360° (c) 360°
15. (a) 20
18. (a) True (c) False (e) False
19. (a) Equal
   (b) $m(\angle 4) = 180° - m(\angle 3)$ (Straight angle)
              $= 180° - [180° - m(\angle 1) - m(\angle 2)]$
              $= m(\angle 1) + m(\angle 2)$
23. 90°
25. They must be supplementary
26. 111°
27. $m(\angle 1) = 60°$, $m(\angle 2) = 30°$, $m(\angle 3) = 110°$
30. If two lines are perpendicular to the same line, then congruent corresponding angles of 90° each are formed, and hence the lines are parallel.
32. 83.5° or 83°30'

33. Answers may vary.
   (a) TO PARALLELOGRAM :L :W :A
       REPEAT 2[FD :L RT 180-:A FD
       :W RT :A]
       END
34. 6
35. No, the union of two rays will always extend infinitely in at least one direction.
38. Sketches may vary, but the possibilities are the empty set, a single point, a segment, a quadrilateral, a triangle, a pentagon, and a hexagon. There are various types of quadrilaterals possible.
39. (a) An octagon (c) Two intersecting segments or lines (e) A rhombus or square
40. (a) All angles must be right angles and all diagonals are the same length. (c) Impossible because all squares are parallelograms.

Problem Set 10-4
1. (a) Quadrilateral pyramid (b) Quadrilateral prism; possibly a trapezoid prism (c) Pentagonal pyramid
2. (a) A, D, R, W (c) $\triangle ARD$, $\triangle AWD$, $\triangle AWR$, $\triangle WDR$
4. (a) 5 (c) 4
5. (a) True (c) True (e) False (g) False
6. 3; Each pair of parallel faces could be considered as bases.
8. (a)   (b)

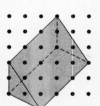 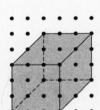

9. (a) Hexagonal pyramid (c) Cube (e) Hexagonal prism
10. (a) iv
11. (a) iv
12. (a) i, ii, and iii
14. Both could be drawings of a quadrilateral pyramid. In part (a) we are directly above the pyramid and in part (b) we are directly below the pyramid.
16. (a)   (b)

 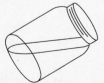

17. **(b)** (4)
18. **(a)**

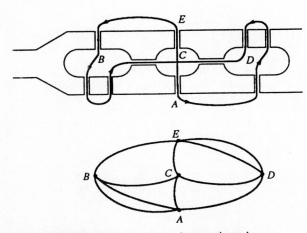

(e)

20.   Pyramid   Prism
   (a) $n+1$ $n+2$
   (b) $n+1$ $2n$
   (c) $2n$ $3n$
   (d) $(n+1) + (n+1) - 2n = 2$
          $(n+2) + 2n - 3n = 2$
21. **(a)** 6   **(c)** 11
23. **(a)** A cone might be described as a many-sided pyramid.
24. **(a)** Yes, if the base is an *n*-gon the number of edges is $3n$. Thus, if $3n = 33$, then $n = 11$ and an 11-gon has exactly 33 edges
25. Parallelogram
26. $m(\angle BCD) = 60°$
27. 140°
28. **(a)** True **(b)** True **(c)** False; The triangle could be equilateral and have three acute angles.

Problem Set 10-5
3.

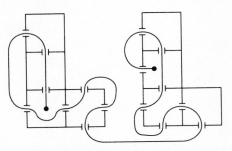

Path: *CEBABCADEDC*; any point can be a starting point.

5. Yes. See figure.

7.

Network	R	V	A	R + V − A
(a)	6	6	10	2
(b)	6	4	8	2
(c)	6	6	10	2
(d)	4	4	6	2
(e)	5	4	7	2
(f)	8	8	14	2
(g)	9	8	15	2
(h)	6	4	8	2
(i)	7	7	12	2
(j)	8	12	18	2

8.

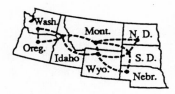

Since all vertices are even, the trip is possible. It makes no difference where she starts.

Problem Set 10-6
1. **(a)** TO RECTANGLE :LENGTH :WIDTH
       PARALLELOGRAM :LENGTH
       :WIDTH 90
     END
4. **(a)** 60° **(c)** 45°
7. Answers may vary.
   (b) TO HONEYCOMB :SIDE
       REPEAT 3 [HEXAGON :SIDE RT 120]
     END
8. Answers may vary.
   TO THIRTY
     FD 100 BK 100 RT 30
     FD 100 BK 100 LT 30
   END
9. Answers may vary.
   (a) TO SEG
       RT 45 FD 50 BK 100
       FD 50 LT 45
     END
12. TO POLYGON :NUM :LEN
     REPEAT :NUM [FD :LEN RT 360/:NUM]
   END

Chapter 10 Test
2. (a) $\overleftrightarrow{AB}$, $\overleftrightarrow{BC}$, and $\overleftrightarrow{AC}$ **(c)** $\overline{AB}$ **(e)** $\overline{AB}$
3. (b) Planes APQ and BPQ **(d)** No. $\overleftrightarrow{PQ}$ and $\overleftrightarrow{AB}$ are skew lines so that no single plane contains them.
6. (a) No. The sum of the measures of two obtuse angles is greater than 180°, which is the sum of the measures of the angles of any triangle.
7. 18°, 36°, 126°.
8. (a) Given any convex n-gon, pick any vertex and draw all possible diagonals from this vertex. This will determine $n - 2$ triangles. Because the sum of the measures of the angles in each triangle is 180°, the sum of the measures of the angles in the n-gon is $(n - 2)180°$.
12. 6
14. (a) 60° **(c)** 120°
16. 48°
17. (a) (i), (ii), and (iv) are traversable.

CHAPTER 11

Problem Set 11-1
1. (c) The side of greater length is opposite the angle of greater measure.
4. 22 triangles
5. (a) Yes; SAS   **(b)** Yes; SSS
(c) No
8. Such a construction tells us that $\triangle BDC \cong \triangle BAC$ by SAS. Therefore, $DB = AB$. By measuring DB, we know AB.
11. (b) Because $\triangle ABC \cong \triangle BCA$, $\angle A \cong \angle B$ and $\angle B \cong \angle C$. Hence $\angle A \cong \angle B \cong \angle C$.
13. Right angles and obtuse angles.
14. (a) F is the midpoint of both diagonals. **(b)** We can show $\triangle ABD \cong \triangle CBD$ by SSS, and then $\angle BDC \cong \angle BDA$ by CPCTC. $\triangle AFD \cong \triangle CFD$ by SAS, so $\overline{AF} \cong \overline{FC}$, thus F is the midpoint of $\overline{AC}$. A similar argument will show $\overline{BF} \cong \overline{FD}$. **(c)** 90° **(d)** From part (b), $\angle AFD \cong \angle CFD$ by CPCTC, but are also supplementary. Hence, $m(\angle AFD) = 90°$. A similar argument is used for $\angle BFA$ and $\angle BFC$.
17. (a) A parallelogram. **(b)** Let $\overline{AB} \cong \overline{CD}$ and $\overline{BC} \cong \overline{AD}$. Prove that $\triangle ABC \cong \triangle CDA$ and conclude that $\overline{BC} \| \overline{AD}$. Similarly show $\triangle ABD \cong \triangle DCB$ and conclude that $\overline{AB} \| \overline{DC}$.
19. Answers may vary.
```
TO EQUITRI :SIDE
  REPEAT 3 [FD :SIDE RT 120]
END
```
22. (c) Add the following:
```
IF NOT (:ANGLE < 180) [PRINT [NO
TRIANGLE IS POSSIBLE.] STOP
```
(In LCSI Logo, use the following line:
```
IF NOT (:ANGLE < 180) [PRINT [NO
TRIANGLE IS POSSIBLE.] STOP].)
```

Problem Set 11-2
3. (a) Yes; ASA **(b)** Yes; AAS **(c)** No, SSA does not assure congruence **(d)** No, AAA does not assure congruence
6. (a) Parallelogram **(c)** None **(e)** Rhombus **(g)** Parallelogram
7. (a) True **(h)** False; A square is both a rectangle and a rhombus. **(j)** True
8. (c) No; any parallelogram with a pair of right angles must have right angles as its other pair of angles and hence be a rectangle.
11. (a) $\overline{OP} \cong \overline{OQ}$
12. (a) $\triangle ABC \cong \triangle ADC$ by SSS. Hence $\angle BAC \cong \angle DAC$ and $\angle BCM \cong \angle DCM$ by CPCTC. Therefore, $\overline{AC}$ bisects $\angle A$ and $\angle C$.
13. (a) The sides opposite congruent angles in an isosceles trapezoid are congruent.
15. (a) Rhombus **(e)** Parallelogram
16. (b) The lengths of the sides of two perpendicular sides of the rectangles.
17. (a) Use the definition of a parallelogram and ASA to prove that $\triangle ABD \cong \triangle CDB$ and $\triangle ADC \cong \triangle CBA$.
19. (a) Answers may vary.
```
TO RHOMBUS :SIDE :ANGLE
  REPEAT 2 [FD :SIDE RT (180-
     :ANGLE) FD :SIDE RT :ANGLE]
END
```
20. Answers may vary.
```
TO ISOSTRI :SIDE :ANGLE
HOME FD :SIDE
RIGHT (2*:ANGLE)
FD :SIDE RIGHT :ANGLE
HOME
END
```
23. (a) Yes; SAS **(c)** No

Problem Set 11-3
6. (d) The lines containing the altitudes meet at a point inside the triangle.
7. (a) The perpendicular bisectors meet at a point inside the triangle.
8. (a) The perpendicular bisector of a chord of a circle contains the center of the circle. **(b)** Choose an arbitrary chord $\overline{AB}$ on the circle with center O. Then $\overline{OA} \cong \overline{OB}$ since both are radii. Construct the angle bisector of $\angle AOB$ and let P be the point of intersection with chord $\overline{AB}$. Then $\triangle AOP \cong \triangle BOP$ by SAS, and $\angle OPB$ is a right angle since it is both congruent and supplementary to $\angle OPA$. Since $\overline{AP} \cong \overline{PB}$ and $\angle OPB$ is a right angle, $\overline{OP}$ is the perpendicular bisector of $\overline{AB}$, and therefore the perpendicular bisector of an arbitrary chord contains point O.
(c) Hint: Construct two non-parallel chords and find their perpendicular bisectors. The intersection of the perpendicular bisectors is the center of the circle.
13. (a) $\overrightarrow{PQ}$ is the perpendicular bisector of $\overline{AB}$. **(b)** Q is on the perpendicular bisector of $\overline{AB}$

because $\overline{AQ} \cong \overline{QB}$. Similarly, P is on the perpendicular bisector of $\overline{AB}$. Because a unique line contains two points, the perpendicular bisector contains $\overline{PQ}$.

17. (a) Since the triangles are congruent, the acute angles formed by the hypotenuse and the line are congruent. Since the corresponding angles are congruent, the hypotenuses are parallel (the line formed by the top of the ruler is the transversal).

20. Answers may vary.
```
TO ALTITUDES
  REPEAT 3 [RT 30 FD 60 RT 90 FD
   110 BK 130 FD 20 LT 90 FD 60
   RT 90]
END
```
24. (b) Yes. (1) $\triangle LYC \cong \triangle UCY$ by SAS. $\overline{LY} \cong \overline{UC}$ is given; $\overline{YC} \cong \overline{CY}$. To show $\angle LYC \cong \angle UCY$, construct $\overline{UV} \| \overline{LY}$. Now $LUVY$ is a parallelogram; $\overline{UV} \cong \overline{LY}$ and, by transitivity, $\overline{UV} \cong \overline{UC}$. $\angle LYC \cong \angle UVC$ (corresponding angles formed by $\overleftrightarrow{LY} \| \overleftrightarrow{UV}$ and transversal $\overleftrightarrow{YC}$). $\angle UVC \cong \angle UCV$ (base angles of isosceles triangle UVC). $\angle LYC \cong \angle UCY$ by transitive property. (2) $\triangle ULY \cong \triangle LUC$ by SAS. $\overline{LY} \cong \overline{UC}$ is given; $\overline{UL} \cong \overline{LU}$. To show $\angle ULY \cong \angle LUC$; use supplementary pairs of angles $\angle ULY$ & $\angle LYC$ and $\angle LUC$ & $\angle UCY$, and $\angle LYC \cong \angle UCY$ from part (1). (3) $\triangle LOY \cong \triangle UOC$ by ASA. $\overline{LY} \cong \overline{UC}$ is given. $\angle YLC \cong \angle CUY$ by CPCTC from part (1). $\angle LYU \cong \angle UCL$ by CPCTC from part (2).

Problem Set 11-4
3. (b) 90° **(d)** An angle whose vertex is on a circle and whose sides intersect the circle in two points determining a diameter is a right angle.
6. (c) Opposite angles are supplementary.
7. Hint: The center of the circle is at the intersection point of the diagonals.
11. (a) Isosceles **(b)** $m(\angle 1) + m(\angle 2) = m(\angle 3)$
(c) $m(\angle 1) = \frac{1}{2}m(\angle 3)$
13. (a) 270°
14. Given that $\overline{AB} \cong \overline{CD}$ (see the figure), prove that $\overline{OM} \cong \overline{ON}$. First prove that $\overline{AM} \cong \overline{MB}$ and $\overline{CN} \cong \overline{ND}$. Then show that $\triangle AOB \cong \triangle COD$. Hence, conclude that $\angle A \cong \angle C$. Now prove that $\triangle AMO \cong \triangle CNO$ (by SAS). Consequently $\overline{OM} \cong \overline{ON}$.

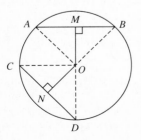

16. Answers may vary. Newer versions of Logo have a FILL primitive.
```
TO FILL.CIRCLE :RADIUS
  REPEAT 360 [FD :RADIUS BK
   :RADIUS RT 1]
END
```
18. $\overline{AB}$
20. If $\angle A$ is not the right angle, the triangles are congruent. If $\angle A$ is the right angle, the triangles are not necessarily congruent.

Problem Set 11-5
1. (a) Yes; AAA **(d)** No **(g)** Yes; sides are proportional and angles are congruent.
5. (c) The triangles are similar if, for example, in $\triangle ABC$ and $\triangle DEF$, we have $AB/DE = AC/DF$ and $\angle A \cong \angle D$.
8. (b) (i) 2/3 (ii) 1/2 (iii) 3/4 (iv) 3/4
12. (b) (1) $AC/AB = CD/CB = AD/AC$
(2) $CB/AB = CD/AC = DB/CB$
(3) $AC/CB = AD/CD = CD/DB$
14. 15 m
16. (a) In $\triangle ABC$ let $\overline{AB} \cong \overline{BC}$ and $\overline{BD}$ be the angle bisector of $\angle B$. Then $\overline{BD} \perp \overline{AC}$ (why?) and we have: $\alpha + \beta = 90°$, $\alpha = \beta$. Hence $\alpha = 45°$ and $2\beta = 90°$. Consequently, the angles of the triangle are 45°, 45°, 90°. Next, suppose the angle bisector is the bisector of one of the base angles. If the measure of each of the base angles is 2α, then $\triangle AEC$ is isosceles if and only if $m(\angle AEC) = 2\alpha$. Then in $\triangle ACE$: $\alpha + 2\alpha + 2\alpha = 180°$ or $5\alpha = 180°$. Hence $\alpha = 36°$ and $2\alpha = 72°$. Now $m(\angle B) = 180° - 2 \cdot 72° = 36°$. Hence the angles of $\triangle ABC$ are 72°, 72°, and 36°.

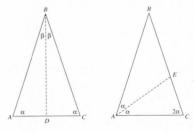

17. 232.62 inches or 19.38 feet
19. $CF = 13$ m $AE = 12$ m
Construct $\overline{BP}$ perpendicular to $\overline{CF}$ as shown. $\triangle CBP \cong \triangle DFE$ by AAS because $\overline{BC} \cong \overline{FD}$ (opposite sides in a rectangle), $\angle DEF \cong \angle CPB$ (right angles), and from $\angle FDE \cong \angle CFD$ (alternate interior angles between the parallels $\overleftrightarrow{CF}$ and $\overleftrightarrow{DE}$ and the transversal $\overleftrightarrow{DF}$) and $\angle CFD \cong \angle BCP$ (alternate interior angles between $\overleftrightarrow{FD} \| \overleftrightarrow{BC}$ and the transversal $\overleftrightarrow{CF}$) it follows that $\angle FDE \cong \angle BCP$. By CPCTC, $\overline{CP} \cong \overline{DE}$ and hence $CP = DE = 4$ m. We have $CF = PF + CP$. Because $ABPF$ is a rectangle, $PF = BA = 9$ m, and

hence $CF = 9 + 4 = 13$ m. Now $\overline{AF} \cong \overline{BP}$ ($ABPF$ is a rectangle), $\overline{FE} \cong \overline{BP}$ (CPCTC in $\triangle CBP$ and $\triangle DFE$). Consequently $\overline{AF} \cong \overline{FE}$. Next show that $\triangle ABF \sim \triangle EFD$ (by AA since $\angle AFB$ and $\angle FDE$ are complements of $\angle DFE$ and each triangle has a right angle.) Consequently $AB/EF = AF/ED$ or $9/EF = AF/4$ or $EF \cdot AF = 36$. Because $\overline{EF} \cong \overline{AF}$, we have $(EF)^2 = 36$ or $EF = 6$. Because $AE = 2EF$, $AE = 12$ m.

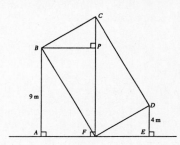

20. Answers may vary.
    (a) TO RECTANGLE :LEN :WID
          REPEAT 2 [FD :LEN RT 90 FD
            :WID RT 90]
        END
        TO SIM.RECT :LEN :WID
          RECTANGLE :LEN*2 :WID*2
        END
    (b) TO SIM.RECTANGLE :LEN :WID
          :SCALE
          RECTANGLE :LEN*:SCALE
            :WID*:SCALE
        END
21. Answers may vary.
    (a) TO TRISECT :LEN
          REPEAT 3 [MARK FD :LEN/3]
        END
        TO MARK
          RT 90 FD 5
          BK 5 LT 90
        END
22. No; the image is two-dimensional while the original person is three-dimensional.

Problem Set 11-6
Answers for the entire section may vary. Procedures are written only as possible answers.
3. TO TRI30 :HYPOT
     DRAW
     FD :HYPOT/2 RT 120
     FD :HYPOT RT 150
     HOME
   END
   (In LCSI Logo, replace DRAW with CLEARSCREEN.)
4. TO STAR :SIDE
     HEXAGON :SIDE
     REPEAT 6 [LT 60 FD :SIDE RT 120
       FD :SIDE]
   END
   TO HEXAGON :SIDE
     REPEAT 6 [FD :SIDE RT 60]
   END

Chapter 11 Test
4. (a) $x = 8$; $y = 5$ (b) $x = 6$
6. $a/b = c/d$ because $a/b = x/y$ and $x/y = c/d$
10. 12 m high
11. (a) (iii) and (iv)
13. 51.2 m
14. (a) True in some cases and false in others: If the diagonals bisect each other, then the quadrilateral is a square. If the diagonals do not bisect each other, then it is not a square. Quadrilateral $ABCD$ is not a square; however its diagonals are perpendicular and congruent.

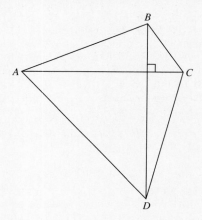

CHAPTER 12

Problem Set 12-1
2. (a) (b)

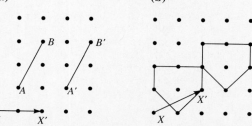

3. (a) (b)

7. (a) **(b)**

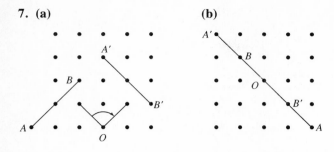

9. Hint: Find the image of the center and one point on the circle.
10. Hint: In (a) and (c) find the images of the vertices.
13. (a) A circle
15. (a) ℓ' is the same as ℓ. **(c)** ℓ' is perpendicular to ℓ.
16. (a) Hint: Find the image m', of line m under a half-turn with center P. The intersection of m' and line l is point A.
18. TO ROTATE :A :SIDE
　　　SQUARE :SIDE
　　　RIGHT :A
　　　SQUARE :SIDE
　　END
　　TO SQUARE :SIDE
　　　REPEAT 4[FORWARD :SIDE RIGHT
　　　　90]
　　END
19. (a) TO TURN.CIRCLE :A
　　　　CIRCLE
　　　　LEFT :A
　　　　CIRCLE
　　　　END
　　　TO CIRCLE
　　　　REPEAT 360[FORWARD 1 RT 1]
　　　　END
To produce the desired transformation, execute TURN.CIRCLE 180.

Problem Set 12-2
4. (a) Yes, there are infinitely many such lines all of which contain the center of the circle. **(c)** Yes, the reflecting line is the line containing the ray. **(e)** Yes, there are two such lines: the perpendicular bisectors of the pairs of parallel sides. **(g)** Yes, there is one such line in a general isosceles triangle: the line that is the perpendicular bisector of the side that is not congruent to the other two. **(i)** There are none. **(k)** Yes, there is one such line; the line that is the perpendicular bisector of the chord determined by the endpoints of the arc. **(m)** Yes, there are two, the diagonals. **(o)** Yes, there are n such lines.
6. The final image is the same as the original.
7. No, the final images are in different locations.
13. (a) The images are the same.

16. Reflect B in the line that contains side 1 to point B', then reflect B' in the line that contains side 2 to point B''. Next reflect B'' in the line that contains side 3, to point B'''. If the ball at A is shot at point B''', it will bounce as desired and hit the ball at B. (Why?)
21. (a) TO EQTRI :SIDE
　　　REPEAT 3[FORWARD :SIDE RIGHT
　　　　120]
　　END
(c) A reflection in a vertical line through the turtle's home.
24. Half-turn about the center of the letter O

Problem Set 12-3
2. (a) A translation from N to M.
(b) A rotation of 75 degrees with center O in a counterclockwise direction. **(e)** A reflection in line n.
3. (a) (2, 1) **(d)** (6, 4)
5. (a) A translation from C to B followed by a size transformation with center B and scale factor 2.
(b) A translation from C to E (which is equivalent to a translation by one unit vertically up followed by a translation 5 units horizontally to the right) followed by a size transformation with center E and scale factor 2.
(c) A 90° counterclockwise rotation about C followed by a size transformation with center C and scale factor 2.

(a)　　　　　　　　　(b)

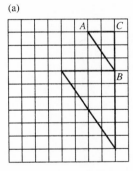 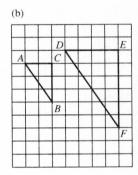

(c)

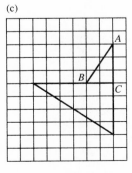

7. Answers may vary.

Problem Set 12-4
1. (a) Line, rotational, and point symmetry **(b)** Line symmetry **(c)** Line symmetry **(d)** Line symmetry
3. (a)   **(b)**

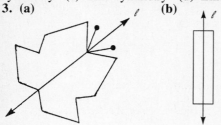

4. (a) (i) 4 (ii) None (iii) 2 (iv) 1
5. (a) 1 vertical **(b)** 1 vertical **(c)** None **(d)** 1 vertical **(e)** 5 lines **(f)** 1 vertical
6. (c) Not possible
8. (a) Yes, a figure with point symmetry has 180 degree rotational symmetry. **(e)** Yes, see part (a).
10. (a) 7 **(b)** 2 **(c)** 7 **(d)** 33
11. (b)
```
TO TURN.SYM :S :N :A
  REPEAT :N[SQUARE :S RIGHT :A]
END
TO SQUARE :S
  REPEAT 4[FORWARD :S RIGHT 90]
END
```
Execute TURN.SYM 50 3 120
12. (a)
```
TO TURN.SY :S :N :A
  REPEAT :N[EQTRI :S RIGHT :A]
END
TO EQTRI :S
  REPEAT 3[FORWARD :S RIGHT 120]
END
```
Execute TURN.SY 50 6 60

Problem Set 12-5
1. (a)   **(b)**

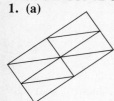 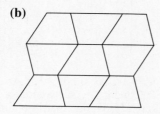

2. (b) Yes
4. (a) The dual is also a tessellation of squares.
5. Hint: Consider figures like a pentagon formed by combining a square and an equilateral triangle, or the figure in Problem 7.
6. (b)
```
TO TESSELTRI
  PENUP BACK 70 PENDOWN
  REPEAT 9[TRIANGLE 20 FORWARD
    20]
  PENUP BACK 180 RIGHT 60
  FORWARD 20 LEFT 60 PENDOWN
  REPEAT 9[TRIANGLE 20 FORWARD
    20]
END
TO TRIANGLE :SIDE
  REPEAT 3[FORWARD :SIDE RIGHT
    120]
END
```

Problem Set 12-6
1.
```
TO WALL3 :XPT :YPT :SIDE
  DRAW
  SETUP :XPT :YPT
  WALLPAPER3 :YPT :SIDE
END
```
(In LCSI Logo, replace DRAW with CLEARSCREEN.)
```
TO SETUP :XPT :YPT
  PENUP
  SETXY :XPT :YPT
  PENDOWN
END
```
(In LCSI Logo, replace SETXY :XPT :YPT with SETPOS (LIST :XPT :YPT.))
```
TO WALLPAPER3 :YPT :SIDE
  TRISTRIP :SIDE
  PENUP
  SETUP (:XPT+:SIDE*(SQRT 3)/2) :YPT
  PENDOWN
  WALLPAPER3 :YPT :SIDE
END
TO TRISTRIP :SIDE
  IF XCOR+:SIDE>120 TOPLEVEL
  IF (ANYOF (XCOR<-120)
    (XCOR+:SIDE*(SQRT 3)/2>120)
    (YCOR<-100)(YCOR+:SIDE>100))
    STOP
  TRIANGLE :SIDE
  FORWARD :SIDE
  RIGHT 60
  TRIANGLE :SIDE
  LEFT 60
  TRISTRIP :SIDE
END
```
(In LCSI Logo, replace IF XCOR+:SIDE >120 TOPLEVEL with IF XCOR+:SIDE>120 [THROW "TOPLEVEL]. Also replace IF (ANYOF (XCOR<-120) (XCOR+:SIDE*(SQRT 3)/2>120) (YCOR<-100) (YCOR+:SIDE>100)) STOP with IF (OR (XCOR<-120) (XCOR+:SIDE*(SQRT 3)/2>120) (YCOR <-100) (YCOR+:SIDE>100)) [THROW "TOPLEVEL].)
```
TO TRIANGLE :SIDE
  REPEAT 3[FORWARD :SIDE RIGHT
    120]
END
```
2. The conditions are to keep the turtle from drawing off the screen. It forces the boundaries to be as follows: $-120 < x < 120$ and $-100 < y < 100$.
5. Yes; once the figures fit between two parallel lines, then one could make a rubber stamp of the parallel lines and the drawings between them and stamp them all across the plane.

7. ```
TO WALL6 :XPT :YPT :SIDE
  DRAW
  SETUP :XPT :YPT
  WALLPAPER6 :YPT :SIDE
END
```
(In LCSI Logo, replace DRAW with CLEARSCREEN.)
```
TO SETUP :XPT :YPT
  PENUP
  SETXY :XPT :YPT
  PENDOWN
END
```
(In LCSI Logo, replace SETXY :XPT :YPT with SETPOS (LIST :XPT :YPT).)
```
  TO WALLPAPER6 :YPT :SIDE
    CHEVRONSTRIP :SIDE
    PENUP
    SETUP (XCOR+:SIDE) :YPT
    PENDOWN
    WALLPAPER6 :YPT :SIDE
END
TO CHEVRONSTRIP :SIDE
  IF XCOR+:SIDE>120 TOPLEVEL
  IF (ANYOF (XCOR-:SIDE<-120)
    (XCOR+:SIDE>120)
    (YCOR-:SIDE*(SQRT 2)/2<-100)
    (YCOR+:SIDE>100)) STOP
  CHEVRON :SIDE
  FORWARD :SIDE
  CHEVRONSTRIP :SIDE
END
```
(In LCSI Logo, replace IF XCOR+:SIDE>120 TOPLEVEL , with IF XCOR+:SIDE>120 [THROW "TOPLEVEL] , and replace IF (ANYOF (XCOR-:SIDE<-120) (XCOR+:SIDE>120) (YCOR-:SIDE*(SQRT 2)/2<-100) (YCOR+:SIDE>100)) STOP with IF (OR (XCOR-:SIDE<-120) (XCOR+:SIDE>120) (YCOR-:SIDE*(SQRT 2)/2<-100) (YCOR+:SIDE>100)) [STOP].)
```
TO CHEVRON :SIDE
  FORWARD :SIDE RIGHT 135
  FORWARD :SIDE*(SQRT 2)/2 LEFT 90
  FORWARD :SIDE*(SQRT 2)/2
    RIGHT 135
  FORWARD :SIDE RIGHT 45
  FORWARD :SIDE*(SQRT 2)/2
    RIGHT 90
  FORWARD :SIDE*(SQRT 2)/2
    RIGHT 45
END
```

## Chapter 12 Test

1. **(a)**    **(b)**

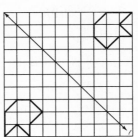 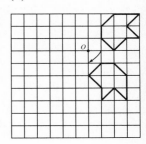

**(c)**

2. Hint: Find the images of the vertices.
3. **(a)** 4 **(b)** 1 **(c)** 1 **(d)** None **(e)** 2 **(f)** 2
4. **(a)** Line and rotational **(c)** Line
5. **(c)** 3
9. Hint: $\overline{BC}$ formed by the midpoints of $\overline{A'B'}$ and $\overline{A'C'}$.
10. In each part half turn about X.
11. **(b)** Answers may vary. Reflection in $\overleftrightarrow{BU}$ followed by a rotation about the center of the hexagon by 60° counterclockwise.
13. If △SER was the image of △HOR under a succession of isometries with a size transformation, then △SER would be similar to △HOR which in turn would imply that ER/OR = SR/HR. However, the last equation is false because $\frac{ER}{OR} = \frac{2}{3}$ and $\frac{SR}{HR} = \frac{5}{5}$. This contradiction implies that △SER is not the image of △HOR under any succession of isometries with size transformation. Notice that in a similar way we could show that △SER is not the image of △OHR under such transformations.

# CHAPTER 13

### Problem Set 13-1

1. **(a)** 9 mm, or 0.9 cm **(c)** 80 mm, or 8 cm **(e)** 7 mm, or 0.7 cm **(g)** 73 mm, or 7.3 cm
2. **(a)** $2\frac{7}{9}$ **(c)** 100
5. **(a)** Centimeters **(c)** Centimeters **(e)** Centimeters **(g)** Centimeters
7. **(a)** 0.35, 350 **(c)** 0.035, 3.5 **(e)** 200, 2000
8. **(a)** 10.00 **(c)** 10.0 **(e)** 195.0 **(g)** 40.0
11. **(a)** 8 cm **(c)** 9 cm
```

12. (a) 1 cm (c) 0.262 km (e) 0.03 m (g) 3500 cm (i) 0.1 cm (k) 1 mm (m) 5130 cm
14. (a) Can be (c) Cannot be
18. (a) Answers vary. (c) 20 squares
19. (a) 6 cm (c) $0.335/\pi$ m
20. (a) 6π cm (c) 4 cm
22. πr
24. (a) 2:1 (c) If $\triangle ABC \sim \triangle A'B'C'$, with $A'B'/AB = r$, then perimeter $\triangle A'B'C'$/perimeter $\triangle ABC = (rAB + rBC + rAC)/(AB + BC + AC) = r(AB + BC + AC)/(AB + BC + AC) = r$.
25. (a) About $9.5 \cdot 10^{12}$ km (c) About $6.9 \cdot 10^8$ k, or about 78,000 years
26. (a) 3096 (c) Approximately Mach 4.04
27. $(50 + 6\pi)$ ft
28. (a) 6 foot longs

Problem Set 13-2
1. (a) cm^2, in.^{2} (c) cm^2, in.^{2} (e) m^2, yd^2
5. (a) 444.4 yd^2 (c) 6400 acres
6. (a) 4900 m^2 (c) 0.98 ha
7. (a) 3 sq. units (c) 2 sq. units (e) 6 sq. units
8. They all check; i.e., $1 + 1/2B - 1 = A$
9. (a) 20 cm^2 (c) 7.5 m^2 (e) 600 cm^2
11. (a) 9 cm^2 (c) $(2\sqrt{21} - 2\sqrt{5})$ cm^2 (e) 84 cm^2
12. (a) (i) 1.95 km^2 (ii) 195 ha
13. (a) True (c) Don't know
14. (a) 75 cm^2
15. $(1/2)ab$
16. (a) $405.11
17. (a) 25π cm^2 (c) $(18/5)\pi$ cm^2 (e) 100 cm^2
18. 1200 tiles
20. (a) (i) 12, 10 (ii) 14, 10 (iii) 62, 22 (c) $2(n + 1)$
21. (a) $24\sqrt{3}$ cm^2
22. (a) 16π cm^2
23. (a) 2π cm^2 (c) 2π cm^2 (e) $(400 - 100\pi)$ cm^2 (g) $(1/8)\pi r^2$
25. (a) About 21.46%
28. (a) 4:1
31. The new figure is a parallelogram that has twice the area of the trapezoid. The area of the parallelogram is $A = (1/2)(AB + DC)h$, where h is the height of the parallelogram. Thus the area of the trapezoid is $(1/2)(AB + DC)h$.
34. $(320 + 64\pi)$ m^2
36. Draw altitudes $\overline{BE}$ and $\overline{DF}$ of triangles BCP and DCP, respectively. $\triangle ABE \cong \triangle CDF$ by AAS. Thus $\overline{BE} \cong \overline{DF}$. Because $\overline{CP}$ is a base of $\triangle BCP$ and $\triangle DCP$, and because the heights are the same, the areas must be equal.
37. (a) 10 (c) 0.35 (e) 8000 m
38. (a) $(2\pi + 4)$ cm

Problem Set 13-3
1. (a) 6 (c) 5a (e) $(s\sqrt{3})/2$ (g) 9 (i) $2\sqrt{2}$ (k) $3\sqrt{3}$
3. (a) No (c) Yes (e) Yes
4. $\sqrt{450}$, or $15\sqrt{2}$ cm
5. (a) $x = 8$, $y = 2\sqrt{3}$
6. $20\sqrt{13}$ km
8. $\sqrt{125}$ mi or about 11.2 mi
10. $6\sqrt{6}$ ft, or about 14.7 ft
11. (a) $37.5\sqrt{3}$ cm^2
13. (a) $(s^2\sqrt{3})/4$
15. 12.5 cm; 15 cm
17. $c/\sqrt{2}$, or $(c\sqrt{2})/2$
20. $\triangle ACD \sim \triangle ABC$; $AC/AB = AD/AC$ implies $b/c = x/b$, which implies $b^2 = cx$. $\triangle BCD \sim \triangle ABC$; $AB/CB = CB/DB$ implies $c/a = a/y$, which implies $a^2 = cy$, $a^2 + b^2 = cx + cy = c(x + y) = cc = c^2$
22. Yes
26. 8 cm
28. $90\sqrt{2}$ ft or about 127.28 ft
30. 0.032 km, 322 cm, 3.2 m, 3.020 mm
31. (a) 33.25 cm^2 (c) 32 m^2
32. (a) 10 cm, 10π cm, 25π cm^2 (c) $\sqrt{17}$ m, $2\sqrt{17}$ m, $2\pi\sqrt{17}$ m

Problem Set 13-4
1. (a) 96 cm^2 (c) 236 cm^2 (e) 24π cm^2 (g) 5600 ft^2 (i) $(32\pi + 16\pi\sqrt{5})$ cm^2
3. 2688π m^2
5. 4:9
6. (a) They have equal lateral surface areas.
9. (a) The surface area is multiplied by 4. (c) The surface area is multiplied by k^2
10. (a) Lateral surface area is tripled. (c) It is 9 times the original area.
12. (a) 44
16. (a) 1.5π m^2
17. (a) 202.5 cm^2
19. 720 ft by 162 ft
21. (a) $100\pi(1 + \sqrt{5})$ cm^2 (c) 2250π cm^2
22. (a) Approximately 42 cm
23. 15.2 cm
25. 375π cm^2
26. (a) 100,000 (c) 500 (e) 1
28. $20\sqrt{5}$ cm
30. Length of side = 25 cm
Diagonal $\overline{BD}$ is 30 cm.

Problem Set 13-5
1. (a) 8000 (c) 0.000675 (e) 7 (g) 0.00857 (i) 345.6
3. (a) 64 cm^3 (c) 216 cm^3 (e) 21π cm^3 (g) $(4000/3)\pi$ cm^3 (i) $(20,000/3)\pi$ ft^3 (k) $(256/3)\pi$ cm^3
4. (a) 2000, 2, 2000 (c) 1500, 1.5, 1500 (e) 0.75, 0.75, 750
5. (a) 200.0 (c) 1.0
6. 1680π mm^3
8. It is multiplied by 8.
9. (a) 2000, 2, 2 (c) 2 dm, 4000, 4
10. $253,500\pi$ L
12. 1.62 L
14. 32.4 L

16. (a) 25π L
19. Larger; The volume of the larger melon is 1.728 times the volume of the smaller, but is only $1\frac{1}{2}$-times as expensive.
21. They're equal.
23. About 21.5%.
25. $(2/3)\sqrt{2/(5\sqrt{5})}$ m^3, or approximately 0.28 m^3
26. (a) 512,000 cm^3
27. Let two square pyramids have sides and heights of s, h, and s_1 and h_1. If pyramids are similar, then $s/s_1 = h/h_1 = r$. Hence $V/V_1 = ((1/3)s^2h)/((1/3)s_1^2h_1) = (s/s_1)^2 \cdot (h/h_1) = r^2r = r^3$.
29. Approximately 45.90 in.^{3}
31. 128 ft^3
32. It won't hold the cream at 10 cm tall; it would have to be 20 cm tall.
34. (a) 15,600 cm^2 (c) $(1649 + (81\sqrt{3})/2)$ m^2, or about 1719.1 m^2
35. (a) 340 cm
37. 62 cm

Problem Set 13-6
1. (a) Kilograms or tons (c) Grams (e) Grams (g) Tons (i) Grams or kilograms
2. (a) Milligrams (c) Milligrams (e) Grams
3. (a) 15 (c) 36 (e) 4.230 (g) 5750 (i) 30 (k) 1.56 (m) 60.8
4. (a) No (c) Yes (e) Yes
5. 16 kg
7. 2¢
9. (a) $^-12°$ C (c) $^-1°$ C (e) 100° C
10. (a) No (c) No (e) No (g) Yes (i) Hot
11. (a) 50° F (c) 86° F (e) 414° F
12. (a) $(20 + 6\pi)$ cm; $(48 + 18\pi)$ cm^2 (c) 50 m; 80 m^2
13. (a) 35 (c) 400,000 (e) 5200
14. (a) Yes (b) No (c) Yes (d) No
15. $\sqrt{61}$ km
16. (a) 12,000π cm^3; 2400π cm^2

Problem Set 13-7
1. (a) It draws a circle five times. The circumference is 1/5 of the circumference of CIRCLE1.
3. (a)
```
TO ARC :S :D
  REPEAT :D [FD :S RT1]
END
```
5. (a)
```
TO CIRCS :RAD
  HT
  CIRCLE :RAD
  ARCRAD :RAD/2 180
  LARCRAD :RAD/2 180
END

TO CIRCLE :R
  VCIRCLE 2*3.14159*:R/360
END

TO VCIRCLE :S
  REPEAT 360 [FD :S RT 1]
END

TO ARCRAD :R :D
  REPEAT :D [FD 2*3.14159*:R/360 RT 1]
END

TO LARCRAD :R :D
  REPEAT :D [FD 2*3.14159*:R/360 LT 1]
END
```
(c)
```
TO SEMIS :R
  ARCRAD :R 180
  RT 90 FD :R*2 RT 90
  ARCRAD :R/2 180
  RT 180
  ARCRAD :R/2 180
END

TO ARCRAD :R :D
  REPEAT :D [FD 2*3.14159*:R/360 RT 1]
END
```
(e)
```
TO FRAME :SIZE
  SQUARE :SIZE
  FD :SIZE/2
  CIRCLE :SIZE/2
END

TO SQUARE :S
  REPEAT 4[FD :S RT 90]
END
```
6.
```
TO SYMBOL :R
  PU LT 90 FD 140 RT 90 PD
  REPEAT 3 [CIRCLE :R PU RT 90 FD 9*:R/4 LT 90 PD]
  PU LT 90 FD :R*45/8 LT 90 FD :R RT 180 PD
  REPEAT 2 [CIRCLE :R PU RT 90 FD 9*:R/4 LT 90 PD]
  HIDETURTLE
END

TO CIRCLE :R
  HT
  VCIRCLE 2*3.14159*:R/360
  ST
END

TO VCIRCLE :S
  REPEAT 360 [FD :S RT 1]
END
```
8.
```
TO DIAMCIRC :R
  REPEAT 360 [FD 2*3.14159*:R/360 RT 1]
  RT 90
  FD 2*:R
END
```

Chapter 13 Test
1. (a) 50,000, 5000, 50 (c) 26,000,000, 260,000, 260
2. (a) Millimeters (c) Millimeters (e) Centimeters
4. 16
5. (a) 8 1/2 cm^2 (c) 7 cm^2
8. (a) $54\sqrt{3}$ cm^2
9. (a) 12π cm^2 (c) 24 cm^2 (e) 178.5 m^2
11. (a) SA = $32(2 + \sqrt{13})$ cm^2. Vol = 128 cm^3.

(c) SA = 100π m²
    Vol = $(500\pi)/3$ m³
(e) SA = 304 m²
    Vol = 320 m³
12. 65π m²
14. (a) Metric tons (c) 1 g (e) 25
   (g) 51,800 (i) 50,000 (k) 25,000 (m) 52.813
15. (a) 16.7 (c) 1089 (e) 486 (g) 60.8 (i) 82.4
17. (a) 6000 kg (c) $r = \sqrt[3]{9/(4\pi)}$ m
18. (a) L (c) g (e) kg (g) mL
19. (a) Unlikely (c) Unlikely (e) Unlikely
20. (a) 2000 (c) 3 (e) 0.0002

CHAPTER 14

Problem Set 14-1

2. (b) (2, ⁻3); answers may vary.
3. (a) I (c) II (e) Between I and II
4. Quadrant I = $\{(x, y) | x > 0 \text{ and } y > 0\}$
   Quadrant II = $\{(x, y) | x < 0 \text{ and } y > 0\}$
   Quadrant III = $\{(x, y) | x < 0 \text{ and } y < 0\}$
   Quadrant IV = $\{(x, y) | x > 0 \text{ and } y < 0\}$
5. $D(4, ⁻2)$
7. (a) $P(3, 4); Q(6, 1)$
8. (a) $P'(2, ⁻2); Q'(2, ⁻5); R'(4, ⁻2)$ (c) $P'(4, 2);$
   $Q'(4, 5); R'(6, 2)$
9. (a) (0, ⁻1), (1, 0), (2, ⁻4), (⁻2, ⁻4), (⁻2, 4),
   (2, 4) (c) (⁻1, 0), (0, 1), (⁻4, 2), (⁻4, ⁻2),
   (4, ⁻2), (4, 2) (e) (0, ⁻3), (1, ⁻4), (2, 0), (⁻2, 0),
   (⁻2, ⁻8), (2, ⁻8)
10. (a) $A'(⁻2, ⁻5); B'(2, ⁻6); C'(5, ⁻1)$
    (c) $A'(2, ⁻5); B'(⁻2, ⁻6); C'(⁻5, ⁻1)$
11. (a) (⁻2, ⁻4)
12. $A'(0, 1); B'(2, 2); C'(1, 3); D'(⁻1, 3); E'(b, a)$
13. (a) $(a - b, 0)$
14. (a) 4 (c) 5 (e) $\sqrt{52}$, or $2\sqrt{13}$ (g) $\sqrt{365}/4$,
    or approx. 4.78 (i) 5
16. The sides have lengths $\sqrt{45}, \sqrt{180}$, and $\sqrt{225}$.
    Since $(\sqrt{45})^2 + (\sqrt{180})^2 = (\sqrt{225})^2$, the triangle is a
    right triangle.
18. $x = 9$ or ⁻7
19. (a) (0, 5) (c) (2, ⁻1.2)
20. (⁻7, 11)
21. (a) (⁻2, 3), (2, 1), (0, 4)
22. (a) $(x - 3)^2 + (y + 2)^2 = 4$ (c) $(x + 1)^2 + y^2 = 4$
23. (a) Exterior (c) Exterior (e) Exterior
    (g) Exterior
24. $x^2 + y^2 = 34$
25. (a) $(x - 4)^2 + (y + 3)^2 = 25$
27. (a)   (c)

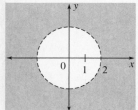
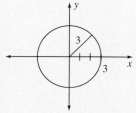

(e) This equation has no graph or the graph is the empty set.
31. If M is equidistant from the vertices, M has coordinates $(a/2, b/2)$. $BM = \sqrt{(a/2)^2 + (b - b/2)^2} = \sqrt{(a/2)^2 + (b/2)^2} = \sqrt{((a^2 + b^2)/4)}$, or $\sqrt{a^2 + b^2}/2$
$AM = \sqrt{(a/2)^2 + (b/2)^2} = \sqrt{((a^2 + b^2)/4)} = \sqrt{a^2 + b^2}/2$.
Therefore, $AM = BM$.
$OM = \sqrt{(a/2)^2 + (b/2)^2} = \sqrt{a^2 + b^2}/2$. Therefore, $OM = AM = BM$.

Problem Set 14-2

1.

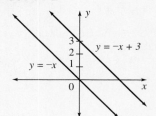

3. (a)   (c)

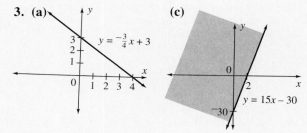

(e)

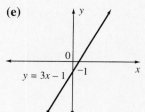

5. (a)

6. (a) $y = \dfrac{1}{3x}$ (c) $y = \dfrac{⁻4x}{3} + 4$ (e) $y = \dfrac{1}{3}x$
7. (a) $\dfrac{1}{3}$ (c) 0 (e) 20,000
8. (a) $y = ⁻x - 1$ (c) $y = 1$ (e) $y = x - \dfrac{1}{2}$
9. (a) $y = -\dfrac{1}{2}x - \dfrac{3}{2}$ (c) $y = ⁻3$

ANSWERS TO SELECTED PROBLEMS   891

10. (a) Parallel (c) Parallel
11. (a) $y = {}^-2x - 1$ (c) $x = {}^-2$ (e) $x = {}^-2$
(g) $y = {}^-x + 1$
12. (a) Perpendicular (c) Perpendicular
15. All three determine a line of slope 3.
17. (a) $y = {}^-3x - 1$ (c) $y = \frac{1}{3}x - \frac{1}{3}$
18. (a) $y = {}^-x$
19. (a) Yes
22. (a) $x = {}^-2$; y is any real number. (c) $x > 0$ and $y < 0$; x and y are real numbers.
24. (a)   (c)

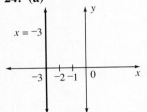

(e)

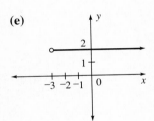

25. (a) $x = 3$ (c) $y = 5$
26. (a)

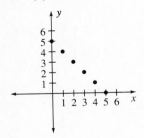

(c)

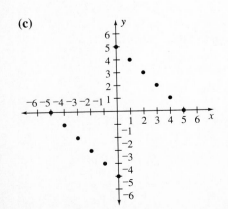

27. (a)

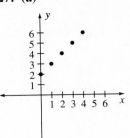

(c)

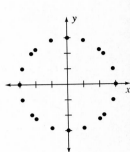

(e)

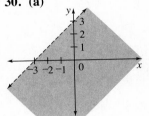

28. (a) $\{(x, y) | {}^-1 \leq x \leq 1 \text{ and } {}^-1 \leq y \leq 1\}$
(c) $\{(x, y) | {}^-1 \leq x \leq 2 \text{ and } {}^-1 \leq y \leq 1\}$
29. (a) $x = 3$ (c) $y = 3$
30. (a)   (c)

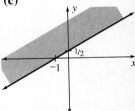

31. (a)

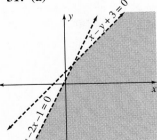

(c)

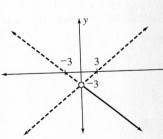

32. (a) $y = {}^-x$ (c) $x = 0$ (e) $y = x$ (g) $y = x + 3$
(i) One is reflection of the other in the x-axis.
33. (a) $y = {}^-x + 1$ (c) $y = x + 1$

34. (a)

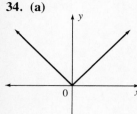

(c)

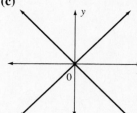

36. (a) $y = {}^-8$
37. (a) Answers vary; $(3, {}^-2)$ and $(7, {}^-2)$
38. There are three possible locations for D; $(4, {}^-8)$, $(12, 0)$ or $(0, 12)$.
39. (a) $\frac{3}{2}$ sq. units **(c)** $\frac{15}{2}$ sq. units **(e)** 3 sq. units

Problem Set 14-3
1. (a) Answers vary; $(0, \frac{-5}{3})$; $(\frac{5}{2}, 0)$; $(1, {}^-1)$; $(2, -\frac{1}{3})$
(c)

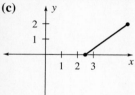

2. (a) $(2, 5)$, unique solution **(c)** $(1, {}^-5)$, unique solution **(e)** $(0, 0)$, unique solution
5. (a) Unique solution **(c)** Infinitely many solutions; same line
7. $y = \frac{1}{3}x - 4$ and $y = -\frac{1}{2}x - 4$
9. (a) $(\frac{17}{7}, \frac{17}{7})$
10. $\frac{55}{72}$ and $-\frac{1}{72}$
12. $133\frac{1}{3}$ lb cashew nut granola and $66\frac{2}{3}$ lb golden granola
14. $20,000 and $60,000, respectively
15. (a) $2000
16. Width = 60 in.; length 75 in.
20. (a) is equation **(c)** is not an equation
21. (a) Slope $-\frac{5}{6}$, y-intercept $\frac{7}{6}$ **(c)** Slope 3.75, y-intercept 1.85
22. (a) $y = \frac{9}{10}x + \frac{17}{5}$ **(c)** $y = {}^-8$
23. (a)

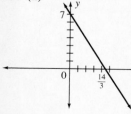

(c)

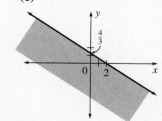

24. (a) $\frac{1}{2}$ **(c)** 0
25. $y = 2$

Problem Set 14-4
1. The answer will vary depending upon the version of Logo used.
3. TO AXES
    SETXY 0 120
    SETXY 0 (-120)
    SETXY 0 0
    SETXY 130 0
    SETXY -130 0
    SETXY 0 0
END
(In LCSI, use SETPOS LIST instead of SETXY.)
5. TO FILL.RECT
    REPEAT 50 [SETY 30 SETY 0 RT 90 FD 1 LT 90]
END
7. TO QUAD :X1 :Y1 :X2 :Y2 :X3 :Y3 :X4 :Y4
    PU SETXY :X1 :Y1 PD
    SETXY :X2 :Y2
    SETXY :X3 :Y3
    SETXY :X4 :Y4
    SETXY :X1 :Y1
END
(In LCSI, use SETPOS LIST instead of SETXY.)
9. Use the QUAD procedure from problem 7 and the following:
TO MEDIAL.QUADS :NUM :X1 :Y1 :X2
    :Y2 :X3 :Y3 :X4 :Y4
    IF :NUM = 0 STOP
    QUAD :X1 :Y1 :X2 :Y2 :X3 :Y3 :X4
    :Y4
    MEDIAL.QUADS :NUM - 1 (:X1 +
        :X2)/2 (:Y1 + :Y2)/2 (:X2 +
        :X3)/2 (:Y2 + :Y3)/2 (:X3 +
        :X4)/2 (:Y3 + :Y4)/2 (:X4 +
        :X1)/2 (:Y4 + :Y1)/2
END
(In LCSI, replace STOP with [STOP] and use SETPOS LIST instead of SETXY.)
11. TO R.ISOS.TRI :LEN
    FD :LEN
    RT 90
    FD :LEN
    RT 135
    FD (SQRT 2)*:LEN
END
13. TO GENCIRC :X :Y :R
    PU
    SETXY :X :Y
    PD
    REPEAT 360 [FD 2*3.14159*:R/360 RT 1]
END
(In LCSI, use SETPOS LIST instead of SETXY.)

Chapter 14 Test
1. 16
3. (a) 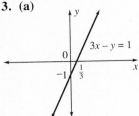   (c)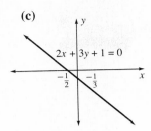

4. (a) $y = -\frac{4}{3}x - \frac{1}{3}$ (c) $y = 3$
5. (a) $y = \frac{4}{3}x + \frac{7}{3}$ (c) $(\frac{3}{4}, 5)$
6. $y = \frac{3}{2}x - \frac{1}{3}$
8. 80 regular and 30 deluxe
9. (a) $(1, {}^-1/2)$
10. (a) 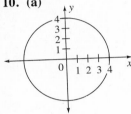   (c)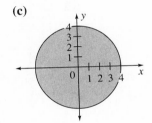

11. $(x + 3)^2 + (y - 4)^2 = 25$
12. (a)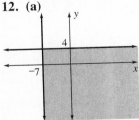

14. 275 freshmen and 500 sophomores
16. (a) $\left(\frac{5}{2}, \frac{3}{2}\right)$
    (c) $\sqrt{(5/2 - 0)^2 + (3/2 - 0)^2} = \sqrt{34}/2$
    $\sqrt{(5/2 - 1)^2 + (3/2 - 4)^2} = \sqrt{34}/2$
    $\sqrt{(5/2 - 5)^2 + (3/2 - 0)^2} = \sqrt{34}/2$
17. $(3/2, 3/2)$
19. $(2/5)\sqrt{5}$
20. (a) $(5, 3)$ (c) $(5, 0, 8)$ (e) 7

APPENDIX I

Problem Set A1
1. (a), (b), (d), (e), and (g)

2. (a) $X \wedge 2 + Y \wedge 2 - 3*Z$ (c) $A + B - C \wedge 2/D$
   (e) $15/(A*(2*B \wedge 2 + 5))$
3. (a) 7 (c) 6 (e) 16
4. (a) 35,200,000 (c) $^-0.000001233$
6. (a)
```
5 REM WE INPUT THE VALUE OF THE
  VARIABLE X
10 INPUT X
20 LET Y = 13*X ∧ 5 - 27/X + 3
30 PRINT "WHEN X = ";X;", Y = "; Y
40 END
```
   (b) (i)
```
RUN
? 1.873
WHEN X = 1.873, Y = 288.247015
```
   (ii)
```
RUN
? 7
WHEN X = 7, Y = 218490.143
```
7. (a)
```
10 REM THIS PROGRAM CONVERTS
   DEGREES FAHRENHEIT
20 REM TO DEGREES CELSIUS
30 PRINT "THE NUMBER OF DEGREES
   FAHRENHEIT IS ";
40 INPUT F
50 LET C = 5/9*(F - 32)
60 PRINT F;" DEGREES
   FAHRENHEIT = ";C;" DEGREES
   CELSIUS"
70 END
```
9. (a)
```
PRINT 100*1.18 ∧ 25
6266.8628
```
   (b)
```
5 REM N IS THE NUMBER OF YEARS
  THE MONEY IS INVESTED
10 INPUT N
15 REM B IS THE BALANCE
20 LET B = 100*1.18 ∧ N
30 PRINT "AFTER ";N;" YEARS,
   THE BALANCE IS $";B
40 END
```
   (The balance in (b) is the same as in (a).)
11. (a) HEY YOU OUT THERE
12. (a)
```
10 FOR N = 1 TO 20
20 PRINT 2*N
30 NEXT N
40 END
```
   (b)
```
10 FOR N = 1 TO 20
20 PRINT 2*N " ";
30 NEXT N
40 END
```
13.
```
5 REM THIS PROGRAM COMPUTES THE
  SUM OF THE SQUARES OF THE
6 REM FIRST 100 POSITIVE
  INTEGERS
7 REM N IS A POSITIVE INTEGER
10 FOR N = 1 TO 100
15 REM Y IS USED TO ACCUMULATE
20 LET Y = Y + N*N
30 NEXT N
40 PRINT Y
50 END
5050
```

17.
```
10 REM THIS PROGRAM CONVERTS
      DEGREES FAHRENHEIT
20 REM TO DEGREES CELSIUS
25 REM F REPRESENTS A NUMBER OF
      DEGREES FAHRENHEIT
26 REM C REPRESENTS A NUMBER OF
      DEGREES CELSIUS
30 PRINT "DEGREE FAHRENHEIT",
      "DEGREE CELSIUS"
40 FOR F = -40 TO 220 STEP 10
50 LET C = 5/9*(F - 32)
60 PRINT F,,C
70 NEXT F
80 END
```

19.
```
10 REM THIS PROGRAM CALCULATES N!
15 REM N IS A NATURAL NUMBER
16 REM T IS USED TO ACCUMULATE
      PRODUCTS
17 REM A IS USED AS A COUNTER
20 PRINT "WHAT IS THE VALUE OF N";
30 INPUT N
40 LET T = 1
50 FOR A = 1 TO N
60 LET T = T*A
70 NEXT A
80 PRINT "N", "N!"
90 PRINT N, T
100 END
```

20.
```
10 REM LAURA'S LOTTERY
11 REM P IS THE PRIZE
12 REM A IS THE AMOUNT ACCUMULATED
13 REM N IS A COUNTER
14 REM P IS THE DIFFERENCE IN A
      AND $10000
15 LET N = 0
20 LET P = 50000
30 LET A = P*(1 + .15)
40 LET N = N + 1
50 LET P = A - 10000
60 IF P <= 0 THEN 80
70 GOTO 30
80 PRINT "YOU'RE OUT OF MONEY."
90 PRINT "IT LASTED ";N - 1;"
      YEARS."
100 END
```

```
YOU'RE OUT OF MONEY.
IT LASTED 9 YEARS.
```

22.
```
10 REM THIS PROGRAM COMPUTES THE
      SUM OF THE
20 REM FIRST 1000 ODD NATURAL
      NUMBERS
21 REM A IS A NATURAL NUMBER
22 REM N DETERMINES ODD NATURAL
      NUMBERS
23 REM Y ACCUMULATES THE SUM OF
      THE ODD NATURAL NUMBERS
24 LET Y = 0
30 FOR A = 1 TO 1000
40 LET N = 2*A - 1
50 LET Y = Y + N
60 NEXT A
70 PRINT "THE SUM OF THE FIRST
      1000 ODD NATURAL ";
80 PRINT "NUMBERS IS ";Y
90 END
```

24.
```
10 REM PLASTIC CARD COMPANY
      BILLS
11 REM O1 IS THE AMOUNT FOR
      FIRST METHOD
12 REM O2 IS THE AMOUNT FOR
      SECOND METHOD
13 REM O3 IS THE AMOUNT FOR
      THIRD METHOD
20 PRINT "WHAT IS THE CUSTOMER
      ID NUMBER";
30 INPUT I
40 PRINT "WHAT IS THE BILL";
50 INPUT B
60 LET O1 = B - 5/100*B
70 LET O2 = B
80 LET O3 = B + 2/100*B
90 PRINT "ID#","OPTION
      1","OPTION 2","OPTION 3"
100 PRINT I,O1,O2,O3
110 END
```

26.
```
10 REM THIS PROGRAM PRINTS
      INTEREST COMPOUNDED DAILY
15 REM N IS A COUNTER
16 REM B IS THE AMOUNT ACCUMULATED
17 LET N = 0
20 LET N = N + 1
30 LET B = 1000*(1 + 5/36500)^N
40 IF B > 5000 THEN 60
50 GOTO 20
60 PRINT "AFTER ";N;" DAYS, THE
      BALANCE EXCEEDS $5000."
70 END
```

```
AFTER 11750 DAYS, THE BALANCE
EXCEEDS $5000.
```

28.
```
10 REM THIS PROGRAM OUTPUTS
      BATTING AVERAGES
11 REM K COUNTS THE PLAYERS
12 REM A IS THE BATTING AVERAGE
13 REM R IS THE BATTING AVERAGE IN
      THOUSANDTHS
15 LET K = 0
20 PRINT "HOW MANY PLAYERS ARE
      THERE?";
30 INPUT N
40 PRINT "PLAYER #","# OF BATS","#
      OF HITS","BATTING AVERGE"
50 LET K = K + 1
60 PRINT "AFTER THE QUESTION MARK,
      TYPE THE # OF AT BATS,"
70 PRINT "THE # OF HITS OF
      PLAYER ";K;" SEPARATED BY
      COMMAS."
80 INPUT B,H
```

ANSWERS TO SELECTED PROBLEMS 895

```
 90 LET A = H/B
100 LET R = INT(1000*A)
105 LET R = R/1000
110 PRINT K,B,H,R
120 IF K < N THEN 50
130 END
```

30.
```
10 REM SUM OF TWO SQUARES
11 REM X AND Y REPRESENT
   POSSIBLE NUMBERS
12 REM Z IS THE SUM OF X^2 AND
   Y^2
20 FOR X = 1 TO 6
30 FOR Y = 1 TO 6
40 LET Z = X^2 + Y^2
50 IF X >= Y THEN 70
60 IF Z < 40 THEN 100
70 NEXT Y
80 NEXT X
90 GOTO 120
100 PRINT Z;"=";X;"^2 +";Y;"^2"
110 GOTO 70
120 END
```

$5 = 1^2 + 2^2$
$10 = 1^2 + 3^2$
$17 = 1^2 + 4^2$
$26 = 1^2 + 5^2$
$37 = 1^2 + 6^2$
$13 = 2^2 + 3^2$
$20 = 2^2 + 4^2$
$29 = 2^2 + 5^2$
$25 = 3^2 + 4^2$
$34 = 3^2 + 5^2$

APPENDIX II

Problem Set AII-1

1. (a)   (c)   (e) 0   (g)

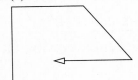
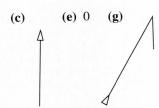

3. (a)   (c)

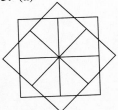
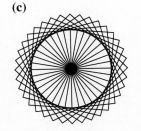

4. Answers may vary.
   (a) TO RECT
       REPEAT 2[FORWARD 30 RIGHT 90
         FORWARD 60 RIGHT 90]
       END
   (c) TO HAT
       REPEAT 2 [FORWARD 60 RIGHT 90
         FORWARD 30 RIGHT 90]
       PENUP LEFT 90 FORWARD 30
       PENDOWN
       REPEAT 2[LEFT 90 FORWARD 6
         LEFT 90 FORWARD 90]
       END
   (d) TO T
       FORWARD 60 LEFT 90
       FORWARD 30 BACK 60
       END
   (e) TO RHOMBUS
       RIGHT 20
       REPEAT 2[FD 40 RT 70 FD 40 RT
       110]
       END

6. Answers may vary
   TO SQUARE1
     REPEAT 4[FORWARD 50 RIGHT 90]
   END
   TO TRIANGLE1
     REPEAT 3[FORWARD 50 RIGHT 120]
   END
   (a) TO SQUARE.PILE
         REPEAT 4[SQUARE1 RIGHT 90]
       END
   (c) TO RECT1
         REPEAT 2[FORWARD 60 RIGHT 90
           FORWARD 30 RIGHT 90]
       END
       TO RECL.SWIRL
         REPEAT 4[RECT1 LEFT 90]
       END
   (e) TO STAR
         RIGHT 30
         REPEAT 4[TRIANGLE1 RIGHT 60
           FORWARD 50 RIGHT 30]
       END

7. Answers may vary.
   (b) TO BUILD.SQR :S
         SQUARE :S
         SQUARE :S + 10
         SQUARE :S + 20
         SQUARE :S + 30
         SQUARE :S + 40
       END
   (d) TO TOWER :S
         SQUARE :S
         FORWARD :S RIGHT 90 FORWARD
           :S/4 LEFT 90
         SQUARE :S/2
         FORWARD :S/2 RIGHT 90 FORWARD
           :S/8 LEFT 90 SQUARE :S/4
       END

9. TO KITE
     LEFT 45
     REPEAT 4[FORWARD 40 RIGHT 90]
     RIGHT 45
     REPEAT 3[BACK 20 K.TAIL RIGHT
       60]
     BACK 20
     END

     TO K.TAIL
       RIGHT 60
       REPEAT 3[FORWARD 10 RIGHT 120]
       LEFT 120
       REPEAT 3[FORWARD 10 LEFT 120]
       END

12. TO RECTANGLES :S
     LEFT 90
     REPEAT 4[RECTANGLE :S/3 :S
       RIGHT 90 FORWARD :S/3]
     END
     TO RECTANGLE :S1 :S2
       REPEAT 2[FORWARD :S1 RIGHT 90
         FORWARD :S2 RIGHT 90]
     END

Problem Set AII-2

3. Answers may vary.
   (a) TO STRETCH :S
       IF :S<5 STOP
       SQUARE :S
       FORWARD :S RIGHT 90
       FORWARD :S LEFT 90
       STRETCH :S − 10
       END
       (In LCSI, replace IF :S<5 STOP *with* IF :S<5 [STOP].*)*
   (c) TO PISA :S :A
       IF :S<5 STOP
       SQUARE :S
       FORWARD :S LEFT :A
       PISA :S*0.75 :A
       END
       (In LCSI, replace IF :S<5 STOP *with* IF :S<5 [STOP].*)*

   (e) TO ROW.HOUSE :S
       IF :S<5 STOP
       HOUSE :S
       SETUP :S
       ROW.HOUSE :S/2
       END
       (In LCSI, replace IF :S<5 STOP *with* IF :S<5 [STOP].*)*

4. TO NEST.TRI :S
     IF :S<10 STOP
     RIGHT 30 TRIANGLE :S
     FD :S/2 RIGHT 30
     NEST.TRI :S/2
     END
     (In LCSI, replace IF :S<10 STOP *with* IF :S<10 [STOP].*)*
   TO TRIANGLE :S
     REPEAT 3[FORWARD :S RIGHT 120]
     END

7. Answers may vary.
   TO SPIN.SQ :S
     IF :S<5 STOP
     SQUARE :S
     RIGHT 20
     SPIN.SQ :S−5
     END
     (In LCSI, replace IF :S<5 STOP *with* IF :S<5 [STOP].*)*
   TO SQUARE :S
     REPEAT 4[FORWARD :S RIGHT 90]
     END

9. Answers may vary.
   TO SQ.TOWER :S
     IF :S<5 STOP
     SQUARE :S FORWARD :S
     SQ.TOWER :S/2
     SQUARE :S FORWARD :S
     END
     (In LCSI, replace IF :S<5 STOP *with* IF :S<5 [STOP].*)*
   TO SQUARE :S
     REPEAT 4[FORWARD :S RIGHT 90]
     END

Index

Abacus, 147
Abscissa, 766
Absolute value, 176
Abuses of statistics, 484
Acre, 699
Acute angle, 510
Acute triangle, 523
Addends, 112
  missing, 117
Addition
  algorithms, 134
  associative property, 115, 180
  BASIC, 815
  clock, 254
  closure, 113, 180
  commutative property, 114, 180
  decimals, 333
  different bases, 161, 162
  grouping property, 118
  identity element, 115, 180
  integers, 177
  mental, 138
  rational numbers, 276, 277, 278
  real numbers, 372
  whole numbers, 111
Addition property
  equality, 198, 281
  greater than, 199
  integers, 177
  rationals, 276
  real numbers, 372
Additive identity
  integers, 180
  rational numbers, 280
  real numbers, 372
  whole numbers, 115
Additive inverse
  decimal numbers, 372
  integers, 180
  rational numbers, 280
  real numbers, 372
Additive property, 106
Adjacent angles, 508
Al-jabr wa'l muqabalah, 201
al-Khowarizmi, Mohammed, 133, 176, 201
Algebra, 203, 305
Algebraic operating system, 38
Algorithm(s), 133
  addition, 134
  cashier's, 141
  division, 129, 130, 295
  Euclidean, 247, 248, 249
  multiplication, 148
  Russian Peasant, 155
  scratch addition, 137
  subtraction, 140
Alternate exterior angles, 532
Alternate interior angles, 532
Altitude
  cone, 548
  triangle, 591
Amicable numbers, 253
And, 60
  in compound statement, 85
Angle, 507
  acute, 510
  adjacent, 508
  alternate exterior, 532

alternate interior, 532
bisector, 589
central, 597
complementary, 530
congruent, 522
corresponding, 532
definition, 507
dihedral, 511
exterior, 520
included, 576
interior, 520, 532
measure, 508
obtuse, 510
right, 510
sides, 507
straight, 510
supplementary, 530
in a triangle, sum of, 533
vertex of, 507
vertical, 530
Angle Angle (AA), 606
Angle Angle Angle (AAA), 605, 606
Angle Angle Side (AAS), 582
Angle Side Angle (ASA), 581
Anleitung zur Algebra, 190
Annuities upon Lives, 475
Apex, 543
Apothem, 707
Applications of functions, 79
Arc, 571
  center, 571
  circle, 571
  length, 692
  major, 572
  minor, 571
  of a network, 554
  semicircle, 572
Archimedes, 692
Are, 700
Area, 698
  circle, 707
  lateral surface, 726
  metric units of, 699
  on geoboards, 700
  parallelogram, 702
  rectangle, 701
  regular n-gon, 707
  sector, 708
  square, 698
  surface, 726
  trapezoid, 704
  triangle, 703
Arithmetic
  clock, 254
  Fundamental Theorem of, 235
  mean, 458
  mental, 138
  modular, 256
  sequence, 7, 8
Arithmetica, 201, 221
Array model, 123
Arrow
  slide, 626
  turn, 630
Ars Magna, 201
Associative property
  addition, 115, 180, 372
  intersection, 62

multiplication, 125, 191, 372
union, 62
At random, 387
Average, 458
  mean, 458
  median, 459
  mode, 460

Babylonian numeration system, 106
BACK (BK), 834
Balance, 364
Balance point, 459
Bar graph, 446
  double, 447
Base
  cone, 548
  cylinder, 548
  of exponent, 107
  five, 161, 162, 163
  number, 158
  parallelogram, 702
  power, 107
  prism, 542
  pyramid, 543
  ten system, 106
  ten blocks, 134
  triangle, 703
  twelve, 161
  unit, 687
BASIC, 815
  arithmetic, 815
  variables, 816
BASIC programming commands, 817
  ABS, 815
  DATA, 825
  END, 817
  FOR-NEXT, 823
  GOTO, 821
  IF-THEN, 820
  INPUT, 817
  INT, 820
  LET, 817
  PRINT, 817
  READ, 825
  REM, 819
  RND, 820
  SQR, 815
  STEP, 823
BASIC system commands, 816, 817
  LIST, 816
  NEW, 817
  RUN, 816
Bernoulli, John, 74
Betweenness of points, 501
Biconditional, 90
Bimodal, 460
Binary operations, 111
Binary system, 160
Bisector
  angle, 589
  perpendicular, 590
Blocks, base-ten, 134
Bobrow, D., 836
Box plot, 464
Brahmagupta, 174
Brain Teasers
  airplane, 432

INDEX

bags of gold, 298
bike sale, 133
birthdays, 408
bridge, 648
calculator display, 214
circumference of Earth, 697
coins in boxes, 111
connect the dots, 516
cut square to rectangle, 714
cyclists, 310
day before yesterday, 36
diagonals of rectangle, 254
dice, 395
elephant-ant, 212
foreign languages, 288
four 4's, 195
four 7's, 340
Grandpa Elmer's birthday, 473
holiday array, 42
hotel, 83
Janna's walk, 318
letter patterns, 17
license plate, 145
Lyndon B. Johnson, 158
Making a Better Beer Glass, 552
Nine digit puzzle, 187
paper cup, 734
popcorn, 755
Primes, 259
Product, 197
racing cars, 470
rotating coins, 635
school committee meeting, 801
shaving soldiers, 58
size of Missoula, 69
social security number, 231
spider and fly, 724
sum of star's angles, 537
toymakers, 615
triangular property, 611
truth and lies, 88
unmagic square, 122
utilities, 529
woman's will, 302
Branching, 819
  conditional, 820

Calculator
  algebraic operating system, 38
  as problem solving tool, 37
  automatic constant, 39
  change-of-sign key, 39, 183
  CLEAR ENTRY key, 37
  CLEAR key, 37
  constant key, 39, 75, 107
  decimal notation, 38
  error indicator, 39
  fraction, 39
  fraction bar key, 39
  integer division key, 40
  logic, 37
  Math Explorer, 39
  percent key, 39
  reciprocal key, 37
  scientific, 37
  simplify key, 39, 271
  square root key, 41
  y to the x power key, 39, 40, 107
Cantor, Georg, 50
Cardano, Girolamo, 174

Cardinal number, 54
Cartesian coordinate system, 764
Cartesian product, 64, 65
  definition of multiplication, 124
  model, 124
Cashier's algorithm, 141
Cavalieri, Bonaventura, 739
Cavalieri's principle, 738, 739
Celsius, Anders, 750
Celsius, degree, 748
Center
  arc, 571
  circle, 526
  size transformation, 649
  turn, 629
Centi-, 687
Centimeter, 687
  cubic, 735
  square, 699
Central angle(s), 597
Central tendency, 457
Certain event, 388
Chain Rule, 94
Charged-field model, 178, 189
Chip model, 177, 189
Chudnovsky, 692
Chord(s), 597
Circle, 526, 597
  area of, 707
  center of, 526, 597
  central angle, 597
  chord of, 597
  circumference, 691
  circumscribing, 599
  concentric, 601
  definition, 526
  diameter, 597
  equation of, 768, 769
  graph, 451
  great, 600
  incribed, 600
  latitude, 600
  longitude, 601
  radius, 597
  secant, 597
  sector, 708
  semicircle, 572, 692
  tangent, 597
  turtle-type, 752
Circular cylinder, 548
Circumference, 691
Circumscribing triangles, 599
Class mark, 450
Classes
  data, 450
  equivalence, 72
Clock arithmetic, 254
Closed
  curve, 517
  curve, simple, 517
  path theorem, 534
  surface, 540
Closure property
  addition, 113, 180
  multiplication, 125, 191
Clusters, 440
Collinear points, 501
Combination, 427, 428
Commutative property
  addition, 114, 180
  intersection, 62
  multiplication, 125, 191
  union, 62

Comparison model, 118
Compass, 570, 571
Complement, 55
  of an angle, 531, 532
  of an event, 390
  set, 55
  set A relative to set B, 61
Complementary angles, 530
Components, 64
Composite(s), 232
Composition of functions, 77
Compound interest, 365
Compound statement, 85
Concentric circles, 601
Conclusion, 89
Concurrent lines, 503
Conditional, 88
Cone
  altitude, 548
  base, 548
  lateral surface, 548
  oblique circular, 548
  right circular, 548
  slant height, 729
  surface area, 728
  vertex, 548
  volume, 740
Congruence, 570
Congruent
  angles, 522, 576
  central angles, 597
  chords, 597
  modulo, 256
  parts, 522
  polygons, 583
  segments, 522
  squares, 570
  triangles, 572
Conjecture, 6
  Goldbach's, 240
Conjunction, 86
Constructions
  angle, 576
  angle bisector, 589
  arc, 571
  circle, 571
  circumscribed circle, 599
  inscribed circle, 600
  inscribed polygon, 598
  line segment, 572
  line segment bisector, 590
  Mira, 592
  paper folding, 592
  parallel lines, 587, 588
  perpendicular bisector, 590
  perpendicular at a point on a line, 591
  perpendicular from point to line, 590
  reflection, 637, 638, 639
  rotations, 639, 630, 631
  translations, 628, 629
  triangles, 575
Contrapositive, 89
CONTROL key, 821
CONTROL-C (CTRL-C), 821, 839
CONTROL-G (CTRL-G), 850
Converse, 89
  of the Pythagorean Theorem, 720
Convex
  polygon, 519
  polyhedron, 544
Coordinate system, 764, 765
Coplanar
  lines, 503

points, 503
Correlation
  negative, 481
  no, 481
  positive, 480
Correspondence, one-to-one, 52
Corresponding angles, 532
Counter(s), 821
  initializing, 822
Counterexample, 6, 62
Counting
  methods of, 424
  numbers, 51
  subsets, 58
Cube, 547
  surface area, 725
  volume, 728
Cubic
  centimeter, 735
  decimeter, 736
  foot, 737
  meter, 735
Curve(s)
  closed, 517
  Jordan Curve Theorem, 518
  polygonal, 517
  simple, 517
Cylinder, 548
  base, 548
  circular 548
  lateral surface, 548
  oblique, 548
  right, 548
  right circular, 548
  surface area, 727
  volume, 738, 739

Data, 439
  graphs, 439
  grouping, 450
DATA, 825
De Mere, Chevalier, 384
De Moivre, Abraham, 475
De Morgan's Laws, 88
Debugging, 842
Decagon, 520
Deci-, 687
decimal number(s), 328
  adding, 333
  additive identity, 372
  additive inverse, 372
  associative property, 372
  closure property, 372
  commutative property, 372
  distributive property, 372
  dividing, 335
  fractional notation, 330, 331
  irrational, 368
  multiplicative identity, 372
  multiplicative inverse, 372
  multiplying, 334
  ordering, 344
  repeating, 341, 342, 343, 344
  representation of fractions, 330, 331
  rounding, 345
  subtracting, 333
  terminating, 332
Decimal point, 328
  floating, 38
Decimal system, 109
Deci-, 687
Deductive system, 498

Degree, 508
  Celsius, 748
  Fahrenheit, 748
  Kelvin, 748
Deka-, 687
Denominator, 266
  least common, 277
  like, 276
  unlike, 277
Denseness property, 309
Descartes, Rene, 4, 74, 764
Deviation, standard, 462, 463
Diagonal, 521
Diagram
  tree, 65, 125
  Venn, 55, 60, 61
Diameter, 597
Difference, 7, 13, 14, 118
  of squares, 193
  set, 61
Digit, 109, 348
Dihedral angle, 511
  edge, 511, 512
  faces, 511, 512
Diophantine equations, 228
Diophantus, 201, 228
Direct reasoning, 93
Disjoint sets, 60
Disjunction, 86
Dispersion, 461
Disquisitiones Arithmeticae, 254
Disraeli, Benjamin, 484
Distance
  between points on a line, 176, 689
  between points in a plane, 766, 767
  formula, 766, 768
  properties, 689
Distributive property
  multiplication over addition, 126, 191
  multiplication over subtraction, 192
  set intersection over set union, 62
  set union over set intersection, 62
Dividend, 129
Divides, 220
Divisibility, 220
  tests for 2, 224
  tests for 3, 226
  tests for 4, 225
  tests for 5, 224
  tests for 6, 227
  tests for 7, 226
  tests for 8, 225
  tests for 9, 226
  tests for 10, 224
  tests for 11, 227
Division
  algorithm, 129, 130, 151
  by powers of ten, 330
  by primes method for LCM, 250
  by zero, 130
  decimals, 335, 336, 337
  definition, 128
  different bases, 164
  estimation, 162
  integers, 193
  mental, 154
  rational numbers, 294
  short, 153
  whole numbers, 128
Divisor, 129, 220
  greatest common (GCD), 244
Dodecahedron, 547
Domain, 74

DRAW, 833
Draw mode, 833
Dual of a tessellation, 669
Duodecimal system, 161
Durer, 147

E-notation, 818
Edge, 542
Edit mode, 838
Effective annual yield, 366
Egyptians, 105
Egyptian numeration system, 105
Element, 51
Elements, The, 498
Elimination method, 796
Ellipsis, 8
Embedded recursion, 850
Empirical probability, 386
Empty set, 54
END, 838
English system, 686
Equal sets, 51
Equal fractions, 273
Equality
  addition property, 198, 281
  cancellation property, 199
  of fractions, 272
  multiplication property, 198
  rational numbers, 273
  real numbers, 373
Equally likely outcomes, 386
Equation(s)
  of a circle, 768, 769
  linear, 781
  point-slope form, 785
  properties of, 198
  slope, 783
  slope-intercept form, 780
  systems of, 794
  vertical lines, 776
Equiangular, 523
  triangle, 524
Equilateral, 523
  triangle, 524
Equivalence
  classes, 72, 269
  logical, 87
  relation, 72
  sets, 53
Equivalent fractions, 269
Eratosthenes, 239
Erlanger Programm, 628
Escher, M., 663
Estimating, 141, 143, 144, 154, 284, 306, 307, 346, 347, 358
Euclid, 240, 498
Euclidean Algorithm, 247, 248
Euler, Leonhard, 74, 190, 547, 549, 692
Euler's formula, 547
Even vertex, 555
Event, 385
  certain, 388
  complementary, 390
  impossible, 388
  mutually exclusive, 388
Expanded form, 106
Expected value, 421
Experiment, 384
  multistage, 396
  two-stage, 396
Experimental probability, 386
Exponents, 107, 318

addition, 319
  properties of, 319, 321, 376
  rational, 375
  subtraction, 320
Exterior angle, 520
  alternate, 532
  polygon, 520, 534
  triangle, 533

Face, 542
  lateral, 542
  value, 106
Factor(s), 107, 125, 220
  scale, 603, 649
  tree, 234
  prime, 234, 245, 249
Factorial, 43, 424
Factorization, 233
Fahrenheit
  degree, 748
  Gabriel Daniel, 750
Fair, 421
Falting, Gerd, 221
Fermat, Pierre de, 220, 221, 384
Feurzeig, W., 836
Fibonacci, 16, 201, 290
  numbers, 16
  sequence, 16
Figurate numbers, 12
Figures, similar, 603, 652
Finger multiplication, 158
Finite set, 54
Five-number summary, 464
Flagging, 827
Flip, 636, 637
  image, 636
  line, 636
Floating decimal point, 38
FOR-NEXT, 823
FORWARD (FD), 834
Fourier, Joseph, 74
Fractions, 266
  algorithm for division of, 295
  equivalence classes, 72, 269
  equivalent, 268, 269
  Fundamental Law of, 269
  proper, 279
  simplifying, 270
  simplest form of, 271
Frequency polygon, 449
Frequency, relative, 386
Frequency table, 444
  grouped, 450
Function(s), 73
  applications, 79
  composition, 77
  domain, 74
  definition, 74
  identity, 79
  operations on, 77
  range, 74
Fundamental Law of Fractions, 269
Fundamental Theorem of Arithmetic, 235
Fundamental Counting Principle, 53

Galois, Evariste, 571
Gaps, 440
Gardner, M., 666
Gauss, Karl Friedrich, 20, 220, 254, 475, 599
General term

arithmetic sequence, 7, 8
  geometric sequence, 10
Geometric probability, 402
Geometric sequence, 10
Geometry, 498
Glide reflection, 643, 644
Goldbach, Christian, 240
Goldbach's conjecture, 240
GOTO, 821
Gram, 747
Graphs, 776
  bar, 446, 447
  circle or pie, 451
  definition, 439
  frequency polygon, 449
  frequency table, 444
  histogram, 445
  inequalities, 789
  lattice, 79
  line, 449
  line plot, 440
  linear inequalities, 789
  misleading, 486, 487, 488
  picto-, 439
  statistical, 439
  stem and leaf plot, 440, 441, 442
Graunt, John, 438, 439
Great circle, 600
Greater than, 57, 113, 199, 302, 303, 304
Greater than or equal to, 113, 200
Greatest common divisor, (GCD), 244
  method for LCM, 250
  simplest form of fractions, 271
Grouped frequency table, 450
Grouping property of addition, 115
Grouping system, 105

Half-closed segment, 502
Half-line, 501
Half-open segment, 502
Half-plane, 507
Half-space, 507
Half-turn, 631
Halmos, Paul, 2
Hardy, G. H., 4
HEADING, 835
Hectare, 700
  heuristics, 3
Hecto-, 687
Hectometer, 688
  square, 699
Hectare, 700
Height
  parallelogram, 702
  slant, 727, 729
Heptagon, 520
Herigone, Pierre, 508
Hermes, 599
Hexagon, 520
  tessellation, 664
HIDETURTLE (HT), 834
Hilbert, D., 228
Hilton, P., 241
Hindu-Arabic system, 109
Hippasus, 369
Histogram, 445
Hoffer, A., 552
HOME, 835
How to Lie with Statistics, 489
How to Solve It, 3, 18
Huff, Darrell, 489
Huygens, C., 384

Hypotenuse, 714, 715
Hypothesis, 89

Icosahedron, 547
Identity function, 79
Identity property
  addition, 115, 180
  multiplication, 125, 191
  set intersection, 62
  set union, 62
IF-THEN, 820
Image, 626
  flip, 636
  turn, 630
Implication, 88
Impossible event, 388
Included angle, 576
Included side, 581, 582
Inclusive or, 61, 86
Index, 375
Indirect reasoning, 93
Inductive reasoning, 6
Inequalities, 56, 305
  graphing, 789
  properties of, 199, 200, 201
Infinite set, 54
Initialize, 822
INPUT, 817
Inscribed, 600
  circle, 600
  polygon, 598
INT, 820
Integer(s), 175
  addition, 177
  additive identity, 180
  additive inverse, 180
  associative property, 180, 191
  closure property, 180, 191
  commutative property, 180, 191
  definition, 175
  distributive property, 191, 192
  division, 193
  multiplication, 188, 189, 190
  multiplicative identity, 191
  negative, 175
  order of operations, 194
  positive, 175
  subtraction, 181, 182, 183
Interest, 363
  compound, 365
  rate, 363
  simple, 363
Interior angle, 520, 532
  alternate, 532
  polygon, 520
Interquartile range (IQR), 465
Intersecting lines, 503
  reflecting, 637
Intersection of sets, 60
  definition, 60
  method for GCD, 245
  method for LCM, 248
Inverse, 89
  additive, 180, 280
  multiplicative, 290
Irrational numbers, 368
Isometry, 628
Isosceles
  trapezoid, 524
  triangle, 524

Jordan Curve Theorem, 518

INDEX

Kelvin, degree, 748
Keys, special, 37
Kilo-, 687
Kilogram, 747
Kilometer, 688
   square, 699
Kite, 524
Klein, Felix, 628
Konigsberg bridge problem, 553

Laboratory activities
   approximation of pi, 693
   area of triangle, 802
   attribute blocks, 69
   base 2, 166
   circles, 597
   coordinates, 775
   congruent cubes, 734
   finger multiplication, 158
   geoboard exercise, 713
   geometry, 516
   hypotenuses, 724
   ISBN numbers, 259
   linoleum print, 648
   mass and temperature, 751
   means, 473
   Mobius strip, 558
   Paper cup toss, 394
   pantograph, 611
   parallelogram/quadrilateral, 587
   perimeters, 697
   polygon construction, 603
   quadrilaterals, 529, 530
   reflected lines, 654
   shape properties, 581
   slopes, 793
   stacking cubes, 746
   spiral primes, 243
   tangram pieces, 562
   three dimensional figures, 552
   Tower of Hanoi, 36
   transformations, 635
   Venn diagram of quadrilaterals, 541
La Disme, 330
La géométrie, 764
Lagrange, Joseph Louis, 74
Lambert, J., 692
Land measure, 699
Lateral face
   prism, 542
   pyramid, 543
Lateral surface
   cone, 548
   cylinder, 548
Latitude, circles of, 600
Lattice
   graph, 79
   multiplication, 150, 164
   point, 792
   polygon, 709
Law of Detachment, 93
Leaf, 440
Least Common Denominator, 277
Least Common Multiple (LCM), 248
   from Least Common Denominator, 277
LEFT (LT), 834
Legs of a right triangle, 714
Leonardo of Pisa, 16, 201, 290
Less than, 56, 57, 199
Less than or equal to, 59, 200
LET, 817

Line(s), 501
   concurrent, 503
   coplanar, 503
   equation, 781
   flip, 636
   graph, 449
   half-line, 501
   intersecting, 503
   parallel, 503
   perpendicular, 510
   plot, 440
   point-slope form of, 785
   reflective, 637
   segment, 501
   skew, 503
   slope-intercept form, 780
   symmetry, 654
Linear equation, 781
   systems of, 794
Linear measure, 689
LIST, 816
Liter, 736
Logic
   calculator, 37, 38
   negation, 83, 84
   quantifiers, 84
Logically equivalent, 87
Logo, 559
   circles, 752
   coordinates, 803
   geometry, 559, 615
   tessellation, 670
Logo commands
   BACK (BK), 834
   DRAW, 833
   END, 838
   FORWARD, 834
   HEADING, 835
   HIDETURTLE (HT), 834
   HOME, 835
   IF, 852
   LEFT (LT), 834
   MAKE, 807
   NODRAW (ND), 834
   OUTPUT (OP), 806
   PENDOWN (PD), 834
   PENUP (PU), 834
   PRINT, 835
   REPEAT, 837
   RIGHT (RT), 834
   SETHEADING (SETH), 835
   SETPOS, 803
   SETX, 803
   SETXY, 803
   SETY, 803
   SHOWTURTLE (ST), 834
   SQRT, 806
   STOP, 852
   TO, 838
   TOWARDS, 807
   XCOR, 807
   YCOR, 807
Longitude, circles of, 601
Loomis, E., 715

Mach number, 695
Magic square, 24, 121
Major arc, 572
MAKE, 807
"Making a Better Beer Glass," 552
Mass, 747
Matijasevič, Yuri, 228

Mean, 458
Measure
   dispersion, 461
   land, 699
   linear, 688
Median, 459
Mega, 688
Melancholia, 147
Member, 51
Mental addition, 138, 337
Mental arithmetic, 138, 357
Mental division, 154, 338
Mental math with percent, 357
Mental multiplication, 150
Mental subtraction, 141, 338
Meter, 687
   cubic, 735
   definition, 687
   square, 699
Method
   elimination, 796
   substitution, 795
Metric measurement
   area, 698
   capacity, 735, 736
   land, 700
   mass, 747
   temperature, 748
   length, 688
   volume, 735
Metric system, 687
Metric ton, 747
Micro, 688
Midpoint formula, 770, 771
Milli-, 687
Milligram, 747
Millimeter
   square, 688
Mind set, 17
Minor arc, 571
Minuend, 118
Minute, 508
Mira, 592
Missing addend, 117
Missing addend model, 117
Missing factor model, 128
Mixed number, 278
Mode, 460
   average, 460
   draw, 833
   edit, 838
   nodraw, 833
   split-screen, 834
Model(s)
   array, 123
   Cartesian product, 124
   Comparison, 118
   integer multiplication, 188, 189, 190
   missing addend, 117
   number-line, 118, 123, 179, 182, 189
   partition, 128
   repeated addition, 122
   take away, 117
Modular arithmetic, 256
Modulo, 256
Modus Ponens, 93
Modus Tollens, 94
Moebius strip, 558, 559
Mouton, Gabriel, 687
Multiple, 220
Multiple, least common (LCM), 248
Multiplication
   algorithm, 148

INDEX

associative property, 125, 191
clock, 255
closure property, 125, 191
commutative property, 125, 191
decimal, 334, 335
different bases, 163
estimation, 162
finger, 158
identity property, 125, 191
integers, 188
lattice, 150, 164
mental, 150
rational numbers, 289, 304
real numbers, 372
whole numbers, 122, 123
zero property, 191
Multiplication properties, 125, 191
divisibility, 221
equality, 198
greater than, 200
inequality, 199, 200, 201
Multiplicative
identity, 125, 191, 290
inverse, 290
property, 109, 290
Multistage experiments, 396
with replacement, 396
without replacement, 397
Mutually exclusive events, 388

n factorial (n!), 424
n-gon, 520
Natural numbers, 19, 51
Negation, 83, 84
Negative
correlation, 481
integers, 175
Nets, 546, 724
Network, 553, 554
arc, 554
traversable, 554
vertex, 554
NEW, 817
Newton, Isaac, 747
Nickel, L., 240
NIM, 43
NODRAW (ND), 834
Noether, E., 191
Noll, C., 240
Nonagon, 520
Noncollinear, 501
Noncoplanar lines, 503
Noncoplanar points, 503
Non-mutually exclusive events, 390
normal
curve, 474
distribution, 474, 478, 479
Notation
E-, 818
set builder, 51
nth
power, 107
root, 375
term, 9, 11
null set, 54
Number(s)
amicable, 253
base five, 161, 162, 163
bases, 106, 158
cardinal, 54
composite, 232
counting, 51

decimal, 328
figurate, 12
integers, 175
irrational, 368
line, 112, 113, 114, 118, 123, 179, 180, 182, 183, 189
line model, 118, 123, 179, 182, 189
mixed, 278
natural, 19, 51
negative, 175
pentagonal, 16
perfect, 253
prime, 232
rational, 266
real numbers, 368, 372
square, 12
theory, 220
triangular, 12
Numerals, 104
Babylonian, 106
Egyptian, 105
Hindu-Arabic, 104, 109
Mayan, 107
Roman, 108
Numeration systems, 104
Babylonian, 106
decimal, 109
Egyptian, 105
Hindu-Arabic, 109
Mayan, 107
other bases, 158
Roman, 108
Numerator, 266

Oblique
circular cone, 548
cylinder, 548
prism, 542
Obtuse, 510
triangle, 523
Octagon, 520
Octahedron, 547
Odd vertex, 555
Odds, 417
against, 418
in favor, 418
One-to-one correspondence, 52
Operations
binary, 111
clock, 254, 255
on functions, 77
order of, 128, 184, 194
on sets, 60
Order of operations, 128, 184, 194
Order relations
decimals, 344
rational numbers, 302
Ordered pairs, 64
Ordinate, 766
Origin, 764
Oughtred, William, 123
Outcome(s), 384
equally likely, 386
Outliers, 440, 467
OUTPUT(OP), 806

Palindromes, 146
Papert, Seymour, 836
Parallel
lines and planes, 503, 506
planes, 506

Parallel lines, 503
slopes of, 786, 787
Parallelogram, 524
area of, 702
rhombus, 524
Partitioning, 132
Pascal, Blaise, 16, 384
Pascal's triangle, 432
Patterns, 4
Patterns model, 182, 188
PENDOWN (PD), 834
Penrose, R., 668
Penrose tilings, 668
Pentagon, 520
Percent, 351
conversion to decimal, 353
definition, 352
use and misuse, 860
Percentile, 477
Perfect numbers, 253
Perimeter, 691
Permutation, 425, 426
Perpendicular
bisector, 590
lines and planes, 511
Perpendicular lines
definition, 510
slope of, 786
Pi, 368, 692
Pictograph, 439
Pie chart, 451
Place value, 106, 109
Plane, 502
curve, 517
half-plane, 507
parallel, 506
parallel to line, 507
perpendicular, 511
perpendicular to lines, 511
properties, 506
symmetry, 659
Plato, 545
Platonic solids, 545
Point(s), 500
betweenness of, 501
collinear, 501
of contact, 597
coplanar, 503
of intersection, 503
properties, 506
of tangency, 597
Point-slope form of a line, 785
Polya, George, 3, 4
Polygons, 518
area of, 707
congruent, 583
convex, 519
diagonal, 521
exterior angle, 520
inscribed, 597
interior angle, 520
lattice, 709
regular, 523
similar, 606
sum of measures of exterior angles, 534
Polygonal curve, 518
Polygonal region, 518
Polyhedron, 542
convex, 544
prism, 542
pyramid, 543
regular, 544, 545
semiregular, 546

types of, 545
Platonic solids, 545
Positive
   correlation, 481
   integers, 175
Prime(s), 232
   factors, 234
   relatively, 245
Prime factorization, 234
   method for GCD, 245
   method for LCM, 249
Prime numbers, 232
Primitives, 834
Principal, 363
Principal square root, 368, 369
Principle, Cavalieri's, 738, 739
PRINT, 817, 835
Prism, 542
   base, 542
   lateral face, 542
   oblique, 542
   right 542
   surface area, 725
   volume, 735
Probability, 384, 387, 389
   definition, 387, 389
   experimental, 386
   geometric, 402
   Multiplication Rule for, 397
   multistage experiments, 396
   properties, 391
   simulations, 408
Problem solving
   steps to, 3
   strategies, 3, 19
   word problems, 205, 206
Problem solving process, explanation, 17
   Carrying Out the Plan, 3, 18
   Devising a Plan, 3, 18
   Looking Back, 3, 19
   Understanding the Problem, 3, 18
Procedure, 838
Proceedings of the St. Petersburg Academy, 549
Product, 125
Program, 815
Programming, top-down, 841
Proper fraction, 279
Properties
   additive, 113
   distance, 689
   equality, 198
   exponents, 319, 320, 321, 376
   grouping, 115
   inequality, 304
   multiplicative, 125
   proportions, 314, 315
   probability, 391
   radicals, 376
   rational numbers, 280, 281, 290, 291
   real numbers, 372
   rhombus, 584
   subtractive, 109
Properties of inequality, 199, 200, 201, 304
Proportion, 312, 313
   properties, 314, 315, 607
Protractor, 508
Ptolemy, 267
Pyramid, 543
   apex, 543
   base, 543
   lateral faces, 543
   surface area, 727, 728

volume, 739
Pythagoras, 369, 714, 716
Pythagorean society, 369, 716
Pythagorean Theorem, 370, 715
   converse, 720
Pythagorean Proposition, 715

Quadrants, 765
Quadrilateral, 520, 584
   kite, 524
   parallelogram, 524, 584
   rectangle, 524, 584
   rhombus, 524, 584
   square, 524, 584
   tessellation, 665
   trapezoid, 524
Quantifiers
   existential, 84
   universal, 84
Quartile, 465
Quotient, 129

Radical(s)
   properties, 376
   sign, 369
Radicand, 369
Radius, 597
Random, 387
Random Digit Table, 409
Range
   of data, 462
   of a function, 74
Ratio, 10, 312
Rational numbers, 266
   addition, 276, 277
   additive identity, 280, 372
   additive inverse, 280, 372
   associative property, 280, 372
   closure property, 280, 372
   commutative property, 280, 372
   definition, 266
   denseness property, 309, 372
   distributive property, 280, 372
   division, 294, 295
   estimation, 284
   fractions, 266
   inverse property, 280, 372
   multiplication, 289
   multiplicative identity, 290
   multiplicative inverse, 290
   ordering properties, 304
   subtraction, 281, 283
   use of, 267
Multiplication Property of Zero, 291

Ray, 501
READ, 825
Real numbers, 368
   addition, 372
   additive inverse, 372
   definition, 368
   denseness, 372
   multiplication, 372
   subtraction, 372
   system, 372
Reasoning
   direct, 93
   indirect, 93
   inductive, 6
   valid, 91

Rectangle, 524
   area of, 701
Recursion, 850
   embedded, 853
   tail-end, 850
Reflection, 636, 637
   glide-, 643, 644
   line, 636
Reflexive property, relations, 71
Regular polygon, 523
Regular polyhedra, 544, 545, 546
Relation, 69
   equivalence, 72
   greater than, 113
   less than, 113
   reflexive property, 71
   symmetric property, 71
   transitive property, 72
Relatively prime, 245
REM, 819
Remainder, 130
REPEAT, 837
Repeated addition model, 122
Repeated subtraction model, 129
Repeating decimals, 341
Repetend, 341
Rhombus, 524, 584
Rice, M., 666
Richelot, F., 599
RIGHT (RT), 834
Right angle, 510
Right circular cone, 548
   surface area, 728
   volume, 740
Right circular cylinder
   surface area, 727
   volume, 738, 739
Right triangle, 523
   hypotenuse, 714, 715
   leg, 714
Right prism, 542
Rigid, 574
Rigid motion, 628
Rise, 782
Rise over run, 782
Robinson, Julia, 228
Roman numerals, 108
Rotation, 629
Rotational symmetry, 657
Rounding, 143, 345
Rudolff, C., 369
Rule, chain, 94
Russian Peasant Algorithm, 155
RND, 820
RSA system, 241
rth percentile, 477
Run, 782
RUN, 816
Russell, Bertrand, 58

Sample space, 385
Scale factor, 603, 649
Scalene triangle, 524
Scattergram, 480
Scientific notation, 331, 348
Scratch addition, 137
Second, 508
Sector, 708
   area of, 708
Segments, 552
   congruent, 522
   half-closed, 502

half-open, 502
line, 501
unit, 112
Semicircle, 572, 692
Semiregular polyhedra, 546
Sequence, 7
  arithmetic, 7, 8
  Fibonacci, 16
  geometric, 10
Set(s), 51
  builder notation, 51
  Cartesian product, 64, 65
  complement, 55
  complement relative to, 61
  difference, 61
  disjoint, 60
  empty, 54
  equal, 51
  equivalent, 53
  finite, 54
  infinite, 54
  intersection, 60
  null, 54
  operations, 60
  properties, 62
  subset, 55
  theory, 50
  union, 50
  universal, 55
  well defined, 51
SETHEADING (SETH), 835
SETPOS, 803
SETX, 803
SETXY, 803
SETY, 803
SHOWTURTLE (ST), 834
Side Angle Side (SAS), 577
Side Side Side (SSS), 574
Sides
  angle, 507
  included, 581, 582
Sieve of Eratosthenes, 239
Significant digits, 348
Similar
  figures, 603, 652
  polygons, 606
  triangles, 603
Simple
  closed curve, 517
  closed surface, 541
  curve, 517
  interest, 363
Simplest form, 271
Simplifying fractions, 270
Simulations, 408, 409
Size transformation, 648, 649
  center of, 649
Skew lines, 503
Slant height
  cone, 729
  pyramid, 727
Slide, 626
  arrow, 626
  line, 626
Slope, 779, 780, 781, 782
  formula, 783
  parallel lines, 786, 787
  perpendicular lines, 786
  rise/run, 781
Slope-intercept form of a line, 780
Solid, 541
  Platonic, 545
Somerville, Mary, 205

Space, 507
  sample, 385
Special keys, calculator, 39
Sphere, 541, 600
  surface area, 730
  volume, 741
Square, 524, 698
  area, 698
  centimeter, 699
  congruent, 571
  kilometer, 699
  magic, 24, 121
  meter, 699
  millimeter, 699
  numbers, 12
  perfect, 12
Square root, 368
Standard deviation, 462, 463
State transparent, 837
Statement, 83
  compound, 85
  contrapositive, 89
  converse, 89
  inverse, 89
  logically equivalent, 87
  negation, 83
Statistics, 438
  abuses of, 484
  graphing, 439
Stems, 440
Stem and leaf plot
  back-to-back, 440
  ordered, 441
STEP, 823
Stevin, Simon, 330
STOP, 852
Straight angle, 510
Strategy, 19
  draw a diagram, 29
  examine related problems, 27
  examine simpler cases, 22
  identify a subgoal, 24
  indirect reasoning, 32, 33
  look for a pattern, 19
  make a table, 8, 9, 21
  guess and check, 31
  work backwards, 28
  write an equation, 28
String variables, 816
Study of Regular Division of the Plane with Reptiles, 663
Subsets, 55
  definition, 55
  number of, 58
  proper, 56
Substitution method, 795
  property, 199
Subtraction
  algorithms, 140
  decimals, 333
  different bases, 162, 163
  integers, 182
  as inverse of addition, 182
  properties of integers, 180, 191
  properties of whole numbers, 120
  rational numbers, 283
  whole numbers, 118
Subtractive property, 109
Subtrahend, 118
Sum, 112
Sum of angle measures
  exterior angles in a polygon, 534
  interior angles in a polygon, 536

interior angles in a triangle, 533, 534
Supplement of an angle, 531
Supplementary angle, 530
Surface area, 725
  lateral, 726
  cube, 725
  right circular cone, 728
  right circular cylinder, 727
  right prism, 725
  right regular pyramid, 727
  sphere, 730
Symmetric property, 71
Symmetry
  line, 654
  line of, 654
  point, 658
  plane, 659
  rotational, 657
  turn, 657
System of linear equations, 794, 795, 796, 797

Table
  frequency, 444
  truth, 85
Tally marks, 105
Take-away model, 117
Tail-end recursion, 850
Tangent, 597
Tautology, 91
Temperature, 748
Terminating decimals, 332
Tessellations, 663
  dual of, 669
  Escher-like, 670
  hexagon, 664
  pentagon, 665
  quadrilateral, 665
  regular, 664
  triangle, 665
Tetrahedron, 547
Theoretical probability, 387
TO, 838
Ton, metric, 747
Top-down programming, 841
Total Turtle Trip Theorem, 533, 534
TOWARDS, 807
Tower of Hanoi, 36
Transformations, 627
  size, 648, 649
Transitive property, 72, 304
Translation, 626, 627
Transparent, state, 837
Transversal, 532
Trapezoid, 524
  area, 704
  isosceles, 524
Traversable, 554
Tree diagram, 65, 125, 415
Trend line, 480
Triangle, 523
  altitude, 591
  acute, 523
  area, 703
  congruent, 572
  equiangular, 523
  equilateral, 524
  isosceles, 524
  obtuse, 523
  right, 523
  scalene, 524
  similar, 603

sum of measures of interior angles, 536
Triangular inequality, 576, 689
Triangular numbers, 12
Truth tables, 85
Turn, 629, 630
   angle or, 629
   arrow, 630
   center of, 629
   half, 631
   symmetry, 657
Turtle, 833
   graphics, 833
   -type circle, 752

Undefined terms, 500
Union, 60
Unit segment, 112
Universe, 55

Valid reasoning, 91
Value
   expected, 421
   face, 106
   place, 106, 109
van Ceulen, Ludolph, 692
van Hiele
   levels, 499
   phases, 499

van Hiele, Pierre, 498, 499
van Hiele-Geldof, Dina, 498, 499
Variable, string, 816
Variance, 462
Vector, 626
Venn, John, 55
Venn diagram, 55
   as a problem-solving tool, 63
   use with quantifiers, 84
Vertex
   angle, 507
   cone, 548
   even, 555
   network, 554
   polygon, 518
   polyhedron, 542
Vertical angles, 530
Volume, 735
   cones, 740
   cube, 738
   metric units of, 735
   pyramids, 739
   right circular cylinder, 738
   right prism, 735
   sphere, 741
von Leibnitz, Gottfried Wilhelm, 123

Weight, 747
Well-defined, 51

Whiskers, 465
Whole numbers
   addition, 111
   additive identity, 115
   associative property (addition), 115
   closure property (addition), 113
   commutative property, 114
   distributive property, 126
   division, 128
   definition, 111
   multiplication, 122, 123
   properties of multiplication, 125
   subtraction, 117
Word problem, 205

x-axis, 764
x-coordinate, 766
x-intercept, 780

y-axis, 764
y-coordinate, 766
y-intercept, 780

Z-score, 477
Zero, 175
Zero multiplication property, 191

The Front Cover Solution

The pairs are:

1 and 5
2 and 8
3 and 4
6 and 9

7 does not have a match.

The Back Cover Solution

The pairs are:

A1 and E6	A4 and E5
B1 and A3	B4 and A6
C1 and E2	C4 and B6
D1 and A2	D4 and D6
E1 and B3	E4 and C6
B2 and C5	C3 and A5
C2 and D5	D3 and B5
D2 and E3	

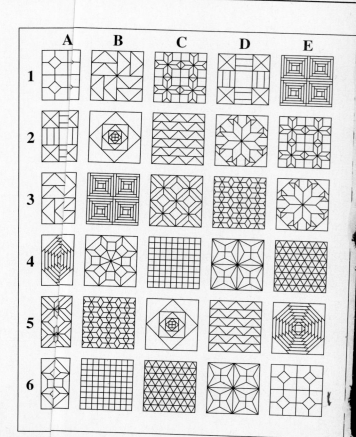

Patchwork Sampler quilt by Margit Echols
© 1983. Photo: Schecter Lee.

STANDARD 4: Mathematical Connections

In grades 5-8, the mathematics curriculum should include the investigation of mathematical connections so that students can—

- see mathematics as an integrated whole;
- explore problems and describe results using graphical, numerical, physical, algebraic, and verbal mathematical models or representations;
- use a mathematical idea to further their understanding of other mathematical ideas;
- apply mathematical thinking and modeling to solve problems that arise in other disciplines, such as art, music, psychology, science, and business;
- value the role of mathematics in our culture and society.

STANDARD 5: Number and Number Relationships

In grades 5-8, the mathematics curriculum should include the continued development of number and number relationships so that students can—

- understand, represent, and use numbers in a variety of equivalent forms (integer, fraction, decimal, percent, exponential, and scientific notation) in real-world and mathematical problem situations;
- develop number sense for whole numbers, fractions, decimals, integers, and rational numbers;
- understand and apply ratios, proportions, and percents in a wide variety of situations;
- investigate relationships among fractions, decimals, and percents;
- represent numerical relationships in one- and two-dimensional graphs.

STANDARD 6: Number Systems and Number Theory

In grades 5-8, the mathematics curriculum should include the study of number systems and number theory so that students can—

- understand and appreciate the need for numbers beyond the whole numbers;
- develop and use order relations for whole numbers, fractions, decimals, integers, and rational numbers;
- extend their understanding of whole number operations to fractions, decimals, integers, and rational numbers;
- understand how the basic arithmetic operations are related to one another;
- develop and apply number theory concepts (e.g., primes, factors, and multiples) in real-world and mathematical problem situations.

STANDARD 7: Computation and Estimation

In grades 5-8, the mathematics curriculum should develop the concepts underlying computation and estimation in various contexts so that students can—

- compute with whole numbers, fractions, decimals, integers, and rational numbers;
- develop, analyze, and explain procedures for computation and techniques for estimation;
- develop, analyze, and explain methods for solving proportions;
- select and use an appropriate method for computing from among mental arithmetic, paper-and-pencil, calculator, and computer methods;
- use computation, estimation, and proportions to solve problems;
- use estimation to check the reasonableness of results.

STANDARD 8: Patterns and Functions

In grades 5-8, the mathematics curriculum should include explorations of patterns and functions so that students can—

- describe, extend, analyze, and create a wide variety of patterns;
- describe and represent relationships with tables, graphs, and rules;
- analyze functional relationships to explain how a change in one quantity results in a change in another;
- use patterns and functions to represent and solve problems.

STANDARD 9: Algebra

In grades 5-8, the mathematics curriculum should include explorations of algebraic concepts and processes so that students can—

- understand the concepts of variable, expression, and equation;
- represent situations and number patterns with tables, graphs, verbal rules, and equations and explore the interrelationships of these representations;
- analyze tables and graphs to identify properties and relationships;
- develop confidence in solving linear equations using concrete, informal, and formal methods;
- investigate inequalities and nonlinear equations informally;
- apply algebraic methods to solve a variety of real-world and mathematical problems.

STANDARD 10: Statistics

In grades 5-8, the mathematics curriculum should include exploration of statistics in real-world situations so students can—

- systematically collect, organize, and describe data;
- construct, read, and interpret tables, charts, and graphs;
- make inferences and convincing arguments that are based on data analysis;
- evaluate arguments that are based on data analysis;
- develop an appreciation for statistical methods as powerful means for decision making.